mwlibrary@guildford.ac.uk

The UK Pesticide Guide 2013

D1464621

Editor: M.A. Lainsbury, BSc(Hons)

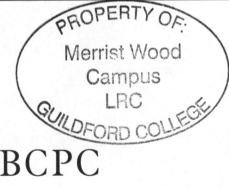
BCPC

www.cabi.org

CABI is one of the world's foremost publishers of databases, books and compendia in agriculture and applied life sciences. It has a worldwide reputation for producing high quality, value-added information, drawing on its links with the scientific community. CABI produces CAB Abstracts – the leading agricultural A&I database; together with the Crop Protection Compendium which provides solutions for identifying, preventing and solving crop health problems. CABI is a not-for-profit organization. For further information, please contact CABI, Nosworthy Way, Wallingford, Oxon OX10 8DE, UK.

Telephone: (01491) 832111
Fax: (01491) 833508
e-mail: enquiries@cabi.org
Web: www.cabi.org

BCPC (British Crop Production Council) promotes the use of good science and technology in understanding and applying effective and sustainable crop production. Its key objectives are to identify developing issues in the science and practice of crop protection and production and provide informed, independent analysis; to publish definitive information for growers, advisors and other stakeholders; to organise conferences and symposia; and to stimulate interest and learning. BCPC is a registered Charity and a Company limited by Guarantee. Further details available from BCPC, 7 Omni Business Centre, Omega Park, Alton, Hampshire GU34 2QD, UK.

Telephone: (01420) 593200
Fax: (01420) 593209
e-mail: md@bcpc.org
Web: www.bcpc.org

ISBN: 978 1 780642 77 2

Typeset and Printed in the United Kingdom by
Hobbs the Printers Ltd, Totton, Hampshire, SO40 3WX
Website: www.hobbs.uk.com

Contents

Disclaimer

Every effort has been made to ensure that the information in this book is correct at the time of going to press, but the Editor and the publishers do not accept liability for any error or omission in the content, or for any loss, damage or other accident arising from the use of the products listed herein. Omission of a product does not necessarily mean that it is not approved or that it is unavailable.

It is essential to follow the instructions on the approved label when handling, storing or using any crop protection product. Approved 'off-label' extensions of use now termed Extension of Authorisation for Minor Use (EAMUs) are undertaken entirely at the risk of the user.

Editor's Note

This is the 26th edition of *The UK Pesticide Guide* and 2012 has been a relatively calm period after the introduction of the changes in the approval process caused by the enactment of EC Regulation 1107/2009. The SDHI fungicides are now firmly established as part of many programmes and have proved to be very effective. New requirements to be met before aerial applications are permitted were introduced in June 2012 and these are explained in Section 5. Some pest control products such as rodenticides containing difenacoum are now approved under the EC Biocide Regulations and these have an approval number with a different format. Specific Off-Label Approvals (SOLAs) are now known as Extension of Authorisation for Minor Use (EAMUs) and companies are encouraged to adopt these and add them to product labels by small extensions to patent protection for each one added. Thus they are no longer 'off-label' in all circumstances.

Products are listed according to their active ingredient profile and while 9 profiles such as asulam and chloropicrin have been deleted, 11 new ones have been added. Some of these new listings are new combinations of existing actives, e.g. dicamba + prosulfuron but some are genuinely new active ingredients. The new ones include Ampelomyces quisqualis strain AQ 10, benzyladenine, Phlebiopsis gigantean, pyriofenone, quinoclamine and Trichoderma asperellum (Strain T34). These are niche products but at least we are still getting new active ingredients approved despite the higher environmental hurdles to be met.

Many of the existing products listed in this publication have been issued with revised expiry dates. Many approvals that were due to expire in 2012 or 2013 have now been granted extensions. However, a number of products have still been lost and so this book can help you identify what is still legal to use and what must be disposed of safely from your pesticide store. This edition also lists hundreds of new EAMUs that are essential for the protection of the less mainstream crops. There is also a new section on biodiversity in Section 5. The way 'Products also Registered', (i.e. approved but not actively marketed) are listed in the index has been revised to make it easier to locate the entry in Section 3.

This book is an annual publication and provides a snapshot of what is approved at the time of publication. If you need more up-to-date information, this book is also available as an online subscription database – plantprotection.co.uk – it is frequently updated, is searchable and contains much more information than can be squeezed into the book. This online resource was completely redesigned in 2012 and was launched at CEREALS in June. It is now very much easier to use and is worth looking at if you are a regular user of the book. New for this year, *The UK Pesticide Guide* will also be available as an eBook from Amazon and Google Play.

Each active ingredient entry carries a classification on its mode of action which is important for managing the risk of resistance. By ensuring that pests and diseases are subjected to a range of different modes of action, both within crops and across years, the risk of reduced pesticide activity is lessened. For people who transport these products, but are not especially concerned about their use, information on the UN Numbers, Transport Code and Packaging Group is also listed. This information is of value to the Emergency Services facing a pesticide emergency.

We express our gratitude to all those people who have provided the information for the compilation of this guide. As always, criticisms and suggestions for improvement are always welcome and in particular, if you notice any errors or omissions in the entries, please let me know via the e-mail address below.

Martin Lainsbury, Editor
ukpg@bcpc.org.

The information in this publication has been selected from official sources, from the suppliers product labels and from the product manuals of the pesticides approved for use in the UK under the Control of Pesticides Regulations 1986 (COPR) or the Plant Protection Products Regulation 1995 (PPPR). The content is based on information received by the editor up to October 2012.

Changes Since 2012 Edition

Pesticides and adjuvants added or deleted since the 2012 edition are listed below. Every effort has been made to ensure the accuracy of this list, but late notifications to the Editor are not included.

Products Added

Products new to this edition are shown here. In addition, products that were listed in the previous edition whose PSD/HSE registration number and/or supplier have changed are included.

Product	Reg. No.	Supplier
Abringo	15607	Belcrop
Actellic Smoke Generator No. 20	15739	Syngenta
Agrichem MCPA 500	15836	AgriChem BV
Answer SX	15632	DuPont
Anthem	15761	Makhteshim
AQ 10	15518	Belchim
Arenales	15814	Agroquimicos
Avoca Premium	15619	Barclay
Barclay Avoca	15549	Barclay
Becki	15732	ChemSource
Beetup Compact SC	15566	United Phosphorus
Betanal Turbo	15505	Bayer CropScience
Betasana Trio SC	15551	United Phosphorus
Better Flowable	15740	Sipcam
Bishop Extra	15824	ChemSource
Blaster Pro	15752	Headland Amenity
Brigida	15658	Agroquimicos
Brogue	15638	AgriGuard
Busa	15805	AgChem Access
Caramba 90	15524	BASF
Casper	15573	Syngenta
Ceriax	15639	BASF
Chubut	15648	Agroquimicos
Clayton Astra	A0724	Clayton
Clayton Concorde	15665	Clayton
Clayton Dent	15546	Clayton
Clayton Dolmen	A0725	Clayton
Clayton Kibo	15822	Clayton
Clayton Krypton WDG	15497	Clayton
Clayton Lithium	15803	Clayton
Clayton Nero	15511	Clayton

Product	Reg. No.	Supplier
Clayton Putter	15827	Clayton
Clayton Union	A0686	Clayton
Cleanrun Pro	15828	Everris Ltd
Clinker	15453	ChemSource
Clovermax	15780	Nufarm UK
Crew	15770	Nufarm UK
Croupier	15643	AgriGuard
Cupari	15883	Agroquimicos
Damocles	15689	AgChem Access
Danadim Progress	15890	Headland
Demeo	15595	Dow
Di Novia	15840	Agroquimicos
Doxstar Pro	15664	Dow
Easel	15548	Nufarm UK
Entail	15671	Headland
Exemptor	15615	Everris Ltd
Exilis	15706	Fine
Forefront T	15568	Dow
Garrison	15657	Pan Agriculture
Globaztar SC	15575	Globachem
Goiana	15778	Agroquimicos
Grazon	15751	Dow
Grazon Pro	15785	Dow
Greenmaster Autumn	15808	Everris Ltd
Greenmaster Extra	15817	Everris Ltd
Greenmaster Mosskiller	15796	Everris Ltd
Helmstar Agrochemical	15070	Rotam
Impede	15601	Headland
Intracrop Evoque	A0754	Intracrop
Intracrop F16	A0752	Intracrop
Jatapu	15776	Agroquimicos
Joules	15730	Nufarm UK
Juliet	15788	ChemSource

Product	Reg. No.	Supplier
Juliet	15947	ChemSource
Juventus	15528	BASF
Kerb Flo 500	15586	Dow
Laminator 75 WG	15667	Interfarm
Landscaper Pro Moss Control + Fertilizer	15795	Everris Ltd
Landscaper Pro Weed Control + Fertilizer	15820	Everris Ltd
Legara	15533	AgChem Access
Mascot Exocet	15598	Rigby Taylor
Maxim 100FS	15683	Syngenta
Meta Flo	15486	ChemSource
Metro 20	15633	AgriGuard
Metro Clear	15624	AgriGuard
Mirador	15557	Makhteshim
MiteStop	15775	ChemSource
Mitron 700 SC	15621	Belcrop
Mogeton	15837	Certis
MyZus	15903	ChemSource
Nebula XL	15555	BASF
Noah	15855	ChemSource
Osiris P	15627	BASF
Palamares	15653	Agroquimicos
Palometa	15833	Agroquimicos
Pan Cradle	15923	Pan Agriculture
Pan Epoflux	15889	Pan Agriculture
Pan Hale	15924	Pan Agriculture
Pan Tepee	15867	Pan Agriculture
Peak	15521	Syngenta
Pendragon	15512	AgriGuard
Pergardo Uni	15734	Fargro
Pexan	15584	BASF
Piper	15786	Nufarm UK
Pirapima	15777	Agroquimicos
Pitch	15485	Nufarm UK
Pitch	15693	Nufarm UK
Plant Trust	15779	Everris Ltd
Polax	15915	Pan Amenity
Pontoon	15821	Agroquimicos
Prop 400	15832	ChemSource
Property 180 SC	15661	Belchim
Proplant	15422	Fargro
Prova	15641	Belcrop
Pure Gold	15559	Pure Amenity
PureFloxy	15473	Pure Amenity
PureProgress	15471	Pure Amenity
PureProteb	15469	Pure Amenity
Quazar	15357	Belchim
Racona i-MIX	15574	Chemtura
Rattler	15522	Nufarm UK
Rattler	15692	Nufarm UK
Reconcile	15379	Becesane
Refine Max SX	15622	DuPont
Renovator Pro	15815	Everris Ltd
Rezist Liquid	14643	Sphere
Rifle	15705	Makhteshim
Rio	15874	ChemSource
Roquat 20	15703	Q-Chem
Roundup Flex	15541	Monsanto
Runner	15629	Landseer
Rustler Pro-Green	15454	ChemSource
Sakarat D Pasta Bait	UK12-0371	Killgerm
Sakarat D Wax Bait	UK12-0370	Killgerm
Sakarat D Whole Wheat	UK12-0301	Killgerm
Samson Extra 6%	15662	Syngenta
Segundo	15710	Agroquimicos
Serenade ASO	15625	BASF
Setanta 50 WP	15642	AgriGuard
Setanta Flo	15791	AgriGuard
Shogi	15736	ChemSource
Shotput	15968	Makhteshim
Shrink	15660	AgChem Access
Simba SX	15503	DuPont
Smartgrass	15628	Interfarm
Snapper	15489	Nufarm UK
Snapper	15695	Nufarm UK
Snicket	15681	AgChem Access
Solitaire	15637	AgriGuard
Solitaire 50 WP	15636	AgriGuard
Spaniel	15530	Pan Amenity
Spudster 360 CS	15711	AgChem Access
Stay Fresh	15760	ChemSource
Sterisoil	15841	ChemSource
T34 Biocontrol	15603	Fargro
Tanaru	15726	Agroquimicos
Tanker	15694	Nufarm UK
Taurus	15508	Certis
Tecka	15831	Agroquimicos
Tercero	15819	Agroquimicos

Product	Reg. No.	Supplier	Product	Reg. No.	Supplier
Tipoff	15728	Globachem	Vortex	15663	BASF
Toppel 100	15696	United Phosphorus	Walia	15787	ChemSource
			Yomp	15609	Becesane
Trilogy	15644	United Phosphorus	Zeal Plus	15850	ChemSource
			Zenith	15649	Becesane
Trimazone	14726	United Phosphorus	Zubarone	15798	AgChem Access
Unite	15758	Dow			

Products Deleted

The appearance of a product name in the following list may not necessarily mean that it is no longer available. In cases of doubt, refer to Section 3 or the supplier.

Product	Reg. No.	Supplier	Product	Reg. No.	Supplier
Agriguard Ethofumesate 200	14129	AgriGuard	Greencrop Rookie	A0515	Greencrop
			Grey Squirrel Bait	H9175	Killgerm
Aliette 80 WG	13130	Certis	Gringo	14467	AgChem Access
Atom	15038	Nufarm UK			
Barclay Gallup Biograde Amenity	12716	Barclay	Mascot Hi-Activ Amenity	15178	Barclay
Becki	14677	ChemSource	MSS Sprout Nip	14184	Whyte
Binder	A0598	Nufarm UK	Agrochemicals		
Bonser	A0601	Nufarm UK	Nobel	15332	ChemSource
Bonser	A0715	Nufarm UK	Pan Tepee	12279	Pan Agriculture
Caesar	13779	AgChem Access	Pitch	15485	Nufarm UK
Caramba 90	12655	BASF	Redstar	14197	AgChem Access
Charger	14107	AgChem Access	Reparto	15452	Syngenta
Charger B	14269	AgChem Access	SHL Granular Feed and Weed	10970	Sinclair
Clenecorn Super	14628	Nufarm UK	SHL Turf Feed and Weed	10963	Sinclair
Conquest	15031	AgChem Access	Shrink	13762	AgChem Access
Cosavet DF	11477	BASF	Tacet	14342	Goldengrass
CS Azoxy	14954	ChemSource	Tempo	13799	Syngenta
Dithane NT Dry Flowable	15721	Interfarm	Tevday	13783	AgChem Access
Dow Shield	10988	Dow	Try-Flex	A0489	Greenaway
EA Clodinafop	15006	European Ag	Walia	14976	ChemSource
Greencrop Fenit	A0602	Greencrop			

Introduction

Purpose

The primary aim of this book is to provide a practical handbook of pesticides, plant growth regulators and adjuvants that the farmer or grower can realistically and legally obtain in the UK and to identify the purposes for which they can be used. It is designed to help in the identification of products appropriate to a particular problem. In addition to uses recommended on product labels, details are provided of those uses which do not appear on product labels but which have been granted Extension of Authorisation for Minor Use (EAMUs) – previously known as SOLAs. As well as identifying the products available, the book provides guidance on how to use them safely and effectively but without giving details of doses, volumes, spray schedules or approved tank mixtures. Sections 5 and 6 provide essential background information on a wide range of pesticide-related issues including: legislation, biodiversity, codes of practice, poisons and treatment of poisoning, products for use in special situations and weed and crop growth stage keys.

While we have tried to cover all other important factors, this book does not provide a full statement of product recommendations. Before using any pesticide product it is essential that the user should read the label carefully and comply strictly with the instructions it contains. **This** *Guide* **does not substitute for the product label.**

Scope

The *Guide* is confined to pesticides registered in the UK for use in arable agriculture, horticulture (including amenity use), forestry and areas in or near water. Within these fields of use there are about 2,800 products in the UK with current approval. The *Guide* is a reference for virtually all of them given in two sections. Section 2 gives full details of the products notified to the editor as available on the market. Products are included in this section only if requested by the supplier and supported by evidence of approval. Section 3 gives brief details of all other registered products with extant approval.

Section 2 lists some 450 individual active ingredients and mixtures. For each entry a list is shown of the available approved products together with the name of the supplier from whom each may be obtained. All types of pesticide covered by Control of Pesticides Regulations or Plant Protection Products Regulations are included. This embraces acaricides, algicides, herbicides, fungicides, insecticides, lumbricides, molluscicides, nematacides and rodenticides together with plant growth regulators used as straw shorteners, sprout inhibitors and for various horticultural purposes. The total number of products included in section 2 (i.e. available in the market in 2013) is about 1,450.

The *Guide* also gives information (in section 4) on around 180 authorised adjuvants which, although not active pesticides themselves, may be added to pesticide products to improve their effectiveness. The *Guide* does not include products solely approved for amateur, home and garden, domestic, food storage, public or animal health uses.

Sources of Information

The information in this edition has been drawn from these authoritative sources:

- approved labels and product manuals received from suppliers of pesticides up to October 2012.
- websites of the Chemicals Regulation Directorate (CRD, www.pesticides.gov.uk) and the Health and Safety Executive (HSE, www.hse.gov.uk).

Criteria for Inclusion

To be included, a pesticide must meet the following conditions:

- it must have extant approval under UK pesticides legislation

- information on the approved uses must have been provided by the supplier
- it must be expected to be on the UK market during the currency of this edition.

When a company changes its name, whether by merger, takeover or joint venture, it is obliged to re-register its products in the name of the new company, and new MAPP numbers are normally assigned. Where stocks of the previously registered product remain legally available, both old and new identities are included in the *Guide* and remain until approval of the former lapses or stocks are notified as exhausted.

Products that have been withdrawn from the market and whose approval will finally lapse during 2013 are identified in each profile. After the indicated date, sale, storage or use of the product bearing that approval number becomes illegal. Where there is a direct replacement product, this is indicated.

The Voluntary Initiative

The Voluntary Initiative (VI) is a programme of measures, agreed by the crop protection industry and related organisations with Government, to minimise the environmental impacts of crop protection products. The programme has provided a framework of practices and principles to help Government achieve its objective of protection of water quality and enhancement of farmland biodiversity. Many of the environmental protection schemes launched under the VI represent current best practice.

The first five-year phase of the programme formally concluded in March 2006, but it is continuing with a rolling two-year review of proposals and targets; current proposals include a new online crop protection management plan and greater involvement in catchment-sensitive farming (see below).

A key element of the VI has been the provision of environmental information on crop protection products. Members of the Crop Protection Association (CPA) have committed to do this by producing Environmental Information Sheets (EISs) for all their marketed professional products.

EISs reinforce and supplement information on a product label by giving specific environmental impact information in a standardised format. They highlight any situations where risk management is essential to ensure environmental protection. Their purpose is to provide user-friendly information to advisers (including those in the amenity sector), farmers and growers on the environmental impact of crop protection products, to allow planning with a better understanding of the practical implications.

Links are provided to give users rapid access to EISs on the VI website: www. voluntaryinitiative.org.uk.

Catchment-sensitive Farming

Catchment-sensitive Farming (CSF) is the government's response to climate change and the new European Water Directive. Rainfall events are likely to become heavier, with a greater risk of soil, nutrients and pesticides ending up in the waterways. The CSF programme is investigating the effects of specific targeted advice on 20 catchment areas perceived as being at risk, and new entries in this Guide can help identify the pesticides that pose the greatest risk of water contamination (see Environmental Safety).

Notes on Contents and Layout

The book consists of six main sections:

1 Crop/Pest Guide
2 Pesticide Profiles
3 Products Also Registered
4 Adjuvants
5 Useful Information
6 Appendices

1 Crop/Pest Guide

This section enables the user to identify which active ingredients are approved for a particular crop/pest combination. The crops are grouped as shown in the Crop Index. For convenience, some crops and pests have been grouped into generic units (for example, 'cereals' are covered as a group, not individually). Indications from the Crop/Pest Guide must always be checked against the specific entry in the pesticide profiles section because a product may not be approved on all crops in the group. Chemicals indicated as having uses in cereals, for example, may be approved for use only on winter wheat and winter barley, not for other cereals. Because of the difference in wording of product labels, it may sometimes be necessary to refer to broader categories of organism in addition to specific organisms (e.g. annual grasses and annual weeds in addition to annual meadow-grass).

2 Pesticide Profiles

Each active ingredient and mixture of ingredients has a separate numbered profile entry. The entries are arranged in alphabetical order using the common names approved by the British Standards Institution and used in *The Pesticide Manual* (15th Edition; BCPC, 2009). Where an active ingredient is available only in mixtures, this is stated. The ingredients of the mixtures are themselves ordered alphabetically, and the entries appear at the correct point for the first named active ingredient.

Below the profile title, a brief phrase describes the main use of the active ingredient(s). This is followed where appropriate by the mode of action code(s) as published by the Fungicide, Herbicide and Insecticide Resistance Action Committees.

Within each profile entry, a table lists the products approved and available on the market in the following style:

Product name	Main supplier	Active ingredient contents	Formulation type	Registration number
1 Broadsword	United Phosphorus	200:85:65 g/l	EC	09140
2 Greengard	SumiAgro Amenity	200:85:65 g/l	EC	11715
3 Nu-Shot	Nufarm UK	200:85:65 g/l	EC	11148

Many of the **product names** are registered trademarks, but no special indication of this is given. Individual products are indexed under their entry number in the Index of Proprietary Names at the back of the book. The **main supplier** indicates the marketing outlet through which the product may be purchased. Full addresses and contact details of suppliers are listed in Appendix 1. For mixtures, the **active ingredient contents** are given in the same order as they appear in the profile heading. The **formulation types** are listed in full in the Key to Abbreviations and Acronyms (Appendix 5). The **registration number** normally refers to the registration with the Chemicals Regulation Directorate (CRD). In cases where the products registered with the Health and Safety Executive are included, HSE numbers are quoted, e.g. H0462 and approved biocides are numbered UK12-XXXX.

Below the product table, a **Uses** section lists all approved uses notified by suppliers to the Editor by October 2012, giving both the principal target organisms and the recommended crops or situations (identified in **bold italics**). Where there is an important condition of use, or where the approval is off-label, this is shown in parentheses as (*off-label*). Numbers in square brackets refer to the numbered products in the table above. Thus, in the example shown above, a use approved for Nu-Shot (product 3) but not for Broadsword or Greengard (products 1 and 2) appears as:

Annual dicotyledons in **rotational grassland** [3]

Any **Extensions of Authorisation for Minor Uses (EAMUs)**, previously known as SOLAs, for products in the profile are detailed below the list of approved uses. Each EAMU has a separate entry and shows the crops to which it applies, the Notice of Approval number, the expiry date (if due to fall during the current edition), and the product reference number in square brackets. EAMUs do not appear on the product label and are undertaken entirely at the risk of the user.

Below the EAMU paragraph, **Notes** are listed under the following headings. Unless otherwise stated, any reference to dose is made in terms of product rather than active ingredient. Where a note refers to a particular product rather than to the entry generally, this is indicated by numbers in square brackets as described above.

Approval information	Information of a general nature about the approval status of the active ingredient or products in the profile is given here. Notes on approval for aerial or ULV application are given.
	Inclusion of the active ingredient in Annex I under EC Directive 1107/2009 is shown, as well as any acceptance from the British Beer and Pub Association (BBPA) for its use on barley or hops for brewing or malting.
	Where a product approval will finally expire in 2013, the expiry date is shown here.
Efficacy	This entry identifies factors important in making the most effective use of the treatment. Certain factors, such as the need to apply chemicals uniformly, the need to use appropriate volume rates, and the need to prevent settling out of the active ingredient in the spray tank, are assumed to be important for all treatments and are not emphasised in individual profiles.
Restrictions	Notes are included in this section where products are subject to the Poisons Law, and where the label warns about organophosphorus and/or anticholinesterase compounds. Factors important in optimising performance and minimising the risk of crop damage, including statutory conditions relating to the maximum permitted number of applications, are listed. Any restrictions on crop varieties that may not be sprayed are mentioned here.
Environmental safety	Where the label specifies an environmental hazard classification, it is noted together with the associated risk phrases. Any other special operator precautions and any conditions concerning withholding livestock from treated areas are specified. Where any of the products in the profile are subject to Category A, Category B or broadcast air-assisted buffer zone restrictions under the LERAP scheme, the relevant classification is shown.
	Other environmental hazards are also noted here, including potential dangers to livestock, game, wildlife, bees and fish. The need to avoid drift onto neighbouring crops and the need to wash out equipment thoroughly after use are important with all pesticide treatments, but may receive special mention here if of particular significance.
Crop-specific information	Instructions about the timing of application or cultivations that are specific to a particular crop, rather than generally applicable, are mentioned here. The latest permitted use and harvest intervals (if any) are shown for individual crops.
Following crops guidance	Any specific label instructions about what may be grown after a treated crop, whether harvested normally or after failure, are shown here.

Hazard classification and safety precautions	The label hazard classifications and precautions are shown using a series of letter and number codes which are explained in Appendix 4. Hazard warnings are now given in full, with the other precautions listed under the following subheadings:

- Risk phrases
- Operator protection
- Environmental protection
- Consumer protection
- Storage and disposal
- Treated seed
- Vertebrate/rodent control products
- Medical advice

This section is given for information only and should not be used for the purpose of making a COSHH assessment without reference to the actual label of the product to be used.

3 Products also Registered

Products with an extant approval for all or part of the year of the edition in which they appear are listed in this section if they have not been requested by their supplier or manufacturer for inclusion in Section 2. Details shown are the same (apart from formulation) as those in the product tables of Section 2 and, where relevant, an approval expiry date is shown. Not all products listed here will be available in the market, but this list, together with the products in Section 2, comprises a comprehensive listing of all approved products in the UK for uses within the scope of the *Guide* for the year in question.

4 Adjuvants

Adjuvants are listed in a table ordered alphabetically by product name. For each product, details are shown of the main supplier, the authorisation number and the mode of action (e.g. wetter, sticker, etc.) as shown on the label. A brief statement of the uses of the adjuvant is given. Protective clothing requirements and label precautions are listed using the codes from Appendix 4.

5 Useful Information

This section summarises legislation covering approval, storage, sale and use of pesticides in the UK. There are brief articles on the broader issues concerning the use of crop protection chemicals, including resistance management. Lists are provided of products approved for use in or near water, in forestry, as seed treatments and for aerial application. Chemicals subject to the Poisons Laws are listed, and there is a summary of first aid measures if pesticide poisoning should be suspected. Finally, this section provides guidance on environmental protection issues and covers the protection of bees and the use of pesticides in or near water.

6 Appendices

Appendix 1	Gives names, addresses, telephone and fax numbers of all companies listed as main suppliers in the pesticide profiles. E-mail and website addresses are also listed where available.
	Where a supplier is no longer trading under that name (usually following a merger or takeover), but products under that name are still listed in the *Guide* because they are still available in the supply chain, a cross-reference indicates the new 'parent' company from which technical or commercial information can be obtained.
Appendix 2	Gives names, addresses, telephone and fax numbers of useful contacts, including the National Poisons Information Service. Website addresses are included where available.

Appendix 3   Gives details of the keys to crop and weed growth stages, including the publication reference for each. The numeric codes are used in the descriptive sections of the pesticide profiles (Section 2).

Appendix 4   Shows the full text for code letters and numbers used in the pesticide profiles (Section 2) to indicate personal protective equipment requirements and label precautions.

Appendix 5   Shows the full text for the formulation abbreviations used in the pesticide profiles (Section 2). Other abbreviations and acronyms used in the *Guide* are also explained here.

Appendix 6   Provides full definitions of officially agreed descriptive phrases for crops or situations used in the pesticide profiles (Section 2) where misunderstandings can occur.

Appendix 7   Shows a list of useful reference publications, which amplify the summarised information provided in this section.

SECTION 1
CROP/PEST GUIDE

Crop/Pest Guide Index

Important note: The Crop/Pest Guide Index refers to pages on which the subject can be located.

Crop/Pest Guide

Important note: For convenience, some crops and pests or targets have been brought together into genetic groups in this guide, e.g. 'cereals', 'annual grasses'. It is essential to check the profile entry in Section 2 *and* the label to ensure that a product is approved for a specific crop/pest combination, e.g. winter wheat/blackgrass.

Arable and vegetable crops

Agricultural herbage - Grass

Crop control	Desiccation	glyphosate
Diseases	Crown rust	propiconazole
	Drechslera leaf spot	propiconazole
	Powdery mildew	propiconazole
	Rhynchosporium	propiconazole
Pests	Aphids	esfenvalerate
	Birds/mammals	aluminium ammonium sulphate
	Flies	chlorpyrifos, cypermethrin, esfenvalerate
	Leatherjackets	chlorpyrifos, methiocarb (*seed admixture*)
	Slugs/snails	methiocarb (*seed admixture*)
Plant growth regulation	Growth control	gibberellins
	Quality/yield control	sulphur
Weeds	Broad-leaved weeds	2,4-D, 2,4-D + dicamba, 2,4-D + dicamba + dichlorprop-P, 2,4-D + dicamba + triclopyr, 2,4-D + MCPA, 2,4-DB, 2,4-DB + MCPA, amidosulfuron, aminopyralid + fluroxypyr, aminopyralid + triclopyr, citronella oil, clopyralid, clopyralid + fluroxypyr + triclopyr, clopyralid + triclopyr, dicamba + MCPA + mecoprop-P, dicamba + mecoprop-P, fluroxypyr, fluroxypyr + triclopyr, glyphosate, glyphosate (*wiper application*), MCPA, mecoprop-P, metsulfuron-methyl, thifensulfuron-methyl
	Crops as weeds	fluroxypyr
	Grass weeds	glyphosate
	Weeds, miscellaneous	aminopyralid + triclopyr, clopyralid + triclopyr, clopyralid + triclopyr (*off-label*), fluroxypyr, glufosinate-ammonium, glyphosate, MCPA
	Woody weeds/scrub	clopyralid + triclopyr

Agricultural herbage - Herbage legumes

Crop control	Desiccation	diquat
Weeds	Broad-leaved weeds	2,4-DB, carbetamide, diquat (*off-label*), isoxaben (*off-label - grown for game cover*), propyzamide
	Crops as weeds	carbetamide
	Grass weeds	carbetamide, diquat (*seed crop*), fluazifop-P-butyl (*off-label*), propyzamide, tri-allate
	Weeds, miscellaneous	diquat (*seed crop*)

Agricultural herbage - Herbage seed

Crop control	Desiccation	diquat
Diseases	Crown rust	propiconazole
	Damping off	thiram (*seed treatment*)
	Disease control	azoxystrobin (*off-label*), difenoconazole (*off-label*), epoxiconazole (*off-label*), maneb (*off-label*), prochloraz + propiconazole (*off-label*), spiroxamine + tebuconazole (*off-label*), tebuconazole (*off-label*)
	Drechslera leaf spot	propiconazole
	Foliar diseases	azoxystrobin (*off-label*), azoxystrobin + cyproconazole (*off-label*), azoxystrobin + fenpropimorph (*off-label*), chlorothalonil + flutriafol (*off-label*), cyproconazole (*off-label*), cyproconazole + quinoxyfen (*off-label*), difenoconazole (*off-label*), dimoxystrobin + epoxiconazole (*off-label*), epoxiconazole (*off-label*), epoxiconazole + fenpropimorph (*off-label*), epoxiconazole + fenpropimorph + kresoxim-methyl (*off-label*), epoxiconazole + kresoxim-methyl (*off-label*), epoxiconazole + kresoxim-methyl + pyraclostrobin (*off-label*), epoxiconazole + pyraclostrobin (*off-label*), famoxadone + flusilazole (*off-label*), fenpropimorph (*off-label*), fenpropimorph + flusilazole (*off-label*), fluquinconazole (*off-label*), fluquinconazole + prochloraz (*off-label*), prothioconazole + tebuconazole (*off-label*), pyraclostrobin (*off-label*), tebuconazole (*off-label*), trifloxystrobin (*off-label*)
	Powdery mildew	cyflufenamid (*off-label*), propiconazole, sulphur (*off-label*)
	Rhynchosporium	propiconazole
	Rust	tebuconazole (*off-label*)
Pests	Aphids	deltamethrin (*off-label*), esfenvalerate (*off-label*), lambda-cyhalothrin (*off-label*), tau-fluvalinate (*off-label*), zeta-cypermethrin (*off-label*)
	Beetles	deltamethrin (*off-label*)
	Caterpillars	deltamethrin (*off-label*)
	Flies	esfenvalerate
	Leafhoppers	deltamethrin (*off-label*)
	Pests, miscellaneous	alpha-cypermethrin (*off-label*), lambda-cyhalothrin (*off-label*), tau-fluvalinate (*off-label*), zeta-cypermethrin (*off-label*)
	Thrips	deltamethrin (*off-label*)
Plant growth regulation	Growth control	trinexapac-ethyl
Weeds	Broad-leaved weeds	amidosulfuron + iodosulfuron-methyl-sodium (*off-label*), bifenox (*off-label*), bromoxynil (*off-label*), carfentrazone-ethyl (*off-label*), carfentrazone-ethyl + flupyrsulfuron-methyl (*off-label*), dicamba + MCPA + mecoprop-P, diflufenican (*off-label*), diflufenican + flufenacet (*off-label*), diflufenican + flurtamone (*off-label*), ethofumesate, florasulam (*off-label*), florasulam + fluroxypyr (*off-label*), iodosulfuron-methyl-sodium (*off-label*), MCPA, mecoprop-P, pendimethalin (*off-label*), propyzamide, sulfosulfuron (*off-label*)

	Crops as weeds	clopyralid (*off-label*), iodosulfuron-methyl-sodium (*off-label - from seed*)
	Grass weeds	clodinafop-propargyl (*off-label*), diflufenican (*off-label*), diflufenican + flufenacet (*off-label*), diflufenican + flurtamone (*off-label*), ethofumesate, fenoxaprop-P-ethyl (*off-label*), pendimethalin (*off-label*), propyzamide, sulfosulfuron (*off-label*)
	Weeds, miscellaneous	MCPA

Brassicas - Brassica seed crops

Diseases	Alternaria	iprodione, iprodione (*pre-storage only*)
	Botrytis	iprodione
	Damping off	thiram (*off-label*)
Weeds	Broad-leaved weeds	carbetamide, propyzamide
	Crops as weeds	carbetamide
	Grass weeds	carbetamide, propyzamide

Brassicas - Brassicas, general

Diseases	Botrytis	Bacillus subtilis (*off-label*), Gliocladium catenulatum (*off-label - moderate control*)
	Disease control	mandipropamid
	Downy mildew	copper oxychloride (*off-label*), dimethomorph (*off-label*), mandipropamid
	Fusarium	Gliocladium catenulatum (*off-label - moderate control*)
	Phytophthora	Gliocladium catenulatum (*off-label - moderate control*)
	Pythium	Gliocladium catenulatum (*off-label - moderate control*)
	Rhizoctonia	Gliocladium catenulatum (*off-label - moderate control*)
Pests	Aphids	deltamethrin (*off-label*), dimethoate (*off-label*), pymetrozine (*off-label*), spirotetramat (*off-label*)
	Flies	Metarhizium anisopliae (*off-label*)
	Pests, miscellaneous	deltamethrin (*off-label*), dimethoate (*off-label*)
	Whiteflies	spirotetramat (*off-label*)
Plant growth regulation	Quality/yield control	sulphur
Weeds	Broad-leaved weeds	pyridate (*off-label*)
	Grass weeds	fluazifop-P-butyl (*off-label*)

Brassicas - Fodder brassicas

Diseases	Alternaria	azoxystrobin, fludioxonil + metalaxyl-M + thiamethoxam (*moderate control*), iprodione (*off-label*)
	Bacterial blight	copper oxychloride (*off-label*)
	Black rot	copper oxychloride (*off-label*)
	Bottom rot	copper oxychloride (*off-label*)
	Damping off	Bacillus subtilis (*off-label*), thiram (*off-label*)
	Disease control	difenoconazole (*off-label*)
	Downy mildew	fludioxonil + metalaxyl-M + thiamethoxam, metalaxyl-M (*off-label*)

	Helminthosporium seedling rot	Bacillus subtilis (*off-label*)
	Phytophthora	Bacillus subtilis (*off-label*), copper oxychloride (*off-label*)
	Powdery mildew	azoxystrobin + difenoconazole
	Pythium	Bacillus subtilis (*off-label*), fludioxonil + metalaxyl-M + thiamethoxam (*reduction*)
	Rhizoctonia	Bacillus subtilis (*off-label*)
	Ring spot	azoxystrobin, difenoconazole (*off-label*)
	Seed-borne diseases	fludioxonil + metalaxyl-M + thiamethoxam
	Spear rot	copper oxychloride (*off-label*)
	Stem canker	copper oxychloride (*off-label*)
	White blister	azoxystrobin, azoxystrobin + difenoconazole
	White rot	Bacillus subtilis (*off-label*)
Pests	Aphids	cypermethrin, dimethoate (*off-label*), esfenvalerate, fatty acids (*off-label*), fludioxonil + metalaxyl-M + thiamethoxam (*early season only*), pirimicarb, pymetrozine (*off-label*)
	Beetles	alpha-cypermethrin, deltamethrin, fludioxonil + metalaxyl-M + thiamethoxam (*reduction of damage*)
	Caterpillars	alpha-cypermethrin, Bacillus thuringiensis (*off-label*), cypermethrin, deltamethrin, esfenvalerate
	Flies	chlorpyrifos (*off-label*)
	Pests, miscellaneous	dimethoate (*off-label*)
	Thrips	fatty acids (*off-label*)
Weeds	Broad-leaved weeds	clomazone (*off-label*), metazachlor (*off-label*), napropamide, pendimethalin (*off-label*), pyridate (*off-label*)
	Crops as weeds	fluazifop-P-butyl (*stockfeed only*)
	Grass weeds	fluazifop-P-butyl (*stockfeed only*), metazachlor (*off-label*), napropamide, pendimethalin (*off-label*), S-metolachlor (*off-label*)

Brassicas - Leaf and flowerhead brassicas

Diseases	Alternaria	azoxystrobin, boscalid + pyraclostrobin, chlorothalonil + metalaxyl-M (*moderate control*), difenoconazole, iprodione, iprodione (*off-label*), prothioconazole, tebuconazole, tebuconazole + trifloxystrobin
	Bacterial blight	copper oxychloride (*off-label*)
	Black rot	copper oxychloride (*off-label*)
	Botrytis	iprodione, iprodione (*off-label*)
	Bottom rot	copper oxychloride (*off-label*)
	Damping off	Bacillus subtilis (*off-label*), thiram (*seed treatment*), tolclofos-methyl
	Disease control	fluopicolide + propamocarb hydrochloride (*off-label*)
	Downy mildew	chlorothalonil + metalaxyl-M, metalaxyl-M (*off-label*)
	Foliar diseases	flusilazole (*off-label*)
	Helminthosporium seedling rot	Bacillus subtilis (*off-label*)
	Light leaf spot	prothioconazole, tebuconazole, tebuconazole + trifloxystrobin

	Phoma leaf spot	tebuconazole + trifloxystrobin
	Phytophthora	Bacillus subtilis (*off-label*), copper oxychloride (*off-label*)
	Powdery mildew	azoxystrobin + difenoconazole, prothioconazole, tebuconazole, tebuconazole + trifloxystrobin
	Pythium	Bacillus subtilis (*off-label*), metalaxyl-M
	Rhizoctonia	Bacillus subtilis (*off-label*), tolclofos-methyl (*off-label*)
	Ring spot	azoxystrobin, boscalid + pyraclostrobin, chlorothalonil + metalaxyl-M (*reduction*), difenoconazole, difenoconazole (*off-label*), prothioconazole, tebuconazole, tebuconazole (*off-label*), tebuconazole + trifloxystrobin
	Spear rot	copper oxychloride (*off-label*)
	Stem canker	copper oxychloride (*off-label*), flusilazole (*off-label*), prothioconazole
	Storage rots	1-methylcyclopropene (*off-label*), metalaxyl-M (*off-label*)
	White blister	azoxystrobin, azoxystrobin + difenoconazole, boscalid + pyraclostrobin, boscalid + pyraclostrobin (*qualified minor use*), chlorothalonil + metalaxyl-M, mancozeb + metalaxyl-M (*off-label*), tebuconazole + trifloxystrobin
	White rot	Bacillus subtilis (*off-label*)
Pests	Aphids	acetamiprid (*off-label*), beta-cyfluthrin, chlorpyrifos, cypermethrin, deltamethrin (*off-label*), dimethoate, dimethoate (*off-label*), esfenvalerate, fatty acids, fatty acids (*off-label*), pirimicarb, pymetrozine, pymetrozine (*off-label*), pymetrozine (*useful levels of control*), pyrethrins, spirotetramat, thiacloprid
	Beetles	alpha-cypermethrin, deltamethrin, pyrethrins
	Birds/mammals	aluminium ammonium sulphate
	Caterpillars	alpha-cypermethrin, Bacillus thuringiensis, Bacillus thuringiensis (*off-label*), beta-cyfluthrin, chlorpyrifos, cypermethrin, deltamethrin, diflubenzuron, esfenvalerate, indoxacarb, indoxacarb (*off-label*), lambda-cyhalothrin, pyrethrins, spinosad
	Cutworms	chlorpyrifos
	Flies	chlorpyrifos, chlorpyrifos (*off-label*)
	Leatherjackets	chlorpyrifos
	Mealybugs	fatty acids
	Pests, miscellaneous	alpha-cypermethrin (*off-label*), deltamethrin (*off-label*), dimethoate, dimethoate (*off-label*)
	Scale insects	fatty acids
	Slugs/snails	metaldehyde, methiocarb
	Spider mites	fatty acids
	Thrips	fatty acids (*off-label*)
	Whiteflies	acetamiprid (*off-label*), chlorpyrifos, cypermethrin, fatty acids, lambda-cyhalothrin, pyrethrins, spirotetramat
Weeds	Broad-leaved weeds	carbetamide, clomazone (*off-label*), clopyralid, dimethenamid-p + pendimethalin (*off-label*), metazachlor, metazachlor (*off-label*), napropamide, napropamide (*off-label*), pendimethalin,

		pendimethalin (*off-label*), propyzamide (*off-label*), pyridate, pyridate (*off-label*)
	Crops as weeds	carbetamide, cycloxydim, pendimethalin, tepraloxydim
	Grass weeds	carbetamide, cycloxydim, dimethenamid-p + pendimethalin (*off-label*), fluazifop-P-butyl (*off-label*), metazachlor, metazachlor (*off-label*), napropamide, napropamide (*off-label*), pendimethalin, pendimethalin (*off-label*), propyzamide (*off-label*), S-metolachlor (*off-label*), tepraloxydim, tepraloxydim (*off-label*)

Brassicas - Mustard

Crop control	Desiccation	diquat (*off-label*), glyphosate
Diseases	Alternaria	fludioxonil + metalaxyl-M + thiamethoxam (*moderate control*)
	Disease control	boscalid (*off-label*), difenoconazole (*off-label*), mandipropamid, prothioconazole (*off-label*), tebuconazole (*off-label*)
	Downy mildew	fludioxonil + metalaxyl-M + thiamethoxam, mandipropamid
	Foliar diseases	cyproconazole (*off-label*), difenoconazole (*off-label*), tebuconazole (*off-label*)
	Pythium	fludioxonil + metalaxyl-M + thiamethoxam (*reduction*)
	Rust	tebuconazole (*off-label*)
	Sclerotinia	azoxystrobin (*off-label*)
	Seed-borne diseases	fludioxonil + metalaxyl-M + thiamethoxam
Pests	Aphids	fludioxonil + metalaxyl-M + thiamethoxam (*early season only*), lambda-cyhalothrin (*off-label*), pirimicarb (*off-label*), spirotetramat (*off-label*), tau-fluvalinate (*off-label*)
	Beetles	beta-cyfluthrin + imidacloprid (*off-label*), beta-cyfluthrin + imidacloprid (*off-label - seed treatment*), deltamethrin, fludioxonil + metalaxyl-M + thiamethoxam (*reduction of damage*), thiacloprid
	Caterpillars	deltamethrin
	Flies	tefluthrin (*off-label - seed treatment*)
	Pests, miscellaneous	lambda-cyhalothrin (*off-label*), tau-fluvalinate (*off-label*)
	Weevils	deltamethrin
	Whiteflies	lambda-cyhalothrin (*off-label*), spirotetramat (*off-label*)
Weeds	Broad-leaved weeds	carbetamide (*off-label*), clopyralid (*off-label*), clopyralid + picloram (*off-label*), glyphosate, metazachlor (*off-label*), metazachlor + quinmerac (*off-label*), napropamide (*off-label*)
	Crops as weeds	glyphosate, propaquizafop (*off-label*)
	Grass weeds	carbetamide (*off-label*), cycloxydim (*off-label*), fluazifop-P-butyl (*off-label*), glyphosate, metazachlor (*off-label*), metazachlor + quinmerac (*off-label*), napropamide (*off-label*), propaquizafop (*off-label*), tepraloxydim (*off-label*)
	Weeds, miscellaneous	glyphosate

Brassicas - Root brassicas

Diseases	Alternaria	azoxystrobin (*off-label*), azoxystrobin + difenoconazole (*off-label*), fenpropimorph (*off-label*), iprodione + thiophanate-methyl (*off-label*), prothioconazole, tebuconazole, tebuconazole (*off-label*)
	Crown rot	fenpropimorph (*off-label*), iprodione + thiophanate-methyl (*off-label*)
	Damping off	thiram (*off-label*), thiram (*off-label - seed treatment*), thiram (*seed treatment*), tolclofos-methyl (*off-label*)
	Disease control	boscalid + pyraclostrobin (*off-label*), tebuconazole (*off-label*)
	Downy mildew	azoxystrobin (*off-label*), metalaxyl-M (*off-label*)
	Foliar diseases	azoxystrobin (*off-label*), flusilazole (*off-label*), tebuconazole (*off-label*), tebuconazole + trifloxystrobin (*off-label*)
	Fungus diseases	tebuconazole (*off-label*)
	Powdery mildew	azoxystrobin (*off-label*), azoxystrobin + difenoconazole (*off-label*), fenpropimorph (*off-label*), prothioconazole, sulphur, sulphur (*off-label*), tebuconazole, tebuconazole (*off-label*)
	Pythium	metalaxyl-M, metalaxyl-M (*off-label*)
	Rhizoctonia	azoxystrobin (*off-label*), tolclofos-methyl (*off-label*)
	Ring spot	tebuconazole (*off-label*)
	Rust	boscalid + pyraclostrobin (*off-label*)
	Sclerotinia	boscalid + pyraclostrobin (*off-label*), cyprodinil + fludioxonil (*off-label*), tebuconazole, tebuconazole (*off-label*)
	Seed-borne diseases	thiram (*off-label - seed treatment*)
	Stem canker	flusilazole (*off-label*)
	White blister	metalaxyl-M (*off-label*)
Pests	Aphids	deltamethrin (*off-label*), dimethoate (*off-label*), esfenvalerate, fatty acids (*off-label*), pirimicarb, pirimicarb (*off-label*), thiacloprid (*off-label*)
	Beetles	deltamethrin, deltamethrin (*off-label*), thiamethoxam (*off-label*)
	Caterpillars	deltamethrin, deltamethrin (*off-label*), esfenvalerate, lambda-cyhalothrin (*off-label*)
	Cutworms	Bacillus thuringiensis (*off-label*), cypermethrin (*off-label*), lambda-cyhalothrin (*off-label*)
	Flies	chlorpyrifos (*off-label*), lambda-cyhalothrin (*off-label*)
	Leafhoppers	deltamethrin (*off-label*)
	Pests, miscellaneous	deltamethrin (*off-label*), dimethoate (*off-label*), lambda-cyhalothrin (*off-label*)
	Thrips	deltamethrin (*off-label*), fatty acids (*off-label*)
	Weevils	lambda-cyhalothrin (*off-label*)
Plant growth regulation	Quality/yield control	sulphur
Weeds	Broad-leaved weeds	clomazone (*off-label*), clopyralid, dimethenamid-p + metazachlor (*off-label*), glyphosate, metamitron (*off-label*), metazachlor, metazachlor (*off-label*), pendimethalin (*off-label*), prosulfocarb (*off-label*), S-metolachlor (*off-label*)

Crops as weeds	cycloxydim, fluazifop-P-butyl (*stockfeed only*), glyphosate, propaquizafop (*off-label*)
Grass weeds	cycloxydim, cycloxydim (*off-label*), fluazifop-P-butyl (*off-label*), fluazifop-P-butyl (*stockfeed only*), glyphosate, metamitron (*off-label*), metazachlor, metazachlor (*off-label*), propaquizafop, propaquizafop (*off-label*), prosulfocarb (*off-label*), S-metolachlor (*off-label*), tepraloxydim (*off-label*)
Weeds, miscellaneous	glyphosate, metamitron (*off-label*)

Brassicas - Salad greens

Diseases		
	Alternaria	azoxystrobin (*off-label - for baby leaf production*), iprodione (*off-label*)
	Aphanomyces	fatty acids (*off-label*)
	Bacterial blight	copper oxychloride (*off-label*)
	Black rot	copper oxychloride (*off-label*)
	Botrytis	cyprodinil + fludioxonil (*off-label*), fenhexamid (*off-label - baby leaf production*)
	Bottom rot	copper oxychloride (*off-label*)
	Damping off	Bacillus subtilis (*off-label*), tolclofos-methyl
	Disease control	dimethomorph + mancozeb (*off-label*), mandipropamid, prochloraz (*off-label*)
	Downy mildew	azoxystrobin (*off-label*), azoxystrobin (*off-label - for baby leaf production*), copper oxychloride (*off-label*), dimethomorph + mancozeb (*off-label*), mandipropamid
	Helminthosporium seedling rot	Bacillus subtilis (*off-label*)
	Phytophthora	Bacillus subtilis (*off-label*), copper oxychloride (*off-label*)
	Pythium	Bacillus subtilis (*off-label*), metalaxyl-M
	Rhizoctonia	Bacillus subtilis (*off-label*), tolclofos-methyl (*off-label*)
	Ring spot	difenoconazole (*off-label*), tebuconazole (*off-label*)
	Spear rot	copper oxychloride (*off-label*)
	Stem canker	copper oxychloride (*off-label*)
	White blister	boscalid + pyraclostrobin (*off-label*)
	White rot	Bacillus subtilis (*off-label*)
Pests	Aphids	acetamiprid (*off-label*), chlorpyrifos, cypermethrin (*off-label*), deltamethrin (*off-label*), dimethoate (*off-label*), esfenvalerate, fatty acids (*off-label*), lambda-cyhalothrin (*off-label*), pirimicarb, pirimicarb (*for baby leaf production*), pirimicarb (*off-label*), pymetrozine (*off-label*), pymetrozine (*off-label - for baby leaf production*), spirotetramat (*off-label*), thiacloprid (*off-label*), thiacloprid (*off-label - baby leaf production*)
	Beetles	deltamethrin (*off-label*), deltamethrin (*off-label - baby leaf production*)
	Birds/mammals	aluminium ammonium sulphate
	Caterpillars	alpha-cypermethrin (*off-label - for baby leaf production*), Bacillus thuringiensis (*off-label*), chlorpyrifos, cypermethrin (*off-label*), deltamethrin (*off-label*), deltamethrin (*off-label - baby leaf*

		production), diflubenzuron (*off-label - for baby leaf production*), esfenvalerate
	Cutworms	chlorpyrifos
	Flies	chlorpyrifos, chlorpyrifos (*off-label*), Metarhizium anisopliae (*off-label*), tefluthrin (*off-label - seed treatment*)
	Leafhoppers	deltamethrin (*off-label*)
	Leatherjackets	chlorpyrifos
	Midges	Metarhizium anisopliae (*off-label*)
	Pests, miscellaneous	alpha-cypermethrin (*off-label*), deltamethrin (*off-label*), dimethoate (*off-label*), lambda-cyhalothrin (*off-label*)
	Thrips	deltamethrin (*off-label*), fatty acids (*off-label*), Metarhizium anisopliae (*off-label*)
	Whiteflies	chlorpyrifos, spirotetramat (*off-label*)
Weeds	Broad-leaved weeds	clomazone (*off-label*), metazachlor (*off-label*), napropamide (*off-label*), pendimethalin (*off-label*), propyzamide (*off-label*)
	Grass weeds	fluazifop-P-butyl (*off-label*), metazachlor (*off-label*), napropamide (*off-label*), pendimethalin (*off-label*), propyzamide (*off-label*), S-metolachlor (*off-label*)

Cereals - Barley

Crop control	Desiccation	diquat, diquat (*off-label*), diquat (*stockfeed only*), glyphosate
Diseases	Covered smut	carboxin + thiram (*seed treatment*), clothianidin + prothioconazole (*seed treatment*), fludioxonil (*seed treatment*), fludioxonil + flutriafol (*seed treatment*), fludioxonil + tefluthrin, fluopyram + prothioconazole + tebuconazole, fluquinconazole + prochloraz (*seed treatment*), fuberidazole + imidacloprid + triadimenol (*seed treatment*), fuberidazole + triadimenol (*seed treatment*), prochloraz + triticonazole (*seed treatment*), prothioconazole (*seed treatment*)
	Ear blight	prothioconazole + tebuconazole
	Eyespot	azoxystrobin + cyproconazole (*reduction*), bixafen + prothioconazole (*reduction of incidence and severity*), boscalid + epoxiconazole (*moderate*), boscalid + epoxiconazole (*moderate control*), boscalid + epoxiconazole + pyraclostrobin (*moderate control*), carbendazim, carbendazim + flusilazole, chlorothalonil + cyproconazole, cyproconazole (*useful reduction*), cyprodinil, cyprodinil + picoxystrobin, epoxiconazole (*reduction*), epoxiconazole + fenpropimorph (*reduction*), epoxiconazole + fenpropimorph + kresoxim-methyl (*reduction*), epoxiconazole + kresoxim-methyl (*reduction*), epoxiconazole + metconazole (*reduction*), epoxiconazole + prochloraz, fluoxastrobin + prothioconazole (*reduction*), fluoxastrobin + prothioconazole + trifloxystrobin (*reduction*), flusilazole, prochloraz, prochloraz + propiconazole, prochloraz + proquinazid + tebuconazole, prochloraz + tebuconazole, prothioconazole, prothioconazole + spiroxamine, prothioconazole + spiroxamine + tebuconazole (*reduction*), prothioconazole + tebuconazole,

prothioconazole + tebuconazole (*reduction*), prothioconazole + trifloxystrobin (*reduction*), prothioconazole + trifloxystrobin (*reduction in severity*)

Foot rot

carboxin + thiram (*seed treatment*), clothianidin + prothioconazole (*seed treatment*), fludioxonil (*seed treatment*), fludioxonil + flutriafol (*seed treatment*), fludioxonil + tefluthrin, fluquinconazole + prochloraz (*seed treatment*), fuberidazole + imidacloprid + triadimenol (*seed treatment*), fuberidazole + imidacloprid + triadimenol (*seed treatment - reduction*), fuberidazole + triadimenol (*seed treatment*), imazalil + ipconazole (*useful protection*), prochloraz + triticonazole (*seed treatment*), prothioconazole (*seed treatment*)

Fusarium

fluopyram + prothioconazole + tebuconazole

Late ear diseases

bixafen + prothioconazole, fluoxastrobin + prothioconazole, fluoxastrobin + prothioconazole + trifloxystrobin, prothioconazole, prothioconazole + spiroxamine + tebuconazole, prothioconazole + tebuconazole

Leaf stripe

carboxin + thiram (*seed treatment*), carboxin + thiram (*seed treatment - reduction*), clothianidin + prothioconazole (*seed treatment*), fludioxonil (*seed treatment - reduction*), fludioxonil + flutriafol (*seed treatment*), fludioxonil + tefluthrin (*partial control*), fluopyram + prothioconazole + tebuconazole, fuberidazole + triadimenol (*seed treatment*), imazalil, imazalil + ipconazole, prochloraz + triticonazole (*seed treatment*), prothioconazole (*seed treatment*), tebuconazole

Loose smut

carboxin + thiram (*seed treatment - reduction*), clothianidin + prothioconazole (*seed treatment*), fludioxonil + flutriafol (*seed treatment*), fluopyram + prothioconazole + tebuconazole, fluquinconazole + prochloraz (*seed treatment*), fuberidazole + imidacloprid + triadimenol (*seed treatment*), fuberidazole + triadimenol (*seed treatment*), imazalil + ipconazole, prochloraz + triticonazole (*seed treatment*), prothioconazole (*seed treatment*), tebuconazole

Net blotch

azoxystrobin, azoxystrobin + chlorothalonil, azoxystrobin + cyproconazole, azoxystrobin + fenpropimorph, bixafen + prothioconazole, boscalid + epoxiconazole, boscalid + epoxiconazole + pyraclostrobin, carboxin + thiram (*seed treatment*), chlorothalonil + cyproconazole, chlorothalonil + cyproconazole + propiconazole (*reduction*), chlorothalonil + picoxystrobin, cyproconazole (*useful reduction*), cyproconazole + picoxystrobin (*moderate control*), cyproconazole + propiconazole, cyprodinil, cyprodinil + isopyrazam, cyprodinil + picoxystrobin, epoxiconazole, epoxiconazole (*moderate control*), epoxiconazole + fenpropimorph, epoxiconazole + fenpropimorph + kresoxim-methyl, epoxiconazole + fenpropimorph + metrafenone, epoxiconazole + fenpropimorph + pyraclostrobin, epoxiconazole + fluxapyroxad , epoxiconazole + fluxapyroxad + pyraclostrobin, epoxiconazole + isopyrazam,

epoxiconazole + kresoxim-methyl, epoxiconazole +
kresoxim-methyl + pyraclostrobin, epoxiconazole +
metconazole, epoxiconazole + metrafenone,
epoxiconazole + prochloraz, epoxiconazole +
pyraclostrobin, famoxadone + flusilazole,
fenpropimorph + flusilazole, fenpropimorph +
pyraclostrobin, fluopyram + prothioconazole +
tebuconazole (*seed borne*), fluoxastrobin +
prothioconazole, fluoxastrobin + prothioconazole +
trifloxystrobin, flusilazole, fuberidazole + imidacloprid
+ triadimenol (*seed treatment - seed-borne only*),
imazalil, maneb, metconazole (*reduction*),
picoxystrobin, prochloraz, prochloraz + proquinazid +
tebuconazole (*moderate control*), prochloraz +
tebuconazole, prothioconazole, prothioconazole +
spiroxamine, prothioconazole + spiroxamine +
tebuconazole, prothioconazole + tebuconazole,
prothioconazole + trifloxystrobin, pyraclostrobin,
spiroxamine + tebuconazole, tebuconazole,
tebuconazole + triadimenol, trifloxystrobin

Powdery mildew   azoxystrobin, azoxystrobin (*moderate control*),
azoxystrobin + cyproconazole, azoxystrobin +
fenpropimorph, bixafen + prothioconazole, boscalid +
epoxiconazole, boscalid + epoxiconazole (*moderate
control*), boscalid + epoxiconazole + pyraclostrobin
(*moderate control*), carbendazim + flusilazole,
chlorothalonil + cyproconazole, chlorothalonil +
cyproconazole + propiconazole, chlorothalonil +
proquinazid, cyflufenamid, cyproconazole,
cyproconazole + propiconazole, cyprodinil, cyprodinil
+ isopyrazam, epoxiconazole, epoxiconazole +
fenpropimorph, epoxiconazole + fenpropimorph +
kresoxim-methyl, epoxiconazole + fenpropimorph +
metrafenone, epoxiconazole + fenpropimorph +
pyraclostrobin, epoxiconazole + fluxapyroxad
(*moderate control*), epoxiconazole + fluxapyroxad +
pyraclostrobin (*moderate control*), epoxiconazole +
kresoxim-methyl, epoxiconazole + metconazole
(*moderate control*), epoxiconazole + metrafenone,
epoxiconazole + prochloraz, epoxiconazole +
prochloraz (*moderate control*), fenpropidin,
fenpropimorph, fenpropimorph + flusilazole,
fenpropimorph + kresoxim-methyl, fenpropimorph +
pyraclostrobin, fenpropimorph + quinoxyfen,
fluoxastrobin + prothioconazole, fluoxastrobin +
prothioconazole + trifloxystrobin, fluquinconazole +
prochloraz, flusilazole, flusilazole (*moderate control*),
flutriafol, fuberidazole + imidacloprid + triadimenol
(*seed treatment*), metconazole, metconazole
(*moderate control*), metrafenone, picoxystrobin,
prochloraz, prochloraz + proquinazid + tebuconazole,
prochloraz + tebuconazole, propiconazole,
proquinazid, prothioconazole, prothioconazole +
spiroxamine, prothioconazole + spiroxamine +
tebuconazole, prothioconazole + tebuconazole,
prothioconazole + trifloxystrobin, quinoxyfen,
spiroxamine, spiroxamine + tebuconazole, sulphur,
tebuconazole, tebuconazole + triadimenol

Ramularia leaf spots   bixafen + prothioconazole, boscalid + epoxiconazole
(*moderate control*), boscalid + epoxiconazole +
pyraclostrobin (*moderate control*), cyprodinil +

isopyrazam, epoxiconazole (*reduction*), epoxiconazole + fluxapyroxad , epoxiconazole + fluxapyroxad + pyraclostrobin, epoxiconazole + isopyrazam, epoxiconazole + metconazole, epoxiconazole + metconazole (*moderate control*)

Rhynchosporium
azoxystrobin, azoxystrobin (*reduction*), azoxystrobin + chlorothalonil, azoxystrobin + chlorothalonil (*moderate control*), azoxystrobin + cyproconazole (*moderate control*), azoxystrobin + fenpropimorph, bixafen + prothioconazole, boscalid + epoxiconazole (*moderate*), boscalid + epoxiconazole (*moderate control*), boscalid + epoxiconazole + pyraclostrobin, carbendazim, chlorothalonil (*moderate control*), chlorothalonil + cyproconazole, chlorothalonil + cyproconazole + propiconazole (*reduction*), chlorothalonil + picoxystrobin, chlorothalonil + propiconazole, chlorothalonil + propiconazole (*moderate control*), chlorothalonil + proquinazid (*suppression only*), cyproconazole, cyproconazole + picoxystrobin (*moderate control*), cyproconazole + propiconazole, cyprodinil, cyprodinil (*moderate control*), cyprodinil + isopyrazam, cyprodinil + picoxystrobin, epoxiconazole, epoxiconazole + fenpropimorph, epoxiconazole + fenpropimorph + kresoxim-methyl, epoxiconazole + fenpropimorph + metrafenone, epoxiconazole + fenpropimorph + pyraclostrobin, epoxiconazole + fluxapyroxad , epoxiconazole + fluxapyroxad + pyraclostrobin, epoxiconazole + isopyrazam, epoxiconazole + kresoxim-methyl, epoxiconazole + kresoxim-methyl + pyraclostrobin, epoxiconazole + metconazole, epoxiconazole + metrafenone, epoxiconazole + prochloraz, epoxiconazole + pyraclostrobin, famoxadone + flusilazole, fenpropidin (*moderate control*), fenpropimorph, fenpropimorph + flusilazole, fenpropimorph + kresoxim-methyl, fenpropimorph + pyraclostrobin, fluoxastrobin + prothioconazole, fluoxastrobin + prothioconazole + trifloxystrobin, fluquinconazole + prochloraz, flusilazole, flusilazole (*moderate control*), flutriafol, folpet (*reduction*), maneb, metconazole, metconazole (*reduction*), picoxystrobin, prochloraz, prochloraz + propiconazole, prochloraz + proquinazid + tebuconazole (*moderate control*), prochloraz + tebuconazole, propiconazole, prothioconazole, prothioconazole + spiroxamine, prothioconazole + spiroxamine + tebuconazole, prothioconazole + tebuconazole, prothioconazole + trifloxystrobin, pyraclostrobin (*moderate*), spiroxamine (*reduction*), spiroxamine + tebuconazole, tebuconazole, tebuconazole + triadimenol, trifloxystrobin

Rust
azoxystrobin, azoxystrobin + chlorothalonil, azoxystrobin + cyproconazole, azoxystrobin + fenpropimorph, bixafen + prothioconazole, boscalid + epoxiconazole, boscalid + epoxiconazole + pyraclostrobin, carbendazim + flusilazole, chlorothalonil + cyproconazole, chlorothalonil + cyproconazole + propiconazole, chlorothalonil + picoxystrobin, chlorothalonil + propiconazole, cyproconazole, cyproconazole + picoxystrobin, cyproconazole + propiconazole, cyprodinil +

isopyrazam, cyprodinil + picoxystrobin, epoxiconazole, epoxiconazole + fenpropimorph, epoxiconazole + fenpropimorph + kresoxim-methyl, epoxiconazole + fenpropimorph + metrafenone, epoxiconazole + fenpropimorph + pyraclostrobin, epoxiconazole + fluxapyroxad , epoxiconazole + fluxapyroxad + pyraclostrobin, epoxiconazole + isopyrazam, epoxiconazole + kresoxim-methyl, epoxiconazole + kresoxim-methyl + pyraclostrobin, epoxiconazole + metconazole, epoxiconazole + metrafenone, epoxiconazole + prochloraz, epoxiconazole + pyraclostrobin, famoxadone + flusilazole, fenpropidin, fenpropidin (*moderate control*), fenpropimorph, fenpropimorph + flusilazole, fenpropimorph + pyraclostrobin, fluoxastrobin + prothioconazole, fluoxastrobin + prothioconazole + trifloxystrobin, fluquinconazole + prochloraz, fluquinconazole + prochloraz (*seed treatment*), flusilazole, flutriafol, fuberidazole + imidacloprid + triadimenol (*seed treatment*), maneb, metconazole, picoxystrobin, prochloraz + proquinazid + tebuconazole, prochloraz + tebuconazole, propiconazole, prothioconazole, prothioconazole + spiroxamine, prothioconazole + spiroxamine + tebuconazole, prothioconazole + tebuconazole, prothioconazole + trifloxystrobin, pyraclostrobin, spiroxamine, spiroxamine + tebuconazole, tebuconazole, tebuconazole + triadimenol, trifloxystrobin

	Seed-borne diseases	fluopyram + prothioconazole + tebuconazole, fluquinconazole, imazalil + ipconazole (*useful protection*), ipconazole
	Snow mould	fludioxonil (*seed treatment*), fludioxonil + flutriafol (*seed treatment*), fludioxonil + tefluthrin, fuberidazole + imidacloprid + triadimenol (*seed treatment - reduction*)
	Soil-borne diseases	ipconazole
	Sooty moulds	maneb, prothioconazole + tebuconazole
	Take-all	azoxystrobin (*reduction*), azoxystrobin + chlorothalonil (*reduction*), azoxystrobin + chlorothalonil (*reduction only*), azoxystrobin + cyproconazole (*reduction*), azoxystrobin + fenpropimorph (*reduction*), fluoxastrobin + prothioconazole (*reduction*), fluquinconazole + prochloraz (*seed treatment - reduction*), silthiofam (*seed treatment*)
Pests	Aphids	alpha-cypermethrin, beta-cyfluthrin, chlorpyrifos, clothianidin + prothioconazole (*seed treatment*), cypermethrin, cypermethrin (*autumn sown*), cypermethrin (*moderate control*), deltamethrin, esfenvalerate, fuberidazole + imidacloprid + triadimenol (*seed treatment*), lambda-cyhalothrin, pirimicarb, tau-fluvalinate, zeta-cypermethrin
	Birds/mammals	aluminium ammonium sulphate
	Flies	alpha-cypermethrin, alpha-cypermethrin (*some control if present at application*), chlorpyrifos, cypermethrin, cypermethrin (*autumn sown*), fludioxonil + tefluthrin

	Leafhoppers	clothianidin + prothioconazole (*seed treatment*)
	Leatherjackets	chlorpyrifos
	Midges	chlorpyrifos
	Slugs/snails	clothianidin + prothioconazole (*seed treatment*)
	Wireworms	clothianidin + prothioconazole (*seed treatment*), fludioxonil + tefluthrin, fuberidazole + imidacloprid + triadimenol (*reduction of damage*)
Plant growth regulation	Growth control	2-chloroethylphosphonic acid, 2-chloroethylphosphonic acid + mepiquat chloride, chlormequat, chlormequat + 2-chloroethylphosphonic acid, chlormequat + 2-chloroethylphosphonic acid + mepiquat chloride, mepiquat chloride + prohexadione-calcium, trinexapac-ethyl
	Quality/yield control	2-chloroethylphosphonic acid + mepiquat chloride (*low lodging situations*), chlormequat, mepiquat chloride + prohexadione-calcium, sulphur
Weeds	Broad-leaved weeds	2,4-D, 2,4-D + MCPA, 2,4-DB, 2,4-DB + MCPA, amidosulfuron, amidosulfuron + iodosulfuron-methyl-sodium, bifenox, bromoxynil, bromoxynil + ioxynil, carfentrazone-ethyl, carfentrazone-ethyl + mecoprop-P, chlorotoluron + diflufenican, chlorotoluron + diflufenican (*off-label*), clopyralid, clopyralid + florasulam + fluroxypyr, dicamba + MCPA + mecoprop-P, dicamba + mecoprop-P, dichlorprop-P + MCPA + mecoprop-P, diflufenican, diflufenican + flufenacet, diflufenican + flufenacet + flurtamone, diflufenican + flupyrsulfuron-methyl, diflufenican + flurtamone, diflufenican + metsulfuron-methyl, diflufenican + pendimethalin, florasulam, florasulam + fluroxypyr, flufenacet + pendimethalin, flupyrsulfuron-methyl, fluroxypyr, fluroxypyr + metsulfuron-methyl, glyphosate, iodosulfuron-methyl-sodium, MCPA, mecoprop-P, metsulfuron-methyl, metsulfuron-methyl + thifensulfuron-methyl, metsulfuron-methyl + tribenuron-methyl, pendimethalin, pendimethalin + picolinafen, picolinafen, prosulfocarb, thifensulfuron-methyl + tribenuron-methyl, tribenuron-methyl
	Crops as weeds	amidosulfuron + iodosulfuron-methyl-sodium, diflufenican, diflufenican + flufenacet, diflufenican + flurtamone, florasulam, florasulam + fluroxypyr, fluroxypyr, glyphosate, iodosulfuron-methyl-sodium, metsulfuron-methyl + tribenuron-methyl, pendimethalin
	Grass weeds	chlorotoluron + diflufenican, chlorotoluron + diflufenican (*off-label*), diflufenican, diflufenican + flufenacet, diflufenican + flufenacet (*off-label*), diflufenican + flufenacet + flurtamone, diflufenican + flupyrsulfuron-methyl, diflufenican + flurtamone, diflufenican + pendimethalin, diquat (*animal feed*), fenoxaprop-P-ethyl, flufenacet + pendimethalin, flupyrsulfuron-methyl, glyphosate, pendimethalin, pendimethalin + picolinafen, pinoxaden, prosulfocarb, prosulfocarb (*off-label*), tralkoxydim, tri-allate
	Weeds, miscellaneous	diquat (*animal feed only*), fluroxypyr, glyphosate

Cereals - Cereals, general

Crop control	Desiccation	diquat, glyphosate, glyphosate (*off-label*), glyphosate (*off-label - for wild bird seed production*)
Diseases	Bunt	fludioxonil (*off-label*)
	Crown rust	boscalid + epoxiconazole, boscalid + epoxiconazole + pyraclostrobin, epoxiconazole, epoxiconazole + fluxapyroxad , epoxiconazole + fluxapyroxad + pyraclostrobin
	Damping off	thiram (*off-label*)
	Disease control	azoxystrobin + chlorothalonil (*off-label*), chlorothalonil (*off-label*), epoxiconazole + pyraclostrobin (*off-label*)
	Ear blight	metconazole (*reduction*)
	Eyespot	boscalid + epoxiconazole (*moderate control*), boscalid + epoxiconazole + pyraclostrobin (*moderate control*), epoxiconazole + fluxapyroxad (*moderate control*), epoxiconazole + fluxapyroxad + pyraclostrobin (*moderate control*), epoxiconazole + metconazole (*reduction*), fluoxastrobin + prothioconazole + trifloxystrobin (*reduction*), prothioconazole + trifloxystrobin
	Foot rot	fludioxonil (*off-label*)
	Late ear diseases	fluoxastrobin + prothioconazole + trifloxystrobin, prothioconazole + trifloxystrobin
	Powdery mildew	azoxystrobin (*moderate control*), boscalid + epoxiconazole (*moderate control*), boscalid + epoxiconazole + pyraclostrobin (*moderate control*), cyflufenamid (*off-label*), epoxiconazole, epoxiconazole + fluxapyroxad (*moderate control*), epoxiconazole + fluxapyroxad + pyraclostrobin (*moderate control*), epoxiconazole + metconazole (*moderate control*), fenpropidin (*off-label*), fluoxastrobin + prothioconazole + trifloxystrobin, metconazole (*moderate control*), prothioconazole + trifloxystrobin
	Rhynchosporium	azoxystrobin (*reduction*), boscalid + epoxiconazole (*moderate control*), boscalid + epoxiconazole + pyraclostrobin, epoxiconazole, epoxiconazole + fluxapyroxad + pyraclostrobin, epoxiconazole + metconazole (*moderate control*), metconazole (*moderate control*)
	Rust	azoxystrobin, boscalid + epoxiconazole, boscalid + epoxiconazole + pyraclostrobin, epoxiconazole, epoxiconazole + fluxapyroxad , epoxiconazole + fluxapyroxad + pyraclostrobin, epoxiconazole + metconazole, fenpropidin (*off-label*), fluoxastrobin + prothioconazole + trifloxystrobin, metconazole, prothioconazole + trifloxystrobin
	Septoria	fludioxonil (*off-label*), fluoxastrobin + prothioconazole + trifloxystrobin, metconazole, prothioconazole + trifloxystrobin
	Snow mould	fludioxonil (*off-label*)
	Take-all	azoxystrobin (*reduction*)
	Tan spot	fluoxastrobin + prothioconazole + trifloxystrobin
Pests	Aphids	beta-cyfluthrin, dimethoate
	Flies	chlorpyrifos (*off-label*)

	Leatherjackets	chlorpyrifos (*off-label*), methiocarb (*reduction*)
	Midges	beta-cyfluthrin, chlorpyrifos (*off-label*)
	Pests, miscellaneous	chlorpyrifos (*off-label*), deltamethrin (*off-label*), lambda-cyhalothrin (*off-label*)
	Slugs/snails	methiocarb
Plant growth regulation	Growth control	chlormequat (*off-label*), trinexapac-ethyl (*off-label*)
Weeds	Broad-leaved weeds	2,4-DB + MCPA, bromoxynil (*off-label*), bromoxynil + prosulfuron (*off-label*), dicamba + MCPA + mecoprop-P (*off-label*), diflufenican (*off-label*), diflufenican + flufenacet (*off-label*), diflufenican + flurtamone (*off-label*), florasulam + fluroxypyr, fluroxypyr (*off-label - grown for wild bird seed production*), isoxaben (*off-label - grown for game cover*), MCPA, mecoprop-P (*off-label*), metsulfuron-methyl + tribenuron-methyl (*off-label*)
	Crops as weeds	florasulam + fluroxypyr
	Grass weeds	diflufenican (*off-label*), diflufenican + flufenacet (*off-label*), diflufenican + flurtamone (*off-label*), glyphosate, prosulfocarb (*off-label*)
	Weeds, miscellaneous	diquat, glyphosate (*off-label*)

Cereals - Maize/sweetcorn

Diseases	Damping off	fludioxonil + metalaxyl M, thiram (*seed treatment*)
	Eyespot	epoxiconazole + pyraclostrobin (*off-label*), flusilazole (*off-label*)
	Foliar diseases	flusilazole (*off-label*), pyraclostrobin (*off-label*)
	Fusarium	fludioxonil + metalaxyl M
	Phytophthora	epoxiconazole + pyraclostrobin (*off-label*)
	Pythium	fludioxonil + metalaxyl M
	Rust	epoxiconazole + pyraclostrobin (*off-label*)
Pests	Aphids	pirimicarb, pirimicarb (*off-label*), pymetrozine (*off-label*)
	Caterpillars	Bacillus thuringiensis (*off-label*), indoxacarb (*off-label*)
	Flies	chlorpyrifos, clothianidin, lambda-cyhalothrin (*off-label*), methiocarb
	Pests, miscellaneous	lambda-cyhalothrin (*off-label*)
	Slugs/snails	methiocarb
	Symphylids	clothianidin (*reduction*)
	Wireworms	clothianidin
Weeds	Bindweeds	dicamba + prosulfuron
	Broad-leaved weeds	bromoxynil, bromoxynil + prosulfuron, bromoxynil + terbuthylazine, bromoxynil + terbuthylazine (*off-label*), clopyralid, dicamba + prosulfuron, dicamba + prosulfuron (*seedlings only*), dimethenamid-p + pendimethalin, dimethenamid-p + pendimethalin (*off-label*), flufenacet + isoxaflutole, flufenacet + isoxaflutole (*off-label*), fluroxypyr, fluroxypyr (*off-label*), isoxaben (*off-label - grown for game cover*), mesotrione, mesotrione (*off-label*), mesotrione + terbuthylazine, mesotrione + terbuthylazine (*off-label*), nicosulfuron, nicosulfuron (*off-label*),

pendimethalin, pendimethalin (*off-label*), pendimethalin + terbuthylazine, prosulfuron, rimsulfuron, S-metolachlor, S-metolachlor (*moderately susceptible*)

Crops as weeds | fluroxypyr, fluroxypyr (*off-label*), mesotrione, nicosulfuron, pendimethalin, rimsulfuron

Grass weeds | bromoxynil + terbuthylazine (*off-label*), dimethenamid-p + pendimethalin, dimethenamid-p + pendimethalin (*off-label*), flufenacet + isoxaflutole, flufenacet + isoxaflutole (*off-label*), mesotrione, mesotrione (*off-label*), mesotrione + terbuthylazine, mesotrione + terbuthylazine (*off-label*), nicosulfuron, nicosulfuron (*off-label*), pendimethalin, pendimethalin + terbuthylazine, S-metolachlor

Weed grasses | nicosulfuron

Weeds, miscellaneous | fluroxypyr

Cereals - Oats

Crop control | Desiccation | diquat, diquat (*off-label*), diquat (*stockfeed only*), glyphosate

Diseases | Covered smut | carboxin + thiram (*seed treatment*), prochloraz + triticonazole (*seed treatment*)

Crown rust | azoxystrobin, azoxystrobin + cyproconazole, azoxystrobin + fenpropimorph, bixafen + prothioconazole, boscalid + epoxiconazole, cyproconazole, cyproconazole + picoxystrobin, epoxiconazole + fenpropimorph + metrafenone, epoxiconazole + fenpropimorph + pyraclostrobin, epoxiconazole + kresoxim-methyl + pyraclostrobin, epoxiconazole + metrafenone, epoxiconazole + pyraclostrobin, epoxiconazole + pyraclostrobin (*qualified minor use*), fenpropimorph + pyraclostrobin, fluoxastrobin + prothioconazole, picoxystrobin, prochloraz + proquinazid + tebuconazole (*qualified minor use recommendation*), propiconazole, prothioconazole, prothioconazole + spiroxamine, prothioconazole + spiroxamine + tebuconazole, prothioconazole + tebuconazole, pyraclostrobin, tebuconazole, tebuconazole (*reduction*), tebuconazole + triadimenol

Ear blight | tebuconazole

Eyespot | bixafen + prothioconazole (*reduction of incidence and severity*), boscalid + epoxiconazole (*moderate control*), epoxiconazole + fenpropimorph + kresoxim-methyl (*reduction*), epoxiconazole + kresoxim-methyl (*reduction*), fluoxastrobin + prothioconazole, fluoxastrobin + prothioconazole (*reduction of incidence and severity*), prothioconazole, prothioconazole + spiroxamine, prothioconazole + spiroxamine + tebuconazole, prothioconazole + tebuconazole

Foot rot | carboxin + thiram (*seed treatment*), clothianidin + prothioconazole (*seed treatment*), difenoconazole + fludioxonil, fludioxonil (*seed treatment*), fludioxonil + flutriafol (*seed treatment*), fludioxonil + tefluthrin, fuberidazole + imidacloprid + triadimenol (*seed*

		treatment - reduction), fuberidazole + triadimenol (*seed treatment*), prochloraz + triticonazole (*seed treatment*), prothioconazole (*seed treatment*)
	Leaf spot	difenoconazole + fludioxonil, fludioxonil (*seed treatment*), fludioxonil + flutriafol (*seed treatment*), fludioxonil + tefluthrin, fuberidazole + imidacloprid + triadimenol (*seed treatment*), fuberidazole + triadimenol (*seed treatment*)
	Loose smut	carboxin + thiram (*seed treatment - reduction*), clothianidin + prothioconazole (*seed treatment*), fuberidazole + imidacloprid + triadimenol (*seed treatment*), fuberidazole + triadimenol (*seed treatment*), prochloraz + triticonazole (*seed treatment*), prothioconazole (*seed treatment*)
	Powdery mildew	azoxystrobin, azoxystrobin (*moderate control*), azoxystrobin + cyproconazole, azoxystrobin + fenpropimorph, bixafen + prothioconazole, boscalid + epoxiconazole, cyflufenamid, cyproconazole, epoxiconazole, epoxiconazole + fenpropimorph, epoxiconazole + fenpropimorph + kresoxim-methyl, epoxiconazole + fenpropimorph + metrafenone, epoxiconazole + fenpropimorph + pyraclostrobin, epoxiconazole + kresoxim-methyl, epoxiconazole + metrafenone, fenpropimorph, fenpropimorph + kresoxim-methyl, fenpropimorph + pyraclostrobin, fenpropimorph + quinoxyfen, fluoxastrobin + prothioconazole, fuberidazole + imidacloprid + triadimenol (*seed treatment*), metrafenone (*evidence of mildew control on oats is limited*), picoxystrobin, prochloraz + proquinazid + tebuconazole, propiconazole, proquinazid, prothioconazole, prothioconazole + spiroxamine, prothioconazole + spiroxamine + tebuconazole, prothioconazole + tebuconazole, quinoxyfen, sulphur, tebuconazole, tebuconazole + triadimenol
	Rust	tebuconazole
	Seed-borne diseases	difenoconazole + fludioxonil
	Septoria	tebuconazole
	Snow mould	fludioxonil (*seed treatment*)
	Sooty moulds	tebuconazole
Pests	Aphids	clothianidin + prothioconazole (*seed treatment*), cypermethrin, cypermethrin (*moderate control*), deltamethrin, fuberidazole + imidacloprid + triadimenol (*seed treatment*), lambda-cyhalothrin, pirimicarb, zeta-cypermethrin
	Birds/mammals	aluminium ammonium sulphate
	Flies	chlorpyrifos, cypermethrin
	Leafhoppers	clothianidin + prothioconazole (*seed treatment*)
	Leatherjackets	chlorpyrifos
	Midges	chlorpyrifos
	Slugs/snails	clothianidin + prothioconazole (*seed treatment*)
	Thrips	chlorpyrifos
	Wireworms	clothianidin + prothioconazole (*seed treatment*), fludioxonil + tefluthrin, fuberidazole + imidacloprid + triadimenol (*reduction of damage*)

Plant growth regulation	Growth control	chlormequat, trinexapac-ethyl
	Quality/yield control	sulphur
Weeds	Broad-leaved weeds	2,4-D, 2,4-D + MCPA, 2,4-DB, amidosulfuron, bromoxynil, bromoxynil + ioxynil, carfentrazone-ethyl, carfentrazone-ethyl + flupyrsulfuron-methyl, carfentrazone-ethyl + mecoprop-P, clopyralid, clopyralid + florasulam + fluroxypyr, dicamba + MCPA + mecoprop-P, dicamba + mecoprop-P, dichlorprop-P + MCPA + mecoprop-P, diflufenican + flupyrsulfuron-methyl (*off-label*), diflufenican + flurtamone (*off-label*), florasulam, florasulam + fluroxypyr, flupyrsulfuron-methyl, flupyrsulfuron-methyl + thifensulfuron-methyl, fluroxypyr, glyphosate, MCPA, mecoprop-P, metsulfuron-methyl, metsulfuron-methyl + thifensulfuron-methyl, metsulfuron-methyl + tribenuron-methyl, thifensulfuron-methyl + tribenuron-methyl, tribenuron-methyl
	Crops as weeds	florasulam, flumioxazine (*off-label*), fluroxypyr, glyphosate, metsulfuron-methyl + tribenuron-methyl
	Grass weeds	carfentrazone-ethyl + flupyrsulfuron-methyl, diflufenican + flufenacet (*off-label*), diflufenican + flupyrsulfuron-methyl (*off-label*), diflufenican + flurtamone (*off-label*), diquat (*animal feed*), flumioxazine (*off-label*), flupyrsulfuron-methyl, flupyrsulfuron-methyl + thifensulfuron-methyl, glyphosate
	Weeds, miscellaneous	diquat (*animal feed only*), fluroxypyr, glyphosate

Cereals - Rye/triticale

Diseases	Blue mould	fuberidazole + triadimenol (*seed treatment*)
	Bunt	clothianidin + prothioconazole (*seed treatment*), fludioxonil (*off-label*), prothioconazole (*seed treatment*)
	Disease control	azoxystrobin + chlorothalonil (*off-label*), chlorothalonil (*off-label*), difenoconazole (*off-label*), tebuconazole (*off-label*)
	Ear blight	epoxiconazole + metconazole (*good reduction*), epoxiconazole + metconazole (*moderate control*), epoxiconazole + prochloraz (*good reduction*), metconazole (*reduction*), tebuconazole (*off-label*), thiophanate-methyl (*reduction*), thiophanate-methyl (*reduction only*)
	Eyespot	bixafen + prothioconazole (*reduction of incidence and severity*), bixafen + prothioconazole + spiroxamine, bixafen + prothioconazole + tebuconazole (*reduction of incidence and severity*), boscalid + epoxiconazole (*moderate control*), boscalid + epoxiconazole + pyraclostrobin (*moderate control*), epoxiconazole + fenpropimorph + kresoxim-methyl (*reduction*), epoxiconazole + fluxapyroxad (*moderate control*), epoxiconazole + fluxapyroxad + pyraclostrobin (*moderate control*), epoxiconazole + kresoxim-methyl (*reduction*), epoxiconazole + metconazole (*reduction*), epoxiconazole + prochloraz, fluoxastrobin + prothioconazole (*reduction*), fluoxastrobin +

prothioconazole + trifloxystrobin (*reduction*), prochloraz, prochloraz + tebuconazole, prothioconazole, prothioconazole + spiroxamine, prothioconazole + spiroxamine + tebuconazole (*reduction*), prothioconazole + tebuconazole, prothioconazole + tebuconazole (*reduction*), prothioconazole + trifloxystrobin

Foliar diseases
azoxystrobin + cyproconazole (*off-label*), carbendazim (*off-label*), chlorothalonil + flutriafol (*off-label*), cyproconazole (*off-label*), cyproconazole + propiconazole (*off-label*), difenoconazole (*off-label*), dimoxystrobin + epoxiconazole (*off-label*), epoxiconazole + kresoxim-methyl + pyraclostrobin (*off-label*), epoxiconazole + pyraclostrobin (*off-label*), famoxadone + flusilazole (*off-label*), fenpropimorph + flusilazole (*off-label*), fluquinconazole (*off-label*), fluquinconazole + prochloraz (*off-label*), prochloraz (*off-label*), pyraclostrobin (*off-label*), tebuconazole (*off-label*), trifloxystrobin (*off-label*)

Foot rot
carboxin + thiram (*seed treatment*), clothianidin + prothioconazole (*seed treatment*), fludioxonil (*off-label*), fuberidazole + triadimenol (*seed treatment*), prochloraz + triticonazole (*seed treatment*), prothioconazole (*seed treatment*)

Fusarium
thiophanate-methyl, thiophanate-methyl (*reduction*)

Late ear diseases
bixafen + fluoxystrobin + prothioconazole, bixafen + prothioconazole, bixafen + prothioconazole + spiroxamine, bixafen + prothioconazole + tebuconazole, fluoxastrobin + prothioconazole + trifloxystrobin, prothioconazole + spiroxamine, prothioconazole + trifloxystrobin

Net blotch
prothioconazole + tebuconazole

Powdery mildew
azoxystrobin, azoxystrobin (*moderate control*), azoxystrobin + cyproconazole, azoxystrobin + fenpropimorph, bixafen + fluoxystrobin + prothioconazole, bixafen + prothioconazole, bixafen + prothioconazole + spiroxamine, bixafen + prothioconazole + tebuconazole, boscalid + epoxiconazole + pyraclostrobin (*moderate control*), cyflufenamid, cyproconazole, epoxiconazole, epoxiconazole + fenpropimorph, epoxiconazole + fenpropimorph + kresoxim-methyl, epoxiconazole + fenpropimorph + metrafenone, epoxiconazole + fluxapyroxad (*moderate control*), epoxiconazole + fluxapyroxad + pyraclostrobin (*moderate control*), epoxiconazole + kresoxim-methyl, epoxiconazole + metconazole (*moderate control*), epoxiconazole + metconazole (*moderate reduction*), epoxiconazole + metrafenone, epoxiconazole + prochloraz (*moderate control*), fenpropidin (*off-label*), fenpropimorph, fenpropimorph + kresoxim-methyl, fenpropimorph + quinoxyfen, fluoxastrobin + prothioconazole, fluoxastrobin + prothioconazole + trifloxystrobin, metconazole (*moderate control*), prochloraz, prochloraz + tebuconazole, propiconazole, proquinazid, prothioconazole, prothioconazole + spiroxamine, prothioconazole + spiroxamine + tebuconazole, prothioconazole + tebuconazole, prothioconazole + trifloxystrobin, quinoxyfen,

	spiroxamine, spiroxamine + tebuconazole, sulphur, sulphur (*off-label*), tebuconazole, tebuconazole (*off-label*), tebuconazole + triadimenol
Rhynchosporium	azoxystrobin, azoxystrobin (*reduction*), azoxystrobin + cyproconazole (*moderate control*), azoxystrobin + fenpropimorph, azoxystrobin + fenpropimorph (*reduction*), bixafen + fluoxystrobin + prothioconazole, bixafen + prothioconazole, bixafen + prothioconazole + spiroxamine, bixafen + prothioconazole + tebuconazole, epoxiconazole, epoxiconazole + fenpropimorph, epoxiconazole + fenpropimorph + kresoxim-methyl, epoxiconazole + fenpropimorph + metrafenone, epoxiconazole + isopyrazam, epoxiconazole + kresoxim-methyl, epoxiconazole + metrafenone, epoxiconazole + prochloraz, fenpropimorph + kresoxim-methyl, fluoxastrobin + prothioconazole, prochloraz, prochloraz + tebuconazole, propiconazole, prothioconazole, prothioconazole + spiroxamine, prothioconazole + spiroxamine + tebuconazole, prothioconazole + tebuconazole, spiroxamine (*reduction*), tebuconazole
Rust	azoxystrobin, azoxystrobin + cyproconazole, azoxystrobin + fenpropimorph, bixafen + fluoxystrobin + prothioconazole, bixafen + prothioconazole, bixafen + prothioconazole + spiroxamine, bixafen + prothioconazole + tebuconazole, boscalid + epoxiconazole, boscalid + epoxiconazole + pyraclostrobin, cyproconazole, epoxiconazole, epoxiconazole + fenpropimorph, epoxiconazole + fenpropimorph + kresoxim-methyl, epoxiconazole + fenpropimorph + metrafenone, epoxiconazole + fluxapyroxad , epoxiconazole + fluxapyroxad + pyraclostrobin, epoxiconazole + isopyrazam, epoxiconazole + kresoxim-methyl, epoxiconazole + metconazole, epoxiconazole + metrafenone, epoxiconazole + prochloraz, fenpropidin (*off-label*), fenpropimorph, fluoxastrobin + prothioconazole, fluoxastrobin + prothioconazole + trifloxystrobin, metconazole, prochloraz + tebuconazole, propiconazole, prothioconazole, prothioconazole + spiroxamine, prothioconazole + spiroxamine + tebuconazole, prothioconazole + tebuconazole, prothioconazole + trifloxystrobin, spiroxamine, spiroxamine + tebuconazole, tebuconazole, tebuconazole (*off-label*), tebuconazole + triadimenol
Seed-borne diseases	fludioxonil + tefluthrin (*off-label*), fluquinconazole (*off-label*), silthiofam (*off-label*)
Septoria	bixafen + fluoxystrobin + prothioconazole, bixafen + prothioconazole, bixafen + prothioconazole + spiroxamine, bixafen + prothioconazole + tebuconazole, boscalid + epoxiconazole, boscalid + epoxiconazole + pyraclostrobin, epoxiconazole, epoxiconazole + fenpropimorph, epoxiconazole + fenpropimorph + kresoxim-methyl, epoxiconazole + fenpropimorph + metrafenone, epoxiconazole + fluxapyroxad , epoxiconazole + fluxapyroxad + pyraclostrobin, epoxiconazole + isopyrazam, epoxiconazole + kresoxim-methyl, epoxiconazole +

metconazole, epoxiconazole + metrafenone, epoxiconazole + prochloraz, fenpropimorph + kresoxim-methyl (*reduction*), fludioxonil (*off-label*), fluoxastrobin + prothioconazole + trifloxystrobin, metconazole, prochloraz, propiconazole, prothioconazole + trifloxystrobin, tebuconazole (*off-label*)

	Snow mould	fludioxonil (*off-label*)
	Stripe smut	difenoconazole + fludioxonil
	Take-all	azoxystrobin (*reduction*), azoxystrobin + fenpropimorph (*reduction*)
	Tan spot	bixafen + fluoxystrobin + prothioconazole, bixafen + prothioconazole, bixafen + prothioconazole + spiroxamine, bixafen + prothioconazole + tebuconazole, fluoxastrobin + prothioconazole + trifloxystrobin
Pests	Aphids	beta-cyfluthrin, clothianidin + prothioconazole (*seed treatment*), cypermethrin, cypermethrin (*moderate control*), deltamethrin (*off-label*), dimethoate, lambda-cyhalothrin (*off-label*), pirimicarb, tau-fluvalinate (*off-label*), zeta-cypermethrin (*off-label*)
	Flies	chlorpyrifos (*off-label*), cypermethrin
	Leafhoppers	clothianidin + prothioconazole (*seed treatment*)
	Leatherjackets	chlorpyrifos (*off-label*)
	Midges	beta-cyfluthrin, chlorpyrifos (*off-label*)
	Pests, miscellaneous	alpha-cypermethrin (*off-label*), chlorpyrifos (*off-label*), deltamethrin (*off-label*), lambda-cyhalothrin (*off-label*), zeta-cypermethrin (*off-label*)
	Slugs/snails	clothianidin + prothioconazole (*seed treatment*)
	Wireworms	clothianidin + prothioconazole (*seed treatment*)
Plant growth regulation	Growth control	2-chloroethylphosphonic acid, 2-chloroethylphosphonic acid + mepiquat chloride, chlormequat, chlormequat (*off-label*), chlormequat + 2-chloroethylphosphonic acid (*off-label*), chlormequat + 2-chloroethylphosphonic acid + mepiquat chloride (*off-label*), chlormequat + imazaquin (*off-label*), mepiquat chloride + prohexadione-calcium, trinexapac-ethyl
	Quality/yield control	mepiquat chloride + prohexadione-calcium
Weeds	Broad-leaved weeds	2,4-D, amidosulfuron, amidosulfuron + iodosulfuron-methyl-sodium, bifenox, bromoxynil (*off-label*), bromoxynil + ioxynil, carfentrazone-ethyl, carfentrazone-ethyl (*off-label*), carfentrazone-ethyl + flupyrsulfuron-methyl, chlorotoluron + diflufenican, clopyralid (*off-label*), dicamba + MCPA + mecoprop-P, dicamba + MCPA + mecoprop-P (*off-label*), dicamba + mecoprop-P, dicamba + mecoprop-P (*off-label*), diflufenican, diflufenican + flufenacet (*off-label*), diflufenican + flurtamone (*off-label*), diflufenican + pendimethalin, florasulam (*off-label*), florasulam + fluroxypyr, florasulam + fluroxypyr (*off-label*), florasulam + pyroxsulam, flupyrsulfuron-methyl, fluroxypyr, iodosulfuron-methyl-sodium, iodosulfuron-methyl-sodium + mesosulfuron-methyl (*off-label*), MCPA, mecoprop-P (*off-label*), metsulfuron-methyl, metsulfuron-methyl + tribenuron-methyl, metsulfuron-methyl + tribenuron-methyl (*off-label*),

		pendimethalin, pendimethalin + pyroxsulam, sulfosulfuron (*off-label*), thifensulfuron-methyl + tribenuron-methyl, tribenuron-methyl
	Crops as weeds	amidosulfuron + iodosulfuron-methyl-sodium, diflufenican, florasulam + fluroxypyr, florasulam + pyroxsulam, fluroxypyr, metsulfuron-methyl + tribenuron-methyl, pendimethalin, pendimethalin + pyroxsulam (*from seed*)
	Grass weeds	carfentrazone-ethyl + flupyrsulfuron-methyl, chlorotoluron + diflufenican, clodinafop-propargyl, diflufenican, diflufenican + flufenacet (*off-label*), diflufenican + flurtamone (*off-label*), diflufenican + pendimethalin, florasulam + pyroxsulam, flupyrsulfuron-methyl, iodosulfuron-methyl-sodium (*from seed*), iodosulfuron-methyl-sodium + mesosulfuron-methyl (*off-label*), pendimethalin, pendimethalin + pyroxsulam, pendimethalin + pyroxsulam (*MS only*), prosulfocarb (*off-label*), sulfosulfuron (*off-label*), tralkoxydim, tri-allate
	Weeds, miscellaneous	fluroxypyr

Cereals - Undersown cereals

Diseases	Powdery mildew	fenpropimorph
	Rhynchosporium	fenpropimorph
	Rust	fenpropimorph
Weeds	Broad-leaved weeds	2,4-D, 2,4-DB + MCPA, dicamba + MCPA + mecoprop-P, dicamba + MCPA + mecoprop-P (*grass only*), dicamba + mecoprop-P, MCPA, MCPA (*red clover or grass*), tribenuron-methyl

Cereals - Wheat

Crop control	Desiccation	glyphosate
Diseases	Blue mould	fuberidazole + imidacloprid + triadimenol (*seed treatment*), fuberidazole + triadimenol (*seed treatment*)
	Bunt	carboxin + thiram (*seed treatment*), clothianidin + prothioconazole (*seed treatment*), difenoconazole + fludioxonil, fludioxonil (*off-label*), fludioxonil (*seed treatment*), fludioxonil + flutriafol (*seed treatment*), fludioxonil + tefluthrin, fluquinconazole (*seed treatment*), fluquinconazole + prochloraz (*seed treatment*), fuberidazole + imidacloprid + triadimenol (*seed treatment*), fuberidazole + triadimenol (*seed treatment*), prochloraz + triticonazole (*off-label - seed treatment*), prochloraz + triticonazole (*seed treatment*), prothioconazole (*seed treatment*)
	Disease control	azoxystrobin (*off-label*), azoxystrobin + chlorothalonil (*off-label*), chlorothalonil (*off-label*), difenoconazole (*off-label*)
	Drechslera leaf spot	fenpropimorph + pyraclostrobin
	Ear blight	boscalid + epoxiconazole (*good reduction*), boscalid + epoxiconazole + pyraclostrobin (*good reduction*), chlorothalonil + tebuconazole, dimoxystrobin + epoxiconazole, epoxiconazole, epoxiconazole (*good*

reduction), epoxiconazole (*reduction*), epoxiconazole + fenpropimorph, epoxiconazole + fenpropimorph + kresoxim-methyl (*reduction*), epoxiconazole + fenpropimorph + metrafenone (*reduction*), epoxiconazole + fenpropimorph + pyraclostrobin (*reduction*), epoxiconazole + fluxapyroxad (*good reduction*), epoxiconazole + fluxapyroxad + pyraclostrobin (*good reduction*), epoxiconazole + kresoxim-methyl (*reduction*), epoxiconazole + metconazole (*good reduction*), epoxiconazole + metconazole (*moderate control*), epoxiconazole + metrafenone (*reduction*), epoxiconazole + prochloraz (*good reduction*), epoxiconazole + pyraclostrobin (*apply during flowering*), epoxiconazole + pyraclostrobin (*reduction*), metconazole (*reduction*), prothioconazole + tebuconazole, prothioconazole + tebuconazole (*moderate control*), spiroxamine + tebuconazole, tebuconazole, tebuconazole + triadimenol, thiophanate-methyl (*reduction*), thiophanate-methyl (*reduction only*)

Eyespot — azoxystrobin + cyproconazole (*reduction*), bixafen + prothioconazole (*reduction of incidence and severity*), bixafen + prothioconazole + spiroxamine, bixafen + prothioconazole + tebuconazole (*reduction of incidence and severity*), boscalid + epoxiconazole (*moderate*), boscalid + epoxiconazole (*moderate control*), boscalid + epoxiconazole + pyraclostrobin (*moderate control*), carbendazim, carbendazim + flusilazole, chlorothalonil + cyproconazole, cyproconazole (*useful reduction*), cyproconazole + quinoxyfen (*reduction*), cyprodinil, cyprodinil + picoxystrobin (*moderate control*), epoxiconazole (*reduction*), epoxiconazole + fenpropimorph (*reduction*), epoxiconazole + fenpropimorph + kresoxim-methyl (*reduction*), epoxiconazole + fluxapyroxad (*moderate control*), epoxiconazole + fluxapyroxad + pyraclostrobin (*moderate control*), epoxiconazole + kresoxim-methyl (*reduction*), epoxiconazole + metconazole (*reduction*), epoxiconazole + prochloraz, fenpropidin + prochloraz + tebuconazole, fluoxastrobin + prothioconazole, fluoxastrobin + prothioconazole (*reduction*), fluoxastrobin + prothioconazole + trifloxystrobin (*reduction*), flusilazole, metrafenone (*reduction*), picoxystrobin (*reduction*), prochloraz, prochloraz + propiconazole, prochloraz + proquinazid + tebuconazole, prochloraz + tebuconazole, prothioconazole, prothioconazole + spiroxamine, prothioconazole + spiroxamine + tebuconazole (*reduction*), prothioconazole + tebuconazole, prothioconazole + tebuconazole (*reduction*), prothioconazole + trifloxystrobin

Foliar diseases — azoxystrobin (*off-label*), azoxystrobin + cyproconazole (*off-label*), azoxystrobin + fenpropimorph (*off-label*), boscalid + epoxiconazole (*off-label*), carbendazim (*off-label*), chlorothalonil + flutriafol (*off-label*), cyproconazole (*off-label*), cyproconazole + propiconazole (*off-label*), difenoconazole (*off-label*), dimoxystrobin + epoxiconazole (*off-label*), epoxiconazole (*off-label*), epoxiconazole + fenpropimorph (*off-label*), epoxiconazole +

fenpropimorph + kresoxim-methyl (*off-label*), epoxiconazole + kresoxim-methyl (*off-label*), epoxiconazole + kresoxim-methyl + pyraclostrobin (*off-label*), epoxiconazole + pyraclostrobin (*off-label*), famoxadone + flusilazole (*off-label*), fenpropimorph (*off-label*), fenpropimorph + flusilazole (*off-label*), fluquinconazole + prochloraz (*off-label*), pyraclostrobin (*off-label*), trifloxystrobin (*off-label*)

Foot rot   carboxin + thiram (*seed treatment*), clothianidin + prothioconazole (*seed treatment*), difenoconazole + fludioxonil, fludioxonil (*off-label*), fludioxonil (*seed treatment*), fludioxonil + flutriafol (*seed treatment*), fludioxonil + tefluthrin, fluoxastrobin + prothioconazole (*reduction*), fluquinconazole + prochloraz (*seed treatment*), fuberidazole + imidacloprid + triadimenol (*seed treatment - reduction*), fuberidazole + triadimenol (*seed treatment*), prochloraz + triticonazole (*off-label - seed treatment*), prochloraz + triticonazole (*seed treatment*), prothioconazole (*seed treatment*)

Fusarium   thiophanate-methyl, thiophanate-methyl (*reduction*)

Late ear diseases   azoxystrobin, azoxystrobin + fenpropimorph (*moderate control*), bixafen + fluoxystrobin + prothioconazole, bixafen + prothioconazole, bixafen + prothioconazole + spiroxamine, bixafen + prothioconazole + tebuconazole, chlorothalonil + flutriafol, cyproconazole, cyproconazole + propiconazole, fenpropidin + prochloraz + tebuconazole, fluoxastrobin + prothioconazole, fluoxastrobin + prothioconazole + trifloxystrobin, picoxystrobin, prochloraz + proquinazid + tebuconazole (*reduction only*), prochloraz + tebuconazole, prothioconazole, prothioconazole + spiroxamine, prothioconazole + spiroxamine + tebuconazole, prothioconazole + tebuconazole, prothioconazole + trifloxystrobin

Leaf spot   folpet (*reduction*)

Loose smut   clothianidin + prothioconazole (*seed treatment*), fludioxonil + flutriafol (*seed treatment*), fuberidazole + imidacloprid + triadimenol (*seed treatment*), fuberidazole + triadimenol (*seed treatment*), prochloraz + triticonazole (*off-label - seed treatment*), prochloraz + triticonazole (*seed treatment*), prothioconazole (*seed treatment*)

Net blotch   metconazole (*reduction*)

Powdery mildew   azoxystrobin, azoxystrobin + cyproconazole, bixafen + fluoxystrobin + prothioconazole, bixafen + prothioconazole, bixafen + prothioconazole + spiroxamine, bixafen + prothioconazole + tebuconazole, boscalid + epoxiconazole, carbendazim + flusilazole, chlorothalonil + cyproconazole, chlorothalonil + cyproconazole + propiconazole (*moderate*), chlorothalonil + flutriafol, chlorothalonil + proquinazid, cyflufenamid, cyproconazole, cyproconazole + propiconazole, cyproconazole + quinoxyfen, cyprodinil (*moderate control*), epoxiconazole, epoxiconazole + fenpropimorph, epoxiconazole + fenpropimorph + metrafenone, epoxiconazole + fluxapyroxad (*moderate control*), epoxiconazole + fluxapyroxad + pyraclostrobin (*moderate control*), epoxiconazole + kresoxim-methyl, epoxiconazole + metconazole (*moderate control*), epoxiconazole + metrafenone, epoxiconazole + prochloraz (*moderate control*), fenpropidin, fenpropidin (*off-label*), fenpropidin + prochloraz + tebuconazole, fenpropimorph, fenpropimorph (*off-label*), fenpropimorph + flusilazole, fenpropimorph + quinoxyfen, fluoxastrobin + prothioconazole, fluoxastrobin + prothioconazole + trifloxystrobin, fluquinconazole, fluquinconazole + prochloraz (*moderate control*), flusilazole, flutriafol, fuberidazole + imidacloprid + triadimenol (*seed treatment*), metconazole (*moderate control*), metrafenone, prochloraz, prochloraz + proquinazid + tebuconazole, prochloraz + tebuconazole, propiconazole, proquinazid, proquinazid (*off-label*), prothioconazole, prothioconazole + spiroxamine, prothioconazole + spiroxamine + tebuconazole, prothioconazole + tebuconazole, prothioconazole + trifloxystrobin, pyriofenone, quinoxyfen, spiroxamine, spiroxamine + tebuconazole, sulphur, tebuconazole, tebuconazole + triadimenol

Rust

azoxystrobin, azoxystrobin + chlorothalonil, azoxystrobin + cyproconazole, azoxystrobin + fenpropimorph, bixafen + fluoxystrobin + prothioconazole, bixafen + prothioconazole, bixafen + prothioconazole + spiroxamine, bixafen + prothioconazole + tebuconazole, boscalid + epoxiconazole, boscalid + epoxiconazole + pyraclostrobin, carbendazim + flusilazole, chlorothalonil + cyproconazole, chlorothalonil + cyproconazole + propiconazole, chlorothalonil + flutriafol, chlorothalonil + picoxystrobin, chlorothalonil + propiconazole, cyproconazole, cyproconazole + picoxystrobin, cyproconazole + propiconazole, cyproconazole + quinoxyfen, cyprodinil + picoxystrobin, difenoconazole, dimoxystrobin + epoxiconazole, epoxiconazole, epoxiconazole + fenpropimorph, epoxiconazole + fenpropimorph + kresoxim-methyl, epoxiconazole + fenpropimorph + metrafenone, epoxiconazole + fenpropimorph + pyraclostrobin, epoxiconazole + fluxapyroxad , epoxiconazole + fluxapyroxad + pyraclostrobin, epoxiconazole + isopyrazam, epoxiconazole + kresoxim-methyl, epoxiconazole + kresoxim-methyl + pyraclostrobin, epoxiconazole + metconazole, epoxiconazole + metrafenone, epoxiconazole + prochloraz, epoxiconazole + pyraclostrobin, famoxadone + flusilazole, fenpropidin (*moderate control*), fenpropidin (*off-label*), fenpropidin + prochloraz + tebuconazole, fenpropimorph, fenpropimorph + flusilazole, fenpropimorph + pyraclostrobin, fluoxastrobin + prothioconazole, fluoxastrobin + prothioconazole + trifloxystrobin, fluquinconazole, fluquinconazole (*seed treatment*), fluquinconazole + prochloraz, fluquinconazole + prochloraz (*seed treatment*), fluquinconazole + prochloraz (*seed treatment - moderate control*), flusilazole, flutriafol, fuberidazole + imidacloprid +

triadimenol (*seed treatment*), mancozeb, mancozeb (*useful control*), maneb, metconazole, picoxystrobin, prochloraz + proquinazid + tebuconazole, prochloraz + tebuconazole, propiconazole, prothioconazole, prothioconazole + spiroxamine, prothioconazole + spiroxamine + tebuconazole, prothioconazole + tebuconazole, prothioconazole + trifloxystrobin, pyraclostrobin, spiroxamine, spiroxamine + tebuconazole, tebuconazole, tebuconazole + triadimenol, trifloxystrobin

Seed-borne diseases
difenoconazole + fludioxonil, fluquinconazole, fluquinconazole (*off-label*), ipconazole, silthiofam (*off-label*)

Septoria
azoxystrobin, azoxystrobin + chlorothalonil, azoxystrobin + cyproconazole, azoxystrobin + fenpropimorph, bixafen + fluoxystrobin + prothioconazole, bixafen + prothioconazole, bixafen + prothioconazole + spiroxamine, bixafen + prothioconazole + tebuconazole, boscalid + epoxiconazole, boscalid + epoxiconazole + pyraclostrobin, carbendazim + flusilazole, carboxin + thiram (*seed treatment*), chlorothalonil, chlorothalonil + cyproconazole, chlorothalonil + cyproconazole + propiconazole, chlorothalonil + flutriafol, chlorothalonil + mancozeb, chlorothalonil + picoxystrobin, chlorothalonil + propiconazole, chlorothalonil + proquinazid (*suppression only*), chlorothalonil + tebuconazole, cyproconazole, cyproconazole + picoxystrobin, cyproconazole + propiconazole, cyproconazole + quinoxyfen, cyprodinil + picoxystrobin, difenoconazole, difenoconazole + fludioxonil, dimoxystrobin + epoxiconazole, epoxiconazole, epoxiconazole + fenpropimorph, epoxiconazole + fenpropimorph + kresoxim-methyl, epoxiconazole + fenpropimorph + metrafenone, epoxiconazole + fenpropimorph + pyraclostrobin, epoxiconazole + fluxapyroxad, epoxiconazole + fluxapyroxad + pyraclostrobin, epoxiconazole + isopyrazam, epoxiconazole + kresoxim-methyl, epoxiconazole + kresoxim-methyl + pyraclostrobin, epoxiconazole + metconazole, epoxiconazole + metrafenone, epoxiconazole + prochloraz, epoxiconazole + pyraclostrobin, famoxadone + flusilazole, fenpropidin + prochloraz + tebuconazole, fenpropimorph + flusilazole, fenpropimorph + kresoxim-methyl (*reduction*), fenpropimorph + pyraclostrobin, fludioxonil (*off-label*), fludioxonil (*seed treatment*), fludioxonil + flutriafol (*seed treatment*), fludioxonil + tefluthrin, fluoxastrobin + prothioconazole, fluoxastrobin + prothioconazole (*reduction*), fluoxastrobin + prothioconazole + trifloxystrobin, fluquinconazole, fluquinconazole (*seed treatment*), fluquinconazole + prochloraz, fluquinconazole + prochloraz (*seed treatment*), fluquinconazole + prochloraz (*seed treatment - moderate control*), flusilazole, flutriafol, fuberidazole + imidacloprid + triadimenol (*seed treatment*), fuberidazole + imidacloprid + triadimenol (*seed treatment - reduction*), mancozeb, mancozeb (*reduction*), maneb, metconazole, picoxystrobin, prochloraz, prochloraz + propiconazole, prochloraz +

proquinazid + tebuconazole, prochloraz + proquinazid + tebuconazole (*moderate control*), prochloraz + tebuconazole, prochloraz + triticonazole (*off-label - seed treatment*), prochloraz + triticonazole (*seed treatment*), propiconazole, prothioconazole, prothioconazole + spiroxamine, prothioconazole + spiroxamine + tebuconazole, prothioconazole + tebuconazole, prothioconazole + trifloxystrobin, pyraclostrobin, spiroxamine + tebuconazole, tebuconazole, tebuconazole + triadimenol, trifloxystrobin

Sharp eyespot	fluoxastrobin + prothioconazole, fluoxastrobin + prothioconazole (*reduction*)
Snow mould	difenoconazole + fludioxonil, fludioxonil (*off-label*), fludioxonil (*seed treatment*), fludioxonil + flutriafol (*seed treatment*), fludioxonil + tefluthrin
Soil-borne diseases	ipconazole
Sooty moulds	boscalid + epoxiconazole, boscalid + epoxiconazole + pyraclostrobin (*good reduction*), chlorothalonil + tebuconazole, epoxiconazole (*reduction*), epoxiconazole + fenpropimorph (*reduction*), epoxiconazole + fenpropimorph + kresoxim-methyl (*reduction*), epoxiconazole + fenpropimorph + metrafenone (*reduction*), epoxiconazole + fluxapyroxad (*reduction*), epoxiconazole + fluxapyroxad + pyraclostrobin (*reduction*), epoxiconazole + kresoxim-methyl (*reduction*), epoxiconazole + metconazole (*good reduction*), epoxiconazole + metconazole (*moderate control*), epoxiconazole + metrafenone (*reduction*), epoxiconazole + prochloraz (*good reduction*), fluoxastrobin + prothioconazole (*reduction*), mancozeb, maneb, propiconazole, prothioconazole + tebuconazole, prothioconazole + tebuconazole (*reduction*), spiroxamine + tebuconazole, tebuconazole, tebuconazole + triadimenol
Take-all	azoxystrobin (*reduction*), azoxystrobin + chlorothalonil (*reduction*), azoxystrobin + chlorothalonil (*reduction only*), azoxystrobin + cyproconazole (*reduction*), azoxystrobin + fenpropimorph (*reduction*), fluoxastrobin + prothioconazole (*reduction*), fluquinconazole (*seed treatment - reduction*), fluquinconazole + prochloraz (*seed treatment - reduction*), silthiofam (*seed treatment*)
Tan spot	bixafen + fluoxystrobin + prothioconazole, bixafen + prothioconazole, bixafen + prothioconazole + spiroxamine, bixafen + prothioconazole + tebuconazole, boscalid + epoxiconazole (*reduction*), boscalid + epoxiconazole (*reduction only*), boscalid + epoxiconazole + pyraclostrobin (*moderate control*), chlorothalonil + picoxystrobin, dimoxystrobin + epoxiconazole, epoxiconazole + fluxapyroxad (*moderate control*), epoxiconazole + fluxapyroxad + pyraclostrobin (*moderate control*), epoxiconazole + metconazole (*moderate control*), epoxiconazole + pyraclostrobin, fluoxastrobin + prothioconazole, fluoxastrobin + prothioconazole + trifloxystrobin,

		prothioconazole, prothioconazole + spiroxamine, prothioconazole + tebuconazole
Pests	Aphids	alpha-cypermethrin, beta-cyfluthrin, chlorpyrifos, clothianidin + prothioconazole (*seed treatment*), cypermethrin, cypermethrin (*autumn sown*), cypermethrin (*moderate control*), deltamethrin, deltamethrin (*off-label*), dimethoate, esfenvalerate, flonicamid, fuberidazole + imidacloprid + triadimenol (*seed treatment*), lambda-cyhalothrin, pirimicarb, tau-fluvalinate, tau-fluvalinate (*off-label*), zeta-cypermethrin, zeta-cypermethrin (*off-label*)
	Beetles	lambda-cyhalothrin
	Birds/mammals	aluminium ammonium sulphate
	Caterpillars	lambda-cyhalothrin
	Flies	alpha-cypermethrin, alpha-cypermethrin (*some control if present at application*), chlorpyrifos, chlorpyrifos (*off-label*), cypermethrin, cypermethrin (*autumn sown*), dimethoate, fludioxonil + tefluthrin, lambda-cyhalothrin
	Leafhoppers	clothianidin + prothioconazole (*seed treatment*)
	Leatherjackets	chlorpyrifos, chlorpyrifos (*off-label*)
	Midges	beta-cyfluthrin, chlorpyrifos, chlorpyrifos (*off-label*), lambda-cyhalothrin, thiacloprid
	Pests, miscellaneous	alpha-cypermethrin (*off-label*), chlorpyrifos (*off-label*), deltamethrin (*off-label*), zeta-cypermethrin (*off-label*)
	Slugs/snails	clothianidin + prothioconazole (*seed treatment*)
	Suckers	lambda-cyhalothrin
	Thrips	chlorpyrifos
	Weevils	lambda-cyhalothrin
	Wireworms	clothianidin + prothioconazole (*seed treatment*), fludioxonil + tefluthrin, fuberidazole + imidacloprid + triadimenol (*reduction of damage*)
Plant growth regulation	Growth control	2-chloroethylphosphonic acid, 2-chloroethylphosphonic acid (*off-label*), 2-chloroethylphosphonic acid + mepiquat chloride, 2-chloroethylphosphonic acid + mepiquat chloride (*off-label*), chlormequat, chlormequat (*off-label*), chlormequat + 2-chloroethylphosphonic acid, chlormequat + 2-chloroethylphosphonic acid (*off-label*), chlormequat + 2-chloroethylphosphonic acid + mepiquat chloride, chlormequat + 2-chloroethylphosphonic acid + mepiquat chloride (*off-label*), chlormequat + imazaquin, chlormequat + mepiquat chloride, mepiquat chloride + prohexadione-calcium, mepiquat chloride + prohexadione-calcium (*off-label*), trinexapac-ethyl, trinexapac-ethyl (*off-label*)
	Quality/yield control	chlormequat + imazaquin, mepiquat chloride + prohexadione-calcium, mepiquat chloride + prohexadione-calcium (*off-label*), sulphur
Weeds	Broad-leaved weeds	2,4-D, 2,4-D + MCPA, 2,4-DB, 2,4-DB + MCPA, amidosulfuron, amidosulfuron + iodosulfuron-methyl-sodium, amidosulfuron + iodosulfuron-methyl-sodium (*off-label*), bifenox, bifenox (*off-label*), bromoxynil, bromoxynil (*off-label*), bromoxynil + ioxynil, carfentrazone-ethyl, carfentrazone-ethyl +

flupyrsulfuron-methyl, carfentrazone-ethyl + flupyrsulfuron-methyl (*off-label*), carfentrazone-ethyl + mecoprop-P, chlorotoluron + diflufenican, chlorotoluron + diflufenican (*off-label*), clodinafop-propargyl + prosulfocarb, clopyralid, clopyralid (*off-label*), clopyralid + florasulam + fluroxypyr, dicamba + MCPA + mecoprop-P, dicamba + MCPA + mecoprop-P (*off-label*), dicamba + mecoprop-P, dicamba + mecoprop-P (*off-label*), dichlorprop-P + MCPA + mecoprop-P, diflufenican, diflufenican + flufenacet, diflufenican + flufenacet (*off-label*), diflufenican + flufenacet + flurtamone, diflufenican + flupyrsulfuron-methyl, diflufenican + flurtamone, diflufenican + iodosulfuron-methyl-sodium + mesosulfuron-methyl, diflufenican + metsulfuron-methyl, diflufenican + pendimethalin, florasulam, florasulam + fluroxypyr, florasulam + fluroxypyr (*off-label*), florasulam + pyroxsulam, flufenacet + pendimethalin, flumioxazine, flupyrsulfuron-methyl, flupyrsulfuron-methyl + pyroxsulam, flupyrsulfuron-methyl + thifensulfuron-methyl, fluroxypyr, fluroxypyr + metsulfuron-methyl, glyphosate, iodosulfuron-methyl-sodium, iodosulfuron-methyl-sodium (*off-label*), iodosulfuron-methyl-sodium + mesosulfuron-methyl, iodosulfuron-methyl-sodium + mesosulfuron-methyl (*off-label*), MCPA, mecoprop-P, mecoprop-P (*off-label*), metsulfuron-methyl, metsulfuron-methyl + thifensulfuron-methyl, metsulfuron-methyl + thifensulfuron-methyl (*off-label*), metsulfuron-methyl + tribenuron-methyl, metsulfuron-methyl + tribenuron-methyl (*off-label*), pendimethalin, pendimethalin (*off-label*), pendimethalin + picolinafen, pendimethalin + pyroxsulam, picolinafen, prosulfocarb, prosulfocarb (*off-label*), sulfosulfuron, thifensulfuron-methyl + tribenuron-methyl, tribenuron-methyl

Crops as weeds   amidosulfuron + iodosulfuron-methyl-sodium, diflufenican, diflufenican + flufenacet, diflufenican + flurtamone, diflufenican + iodosulfuron-methyl-sodium + mesosulfuron-methyl, florasulam, florasulam + fluroxypyr, florasulam + pyroxsulam, flumioxazine, flupyrsulfuron-methyl + pyroxsulam (*from seed*), fluroxypyr, glyphosate, iodosulfuron-methyl-sodium (*off-label - from seed*), metsulfuron-methyl + tribenuron-methyl, pendimethalin, pendimethalin + pyroxsulam (*from seed*)

Grass weeds   carfentrazone-ethyl + flupyrsulfuron-methyl, chlorotoluron + diflufenican, chlorotoluron + diflufenican (*off-label*), clodinafop-propargyl, clodinafop-propargyl + pinoxaden, clodinafop-propargyl + pinoxaden (*from seed*), clodinafop-propargyl + prosulfocarb, diflufenican, diflufenican + flufenacet, diflufenican + flufenacet (*off-label*), diflufenican + flufenacet + flurtamone, diflufenican + flufenacet + flurtamone (*moderately susceptible*), diflufenican + flupyrsulfuron-methyl, diflufenican + flurtamone, diflufenican + iodosulfuron-methyl-sodium + mesosulfuron-methyl, diflufenican + pendimethalin, fenoxaprop-P-ethyl, florasulam + pyroxsulam, flufenacet + pendimethalin, flumioxazine, flupyrsulfuron-methyl, flupyrsulfuron-methyl + pyroxsulam, flupyrsulfuron-methyl + pyroxsulam (*moderately susceptible*), flupyrsulfuron-methyl + thifensulfuron-methyl, glyphosate, iodosulfuron-methyl-sodium (*from seed*), iodosulfuron-methyl-sodium + mesosulfuron-methyl, iodosulfuron-methyl-sodium + mesosulfuron-methyl (*off-label*), pendimethalin, pendimethalin (*off-label*), pendimethalin + picolinafen, pendimethalin + pyroxsulam, pendimethalin + pyroxsulam (*MS only*), pinoxaden, propoxycarbazone-sodium, prosulfocarb, prosulfocarb (*off-label*), sulfosulfuron, sulfosulfuron (*moderate control of barren brome*), sulfosulfuron (*moderate control only*), tralkoxydim, tri-allate, tri-allate (*off-label*)

Weeds, miscellaneous | fluroxypyr, glyphosate

Edible fungi - Mushrooms

Diseases | Bacterial blotch | sodium hypochlorite (commodity substance)
Pests | Flies | deltamethrin (*off-label*), diflubenzuron (*off-label - other than mushrooms*), Metarhizium anisopliae (*off-label*)

Fruiting vegetables - Aubergines

Diseases | Botrytis | azoxystrobin (*off-label*)
| Damping off | Bacillus subtilis (*off-label*)
| Didymella stem rot | azoxystrobin (*off-label*)
| Helminthosporium seedling rot | Bacillus subtilis (*off-label*)
| Phytophthora | azoxystrobin (*off-label*), Bacillus subtilis (*off-label*)
| Powdery mildew | azoxystrobin (*off-label*), sulphur (*off-label*)
| Pythium | Bacillus subtilis (*off-label*)
| Rhizoctonia | Bacillus subtilis (*off-label*)
| White rot | Bacillus subtilis (*off-label*)
Pests | Aphids | fatty acids (*off-label*)
| Thrips | fatty acids (*off-label*)
| Whiteflies | Lecanicillium lecanii

Fruiting vegetables - Cucurbits

Diseases | Botrytis | mepanipyrim (*off-label*)
| Damping off | Bacillus subtilis (*off-label*)
| Disease control | mancozeb
| Helminthosporium seedling rot | Bacillus subtilis (*off-label*)
| Phytophthora | Bacillus subtilis (*off-label*)
| Powdery mildew | azoxystrobin (*off-label*), bupirimate (*off-label*), bupirimate (*outdoor only*), myclobutanil (*off-label*)
| Pythium | Bacillus subtilis (*off-label*)
| Rhizoctonia | Bacillus subtilis (*off-label*)
| White rot | Bacillus subtilis (*off-label*)

Pests	Aphids	fatty acids, fatty acids (*off-label*), pirimicarb, pirimicarb (*off-label*)
	Mealybugs	fatty acids
	Pests, miscellaneous	thiacloprid (*off-label*)
	Scale insects	fatty acids
	Spider mites	fatty acids
	Thrips	fatty acids (*off-label*)
	Whiteflies	fatty acids
Weeds	Broad-leaved weeds	isoxaben (*off-label*), propyzamide (*off-label*)
	Grass weeds	propyzamide (*off-label*)

Fruiting vegetables - Peppers

Diseases	Damping off	Bacillus subtilis (*off-label*)
	Helminthosporium seedling rot	Bacillus subtilis (*off-label*)
	Phytophthora	Bacillus subtilis (*off-label*)
	Pythium	Bacillus subtilis (*off-label*)
	Rhizoctonia	Bacillus subtilis (*off-label*)
	White rot	Bacillus subtilis (*off-label*)
Pests	Aphids	fatty acids, pirimicarb
	Mealybugs	fatty acids
	Scale insects	fatty acids
	Spider mites	fatty acids
	Whiteflies	fatty acids

Fruiting vegetables - Tomatoes

Diseases	Botrytis	fenhexamid (*off-label*)
	Damping off	Bacillus subtilis (*off-label*), copper oxychloride
	Foot rot	copper oxychloride
	Helminthosporium seedling rot	Bacillus subtilis (*off-label*)
	Phytophthora	Bacillus subtilis (*off-label*), copper oxychloride, mancozeb
	Pythium	Bacillus subtilis (*off-label*)
	Rhizoctonia	Bacillus subtilis (*off-label*)
	White rot	Bacillus subtilis (*off-label*)
Pests	Aphids	fatty acids, pirimicarb, pyrethrins
	Beetles	pyrethrins
	Caterpillars	pyrethrins
	Mealybugs	fatty acids
	Scale insects	fatty acids
	Spider mites	fatty acids
	Whiteflies	fatty acids, pyrethrins

Herb crops - Herbs

Crop control	Desiccation	diquat (*off-label*)
Diseases	Botrytis	Bacillus subtilis (*off-label*), cyprodinil + fludioxonil (*off-label*), fenhexamid (*off-label*), Gliocladium catenulatum (*off-label - moderate control*)
	Damping off	Bacillus subtilis (*off-label*)
	Disease control	dimethomorph + mancozeb (*off-label*), mancozeb + metalaxyl-M (*off-label*), mandipropamid, prochloraz (*off-label*)
	Downy mildew	copper oxychloride (*off-label*), dimethomorph (*off-label*), dimethomorph + mancozeb (*off-label*), mandipropamid, metalaxyl-M (*off-label*)
	Foliar diseases	difenoconazole (*off-label*), tebuconazole (*off-label*)
	Fusarium	Gliocladium catenulatum (*off-label - moderate control*)
	Helminthosporium seedling rot	Bacillus subtilis (*off-label*)
	Phytophthora	Bacillus subtilis (*off-label*), Gliocladium catenulatum (*off-label - moderate control*)
	Powdery mildew	azoxystrobin (*off-label*), tebuconazole (*off-label*)
	Pythium	Bacillus subtilis (*off-label*), Gliocladium catenulatum (*off-label - moderate control*), metalaxyl-M (*off-label*)
	Rhizoctonia	Bacillus subtilis (*off-label*), Gliocladium catenulatum (*off-label - moderate control*)
	Ring spot	azoxystrobin (*off-label*)
	Rust	azoxystrobin (*off-label*), azoxystrobin + difenoconazole (*off-label*), tebuconazole (*off-label*)
	Sclerotinia	cyprodinil + fludioxonil (*off-label*)
	Seed-borne diseases	thiram (*seed soak*)
	Septoria	azoxystrobin + difenoconazole (*off-label*)
	White blister	boscalid + pyraclostrobin (*off-label*)
	White rot	Bacillus subtilis (*off-label*)
Pests	Aphids	acetamiprid (*off-label*), cypermethrin (*off-label*), deltamethrin (*off-label*), fatty acids (*off-label*), lambda-cyhalothrin (*off-label*), Metarhizium anisopliae (*off-label*), pirimicarb (*off-label*), spirotetramat (*off-label*)
	Beetles	deltamethrin (*off-label*), deltamethrin (*off-label - baby leaf production*), lambda-cyhalothrin (*off-label*)
	Caterpillars	Bacillus thuringiensis (*off-label*), cypermethrin (*off-label*), deltamethrin (*off-label*), deltamethrin (*off-label - baby leaf production*), diflubenzuron (*off-label*)
	Cutworms	lambda-cyhalothrin (*off-label*)
	Flies	lambda-cyhalothrin (*off-label*), Metarhizium anisopliae (*off-label*), tefluthrin (*off-label - seed treatment*)
	Leafhoppers	deltamethrin (*off-label*)
	Midges	Metarhizium anisopliae (*off-label*)
	Pests, miscellaneous	deltamethrin (*off-label*), lambda-cyhalothrin (*off-label*), spinosad
	Thrips	deltamethrin (*off-label*), fatty acids (*off-label*), Metarhizium anisopliae (*off-label*), pirimicarb (*off-label*)
	Whiteflies	spirotetramat (*off-label*)

Weeds	Broad-leaved weeds	bentazone (*off-label*), chloridazon (*off-label*), chloridazon + quinmerac (*off-label*), chlorpropham (*off-label*), clomazone (*off-label*), clopyralid (*off-label*), dimethenamid-p + pendimethalin (*off-label*), ioxynil (*off-label*), lenacil (*off-label*), linuron, linuron (*off-label*), metamitron (*off-label*), metazachlor (*off-label*), metazachlor + quinmerac (*off-label*), napropamide (*off-label*), pendimethalin (*off-label*), phenmedipham (*off-label*), propyzamide (*off-label*), prosulfocarb (*off-label*), pyridate (*off-label*), S-metolachlor (*off-label*)
	Crops as weeds	propaquizafop (*off-label*)
	Grass weeds	chloridazon (*off-label*), chlorpropham (*off-label*), dimethenamid-p + pendimethalin (*off-label*), fluazifop-P-butyl (*off-label*), lenacil (*off-label*), linuron, linuron (*off-label*), metamitron (*off-label*), metazachlor (*off-label*), metazachlor + quinmerac (*off-label*), napropamide (*off-label*), pendimethalin (*off-label*), propaquizafop (*off-label*), propyzamide (*off-label*), prosulfocarb (*off-label*), S-metolachlor (*off-label*), tepraloxydim (*off-label*)
	Weeds, miscellaneous	diquat (*off-label*), glyphosate (*off-label*)

Leafy vegetables - Endives

Diseases	Botrytis	cyprodinil + fludioxonil (*off-label*), fenhexamid (*off-label*), Gliocladium catenulatum (*off-label - moderate control*)
	Damping off	thiram (*off-label*)
	Disease control	dimethomorph + mancozeb, dimethomorph + mancozeb (*off-label*), mandipropamid, prochloraz (*off-label*)
	Downy mildew	boscalid + pyraclostrobin (*off-label*), copper oxychloride (*off-label*), dimethomorph + mancozeb (*off-label*), mandipropamid
	Fusarium	Gliocladium catenulatum (*off-label - moderate control*)
	Phytophthora	Gliocladium catenulatum (*off-label - moderate control*)
	Powdery mildew	azoxystrobin (*off-label*)
	Pythium	Gliocladium catenulatum (*off-label - moderate control*)
	Rhizoctonia	Gliocladium catenulatum (*off-label - moderate control*)
	Rust	azoxystrobin (*off-label*)
Pests	Aphids	acetamiprid (*off-label*), cypermethrin (*off-label*), deltamethrin (*off-label*), fatty acids (*off-label*), lambda-cyhalothrin (*off-label*), pirimicarb (*off-label*), pymetrozine (*off-label*), spirotetramat (*off-label*)
	Caterpillars	Bacillus thuringiensis (*off-label*), cypermethrin (*off-label*), diflubenzuron (*off-label*)
	Pests, miscellaneous	deltamethrin (*off-label*), lambda-cyhalothrin (*off-label*), spinosad
	Thrips	fatty acids (*off-label*), pirimicarb (*off-label*)
	Whiteflies	spirotetramat (*off-label*)
Weeds	Broad-leaved weeds	chlorpropham, pendimethalin (*off-label*), propyzamide (*off-label*), S-metolachlor (*off-label*)
	Grass weeds	chlorpropham, pendimethalin (*off-label*), propyzamide (*off-label*), S-metolachlor (*off-label*)

Leafy vegetables - Lettuce

Diseases	Alternaria	iprodione
	Botrytis	boscalid + pyraclostrobin, boscalid + pyraclostrobin (*off-label*), cyprodinil + fludioxonil (*off-label*), fenhexamid (*off-label*), Gliocladium catenulatum (*off-label - moderate control*), iprodione
	Bottom rot	boscalid + pyraclostrobin
	Damping off	Bacillus subtilis (*off-label*), thiram (*off-label*)
	Disease control	boscalid + pyraclostrobin (*off-label*), dimethomorph + mancozeb (*off-label*), mancozeb + metalaxyl-M (*off-label*), mandipropamid, prochloraz (*off-label*)
	Downy mildew	azoxystrobin (*off-label*), copper oxychloride (*off-label*), dimethomorph (*off-label*), dimethomorph + mancozeb (*off-label*), fosetyl-aluminium + propamocarb hydrochloride, mancozeb (*off-label*), mandipropamid
	Fusarium	Gliocladium catenulatum (*off-label - moderate control*)
	Helminthosporium seedling rot	Bacillus subtilis (*off-label*)
	Phytophthora	Bacillus subtilis (*off-label*), Gliocladium catenulatum (*off-label - moderate control*)
	Powdery mildew	azoxystrobin (*off-label*)
	Pythium	Bacillus subtilis (*off-label*), fosetyl-aluminium + propamocarb hydrochloride, Gliocladium catenulatum (*off-label - moderate control*)
	Rhizoctonia	Bacillus subtilis (*off-label*), Gliocladium catenulatum (*off-label - moderate control*)
	Rust	azoxystrobin (*off-label*)
	Sclerotinia	azoxystrobin (*off-label*)
	Soft rot	boscalid + pyraclostrobin
	White rot	Bacillus subtilis (*off-label*)
Pests	Aphids	acetamiprid (*off-label*), cypermethrin (*off-label*), deltamethrin, deltamethrin (*off-label*), dimethoate, fatty acids, fatty acids (*off-label*), lambda-cyhalothrin (*off-label*), pirimicarb, pirimicarb (*off-label*), pirimicarb (*outdoor crops*), pymetrozine (*off-label*), pyrethrins, spirotetramat, spirotetramat (*off-label*)
	Beetles	deltamethrin
	Caterpillars	Bacillus thuringiensis (*off-label*), cypermethrin (*off-label*), deltamethrin, diflubenzuron (*off-label*), pyrethrins
	Cutworms	deltamethrin, lambda-cyhalothrin
	Leafhoppers	deltamethrin
	Mealybugs	fatty acids
	Pests, miscellaneous	deltamethrin (*off-label*), lambda-cyhalothrin (*off-label*), spinosad
	Scale insects	fatty acids
	Slugs/snails	methiocarb
	Spider mites	fatty acids
	Thrips	deltamethrin, fatty acids (*off-label*)
	Whiteflies	fatty acids, spirotetramat (*off-label*)

Weeds	Broad-leaved weeds	chlorpropham, dimethenamid-p + pendimethalin (*off-label*), napropamide (*off-label*), pendimethalin (*off-label*), propyzamide, propyzamide (*off-label*), propyzamide (*outdoor crops*), S-metolachlor (*off-label*)
	Grass weeds	chlorpropham, dimethenamid-p + pendimethalin (*off-label*), napropamide (*off-label*), pendimethalin (*off-label*), propyzamide, propyzamide (*off-label*), propyzamide (*outdoor crops*), S-metolachlor (*off-label*)

Leafy vegetables - Spinach

Diseases	Botrytis	cyprodinil + fludioxonil (*off-label*), Gliocladium catenulatum (*off-label - moderate control*)
	Damping off	Bacillus subtilis (*off-label*)
	Disease control	mandipropamid
	Downy mildew	boscalid + pyraclostrobin (*off-label*), copper oxychloride (*off-label*), dimethomorph (*off-label*), mandipropamid, metalaxyl-M (*off-label*)
	Fusarium	Gliocladium catenulatum (*off-label - moderate control*)
	Helminthosporium seedling rot	Bacillus subtilis (*off-label*)
	Phytophthora	Bacillus subtilis (*off-label*), Gliocladium catenulatum (*off-label - moderate control*)
	Pythium	Bacillus subtilis (*off-label*), Gliocladium catenulatum (*off-label - moderate control*), metalaxyl-M (*off-label*)
	Rhizoctonia	Bacillus subtilis (*off-label*), Gliocladium catenulatum (*off-label - moderate control*)
	White rot	Bacillus subtilis (*off-label*)
	White tip	spirotetramat (*off-label*)
Pests	Aphids	acetamiprid (*off-label*), cypermethrin (*off-label*), fatty acids (*off-label*), pirimicarb (*off-label*), pymetrozine (*off-label*), spirotetramat (*off-label*)
	Caterpillars	Bacillus thuringiensis (*off-label*), cypermethrin (*off-label*), diflubenzuron (*off-label*)
	Flies	tefluthrin (*off-label - seed treatment*)
	Slugs/snails	methiocarb
	Thrips	fatty acids (*off-label*)
Weeds	Broad-leaved weeds	chloridazon (*off-label*), chloridazon + quinmerac (*off-label*), chlorpropham (*off-label*), clopyralid (*off-label*), lenacil (*off-label*), phenmedipham (*off-label*)
	Grass weeds	chloridazon (*off-label*), chlorpropham (*off-label*), lenacil (*off-label*)

Leafy vegetables - Watercress

Diseases	Bacterial blight	copper oxychloride (*off-label*)
	Bottom rot	copper oxychloride (*off-label*)
	Damping off	metalaxyl-M (*off-label*)
	Downy mildew	metalaxyl-M (*off-label*)
	Phytophthora	copper oxychloride (*off-label*)
	Pythium	copper oxychloride (*off-label - during propagation*)
	Rhizoctonia	copper oxychloride (*off-label - during propagation*)
	Spear rot	copper oxychloride (*off-label*)

	Stem canker	copper oxychloride (*off-label*)
Pests	Caterpillars	Bacillus thuringiensis (*off-label*)

Legumes - Beans (Phaseolus)

Crop control	Desiccation	diquat (*off-label*)
Diseases	Alternaria	iprodione (*off-label*)
	Botrytis	azoxystrobin (*off-label*), boscalid + pyraclostrobin (*off-label*), cyprodinil + fludioxonil (*moderate control*), iprodione (*off-label*)
	Damping off	thiram (*seed treatment*)
	Rust	tebuconazole (*off-label*)
	Sclerotinia	cyprodinil + fludioxonil
Pests	Aphids	fatty acids, pirimicarb
	Caterpillars	Bacillus thuringiensis (*off-label*), lambda-cyhalothrin (*off-label*)
	Flies	chlorpyrifos (*off-label*), chlorpyrifos (*off-label - harvested as a dry pulse*), chlorpyrifos (*off-label - harvested dry as a pulse*)
	Mealybugs	fatty acids
	Pests, miscellaneous	lambda-cyhalothrin (*off-label*)
	Scale insects	fatty acids
	Spider mites	fatty acids
	Whiteflies	fatty acids
Weeds	Broad-leaved weeds	bentazone, clomazone (*off-label*), linuron (*off-label*), pendimethalin, pendimethalin (*off-label*), S-metolachlor (*off-label*)
	Crops as weeds	cycloxydim
	Grass weeds	cycloxydim, cycloxydim (*off-label*), fluazifop-P-butyl (*off-label*), linuron (*off-label*), pendimethalin (*off-label*), S-metolachlor (*off-label*)

Legumes - Beans (Vicia)

Crop control	Desiccation	diquat, glufosinate-ammonium, glyphosate
Diseases	Ascochyta	azoxystrobin (*off-label*)
	Botrytis	azoxystrobin (*off-label*), boscalid + pyraclostrobin (*off-label*), cyprodinil + fludioxonil (*moderate control*)
	Chocolate spot	boscalid + pyraclostrobin (*moderate*), chlorothalonil + cyproconazole, chlorothalonil + pyrimethanil, cyproconazole, iprodione + thiophanate-methyl, tebuconazole
	Damping off	thiram (*seed treatment*)
	Disease control	metconazole (*off-label*)
	Downy mildew	chlorothalonil + metalaxyl-M, cymoxanil + fludioxonil + metalaxyl-M (*off-label - seed treatment*)
	Rust	azoxystrobin, boscalid + pyraclostrobin, chlorothalonil + cyproconazole, cyproconazole, metconazole, tebuconazole, tebuconazole (*off-label*)
	Sclerotinia	cyprodinil + fludioxonil

Pests	Aphids	cypermethrin, cypermethrin (*off-label*), esfenvalerate, fatty acids, fatty acids (*off-label*), lambda-cyhalothrin, pirimicarb, zeta-cypermethrin (*off-label*)
	Beetles	lambda-cyhalothrin, lambda-cyhalothrin (*off-label*)
	Birds/mammals	aluminium ammonium sulphate
	Caterpillars	Bacillus thuringiensis (*off-label*), lambda-cyhalothrin
	Mealybugs	fatty acids
	Pests, miscellaneous	lambda-cyhalothrin (*off-label*), zeta-cypermethrin (*off-label*)
	Scale insects	fatty acids
	Spider mites	fatty acids
	Thrips	fatty acids (*off-label*), lambda-cyhalothrin (*off-label*)
	Weevils	alpha-cypermethrin, cypermethrin, deltamethrin, esfenvalerate, lambda-cyhalothrin, zeta-cypermethrin
	Whiteflies	fatty acids
Weeds	Broad-leaved weeds	bentazone, carbetamide, clomazone, clomazone (*off-label*), clomazone + linuron, glyphosate, imazamox + pendimethalin, imazamox + pendimethalin (*off-label*), isoxaben + terbuthylazine, isoxaben + terbuthylazine (*off-label*), linuron, pendimethalin (*off-label*), propyzamide, prosulfocarb (*off-label*)
	Crops as weeds	carbetamide, cycloxydim, fluazifop-P-butyl, glyphosate, pendimethalin (*off-label*), propyzamide, quizalofop-P-ethyl, quizalofop-P-tefuryl, tepraloxydim
	Grass weeds	carbetamide, clomazone + linuron, cycloxydim, cycloxydim (*off-label*), diquat, fluazifop-P-butyl, fluazifop-P-butyl (*off-label*), glyphosate, isoxaben + terbuthylazine (*off-label*), linuron, pendimethalin (*off-label*), propaquizafop, propyzamide, prosulfocarb (*off-label*), quizalofop-P-ethyl, quizalofop-P-tefuryl, tepraloxydim, tepraloxydim (*off-label*), tri-allate, tri-allate (*off-label*)
	Weeds, miscellaneous	diquat, glyphosate

Legumes - Forage legumes, general

Weeds	Broad-leaved weeds	pendimethalin (*off-label*)
	Grass weeds	pendimethalin (*off-label*)

Legumes - Lupins

Crop control	Desiccation	diquat (*off-label*), glyphosate (*off-label*)
Diseases	Ascochyta	azoxystrobin (*off-label*), metconazole (*qualified minor use*)
	Botrytis	cyprodinil + fludioxonil (*off-label*), metconazole (*qualified minor use*)
	Damping off	cymoxanil + fludioxonil + metalaxyl-M (*off-label*), thiram (*off-label - seed treatment*)
	Disease control	azoxystrobin (*off-label*)
	Downy mildew	cymoxanil + fludioxonil + metalaxyl-M (*off-label*)
	Pythium	cymoxanil + fludioxonil + metalaxyl-M (*off-label*)
	Rust	azoxystrobin (*off-label*), metconazole (*qualified minor use*)

	Seed-borne diseases	thiram (*off-label - seed treatment*)
Pests	Aphids	deltamethrin (*off-label*), fatty acids (*off-label*), lambda-cyhalothrin (*off-label*), zeta-cypermethrin (*off-label*)
	Pests, miscellaneous	alpha-cypermethrin (*off-label*), deltamethrin (*off-label*), lambda-cyhalothrin (*off-label*), zeta-cypermethrin (*off-label*)
	Thrips	fatty acids (*off-label*)
Weeds	Broad-leaved weeds	carbetamide (*off-label*), clomazone (*off-label*), isoxaben + terbuthylazine (*off-label*), pendimethalin (*off-label*), pyridate (*off-label*)
	Crops as weeds	propaquizafop (*off-label*)
	Grass weeds	carbetamide (*off-label*), cycloxydim (*off-label*), fluazifop-P-butyl (*off-label*), isoxaben + terbuthylazine (*off-label*), pendimethalin (*off-label*), propaquizafop (*off-label*), tepraloxydim (*off-label*), tri-allate (*off-label*)
	Weeds, miscellaneous	glyphosate (*off-label*)

Legumes - Peas

Crop control	Desiccation	diquat, glufosinate-ammonium, glufosinate-ammonium (*not for seed*), glyphosate
Diseases	Alternaria	iprodione (*off-label*)
	Ascochyta	azoxystrobin, boscalid + pyraclostrobin (*moderate control only*), chlorothalonil + pyrimethanil, cymoxanil + fludioxonil + metalaxyl-M (*seed treatment*), cyprodinil + fludioxonil, metconazole (*reduction*)
	Botrytis	chlorothalonil + cyproconazole, chlorothalonil + pyrimethanil, cyprodinil + fludioxonil (*moderate control*), iprodione (*off-label*), metconazole (*reduction*)
	Damping off	cymoxanil + fludioxonil + metalaxyl-M (*seed treatment*), thiram (*seed treatment*)
	Downy mildew	cymoxanil + fludioxonil + metalaxyl-M (*seed treatment*)
	Mycosphaerella	chlorothalonil + pyrimethanil, cyprodinil + fludioxonil, metconazole (*reduction*)
	Powdery mildew	sulphur (*off-label*)
	Pythium	cymoxanil + fludioxonil + metalaxyl-M (*seed treatment*)
	Rust	metconazole
	Sclerotinia	cyprodinil + fludioxonil
Pests	Aphids	alpha-cypermethrin, alpha-cypermethrin (*reduction*), cypermethrin, esfenvalerate, fatty acids, lambda-cyhalothrin, pirimicarb, thiacloprid, zeta-cypermethrin
	Beetles	lambda-cyhalothrin
	Birds/mammals	aluminium ammonium sulphate
	Caterpillars	alpha-cypermethrin, cypermethrin, deltamethrin, lambda-cyhalothrin, zeta-cypermethrin
	Mealybugs	fatty acids
	Midges	deltamethrin, lambda-cyhalothrin, thiacloprid
	Scale insects	fatty acids
	Spider mites	fatty acids

	Weevils	alpha-cypermethrin, cypermethrin, deltamethrin, esfenvalerate, lambda-cyhalothrin, zeta-cypermethrin
	Whiteflies	fatty acids
Weeds	Broad-leaved weeds	bentazone, clomazone, clomazone + linuron, flumioxazine (*off-label*), glyphosate, imazamox + pendimethalin, isoxaben + terbuthylazine, linuron, MCPB, pendimethalin, pendimethalin (*off-label*)
	Crops as weeds	cycloxydim, fluazifop-P-butyl, flumioxazine (*off-label*), glyphosate, pendimethalin, quizalofop-P-ethyl, quizalofop-P-tefuryl, tepraloxydim
	Grass weeds	clomazone + linuron, cycloxydim, cycloxydim (*off-label*), diquat (*dry harvested*), fluazifop-P-butyl, glyphosate, linuron, pendimethalin, pendimethalin (*off-label*), propaquizafop, quizalofop-P-ethyl, quizalofop-P-tefuryl, tepraloxydim, tri-allate
	Weeds, miscellaneous	diquat (*dry harvested only*), glyphosate

Miscellaneous arable - Industrial crops

Crop control	Desiccation	glyphosate (*off-label*), glyphosate (*off-label - for wild bird seed production*)
Pests	Aphids	lambda-cyhalothrin (*off-label*)
	Pests, miscellaneous	lambda-cyhalothrin (*off-label*)
	Thrips	lambda-cyhalothrin (*off-label*)
	Whiteflies	lambda-cyhalothrin (*off-label*)
Weeds	Broad-leaved weeds	bromoxynil + ioxynil (*off-label*), clopyralid (*off-label*), flufenacet + isoxaflutole (*off-label*), fluroxypyr (*off-label*), MCPA (*off-label*), mecoprop-P (*off-label*), metsulfuron-methyl (*off-label*), metsulfuron-methyl + thifensulfuron-methyl (*off-label*), pendimethalin (*off-label*), tribenuron-methyl (*off-label*)
	Grass weeds	flufenacet + isoxaflutole (*off-label*), propoxycarbazone-sodium (*off-label*)
	Weeds, miscellaneous	glyphosate (*off-label*)

Miscellaneous arable - Miscellaneous arable crops

Crop control	Desiccation	glyphosate (*off-label*), glyphosate (*off-label - for wild bird seed production*)
	Miscellaneous non-selective situations	benzoic acid
Diseases	Botrytis	Bacillus subtilis (*off-label*), Gliocladium catenulatum (*moderate control*), iprodione + thiophanate-methyl (*off-label*)
	Didymella stem rot	Gliocladium catenulatum (*moderate control*)
	Disease control	azoxystrobin (*off-label*), boscalid (*off-label*), boscalid + pyraclostrobin (*off-label*), chlorothalonil (*off-label*), epoxiconazole (*off-label*), potassium bicarbonate (commodity substance)
	Downy mildew	boscalid + pyraclostrobin (*off-label*)
	Fusarium	Gliocladium catenulatum (*moderate control*)
	Phytophthora	Gliocladium catenulatum (*moderate control*)

	Powdery mildew	kresoxim-methyl (*off-label*)
	Pythium	Gliocladium catenulatum (*moderate control*)
	Rhizoctonia	Gliocladium catenulatum (*moderate control*)
	Sclerotinia	Coniothyrium minitans
Pests	Aphids	lambda-cyhalothrin (*off-label*), maltodextrin, pyrethrins, thiacloprid (*off-label*)
	Birds/mammals	aluminium ammonium sulphate
	Caterpillars	indoxacarb (*off-label*), pyrethrins
	Pests, miscellaneous	indoxacarb (*off-label*), lambda-cyhalothrin (*off-label*)
	Slugs/snails	ferric phosphate, metaldehyde, metaldehyde (*excluding potato and cauliflower*), metaldehyde (*excluding potatoes*)
	Spider mites	maltodextrin, pyrethrins
	Thrips	lambda-cyhalothrin (*off-label*), pyrethrins
	Whiteflies	lambda-cyhalothrin (*off-label*), maltodextrin
Plant growth regulation	Growth control	trinexapac-ethyl (*off-label*)
Weeds	Broad-leaved weeds	bentazone (*off-label*), bromoxynil (*off-label*), bromoxynil + ioxynil (*off-label*), bromoxynil + prosulfuron (*off-label*), carfentrazone-ethyl (*before planting*), clomazone (*off-label*), clopyralid (*off-label*), dicamba + mecoprop-P (*off-label*), diflufenican (*off-label*), diquat, diquat (*around*), florasulam (*off-label*), florasulam + fluroxypyr (*off-label*), flufenacet + isoxaflutole (*off-label*), flufenacet + pendimethalin (*off-label*), fluroxypyr (*off-label*), isoxaben (*off-label - grown for game cover*), linuron (*off-label*), MCPA (*off-label*), MCPB (*off-label*), mecoprop-P (*off-label*), mesotrione (*off-label*), metsulfuron-methyl (*off-label*), pendimethalin (*off-label*), propyzamide (*off-label*), prosulfuron (*off-label*), pyridate (*off-label*), sulfosulfuron (*off-label*), thifensulfuron-methyl + tribenuron-methyl (*off-label*)
	Crops as weeds	carfentrazone-ethyl (*before planting*), quizalofop-P-tefuryl
	Grass weeds	diquat, diquat (*around only*), flufenacet + isoxaflutole (*off-label*), flufenacet + pendimethalin (*off-label*), linuron (*off-label*), mesotrione (*off-label*), pinoxaden (*off-label*), propyzamide (*off-label*), quizalofop-P-tefuryl, sulfosulfuron (*off-label*), tepraloxydim (*off-label*), tri-allate (*off-label - for wild bird seed production*)
	Weeds, miscellaneous	diquat (*around only*), glyphosate, glyphosate (*before planting*), glyphosate (*off-label*), glyphosate (*pre-sowing/planting*)

Miscellaneous arable - Miscellaneous arable situations

Pests	Slugs/snails	metaldehyde
Weeds	Broad-leaved weeds	amitrole, citronella oil, glufosinate-ammonium, glufosinate-ammonium (*uncropped*), glyphosate
	Crops as weeds	amitrole (*barley stubble*), fluazifop-P-butyl, glyphosate, tepraloxydim

Grass weeds		amitrole, fluazifop-P-butyl, glufosinate-ammonium, glufosinate-ammonium (*uncropped*), glyphosate, tepraloxydim
Weeds, miscellaneous		2,4-D + dicamba + triclopyr, cycloxydim, fluazifop-P-butyl, glufosinate-ammonium, glyphosate, glyphosate (*wiper application*), metsulfuron-methyl, tepraloxydim, thifensulfuron-methyl

Miscellaneous field vegetables - All vegetables

Plant growth regulation	Quality/yield control	sulphur
Weeds	Broad-leaved weeds	glufosinate-ammonium
	Grass weeds	glufosinate-ammonium
	Weeds, miscellaneous	glufosinate-ammonium

Oilseed crops - Linseed/flax

Crop control	Desiccation	diquat, glufosinate-ammonium, glyphosate
Diseases	Botrytis	tebuconazole (*reduction*)
	Disease control	boscalid (*off-label*), difenoconazole (*off-label*), metconazole (*off-label*), prochloraz (*off-label*)
	Downy mildew	fludioxonil + metalaxyl-M + thiamethoxam (*off-label*)
	Foliar diseases	cyproconazole (*off-label*), difenoconazole (*off-label*)
	Powdery mildew	tebuconazole
	Pythium	fludioxonil + metalaxyl-M + thiamethoxam (*off-label*)
	Seed-borne diseases	prochloraz (*seed treatment*)
Pests	Aphids	lambda-cyhalothrin (*off-label*), pirimicarb (*off-label*), tau-fluvalinate (*off-label*)
	Beetles	beta-cyfluthrin + imidacloprid (*off-label*), beta-cyfluthrin + imidacloprid (*off-label - seed treatment*), cypermethrin (*off-label*), zeta-cypermethrin
	Pests, miscellaneous	alpha-cypermethrin (*off-label*), lambda-cyhalothrin (*off-label*)
	Whiteflies	lambda-cyhalothrin (*off-label*)
Weeds	Broad-leaved weeds	amidosulfuron, amidosulfuron + iodosulfuron-methyl-sodium (*off-label*), bentazone, bromoxynil, bromoxynil (*off-label*), carbetamide (*off-label*), clopyralid + picloram (*off-label*), flupyrsulfuron-methyl (*off-label*), glyphosate, mesotrione (*off-label*), metazachlor (*off-label*), metazachlor + quinmerac (*off-label*), metsulfuron-methyl, napropamide (*off-label*), prosulfocarb (*off-label*)
	Crops as weeds	amidosulfuron + iodosulfuron-methyl-sodium (*off-label*), cycloxydim, fluazifop-P-butyl, glyphosate, mesotrione (*off-label*), propaquizafop (*off-label*), quizalofop-P-ethyl, quizalofop-P-tefuryl, tepraloxydim
	Grass weeds	carbetamide (*off-label*), cycloxydim, diquat, fluazifop-P-butyl, fluazifop-P-butyl (*off-label*), flupyrsulfuron-methyl (*off-label*), glyphosate, mesotrione (*off-label*), metazachlor (*off-label*), metazachlor + quinmerac (*off-label*), napropamide (*off-label*), propaquizafop, propaquizafop (*off-label*), prosulfocarb (*off-label*),

quizalofop-P-ethyl, quizalofop-P-tefuryl, tepraloxydim, tri-allate (*off-label*)

	Weeds, miscellaneous	diquat, glyphosate

Oilseed crops - Miscellaneous oilseeds

Crop control	Desiccation	diquat (*off-label*), glyphosate (*off-label*)
Diseases	Alternaria	boscalid + pyraclostrobin (*off-label*)
	Botrytis	boscalid (*off-label*), boscalid + pyraclostrobin (*off-label*)
	Damping off	cymoxanil + fludioxonil + metalaxyl-M (*off-label - seed treatment*), thiram (*off-label - seed treatment*)
	Disease control	azoxystrobin + chlorothalonil (*off-label*), boscalid (*off-label*), chlorothalonil + metalaxyl-M (*off-label*), difenoconazole (*off-label*), dimethomorph + mancozeb (*off-label*), propiconazole (*off-label*), tebuconazole (*off-label*)
	Downy mildew	chlorothalonil (*off-label*), cymoxanil + fludioxonil + metalaxyl-M (*off-label - seed treatment*), dimethomorph + mancozeb (*off-label*), mancozeb + metalaxyl-M (*off-label*)
	Fire	boscalid + pyraclostrobin (*off-label*)
	Foliar diseases	azoxystrobin (*off-label*), cyproconazole (*off-label*), difenoconazole (*off-label*), tebuconazole (*off-label*)
	Powdery mildew	azoxystrobin (*off-label*), sulphur (*off-label*)
	Pythium	cymoxanil + fludioxonil + metalaxyl-M (*off-label - seed treatment*), fludioxonil + metalaxyl-M + thiamethoxam (*off-label*), thiram (*off-label - seed treatment*)
	Rust	tebuconazole (*off-label*)
	Sclerotinia	azoxystrobin (*off-label*), boscalid (*off-label*)
	Seed-borne diseases	prochloraz (*off-label - seed treatment*)
Pests	Aphids	deltamethrin (*off-label*), lambda-cyhalothrin (*off-label*), pirimicarb (*off-label*), tau-fluvalinate (*off-label*)
	Beetles	beta-cyfluthrin + clothianidin (*off-label*), beta-cyfluthrin + imidacloprid (*off-label*), beta-cyfluthrin + imidacloprid (*off-label - seed treatment*), deltamethrin (*off-label*), lambda-cyhalothrin (*off-label*)
	Caterpillars	deltamethrin (*off-label*)
	Leafhoppers	deltamethrin (*off-label*)
	Pests, miscellaneous	alpha-cypermethrin (*off-label*), lambda-cyhalothrin (*off-label*)
	Thrips	deltamethrin (*off-label*)
	Whiteflies	lambda-cyhalothrin (*off-label*)
Weeds	Broad-leaved weeds	bifenox (*off-label*), carbetamide (*off-label*), clomazone (*off-label*), clopyralid (*off-label*), clopyralid + picloram (*off-label*), fluroxypyr (*off-label*), mesotrione (*off-label*), metazachlor (*off-label*), metazachlor + quinmerac (*off-label*), napropamide (*off-label*), pendimethalin (*off-label*)
	Crops as weeds	fluroxypyr (*off-label*), propaquizafop (*off-label*)
	Grass weeds	carbetamide (*off-label*), cycloxydim (*off-label*), fluazifop-P-butyl (*off-label*), metazachlor (*off-label*),

		metazachlor + quinmerac (*off-label*), napropamide (*off-label*), pendimethalin (*off-label*), propaquizafop (*off-label*), prosulfocarb (*off-label*), tepraloxydim (*off-label*)
	Weeds, miscellaneous	diquat (*off-label*), glyphosate (*off-label*)

Oilseed crops - Oilseed rape

Crop control	Desiccation	diquat, glufosinate-ammonium, glyphosate
Diseases	Alternaria	azoxystrobin, azoxystrobin + cyproconazole, boscalid, difenoconazole, fludioxonil + metalaxyl-M + thiamethoxam (*moderate control*), iprodione, iprodione + thiophanate-methyl, metconazole, prochloraz, tebuconazole
	Black scurf and stem canker	bixafen + prothioconazole + tebuconazole, difenoconazole, iprodione + thiophanate-methyl, prothioconazole + tebuconazole, tebuconazole
	Botrytis	iprodione, iprodione + thiophanate-methyl, prochloraz
	Damping off	thiram (*seed treatment*)
	Downy mildew	fludioxonil + metalaxyl-M + thiamethoxam
	Ear blight	thiophanate-methyl (*reduction only*)
	Fusarium	thiophanate-methyl
	Light leaf spot	bixafen + prothioconazole + tebuconazole, carbendazim, carbendazim + flusilazole, cyproconazole, difenoconazole, famoxadone + flusilazole, flusilazole, iprodione + thiophanate-methyl, metconazole, metconazole (*reduction*), prochloraz, prochloraz + propiconazole, propiconazole (*reduction*), prothioconazole + tebuconazole, prothioconazole + tebuconazole (*moderate control*), prothioconazole + tebuconazole (*moderate control only*), tebuconazole
	Phoma leaf spot	carbendazim + flusilazole, cyproconazole, prothioconazole + tebuconazole
	Powdery mildew	sulphur
	Pythium	fludioxonil + metalaxyl-M + thiamethoxam (*reduction*)
	Ring spot	tebuconazole (*reduction*)
	Sclerotinia	azoxystrobin, azoxystrobin + cyproconazole, bixafen + prothioconazole + tebuconazole, boscalid, boscalid + metconazole, iprodione (*moderate control*), iprodione + thiophanate-methyl, picoxystrobin, prochloraz, prochloraz + tebuconazole, prothioconazole, prothioconazole + tebuconazole, tebuconazole, thiophanate-methyl
	Seed-borne diseases	fludioxonil + metalaxyl-M + thiamethoxam
	Stem canker	bixafen + prothioconazole + tebuconazole, carbendazim + flusilazole (*reduction*), famoxadone + flusilazole (*suppression*), metconazole (*reduction*), prochloraz, prochloraz + propiconazole, prochloraz + thiram (*seed treatment*), prothioconazole + tebuconazole, tebuconazole
	White leaf spot	prochloraz
Pests	Aphids	acetamiprid, alpha-cypermethrin, beta-cyfluthrin + clothianidin, cypermethrin, deltamethrin, fludioxonil +

		metalaxyl-M + thiamethoxam (*early season only*), lambda-cyhalothrin, pirimicarb, tau-fluvalinate
	Beetles	alpha-cypermethrin, beta-cyfluthrin, beta-cyfluthrin + clothianidin, beta-cyfluthrin + imidacloprid (*seed treatment*), cypermethrin, deltamethrin, fludioxonil + metalaxyl-M + thiamethoxam (*reduction of damage*), indoxacarb, lambda-cyhalothrin, pymetrozine, tau-fluvalinate, thiacloprid, zeta-cypermethrin
	Birds/mammals	aluminium ammonium sulphate
	Caterpillars	lambda-cyhalothrin
	Flies	beta-cyfluthrin + clothianidin, beta-cyfluthrin + clothianidin (*reduction of feeding activity only*)
	Midges	alpha-cypermethrin, beta-cyfluthrin, cypermethrin, cypermethrin (*some coincidental control only*), lambda-cyhalothrin, zeta-cypermethrin
	Slugs/snails	methiocarb
	Weevils	alpha-cypermethrin, beta-cyfluthrin, cypermethrin, deltamethrin, lambda-cyhalothrin, zeta-cypermethrin
Plant growth regulation	Growth control	metconazole, tebuconazole
	Quality/yield control	sulphur
Weeds	Broad-leaved weeds	bifenox (*off-label*), carbetamide, clomazone, clomazone + metazachlor, clopyralid, clopyralid + picloram, clopyralid + picloram (*off-label*), dimethenamid-p + metazachlor, dimethenamid-p + metazachlor + quinmerac, glyphosate, imazamox + metazachlor, metazachlor, metazachlor + quinmerac, napropamide, propyzamide, pyridate (*off-label*)
	Crops as weeds	carbetamide, cycloxydim, fluazifop-P-butyl, glyphosate, propyzamide, quizalofop-P-ethyl, quizalofop-P-tefuryl, tepraloxydim
	Grass weeds	carbetamide, clomazone + metazachlor, cycloxydim, dimethenamid-p + metazachlor + quinmerac, diquat, fluazifop-P-butyl, glyphosate, imazamox + metazachlor, metazachlor, metazachlor + quinmerac, napropamide, propaquizafop, propyzamide, quizalofop-P-ethyl, quizalofop-P-tefuryl, tepraloxydim, tri-allate (*off-label*)
	Weeds, miscellaneous	diquat, glyphosate

Oilseed crops - Soya

Diseases	Damping off	thiram (*seed treatment - qualified minor use*)
Weeds	Broad-leaved weeds	bentazone (*off-label*), clomazone (*off-label*), thifensulfuron-methyl (*off-label*)

Oilseed crops - Sunflowers

Crop control	Desiccation	diquat (*off-label*)
Pests	Slugs/snails	methiocarb
Plant growth regulation	Growth control	daminozide
Weeds	Broad-leaved weeds	isoxaben (*off-label - grown for game cover*), pendimethalin

Crops as weeds	pendimethalin
Grass weeds	pendimethalin

Root and tuber crops - Beet crops

Diseases	Black leg	hymexazol (*seed treatment*)
	Botrytis	cyprodinil + fludioxonil (*off-label*), Gliocladium catenulatum (*off-label - moderate control*), iprodione (*off-label*)
	Cercospora leaf spot	azoxystrobin + cyproconazole, cyproconazole + trifloxystrobin, epoxiconazole + pyraclostrobin
	Damping off	Bacillus subtilis (*off-label*), cymoxanil + fludioxonil + metalaxyl-M (*off-label - seed treatment*)
	Disease control	boscalid + pyraclostrobin (*off-label*), flusilazole (*off-label*), mandipropamid
	Downy mildew	boscalid + pyraclostrobin (*off-label*), copper oxychloride (*off-label*), dimethomorph (*off-label*), fosetyl-aluminium + propamocarb hydrochloride (*off-label*), mandipropamid, metalaxyl-M (*off-label*)
	Foliar diseases	cyproconazole (*off-label*), epoxiconazole + pyraclostrobin (*off-label*), flusilazole (*off-label*)
	Fusarium	Gliocladium catenulatum (*off-label - moderate control*)
	Helminthosporium seedling rot	Bacillus subtilis (*off-label*)
	Phytophthora	Bacillus subtilis (*off-label*), Gliocladium catenulatum (*off-label - moderate control*)
	Powdery mildew	azoxystrobin + cyproconazole, carbendazim + flusilazole, cyproconazole, cyproconazole + trifloxystrobin, difenoconazole + fenpropidin, epoxiconazole + pyraclostrobin, fenpropimorph (*off-label*), flusilazole, quinoxyfen, sulphur, sulphur (*off-label*)
	Pythium	Bacillus subtilis (*off-label*), Gliocladium catenulatum (*off-label - moderate control*), metalaxyl-M, metalaxyl-M (*off-label*)
	Ramularia leaf spots	azoxystrobin + cyproconazole, cyproconazole, cyproconazole + trifloxystrobin, difenoconazole + fenpropidin, epoxiconazole + pyraclostrobin, propiconazole (*reduction*)
	Rhizoctonia	azoxystrobin (*off-label*), Bacillus subtilis (*off-label*), Gliocladium catenulatum (*off-label - moderate control*)
	Root malformation disorder	azoxystrobin (*off-label*), metalaxyl-M (*off-label*)
	Rust	azoxystrobin + cyproconazole, carbendazim + flusilazole, cyproconazole + trifloxystrobin, difenoconazole + fenpropidin, epoxiconazole + pyraclostrobin, fenpropimorph (*off-label*), flusilazole, propiconazole
	Sclerotinia	cyprodinil + fludioxonil (*off-label*)
	Seed-borne diseases	thiram (*seed soak*)
	White rot	Bacillus subtilis (*off-label*)
Pests	Aphids	acetamiprid (*off-label*), beta-cyfluthrin, beta-cyfluthrin + clothianidin (*seed treatment*), cypermethrin, cypermethrin (*off-label*), deltamethrin (*off-label*), dimethoate (*excluding Myzus persicae*), fatty acids

		(*off-label*), imidacloprid, lambda-cyhalothrin, lambda-cyhalothrin (*off-label*), oxamyl, oxamyl (*off-label*), pirimicarb, pirimicarb (*off-label*), pymetrozine (*off-label*), spirotetramat (*off-label*), thiacloprid (*off-label*), zeta-cypermethrin (*off-label*)
	Beetles	chlorpyrifos, chlorpyrifos (*off-label*), deltamethrin, imidacloprid, lambda-cyhalothrin, lambda-cyhalothrin (*off-label*), oxamyl, oxamyl (*off-label*), tefluthrin (*seed treatment*), thiamethoxam (*seed treatment*)
	Birds/mammals	aluminium ammonium sulphate
	Caterpillars	Bacillus thuringiensis (*off-label*), cypermethrin, cypermethrin (*off-label*), lambda-cyhalothrin, lambda-cyhalothrin (*off-label*)
	Cutworms	Bacillus thuringiensis (*off-label*), cypermethrin, lambda-cyhalothrin, lambda-cyhalothrin (*off-label*), methiocarb (*reduction*), zeta-cypermethrin
	Flies	oxamyl, oxamyl (*off-label*)
	Free-living nematodes	oxamyl, oxamyl (*off-label*)
	Leaf miners	dimethoate, imidacloprid, lambda-cyhalothrin, lambda-cyhalothrin (*off-label*), thiamethoxam (*seed treatment*)
	Leatherjackets	chlorpyrifos, chlorpyrifos (*off-label*), methiocarb (*reduction*)
	Millipedes	imidacloprid, methiocarb (*reduction*), oxamyl, oxamyl (*off-label*), tefluthrin (*seed treatment*), thiamethoxam (*seed treatment*)
	Pests, miscellaneous	chlorpyrifos (*off-label*), deltamethrin (*off-label*), lambda-cyhalothrin (*off-label*), zeta-cypermethrin (*off-label*)
	Slugs/snails	methiocarb
	Springtails	imidacloprid, tefluthrin (*seed treatment*), thiamethoxam (*seed treatment*)
	Symphylids	imidacloprid, tefluthrin (*seed treatment*), thiamethoxam (*seed treatment*)
	Thrips	fatty acids (*off-label*)
	Weevils	lambda-cyhalothrin
	Whiteflies	spirotetramat (*off-label*)
	Wireworms	thiamethoxam (*seed treatment - reduction*)
Plant growth regulation	Quality/yield control	sulphur
Weeds	Broad-leaved weeds	carbetamide, chloridazon, chloridazon (*off-label*), chloridazon + ethofumesate, chloridazon + metamitron, chloridazon + metamitron (*off-label*), chloridazon + quinmerac, chloridazon + quinmerac (*off-label*), chlorpropham (*off-label*), chlorpropham + metamitron, clomazone (*off-label*), clopyralid, clopyralid (*off-label*), desmedipham + ethofumesate + lenacil + phenmedipham, desmedipham + ethofumesate + phenmedipham, desmedipham + phenmedipham, diquat, ethofumesate, ethofumesate + metamitron, ethofumesate + metamitron + phenmedipham, ethofumesate + phenmedipham, glufosinate-ammonium, glyphosate, lenacil, lenacil (*off-label*), lenacil + triflusulfuron-methyl, lenacil + triflusulfuron-methyl (*off-label*),

		metamitron, phenmedipham, phenmedipham (*off-label*), propyzamide, S-metolachlor (*off-label*), triflusulfuron-methyl, triflusulfuron-methyl (*off-label*)
	Crops as weeds	carbetamide, cycloxydim, fluazifop-P-butyl, glyphosate, glyphosate (*wiper application*), lenacil + triflusulfuron-methyl, propaquizafop (*off-label*), propyzamide, quizalofop-P-ethyl, quizalofop-P-tefuryl, tepraloxydim
	Grass weeds	carbetamide, chloridazon, chloridazon (*off-label*), chloridazon + ethofumesate, chloridazon + metamitron, chloridazon + metamitron (*off-label*), chloridazon + quinmerac, chlorpropham (*off-label*), chlorpropham + metamitron, cycloxydim, desmedipham + ethofumesate + lenacil + phenmedipham, desmedipham + ethofumesate + phenmedipham, diquat, ethofumesate, ethofumesate + metamitron, ethofumesate + metamitron + phenmedipham, ethofumesate + phenmedipham, fluazifop-P-butyl, fluazifop-P-butyl (*off-label*), glufosinate-ammonium, glyphosate, lenacil, lenacil (*off-label*), metamitron, propaquizafop, propaquizafop (*off-label*), propyzamide, quizalofop-P-ethyl, quizalofop-P-tefuryl, S-metolachlor (*off-label*), tepraloxydim, tepraloxydim (*off-label*), tri-allate
	Weeds, miscellaneous	diquat, glufosinate-ammonium, glyphosate

Root and tuber crops - Carrots/parsnips

Diseases	Alternaria	azoxystrobin, azoxystrobin (*off-label*), azoxystrobin + difenoconazole, azoxystrobin + difenoconazole (*off-label*), boscalid + pyraclostrobin (*moderate*), cyprodinil + fludioxonil (*moderate control*), fenpropimorph (*off-label*), iprodione + thiophanate-methyl (*off-label*), prothioconazole, tebuconazole, tebuconazole + trifloxystrobin
	Botrytis	cyprodinil + fludioxonil (*moderate control*), Gliocladium catenulatum (*off-label - moderate control*)
	Cavity spot	metalaxyl-M, metalaxyl-M (*off-label*), metalaxyl-M (*off-label - reduction*), metalaxyl-M (*reduction only*)
	Crown rot	fenpropimorph (*off-label*), iprodione + thiophanate-methyl (*off-label*)
	Damping off	Bacillus subtilis (*off-label*)
	Disease control	mancozeb
	Foliar diseases	tebuconazole + trifloxystrobin (*off-label*)
	Fusarium	Gliocladium catenulatum (*off-label - moderate control*)
	Helminthosporium seedling rot	Bacillus subtilis (*off-label*)
	Phytophthora	Bacillus subtilis (*off-label*), Gliocladium catenulatum (*off-label - moderate control*)
	Powdery mildew	azoxystrobin, azoxystrobin (*off-label*), azoxystrobin + difenoconazole, azoxystrobin + difenoconazole (*off-label*), boscalid + pyraclostrobin, fenpropimorph (*off-label*), prothioconazole, sulphur (*off-label*), tebuconazole, tebuconazole + trifloxystrobin

	Pythium	Bacillus subtilis (*off-label*), cymoxanil + fludioxonil + metalaxyl-M (*off-label - seed treatment*), Gliocladium catenulatum (*off-label - moderate control*)
	Rhizoctonia	Bacillus subtilis (*off-label*), Gliocladium catenulatum (*off-label - moderate control*)
	Sclerotinia	boscalid + pyraclostrobin (*moderate*), cyprodinil + fludioxonil (*moderate control*), cyprodinil + fludioxonil (*off-label*), prothioconazole, tebuconazole, tebuconazole + trifloxystrobin
	Seed-borne diseases	thiram (*seed soak*)
	White rot	Bacillus subtilis (*off-label*)
Pests	Aphids	deltamethrin (*off-label*), fatty acids (*off-label*), pirimicarb, thiacloprid
	Beetles	deltamethrin (*off-label*)
	Birds/mammals	aluminium ammonium sulphate
	Caterpillars	deltamethrin (*off-label*)
	Cutworms	Bacillus thuringiensis (*off-label*), cypermethrin (*off-label*), lambda-cyhalothrin
	Flies	lambda-cyhalothrin (*off-label*), tefluthrin (*off-label - seed treatment*)
	Leafhoppers	deltamethrin (*off-label*)
	Pests, miscellaneous	deltamethrin (*off-label*), lambda-cyhalothrin (*off-label*)
	Stem nematodes	oxamyl
	Thrips	deltamethrin (*off-label*), fatty acids (*off-label*)
Plant growth regulation	Growth control	maleic hydrazide (*off-label*)
Weeds	Broad-leaved weeds	clomazone, flumioxazine (*off-label*), isoxaben (*off-label*), linuron, metamitron (*off-label*), metribuzin (*off-label*), pendimethalin, pendimethalin (*off-label*), prosulfocarb (*off-label*)
	Crops as weeds	cycloxydim, fluazifop-P-butyl, flumioxazine (*off-label*), metribuzin (*off-label*), pendimethalin, tepraloxydim
	Grass weeds	cycloxydim, fluazifop-P-butyl, fluazifop-P-butyl (*off-label*), linuron, metamitron (*off-label*), metribuzin (*off-label*), pendimethalin, propaquizafop, prosulfocarb (*off-label*), tepraloxydim, tepraloxydim (*off-label*)
	Weeds, miscellaneous	metamitron (*off-label*)

Root and tuber crops - Miscellaneous root crops

Crop control	Desiccation	diquat (*off-label*)
Diseases	Alternaria	azoxystrobin (*off-label*), azoxystrobin + difenoconazole (*off-label*), fenpropimorph (*off-label*)
	Botrytis	chlorothalonil (*off-label*), cyprodinil + fludioxonil (*moderate control*), Gliocladium catenulatum (*off-label - moderate control*)
	Crown rot	fenpropimorph (*off-label*)
	Damping off	Bacillus subtilis (*off-label*)
	Disease control	tebuconazole (*off-label*)
	Foliar diseases	tebuconazole (*off-label*), tebuconazole + trifloxystrobin (*off-label*)
	Fusarium	Gliocladium catenulatum (*off-label - moderate control*)

	Helminthosporium seedling rot	Bacillus subtilis (*off-label*)
	Leaf spot	boscalid + pyraclostrobin (*off-label*), chlorothalonil (*off-label*)
	Phytophthora	Bacillus subtilis (*off-label*), difenoconazole (*off-label*), Gliocladium catenulatum (*off-label - moderate control*)
	Powdery mildew	azoxystrobin (*off-label*), azoxystrobin + difenoconazole (*off-label*), boscalid + pyraclostrobin (*off-label*), fenpropimorph (*off-label*)
	Pythium	Bacillus subtilis (*off-label*), Gliocladium catenulatum (*off-label - moderate control*)
	Rhizoctonia	Bacillus subtilis (*off-label*), Gliocladium catenulatum (*off-label - moderate control*)
	Rust	azoxystrobin (*off-label*), boscalid + pyraclostrobin (*off-label*), tebuconazole (*off-label*)
	Sclerotinia	azoxystrobin (*off-label*), boscalid + pyraclostrobin (*off-label*), cyprodinil + fludioxonil (*off-label*)
	Septoria	difenoconazole (*off-label*)
	White rot	Bacillus subtilis (*off-label*)
Pests	Aphids	deltamethrin (*off-label*), dimethoate (*off-label*), fatty acids (*off-label*), pirimicarb (*off-label*), pymetrozine (*off-label*), thiacloprid (*off-label*)
	Cutworms	Bacillus thuringiensis (*off-label*), cypermethrin (*off-label*), lambda-cyhalothrin (*off-label*)
	Flies	lambda-cyhalothrin (*off-label*)
	Pests, miscellaneous	deltamethrin (*off-label*), dimethoate (*off-label*), lambda-cyhalothrin (*off-label*)
	Thrips	fatty acids (*off-label*)
	Weevils	lambda-cyhalothrin (*off-label*)
Weeds	Broad-leaved weeds	clomazone (*off-label*), linuron, linuron (*off-label*), metribuzin (*off-label*), pendimethalin (*off-label*), propyzamide (*off-label*), prosulfocarb (*off-label*), S-metolachlor (*off-label*)
	Crops as weeds	metribuzin (*off-label*)
	Grass weeds	fluazifop-P-butyl (*off-label*), linuron (*off-label*), metribuzin (*off-label*), pendimethalin (*off-label*), propyzamide (*off-label*), prosulfocarb (*off-label*), S-metolachlor (*off-label*)

Root and tuber crops - Potatoes

Crop control	Desiccation	carfentrazone-ethyl, diquat, glufosinate-ammonium, pyraflufen-ethyl
Diseases	Alternaria	azoxystrobin + chlorothalonil (*off-label*), fenamidone + propamocarb hydrochloride, mancozeb
	Black dot	azoxystrobin, fludioxonil (*some reduction*)
	Black scurf and stem canker	azoxystrobin, fludioxonil, flutolanil (*tuber treatment*), imazalil + pencycuron (*tuber treatment*), imazalil + pencycuron (*tuber treatment - reduction*), pencycuron (*tuber treatment*), tolclofos-methyl (*tuber treatment*), tolclofos-methyl (*tuber treatment only to be used with automatic planters*)
	Blight	ametoctradin + dimethomorph (*reduction*), amisulbrom

	Damping off	Bacillus subtilis (*off-label*)
	Dry rot	imazalil, thiabendazole, thiabendazole (*tuber treatment - post-harvest*)
	Gangrene	imazalil, thiabendazole, thiabendazole (*tuber treatment - post-harvest*)
	Helminthosporium seedling rot	Bacillus subtilis (*off-label*)
	Phytophthora	ametoctradin + dimethomorph, amisulbrom, Bacillus subtilis (*off-label*), benthiavalicarb-isopropyl + mancozeb, boscalid + pyraclostrobin (*off-label*), chlorothalonil + propamocarb hydrochloride, copper oxychloride, cyazofamid, cymoxanil, cymoxanil + famoxadone, cymoxanil + mancozeb, cymoxanil + propamocarb, cymoxanil + propamocarb (*off-label*), dimethomorph, dimethomorph + mancozeb, fenamidone + propamocarb hydrochloride, fluazinam, fluazinam + metalaxyl-M, fluopicolide + propamocarb hydrochloride, mancozeb, mancozeb + metalaxyl-M, mancozeb + zoxamide, mandipropamid, maneb
	Powdery scab	fluazinam (*off-label*)
	Pythium	Bacillus subtilis (*off-label*)
	Rhizoctonia	Bacillus subtilis (*off-label*), tolclofos-methyl (*chitted seed treatment*)
	Scab	fluazinam (*off-label*)
	Silver scurf	fludioxonil (*reduction*), imazalil, imazalil + pencycuron (*tuber treatment - reduction*), thiabendazole, thiabendazole (*tuber treatment - post-harvest*)
	Skin spot	imazalil, thiabendazole, thiabendazole (*tuber treatment - post-harvest*)
	White rot	Bacillus subtilis (*off-label*)
Pests	Aphids	acetamiprid, cypermethrin, esfenvalerate, flonicamid, lambda-cyhalothrin, oxamyl, pirimicarb, pymetrozine, thiacloprid, thiamethoxam
	Beetles	lambda-cyhalothrin
	Caterpillars	cypermethrin, lambda-cyhalothrin
	Cutworms	chlorpyrifos, cypermethrin, lambda-cyhalothrin (*if present at time of application*), zeta-cypermethrin
	Cyst nematodes	fosthiazate, oxamyl
	Free-living nematodes	fosthiazate (*reduction*), oxamyl
	Leatherjackets	methiocarb (*reduction*)
	Slugs/snails	metaldehyde, methiocarb
	Weevils	lambda-cyhalothrin
	Wireworms	fosthiazate (*reduction*)
Plant growth regulation	Growth control	chlorpropham, chlorpropham (*thermal fog*), ethylene, maleic hydrazide
	Quality/yield control	sulphur
Weeds	Broad-leaved weeds	bentazone, carfentrazone-ethyl, clomazone, clomazone + linuron, clomazone + metribuzin, diquat, flufenacet + metribuzin, glufosinate-ammonium, linuron, metribuzin, pendimethalin, prosulfocarb, pyraflufen-ethyl, rimsulfuron
	Crops as weeds	carfentrazone-ethyl, cycloxydim, metribuzin, pendimethalin, quizalofop-P-tefuryl, rimsulfuron

| | Grass weeds | clomazone + linuron, clomazone + metribuzin, cycloxydim, diquat, diquat (*weed control and dessication*), flufenacet + metribuzin, glufosinate-ammonium, linuron, metribuzin, pendimethalin, propaquizafop, prosulfocarb, quizalofop-P-tefuryl |
| | Weeds, miscellaneous | diquat, diquat (*weed control or dessication*), glufosinate-ammonium (*not for seed*), glyphosate |

Stem and bulb vegetables - Asparagus

Diseases	Botrytis	azoxystrobin + chlorothalonil (*qualified minor use recommendation*), boscalid + pyraclostrobin (*off-label*), cyprodinil + fludioxonil (*off-label*)
	Downy mildew	metalaxyl-M (*off-label*)
	Rust	azoxystrobin, azoxystrobin + chlorothalonil (*moderate control only*), azoxystrobin + difenoconazole (*off-label*), boscalid + pyraclostrobin (*off-label*), difenoconazole (*off-label*)
	Stemphylium	azoxystrobin, azoxystrobin + chlorothalonil (*qualified minor use recommendation*), azoxystrobin + difenoconazole (*off-label*)
Pests	Aphids	fatty acids (*off-label*)
	Beetles	cypermethrin (*off-label*), thiacloprid (*off-label*)
	Thrips	fatty acids (*off-label*)
Weeds	Broad-leaved weeds	clomazone (*off-label*), clopyralid (*off-label*), isoxaben (*off-label*), linuron, mesotrione (*off-label*), metamitron (*off-label*), metribuzin (*off-label*), pendimethalin (*off-label*), pyridate (*off-label*), triclopyr (*off-label*)
	Crops as weeds	metribuzin (*off-label*)
	Grass weeds	fluazifop-P-butyl (*off-label*), metamitron (*off-label*), metribuzin (*off-label*), pendimethalin (*off-label*)
	Weeds, miscellaneous	glyphosate, glyphosate (*off-label*), metamitron (*off-label*)

Stem and bulb vegetables - Celery/chicory

Diseases	Botrytis	azoxystrobin (*off-label*), cyprodinil + fludioxonil (*off-label*)
	Damping off	Bacillus subtilis (*off-label*), thiram (*off-label*)
	Disease control	dimethomorph + mancozeb (*off-label*), mandipropamid, prochloraz (*off-label*)
	Downy mildew	copper oxychloride (*off-label*), dimethomorph + mancozeb (*off-label*), mandipropamid
	Helminthosporium seedling rot	Bacillus subtilis (*off-label*)
	Phytophthora	Bacillus subtilis (*off-label*), difenoconazole (*off-label*)
	Powdery mildew	azoxystrobin (*off-label*)
	Pythium	Bacillus subtilis (*off-label*)
	Rhizoctonia	azoxystrobin (*off-label*), Bacillus subtilis (*off-label*)
	Rust	azoxystrobin (*off-label*)
	Sclerotinia	azoxystrobin (*off-label*)
	Seed-borne diseases	thiram (*seed soak*)
	Septoria	copper oxychloride

	White rot	Bacillus subtilis (*off-label*)
Pests	Aphids	acetamiprid (*off-label*), cypermethrin (*off-label*), deltamethrin (*off-label*), fatty acids (*off-label*), pirimicarb (*off-label*), pymetrozine (*off-label*)
	Beetles	deltamethrin (*off-label*)
	Caterpillars	Bacillus thuringiensis (*off-label*), cypermethrin (*off-label*), deltamethrin (*off-label*), diflubenzuron (*off-label*), lambda-cyhalothrin (*off-label*)
	Cutworms	lambda-cyhalothrin (*off-label*)
	Flies	lambda-cyhalothrin (*off-label*)
	Leafhoppers	deltamethrin (*off-label*)
	Pests, miscellaneous	deltamethrin (*off-label*), lambda-cyhalothrin (*off-label*), spinosad
	Thrips	deltamethrin (*off-label*), fatty acids (*off-label*)
Plant growth regulation	Quality/yield control	gibberellins
Weeds	Broad-leaved weeds	clomazone (*off-label*), metamitron (*off-label*), pendimethalin (*off-label*), propyzamide (*off-label*), S-metolachlor (*off-label*), triflusulfuron-methyl (*off-label*)
	Crops as weeds	propaquizafop (*off-label*)
	Grass weeds	fluazifop-P-butyl (*off-label*), metamitron (*off-label*), pendimethalin (*off-label*), propaquizafop (*off-label*), propyzamide (*off-label*), prosulfocarb (*off-label*), S-metolachlor (*off-label*)

Stem and bulb vegetables - Globe artichokes/cardoons

Diseases	Downy mildew	azoxystrobin (*off-label*)
	Powdery mildew	azoxystrobin (*off-label*), myclobutanil (*off-label*)
Weeds	Broad-leaved weeds	metamitron (*off-label*)
	Grass weeds	metamitron (*off-label*)

Stem and bulb vegetables - Onions/leeks/garlic

Diseases	Alternaria	iprodione, iprodione (*off-label*)
	Bacterial blight	copper oxychloride (*off-label*)
	Blight	fluoxastrobin + prothioconazole (*control*)
	Botrytis	azoxystrobin + chlorothalonil, cyprodinil + fludioxonil (*off-label*), fluoxastrobin + prothioconazole (*useful reduction*), iprodione, iprodione (*off-label*)
	Bottom rot	copper oxychloride (*off-label*)
	Cladosporium	fluoxastrobin + prothioconazole (*useful reduction*)
	Collar rot	iprodione, iprodione (*off-label*)
	Damping off	Bacillus subtilis (*off-label*), thiram (*off-label - seed treatment*), thiram (*seed treatment*)
	Disease control	azoxystrobin (*off-label*), dimethomorph + mancozeb, dimethomorph + mancozeb (*off-label*), mancozeb
	Downy mildew	azoxystrobin, azoxystrobin (*off-label*), azoxystrobin + chlorothalonil, benthiavalicarb-isopropyl + mancozeb (*off-label*), dimethomorph + mancozeb, dimethomorph + mancozeb (*off-label*), fluoxastrobin + prothioconazole (*control*), mancozeb, mancozeb + metalaxyl-M (*off-label*), mancozeb + metalaxyl-M

		(*useful control*), maneb, metalaxyl-M (*off-label*), propiconazole (*off-label*)
	Fusarium	boscalid + pyraclostrobin (*off-label*)
	Helminthosporium seedling rot	Bacillus subtilis (*off-label*)
	Phytophthora	Bacillus subtilis (*off-label*), copper oxychloride (*off-label*)
	Purple blotch	azoxystrobin, azoxystrobin + difenoconazole (*moderate control*), prothioconazole, tebuconazole + trifloxystrobin
	Pythium	Bacillus subtilis (*off-label*), metalaxyl-M, metalaxyl-M (*off-label*)
	Rhizoctonia	Bacillus subtilis (*off-label*)
	Rhynchosporium	prothioconazole (*useful reduction*)
	Rust	azoxystrobin, azoxystrobin + difenoconazole, fenpropimorph, fluoxastrobin + prothioconazole (*useful reduction*), prothioconazole, tebuconazole, tebuconazole + trifloxystrobin
	Seed-borne diseases	thiram (*off-label - seed treatment*)
	Spear rot	copper oxychloride (*off-label*)
	Stem canker	copper oxychloride (*off-label*)
	Stemphylium	prothioconazole (*useful reduction*)
	Storage rots	copper oxychloride (*off-label*)
	White rot	Bacillus subtilis (*off-label*), boscalid + pyraclostrobin (*off-label*), tebuconazole (*off-label*)
	White tip	azoxystrobin + difenoconazole (*qualified minor use*), boscalid + pyraclostrobin (*off-label*), dimethomorph + mancozeb (*reduction*), tebuconazole + trifloxystrobin
Pests	Aphids	deltamethrin (*off-label*), fatty acids (*off-label*), lambda-cyhalothrin (*off-label*)
	Beetles	deltamethrin (*off-label*)
	Caterpillars	deltamethrin (*off-label*)
	Cutworms	Bacillus thuringiensis (*off-label*), chlorpyrifos
	Flies	tefluthrin (*off-label*)
	Free-living nematodes	oxamyl (*off-label*)
	Leafhoppers	deltamethrin (*off-label*)
	Pests, miscellaneous	abamectin (*off-label*), chlorpyrifos (*off-label*), deltamethrin (*off-label*), lambda-cyhalothrin (*off-label*), oxamyl (*off-label*)
	Stem nematodes	oxamyl (*off-label*)
	Thrips	abamectin (*off-label*), chlorpyrifos (*off-label*), deltamethrin (*off-label*), fatty acids (*off-label*), lambda-cyhalothrin (*off-label*), spinosad
Plant growth regulation	Growth control	maleic hydrazide, maleic hydrazide (*off-label*)
Weeds	Broad-leaved weeds	bentazone (*off-label*), bromoxynil, chloridazon (*off-label*), chlorpropham, chlorpropham (*off-label*), dimethenamid-p + metazachlor (*off-label*), dimethenamid-p + pendimethalin (*off-label*), flumioxazine (*off-label*), fluroxypyr (*off-label*), glyphosate, ioxynil, ioxynil (*off-label*), linuron (*off-label*), metazachlor (*off-label*), pendimethalin,

		pendimethalin (*off-label*), prosulfocarb (*off-label*), pyridate, pyridate (*off-label*), S-metolachlor (*off-label*)
Crops as weeds		cycloxydim, fluazifop-P-butyl, flumioxazine (*off-label*), fluroxypyr (*off-label*), glyphosate, pendimethalin, propaquizafop (*off-label*), prosulfocarb (*off-label*), tepraloxydim
Grass weeds		chloridazon (*off-label*), chlorpropham, cycloxydim, cycloxydim (*off-label*), dimethenamid-p + metazachlor (*off-label*), dimethenamid-p + pendimethalin (*off-label*), fluazifop-P-butyl, fluazifop-P-butyl (*off-label*), glyphosate, linuron (*off-label*), metazachlor (*off-label*), pendimethalin, pendimethalin (*off-label*), propaquizafop, propaquizafop (*off-label*), prosulfocarb (*off-label*), S-metolachlor (*off-label*), tepraloxydim, tepraloxydim (*off-label*)
Weeds, miscellaneous		glyphosate

Stem and bulb vegetables - Vegetables

Diseases	Botrytis	Bacillus subtilis (*off-label*), Gliocladium catenulatum (*off-label - moderate control*)
	Downy mildew	azoxystrobin (*off-label*)
	Fusarium	Gliocladium catenulatum (*off-label - moderate control*)
	Phytophthora	Gliocladium catenulatum (*off-label - moderate control*)
	Pythium	Gliocladium catenulatum (*off-label - moderate control*)
	Rhizoctonia	Gliocladium catenulatum (*off-label - moderate control*)
Pests	Aphids	fatty acids (*off-label*), Metarhizium anisopliae (*off-label*)
	Flies	Metarhizium anisopliae (*off-label*)
	Thrips	fatty acids (*off-label*), Metarhizium anisopliae (*off-label*)
Weeds	Grass weeds	fluazifop-P-butyl (*off-label*)

Flowers and ornamentals

Flowers - Bedding plants, general

Diseases	Botrytis	thiram
	Powdery mildew	bupirimate
	Rust	azoxystrobin (*off-label - in pots*), thiram
Pests	Aphids	imidacloprid, thiacloprid
	Beetles	thiacloprid
	Flies	imidacloprid
	Weevils	imidacloprid, thiacloprid
	Whiteflies	imidacloprid, thiacloprid
Plant growth regulation	Flowering control	paclobutrazol
	Growth control	chlormequat, daminozide, paclobutrazol

Flowers - Bulbs/corms

Diseases	Basal stem rot	chlorothalonil (*off-label*), thiabendazole (*off-label*)
	Botrytis	Bacillus subtilis (*off-label*), chlorothalonil (*off-label*), Gliocladium catenulatum (*off-label - moderate control*), mancozeb, tebuconazole (*off-label*), tebuconazole (*off-label - for galanthamine production*), thiabendazole (*off-label*), thiram
	Crown rot	metalaxyl-M (*off-label*)
	Downy mildew	metalaxyl-M (*off-label*)
	Fusarium	Gliocladium catenulatum (*off-label - moderate control*), thiabendazole, thiabendazole (*post-lifting or pre-planting dip*), thiabendazole (*post-lifting spray*)
	Phytophthora	Gliocladium catenulatum (*off-label - moderate control*)
	Powdery mildew	bupirimate
	Pythium	Gliocladium catenulatum (*off-label - moderate control*)
	Rhizoctonia	Gliocladium catenulatum (*off-label - moderate control*)
Pests	Flies	chlorpyrifos (*off-label*)
Plant growth regulation	Flowering control	paclobutrazol
	Growth control	chlormequat, paclobutrazol
Weeds	Broad-leaved weeds	bentazone, carfentrazone-ethyl (*off-label*), pendimethalin (*off-label*)
	Crops as weeds	cycloxydim
	Grass weeds	cycloxydim, pendimethalin (*off-label*), S-metolachlor (*off-label*)

Flowers - Miscellaneous flowers

Diseases	Black spot	captan, kresoxim-methyl, myclobutanil
	Crown rot	fenpropimorph (*off-label*)
	Disease control	tebuconazole (*off-label*)
	Foliar diseases	tebuconazole (*off-label*)
	Powdery mildew	bupirimate, fenpropimorph (*off-label*), kresoxim-methyl, myclobutanil
	Rust	myclobutanil, penconazole, tebuconazole (*off-label*)
Pests	Aphids	cypermethrin (*off-label*), deltamethrin (*off-label*), imidacloprid, pirimicarb (*off-label*)
	Beetles	deltamethrin (*off-label*)
	Caterpillars	cypermethrin (*off-label*), deltamethrin (*off-label*)
	Flies	imidacloprid, lambda-cyhalothrin (*off-label*)
	Leafhoppers	deltamethrin (*off-label*)
	Pests, miscellaneous	deltamethrin (*off-label*), lambda-cyhalothrin (*off-label*)
	Thrips	deltamethrin (*off-label*)
	Weevils	imidacloprid
	Whiteflies	imidacloprid
Plant growth regulation	Flowering control	paclobutrazol
	Growth control	paclobutrazol
Weeds	Broad-leaved weeds	linuron (*off-label*), metribuzin (*off-label*), pendimethalin (*off-label*), propyzamide
	Crops as weeds	metribuzin (*off-label*), propaquizafop (*off-label*)

| | Grass weeds | cycloxydim (*off-label*), fluazifop-P-butyl (*off-label*), linuron (*off-label*), metribuzin (*off-label*), pendimethalin (*off-label*), propaquizafop (*off-label*), propyzamide, tepraloxydim (*off-label*) |

Flowers - Pot plants

Diseases	Rust	azoxystrobin (*off-label*)
Pests	Aphids	deltamethrin, imidacloprid, thiacloprid
	Beetles	thiacloprid
	Caterpillars	deltamethrin
	Flies	imidacloprid
	Mealybugs	deltamethrin, petroleum oil
	Scale insects	deltamethrin, petroleum oil
	Spider mites	petroleum oil
	Weevils	imidacloprid, thiacloprid
	Whiteflies	deltamethrin, imidacloprid, thiacloprid
Plant growth regulation	Flowering control	paclobutrazol
	Fruiting control	paclobutrazol
	Growth control	chlormequat, daminozide, paclobutrazol

Flowers - Protected bulbs/corms

Pests	Aphids	pirimicarb

Flowers - Protected flowers

Diseases	Black spot	kresoxim-methyl
	Crown rot	metalaxyl-M (*off-label*)
	Disease control	difenoconazole (*off-label*)
	Downy mildew	metalaxyl-M (*off-label*)
	Powdery mildew	kresoxim-methyl
	Ring spot	difenoconazole (*off-label - ground grown (or pot grown - 20050159)*), difenoconazole (*off-label - ground grown (or pot grown - SOLA 20050159)*)
	Rust	azoxystrobin (*off-label*), difenoconazole (*off-label*), difenoconazole (*off-label - ground grown (or pot grown - 20050159)*), difenoconazole (*off-label - ground grown (or pot grown - SOLA 20050159)*)
Pests	Aphids	pirimicarb
	Spider mites	tebufenpyrad

Miscellaneous flowers and ornamentals - Miscellaneous uses

Diseases	Botrytis	boscalid + pyraclostrobin (*off-label*)
	Disease control	boscalid + pyraclostrobin (*off-label*), potassium bicarbonate (commodity substance)
	Leaf spot	boscalid + pyraclostrobin (*off-label*)
	Powdery mildew	boscalid + pyraclostrobin (*off-label*), physical pest control
	Rust	boscalid + pyraclostrobin (*off-label*)

	Sclerotinia	Coniothyrium minitans
Pests	Aphids	physical pest control, pyrethrins
	Birds/mammals	aluminium ammonium sulphate
	Caterpillars	diflubenzuron, pyrethrins
	Mealybugs	physical pest control
	Scale insects	physical pest control
	Slugs/snails	ferric phosphate, metaldehyde, metaldehyde (*do not apply to cauliflowers*), methiocarb (*outdoor only*)
	Spider mites	physical pest control, pyrethrins
	Suckers	physical pest control
	Thrips	pyrethrins
	Whiteflies	physical pest control
Weeds	Broad-leaved weeds	carfentrazone-ethyl (*before planting*), diquat, diquat (*around*), propyzamide
	Crops as weeds	carfentrazone-ethyl (*before planting*)
	Grass weeds	diquat, diquat (*around only*), glyphosate, propyzamide
	Weeds, miscellaneous	diquat (*around only*), glyphosate, glyphosate (*before planting*), glyphosate (*pre-sowing/planting*)

Ornamentals - Nursery stock

Diseases	Alternaria	iprodione
	Black spot	myclobutanil
	Botrytis	Bacillus subtilis (*off-label*), boscalid + pyraclostrobin (*off-label*), cyprodinil + fludioxonil, Gliocladium catenulatum (*off-label - moderate control*), iprodione, iprodione (*off-label*), pyrimethanil (*off-label*), thiram, thiram (*except Hydrangea*)
	Disease control	azoxystrobin (*off-label*), chlorothalonil (*off-label*), chlorothalonil + metalaxyl-M (*off-label*), dimethomorph + mancozeb (*off-label*), fosetyl-aluminium, iprodione (*off-label*), mancozeb, mancozeb (*off-label*), mancozeb + metalaxyl-M (*off-label*), mandipropamid (*off-label*), metrafenone (*off-label*)
	Downy mildew	benthiavalicarb-isopropyl + mancozeb (*off-label*), dimethomorph (*off-label*), dimethomorph + mancozeb (*off-label*), mandipropamid (*off-label*)
	Foliar diseases	picoxystrobin (*off-label*), pyraclostrobin (*off-label*), thiamethoxam (*off-label*), trifloxystrobin (*off-label*)
	Foot rot	tolclofos-methyl, Trichoderma asperellum (Strain T34), Trichoderma asperellum (Strain T34) (*off-label*)
	Fungus diseases	prochloraz
	Fusarium	Gliocladium catenulatum (*off-label - moderate control*), thiophanate-methyl (*off-label*)
	Phytophthora	fenamidone + fosetyl-aluminium, Gliocladium catenulatum (*off-label - moderate control*), metalaxyl-M
	Powdery mildew	Ampelomyces quisqualis strain AQ10 (*off-label*), cyflufenamid (*off-label*), metrafenone, myclobutanil, physical pest control, quinoxyfen (*off-label*)
	Pythium	fenamidone + fosetyl-aluminium, fosetyl-aluminium + propamocarb hydrochloride (*off-label*), Gliocladium

		catenulatum (*off-label - moderate control*), metalaxyl-M, metalaxyl-M (*off-label*)
	Rhizoctonia	Gliocladium catenulatum (*off-label - moderate control*)
	Root diseases	thiophanate-methyl (*off-label*)
	Root rot	tolclofos-methyl
	Rust	myclobutanil, propiconazole (*off-label*)
	Seed-borne diseases	iprodione
Pests	Aphids	acetamiprid, Beauveria bassiana, deltamethrin, dimethoate, esfenvalerate, flonicamid (*off-label*), imidacloprid, imidacloprid (*container grown*), lambda-cyhalothrin (*off-label*), physical pest control, pirimicarb, pymetrozine, pymetrozine (*off-label*), pyrethrins, thiacloprid, thiacloprid (*off-label*)
	Beetles	Beauveria bassiana, thiacloprid
	Capsid bugs	deltamethrin
	Caterpillars	Bacillus thuringiensis, deltamethrin, diflubenzuron, indoxacarb, indoxacarb (*off-label*), pyrethrins, spinosad (*off-label*)
	Flies	imidacloprid, imidacloprid (*container grown*), Metarhizium anisopliae (*off-label*)
	Free-living nematodes	fosthiazate (*off-label*)
	Leaf miners	dimethoate, oxamyl (*off-label*), thiacloprid (*off-label*)
	Mealybugs	deltamethrin, petroleum oil, physical pest control
	Midges	Metarhizium anisopliae (*off-label*)
	Pests, miscellaneous	esfenvalerate, indoxacarb (*off-label*), lambda-cyhalothrin (*off-label*), methoxyfenozide (*off-label*), spinosad (*off-label*), spirotetramat (*off-label*), thiacloprid (*off-label*)
	Scale insects	deltamethrin, petroleum oil, physical pest control
	Spider mites	abamectin, dimethoate, etoxazole (*off-label*), petroleum oil, physical pest control, spirodiclofen (*off-label*), spiromesifen (*off-label*)
	Suckers	physical pest control
	Thrips	abamectin, deltamethrin, deltamethrin (*off-label*), Metarhizium anisopliae (*off-label*), spinosad, thiacloprid (*off-label*)
	Weevils	imidacloprid, imidacloprid (*container grown*), Metarhizium anisopliae, thiacloprid
	Whiteflies	acetamiprid, Beauveria bassiana, deltamethrin, flonicamid (*off-label*), imidacloprid, imidacloprid (*container grown*), Lecanicillium lecanii, physical pest control, pymetrozine (*off-label*), spiromesifen (*off-label*), thiacloprid, thiacloprid (*off-label*)
Plant growth regulation	Growth control	1-naphthylacetic acid, 2-chloroethylphosphonic acid (*off-label*), 4-indol-3-ylbutyric acid, chlormequat, daminozide, gibberellins (*off-label*), prohexadione-calcium (*off-label*), trinexapac-ethyl (*off-label*)
Weeds	Broad-leaved weeds	bentazone (*off-label*), chlorpropham, clopyralid, diflufenican (*off-label*), diquat, florasulam (*off-label*), florasulam + fluroxypyr (*off-label*), flufenacet + isoxaflutole (*off-label*), flumioxazine (*off-label*), fluroxypyr (*off-label*), glufosinate-ammonium, imazamox + pendimethalin (*off-label*), isoxaben, isoxaben (*off-label*), linuron, linuron (*off-label*),

		metazachlor, metsulfuron-methyl (*off-label*), napropamide, oxadiazon, pendimethalin (*off-label*), rimsulfuron (*off-label*), S-metolachlor (*off-label*)
	Crops as weeds	flumioxazine (*off-label*)
	Grass weeds	chlorpropham, diquat, fluazifop-P-butyl (*off-label*), flufenacet + isoxaflutole (*off-label*), glufosinate-ammonium, linuron (*off-label*), maleic hydrazide (*off-label*), metazachlor, napropamide, oxadiazon, pendimethalin (*off-label*), quizalofop-P-tefuryl (*off-label*), S-metolachlor (*off-label*), tepraloxydim (*off-label*)
	Mosses	maleic hydrazide + pelargonic acid
	Weeds, miscellaneous	carfentrazone-ethyl (*off-label*), diquat, glyphosate, glyphosate (*off-label*), glyphosate + pyraflufen-ethyl, maleic hydrazide + pelargonic acid

Ornamentals - Trees and shrubs

Diseases	Botrytis	boscalid + pyraclostrobin (*off-label*)
	Disease control	mancozeb
	Downy mildew	benthiavalicarb-isopropyl + mancozeb (*off-label*)
	Fungus diseases	prochloraz
	Leaf spot	boscalid + pyraclostrobin (*off-label*)
	Phytophthora	Bacillus subtilis (*off-label*), metalaxyl-M (*off-label*)
	Powdery mildew	boscalid + pyraclostrobin (*off-label*), cyflufenamid (*off-label*), penconazole
	Rust	boscalid + pyraclostrobin (*off-label*)
	Scab	penconazole
Pests	Aphids	deltamethrin, lambda-cyhalothrin (*off-label*)
	Birds/mammals	aluminium ammonium sulphate, warfarin (*nuts*)
	Capsid bugs	deltamethrin
	Caterpillars	Bacillus thuringiensis, deltamethrin, diflubenzuron
	Flies	Metarhizium anisopliae (*off-label*)
	Mealybugs	deltamethrin
	Midges	Metarhizium anisopliae (*off-label*)
	Pests, miscellaneous	lambda-cyhalothrin (*off-label*)
	Scale insects	deltamethrin
	Slugs/snails	ferric phosphate
	Thrips	deltamethrin, Metarhizium anisopliae (*off-label*)
	Weevils	Metarhizium anisopliae (*off-label*)
	Whiteflies	deltamethrin
Plant growth regulation	Growth control	glyphosate, maleic hydrazide (*off-label*)
Weeds	Broad-leaved weeds	clopyralid (*off-label*), isoxaben, metazachlor, oxadiazon (*off-label*), propyzamide
	Grass weeds	cycloxydim (*off-label*), metazachlor, oxadiazon (*off-label*), propyzamide, quizalofop-P-tefuryl (*off-label*)
	Mosses	maleic hydrazide + pelargonic acid
	Weeds, miscellaneous	glyphosate, glyphosate (*wiper application*), glyphosate + pyraflufen-ethyl, glyphosate + sulfosulfuron, maleic hydrazide + pelargonic acid, propyzamide

Woody weeds/scrub   glyphosate

Ornamentals - Woody ornamentals

Diseases	Fungus diseases	prochloraz
Pests	Aphids	fatty acids
	Mealybugs	fatty acids
	Scale insects	fatty acids
	Spider mites	fatty acids
	Whiteflies	fatty acids
Weeds	Bindweeds	oxadiazon
	Broad-leaved weeds	glufosinate-ammonium, napropamide, oxadiazon, propyzamide
	Grass weeds	glufosinate-ammonium, napropamide, oxadiazon, propyzamide
	Weeds, miscellaneous	glyphosate

Forestry

Forest nurseries, general - Forest nurseries

Crop control	Chemical stripping/ thinning	glyphosate
Diseases	Botrytis	boscalid + pyraclostrobin (*off-label*), cyprodinil + fludioxonil, fenhexamid (*off-label*), iprodione (*off-label*), mepanipyrim (*off-label*), pyrimethanil (*off-label*)
	Damping off	fosetyl-aluminium + propamocarb hydrochloride (*off-label*)
	Disease control	azoxystrobin (*off-label*), benthiavalicarb-isopropyl + mancozeb (*off-label*), chlorothalonil (*off-label*), chlorothalonil + metalaxyl-M (*off-label*), cyprodinil (*off-label*), dimethomorph + mancozeb (*off-label*), mancozeb (*off-label*), mancozeb + metalaxyl-M (*off-label*), propiconazole (*off-label*), pyrimethanil (*off-label*)
	Downy mildew	dimethomorph + mancozeb (*off-label*), fosetyl-aluminium + propamocarb hydrochloride (*off-label*)
	Foliar diseases	fluoxastrobin + prothioconazole (*off-label*)
	Foot rot	Trichoderma asperellum (Strain T34) (*off-label*)
	Phytophthora	Bacillus subtilis (*off-label*), metalaxyl-M (*off-label*)
	Powdery mildew	Ampelomyces quisqualis strain AQ10 (*off-label*), cyflufenamid (*off-label*), metrafenone (*off-label*), proquinazid (*off-label*), quinoxyfen (*off-label*)
	Root rot	Phlebiopsis gigantea
Pests	Aphids	acetamiprid (*off-label*), lambda-cyhalothrin (*off-label*), pirimicarb, pymetrozine (*off-label*)
	Birds/mammals	aluminium ammonium sulphate
	Caterpillars	indoxacarb (*off-label*)
	Pests, miscellaneous	dimethoate (*off-label*), indoxacarb (*off-label*), lambda-cyhalothrin (*off-label*), spirotetramat (*off-label*), thiacloprid (*off-label*)

	Spider mites	bifenazate (*off-label*)
	Weevils	alpha-cypermethrin (*off-label*)
	Whiteflies	pymetrozine (*off-label*)
Plant growth regulation	Growth control	trinexapac-ethyl (*off-label*)
Weeds	Broad-leaved weeds	carfentrazone-ethyl (*off-label*), chlorpropham (*off-label*), clopyralid (*off-label*), diflufenican (*off-label*), florasulam (*off-label*), flumioxazine (*off-label*), fluroxypyr (*off-label*), linuron, metazachlor, metsulfuron-methyl (*off-label*), napropamide, pendimethalin (*off-label*), propyzamide, propyzamide (*off-label*), rimsulfuron (*off-label*), S-metolachlor (*off-label*), tribenuron-methyl (*off-label*)
	Crops as weeds	flumioxazine (*off-label*)
	Grass weeds	chlorpropham (*off-label*), ethofumesate (*off-label*), metazachlor, napropamide, pendimethalin (*off-label*), propaquizafop, propyzamide, propyzamide (*off-label*), S-metolachlor (*off-label*), tepraloxydim (*off-label*)
	Weeds, miscellaneous	diquat (*off-label*), glyphosate
	Woody weeds/scrub	glyphosate

Forestry plantations, general - Forestry plantations

Crop control	Chemical stripping/ thinning	glyphosate, glyphosate (*stump treatment*)
Diseases	Root rot	Phlebiopsis gigantea
Pests	Birds/mammals	aluminium ammonium sulphate, warfarin
	Caterpillars	diflubenzuron
	Weevils	acetamiprid (*off-label*), alpha-cypermethrin (*off-label*), cypermethrin
Plant growth regulation	Growth control	gibberellins (*off-label*), glyphosate
Weeds	Aquatic weeds	glyphosate, propyzamide
	Broad-leaved weeds	2,4-D + dicamba + triclopyr, glufosinate-ammonium, glyphosate, glyphosate (*wiper application*), isoxaben, metamitron (*off-label*), metazachlor, napropamide (*off-label*), pendimethalin (*off-label*), propyzamide, triclopyr
	Grass weeds	cycloxydim, glufosinate-ammonium, glyphosate, metamitron (*off-label*), metazachlor, napropamide (*off-label*), propaquizafop, propyzamide
	Weeds, miscellaneous	carfentrazone-ethyl (*off-label*), glufosinate-ammonium, glyphosate, metamitron (*off-label*), propyzamide
	Woody weeds/scrub	2,4-D + dicamba + triclopyr, glyphosate, triclopyr

Miscellaneous forestry situations - Cut logs/timber

| Pests | Beetles | chlorpyrifos, cypermethrin |
| Weeds | Grass weeds | propaquizafop |

Woodland on farms - Woodland

Crop control	Chemical stripping/ thinning	glyphosate
Diseases	Disease control	mancozeb (*off-label*)
	Root rot	Phlebiopsis gigantea
Pests	Aphids	lambda-cyhalothrin (*off-label*)
	Beetles	lambda-cyhalothrin (*off-label*)
	Pests, miscellaneous	lambda-cyhalothrin (*off-label*)
	Sawflies	lambda-cyhalothrin (*off-label*)
	Thrips	lambda-cyhalothrin (*off-label*)
	Whiteflies	lambda-cyhalothrin (*off-label*)
Plant growth regulation	Growth control	gibberellins (*do not exceed 2.5 g per 5 litres when used as a seed soak*), gibberellins (*off-label*)
	Plant growth regulation, miscellaneous	gibberellins (*qualified minor use*)
Weeds	Broad-leaved weeds	2,4-D + dicamba + triclopyr, clopyralid (*off-label*), florasulam (*off-label*), flumioxazine (*off-label*), fluroxypyr (*off-label*), lenacil (*off-label*), MCPA (*off-label*), metazachlor, metsulfuron-methyl (*off-label*), napropamide (*off-label*), pendimethalin (*off-label*), propyzamide, tribenuron-methyl (*off-label*)
	Crops as weeds	fluazifop-P-butyl, flumioxazine (*off-label*)
	Grass weeds	cycloxydim, fluazifop-P-butyl, lenacil (*off-label*), metazachlor, napropamide (*off-label*), pendimethalin (*off-label*), propaquizafop, propyzamide
	Weeds, miscellaneous	amitrole (*off-label*), glyphosate, glyphosate (*off-label*)
	Woody weeds/scrub	2,4-D + dicamba + triclopyr, glyphosate

Fruit and hops

All bush fruit - All currants

Diseases	Botrytis	Bacillus subtilis (*off-label*), boscalid + pyraclostrobin, boscalid + pyraclostrobin (*off-label*), cyprodinil + fludioxonil (*qualified minor use recommendation*), fenhexamid, Gliocladium catenulatum (*off-label - moderate control*), pyrimethanil (*off-label*)
	Damping off	Bacillus subtilis (*off-label*)
	Disease control	sulphur (*off-label*)
	Downy mildew	copper oxychloride (*off-label*)
	Fusarium	Gliocladium catenulatum (*off-label - moderate control*)
	Helminthosporium seedling rot	Bacillus subtilis (*off-label*)
	Leaf spot	boscalid + pyraclostrobin (*moderate control only*), boscalid + pyraclostrobin (*off-label*), dodine, dodine (*off-label*)
	Phytophthora	Bacillus subtilis (*off-label*), Gliocladium catenulatum (*off-label - moderate control*)
	Powdery mildew	boscalid + pyraclostrobin (*moderate control only*), boscalid + pyraclostrobin (*off-label*), bupirimate,

		bupirimate (*off-label*), fenpropimorph (*off-label*), kresoxim-methyl, kresoxim-methyl (*off-label*), myclobutanil, myclobutanil (*off-label*), penconazole, penconazole (*off-label*), quinoxyfen (*off-label*)
	Pythium	Bacillus subtilis (*off-label*), Gliocladium catenulatum (*off-label - moderate control*)
	Rhizoctonia	Bacillus subtilis (*off-label*), Gliocladium catenulatum (*off-label - moderate control*)
	Rust	copper oxychloride
	White rot	Bacillus subtilis (*off-label*)
Pests	Aphids	chlorpyrifos, fatty acids (*off-label*), pirimicarb, pymetrozine (*off-label*), spirotetramat (*off-label*)
	Capsid bugs	chlorpyrifos
	Caterpillars	Bacillus thuringiensis (*off-label*), chlorpyrifos, diflubenzuron, diflubenzuron (*off-label*), spinosad (*off-label*)
	Flies	Metarhizium anisopliae (*off-label*)
	Gall, rust and leaf & bud mites	sulphur, tebufenpyrad (*off-label*)
	Leatherjackets	Metarhizium anisopliae (*off-label*)
	Midges	lambda-cyhalothrin (*off-label*), Metarhizium anisopliae (*off-label*)
	Pests, miscellaneous	lambda-cyhalothrin (*off-label*), spinosad (*off-label*), thiacloprid (*off-label*)
	Sawflies	lambda-cyhalothrin (*off-label*), spinosad (*off-label*)
	Spider mites	chlorpyrifos, chlorpyrifos (*non OP resistant strains only*), clofentezine (*off-label*)
	Thrips	fatty acids (*off-label*), Metarhizium anisopliae (*off-label*)
	Weevils	Metarhizium anisopliae
Weeds	Bindweeds	oxadiazon, oxadiazon (*off-label*)
	Broad-leaved weeds	flufenacet + metribuzin (*off-label*), glufosinate-ammonium, isoxaben, isoxaben (*off-label*), lenacil (*off-label*), napropamide, napropamide (*off-label*), oxadiazon, oxadiazon (*off-label*), pendimethalin, pendimethalin (*off-label*), propyzamide
	Crops as weeds	fluazifop-P-butyl, pendimethalin
	Grass weeds	fluazifop-P-butyl, fluazifop-P-butyl (*off-label*), flufenacet + metribuzin (*off-label*), glufosinate-ammonium, lenacil (*off-label*), napropamide, napropamide (*off-label*), oxadiazon, oxadiazon (*off-label*), pendimethalin, pendimethalin (*off-label*), propyzamide
	Weeds, miscellaneous	glufosinate-ammonium, glyphosate (*off-label*)

All bush fruit - All protected bush fruit

Diseases	Anthracnose	boscalid + pyraclostrobin (*off-label*)
	Botrytis	boscalid + pyraclostrobin (*off-label*)
	Leaf spot	boscalid + pyraclostrobin (*off-label*)
	Powdery mildew	Ampelomyces quisqualis strain AQ10 (*off-label*)
Pests	Aphids	pymetrozine (*off-label*), pyrethrins

	Caterpillars	Bacillus thuringiensis (*off-label*), pyrethrins, thiacloprid (*off-label*)
	Midges	thiacloprid (*off-label*)
	Pests, miscellaneous	thiacloprid (*off-label*)
	Whiteflies	Lecanicillium lecanii (*off-label*)

All bush fruit - All vines

Crop control	Desiccation	carfentrazone-ethyl (*off-label*)
Diseases	Botrytis	Bacillus subtilis (*off-label*), fenhexamid, Gliocladium catenulatum (*off-label - moderate control*)
	Disease control	mancozeb
	Downy mildew	benthiavalicarb-isopropyl + mancozeb (*off-label*), copper oxychloride
	Fusarium	Gliocladium catenulatum (*off-label - moderate control*)
	Phytophthora	Gliocladium catenulatum (*off-label - moderate control*)
	Powdery mildew	Ampelomyces quisqualis strain AQ10 (*off-label*), fenbuconazole (*off-label*), kresoxim-methyl (*off-label*), meptyldinocap, myclobutanil (*off-label*), sulphur, tebuconazole + trifloxystrobin (*off-label*)
	Pythium	Gliocladium catenulatum (*off-label - moderate control*)
	Rhizoctonia	Gliocladium catenulatum (*off-label - moderate control*)
Pests	Aphids	spirotetramat (*off-label*)
	Caterpillars	Bacillus thuringiensis (*off-label*)
	Flies	Metarhizium anisopliae (*off-label*)
	Leatherjackets	Metarhizium anisopliae (*off-label*)
	Mealybugs	petroleum oil
	Midges	Metarhizium anisopliae (*off-label*)
	Scale insects	petroleum oil
	Spider mites	petroleum oil
	Thrips	Metarhizium anisopliae (*off-label*)
	Whiteflies	Lecanicillium lecanii (*off-label*)
Weeds	Bindweeds	oxadiazon
	Broad-leaved weeds	glufosinate-ammonium, isoxaben, oxadiazon, propyzamide (*off-label*)
	Grass weeds	glufosinate-ammonium, oxadiazon, propyzamide (*off-label*)
	Weeds, miscellaneous	glufosinate-ammonium, glyphosate (*off-label*)

All bush fruit - Bilberries/blueberries/cranberries

Diseases	Anthracnose	boscalid + pyraclostrobin (*off-label*)
	Botrytis	Bacillus subtilis (*off-label*), boscalid + pyraclostrobin (*off-label*), cyprodinil + fludioxonil (*qualified minor use recommendation*), fenhexamid (*off-label*), Gliocladium catenulatum (*off-label - moderate control*), pyrimethanil (*off-label*)
	Damping off	Bacillus subtilis (*off-label*)
	Disease control	sulphur (*off-label*)
	Downy mildew	copper oxychloride (*off-label*)

	Fusarium	Gliocladium catenulatum (*off-label - moderate control*)
	Helminthosporium seedling rot	Bacillus subtilis (*off-label*)
	Leaf spot	boscalid + pyraclostrobin (*off-label*), dodine (*off-label*)
	Phytophthora	Bacillus subtilis (*off-label*), Gliocladium catenulatum (*off-label - moderate control*)
	Powdery mildew	fenpropimorph (*off-label*), penconazole (*off-label*), quinoxyfen (*off-label*)
	Pythium	Bacillus subtilis (*off-label*), Gliocladium catenulatum (*off-label - moderate control*)
	Rhizoctonia	Bacillus subtilis (*off-label*), Gliocladium catenulatum (*off-label - moderate control*)
	White rot	Bacillus subtilis (*off-label*)
Pests	Aphids	fatty acids (*off-label*), pirimicarb (*off-label*), pymetrozine (*off-label*), spirotetramat (*off-label*)
	Caterpillars	Bacillus thuringiensis (*off-label*), diflubenzuron (*off-label*)
	Flies	Metarhizium anisopliae (*off-label*)
	Gall, rust and leaf & bud mites	tebufenpyrad (*off-label*)
	Leatherjackets	Metarhizium anisopliae (*off-label*)
	Midges	Metarhizium anisopliae (*off-label*)
	Pests, miscellaneous	thiacloprid (*off-label*)
	Thrips	Metarhizium anisopliae (*off-label*)
	Weevils	Metarhizium anisopliae
Weeds	Bindweeds	oxadiazon (*off-label*)
	Broad-leaved weeds	flufenacet + metribuzin (*off-label*), isoxaben (*off-label*), lenacil (*off-label*), napropamide (*off-label*), oxadiazon (*off-label*), pendimethalin (*off-label*), propyzamide (*off-label*)
	Grass weeds	fluazifop-P-butyl (*off-label*), flufenacet + metribuzin (*off-label*), lenacil (*off-label*), napropamide (*off-label*), oxadiazon (*off-label*), pendimethalin (*off-label*), propyzamide (*off-label*)
	Weeds, miscellaneous	glufosinate-ammonium, glyphosate (*off-label*)

All bush fruit - Bush fruit, general

Diseases	Phytophthora	fluazinam (*off-label*)
Pests	Aphids	pyrethrins
	Birds/mammals	aluminium ammonium sulphate
	Caterpillars	pyrethrins
	Mealybugs	petroleum oil
	Scale insects	petroleum oil
	Spider mites	petroleum oil
Plant growth regulation	Quality/yield control	sulphur

All bush fruit - Gooseberries

Diseases	Anthracnose	boscalid + pyraclostrobin (*off-label*)
	Botrytis	Bacillus subtilis (*off-label*), boscalid + pyraclostrobin (*off-label*), cyprodinil + fludioxonil (*qualified minor use recommendation*), fenhexamid, Gliocladium catenulatum (*off-label - moderate control*), pyrimethanil (*off-label*)
	Damping off	Bacillus subtilis (*off-label*)
	Downy mildew	copper oxychloride (*off-label*)
	Fusarium	Gliocladium catenulatum (*off-label - moderate control*)
	Helminthosporium seedling rot	Bacillus subtilis (*off-label*)
	Leaf spot	boscalid + pyraclostrobin (*off-label*), dodine (*off-label*)
	Phytophthora	Bacillus subtilis (*off-label*), Gliocladium catenulatum (*off-label - moderate control*)
	Powdery mildew	bupirimate, fenpropimorph (*off-label*), kresoxim-methyl (*off-label*), myclobutanil, penconazole (*off-label*), quinoxyfen (*off-label*), sulphur
	Pythium	Bacillus subtilis (*off-label*), Gliocladium catenulatum (*off-label - moderate control*)
	Rhizoctonia	Bacillus subtilis (*off-label*), Gliocladium catenulatum (*off-label - moderate control*)
	White rot	Bacillus subtilis (*off-label*)
Pests	Aphids	chlorpyrifos, fatty acids (*off-label*), pirimicarb, pymetrozine (*off-label*), spirotetramat (*off-label*)
	Capsid bugs	chlorpyrifos
	Caterpillars	Bacillus thuringiensis (*off-label*), chlorpyrifos, diflubenzuron (*off-label*)
	Gall, rust and leaf & bud mites	tebufenpyrad (*off-label*)
	Midges	lambda-cyhalothrin (*off-label*)
	Pests, miscellaneous	lambda-cyhalothrin (*off-label*), thiacloprid (*off-label*)
	Sawflies	lambda-cyhalothrin (*off-label*)
	Spider mites	chlorpyrifos, chlorpyrifos (*non OP resistant strains only*)
	Thrips	fatty acids (*off-label*)
	Weevils	Metarhizium anisopliae
Weeds	Bindweeds	oxadiazon
	Broad-leaved weeds	flufenacet + metribuzin (*off-label*), isoxaben, lenacil (*off-label*), napropamide, oxadiazon, pendimethalin, propyzamide
	Crops as weeds	fluazifop-P-butyl, pendimethalin
	Grass weeds	fluazifop-P-butyl, flufenacet + metribuzin (*off-label*), lenacil (*off-label*), napropamide, oxadiazon, pendimethalin, propyzamide
	Weeds, miscellaneous	glyphosate (*off-label*)

All bush fruit - Miscellaneous bush fruit

Diseases	Alternaria	iprodione (*off-label*)
	Botrytis	cyprodinil + fludioxonil (*off-label*), fenhexamid, iprodione (*off-label*), pyrimethanil (*off-label*)
	Downy mildew	cymoxanil (*off-label*), metalaxyl-M (*off-label*)
	Powdery mildew	proquinazid (*off-label*), sulphur
Pests	Capsid bugs	lambda-cyhalothrin (*off-label*)
	Caterpillars	indoxacarb (*off-label*)
	Pests, miscellaneous	lambda-cyhalothrin (*off-label*), spinosad (*off-label*)
	Wasps	lambda-cyhalothrin (*off-label*)
Plant growth regulation	Growth control	prohexadione-calcium (*off-label*)
Weeds	Grass weeds	fluazifop-P-butyl (*off-label*)
	Weeds, miscellaneous	glyphosate (*off-label*)

Cane fruit - All outdoor cane fruit

Crop control	Desiccation	carfentrazone-ethyl (*off-label*)
Diseases	Alternaria	iprodione
	Botrytis	Bacillus subtilis (*off-label*), fenhexamid, Gliocladium catenulatum (*off-label - moderate control*), iprodione, pyrimethanil (*off-label*)
	Botrytis fruit rot	cyprodinil + fludioxonil
	Cane blight	boscalid + pyraclostrobin (*off-label*), tebuconazole (*off-label*)
	Cane spot	copper oxychloride, copper oxychloride (*off-label*)
	Damping off	Bacillus subtilis (*off-label*)
	Downy mildew	metalaxyl-M (*off-label*)
	Fusarium	Gliocladium catenulatum (*off-label - moderate control*)
	Helminthosporium seedling rot	Bacillus subtilis (*off-label*)
	Phytophthora	Bacillus subtilis (*off-label*), fluazinam (*off-label*), Gliocladium catenulatum (*off-label - moderate control*), tebuconazole (*off-label*)
	Powdery mildew	azoxystrobin (*off-label*), bupirimate (*outdoor only*), fenpropimorph (*off-label*)
	Purple blotch	boscalid + pyraclostrobin (*off-label*), copper oxychloride
	Pythium	Bacillus subtilis (*off-label*), Gliocladium catenulatum (*off-label - moderate control*)
	Rhizoctonia	Bacillus subtilis (*off-label*), Gliocladium catenulatum (*off-label - moderate control*)
	Root rot	dimethomorph, dimethomorph (*moderate control*), fluazinam (*off-label*)
	White rot	Bacillus subtilis (*off-label*)
Pests	Aphids	chlorpyrifos, chlorpyrifos (*off-label*), deltamethrin (*off-label*), fatty acids (*off-label*), pirimicarb, pirimicarb (*off-label*), pymetrozine (*off-label*), pyrethrins
	Beetles	chlorpyrifos, deltamethrin
	Birds/mammals	aluminium ammonium sulphate

	Capsid bugs	lambda-cyhalothrin (*off-label*)
	Caterpillars	Bacillus thuringiensis, Bacillus thuringiensis (*off-label*), diflubenzuron (*off-label*), pyrethrins
	Flies	Metarhizium anisopliae (*off-label*)
	Gall, rust and leaf & bud mites	tebufenpyrad (*off-label*)
	Leatherjackets	Metarhizium anisopliae (*off-label*)
	Mealybugs	petroleum oil
	Midges	chlorpyrifos, Metarhizium anisopliae (*off-label*)
	Pests, miscellaneous	chlorpyrifos (*off-label*), deltamethrin (*off-label*), lambda-cyhalothrin (*off-label*), spinosad (*off-label*), thiacloprid (*off-label*)
	Scale insects	petroleum oil
	Slugs/snails	metaldehyde
	Spider mites	chlorpyrifos, chlorpyrifos (*non OP resistant strains only*), chlorpyrifos (*off-label*), clofentezine (*off-label*), petroleum oil, tebufenpyrad (*off-label*)
	Thrips	fatty acids (*off-label*), Metarhizium anisopliae (*off-label*), spinosad (*off-label*)
	Weevils	lambda-cyhalothrin (*off-label*), Metarhizium anisopliae
Plant growth regulation	Growth control	1-naphthylacetic acid, 1-naphthylacetic acid (*off-label*)
	Quality/yield control	sulphur
Weeds	Bindweeds	oxadiazon, oxadiazon (*off-label*)
	Broad-leaved weeds	glufosinate-ammonium, isoxaben, isoxaben (*off-label*), lenacil (*off-label*), napropamide, napropamide (*off-label*), oxadiazon, oxadiazon (*off-label*), pendimethalin, propyzamide, propyzamide (*England only*)
	Crops as weeds	fluazifop-P-butyl, pendimethalin
	Grass weeds	fluazifop-P-butyl, fluazifop-P-butyl (*off-label*), glufosinate-ammonium, napropamide, napropamide (*off-label*), oxadiazon, oxadiazon (*off-label*), pendimethalin, propyzamide, propyzamide (*England only*)
	Weeds, miscellaneous	glufosinate-ammonium

Cane fruit - All protected cane fruit

Crop control	Desiccation	carfentrazone-ethyl (*off-label*)
Diseases	Botrytis	pyrimethanil (*off-label*)
	Cane blight	boscalid + pyraclostrobin (*off-label*)
	Phytophthora	fluazinam (*off-label*)
	Powdery mildew	azoxystrobin (*off-label*), myclobutanil (*off-label*)
	Purple blotch	boscalid + pyraclostrobin (*off-label*)
	Root rot	dimethomorph (*moderate control*), fluazinam (*off-label*)
	Rust	boscalid + pyraclostrobin (*off-label*)
Pests	Aphids	pymetrozine (*off-label*), pyrethrins
	Beetles	thiacloprid (*off-label*)
	Capsid bugs	thiacloprid (*off-label*)

	Caterpillars	pyrethrins
	Gall, rust and leaf & bud mites	abamectin (*off-label*), tebufenpyrad (*off-label*)
	Mites	abamectin (*off-label*)
	Pests, miscellaneous	thiacloprid (*off-label*)
	Spider mites	abamectin (*off-label*), clofentezine (*off-label*), tebufenpyrad (*off-label*)
	Whiteflies	Lecanicillium lecanii (*off-label*)

Hops, general - Hops

Crop control	Chemical stripping/ thinning	diquat
	Desiccation	diquat (*off-label*)
Diseases	Botrytis	Bacillus subtilis (*off-label*), fenhexamid (*off-label*), Gliocladium catenulatum (*off-label - moderate control*), iprodione (*off-label*)
	Disease control	benthiavalicarb-isopropyl + mancozeb (*off-label*), boscalid + pyraclostrobin (*off-label*), cyazofamid (*off-label*), mancozeb (*off-label*), mancozeb + metalaxyl-M (*off-label*), propiconazole (*off-label*), thiophanate-methyl (*off-label - do not harvest for human or animal consumption (including idling) within 12 months of treatment.*)
	Downy mildew	copper oxychloride, cymoxanil (*off-label*), metalaxyl-M (*off-label*)
	Fusarium	Gliocladium catenulatum (*off-label - moderate control*)
	Phytophthora	Gliocladium catenulatum (*off-label - moderate control*)
	Powdery mildew	bupirimate, cyflufenamid (*off-label*), fenpropimorph (*off-label*), kresoxim-methyl (*off-label*), myclobutanil (*off-label*), penconazole, quinoxyfen (*off-label*), sulphur
	Pythium	Gliocladium catenulatum (*off-label - moderate control*)
	Rhizoctonia	Gliocladium catenulatum (*off-label - moderate control*)
Pests	Aphids	acetamiprid (*off-label*), deltamethrin, flonicamid (*off-label*), lambda-cyhalothrin (*off-label*), pymetrozine (*off-label*), spirotetramat (*off-label*), tebufenpyrad
	Caterpillars	indoxacarb (*off-label*), spinosad (*off-label*)
	Free-living nematodes	fosthiazate (*off-label*)
	Mealybugs	petroleum oil
	Pests, miscellaneous	indoxacarb (*off-label*), lambda-cyhalothrin (*off-label*), methoxyfenozide (*off-label*), spinosad (*off-label*), thiacloprid (*off-label*)
	Scale insects	petroleum oil
	Spider mites	abamectin (*off-label*), petroleum oil, spirodiclofen (*off-label*), tebufenpyrad
Plant growth regulation	Growth control	diquat, prohexadione-calcium (*off-label*)
Weeds	Bindweeds	oxadiazon
	Broad-leaved weeds	bentazone (*off-label*), clopyralid (*off-label*), flufenacet + isoxaflutole (*off-label*), flumioxazine (*off-label*), imazamox + pendimethalin (*off-label*), isoxaben, oxadiazon, pendimethalin (*off-label*), propyzamide (*off-label*)

	Crops as weeds	fluazifop-P-butyl, flumioxazine (*off-label*)
	Grass weeds	diquat, fluazifop-P-butyl, flufenacet + isoxaflutole (*off-label*), maleic hydrazide (*off-label*), oxadiazon, propyzamide (*off-label*), tepraloxydim (*off-label*)
	Weeds, miscellaneous	diquat, glyphosate (*off-label*)

Miscellaneous fruit situations - Fruit crops, general

Diseases	Botrytis	Bacillus subtilis (*off-label*), fenhexamid (*off-label*), Gliocladium catenulatum (*off-label - moderate control*), iprodione (*off-label*), pyrimethanil (*off-label*)
	Damping off	fosetyl-aluminium + propamocarb hydrochloride (*off-label*)
	Disease control	azoxystrobin (*off-label*), benthiavalicarb-isopropyl + mancozeb (*off-label*), boscalid + pyraclostrobin (*off-label*), mancozeb (*off-label*), mancozeb + metalaxyl-M (*off-label*), propiconazole (*off-label*), pyrimethanil (*off-label*), thiophanate-methyl (*off-label*)
	Downy mildew	fosetyl-aluminium + propamocarb hydrochloride (*off-label*), metalaxyl-M (*off-label*)
	Fusarium	Gliocladium catenulatum (*off-label - moderate control*)
	Phytophthora	Gliocladium catenulatum (*off-label - moderate control*)
	Powdery mildew	cyflufenamid (*off-label*), kresoxim-methyl (*off-label*), proquinazid (*off-label*)
	Pythium	Gliocladium catenulatum (*off-label - moderate control*)
	Rhizoctonia	Gliocladium catenulatum (*off-label - moderate control*)
Pests	Aphids	acetamiprid (*off-label*), fatty acids, lambda-cyhalothrin (*off-label*), pymetrozine (*off-label*)
	Caterpillars	Bacillus thuringiensis (*off-label*), indoxacarb (*off-label*), spinosad (*off-label*)
	Flies	Metarhizium anisopliae (*off-label*)
	Free-living nematodes	fosthiazate (*off-label*)
	Mealybugs	fatty acids, petroleum oil
	Midges	Metarhizium anisopliae (*off-label*)
	Pests, miscellaneous	lambda-cyhalothrin (*off-label*), methoxyfenozide (*off-label*), spinosad (*off-label*), thiacloprid (*off-label*)
	Scale insects	fatty acids, petroleum oil
	Spider mites	fatty acids, petroleum oil, spirodiclofen (*off-label*)
	Thrips	Metarhizium anisopliae (*off-label*)
	Whiteflies	fatty acids, pymetrozine (*off-label*)
Plant growth regulation	Growth control	prohexadione-calcium (*off-label*)
Weeds	Broad-leaved weeds	carfentrazone-ethyl (*off-label*), clopyralid (*off-label*), flufenacet + isoxaflutole (*off-label*), flumioxazine (*off-label*), imazamox + pendimethalin (*off-label*), pendimethalin (*off-label*), propyzamide (*off-label*)
	Crops as weeds	flumioxazine (*off-label*)
	Grass weeds	fluazifop-P-butyl (*off-label*), flufenacet + isoxaflutole (*off-label*), maleic hydrazide (*off-label*), propyzamide (*off-label*), tepraloxydim (*off-label*)
	Weeds, miscellaneous	glyphosate (*off-label*)

Miscellaneous fruit situations - Fruit nursery stock

Weeds	Broad-leaved weeds	metazachlor
	Grass weeds	metazachlor

Other fruit - All outdoor strawberries

Diseases	Alternaria	iprodione
	Black spot	azoxystrobin (*off-label*), boscalid + pyraclostrobin, cyprodinil + fludioxonil (*qualified minor use*)
	Botrytis	Bacillus subtilis (*off-label*), boscalid + pyraclostrobin, captan, chlorothalonil (*off-label*), cyprodinil + fludioxonil, fenhexamid, Gliocladium catenulatum (*moderate control*), iprodione, mepanipyrim, pyrimethanil, thiram
	Collar rot	fenamidone + fosetyl-aluminium (*moderate control from foliar spray*)
	Crown rot	dimethomorph (*moderate control*), fenamidone + fosetyl-aluminium (*moderate control from foliar spray*)
	Didymella stem rot	Gliocladium catenulatum (*moderate control*)
	Fusarium	Gliocladium catenulatum (*moderate control*)
	Leaf spot	chlorothalonil (*off-label*)
	Phytophthora	fenamidone + fosetyl-aluminium, Gliocladium catenulatum (*moderate control*)
	Powdery mildew	boscalid + pyraclostrobin, bupirimate, fenpropimorph (*off-label*), kresoxim-methyl, myclobutanil, penconazole (*off-label*), quinoxyfen (*off-label*), sulphur
	Pythium	fenamidone + fosetyl-aluminium, Gliocladium catenulatum (*moderate control*)
	Red core	fenamidone + fosetyl-aluminium
	Rhizoctonia	Gliocladium catenulatum (*moderate control*)
Pests	Aphids	chlorpyrifos, fatty acids (*off-label*), pirimicarb, pymetrozine (*off-label*)
	Beetles	methiocarb
	Birds/mammals	aluminium ammonium sulphate
	Capsid bugs	thiacloprid (*off-label*)
	Caterpillars	Bacillus thuringiensis, chlorpyrifos
	Pests, miscellaneous	lambda-cyhalothrin (*off-label*)
	Slugs/snails	methiocarb
	Spider mites	abamectin (*off-label*), chlorpyrifos, chlorpyrifos (*non OP resistant strains only*), clofentezine (*off-label*), fenpyroximate (*off-label*), spirodiclofen (*off-label*), tebufenpyrad
	Tarsonemid mites	abamectin (*off-label*), abamectin (*off-label - in propagation*)
	Weevils	chlorpyrifos, chlorpyrifos (*non-protected crops only*), Metarhizium anisopliae
Weeds	Broad-leaved weeds	glufosinate-ammonium, isoxaben, lenacil (*off-label*), metamitron (*off-label*), napropamide, pendimethalin, phenmedipham, propyzamide
	Crops as weeds	cycloxydim, fluazifop-P-butyl, pendimethalin

| Grass weeds | cycloxydim, fluazifop-P-butyl, glufosinate-ammonium, metamitron (*off-label*), napropamide, pendimethalin, propyzamide, S-metolachlor (*off-label*) |
| Weeds, miscellaneous | glufosinate-ammonium |

Other fruit - Protected miscellaneous fruit

Diseases	Alternaria	iprodione
	Black spot	azoxystrobin (*off-label*), boscalid + pyraclostrobin, cyprodinil + fludioxonil (*qualified minor use*)
	Botrytis	Bacillus subtilis (*reduction of damage to fruit*), boscalid + pyraclostrobin, cyprodinil + fludioxonil, iprodione, iprodione (*off-label*), mepanipyrim, pyrimethanil, thiram
	Botrytis fruit rot	Bacillus subtilis (*reduction of damage to fruit*)
	Crown rot	dimethomorph (*moderate control*)
	Damping off	fosetyl-aluminium + propamocarb hydrochloride (*off-label*)
	Disease control	boscalid + pyraclostrobin (*off-label*), mancozeb + metalaxyl-M (*off-label*)
	Downy mildew	fosetyl-aluminium + propamocarb hydrochloride (*off-label*)
	Powdery mildew	Ampelomyces quisqualis strain AQ10, boscalid + pyraclostrobin, bupirimate, kresoxim-methyl, meptyldinocap, penconazole (*off-label*), quinoxyfen (*off-label*)
Pests	Aphids	acetamiprid (*off-label*), deltamethrin (*off-label*), pymetrozine (*off-label*)
	Beetles	deltamethrin (*off-label*)
	Capsid bugs	thiacloprid (*off-label*)
	Caterpillars	Bacillus thuringiensis (*off-label*), deltamethrin (*off-label*), indoxacarb (*off-label*)
	Leafhoppers	deltamethrin (*off-label*)
	Pests, miscellaneous	deltamethrin (*off-label*), indoxacarb (*off-label*), lambda-cyhalothrin (*off-label*), thiacloprid (*off-label*)
	Spider mites	abamectin (*off-label*), bifenazate, bifenazate (*off-label*), clofentezine (*off-label*), etoxazole (*off-label*), fenpyroximate (*off-label*), spirodiclofen (*off-label*), spiromesifen (*off-label*)
	Tarsonemid mites	abamectin (*off-label*)
	Thrips	deltamethrin (*off-label*)
	Whiteflies	Lecanicillium lecanii (*off-label*), pymetrozine (*off-label*), spiromesifen (*off-label*)
Plant growth regulation	Quality/yield control	gibberellins

Other fruit - Rhubarb

Diseases	Bacterial blight	copper oxychloride (*off-label*)
	Damping off	thiram (*off-label - seed treatment*)
	Downy mildew	mancozeb + metalaxyl-M (*off-label*)
	Foliar diseases	copper oxychloride (*off-label*)

	Phytophthora	difenoconazole (*off-label*)
Pests	Aphids	deltamethrin (*off-label*), fatty acids (*off-label*), pirimicarb (*off-label*)
	Beetles	deltamethrin (*off-label*)
	Caterpillars	Bacillus thuringiensis (*off-label*), deltamethrin (*off-label*)
	Leafhoppers	deltamethrin (*off-label*)
	Pests, miscellaneous	deltamethrin (*off-label*)
	Thrips	deltamethrin (*off-label*), fatty acids (*off-label*)
Plant growth regulation	Quality/yield control	gibberellins
Weeds	Broad-leaved weeds	clomazone (*off-label*), metamitron (*off-label*), pendimethalin (*off-label*), propyzamide, propyzamide (*outdoor*)
	Crops as weeds	propaquizafop (*off-label*)
	Grass weeds	metamitron (*off-label*), pendimethalin (*off-label*), propaquizafop (*off-label*), propyzamide, propyzamide (*outdoor*)
	Weeds, miscellaneous	glyphosate (*off-label*)

Tree fruit - All nuts

Diseases	Bacterial blight	copper oxychloride (*off-label*)
	Bacterial canker	copper oxychloride (*off-label*)
	Bottom rot	copper oxychloride (*off-label*)
	Phytophthora	copper oxychloride (*off-label*)
	Spear rot	copper oxychloride (*off-label*)
	Stem canker	copper oxychloride (*off-label*), tebuconazole (*off-label*)
Pests	Aphids	lambda-cyhalothrin (*off-label*)
	Caterpillars	Bacillus thuringiensis (*off-label*), diflubenzuron (*off-label*)
	Mites	tebufenpyrad (*off-label*), thiacloprid (*off-label*)
	Pests, miscellaneous	lambda-cyhalothrin (*off-label*)
Weeds	Bindweeds	oxadiazon (*off-label*)
	Broad-leaved weeds	dicamba + MCPA + mecoprop-P (*off-label*), fluroxypyr (*off-label*), glufosinate-ammonium, isoxaben (*off-label*), metazachlor, oxadiazon (*off-label*), pendimethalin (*off-label*), propyzamide (*off-label*)
	Grass weeds	fluazifop-P-butyl (*off-label*), glufosinate-ammonium, metazachlor, oxadiazon (*off-label*), pendimethalin (*off-label*), propyzamide (*off-label*)
	Weeds, miscellaneous	glufosinate-ammonium, glyphosate (*off-label*)

Tree fruit - All pome fruit

Crop control	Sucker/shoot control	glyphosate
Diseases	Alternaria	cyprodinil + fludioxonil
	Botrytis	iprodione (*off-label*), pyrimethanil (*off-label*), thiram
	Botrytis fruit rot	cyprodinil + fludioxonil, thiram
	Collar rot	copper oxychloride, copper oxychloride (*off-label*)

	Disease control	mancozeb
	Fusarium	cyprodinil + fludioxonil
	Leaf curl	copper oxychloride (*off-label*)
	Penicillium rot	cyprodinil + fludioxonil
	Phytophthora	mancozeb + metalaxyl-M (*off-label*)
	Powdery mildew	boscalid + pyraclostrobin, bupirimate, dithianon + pyraclostrobin, fenbuconazole (*reduction*), kresoxim-methyl (*off-label*), kresoxim-methyl (*reduction*), meptyldinocap (*off-label*), myclobutanil, myclobutanil (*off-label*), penconazole, sulphur, sulphur (*off-label*)
	Scab	boscalid + pyraclostrobin, captan, captan (*off-label*), dithianon, dithianon (*off-label*), dithianon + pyraclostrobin, dodine, dodine (*off-label*), fenbuconazole, kresoxim-methyl, mancozeb, maneb, myclobutanil, pyrimethanil, pyrimethanil (*off-label*), sulphur, sulphur (*off-label*), thiram
	Stem canker	copper oxychloride, copper oxychloride (*off-label*), tebuconazole (*off-label*)
	Storage rots	captan, cyprodinil + fludioxonil, thiram
Pests	Aphids	acetamiprid, chlorpyrifos, chlorpyrifos (*off-label*), cypermethrin, deltamethrin, flonicamid, pirimicarb, thiacloprid
	Capsid bugs	chlorpyrifos, chlorpyrifos (*off-label*), cypermethrin, deltamethrin
	Caterpillars	Bacillus thuringiensis (*off-label*), chlorantraniliprole, chlorpyrifos, chlorpyrifos (*off-label*), Cydia pomonella GV, cypermethrin, deltamethrin, diflubenzuron, diflubenzuron (*off-label*), dodeca-8,10-dienyl acetate, fenoxycarb, indoxacarb, methoxyfenozide, spinosad
	Gall, rust and leaf & bud mites	diflubenzuron, spirodiclofen
	Midges	thiacloprid (*off-label*)
	Pests, miscellaneous	abamectin (*off-label*), chlorpyrifos (*off-label*), thiacloprid (*off-label*)
	Sawflies	chlorpyrifos, cypermethrin, deltamethrin
	Scale insects	spirodiclofen
	Spider mites	chlorpyrifos, chlorpyrifos (*non OP resistant strains only*), clofentezine, fenpyroximate, spirodiclofen, tebufenpyrad
	Suckers	abamectin (*effective against larval stages but little activity against adults*), chlorpyrifos, cypermethrin, deltamethrin, diflubenzuron, lambda-cyhalothrin, spirodiclofen
	Weevils	chlorpyrifos
	Whiteflies	acetamiprid
Plant growth regulation	Fruiting control	1-methylcyclopropene (*off-label - post harvest only*), 1-methylcyclopropene (*post-harvest use*), benzyladenine, gibberellins, paclobutrazol, paclobutrazol (*off-label*)
	Growth control	1-naphthylacetic acid, 2-chloroethylphosphonic acid (*off-label*), gibberellins, gibberellins (*off-label*), paclobutrazol, paclobutrazol (*off-label*), prohexadione-calcium, prohexadione-calcium (*off-label*)
	Quality/yield control	gibberellins, gibberellins (*off-label*)

Weeds	Bindweeds	oxadiazon
	Broad-leaved weeds	2,4-D, amitrole, clopyralid (*off-label*), dicamba + MCPA + mecoprop-P, fluroxypyr (*off-label*), glufosinate-ammonium, isoxaben, metazachlor, oxadiazon, pendimethalin, pendimethalin (*off-label*), propyzamide, propyzamide (*off-label*)
	Crops as weeds	pendimethalin
	Grass weeds	amitrole, fluazifop-P-butyl (*off-label*), glufosinate-ammonium, glyphosate, metazachlor, oxadiazon, pendimethalin, pendimethalin (*off-label*), propyzamide, propyzamide (*off-label*)
	Weeds, miscellaneous	amitrole, amitrole (*off-label*), glufosinate-ammonium, glyphosate, glyphosate (*off-label*)

Tree fruit - All stone fruit

Crop control	Sucker/shoot control	glyphosate
Diseases	Bacterial canker	copper oxychloride
	Blossom wilt	boscalid + pyraclostrobin (*off-label*), cyprodinil + fludioxonil (*off-label*), fenbuconazole (*off-label*), myclobutanil (*off-label*)
	Botrytis	cyprodinil + fludioxonil (*off-label*), fenhexamid (*off-label*)
	Leaf curl	copper oxychloride, copper oxychloride (*off-label*)
	Powdery mildew	myclobutanil (*off-label*)
	Rust	myclobutanil (*off-label*)
	Sclerotinia	boscalid + pyraclostrobin (*off-label*), fenbuconazole (*off-label*), myclobutanil (*off-label*)
Pests	Aphids	acetamiprid, chlorpyrifos, chlorpyrifos (*Non OP resistant strains only*), deltamethrin, pirimicarb, pirimicarb (*off-label*), tebufenpyrad (*off-label*), thiacloprid (*off-label*)
	Caterpillars	Bacillus thuringiensis (*off-label*), chlorpyrifos, deltamethrin, diflubenzuron, indoxacarb (*off-label*)
	Gall, rust and leaf & bud mites	diflubenzuron
	Pests, miscellaneous	thiacloprid (*off-label*), thiacloprid (*off-label - under temporary protective rain covers*)
	Sawflies	deltamethrin
	Spider mites	chlorpyrifos, chlorpyrifos (*non OP resistant strains only*), clofentezine, fenpyroximate (*off-label*)
	Whiteflies	acetamiprid
Plant growth regulation	Growth control	1-naphthylacetic acid, paclobutrazol (*off-label*)
	Quality/yield control	gibberellins (*off-label*)
Weeds	Broad-leaved weeds	glufosinate-ammonium, isoxaben, metazachlor, pendimethalin, propyzamide, propyzamide (*off-label*)
	Crops as weeds	pendimethalin
	Grass weeds	fluazifop-P-butyl (*off-label*), glufosinate-ammonium, metazachlor, pendimethalin, propyzamide, propyzamide (*off-label*)
	Weeds, miscellaneous	amitrole (*off-label*), glufosinate-ammonium, glyphosate, glyphosate (*off-label*)

Tree fruit - Miscellaneous top fruit

Pests	Birds/mammals	aluminium ammonium sulphate
Plant growth regulation	Quality/yield control	sulphur

Tree fruit - Protected fruit

Diseases	Powdery mildew	Ampelomyces quisqualis strain AQ10 (*off-label*)
Pests	Caterpillars	Bacillus thuringiensis (*off-label*)

Grain/crop store uses

Stored seed - Stored grain/rape/linseed

Pests	Birds/mammals	aluminium ammonium sulphate
	Food/grain storage pests	chlorpyrifos-methyl, d-phenothrin + tetramethrin, pirimiphos-methyl
	Mites	physical pest control, pirimiphos-methyl
	Pests, miscellaneous	aluminium phosphide, deltamethrin, deltamethrin (*off-label*), magnesium phosphide, pirimiphos-methyl

Grass

Turf/amenity grass - Amenity grassland

Diseases	Anthracnose	azoxystrobin, azoxystrobin + propiconazole (*moderate control*), chlorothalonil + fludioxonil + propiconazole (*reduction*), fludioxonil (*reduction*), propiconazole (*qualified minor use*), tebuconazole + trifloxystrobin
	Brown patch	azoxystrobin, chlorothalonil + fludioxonil + propiconazole, iprodione, propiconazole (*qualified minor use*)
	Crown rust	azoxystrobin
	Dollar spot	azoxystrobin + propiconazole, chlorothalonil + fludioxonil + propiconazole, iprodione, propiconazole, tebuconazole + trifloxystrobin
	Drechslera leaf spot	fludioxonil (*useful levels of control*)
	Fairy rings	azoxystrobin
	Fusarium	azoxystrobin, azoxystrobin + propiconazole, chlorothalonil + fludioxonil + propiconazole, fludioxonil, iprodione, propiconazole, tebuconazole + trifloxystrobin, trifloxystrobin
	Melting out	azoxystrobin, iprodione, tebuconazole + trifloxystrobin
	Red thread	iprodione, tebuconazole + trifloxystrobin, trifloxystrobin
	Rust	tebuconazole + trifloxystrobin
	Seed-borne diseases	azoxystrobin + propiconazole, fludioxonil
	Snow mould	iprodione
	Take-all patch	azoxystrobin

Pests	Birds/mammals	aluminium ammonium sulphate, aluminium ammonium sulphate (*anti-fouling*)
	Flies	chlorpyrifos
	Leatherjackets	chlorpyrifos
	Slugs/snails	metaldehyde
Plant growth regulation	Growth control	trinexapac-ethyl
Weeds	Broad-leaved weeds	2,4-D, 2,4-D + dicamba, 2,4-D + dicamba + dichlorprop-P, 2,4-D + dicamba + MCPA + mecoprop-P, aminopyralid + fluroxypyr, aminopyralid + triclopyr, clopyralid + florasulam + fluroxypyr, clopyralid + triclopyr, dicamba + MCPA + mecoprop-P, ethofumesate, florasulam + fluroxypyr, fluroxypyr + triclopyr, mecoprop-P
	Crops as weeds	2,4-D + dicamba + MCPA + mecoprop-P, florasulam + fluroxypyr, pinoxaden (*off-label - reduction*), pinoxaden (*reduction only*)
	Grass weeds	cycloxydim (*off-label*), ethofumesate, maleic hydrazide (*off-label*)
	Mosses	carfentrazone-ethyl + mecoprop-P, ferrous sulphate
	Weeds, miscellaneous	aminopyralid + triclopyr, carfentrazone-ethyl + mecoprop-P, florasulam + fluroxypyr, glyphosate, glyphosate (*wiper application*)
	Woody weeds/scrub	clopyralid + triclopyr

Turf/amenity grass - Managed amenity turf

Crop control	Miscellaneous non-selective situations	glufosinate-ammonium
Diseases	Anthracnose	azoxystrobin, azoxystrobin + propiconazole (*moderate control*), chlorothalonil + fludioxonil + propiconazole (*reduction*), fludioxonil (*reduction*), iprodione, propiconazole (*qualified minor use*), tebuconazole + trifloxystrobin
	Brown patch	azoxystrobin, chlorothalonil + fludioxonil + propiconazole, iprodione, propiconazole (*qualified minor use*)
	Crown rust	azoxystrobin
	Disease control	carbendazim, iprodione
	Dollar spot	azoxystrobin + propiconazole, chlorothalonil + fludioxonil + propiconazole, iprodione, propiconazole, pyraclostrobin (*useful reduction*), tebuconazole + trifloxystrobin
	Drechslera leaf spot	fludioxonil (*useful levels of control*)
	Fairy rings	azoxystrobin
	Fusarium	azoxystrobin, azoxystrobin + propiconazole, chlorothalonil + fludioxonil + propiconazole, fludioxonil, iprodione, myclobutanil, prochloraz + tebuconazole, propiconazole, pyraclostrobin (*moderate control*), pyraclostrobin (*moderate control only*), tebuconazole + trifloxystrobin, trifloxystrobin
	Melting out	azoxystrobin, iprodione, tebuconazole + trifloxystrobin
	Red thread	iprodione, pyraclostrobin, tebuconazole + trifloxystrobin, trifloxystrobin

	Rust	iprodione, tebuconazole + trifloxystrobin
	Seed-borne diseases	azoxystrobin + propiconazole, fludioxonil
	Snow mould	iprodione
	Take-all patch	azoxystrobin
Pests	Aphids	esfenvalerate
	Birds/mammals	aluminium ammonium sulphate, aluminium ammonium sulphate (*anti-fouling*)
	Chafers	imidacloprid
	Earthworms	carbendazim
	Flies	chlorpyrifos, esfenvalerate
	Leatherjackets	chlorpyrifos, imidacloprid
	Pests, miscellaneous	imidacloprid
	Slugs/snails	metaldehyde
Plant growth regulation	Growth control	trinexapac-ethyl
	Quality/yield control	sulphur
Weeds	Broad-leaved weeds	2,4-D, 2,4-D + dicamba, 2,4-D + dicamba + fluroxypyr, 2,4-D + dicamba + MCPA + mecoprop-P, 2,4-D + florasulam, clopyralid + 2,4-D + MCPA, clopyralid + diflufenican + MCPA, clopyralid + florasulam + fluroxypyr, clopyralid + fluroxypyr + MCPA, dicamba + dichlorprop-P + MCPA, dicamba + MCPA + mecoprop-P, dichlorprop-P + ferrous sulphate + MCPA, ethofumesate (*off-label*), ferrous sulphate + MCPA + mecoprop-P, florasulam + fluroxypyr, MCPA + mecoprop-P, mecoprop-P
	Crops as weeds	2,4-D + dicamba + fluroxypyr, 2,4-D + dicamba + MCPA + mecoprop-P, 2,4-D + florasulam, florasulam + fluroxypyr, pinoxaden (*off-label - reduction*), pinoxaden (*reduction only*)
	Grass weeds	amitrole (*off-label - on amitrole resistant turf*), ethofumesate (*off-label*), tepraloxydim (*off-label*)
	Mosses	carfentrazone-ethyl + mecoprop-P, dichlorprop-P + ferrous sulphate + MCPA, ferrous sulphate, ferrous sulphate + MCPA + mecoprop-P, quinoclamine (*moderate control*)
	Weeds, miscellaneous	2,4-D + dicamba, carfentrazone-ethyl + mecoprop-P, florasulam + fluroxypyr, glufosinate-ammonium, glyphosate (*pre-establishment*), glyphosate (*pre-establishment only*)

Non-crop pest control

Farm buildings/yards - Farm buildings

Pests	Birds/mammals	aluminium ammonium sulphate, bromadiolone, difenacoum, warfarin
	Flies	diflubenzuron, d-phenothrin + tetramethrin, tetramethrin
	Food/grain storage pests	alpha-cypermethrin
	Pests, miscellaneous	alpha-cypermethrin, cypermethrin, pyrethrins
	Wasps	d-phenothrin + tetramethrin

| Weeds | Weeds, miscellaneous | glyphosate |

Farm buildings/yards - Farmyards

| Pests | Birds/mammals | bromadiolone, difenacoum |

Farmland pest control - Farmland situations

| Crop control | Desiccation | diquat (*off-label*) |
| Pests | Birds/mammals | aluminium phosphide |

Miscellaneous non-crop pest control - Manure/rubbish

| Pests | Flies | diflubenzuron |

Miscellaneous non-crop pest control - Miscellaneous pest control situations

| Pests | Birds/mammals | carbon dioxide (commodity substance) |

Protected salad and vegetable crops

Protected brassicas - Protected brassica vegetables

Diseases	Bacterial blight	copper oxychloride (*off-label*)
	Black rot	copper oxychloride (*off-label*)
	Bottom rot	copper oxychloride (*off-label*)
	Damping off	fosetyl-aluminium + propamocarb hydrochloride, propamocarb hydrochloride
	Disease control	prochloraz (*off-label*)
	Downy mildew	fosetyl-aluminium + propamocarb hydrochloride, propamocarb hydrochloride
	Fungus diseases	propamocarb hydrochloride
	Phytophthora	copper oxychloride (*off-label*), propamocarb hydrochloride
	Pythium	propamocarb hydrochloride
	Rhizoctonia	azoxystrobin (*off-label*)
	Root rot	propamocarb hydrochloride
	Spear rot	copper oxychloride (*off-label*)
	Stem canker	copper oxychloride (*off-label*)
	White blister	boscalid + pyraclostrobin (*off-label*)
Pests	Aphids	acetamiprid (*off-label*), cypermethrin (*off-label*), dimethoate (*off-label*), pirimicarb (*off-label*), pyrethrins, spirotetramat (*off-label*)
	Beetles	deltamethrin (*off-label*), pyrethrins
	Caterpillars	Bacillus thuringiensis (*off-label*), cypermethrin (*off-label*), deltamethrin (*off-label*), pyrethrins
	Mites	abamectin (*off-label*)
	Pests, miscellaneous	dimethoate (*off-label*)
	Whiteflies	pyrethrins, spirotetramat (*off-label*)

Protected brassicas - Protected salad brassicas

Diseases	Bacterial blight	copper oxychloride (*off-label*)
	Bottom rot	copper oxychloride (*off-label*)
	Damping off	fosetyl-aluminium + propamocarb hydrochloride
	Downy mildew	copper oxychloride (*off-label*), fosetyl-aluminium + propamocarb hydrochloride
	Phytophthora	copper oxychloride (*off-label*)
	Ring spot	difenoconazole (*off-label*)
	Spear rot	copper oxychloride (*off-label*)
	Stem canker	copper oxychloride (*off-label*)
Pests	Aphids	cypermethrin (*off-label*), deltamethrin (*off-label*), dimethoate (*off-label*), lambda-cyhalothrin (*off-label*), pymetrozine (*off-label*), pymetrozine (*off-label - for baby leaf production*)
	Beetles	deltamethrin (*off-label*), lambda-cyhalothrin (*off-label*)
	Caterpillars	Bacillus thuringiensis (*off-label*), cypermethrin (*off-label*), deltamethrin (*off-label*), lambda-cyhalothrin (*off-label*)
	Leafhoppers	deltamethrin (*off-label*)
	Pests, miscellaneous	deltamethrin (*off-label*), dimethoate (*off-label*), lambda-cyhalothrin (*off-label*)
	Thrips	deltamethrin (*off-label*)
	Whiteflies	Lecanicillium lecanii (*off-label*)

Protected crops, general - All protected crops

Diseases	Sclerotinia	Coniothyrium minitans
Pests	Aphids	Beauveria bassiana, pyrethrins
	Beetles	Beauveria bassiana, pyrethrins
	Caterpillars	indoxacarb (*off-label*), pyrethrins
	Mealybugs	pyrethrins
	Pests, miscellaneous	lambda-cyhalothrin (*off-label*)
	Scale insects	pyrethrins
	Slugs/snails	ferric phosphate, metaldehyde
	Spider mites	pyrethrins
	Thrips	pyrethrins
	Whiteflies	Beauveria bassiana, pyrethrins

Protected crops, general - Glasshouses

Pests	Pests, miscellaneous	pyrethrins

Protected crops, general - Protected vegetables, general

Diseases	Powdery mildew	Ampelomyces quisqualis strain AQ10 (*off-label*)
Pests	Aphids	pirimicarb (*off-label*), pymetrozine (*off-label*)
	Whiteflies	Lecanicillium lecanii (*off-label*), pymetrozine (*off-label*)

Protected crops, general - Soils and compost

Diseases	Root rot	metam-sodium
	Sclerotinia	metam-sodium
	Soil-borne diseases	dazomet
Pests	Cyst nematodes	metam-sodium
	Millipedes	metam-sodium
	Pests, miscellaneous	dimethoate (*off-label*)
	Root-knot nematodes	metam-sodium
	Soil pests	dazomet
	Symphylids	metam-sodium
	Wireworms	metam-sodium
Weeds	Weeds, miscellaneous	metam-sodium

Protected fruiting vegetables - Protected aubergines

Diseases	Botrytis	azoxystrobin (*off-label*), boscalid + pyraclostrobin (*off-label*), cyprodinil + fludioxonil (*off-label*), fenhexamid (*off-label*), iprodione (*off-label*), pyrimethanil (*off-label*)
	Damping off	fosetyl-aluminium + propamocarb hydrochloride (*off-label*)
	Downy mildew	azoxystrobin (*off-label*)
	Phytophthora	azoxystrobin (*off-label*)
	Powdery mildew	Ampelomyces quisqualis strain AQ10, azoxystrobin (*off-label*), myclobutanil (*off-label*), sulphur (*off-label*)
	Septoria	azoxystrobin (*off-label*)
Pests	Aphids	acetamiprid, deltamethrin (*off-label*), lambda-cyhalothrin (*off-label*), pirimicarb (*off-label*), pymetrozine (*off-label*)
	Caterpillars	Bacillus thuringiensis (*off-label*), indoxacarb
	Leaf miners	oxamyl (*off-label*), thiacloprid (*off-label*)
	Mites	etoxazole
	Pests, miscellaneous	deltamethrin (*off-label*), lambda-cyhalothrin (*off-label*)
	Spider mites	abamectin (*off-label*), spiromesifen (*off-label*)
	Thrips	abamectin (*off-label*), deltamethrin (*off-label*), spinosad, thiacloprid (*off-label*)
	Whiteflies	acetamiprid, pymetrozine (*off-label*), spiromesifen (*off-label*), thiacloprid (*off-label*)

Protected fruiting vegetables - Protected cucurbits

Diseases	Botrytis	azoxystrobin (*off-label*), boscalid + pyraclostrobin (*off-label*), cyprodinil + fludioxonil (*off-label*), fenhexamid (*off-label*)
	Damping off	fosetyl-aluminium + propamocarb hydrochloride, fosetyl-aluminium + propamocarb hydrochloride (*off-label*)
	Downy mildew	azoxystrobin (*off-label*), fosetyl-aluminium + propamocarb hydrochloride, metalaxyl-M (*off-label*)
	Fusarium	cyprodinil + fludioxonil (*off-label*)
	Phytophthora	azoxystrobin (*off-label*)

	Powdery mildew	Ampelomyces quisqualis strain AQ10, Ampelomyces quisqualis strain AQ10 (*off-label*), azoxystrobin (*off-label*), bupirimate, myclobutanil (*off-label*), sulphur (*off-label*)
	Pythium	fosetyl-aluminium + propamocarb hydrochloride
	Septoria	azoxystrobin (*off-label*)
Pests	Aphids	acetamiprid, deltamethrin, deltamethrin (*off-label*), lambda-cyhalothrin (*off-label*), pirimicarb (*off-label*), pymetrozine, pymetrozine (*off-label*), pyrethrins (*off-label*)
	Caterpillars	Bacillus thuringiensis, Bacillus thuringiensis (*off-label*), deltamethrin, indoxacarb
	Leaf miners	thiacloprid (*off-label*)
	Mealybugs	deltamethrin, petroleum oil
	Pests, miscellaneous	deltamethrin (*off-label*), lambda-cyhalothrin (*off-label*)
	Scale insects	deltamethrin, petroleum oil
	Spider mites	abamectin, abamectin (*off-label*), petroleum oil, spiromesifen (*off-label*)
	Thrips	abamectin, deltamethrin, deltamethrin (*off-label*), spinosad, thiacloprid (*off-label*)
	Whiteflies	acetamiprid, deltamethrin, Lecanicillium lecanii, Lecanicillium lecanii (*off-label*), pymetrozine (*off-label*), spiromesifen (*off-label*), thiacloprid (*off-label*)
Weeds	Broad-leaved weeds	isoxaben (*off-label*)

Protected fruiting vegetables - Protected tomatoes

Diseases	Alternaria	iprodione
	Botrytis	azoxystrobin (*off-label*), boscalid + pyraclostrobin (*off-label*), cyprodinil + fludioxonil (*off-label*), fenhexamid (*off-label*), iprodione, pyrimethanil (*off-label*)
	Damping off	copper oxychloride
	Didymella stem rot	azoxystrobin (*off-label*)
	Disease control	maneb
	Downy mildew	azoxystrobin (*off-label*)
	Foot rot	copper oxychloride
	Phytophthora	azoxystrobin (*off-label*), copper oxychloride
	Powdery mildew	Ampelomyces quisqualis strain AQ10, azoxystrobin (*off-label*), bupirimate (*off-label*), myclobutanil (*off-label*), sulphur (*off-label*)
	Pythium	fosetyl-aluminium + propamocarb hydrochloride
	Septoria	azoxystrobin (*off-label*)
	Verticillium wilt	thiophanate-methyl (*off-label*)
Pests	Aphids	acetamiprid, deltamethrin, fatty acids, lambda-cyhalothrin (*off-label*), pirimicarb, pymetrozine (*off-label*), pyrethrins
	Beetles	pyrethrins
	Caterpillars	Bacillus thuringiensis, deltamethrin, indoxacarb, pyrethrins
	Leaf miners	abamectin, abamectin (*off-label*), oxamyl (*off-label*), thiacloprid (*off-label*)

	Mealybugs	deltamethrin, fatty acids, petroleum oil, pyrethrins, pyrethrins (*off-label*)
	Mirid bugs	pyrethrins, pyrethrins (*off-label*)
	Mites	etoxazole
	Pests, miscellaneous	lambda-cyhalothrin (*off-label*)
	Scale insects	deltamethrin, fatty acids, petroleum oil
	Spider mites	abamectin, fatty acids, petroleum oil
	Thrips	deltamethrin, spinosad, thiacloprid (*off-label*)
	Whiteflies	acetamiprid, deltamethrin, fatty acids, Lecanicillium lecanii, pymetrozine (*off-label*), pyrethrins, spiromesifen, thiacloprid (*off-label*)
Plant growth regulation	Growth control	2-chloroethylphosphonic acid (*off-label*)

Protected herb crops - Protected herbs

Diseases	Alternaria	iprodione (*off-label*)
	Botrytis	fenhexamid (*off-label*), iprodione (*off-label*), pyrimethanil (*off-label*)
	Damping off	fosetyl-aluminium + propamocarb hydrochloride (*off-label*)
	Disease control	mandipropamid, prochloraz (*off-label*)
	Downy mildew	copper oxychloride (*off-label*), dimethomorph (*off-label*), fosetyl-aluminium + propamocarb hydrochloride (*off-label*), mancozeb + metalaxyl-M (*off-label*), mandipropamid, metalaxyl-M (*off-label*)
	Powdery mildew	Ampelomyces quisqualis strain AQ10 (*off-label*), sulphur (*off-label*)
	Rhizoctonia	azoxystrobin (*off-label*)
	White blister	boscalid + pyraclostrobin (*off-label*)
Pests	Aphids	acetamiprid (*off-label*), cypermethrin (*off-label*), deltamethrin (*off-label*), fatty acids (*off-label*), pirimicarb (*off-label*), pymetrozine (*off-label*), pyrethrins (*off-label*), spirotetramat (*off-label*), thiacloprid (*off-label*)
	Beetles	deltamethrin (*off-label*)
	Caterpillars	Bacillus thuringiensis (*off-label*), cypermethrin (*off-label*), deltamethrin (*off-label*)
	Leafhoppers	deltamethrin (*off-label*)
	Mites	abamectin (*off-label*)
	Pests, miscellaneous	abamectin (*off-label*), deltamethrin (*off-label*), spinosad (*off-label*)
	Thrips	deltamethrin (*off-label*), fatty acids (*off-label*), pirimicarb (*off-label*)
	Whiteflies	Lecanicillium lecanii (*off-label*), pymetrozine (*off-label*), spirotetramat (*off-label*)
Weeds	Broad-leaved weeds	propyzamide (*off-label*)
	Grass weeds	propyzamide (*off-label*)

Protected leafy vegetables - Mustard and cress

Diseases	Botrytis	cyprodinil + fludioxonil (*off-label*), Gliocladium catenulatum (*off-label - moderate control*)
	Damping off	thiram (*off-label*)
	Disease control	dimethomorph + mancozeb (*off-label*), mandipropamid
	Downy mildew	copper oxychloride (*off-label*), dimethomorph + mancozeb (*off-label*), mandipropamid
	Fusarium	Gliocladium catenulatum (*off-label - moderate control*)
	Phytophthora	Gliocladium catenulatum (*off-label - moderate control*)
	Powdery mildew	azoxystrobin (*off-label*)
	Pythium	Gliocladium catenulatum (*off-label - moderate control*)
	Rhizoctonia	Gliocladium catenulatum (*off-label - moderate control*)
	Rust	azoxystrobin (*off-label*)
Pests	Aphids	acetamiprid (*off-label*), fatty acids (*off-label*), spirotetramat (*off-label*)
	Pests, miscellaneous	spinosad
	Thrips	fatty acids (*off-label*)
	Whiteflies	spirotetramat (*off-label*)
Weeds	Broad-leaved weeds	pendimethalin (*off-label*), propyzamide (*off-label*)
	Grass weeds	pendimethalin (*off-label*), propyzamide (*off-label*)

Protected leafy vegetables - Protected leafy vegetables

Diseases	Alternaria	iprodione
	Botrytis	boscalid + pyraclostrobin, iprodione, pyrimethanil (*off-label*)
	Bottom rot	boscalid + pyraclostrobin, tolclofos-methyl
	Disease control	mandipropamid, prochloraz (*off-label*)
	Downy mildew	boscalid + pyraclostrobin (*off-label*), copper oxychloride (*off-label*), dimethomorph (*off-label*), fosetyl-aluminium + propamocarb hydrochloride, mandipropamid
	Pythium	fosetyl-aluminium + propamocarb hydrochloride
	Rhizoctonia	azoxystrobin (*off-label*)
	Ring spot	difenoconazole (*off-label*)
	Soft rot	boscalid + pyraclostrobin
Pests	Aphids	acetamiprid (*off-label*), cypermethrin (*off-label*), deltamethrin (*off-label*), lambda-cyhalothrin (*off-label*), pirimicarb, pirimicarb (*off-label*), pymetrozine (*off-label*), pyrethrins, spirotetramat (*off-label*), thiacloprid (*off-label*)
	Beetles	deltamethrin (*off-label*), lambda-cyhalothrin (*off-label*)
	Caterpillars	Bacillus thuringiensis (*off-label*), cypermethrin (*off-label*), deltamethrin (*off-label*), lambda-cyhalothrin (*off-label*), pyrethrins
	Leafhoppers	deltamethrin (*off-label*)
	Mites	abamectin (*off-label*)
	Pests, miscellaneous	abamectin (*off-label*), lambda-cyhalothrin (*off-label*)
	Thrips	deltamethrin (*off-label*), pirimicarb (*off-label*)
	Whiteflies	Lecanicillium lecanii, spirotetramat (*off-label*)

Weeds	Broad-leaved weeds	propyzamide (*off-label*)
	Grass weeds	propyzamide (*off-label*)

Protected leafy vegetables - Protected spinach

Diseases	Disease control	mandipropamid
	Downy mildew	copper oxychloride (*off-label*), mandipropamid, metalaxyl-M (*off-label*)
Pests	Aphids	acetamiprid (*off-label*), cypermethrin (*off-label*), deltamethrin (*off-label*), dimethoate (*off-label*), pirimicarb (*off-label*), spirotetramat (*off-label*)
	Beetles	deltamethrin (*off-label*)
	Caterpillars	Bacillus thuringiensis (*off-label*), cypermethrin (*off-label*), deltamethrin (*off-label*)
	Leafhoppers	deltamethrin (*off-label*)
	Pests, miscellaneous	dimethoate (*off-label*)
	Thrips	deltamethrin (*off-label*)
	Whiteflies	spirotetramat (*off-label*)

Protected legumes - Protected peas and beans

Diseases	Damping off	fosetyl-aluminium + propamocarb hydrochloride (*off-label*)
Pests	Whiteflies	Lecanicillium lecanii, Lecanicillium lecanii (*off-label*)

Protected root and tuber vegetables - Protected carrots/parsnips/celeriac

Pests	Aphids	dimethoate (*off-label*)
	Pests, miscellaneous	dimethoate (*off-label*)
Weeds	Broad-leaved weeds	isoxaben (*off-label - temporary protection*)

Protected root and tuber vegetables - Protected root brassicas

Diseases	Damping off	fosetyl-aluminium + propamocarb hydrochloride
	Disease control	boscalid + pyraclostrobin (*off-label*)
	Downy mildew	fosetyl-aluminium + propamocarb hydrochloride
	Rhizoctonia	tolclofos-methyl (*off-label*)
Pests	Aphids	pirimicarb (*off-label*)

Protected stem and bulb vegetables - Protected celery/chicory

Diseases	Botrytis	azoxystrobin (*off-label*)
	Disease control	mandipropamid, prochloraz (*off-label*)
	Downy mildew	copper oxychloride (*off-label*), mandipropamid
	Phytophthora	azoxystrobin (*off-label - for forcing*)
	Powdery mildew	azoxystrobin (*off-label*)
	Rhizoctonia	azoxystrobin (*off-label*), tolclofos-methyl (*off-label*)
	Rust	azoxystrobin (*off-label*)
	Sclerotinia	azoxystrobin (*off-label*)

Pests	Aphids	cypermethrin (*off-label*), deltamethrin (*off-label*), pirimicarb (*off-label*), pymetrozine (*off-label*)
	Beetles	deltamethrin (*off-label*)
	Caterpillars	Bacillus thuringiensis (*off-label*), cypermethrin (*off-label*), deltamethrin (*off-label*)
	Leafhoppers	deltamethrin (*off-label*)
	Mites	abamectin (*off-label*)
	Pests, miscellaneous	abamectin (*off-label*), deltamethrin (*off-label*)
	Thrips	deltamethrin (*off-label*)

Protected stem and bulb vegetables - Protected onions/leeks/garlic

Pests	Aphids	deltamethrin (*off-label*)
	Beetles	deltamethrin (*off-label*)
	Caterpillars	deltamethrin (*off-label*)
	Leafhoppers	deltamethrin (*off-label*)
	Pests, miscellaneous	deltamethrin (*off-label*)
	Thrips	deltamethrin (*off-label*)

Total vegetation control

Aquatic situations, general - Aquatic situations

Weeds	Aquatic weeds	glyphosate
	Broad-leaved weeds	2,4-D
	Grass weeds	glyphosate
	Weeds, miscellaneous	glyphosate

Non-crop areas, general - Miscellaneous non-crop situations

Weeds	Broad-leaved weeds	triclopyr
	Weeds, miscellaneous	flazasulfuron, glyphosate
	Woody weeds/scrub	glyphosate, triclopyr

Non-crop areas, general - Non-crop farm areas

Weeds	Broad-leaved weeds	2,4-D + dicamba + triclopyr, dicamba + MCPA + mecoprop-P, glufosinate-ammonium, metsulfuron-methyl, picloram, tribenuron-methyl, triclopyr
	Grass weeds	glufosinate-ammonium, glyphosate
	Weeds, miscellaneous	glufosinate-ammonium, glyphosate, glyphosate + pyraflufen-ethyl, picloram
	Woody weeds/scrub	glyphosate, picloram, triclopyr

Non-crop areas, general - Paths/roads etc

Pests	Birds/mammals	aluminium ammonium sulphate, aluminium ammonium sulphate (*anti-fouling*)
	Slugs/snails	metaldehyde
Weeds	Broad-leaved weeds	2,4-D + dicamba + triclopyr, picloram
	Grass weeds	glyphosate, quizalofop-P-tefuryl (*off-label*)
	Mosses	maleic hydrazide + pelargonic acid
	Weeds, miscellaneous	acetic acid, diflufenican + glyphosate, flazasulfuron, glyphosate, glyphosate + pyraflufen-ethyl, glyphosate + sulfosulfuron, maleic hydrazide + pelargonic acid, picloram
	Woody weeds/scrub	2,4-D + dicamba + triclopyr, glyphosate, picloram

SECTION 2
PESTICIDE PROFILES

1 abamectin

A selective acaricide and insecticide for use in ornamentals and other protected crops
IRAC mode of action code: 6

Products

1	Agrimec	Syngenta	18 g/l	EC	14491
2	Clayton Abba	Clayton	18 g/l	EC	13808
3	Dynamec	Syngenta Bioline	18 g/l	EC	13331

Uses

- Insect control in **apples** *(off-label)*, **crab apples** *(off-label)*, **quinces** *(off-label)* [1]; **protected chives** *(off-label)*, **protected cress** *(off-label)*, **protected frise** *(off-label)*, **protected herbs (see appendix 6)** *(off-label)*, **protected lamb's lettuce** *(off-label)*, **protected lettuce** *(off-label)*, **protected parsley** *(off-label)*, **protected radicchio** *(off-label)*, **protected scarole** *(off-label)* [3]
- Insect pests in **salad onions** *(off-label)* [1]
- Leaf and bud mite in **protected blackberries** *(off-label)*, **protected raspberries** *(off-label)* [3]
- Leaf miner in **protected cherry tomatoes** *(off-label)* [3]; **protected tomatoes** [2, 3]
- Mites in **protected blackberries** *(off-label)*, **protected cress** *(off-label)*, **protected frise** *(off-label)*, **protected herbs (see appendix 6)** *(off-label)*, **protected lamb's lettuce** *(off-label)*, **protected leaf brassicas** *(off-label)*, **protected lettuce** *(off-label)*, **protected radicchio** *(off-label)*, **protected raspberries** *(off-label)*, **protected scarole** *(off-label)* [2]
- Pear sucker in **pears** *(effective against larval stages but little activity against adults)* [1]
- Red spider mites in **hops** *(off-label)* [1]; **protected blackberries** *(off-label)*, **protected cayenne peppers** *(off-label)*, **protected peppers** *(off-label)*, **protected raspberries** *(off-label)* [3]; **protected strawberries** *(off-label)* [2, 3]; **strawberries** *(off-label)* [2]
- Spider mites in **hops** *(off-label)* [1]
- Tarsonemid mites in **protected strawberries** *(off-label)*, **strawberries** *(off-label)* [2]; **strawberries** *(off-label - in propagation)* [3]
- Thrips in **salad onions** *(off-label)* [1]
- Two-spotted spider mite in **ornamental plant production**, **protected cucumbers**, **protected ornamentals**, **protected tomatoes** [2, 3]; **protected aubergines** *(off-label)* [3]
- Western flower thrips in **ornamental plant production**, **protected cucumbers**, **protected ornamentals** [2, 3]; **protected aubergines** *(off-label)* [3]

Extension of Authorisation for Minor Use (EAMUs)

- **apples** *20111164* [1]
- **crab apples** *20111164* [1]
- **hops** *20091581* [1]
- **protected aubergines** *20070421* [3]
- **protected blackberries** *20101940* [2], *20072290* [3]
- **protected cayenne peppers** *20070422* [3]
- **protected cherry tomatoes** *20070422* [3]
- **protected chives** *20070430* [3]
- **protected cress** *20101939* [2], *20070430* [3]
- **protected frise** *20101939* [2], *20070430* [3]
- **protected herbs (see appendix 6)** *20101939* [2], *20070430* [3]
- **protected lamb's lettuce** *20101939* [2], *20070430* [3]
- **protected leaf brassicas** *20101939* [2]
- **protected lettuce** *20101939* [2], *20070430* [3]
- **protected parsley** *20070430* [3]
- **protected peppers** *20070422* [3]
- **protected radicchio** *20101939* [2], *20070430* [3]
- **protected raspberries** *20101940* [2], *20072290* [3]
- **protected scarole** *20101939* [2], *20070430* [3]
- **protected strawberries** *20090342* [2], *20070423* [3]
- **quinces** *20111164* [1]
- **salad onions** *20112505* [1]
- **strawberries** *20090342* [2], *(in propagation) 20070423* [3]

SECTION 2

SEE SECTION 3 FOR PRODUCTS ALSO REGISTERED

Approval information
- Abamectin included in Annex I under EC Regulation 1107/2009

Efficacy guidance
- Abamectin controls adults and immature stages of the two-spotted spider mite, the larval stages of leaf miners and the nymphs of Western Flower Thrips, plus a useful reduction in adults
- Treat at first sign of infestation. Repeat sprays may be required
- For effective control total cover of all plant surfaces is essential, but avoid run-off
- Target pests quickly become immobilised but 3-5 d may be required for maximum mortality
- Indoor applications should be made through a hydraulic nozzle applicator or a knapsack applicator. Outdoors suitable high volume hydraulic nozzle applicators should be used
- Limited data shows abamectin only slightly harmful to Anthocorid bugs and so compatible with biological control systems in which Anthocorid bugs are important [1]

Restrictions
- Number of treatments 6 on protected tomatoes and cucumbers (only 4 of which can be made when flowers or fruit present); not restricted on flowers but rotation with other products advised
- Maximum concentration must not exceed 50 ml per 100 l water
- Do not mix with wetters, stickers or other adjuvants
- Do not use on ferns (*Adiantum* spp) or Shasta daisies
- Do not treat protected crops which are in flower or have set fruit between 1 Nov and 28 Feb
- For use on all varieties of pears [1]
- Do not treat cherry tomatoes (but protected cherry tomatoes may be treated off-label)
- Consult manufacturer for list of plant varieties tested for safety
- There is insufficient evidence to support product compatibility with integrated and biological pest control programmes
- Unprotected persons must be kept out of treated areas until the spray has dried

Crop-specific information
- HI 3 d for protected edible crops
- On tomato or cucumber crops that are in flower, or have started to set fruit, treat only between 1 Mar and 31 Oct. Seedling tomatoes or cucumbers that have not started to flower or set fruit may be treated at any time
- Some spotting or staining may occur on carnation, kalanchoe and begonia foliage

Environmental safety
- Dangerous for the environment
- Very toxic to aquatic organisms
- High risk to bees. Do not apply to crops in flower or to those in which bees are actively foraging. Do not apply when flowering weeds are present
- Keep in original container, tightly closed, in a safe place, under lock and key
- Where bumble bees are used in tomatoes as pollinators, keep them out for 24 h after treatment
- Broadcast air-assisted LERAP [1] (5 m)

Hazard classification and safety precautions
Hazard Harmful, Dangerous for the environment
Transport code 9
Packaging group III
UN Number 3082
Risk phrases R22a, R50, R53a
Operator protection A, C, D, H, K, M; U02a, U05a [1-3]; U20a [2, 3]; U20c [1]
Environmental protection E02a [2, 3] (until spray has dried); E12a, E15b, E34, E38 [1-3]; E12e [2, 3]; E12f, E22c [1]; E17b [1] (5 m)
Storage and disposal D01, D02, D05, D09b, D10c, D12a
Medical advice M04a

2   acetamiprid

A neonicotinoid insecticide for use in top fruit and horticulture
IRAC mode of action code: 4A

Products

1	Gazelle SG	Certis	20% w/w	SG	13725
2	Insyst	Certis	20% w/w	SP	13414

Uses

* Aphids in *apples*, *chard* (off-label), *cherries*, *cress* (off-label), *forest nurseries* (off-label), *frise* (off-label), *herbs (see appendix 6)* (off-label), *hops* (off-label), *lamb's lettuce* (off-label), *leaf brassicas* (off-label), *lettuce* (off-label), *ornamental plant production*, *parsley* (off-label), *pears*, *plums*, *protected aubergines*, *protected cress* (off-label), *protected herbs (see appendix 6)* (off-label), *protected hops* (off-label), *protected lamb's lettuce* (off-label), *protected leaf brassicas* (off-label), *protected lettuce* (off-label), *protected ornamentals*, *protected parsley* (off-label), *protected peppers*, *protected rocket* (off-label), *protected soft fruit* (off-label), *protected spinach* (off-label), *protected tomatoes*, *radicchio* (off-label), *rocket* (off-label), *scarole* (off-label), *soft fruit* (off-label), *spinach* (off-label), *spinach beet* (off-label), *top fruit* (off-label) [1]; *brussels sprouts* (off-label), *seed potatoes*, *spring oilseed rape*, *ware potatoes*, *winter oilseed rape* [2]
* Large pine weevil in *forest* (off-label) [1]
* Whitefly in *apples*, *cherries*, *ornamental plant production*, *pears*, *plums*, *protected aubergines*, *protected ornamentals*, *protected peppers*, *protected tomatoes* [1]; *brussels sprouts* (off-label) [2]

Extension of Authorisation for Minor Use (EAMUs)

* *brussels sprouts* 20072866 [2]
* *chard* 20113144 [1]
* *cress* 20101101 [1]
* *forest* 20121068 [1]
* *forest nurseries* 20111313 [1]
* *frise* 20111994 [1]
* *herbs (see appendix 6)* 20101101 [1]
* *hops* 20082857 [1]
* *lamb's lettuce* 20111994 [1]
* *leaf brassicas* 20111994 [1]
* *lettuce* 20111994 [1]
* *parsley* 20111994 [1]
* *protected cress* 20101101 [1]
* *protected herbs (see appendix 6)* 20101101 [1]
* *protected hops* 20082857 [1]
* *protected lamb's lettuce* 20111994 [1]
* *protected leaf brassicas* 20111994 [1]
* *protected lettuce* 20111994 [1]
* *protected parsley* 20111994 [1]
* *protected rocket* 20111994 [1]
* *protected soft fruit* 20082857 [1]
* *protected spinach* 20101101 [1]
* *radicchio* 20111994 [1]
* *rocket* 20111994 [1]
* *scarole* 20111994 [1]
* *soft fruit* 20082857 [1]
* *spinach* 20101101 [1]
* *spinach beet* 20113144 [1]
* *top fruit* 20082857 [1]

Approval information

* Acetamiprid included in Annex I under EC Regulation 1107/2009

SEE SECTION 3 FOR PRODUCTS ALSO REGISTERED

Efficacy guidance
- Best results obtained from application at the first sign of pest attack or when appropriate thresholds are reached
- Thorough coverage of foliage is essential to ensure best control. Acetamiprid has contact, systemic and translaminar activity

Restrictions
- Maximum number of treatments 1 per yr for cherries, ware potatoes; 2 per yr or crop for apples, pears, plums, ornamental plant production, protected crops, seed potatoes
- Do not use more than two applications of any neonicotinoid insecticide (e.g. acetamiprid, clothianidin, imidacloprid, thiacloprid) on any crop. Previous soil or seed treatment with a neonicotinoid counts as one such treatment
- 2 treatments are permitted on seed potatoes but must not be used consecutively [2]

Crop-specific information
- HI 3 d for protected crops; 14 d for apples, cherries, pears, plums, potatoes

Environmental safety
- Harmful to aquatic organisms
- Acetamiprid is slightly toxic to predatory mites and generally slightly toxic to other beneficials
- Broadcast air-assisted LERAP [1] (18 m); LERAP Category B [1, 2]

Hazard classification and safety precautions
Hazard Dangerous for the environment
UN Number N/C
Risk phrases R22a [1]; R52, R53a [1, 2]
Operator protection A [1, 2]; H [1]; U02a, U19a [1, 2]; U20a [2]; U20b [1]
Environmental protection E15b, E16a, E16b [1, 2]; E17b [1] (18 m); E38 [2]
Storage and disposal D05 [1]; D09a, D10b [1, 2]; D12a [2]

---

3   acetic acid

A non-selective herbicide for non-crop situations

Products

Natural Weed & Moss Spray No 1	Headland Amenity	240 g/l	SL	14509

Uses
- General weed control in **hard surfaces**, **natural surfaces not intended to bear vegetation**, **permeable surfaces overlying soil**

Approval information
- Acetic acid is included in Annex 1 under EC Regulation 1107/2009
- Approval expiry 31 Jul 2013 [1]

Efficacy guidance
- Best results obtained from treatment of young tender weeds less than 10 cm high
- Treat in spring and repeat as necessary throughout the growing season
- Treat survivors as soon as fresh growth is seen
- Ensure complete coverage of foliage to the point of run-off
- Rainfall after treatment may reduce efficacy

Restrictions
- No restriction on number of treatments

Following crops guidance
- There is no residual activity in the soil and sowing or planting may take place as soon as treated weeds have died

Environmental safety
- Harmful to aquatic organisms
- High risk to bees

- Do not apply to crops in flower or to those in which bees are actively foraging. Do not apply when flowering weeds are present
- Keep people and animals off treated dense weed patches until spray has dried. This is not necessary for areas with occasional low growing prostrate weeds such as on pathways

Hazard classification and safety precautions
Hazard Irritant
UN Number N/C
Risk phrases R36, R37, R38, R52
Operator protection A, C, D, H; U05a, U09a, U19a, U20b
Environmental protection E12a, E12e, E15a
Storage and disposal D01, D02, D09a, D12a

4   alpha-cypermethrin

A contact and ingested pyrethroid insecticide for use in arable crops and agricultural buildings
IRAC mode of action code: 3

Products

1	A-Cyper 100EC	Goldengrass	100 g/l	EC	14175
2	Antec Durakil 1.5 SC	Antec Biosentry	15 g/l	SC	H7560
3	Antec Durakil 6SC	Antec Biosentry	60 g/l	SC	H7559
4	Contest	BASF	15% w/w	WG	10216

Uses

- Aphids in *spring barley, spring oilseed rape, spring wheat, winter barley, winter oilseed rape, winter wheat* [1]
- Brassica pod midge in *spring oilseed rape* [1]; *winter oilseed rape* [1, 4]
- Cabbage seed weevil in *spring oilseed rape, winter oilseed rape* [1, 4]
- Cabbage stem flea beetle in *spring oilseed rape* [1]; *winter oilseed rape* [1, 4]
- Cabbage stem weevil in *spring oilseed rape, winter oilseed rape* [1]
- Cabbage white butterfly in *salad brassicas (off-label - for baby leaf production)* [4]
- Caterpillars in *broccoli, brussels sprouts, cabbages, calabrese, cauliflowers, kale* [1, 4]
- Cereal aphid in *spring barley, spring wheat, winter barley, winter wheat* [4]
- Diamond-back moth in *salad brassicas (off-label - for baby leaf production)* [4]
- Flea beetle in *broccoli, brussels sprouts, cabbages, calabrese, cauliflowers, kale* [1, 4]
- Insect pests in *borage for oilseed production (off-label), chinese cabbage (off-label), choi sum (off-label), collards (off-label), durum wheat (off-label), grass seed crops (off-label), linseed (off-label), lupins (off-label), pak choi (off-label), spring rye (off-label), triticale (off-label), winter rye (off-label)* [4]
- Lesser mealworm in *poultry houses* [2, 3]
- Pea and bean weevil in *broad beans, combining peas, spring field beans, vining peas, winter field beans* [1, 4]
- Pea aphid in *combining peas, vining peas* [1]; *combining peas (reduction), vining peas (reduction)* [4]
- Pea moth in *combining peas, vining peas* [1, 4]
- Pine weevil in *forest (off-label), forest nurseries (off-label)* [4]
- Pollen beetle in *spring oilseed rape, winter oilseed rape* [1, 4]
- Poultry red mite in *poultry houses* [2, 3]
- Rape winter stem weevil in *winter oilseed rape* [4]
- Small white butterfly in *salad brassicas (off-label - for baby leaf production)* [4]
- Yellow cereal fly in *spring barley (some control if present at application), spring wheat (some control if present at application), winter barley (some control if present at application), winter wheat (some control if present at application)* [1]; *winter barley, winter wheat* [4]

Extension of Authorisation for Minor Use (EAMUs)

- *borage for oilseed production 20052268* [4]
- *chinese cabbage 20052265* [4]
- *choi sum 20052265* [4]

SEE SECTION 3 FOR PRODUCTS ALSO REGISTERED

- **collards** *20052265* [4]
- **durum wheat** *20052269* [4]
- **forest** *20062391* [4]
- **forest nurseries** *20062391* [4]
- **grass seed crops** *20052269* [4]
- **linseed** *20052268* [4]
- **lupins** *20052267* [4]
- **pak choi** *20052265* [4]
- **salad brassicas** *(for baby leaf production) 20011221* [4]
- **spring rye** *20052269* [4]
- **triticale** *20052269* [4]
- **winter rye** *20052269* [4]

Approval information
- Alpha-cypermethrin included in Annex I under EC Regulation 1107/2009
- Accepted by BBPA for use on malting barley

Efficacy guidance
- For cabbage stem flea beetle control spray oilseed rape when adult or larval damage first seen and about 1 mth later [4]
- For flowering pests on oilseed rape apply at any time during flowering, against pollen beetle best results achieved at green to yellow bud stage (GS 3.3-3.7), against seed weevil between 20 pods set and 80% petal fall (GS 4.7-5.8) [4]
- Spray cereals in autumn for control of cereal aphids, in spring/summer for grain aphids. (See label for details) [4]
- For flea beetle, caterpillar and cabbage aphid control on brassicas apply when the pest or damage first seen or as a preventive spray. Repeat if necessary [4]
- For pea and bean weevil control in peas and beans apply when pest attack first seen and repeat as necessary [4]
- For lesser mealworm control in poultry houses apply a coarse, low-pressure spray as routine treatment after clean-out and before each new crop. Spray vertical surfaces and ensure an overlap onto ceilings. It is not necessary to treat the floor [2, 3]
- Use the highest recommended concentration in poultry houses where extreme residual action is required or surfaces are dirty or highly absorbent [2, 3]

Restrictions
- Maximum number of treatments 4 per crop on edible brassicas, 3 per crop on winter oilseed rape, peas, 2 on beans, spring oilseed rape (only 1 after yellow bud stage - GS 3,7) [4]
- Maximum number of applications in animal husbandry use 2 when used in premises that are occupied by poultry [2, 3]
- Apply up to 2 sprays on cereals in autumn and spring, 1 in summer between 1 Apr and 31 Aug. See label for details of rates and maximum total dose [4]
- Only 1 aphicide treatment may be applied in cereals between 1 Apr and 31 Aug and spray volume must not be reduced in this period [4]
- Do not apply to a cereal crop if any product containing a pyrethroid or dimethoate has been applied after the start of ear emergence (GS 51) [4]

Crop-specific information
- Latest use: before the end of flowering for oilseed rape; before 31 Aug for cereals [4]
- HI vining peas 1 d; brassicas 7 d; combining peas, broad beans, field beans 11 d [4]
- For summer cereal application do not spray within 6 m from edge of crop and do not reduce volume when used after 31 Mar [4]

Environmental safety
- Dangerous for the environment [4]
- Very toxic to aquatic organisms [4]
- Extremely dangerous to fish or other aquatic life. Do not contaminate surface waters or ditches with chemical or used container [2, 3]
- Dangerous to bees [2, 3]
- Where possible spray oilseed rape crops in the late evening or early morning or in dull weather [4]
- Do not spray within 6 m of the edge of a cereal crop after 31 Mar in yr of harvest [4]

FOR FULL CONDITIONS OF USE ALWAYS READ THE PRODUCT LABEL

- Do not apply directly to poultry; collect eggs before application [2, 3]
- LERAP Category A [1, 4]

Hazard classification and safety precautions

Hazard Harmful, Dangerous for the environment [1, 4]; Flammable [1]

Transport code 3 [1]; 9 [2-4]

Packaging group III

UN Number 1993 [1]; 3077 [4]; 3082 [2, 3]

Risk phrases R20, R22b, R66 [1]; R22a, R48, R50, R53a [1, 4]

Operator protection A, H [1-4]; C [1]; U02a, U19a [1-3]; U02b, U09a [2, 3]; U04a, U11, U14, U20a [1]; U05a, U10 [1, 4]; U20b [2-4]

Environmental protection E02a [2, 3] (until dry); E05a, E12c, E13a [2, 3]; E15a [1-4]; E16c, E16d [1, 4]; E34 [1]; E38 [4]

Consumer protection C05, C06, C07, C09, C11 [2, 3]

Storage and disposal D01, D02, D05, D10b [1, 4]; D09a, D12a [1-4]; D11a [2, 3]

Medical advice M04a, M05b [1]; M05a [4]

5   aluminium ammonium sulphate

An inorganic bird and animal repellent

Products

1	Curb Crop Spray Powder	Sphere	88% w/w	WP	02480
2	Liquid Curb Crop Spray	Sphere	83 g/l	SC	03164
3	Rezist	Sphere	88% w/w	WP	08576
4	Rezist Liquid	Sphere	83 g/l	SC	14643
5	Sphere ASBO	Sphere	83 g/l	SC	15064

Uses

- Animal repellent in **agricultural premises, all top fruit, broad beans, bush fruit, cane fruit, carrots, flowerhead brassicas, forest nursery beds, forestry plantations, grain stores, leaf brassicas, peas, spring barley, spring field beans, spring oats, spring oilseed rape, spring wheat, strawberries, sugar beet, winter barley, winter field beans, winter oats, winter oilseed rape, winter wheat** [1, 2]; **all edible crops (outdoor), all non-edible crops (outdoor), amenity grassland, forest** [5]; **amenity vegetation** [1]; **hard surfaces, managed amenity turf** [4, 5]; **permanent grassland** [1, 2, 4]
- Bird repellent in **agricultural premises, all top fruit, broad beans, bush fruit, cane fruit, carrots, flowerhead brassicas, forest nursery beds, forestry plantations, grain stores, leaf brassicas, peas, spring barley, spring field beans, spring oats, spring oilseed rape, spring wheat, strawberries, sugar beet, winter barley, winter field beans, winter oats, winter oilseed rape, winter wheat** [1, 2]; **all edible crops (outdoor), all non-edible crops (outdoor), amenity grassland, forest** [5]; **amenity vegetation** [1]; **hard surfaces, managed amenity turf** [4, 5]; **permanent grassland** [1, 2, 4]
- Dogs in **amenity grassland** (anti-fouling), **hard surfaces** (anti-fouling), **managed amenity turf** (anti-fouling) [3]
- Moles in **amenity grassland, hard surfaces, managed amenity turf** [3]

Approval information

- aluminium ammonium sulphate is included in Annex 1 under EC Regulation 1107/2009

Efficacy guidance

- Apply as overall spray to growing crops before damage starts or mix powder with seed depending on type of protection required
- Spray deposit protects growth present at spraying but gives little protection to new growth
- Product must be sprayed onto dry foliage to be effective and must dry completely before dew or frost forms. In winter this may require some wind

Crop-specific information

- Latest use: no restriction

SEE SECTION 3 FOR PRODUCTS ALSO REGISTERED

Hazard classification and safety precautions
 UN Number N/C
 Operator protection U05a [1, 2, 5]; U15 [4]; U20a [1-5]
 Environmental protection E15a [1-5]; E19b [1, 2, 4, 5]
 Storage and disposal D01, D02, D05 [1, 2, 4, 5]; D09a [3, 4]; D10b [4]; D11a [1-5]; D12a [1, 2, 5]
 Medical advice M03 [1, 2, 4, 5]

6 aluminium phosphide

A phosphine generating compound used against vertebrates and grain store pests. Under new regulations any person wishing to purchase or use Aluminium Phosphide must be suitably qualified and licensed by January 2015. A new licensing body has been set up.
IRAC mode of action code: 24A

Products

1	Degesch Fumigation Tablets	Rentokil	56% w/w	GE	09313
2	Phostoxin	Rentokil	56% w/w	GE	09315

Uses
- Insect pests in **stored grain** [1]
- Moles in **farmland** [2]
- Rabbits in **farmland** [2]
- Rats in **farmland** [2]

Approval information
- Aluminium phosphide is included in Annex 1 under EC Regulation 1107/2009
- Accepted by BBPA for use in stores for malting barley

Efficacy guidance
- Product releases poisonous hydrogen phosphide gas in contact with moisture
- Place fumigation tablets in grain stores as directed [1]
- Place pellets in burrows or runs and seal hole by heeling in or covering with turf. Do not cover pellets with soil. Inspect daily and treat any new or re-opened holes [2]

Restrictions
- Aluminium phosphide is subject to the Poisons Rules 1982 and the Poisons Act 1972. See Section 5 for more information
- Only to be used by operators instructed or trained in the use of aluminium phosphide and familiar with the precautionary measures to be taken. See label and HSE Guidance Notes for full precautions
- Only open container outdoors [2] and for immediate use. Keep away from liquid or water as this causes immediate release of gas. Do not use in wet weather
- Do not use within 3 m of human or animal habitation. Before application ensure that no humans or domestic animals are in adjacent buildings or structures. Allow a minimum airing-off period of 4 h before re-admission

Environmental safety
- Product liberates very toxic, highly flammable gas
- Dangerous to fish or other aquatic life. Do not contaminate surface waters or ditches with chemical or used container [1, 2]
- Prevent access by livestock, pets and other non-target mammals and birds to buildings under fumigation and ventilation [1]
- Pellets must never be placed or allowed to remain on ground surface
- Do not use adjacent to watercourses
- Take particular care to avoid gassing non-target animals, especially those protected under the Wildlife and Countryside Act (e.g. badgers, polecat, reptiles, natterjack toads, most birds). Do not use in burrows where there is evidence of badger or fox activity, or when burrows might be occupied by birds
- Dust remaining after decomposition is harmless and of no environmental hazard
- Keep in original container, tightly closed, in a safe place, under lock and key

FOR FULL CONDITIONS OF USE ALWAYS READ THE PRODUCT LABEL

- Dispose of empty containers as directed on label

Hazard classification and safety precautions
    Hazard Very toxic, Highly flammable, Dangerous for the environment [1, 2]; Harmful [2]
    Transport code 4.3
    Packaging group I
    UN Number 1397
    Risk phrases R21, R26, R28, R50
    Operator protection A, D, H; U01, U07, U13, U19a, U20a [1, 2]; U05a [2]; U05b, U18 [1]
    Environmental protection E02a [1] (4 h min); E02a [2] (4 h); E02b [1]; E13b, E34 [1, 2]
    Storage and disposal D01, D02, D07, D09b [1, 2]; D11b [2]; D11c [1]
    Vertebrate/rodent control products V04a [2]
    Medical advice M04a

7   ametoctradin

A potato blight fungicide available only in mixtures
FRAC mode of action code: 45

See also ametoctradin + dimethomorph
        ametoctradin + mancozeb

8   ametoctradin + dimethomorph

A systemic and protectant fungicide for potato blight control
FRAC mode of action code: 40 + 45

Products

Zampro DM	BASF	300:225 g/l	SC	15013

Uses
- Blight in *potatoes*
- Tuber blight in *potatoes* (reduction)

Approval information
- Ametoctradin not yet included in Annex 1 under EC Regulation 1107/2009. Dimethomorph listed in Annex 1

Efficacy guidance
- Application to dry foliage is rainfast within 1 hour of drying on the leaf.
- Will give effective control of phenylamide-resistant strains of blight

Crop-specific information
- HI 7 days for potatoes

Hazard classification and safety precautions
    Hazard Harmful, Dangerous for the environment
    UN Number N/C
    Risk phrases R22a, R52, R53a
    Operator protection U05a
    Environmental protection E15b, E34, E38
    Storage and disposal D01, D02, D09a, D10c
    Medical advice M03, M05a

SEE SECTION 3 FOR PRODUCTS ALSO REGISTERED

9 amidosulfuron

A post-emergence sulfonylurea herbicide for cleavers and other broad-leaved weed control in cereals

HRAC mode of action code: B

Products

1	Eagle	Bayer CropScience	75% w/w	WG	07318
2	Gauntlet	ChemSource	75% w/w	WG	15018
3	Squire Ultra	Interfarm	75% w/w	WG	14401

Uses

- Annual dicotyledons in *durum wheat*, *linseed*, *spring barley*, *spring oats*, *spring rye*, *spring wheat*, *triticale*, *winter barley*, *winter oats*, *winter rye*, *winter wheat* [1, 2]; *grassland* [3]
- Charlock in *durum wheat* [1, 2]
- Cleavers in *durum wheat*, *linseed*, *spring barley*, *spring oats*, *spring rye*, *spring wheat*, *triticale*, *winter barley*, *winter oats*, *winter rye*, *winter wheat* [1, 2]
- Docks in *grassland* [3]

Approval information

- Accepted by BBPA for use on malting barley
- Amidosulfuron included in Annex I under EC Regulation 1107/2009

Efficacy guidance

- For best results apply in spring (from 1 Feb) in warm weather when soil moist and weeds growing actively. When used in grassland following cutting or grazing, docks should be allowed to regrow before treatment
- Weed kill is slow, especially under cool, dry conditions. Weeds may sometimes only be stunted but will have little or no competitive effect on crop
- May be used on all soil types unless certain sequences are used on linseed. See label
- Spray is rainfast after 1 h
- Cleavers controlled from emergence to flower bud stage. If present at application charlock (up to flower bud), shepherds purse (up to flower bud) and field forget-me-not (up to 6 leaves) will also be controlled
- Amidosulfuron is a member of the ALS-inhibitor group of herbicides and products should be used in a planned Resistance Management strategy. See Section 5 for more information

Restrictions

- Maximum number of treatments 1 per crop [1]
- Use after 1 Feb and do not apply to rotational grass after 30 Jun, or to permanent grassland after 15 Oct [3]
- Do not apply to crops undersown or due to be undersown with clover or alfalfa [1]
- Do not spray crops under stress, suffering drought, waterlogged, grazed, lacking nutrients or if soil compacted
- Do not spray if frost expected
- Do not roll or harrow within 1 wk of spraying
- Specific restrictions apply to use in sequence or tank mixture with other sulfonylurea or ALS-inhibiting herbicides. See label for details. There are no recommendations for mixtures with metsulfuron-methyl products on linseed
- Certain mixtures with fungicides are expressly forbidden. See label for details

Crop-specific information

- Latest use: before first spikelets just visible (GS 51) for cereals; before flower buds visible for linseed; 15 Oct for grassland
- Broadcast cereal crops should be sprayed post-emergence after plants have a well established root system

Following crops guidance

- If a treated crop fails cereals may be sown after 15 d and thorough cultivation

- After normal harvest of a treated crop only cereals, winter oilseed rape, mustard, turnips, winter field beans or vetches may be sown in the same year as treatment and these must be preceded by ploughing or thorough cultivation
- Only cereals may be sown within 12 mth of application to grassland [3]
- Cereals or potatoes must be sown as the following crop after use of permitted mixtures or sequences with other sulfonylurea herbicides in cereals. Only cereals may be sown after the use of such sequences in linseed

Environmental safety
- Take care to wash out sprayers thoroughly. See label for details
- Avoid drift onto neighbouring broad-leaved plants or onto surface waters or ditches

Hazard classification and safety precautions
Hazard Dangerous for the environment [3]
Transport code 9
Packaging group III
UN Number 3077
Risk phrases R50, R53a [3]
Operator protection U20a [1, 2]; U20b [3]
Environmental protection E07b [3] (1 week); E15a [1-3]; E38, E41 [3]
Storage and disposal D10a [1-3]; D12a [3]

10   amidosulfuron + iodosulfuron-methyl-sodium

A post-emergence sulfonylurea herbicide mixture for cereals
HRAC mode of action code: B + B

See also iodosulfuron-methyl-sodium

Products
1	Chekker	Bayer CropScience	12.5:1.25% w/w	WG	10955
2	Sekator	Interfarm	5:1.25% w/w	WG	14746

Uses
- Annual dicotyledons in **durum wheat** *(off-label)*, **grass seed crops** *(off-label)*, **linseed** *(off-label)* [1]; **spring barley, spring rye, spring wheat, triticale, winter barley, winter rye, winter wheat** [1, 2]
- Chickweed in **linseed** *(off-label)* [1]; **spring barley, spring rye, spring wheat, triticale, winter barley, winter rye, winter wheat** [1, 2]
- Cleavers in **linseed** *(off-label)* [1]; **spring barley, spring rye, spring wheat, triticale, winter barley, winter rye, winter wheat** [1, 2]
- Mayweeds in **linseed** *(off-label)* [1]; **spring barley, spring rye, spring wheat, triticale, winter barley, winter rye, winter wheat** [1, 2]
- Volunteer oilseed rape in **linseed** *(off-label)* [1]; **spring barley, spring rye, spring wheat, triticale, winter barley, winter rye, winter wheat** [1, 2]

Extension of Authorisation for Minor Use (EAMUs)
- **durum wheat** *20060858* [1]
- **grass seed crops** *20060858* [1]
- **linseed** *20101125* [1]

Approval information
- Amidosulfuron and iodosulfuron-methyl-sodium included in Annex I under EC Regulation 1107/2009
- Accepted by BBPA for use on malting barley

Efficacy guidance
- Best results obtained from treatment in warm weather when soil is moist and the weeds are growing actively
- Weeds must be present at application to be controlled
- Dry conditions resulting in moisture stress may reduce effectiveness

SEE SECTION 3 FOR PRODUCTS ALSO REGISTERED

- Weed control is slow especially under cool dry conditions
- Occasionally weeds may only be stunted but they will normally have little or no competitive effect on the crop
- Amidosulfuron and iodosulfuron are members of the ALS-inhibitor group of herbicides and products should be used in a planned Resistance Management strategy. See Section 5 for more information

Restrictions
- Maximum number of treatments 1 per crop
- Must only be applied between 1 Feb in yr of harvest and specified latest time of application
- Do not apply to crops undersown or to be undersown with grass, clover or alfalfa
- Do not roll or harrow within 1 wk of spraying
- Do not spray crops under stress from any cause or if the soil is compacted
- Do not spray if rain or frost expected
- Do not apply in mixture or in sequence with any other ALS inhibitor

Crop-specific information
- Latest use: before first spikelet of inflorescence just visible (GS 51)
- Treat drilled crops after the 2-leaf stage; treat broadcast crops after the plants have a well-established root system
- Applications to spring barley may cause transient crop yellowing

Following crops guidance
- Cereals, winter oilseed rape and winter field beans may be sown in the same yr as treatment provided they are preceded by ploughing or thorough cultivation. Any crop may be sown in the spring of the yr following treatment
- A minimum of 3 mth must elapse between treatment and sowing winter oilseed rape

Environmental safety
- Dangerous for the environment
- Toxic to aquatic organisms
- Take extreme care to avoid damage by drift onto broad-leaved plants outside the target area or onto ponds, waterways and ditches
- Observe carefully label instructions for sprayer cleaning
- LERAP Category B

Hazard classification and safety precautions
    Hazard Irritant, Dangerous for the environment
    Transport code 9
    Packaging group III
    UN Number 3077
    Risk phrases R36, R51, R53a
    Operator protection A, C, H; U05a, U08, U11, U14, U15, U20b
    Environmental protection E15a, E16a, E16b, E38
    Storage and disposal D01, D02, D10a, D12a

11   aminopyralid

A pyridine carboxylic acid herbicide available only in mixtures
HRAC mode of action code: O

12   aminopyralid + fluroxypyr

A foliar acting herbicide mixture for use in grassland
HRAC mode of action code: O + O

See also fluroxypyr

Products

1	Forefront	Dow	30:100 g/l	EO	14701

FOR FULL CONDITIONS OF USE ALWAYS READ THE PRODUCT LABEL

Products – continued

2	Mileway	Dow	30:100 g/l	EW	14702
3	Synero	Dow	30:100 g/l	EW	14708

Uses

- Buttercups in *amenity grassland* [2, 3]; *permanent grassland, rotational grass* [1]
- Chickweed in *amenity grassland* [2, 3]; *permanent grassland, rotational grass* [1]
- Dandelions in *amenity grassland* [2, 3]; *permanent grassland, rotational grass* [1]
- Docks in *amenity grassland* [2, 3]; *permanent grassland, rotational grass* [1]
- Stinging nettle in *amenity grassland* [2, 3]; *permanent grassland, rotational grass* [1]
- Thistles in *amenity grassland* [2, 3]; *permanent grassland, rotational grass* [1]

Approval information

- Fluroxypyr included in Annex I under EC Regulation 1107/2009 while aminopyralid has not yet been included.

Efficacy guidance

- For best results and to avoid crop check, grass and weeds must be growing actively
- Allow 2-3 wk after cutting for hay or silage for sufficient regrowth to occur before spraying and leave 7 d afterwards to allow maximum translocation
- Where there is a high reservoir of weed seed or a historically high weed population a programmed approach may be needed involving a second treatment in the following yr
- Control may be reduced if rain falls within 1 h of spraying

Restrictions

- Maximum number of treatments 1 per yr.
- Do not apply to leys less than 1 year old.
- Do not apply by hand-held equipment.
- Do not use on grassland that will be used for animal feed, bedding, composting or mulching within 1 calender year of application.
- Do not use on grassland that will be grazed by animals other than cattle or sheep.
- Use of an antifoam is compulsory
- Do not use any treated plant material, or manure from animals fed on treated crops, for composting or mulching
- Do not use on crops grown for seed
- Manure from animals fed on pasture or silage treated with [1] should not leave the farm

Crop-specific information

- Treatment will kill clover
- Treatment may occasionally cause transient yellowing of the sward which is quickly outgrown
- Late treatments may lead to a slight transient leaning of grass that does not affect yield

Following crops guidance

- Do not drill clover or other legumes within 4 mth of treatment, or potatoes in the spring following treatment in the previous autumn
- In the event of failure of newly seeded treated grassland, grass may be re-seeded immediately, or wheat may be sown provided 4 mth have elapsed since application
- Residues in plant tissues, including manure, may affect succeeding susceptible crops of peas, beans, other legumes, carrots, other Umbelliferae, potatoes, tomatoes, lettuce and other Compositae. These crops should not be sown within 3 mth of ploughing up treated grassland

Environmental safety

- Dangerous for the environment
- Toxic to aquatic organisms
- Keep livestock out of treated areas for at least 7 d following treatment and until poisonous weeds, such as ragwort, have died down and become unpalatable
- Avoid damage by drift onto susceptible crops, non-target plants or waterways

Hazard classification and safety precautions

Hazard Irritant, Dangerous for the environment
Transport code 9
Packaging group III

SEE SECTION 3 FOR PRODUCTS ALSO REGISTERED

UN Number 3082
Risk phrases R38, R41, R51, R53a, R67
Operator protection A, C; U05a, U08, U11, U14, U23a
Environmental protection E07a, E15b, E38
Consumer protection C01
Storage and disposal D01, D02, D05, D09a, D10b, D12a

13   aminopyralid + triclopyr

A foliar acting herbicide mixture for broad-leaved weed control in grassland
HRAC mode of action code: O + O

See also triclopyr

Products

1	Forefront T	Dow	30:240g/l	EW	15568
2	Pharaoh	Dow	30:240g/l	EW	14731
3	Speedline	Dow	30:240g/l	EW	15391

Uses

- Buttercups in *amenity grassland* [1, 3]; *grassland* [1, 2]
- Common nettle in *amenity grassland* [1, 3]; *grassland* [1, 2]
- Dandelions in *amenity grassland* [1, 3]; *grassland* [1, 2]
- Docks in *amenity grassland* [1, 3]; *grassland* [1, 2]
- Thistles in *amenity grassland* [1, 3]; *grassland* [1, 2]

Approval information

- Triclopyr included in Annex I under EC Regulation 1107/2009 while aminopyralid has not yet been included.

Efficacy guidance

- For best results and to avoid crop check, grass and weeds must be growing actively
- Use adeqate water volume to ensure good weed coverage. Increase water volume if necessary where the weed population is high and where the grass is dense
- Allow 2-3 wk after cutting for hay or silage for sufficient regrowth to occur before spraying and leave 7 d afterwards to allow maximum translocation
- Where there is a high reservoir of weed seed or a historically high weed population a programmed approach may be needed involving a second treatment in the following yr
- Control may be reduced if rain falls within 1 h of spraying

Restrictions

- Maximum number of treatments 1 per yr.
- Do not apply to leys less than 1 year old.
- Do not apply by hand-held equipment.
- Do not use on grassland that will be used for animal feed, bedding, composting or mulching within 1 calender year of application.
- Do not use on grassland that will be grazed by animals other than cattle or sheep.
- Use of an antifoam is compulsory
- Do not use any treated plant material, or manure from animals fed on treated crops, for composting or mulching
- Do not use on crops grown for seed
- Manure from animals fed on pasture or silage from treated crops should not leave the farm

Crop-specific information

- Latest use: 7 d before grazing or harvest for grassland
- Late applications may lead to transient leaning of grass which does not affect final yield

Following crops guidance

- Ensure that all plant remains of a treated crop have completely decayed before planting susceptible crops such as peas, beans and other legumes, sugar beet, carrots and other Umbelliferae, potatoes and tomatoes, lettuce and other Compositae

FOR FULL CONDITIONS OF USE ALWAYS READ THE PRODUCT LABEL

- Do not plant potatoes, sugar beet, vegetables, beans or other leguminous crops in the calendar yr following application

Environmental safety
- Dangerous for the environment
- Toxic to aquatic organisms
- Keep livestock out of treated areas for at least 7 d after treatment or until foliage of any poisonous weeds such as ragwort has died and become unpalatable
- To protect groundwater do not apply to grass leys less than 1 yr old
- Take extreme care to avoid drift onto susceptible crops, non-target plants or waterways. All conifers, especially pine and larch, are very sensitive and may be damaged by vapour drift in hot conditions
- LERAP Category B

Hazard classification and safety precautions
Hazard Irritant, Dangerous for the environment
Transport code 9
Packaging group III
UN Number 3082
Risk phrases R43, R51, R53a
Operator protection A, H; U05a, U08, U14, U23a
Environmental protection E06a (7 days); E15b, E15c, E16a, E38
Consumer protection C01
Storage and disposal D01, D02, D09a, D10b, D12a

14   amisulbrom

A fungicide for use in potatoes
FRAC mode of action code: C4

Products

Shinkon	Procam	200 g/l	SC	13722

Uses
- Late blight in *potatoes*
- Tuber blight in *potatoes*

Approval information
- Amisulbrom is pending inclusion in Annex 1 under EC Regulation 1107/2009

Efficacy guidance
- Make no more than 3 consecutive applications before switching to a blight fungicide with a different mode of action to protect against the risk of resistance.
- Do not use if blight is already visible on 1% of leaves.

Crop-specific information
- Rainfast within 3 hours of application but apply to dry foliage.

Following crops guidance
- Plough before planting new crops in the same soil.

Environmental safety
- LERAP Category B

Hazard classification and safety precautions
Hazard Dangerous for the environment
Risk phrases R50, R53a
Operator protection A, C; U05a, U15, U20b
Environmental protection E13b, E15a, E16a, E34, E38
Storage and disposal D01, D02, D05, D09a, D10b, D12a

SEE SECTION 3 FOR PRODUCTS ALSO REGISTERED

15 amitrole

A translocated, foliar-acting, non-selective triazole herbicide
HRAC mode of action code: F3

Products

Weedazol-TL	Nufarm UK	225 g/l	SL	11968

Uses

- Annual and perennial weeds in **apricots** *(off-label)*, **cherries** *(off-label)*, **peaches** *(off-label)*, **plums** *(off-label)*, **quinces** *(off-label)*
- Annual dicotyledons in **fallows, headlands, stubbles**
- Annual grasses in **fallows, headlands, stubbles**
- Barren brome in **apple orchards, pear orchards**
- Couch in **apple orchards, fallows, headlands, pear orchards, stubbles**
- Creeping thistle in **fallows, stubbles**
- Docks in **fallows, headlands, stubbles**
- General weed control in **apple orchards, farm woodland** *(off-label)*, **pear orchards**
- Grass weeds in **managed amenity turf** *(off-label - on amitrole resistant turf)*
- Perennial dicotyledons in **apple orchards, fallows, headlands, pear orchards**
- Perennial grasses in **apple orchards, fallows, headlands, pear orchards**
- Volunteer potatoes in **stubbles** *(barley stubble)*

Extension of Authorisation for Minor Use (EAMUs)

- **apricots** *20051981*
- **cherries** *20051981*
- **farm woodland** *20051980*
- **managed amenity turf** *(on amitrole resistant turf) 20051979*
- **peaches** *20051981*
- **plums** *20051981*
- **quinces** *20051981*

Approval information

- Amitrole included in Annex I under EC Regulation 1107/2009

Efficacy guidance

- In non-crop land may be applied at any time from Apr to Oct. Best results achieved in spring or early summer when weeds growing actively. For coltsfoot, hogweed and horsetail summer and autumn applications are preferred
- Uptake is via foliage and heavy rain immediately after application will reduce efficacy. Amitrole is less affected by drought than some residual herbicides and remains effective for up to 2 mth
- Applications made in summer may not give complete control of couch if past the shooting stage or not actively growing
- Effective crop competition and efficient ploughing are essential for good couch control

Restrictions

- Maximum number of treatments 1 per yr
- Keep off suckers or foliage of desirable trees or shrubs
- Do not spray areas into which the roots of adjacent trees or shrubs extend
- Application to land intended for spring barley should be in the preceding autumn, not the spring
- Do not spray on sloping ground when rain imminent and run-off may occur
- Do not spray if foliage is wet or rain imminent
- Do not use in low temperatures or in drought
- Do not mix product with acids

Crop-specific information

- Latest use: before end Jun or after harvest for apple and pear orchards; end Oct for headlands; end Oct and at least 2 wk before cultivation and drilling for stubbles and fallows

Following crops guidance

- Amitrole breaks down fairly quickly in medium and heavy soils and 3 wk should be allowed between application and sowing or planting. On sandy soils the interval should be 6 wk

FOR FULL CONDITIONS OF USE ALWAYS READ THE PRODUCT LABEL

Environmental safety
- Harmful to aquatic organisms
- Keep livestock out of treated areas for at least two weeks following treatment and until poisonous weeds, such as ragwort, have died down and become unpalatable
- Harmful to fish or other aquatic life. Do not contaminate surface waters or ditches with chemical or used container

Hazard classification and safety precautions
  Hazard Harmful
  UN Number N/C
  Risk phrases R48, R52, R53a, R63
  Operator protection A, C; U05a, U08, U19a, U20b
  Environmental protection E07a, E15a
  Storage and disposal D01, D02, D09a, D10b

16   Ampelomyces quisqualis strain AQ10

A biological fungicide that combats powdery mildew in fruit and ornamentals

Products

AQ 10	Belchim	58% w/w	WG	15518

Uses
- Powdery mildew in *protected apple* (off-label), *protected aubergines*, *protected blueberry* (off-label), *protected chilli peppers* (off-label), *protected courgettes*, *protected crabapple* (off-label), *protected cucumbers*, *protected forest nurseries* (off-label), *protected gooseberries* (off-label), *protected herbs (see appendix 6)* (off-label), *protected marrows* (off-label), *protected melons*, *protected ornamentals* (off-label), *protected pear* (off-label), *protected peppers*, *protected pumpkins*, *protected quince* (off-label), *protected redcurrants* (off-label), *protected squashes*, *protected strawberries*, *protected table grapes* (off-label), *protected tomatoes*, *protected watermelon* (off-label), *protected whitecurrants* (off-label), *protected wine grapes* (off-label)

Extension of Authorisation for Minor Use (EAMUs)
- *protected apple* 20121324
- *protected blueberry* 20121324
- *protected chilli peppers* 20121324
- *protected crabapple* 20121324
- *protected forest nurseries* 20121324
- *protected gooseberries* 20121324
- *protected herbs (see appendix 6)* 20121324
- *protected marrows* 20121324
- *protected ornamentals* 20121324
- *protected pear* 20121324
- *protected quince* 20121324
- *protected redcurrants* 20121324
- *protected table grapes* 20121324
- *protected watermelon* 20121324
- *protected whitecurrants* 20121324
- *protected wine grapes* 20121324

Approval information
- Ampelomyces quisqualis is included in Annex 1 under EC Regulation 1107/2009

Hazard classification and safety precautions
  UN Number N/C
  Operator protection A, D, H; U14, U20b
  Environmental protection E15b, E34
  Storage and disposal D01, D02, D05, D09a, D10a, D16, D20
  Medical advice M03

SEE SECTION 3 FOR PRODUCTS ALSO REGISTERED

17   asulam

A translocated carbamate herbicide for control of docks and bracken but all approvals expired in 2012
HRAC mode of action code: I

18   azoxystrobin

A systemic translaminar and protectant strobilurin fungicide for a wide range of crops
FRAC mode of action code: 11

Products

1	Amistar	Syngenta	250 g/l	SC	10443
2	Aubrac	AgChem Access	250 g/l	SC	13483
3	Azzox	Goldengrass	250 g/l	SC	14296
4	Clayton Belfry	Clayton	250 g/l	SC	12886
5	Clayton Putter	Clayton	50% w/w	WG	15827
6	EA Azoxystrobin	European Ag	250 g/l	SC	15048
7	Globaztar SC	Globachem	250 g/l	SC	15575
8	Harness	ChemSource	50% w/w	WG	14803
9	Heritage	Syngenta	50% w/w	WG	13536
10	Heritage Maxx	Syngenta	95 g/l	DC	14787
11	Life Scientific Azoxystrobin	Life Scientific	250 g/l	SC	15341
12	Mirador	Makhteshim	250 g/l	SC	15557
13	Panama	Pan Amenity	50% w/w	SG	13950
14	PureAzoxy	Pure Amenity	50% w/w	WG	15136
15	Reconcile	Becesane	250 g/l	SC	15379
16	Standon Azoxystrobin	Standon	250 g/l	SC	09515
17	Tazer	Nufarm UK	250 g/l	SC	15495

Uses

- Alternaria in *broccoli*, *brussels sprouts*, *cabbages*, *calabrese*, *carrots*, *cauliflowers*, *collards*, *kale*, *spring oilseed rape*, *winter oilseed rape* [1-4, 6, 15]; *chicory root* (off-label), *horseradish* (off-label), *parsnips* (off-label), *salad brassicas* (off-label - for baby leaf production) [1]
- Anthracnose in *amenity grassland*, *managed amenity turf* [5, 8-10, 13, 14]
- Ascochyta in *broad beans* (off-label), *lupins* (off-label) [1]; *combining peas*, *vining peas* [1-4, 6, 11, 15]
- Black dot in *potatoes* [1-4, 6, 15]
- Black scurf and stem canker in *potatoes* [1-4, 6, 15]
- Black spot in *protected strawberries* (off-label), *strawberries* (off-label) [1, 11]
- Blight in *protected aubergines* (off-label), *protected courgettes* (off-label), *protected cucumbers* (off-label), *protected gherkins* (off-label), *protected tomatoes* (off-label) [11]
- Botrytis in *broad beans* (off-label), *celery (outdoor)* (off-label), *dwarf beans* (off-label), *navy beans* (off-label), *protected celery* (off-label), *runner beans* (off-label) [1, 11]; *protected aubergines* (off-label), *protected courgettes* (off-label), *protected cucumbers* (off-label), *protected gherkins* (off-label), *protected tomatoes* (off-label) [11]
- Brown patch in *amenity grassland*, *managed amenity turf* [5, 8-10, 13, 14]
- Brown rust in *rye* [7, 12, 17]; *spring barley*, *spring wheat*, *winter barley*, *winter wheat* [1-4, 6, 7, 11, 12, 15-17]; *spring rye*, *winter rye* [1-4, 6, 11, 15]; *triticale* [1-4, 6, 7, 11, 12, 15, 17]
- Crown rust in *amenity grassland*, *managed amenity turf* [5, 8-10, 13, 14]; *spring oats*, *winter oats* [1-4, 6, 7, 12, 15]
- Dark leaf spot in *spring oilseed rape*, *winter oilseed rape* [12]
- Didymella in *inert substrate aubergines* (off-label), *inert substrate tomatoes* (off-label) [1]
- Didymella stem rot in *aubergines* (off-label), *protected tomatoes* (off-label) [1]
- Disease control in *all edible seed crops grown outdoors* (off-label), *all non-edible seed crops grown outdoors* (off-label), *forest nurseries* (off-label), *ornamental plant production* (off-

label), **protected forest nurseries** *(off-label),* **protected ornamentals** *(off-label),* **protected soft fruit** *(off-label),* **soft fruit** *(off-label)* [1, 2, 11]; **durum wheat** *(off-label),* **garlic** *(off-label),* **lupins** *(off-label),* **shallots** *(off-label)* [11]; **grass seed crops** *(off-label)* [11, 16]

- Downy mildew in **artichokes** *(off-label),* **garlic** *(off-label),* **inert substrate courgettes** *(off-label),* **inert substrate cucumbers** *(off-label),* **inert substrate gherkins** *(off-label),* **salad brassicas** *(off-label - for baby leaf production),* **shallots** *(off-label)* [1]; **bulb onions** [1-4, 6, 15]; **globe artichoke** *(off-label),* **leaf brassicas** *(off-label),* **lettuce** *(off-label),* **protected aubergines** *(off-label),* **protected tomatoes** *(off-label),* **radishes** *(off-label)* [11]; **protected courgettes** *(off-label),* **protected cucumbers** *(off-label),* **protected gherkins** *(off-label),* **protected marrows** *(off-label),* **protected melons** *(off-label),* **protected pumpkins** *(off-label),* **protected squashes** *(off-label),* **salad onions** *(off-label)* [1, 11]
- Ear diseases in **spring wheat**, **winter wheat** [7, 12, 17]
- Fairy rings in **amenity grassland**, **managed amenity turf** [5, 8-10, 13, 14]
- Foliar disease control in **crambe** *(off-label),* **durum wheat** *(off-label),* **grass seed crops** *(off-label),* **radishes** *(off-label)* [1]
- Fusarium patch in **amenity grassland**, **managed amenity turf** [5, 8-10, 13, 14]
- Glume blotch in **spring wheat**, **winter wheat** [1-4, 6, 7, 11, 12, 15-17]
- Grey mould in **aubergines** *(off-label),* **inert substrate aubergines** *(off-label),* **inert substrate tomatoes** *(off-label)* [1]; **protected aubergines** *(off-label),* **protected courgettes** *(off-label),* **protected cucumbers** *(off-label),* **protected gherkins** *(off-label)* [11]; **protected tomatoes** *(off-label)* [1, 11]
- Late blight in **aubergines** *(off-label),* **inert substrate aubergines** *(off-label),* **inert substrate tomatoes** *(off-label),* **protected tomatoes** *(off-label)* [1]
- Late ear diseases in **spring wheat**, **winter wheat** [1-4, 6, 11, 15, 16]
- Melting out in **amenity grassland**, **managed amenity turf** [5, 8-10, 13, 14]
- Net blotch in **spring barley**, **winter barley** [1-4, 6, 7, 11, 12, 15-17]
- Phytophthora in **protected chicory** *(off-label - for forcing)* [1]
- Powdery mildew in **artichokes** *(off-label),* **aubergines** *(off-label),* **chicory root** *(off-label),* **chives** *(off-label),* **inert substrate aubergines** *(off-label),* **inert substrate courgettes** *(off-label),* **inert substrate cucumbers** *(off-label),* **inert substrate gherkins** *(off-label),* **inert substrate tomatoes** *(off-label),* **parsley** *(off-label),* **protected marrows** *(off-label),* **protected melons** *(off-label),* **protected pumpkins** *(off-label),* **protected squashes** *(off-label)* [1]; **blackberries** *(off-label),* **courgettes** *(off-label),* **herbs (see appendix 6)** *(off-label),* **horseradish** *(off-label),* **parsnips** *(off-label),* **poppies for morphine production** *(off-label),* **protected blackberries** *(off-label),* **protected cayenne peppers** *(off-label),* **protected courgettes** *(off-label),* **protected cucumbers** *(off-label),* **protected gherkins** *(off-label),* **protected peppers** *(off-label),* **protected raspberries** *(off-label),* **protected tomatoes** *(off-label),* **raspberries** *(off-label)* [1, 11]; **carrots, spring oats, winter oats** [1-4, 6, 15]; **chicory** *(off-label),* **cress** *(off-label),* **frise** *(off-label),* **lamb's lettuce** *(off-label),* **protected aubergines** *(off-label),* **protected chicory** *(off-label)* [11]; **rye** *(moderate control),* **spring barley** *(moderate control),* **triticale** *(moderate control),* **winter barley** *(moderate control)* [7, 12, 17]; **spring barley, winter barley** [1-4, 6, 11, 15, 16]; **spring oats** *(moderate control),* **winter oats** *(moderate control)* [7, 12]; **spring wheat, winter wheat** [16]; **triticale, winter rye** [1-4, 6, 11, 15]
- Purple blotch in **leeks** [1-4, 6, 15]
- Rhizoctonia in **celery (outdoor)** *(off-label),* **protected celery** *(off-label),* **protected leaf brassicas** *(off-label)* [1]; **protected herbs (see appendix 6)** *(off-label),* **protected lettuce** *(off-label),* **protected radicchio** *(off-label),* **protected scarole** *(off-label)* [1, 11]; **red beet** *(off-label)* [2, 11]; **swedes** *(off-label),* **turnips** *(off-label)* [1, 2, 11]
- Rhynchosporium in **rye** *(reduction),* **spring barley** *(reduction),* **triticale** *(reduction),* **winter barley** *(reduction)* [7, 12, 17]; **spring barley, winter barley** [1-4, 6, 11, 15, 16]; **spring rye, triticale, winter rye** [1-4, 6, 11, 15]
- Ring spot in **broccoli, brussels sprouts, cabbages, calabrese, cauliflowers, collards, kale** [1-4, 6, 15]; **chives** *(off-label),* **herbs (see appendix 6)** *(off-label),* **parsley** *(off-label)* [1]
- Root malformation disorder in **red beet** *(off-label)* [1]
- Rust in **asparagus** [1-4, 6, 11, 15]; **chicory** *(off-label),* **cress** *(off-label),* **frise** *(off-label),* **lamb's lettuce** *(off-label),* **pot chrysanthemums** *(off-label),* **protected chicory** *(off-label),* **protected chrysanthemums** *(off-label)* [11]; **chicory root** *(off-label),* **chives** *(off-label),* **lupins** *(off-label),* **parsley** *(off-label)* [1]; **herbs (see appendix 6)** *(off-label)* [1, 11]; **leeks, spring field beans, winter field beans** [1-4, 6, 15]

SEE SECTION 3 FOR PRODUCTS ALSO REGISTERED

- Sclerotinia in *celeriac (off-label)*, *celery (outdoor) (off-label)*, *mustard (off-label)*, *protected celery (off-label)* [1, 11]; *crambe (off-label)* [11]; *lettuce (off-label)* [1]
- Sclerotinia stem rot in *spring oilseed rape*, *winter oilseed rape* [1-4, 6, 12, 15]
- Septoria leaf blotch in *protected aubergines (off-label)*, *protected courgettes (off-label)*, *protected cucumbers (off-label)*, *protected gherkins (off-label)*, *protected tomatoes (off-label)* [11]; *spring wheat*, *winter wheat* [1-4, 6, 7, 11, 12, 15-17]
- Stemphylium in *asparagus* [1-4, 6, 11, 15]
- Take-all in *rye (reduction)*, *triticale (reduction)* [7, 12, 17]; *spring barley (reduction)*, *spring wheat (reduction)*, *winter barley (reduction)*, *winter wheat (reduction)* [1-4, 6, 7, 11, 12, 15, 17]
- Take-all patch in *amenity grassland*, *managed amenity turf* [5, 8-10, 13, 14]
- White blister in *broccoli*, *brussels sprouts*, *cabbages*, *calabrese*, *cauliflowers*, *collards*, *kale* [1-4, 6, 15]
- White rust in *chrysanthemums (off-label - in pots)* [1]; *pot chrysanthemums (off-label)* [11]; *protected chrysanthemums (off-label)* [1, 11]
- Yellow rust in *spring wheat*, *winter wheat* [1-4, 6, 7, 11, 12, 15-17]

Extension of Authorisation for Minor Use (EAMUs)
- *all edible seed crops grown outdoors* 20090443 [1], 20100372 [2], 20111418 [11]
- *all non-edible seed crops grown outdoors* 20090443 [1], 20100372 [2], 20111418 [11]
- *artichokes* 20051814 [1]
- *aubergines* 20021533 [1]
- *blackberries* 20030365 [1], 20111403 [11]
- *broad beans* 20032311 [1], 20111407 [11]
- *celeriac* 20031862 [1], 20111406 [11]
- *celery (outdoor)* 20011041 [1], 20111391 [11]
- *chicory* 20111390 [11]
- *chicory root* 20051813 [1]
- *chives* 20021293 [1]
- *chrysanthemums (in pots)* 20011684 [1]
- *courgettes* 20051985 [1], 20111410 [11]
- *crambe* 20081341 [1], 20111416 [11]
- *cress* 20111398 [11]
- *durum wheat* 20061722 [1], 20111412 [11]
- *dwarf beans* 20032311 [1], 20111407 [11]
- *forest nurseries* 20090443 [1], 20100372 [2], 20111418 [11]
- *frise* 20111398 [11]
- *garlic* 20061724 [1], 20111414 [11]
- *globe artichoke* 20111430 [11]
- *grass seed crops* 20061722 [1], 20111412 [11], 20121539 expires 31 Dec 2013 [16]
- *herbs (see appendix 6)* 20021293 [1], 20111398 [11]
- *horseradish* 20061721 [1], 20111411 [11]
- *inert substrate aubergines* 20011685 [1]
- *inert substrate courgettes* 20011685 [1]
- *inert substrate cucumbers* 20011685 [1]
- *inert substrate gherkins* 20011685 [1]
- *inert substrate tomatoes* 20011685 [1]
- *lamb's lettuce* 20111398 [11]
- *leaf brassicas* 20111395 [11]
- *lettuce* 20011465 [1], 20111395 [11]
- *lupins* 20061723 [1], 20111413 [11]
- *mustard* 20111165 [1], 20111419 [11]
- *navy beans* 20032311 [1], 20111407 [11]
- *ornamental plant production* 20090443 [1], 20100372 [2], 20111418 [11]
- *parsley* 20021293 [1]
- *parsnips* 20061721 [1], 20111411 [11]
- *poppies for morphine production* 20031137 [1], 20111405 [11]
- *pot chrysanthemums* 20111396 [11]
- *protected aubergines* 20111397 [11], 20111401 [11]
- *protected blackberries* 20051194 [1], 20111409 [11]

FOR FULL CONDITIONS OF USE ALWAYS READ THE PRODUCT LABEL

- *protected cayenne peppers* 20021295 [1], 20111400 [11]
- *protected celery* 20011041 [1], 20111391 [11]
- *protected chicory* (for forcing) 20051813 [1], 20111390 [11]
- *protected chrysanthemums* 20011684 [1], 20111396 [11]
- *protected courgettes* 20021533 [1], 20111397 [11], 20111401 [11]
- *protected cucumbers* 20021533 [1], 20111397 [11], 20111401 [11]
- *protected forest nurseries* 20090443 [1], 20100372 [2], 20111418 [11]
- *protected gherkins* 20021533 [1], 20111397 [11], 20111401 [11]
- *protected herbs (see appendix 6)* 20120876 [1], 20111404 [11]
- *protected leaf brassicas* 20120876 [1]
- *protected lettuce* 20120876 [1], 20111404 [11]
- *protected marrows* 20070371 [1], 20111415 [11]
- *protected melons* 20070371 [1], 20111415 [11]
- *protected ornamentals* 20090443 [1], 20100372 [2], 20111418 [11]
- *protected peppers* 20021295 [1], 20111400 [11]
- *protected pumpkins* 20070371 [1], 20111415 [11]
- *protected radicchio* 20120876 [1], 20111404 [11]
- *protected raspberries* 20051194 [1], 20111409 [11]
- *protected scarole* 20120876 [1], 20111404 [11]
- *protected soft fruit* 20090443 [1], 20100372 [2], 20111418 [11]
- *protected squashes* 20070371 [1], 20111415 [11]
- *protected strawberries* 20021294 [1], 20111399 [11]
- *protected tomatoes* 20021533 [1], 20111397 [11], 20111401 [11]
- *radishes* 20081448 [1], 20111417 [11]
- *raspberries* 20030365 [1], 20111403 [11]
- *red beet* 20040614 [1], 20100371 [2], 20111408 [11]
- *runner beans* 20032311 [1], 20111407 [11]
- *salad brassicas* (for baby leaf production) 20011465 [1]
- *salad onions* 20021687 [1], 20111402 [11]
- *shallots* 20061724 [1], 20111414 [11]
- *soft fruit* 20090443 [1], 20100372 [2], 20111418 [11]
- *strawberries* 20021294 [1], 20111399 [11]
- *swedes* 20040614 [1], 20100371 [2], 20111408 [11]
- *turnips* 20040614 [1], 20100371 [2], 20111408 [11]

Approval information
- Azoxystrobin acid included in Annex I under EC Regulation 1107/2009
- Accepted by BBPA for use on malting barley
- Approval expiry 31 Dec 2013 [16]

Efficacy guidance
- Best results obtained from use as a protectant or during early stages of disease establishment or when a predictive assessment indicates a risk of disease development
- Azoxystrobin inhibits fungal respiration and should always be used in mixture with fungicides with other modes of action
- Treatment under poor growing conditions may give less reliable results
- For good control of *Fusarium* patch in amenity turf and grass repeat treatment at minimum intervals of 2 wk
- Azoxystrobin is a member of the QoI cross resistance group. Product should be used preventatively and not relied on for its curative potential
- Use product in cereals as part of an Integrated Crop Management strategy incorporating other methods of control, including where appropriate other fungicides with a different mode of action. Do not apply more than two foliar applications of QoI containing products to any cereal crop
- There is a significant risk of widespread resistance occurring in *Septoria tritici* populations in UK. Failure to follow resistance management action may result in reduced levels of disease control
- On cereal crops product must always be used in mixture with another product, recommended for control of the same target disease, that contains a fungicide from a different cross resistance group and is applied at a dose that will give robust control
- Strains of barley powdery mildew resistant to QoI's are common in the UK

SEE SECTION 3 FOR PRODUCTS ALSO REGISTERED

Restrictions
- Maximum number of treatments 1 per crop for potatoes; 2 per crop for brassicas, peas, cereals, oilseed rape; 4 per crop or yr for onions, carrots, leeks, amenity turf [1, 4]
- Maximum total dose ranges from 2-4 times the single full dose depending on crop and product. See labels for details
- On turf the maximum number of treatments is 4 per yr but they must not exceed one third of the total number of fungicide treatments applied
- Do not use where there is risk of spray drift onto neighbouring apple crops
- The same spray equipment should not be used to treat apples

Crop-specific information
- Latest use: at planting for potatoes; grain watery ripe (GS 71) for cereals; before senescence for asparagus
- HI: 10 d for carrots;14 d for broccoli, Brussels sprouts, bulb onions, cabbages, calabrese, cauliflowers, collards, kale, vining peas; 21 d for leeks, spring oilseed rape, winter oilseed rape; 36 d for combining peas, 35 d for field beans [1, 4]
- In cereals control of established infections can be improved by appropriate tank mixtures or application as part of a programme. Always use in mixture with another product from a different cross-resistance group
- In turf use product at full dose rate in a disease control programme, alternating with fungicides of different modes of action
- In potatoes when used as incorporated treatment apply overall to the entire area to be planted, incorporate to 15 cm and plant on the same day. In-furrow spray should be directed at the furrow and not the seed tubers
- Applications to brassica crops must only be made to a developed leaf canopy and not before growth stages specified on the label
- Heavy disease pressure in brassicae and oilseed rape may require a second treatment
- All crops should be treated when not under stress. Check leaf wax on peas if necessary
- Consult processor before treating any crops for processing
- Treat asparagus after the harvest season. Where a new bed is established do not treat within 3 wk of transplanting out the crowns
- Do not apply to turf when ground is frozen or during drought

Environmental safety
- Dangerous for the environment
- Very toxic to aquatic organisms
- Avoid spray drift onto surrounding areas or crops, especially apples, plums or privet
- LERAP Category B [1-4, 6, 15] (potatoes only)

Hazard classification and safety precautions
   Hazard Harmful [7, 11, 17]; Dangerous for the environment [1-12, 14-17]
   Transport code 9
   Packaging group III
   UN Number 3077 [5, 8, 9, 13, 14]; 3082 [1-4, 6, 7, 10-12, 15-17]
   Risk phrases R20 [7, 11]; R50 [1-12, 14-16]; R51 [17]; R53a [1-12, 14-17]
   Operator protection A [1-3]; A [4] (during treatment of potatoes); A [6, 7, 10-12, 15, 17]; U05a [1-12, 14, 15, 17]; U09a, U19a [1-4, 6, 7, 11, 12, 15-17]; U20b [1-12, 14-17]
   Environmental protection E15a [1-4, 6, 15, 16]; E15b [5, 7-12, 14, 17]; E16a, E16b [1-4, 6, 15] (potatoes only); E23 [11]; E38 [1-12, 14, 15, 17]
   Storage and disposal D01, D02, D12a [1-12, 14, 15, 17]; D03 [5, 8-10, 14]; D05 [1-11, 14-16]; D09a [1-12, 14-17]; D10b [4, 5, 8-10, 14, 16]; D10c [1-3, 6, 7, 11, 12, 15, 17]
   Medical advice M03 [11]; M05a [5, 8-10, 14]

19   azoxystrobin + chlorothalonil

A preventative and systemic fungicide mixture for cereals
FRAC mode of action code: 11 + M5

See also chlorothalonil

Products

1	Amistar Opti	Syngenta	100:500 g/l	SC	14582
2	Curator	Syngenta	80:400 g/l	SC	14955
3	Olympus	Syngenta	80:400 g/l	SC	13797

Uses

- Alternaria in **potatoes** *(off-label)* [3]
- Botrytis in **asparagus** *(qualified minor use recommendation)*, **bulb onions**, **garlic**, **shallots** [3]
- Brown rust in **spring barley**, **spring wheat**, **winter barley**, **winter wheat** [1, 2]
- Disease control in **durum wheat** *(off-label)*, **poppies for morphine production** *(off-label)*, **rye** *(off-label)*, **triticale** *(off-label)* [1]
- Downy mildew in **bulb onions**, **garlic**, **shallots** [3]
- Glume blotch in **spring wheat**, **winter wheat** [1, 2]
- Net blotch in **spring barley**, **winter barley** [1, 2]
- Rhynchosporium in **spring barley**, **winter barley** [2]; **spring barley** *(moderate control)*, **winter barley** *(moderate control)* [1]
- Rust in **asparagus** *(moderate control only)* [3]
- Septoria leaf blotch in **spring wheat**, **winter wheat** [1, 2]
- Stemphylium in **asparagus** *(qualified minor use recommendation)* [3]
- Take-all in **spring barley** *(reduction)*, **spring wheat** *(reduction)*, **winter barley** *(reduction)* [1]; **spring barley** *(reduction only)*, **spring wheat** *(reduction only)*, **winter barley** *(reduction only)* [2]; **winter wheat** *(reduction)* [1, 2]
- Yellow rust in **spring barley**, **winter barley** [2]; **spring wheat**, **winter wheat** [1, 2]

Extension of Authorisation for Minor Use (EAMUs)

- **durum wheat** *20120472* [1]
- **poppies for morphine production** *20120473* [1]
- **potatoes** *20090711* [3]
- **rye** *20120472* [1]
- **triticale** *20120472* [1]

Approval information

- Azoxystrobin and chlorothalonil included in Annex I under EC Regulation 1107/2009
- Accepted by BBPA for use on malting barley

Efficacy guidance

- Best results obtained from applications made as a protectant treatment or in earliest stages of disease development. Further applications may be needed if disease attack is prolonged
- Reduction of barley spotting occurs when used as part of a programme with other fungicides
- Control of *Septoria* and rust diseases may be improved by mixture with a triazole fungicide
- Azoxystrobin is a member of the QoI cross resistance group. Product should be used preventatively and not relied on for its curative potential
- Use product in cereals as part of an Integrated Crop Management strategy incorporating other methods of control, including where appropriate other fungicides with a different mode of action. Do not apply more than two foliar applications of QoI containing products to any cereal crop
- There is a significant risk of widespread resistance occurring in *Septoria tritici* populations in UK. Failure to follow resistance management action may result in reduced levels of disease control

Restrictions

- Maximum number of treatments 2 per crop
- Maximum total dose on barley equivalent to one full dose treatment
- Do not use where there is risk of spray drift onto neighbouring apple crops
- The same spray equipment should not be used to treat apples

SEE SECTION 3 FOR PRODUCTS ALSO REGISTERED

Crop-specific information
- Latest use: before beginning of heading (GS 51) for barley; before caryopsis watery ripe (GS 71) for wheat

Environmental safety
- Dangerous for the environment
- Very toxic to aquatic organisms
- LERAP Category B

Hazard classification and safety precautions
Hazard Toxic [1]; Harmful [2, 3]; Dangerous for the environment [1-3]
Transport code 9
Packaging group III
UN Number 3082
Risk phrases R20 [2, 3]; R22a, R23 [1]; R37, R40, R41, R43, R50, R53a [1-3]
Operator protection A, C, H; U02a, U05a, U09a, U11, U14, U15, U19a [1-3]; U20a [2]; U20b [1, 3]
Environmental protection E15a [1, 3]; E15b [2]; E16a, E34, E38 [1-3]
Storage and disposal D01, D02, D09a, D10c, D12a [1-3]; D05 [1, 3]
Medical advice M04a [1, 3]

20   azoxystrobin + cyproconazole

A contact and systemic broad spectrum fungicide mixture for cereals, beet crops and oilseed rape
FRAC mode of action code: 11 + 3

See also cyproconazole

Products

Priori Xtra	Syngenta	200:80 g/l	SC	11518

Uses
- Alternaria in *spring oilseed rape*, *winter oilseed rape*
- Brown rust in *spring barley*, *spring rye*, *spring wheat*, *winter barley*, *winter rye*, *winter wheat*
- Cercospora leaf spot in *fodder beet*, *sugar beet*
- Crown rust in *spring oats*, *winter oats*
- Eyespot in *spring barley* (reduction), *spring wheat* (reduction), *winter barley* (reduction), *winter wheat* (reduction)
- Foliar disease control in *durum wheat* (off-label), *grass seed crops* (off-label), *triticale* (off-label)
- Glume blotch in *spring wheat*, *winter wheat*
- Net blotch in *spring barley*, *winter barley*
- Powdery mildew in *fodder beet*, *spring barley*, *spring oats*, *spring rye*, *spring wheat*, *sugar beet*, *winter barley*, *winter oats*, *winter rye*, *winter wheat*
- Ramularia leaf spots in *fodder beet*, *sugar beet*
- Rhynchosporium in *spring barley* (moderate control), *spring rye* (moderate control), *winter barley* (moderate control), *winter rye* (moderate control)
- Rust in *fodder beet*, *sugar beet*
- Sclerotinia stem rot in *spring oilseed rape*, *winter oilseed rape*
- Septoria leaf blotch in *spring wheat*, *winter wheat*
- Take-all in *spring barley* (reduction), *spring wheat* (reduction), *winter barley* (reduction), *winter wheat* (reduction)
- Yellow rust in *spring wheat*, *winter wheat*

Extension of Authorisation for Minor Use (EAMUs)
- *durum wheat* 20052891
- *grass seed crops* 20052891
- *triticale* 20052891

Approval information
- Azoxystrobin and cyproconazole included in Annex I under EC Regulation 1107/2009
- Accepted by BBPA for use on malting barley

Efficacy guidance
- Best results obtained from treatment during the early stages of disease development
- A second application may be needed if disease attack is prolonged
- Azoxystrobin is a member of the QoI cross resistance group. Product should be used preventatively and not relied on for its curative potential
- Use product as part of an Integrated Crop Management strategy incorporating other methods of control, including where appropriate other fungicides with a different mode of action. Do not apply more than two foliar applications of QoI containing products to any cereal crop
- There is a significant risk of widespread resistance occurring in *Septoria tritici* populations in UK. Failure to follow resistance management action may result in reduced levels of disease control
- Strains of wheat and barley powdery mildew resistant to QoI's are common in the UK. Control of wheat powdery mildew can only be relied upon from the triazole component
- Where specific control of wheat mildew is required this should be achieved through a programme of measures including products recommended for the control of mildew that contain a fungicide from a different cross-resistance group and applied at a dose that will give robust control
- Cyproconazole is a DMI fungicide. Resistance to some DMI fungicides has been identified in Septoria leaf blotch which may seriously affect performance of some products. For further advice contact a specialist advisor and visit the Fungicide Resistance Action Group (FRAG)-UK website

Restrictions
- Maximum total dose equivalent to two full dose treatments
- Do not use where there is risk of spray drift onto neighbouring apple crops
- The same spray equipment should not be used to treat apples

Crop-specific information
- Latest use: up to and including anthesis complete (GS 69) for rye and wheat; up to and including emergence of ear complete (GS 59) for barley and oats; BBCH79 (nearly all pods at final size) or 30 days before harvest, whichever is sooner for oilseed rape

Environmental safety
- Dangerous for the environment
- Very toxic to aquatic organisms

Hazard classification and safety precautions
Hazard Harmful, Dangerous for the environment
Transport code 9
Packaging group III
UN Number 3082
Risk phrases R22a, R50, R53a, R63
Operator protection A; U05a, U09a, U19a, U20b
Environmental protection E15b, E34, E38
Storage and disposal D01, D02, D05, D09a, D10c, D12a
Medical advice M03

21   azoxystrobin + difenoconazole

A broad spectrum fungicide mixture for field crops
FRAC mode of action code: 11 + 3

See also difenoconazole

Products

Amistar Top	Syngenta	200:125 g/l	SC	12761

SECTION 2

Uses

- Alternaria blight in **carrots**, **horseradish** *(off-label)*, **parsley root** *(off-label)*, **parsnips** *(off-label)*, **salsify** *(off-label)*
- Powdery mildew in **broccoli**, **brussels sprouts**, **cabbages**, **calabrese**, **carrots**, **cauliflowers**, **collards**, **horseradish** *(off-label)*, **kale**, **parsley root** *(off-label)*, **parsnips** *(off-label)*, **salsify** *(off-label)*
- Purple blotch in **leeks** *(moderate control)*
- Rust in **asparagus** *(off-label)*, **herbs (see appendix 6)** *(off-label)*, **leeks**
- Septoria leaf blotch in **herbs (see appendix 6)** *(off-label)*
- Stemphylium in **asparagus** *(off-label)*
- White blister in **broccoli**, **brussels sprouts**, **cabbages**, **calabrese**, **cauliflowers**, **collards**, **kale**
- White tip in **leeks** *(qualified minor use)*

Extension of Authorisation for Minor Use (EAMUs)

- **asparagus** *20070831*
- **herbs (see appendix 6)** *20121838*
- **horseradish** *20061476*
- **parsley root** *20061476*
- **parsnips** *20061476*
- **salsify** *20061476*

Approval information

- Azoxystrobin and difenoconazole included in Annex I under EC Regulation 1107/2009

Efficacy guidance

- Best results obtained from applications made in the earliest stages of disease development or as a protectant treatment following a disease risk assessment
- Ensure the crop is free from any stress caused by environmental or agronomic effects
- Azoxystrobin is a member of the QoI cross resistance group. Product should be used preventatively and not relied on for its curative potential
- Use as part of an Integrated Crop Management strategy incorporating other methods of control, including where appropriate other fungicides with a different mode of action. Do not apply more than two foliar applications of QoI containing products

Restrictions

- Maximum number of treatments 2 per crop
- Do not apply where there is a risk of spray drift onto neighbouring apple crops
- Consult processors before treating a crop destined for processing

Crop-specific information

- HI 14 d for carrots; 21 d for brassicas, leeks
- Minimum spray interval of 14 d must be observed on brassicas

Environmental safety

- Dangerous for the environment
- Very toxic to aquatic organisms

Hazard classification and safety precautions

Hazard Irritant, Dangerous for the environment
Transport code 9
Packaging group III
UN Number 3082
Risk phrases R20, R43, R50, R53a
Operator protection A, H; U05a, U09a, U19a, U20b
Environmental protection E15b, E38
Storage and disposal D01, D02, D05, D09a, D10c, D12a

FOR FULL CONDITIONS OF USE ALWAYS READ THE PRODUCT LABEL

SECTION 2

22   azoxystrobin + fenpropimorph

A protectant and eradicant fungicide mixture for cereals
FRAC mode of action code: 11 + 5

See also fenpropimorph

Products

1 Amistar Pro	Syngenta	100:280 g/l	SE	10513
2 Aspect	Syngenta	100:280 g/l	SE	10516

Uses

- Brown rust in **spring barley**, **spring rye**, **spring wheat**, **triticale**, **winter barley**, **winter rye**, **winter wheat** [1, 2]
- Crown rust in **spring oats**, **winter oats** [1, 2]
- Foliar disease control in **durum wheat** *(off-label)*, **grass seed crops** *(off-label)* [1]
- Glume blotch in **spring wheat**, **winter wheat** [1, 2]
- Late ear diseases in **spring wheat** *(moderate control)*, **winter wheat** *(moderate control)* [1, 2]
- Net blotch in **spring barley**, **winter barley** [1, 2]
- Powdery mildew in **spring barley**, **spring oats**, **spring rye**, **triticale**, **winter barley**, **winter oats**, **winter rye** [1, 2]
- Rhynchosporium in **spring barley**, **spring rye**, **triticale** *(reduction)*, **winter barley**, **winter rye** [1, 2]
- Septoria leaf blotch in **spring wheat**, **winter wheat** [1, 2]
- Take-all in **spring barley** *(reduction)*, **spring rye** *(reduction)*, **spring wheat** *(reduction)*, **triticale** *(reduction)*, **winter barley** *(reduction)*, **winter rye** *(reduction)*, **winter wheat** *(reduction)* [1, 2]
- Yellow rust in **spring wheat**, **winter wheat** [1, 2]

Extension of Authorisation for Minor Use (EAMUs)

- **durum wheat** *20060962* [1]
- **grass seed crops** *20060962* [1]

Approval information

- Azoxystrobin and fenpropimorph included in Annex I under EC Regulation 1107/2009
- Accepted by BBPA for use on malting barley

Efficacy guidance

- Best results obtained from application before infection following a disease risk assessment, or when disease first seen in crop
- Results may be less reliable when used on crops under stress
- Treatments for protection against ear disease should be made at ear emergence
- Azoxystrobin is a member of the QoI cross resistance group. Product should be used preventatively and not relied on for its curative potential
- Use product as part of an Integrated Crop Management strategy incorporating other methods of control, including where appropriate other fungicides with a different mode of action. Do not apply more than two foliar applications of QoI containing products to any cereal crop
- There is a significant risk of widespread resistance occurring in *Septoria tritici* populations in UK. Failure to follow resistance management action may result in reduced levels of disease control
- Strains of barley powdery mildew resistant to QoI's are common in the UK
- In wheat product must always be used in mixture with another product, recommended for control of the same target disease, that contains a fungicide from a different cross resistance group and is applied at a dose that will give robust control

Restrictions

- Maximum total dose equivalent to 2 full dose treatments
- Do not use where there is risk of spray drift onto neighbouring apple crops
- The same spray equipment should not be used to treat apples

Crop-specific information

- Latest use: before early milk stage (GS 73) for wheat, barley; up to and including grain watery ripe stage (GS 71) for oats, rye, triticale
- HI 5 wk

SEE SECTION 3 FOR PRODUCTS ALSO REGISTERED

Environmental safety
- Dangerous for the environment
- Very toxic to aquatic organisms

Hazard classification and safety precautions

Hazard Harmful, Dangerous for the environment
Transport code 9
Packaging group III
UN Number 3082
Risk phrases R20, R38, R43, R50, R53a, R63
Operator protection A, H; U02a, U05a, U09a, U14, U15, U19a, U20b
Environmental protection E15a, E38
Storage and disposal D01, D02, D05, D09a, D10c, D12a
Medical advice M03

23   azoxystrobin + propiconazole

A contact and systemic broad spectrum fungicide mixture for use on grass
FRAC mode of action code: 11 + 3

Products

1 Headway	Syngenta	62.5:104 g/l	EC	14396
2 PureProgress	Pure Amenity	62.5:104 g/l	EC	15471

Uses
- Anthracnose in **amenity grassland** *(moderate control)*, **managed amenity turf** *(moderate control)*
- Dollar spot in **amenity grassland**, **managed amenity turf**
- Fusarium diseases in **amenity grassland**, **managed amenity turf**
- Microdochium nivale in **amenity grassland**, **managed amenity turf**

Approval information
- Azoxystrobin and propiconazole included in Annex I under EC Regulation 1107/2009

Environmental safety
- LERAP Category B

Hazard classification and safety precautions

Hazard Dangerous for the environment
Transport code 9
Packaging group III
UN Number 3082
Risk phrases R50, R53a
Operator protection A; U05a, U09a, U19a, U20a
Environmental protection E15b, E16a, E16b, E34, E38
Storage and disposal D01, D02, D05, D09a, D10c, D12a

24   Bacillus subtilis

A bacterial fungicide for the control of Botrytis cinerea
FRAC mode of action code: 44

Products

Serenade ASO	BASF	13.96 g/l	SC	15625

Uses
- Botrytis in **bilberries** *(off-label)*, **blackcurrants** *(off-label)*, **blueberries** *(off-label)*, **bulb vegetables** *(off-label)*, **canary grass** *(off-label)*, **cane fruit** *(off-label)*, **cranberries** *(off-label)*, **figs** *(off-label)*, **fruiting vegetables** *(off-label)*, **gooseberries** *(off-label)*, **herbs (see appendix 6)** *(off-label)*, **hops** *(off-label)*, **leafy vegetables** *(off-label)*, **legumes** *(off-label)*, **ornamental plant production** *(off-label)*, **redcurrants** *(off-label)*, **ribes hybrids** *(off-label)*, **root & tuber crops** *(off-*

label), **stem vegetables** *(off-label)*, **strawberries** *(off-label)*, **table grapes** *(off-label)*, **top fruit** *(off-label)*, **vegetable brassicas** *(off-label)*, **wine grapes** *(off-label)*
- Botrytis fruit rot in **protected strawberries** *(reduction of damage to fruit)*
- Damping off in **aubergines** *(off-label)*, **blackberries** *(off-label)*, **blackcurrants** *(off-label)*, **blueberries** *(off-label)*, **broccoli** *(off-label)*, **brussels sprouts** *(off-label)*, **bulb onions** *(off-label)*, **cabbages** *(off-label)*, **calabrese** *(off-label)*, **carrots** *(off-label)*, **cauliflowers** *(off-label)*, **celeriac** *(off-label)*, **celery (outdoor)** *(off-label)*, **chard** *(off-label)*, **collards** *(off-label)*, **courgettes** *(off-label)*, **cranberries** *(off-label)*, **cucumbers** *(off-label)*, **garlic** *(off-label)*, **gooseberries** *(off-label)*, **herbs (see appendix 6)** *(off-label)*, **kale** *(off-label)*, **leaf brassicas** *(off-label)*, **leeks** *(off-label)*, **lettuce** *(off-label)*, **marrows** *(off-label)*, **parsnips** *(off-label)*, **peppers** *(off-label)*, **potatoes** *(off-label)*, **pumpkins** *(off-label)*, **raspberries** *(off-label)*, **redcurrants** *(off-label)*, **ribes hybrids** *(off-label)*, **rubus hybrids** *(off-label)*, **salad onions** *(off-label)*, **shallots** *(off-label)*, **spinach** *(off-label)*, **squashes** *(off-label)*, **tomatoes (outdoor)** *(off-label)*, **whitecurrants** *(off-label)*
- Grey mould in **protected strawberries** *(reduction of damage to fruit)*
- Helminthosporium seedling rot in **aubergines** *(off-label)*, **blackberries** *(off-label)*, **blackcurrants** *(off-label)*, **blueberries** *(off-label)*, **broccoli** *(off-label)*, **brussels sprouts** *(off-label)*, **bulb onions** *(off-label)*, **cabbages** *(off-label)*, **calabrese** *(off-label)*, **carrots** *(off-label)*, **cauliflowers** *(off-label)*, **celeriac** *(off-label)*, **celery (outdoor)** *(off-label)*, **chard** *(off-label)*, **collards** *(off-label)*, **courgettes** *(off-label)*, **cranberries** *(off-label)*, **cucumbers** *(off-label)*, **garlic** *(off-label)*, **gooseberries** *(off-label)*, **herbs (see appendix 6)** *(off-label)*, **kale** *(off-label)*, **leaf brassicas** *(off-label)*, **leeks** *(off-label)*, **lettuce** *(off-label)*, **marrows** *(off-label)*, **parsnips** *(off-label)*, **peppers** *(off-label)*, **potatoes** *(off-label)*, **pumpkins** *(off-label)*, **raspberries** *(off-label)*, **redcurrants** *(off-label)*, **ribes hybrids** *(off-label)*, **rubus hybrids** *(off-label)*, **salad onions** *(off-label)*, **shallots** *(off-label)*, **spinach** *(off-label)*, **squashes** *(off-label)*, **tomatoes (outdoor)** *(off-label)*, **whitecurrants** *(off-label)*
- Phytophthora in **amenity vegetation** *(off-label)*, **aubergines** *(off-label)*, **blackberries** *(off-label)*, **blackcurrants** *(off-label)*, **blueberries** *(off-label)*, **broccoli** *(off-label)*, **brussels sprouts** *(off-label)*, **bulb onions** *(off-label)*, **cabbages** *(off-label)*, **calabrese** *(off-label)*, **carrots** *(off-label)*, **cauliflowers** *(off-label)*, **celeriac** *(off-label)*, **celery (outdoor)** *(off-label)*, **chard** *(off-label)*, **collards** *(off-label)*, **courgettes** *(off-label)*, **cranberries** *(off-label)*, **cucumbers** *(off-label)*, **forest nurseries** *(off-label)*, **garlic** *(off-label)*, **gooseberries** *(off-label)*, **herbs (see appendix 6)** *(off-label)*, **kale** *(off-label)*, **leaf brassicas** *(off-label)*, **leeks** *(off-label)*, **lettuce** *(off-label)*, **marrows** *(off-label)*, **parsnips** *(off-label)*, **peppers** *(off-label)*, **potatoes** *(off-label)*, **pumpkins** *(off-label)*, **raspberries** *(off-label)*, **redcurrants** *(off-label)*, **ribes hybrids** *(off-label)*, **rubus hybrids** *(off-label)*, **salad onions** *(off-label)*, **shallots** *(off-label)*, **spinach** *(off-label)*, **squashes** *(off-label)*, **tomatoes (outdoor)** *(off-label)*, **whitecurrants** *(off-label)*
- Pythium in **aubergines** *(off-label)*, **blackberries** *(off-label)*, **blackcurrants** *(off-label)*, **blueberries** *(off-label)*, **broccoli** *(off-label)*, **brussels sprouts** *(off-label)*, **bulb onions** *(off-label)*, **cabbages** *(off-label)*, **calabrese** *(off-label)*, **carrots** *(off-label)*, **cauliflowers** *(off-label)*, **celeriac** *(off-label)*, **celery (outdoor)** *(off-label)*, **chard** *(off-label)*, **collards** *(off-label)*, **courgettes** *(off-label)*, **cranberries** *(off-label)*, **cucumbers** *(off-label)*, **garlic** *(off-label)*, **gooseberries** *(off-label)*, **herbs (see appendix 6)** *(off-label)*, **kale** *(off-label)*, **leaf brassicas** *(off-label)*, **leeks** *(off-label)*, **lettuce** *(off-label)*, **marrows** *(off-label)*, **parsnips** *(off-label)*, **peppers** *(off-label)*, **potatoes** *(off-label)*, **pumpkins** *(off-label)*, **raspberries** *(off-label)*, **redcurrants** *(off-label)*, **ribes hybrids** *(off-label)*, **rubus hybrids** *(off-label)*, **salad onions** *(off-label)*, **shallots** *(off-label)*, **spinach** *(off-label)*, **squashes** *(off-label)*, **tomatoes (outdoor)** *(off-label)*, **whitecurrants** *(off-label)*
- Rhizoctonia in **aubergines** *(off-label)*, **blackberries** *(off-label)*, **blackcurrants** *(off-label)*, **blueberries** *(off-label)*, **broccoli** *(off-label)*, **brussels sprouts** *(off-label)*, **bulb onions** *(off-label)*, **cabbages** *(off-label)*, **calabrese** *(off-label)*, **carrots** *(off-label)*, **cauliflowers** *(off-label)*, **celeriac** *(off-label)*, **celery (outdoor)** *(off-label)*, **chard** *(off-label)*, **collards** *(off-label)*, **courgettes** *(off-label)*, **cranberries** *(off-label)*, **cucumbers** *(off-label)*, **garlic** *(off-label)*, **gooseberries** *(off-label)*, **herbs (see appendix 6)** *(off-label)*, **kale** *(off-label)*, **leaf brassicas** *(off-label)*, **leeks** *(off-label)*, **lettuce** *(off-label)*, **marrows** *(off-label)*, **parsnips** *(off-label)*, **peppers** *(off-label)*, **potatoes** *(off-label)*, **pumpkins** *(off-label)*, **raspberries** *(off-label)*, **redcurrants** *(off-label)*, **ribes hybrids** *(off-label)*, **rubus hybrids** *(off-label)*, **salad onions** *(off-label)*, **shallots** *(off-label)*, **spinach** *(off-label)*, **squashes** *(off-label)*, **tomatoes (outdoor)** *(off-label)*, **whitecurrants** *(off-label)*
- White rot in **aubergines** *(off-label)*, **blackberries** *(off-label)*, **blackcurrants** *(off-label)*, **blueberries** *(off-label)*, **broccoli** *(off-label)*, **brussels sprouts** *(off-label)*, **bulb onions** *(off-label)*, **cabbages** *(off-label)*, **calabrese** *(off-label)*, **carrots** *(off-label)*, **cauliflowers** *(off-label)*, **celeriac** *(off-label)*, **celery (outdoor)** *(off-label)*, **chard** *(off-label)*, **collards** *(off-label)*, **courgettes** *(off-*

label), **cranberries** *(off-label)*, **cucumbers** *(off-label)*, **garlic** *(off-label)*, **gooseberries** *(off-label)*, **herbs (see appendix 6)** *(off-label)*, **kale** *(off-label)*, **leaf brassicas** *(off-label)*, **leeks** *(off-label)*, **lettuce** *(off-label)*, **marrows** *(off-label)*, **parsnips** *(off-label)*, **peppers** *(off-label)*, **potatoes** *(off-label)*, **pumpkins** *(off-label)*, **raspberries** *(off-label)*, **redcurrants** *(off-label)*, **ribes hybrids** *(off-label)*, **rubus hybrids** *(off-label)*, **salad onions** *(off-label)*, **shallots** *(off-label)*, **spinach** *(off-label)*, **squashes** *(off-label)*, **tomatoes (outdoor)** *(off-label)*, **whitecurrants** *(off-label)*

Extension of Authorisation for Minor Use (EAMUs)

- *amenity vegetation 20120500*
- *aubergines 20120663*
- *bilberries 20120475*
- *blackberries 20120663*
- *blackcurrants 20120475, 20120663*
- *blueberries 20120475, 20120663*
- *broccoli 20120663*
- *brussels sprouts 20120663*
- *bulb onions 20120663*
- *bulb vegetables 20120475*
- *cabbages 20120663*
- *calabrese 20120663*
- *canary grass 20120475*
- *cane fruit 20120475*
- *carrots 20120663*
- *cauliflowers 20120663*
- *celeriac 20120663*
- *celery (outdoor) 20120663*
- *chard 20120663*
- *collards 20120663*
- *courgettes 20120663*
- *cranberries 20120475, 20120663*
- *cucumbers 20120663*
- *figs 20120475*
- *forest nurseries 20120500*
- *fruiting vegetables 20120475*
- *garlic 20120663*
- *gooseberries 20120475, 20120663*
- *herbs (see appendix 6) 20120475, 20120663*
- *hops 20120475*
- *kale 20120663*
- *leaf brassicas 20120663*
- *leafy vegetables 20120475*
- *leeks 20120663*
- *legumes 20120475*
- *lettuce 20120663*
- *marrows 20120663*
- *ornamental plant production 20120475*
- *parsnips 20120663*
- *peppers 20120663*
- *potatoes 20120663*
- *pumpkins 20120663*
- *raspberries 20120663*
- *redcurrants 20120475, 20120663*
- *ribes hybrids 20120475, 20120663*
- *root & tuber crops 20120475*
- *rubus hybrids 20120663*
- *salad onions 20120663*
- *shallots 20120663*
- *spinach 20120663*
- *squashes 20120663*

FOR FULL CONDITIONS OF USE ALWAYS READ THE PRODUCT LABEL

- **stem vegetables** *20120475*
- **strawberries** *20120475*
- **table grapes** *20120475*
- **tomatoes (outdoor)** *20120663*
- **top fruit** *20120475*
- **vegetable brassicas** *20120475*
- **whitecurrants** *20120663*
- **wine grapes** *20120475*

Approval information
- Bacillus subtilis included in Annex I under EC Regulation 1107/2009

Efficacy guidance
- Alternating applications with fungicides using a different mode of action is recommended for resistance management.
- Do not apply using irrigation equipment.
- For maximum effectiveness, start applications before disease development.
- Apply in a minimum water volume of 400 l/ha.

Restrictions
- Consult processor before using on crops grown for processing

Hazard classification and safety precautions
Hazard Irritant
UN Number N/C
Risk phrases R43
Operator protection A, H; U05a, U11, U14, U20b
Environmental protection E15b
Storage and disposal D01, D02, D05, D10a, D16

25   Bacillus thuringiensis

A bacterial insecticide for control of caterpillars
IRAC mode of action code: 11

Products

Dipel DF	Interfarm	32000 IU/mg	WG	14119

Uses
- Caterpillars in **amenity vegetation**, **apples** *(off-label)*, **apricots** *(off-label)*, **blackberries** *(off-label)*, **borage** *(off-label)*, **broad beans** *(off-label)*, **broccoli**, **brussels sprouts**, **cabbages**, **calabrese** *(off-label)*, **cauliflowers**, **celery (outdoor)** *(off-label)*, **cherries** *(off-label)*, **chinese cabbage** *(off-label)*, **chives** *(off-label)*, **choi sum** *(off-label)*, **cob nuts** *(off-label)*, **collards** *(off-label)*, **crab apples** *(off-label)*, **dwarf beans** *(off-label)*, **filberts** *(off-label)*, **french beans** *(off-label)*, **frise** *(off-label)*, **hazel nuts** *(off-label)*, **herbs (see appendix 6)** *(off-label)*, **kale** *(off-label)*, **kiwi fruit** *(off-label)*, **leaf spinach** *(off-label)*, **lettuce** *(off-label)*, **olives** *(off-label)*, **ornamental plant production**, **pak choi** *(off-label)*, **parsley** *(off-label)*, **pears** *(off-label)*, **plums** *(off-label)*, **protected apricots** *(off-label)*, **protected aubergines** *(off-label)*, **protected blueberry** *(off-label)*, **protected broccoli** *(off-label)*, **protected brussels sprouts** *(off-label)*, **protected cabbages** *(off-label)*, **protected calabrese** *(off-label)*, **protected cauliflowers** *(off-label)*, **protected cayenne peppers** *(off-label)*, **protected celery** *(off-label)*, **protected chestnuts** *(off-label)*, **protected chinese cabbage** *(off-label)*, **protected chives** *(off-label)*, **protected choi sum** *(off-label)*, **protected collards** *(off-label)*, **protected cucumbers**, **protected frise** *(off-label)*, **protected hazelnuts** *(off-label)*, **protected herbs (see appendix 6)** *(off-label)*, **protected kale** *(off-label)*, **protected kiwi fruit** *(off-label)*, **protected lettuce** *(off-label)*, **protected olives** *(off-label)*, **protected ornamentals**, **protected parsley** *(off-label)*, **protected peppers**, **protected plums** *(off-label)*, **protected quince** *(off-label)*, **protected rhubarb** *(off-label)*, **protected salad brassicas** *(off-label)*, **protected scarole** *(off-label)*, **protected spinach** *(off-label)*, **protected tomatoes**, **protected walnuts** *(off-label)*, **protected watercress** *(off-label)*, **protected wine grapes** *(off-label)*, **quinces** *(off-label)*, **radicchio** *(off-label)*, **raspberries**, **rhubarb** *(off-label)*,

SEE SECTION 3 FOR PRODUCTS ALSO REGISTERED

rubus hybrids (off-label), **runner beans** (off-label), **salad brassicas** (off-label), **scarole** (off-label), **spinach** (off-label), **spring cabbage** (off-label), **strawberries**, **sweetcorn** (off-label), **walnuts** (off-label), **watercress** (off-label), **wine grapes** (off-label)

- Cutworms in **bulb onions** (off-label), **carrots** (off-label), **garlic** (off-label), **horseradish** (off-label), **leeks** (off-label), **parsley root** (off-label), **parsnips** (off-label), **red beet** (off-label), **salsify** (off-label), **shallots** (off-label)
- Silver Y moth in **broad beans** (off-label), **dwarf beans** (off-label), **french beans** (off-label), **runner beans** (off-label), **sweetcorn** (off-label)
- Winter moth in **bilberries** (off-label), **blackcurrants** (off-label), **blueberries** (off-label), **cranberries** (off-label), **gooseberries** (off-label), **redcurrants** (off-label), **vaccinium spp.** (off-label), **whitecurrants** (off-label)

Extension of Authorisation for Minor Use (EAMUs)

- **apples** *20091070*
- **apricots** *20102882*
- **bilberries** *20091069*
- **blackberries** *20091058*
- **blackcurrants** *20091069*
- **blueberries** *20091069*
- **borage** *20091070*
- **broad beans** *20091067*
- **bulb onions** *20091065*
- **calabrese** *20091064*
- **carrots** *20091068*
- **celery (outdoor)** *20091064*
- **cherries** *20091070*
- **chinese cabbage** *20091064*
- **chives** *20091070*
- **choi sum** *20091064*
- **cob nuts** *20102882*
- **collards** *20091064*
- **crab apples** *20091070*
- **cranberries** *20091069*
- **dwarf beans** *20091067*
- **filberts** *20102882*
- **french beans** *20091067*
- **frise** *20091070*
- **garlic** *20091065*
- **gooseberries** *20091069*
- **hazel nuts** *20102882*
- **herbs (see appendix 6)** *20091070*
- **horseradish** *20091068*
- **kale** *20091064*
- **kiwi fruit** *20102882*
- **leaf spinach** *20091070*
- **leeks** *20091062*
- **lettuce** *20091070*
- **olives** *20102882*
- **pak choi** *20091064*
- **parsley** *20091070*
- **parsley root** *20091068*
- **parsnips** *20091068*
- **pears** *20091070*
- **plums** *20102882*
- **protected apricots** *20102882*
- **protected aubergines** *20091059*
- **protected blueberry** *20102882*
- **protected broccoli** *20091064*
- **protected brussels sprouts** *20091064*

FOR FULL CONDITIONS OF USE ALWAYS READ THE PRODUCT LABEL

- *protected cabbages* 20091064
- *protected calabrese* 20091064
- *protected cauliflowers* 20091064
- *protected cayenne peppers* 20091060
- *protected celery* 20091064
- *protected chestnuts* 20102882
- *protected chinese cabbage* 20091064
- *protected chives* 20091070
- *protected choi sum* 20091064
- *protected collards* 20091064
- *protected frise* 20091070
- *protected hazelnuts* 20102882
- *protected herbs (see appendix 6)* 20091070
- *protected kale* 20091064
- *protected kiwi fruit* 20102882
- *protected lettuce* 20091070
- *protected olives* 20102882
- *protected parsley* 20091070
- *protected plums* 20102882
- *protected quince* 20102882
- *protected rhubarb* 20091064
- *protected salad brassicas* 20091070
- *protected scarole* 20091070
- *protected spinach* 20091070
- *protected walnuts* 20102882
- *protected watercress* 20091063
- *protected wine grapes* 20102882
- *quinces* 20102882
- *radicchio* 20091070
- *red beet* 20091061
- *redcurrants* 20091069
- *rhubarb* 20091064
- *rubus hybrids* 20091058
- *runner beans* 20091067
- *salad brassicas* 20091070
- *salsify* 20091068
- *scarole* 20091070
- *shallots* 20091065
- *spinach* 20091070
- *spring cabbage* 20091064
- *sweetcorn* 20091057
- *vaccinium spp.* 20091069
- *walnuts* 20102882
- *watercress* 20091063
- *whitecurrants* 20091069
- *wine grapes* 20102882

Approval information
- Bacillus thuringiensis included in Annex I under EC Regulation 1107/2009

Efficacy guidance
- Pest control achieved by ingestion by caterpillars of the treated plant vegetation. Caterpillars cease feeding and die in 1-3 d
- Apply as soon as larvae appear on crop and repeat every 7-10 d until the end of the hatching period
- Good coverage is essential, especially of undersides of leaves. Spray onto dry foliage and do not apply if rain expected within 6 h

Restrictions
- No restriction on number of treatments on edible crops

SEE SECTION 3 FOR PRODUCTS ALSO REGISTERED

- Apply spray mixture as soon as possible after preparation

Crop-specific information
- HI zero

Environmental safety
- Store out of direct sunlight

Hazard classification and safety precautions
 UN Number N/C
 Operator protection A, C, D; U05a, U15, U19a, U20c
 Environmental protection E15a
 Storage and disposal D01, D02, D05, D09a, D11a

26   Beauveria bassiana

It is an entomopathogenic fungus causing white muscardine disease. It can be used as a biological insecticide to control a number of pests such as termites, whitefly and some beetles

Products

Naturalis-L	Belchim	7.16% w/w	OD	14655

Uses
- Aphids in *protected edible crops*, *protected ornamentals*
- Beetles in *protected edible crops*, *protected ornamentals*
- Whitefly in *protected edible crops*, *protected ornamentals*

Approval information
- Beauveria bassiana included in Annex 1 under EC Regulation 1107/2009

Hazard classification and safety precautions
 UN Number N/C
 Operator protection A, D, H; U05a, U11, U14, U15, U16b, U19a, U20b
 Environmental protection E15b, E34
 Storage and disposal D01, D02, D05, D09a, D10a
 Medical advice M03, M04a

27   benalaxyl

A phenylamide (acylalanine) fungicide available only in mixtures
FRAC mode of action code: 4

28   bentazone

A post-emergence contact benzothiadiazinone herbicide
HRAC mode of action code: C3

Products

1	Basagran SG	BASF	87% w/w	SG	08360
2	Benta 480 SL	Sharda	480 g/l	SL	14940
3	Bentazone 480	Goldengrass	480 g/l	SL	14423
4	Clayton Dent	Clayton	480 g/l	SL	15546
5	Mac-Bentazone 480 SL	AgChem Access	480 g/l	SL	13598
6	Tanaru	Agroquimicos	480 g/l	SL	15726
7	Troy 480	AgriChem BV	480 g/l	SL	12341
8	UPL B Zone	United Phosphorus	480 g/l	SL	14808

Uses
- Annual dicotyledons in *broad beans*, *linseed*, *narcissi*, *potatoes*, *runner beans*, *spring field beans*, *winter field beans* [1-8]; *bulb onions* (off-label), *chives* (off-label), *garlic* (off-label),

FOR FULL CONDITIONS OF USE ALWAYS READ THE PRODUCT LABEL

leeks (off-label), *salad onions* (off-label), *shallots* (off-label), *soya beans* (off-label) [1]; *combining peas, dwarf beans, vining peas* [6, 7]; *french beans* [1-5, 8]; *navy beans* [1, 3, 5-8]; *peas* [1-5]

- Chickweed in *game cover* (off-label), *hops* (off-label), *ornamental plant production* (off-label) [1]
- Cleavers in *game cover* (off-label), *hops* (off-label), *ornamental plant production* (off-label) [1]
- Common storksbill in *leeks* (off-label) [1]
- Fool's parsley in *leeks* (off-label) [1]
- Groundsel in *game cover* (off-label), *hops* (off-label), *ornamental plant production* (off-label) [1]
- Mayweeds in *bulb onions* (off-label), *game cover* (off-label), *garlic* (off-label), *hops* (off-label), *ornamental plant production* (off-label), *shallots* (off-label) [1]

Extension of Authorisation for Minor Use (EAMUs)
- *bulb onions* 20061631 [1]
- *chives* 20102994 [1]
- *game cover* 20082819 [1]
- *garlic* 20061631 [1]
- *hops* 20082819 [1]
- *leeks* 20061630 [1]
- *ornamental plant production* 20082819 [1]
- *salad onions* 20102994 [1]
- *shallots* 20061631 [1]
- *soya beans* 20061629 [1]

Approval information
- Bentazone included in Annex I under EC Regulation 1107/2009

Efficacy guidance
- Most effective control obtained when weeds are growing actively and less than 5 cm high or across. Good spray cover is essential
- The addition of specified adjuvant oils is recommended for use on some crops to improve fat hen control. Do not use under hot or humid conditions. See label for details
- Split dose application may be made in all recommended crops except peas and generally gives better weed control. See label for details

Restrictions
- Maximum number of treatments normally 2 per crop but check label
- Crops must be treated at correct stage of growth to avoid danger of scorch. See label for details
- Not all varieties of recommended crops are fully tolerant. Use only on tolerant varieties named in label. Do not use on forage or mange-tout varieties of peas
- Do not use on crops which have been affected by drought, waterlogging, frost or other stress conditions
- Do not apply insecticides within 7 d of treatment
- Leave 14 d after using a post-emergence grass herbicide and carry out a leaf wax test where relevant or 7 d where treatment precedes the grass herbicide
- Do not spray at temperatures above 21°C. Delay spraying until evening if necessary
- Do not apply if rain or frost expected, if unseasonably cold, if foliage wet or in drought
- A minimum of 6 h (preferably 12 h) free from rain is required after application
- May be used on selected varieties of maincrop and second early potatoes (see label for details), not on seed crops or first earlies

Crop-specific information
- Latest use: before shoots exceed 15 cm high for potatoes and spring field beans (or 6-7 leaf pairs); 4 leaf pairs (6 pairs or 15 cm high with split dose) for broad beans; before flower buds visible for French, navy, runner and winter field beans, and linseed; before flower buds can be found enclosed in terminal shoot for peas
- Best results in narcissi obtained by using a suitable pre-emergence herbicide first
- Consult processor before using on crops for processing
- A satisfactory wax test must be carried out before use on peas
- Do not treat narcissi during flower bud initiation

SEE SECTION 3 FOR PRODUCTS ALSO REGISTERED

Environmental safety
- Dangerous for the environment
- Harmful to aquatic organisms
- Some pesticides pose a greater threat of contamination of water than others and bentazone is one of these pesticides. Take special care when applying bentazone near water and do not apply if heavy rain is forecast

Hazard classification and safety precautions
    Hazard Harmful [1-5, 8]; Irritant [2, 4, 6, 7]
    Packaging group III [1]
    UN Number 2588 [1]; N/C [2-8]
    Risk phrases R22a [1-5, 8]; R36 [6, 7]; R41 [1, 5]; R43 [1-8]; R52, R53a [1, 3, 5, 8]
    Operator protection A [1-8]; C [1, 3, 5-7]; H [6-8]; U05a, U08, U19a [1-8]; U11, U20b [1, 3, 5-8]; U14 [1, 3, 5, 8]; U20a [2, 4]
    Environmental protection E15a [2, 4]; E15b, E34 [1, 3, 5-8]
    Storage and disposal D01, D02, D09a [1-8]; D10b [1, 3, 5, 8]; D10c [2, 4, 6, 7]
    Medical advice M03 [1, 3, 5-8]; M05a [1, 3, 5, 8]

29   benthiavalicarb-isopropyl

An amino acid amide carbamate fungicide available only in mixtures
FRAC mode of action code: 40

30   benthiavalicarb-isopropyl + mancozeb

A fungicide mixture for potatoes
FRAC mode of action code: 40 + M3

See also mancozeb

Products

1	Hobby	ChemSource	1.75:70% w/w	WG	15021
2	Valbon	Certis	1.75:70% w/w	WG	14868

Uses
- Blight in **potatoes** [1, 2]
- Disease control in **forest nurseries** *(off-label)*, **hops** *(off-label)*, **soft fruit** *(off-label)* [2]
- Downy mildew in **amenity vegetation** *(off-label)*, **bulb onions** *(off-label)*, **garlic** *(off-label)*, **ornamental plant production** *(off-label)*, **shallots** *(off-label)*, **table grapes** *(off-label)*, **wine grapes** *(off-label)* [2]

Extension of Authorisation for Minor Use (EAMUs)
- **amenity vegetation** *20101513* [2]
- **bulb onions** *20101518* [2]
- **forest nurseries** *20111123* [2]
- **garlic** *20101518* [2]
- **hops** *20111123* [2]
- **ornamental plant production** *20101513* [2]
- **shallots** *20101518* [2]
- **soft fruit** *20111123* [2]
- **table grapes** *20101973* [2]
- **wine grapes** *20101973* [2]

Approval information
- Benthiavalicarb-isopropyl and mancozeb included in Annex I under EC Regulation 1107/2009

Efficacy guidance
- Apply as a protectant spray commencing before blight enters the crop irrespective of growth stage following a blight warning or where there is a local source of infection

FOR FULL CONDITIONS OF USE ALWAYS READ THE PRODUCT LABEL

- In the absence of a warning commence the spray programme before the foliage meets in the rows
- Repeat treatment every 7-10 d depending on disease pressure
- Increase water volume as necessary to ensure thorough coverage of the plant
- Use of air assisted sprayers or drop legs may help to improve coverage

Restrictions
- Maximum number of treatments 6 per crop
- Do not use more than three consecutive treatments. Include products with a different mode of action in the programme

Crop-specific information
- HI 7 d for potatoes

Environmental safety
- Very toxic to aquatic organisms
- Risk to non-target insects or other arthropods. Avoid spraying within 6 m of field boundary
- LERAP Category B

Hazard classification and safety precautions
Hazard Irritant
Transport code 9
Packaging group III
UN Number 3077
Risk phrases R40, R43, R50, R53a
Operator protection A, H [1, 2]; D [2]; U14
Environmental protection E15a, E16a, E22c
Storage and disposal D01, D02, D05, D09a, D11a, D12b

31   benzoic acid

An organic horticultural disinfectant

Products

Menno Florades	Fargro	90 g/kg	SL	13985

Uses
- Bacteria in *all protected non-edible crops*
- Fungal spores in *all protected non-edible crops*
- Viroids in *all protected non-edible crops*
- Virus in *all protected non-edible crops*

Approval information
- Benzoic acid is included in Annex 1 under EC Regulation 1107/2009

Hazard classification and safety precautions
Hazard Irritant, Flammable
Transport code 3
Packaging group III
UN Number 1987
Risk phrases R41, R67
Operator protection A, C, H; U05a, U11, U13, U14, U15
Environmental protection E15a
Storage and disposal D01, D02, D09a, D10a
Medical advice M05a

SEE SECTION 3 FOR PRODUCTS ALSO REGISTERED

32   benzyladenine

A cytokinin plant growth regulator for use in apples and pears

Products

Exilis	Fine	20 g/l	SL	15706

Uses
* Fruit thinning in *apples*, *pears*

Approval information
* Benzyladenine is included in Annex 1 under EC Directive 1107/2009

Efficacy guidance
* A post-bloom thinner for apples & pears with excessive blossom.
* Apply when the temperature wil exceed 15 deg C on day of application but note that temperatures above 28 deg C may result in excessive thinning.

Restrictions
* Use in pears is a Qualified Minor Use reccomendation. Evidence of efficacy and crop safety in pears is limited to the pear variety 'Conference' .

Hazard classification and safety precautions
   UN Number N/C
   Risk phrases R52, R53a
   Operator protection A; U02a, U05a, U09a, U19a, U20b
   Environmental protection E15b, E38
   Storage and disposal D02, D03, D05, D09a, D10b, D12a
   Medical advice M05a

33   beta-cyfluthrin

A non-systemic pyrethroid insecticide for insect control in cereals, oilseed rape, cabbages, cauliflowers and sugar beet
IRAC mode of action code: 3

Products

Gandalf	Makhteshim	25 g/l	EC	12865

Uses
* Aphids in *oats*, *rye*, *spring barley*, *spring wheat*, *sugar beet*, *triticale*, *winter barley*, *winter wheat*
* Brassica pod midge in *oilseed rape*
* Cabbage moth in *cabbages*, *cauliflowers*
* Cabbage seed weevil in *oilseed rape*
* Cabbage stem flea beetle in *oilseed rape*
* Cabbage stem weevil in *oilseed rape*
* Peach-potato aphid in *cabbages*, *cauliflowers*
* Wheat-blossom midge in *rye*, *spring wheat*, *triticale*, *winter wheat*

Approval information
* Beta-cyfluthrin included in Annex I under EC Regulation 1107/2009
* Approval expiry 31 Dec 2013 [1]

Hazard classification and safety precautions
   Hazard Harmful, Dangerous for the environment
   Transport code 9
   Packaging group III
   UN Number 3082
   Risk phrases R20, R22a, R22b, R36, R38, R50, R53a, R66, R67
   Operator protection A, C, H; U04a, U05a, U10, U12, U14, U19a, U20a

FOR FULL CONDITIONS OF USE ALWAYS READ THE PRODUCT LABEL

Environmental protection E16e, E34, E38
Storage and disposal D01, D02, D09a, D10a, D12a
Medical advice M03, M05b

34 beta-cyfluthrin + clothianidin

An insecticidal seed treatment mixture for beet
IRAC mode of action code: 3 + 4A

See also clothianidin

Products
1	Modesto	Bayer CropScience	80:400 g/l	FS	14029
2	Poncho Beta	Bayer CropScience	53:400 g/l	FS	12076

Uses
- Aphids in **winter oilseed rape** [1]
- Beet virus yellows vectors in **fodder beet** *(seed treatment)*, **sugar beet** *(seed treatment)* [2]
- Cabbage root fly in **winter oilseed rape** *(reduction of feeding activity only)* [1]
- Cabbage stem flea beetle in **winter oilseed rape** [1]
- Flea beetle in **poppies for morphine production** *(off-label)* [1]
- Turnip sawfly larvae in **winter oilseed rape** [1]

Extension of Authorisation for Minor Use (EAMUs)
- *poppies for morphine production* 20090479 expires 31 Dec 2013 [1]

Approval information
- Beta-cyfluthrin and clothianidin included in Annex I under EC Regulation 1107/2009
- Approval expiry 31 Dec 2013 [1, 2]

Efficacy guidance
- In addition to control of aphid virus vectors, product improves crop establishment by reducing damage caused by symphylids, springtails, millipedes, wireworms and leatherjackets
- Additional control measures should be taken where very high populations of soil pests are present
- Product reduces direct feeding damage by foliar pests such as pygmy mangold beetle, beet flea beetle, mangold fly
- Product is not active against nematodes
- Treatment does not alter physical characteristics of pelleted seed and no change to standard drill settings should be necessary

Restrictions
- Maximum number of treatments: 1 per seed batch
- Product must be co-applied with a colouring dye
- Product must only be applied as a coating to pelleted seed using special treatment machinery

Crop-specific information
- Latest use: pre-drilling

Environmental safety
- Extremely dangerous to fish or other aquatic life. Do not contaminate surface waters or ditches with chemical or used container

Hazard classification and safety precautions
Hazard Harmful [1, 2]; Dangerous for the environment [1]
Transport code 9
Packaging group III
UN Number 3082
Risk phrases R22a [1, 2]; R50, R53a [1]
Operator protection A, H; U04a, U05a, U07, U13, U14 [1, 2]; U20b, U24 [2]; U20c [1]
Environmental protection E03, E15a, E38 [1]; E13a [2]; E34, E36a [1, 2]
Storage and disposal D01, D02, D09a, D14 [1, 2]; D05 [2]; D12a [1]

SEE SECTION 3 FOR PRODUCTS ALSO REGISTERED

Treated seed S02, S04b, S07 [1, 2]; S03, S04a, S05, S06a, S06b [2]
Medical advice M03

35   beta-cyfluthrin + imidacloprid

An insecticide mixture for seed treatment of oilseed rape
IRAC mode of action code: 3 + 4A

See also imidacloprid

Products

1	Chinook Blue	Bayer CropScience	100:100 g/l	LS	11262
2	Chinook Colourless	Bayer CropScience	100:100 g/l	LS	11206

Uses

- Cabbage stem flea beetle in **evening primrose** *(off-label)*, **honesty** *(off-label)*, **linseed** *(off-label)*, **mustard** *(off-label)* [2]; **evening primrose** *(off-label - seed treatment)*, **honesty** *(off-label - seed treatment)*, **linseed** *(off-label - seed treatment)*, **mustard** *(off-label - seed treatment)*, **winter oilseed rape** *(seed treatment)* [1, 2]; **spring oilseed rape** *(seed treatment)* [1]
- Flea beetle in **evening primrose** *(off-label)*, **honesty** *(off-label)*, **linseed** *(off-label)*, **mustard** *(off-label)* [2]; **evening primrose** *(off-label - seed treatment)*, **honesty** *(off-label - seed treatment)*, **linseed** *(off-label - seed treatment)*, **mustard** *(off-label - seed treatment)* [1, 2]; **spring oilseed rape** *(seed treatment)*, **winter oilseed rape** *(seed treatment)* [1]

Extension of Authorisation for Minor Use (EAMUs)

- **evening primrose** *(seed treatment)* 20052262 [1], *(seed treatment)* 20052263 [2], 20113181 [2]
- **honesty** *(seed treatment)* 20052262 [1], *(seed treatment)* 20052263 [2], 20113181 [2]
- **linseed** *(seed treatment)* 20052262 [1], *(seed treatment)* 20052263 [2], 20113181 [2]
- **mustard** *(seed treatment)* 20052262 [1], *(seed treatment)* 20052263 [2], 20113181 [2]

Approval information

- Beta-cyfluthrin and imidacloprid included in Annex I under EC Regulation 1107/2009

Efficacy guidance

- Drill treated seed as soon as possible after treatment
- Product will reduce damage by early attacks of flea beetle and may also reduce subsequent larval damage. However a follow-up spray treatment may be required if pest activity is heavy and prolonged

Restrictions

- Maximum number of treatments 1 per batch of seed
- Do not use on seed with more than 9% moisture content, on sprouted seed or on cracked, split or otherwise damaged seed
- Product must be co-applied with a colouring agent [2]
- Must be applied using seed treatment machine recommended by manufacturer
- Allow treated seed to dry before packaging. Drying can be assisted by subsequent application of talc
- Product can cause a transient tingling or numbing sensation to exposed skin. Avoid skin contact with the product, treated seed or dust throughout all operations in the treatment plant and during drilling

Crop-specific information

- Latest use: before drilling

Environmental safety

- Dangerous for the environment
- Toxic to aquatic organisms
- Extremely dangerous to fish or other aquatic life. Do not contaminate surface waters or ditches with chemical or used container
- Do not use treated seed as food or feed

FOR FULL CONDITIONS OF USE ALWAYS READ THE PRODUCT LABEL

- Dangerous to birds, game and other wildlife. Treated seed should not be left on the soil surface. Bury spillages
- Do not broadcast treated seed

Hazard classification and safety precautions

Hazard Harmful, Dangerous for the environment
Transport code 6.1
Packaging group III
UN Number 3352
Risk phrases R22a, R51, R53a
Operator protection A, H; U04a, U05a, U07, U20b
Environmental protection E13a, E34, E36a, E38
Storage and disposal D01, D02, D05, D09a, D12a, D14
Treated seed S01, S02, S03, S04c, S05, S06a, S06b, S07, S08
Medical advice M03

36   bifenazate

A bifenazate acaricide for use on protected strawberries
IRAC mode of action code: 25

Products

Floramite 240 SC	Certis	240 g/l	SC	13686

Uses

- Spider mites in **protected forest nurseries** *(off-label)*, **protected hops** *(off-label)*
- Two-spotted spider mite in **protected strawberries**

Extension of Authorisation for Minor Use (EAMUs)

- **protected forest nurseries** *20082851*
- **protected hops** *20082851*

Approval information

- Bifenazate included in Annex I under EC Regulation 1107/2009

Efficacy guidance

- Bifenazate acts by contact knockdown after approximately 4 d and a subsequent period of residual control
- All mobile stages of mites are controlled with occasionally some ovicidal activity
- Best results obtained from a programme of two treatments started as soon as first spider mites are seen
- To minimise possible development of resistance use in a planned Resistance Management strategy. Alternate at least two products with different modes of action between treatment programmes of bifenazate. See Section 5 for more information

Restrictions

- Maximum number of treatments 2 per yr for protected strawberries
- After use in protected environments avoid re-entry for at least 2 h while full ventilation is carried out allowing spray to dry on leaves
- Personnel working among treated crops should wear suitable long-sleeved garments and gloves for 14 d after treatment
- Do not apply as a low volume spray

Crop-specific information

- HI: 7 d for protected strawberries
- Carry out small scale tolerance test on the strawberry cultivar before large scale use

Environmental safety

- Dangerous for the environment
- Toxic to aquatic organisms

SEE SECTION 3 FOR PRODUCTS ALSO REGISTERED

SECTION 2

- Harmful to non-target predatory mites. Avoid spray drift onto field margins, hedges, ditches, surface water and neighbouring crops

Hazard classification and safety precautions
Hazard Irritant, Dangerous for the environment
Transport code 9
Packaging group III
UN Number 3082
Risk phrases R43, R51, R53a
Operator protection A, H; U02a, U04a, U05a, U10, U14, U15, U19a, U20b
Environmental protection E15b, E38
Storage and disposal D01, D02, D05, D09a, D10b, D12b
Medical advice M03

37 bifenox

A diphenyl ether herbicide for use in cereals and (off-label) in oilseed rape.
HRAC mode of action code: E

Products

1	Clayton Belstone	Clayton	480 g/l	SC	14033
2	Fox	Makhteshim	480 g/l	SC	11981

Uses
- Annual dicotyledons in *canary flower (echium spp.)* (off-label), *triticale, winter barley, winter rye, winter wheat* [1]; *durum wheat* (off-label), *grass seed crops* (off-label) [1, 2]
- Cleavers in *triticale, winter barley, winter rye, winter wheat* [2]
- Field pansy in *triticale, winter barley, winter rye, winter wheat* [2]
- Field speedwell in *triticale, winter barley, winter rye, winter wheat* [2]
- Forget-me-not in *triticale, winter barley, winter rye, winter wheat* [2]
- Geranium species in *spring oilseed rape* (off-label), *winter oilseed rape* (off-label) [2]
- Ivy-leaved speedwell in *triticale, winter barley, winter rye, winter wheat* [2]
- Poppies in *triticale, winter barley, winter rye, winter wheat* [2]
- Red dead-nettle in *triticale, winter barley, winter rye, winter wheat* [2]

Extension of Authorisation for Minor Use (EAMUs)
- *canary flower (echium spp.)* 20121422 [1]
- *durum wheat* 20121421 [1], 20072364 [2]
- *grass seed crops* 20121421 [1], 20072364 [2]
- *spring oilseed rape* 20060321 [2]
- *winter oilseed rape* 20060321 [2]

Approval information
- Accepted by BBPA for use on malting barley
- Bifenox included in Annex I under EC Regulation 1107/2009

Efficacy guidance
- Bifenox is absorbed by foliage and emerging roots of susceptible species
- Best results obtained when weeds are growing actively with adequate soil moisture

Restrictions
- Maximum number of treatments: one per crop
- Do not apply to crops suffering from stress from whatever cause
- Do not apply if the crop is wet or if rain or frost is expected
- Avoid drift onto broad-leaved plants outside the target area

Crop-specific information
- Latest use: before 2nd node detectable (GS 32) for all crops

Environmental safety
- Dangerous for the environment

FOR FULL CONDITIONS OF USE ALWAYS READ THE PRODUCT LABEL

- Very toxic to aquatic organisms
- Do not empty into drains

Hazard classification and safety precautions
  Hazard Dangerous for the environment
  Transport code 9
  Packaging group III
  UN Number 3082
  Risk phrases R50, R53a
  Operator protection A; U05a, U08, U20b
  Environmental protection E15a, E19b, E34, E38
  Storage and disposal D01, D02, D09a, D12a
  Medical advice M03

38   bifenthrin

A contact and residual pyrethroid acaricide/insecticide for use in agricultural and horticultural crops but all approvals expired in 2011
IRAC mode of action code: 3

39   bixafen

A succinate dehydrogenase inhibitor (SDHI) fungicide available only in mixtures
FRAC mode of action code: 7

40   bixafen + fluoxystrobin + prothioconazole

An SDHI + strobilurin + triazole mixture for disease control in cereals
FRAC mode of action code: 7 + 11 + 3

Products
  Variano Xpro   Bayer CropScience   40:50:100 g/l   EC   15218

Uses
- Brown rust in *spring wheat, triticale, winter rye, winter wheat*
- Ear diseases in *spring wheat, triticale, winter rye, winter wheat*
- Glume blotch in *spring wheat, triticale, winter rye, winter wheat*
- Leaf blotch in *triticale, winter rye*
- Powdery mildew in *spring wheat, triticale, winter rye, winter wheat*
- Rhynchosporium in *winter rye*
- Tan spot in *spring wheat, triticale, winter rye, winter wheat*
- Yellow rust in *spring wheat, triticale, winter rye, winter wheat*

Approval information
- Prothioconazole is included in Annex 1 under EC Regulation 1107/2009 but bixafen and fluroxystrobin have not yet been included.

Environmental safety
- LERAP Category B

Hazard classification and safety precautions
  Hazard Irritant, Dangerous for the environment
  Transport code 9
  Packaging group III
  UN Number 3082
  Risk phrases R43, R51, R53a
  Operator protection A, H; U05b, U08, U14, U20b
  Environmental protection E15a, E16a, E16b, E34, E38

SEE SECTION 3 FOR PRODUCTS ALSO REGISTERED

Storage and disposal D01, D02, D05, D09a, D10b, D12a
Medical advice M03

41   bixafen + prothioconazole

Succinate dehydrogenase inhibitor (SDHI) + triazole fungicide mixture for disease control in cereals
FRAC mode of action code: 3 + 7

Products

1	Aviator 235 Xpro	Bayer CropScience	75:160 g/l	EC	15026
2	Siltra Xpro	Bayer CropScience	60:200 g/l	EC	15082

Uses

*   Brown rust in *spring barley, winter barley* [2]; *spring wheat, triticale, winter rye, winter wheat* [1]
*   Crown rust in *spring oats, winter oats* [2]
*   Ear diseases in *spring barley, winter barley* [2]; *spring wheat, triticale, winter wheat* [1]
*   Eyespot in *spring barley* (reduction of incidence and severity), *spring oats* (reduction of incidence and severity), *winter barley* (reduction of incidence and severity), *winter oats* (reduction of incidence and severity) [2]; *spring wheat* (reduction of incidence and severity), *triticale* (reduction of incidence and severity), *winter wheat* (reduction of incidence and severity) [1]
*   Glume blotch in *spring wheat, triticale, winter wheat* [1]
*   Net blotch in *spring barley, winter barley* [2]
*   Powdery mildew in *spring barley, spring oats, winter barley, winter oats* [2]; *spring wheat, triticale, winter rye, winter wheat* [1]
*   Ramularia leaf spots in *spring barley, winter barley* [2]
*   Rhynchosporium in *spring barley, winter barley* [2]; *winter rye* [1]
*   Septoria leaf blotch in *spring wheat, triticale, winter wheat* [1]
*   Tan spot in *spring wheat, triticale, winter wheat* [1]
*   Yellow rust in *spring barley, winter barley* [2]; *spring wheat, triticale, winter wheat* [1]

Approval information

*   Prothioconazole included in Annex I under EC Regulation 1107/2009 but bixafen not yet included
*   Accepted by BBPA for use on malting barley

Efficacy guidance

*   Resistance to some DMI fungicides has been identified in Septoria leaf blotch (*Mycosphaerella graminicola*) which may seriously affect the performance of some products. For further advice on resistance management in DMIs contact your agronomist or specialist advisor, and visit the FRAG-UK website.

Environmental safety

*   LERAP Category B

Hazard classification and safety precautions

Hazard Harmful, Dangerous for the environment
Transport code 9
Packaging group III
UN Number 3082
Risk phrases R22a, R43, R51, R53a, R63
Operator protection U05a, U09a, U20b
Environmental protection E15b, E16a, E16b, E34, E38
Storage and disposal D01, D02, D09a, D10b, D12a

42   bixafen + prothioconazole + spiroxamine

A broad-spectrum fungicide mixture for disease control in cereals
FRAC mode of action code: 3 + 5 + 7

Products

Boogie Xpro	Bayer CropScience	50:100:250 g/l	EC	15061

Uses

- Brown rust in *spring wheat*, *triticale*, *winter rye*, *winter wheat*
- Ear diseases in *spring wheat*, *triticale*, *winter rye*, *winter wheat*
- Eyespot in *spring wheat*, *triticale*, *winter rye*, *winter wheat*
- Glume blotch in *spring wheat*, *triticale*, *winter rye*, *winter wheat*
- Powdery mildew in *spring wheat*, *triticale*, *winter rye*, *winter wheat*
- Rhynchosporium in *triticale*, *winter rye*
- Septoria leaf blotch in *spring wheat*, *triticale*, *winter rye*, *winter wheat*
- Tan spot in *spring wheat*, *triticale*, *winter rye*, *winter wheat*
- Yellow rust in *spring wheat*, *triticale*, *winter rye*, *winter wheat*

Approval information

- Prothioconazole and spiroxamine included in Annex I under EC Regulation 1107/2009 but bixafen not yet included

Efficacy guidance

- Resistance to some DMI fungicides has been identified in Septoria leaf blotch (*Mycosphaerella graminicola*) which may seriously affect the performance of some products. For further advice on resistance management in DMIs contact your agronomist or specialist advisor, and visit the FRAG-UK website.

Restrictions

- No more than two applications of SDH inhibitors must be applied to the same cereal crop.

Environmental safety

- LERAP Category B

Hazard classification and safety precautions

    Hazard Harmful, Dangerous for the environment
    Risk phrases R20, R22a, R41, R50, R53a
    Operator protection A, C, H; U05a, U09b, U11, U20a, U26
    Environmental protection E15b, E16a, E16b, E34, E38
    Storage and disposal D01, D02, D05, D09a, D10b, D12a
    Medical advice M03

43   bixafen + prothioconazole + tebuconazole

A succinate dehydrogenase inhibitor (SDHI) + triazole fungicide mixture for disease control in cereals
FRAC mode of action code: 3 + 7

Products

1	Skyway 285 Xpro	Bayer CropScience	75:110:100 g/l	EC	15028
2	Sparticus Xpro	Bayer CropScience	75:110:90 g/l	EC	15162

Uses

- Brown rust in *spring wheat*, *triticale*, *winter rye*, *winter wheat*
- Ear diseases in *spring wheat*, *triticale*, *winter wheat*
- Eyespot in *spring wheat* (reduction of incidence and severity), *triticale* (reduction of incidence and severity), *winter wheat* (reduction of incidence and severity)
- Glume blotch in *spring wheat*, *triticale*, *winter wheat*
- Light leaf spot in *winter oilseed rape*
- Phoma in *winter oilseed rape*

SEE SECTION 3 FOR PRODUCTS ALSO REGISTERED

- Powdery mildew in *spring wheat*, *triticale*, *winter rye*, *winter wheat*
- Rhynchosporium in *winter rye*
- Sclerotinia in *winter oilseed rape*
- Septoria leaf blotch in *spring wheat*, *triticale*, *winter wheat*
- Stem canker in *winter oilseed rape*
- Tan spot in *spring wheat*, *triticale*, *winter wheat*
- Yellow rust in *spring wheat*, *triticale*, *winter wheat*

Approval information
- Prothioconazole and tebuconazole included in Annex I under EC Regulation 1107/2009 but bixafen not yet included

Efficacy guidance
- Resistance to some DMI fungicides has been identified in Septoria leaf blotch (*Mycosphaerella graminicola*) which may seriously affect the performance of some products. For further advice on resistance management in DMIs contact your agronomist or specialist advisor, and visit the FRAG-UK website.

Environmental safety
- LERAP Category B

Hazard classification and safety precautions
Hazard Harmful, Dangerous for the environment
Transport code 9
Packaging group III
UN Number 3082
Risk phrases R22a, R43, R51, R53a, R63
Operator protection A, H; U05a, U09a, U20b
Environmental protection E15b, E16a, E16b, E34, E38
Storage and disposal D01, D02, D09a, D10b, D12a

44   boscalid

A translocated and translaminar anilide fungicide
FRAC mode of action code: 7

Products

Filan	BASF	50% w/w	WG	11449

Uses
- Alternaria in *spring oilseed rape*, *winter oilseed rape*
- Botrytis in *poppies for morphine production* (off-label)
- Disease control in *all edible seed crops grown outdoors* (off-label), *all non-edible seed crops grown outdoors* (off-label), *borage for oilseed production* (off-label), *canary flower (echium spp.)* (off-label), *evening primrose* (off-label), *honesty* (off-label), *linseed* (off-label), *mustard* (off-label)
- Sclerotinia in *poppies for morphine production* (off-label)
- Sclerotinia stem rot in *spring oilseed rape*, *winter oilseed rape*

Extension of Authorisation for Minor Use (EAMUs)
- *all edible seed crops grown outdoors* 20093231
- *all non-edible seed crops grown outdoors* 20093231
- *borage for oilseed production* 20093229
- *canary flower (echium spp.)* 20093229
- *evening primrose* 20093229
- *honesty* 20093229
- *linseed* 20093229
- *mustard* 20093229
- *poppies for morphine production* 20093222

Approval information
- Boscalid included in Annex I under EC Regulation 1107/2009
- Accepted by BBPA on malting barley before ear emergence only

Efficacy guidance
- For best results apply as a protectant spray before symptoms are visible
- Applications against *Sclerotinia* should be made in high disease risk situations at early to full flower
- *Sclerotinia* control may be reduced when high risk conditions occur after flowering, leading to secondary disease spread
- Applications against *Alternaria* should be made at full flowering. Later applications may result in reduced levels of control
- Ensure adequate spray penetration and good coverage

Restrictions
- Maximum total dose equivalent to two full dose treatments
- Do no treat crops to be used for seed production later than full flowering stage
- Avoid spray drift onto neighbouring crops

Crop-specific information
- Latest use: up to and including when 50% of pods have reached final size (GS 75) for oilseed rape

Environmental safety
- Dangerous for the environment
- Toxic to aquatic organisms
- Product represents minimal hazard to bees when used as directed. However local bee-keepers should be notified if crops are to be sprayed when in flower

Hazard classification and safety precautions
Hazard Dangerous for the environment
Transport code 9
Packaging group III
UN Number 3077
Risk phrases R51, R53a
Operator protection U05a, U20c
Environmental protection E15a, E38
Storage and disposal D01, D02, D05, D09a, D10c, D12a

45   boscalid + epoxiconazole

A broad spectrum fungicide mixture for cereals
FRAC mode of action code: 7 + 3

See also epoxiconazole

Products
1	Chord	BASF	210:75 g/l	SC	13928
2	Deuce	BASF	233:67 g/l	SC	13240
3	Enterprise	BASF	140:50 g/l	OD	15228
4	Injun	Goldengrass	233:67 g/l	SC	14305
5	Kingdom	BASF	140:50 g/l	OD	15240
6	Tracker	BASF	233:67 g/l	SC	12295
7	Whistle	BASF	210:75 g/l	SC	14757

Uses
- Brown rust in *rye*, *triticale* [3, 5]; *spring barley*, *spring wheat*, *winter barley*, *winter wheat* [1-7]
- Crown rust in *oats* [3, 5]; *spring oats*, *winter oats* [1, 2, 4, 6, 7]
- Eyespot in *rye* (moderate control), *triticale* (moderate control) [3, 5]; *spring barley* (moderate), *spring wheat* (moderate), *winter barley* (moderate), *winter wheat* (moderate) [2, 4, 6]; *spring*

barley *(moderate control)*, **spring wheat** *(moderate control)*, **winter barley** *(moderate control)*, **winter wheat** *(moderate control)* [1, 3, 5, 7]; **spring oats** *(moderate control)*, **winter oats** *(moderate control)* [1, 7]

- Foliar disease control in **durum wheat** *(off-label)* [6]
- Fusarium ear blight in **spring wheat** *(good reduction)*, **winter wheat** *(good reduction)* [1, 3, 5, 7]
- Glume blotch in **spring wheat**, **winter wheat** [1, 3, 5, 7]; **triticale** [3, 5]
- Net blotch in **spring barley**, **winter barley** [1-7]
- Powdery mildew in **oats** *(moderate control)*, **rye** *(moderate control)*, **spring barley** *(moderate control)*, **winter barley** *(moderate control)* [3, 5]; **spring barley**, **spring oats**, **spring wheat**, **winter barley**, **winter oats**, **winter wheat** [1, 7]
- Ramularia leaf spots in **spring barley** *(moderate control)*, **winter barley** *(moderate control)* [3, 5]
- Rhynchosporium in **rye** *(moderate control)* [3, 5]; **spring barley** *(moderate)*, **winter barley** *(moderate)* [2, 4, 6]; **spring barley** *(moderate control)*, **winter barley** *(moderate control)* [1, 3, 5, 7]
- Septoria leaf blotch in **spring wheat**, **winter wheat** [1-7]; **triticale** [3, 5]
- Sooty moulds in **spring wheat**, **winter wheat** [1, 3, 5, 7]
- Tan spot in **spring wheat** *(reduction)*, **winter wheat** *(reduction)* [3, 5]; **spring wheat** *(reduction only)*, **winter wheat** *(reduction only)* [1, 7]
- Yellow rust in **rye**, **triticale** [3, 5]; **spring barley**, **spring wheat**, **winter barley**, **winter wheat** [1, 3, 5, 7]

Extension of Authorisation for Minor Use (EAMUs)
- **durum wheat** *20070462* [6]

Approval information
- Boscalid and epoxiconazole included in Annex I under EC Regulation 1107/2009
- Accepted by BBPA for use on malting barley (before ear emergence only)

Efficacy guidance
- Apply at the start of foliar or stem based disease attack
- Optimum effect against eyespot achieved by spraying between leaf-sheath erect and second node detectable stages (GS 30-32)
- Epoxiconazole is a DMI fungicide. Resistance to some DMI fungicides has been identified in Septoria leaf blotch which may seriously affect performance of some products. For further advice contact a specialist advisor and visit the Fungicide Resistance Action Group (FRAG)-UK website

Restrictions
- Maximum total dose equivalent to two full dose treatments [2, 6]

Crop-specific information
- Latest use: before cereal ear emergence

Environmental safety
- Dangerous for the environment
- Toxic to aquatic organisms
- Avoid drift onto neighbouring crops. May cause damage to broad-leaved plant species
- LERAP Category B

Hazard classification and safety precautions
Hazard Harmful, Dangerous for the environment
Transport code 9 [2-6]
Packaging group III [2-6]
UN Number 3082 [2-6]; N/C [1, 7]
Risk phrases R36, R51 [1, 2, 4, 6, 7]; R38, R41, R43, R50 [3, 5]; R40, R53a, R62, R63 [1-7]
Operator protection A, C [1-7]; H [1, 2, 6, 7]; U05a [1-7]; U14, U20a [3, 5]; U20b [1, 2, 4, 6, 7]
Environmental protection E15a [1, 2, 4, 6, 7]; E15b [3, 5]; E16a, E38 [1-7]; E34 [1, 7]
Storage and disposal D01, D02, D10c, D12a [1-7]; D05 [3, 5]; D08 [1, 2, 4, 6, 7]; D09a [1, 3, 5, 7]
Medical advice M03 [1-7]; M05a [3, 5]

FOR FULL CONDITIONS OF USE ALWAYS READ THE PRODUCT LABEL

46   boscalid + epoxiconazole + pyraclostrobin

A systemic and protectant fungicide mixture for diease control in cereals
FRAC mode of action code: 3 + 7 + 11

Products

Nebula XL	BASF	140:50:60 g/l	OD	15555

Uses
- Brown rust in **rye**, **spring wheat**, **triticale**, **winter barley**, **winter wheat**
- Crown rust in **oats**
- Eyespot in **rye** *(moderate control)*, **spring wheat** *(moderate control)*, **triticale** *(moderate control)*, **winter barley** *(moderate control)*, **winter wheat** *(moderate control)*
- Fusarium ear blight in **spring wheat** *(good reduction)*, **winter wheat** *(good reduction)*
- Glume blotch in **spring wheat**, **triticale**, **winter wheat**
- Net blotch in **winter barley**
- Powdery mildew in **oats** *(moderate control)*, **rye** *(moderate control)*, **triticale** *(moderate control)*, **winter barley** *(moderate control)*
- Ramularia leaf spots in **winter barley** *(moderate control)*
- Rhynchosporium in **rye**, **winter barley**
- Septoria leaf blotch in **spring wheat**, **triticale**, **winter wheat**
- Sooty moulds in **spring wheat** *(good reduction)*, **winter wheat** *(good reduction)*
- Tan spot in **spring wheat** *(moderate control)*, **winter wheat** *(moderate control)*
- Yellow rust in **rye**, **spring wheat**, **triticale**, **winter barley**, **winter wheat**

Approval information
- Boscalid, epoxiconazole and pyraclostrobin included in Annex I under EC Regulation 1107/2009

Following crops guidance
- Onions, oats, sugar beet, oilseed rape, cabbages, carrots, sunflowers, barley, lettuce, ryegrass, dwarf french beans, peas, potatoes, clover, wheat, field beans and maize may be sown as the following crop.

Environmental safety
- LERAP Category B

Hazard classification and safety precautions
   Hazard Harmful, Dangerous for the environment
   Transport code 9
   Packaging group III
   UN Number 3082
   Risk phrases R20, R38, R40, R43, R50, R53a, R62, R63
   Operator protection A, H; U05a, U14, U20a
   Environmental protection E15b, E16a, E34, E38
   Storage and disposal D01, D02, D05, D09a, D10c, D12a
   Medical advice M03, M05a

47   boscalid + metconazole

An anilide and triazole fungicide mixture for disease control in oilseed rape
FRAC mode of action code: 3 + 7

Products

1	Highgate	BASF	133:60 g/l	SC	15251
2	Tectura	BASF	133:60 g/l	SC	15232

Uses
- Sclerotinia stem rot in **oilseed rape**

Approval information
- Boscalid and metconazole included in Annex I under EC Regulation 1107/2009

SEE SECTION 3 FOR PRODUCTS ALSO REGISTERED

Following crops guidance
- Any crop can follow normally-harvested oilseed rape treated with [1] or [2]

Environmental safety
- LERAP Category B

Hazard classification and safety precautions
    Hazard Harmful
    UN Number N/C
    Risk phrases R52, R53a, R63
    Operator protection A, H; U05a, U20a
    Environmental protection E15b, E16a, E16b, E34, E38
    Storage and disposal D01, D02, D05, D09a, D10c, D12a
    Medical advice M03, M05a

---

48   boscalid + pyraclostrobin

A protectant and systemic fungicide mixture
FRAC mode of action code: 7 + 11

See also pyraclostrobin

Products

1	Bellis	BASF	25.2:12.8% w/w	WG	12522
2	Signum	BASF	26.7:6.7% w/w	WG	11450

Uses
- Alternaria in **cabbages, calabrese, carrots** *(moderate)*, **cauliflowers, poppies for morphine production** *(off-label)* [2]
- American gooseberry mildew in **blackcurrants** *(moderate control only)* [2]
- Anthracnose in **blueberries** *(off-label)*, **gooseberries** *(off-label)*, **protected blueberry** *(off-label)*, **protected gooseberries** *(off-label)* [2]
- Black spot in **protected strawberries, strawberries** [2]
- Blight in **potatoes** *(off-label)* [2]
- Blossom wilt in **cherries** *(off-label)*, **mirabelles** *(off-label)*, **plums** *(off-label)*, **protected cherries** *(off-label)*, **protected mirabelles** *(off-label)*, **protected plums** *(off-label)* [2]
- Botrytis in **amenity vegetation** *(off-label)*, **asparagus** *(off-label)*, **beans without pods (fresh)** *(off-label)*, **blueberries** *(off-label)*, **broad beans** *(off-label)*, **forest nurseries** *(off-label)*, **gooseberries** *(off-label)*, **interior landscapes** *(off-label)*, **lamb's lettuce** *(off-label)*, **ornamental plant production** *(off-label)*, **poppies for morphine production** *(off-label)*, **protected aubergines** *(off-label)*, **protected blueberry** *(off-label)*, **protected forest nurseries** *(off-label)*, **protected gooseberries** *(off-label)*, **protected ornamentals** *(off-label)*, **protected peppers** *(off-label)*, **protected tomatoes** *(off-label)*, **redcurrants** *(off-label)*, **whitecurrants** *(off-label)* [2]
- Bottom rot in **lettuce, protected lettuce** [2]
- Brown rot in **apricots** *(off-label)*, **nectarines** *(off-label)*, **peaches** *(off-label)* [2]
- Cane blight in **blackberries** *(off-label)*, **protected blackberries** *(off-label)*, **raspberries** *(off-label)* [2]
- Chocolate spot in **spring field beans** *(moderate)*, **winter field beans** *(moderate)* [2]
- Disease control in **all edible crops (outdoor and protected)** *(off-label)*, **all non-edible crops (outdoor)** *(off-label)*, **all protected non-edible crops** *(off-label)*, **beetroot** *(off-label)*, **kohlrabi** *(off-label)*, **lamb's lettuce** *(off-label)*, **protected hops** *(off-label)*, **protected radishes** *(off-label)*, **protected soft fruit** *(off-label)*, **protected top fruit** *(off-label)*, **radishes** *(off-label)*, **red beet** *(off-label)*, **soft fruit** *(off-label)*, **top fruit** *(off-label)* [2]; **hops** *(off-label)* [1, 2]
- Downy mildew in **bulls blood** *(off-label)*, **chard** *(off-label)*, **endives** *(off-label)*, **protected endives** *(off-label)*, **spinach** *(off-label)*, **spinach beet** *(off-label)* [2]
- Fusarium in **bullb onion sets** *(off-label)* [2]
- Grey mould in **blackcurrants, lettuce, protected lettuce, protected strawberries, strawberries** [2]
- Late blight in **potatoes** *(off-label)* [2]

FOR FULL CONDITIONS OF USE ALWAYS READ THE PRODUCT LABEL

- Leaf and pod spot in **combining peas** *(moderate control only)*, **vining peas** *(moderate control only)* [2]
- Leaf spot in **amenity vegetation** *(off-label)*, **black salsify** *(off-label)*, **blackcurrants** *(moderate control only)*, **blueberries** *(off-label)*, **gooseberries** *(off-label)*, **interior landscapes** *(off-label)*, **parsley root** *(off-label)*, **protected blueberry** *(off-label)*, **protected gooseberries** *(off-label)*, **redcurrants** *(off-label)*, **salsify** *(off-label)*, **whitecurrants** *(off-label)* [2]
- Poppy fire in **poppies for morphine production** *(off-label)* [2]
- Powdery mildew in **amenity vegetation** *(off-label)*, **black salsify** *(off-label)*, **carrots**, **interior landscapes** *(off-label)*, **parsley root** *(off-label)*, **protected strawberries**, **redcurrants** *(off-label)*, **salsify** *(off-label)*, **strawberries**, **whitecurrants** *(off-label)* [2]; **apples** [1]
- Purple blotch in **blackberries** *(off-label)*, **protected blackberries** *(off-label)*, **raspberries** *(off-label)* [2]
- Ring spot in **broccoli**, **brussels sprouts**, **cabbages**, **calabrese**, **cauliflowers** [2]
- Rust in **amenity vegetation** *(off-label)*, **asparagus** *(off-label)*, **black salsify** *(off-label)*, **interior landscapes** *(off-label)*, **kohlrabi** *(off-label)*, **parsley root** *(off-label)*, **protected blackberries** *(off-label)*, **salsify** *(off-label)*, **spring field beans**, **winter field beans** [2]
- Scab in **apples**, **pears** [1]
- Sclerotinia in **black salsify** *(off-label)*, **carrots** *(moderate)*, **horseradish** *(off-label)*, **parsley root** *(off-label)*, **salsify** *(off-label)* [2]
- Soft rot in **lettuce**, **protected lettuce** [2]
- White blister in **broccoli**, **brussels sprouts**, **cabbages** *(qualified minor use)*, **calabrese** *(qualified minor use)*, **herbs (see appendix 6)** *(off-label)*, **leaf brassicas** *(off-label)*, **protected herbs (see appendix 6)** *(off-label)*, **protected leaf brassicas** *(off-label)* [2]
- White rot in **bulb onions** *(off-label)*, **garlic** *(off-label)*, **salad onions** *(off-label)*, **shallots** *(off-label)* [2]
- White rust in **kohlrabi** *(off-label)* [2]
- White tip in **leeks** *(off-label)* [2]

Extension of Authorisation for Minor Use (EAMUs)
- **all edible crops (outdoor and protected)** *20102111* [2]
- **all non-edible crops (outdoor)** *20102111* [2]
- **all protected non-edible crops** *20102111* [2]
- **amenity vegetation** *20121043* [2], *20122317* [2]
- **apricots** *20121721* [2]
- **asparagus** *20102105* [2]
- **beans without pods (fresh)** *20121009* [2]
- **beetroot** *20121717* [2]
- **black salsify** *20121720* [2]
- **blackberries** *20102110* [2]
- **blueberries** *20121722* [2]
- **broad beans** *20121009* [2]
- **bulb onions** *20102108* [2]
- **bullb onion sets** *20103122* [2]
- **bulls blood** *20102136* [2]
- **chard** *20102136* [2]
- **cherries** *20102109* [2]
- **endives** *20102136* [2]
- **forest nurseries** *20102119* [2]
- **garlic** *20102108* [2]
- **gooseberries** *20121722* [2]
- **herbs (see appendix 6)** *20102115* [2]
- **hops** *20112732* [1], *20102111* [2]
- **horseradish** *20093375* [2]
- **interior landscapes** *20121043* [2], *20122317* [2]
- **kohlrabi** *20121719* [2]
- **lamb's lettuce** *20121718* [2]
- **leaf brassicas** *20102115* [2]
- **leeks** *20102134* [2]
- **mirabelles** *20102109* [2]

SEE SECTION 3 FOR PRODUCTS ALSO REGISTERED

- **nectarines** *20121721* [2]
- **ornamental plant production** *20091842 expires 30 Sep 2013* [2], *20122141* [2]
- **parsley root** *20121720* [2]
- **peaches** *20121721* [2]
- **plums** *20102109* [2]
- **poppies for morphine production** *20101233* [2]
- **potatoes** *20110394* [2]
- **protected aubergines** *20120427* [2]
- **protected blackberries** *20102102* [2]
- **protected blueberry** *20121722* [2]
- **protected cherries** *20102109* [2]
- **protected endives** *20102136* [2]
- **protected forest nurseries** *20102119* [2]
- **protected gooseberries** *20121722* [2]
- **protected herbs (see appendix 6)** *20102115* [2]
- **protected hops** *20102111* [2]
- **protected leaf brassicas** *20102115* [2]
- **protected mirabelles** *20102109* [2]
- **protected ornamentals** *20091842 expires 30 Sep 2013* [2], *20122141* [2]
- **protected peppers** *20120427* [2]
- **protected plums** *20102109* [2]
- **protected radishes** *20121717* [2]
- **protected soft fruit** *20102111* [2]
- **protected tomatoes** *20120427* [2]
- **protected top fruit** *20102111* [2]
- **radishes** *20121717* [2]
- **raspberries** *20102110* [2]
- **red beet** *20121717* [2]
- **redcurrants** *20102114* [2]
- **salad onions** *20102107* [2]
- **salsify** *20121720* [2]
- **shallots** *20102108* [2]
- **soft fruit** *20102111* [2]
- **spinach** *20102136* [2]
- **spinach beet** *20102136* [2]
- **top fruit** *20102111* [2]
- **whitecurrants** *20102114* [2]

Approval information
- Boscalid and pyraclostrobin included in Annex I under EC Regulation 1107/2009
- Accepted by BBPA for use on malting barley (before ear emergence only)

Efficacy guidance
- On brassicas apply as a protectant spray or at the first sign of disease and repeat at 3-4 wk intervals depending on disease pressure [2]
- Ensure adequate spray penetration and coverage by increasing water volume in dense crops [2]
- For best results on strawberries apply as a protectant spray at the white bud stage. Applications should be made in sequence with other products as part of a fungicide spray programme during flowering at 7-10 day intervals [2]
- On carrots and field beans apply as a protectant spray or at the first sign of disease with a repeat treatment if needed, as directed on the label [2]
- On lettuce apply as a protectant spray 1-2 wk after planting [2]
- Optimum results on apples and pears obtained from a protectant treatment from bud burst [1]
- Application as the final 2 sprays on apples and pears gives reduction in storage rots [1]
- Pyraclostrobin is a member of the QoI cross-resistance group of fungicides and should be used in programmes with fungicides with a different mode of action

Restrictions
- Maximum total dose equivalent to three full dose treatments on brassicas; two full dose treatments on all other field crops [2]

FOR FULL CONDITIONS OF USE ALWAYS READ THE PRODUCT LABEL

- Maximum number of treatments (including other QoI treatments) on apples and pears 4 per yr if total number of applications is 12 or more, or 3 per yr if the total number is fewer than 12 [1]
- Do not use more than 2 consecutive treatments on apples and pears, and these must be separated by a minimum of 2 applications of a fungicide with a different mode of action [1]
- Use a maximum of three applications per yr on brassicas and no more than two per yr on other field crops [2]
- Do not use consecutive treatments; apply in alternation with fungicides from a different cross resistance group and effective against the target diseases [2]
- Do not apply more than 6 kg/ha to the same area of land per yr [2]
- Consult processor before use on crops for processing
- Applications to lettuce and protected salad crops may only be made between 1 Apr and 31 Oct [2]

Crop-specific information
- HI 21 d for field beans; 14 d for brassica crops, carrots, lettuce; 7 d for apples, pears; 3 d for strawberries

Environmental safety
- Dangerous for the environment
- Very toxic to aquatic organisms
- Broadcast air-assisted LERAP [1] (40 m); LERAP Category B [2]

Hazard classification and safety precautions
Hazard Harmful, Dangerous for the environment
Transport code 9
Packaging group III
UN Number 3077
Risk phrases R22a, R50, R53a
Operator protection A; U05a, U20c [1, 2]; U13 [1]
Environmental protection E15a, E38 [1, 2]; E16a [2]; E17b [1] (40 m); E34 [1]
Storage and disposal D01, D02, D10c, D12a [1, 2]; D08 [1]; D09a [2]
Medical advice M03 [1]; M05a [1, 2]

49   bromadiolone

An anti-coagulant coumarin-derivative rodenticide

Products

1	Bromag	Killgerm	0.005% w/w	RB	H8285
2	Bromag Fresh Bait	Killgerm	0.005% w/w	RB	H8557
3	Endorats Rat Killer	Irish Drugs	0.005% w/w	AB	H9343
4	Endorats Rat Killer Blocks	Irish Drugs	0.005% w/w	BB	H9336
5	Sakarat Bromabait	Killgerm	0.005% w/w	RB	H7902

Uses
- Mice in *farm buildings*, *farmyards* [5]; *farm buildings/yards* [1, 2]
- Rats in *farm buildings* [3-5]; *farm buildings/yards* [1, 2]; *farmyards* [5]

Approval information
- Bromadiolone included in Annex I under EC Directive 1107/2009

Efficacy guidance
- Ready-to-use baits are formulated on a mould-resistant, whole-wheat base
- Use in baiting programme. Place baits in protected situations, sufficient for continuous feeding between treatments
- Chemical is effective against warfarin- and coumatetralyl-resistant rats and mice and does not induce bait shyness
- Use bait bags where loose baiting inconvenient (eg behind ricks, silage clamps etc)
- The resistance status of the rodent population should be assessed when considering the choice of product to use

SEE SECTION 3 FOR PRODUCTS ALSO REGISTERED

Restrictions
- For use only by professional operators

Environmental safety
- Access to baits by children, birds and animals, particularly cats, dogs, pigs and poultry, must be prevented
- Baits must not be placed where food, feed or water could become contaminated
- Remains of bait and bait containers must be removed after treatment and burned or buried
- Rodent bodies must be searched for and burned or buried. They must not be placed in refuse bins or on rubbish tips
- Take extreme care to prevent domestic animals having access to the bait

Hazard classification and safety precautions
Operator protection A [1]; U13 [1-5]; U20b [1, 2, 5]
Storage and disposal D05, D07, D11a [1, 2, 5]; D09a [1-5]
Treated seed S06a [3, 4]
Vertebrate/rodent control products V01a, V03a, V04a [3, 4]; V01b, V03b, V04b [1, 2, 5]; V02 [1-5]
Medical advice M03 [1, 2, 5]

50   bromoxynil

A contact acting HBN herbicide
HRAC mode of action code: C3

Products

1	Butryflow	Nufarm UK	402 g/l	SC	14056
2	Flagon 400 EC	Makhteshim	400 g/l	EC	14921

Uses
- Annual dicotyledons in **bulb onions**, **forage maize**, **grain maize**, **linseed**, **millet** *(off-label)*, **shallots**, **sweetcorn** [1]; **durum wheat** *(off-label)*, **game cover** *(off-label)*, **grass seed crops** *(off-label)*, **linseed** *(off-label)*, **spring barley**, **spring oats**, **spring rye** *(off-label)*, **spring wheat**, **triticale** *(off-label)*, **winter barley**, **winter oats**, **winter rye** *(off-label)*, **winter wheat** [2]
- Black bindweed in **millet** *(off-label)* [1]
- Black nightshade in **millet** *(off-label)* [1]
- Chickweed in **millet** *(off-label)* [1]
- Fat hen in **durum wheat** *(off-label)*, **game cover** *(off-label)*, **grass seed crops** *(off-label)*, **linseed** *(off-label)*, **spring barley**, **spring oats**, **spring rye** *(off-label)*, **spring wheat**, **triticale** *(off-label)*, **winter barley**, **winter oats**, **winter rye** *(off-label)*, **winter wheat** [2]; **millet** *(off-label)* [1]
- Redshank in **millet** *(off-label)* [1]

Extension of Authorisation for Minor Use (EAMUs)
- **durum wheat** *20102116* [2]
- **game cover** *20102116* [2]
- **grass seed crops** *20102116* [2]
- **linseed** *20110874* [2]
- **millet** *20120593* [1]
- **spring rye** *20102116* [2]
- **triticale** *20102116* [2]
- **winter rye** *20102116* [2]

Approval information
- Bromoxynil included in Annex I under EC Regulation 1107/2009
- Accepted by BBPA for use on malting barley

Efficacy guidance
- Spray when main weed flush has germinated and the largest are at the 4 leaf stage
- Weed control can be enhanced by using a split treatment spraying each application when the weeds are seedling to 2 true leaves.

FOR FULL CONDITIONS OF USE ALWAYS READ THE PRODUCT LABEL

Restrictions

- Maximum number of treatments 1 per crop or yr or maximum total dose equivalent to one full dose treatment
- Do not apply with oils or other adjuvants
- Do not apply using hand-held equipment or at concentrations higher than those recommended
- Do not apply during frosty weather, drought, when soil is waterlogged, when rain expected within 4 h or to crops under any stress
- Take particular care to avoid drift onto neighbouring susceptible crops or open water surfaces

Crop-specific information

- Latest use: before 10 fully expanded leaf stage of crop for maize, sweetcorn; before crop 20 cm tall and before flower buds visible for linseed; before 2nd node detectable (GS 32) for cereals
- Foliar scorch, which rapidly disappears without affecting growth, will occur if treatment made in hot weather or during rapid growth

Environmental safety

- Very toxic to aquatic organisms
- Keep livestock out of treated areas for at least 6 wk after treatment
- High risk to bees. Do not apply to crops in flower or to those in which bees are actively foraging
- Do not contaminate surface waters or ditches with chemical or used container
- LERAP Category B

Hazard classification and safety precautions

Hazard Harmful, Dangerous for the environment
Transport code 3 [2]; 9 [1]
Packaging group III
UN Number 1993 [2]; 3082 [1]
Risk phrases R20, R22a, R43, R50, R53a, R63 [1, 2]; R22b, R36, R37, R38 [2]
Operator protection A, H [1, 2]; C [2]; K [1]; U02a, U03, U08, U10, U11, U13, U19a, U20a [2]; U05a, U14, U15, U23a [1, 2]
Environmental protection E06a [2] (6 wk); E12d [2]; E15a, E16a, E34, E38 [1, 2]
Consumer protection C02a [2] (6 wk)
Storage and disposal D01, D02, D05, D09a, D12a [1, 2]; D10b [2]; D10c [1]
Medical advice M03 [1]; M04a, M05b [2]

---

51   bromoxynil + ioxynil

A contact acting post-emergence HBN herbicide for cereals
HRAC mode of action code: C3 + C3

See also ioxynil

Products

1	Mextrol-Biox	Nufarm UK	200:200 g/l	EC	14697
2	Oxytril CM	Bayer CropScience	200:200 g/l	EC	14511
3	Stellox	Nufarm UK	200:200 g/l	EC	14696

Uses

- Annual dicotyledons in *game cover* (off-label), *miscanthus* (off-label) [1, 2]; *spring barley*, *spring oats*, *spring rye*, *spring wheat*, *triticale*, *winter barley*, *winter oats*, *winter rye*, *winter wheat* [1-3]

Extension of Authorisation for Minor Use (EAMUs)

- *game cover* 20111126 [1], 20111124 [2]
- *miscanthus* 20111126 [1], 20111124 [2]

Approval information

- Bromoxynil and ioxynil included in Annex I under EC Regulation 1107/2009
- Accepted by BBPA for use on malting barley

Efficacy guidance

- Best results achieved on young weeds growing actively in a highly competitive crop

---

SEE SECTION 3 FOR PRODUCTS ALSO REGISTERED

- Do not apply during periods of drought or when rain imminent (some labels say 'if likely within 4 or 6 h')
- Recommended for tank mixture with hormone herbicides to extend weed spectrum. See labels for details

Restrictions
- Maximum number of treatments 1 per crop
- Do not spray crops stressed by drought, waterlogging or other factors
- Do not roll or harrow for several days before or after spraying. Number of days specified varies with product. See label for details
- Do not apply by hand-held equipment or at concentrations higher than those recommended

Crop-specific information
- Latest use: before 2nd node detectable stage (GS 32)
- HI (animal consumption) 6 wk
- Apply to winter or spring cereals from 1-2 fully expanded leaf stage, but before second node detectable (GS 32)
- Spray oats in spring when danger of frost past. Do not spray winter oats in autumn

Environmental safety
- Dangerous for the environment
- Very toxic to aquatic organisms
- Keep livestock out of treated areas for at least 6 wk after treatment and until poisonous weeds, such as ragwort, have died down and become unpalatable
- LERAP Category B

Hazard classification and safety precautions
    Hazard Harmful, Dangerous for the environment
    Transport code 9
    Packaging group III
    UN Number 3082
    Risk phrases R22a, R22b, R43, R50, R53a, R63, R67 [1-3]; R36, R38 [1, 3]; R66 [2]
    Operator protection A, C [1-3]; H [1, 2]; U05a, U08, U13, U19a, U20b, U23a
    Environmental protection E07a, E12d, E12e, E13b, E16a, E34, E38
    Consumer protection C02a (14 d)
    Storage and disposal D01, D02, D09a, D11a, D12a
    Medical advice M03, M05b

---

52   bromoxynil + prosulfuron

A contact and residual herbicide mixture for maize
HRAC mode of action code: C3 + B

See also prosulfuron

Products

Jester	Syngenta	60:3% w/w	WG	13500

Uses
- Annual dicotyledons in **forage maize**, **game cover** *(off-label)*, **grain maize**, **millet** *(off-label)*
- Black bindweed in **forage maize**, **grain maize**
- Chickweed in **forage maize**, **grain maize**
- Fat hen in **millet** *(off-label)*
- Hemp-nettle in **forage maize**, **grain maize**
- Knotgrass in **forage maize**, **grain maize**
- Mayweeds in **forage maize**, **grain maize**

Extension of Authorisation for Minor Use (EAMUs)
- *game cover* 20100322
- *millet* 20101803

FOR FULL CONDITIONS OF USE ALWAYS READ THE PRODUCT LABEL

SECTION 2

Approval information
- Bromoxynil and prosulfuron included in Annex I under EC Regulation 1107/2009

Efficacy guidance
- Prosulfuron is a member of the ALS-inhibitor group of herbicides and products should be used in a planned Resistance Management strategy. See Section 5 for more information

Restrictions
- Maximum total dose equivalent to one full dose treatment
- Do not treat crops grown for seed production
- Consult before use on crops intended for processing
- Do not use in frosty weather or on crops under stress
- Do not tank mix with organophosphate insecticides or apply in tank mix or sequence with any other sulfonylurea
- Do not apply by knapsack sprayer or in volumes less than those recommended
- Take care to wash out sprayers thoroughly. See label for details

Crop-specific information
- Latest use: before 7 crop leaves unfolded for maize
- Apply post-emergence up to when the crop has six unfolded leaves
- Product must be used with a non-ionic wetter that is not an organosilicone
- Apply in cool conditions, or during the evening, to avoid scorch

Following crops guidance
- Winter or spring cereals, winter or spring beans, spring sown peas or oilseed rape may be sown following normal harvest of a treated crop. Mould-board ploughing to 20 cm is recommended in some cases

Environmental safety
- Dangerous for the environment
- Very toxic to aquatic organisms
- Take special care to avoid drift outside the target area

Hazard classification and safety precautions
Hazard Toxic, Dangerous for the environment
Transport code 6.1
Packaging group III
UN Number 2588
Risk phrases R22a, R23, R36, R43, R50, R53a, R63
Operator protection A, D, H; U05a, U11, U15, U19a, U20c, U23a
Environmental protection E15b, E34, E38
Storage and disposal D01, D02, D05, D09a, D10c, D12a
Medical advice M04a

53   bromoxynil + terbuthylazine

A post-emergence contact and residual herbicide for maize
HRAC mode of action code: C3 + C1

See also terbuthylazine

Products

Templar	Makhteshim	200:300 g/l	SC	10254

Uses
- Annual dicotyledons in **forage maize**, **forage maize (under plastic mulches)** *(off-label)*, **grain maize under plastic mulches** *(off-label)*, **sweetcorn** *(off-label)*, **sweetcorn under plastic mulches** *(off-label)*
- Annual grasses in **forage maize (under plastic mulches)** *(off-label)*, **grain maize under plastic mulches** *(off-label)*, **sweetcorn under plastic mulches** *(off-label)*

SEE SECTION 3 FOR PRODUCTS ALSO REGISTERED

Extension of Authorisation for Minor Use (EAMUs)
- *forage maize (under plastic mulches)* 20091844
- *grain maize under plastic mulches* 20091844
- *sweetcorn* 20072179
- *sweetcorn under plastic mulches* 20091844

Approval information
- Bromoxynil and terbuthylazine included in Annex I under EC Regulation 1107/2009

Efficacy guidance
- Best results obtained from treatment when main flush of weeds has germinated and largest at 4 leaf stage
- Control can be enhanced by use of split dose treatment but second spray must be applied before crop canopy covers ground
- Soils should be moist at time of treatment and should not be disturbed afterwards
- Residual activity may be reduced on soils with high OM content, or where high amounts of slurry or manure have been applied

Restrictions
- Maximum total dose equivalent to one full dose treatment
- Foliar scorch may occur after treatment in hot conditions or when growth particularly rapid. In such conditions spray in the evening
- Do not apply during frosty weather, drought, when soil is waterlogged or when rain expected within 12 hr
- Do not treat crops under any stress whatever
- Product should be used only where damaging populations of competitive weeds have emerged. Yield may be reduced when used in absence of significant weed populations
- Succeeding crops may not be planted until the spring following application. Mould board plough to 150 mm before doing so

Crop-specific information
- Apply before 9 fully expanded leaf stage

Following crops guidance
- Land should be mould board ploughed to a depth of 150mm before any succeeding crop is planted. Succeeding crops may only be planted in the spring following application

Environmental safety
- Dangerous for the environment
- Very toxic to aquatic organisms
- Keep livestock out of treated areas for at least 6 wk after treatment
- Harmful to bees. Do not apply to crops in flower or to those in which bees are actively foraging. Do not apply when flowering weeds are present
- Do not apply by hand-held equipment or at concentrations higher than those recommended

Hazard classification and safety precautions
Hazard Harmful, Dangerous for the environment
Transport code 9
Packaging group III
UN Number 3082
Risk phrases R20, R22a, R50, R53a
Operator protection A, C, H; U02a, U05a, U08, U13, U14, U15, U19a, U20a, U23a
Environmental protection E06a (6 wk); E12d, E12e, E13b, E38
Storage and disposal D01, D02, D12a
Medical advice M04a

FOR FULL CONDITIONS OF USE ALWAYS READ THE PRODUCT LABEL

54   bupirimate

A systemic aminopyrimidinol fungicide active against powdery mildew
FRAC mode of action code: 8

Products

1 Clayton Concorde	Clayton	250 g/l	EC	15665
2 Jatapu	Agroquimicos	250 g/l	EC	15776
3 Nimrod	Makhteshim	250 g/l	EC	13046

Uses
- Powdery mildew in **apples**, **begonias**, **blackcurrants**, **chrysanthemums**, **courgettes** (outdoor only), **gooseberries**, **hops**, **marrows** (outdoor only), **pears**, **protected cucumbers**, **protected strawberries**, **raspberries** (outdoor only), **roses**, **strawberries** [1-3]; **protected tomatoes** (off-label), **pumpkins** (off-label), **redcurrants** (off-label), **squashes** (off-label), **whitecurrants** (off-label) [3]

Extension of Authorisation for Minor Use (EAMUs)
- **protected tomatoes** 20070997 [3]
- **pumpkins** 20070994 [3]
- **redcurrants** 20082082 [3]
- **squashes** 20070994 [3]
- **whitecurrants** 20082082 [3]

Approval information
- Bupirimate included in Annex 1 under EC Regulation 1107/2009
- Accepted by BBPA for use on hops

Efficacy guidance
- On apples during periods that favour disease development lower doses applied weekly give better results than higher rates fortnightly
- Not effective in protected crops against strains of mildew resistant to bupirimate

Restrictions
- Maximum number of treatments or maximum total dose depends on crop and dose (see label for details)

Crop-specific information
- HI: 1 d for apples, pears, strawberries; 2 d for cucurbits; 7 d for blackcurrants; 8 d for raspberries; 14 d for gooseberries, hops
- Apply before or at first signs of disease and repeat at 7-14 d intervals. Timing and maximum dose vary with crop. See label for details
- With apples, hops and ornamentals cultivars may vary in sensitivity to spray. See label for details
- If necessary to spray cucurbits in winter or early spring spray a few plants 10-14 d before spraying whole crop to test for likelihood of leaf spotting problem
- On roses some leaf puckering may occur on young soft growth in early spring or under low light intensity. Avoid use of high rates or wetter on such growth
- Never spray flowering begonias (or buds showing colour) as this can scorch petals
- Do not mix with other chemicals for application to begonias, cucumbers or gerberas

Environmental safety
- Dangerous for the environment
- Toxic to aquatic organisms
- Flammable
- Product has negligible effect on *Phytoseiulus* and *Encarsia* and may be used in conjunction with biological control of red spider mite

Hazard classification and safety precautions
  Hazard Irritant, Flammable, Dangerous for the environment
  Transport code 3
  Packaging group III
  UN Number 1993

SEE SECTION 3 FOR PRODUCTS ALSO REGISTERED

Risk phrases R22b, R38, R51, R53a
Operator protection A, C; U05a, U20b
Environmental protection E15a
Storage and disposal D01, D02, D09a, D10c
Medical advice M05b

55   captan

A protectant phthalimide fungicide with horticultural uses
FRAC mode of action code: M4

Products

Alpha Captan 80 WDG	Makhteshim	80% w/w	WG	07096

Uses
- Black spot in **roses**
- Botrytis in **strawberries**
- Gloeosporium rot in **apples**
- Scab in **apples**, **pears**, **quinces** *(off-label)*

Extension of Authorisation for Minor Use (EAMUs)
- *quinces 20091388 expires 31 Mar 2013*

Approval information
- Captan included in Annex I under EC Regulation 1107/2009

Restrictions
- Maximum number of treatments 12 per yr on apples and pears as pre-harvest sprays
- Do not use on apple cultivars Bramley, Monarch, Winston, King Edward, Spartan, Kidd's Orange or Red Delicious or on pear cultivar D'Anjou
- Do not mix with alkaline materials or oils
- Do not use on fruit for processing
- Powered visor respirator with hood and neck cape must be used when handling concentrate

Crop-specific information
- HI apples, pears 14 d; strawberries 7 d
- For control of scab apply at bud burst and repeat at 10-14 d intervals until danger of scab infection ceased
- For suppression of fruit storage rots apply from late Jul and repeat at 2-3 wk intervals
- For black spot control in roses apply after pruning with 3 further applications at 14 d intervals or spray when spots appear and repeat at 7-10 d intervals
- For grey mould in strawberries spray at first open flower and repeat every 7-10 d
- Do not leave diluted material for more than 2 h. Agitate well before and during spraying

Environmental safety
- Dangerous for the environment
- Very toxic to aquatic organisms

Hazard classification and safety precautions
Hazard Harmful, Dangerous for the environment
Transport code 9
Packaging group III
UN Number 3077
Risk phrases R36, R40, R41, R43, R50, R53a
Operator protection A, D, E, H; U05a, U09a, U11, U19a, U20c
Environmental protection E15a
Storage and disposal D01, D02, D09a, D11a, D12a

56   carbendazim

A systemic benzimidazole fungicide with curative and protectant activity
FRAC mode of action code: 1

Products

1 Caste Off	Nufarm UK	500 g/l	SC	12832
2 Delsene 50 Flo	Nufarm UK	500 g/l	SC	11452
3 Mascot Systemic	Rigby Taylor	500 g/l	SC	14488
4 Ringer	Barclay	500 g/l	SC	13289

Uses

- Disease control in **managed amenity turf** [3]
- Eyespot in **spring barley, winter barley, winter wheat** [2]
- Foliar disease control in **durum wheat** *(off-label)*, **spring rye** *(off-label)*, **triticale** *(off-label)*, **winter rye** *(off-label)* [2]
- Light leaf spot in **spring oilseed rape, winter oilseed rape** [2]
- Rhynchosporium in **spring barley, winter barley** [2]
- Wormcast formation in **managed amenity turf** [1, 4]

Extension of Authorisation for Minor Use (EAMUs)

- **durum wheat** *20073118* [2]
- **spring rye** *20073118* [2]
- **triticale** *20073118* [2]
- **winter rye** *20073118* [2]

Approval information

- Carbendazim included in Annex I under EC Regulation 1107/2009
- Accepted by BBPA for use on malting barley

Efficacy guidance

- Products vary in the diseases listed as controlled for several crops. Labels must be consulted for full details and for rates and timings
- Mostly applied as spray or drench. Spray treatments normally applied at first sign of disease and repeated after 1 mth if required
- Apply as drench rather than spray where red spider mite predators are being used
- Do not apply during drought conditions. Rain or irrigation after treatment may improve control
- Leave 4 mth intervals where used only for worm cast control [4]
- To delay appearance of resistant strains alternate treatment with non-MBC fungicide. Eyespot in cereals and *Botrytis cinerea* in many crops are now widely resistant. To avoid the development of resistance, a maximum of 2 applications of any MBC product (thiophanate-methyl or carbendazim) are allowed in any one crop. Avoid using MBC fungicides alone.

Restrictions

- Maximum number of treatments (including applications of any product containing benomyl, carbendazim or thiophanate-methyl) varies with crop treated and product used - see labels for details
- Do not treat crops or turf suffering from drought or other physical or chemical stress
- Consult processors before using on crops for processing
- Not compatible with alkaline products such as lime sulphur

Crop-specific information

- Latest use: before 2nd node detectable (GS32) for cereals
- HI 2 d for inert substrate cucumbers, protected tomatoes; 7-14 d (depends on dose) for apples, pears; 14 d for mushrooms, protected celery; 21 d for oilseed rape, dwarf beans, lettuce, 6 mth for cane fruit
- On turf apply as preventative treatment in spring or autumn during periods of high disease risk [4]
- Apply as a drench to control soil-borne diseases in cucumbers and tomatoes and as a pre-planting dip treatment for bulbs

Environmental safety

- Dangerous for the environment

SEE SECTION 3 FOR PRODUCTS ALSO REGISTERED

- Very toxic to aquatic organisms
- After use dipping suspension must not be discharged directly into ditches or drains

Hazard classification and safety precautions
 Hazard Toxic, Dangerous for the environment [1, 2, 4]
 Transport code 9
 Packaging group III
 UN Number 3082
 Risk phrases R46 [2, 4]; R50, R53a, R60, R61 [1, 2, 4]; R60b [1, 2]; R68 [4]
 Operator protection A, C, H, M [1, 2, 4]; U14 [4]; U19a [1, 4]; U20b [2, 4]; U20c [1]; U23a [2]
 Environmental protection E13c [2]; E14a, E34 [4]; E15a [1]; E38 [1, 2, 4]
 Storage and disposal D01, D02, D11a [1, 2]; D05, D09a, D12a [1, 2, 4]; D08, D10a [4]
 Medical advice M04a [4]

57   carbendazim + flusilazole

A broad-spectrum systemic and protectant fungicide for cereals, oilseed rape and sugar beet
FRAC mode of action code: 1 + 3

See also flusilazole

Products

1	Contrast	DuPont	125:250 g/l	SC	13107
2	Harvesan	DuPont	125:250 g/l	SC	13106
3	Punch C	DuPont	125:250 g/l	SC	13023

Uses
- Brown rust in *spring barley, spring wheat, winter barley, winter wheat*
- Canker in *spring oilseed rape* (reduction), *winter oilseed rape* (reduction)
- Eyespot in *spring barley, spring wheat, winter barley, winter wheat*
- Glume blotch in *spring wheat, winter wheat*
- Light leaf spot in *spring oilseed rape, winter oilseed rape*
- Phoma leaf spot in *spring oilseed rape, winter oilseed rape*
- Powdery mildew in *spring barley, spring wheat, sugar beet, winter barley, winter wheat*
- Rust in *sugar beet*
- Septoria leaf blotch in *spring wheat, winter wheat*
- Yellow rust in *spring barley, spring wheat, winter barley, winter wheat*

Approval information
- Carbendazim included in Annex I under EC Regulation 1107/2009 and flusilazole has been reinstated in Annex 1 under EC Regulation 1107/2009 following an appeal
- Accepted by BBPA for use on malting barley

Efficacy guidance
- Apply at early stage of disease development or in routine preventive programme
- Most effective timing of treatment on cereals varies with disease. See label for details
- Higher rate active against both MBC-sensitive and MBC-resistant eyespot
- Rain occurring within 2 h of spraying may reduce effectiveness
- To prevent build-up of resistant strains of cereal mildew tank mix with approved morpholine fungicide
- Flusilazole is a DMI fungicide. Resistance to some DMI fungicides has been identified in Septoria leaf blotch which may seriously affect performance of some products. For further advice contact a specialist advisor and visit the Fungicide Resistance Action Group (FRAG)-UK website

Restrictions
- Maximum number of treatments 1 per crop on sugar beet; 2 per crop on cereals and oilseed rape
- Do not apply to crops under stress or during frosty weather

Crop-specific information

- Latest use on cereals varies with crop and dose used - see label for details; before first flower opened stage for oilseed rape
- HI 7 wk for sugar beet
- Treat oilseed rape in autumn when leaf lesions first appear and spring from the start of stem extension when disease appears

Environmental safety

- Dangerous for the environment
- Toxic to aquatic organisms
- Dangerous to fish or other aquatic life. Do not contaminate surface waters or ditches with chemical or used container

Hazard classification and safety precautions

Hazard Toxic, Dangerous for the environment
Transport code 9
Packaging group III
UN Number 3082
Risk phrases R22a, R40, R41, R43, R46, R51, R53a, R60, R61
Operator protection A, C, H, M; U05a, U11, U19a, U20b
Environmental protection E13b, E34, E38
Storage and disposal D01, D02, D05, D09a, D10b, D12a
Medical advice M04a

58   carbetamide

A residual pre- and post-emergence carbamate herbicide for a range of field crops
HRAC mode of action code: K2

Products

Crawler	Makhteshim	60% w/w	WG	11840

Uses

- Annual dicotyledons in *cabbage seed crops, collards, evening primrose* (off-label), *fodder rape seed crops, honesty* (off-label), *kale seed crops, linseed* (off-label), *lucerne, lupins* (off-label), *mustard* (off-label), *red clover, sainfoin, spring cabbage, sugar beet seed crops, swede seed crops, turnip seed crops, white clover, winter field beans, winter oilseed rape*
- Annual grasses in *cabbage seed crops, collards, evening primrose* (off-label), *fodder rape seed crops, honesty* (off-label), *kale seed crops, linseed* (off-label), *lucerne, lupins* (off-label), *mustard* (off-label), *red clover, sainfoin, spring cabbage, sugar beet seed crops, swede seed crops, turnip seed crops, white clover, winter field beans, winter oilseed rape*
- Volunteer cereals in *cabbage seed crops, collards, fodder rape seed crops, kale seed crops, lucerne, red clover, sainfoin, spring cabbage, sugar beet seed crops, swede seed crops, turnip seed crops, white clover, winter field beans, winter oilseed rape*

Extension of Authorisation for Minor Use (EAMUs)

- *evening primrose* 20061236
- *honesty* 20061236
- *linseed* 20061236
- *lupins* 20061237
- *mustard* 20061236

Approval information

- Carbetamide included in Annex I under EC Regulation 1107/2009

Efficacy guidance

- Best results pre- or early post-emergence of weeds under cool, moist conditions. Adequate soil moisture is essential
- Dicotyledons controlled include chickweed, cleavers and speedwell
- Weed growth stops rapidly after treatment but full effects may take 6-8 wk to develop
- Various tank mixes are recommended to broaden the weed spectrum. See label for details

SEE SECTION 3 FOR PRODUCTS ALSO REGISTERED

- Always follow WRAG guidelines for preventing and managing herbicide resistant weeds. See Section 5 for more information

Restrictions
- Maximum number of treatments 1 per crop for all crops
- Do not treat any crop on waterlogged soil
- Do not use on soils with more than 10% organic matter as residual activity is impaired
- Do not apply during prolonged periods of cold weather when weeds are dormant
- Do not tank-mix with prothioconazole or liquid manganese

Crop-specific information
- HI 6 wk for all crops
- Apply to brassicas from mid-Oct to end-Feb provided crop has at least 4 true leaves (spring cabbage, spring greens), 3-4 true leaves (seed crops, oilseed rape)
- Apply to established lucerne or sainfoin from Nov to end-Feb
- Apply to established red or white clover from Feb to mid-Mar

Following crops guidance
- Succeeding crops may be sown 2 wk after treatment for brassicas, field beans, 8 wk after treatment for peas, runner beans, 16 wk after treatment for cereals, maize
- Ploughing is not necessary before sowing subsequent crops

Environmental safety
- Do not graze crops for at least 6 wk after treatment
- Some pesticides pose a greater threat of contamination of water than others and carbetamide is one of these pesticides. Take special care when applying carbetamide near water and do not apply if heavy rain is forecast

Hazard classification and safety precautions
    Hazard Dangerous for the environment
    Transport code 9
    Packaging group III
    UN Number 3077
    Risk phrases R51, R53a
    Operator protection U20c
    Environmental protection E15a, E38
    Storage and disposal D01, D02, D09a, D11a

59   carbon dioxide (commodity substance)

A gas for the control of trapped rodents and other vertebrates. Approval valid until 31/8/2019

Products
carbon dioxide	various	99.9%	GA	-

Uses
- Birds in *traps*
- Mice in *traps*
- Rats in *traps*

Approval information
- Carbon dioxide included in Annex 1 under EC Regulation 1107/2009

Efficacy guidance
- Use to destroy trapped rodent pests
- Use to control birds covered by general licences issued by the Agriculture and Environment Departments under Section 16(1) of the Wildlife and Countryside Act (1981) for the control of opportunistic bird species, where birds have been trapped or stupefied with alphachloralose/ seconal

Restrictions
- Operators must wear self-contained breathing apparatus when carbon dioxide levels are greater than 0.5% v/v
- Operators must be suitably trained and competent

Environmental safety
- Unprotected persons and non-target animals must be excluded from the treatment enclosures and surrounding areas unless the carbon dioxide levels are below 0.5% v/v

Hazard classification and safety precautions
Operator protection G

60   carboxin

A carboxamide fungicide available only in mixtures
FRAC mode of action code: 7

61   carboxin + thiram

A fungicide seed dressing for cereals
FRAC mode of action code: 7 + M3

See also thiram

Products

Anchor	Chemtura	200:200 g/l	FS	08684

Uses
- Bunt in **winter wheat** *(seed treatment)*
- Covered smut in **spring barley** *(seed treatment)*, **spring oats** *(seed treatment)*, **winter barley** *(seed treatment)*, **winter oats** *(seed treatment)*
- Fusarium foot rot and seedling blight in **spring barley** *(seed treatment)*, **spring oats** *(seed treatment)*, **spring rye** *(seed treatment)*, **spring wheat** *(seed treatment)*, **triticale** *(seed treatment)*, **winter barley** *(seed treatment)*, **winter oats** *(seed treatment)*, **winter rye** *(seed treatment)*, **winter wheat** *(seed treatment)*
- Leaf stripe in **spring barley** *(seed treatment)*, **winter barley** *(seed treatment - reduction)*
- Loose smut in **spring barley** *(seed treatment - reduction)*, **spring oats** *(seed treatment - reduction)*, **winter barley** *(seed treatment - reduction)*, **winter oats** *(seed treatment - reduction)*
- Net blotch in **spring barley** *(seed treatment)*
- Septoria seedling blight in **spring wheat** *(seed treatment)*, **winter wheat** *(seed treatment)*
- Stinking smut in **spring wheat** *(seed treatment)*

Approval information
- Carboxin and thiram included in Annex I under EC Regulation 1107/2009
- Accepted by BBPA for use on malting barley

Efficacy guidance
- Apply through suitable liquid flowable seed treating equipment of the batch treatment or continuous flow type where a secondary mixing auger is fitted
- Drill flow may be affected by treatment. Always re-calibrate seed drill before use

Restrictions
- Maximum number of treatments 1 per batch of seed
- Do not treat seed with moisture content above 16%
- Do not apply to cracked, split or sprouted seed
- Do not store treated seed from one season to the next

Crop-specific information
- Latest use: pre-drilling

SEE SECTION 3 FOR PRODUCTS ALSO REGISTERED

Environmental safety

- Dangerous for the environment
- Very toxic to aquatic organisms
- Do not use treated seed as food or feed
- Treated seed harmful to game and wildlife

Hazard classification and safety precautions

Hazard Harmful, Dangerous for the environment
Transport code 9
Packaging group III
UN Number 3082
Risk phrases R22a, R48, R50, R53a
Operator protection A, D, H; U05a, U09a, U20c
Environmental protection E15a, E34, E38
Storage and disposal D01, D02, D05, D06a, D09a, D10a
Treated seed S01, S02, S04b, S05, S06a, S07
Medical advice M03

62   carfentrazone-ethyl

A triazolinone contact herbicide
HRAC mode of action code: E

Products

1	Aurora 50 WG	Belchim	50% w/w	WG	11613
2	Headlite	AgChem Access	60 g/l	ME	13624
3	Shark	Belchim	60 g/l	ME	12762
4	Shylock	AgChem Access	60 g/l	ME	15405
5	Spotlight Plus	Belchim	60 g/l	ME	12436

Uses

- Annual and perennial weeds in **forest** *(off-label)*, **ornamental plant production** *(off-label)*, **protected forest** *(off-label)*, **protected ornamentals** *(off-label)* [3]
- Annual dicotyledons in **all edible crops (outdoor)** *(before planting)*, **all non-edible crops (outdoor)** *(before planting)*, **potatoes** [3, 4]; **forest nurseries** *(off-label)*, **protected forest nurseries** *(off-label)*, **protected soft fruit** *(off-label)*, **soft fruit** *(off-label)* [3]; **grass seed crops** *(off-label)*, **spring rye** *(off-label)*, **winter rye** *(off-label)* [1]; **narcissi** *(off-label)* [5]
- Black bindweed in **potatoes** [3, 4]
- Cleavers in **durum wheat**, **spring barley**, **spring oats**, **spring wheat**, **triticale**, **winter barley**, **winter oats**, **winter wheat** [1]; **potatoes** [3, 4]
- Desiccation in **blackberries** *(off-label)*, **protected blackberries** *(off-label)*, **protected raspberries** *(off-label)*, **protected rubus hybrids** *(off-label)*, **raspberries** *(off-label)*, **rubus hybrids** *(off-label)*, **wine grapes** *(off-label)* [3]
- Fat hen in **potatoes** [3, 4]
- Haulm destruction in **seed potatoes**, **ware potatoes** [2, 5]
- Ivy-leaved speedwell in **durum wheat**, **spring barley**, **spring oats**, **spring wheat**, **triticale**, **winter barley**, **winter oats**, **winter wheat** [1]; **potatoes** [3, 4]
- Knotgrass in **potatoes** [3, 4]
- Redshank in **potatoes** [3, 4]
- Volunteer oilseed rape in **all edible crops (outdoor)** *(before planting)*, **all non-edible crops (outdoor)** *(before planting)*, **potatoes** [3, 4]

Extension of Authorisation for Minor Use (EAMUs)

- **blackberries** *20080551 expires 30 Sep 2013* [3]
- **forest** *20080552 expires 30 Sep 2013* [3]
- **forest nurseries** *20082922 expires 30 Sep 2013* [3]
- **grass seed crops** *20060954 expires 30 Sep 2013* [1]
- **narcissi** *20091003 expires 30 Sep 2013* [5]
- **ornamental plant production** *20080552 expires 30 Sep 2013* [3]
- **protected blackberries** *20080551 expires 30 Sep 2013* [3]

FOR FULL CONDITIONS OF USE ALWAYS READ THE PRODUCT LABEL

- **protected forest** *20080552 expires 30 Sep 2013* [3]
- **protected forest nurseries** *20082922 expires 30 Sep 2013* [3]
- **protected ornamentals** *20080552 expires 30 Sep 2013* [3]
- **protected raspberries** *20080551 expires 30 Sep 2013* [3]
- **protected rubus hybrids** *20080551 expires 30 Sep 2013* [3]
- **protected soft fruit** *20082922 expires 30 Sep 2013* [3]
- **raspberries** *20080551 expires 30 Sep 2013* [3]
- **rubus hybrids** *20080551 expires 30 Sep 2013* [3]
- **soft fruit** *20082922 expires 30 Sep 2013* [3]
- **spring rye** *20060954 expires 30 Sep 2013* [1]
- **wine grapes** *20102957 expires 30 Sep 2013* [3]
- **winter rye** *20060954 expires 30 Sep 2013* [1]

Approval information
- Carfentrazone-ethyl included in Annex I under EC Regulation 1107/2009
- Accepted by BBPA for use on malting barley and hops
- Approval expiry 30 Sep 2013 [1-5]

Efficacy guidance
- Best weed control results achieved from good spray cover applied to small actively growing weeds [1, 3]
- Carfentrazone-ethyl acts by contact only; see label for optimum timing on specified weeds. Weeds emerging after application will not be controlled [1, 3]
- For weed control in cereals use as two spray programme with one application in autumn and one in spring [1]
- Efficacy of haulm destruction will be reduced where flailed haulm covers the stems at application [5]
- For potato crops with very dense vigorous haulm or where regrowth occurs following a single application a second application may be necessary to achieve satisfactory desiccation. A minimum interval between applications of 7 d should be observed to achieve optimum performance [5]

Restrictions
- Maximum number of treatments 2 per crop for cereals (1 in Autumn and 1 in Spring) [1]; 1 per crop for weed control in potatoes [3]; 1 per yr for pre-planting treatments [3]
- Maximum total dose for potato haulm destruction equivalent to 1.6 full dose treatments [5]
- Do not treat cereal crops under stress from drought, waterlogging, cold, pests, diseases, nutrient or lime deficiency or any factors reducing plant growth [1]
- Do not treat cereals undersown with clover or other legumes [1]
- Allow at least 2-3 wk between application and lifting potatoes to allow skins to set if potatoes are to be stored [5]
- Follow label instructions for sprayer cleaning
- Do not apply through knapsack sprayers
- Contact processor before using a split dose on potatoes for processing

Crop-specific information
- Latest use: 1 mth before planting edible and non-edible crops; before 3rd node detectable (GS 33) on cereals [1]
- HI 7 d for potatoes [5]
- For weed control in cereals treat from 2 leaf stage [1]
- When used as a treatment prior to planting a subsequent crop apply before weeds exceed maximum sizes indicated in the label [3]
- If treated potato tubers are to be stored then allow at least 2-3 wk between the final application and lifting to allow skins to set [5]

Following crops guidance
- No restrictions apply on the planting of succeeding crops 1 mth after application to potatoes for haulm destruction or as a pre-planting treatment, or 3 mth after application to cereals for weed control
- In the event of failure of a treated cereal crop, all cereals, ryegrass, maize, oilseed rape, peas, sunflowers, *Phacelia*, vetches, carrots or onions may be planted within 1 mth of treatment

SEE SECTION 3 FOR PRODUCTS ALSO REGISTERED

Environmental safety
- Dangerous for the environment
- Very toxic to aquatic organisms
- Some non-target crops are sensitive. Avoid drift on to broad-leaved plants outside the treated area, or onto ponds waterways or ditches

Hazard classification and safety precautions
    Hazard Irritant, Dangerous for the environment
    Transport code 9
    Packaging group III
    UN Number 3077 [1]; 3082 [2-5]
    Risk phrases R38 [2, 5]; R43, R50, R53a [1-5]
    Operator protection A, H; U05a, U14 [1-5]; U08, U13 [1]; U20a [2-5]
    Environmental protection E15a [1, 3, 4]; E15b [2, 5]; E34 [1]; E38 [1-5]
    Storage and disposal D01, D02, D09a, D12a [1-5]; D10b [1]; D10c [2-5]
    Medical advice M05a [1]

---

63   carfentrazone-ethyl + flupyrsulfuron-methyl

A foliar and residual acting herbicide for cereals
HRAC mode of action code: E + B

See also flupyrsulfuron-methyl

Products

Lexus Class	DuPont	33.3:16.7% w/w	WG	10809

Uses
- Annual dicotyledons in **durum wheat** *(off-label)*, **grass seed crops** *(off-label)*, **triticale**, **winter oats**, **winter rye**, **winter wheat**
- Blackgrass in **triticale**, **winter oats**, **winter rye**, **winter wheat**

Extension of Authorisation for Minor Use (EAMUs)
- **durum wheat** 20080895
- **grass seed crops** 20080895

Approval information
- Carfentrazone-ethyl and flupyrsulfuron-methyl included in Annex I under EC Regulation 1107/2009

Efficacy guidance
- Best results obtained when applied to small actively growing weeds
- Good spray cover of weeds must be obtained
- Increased degradation of active ingredient in high soil temperatures reduces residual activity
- Weed control may be reduced in dry soil conditions but susceptible weeds germinating soon after treatment will be controlled if adequate soil moisture present
- Symptoms may not be apparent on some weeds for up to 4 wk, depending on weather conditions
- Blackgrass should be treated from 1 leaf but before first node
- Flupyrsulfuron-methyl is a member of the ALS-inhibitor group of herbicides. To avoid the build up of resistance do not use any product containing an ALS-inhibitor herbicide with claims for control of grass weeds more than once on any crop
- Use these products as part of a resistance management strategy that includes cultural methods of control and does not use ALS inhibitors as the sole chemical method of grass weed control

Restrictions
- Maximum number of treatments 1 per crop
- Do not use on barley, on crops undersown with grasses or legumes, or any other broad-leaved crop
- Do not apply within 7 d of rolling

- Do not treat any crop suffering from drought, waterlogging, pest or disease attack, nutrient deficiency, or any other stress factors
- Specific restrictions apply to use in sequence or tank mixture with other sulfonylurea or ALS-inhibiting herbicides. See label for details

Crop-specific information
- Latest use: before 31 Dec in yr of sowing for winter oats, winter rye, triticale; before 1st node detectable (GS 31) for winter wheat
- Slight chlorosis and stunting may occur in certain conditions. Recovery is rapid and yield not affected

Following crops guidance
- Only cereals, oilseed rape, field beans, clover or grass may be sown in the yr of harvest of a treated crop
- In the event of crop failure only winter or spring wheat may be sown 1-3 mth after treatment. Land should be ploughed and cultivated to 15 cm minimum before resowing

Environmental safety
- Dangerous for the environment
- Very toxic to aquatic organisms
- Take extreme care to avoid damage by drift outside the target area or onto surface waters or ditches
- Spraying equipment should not be drained or flushed onto land planted, or to be planted, with trees or crops other than cereals and should be thoroughly cleansed after use - see label for instructions

Hazard classification and safety precautions
Hazard Irritant, Dangerous for the environment
Transport code 9
Packaging group III
UN Number 3077
Risk phrases R43, R50, R53a
Operator protection A, H; U05a, U08, U14, U19a, U20b
Environmental protection E15b, E38
Storage and disposal D01, D02, D09a, D10b, D11a, D12a

64   carfentrazone-ethyl + mecoprop-P

A foliar applied herbicide for cereals
HRAC mode of action code: E + O

See also mecoprop-P

Products
1	Jewel	Everris Ltd	1.5:60% w/w	WG	14327
2	Pan Glory	Pan Amenity	1.5:60% w/w	WG	14487
3	Platform S	Belchim	1.5:60% w/w	WG	14380

Uses
- Annual and perennial weeds in *amenity grassland, managed amenity turf* [1, 2]
- Charlock in *spring barley, spring oats, spring wheat, winter barley, winter oats, winter wheat* [3]
- Chickweed in *spring barley, spring oats, spring wheat, winter barley, winter oats, winter wheat* [3]
- Cleavers in *spring barley, spring oats, spring wheat, winter barley, winter oats, winter wheat* [3]
- Field speedwell in *spring barley, spring oats, spring wheat, winter barley, winter oats, winter wheat* [3]
- Ivy-leaved speedwell in *spring barley, spring oats, spring wheat, winter barley, winter oats, winter wheat* [3]
- Moss in *amenity grassland, managed amenity turf* [1, 2]

SEE SECTION 3 FOR PRODUCTS ALSO REGISTERED

- Red dead-nettle in **spring barley**, **spring oats**, **spring wheat**, **winter barley**, **winter oats**, **winter wheat** [3]

Approval information
- Carfentrazone-ethyl and mecoprop-P included in Annex I under EC Regulation 1107/2009
- Accepted by BBPA for use on malting barley

Efficacy guidance
- Best results obtained when weeds have germinated and growing vigorously in warm moist conditions
- Treatment of large weeds and poor spray coverage may result in reduced weed control

Restrictions
- Maximum number of treatments 2 per crop. The total amount of mecoprop-P applied in a single yr must not exceed the maximum total dose approved for any single product for the crop per situation
- Do not treat crops suffering from stress from any cause
- Do not treat crops undersown or to be undersown

Crop-specific information
- Latest use: before 3rd node detectable (GS 33)
- Can be used on all varieties of wheat and barley in autumn or spring from the beginning of tillering
- Early sown crops may be prone to damage if treated after period of rapid growth in autumn

Following crops guidance
- In the event of crop failure, any cereal, maize, oilseed rape, peas, vetches or sunflowers may be sown 1 mth after a spring treatment. Any crop may be planted 3 mth after treatment

Environmental safety
- Dangerous for the environment
- Very toxic to aquatic organisms
- Keep livestock out of treated areas for at least two weeks following treatment and until poisonous weeds, such as ragwort, have died down and become unpalatable
- Some pesticides pose a greater threat of contamination of water than others and mecoprop-P is one of these pesticides. Take special care when applying mecoprop-P near water and do not apply if heavy rain is forecast

Hazard classification and safety precautions
    Hazard Harmful, Dangerous for the environment
    Transport code 9
    Packaging group III
    UN Number 3077
    Risk phrases R22a, R41, R43, R50, R53a
    Operator protection A, C, H [1-3]; M [3]; U05a, U11, U13, U14, U20b [1-3]; U08 [3]; U09a [1, 2]
    Environmental protection E07a, E34, E38 [1-3]; E15a [3]; E15b [1, 2]
    Storage and disposal D01, D02, D09a, D10b [1-3]; D12a [3]
    Medical advice M05a

65   chlorantraniliprole

An ingested and contact insecticide for insect pest control in apples and pears and, in Ireland only, for colorado beetle control in potatoes.
IRAC mode of action code: 28

Products

Coragen	DuPont	200 g/l	SC	14930

Uses
- Codling moth in **apples**, **pears**

Approval information
- Chlorantraniliprole is pending inclusion in Annex 1 under EC Regulation 1107/2009

Efficacy guidance
- Can be used as part of an Integrated Pest Management programme.
- For best fruit protection in apples and pears, apply before egg hatch.
- Best applied early morning or late evening to avoid applications when bees may be present.

Restrictions
- Maximum number of applications is two per year

Crop-specific information
- Latest time of application is 14 days before harvest

Environmental safety
- To protect bees and pollinating insects, do not apply to crops when in flower. Do not apply when bees are actively foraging or when flowering plants are present.
- Broadcast air-assisted LERAP (10 m)

Hazard classification and safety precautions
Hazard Dangerous for the environment
Transport code 9
Packaging group III
UN Number 3082
Risk phrases R50, R53a
Environmental protection E15b, E22b, E38; E17b (10 m)
Storage and disposal D01, D02, D05, D09a, D12a

66   chloridazon

A residual pyridazinone herbicide for beet crops
HRAC mode of action code: C1

Products

1	Better DF	Sipcam	65% w/w	SG	06250
2	Better Flowable	Sipcam	430 g/l	SC	04924
3	Better Flowable	Sipcam	430 g/l	SC	15740
4	Goiana	Agroquimicos	65% w/w	WG	15778
5	Parador	United Phosphorus	430 g/l	SC	12372
6	Pyramin DF	BASF	65% w/w	WG	03438
7	Takron	BASF	430 g/l	SC	11627

Uses
- Annual dicotyledons in *beet leaves* (off-label), *chard* (off-label), *herbs (see appendix 6)* (off-label), *spinach* (off-label), *spinach beet* (off-label) [6]; *bulb onions* (off-label), *garlic* (off-label), *leeks* (off-label), *salad onions* (off-label), *shallots* (off-label) [1, 6]; *fodder beet*, *mangels*, *sugar beet* [1-7]
- Annual meadow grass in *beet leaves* (off-label), *bulb onions* (off-label), *chard* (off-label), *garlic* (off-label), *herbs (see appendix 6)* (off-label), *leeks* (off-label), *salad onions* (off-label), *shallots* (off-label), *spinach* (off-label), *spinach beet* (off-label) [6]; *fodder beet*, *mangels*, *sugar beet* [1-7]

Extension of Authorisation for Minor Use (EAMUs)
- *beet leaves* 20121682 [6]
- *bulb onions* 20121681 [1], 20121683 [6]
- *chard* 20121682 [6]
- *garlic* 20121681 [1], 20121683 [6]
- *herbs (see appendix 6)* 20121682 [6]
- *leeks* 20121681 [1], 20121683 [6]
- *salad onions* 20121681 [1], 20121683 [6]
- *shallots* 20121681 [1], 20121683 [6]

SEE SECTION 3 FOR PRODUCTS ALSO REGISTERED

- **spinach** *20121682* [6]
- **spinach beet** *20121682* [6]

Approval information
- Chloridazon included in Annex I under EC Regulation 1107/2009

Efficacy guidance
- Absorbed by roots of germinating weeds and best results achieved pre-emergence of weeds or crop when soil moist and adequate rain falls after application
- Application rate for some products depends on soil type. Check label for details

Restrictions
- Maximum number of treatments generally 1 per crop (pre-emergence) for fodder beet and mangels; 1 (pre-emergence) + 3 (post-emergence) per crop for sugar beet, but labels vary slightly. Check labels for maximum total dose for the crop to be treated
- Maximum total dose for onions and leeks equivalent to one full dose recommended for these crops [6]
- Do not use on Coarse Sands, Sands or Fine Sands or where organic matter exceeds 5%
- A maximum total dose of 2.6 kg/ha chloridazon may only be applied every third year on the same field

Crop-specific information
- Latest use: pre-emergence for fodder beet and mangels; normally before leaves of crop meet in row for sugar beet, but labels vary slightly; up to and including second true leaf stage for onions and leeks [6]
- Where used pre-emergence spray as soon as possible after drilling in mid-Mar to mid-Apr on fine, firm, clod-free seedbed
- Where crop drilled after mid-Apr or soil dry apply pre-drilling and incorporate to 2.5 cm immediately afterwards
- Various tank mixes recommended on sugar beet for pre- and post-emergence use and as repeated low dose treatments. See label for details
- Crop vigour may be reduced by treatment of crops growing under unfavourable conditions including poor tilth, drilling at incorrect depth, soil capping, physical damage, pest or disease damage, excess seed dressing, trace-element deficiency or a sudden rise in temperature after a cold spell

Following crops guidance
- Winter cereals or any spring sown crop may follow a treated beet crop harvested at the normal time and after ploughing to at least 150 mm
- In the event of failure of a treated crop only a beet crop or maize may be drilled, after cultivation

Environmental safety
- Dangerous for the environment
- Very toxic to aquatic organisms

Hazard classification and safety precautions
    Hazard Harmful [1, 4, 6]; Irritant [2, 3, 5, 7]; Dangerous for the environment [1-7]
    Transport code 9 [4-7]
    Packaging group III [4-7]
    UN Number 3077 [4, 6]; 3082 [5, 7]; N/C [1-3]
    Risk phrases R20 [4, 6]; R22a [1, 4, 6]; R43 [2-7]; R50, R53a [1-7]
    Operator protection A [4-7]; U05a, U08, U20b [1-7]; U14 [2-7]; U19a [2, 3, 5, 7]
    Environmental protection E13b [5]; E15a, E38 [1-7]
    Storage and disposal D01, D02, D09a, D12a [1-7]; D10b [2, 3, 5, 7]; D10c [5]; D11a [1, 4, 6]
    Medical advice M05a

FOR FULL CONDITIONS OF USE ALWAYS READ THE PRODUCT LABEL

67   chloridazon + ethofumesate

A contact and residual herbicide for beet crops
HRAC mode of action code: C1 + N

See also ethofumesate

SECTION 2

Products

Magnum	BASF	275:170 g/l	SC	11727

Uses

- Annual dicotyledons in **fodder beet**, **sugar beet**
- Annual meadow grass in **fodder beet**, **sugar beet**

Approval information

- Chloridazon and ethofumesate included in Annex I under EC Regulation 1107/2009

Efficacy guidance

- Best results achieved on a fine firm clod-free seedbed when soil moist and adequate rain falls after spraying. Efficacy and crop safety may be reduced if heavy rain falls just after incorporation
- Effectiveness may be reduced under conditions of low pH
- May be applied by conventional or repeat low dose method. See label for details

Restrictions

- Maximum number of treatments 1 pre-emergence for fodder beet; 1 pre-emergence plus 3 post-emergence for sugar beet
- May be used on soil classes Loamy Sand - Silty Clay Loam. Additional restrictions apply for some tank mixtures
- Do not treat beet post-emergence with recommended tank mixtures when temperature is, or is likely to be, above 21°C on day of spraying or under conditions of high light intensity
- If a mixture with phenmedipham is applied to a crop previously treated with a pre-emergence herbicide and the crop is suffering from stress from whatever cause, no further such post-emergence applications may be made
- A maximum total dose of 2.6 kg/ha chloridazon may only be applied every third year on the same field.
- The maximum total dose must not exceed 1.0 kg ethofumesate per hectare in any three year period

Crop-specific information

- Latest use: pre-crop emergence for fodder beet; before crop leaves meet between rows for sugar beet
- Apply up to cotyledon stage of weeds
- Crop vigour may be reduced by treatment of crops growing under unfavourable conditions including poor tilth, drilling at incorrect depth, soil capping, physical damage, excess nitrogen, excess seed dressing, trace element deficiency or a sudden rise in temperature after a cold spell. Frost after pre-emergence treatment may check crop growth

Following crops guidance

- In the event of crop failure only sugar beet, fodder beet or mangels may be re-drilled
- Any crop may be sown 3 mth after spraying following ploughing to 15 cm

Environmental safety

- Dangerous for the environment
- Very toxic to aquatic organisms

Hazard classification and safety precautions

Hazard Harmful, Dangerous for the environment
Transport code 9
Packaging group III
UN Number 3082
Risk phrases R22a, R50, R53a
Operator protection A, H; U05a, U08, U19a, U20a

SEE SECTION 3 FOR PRODUCTS ALSO REGISTERED

Environmental protection E15a, E38
Storage and disposal D01, D02, D09a, D10b, D12a
Medical advice M05a

68   chloridazon + metamitron

A contact and residual herbicide mixture for sugar beet
HRAC mode of action code: C1 + C1

See also metamitron

Products

Volcan Combi	Sipcam	300:280 g/l	SC	10256

Uses
- Annual dicotyledons in **fodder beet** *(off-label)*, **sugar beet**
- Annual meadow grass in **fodder beet** *(off-label)*, **sugar beet**

Extension of Authorisation for Minor Use (EAMUs)
- **fodder beet** *20101104*

Approval information
- Chloridazon and metamitron included in Annex I under EC Regulation 1107/2009

Efficacy guidance
- Best results achieved from a sequential programmme of sprays
- Pre-emergence use improves efficacy of post-emergence programme. Best results obtained from application to a moist seed bed
- First post-emergence treatment should be made when weeds at early cotyledon stage and subsequent applications made when new weed flushes reach this stage
- Weeds surviving an earlier treatment should be treated again after 7-10 d even if no new weeds have appeared

Restrictions
- Maximum number of treatments 1 pre-emergence followed by 3 post-emergence
- Take advice if light soils have a high proportion of stones

Crop-specific information
- Latest use: when leaves of crop meet in rows
- Tolerance of crops growing under stress from any cause may be reduced
- Crops treated pre-emergence and subsequently subjected to frost may be checked and recovery may not be complete

Following crops guidance
- After the last application only sugar beet or mangels may be sown within 4 mth; cereals may be sown after 16 wk.   Land should be mouldboard ploughed to 15 cm and thoroughly cultivated before any succeeding crop

Environmental safety
- Harmful to fish or other aquatic life. Do not contaminate surface waters or ditches with chemical or used container

Hazard classification and safety precautions
Hazard Irritant
UN Number N/C
Risk phrases R36, R38
Operator protection U08, U14, U15, U20a
Environmental protection E13c, E34
Storage and disposal D09a, D10b

FOR FULL CONDITIONS OF USE ALWAYS READ THE PRODUCT LABEL

69   chloridazon + quinmerac

A herbicide mixture for use in beet crops
HRAC mode of action code: C1 + O

See also quinmerac

Products
Fiesta T   BASF   360:60 g/l   SC   11734

Uses
- Annual dicotyledons in **beet leaves** *(off-label)*, **chard** *(off-label)*, **fodder beet**, **herbs (see appendix 6)** *(off-label)*, **mangels**, **spinach** *(off-label)*, **spinach beet** *(off-label)*, **sugar beet**
- Annual meadow grass in **fodder beet**, **mangels**, **sugar beet**
- Chickweed in **fodder beet**, **mangels**, **sugar beet**
- Cleavers in **beet leaves** *(off-label)*, **chard** *(off-label)*, **fodder beet**, **herbs (see appendix 6)** *(off-label)*, **mangels**, **spinach** *(off-label)*, **spinach beet** *(off-label)*, **sugar beet**
- Field speedwell in **fodder beet**, **mangels**, **sugar beet**
- Ivy-leaved speedwell in **fodder beet**, **mangels**, **sugar beet**
- Mayweeds in **fodder beet**, **mangels**, **sugar beet**
- Poppies in **fodder beet**, **mangels**, **sugar beet**

Extension of Authorisation for Minor Use (EAMUs)
- **beet leaves** *20101086*
- **chard** *20101086*
- **herbs (see appendix 6)** *20101086*
- **spinach** *20101086*
- **spinach beet** *20101086*

Approval information
- Chloridazon and quinmerac included in Annex I under EC Regulation 1107/2009

Efficacy guidance
- Best results obtained when adequate soil moisture is present at application and afterwards to form an active herbicidal layer in the soil
- A programme of pre-emergence treatment followed by post-emergence application(s) in mixture with a contact herbicide optimises weed control and is essential for some difficult weed species such as cleavers
- Treatment pre-emergence only may not provide sufficient residual activity to give season-long weed control
- Effectiveness may be reduced under conditions of low pH
- Always follow WRAG guidelines for preventing and managing herbicide resistant weeds. See Section 5 for more information

Restrictions
- Maximum total dose on sugar beet equivalent to one treatment pre-emergence plus two treatments post-emergence, all at maximum individual dose; maximum total dose on fodder beet and mangels equivalent to one full dose pre-emergence treatment
- Heavy rain shortly after treatment may check crop growth particularly when it leaves water standing in surface depressions
- Treatment of stressed crops or those growing in unfavourable conditions may depress crop vigour and possibly reduce stand
- Do not use on Sands or soils of high organic matter content
- Where rates of nitrogen higher than those generally recommended are considered necessary, apply at least 3 wk before drilling
- A maximum total dose of 2.6 kg/ha chloridazon may only be applied every third year on the same field

Crop-specific information
- Latest use: before plants meet between the rows for sugar beet; pre-emergence for fodder beet, mangels

SEE SECTION 3 FOR PRODUCTS ALSO REGISTERED

Following crops guidance
- In the spring following normal harvest of a treated crop sow only cereals, beet or mangel crops, potatoes or field beans, after ploughing
- In the event of failure of a treated crop only a beet crop may be drilled, after cultivation

Environmental safety
- Dangerous for the environment
- Toxic to aquatic organisms
- To reduce movement to groundwater do not apply to dry soil or when heavy rain is forecast

Hazard classification and safety precautions
    Hazard Irritant, Dangerous for the environment
    Transport code 9
    Packaging group III
    UN Number 3082
    Risk phrases R43, R51, R53a
    Operator protection A; U05a, U08, U14, U19a, U20b
    Environmental protection E15a, E38
    Storage and disposal D01, D02, D08, D09a, D10c, D12a
    Medical advice M03, M05a

70   chlormequat

A plant-growth regulator for reducing stem growth and lodging

Products

1	3C Chlormequat 720	BASF	720 g/l	SL	13973
2	Adjust	Taminco	620 g/l	SL	13961
3	Agrovista 3 See 750	Agrovista	750 g/l	SL	14797
4	Barleyquat B	Taminco	620 g/l	SL	13962
5	Belcocel	Taminco	720 g/l	SL	11881
6	Bettaquat B	Taminco	620 g/l	SL	13965
7	CCC 720	European Ag	720 g/l	SL	14891
8	Clayton Coldstream	Clayton	400 g/l	SL	12724
9	CS Chlormequat	ChemSource	700 g/l	SL	14946
10	Fargro Chlormequat	Fargro	460 g/l	SL	02600
11	Hive	Nufarm UK	730 g/l	SL	11392
12	K2	Taminco	620 g/l	SL	13964
13	Mandops Chlormequat 700	Taminco	700 g/l	SL	13970
14	Manipulator	Taminco	620 g/l	SL	13963
15	Mirquat	Nufarm UK	730 g/l	SL	11406
16	New 5C Cycocel	BASF	645 g/l	SL	01482
17	New 5C Quintacel	Nufarm UK	645 g/l	SL	12074
18	Selon	Taminco	620 g/l	SL	13966
19	Sigma PCT	Nufarm UK	460 g/l	SL	11209
20	Stabilan 700	Nufarm UK	700 g/l	SL	11393
21	Stabilan 750	Nufarm UK	75	ZZ	09303
22	Standon Girder 720	Standon	720 g/l	SL	14251

Uses
- Growth regulation in ***durum wheat*** *(off-label)* [2, 3, 5, 6, 11, 12, 14, 16, 20]; ***ornamental plant production*** [3, 21]; ***rye*** *(off-label)* [3]; ***spring barley*** [2, 4, 12-14, 18]; ***spring durum wheat*** *(off-label)*, ***spring triticale*** *(off-label)* [12]; ***spring oats***, ***winter oats*** [2, 4, 5, 12-14, 18, 22]; ***spring rye*** *(off-label)*, ***winter rye*** *(off-label)* [2, 4-6, 12-14]; ***spring wheat***, ***winter wheat*** [2, 5, 6, 12-14, 18, 22]; ***triticale*** [9, 11, 20, 21]; ***triticale*** *(off-label)* [2-4, 6, 11, 12, 14, 20]; ***winter barley*** [2, 5, 12-14, 18, 22]; ***winter rye*** [22]
- Increasing yield in ***winter barley*** [1, 4, 7, 16, 17]
- Lodging control in ***durum wheat*** *(off-label)* [2, 5, 6, 12, 14]; ***spring barley*** [2, 4, 12-14, 18]; ***spring durum wheat*** *(off-label)*, ***spring triticale*** *(off-label)* [12]; ***spring oats***, ***winter oats*** [1-5,

7-9, 11-18, 20, 21]; *spring rye* [1, 3, 7, 16, 17]; *spring rye* *(off-label)*, *winter rye* *(off-label)* [2, 4-6, 12-14]; *spring wheat*, *winter wheat* [1-3, 5-9, 11-21]; *triticale* [1, 3, 7, 9, 11, 15-17, 20, 21]; *triticale* *(off-label)* [2, 4, 6, 12, 14]; *winter barley* [1-5, 7-9, 11-15, 18-21]; *winter rye* [1, 3, 7-9, 11, 15-17, 20, 21]
- Stem shortening in *bedding plants* [9-11, 15, 20]; *camellias*, *hibiscus trionum*, *lilies* [9-11, 15, 20, 21]; *geraniums*, *poinsettias* [1, 7, 9-11, 15-17, 20, 21]

Extension of Authorisation for Minor Use (EAMUs)
- *durum wheat* *20081166* [2], *20122274* [3], *20060952* [5], *20081173* [6], *20122275* [11], *20081175* [12], *20081179* [14], *20061298* [16], *20122276* [20]
- *rye* *20122274* [3]
- *spring durum wheat* *20081175* [12]
- *spring rye* *20081166* [2], *20081170* [4], *20060952* [5], *20081173* [6], *20081175* [12], *20120865* [13], *20081179* [14]
- *spring triticale* *20081175* [12]
- *triticale* *20081166* [2], *20122274* [3], *20081170* [4], *20081173* [6], *20122275* [11], *20081175* [12], *20081179* [14], *20122276* [20]
- *winter rye* *20081166* [2], *20081170* [4], *20060952* [5], *20081173* [6], *20081175* [12], *20120865* [13], *20081179* [14]

Approval information
- Chlormequat included in Annex 1 under EC Regulation 1107/2009
- Approved for aerial application on wheat and oats [11, 15, 19]; See Section 5 for more information
- Accepted by BBPA for use on malting barley

Efficacy guidance
- Most effective results on cereals normally achieved from application from Apr onwards, on wheat and rye from leaf sheath erect to first node detectable (GS 30-31), on oats at second node detectable (GS 32), on winter barley from mid-tillering to leaf sheath erect (GS 25-30). However, recommendations vary with product. See label for details (n.b. [10] only approved on flowers)
- Influence on growth varies with crop and growth stage. Risk of lodging reduced by application at early stem extension. Root development and yield can be improved by earlier treatment
- Results on barley can be variable
- In tank mixes with other pesticides on cereals optimum timing for herbicide action may differ from that for growth reduction. See label for details of tank mix recommendations
- Most products recommended for use on oats require addition of approved non-ionic wetter. Check label
- Some products are formulated with trace elements to help compensate for increased demand during rapid growth

Restrictions
- Maximum number of treatments or maximum total dose varies with crop and product and whether split dose treatments are recommended. Check labels
- Do not use on very late sown spring wheat or oats or on crops under stress (n.b. [10] only approved on flowers)
- Mixtures with liquid nitrogen fertilizers may cause scorch and are specifically excluded on some labels
- Do not use on soils of low fertility unless such crops regularly receive adequate dressings of nitrogen
- At least 6 h, preferably 24 h, required before rain for maximum effectiveness. Do not apply to wet crops
- Check labels for tank mixtures known to be incompatible
- Not to be used on food crops [10]

Crop-specific information
- Latest use varies with crop and product. See label for details
- May be used on cereals undersown with grass or clovers (n.b. [10] only approved on flowers)
- Ornamentals to be treated must be well established and growing vigorously. Do not treat in strong sunlight or when temperatures are likely to fall below 10°C

SEE SECTION 3 FOR PRODUCTS ALSO REGISTERED

SECTION 2

- Temporary yellow spotting may occur on poinsettias. It can be minimised by use of a non-ionic wetting agent - see label

Environmental safety
- Harmful to aquatic organisms [10, 11, 15, 20]
- Wash equipment thoroughly with water and wetting agent immediately after use and spray out. Spray out again before storing or using for another product. Traces can cause harm to susceptible crops sprayed later
- Do not use straw from treated cereals as horticultural growth medium or mulch [16]

Hazard classification and safety precautions
Hazard Harmful [1-13, 15-22]
Transport code 8 [1, 3, 5-11, 15, 17, 19-22]
Packaging group III [1, 3, 5-11, 15, 17, 19-22]
UN Number 1760 [1, 3, 5-11, 15, 17, 19-22]
Risk phrases R21 [3, 8, 22]; R22a [1-13, 15-22]; R36 [13]; R52 [2, 4, 9-11, 13, 15, 18, 20-22]; R53a [2, 4, 9-11, 15, 18, 20-22]
Operator protection A [1-13, 15-22]; C [13]; U05a, U19a [1-13, 15-22]; U08 [1-8, 10, 12, 13, 16, 18, 19]; U09a [9, 11, 15, 17, 20-22]; U13 [10]; U20a [17]; U20b [1-13, 15, 16, 18-22]
Environmental protection E15a, E34 [1-13, 15-22]
Storage and disposal D01, D02, D09a [1-13, 15-22]; D05 [3, 8-11, 15, 17, 20, 21]; D10a [9, 20-22]; D10b [2-6, 8, 10-13, 15, 18, 19]; D10c [1, 7, 16, 17]; D12a [8]
Medical advice M03 [1-13, 15-22]; M05a [1, 7, 9-11, 15-17, 20, 21]

71   chlormequat + 2-chloroethylphosphonic acid

A plant growth regulator for use in cereals

See also 2-chloroethylphosphonic acid

Products

1	Greencrop Tycoon	Greencrop	305:155 g/l	SL	09571
2	Strate	Bayer CropScience	360:180 g/l	SL	10020
3	Upgrade	Bayer CropScience	360:180 g/l	SL	10029

Uses
- Growth regulation in **durum wheat** *(off-label)*, **spring rye** *(off-label)*, **triticale** *(off-label)*, **winter rye** *(off-label)* [2, 3]
- Lodging control in **spring barley**, **winter barley**, **winter wheat** [1-3]

Extension of Authorisation for Minor Use (EAMUs)
- **durum wheat** *20061541* [2], *20063100* [3]
- **spring rye** *20061541* [2], *20063100* [3]
- **triticale** *20061541* [2], *20063100* [3]
- **winter rye** *20061541* [2], *20063100* [3]

Approval information
- Chlormequat and 2-chloroethylphosphonic acid (ethephon) included in Annex I under EC Regulation 1107/2009
- Accepted by BBPA for use on malting barley
- In 2006 CRD required that all products containing 2-chloroethylphosphonic acid should carry the following warning in the main area of the container label: "2-chloroethylphosphonic acid is an anticholinesterase organophosphate. Handle with care"

Efficacy guidance
- Best results obtained when crops growing vigorously
- Recommended dose varies with growth stage. See labels for details and recommendations for use of sequential treatments

FOR FULL CONDITIONS OF USE ALWAYS READ THE PRODUCT LABEL

Restrictions

- 2-chloroethylphosphonic acid is an anticholinesterase organophosphorus compound. Do not use if under medical advice not to work with such compounds
- Maximum number of treatments 1 per crop; maximum total dose equivalent to one full dose treatment
- Product must always be used with specified non-ionic wetter - see labels
- Do not use on any crop in sequence with any other product containing 2-chloroethylphosphonic acid
- Do not spray when crop wet or rain imminent
- Do not spray during cold weather or periods of night frost, when soil is very dry, when crop diseased or suffering pest damage, nutrient deficiency or herbicide stress
- If used on seed crops grown for certification inform seed merchant beforehand
- Do not use on wheat variety Moulin or on any winter varieties sown in spring [1]
- Do not use on spring barley variety Triumph [1]
- Do not treat barley on soils with more than 10% organic matter [1]
- Do not use in programme with any other product containing 2-chloroethylphosphonic acid [1]
- Only crops growing under conditions of high fertility should be treated

Crop-specific information

- Latest use: before flag leaf ligule/collar just visible (GS 39) or 1st spikelet visible (GS 51) for wheat or barley at top dose; or before flag leaf sheath opening (GS 47) for winter wheat at reduced dose
- Apply before lodging has started

Environmental safety

- Dangerous for the environment [2, 3]
- Harmful to aquatic organisms [1]
- Harmful to fish or other aquatic life. Do not contaminate surface waters or ditches with chemical or used container [2, 3]
- Do not use straw from treated cereals as a horticultural growth medium or as a mulch [1]

Hazard classification and safety precautions

    Hazard Harmful [1-3]; Dangerous for the environment [2, 3]
    Transport code 8
    Packaging group III
    UN Number 3265
    Risk phrases R22a, R37 [1-3]; R41 [2, 3]; R52 [1]
    Operator protection A [1-3]; C [2, 3]; U05a, U08, U19a, U20b [1-3]; U11, U13 [2, 3]
    Environmental protection E13c [2, 3]; E15a [1]; E34 [1-3]
    Storage and disposal D01, D02, D09a, D10b [1-3]; D05 [2, 3]
    Medical advice M01, M03 [1-3]; M05a [1]

---

72 chlormequat + 2-chloroethylphosphonic acid + mepiquat chloride

A plant growth regulator for reducing lodging in cereals

See also 2-chloroethylphosphonic acid
*   mepiquat chloride*

Products

Cyclade	BASF	230:155:75 g/l	SL	08958

Uses

- Growth regulation in **durum wheat** *(off-label)*, **spring rye** *(off-label)*, **triticale** *(off-label)*, **winter rye** *(off-label)*
- Lodging control in **spring barley**, **winter barley**, **winter wheat**

Extension of Authorisation for Minor Use (EAMUs)

- **durum wheat** *20063098*
- **spring rye** *20063098*

SEE SECTION 3 FOR PRODUCTS ALSO REGISTERED

- **triticale** *20063098*
- **winter rye** *20063098*

Approval information
- Chlormequat, 2-chloroethylphosphonic acid (ethephon) and mepiquat chloride included in Annex I under EC Regulation 1107/2009
- Accepted by BBPA for use on malting barley
- In 2006 CRD required that all products containing 2-chloroethylphosphonic acid should carry the following warning in the main area of the container label: "2-chloroethylphosphonic acid is an anticholinesterase organophosphate. Handle with care"

Efficacy guidance
- Best results achieved in a vigorous, actively growing crop with adequate fertility and moisture
- Optimum timing on all crops is from second node detectable stage (GS 32)
- Recommended for use as part of an intensive growing system which includes provision for optimum fertilizer treatment and disease control

Restrictions
- 2-chloroethylphosphonic acid is an anticholinesterase organophosphorus compound. Do not use if under medical advice not to work with such compounds
- Maximum number of treatments 1 per crop
- Maximum total dose depends on spraying regime adopted. See label
- Must be used with a non-ionic wetting agent
- Do not apply to stressed crops or those on soils of low fertility unless receiving adequate dressings of fertilizer
- Do not apply in temperatures above 21°C or if crop is wet or if rain expected
- Do not treat variety Moulin nor any winter varieties sown in spring
- Do not use in a programme with any other product containing 2-chloroethylphosphonic acid
- Do not apply to barley on soils with more than 10% organic matter (winter wheat may be treated)
- Notify seed merchant in advance if use on a seed crop is proposed

Crop-specific information
- Latest use: before first spikelet of ear visible (GS 51) using reduced dose on winter barley; before flag leaf sheath opening (GS 47) using reduced dose on winter wheat; before flag leaf just visible on spring barley
- May be applied to crops undersown with grasses or clovers
- Treatment may cause some delay in ear emergence

Environmental safety
- Harmful to aquatic organisms
- Do not use straw from treated crops as a horticultural growth medium

Hazard classification and safety precautions
Hazard Harmful
Transport code 8
Packaging group III
UN Number 3265
Risk phrases R22a, R37, R52
Operator protection A; U05a, U08, U19a, U20b
Environmental protection E15a, E34
Storage and disposal D01, D02, D09a, D10c
Medical advice M01, M03, M05a

73   chlormequat + imazaquin

A plant growth regulator mixture for winter wheat

See also imazaquin

Products
1	Meteor	BASF	368:0.8 g/l	SL	10403

Products – continued

2	Standon Imazaquin 5C	Standon	368:0.8 g/l	SL	08813
3	Upright	BASF	368:0.8 g/l	SL	10404

Uses
- Growth regulation in **spring rye** *(off-label)*, **triticale** *(off-label)*, **winter rye** *(off-label)* [1]
- Increasing yield in **winter wheat** [1-3]
- Lodging control in **winter wheat** [1-3]

Extension of Authorisation for Minor Use (EAMUs)
- **spring rye** *20063099* [1]
- **triticale** *20063099* [1]
- **winter rye** *20063099* [1]

Approval information
- Chlormequat and imazaquin included in Annex I under EC Regulation 1107/2009

Efficacy guidance
- Apply to crops during good growing conditions or to those at risk from lodging
- On soils of low fertility, best results obtained where adequate nitrogen fertilizer used

Restrictions
- Maximum number of applications 1 per crop (2 per crop at split dose)
- Do not treat durum wheat
- Do not apply to undersown crops
- Do not apply when crop wet or rain imminent

Crop-specific information
- Latest use: before second node detectable (GS 31)
- Apply as single dose from leaf sheath lengthening up to and including 1st node detectable or as split dose, the first from tillers formed to leaf sheath lengthening, the second from leaf sheath erect up to and including 1st node detectable

Environmental safety
- Dangerous for the environment
- Toxic to aquatic organisms
- Do not use straw from treated cereals as horticultural growth medium or mulch

Hazard classification and safety precautions
Hazard Harmful [1-3]; Dangerous for the environment [2]
Transport code 8
Packaging group III
UN Number 1760
Risk phrases R20, R52 [1, 3]; R22a, R36, R53a [1-3]; R51 [2]
Operator protection A, C; U05a, U08, U13, U19a, U20b [1-3]; U11, U15 [1, 3]
Environmental protection E15a, E34
Storage and disposal D01, D02, D05, D09a, D10b [1-3]; D06c, D12a [1, 3]
Medical advice M03 [1-3]; M05a [1, 3]

74   chlormequat + mepiquat chloride

A plant growth regulator for reducing lodging in wheat

See also mepiquat chloride

Products

Stronghold	BASF	345:115 g/l	SL	09134

Uses
- Lodging control in **winter wheat**

Approval information
- Chlormequat and mepiquat chloride included in Annex I under EC Regulation 1107/2009

SEE SECTION 3 FOR PRODUCTS ALSO REGISTERED

SECTION 2

Efficacy guidance
- Optimum timing is when leaf sheaths erect (GS 30)
- Benefit will vary according to crop and stage of growth at application

Restrictions
- Maximum total dose equivalent to one full dose treatment
- Do not apply to stressed crops or those on soils of low fertility unless receiving adequate dressings of fertilizer
- Do not treat crops where significant foot diseases, especially take-all, are expected
- Do not treat crops on soils of low fertility
- Do not apply in temperatures above 21°C or if crop is wet or if rain expected
- Do not treat any winter varieties sown in spring
- Notify seed merchant in advance if use on a seed crop is proposed

Crop-specific information
- Latest use: before 3rd node detectable (GS 33)
- Apply during good growing conditions at the correct timings - see label
- May be applied to crops undersown with grasses or clovers
- Treatment may cause some delay in ear emergence
- Mixtures with liquid fertilizers may cause scorching in some circumstances

Environmental safety
- Do not use straw from treated crops as a horticultural growth medium or mulch

Hazard classification and safety precautions
> **Hazard** Harmful
> **Transport code** 8
> **Packaging group** III
> **UN Number** 1760
> **Risk phrases** R22a
> **Operator protection** A; U05a, U08, U19a, U20b
> **Environmental protection** E15a, E34
> **Storage and disposal** D01, D02, D09a, D10c
> **Medical advice** M03, M05a

75   2-chloroethylphosphonic acid

A plant growth regulator for cereals and various horticultural crops

See also chlormequat + 2-chloroethylphosphonic acid
chlormequat + 2-chloroethylphosphonic acid + imazaquin
chlormequat + 2-chloroethylphosphonic acid + mepiquat chloride

Products

1	Becki	ChemSource	480 g/l	SL	15732
2	Cerone	Bayer CropScience	480 g/l	SL	15087

Uses
- Growth regulation in **apples** *(off-label)*, **durum wheat** *(off-label)*, **ornamental plant production** *(off-label)*, **protected tomatoes** *(off-label)* [2]
- Lodging control in **spring barley**, **triticale**, **winter barley**, **winter rye**, **winter wheat** [1, 2]

Extension of Authorisation for Minor Use (EAMUs)
- **apples** *20122363* [2]
- **durum wheat** *20122362* [2]
- **ornamental plant production** *20122364* [2]
- **protected tomatoes** *20122365* [2]

Approval information
- 2-chloroethylphosphonic acid (ethephon) included in Annex I under EC Regulation 1107/2009

FOR FULL CONDITIONS OF USE ALWAYS READ THE PRODUCT LABEL

- In 2006 CRD required that all products containing this active ingredient should carry the following warning in the main area of the container label: "2-chloroethylphosphonic acid is an anticholinesterase organophosphate. Handle with care"
- Accepted by BBPA for use on malting barley

Efficacy guidance
- Best results achieved on crops growing vigorously under conditions of high fertility
- Optimum timing varies between crops and products. See labels for details
- Do not spray crops when wet or if rain imminent

Restrictions
- 2-chloroethylphosphonic acid is an anticholinesterase organophosphorus compound. Do not use if under medical advice not to work with such compounds
- Maximum number of treatments 1 per crop or yr
- Do not spray crops suffering from stress caused by any factor, during cold weather or period of night frost nor when soil very dry
- Do not apply to cereals within 10 d of herbicide or liquid fertilizer application
- Do not spray wheat or triticale where the leaf sheaths have split and the ear is visible

Crop-specific information
- Latest use: before 1st spikelet visible (GS 51) for spring barley, winter barley, winter rye; before flag leaf sheath opening (GS 47) for triticale, winter wheat
- HI cider apples, tomatoes 5 d

Environmental safety
- Harmful to aquatic organisms
- Avoid accidental deposits on painted objects such as cars, trucks, aircraft

Hazard classification and safety precautions
Hazard Harmful [2]; Irritant [1]
Transport code 8
Packaging group III
UN Number 3265
Risk phrases R20 [2]; R37, R38 [1]; R41, R52, R53a [1, 2]
Operator protection A, C; U05a, U08, U11, U13, U20b [1, 2]; U19c [2]
Environmental protection E13c, E38
Storage and disposal D01, D02, D09a, D10b, D12a
Medical advice M01

---

76   2-chloroethylphosphonic acid + mepiquat chloride

A plant growth regulator for reducing lodging in cereals

See also mepiquat chloride

Products
1 Clayton Mepiquat	Clayton	155:305 g/l	SL	12360
2 Standon Mepiquat Plus	Standon	155:305 g/l	SL	09373
3 Terpal	BASF	155:305 g/l	SL	02103

Uses
- Growth regulation in **durum wheat** *(off-label)* [3]
- Increasing yield in **winter barley** *(low lodging situations)* [1, 3]
- Lodging control in **spring barley** [1, 3]; **triticale, winter barley, winter rye, winter wheat** [1-3]

Extension of Authorisation for Minor Use (EAMUs)
- **durum wheat** 20063104 [3]

Approval information
- 2-chloroethylphosphonic acid (ethephon) and mepiquat chloride included in Annex I under EC Regulation 1107/2009

SEE SECTION 3 FOR PRODUCTS ALSO REGISTERED

- In 2006 CRD required that all products containing 2-chloroethylphosphonic acid should carry the following warning in the main area of the container label: "2-chloroethylphosphonic acid is an anticholinesterase organophosphate. Handle with care"
- Accepted by BBPA for use on malting barley

Efficacy guidance
- Best results achieved on crops growing vigorously under conditions of high fertility
- Recommended dose and timing vary with crop, cultivar, growing conditions, previous treatment and desired degree of lodging control. See label for details
- May be applied to crops undersown with grass or clovers
- Do not apply to crops if wet or rain expected as efficacy will be impaired

Restrictions
- 2-chloroethylphosphonic acid is an anticholinesterase organophosphorus compound. Do not use if under medical advice not to work with such compounds
- Maximum number of treatments 2 per crop
- Add an authorised non-ionic wetter to spray solution. See label for recommended product and rate
- Do not treat crops damaged by herbicides or stressed by drought, waterlogging etc
- Do not treat crops on soils of low fertility unless adequately fertilized
- Do not apply to winter cultivars sown in spring or treat winter barley, triticale or winter rye on soils with more than 10% organic matter (winter wheat may be treated)
- Do not apply at temperatures above 21°C

Crop-specific information
- Latest use: before ear visible (GS 49) for winter barley, spring barley, winter wheat and triticale; flag leaf just visible (GS 37) for winter rye
- Late tillering may be increased with crops subject to moisture stress and may reduce quality of malting barley

Environmental safety
- Do not use straw from treated cereals as a mulch or growing medium

Hazard classification and safety precautions
Hazard Harmful
Transport code 8
Packaging group III
UN Number 3265
Risk phrases R22a, R37, R53a [1-3]; R36 [2, 3]; R41, R53b [1]
Operator protection A, C [3]; U05a, U11 [1]; U20b [1-3]
Environmental protection E15a
Storage and disposal D01, D02, D09a [1-3]; D05, D10b, D12a [1]; D08, D10c [2, 3]
Medical advice M01 [1-3]; M05a [2, 3]

---

77   chlorothalonil

A protectant chlorophenyl fungicide for use in many crops and turf
FRAC mode of action code: M5

See also azoxystrobin + chlorothalonil
carbendazim + chlorothalonil

Products

1	Abringo	Belcrop	500 g/l	SC	15607
2	Acclaim	European Ag	500 g/l	SC	14905
3	Barclay Avoca	Barclay	500 g/l	SC	15549
4	Bravo 500	Syngenta	500 g/l	SC	14548
5	Busa	AgChem Access	500 g/l	SC	15805
6	CTL 500	Goldengrass	500 g/l	SC	14812
7	Damocles	AgChem Access	500 g/l	SC	15689
8	Hail	Becesane	500 g/l	SC	15378

FOR FULL CONDITIONS OF USE ALWAYS READ THE PRODUCT LABEL

Products – continued

9	Joules	Nufarm UK	500 g/l	SC	15730
10	Juliet	ChemSource	500 g/l	SC	15788
11	Juliet	ChemSource	500 g/l	SC	15947
12	Life Scientific Chlorothalonil	Life Scientific	500 g/l	SC	15369
13	Piper	Nufarm UK	500 g/l	SC	15786
14	Rio	ChemSource	500 g/l	SC	15874
15	Supreme	AgriGuard	500 g/l	SC	15395

Uses

- Basal stem rot in *narcissi* *(off-label)* [4, 12]
- Botrytis in *celeriac* *(off-label)*, *strawberries* *(off-label)* [4]
- Disease control in *all edible seed crops grown outdoors* *(off-label)*, *all non-edible seed crops grown outdoors* *(off-label)*, *durum wheat* *(off-label)*, *forest nurseries* *(off-label)*, *ornamental plant production* *(off-label)*, *rye* *(off-label)*, *triticale* *(off-label)* [4, 12]
- Downy mildew in *poppies for morphine production* *(off-label)* [4]
- Glume blotch in *spring wheat*, *winter wheat* [1-15]
- Leaf spot in *celeriac* *(off-label)*, *strawberries* *(off-label)* [4]
- Neck rot in *narcissi* *(off-label)* [4, 12]
- Rhynchosporium in *spring barley* *(moderate control)*, *winter barley* *(moderate control)* [1-15]
- Septoria leaf blotch in *spring wheat*, *winter wheat* [1-15]

Extension of Authorisation for Minor Use (EAMUs)

- *all edible seed crops grown outdoors* *20111130* [4], *20120380* [12]
- *all non-edible seed crops grown outdoors* *20111130* [4], *20120380* [12]
- *celeriac* *20110948* [4]
- *durum wheat* *20110944* [4], *20120381* [12]
- *forest nurseries* *20111130* [4], *20120380* [12]
- *narcissi* *20110943* [4], *20120379* [12]
- *ornamental plant production* *20111130* [4], *20120380* [12]
- *poppies for morphine production* *20110947* [4]
- *rye* *20110944* [4], *20120381* [12]
- *strawberries* *20110949* [4]
- *triticale* *20110944* [4], *20120381* [12]

Approval information

- Chlorothalonil included in Annex I under EC Regulation 1107/2009. However a Standing Committee decision in July 2007 to reduce to the limit of determination the MRL set for chlorothalonil in blackberries and raspberries led to revocation of approvals for use on these crops
- Accepted by BBPA for use on malting barley and hops

Efficacy guidance

- For some crops products differ in diseases listed as controlled. See label for details and for application rates, timing and number of sprays
- Apply as protective spray or as soon as disease appears and repeat as directed
- On cereals activity against Septoria may be reduced where serious mildew or rust present. In such conditions mix with suitable mildew or rust fungicide
- May be used for preventive and curative treatment of turf but treatment at a late stage of disease development may not be so successful and can leave bare patches of soil requiring renovation.

Restrictions

- Maximum number of treatments and maximum total doses vary with crop and product - see labels for details
- Operators of vehicle mounted equipment must use a vehicle fitted with a cab and a forced air filtration unit with a pesticide filter complying with HSE Guidance Note PM74 or to an equivalent or higher standard when making broadcast or air-assisted applications

Crop-specific information

- Latest time of application to cereals varies with product. Consult label

SEE SECTION 3 FOR PRODUCTS ALSO REGISTERED

- Latest use for winter oilseed rape before flowering; end Aug in yr of harvest for blackcurrants, gooseberries, redcurrants,
- HI 8 wk for field beans; 6 wk for combining peas; 28 d or before 31 Aug for post-harvest treatment for blackcurrants, gooseberries, redcurrants; 14 d for onions, strawberries; 10 d for hops; 7-14 d for broccoli, Brussels sprouts, cabbages, cauliflowers, celery, onions, potatoes; 3 d for cane fruit; 12-48 h for protected cucumbers, protected tomatoes
- For Botrytis control in strawberries important to start spraying early in flowering period and repeat at least 3 times at 10 d intervals
- On strawberries some scorching of calyx may occur with protected crops

Environmental safety
- Dangerous for the environment
- Very toxic to aquatic organisms
- Do not spray from the air within 250 m horizontal distance of surface waters or ditches
- Broadcast air-assisted LERAP (18 m); LERAP Category B

Hazard classification and safety precautions
Hazard Harmful, Dangerous for the environment
Transport code 6.1 [9, 13]; 9 [1-8, 10-12, 14, 15]
Packaging group II [9, 13]; III [1-8, 10-12, 14, 15]
UN Number 2902 [9, 13]; 3082 [1-8, 10-12, 14, 15]
Risk phrases R20, R36, R37, R40, R43, R50, R53a
Operator protection A, H [1-15]; C [1, 2, 4-15]; J, M [1, 2, 4, 5, 7, 8, 10-12, 14]; U02a, U05a, U09a, U19a, U20a
Environmental protection E15a, E16a, E16b, E18, E38; E17b (18 m)
Storage and disposal D01, D02, D05, D09a, D10c, D12a
Medical advice M03 [3]

78   chlorothalonil + cyproconazole

A systemic protectant and curative fungicide mixture for cereals, combining peas and some products can also be used in field beans
FRAC mode of action code: M5 + 3

See also cyproconazole

Products

1	Alto Elite	Syngenta	375:40 g/l	SC	08467
2	Octolan	Syngenta	375:40 g/l	SC	11675
3	SAN 703	Syngenta	375:40 g/l	SC	11676

Uses
- Brown rust in *spring barley*, *winter barley*, *winter wheat* [1-3]
- Chocolate spot in *spring field beans*, *winter field beans* [1-3]
- Eyespot in *winter barley*, *winter wheat* [1-3]
- Glume blotch in *winter wheat* [1]
- Grey mould in *combining peas* [1-3]
- Net blotch in *spring barley*, *winter barley* [1-3]
- Powdery mildew in *spring barley*, *winter barley*, *winter wheat* [1-3]
- Rhynchosporium in *spring barley*, *winter barley* [1-3]
- Rust in *spring field beans*, *winter field beans* [1]
- Septoria leaf blotch in *winter wheat* [1-3]
- Yellow rust in *spring barley*, *winter barley*, *winter wheat* [1-3]

Approval information
- Chlorothalonil and cyproconazole included in Annex I under EC Regulation 1107/2009
- Accepted by BBPA for use on malting barley

Efficacy guidance
- Apply at first signs of infection or as soon as disease becomes active
- A repeat application may be made if re-infection occurs

FOR FULL CONDITIONS OF USE ALWAYS READ THE PRODUCT LABEL

- For established mildew tank-mix with an approved mildewicide
- When applied prior to third node detectable (GS 33) a useful reduction of eyespot will be obtained
- Cyproconazole is a DMI fungicide. Resistance to some DMI fungicides has been identified in Septoria leaf blotch which may seriously affect performance of some products. For further advice contact a specialist advisor and visit the Fungicide Resistance Action Group (FRAG)-UK website

Restrictions
- Maximum total dose equivalent to 2 full dose treatments
- Do not apply at concentrations higher than recommended

Crop-specific information
- Latest use: before beginning of anthesis (GS 60) for barley; before caryopsis watery ripe (GS 71) for wheat
- HI peas, field beans 6 wk
- If applied to winter wheat in spring at GS 30-33 straw shortening may occur but yield is not reduced

Environmental safety
- Dangerous for the environment
- Very toxic to aquatic organisms
- LERAP Category B

Hazard classification and safety precautions
Hazard Harmful, Dangerous for the environment
Transport code 9
Packaging group III
UN Number 3082
Risk phrases R20, R37, R40, R41, R43, R50, R53a
Operator protection A, C, H, M; U05a, U19a [1-3]; U11, U20a [1]; U12, U20b [2, 3]
Environmental protection E15a, E16b [1]; E15b [2, 3]; E16a, E38 [1-3]
Storage and disposal D01, D02, D09a, D10c, D12a [1-3]; D05 [1]
Medical advice M04a

79   chlorothalonil + cyproconazole + propiconazole

A broad-spectrum fungicide mixture for cereals
FRAC mode of action code: M5 + 3 + 3

See also cyproconazole
propiconazole

Products

Cherokee	Syngenta	375:50:62.5 g/l	SE	13251

Uses
- Brown rust in **spring barley**, **winter barley**, **winter wheat**
- Net blotch in **spring barley** *(reduction)*, **winter barley** *(reduction)*
- Powdery mildew in **spring barley**, **winter barley**, **winter wheat** *(moderate)*
- Rhynchosporium in **spring barley** *(reduction)*, **winter barley** *(reduction)*
- Septoria leaf blotch in **winter wheat**
- Yellow rust in **winter wheat**

Approval information
- Chlorothalonil, propiconazole and cyproconazole included in Annex I under EC Regulation 1107/2009

Efficacy guidance
- Best results obtained from treatment at the early stages of disease development

SEE SECTION 3 FOR PRODUCTS ALSO REGISTERED

- Cyproconazole and propiconazole are DMI fungicides. Resistance to some DMI fungicides has been identified in Septoria leaf blotch which may seriously affect performance of some products. For further advice contact a specialist advisor and visit the Fungicide Resistance Action Group (FRAG)-UK website
- Product should be used as part of an Integrated Crop Management strategy incorporating other methods of control and including, where appropriate, other fungicides with a different mode of action
- Product should be used preventatively and not relied on for its curative potential
- Always follow FRAG guidelines for resistance management. See Section 5 for more information

Restrictions
- Maximum total dose equivalent to two full dose treatments on wheat and barley
- Do not treat crops under stress
- Spray solution must be used as soon as possible after mixing

Crop-specific information
- Latest use: before caryopsis watery ripe (GS 71) for winter wheat; before first spikelet just visible (GS 51) for barley
- When applied to winter wheat in the spring straw shortening may occur which causes no loss of yield

Environmental safety
- Dangerous for the environment
- Very toxic to aquatic organisms
- LERAP Category B

Hazard classification and safety precautions
Hazard Harmful, Dangerous for the environment
Transport code 9
Packaging group III
UN Number 3082
Risk phrases R20, R36, R37, R40, R43, R50, R53a
Operator protection A, C, H; U02a, U05a, U09a, U10, U11, U14, U15, U19a, U20b
Environmental protection E15b, E16a, E16b, E34, E38
Storage and disposal D01, D02, D05, D07, D09a, D10c, D12a
Medical advice M04a

80 chlorothalonil + fludioxonil + propiconazole

A fungicide treatment for amenity grassland
FRAC mode of action code: M5 + 12 + 3

See also fludioxonil
propiconazole

Products

Instrata	Syngenta	362:14.5:56.9 g/l	SE	14154

Uses
- Anthracnose in **amenity grassland** *(reduction)*, **managed amenity turf** *(reduction)*
- Brown patch in **amenity grassland**, **managed amenity turf**
- Dollar spot in **amenity grassland**, **managed amenity turf**
- Fusarium patch in **amenity grassland**, **managed amenity turf**

Approval information
- Chlorothalonil, fludioxonil and propiconazole included in Annex I under EC Regulation 1107/2009

Environmental safety
- LERAP Category B

Hazard classification and safety precautions
Hazard Toxic, Dangerous for the environment

FOR FULL CONDITIONS OF USE ALWAYS READ THE PRODUCT LABEL

Transport code 9
Packaging group III
UN Number 3082
Risk phrases R23, R36, R40, R43, R50, R53a
Operator protection A, C, H; U02a, U05a, U09a, U10, U11, U19a, U19c, U20b
Environmental protection E15b, E16a, E34, E38
Storage and disposal D01, D02, D09a, D12a
Medical advice M04a

81   chlorothalonil + flutriafol

A systemic eradicant and protectant fungicide for winter wheat
FRAC mode of action code: M5 + 3

See also flutriafol

Products

1	Impact Excel	Headland	300:47 g/l	SC	11547
2	Prospa	Headland	300:47 g/l	SC	11548

Uses

- Brown rust in **winter wheat**
- Foliar disease control in **durum wheat** *(off-label)*, **grass seed crops** *(off-label)*, **spring rye** *(off-label)*, **triticale** *(off-label)*, **winter rye** *(off-label)*
- Late ear diseases in **winter wheat**
- Powdery mildew in **winter wheat**
- Septoria leaf blotch in **winter wheat**
- Yellow rust in **winter wheat**

Extension of Authorisation for Minor Use (EAMUs)

- **durum wheat** *20052890* [1], *20052889* [2]
- **grass seed crops** *20052890* [1], *20052889* [2]
- **spring rye** *20052890* [1], *20052889* [2]
- **triticale** *20052890* [1], *20052889* [2]
- **winter rye** *20052890* [1], *20052889* [2]

Approval information

- Chlorothalonil and flutriafol included in Annex I under EC Regulation 1107/2009

Efficacy guidance

- Generally disease control and yield benefit will be optimised when application is made at an early stage of disease development
- Apply as soon as disease is seen establishing in the crop

Restrictions

- Maximum number of treatments 2 per crop on winter wheat

Crop-specific information

- Latest use: before early milk stage (GS 73)
- On certain cultivars with erect leaves high transpiration can result in flag leaf tip scorch. This may be increased by treatment but does not affect yield

Environmental safety

- Dangerous for the environment [1, 2]
- Very toxic to aquatic organisms [1, 2]
- Dangerous to fish or other aquatic life. Do not contaminate surface waters or ditches with chemical or used container [2]
- LERAP Category B

Hazard classification and safety precautions

Hazard Toxic, Dangerous for the environment
Transport code 6.1

SEE SECTION 3 FOR PRODUCTS ALSO REGISTERED

Packaging group II
UN Number 2902
Risk phrases R23, R40, R41, R43, R50, R53a
Operator protection A, C, H; U02a, U05a, U09a, U11, U19a, U20b
Environmental protection E13b [2]; E15a [1]; E16a, E38 [1, 2]
Storage and disposal D01, D02, D09a, D10b [1, 2]; D05 [2]

82   chlorothalonil + mancozeb

A multi-site protectant fungicide mixture for blight control in winter wheat
FRAC mode of action code: M5 + M3

See also mancozeb

Products

Guru	Interfarm	286:194 g/l	SC	15268

Uses

- Glume blotch in **winter wheat**
- Septoria leaf blotch in **winter wheat**

Approval information

- Chlorothalonil and mancozeb included in Annex I under EC Regulation 1107/2009

Efficacy guidance

- Best results on winter wheat achieved from protective applications to the flag leaf. If disease is already present on the lower leaves treat as soon as the flag leaf is just visible (GS 37).
- Activity on Septoria may be reduced in the presence of a severe mildew infection.

Restrictions

- Maximum total dose on winter wheat equivalent to 1.5 x full dose treatment.
- Do not treat crops under stress from frost, drought, water-logging, trace element deficiency or pest attack.

Crop-specific information

- Latest use: before grain watery ripe (GS 71)

Environmental safety

- Dangerous for the environment
- Very toxic to aquatic organisms
- LERAP Category B

Hazard classification and safety precautions

Hazard Irritant, Dangerous for the environment
Transport code 9
Packaging group III
UN Number 3082
Risk phrases R37, R40, R43, R50, R53a
Operator protection A, C, H, M; U02a, U05a, U08, U10, U11, U13, U14, U19a, U20b
Environmental protection E16a, E16b, E38
Storage and disposal D01, D02, D05, D09a, D10b, D12a

83   chlorothalonil + metalaxyl-M

A systemic and protectant fungicide for various crops
FRAC mode of action code: M5 + 4

See also metalaxyl-M

Products

Folio Gold	Syngenta	500:37.5 g/l	SC	14368

Uses

- Alternaria in **brussels sprouts** *(moderate control)*, **cauliflowers** *(moderate control)*
- Disease control in **forest nurseries** *(off-label)*, **ornamental plant production** *(off-label)*, **poppies for morphine production** *(off-label)*, **protected ornamentals** *(off-label)*
- Downy mildew in **broad beans**, **brussels sprouts**, **cauliflowers**, **spring field beans**, **winter field beans**
- Ring spot in **brussels sprouts** *(reduction)*, **cauliflowers** *(reduction)*
- White blister in **brussels sprouts**, **cauliflowers**

Extension of Authorisation for Minor Use (EAMUs)

- **forest nurseries** *20111128*
- **ornamental plant production** *20111128*
- **poppies for morphine production** *20090251*
- **protected ornamentals** *20120032*

Approval information

- Chlorothalonil and metalaxyl-M included in Annex I under EC Regulation 1107/2009

Efficacy guidance

- Apply at first signs of disease or when weather conditions favourable for disease pressure
- Best results obtained when used in a full and well-timed programme. Repeat treatment at 14-21 d intervals if necessary
- Treatment of established disease will be less effective
- Evidence of effectiveness in bulb onions, shallots and leeks is limited

Restrictions

- Maximum total dose equivalent to 2 full doses on broad beans, field beans, cauliflowers, calabrese; 3 full doses on Brussels sprouts, leeks, onions and shallots

Crop-specific information

- HI 14 d for all crops

Environmental safety

- Dangerous for the environment
- Very toxic to aquatic organisms
- LERAP Category B

Hazard classification and safety precautions

Hazard Harmful, Dangerous for the environment
Transport code 9
Packaging group III
UN Number 3082
Risk phrases R20, R36, R37, R38, R40, R43, R50, R53a
Operator protection A, C, H; U05a, U09a, U15, U20a
Environmental protection E15a, E16a, E38
Consumer protection C02a (14 d)
Storage and disposal D01, D02, D05, D09a, D10c, D12a

SEE SECTION 3 FOR PRODUCTS ALSO REGISTERED

84   chlorothalonil + picoxystrobin

A broad spectrum fungicide mixture for cereals
FRAC mode of action code: M5 + 11

See also picoxystrobin

Products

1	Credo	DuPont	500:100 g/l	SC	14577
2	Zimbrail	DuPont	500:100 g/l	SC	14735

Uses

- Brown rust in **spring barley**, **spring wheat**, **winter barley**, **winter wheat**
- Glume blotch in **spring wheat**, **winter wheat**
- Net blotch in **spring barley**, **winter barley**
- Rhynchosporium in **spring barley**, **winter barley**
- Septoria leaf blotch in **spring wheat**, **winter wheat**
- Tan spot in **spring wheat**, **winter wheat**
- Yellow rust in **spring wheat**, **winter wheat**

Approval information

- Chlorothalonil and picoxystrobin included in Annex I under EC Regulation 1107/2009
- Accepted by BBPA for use on malting barley
- Approval expiry 31 Dec 2013 [1, 2]

Efficacy guidance

- Best results obtained as a protectant treatment, or when disease first seen in crop, applied in good growing conditions with adequate soil moisture
- Disease control for 4-6 wk normally achieved during stem elongation
- Results may be less reliable when used on crops under stress
- Treatments for protection against ear disease should be made at ear emergence
- Picoxystrobin is a member of the QoI cross resistance group. Product should be used preventatively and not relied on for its curative potential
- Use product as part of an Integrated Crop Management strategy incorporating other methods of control, including where appropriate other fungicides with a different mode of action. Do not apply more than two foliar applications of QoI containing products to any cereal crop
- Control of many diseases may be improved by use in mixture with an appropriate triazole
- There is a significant risk of widespread resistance occurring in *Septoria tritici* populations in UK. Failure to follow resistance management action may result in reduced levels of disease control
- Strains of barley powdery mildew resistant to QoI's are common in the UK

Restrictions

- Maximum number of treatments 2 per crop of wheat or barley

Crop-specific information

- Latest use: before first spikelet just visible (GS 51) for barley; before grain watery ripe (GS 71) for wheat

Environmental safety

- Dangerous for the environment
- Very toxic to aquatic organisms
- LERAP Category B

Hazard classification and safety precautions

Hazard Harmful, Dangerous for the environment
Transport code 9
Packaging group III
UN Number 3082
Risk phrases R20, R37, R40, R43, R50, R53a
Operator protection A, H; U05a, U09a, U14, U15, U19a, U20a
Environmental protection E15a, E16a, E16b, E38
Storage and disposal D01, D02, D05, D09a, D10c, D12a

FOR FULL CONDITIONS OF USE ALWAYS READ THE PRODUCT LABEL

85   chlorothalonil + propamocarb hydrochloride

A contact and systemic fungicide mixture for blight control in potatoes
FRAC mode of action code: M5 + 28

See also propamocarb hydrochloride

Products

1 Ledgend	ChemSource	375:375 g/l	SC	15024
2 Pan Magician	Pan Agriculture	375:375 g/l	SC	11992

Uses
- Blight in **potatoes**

Approval information
- Chlorothalonil and propamocarb hydrochloride included in Annex I under EC Regulation 1107/2009
- Approval expiry 31 Mar 2013 [1, 2]

Efficacy guidance
- Commence treatment early in the season as soon as there is risk of infection
- In the absence of a blight warning treatment should start just before potatoes meet along the row
- Use only as a protectant. Stop use when blight readily visible (1% leaf area destroyed)
- Repeat sprays at 10-14 d intervals depending on blight infection risk. See label for details
- Complete blight spray programme after end Aug up to haulm destruction with protectant fungicides

Restrictions
- Maximum total dose equivalent to 6 full dose treatments on potatoes
- Apply to dry foliage. Do not apply if rainfall or irrigation imminent

Crop-specific information
- HI 7 d for potatoes

Environmental safety
- Dangerous for the environment
- Very toxic to aquatic organisms
- LERAP Category B

Hazard classification and safety precautions
Hazard Harmful, Dangerous for the environment
Transport code 9
Packaging group III
UN Number 3082
Risk phrases R20, R40, R41, R43, R50, R53a
Operator protection A, C, H; U05a, U08, U11, U14, U19a, U20a
Environmental protection E15a, E16a, E16b, E34, E38
Storage and disposal D01, D02, D05, D09a, D10b, D12a

86   chlorothalonil + propiconazole

A systemic and protectant fungicide for winter wheat and barley
FRAC mode of action code: M5 + 3

See also propiconazole

Products

1 Avoca Premium	Barclay	250:62.5 g/l	SC	15619
2 Prairie	Syngenta	250:62.5 g/l	SC	13994

SEE SECTION 3 FOR PRODUCTS ALSO REGISTERED

SECTION 2

Uses

- Brown rust in **spring barley**, **spring wheat**, **winter barley**, **winter wheat** [1, 2]
- Glume blotch in **spring wheat**, **winter wheat** [1, 2]
- Rhynchosporium in **spring barley**, **winter barley** [2]; **spring barley** *(moderate control)*, **winter barley** *(moderate control)* [1]
- Septoria leaf blotch in **spring wheat**, **winter wheat** [1, 2]
- Yellow rust in **spring wheat**, **winter wheat** [1, 2]

Approval information

- Chlorothalonil and propiconazole included in Annex I under EC Regulation 1107/2009
- Accepted by BBPA for use on malting barley

Efficacy guidance

- On wheat apply from start of flag leaf emergence up to and including when ears just fully emerged (GS 37-59), on barley at any time to ears fully emerged (GS 59)
- Best results achieved from early treatment, especially if weather wet, or as soon as disease develops

Restrictions

- Maximum number of treatments 2 per crop or 1 per crop if other propiconazole based fungicide used in programme

Crop-specific information

- Latest use: up to and including emergence of ear just complete (GS 59).
- HI 42 d

Environmental safety

- Irritating to eyes, skin and respiratory system
- Risk of serious damage to eyes
- Dangerous to fish or other aquatic life. Do not contaminate surface waters or ditches with chemical or used container
- LERAP Category B

Hazard classification and safety precautions

Hazard Harmful, Dangerous for the environment
Transport code 9
Packaging group III
UN Number 3082
Risk phrases R20, R36, R37, R40, R43, R50, R53a
Operator protection A, C, H; U02a, U05a, U11, U20b
Environmental protection E15b, E16a, E16b, E38
Storage and disposal D01, D02, D09a, D10c, D12a

87 chlorothalonil + proquinazid

A chlorophenyl and quinazolinone fungicide mixture for powdery mildew control in cereals
FRAC mode of action code: M5 + U7

Products

Fielder SE	DuPont	500:25 g/l	SE	15120

Uses

- Powdery mildew in **spring barley**, **spring wheat**, **winter barley**, **winter wheat**
- Rhynchosporium in **spring barley** *(suppression only)*, **winter barley** *(suppression only)*
- Septoria leaf blotch in **spring wheat** *(suppression only)*, **winter wheat** *(suppression only)*

Approval information

- Chlorothalonil and proquinazid included in Annex 1 under EC Regulation 1107/2009

Efficacy guidance

- If disease has actively spread to new growth, application should be in tank-mixture with a fungicide that has an alternative mode of action with curative activity.

FOR FULL CONDITIONS OF USE ALWAYS READ THE PRODUCT LABEL

- If two applications are used consecutively in a crop, the second application should be in tank-mixture with a fungicide with an alternative mode of action against mildew, i.e. not a quinazolinone or quinoline fungicide.
- Should be considered as a member of the quinoline fungicides for resistance management.

Restrictions
- Maximum number of applications: 2 per crop.

Crop-specific information
- Latest use in wheat is before full flowering (GS65)
- Latest use in barley is before first awms visible

Environmental safety
- LERAP Category B

Hazard classification and safety precautions
Hazard Harmful, Dangerous for the environment
Transport code 9
Packaging group III
UN Number 3082
Risk phrases R20, R37, R40, R41, R43, R50, R53a
Operator protection A, C, H; U05a, U09a, U12, U14, U15, U20a
Environmental protection E15a, E15b, E16a, E38, E40b
Storage and disposal D01, D02, D05, D09a, D10c, D12a
Medical advice M05a

88   chlorothalonil + pyrimethanil

A fungicide mixture for use in combining peas
FRAC mode of action code: M5 + 9

See also pyrimethanil

Products

Walabi	BASF	375:150 g/l	SC	12265

Uses
- Ascochyta in **combining peas**
- Botrytis in **combining peas**
- Chocolate spot in **spring field beans**, **winter field beans**
- Mycosphaerella in **combining peas**

Approval information
- Chlorothalonil and pyrimethanil included in Annex I under EC Regulation 1107/2009

Efficacy guidance
- Use only as a protectant spray in a two spray programme
- For best results ensure good foliar cover

Restrictions
- Maximum total dose equivalent to two full dose treatments
- Consult processors before use on crops for processing

Crop-specific information
- HI 6 wk for combining peas; 8 wk for field beans

Environmental safety
- Dangerous for the environment
- Very toxic to aquatic organisms
- LERAP Category B

Hazard classification and safety precautions
Hazard Harmful, Dangerous for the environment

SEE SECTION 3 FOR PRODUCTS ALSO REGISTERED

Transport code 9
Packaging group III
UN Number 3082
Risk phrases R36, R40, R43, R50, R53a
Operator protection A, C, H; U02a, U11
Environmental protection E15a, E16a
Storage and disposal D01, D02, D05, D09a, D10c, D12a
Medical advice M03

89   chlorothalonil + tebuconazole

A protectant and systemic fungicide mixture for winter wheat
FRAC mode of action code: M5 + 3

See also tebuconazole

Products

1 Fezan Plus	Sipcam	166:60 g/l	SC	15490
2 Pentangle	Nufarm UK	500:180 g/l	SC	13746
3 Timpani	Nufarm UK	250:90 g/l	SC	14651

Uses
- Fusarium ear blight in *winter wheat* [1]
- Septoria leaf blotch in *winter wheat* [1-3]
- Sooty moulds in *winter wheat* [1]

Approval information
- Chlorothalonil and tebuconazole included in Annex I under EC Regulation 1107/2009

Efficacy guidance
- Latest time of application before anthesis (GS61)
- Maximum individual dose 1.0 l/ha. Where disease pressure is severe, a second application may be required but applications after flag leaf visible (GS39) may only result in a reduction of disease severity

Environmental safety
- LERAP Category B

Hazard classification and safety precautions
Hazard Very toxic [2, 3]; Harmful [1]; Dangerous for the environment [1-3]
Transport code 6.1 [2]; 9 [1, 3]
Packaging group II [2]; III [1, 3]
UN Number 2902 [2]; 3082 [1, 3]
Risk phrases R20 [1]; R26, R37, R41, R43 [2, 3]; R40, R53a, R63 [1-3]; R50 [1, 2]; R51 [3]
Operator protection A, H [1-3]; C [2, 3]; U02a, U09a, U10, U11, U13, U14, U15, U19b [2, 3]; U05a, U19a [1-3]
Environmental protection E15a, E16b [2, 3]; E15b [1]; E16a, E34, E38 [1-3]
Storage and disposal D01, D02, D09a, D12a [1-3]; D05, D10c [1]; D10b [2, 3]
Medical advice M03 [1]; M04a [2, 3]

90   chlorotoluron

A contact and residual urea herbicide for cereals but all approvals expired 2011
HRAC mode of action code: C2

91   chlorotoluron + diflufenican

A soil and foliar acting herbicide mixture for winter cereals
HRAC mode of action code: C2 + F1

See also diflufenican

Products

1	Dicurane Surpass	Makhteshim	400:25 g/l	SC	14544
2	Hekla	Makhteshim	450:15 g/l	SC	14298
3	Kula	Makhteshim	450:10 g/l	SC	14299
4	Tawny	Makhteshim	400:25 g/l	SC	14038

Uses

- Annual dicotyledons in **durum wheat**, **triticale**, **winter barley**, **winter wheat** [1-4]; **spring barley** *(off-label)*, **spring wheat** *(off-label)* [2-4]
- Annual grasses in **durum wheat**, **spring barley** *(off-label)*, **spring wheat** *(off-label)*, **triticale**, **winter barley**, **winter wheat** [2-4]
- Annual meadow grass in **durum wheat**, **spring barley** *(off-label)*, **spring wheat** *(off-label)*, **triticale**, **winter barley**, **winter wheat** [2, 3]
- Wild oats in **spring barley** *(off-label)*, **spring wheat** *(off-label)* [4]

Extension of Authorisation for Minor Use (EAMUs)

- **spring barley** *20121493* [2], *20121494* [3], *20121508* [4]
- **spring wheat** *20121493* [2], *20121494* [3], *20121508* [4]

Approval information

- Chlorotoluron and diflufenican included in Annex I under EC Regulation 1107/2009

Efficacy guidance

- Best results obtained in crops sown in a fine firm seedbed with clods no greater than fist size and adequate moisture present at spraying. Good even spray coverage of the soil is essential
- Good weed control depends on burial and dispersal of any trash or straw before drilling
- Weed control may be reduced in prolonged periods of below average rainfall
- In direct drilled crops soil surface should be cultivated and the drill slots closed before spraying
- Spring germinating weeds will not be controlled
- Fluffy seedbeds should be rolled before treatment
- Harrowing a treated crop may reduce the level of weed control

Restrictions

- Maximum number of treatments 1 per crop for wheat, barley
- Do not apply if heavy rain is expected within 4 h or if crop is stressed
- Do not roll for 7 d before or after application. Do not roll autumn crops until spring
- Do not treat undersown crops, or those to be undersown
- Do not apply on soils with more than 10% organic matter
- Treat only listed varieties of wheat and barley

Crop-specific information

- Latest use: before end Feb for winter barley, winter wheat
- Transient leaf discolouration occasionally occurs after treatment of wheat or barley
- Treatment of wheat or barley on stony or gravelly soils may cause damage especially if heavy rain falls after application
- Early sown crops of wheat or barley may be damaged if treated during a period of rapid autumn growth

Following crops guidance

- Any crop may be drilled or planted following normal harvest of a treated crop provided the soil is ploughed to 150 mm
- In the event of failure of a treated crop only winter wheat or winter barley may be drilled immediately after ploughing. A period of 12 wk must elapse after ploughing before spring sowing of wheat, barley, oilseed rape, peas, beans, sugar beet, potatoes, carrots, edible brassicas or onions

SECTION 2

SEE SECTION 3 FOR PRODUCTS ALSO REGISTERED

- Successive treatments with any products containing diflufenican can lead to soil build-up and inversion ploughing to 150 mm must precede sowing any following non-cereal crop. Even where ploughing occurs some crops may be damaged

Environmental safety
- Dangerous for the environment
- Very toxic to aquatic organisms
- Do not apply to dry, cracked or waterlogged soils
- Some pesticides pose a greater threat of contamination of water than others and chlorotoluron is one of these pesticides. Take special care when applying chlorotoluron near water and do not apply if heavy rain is forecast
- LERAP Category B

Hazard classification and safety precautions
Hazard Harmful, Dangerous for the environment
Transport code 9
Packaging group III
UN Number 3082
Risk phrases R40, R50, R53a, R63
Operator protection A, H [1-4]; C [4]; U05a [1-4]; U08, U13, U19a, U20a [2, 3]; U14, U15 [1, 4]
Environmental protection E15a, E34 [2, 3]; E16a, E38 [1-4]; E16b [1, 4]
Storage and disposal D01, D02, D12a [1-4]; D05, D09a, D10b [2, 3]
Medical advice M04a [1, 4]

---

92 chlorpropham

A residual carbamate herbicide and potato sprout suppressant
HRAC mode of action code: K2

See also chloridazon + chlorpropham + metamitron

Products

1	BL 500	United Phosphorus	500 g/l	HN	14387
2	Cleancrop Amigo 2	AgriChem BV	400 g/l	EC	15419
3	Gro-Stop 100	Certis	300 g/l	HN	14182
4	Gro-Stop Fog	Certis	300 g/l	HN	14183
5	Gro-Stop Ready	Certis	120 g/l	EW	15109
6	Gro-Stop Solid	Certis	100% w/w	BR	14103
7	Intruder	AgriChem BV	400 g/l	EC	15076
8	MSS CIPC 50 M	United Phosphorus	500 g/l	HN	14388
9	Pro-Long	United Phosphorus	500 g/l	HN	14389

Uses
- Annual dicotyledons in **bulb onions**, **leeks** *(off-label)* [2]; **chard** *(off-label)*, **endives**, **forest nurseries** *(off-label)*, **herbs (see appendix 6)** *(off-label)*, **lettuce**, **ornamental plant production**, **salad onions**, **shallots**, **spinach** *(off-label)* [2, 7]; **onions** [7]
- Annual meadow grass in **bulb onions** [2]; **chard** *(off-label)*, **endives**, **forest nurseries** *(off-label)*, **herbs (see appendix 6)** *(off-label)*, **lettuce**, **ornamental plant production**, **salad onions**, **shallots**, **spinach** *(off-label)* [2, 7]; **onions** [7]
- Chickweed in **leeks** *(off-label)* [2]
- Polygonums in **leeks** *(off-label)* [2]
- Small nettle in **leeks** *(off-label)* [2]
- Sprout suppression in **potatoes** [5]; **ware potatoes** [3, 8]; **ware potatoes** *(thermal fog)* [1, 4, 6, 9]
- Wild oats in **bulb onions**, **chard** *(off-label)*, **endives**, **forest nurseries** *(off-label)*, **herbs (see appendix 6)** *(off-label)*, **lettuce**, **ornamental plant production**, **salad onions**, **shallots**, **spinach** *(off-label)* [2]

Extension of Authorisation for Minor Use (EAMUs)
- **chard** 20120728 [2], 20110596 [7]

FOR FULL CONDITIONS OF USE ALWAYS READ THE PRODUCT LABEL

- *forest nurseries* 20120727 [2], 20111127 [7]
- *herbs (see appendix 6)* 20120728 [2], 20110596 [7]
- *leeks* 20121841 [2]
- *spinach* 20120728 [2], 20110596 [7]

Approval information
- Chlorpropham included in Annex I under EC Regulation 1107/2009
- Some products are formulated for application by thermal fogging. See labels for details

Efficacy guidance
- For sprout suppression apply with suitable fogging or rotary atomiser equipment over dry tubers before sprouting commences. Repeat applications may be needed. See labels for details
- Best results on potatoes obtained in purpose-built box stores with suitable forced draft ventilation. Potatoes in bulk stores should not be stacked more than 3 m high
- Blockage of air spaces between tubers prevents circulation of vapour and consequent loss of efficacy
- It is important to treat potatoes before the eyes open to obtain best results
- Effectiveness of fogging reduced in non-dedicated stores without proper insulation and temperature controls. Best results obtained at 5-10°C and 75-80% humidity

Restrictions
- Only clean, mature, disease-free potatoes should be treated for sprout suppression. Use of chlorpropham can inhibit tuber wound healing and the severity of skin spot infection in store may be increased if damaged tubers are treated
- Do not fog potatoes with a high level of skin spot
- Do not use on potatoes for seed. Do not handle, dry or store seed potatoes or any other seed or bulbs in boxes or buildings in which potatoes are being or have been treated
- Do not remove treated potatoes from store for sale or processing for at least 21 d after application

Crop-specific information
- Cure potatoes according to label instructions before treatment and allow 2 days - 4 wks between completion of loading into store and first treatment - check labels.

Following crops guidance
- There is a risk of damage to seed potatoes which are handled or stored in boxes or buildings previously treated with chlorpropham

Environmental safety
- Dangerous for the environment
- Toxic to aquatic organisms
- Keep unprotected persons out of treated stores for at least 24 h after application
- Keep in original container, tightly closed, in a safe place, under lock and key

Hazard classification and safety precautions
Hazard Toxic, Highly flammable [1, 8, 9]; Harmful [2-5, 7]; Dangerous for the environment [1-9]
Transport code 6.1 [1, 3-5, 8, 9]; 9 [2, 6]
Packaging group III [1-6, 8, 9]
UN Number 2902 [1, 3-5, 8, 9]; 3077 [6]; 3082 [2]
Risk phrases R22a, R22b, R38, R67 [2, 7]; R23, R24 [1, 8, 9]; R25 [1, 9]; R36 [1, 2, 7-9]; R39 [1, 5, 8, 9]; R40 [2-5, 7]; R42 [3, 4]; R43 [2-4, 7]; R48 [2, 5, 7]; R51, R53a [1-9]
Operator protection A, H [1-9]; C [1, 2, 7-9]; D, E, M [1, 3, 4, 6, 8, 9]; J [1, 6, 8, 9]; U01, U19b [6]; U02a [2-4, 6, 7]; U04a, U15, U20a [3, 4]; U05a, U19a [1-4, 6-9]; U08 [1, 3, 4, 6, 8, 9]; U09a, U11, U16b [2, 7]; U14 [3-5]; U20b [1, 2, 6-9]
Environmental protection E02a [1, 8, 9] (24 h); E15a [1, 3, 4, 8, 9]; E15b, E38 [2, 5-7]; E22c [2, 7]; E34 [1-9]
Consumer protection C02b, C12 [3, 4]; C02c [1, 8, 9]
Storage and disposal D01, D02 [1-9]; D05 [1, 3-6, 8, 9]; D06b [1, 8]; D07 [2-4, 6, 7]; D09a, D10c [2, 7]; D09b [1, 3, 4, 8, 9]; D10a, D12b [1, 8, 9]; D11a [3-6]; D12a [2-7]
Medical advice M03, M05b [2, 7]; M04a [1, 3, 4, 6, 8, 9]

SEE SECTION 3 FOR PRODUCTS ALSO REGISTERED

93   chlorpropham + metamitron

A contact and residual herbicide mixture for beet crops
HRAC mode of action code: K2 + C1

See also metamitron

Products

Newgold	United Phosphorus	24.5:280 g/l	SE	15338

Uses

- Annual dicotyledons in **fodder beet**, **mangels**, **sugar beet**
- Grass weeds in **fodder beet**, **mangels**, **sugar beet**

Approval information

- Chlorpropham and metamitron included in Annex I under EC Regulation 1107/2009

Efficacy guidance

- May be used pre-emergence alone or post-emergence in tank mixture or with an authorised adjuvant oil
- Use as part of a spray programme. For optimum efficacy a full programme of pre- and post-emergence sprays is recommended
- Ideally one pre-emergence application should be followed by a repeat overall low dose programme on mineral soils. On organic soils a programme of up to five post-emergence sprays is likely to be most effective
- Product combines residual and contact activity. Adequate soil moisture and a fine firm seedbed are essential for optimum residual activity

Restrictions

- Maximum number of treatments one pre-emergence plus three post-emergence or, where no pre-emergence treatment has been applied, up to five post-emergence treatments with an adjuvant oil

Crop-specific information

- Latest use: before crop foliage meets across the rows

Following crops guidance

- Only sugar beet, fodder beet or mangels should be sown as following crops within 4 mth of treatment. After normal harvest of a treated crop land should be mouldboard ploughed to 15 cm after which any spring crops may be sown or planted

Environmental safety

- Dangerous for the environment
- Very toxic to aquatic organisms

Hazard classification and safety precautions

Hazard Dangerous for the environment
Transport code 9
Packaging group III
UN Number 3082
Risk phrases R22a, R36, R38, R43, R50, R53a
Operator protection A, C; U05a, U08, U11, U14, U15, U19a, U20b
Environmental protection E15a, E34
Storage and disposal D01, D02, D09a, D10c
Medical advice M03

94   chlorpyrifos

A contact and ingested organophosphorus insecticide and acaricide
IRAC mode of action code: 1B

Products

1	Ballad	Headland	480 g/l	EC	11659
2	Chlobber	AgChem Access	480 g/l	EC	13723
3	Clayton Pontoon 480EC	Clayton	480 g/l	EC	14555
4	Cyren	Headland	480 g/l	EC	11028
5	Dursban WG	Dow	75% w/w	WG	09153
6	EA Pyrifos	European Ag	480 g/l	EC	15005
7	Equity	Dow	480 g/l	EC	12465
8	Govern	Dow	75% w/w	WG	13223
9	Parapet	Dow	75 % w/w	WG	12773
10	Pyrinex 48 EC	Makhteshim	480 g/l	EC	13534

Uses

- Ambrosia beetle in *cut logs* [1-10]
- Aphids in *apples*, *broccoli*, *cabbages*, *calabrese*, *cauliflowers*, *chinese cabbage*, *gooseberries*, *pears*, *plums*, *raspberries*, *strawberries* [1-10]; *blackberries* (off-label) [7, 9]; *blackcurrants* [2, 3, 5, 8-10]; *brussels sprouts* [1, 2, 4, 10]; *currants* [1, 4, 6, 7]; *quinces* (off-label) [7]; *redcurrants*, *whitecurrants* [5, 8, 9]
- Apple blossom weevil in *apples* [1-10]
- Apple sucker in *apples* [3, 5, 8, 9]
- Bean seed fly in *green beans* (off-label) [8]; *green beans* (off-label - harvested as a dry pulse) [4, 5]
- Cabbage root fly in *broccoli*, *brussels sprouts*, *cabbages*, *calabrese*, *cauliflowers* [1-10]; *chinese cabbage* [1, 2, 4-10]; *chinese cabbage* (off-label), *green beans harvested dry as a pulse* (off-label) [10]; *choi sum* (off-label), *collards* (off-label), *kale* (off-label), *kohlrabi* (off-label), *mooli* (off-label), *pak choi* (off-label), *radishes* (off-label) [4, 5, 8-10]; *green beans* (off-label - harvested dry as a pulse) [9]; *tatsoi* (off-label) [4, 10]
- Capsids in *apples*, *gooseberries* [1-10]; *blackcurrants* [2, 3, 5, 8-10]; *currants* [1, 4, 6, 7]; *pears* [1-4, 6-10]; *quinces* (off-label) [7]; *redcurrants*, *whitecurrants* [5, 8, 9]
- Caterpillars in *apples*, *pears* [1, 2, 4, 6, 7, 10]; *blackcurrants* [2, 3, 5, 8-10]; *broccoli*, *cabbages*, *calabrese*, *cauliflowers*, *chinese cabbage*, *gooseberries* [1-10]; *brussels sprouts* [1, 2, 4, 10]; *currants* [1, 4, 6, 7]; *quinces* (off-label) [7]; *redcurrants*, *whitecurrants* [5, 8, 9]
- Cereal aphid in *spring barley*, *spring wheat*, *winter barley*, *winter wheat* [3]
- Codling moth in *apples* [1-10]; *pears* [1-4, 6-10]
- Common green capsid in *apples* [5]
- Cutworms in *broccoli*, *cabbages*, *calabrese*, *cauliflowers*, *chinese cabbage* [2, 3, 5-10]; *brussels sprouts* [2, 10]; *bulb onions* [1-10]; *potatoes grown for seed* [3]; *seed potatoes* [1, 2, 5-10]
- Damson-hop aphid in *plums* [2, 5, 8-10]; *plums* (Non OP resistant strains only) [3]
- Elm bark beetle in *cut logs* [1, 4]
- Frit fly in *amenity grassland* [1, 4, 5]; *durum wheat* (off-label), *triticale* (off-label) [7, 9]; *forage maize*, *spring oats*, *spring wheat*, *winter oats*, *winter wheat* [1-10]; *grassland* [3]; *managed amenity turf* [1-5, 10]; *permanent grassland* [1, 2, 4-10]; *rotational grass* [1, 2, 4, 5, 8-10]; *rye* (off-label) [7]; *spring barley*, *winter barley* [3, 6, 7]; *spring rye* (off-label), *winter rye* (off-label) [9]
- Insect pests in *blackberries* (off-label) [4, 8, 10]; *durum wheat* (off-label), *triticale* (off-label) [5, 8, 10]; *fodder beet* (off-label) [5, 8-10]; *leeks* (off-label), *rye* (off-label) [5]; *quinces* (off-label) [7]; *spring rye* (off-label), *winter rye* (off-label) [8, 10]
- Larch shoot beetle in *cut logs* [1-10]
- Leatherjackets in *amenity grassland* [1, 4, 5]; *broccoli*, *cabbages*, *calabrese*, *cauliflowers*, *chinese cabbage* [1, 4-9]; *brussels sprouts* [1, 4]; *durum wheat* (off-label), *triticale* (off-label) [7, 9]; *fodder beet* (off-label), *rye* (off-label) [7]; *grassland* [3]; *managed amenity turf* [1-5, 10]; *permanent grassland* [1, 2, 4-10]; *rotational grass* [1, 2, 4, 5, 8-10]; *spring barley, spring*

SEE SECTION 3 FOR PRODUCTS ALSO REGISTERED

oats, spring wheat, sugar beet, winter barley, winter oats, winter wheat [1-10]; *spring rye* *(off-label)*, *winter rye* *(off-label)* [9]
- Mealy aphid in *plums* [2, 10]
- Mealy plum aphid in *plums* [3, 5]
- Midges in *raspberries* [3]
- Narcissus fly in *bulbs/corms* *(off-label)* [10]; *flower bulbs* *(off-label)* [4]
- Pear sucker in *pears* [3, 5, 8, 9]
- Pine shoot beetle in *cut logs* [1-10]
- Pygmy mangold beetle in *fodder beet* *(off-label)* [7]; *sugar beet* [1, 4-9]
- Raspberry beetle in *raspberries* [1-10]
- Raspberry cane midge in *raspberries* [1, 2, 4-10]
- Red spider mites in *apples, gooseberries, pears, plums, raspberries, strawberries* [1, 2, 4-10]; *apples* *(non OP resistant strains only)*, *blackcurrants* *(non OP resistant strains only)*, *gooseberries* *(non OP resistant strains only)*, *pears* *(non OP resistant strains only)*, *plums* *(non OP resistant strains only)*, *raspberries* *(non OP resistant strains only)*, *strawberries* *(non OP resistant strains only)* [3]; *blackberries* *(off-label)* [7]; *blackcurrants* [2, 5, 8-10]; *currants* [1, 4, 6, 7]; *redcurrants, whitecurrants* [5, 8, 9]
- Sawflies in *apples* [1-10]; *pears* [6, 7]
- Strawberry blossom weevil in *strawberries* [5-7]
- Suckers in *apples, pears* [1, 2, 4, 6, 7, 10]
- Summer-fruit tortrix moth in *apples* [3, 8, 9]
- Thrips in *leeks* *(off-label)* [10]; *spring oats, spring wheat, winter oats, winter wheat* [1, 4]
- Tortrix moths in *apples, plums, strawberries* [1-10]; *pears* [1-4, 6-10]
- Vine weevil in *strawberries* [1, 2, 4-10]; *strawberries* *(non-protected crops only)* [3]
- Wheat bulb fly in *durum wheat* *(off-label)*, *triticale* *(off-label)* [7, 9]; *rye* *(off-label)* [7]; *spring barley, winter barley* [3, 5-9]; *spring oats, spring wheat, winter oats, winter wheat* [1-10]; *spring rye* *(off-label)*, *winter rye* *(off-label)* [9]
- Wheat-blossom midge in *durum wheat* *(off-label)*, *triticale* *(off-label)* [7, 9]; *rye* *(off-label)* [7]; *spring barley, winter barley* [1, 4]; *spring oats, winter oats* [1, 4-10]; *spring rye* *(off-label)*, *winter rye* *(off-label)* [9]; *spring wheat, winter wheat* [1, 3-10]
- Whitefly in *broccoli, cabbages, calabrese, cauliflowers, chinese cabbage* [1, 4-9]; *brussels sprouts* [1, 4]
- Winter moth in *apples, plums* [1-10]; *pears* [1-4, 6-10]
- Woolly aphid in *apples* [1-10]; *pears* [1, 4, 6, 7]

Extension of Authorisation for Minor Use (EAMUs)
- *blackberries* *20052977* [4], *20052964* [7], *20100034* [8], *20061585* [9], *20100044* [10]
- *bulbs/corms* *20100043* [10]
- *chinese cabbage* *20100038* [10]
- *choi sum* *20050237* [4], *20031390* [5], *20100036* [8], *20061587* [9], *20100038* [10]
- *collards* *20050237* [4], *20031390* [5], *20100036* [8], *20061587* [9], *20100038* [10]
- *durum wheat* *20052978* [5], *20052966* [7], *20100037* [8], *20061583* [9], *20100042* [10]
- *flower bulbs* *20050236* [4]
- *fodder beet* *20091120* [5], *20052965* [7], *20091121 expires 31 Dec 2013* [8], *20100035* [8], *20091122* [9], *20100045* [10]
- *green beans* *(harvested as a dry pulse) 20050235* [4], *(harvested as a dry pulse) 20031391* [5], *20100033* [8], *(harvested dry as a pulse) 20061586* [9]
- *green beans harvested dry as a pulse* *20100040* [10]
- *kale* *20050237* [4], *20031390* [5], *20100036* [8], *20061587* [9], *20100038* [10]
- *kohlrabi* *20050235* [4], *20031391* [5], *20100033* [8], *20061586* [9], *20100040* [10]
- *leeks* *20081469* [5], *20100039* [10]
- *mooli* *20050235* [4], *20031391* [5], *20100033* [8], *20061586* [9], *20100040* [10]
- *pak choi* *20050237* [4], *20031390* [5], *20100036* [8], *20061587* [9], *20100038* [10]
- *quinces* *20052974* [7]
- *radishes* *20050235* [4], *20031391* [5], *20100033* [8], *20061586* [9], *20100040* [10]
- *rye* *20052978* [5], *20052966* [7]
- *spring rye* *20100037* [8], *20061583* [9], *20100042* [10]
- *tatsoi* *20050237* [4], *20100038* [10]
- *triticale* *20052978* [5], *20052966* [7], *20100037* [8], *20061583* [9], *20100042* [10]

FOR FULL CONDITIONS OF USE ALWAYS READ THE PRODUCT LABEL

- ***winter rye*** *20100037* [8], *20061583* [9], *20100042* [10]

Approval information
- Chlorpyrifos included in Annex I under EC Regulation 1107/2009
- Accepted by BBPA for use on malting barley
- Following the environmental review of chlorpyrifos uses on ware potatoes, carrots, conifers (drench), forestry trees (dip) and cereals (for aphids and yellow cereal fly) were revoked in September 2005. In addition the maximum permitted number of treatments, or the maximum total dose, on several other crops was reduced.
- In 2006 CRD required that all products containing this active ingredient should carry the following warning in the main area of the container label: "Chlorpyrifos is an anticholinesterase organophosphate. Handle with care"

Efficacy guidance
- Use Low Drift Nozzles for ALL applications of chlorpyrifos to help maintain the approval of this active ingredient. It is important to minimise the effects in non-target areas.
- Brassicas raised in plant-raising beds may require retreatment at transplanting
- Activity on field crops may be reduced when soil temperature below 5°C or on organic soils
- In dry conditions the effect of granules applied as a surface band may be reduced
- Where pear suckers resistant to chlorpyrifos occur control is unlikely to be satisfactory
- Efficacy against frit fly and leatherjackets in grass may be reduced if applied during periods of frost
- Thorough incorporation by hand or mixing machine is essential when used in potting media for ornamental plant production

Restrictions
- Contains an anticholinesterase organophosphate compound. Do not use if under medical advice not to work with such compounds
- Maximum number of treatments and timing vary with crop, product and pest. See label for details
- Do not apply to sugar beet under stress or within 4 d of applying a herbicide
- Once incorporated, treated growing media must be used within 30 d
- Do not use product in growing media used for aquatic or semi-aquatic plants, and not for propagation of any edible crop
- Do not graze lactating cows on treated pasture for 14 d after treatment

Crop-specific information
- Latest use 4d after transplanting edible brassica crops; end Jul for sugar beet; varies for other crops. See individual labels
- HI range from 7 d to 6 wk depending on crop. See labels
- For turf pests apply from Nov where high larval populations detected or damage seen
- On lettuce apply only to strong well developed plants when damage first seen, or on professional advice
- In apples use pre-blossom up to pink/white bud and post-blossom after petal fall
- Test tolerance of glasshouse ornamentals before widescale use for propagating unrooted cuttings, or potting unusual plants and new species, and when using any media with a high content of non-peat ingredients

Environmental safety
- Dangerous for the environment
- Very toxic to aquatic organisms
- Flammable [7]
- Keep livestock out of treated areas for at least 14 d after treatment [1, 4, 5, 7]
- High risk to bees. Do not apply to crops in flower or to those in which bees are actively foraging. Do not apply when flowering weeds are present [1, 4, 5, 7-9]
- Chlorpyrifos is dangerous to some beneficial arthropods, especially parasitoid wasps, ground beetles, rove beetles and hoverflies. Environmental protection advice is not to spray summer applications within 20 m of the edge of the growing crop
- Broadcast air-assisted LERAP [2, 3, 5-10] (18 m); LERAP Category A [1-10]

SEE SECTION 3 FOR PRODUCTS ALSO REGISTERED

Hazard classification and safety precautions

Hazard Harmful, Dangerous for the environment [1-10]; Flammable [3, 6, 7, 10]

Transport code 3 [3, 10]; 6.1 [1, 4, 6, 7]; 9 [5, 8, 9]

Packaging group III [1, 3-10]

UN Number 1993 [3, 10]; 3017 [6, 7]; 3018 [1, 4]; 3077 [5, 8, 9]

Risk phrases R20, R22b, R38 [1-4, 6, 7, 10]; R22a, R50, R53a [1-10]; R36 [2, 3, 6, 7, 10]; R37 [3, 6, 7]; R40 [10]; R41, R43 [2, 10]

Operator protection A [1-10]; C, H [1-4, 10]; P [6, 7]; U02a [2, 10]; U05a [1-4, 6, 7, 10]; U08, U19a, U20b [1-10]; U11 [3, 6, 7]; U14 [1, 2, 4, 10]; U15 [10]

Environmental protection E06a [1-7, 10] (14 d); E12a [1, 4, 8, 9]; E12c [2, 3, 5-7, 10]; E12e, E16c, E16d, E34 [1-10]; E13a [1, 2, 4, 10]; E15a [3, 5-9]; E17a [1, 4] (18 m); E17b [2, 3, 5-10] (18 m); E38 [2, 3, 5-9]

Storage and disposal D01, D10b [1-10]; D02 [2, 3, 5-7, 10]; D05 [1, 2, 4]; D09a [1-7, 10]; D09c [8, 9]; D12a [2, 3, 5-9]; D12b [1, 4]

Medical advice M01 [3, 5-10]; M03 [1, 3-10]; M05b [1-4, 6, 7, 10]

95   chlorpyrifos-methyl

An organophosphorus insecticide and acaricide for grain store use
IRAC mode of action code: 1B

Products

1 Garrison	Pan Agriculture	225 g/l	EC	15657
2 Reldan 22	Dow	225 g/l	EC	12404

Uses

- Grain storage pests in **stored grain**
- Pre-harvest hygiene in **grain stores**

Approval information

- Chlorpyrifos-methyl included in Annex I under EC Regulation 1107/2009
- Accepted by BBPA for use in stores for malting barley
- In 2006 CRD required that all products containing this active ingredient should carry the following warning in the main area of the container label: "Chlorpyrifos-methyl is an anticholinesterase organophosphate. Handle with care"

Efficacy guidance

- Apply to grain after drying to moisture content below 14%, cooling and cleaning
- Insecticide may become depleted at grain surface if grain is being cooled by continuous extraction of air from the base leading to reduced control of grain store pests especially mites
- Resistance to organophosphorus compounds sometimes occurs in insect and mite pests of stored products

Restrictions

- Contains an anticholinesterase organophosphorus compound. Do not use if under medical advice not to work with such compounds
- Maximum number of treatments 1 per batch or 1 per store, prior to storage
- Only treat grain in good condition
- Do not treat grain intended for sowing
- Minimum of 90 d must elapse between treatment of grain and removal from store for consumption or processing

Crop-specific information

- May be applied pre-harvest to surfaces of empty store and grain handling machinery and as admixture with grain
- May be used in wheat, barley, oats, rye or triticale stores

Environmental safety

- Dangerous for the environment
- Very toxic to aquatic organisms

FOR FULL CONDITIONS OF USE ALWAYS READ THE PRODUCT LABEL

Hazard classification and safety precautions
  Hazard Harmful, Dangerous for the environment
  Transport code 9 [1]; M6 [2]
  Packaging group III
  UN Number 3082
  Risk phrases R22b, R41, R50, R53a
  Operator protection A, C, H, M; U05a, U08, U19a, U20b
  Environmental protection E15a, E34 [1, 2]; E38 [1]
  Storage and disposal D01, D02, D05, D09a, D10b [1, 2]; D12a [1]
  Medical advice M01, M03, M05b

96   citronella oil

A natural plant extract herbicide

Products

Barrier H	Barrier	22.9% w/w	EW	10136

Uses
  • Ragwort in *land temporarily removed from production*, *permanent grassland*

Approval information
  • Plant oils such as citronella oil are included in Annex 1 under EC Regulation 1107/2009

Efficacy guidance
  • Best results obtained from spot treatment of ragwort in the rosette stage, during dry still conditions
  • Aerial growth of ragwort is rapidly destroyed. Longer term control depends on overall management strategy
  • Check for regrowth after 28 d and re-apply as necessary

Crop-specific information
  • Contact with grasses will result in transient scorch which is outgrown in good growing conditions

Environmental safety
  • Apply away from bees
  • Harmful to fish or other aquatic life. Do not contaminate surface waters or ditches with chemical or used container
  • Keep livestock out of treated areas for at least 2 wk and until foliage of any poisonous weeds such as ragwort has died and become unpalatable

Hazard classification and safety precautions
  Hazard Irritant
  Risk phrases R36, R38
  Operator protection A, C; U05a, U20c
  Environmental protection E07b (2 wk); E12g, E13c
  Storage and disposal D01, D02, D05, D09a, D11a

97   clodinafop-propargyl

A foliar acting herbicide for annual grass weed control in wheat, triticale and rye
HRAC mode of action code: A

Products

1	Chubut	Agroquimicos	240 g/l	EC	15648
2	Sword	Makhteshim	240 g/l	EC	15099
3	Topik	Syngenta	240 g/l	EC	15123
4	Tuli	AgChem Access	240 g/l	EC	15714
5	Viscount	Syngenta	240 g/l	EC	15149

Uses

- Blackgrass in *durum wheat, spring rye, spring wheat, triticale, winter rye, winter wheat* [1-5]; *grass seed crops* (off-label) [3]
- Rough meadow grass in *durum wheat, spring rye, spring wheat, triticale, winter rye, winter wheat* [1-5]; *grass seed crops* (off-label) [3]
- Wild oats in *durum wheat, spring rye, spring wheat, triticale, winter rye, winter wheat* [1-5]; *grass seed crops* (off-label) [3]

Extension of Authorisation for Minor Use (EAMUs)

- *grass seed crops* 20102871 [3]

Approval information

- Clodinafop-propargyl included in Annex I under EC Regulation 1107/2009

Efficacy guidance

- Spray when majority of weeds have germinated but before competition reduces yield
- Products contain a herbicide safener (cloquintocet-mexyl) that improves crop tolerance to clodinafop-propargyl
- Optimum control achieved when all grass weeds emerged. Wait for delayed germination on dry or cloddy seedbed
- A mineral oil additive is recommended to give more consistent control of very high blackgrass populations or for late season treatments. See label for details
- Weed control not affected by soil type, organic matter or straw residues
- Control may be reduced if rain falls within 1 h of treatment
- Clodinafop-propargyl is an ACCase inhibitor herbicide. To avoid the build up of resistance do not apply products containing an ACCase inhibitor herbicide more than twice to any crop. In addition do not use any product containing clodinafop-propargyl in mixture or sequence with any other product containing the same ingredient
- Use these products as part of a resistance management strategy that includes cultural methods of control and does not use ACCase inhibitors as the sole chemical method of grass weed control
- Applying a second product containing an ACCase inhibitor to a crop will increase the risk of resistance development; only use a second ACCase inhibitor to control different weeds at a different timing
- Always follow WRAG guidelines for preventing and managing herbicide resistant weeds. See Section 5 for more information

Restrictions

- Maximum number of treatments 1 per crop
- Do not use on barley or oats
- Do not treat crops under stress or suffering from waterlogging, pest attack, disease or frost
- Do not treat crops undersown with grass mixtures
- Do not mix with products containing MCPA, mecoprop-P, 2,4-D or 2,4-DB
- MCPA, mecoprop, 2,4-D or 2,4-DB should not be applied within 21 d before, or 7 d after, treatment

Crop-specific information

- Latest use: before second node detectable stage (GS 32) for durum wheat, triticale, rye; before flag leaf sheath extending (GS 41) for wheat
- Spray in autumn, winter or spring from 1 true leaf stage (GS 11) to before second node detectable (GS 32) on durum, rye, triticale; before flag leaf sheath extends (GS 41) on wheat

Following crops guidance

- Any broad leaved crop or cereal (except oats) may be sown after failure of a treated crop provided that at least 3 wk have elapsed between application and drilling a cereal
- After normal harvest of a treated crop any broad leaved crop or wheat, durum wheat, rye, triticale or barley should be sown. Oats and grass should not be sown until the following spring

Environmental safety

- Dangerous for the environment
- Very toxic to aquatic organisms

FOR FULL CONDITIONS OF USE ALWAYS READ THE PRODUCT LABEL

Hazard classification and safety precautions

Hazard Dangerous for the environment
Transport code 9
Packaging group III
UN Number 3082
Risk phrases R50, R53a
Operator protection A, C, H, K; U02a, U05a, U09a, U20a
Environmental protection E15a, E38
Storage and disposal D01, D02, D05, D09a, D10c, D12a

98   clodinafop-propargyl + pinoxaden

A foliar acting herbicide mixture for winter wheat
HRAC mode of action code: A + A

See also pinoxaden

Products

Traxos	Syngenta	100:100 g/l	EC	12742

Uses

* Italian ryegrass in **winter wheat**
* Perennial ryegrass in **winter wheat** *(from seed)*
* Wild oats in **winter wheat**

Approval information

* Clodinafop-propargyl included in Annex I under EC Regulation 1107/2009
* Approved for use on crops for brewing by BBPA

Efficacy guidance

* Products contain a herbicide safener (cloquintocet-mexyl) that improves crop tolerance to clodinafop-propargyl
* Best results obtained from treatment when all grass weeds have emerged. There is no residual activity
* Broad-leaved weeds are not controlled
* Treat before emerged weed competition reduces yield
* For ryegrass control use as part of a programme including other products with activity against ryegrass
* For blackgrass control must be used as part of an integrated control strategy
* Grass weed control may be reduced if rain falls within 1 hr of application
* Clodinafop-propargyl and pinoxaden are ACCase inhibitor herbicides. To avoid the build up of resistance do not apply products containing an ACCase inhibitor herbicide more than twice to any crop. In addition do not use any product containing clodinafop-propargyl or pinoxaden in mixture or sequence with any other product containing the same ingredient
* Use as part of a resistance management strategy that includes cultural methods of control and does not use ACCase inhibitors as the sole chemical method of grass weed control
* Applying a second product containing an ACCase inhibitor to a crop will increase the risk of resistance development; only use a second ACCase inhibitor to control different weeds at a different timing
* Always follow WRAG guidelines for preventing and managing herbicide resistant weeds. See Section 5 for more information

Restrictions

* Maximum number of treatments 1 per crop
* Product must always be used with specified adjuvant. See label
* Do not use on other cereals
* Do not treat crops under stress from any cause
* Do not treat crops undersown with grass or grass mixtures
* Avoid use of hormone-containing herbicides in mixture or in sequence. See label for restrictions on tank mixes

SEE SECTION 3 FOR PRODUCTS ALSO REGISTERED

Crop-specific information
- Latest use: before flag leaf sheath extended (before GS 41) for winter wheat
- All varieties of winter wheat may be treated

Following crops guidance
- There are no restrictions on succeeding crops in a normal rotation
- In the event of failure of a treated crop ryegrass, maize, oats or any broad-leaved crop may be planted after a minimum interval of 4 wk from application

Environmental safety
- Dangerous for the environment
- Toxic to aquatic organisms

Hazard classification and safety precautions
> **Hazard** Irritant, Dangerous for the environment
> **Transport code** 9
> **Packaging group** III
> **UN Number** 3082
> **Risk phrases** R36, R38, R43, R51, R53a
> **Operator protection** A, C, H, K; U05a, U09a, U20b
> **Environmental protection** E15b, E38
> **Storage and disposal** D01, D02, D05, D09a, D10c, D11a, D12a

99   clodinafop-propargyl + prosulfocarb

A foliar acting herbicide mixture for grass and broad-leaved weed control in winter wheat
HRAC mode of action code: A + N

See also prosulfocarb

Products

Auxiliary	BASF	10:800 g/l	EC	14576

Uses
- Annual dicotyledons in **winter wheat**
- Annual grasses in **winter wheat**

Approval information
- Clodinafop-propargyl and prosulfocarb included in Annex I under EC Regulation 1107/2009

Efficacy guidance
- To avoid the build up of resistance do not apply products containing an ACCase inhibitor herbicide more than twice to any crop. In addition, do not use this product in mixture or sequence with any other product containing clodinafop-propargyl.

Restrictions
- Do not use on barley or oats
- Do not spray crops undersown with grass mixtures
- When [1] is applied first, leave 7 days before applying hormone herbicides. If mecoprop-p or 2,4-DB containing products are applied first, leave 14 days before [1] is applied. If MCPA or 2,4-D containing products are applied first, leave 21 days before [1] is applied.
- The cereal seed must be covered by 3 cm of soil and for best results apply to a firm, moist seedbed free of clods.

Crop-specific information
- Only one application per crop
- Latest use before 5 tiller stage (GS25)

Environmental safety
- LERAP Category B

Hazard classification and safety precautions
> **Hazard** Harmful, Dangerous for the environment

FOR FULL CONDITIONS OF USE ALWAYS READ THE PRODUCT LABEL

Transport code 9
Packaging group III
UN Number 3082
Risk phrases R22b, R51, R53a
Operator protection A, H; U02a, U05a, U08, U20c
Environmental protection E15b, E16a, E16b, E38
Storage and disposal D01, D02, D05, D09a, D10c, D12a
Medical advice M05b

100   clofentezine

A selective ovicidal tetrazine acaricide for use in top fruit
IRAC mode of action code: 10A

Products

Apollo 50 SC	Makhteshim	500 g/l	SC	10590

Uses

- Red spider mites in *apples, cherries, pears, plums*
- Spider mites in *blackcurrants* (off-label), *protected blackberries* (off-label), *protected raspberries* (off-label), *protected strawberries* (off-label), *raspberries* (off-label), *strawberries* (off-label)

Extension of Authorisation for Minor Use (EAMUs)

- *blackcurrants* 20012269
- *protected blackberries* 20012268
- *protected raspberries* 20012268
- *protected strawberries* 20012271
- *raspberries* 20012269
- *strawberries* 20012269

Approval information

- Clofentezine included in Annex I under EC Regulation 1107/2009

Efficacy guidance

- Acts on eggs and early motile stages of mites. For effective control total cover of plants is essential, particular care being needed to cover undersides of leaves

Restrictions

- Maximum number of treatments 1 per yr for apples, pears, cherries, plums

Crop-specific information

- HI apples, pears 28 d; cherries, plums 8 wk
- For red spider mite control spray apples and pears between bud burst and pink bud, plums and cherries between white bud and first flower. Rust mite is also suppressed
- On established infestations apply in conjunction with an adult acaricide

Environmental safety

- Harmful to aquatic organisms
- Product safe on predatory mites, bees and other predatory insects

Hazard classification and safety precautions

UN Number N/C
Risk phrases R52
Operator protection U05a, U08, U20b
Environmental protection E15a
Storage and disposal D01, D02, D05, D09a, D11a

SEE SECTION 3 FOR PRODUCTS ALSO REGISTERED

101   clomazone

An isoxazolidinone residual herbicide for oilseed rape, field beans, combining and vining peas
HRAC mode of action code: F3

Products

1	Centium 360 CS	Belchim	360 g/l	CS	13846
2	Cirrus CS	Belchim	371 g/l	CS	13860
3	Clayton Chrome	Clayton	360 g/l	CS	15264
4	Clayton Surat	Clayton	360 g/l	CS	15304
5	Clomaz 36 CS	Goldengrass	360 g/l	CS	14938
6	Concept	ChemSource	360 g/l	CS	14642
7	Fiddle	Goldengrass	360 g/l	CS	14451
8	Gadwall	AgChem Access	360 g/l	CS	14522
9	Gamit 36 CS	Belchim	360 g/l	CS	13861
10	Pan Fillip	Pan Agriculture	360 g/l	CS	15410
11	Spudster 360 CS	AgChem Access	360 g/l	CS	15711

Uses

- Annual dicotyledons in *asparagus* (off-label), *broad beans* (off-label), *broccoli* (off-label), *brussels sprouts* (off-label), *cabbages* (off-label), *calabrese* (off-label), *cauliflowers* (off-label), *celeriac* (off-label), *celery (outdoor)* (off-label), *chinese cabbage* (off-label), *choi sum* (off-label), *collards* (off-label), *fennel* (off-label), *french beans* (off-label), *game cover* (off-label), *herbs (see appendix 6)* (off-label), *kale* (off-label), *lupins* (off-label), *pak choi* (off-label), *rhubarb* (off-label), *runner beans* (off-label), *soya beans* (off-label), *spinach* (off-label), *spring cabbage* (off-label), *swedes* (off-label), *sweet potato* (off-label) [9]; *poppies for morphine production* (off-label) [1, 9]
- Chickweed in *carrots*, *potatoes* [4, 5, 9, 11]; *combining peas*, *spring field beans*, *spring oilseed rape*, *vining peas*, *winter field beans*, *winter oilseed rape* [1-3, 6-8, 10]
- Cleavers in *carrots*, *potatoes* [4, 5, 9, 11]; *combining peas*, *spring field beans*, *spring oilseed rape*, *vining peas*, *winter field beans*, *winter oilseed rape* [1-3, 6-8, 10]
- Fool's parsley in *carrots*, *potatoes* [4, 5, 9, 11]; *combining peas*, *spring field beans*, *spring oilseed rape*, *vining peas* [1-3, 6-8, 10]
- Red dead-nettle in *carrots*, *potatoes* [4, 5, 9, 11]; *combining peas*, *spring field beans*, *spring oilseed rape*, *vining peas*, *winter field beans*, *winter oilseed rape* [1-3, 6-8, 10]
- Shepherd's purse in *carrots*, *potatoes* [4, 5, 9, 11]; *combining peas*, *spring field beans*, *spring oilseed rape*, *vining peas*, *winter field beans*, *winter oilseed rape* [1-3, 6-8, 10]

Extension of Authorisation for Minor Use (EAMUs)

- *asparagus* 20121270 [9]
- *broad beans* 20121268 [9]
- *broccoli* 20121279 [9]
- *brussels sprouts* 20121279 [9]
- *cabbages* 20121279 [9]
- *calabrese* 20121279 [9]
- *cauliflowers* 20121279 [9]
- *celeriac* 20121278 [9]
- *celery (outdoor)* 20121276 [9]
- *chinese cabbage* 20121279 [9]
- *choi sum* 20121279 [9]
- *collards* 20121279 [9]
- *fennel* 20121272 [9]
- *french beans* 20121274 [9]
- *game cover* 20121267 [9]
- *herbs (see appendix 6)* 20121265 [9]
- *kale* 20121279 [9]
- *lupins* 20121264 [9]
- *pak choi* 20121279 [9]
- *poppies for morphine production* 20121262 [1], 20121266 [9]
- *rhubarb* 20121277 [9]

FOR FULL CONDITIONS OF USE ALWAYS READ THE PRODUCT LABEL

- **runner beans** *20121273* [9]
- **soya beans** *20121359* [9]
- **spinach** *20121275* [9]
- **spring cabbage** *20121279* [9]
- **swedes** *20121269* [9]
- **sweet potato** *20121271* [9]

Approval information
- Clomazone included in Annex I under EC Regulation 1107/2009

Efficacy guidance
- Best results obtained from application as soon as possible after sowing crop and before emergence of crop or weeds
- Uptake is via roots and shoots. Seedbeds should be firm, level and free from clods. Loose puffy seedbeds should be consolidated before spraying
- Efficacy is reduced on organic soils, on dry cloddy seedbeds and if prolonged dry weather follows application
- Clomazone acts by inhibiting synthesis of chlorophyll pigments. Susceptible weeds emerge but are chlorotic and die shortly afterwards
- Season-long control of weeds may not be achieved
- Always follow WRAG guidelines for preventing and managing herbicide resistant weeds. See Section 5 for more information

Restrictions
- Maximum number of treatments one per crop
- Crops must be covered by a minimum of 20 mm settled soil. Do not apply to broadcast crops. Direct-drilled crops should be harrowed across the slits to cover seed before spraying
- Do not use on compacted soils or soils of poor structure that may be liable to waterlogging
- Do not use on Sands or Very Light soils or those with more than 10% organic matter
- Do not treat two consecutive crops of carrots with clomazone in one calendar yr
- Consult manufacturer or your advisor before use on potato seed crops
- Do not overlap spray swaths. Crop plants emerged at time of treatment may be severely damaged

Crop-specific information
- Latest use: pre-emergence of crop
- Severe, but normally transient, crop damage may occur in overlaps on field beans
- Some transient crop bleaching may occur under certain climatic conditions and can be severe where heavy rain follows application. This is normally rapidly outgrown and has no effect on final crop yield. Overlapping spray swaths may cause severe damage to field beans

Following crops guidance
- Following normal harvest of a spring or autumn treated crop, cereals, oilseed rape, field beans, combining peas, potatoes, maize, turnips, linseed or sugar beet may be sown
- In the event of failure of an autumn treated crop, winter cereals or winter beans may be sown in the autumn if 6 wk have elapsed since treatment. In the spring following crop failure combining peas, field beans or potatoes may be sown if 6 wk have elapsed since treatment, and spring cereals, maize, turnips, onions, carrots or linseed may be sown if 7 mth have elapsed since treatment
- In the event of a failure of a spring treated crop a wide range of crops may be sown provided intervals of 6-9 wk have elapsed since treatment. See label for details
- Prior to resowing any listed replacement crop the soil should be ploughed and cultivated to 15 cm

Environmental safety
- Take extreme care to avoid drift outside the target area, or on to ponds, waterways or ditches as considerable damage may occur. Apply using a coarse quality spray

Hazard classification and safety precautions
  Transport code 9 [10]
  Packaging group III [10]
  UN Number 3082 [10]; N/C [1-9, 11]

SEE SECTION 3 FOR PRODUCTS ALSO REGISTERED

Operator protection A; U05a, U14, U20b
Environmental protection E15b
Storage and disposal D01, D02, D09a, D10b

102   clomazone + linuron

An isoxazolidinone and urea residual herbicide for peas, field beans and potatoes
HRAC mode of action code: F3 + C2

See also linuron

Products

1 Lingo	Belchim	45:250 g/l	CC	14075
2 Linzone	Belchim	45:250g/l	CC	14688

Uses
- Annual dicotyledons in **combining peas, potatoes, spring field beans, vining peas, winter field beans**
- Annual meadow grass in **combining peas, potatoes, spring field beans, vining peas, winter field beans**

Approval information
- Clomazone and linuron included in Annex I under EC Regulation 1107/2009

Restrictions
- Do not apply to broadcast crops
- Do not apply on light soils if heavy rain is expected within 24 hours of application
- Transient crop bleaching may occur under some conditions. This is normally outgrown with no effect on final yield

Following crops guidance
- Following normal harvest of an autumn treated crop, cereals, oilseed rape, field beans, combining peas, potatoes, maize, turnip, linseed and sugar beet may be sown. Prior to planting, soil should be ploughed and cultivated to a minimum depth of 15 cm.
- Following normal harvest of a spring treated crop, cereals, oilseed rape, field beans, combining peas, potatoes, maize, linseed, sugar beet and turnip may be sown. Prior to planting, soil should be ploughed and cultivated to a minimum depth of 15 cm. Do not sow or plant lettuce in the same season as an application of [1]

Hazard classification and safety precautions
Hazard Toxic, Dangerous for the environment
Transport code 9
Packaging group III
UN Number 3082
Risk phrases R22a, R40, R48, R50, R53a, R61, R62
Operator protection A, H; U02a, U05a, U13, U14, U15, U19a
Environmental protection E15a, E23, E34, E38
Storage and disposal D01, D02, D12a
Medical advice M04a

103   clomazone + metazachlor

A mixture of isoxazolidinone and residual anilide herbicides for weed control in oilseed rape
HRAC mode of action code: F3 + K3

See also metazachlor

Products

Nimbus CS	BASF	33.3:250	SC	14092

FOR FULL CONDITIONS OF USE ALWAYS READ THE PRODUCT LABEL

Uses

- Annual dicotyledons in *winter oilseed rape*
- Annual grasses in *winter oilseed rape*
- Cleavers in *winter oilseed rape*

Approval information

- Clomazone and metazachlor included in Annex I under EC Regulation 1107/2009

Restrictions

- Applications shall be limited to a total dose of not more than 1.0 kg metazachlor/ha in a three year period on the same field.

Environmental safety

- LERAP Category B

Hazard classification and safety precautions

Hazard Irritant, Dangerous for the environment
Transport code 9
Packaging group III
UN Number 3082
Risk phrases R43, R50, R53a
Operator protection A, H; U05a, U08, U14, U19a, U20b
Environmental protection E07a, E15b, E16a, E38
Storage and disposal D01, D02, D08, D09a, D10c, D12a
Medical advice M05a

---

104   clomazone + metribuzin

A mixture of isoxazolidinone and triazinone residual herbicides for use in potatoes
HRAC mode of action code: F3 + C1

See also metribuzin

Products

Metric	Belchim	60:233 g/l	CC	14214

Uses

- Annual dicotyledons in *potatoes*
- Annual meadow grass in *potatoes*

Approval information

- Clomazone and metribuzin included in Annex I under EC Regulation 1107/2009

Restrictions

- For use on specified varieties of potato only
- Safety to daughter tubers has not been tested; consult manufacturer before treating seed crops

Environmental safety

- LERAP Category B

Hazard classification and safety precautions

Hazard Dangerous for the environment
Transport code 9
Packaging group III
UN Number 3082
Risk phrases R50, R53a
Operator protection A, H; U05a, U19a, U20b
Environmental protection E15b, E16a, E16b, E34, E38
Storage and disposal D01, D02, D09a, D10b, D12a
Medical advice M05a

SEE SECTION 3 FOR PRODUCTS ALSO REGISTERED

105   clopyralid

A foliar translocated picolinic herbicide for a wide range of crops
HRAC mode of action code: O

See also bromoxynil + clopyralid

Products

1	Dow Shield 400	Dow	400 g/l	SL	14984
2	Glopyr 400	Globachem	400 g/l	SL	15009
3	Vivendi 200	AgriChem BV	200 g/l	SL	15017

Uses

- Annual dicotyledons in **brussels sprouts, forage maize** [1]; **fodder beet, grassland, mangels, ornamental plant production, red beet, spring barley, spring oats, spring oilseed rape, spring wheat, sugar beet, swedes, turnips, winter barley, winter oats, winter oilseed rape, winter wheat** [1-3]
- Clovers in **grass seed crops** *(off-label)* [1]
- Corn marigold in **brussels sprouts, forage maize** [1]; **fodder beet, grassland, mangels, ornamental plant production, red beet, spring barley, spring oats, spring oilseed rape, spring wheat, sugar beet, swedes, turnips, winter barley, winter oats, winter oilseed rape, winter wheat** [1, 2]
- Creeping thistle in **brussels sprouts, forage maize** [1]; **fodder beet, grassland, mangels, ornamental plant production, red beet, spring barley, spring oats, spring oilseed rape, spring wheat, sugar beet, swedes, turnips, winter barley, winter oats, winter oilseed rape, winter wheat** [1, 2]
- Groundsel in **all edible seed crops grown outdoors** *(off-label)*, **all non-edible seed crops grown outdoors** *(off-label)*, **chard** *(off-label)*, **corn gromwell** *(off-label)*, **farm forestry** *(off-label)*, **hemp grown for fibre production** *(off-label)*, **hops** *(off-label)*, **miscanthus** *(off-label)*, **outdoor leaf herbs** *(off-label)*, **spinach** *(off-label)*, **spinach beet** *(off-label)*, **top fruit** *(off-label)* [1]; **forest nurseries** *(off-label)*, **game cover** *(off-label)* [1, 3]
- Mayweeds in **all edible seed crops grown outdoors** *(off-label)*, **all non-edible seed crops grown outdoors** *(off-label)*, **apples** *(off-label)*, **borage for oilseed production** *(off-label)*, **brussels sprouts, canary flower (echium spp.)** *(off-label)*, **chard** *(off-label)*, **corn gromwell** *(off-label)*, **crab apples** *(off-label)*, **durum wheat** *(off-label)*, **evening primrose** *(off-label)*, **farm forestry** *(off-label)*, **forage maize, hemp grown for fibre production** *(off-label)*, **honesty** *(off-label)*, **hops** *(off-label)*, **miscanthus** *(off-label)*, **mustard** *(off-label)*, **outdoor leaf herbs** *(off-label)*, **pears** *(off-label)*, **poppies for morphine production** *(off-label)*, **spinach** *(off-label)*, **spinach beet** *(off-label)*, **spring rye** *(off-label)*, **top fruit** *(off-label)*, **triticale** *(off-label)*, **winter rye** *(off-label)* [1]; **fodder beet, grassland, mangels, ornamental plant production, red beet, spring barley, spring oats, spring oilseed rape, spring wheat, sugar beet, swedes, turnips, winter barley, winter oats, winter oilseed rape, winter wheat** [1-3]; **forest nurseries** *(off-label)*, **game cover** *(off-label)* [1, 3]
- Sowthistle in **poppies for morphine production** *(off-label)* [1]
- Thistles in **apples** *(off-label)*, **asparagus** *(off-label)*, **borage for oilseed production** *(off-label)*, **canary flower (echium spp.)** *(off-label)*, **corn gromwell** *(off-label)*, **crab apples** *(off-label)*, **durum wheat** *(off-label)*, **evening primrose** *(off-label)*, **honesty** *(off-label)*, **mustard** *(off-label)*, **pears** *(off-label)*, **poppies for morphine production** *(off-label)*, **spring rye** *(off-label)*, **trees** *(off-label)*, **triticale** *(off-label)*, **winter rye** *(off-label)* [1]; **fodder beet, forest nurseries** *(off-label)*, **game cover** *(off-label)*, **grassland, mangels, ornamental plant production, red beet, spring barley, spring oats, spring oilseed rape, spring wheat, sugar beet, swedes, turnips, winter barley, winter oats, winter oilseed rape, winter wheat** [3]

Extension of Authorisation for Minor Use (EAMUs)

- **all edible seed crops grown outdoors** *20122041* [1]
- **all non-edible seed crops grown outdoors** *20122041* [1]
- **apples** *20102080* [1]
- **asparagus** *20102079* [1]
- **borage for oilseed production** *20102086* [1]
- **canary flower (echium spp.)** *20102086* [1]

FOR FULL CONDITIONS OF USE ALWAYS READ THE PRODUCT LABEL

- **chard** *20102081* [1]
- **corn gromwell** *20121046* [1]
- **crab apples** *20102080* [1]
- **durum wheat** *20102085* [1]
- **evening primrose** *20102086* [1]
- **farm forestry** *20122041* [1]
- **forest nurseries** *20122041* [1], *20122043* [3]
- **game cover** *20122041* [1], *20122043* [3]
- **grass seed crops** *20102084* [1]
- **hemp grown for fibre production** *20122041* [1]
- **honesty** *20102086* [1]
- **hops** *20122041* [1]
- **miscanthus** *20122041* [1]
- **mustard** *20102086* [1]
- **outdoor leaf herbs** *20113236* [1]
- **pears** *20102080* [1]
- **poppies for morphine production** *20102082* [1]
- **spinach** *20102081* [1]
- **spinach beet** *20102081* [1]
- **spring rye** *20102085* [1]
- **top fruit** *20122041* [1]
- **trees** *20102083* [1]
- **triticale** *20102085* [1]
- **winter rye** *20102085* [1]

Approval information
- Clopyralid included in Annex I under EC Regulation 1107/2009
- Accepted by BBPA for use on malting barley

Efficacy guidance
- Best results achieved by application to young actively growing weed seedlings. Treat creeping thistle at rosette stage and repeat 3-4 wk later as directed
- High activity on weeds of Compositae family. For most crops recommended for use in tank mixes. See label for details

Restrictions
- Maximum total dose varies between the equivalent of one and two full dose treatments, depending on the crop treated. See labels for details
- Do not apply to cereals later than the second node detectable stage (GS 32)
- Do not apply when crop damp or when rain expected within 6 h
- Do not use straw from treated cereals in compost or any other form for glasshouse crops. Straw may be used for strawing down strawberries
- Straw from treated grass seed crops or linseed should be baled and carted away. If incorporated do not plant winter beans in same year
- Do not use on onions at temperatures above 20˚C or when under stress
- Do not treat maiden strawberries or runner beds or apply to early leaf growth during blossom period or within 4 wk of picking. Aug or early Sep sprays may reduce yield

Crop-specific information
- Latest use: 7 d before cutting grass for hay or silage; before 3rd node detectable (GS 33) for cereals; before flower buds visible from above for oilseed rape, linseed
- HI grassland 7 d; apples, pears, strawberries 4 wk; maize, sweetcorn, onions, Brussels sprouts, broccoli, cabbage, cauliflowers, calabrese, kale, fodder rape, oilseed rape, swedes, turnips, sugar beet, red beet, fodder beet, mangels, sage, honesty 6 wk
- Timing of application varies with weed problem, crop and other ingredients of tank mixes. See labels for details
- Apply as directed spray in woody ornamentals, avoiding leaves, buds and green stems. Do not apply in root zone of families Compositae or Leguminosae

SEE SECTION 3 FOR PRODUCTS ALSO REGISTERED

Following crops guidance
- Do not plant susceptible autumn-sown crops in same year as treatment. Do not apply later than Jul where susceptible crops are to be planted in spring. See label for details

Environmental safety
- Harmful to aquatic organisms [1, 2]
- Wash spray equipment thoroughly with water and detergent immediately after use. Traces of product can damage susceptible plants sprayed later
- Keep livestock out of treated areas for at least 7 d and until foliage of any poisonous weeds such as ragwort has died and become unpalatable
- Some pesticides pose a greater threat of contamination of water than others and clopyralid is one of these pesticides. Take special care when applying clopyralid near water and do not apply if heavy rain is forecast

Hazard classification and safety precautions
    UN Number N/C [1, 2]
    Risk phrases R52, R53a [1, 2]
    Operator protection A, C; U05a [3]; U08, U19a, U20b [1-3]
    Environmental protection E07a [1, 2]; E15a [1-3]; E34 [3]
    Storage and disposal D01, D05, D09a, D10b [1-3]; D02 [3]; D12a [1, 2]

106   clopyralid + 2,4-D + MCPA

A translocated herbicide mixture for managed amenity turf
HRAC mode of action code: O + O + O

See also 2,4-D
    MCPA

Products

Esteem	Vitax	35:150:175 g/l	SL	15181

Uses
- Annual dicotyledons in *managed amenity turf*
- Perennial dicotyledons in *managed amenity turf*

Approval information
- Clopyralid, 2,4-D and MCPA included in Annex I under EC Regulation 1107/2009

Efficacy guidance
- Apply when soil is moist and weeds growing actively, normally between Apr and Oct
- Ensure sufficient leaf area for uptake and select appropriate water volume to achieve good spray coverage
- To allow maximum translocation do not cut grass for 3 d after treatment

Restrictions
- Maximum number of treatments on managed amenity turf 1 per yr
- Do not spray in periods of drought unless irrigation is applied
- Do not apply if night temrperatures are low, if ground frost is imminent or in periods of prolonged cold weather
- Do not treat turf less than 1 yr old
- Do not use any treated plant materials for composting or mulching
- Do not mow for 3 d before or after treatment

Crop-specific information
- Consult manufacturer or carry out small scale safety test on turf grass cultivars not previously treated

Following crops guidance
- If reseeding of treated turf is required allow at least 6 wk after treatment before drilling grasses into the sward

FOR FULL CONDITIONS OF USE ALWAYS READ THE PRODUCT LABEL

- Where treated land is subsequently to be sown with broad-leaved plants allow an interval of at least 6 mth

Environmental safety
- Dangerous for the environment
- Harmful to aquatic organisms
- Wash spray equipment thoroughly with water and detergent immediately after use. Traces of product can damage susceptible plants sprayed later
- Do not spray outside the target area and avoid drift onto non-target plants
- Some pesticides pose a greater threat of contamination of water than others and clopyralid is one of these pesticides. Take special care when applying clopyralid near water and do not apply if heavy rain is forecast

Hazard classification and safety precautions
>**Hazard** Harmful, Dangerous for the environment
>**Transport code** M6
>**Packaging group** III
>**UN Number** 3082
>**Risk phrases** R22a, R37, R41, R52, R53a
>**Operator protection** A, C, H; U05a, U11, U15
>**Environmental protection** E15a, E34
>**Storage and disposal** D01, D02, D12a

107   clopyralid + diflufenican + MCPA

A selective herbicide for use in managed amenity turf
HRAC mode of action code: O + F1 + O

See also diflufenican
*   MCPA*

Products

Spearhead	Bayer Environ.	20:15:300 g/l	SL	09941

Uses
- Annual dicotyledons in *managed amenity turf*
- Perennial dicotyledons in *managed amenity turf*

Approval information
- Clopyralid, diflufenican and MCPA included in Annex I under EC Regulation 1107/2009

Efficacy guidance
- Best results achieved by application when grass and weeds are actively growing
- Treatment during early part of the season is recommended, but not during drought

Restrictions
- Maximum number of treatments 1 per yr
- Only use when sward is satisfactorily established and regular mowing has begun
- Turf sown in spring or early summer may be ready for treatment after 2 mth. Later sown turf should not be sprayed until growth is resumed in the following spring
- Avoid mowing within 3-4 d before or after treatment
- Do not use cuttings from treated area as a mulch for any crop

Environmental safety
- Extremely dangerous to fish or other aquatic life. Do not contaminate surface waters or ditches with chemical or used container
- Keep livestock out of treated areas/away from treated water for at least 1 wk and until foliage of any poisonous weeds, such as ragwort, has died and become unpalatable
- Avoid drift. Small amounts of spray can cause serious injury to herbaceous plants, vegetables, fruit and glasshouse crops
- Keep livestock out of treated areas

SEE SECTION 3 FOR PRODUCTS ALSO REGISTERED

- Some pesticides pose a greater threat of contamination of water than others and clopyralid is one of these pesticides. Take special care when applying clopyralid near water and do not apply if heavy rain is forecast
- LERAP Category B

Hazard classification and safety precautions
 Hazard Harmful, Irritant
 Transport code 9
 Packaging group III
 UN Number 3082
 Risk phrases R20, R21, R22a, R36, R38
 Operator protection A, C; U02a, U05a, U08, U19a, U20b
 Environmental protection E06a (unstipulated period); E13a, E16a, E16b, E34
 Storage and disposal D01, D02, D05, D09a, D10b
 Medical advice M03

108   clopyralid + florasulam + fluroxypyr

A translocated herbicide mixture for cereals
HRAC mode of action code: O + B + O

See also florasulam
*   fluroxypyr*

Products

1 Galaxy	Dow	80:2.5:100 g/l	EC	14085
2 Polax	Pan Amenity	80:2.5;100 g/l	EC	15915
3 Praxys	Dow	80:2.5;100 g/l	EC	13912

Uses
- Annual dicotyledons in *amenity grassland*, *lawns*, *managed amenity turf* [2, 3]; *spring barley*, *spring oats*, *spring wheat*, *winter barley*, *winter oats*, *winter wheat* [1]
- Chickweed in *spring barley*, *spring oats*, *spring wheat*, *winter barley*, *winter oats*, *winter wheat* [1]
- Cleavers in *spring barley*, *spring oats*, *spring wheat*, *winter barley*, *winter oats*, *winter wheat* [1]

Approval information
- Clopyralid, florasulam and fluroxypyr are all included in Annex I under EC Regulation 1107/2009

Efficacy guidance
- Best results obtained when weeds are small and actively growing
- Effectiveness may be reduced when soil is very dry
- Use adequate water volume to achieve complete spray coverage of the weeds
- Florasulam is a member of the ALS-inhibitor group of herbicides

Restrictions
- Maximum number of treatments 1 per yr for all crops
- Do not spray when crops are under stress from any cause
- Do not apply through CDA applicators
- Do not roll or harrow for 7 d before or after application
- Do not use any treated plant material for composting or mulching
- Do not use manure from animals fed on treated crops for composting
- Specific restrictions apply to use in sequence or tank mixture with other sulfonylurea or ALS-inhibiting herbicides. See label for details

Crop-specific information
- Latest use: before second node detectable for oats; before third node detectable for spring barley and spring wheat; before flag leaf detectable in winter barley and winter wheat.

FOR FULL CONDITIONS OF USE ALWAYS READ THE PRODUCT LABEL

Following crops guidance
- Where residues of a treated crop have not completely decayed by the time of planting a succeeding crop, avoid planting peas, beans and other legumes, carrots and other Umbelliferae, potatoes, lettuce and other Compositae, glasshouse and protected crops
- Where the product has been used in mixture with certain named products (see label) only cereals or grass may be sown in the autumn following harvest. Otherwise cereals, oilseed rape, grass or vegetable brassicas as transplants may be sown as a following crop in the same calendar yr as treatment. Oilseed rape may show some temporary reduction of vigour after a dry summer, but yields are not affected
- In addition to the above, field beans, linseed, peas, sugar beet, potatoes, maize, clover (for use in grass/clover mixtures) or carrots may be sown in the calendar yr following treatment
- In the event of failure of a treated crop in spring, only spring wheat, spring barley, spring oats, maize or ryegrass may be sown

Environmental safety
- Dangerous for the environment
- Very toxic to aquatic organisms
- Take extreme care to avoid drift outside the target area
- Some pesticides pose a greater threat of contamination of water than others and clopyralid is one of these pesticides. Take special care when applying clopyralid near water and do not apply if heavy rain is forecast

Hazard classification and safety precautions
Hazard Harmful, Dangerous for the environment
Transport code 9
Packaging group III
UN Number 3082
Risk phrases R20, R36, R38, R50, R53a
Operator protection A, H [1-3]; C [1]; U05a, U11, U19a, U20a
Environmental protection E15b, E34, E38
Storage and disposal D01, D02, D09a, D10b, D12a [1-3]; D05 [2, 3]
Medical advice M03 [1-3]; M05a [1]

109   clopyralid + fluroxypyr + MCPA

A translocated herbicide mixture for use in sports and amenity turf
HRAC mode of action code: O + O + O

See also fluroxypyr
MCPA

Products
Greenor	Rigby Taylor	20:40:200 g/l	ME	15204

Uses
- Annual dicotyledons in **managed amenity turf**

Approval information
- Clopyralid, fluroxypyr and MCPA included in Annex I under EC Regulation 1107/2009

Efficacy guidance
- Best results achieved when weeds actively growing and turf grass competitive
- Treatment should normally be between Apr-Sep when the soil is moist
- Do not apply during drought unless irrigation is applied
- Allow 3 d before or after mowing established turf to ensure sufficient weed leaf surface present to allow uptake and movement

Restrictions
- Maximum number of treatments 2 per yr
- Do not treat grass under stress from frost, drought, waterlogging, trace element deficiency, disease or pest attack

- Do not treat if night temperatures are low, when frost is imminent or during prolonged cold weather

Crop-specific information
- Treat young turf only in spring when at least 2 mth have elapsed since sowing
- Allow 5 d after mowing young turf before treatment
- Product selective on a number of turf grass species (see label) but consultation or testing recommended before treatment of any cultivar

Environmental safety
- Dangerous for the environment
- Very toxic to aquatic organisms
- Wash spray equipment thoroughly with water and detergent immediately after use. Traces of product can damage susceptible plants sprayed later
- Some pesticides pose a greater threat of contamination of water than others and clopyralid is one of these pesticides. Take special care when applying clopyralid near water and do not apply if heavy rain is forecast
- LERAP Category B

Hazard classification and safety precautions
Hazard Irritant, Dangerous for the environment
Transport code 9
Packaging group III
UN Number 3082
Risk phrases R36, R43, R50, R53a
Operator protection A, C; U05a, U08, U14, U19a, U20b
Environmental protection E15a, E16a, E38
Storage and disposal D01, D02, D05, D09a, D10b, D12a

110   clopyralid + fluroxypyr + triclopyr

A foliar acting herbicide mixture for grassland
HRAC mode of action code: O + O + O

See also fluroxypyr
triclopyr

Products

Pastor	Dow	50:75:100 g/l	EC	11168

Uses
- Annual dicotyledons in **rotational grass**
- Docks in **permanent grassland**, **rotational grass**
- Stinging nettle in **permanent grassland**
- Thistles in **permanent grassland**

Approval information
- Clopyralid, fluroxypyr and triclopyr included in Annex I under EC Regulation 1107/2009

Efficacy guidance
- Apply in spring or autumn depending on weeds present or as a split treatment in spring followed by autumn
- Treatment must be made when weeds and grass are actively growing
- It is important to ensure sufficient leaf area is present for uptake especially on established docks and thistles
- On large well established docks and where there is a large soil seed reservoir further treatment in the following year may be needed
- To allow maximum translocation in the weeds do not cut grass for 4 wk after treatment

Restrictions
- Maximum number of treatments 1 per yr at full dose or 2 per yr at half dose

FOR FULL CONDITIONS OF USE ALWAYS READ THE PRODUCT LABEL

- Do not spray in drought, very hot or very cold weather
- Do not treat sports or amenity turf
- Product kills or severely checks clover and should not be used where clover is an important constituent of the sward
- Do not roll or harrow 10 d before or 7 d after treatment
- Do not allow spray or drift to reach other crops, amenity plantings, gardens, ponds, lakes or water courses
- Do not use on newly-sown leys
- Only apply between 1st March & 31st Oct.

Crop-specific information
- HI 7 d
- Application during active growth ensures minimal check to grass. Newly sown grass may be treated from the third leaf visible stage
- Product may be used in established grassland which is under non-rotational setaside arrangements
- Very occasionally some yellowing of the sward may occur after treatment but is quickly outgrown

Following crops guidance
- Residues in incompletely decayed plant tissue may affect succeeding susceptible crops such as peas, beans and other legumes, carrots and related crops, potatoes, tomatoes, lettuce
- Do not plant susceptible autumn-sown crops in the same yr as treatment with product. Spring sown crops may follow if treatment was before end July in the previous yr

Environmental safety
- Dangerous for the environment
- Toxic to aquatic organisms
- Keep livestock out of treated areas for at least 7 d following treatment and until foliage of poisonous weeds such as ragwort has died and become unpalatable
- Wash spray equipment thoroughly with water and detergent immediately after use. Traces of product can damage susceptible plants sprayed later
- Some pesticides pose a greater threat of contamination of water than others and clopyralid is one of these pesticides. Take special care when applying clopyralid near water and do not apply if heavy rain is forecast
- LERAP Category B

Hazard classification and safety precautions
Hazard Harmful, Flammable, Dangerous for the environment
Transport code 3
Packaging group III
UN Number 1993
Risk phrases R22b, R37, R38, R41, R43, R51, R53a, R67
Operator protection A, C; U02a, U05a, U08, U11, U14, U19a, U20b
Environmental protection E07a, E15a, E16a, E34, E38
Consumer protection C01
Storage and disposal D01, D02, D05, D09a, D10b, D12a
Medical advice M05b

111 clopyralid + picloram

A post-emergence herbicide mixture for oilseed rape
HRAC mode of action code: O + O

See also picloram

Products

1 EA Cloporam	European Ag	267:67 g/l	SL	14828
2 Galera	Dow	267:67 g/l	SL	11961
3 Legara	AgChem Access	267:67 g/l	SL	15533

SEE SECTION 3 FOR PRODUCTS ALSO REGISTERED

Products – continued

4	Piccant	Goldengrass	267:67 g/l	SL	14294
5	Renegade	ChemSource	267:67 g/l	SL	14507

Uses
- Cleavers in *evening primrose* (off-label), *honesty* (off-label), *linseed* (off-label), *mustard* (off-label), *spring oilseed rape* (off-label) [2]; *winter oilseed rape* [1-5]
- Mayweeds in *evening primrose* (off-label), *honesty* (off-label), *linseed* (off-label), *mustard* (off-label), *spring oilseed rape* (off-label) [2]; *winter oilseed rape* [1-5]

Extension of Authorisation for Minor Use (EAMUs)
- *evening primrose* 20060960 [2]
- *honesty* 20060960 [2]
- *linseed* 20060960 [2]
- *mustard* 20060960 [2]
- *spring oilseed rape* 20061437 [2]

Approval information
- Clopyralid and picloram included in Annex I under EC Regulation 1107/2009
- Approval expiry 31 May 2013 [5]

Efficacy guidance
- Best results obtained from treatment when weeds are small and actively growing
- Cleavers that germinate after treatment will not be controlled

Restrictions
- Maximum total dose equivalent to one full dose treatment
- Do not treat crops under stress from cold, drought, pest damage, nutrient deficiency or any other cause
- Do not roll or harrow for 7 d before or after spraying
- Do not apply through CDA applicators
- Do not use any treated plant material for composting or mulching
- Do not use manure from animals fed on treated crops for composting
- Chop and incorporate all treated plant remains in early autumn, or as soon as possible after harvest, to release any residues into the soil. Ensure that all treated plant remains have completely decayed before planting susceptible crops

Crop-specific information
- Latest use: before flower buds visible above crop canopy for winter oilseed rape

Following crops guidance
- Any crop may be sown in the calendar yr following treatment
- Ploughing or thorough cultivation should be carried out before planting leguminous crops
- Do not attempt to plant peas, beans, other legumes, carrots, other umbelliferous crops, potatoes, lettuce, other Compositae, or any glasshouse or protected crops if treated crop remains have not fully decayed by the time of planting
- In the event of failure of an autumn treated crop only oilseed rape, wheat, barley, oats, maize or ryegrass may be sown in the spring and only after ploughing or thorough cultivation

Environmental safety
- Dangerous for the environment
- Toxic to aquatic organisms
- Take extreme care to avoid drift onto crops and non-target plants outside the target area
- Some pesticides pose a greater threat of contamination of water than others and clopyralid is one of these pesticides. Take special care when applying clopyralid near water and do not apply if heavy rain is forecast

Hazard classification and safety precautions
  Hazard Dangerous for the environment
  UN Number N/C
  Risk phrases R51, R53a
  Operator protection A, C; U05a
  Environmental protection E15a, E34

FOR FULL CONDITIONS OF USE ALWAYS READ THE PRODUCT LABEL

Storage and disposal D01, D02, D05, D07, D09a

112　clopyralid + triclopyr

A perennial and woody weed herbicide for use in grassland
HRAC mode of action code: O + O

See also triclopyr

Products

1 Blaster	Headland Amenity	60:240 g/l	EC	13267
2 Blaster Pro	Headland Amenity	60:240 g/l	EC	15752
3 Grazon	Dow	60:240 g/l	EC	15751
4 Grazon 90	Dow	60:240 g/l	EC	13117
5 Grazon Pro	Dow	60:240 g/l	EC	15785
6 Headland Flail	Headland	60:240 g/l	EC	15191
7 Thistlex	Dow	200:200 g/l	EC	11533

Uses

- Annual and perennial weeds in **established grassland** *(off-label)* [4]; **grassland** [3-6]
- Brambles in **amenity grassland** [1, 2]; **grassland** [6]
- Broom in **amenity grassland** [1, 2]; **grassland** [6]
- Creeping thistle in **grassland** [6]; **permanent grassland**, **rotational grass** [7]
- Docks in **amenity grassland** [1, 2]; **grassland** [6]
- Gorse in **amenity grassland** [1, 2]; **grassland** [6]
- Perennial dicotyledons in **amenity grassland** [1, 2]
- Stinging nettle in **amenity grassland** [1, 2]
- Thistles in **amenity grassland** [1, 2]

Extension of Authorisation for Minor Use (EAMUs)

- **established grassland** 20121533 expires 30 Nov 2013 [4]

Approval information

- Clopyralid and triclopyr included in Annex I under EC Regulation 1107/2009

Efficacy guidance

- Must be applied to actively growing weeds
- Correct timing crucial for good control. Spray stinging nettle before flowering, docks in rosette stage in spring, creeping thistle before flower stems 15-20 cm high, brambles, broom and gorse in Jun-Aug
- Allow 2-3 wk regrowth after grazing or mowing before spraying perennial weeds
- Where there is a large reservoir of weed seed in the soil further treatment in the following yr may be needed

Restrictions

- Maximum number of treatments 1 per yr
- Only use on permanent pasture or rotational grassland established for at least 1 yr
- Do not apply where clover is an important constituent of sward
- Do not roll or harrow within 10 d before or 7 d after spraying
- Do not cut grass for 21 d before or 28 d after spraying
- Do not use any treated plant material for composting or mulching, and do not use manure for composting from animals fed on treated crops
- Do not apply by hand-held rotary atomiser equipment
- Do not allow drift onto other crops, amenity plantings or gardens, ponds, lakes or water courses. All conifers, especially pine and larch, are very sensitive

Crop-specific information

- Latest use: 7 d before grazing or cutting grass
- Some transient yellowing of treated swards may occur but is quickly outgrown

Following crops guidance

- Residues in plant tissues which have not completely decayed may affect succeeding susceptible crops such as peas, beans, other legumes, carrots, parsnips, potatoes, tomatoes, lettuce, glasshouse and protected crops
- Do not plant susceptible autumn-sown crops (eg winter beans) in same year as treatment and allow at least 9 mth from treatment before planting a susceptible crop in the following yr
- Do not direct drill kale, swedes, turnips, grass or grass mixtures within 6 wk of spraying
- Do not spray after end Jul where susceptible crops are to be planted in the next spring

Environmental safety

- Dangerous for the environment
- Very toxic to aquatic organisms
- Keep livestock out of treated areas for at least 7 d after spraying and until foliage of any poisonous weeds such as ragwort or buttercup has died down and become unpalatable
- Some pesticides pose a greater threat of contamination of water than others and clopyralid is one of these pesticides. Take special care when applying clopyralid near water and do not apply if heavy rain is forecast
- LERAP Category B [2, 3, 5]

Hazard classification and safety precautions

Hazard Harmful [1-6]; Irritant [7]; Flammable [6]; Dangerous for the environment [1-7]
Transport code 3 [1-6]
Packaging group III [1-6]
UN Number 1993 [1-6]; N/C [7]
Risk phrases R22a, R38, R43 [1-6]; R36, R37, R51 [6]; R41 [1-5, 7]; R50 [1-5]; R52 [7]; R53a [1-7]
Operator protection A, C [1-7]; H, M [1-6]; U02a, U05a, U11, U20b [1-7]; U08, U14, U19a, U23b [1-6]; U15, U23a [7]
Environmental protection E07a, E15a [1-7]; E16a [2, 3, 5]; E23, E34, E38 [1-6]
Consumer protection C01 [1-6]
Storage and disposal D01, D02, D05, D09a, D10b, D12a
Medical advice M03 [7]

113　clothianidin

A nitromethylene neonicotinoid insecticide
IRAC mode of action code: 4A

See also beta-cyfluthrin + clothianidin

Products

Poncho	Bayer CropScience	600 g/l	SC	13910

Uses

- Frit fly in **forage maize, grain maize, sweetcorn**
- Symphylids in **forage maize** *(reduction)*, **grain maize** *(reduction)*, **sweetcorn** *(reduction)*
- Wireworm in **forage maize, grain maize, sweetcorn**

Approval information

- Clothianidin included in Annex I under EC Regulation 1107/2009
- Accepted by BBPA for use on malting barley

Efficacy guidance

- May only be used in conjunction with manufacturer's approved seed treatment application equipment or by following procedures given in the operating instructions
- Treated seed should preferably be drilled in the same season
- Evenness of seed cover improved by simultaneous application with a small volume (1.5-3 litres/ tonne) of water
- Calibrate drill for treated seed and drill at 2.5-4 cm into firm, well prepared seedbed
- Use minimum 125 kg treated seed per ha

FOR FULL CONDITIONS OF USE ALWAYS READ THE PRODUCT LABEL

- Seed should be drilled to a depth of 40 mm into a well prepared and firm seedbed. If seed is present on the surface, or if spills have occurred, the field should be harrowed and rolled if conditions are appropriate
- When aphid activity is unusually late, or is heavy and prolonged in areas of high risk, and mild weather predominates, follow-up foliar aphicide may be required
- Incidental suppression of leafhoppers in early spring (Mar/Apr) may be achieved but if a specific attack develops an additional foliar insecticide may be required

Restrictions
- Maximum number of seed treatments one per batch
- Product must be fully re-dispersed and homogeneous before use
- Do not use on seed with more than 16% moisture content, or on sprouted, cracked or skinned seed
- All seed batches should be tested to ensure they are suitable for treatment

Crop-specific information
- Latest use: pre-drilling

Environmental safety
- Harmful to game and wildlife. Treated seed should not be left on the surface. Bury spillages.

Hazard classification and safety precautions
 Hazard Harmful, Dangerous for the environment
 UN Number N/C
 Risk phrases R22a, R50, R53a
 Operator protection A, H; U04a, U05a, U13, U20c
 Environmental protection E03, E15b, E34, E38
 Storage and disposal D01, D02, D09a, D10d, D12a, D14
 Treated seed S04d, S06a
 Medical advice M03

114 clothianidin + prothioconazole

A combined fungicide and insecticide seed dressing for cereals
IRAC mode of action code: 4A

See also prothioconazole

Products

Redigo Deter	Bayer CropScience	250:50 g/l	FS	12423

Uses
- Bunt in **durum wheat** *(seed treatment)*, **triticale** *(seed treatment)*, **winter rye** *(seed treatment)*, **winter wheat** *(seed treatment)*
- Covered smut in **winter barley** *(seed treatment)*
- Fusarium foot rot and seedling blight in **durum wheat** *(seed treatment)*, **triticale** *(seed treatment)*, **winter barley** *(seed treatment)*, **winter oats** *(seed treatment)*, **winter rye** *(seed treatment)*, **winter wheat** *(seed treatment)*
- Leaf stripe in **winter barley** *(seed treatment)*
- Leafhoppers in **durum wheat** *(seed treatment)*, **triticale** *(seed treatment)*, **winter barley** *(seed treatment)*, **winter oats** *(seed treatment)*, **winter rye** *(seed treatment)*, **winter wheat** *(seed treatment)*
- Loose smut in **durum wheat** *(seed treatment)*, **winter barley** *(seed treatment)*, **winter oats** *(seed treatment)*, **winter wheat** *(seed treatment)*
- Slugs in **durum wheat** *(seed treatment)*, **triticale** *(seed treatment)*, **winter barley** *(seed treatment)*, **winter oats** *(seed treatment)*, **winter rye** *(seed treatment)*, **winter wheat** *(seed treatment)*
- Virus vectors in **durum wheat** *(seed treatment)*, **triticale** *(seed treatment)*, **winter barley** *(seed treatment)*, **winter oats** *(seed treatment)*, **winter rye** *(seed treatment)*, **winter wheat** *(seed treatment)*

SEE SECTION 3 FOR PRODUCTS ALSO REGISTERED

- Wireworm in **durum wheat** *(seed treatment)*, **triticale** *(seed treatment)*, **winter barley** *(seed treatment)*, **winter oats** *(seed treatment)*, **winter rye** *(seed treatment)*, **winter wheat** *(seed treatment)*

Approval information
- Clothianidin and prothioconazole included in Annex I under EC Regulation 1107/2009
- Accepted by BBPA for use on malting barley

Efficacy guidance
- May only be used in conjunction with manufacturer's approved seed treatment application equipment or by following procedures given in the operating instructions
- Treated wheat seed should preferably be drilled in the same season. Treated seed of other cereals must be used in the same season
- Evenness of seed cover improved by simultaneous application with a small volume (1.5-3 litres/tonne) of water
- Calibrate drill for treated seed and drill at 2.5-4 cm into firm, well prepared seedbed
- Use minimum 125 kg treated seed per ha
- When aphid activity unusually late, or is heavy and prolonged in areas of high risk, and mild weather predominates, follow-up foliar aphicide may be required
- Incidental suppression of leafhoppers in early spring (Mar/Apr) may be achieved but if a specific attack develops an additional foliar insecticide may be required
- Protection against slugs only applies to germinating seeds and not to aerial parts after emergence
- Where very high populations of aphids, wireworms or slugs occur additional specific control measures may be needed

Restrictions
- Maximum number of seed treatments one per batch for all cereals
- Product must be fully re-dispersed and homogeneous before use
- Do not use on seed with more than 16% moisture content, or on sprouted, cracked or skinned seed
- All seed batches should be tested to ensure they are suitable for treatment

Crop-specific information
- Latest use: pre-drilling of all cereals

Environmental safety
- Harmful to aquatic organisms

Hazard classification and safety precautions
  Hazard Irritant
  UN Number N/C
  Risk phrases R43, R52, R53a
  Operator protection A, H; U14, U19a, U20b
  Environmental protection E15a, E34, E38
  Storage and disposal D01, D02, D09a, D11a, D12a
  Treated seed S01, S02, S03, S04b, S05, S06a, S07, S08

115   Coniothyrium minitans

A fungal parasite of sclerotia in soil

Products

Contans WG	Belchim	1000 IU/mg	WG	12616

Uses
- Sclerotinia in **all edible crops (outdoor), all non-edible crops (outdoor), protected crops**

Approval information
- Coniothyrium minitans included in Annex 1 under EC Regulation 1107/2009

Efficacy guidance

- *C.minitans* is a soil acting biological fungicide with specific action against the resting bodies (sclerotia) of *Sclerotinia sclerotiorum* and *S.minor*
- Treat 3 mth before disease protection is required to allow time for the infective sclerotia in the soil to be reduced
- A post-harvest treatment of soil and debris prevents further contamination of the soil with sclerotia produced by the previous crop
- For best results the soil should be moist and the temperature between 12-20°C
- Application should be followed by soil incorporation into the surface layer to 10 cm with a rotovator or rotary harrow. Thorough spray coverage of the soil is essential to ensure uniform distribution
- If soil temperature drops below 0°C or rises above 27°C fungicidal activity is suspended but restarts when soil temperature returns within this range
- In glasshouses untreated areas at the margins should be covered by a film to avoid spread of spores from untreated sclerotia

Restrictions

- Maximum number of treatments 1 per crop or situation
- Do not plough or cultivate after treatment
- Do not mix with other pesticides, acids, alkalines or any product that attacks organic material
- Store product under cool dry conditions away from direct sunlight and away from heat sources

Crop-specific information

- Latest use: pre-planting
- HI: zero

Following crops guidance

- There are no restrictions on following crops and no waiting interval is specified

Hazard classification and safety precautions

Hazard Harmful
UN Number N/C
Risk phrases R42
Operator protection A, C, D; U05a, U19a
Environmental protection E15b, E34
Storage and disposal D01, D02, D12a
Medical advice M03

116   copper oxychloride

A protectant copper fungicide and bactericide
FRAC mode of action code: M1

Products

1	Cuprokylt	Unicrop	50% w/w (copper)	WP	00604
2	Cuprokylt FL	Unicrop	270 g/l (copper)	SC	08299
3	Headland Inorganic Liquid Copper	Headland	256 g/l (copper)	SC	13009

Uses

- Bacterial canker in **cherries** [1, 3]; **cob nuts** *(off-label)*, **hazel nuts** *(off-label)*, **walnuts** *(off-label)* [2]; **plums** [1, 2]
- Bacterial rot in **broccoli** *(off-label)*, **brussels sprouts** *(off-label)*, **cabbages** *(off-label)*, **calabrese** *(off-label)*, **cauliflowers** *(off-label)*, **chinese cabbage** *(off-label)*, **choi sum** *(off-label)*, **cob nuts** *(off-label)*, **collards** *(off-label)*, **hazel nuts** *(off-label)*, **kale** *(off-label)*, **pak choi** *(off-label)*, **protected broccoli** *(off-label)*, **protected brussels sprouts** *(off-label)*, **protected cabbages** *(off-label)*, **protected calabrese** *(off-label)*, **protected cauliflowers** *(off-label)*, **protected chinese cabbage** *(off-label)*, **protected choi sum** *(off-label)*, **protected collards** *(off-label)*, **tatsoi** *(off-label)*, **walnuts** *(off-label)*, **watercress** *(off-label)* [3]; **bulb onions** *(off-label)*, **garlic** *(off-label)*, **leeks** *(off-label)*, **salad onions** *(off-label)*, **shallots** *(off-label)* [1, 3]; **rhubarb** *(off-label)* [1, 2]

SEE SECTION 3 FOR PRODUCTS ALSO REGISTERED

- Black rot in **broccoli** *(off-label)*, **brussels sprouts** *(off-label)*, **cabbages** *(off-label)*, **calabrese** *(off-label)*, **cauliflowers** *(off-label)*, **chinese cabbage** *(off-label)*, **collards** *(off-label)*, **kale** *(off-label)* [1]; **protected brassica seedlings** *(off-label)* [2]
- Blight in **broccoli** *(off-label)*, **brussels sprouts** *(off-label)*, **bulb onions** *(off-label)*, **cabbages** *(off-label)*, **calabrese** *(off-label)*, **cauliflowers** *(off-label)*, **chinese cabbage** *(off-label)*, **choi sum** *(off-label)*, **collards** *(off-label)*, **garlic** *(off-label)*, **kale** *(off-label)*, **leeks** *(off-label)*, **pak choi** *(off-label)*, **protected broccoli** *(off-label)*, **protected brussels sprouts** *(off-label)*, **protected cabbages** *(off-label)*, **protected calabrese** *(off-label)*, **protected cauliflowers** *(off-label)*, **protected chinese cabbage** *(off-label)*, **protected choi sum** *(off-label)*, **protected collards** *(off-label)*, **salad onions** *(off-label)*, **shallots** *(off-label)*, **tatsoi** *(off-label)*, **watercress** *(off-label)* [3]; **cob nuts** *(off-label)*, **hazel nuts** *(off-label)*, **walnuts** *(off-label)* [2, 3]; **potatoes, tomatoes (outdoor)** [1-3]; **protected tomatoes** [1, 2]
- Buck-eye rot in **protected tomatoes, tomatoes (outdoor)** [1-3]
- Cane spot in **blackberries** *(off-label)*, **loganberries, raspberries, rubus hybrids** *(off-label)* [1, 2]
- Canker in **apples, pears** [1-3]; **apples** [2]; **broccoli** *(off-label)*, **brussels sprouts** *(off-label)*, **bulb onions** *(off-label)*, **cabbages** *(off-label)*, **calabrese** *(off-label)*, **cauliflowers** *(off-label)*, **chinese cabbage** *(off-label)*, **choi sum** *(off-label)*, **cob nuts** *(off-label)*, **collards** *(off-label)*, **garlic** *(off-label)*, **hazel nuts** *(off-label)*, **kale** *(off-label)*, **leeks** *(off-label)*, **pak choi** *(off-label)*, **protected broccoli** *(off-label)*, **protected brussels sprouts** *(off-label)*, **protected cabbages** *(off-label)*, **protected calabrese** *(off-label)*, **protected cauliflowers** *(off-label)*, **protected chinese cabbage** *(off-label)*, **protected choi sum** *(off-label)*, **protected collards** *(off-label)*, **salad onions** *(off-label)*, **shallots** *(off-label)*, **tatsoi** *(off-label)*, **walnuts** *(off-label)*, **watercress** *(off-label)* [3]
- Celery leaf spot in **celery (outdoor)** [1-3]
- Collar rot in **apples, apples** *(off-label)* [2]
- Damping off in **protected tomatoes, tomatoes (outdoor)** [1, 2]
- Downy mildew in **bilberries** *(off-label)*, **blueberries** *(off-label)*, **cranberries** *(off-label)*, **gooseberries** *(off-label)*, **redcurrants** *(off-label)*, **whitecurrants** *(off-label)* [1, 2]; **chard** *(off-label)*, **cress** *(off-label)*, **frise** *(off-label)*, **herbs (see appendix 6)** *(off-label)*, **lamb's lettuce** *(off-label)*, **protected chard** *(off-label)*, **protected cress** *(off-label)*, **protected frise** *(off-label)*, **protected herbs (see appendix 6)** *(off-label)*, **protected lamb's lettuce** *(off-label)*, **protected radicchio** *(off-label)*, **protected salad brassicas** *(off-label)*, **protected scarole** *(off-label)*, **protected spinach** *(off-label)*, **protected spinach beet** *(off-label)*, **radicchio** *(off-label)*, **salad brassicas** *(off-label)*, **scarole** *(off-label)*, **spinach** *(off-label)*, **spinach beet** *(off-label)*, **wine grapes** [3]; **hops** [1-3]
- Foliar disease control in **rhubarb** *(off-label)* [3]
- Foot rot in **protected tomatoes, tomatoes (outdoor)** [1, 2]
- Leaf curl in **apricots** *(off-label)*, **peaches** [1, 2]; **quinces** *(off-label)* [1]
- Pseudomonas storage rots in **bulb onions** *(off-label)*, **garlic** *(off-label)*, **leeks** *(off-label)*, **salad onions** *(off-label)*, **shallots** *(off-label)* [1]
- Purple blotch in **blackberries** [3]
- Pythium in **watercress** *(off-label - during propagation)* [2]
- Rhizoctonia in **watercress** *(off-label - during propagation)* [2]
- Rhizoctonia rot in **broccoli** *(off-label)*, **brussels sprouts** *(off-label)*, **cabbages** *(off-label)*, **calabrese** *(off-label)*, **cauliflowers** *(off-label)*, **chinese cabbage** *(off-label)*, **choi sum** *(off-label)*, **cob nuts** *(off-label)*, **collards** *(off-label)*, **garlic** *(off-label)*, **hazel nuts** *(off-label)*, **kale** *(off-label)*, **leeks** *(off-label)*, **pak choi** *(off-label)*, **protected broccoli** *(off-label)*, **protected brussels sprouts** *(off-label)*, **protected cabbages** *(off-label)*, **protected calabrese** *(off-label)*, **protected cauliflowers** *(off-label)*, **protected chinese cabbage** *(off-label)*, **protected choi sum** *(off-label)*, **protected collards** *(off-label)*, **salad onions** *(off-label)*, **shallots** *(off-label)*, **tatsoi** *(off-label)*, **walnuts** *(off-label)*, **watercress** *(off-label)* [3]
- Rust in **blackcurrants** [1-3]
- Spear rot in **broccoli** *(off-label)*, **brussels sprouts** *(off-label)*, **cabbages** *(off-label)*, **calabrese** *(off-label)*, **cauliflowers** *(off-label)*, **chinese cabbage** *(off-label)*, **collards** *(off-label)*, **kale** *(off-label)* [1, 3]; **bulb onions** *(off-label)*, **choi sum** *(off-label)*, **cob nuts** *(off-label)*, **garlic** *(off-label)*, **hazel nuts** *(off-label)*, **leeks** *(off-label)*, **pak choi** *(off-label)*, **protected broccoli** *(off-label)*, **protected brussels sprouts** *(off-label)*, **protected cabbages** *(off-label)*, **protected calabrese** *(off-label)*, **protected cauliflowers** *(off-label)*, **protected chinese cabbage** *(off-label)*, **protected choi sum** *(off-label)*, **protected collards** *(off-label)*, **salad onions** *(off-label)*, **shallots** *(off-label)*, **tatsoi** *(off-label)*, **walnuts** *(off-label)*, **watercress** *(off-label)* [3]; **protected brassica seedlings** *(off-label)* [2]

Extension of Authorisation for Minor Use (EAMUs)

- **apples** *982336* [2]
- **apricots** *20063134* [1], *20063140* [2]
- **bilberries** *20063131* [1], *20063137* [2]
- **blackberries** *20063132* [1], *20063139* [2]
- **blueberries** *20063131* [1], *20063137* [2]
- **broccoli** *20010115* [1], *20080156* [3]
- **brussels sprouts** *20010115* [1], *20080156* [3]
- **bulb onions** *991127* [1], *20080156* [3]
- **cabbages** *20010115* [1], *20080156* [3]
- **calabrese** *20010115* [1], *20080156* [3]
- **cauliflowers** *20010115* [1], *20080156* [3]
- **chard** *20080157* [3]
- **chinese cabbage** *20010115* [1], *20080156* [3]
- **choi sum** *20080156* [3]
- **cob nuts** *990385* [2], *20080156* [3]
- **collards** *20010115* [1], *20080156* [3]
- **cranberries** *20063131* [1], *20063137* [2]
- **cress** *20080157* [3]
- **frise** *20080157* [3]
- **garlic** *991127* [1], *20080156* [3]
- **gooseberries** *20063131* [1], *20063137* [2]
- **hazel nuts** *990385* [2], *20080156* [3]
- **herbs (see appendix 6)** *20080157* [3]
- **kale** *20010115* [1], *20080156* [3]
- **lamb's lettuce** *20080157* [3]
- **leeks** *991127* [1], *20080156* [3]
- **pak choi** *20080156* [3]
- **protected brassica seedlings** *20010117* [2]
- **protected broccoli** *20080156* [3]
- **protected brussels sprouts** *20080156* [3]
- **protected cabbages** *20080156* [3]
- **protected calabrese** *20080156* [3]
- **protected cauliflowers** *20080156* [3]
- **protected chard** *20080157* [3]
- **protected chinese cabbage** *20080156* [3]
- **protected choi sum** *20080156* [3]
- **protected collards** *20080156* [3]
- **protected cress** *20080157* [3]
- **protected frise** *20080157* [3]
- **protected herbs (see appendix 6)** *20080157* [3]
- **protected lamb's lettuce** *20080157* [3]
- **protected radicchio** *20080157* [3]
- **protected salad brassicas** *20080157* [3]
- **protected scarole** *20080157* [3]
- **protected spinach** *20080157* [3]
- **protected spinach beet** *20080157* [3]
- **quinces** *20063133* [1]
- **radicchio** *20080157* [3]
- **redcurrants** *20063131* [1], *20063137* [2]
- **rhubarb** *20063135* [1], *20063138* [2], *20080158* [3]
- **rubus hybrids** *20063132* [1], *20063139* [2]
- **salad brassicas** *20080157* [3]
- **salad onions** *991127* [1], *20080156* [3]
- **scarole** *20080157* [3]
- **shallots** *991127* [1], *20080156* [3]
- **spinach** *20080157* [3]
- **spinach beet** *20080157* [3]
- **tatsoi** *20080156* [3]

SEE SECTION 3 FOR PRODUCTS ALSO REGISTERED

- **walnuts** *990385* [2], *20080156* [3]
- **watercress** *(during propagation) 20001538* [2], *20080156* [3]
- **whitecurrants** *20063131* [1], *20063137* [2]

Approval information
- Copper oxychloride included in Annex 1 under EC Regulation 1107/2009
- Approved for aerial application on potatoes [1]. See Section 5 for more information
- Accepted by BBPA for use on hops

Efficacy guidance
- Spray crops at high volume when foliage dry but avoid run off. Do not spray if rain expected soon
- Spray interval commonly 10-14 d but varies with crop, see label for details
- If buck-eye rot occurs, spray soil surface and lower parts of tomato plants to protect unaffected fruit [1, 2]
- A follow-up spray in the following spring should be made to top fruit severely infected with bacterial canker

Restrictions
- Maximum number of treatments 3 per crop for apples, blackberries, grapevines, pears; 2 per crop for watercress. Not specified for other crops
- Some peach cultivars are sensitive to copper. Treat non-sensitive varieties only [1, 2]

Crop-specific information
- HI hops 7 d
- Slight damage may occur to leaves of cherries and plums

Environmental safety
- Dangerous for the environment
- Very toxic to aquatic organisms
- Keep all livestock out of treated areas for at least 3 wk

Hazard classification and safety precautions
Hazard Harmful [2]; Dangerous for the environment [1-3]
Transport code 6.1 [2]; 9 [1, 3]
Packaging group III
UN Number 3010 [2]; 3077 [1]; 3082 [3]
Risk phrases R22a, R50 [2]; R51 [1, 3]; R53a [1-3]
Operator protection A [3]; U20a [1, 3]; U20c [2]
Environmental protection E06a [1-3] (3 wk); E13c, E34, E38 [1, 3]; E15a [2]
Storage and disposal D01, D09a [1-3]; D05, D10c, D12a [2]; D10b [1, 3]

117   cyazofamid

A cyanoimidazole sulfonamide protectant fungicide for potatoes
FRAC mode of action code: 21

Products

1	Linford	AgChem Access	400 g/l	SC	13824
2	Ranman Top	Belchim	160 g/l	SC	14753
3	Ranman Twinpack	Belchim	400 g/l	KL	11851
4	Swallow	AgChem Access	400 g/l	SC	14078

Uses
- Blight in **potatoes** [1-4]
- Disease control in **hops** *(off-label)* [3]

Extension of Authorisation for Minor Use (EAMUs)
- **hops** *20082918* [3]

Approval information
- Cyazofamid included in Annex I under EC Regulation 1107/2009
- Approval expiry 30 Jun 2013 [1, 4]

FOR FULL CONDITIONS OF USE ALWAYS READ THE PRODUCT LABEL

- Approval expiry 07 Dec 2013 [2]

Efficacy guidance
- Apply as a protectant treatment before blight enters the crop and repeat every 7-10 d depending on severity of disease pressure
- Commence spray programme immediately the risk of blight in the locality occurs, usually when the crop meets along the rows
- Product must always be used with organosilicone adjuvant provided in the twin pack
- To minimise the chance of development of resistance no more than three applications should be made consecutively (out of a permissible total of six) in the blight control programme. For more information on Resistance Management see Section 5

Restrictions
- Maximum number of treatments 6 per crop (no more than three of which should be consecutive)
- Mixed product must not be allowed to stand overnight
- Consult processor before using on crops intended for processing
- Do not allow diluted product to stand overnight [2]

Crop-specific information
- HI 7 d

Environmental safety
- Dangerous for the environment
- Very toxic to aquatic organisms
- Do not empty into drains

Hazard classification and safety precautions
> **Hazard** Harmful [1, 3, 4] (adjuvant); Irritant [2]; Dangerous for the environment [1-4]
> **Transport code** 9 [2, 3]
> **Packaging group** III [2, 3]
> **UN Number** 3082 [2, 3]
> **Risk phrases** R20, R36, R48 [1, 3, 4] (adjuvant); R41, R53a [1-4]; R50 [1, 3, 4]; R51 [2]
> **Operator protection** A, C; U02a, U05a, U11, U15 [1-4]; U19a [1, 3, 4] (adjuvant); U20a [1, 3, 4]; U20b [2]
> **Environmental protection** E15a, E34 [1, 3, 4]; E15b [2]; E19b, E38 [1-4]
> **Storage and disposal** D01, D02, D09a, D10c, D12a [1-4]; D05, D12b [2]
> **Medical advice** M03

118 cycloxydim

A translocated post-emergence cyclohexanedione oxime herbicide for grass weed control
HRAC mode of action code: A

Products

Laser	BASF	200 g/l	EC	12930

Uses
- Annual grasses in *amenity grassland (off-label)*, *amenity vegetation (off-label)*, *borage for oilseed production (off-label)*, *broad beans (off-label)*, *brussels sprouts*, *bulb onions*, *cabbages*, *canary flower (echium spp.) (off-label)*, *carrots*, *cauliflowers*, *combining peas*, *dwarf beans*, *early potatoes*, *edible podded peas (off-label)*, *evening primrose (off-label)*, *farm forestry*, *flower bulbs*, *fodder beet*, *forest*, *garlic (off-label)*, *honesty (off-label)*, *horseradish (off-label)*, *leeks*, *linseed*, *lupins (off-label)*, *maincrop potatoes*, *mallow (althaea spp.) (off-label)*, *mangels*, *mustard (off-label)*, *parsnips*, *runner beans (off-label)*, *salad onions*, *shallots (off-label)*, *spring field beans*, *spring oilseed rape*, *strawberries*, *sugar beet*, *swedes*, *turnips (off-label)*, *vining peas*, *winter field beans*, *winter oilseed rape*
- Black bent in *brussels sprouts*, *bulb onions*, *cabbages*, *carrots*, *cauliflowers*, *combining peas*, *dwarf beans*, *early potatoes*, *flower bulbs*, *fodder beet*, *leeks*, *linseed*, *maincrop potatoes*, *mangels*, *parsnips*, *salad onions*, *spring field beans*, *spring oilseed rape*, *strawberries*, *sugar beet*, *swedes*, *vining peas*, *winter field beans*, *winter oilseed rape*

SEE SECTION 3 FOR PRODUCTS ALSO REGISTERED

- Blackgrass in **brussels sprouts, bulb onions, cabbages, carrots, cauliflowers, combining peas, dwarf beans, early potatoes, flower bulbs, fodder beet, leeks, linseed, maincrop potatoes, mangels, parsnips, salad onions, spring field beans, spring oilseed rape, strawberries, sugar beet, swedes, vining peas, winter field beans, winter oilseed rape**
- Couch in **brussels sprouts, bulb onions, cabbages, carrots, cauliflowers, combining peas, dwarf beans, early potatoes, flower bulbs, fodder beet, leeks, linseed, maincrop potatoes, mangels, parsnips, salad onions, spring field beans, spring oilseed rape, strawberries, sugar beet, swedes, vining peas, winter field beans, winter oilseed rape**
- Creeping bent in **brussels sprouts, bulb onions, cabbages, carrots, cauliflowers, combining peas, dwarf beans, early potatoes, flower bulbs, fodder beet, leeks, linseed, maincrop potatoes, mangels, parsnips, salad onions, spring field beans, spring oilseed rape, strawberries, sugar beet, swedes, vining peas, winter field beans, winter oilseed rape**
- Green cover in **land temporarily removed from production**
- Onion couch in **brussels sprouts, bulb onions, cabbages, carrots, cauliflowers, combining peas, dwarf beans, early potatoes, flower bulbs, fodder beet, leeks, linseed, maincrop potatoes, mangels, parsnips, salad onions, spring field beans, spring oilseed rape, strawberries, sugar beet, swedes, vining peas, winter field beans, winter oilseed rape**
- Perennial grasses in **amenity vegetation** (off-label), **borage for oilseed production** (off-label), **broad beans** (off-label), **canary flower (echium spp.)** (off-label), **edible podded peas** (off-label), **evening primrose** (off-label), **farm forestry, forest, garlic** (off-label), **honesty** (off-label), **horseradish** (off-label), **lupins** (off-label), **mallow (althaea spp.)** (off-label), **mustard** (off-label), **runner beans** (off-label), **shallots** (off-label), **turnips** (off-label)
- Volunteer cereals in **brussels sprouts, bulb onions, cabbages, carrots, cauliflowers, combining peas, dwarf beans, early potatoes, flower bulbs, fodder beet, leeks, linseed, maincrop potatoes, mangels, parsnips, salad onions, spring field beans, spring oilseed rape, strawberries, sugar beet, swedes, vining peas, winter field beans, winter oilseed rape**
- Wild oats in **brussels sprouts, bulb onions, cabbages, carrots, cauliflowers, combining peas, dwarf beans, early potatoes, flower bulbs, fodder beet, leeks, linseed, maincrop potatoes, mangels, parsnips, salad onions, spring field beans, spring oilseed rape, strawberries, sugar beet, swedes, vining peas, winter field beans, winter oilseed rape**

Extension of Authorisation for Minor Use (EAMUs)
- **amenity grassland** 20113145
- **amenity vegetation** 20081092
- **borage for oilseed production** 20081094
- **broad beans** 20081136
- **canary flower (echium spp.)** 20081094
- **edible podded peas** 20081096
- **evening primrose** 20081094
- **garlic** 20081097
- **honesty** 20081094
- **horseradish** 20081098
- **lupins** 20081093
- **mallow (althaea spp.)** 20081098
- **mustard** 20081094
- **runner beans** 20081096
- **shallots** 20081097
- **turnips** 20081095

Approval information
- Cycloxydim included in Annex 1 under EC Regulation 1107/2009

Efficacy guidance
- Best results achieved when weeds small and have not begun to compete with crop. Effectiveness reduced by drought, cool conditions or stress. Weeds emerging after application are not controlled
- Foliage death usually complete after 3-4 wk but longer under cool conditions, especially late treatments to winter oilseed rape

FOR FULL CONDITIONS OF USE ALWAYS READ THE PRODUCT LABEL

- Perennial grasses should have sufficient foliage to absorb spray and should not be cultivated for at least 14 d after treatment
- On established couch pre-planting cultivation recommended to fragment rhizomes and encourage uniform emergence
- Split applications to volunteer wheat and barley at GS 12-14 will often give adequate control in winter oilseed rape. See label for details
- Apply to dry foliage when rain not expected for at least 2 h
- Cycloxydim is an ACCase inhibitor herbicide. To avoid the build up of resistance do not apply products containing an ACCase inhibitor herbicide more than twice to any crop. In addition do not use any product containing cycloxydim in mixture or sequence with any other product containing the same ingredient
- Use these products as part of a resistance management strategy that includes cultural methods of control and does not use ACCase inhibitors as the sole chemical method of grass weed control
- Applying a second product containing an ACCase inhibitor to a crop will increase the risk of resistance development; only use a second ACCase inhibitor to control different weeds at a different timing
- Always follow WRAG guidelines for preventing and managing herbicide resistant weeds. See Section 5 for more information

Restrictions
- Maximum number of treatments 1 per crop or yr in most situations. See label
- Must be used with authorised adjuvant oil. See label
- Do not apply to crops damaged or stressed by adverse weather, pest or disease attack or other pesticide treatment
- Prevent drift onto other crops, especially cereals and grass

Crop-specific information
- HI cabbage, cauliflower, calabrese, salad onions 4 wk; peas, dwarf beans 5 wk; bulb onions, carrots, parsnips, strawberries 6 wk; sugar and fodder beet, leeks, mangels, potatoes, field beans, swedes, Brussels sprouts, winter field beans 8 wk; oilseed rape, soya beans, linseed 12 wk
- Recommended time of application varies with crop. See label for details
- On peas a crystal violet wax test should be done if leaf wax likely to have been affected by weather conditions or other chemical treatment. The wax test is essential if other products are to be sprayed before or after treatment
- May be used on ornamental bulbs when crop 5-10 cm tall. Product has been used on tulips, narcissi, hyacinths and irises but some subjects may be more sensitive and growers advised to check tolerance on small number of plants before treating the rest of the crop
- May be applied to land temporarily removed from production where the green cover is made up predominantly of tolerant crops listed on label. Use on industrial crops of linseed and oilseed rape on land temporarily removed from production also permitted.

Following crops guidance
- Guideline intervals for sowing succeeding crops after failed treated crop: field beans, peas, sugar beet, rape, kale, swedes, radish, white clover, lucerne 1 wk; dwarf French beans 4 wk; wheat, barley, maize 8 wk
- Oats should not be sown after failure of a treated crop

Environmental safety
- Dangerous for the environment
- Toxic to aquatic organisms
- Harmful to fish or other aquatic life. Do not contaminate surface waters or ditches with chemical or used container

Hazard classification and safety precautions
Hazard Harmful, Dangerous for the environment
Transport code 9
Packaging group III
UN Number 3082
Risk phrases R22b, R38, R41, R51, R53a

SEE SECTION 3 FOR PRODUCTS ALSO REGISTERED

Operator protection A, C; U05a, U08, U20b
Environmental protection E15a, E38
Storage and disposal D01, D02, D09a, D10c, D12a
Medical advice M05b

119   Cydia pomonella GV

An baculovirus insecticide for codling moth control

Products

1 Carpovirusine	Fargro	1.0 x10^13 GV/l	SC	15243
2 Cyd-X	Certis	10 g/l granulovirus	SC	13535

Uses
* Codling moth in *apples* [1, 2]; *pears* [2]

Approval information
* Cydia pomonella included in Annex I under EC Regulation 1107/2009

Efficacy guidance
* Optimum results require a precise spraying schedule commencing soon after egg laying and before first hatch
* First generation of codling moth normally occurs in first 2 wk of Jun in UK
* Apply in sufficient water to achieve good coverage of whole tree
* Normally 3 applications at intervals of 8 sunny days are needed for each codling moth generation
* Fruit damage is reduced significantly in the first yr and will reduce the population of codling moth in the following yr

Restrictions
* Maximum number of treatments 3 per season
* Do not tank mix with copper or any products with a pH lower than 5 or higher than 8
* Do not use as the exclusive measure for codling moth control. Adopt an anti-resistance strategy. See Section 5 for more information
* Consult processor before use on crops for processing

Crop-specific information
* HI for apples 3 days [1]
* Evidence of efficacy and crop safety on pears is limited [2]

Environmental safety
* Product has no adverse effect on predatory insects and is fully compatible with IPM programmes

Hazard classification and safety precautions
Hazard Harmful [2]; Irritant [1]
UN Number N/C [1]
Risk phrases R42, R43 [2]
Operator protection A, D, H; U05a [1]; U14, U19a [1, 2]
Environmental protection E15a [2]; E15b, E34 [1]
Storage and disposal D01, D02, D05, D09a, D10b [1]; D12a [1, 2]
Medical advice M05a [1]

120   cyflufenamid

An amidoxime fungicide for cereals
FRAC mode of action code: U6

Products

Cyflamid	Certis	50 g/l	EW	12403

Uses

- Powdery mildew in *amenity vegetation* (off-label), *durum wheat*, *forest nurseries* (off-label), *grass seed crops* (off-label), *hops* (off-label), *ornamental plant production* (off-label), *rye* (off-label), *soft fruit* (off-label), *spring barley*, *spring oats*, *spring wheat*, *triticale*, *winter barley*, *winter oats*, *winter rye*, *winter wheat*

Extension of Authorisation for Minor Use (EAMUs)

- *amenity vegetation* 20070512
- *forest nurseries* 20082915
- *grass seed crops* 20060602
- *hops* 20082915
- *ornamental plant production* 20070512
- *rye* 20060602
- *soft fruit* 20082915

Approval information

- Cyflufenamid included in Annex 1 under EC Regulation 1107/2009
- Accepted by BBPA on malting barley

Efficacy guidance

- Best results obtained from treatment at the first visible signs of infection by powdery mildew
- Sustained disease pressure may require a second treatment
- Disease spectrum may be broadened by appropriate tank mixtures. See label
- Product is rainfast within 1 h
- Must be used as part of an Integrated Crop Management programme that includes alternating use, or mixture, with fungicides with a different mode of action effective against powdery mildew

Restrictions

- Maximum number of treatments 2 per crop on all recommended cereals
- Apply only in the spring
- Do not apply to crops under stress from drought, waterlogging, cold, pests or diseases, lime or nutrient deficiency or other factors affecting crop growth

Crop-specific information

- Latest use: before start of flowering of cereal crops

Following crops guidance

- No restrictions

Environmental safety

- Dangerous for the environment
- Very toxic to aquatic organisms
- Avoid drift onto ponds, waterways or ditches

Hazard classification and safety precautions

Hazard Dangerous for the environment
Transport code 9
Packaging group III
UN Number 3082
Risk phrases R50, R53a
Operator protection A; U05a
Environmental protection E15b, E34
Storage and disposal D01, D02, D05, D10c, D12b

SEE SECTION 3 FOR PRODUCTS ALSO REGISTERED

121   cymoxanil

A cyanoacetamide oxime fungicide for potatoes
FRAC mode of action code: 27

See also chlorothalonil + cymoxanil
cyazofamid + cymoxanil

Products

1	Cymbal	Belchim	45% w/w	WG	13312
2	Drum	Belchim	45% w/w	WG	13853
3	Harpoon	Headland	60% w/w	WG	14278
4	Option	DuPont	60% w/w	WG	11834
5	Sipcam C50 WG	Sipcam	50% w/w	WG	14650
6	Tabour	ChemSource	45% w/w	WG	15432

Uses

- Blight in **early potatoes** [5]; **potatoes** [1-6]
- Downy mildew in **grapevines** *(off-label)*, **hops** *(off-label)* [4]

Extension of Authorisation for Minor Use (EAMUs)

- **grapevines** *20111826* [4]
- **hops** *20102250* [4]

Approval information

- Cymoxanil included in Annex I under EC Regulation 1107/2009
- Accepted by BBPA for use on hops

Efficacy guidance

- Product to be used in mixture with specified mixture partners in order to combine systemic and protective activity. See labels for details
- Commence spray programme as soon as weather conditions favourable for disease development occur and before infection appears. At latest the first treatment should be made as the foliage meets along the rows
- Repeat treatments at 7-14 day intervals according to disease incidence and weather conditions
- Treat irrigated crops as soon as possible after irrigation and repeat at 10 d intervals

Restrictions

- Minimum spray interval 7 d
- Do not allow packs to become wet during storage and do not keep opened packs from one season to another
- Cymoxanil must be used in tank mixture with other fungicides

Crop-specific information

- HI for potatoes 14 d [1]
- Consult processor before treating potatoes grown for processing

Environmental safety

- Dangerous for the environment
- Very toxic to aquatic organisms

Hazard classification and safety precautions

Hazard Harmful [3-5]; Irritant [1, 2, 6]; Dangerous for the environment [1-4, 6]
Transport code 9 [1-4, 6]
Packaging group III [1-4, 6]
UN Number 3077 [1-4, 6]; N/C [5]
Risk phrases R22a [3-5]; R43 [1, 2, 6]; R50, R53a [1-4, 6]
Operator protection A [1, 3-5]; H [1]; U05a [1-6]; U08 [3-5]; U09a, U14 [1, 2, 6]; U19a [3, 4]; U20a [1]; U20b [2-6]
Environmental protection E13a, E34 [3, 4]; E13c [5]; E15a [1, 2, 6]; E38 [1, 2, 4, 6]
Storage and disposal D01, D02, D09a, D11a [1-6]; D12a [1, 2, 4, 6]
Medical advice M03 [3-5]

FOR FULL CONDITIONS OF USE ALWAYS READ THE PRODUCT LABEL

122   cymoxanil + famoxadone

A preventative and curative fungicide mixture for potatoes
FRAC mode of action code: 27 + 11

See also famoxadone

Products

1 Santos	ChemSource	25:25% w/w	WG	15047
2 Tanos	DuPont	25:25% w/w	WG	10677

Uses
* Blight in **potatoes**

Approval information
* Cymoxanil and famoxadone included in Annex I under EC Regulation 1107/2009

Efficacy guidance
* Commence spray programme before infection appears as soon as weather conditions favourable for disease development to occur. At latest the first treatment should be made as the foliage meets along the rows
* Repeat treatments at 7-10 day intervals according to disease incidence and weather conditions
* Reduce spray interval if conditions are conducive to the spread of blight
* Spray as soon as possible after irrigation
* Increase water volume in dense crops
* Product combines active substances with different modes of action and is effective against strains of potato blight that are insensitive to phenylamide fungicides
* To minimise the likelihood of development of resistance to QoI fungicides these products should be used in a planned Resistance Management strategy. See Section 5 for more information

Restrictions
* Maximum number of treatments 6 per crop
* Consult processors before using on crops for processing
* Do not apply more than 3 consecutive treatments containing QoI active fungicides

Crop-specific information
* HI 14 d for potatoes

Environmental safety
* Dangerous for the environment
* Very toxic to aquatic organisms
* Dangerous to fish or other aquatic life. Do not contaminate surface waters or ditches with chemical or used container
* LERAP Category B

Hazard classification and safety precautions
  Hazard Harmful, Dangerous for the environment
  Transport code 9
  Packaging group III
  UN Number 3077
  Risk phrases R22a, R50, R53a
  Operator protection A, D; U05a, U20b
  Environmental protection E13b, E16a, E16b, E34, E38
  Storage and disposal D01, D02, D09a, D10b, D12a
  Medical advice M03

SEE SECTION 3 FOR PRODUCTS ALSO REGISTERED

123   cymoxanil + fludioxonil + metalaxyl-M

A fungicide seed dressing for peas
FRAC mode of action code: 27 + 12 + 4

See also fludioxonil
      metalaxyl-M

Products

Wakil XL	Syngenta	10:5:17.5% w/w	WS	10562

Uses

- Ascochyta in **combining peas** *(seed treatment)*, **vining peas** *(seed treatment)*
- Damping off in **combining peas** *(seed treatment)*, **lupins** *(off-label)*, **poppies for morphine production** *(off-label - seed treatment)*, **red beet** *(off-label - seed treatment)*, **vining peas** *(seed treatment)*
- Downy mildew in **broad beans** *(off-label - seed treatment)*, **combining peas** *(seed treatment)*, **lupins** *(off-label)*, **poppies for morphine production** *(off-label - seed treatment)*, **spring field beans** *(off-label - seed treatment)*, **vining peas** *(seed treatment)*, **winter field beans** *(off-label - seed treatment)*
- Pythium in **carrots** *(off-label - seed treatment)*, **combining peas** *(seed treatment)*, **lupins** *(off-label)*, **parsnips** *(off-label - seed treatment)*, **poppies for morphine production** *(off-label - seed treatment)*, **vining peas** *(seed treatment)*

Extension of Authorisation for Minor Use (EAMUs)

- **broad beans** *(seed treatment)* 20023932
- **carrots** *(seed treatment)* 20021191
- **lupins** 20060614
- **parsnips** *(seed treatment)* 20021191
- **poppies for morphine production** *(seed treatment)* 20040683
- **red beet** *(seed treatment)* 20032313
- **spring field beans** *(seed treatment)* 20023932
- **winter field beans** *(seed treatment)* 20023932

Approval information

- Cymoxanil, fludioxonil and metalaxyl-M included in Annex I under EC Regulation 1107/2009

Efficacy guidance

- Apply through continuous flow seed treaters which should be calibrated before use

Restrictions

- Max number of treatments 1 per seed batch
- Ensure moisture content of treated seed satisfactory and store in a dry place
- Check calibration of seed drill with treated seed before drilling and sow as soon as possible after treatment
- Consult before using on crops for processing

Crop-specific information

- Latest use: pre-drilling

Environmental safety

- Harmful to aquatic organisms
- Do not use treated seed as food or feed

Hazard classification and safety precautions

    UN Number N/C
    Risk phrases R52, R53a
    Operator protection A, D, H; U02a, U05a, U20b
    Environmental protection E03, E15a
    Storage and disposal D01, D02, D05, D07, D09a, D11a, D12a
    Treated seed S02, S04b, S05, S07

124   cymoxanil + mancozeb

A protectant and systemic fungicide for potato blight control
FRAC mode of action code: 27 + M3

See also mancozeb

Products

1	Astral	AgChem Access	4.5:68% w/w	WP	14049
2	Clayton Krypton	Clayton	4.5:68% w/w	WP	09398
3	Clayton Krypton MZ	Clayton	4.5:68% w/w	WP	14010
4	Clayton Krypton WDG	Clayton	5:68% w/w	WG	15497
5	Curzate M WG	DuPont	4.5:68% w/w	WG	11901
6	Cyman	ChemSource	4.5:68% w/w	WP	15033
7	Globe	Sipcam	6:70% w/w	WP	10339
8	Matilda	Nufarm UK	4.5:68% w/w	WP	12006
9	Nautile DG	United Phosphorus	5:68% w/w	WG	14692
10	Pillar WG	ChemSource	5:68% w/w	WG	15334
11	Video DG	United Phosphorus	5:68% w/w	WG	14850

Uses

- Blight in **potatoes**

Approval information

- Cymoxanil and mancozeb included in Annex I under EC Regulation 1107/2009

Efficacy guidance

- Apply immediately after blight warning or as soon as local conditions dictate and repeat at 7-14 d intervals until haulm dies down or is burnt off
- Spray interval should not be more than 7-10 d in irrigated crops (see product label). Apply treatment after irrigation
- To minimise the likelihood of development of resistance these products should be used in a planned Resistance Management strategy. See Section 5 for more information

Restrictions

- Maximum number of treatments 6 per crop [7]. Not specified for other products
- Do not apply at less than 7 d intervals
- Do not allow packs to become wet during storage

Crop-specific information

- HI 7 d [7]
- Destroy and remove any haulm that remains after harvest of early varieties to reduce blight pressure on neighbouring maincrop potatoes

Environmental safety

- Dangerous for the environment
- Very toxic to aquatic organisms
- Keep product away from fire or sparks

Hazard classification and safety precautions

Hazard Harmful [2, 5, 6, 8]; Irritant [1, 3, 4, 7, 9-11]; Dangerous for the environment [1-6, 8-11]
Transport code 9
Packaging group III
UN Number 3077
Risk phrases R36 [1-3, 5, 6, 8]; R37, R50, R53a [1-6, 8-11]; R38 [3, 8]; R42 [1, 2, 5, 6]; R43 [1-11]; R63 [8]
Operator protection A [1-11]; C [1-3, 5-8]; H [4, 7, 9-11]; U02a [7]; U05a, U08, U20b [1-11]; U11 [1-4, 6, 8-11]; U14 [1, 4-6, 9-11]; U19a [1-6, 8-11]
Environmental protection E13b [7]; E13c [1, 5, 6, 8]; E15a [1-7, 9-11]; E38 [1, 4-6, 9-11]
Storage and disposal D01, D02, D09a [1-11]; D06b [4, 9-11]; D11a [1-3, 5-8]; D12a [1, 4, 6, 9-11]
Medical advice M04a, M04b [2, 3]

SECTION 2

SEE SECTION 3 FOR PRODUCTS ALSO REGISTERED

125   cymoxanil + propamocarb

A cyanoacetamide oxime fungicide + a carbamate fungicide for potatoes
FRAC mode of action code: 27 + 28

Products

1	Axidor	Belchim	50:400 g/l	SC	15100
2	Proxanil	Sipcam	50:400 g/l	SC	13945

Uses

*   Blight in **potatoes** [1, 2]; **potatoes** *(off-label)* [1]

Extension of Authorisation for Minor Use (EAMUs)

*   **potatoes** *20121782* [1]

Approval information

*   Cymoxanil and propamocarb hydrochloride included in Annex I under EC Regulation 1107/2009

Hazard classification and safety precautions
Hazard Irritant
UN Number N/C
Risk phrases R43, R52, R53a
Operator protection A, H; U05a, U08, U14, U20b
Environmental protection E15a, E38
Storage and disposal D01, D02, D09a, D10a

126   cypermethrin

A contact and stomach acting pyrethroid insecticide
IRAC mode of action code: 3

Products

1	Cypermethrin Lacquer	Killgerm	7 g/l	LA	H3164
2	Forester	Fargro	100 g/l	EW	13164
3	Permasect C	Nufarm UK	100 g/l	EC	13158
4	Syper 100	Goldengrass	100 g/l	EC	14194
5	Toppel 100	United Phosphorus	100 g/l	EC	15696

Uses

*   Aphids in **apples**, **vining peas** [3, 5]; **beet leaves** *(off-label)*, **broad beans** *(off-label)*, **chinese cabbage** *(off-label)*, **frise** *(off-label)*, **herbs (see appendix 6)** *(off-label)*, **lamb's lettuce** *(off-label)*, **leaf brassicas** *(off-label)*, **mallow (althaea spp.)** *(off-label)*, **pears**, **protected beet leaves** *(off-label)*, **protected chinese cabbage** *(off-label)*, **protected frise** *(off-label)*, **protected herbs (see appendix 6)** *(off-label)*, **protected leaf brassicas** *(off-label)*, **protected radicchio** *(off-label)*, **protected scarole** *(off-label)*, **protected spinach** *(off-label)*, **protected spinach beet** *(off-label)*, **radicchio** *(off-label)*, **scarole** *(off-label)*, **spinach** *(off-label)*, **spinach beet** *(off-label)*, **spring barley** *(autumn sown)*, **spring rye**, **spring wheat** *(autumn sown)* [5]; **broccoli**, **brussels sprouts**, **cabbages**, **calabrese**, **cauliflowers**, **spring oats**, **triticale**, **winter barley**, **winter oats**, **winter rye**, **winter wheat** [4, 5]; **fodder beet**, **kale**, **mangels**, **potatoes**, **red beet**, **spring barley**, **spring field beans**, **spring oilseed rape**, **spring wheat**, **sugar beet**, **winter field beans**, **winter oilseed rape** [4]; **spring barley** *(moderate control)*, **spring wheat** *(moderate control)*, **winter barley** *(moderate control)*, **winter oats** *(moderate control)*, **winter rye** *(moderate control)*, **winter wheat** *(moderate control)* [3]
*   Asparagus beetle in **asparagus** *(off-label)* [5]
*   Barley yellow dwarf virus vectors in **spring barley** *(autumn sown)*, **spring wheat** *(autumn sown)*, **triticale**, **winter barley**, **winter oats**, **winter rye**, **winter wheat** [3, 5]; **spring oats**, **spring rye** [5]
*   Beetles in **cut logs** [2]
*   Bladder pod midge in **spring oilseed rape**, **winter oilseed rape** [5]
*   Blossom beetle in **spring oilseed rape**, **winter oilseed rape** [3]

FOR FULL CONDITIONS OF USE ALWAYS READ THE PRODUCT LABEL

SECTION 2

- Brassica pod midge in **spring oilseed rape** *(some coincidental control only)*, **winter oilseed rape** *(some coincidental control only)* [4]
- Cabbage seed weevil in **spring oilseed rape**, **winter oilseed rape** [3, 4]
- Cabbage stem flea beetle in **spring oilseed rape** [4]; **winter oilseed rape** [3-5]
- Capsids in **apples** [5]; **pears** [3, 5]
- Caterpillars in **apples**, **beet leaves** *(off-label)*, **chinese cabbage** *(off-label)*, **frise** *(off-label)*, **herbs (see appendix 6)** *(off-label)*, **lamb's lettuce** *(off-label)*, **leaf brassicas** *(off-label)*, **mallow (althaea spp.)** *(off-label)*, **pears**, **protected beet leaves** *(off-label)*, **protected chinese cabbage** *(off-label)*, **protected frise** *(off-label)*, **protected herbs (see appendix 6)** *(off-label)*, **protected leaf brassicas** *(off-label)*, **protected radicchio** *(off-label)*, **protected scarole** *(off-label)*, **protected spinach** *(off-label)*, **protected spinach beet** *(off-label)*, **radicchio** *(off-label)*, **scarole** *(off-label)*, **spinach** *(off-label)*, **spinach beet** *(off-label)* [5]; **broccoli**, **brussels sprouts**, **cabbages**, **calabrese**, **cauliflowers** [3-5]; **fodder beet**, **kale**, **mangels**, **potatoes**, **red beet**, **sugar beet** [3, 4]
- Codling moth in **apples** [3, 5]
- Cutworms in **carrots** *(off-label)*, **horseradish** *(off-label)*, **parsley root** *(off-label)*, **parsnips** *(off-label)* [5]; **fodder beet**, **mangels**, **potatoes**, **red beet**, **sugar beet** [3-5]
- Flea beetle in **spring linseed** *(off-label)*, **winter linseed** *(off-label)* [5]
- Frit fly in **grassland** [3, 4]; **permanent grassland** [5]; **rotational grass** [3, 5]
- Insect pests in **farm buildings** [1]
- Large pine weevil in **forest** [2]
- Pea and bean weevil in **spring field beans**, **vining peas**, **winter field beans** [3-5]
- Pea moth in **vining peas** [3-5]
- Pod midge in **spring oilseed rape**, **winter oilseed rape** [3]
- Pollen beetle in **spring oilseed rape**, **winter oilseed rape** [3-5]
- Rape winter stem weevil in **winter oilseed rape** [5]
- Sawflies in **apples** [5]
- Seed weevil in **spring oilseed rape**, **winter oilseed rape** [5]
- Suckers in **apples** [3]; **pears** [5]
- Tortrix moths in **apples** [3, 5]; **pears** [3]
- Whitefly in **broccoli**, **brussels sprouts**, **cabbages**, **calabrese**, **cauliflowers** [5]
- Winter moth in **apples**, **pears** [3]
- Yellow cereal fly in **spring barley** *(autumn sown)*, **spring wheat** *(autumn sown)*, **triticale**, **winter barley**, **winter oats**, **winter rye** [3, 5]; **spring oats**, **spring rye** [5]; **winter wheat** [3-5]

Extension of Authorisation for Minor Use (EAMUs)
- **asparagus** *20121063* [5]
- **beet leaves** *20121064* [5]
- **broad beans** *20121067* [5]
- **carrots** *20121062* [5]
- **chinese cabbage** *20121064* [5]
- **frise** *20121066* [5]
- **herbs (see appendix 6)** *20121066* [5]
- **horseradish** *20121062* [5]
- **lamb's lettuce** *20121066* [5]
- **leaf brassicas** *20121066* [5]
- **mallow (althaea spp.)** *20121066* [5]
- **parsley root** *20121062* [5]
- **parsnips** *20121062* [5]
- **protected beet leaves** *20121064* [5]
- **protected chinese cabbage** *20121064* [5]
- **protected frise** *20121066* [5]
- **protected herbs (see appendix 6)** *20121066* [5]
- **protected leaf brassicas** *20121066* [5]
- **protected radicchio** *20121066* [5]
- **protected scarole** *20121066* [5]
- **protected spinach** *20121064* [5]
- **protected spinach beet** *20121066* [5]
- **radicchio** *20121066* [5]

SEE SECTION 3 FOR PRODUCTS ALSO REGISTERED

- **scarole** *20121066* [5]
- **spinach** *20121064* [5]
- **spinach beet** *20121066* [5]
- **spring linseed** *20121065* [5]
- **winter linseed** *20121065* [5]

Approval information
- Cypermethrin included in Annex I under EC Regulation 1107/2009
- Accepted by BBPA for use on malting barley and hops

Efficacy guidance
- Products combine rapid action, good persistence, and high activity on Lepidoptera.
- As effect is mainly via contact, good coverage is essential for effective action. Spray volume should be increased on dense crops [3, 5]
- A repeat spray after 10-14 d is needed for some pests of outdoor crops, several sprays at shorter intervals for whitefly and other glasshouse pests [3, 5]
- Rates and timing of sprays vary with crop and pest. See label for details [2, 3, 5]
- Where aphids in hops, pear suckers or glasshouse whitefly resistant to cypermethrin occur control is unlikely to be satisfactory [3, 5]
- Apply lacquer undiluted in a 5 cm wide band at strategic places. Ensure surfaces are clean and dry before application [1]
- Lacquer normally remains active for 12 mth [1]
- Temporary loss of activity for about 2 d will occur if lacquer surface is washed, until cypermethrin has migrated to the surface layer [1]

Restrictions
- Maximum number of treatments varies with crop and product. See label or approval notice for details [2, 3, 5]
- Apply by knapsack or hand-held sprayer [2]

Crop-specific information
- HI vining peas 7 d; other crops 0 d
- Test spray sample of new or unusual ornamentals or trees before committing whole batches [2]
- Post-planting forestry applications should be made before damage is seen or at the onset of damage during the first 2 yrs after transplanting [2]

Environmental safety
- Dangerous for the environment [2, 3, 5]
- Very toxic to aquatic organisms
- High risk to bees. Do not apply to crops in flower, or to those in which bees are actively foraging, except as directed. Do not apply when flowering weeds are present
- Flammable [1, 3, 5]
- Do not spray cereals after 31 Mar within 6 m of the edge of the growing crop [2, 3, 5]
- LERAP A [3, 5]
- Broadcast air-assisted LERAP [3-5] (18 m); LERAP Category A [2-5]; LERAP Category B [4]

Hazard classification and safety precautions
Hazard Harmful [1-5]; Flammable [1, 3-5]; Dangerous for the environment [2-5]
Transport code 3 [3-5]; 9 [2]
Packaging group III [2-5]
UN Number 1993 [3, 5]; 3082 [2]; 3295 [4]
Risk phrases R20, R21, R52 [1]; R22a, R50 [2-5]; R22b, R37, R66, R67 [3-5]; R36 [3, 5]; R38 [2, 3, 5]; R41 [5]; R43 [2]; R53a [1-5]
Operator protection A [1-5]; C [1, 3-5]; H, M [2]; P [1]; U02a, U10 [2-4]; U04a [1-4]; U05a [1-5]; U08 [5]; U09a, U15 [1]; U11 [3, 4]; U14 [2]; U16b [1, 2]; U19a [2-5]; U20a [4]; U20b [3, 5]
Environmental protection E02a [1] (48 h); E13a, E16h [5]; E15a [1-4]; E16a [4]; E16c [3, 5]; E16d, E34 [2-5]; E17b [3-5] (18 m); E38 [2]
Consumer protection C08, C09, C11, C12 [1]
Storage and disposal D01, D02, D09a [1-5]; D05, D10b [2-5]; D12a [2, 3, 5]; D12b [1]
Medical advice M03 [2]; M04a [1, 3, 4]; M05b [3-5]

FOR FULL CONDITIONS OF USE ALWAYS READ THE PRODUCT LABEL

127   cyproconazole

A contact and systemic triazole fungicide
FRAC mode of action code: 3

See also azoxystrobin + cyproconazole
chlorothalonil + cyproconazole
chlorothalonil + cyproconazole + propiconazole
cyproconazole + picoxystrobin

Products

Centaur	Bayer CropScience	200 g/l	EC	13852

Uses
- Brown rust in **spring barley, spring rye, winter barley, winter rye, winter wheat**
- Chocolate spot in **spring field beans, winter field beans**
- Crown rust in **spring oats, winter oats**
- Ear diseases in **winter wheat**
- Eyespot in **winter barley** *(useful reduction),* **winter wheat** *(useful reduction)*
- Foliar disease control in **borage for oilseed production** *(off-label),* **canary flower (echium spp.)** *(off-label),* **durum wheat** *(off-label),* **fodder beet** *(off-label),* **grass seed crops** *(off-label),* **linseed** *(off-label),* **mustard** *(off-label),* **red beet** *(off-label),* **triticale** *(off-label)*
- Glume blotch in **winter wheat**
- Light leaf spot in **winter oilseed rape**
- Net blotch in **spring barley** *(useful reduction),* **winter barley** *(useful reduction)*
- Phoma leaf spot in **winter oilseed rape**
- Powdery mildew in **spring barley, spring oats, spring rye, sugar beet, winter barley, winter oats, winter rye, winter wheat**
- Ramularia leaf spots in **sugar beet**
- Rhynchosporium in **spring barley, winter barley**
- Rust in **spring field beans, winter field beans**
- Septoria leaf blotch in **winter wheat**
- Yellow rust in **spring barley, winter barley, winter wheat**

Extension of Authorisation for Minor Use (EAMUs)
- **borage for oilseed production** *20081099*
- **canary flower (echium spp.)** *20081099*
- **durum wheat** *20081101*
- **fodder beet** *20081100*
- **grass seed crops** *20081101*
- **linseed** *20081099*
- **mustard** *20081099*
- **red beet** *20081100*
- **triticale** *20081101*

Approval information
- Cyproconazole included in Annex I under EC Regulation 1107/2009
- Accepted by BBPA for use on malting barley

Efficacy guidance
- Apply at start of disease development or as preventive treatment and repeat as necessary
- Most effective time of treatment varies with disease and use of tank mixes may be desirable. See label for details
- Product alone gives useful reduction of cereal eyespot. Where high infections probable a tank mix with prochloraz recommended
- On oilseed rape a two spray autumn/spring programme recommended for high risk situations and on susceptible varieties
- Cyproconazole is a DMI fungicide. Resistance to some DMI fungicides has been identified in Septoria leaf blotch which may seriously affect performance of some products. For further advice contact a specialist advisor and visit the Fungicide Resistance Action Group (FRAG)-UK website

SEE SECTION 3 FOR PRODUCTS ALSO REGISTERED

SECTION 2

Restrictions
- Maximum total dose equivalent to 1-3 full dose treatments depending on crop treated. See labels for details

Crop-specific information
- HI 6 wk for field beans; 14 d for sugar beet, red beet, leeks
- Latest use: completion of flowering for rye, wheat; emergence of ear complete (GS 59) for barley, oats; before lowest pods more than 2 cm long (GS 5,1) for winter oilseed rape
- Application to winter wheat in spring between the start of stem elongation and the third node detectable stage (GS 30-33) may cause straw shortening, but does not cause loss of yield

Environmental safety
- Dangerous for the environment
- Toxic to aquatic organisms
- Dangerous to fish or other aquatic life. Do not contaminate surface waters or ditches with chemical or used container

Hazard classification and safety precautions
    Hazard Harmful, Dangerous for the environment
    Transport code 9
    Packaging group III
    UN Number 3082
    Risk phrases R36, R37, R43, R51, R53a, R63
    Operator protection A, C, H; U05a, U08, U20a
    Environmental protection E15a, E38
    Storage and disposal D01, D02, D09a, D10a, D12a
    Medical advice M03

128   cyproconazole + picoxystrobin

A broad spectrum triazole + strobilurin fungicide mixture for use in cereals
FRAC mode of action code: 3 + 11

See also cyproconazole

Products

Furlong	DuPont	80:200 g/l	SC	13818

Uses
- Brown rust in **spring barley**, **spring wheat**, **winter barley**, **winter wheat**
- Crown rust in **spring oats**, **winter oats**
- Glume blotch in **spring wheat**, **winter wheat**
- Net blotch in **spring barley** *(moderate control)*, **winter barley** *(moderate control)*
- Rhynchosporium in **spring barley** *(moderate control)*, **winter barley** *(moderate control)*
- Septoria leaf blotch in **spring wheat**, **winter wheat**
- Yellow rust in **spring wheat**, **winter wheat**

Approval information
- Picoxystrobin and cyproconazole included in Annex I under EC Regulation 1107/2009
- Accepted by BBPA for use on malting barley

Restrictions
- To reduce the risk of resistance developing in target diseases the total number of applications of products containing QoI fungicides made to any cereal crop must not exceed two.
- Maximum number of applications is 2 per crop

Hazard classification and safety precautions
    Hazard Harmful, Dangerous for the environment
    Transport code 9
    Packaging group III
    UN Number 3082
    Risk phrases R50, R53a, R63

FOR FULL CONDITIONS OF USE ALWAYS READ THE PRODUCT LABEL

Operator protection A, H
Environmental protection E15a, E34, E38
Storage and disposal D01, D02, D05, D09a, D12a

129 cyproconazole + propiconazole

A broad-spectrum mixture of conazole fungicides for wheat and barley
FRAC mode of action code: 3 + 3

See also propiconazole

Products

1	Alto Xtra	Syngenta	160:250 g/l	EC	14437
2	Menara	Syngenta	160:250 g/l	EC	14398

Uses

- Brown rust in **spring barley**, **winter barley**, **winter wheat** [1, 2]
- Foliar disease control in **durum wheat** *(off-label)*, **spring rye** *(off-label)*, **triticale** *(off-label)*, **winter rye** *(off-label)* [2]
- Glume blotch in **winter wheat** [1, 2]
- Late ear diseases in **winter wheat** [1, 2]
- Net blotch in **spring barley**, **winter barley** [1, 2]
- Powdery mildew in **spring barley**, **winter barley** [1, 2]; **winter wheat** [1]
- Rhynchosporium in **spring barley**, **winter barley** [1, 2]
- Septoria leaf blotch in **winter wheat** [1, 2]
- Yellow rust in **spring barley**, **winter barley** [1]; **winter wheat** [1, 2]

Extension of Authorisation for Minor Use (EAMUs)

- **durum wheat** *20090685* [2]
- **spring rye** *20090685* [2]
- **triticale** *20090685* [2]
- **winter rye** *20090685* [2]

Approval information

- Propiconazole and cyproconazole included in Annex I under EC Regulation 1107/2009
- Accepted by BBPA for use on malting barley

Efficacy guidance

- Treatment should be made at first sign of disease
- Useful reduction of eyespot when applied in spring during stem extension (GS 30-32)
- Cyproconazole and propiconazole are DMI fungicides. Resistance to some DMI fungicides has been identified in Septoria leaf blotch which may seriously affect performance of some products. For further advice contact a specialist advisor and visit the Fungicide Resistance Action Group (FRAG)-UK website

Restrictions

- Maximum total dose equivalent to one full dose treatment for barley; two full dose treatments for winter wheat

Crop-specific information

- Latest use: before beginning of anthesis (GS 60) for barley; before grain watery ripe (GS 71) for wheat
- Application to winter wheat in spring may cause straw shortening but does not cause loss of yield

Environmental safety

- Dangerous for the environment
- Very toxic to aquatic organisms

Hazard classification and safety precautions

Hazard Harmful, Dangerous for the environment
Transport code 9

SEE SECTION 3 FOR PRODUCTS ALSO REGISTERED

Packaging group III
UN Number 3082
Risk phrases R50, R53a, R63
Operator protection A, C, H; U05a, U08, U20b
Environmental protection E15a, E38
Storage and disposal D01, D02, D05, D07, D09a, D10c, D12a

130   cyproconazole + quinoxyfen

A contact and systemic fungicide mixture for cereals
FRAC mode of action code: 3 + 13

See also quinoxyfen

Products

Excelsior	Dow	80:75 g/l	SC	13378

Uses
- Brown rust in *winter wheat*
- Eyespot in *winter wheat* *(reduction)*
- Foliar disease control in *grass seed crops* *(off-label)*
- Glume blotch in *winter wheat*
- Powdery mildew in *winter wheat*
- Septoria leaf blotch in *winter wheat*
- Yellow rust in *winter wheat*

Extension of Authorisation for Minor Use (EAMUs)
- *grass seed crops* 20090887 expires 31 Mar 2013

Approval information
- Quinoxyfen and cyproconazole included in Annex I under EC Regulation 1107/2009
- Accepted by BBPA for use on malting barley
- Approval expiry 31 Mar 2013 [1]

Efficacy guidance
- Best results are achieved when disease is first detected and seen to be active. A second treatment may be needed under prolonged disease pressure
- Treatment for Septoria control should be made as soon as possible after weather that favours infection to prevent spread to the second and flag leaves
- Useful reduction of eyespot when applied between GS 30-34 to control other diseases
- Systemic activity may be reduced under conditions of severe drought stress

Restrictions
- Maximum number of treatments equivalent to a total of two full doses
- Application to winter wheat in spring may cause straw shortening but does not cause loss of yield

Crop-specific information
- Latest use: before ear emergence for wheat

Environmental safety
- Dangerous for the environment
- Very toxic to aquatic organisms
- LERAP Category B

Hazard classification and safety precautions
Hazard Harmful, Dangerous for the environment
Transport code 9
Packaging group III
UN Number 3082
Risk phrases R43, R50, R53a, R63
Operator protection A, C, H; U05a, U20b

FOR FULL CONDITIONS OF USE ALWAYS READ THE PRODUCT LABEL

Environmental protection E15a, E16a, E16b, E38
Storage and disposal D01, D02, D09a, D10b, D12a

131   cyproconazole + trifloxystrobin

A conazole and strobilurin fungicide mixture for disease control in sugar beet
FRAC mode of action code: 3 + 11

See also trifloxystrobin

Products

Escolta	Bayer CropScience	160:375 g/l	SC	13923

Uses
- Cercospora leaf spot in **sugar beet**
- Powdery mildew in **sugar beet**
- Ramularia leaf spots in **sugar beet**
- Rust in **sugar beet**

Approval information
- Cyproconazole and trifloxystrobin included in Annex I under EC Regulation 1107/2009
- Accepted by BBPA for use on malting barley

Efficacy guidance
- Best results obtained from treatment at early stages of disease development. Further treatment may be needed if disease attack is prolonged
- Cyproconazole is a DMI fungicide. Resistance to some DMI fungicides has been identified in Septoria leaf blotch which may seriously affect performance of some products. For further advice contact a specialist advisor and visit the Fungicide Resistance Action Group (FRAG)-UK website
- Trifloxystrobin is a member of the QoI cross resistance group. Product should be used preventatively and not relied on for its curative potential
- Use product as part of an Integrated Crop Management strategy incorporating other methods of control, including where appropriate other fungicides with a different mode of action. Do not apply more than two foliar applications of QoI containing products to any cereal crop
- There is a significant risk of widespread resistance occurring in *Septoria tritici* populations in UK. Failure to follow resistance management action may result in reduced levels of disease control
- Strains of barley powdery mildew resistant to QoI's are common in the UK

Restrictions
- Maximum total dose per crop equivalent to two full dose treatments

Crop-specific information
- HI 35 d

Environmental safety
- Dangerous for the environment
- Very toxic to aquatic organisms

Hazard classification and safety precautions
   Hazard Harmful, Dangerous for the environment
   Transport code 9
   Packaging group III
   UN Number 3082
   Risk phrases R50, R53a, R63
   Operator protection A; U02a, U05a, U09a, U19a, U20a
   Environmental protection E15b, E38
   Storage and disposal D01, D02, D05, D09a, D10b, D12a
   Medical advice M03

SEE SECTION 3 FOR PRODUCTS ALSO REGISTERED

132   cyprodinil

An anilinopyrimidine systemic broad spectrum fungicide for cereals
FRAC mode of action code: 9

See also cyproconazole + cyprodinil

Products

1 Kayak	Syngenta	300 g/l	EC	14847
2 Unix	Syngenta	75% w/w	WG	14846

Uses
- Disease control in **forest nurseries** *(off-label)* [1]
- Eyespot in **spring barley**, **winter barley** [1, 2]; **spring wheat**, **winter wheat** [2]
- Net blotch in **spring barley**, **winter barley** [1, 2]
- Powdery mildew in **spring barley**, **winter barley** [1, 2]; **spring wheat** *(moderate control)*, **winter wheat** *(moderate control)* [2]
- Rhynchosporium in **spring barley**, **winter barley** [1]; **spring barley** *(moderate control)*, **winter barley** *(moderate control)* [2]

Extension of Authorisation for Minor Use (EAMUs)
- **forest nurseries** *20122056* [1]

Approval information
- Cyprodinil included in Annex I under EC Regulation 1107/2009
- Accepted by BBPA for use on malting barley

Efficacy guidance
- Best results obtained from treatment at early stages of disease development
- For best control of eyespot spray before or during the period of stem extension in spring. Control may be reduced if very dry conditions follow treatment

Restrictions
- Maximum number of treatments 2 per crop on wheat and barley [1]

Crop-specific information
- Latest use: before first spikelet of inflorescence just visible stage for barley, before early milk stage for wheat [2]; up to and including first awns visible stage (GS49) in barley [1]

Environmental safety
- Dangerous for the environment
- Very toxic to aquatic organisms
- LERAP Category B

Hazard classification and safety precautions
> **Hazard** Irritant [1]; Dangerous for the environment [1, 2]
> **Transport code** 9
> **Packaging group** III
> **UN Number** 3077 [2]; 3082 [1]
> **Risk phrases** R38, R43 [1]; R50, R53a [1, 2]
> **Operator protection** A, H; U05a, U20a [1, 2]; U09a [1]
> **Environmental protection** E15a [2]; E15b, E34 [1]; E16a, E38 [1, 2]
> **Storage and disposal** D01, D02, D05, D07, D09a, D12a [1, 2]; D10c [1]

133   cyprodinil + fludioxonil

A broad-spectrum fungicide mixture for legumes and strawberries
FRAC mode of action code: 9 + 12

See also fludioxonil

Products

Switch	Syngenta	37.5:25 % w/w	WG	15129

Uses

- Alternaria in **apples**, **crab apples**, **pears**, **quinces**
- Alternaria blight in **carrots** *(moderate control)*
- Ascochyta in **mange-tout peas**, **sugar snap peas**, **vining peas**
- Black spot in **protected strawberries** *(qualified minor use)*, **strawberries** *(qualified minor use)*
- Blossom wilt in **apricots** *(off-label)*, **cherries** *(off-label)*, **peaches** *(off-label)*, **plums** *(off-label)*
- Botrytis in **apricots** *(off-label)*, **asparagus** *(off-label)*, **beet leaves** *(off-label)*, **broad beans** *(moderate control)*, **broad beans (dry-harvested)** *(moderate control)*, **bulb onions** *(off-label)*, **carrots** *(moderate control)*, **celeriac** *(moderate control)*, **chard** *(off-label)*, **cherries** *(off-label)*, **chicory** *(off-label)*, **combining peas** *(moderate control)*, **cress** *(off-label)*, **forest nurseries**, **french beans (dry-harvested)** *(moderate control)*, **frise** *(off-label)*, **garlic** *(off-label)*, **grapevines** *(off-label)*, **green beans** *(moderate control)*, **lamb's lettuce** *(off-label)*, **leaf brassicas** *(off-label)*, **lettuce** *(off-label)*, **lupins** *(off-label)*, **mange-tout peas** *(moderate control)*, **ornamental plant production**, **peaches** *(off-label)*, **plums** *(off-label)*, **protected aubergines** *(off-label)*, **protected cucumbers** *(off-label)*, **protected forest nurseries**, **protected ornamentals**, **protected peppers** *(off-label)*, **protected strawberries**, **protected tomatoes** *(off-label)*, **radicchio** *(off-label)*, **rocket** *(off-label)*, **runner beans** *(moderate control)*, **salad onions** *(off-label)*, **scarole** *(off-label)*, **shallots** *(off-label)*, **spinach** *(off-label)*, **strawberries**, **sugar snap peas** *(moderate control)*, **vining peas** *(moderate control)*
- Botrytis fruit rot in **apples**, **blackberries**, **crab apples**, **pears**, **quinces**, **raspberries**
- Botrytis root rot in **bilberries** *(qualified minor use recommendation)*, **blackcurrants** *(qualified minor use recommendation)*, **blueberries** *(qualified minor use recommendation)*, **cranberries** *(qualified minor use recommendation)*, **gooseberries** *(qualified minor use recommendation)*, **redcurrants** *(qualified minor use recommendation)*, **whitecurrants** *(qualified minor use recommendation)*
- Fusarium in **protected cucumbers** *(off-label)*
- Fusarium diseases in **apples**, **crab apples**, **pears**, **quinces**
- Gloeosporium rot in **apples**, **crab apples**, **pears**, **quinces**
- Mycosphaerella in **mange-tout peas**, **sugar snap peas**, **vining peas**
- Penicillium rot in **apples**, **crab apples**, **pears**, **quinces**
- Sclerotinia in **beetroot** *(off-label)*, **broad beans**, **broad beans (dry-harvested)**, **carrots** *(moderate control)*, **combining peas**, **french beans (dry-harvested)**, **ginger** *(off-label)*, **green beans**, **herbs for medicinal uses (see appendix 6)** *(off-label)*, **horseradish** *(off-label)*, **mange-tout peas**, **parsley root** *(off-label)*, **parsnips** *(off-label)*, **runner beans**, **salsify** *(off-label)*, **spice roots** *(off-label)*, **sugar snap peas**, **tumeric** *(off-label)*, **vining peas**

Extension of Authorisation for Minor Use (EAMUs)

- **apricots** *20103092*
- **asparagus** *20103173*
- **beet leaves** *20111317*
- **beetroot** *20103087*
- **bulb onions** *20122285*
- **chard** *20111317*
- **cherries** *20103092*
- **chicory** *20122285*
- **cress** *20111317*
- **frise** *20111317*
- **garlic** *20122285*
- **ginger** *20103087*
- **grapevines** *20112098*

SECTION 2

SEE SECTION 3 FOR PRODUCTS ALSO REGISTERED

- *herbs for medicinal uses (see appendix 6)* 20103087
- *horseradish* 20103087
- *lamb's lettuce* 20111317
- *leaf brassicas* 20111317
- *lettuce* 20111317
- *lupins* 20103170
- *parsley root* 20103087
- *parsnips* 20103087
- *peaches* 20103092
- *plums* 20103092
- *protected aubergines* 20103172
- *protected cucumbers* 20103171
- *protected peppers* 20103172
- *protected tomatoes* 20110302
- *radicchio* 20111317
- *rocket* 20111317
- *salad onions* 20103174
- *salsify* 20103087
- *scarole* 20111317
- *shallots* 20122285
- *spice roots* 20103087
- *spinach* 20111317
- *tumeric* 20103087

Approval information
- Cyprodinil and fludioxonil included in Annex I under EC Regulation 1107/2009

Efficacy guidance
- Best results obtained from application at the earliest stage of disease development or as a protective treatment following disease risk assessment
- Subsequent treatments may follow after a minimum of 10 d if disease pressure remains high
- To minimise the likelihood of development of resistance product should be used in a planned Resistance Management strategy. See Section 5 for more information

Restrictions
- Maximum number of treatments 2 per crop
- Consult processor before treating any crop for processing

Crop-specific information
- HI broad beans, green beans, mange-tout peas, runner beans, sugar snap peas, vining peas 14 d; protected strawberries, strawberries 3 d
- First signs of disease infection likely to be seen from early flowering for all crops
- Ensure peas are free from any stress before treatment. If necessary check wax with a crystal violet test

Environmental safety
- Dangerous for the environment
- Very toxic to aquatic organisms
- LERAP Category B

Hazard classification and safety precautions
Hazard Dangerous for the environment
Transport code 9
Packaging group III
UN Number 3077
Risk phrases R50, R53a
Operator protection A, H; U05a, U20a
Environmental protection E15b, E16a, E16b, E38
Storage and disposal D01, D02, D05, D07, D09a, D10c, D12a

FOR FULL CONDITIONS OF USE ALWAYS READ THE PRODUCT LABEL

134   cyprodinil + isopyrazam

A fungicide mixture for disease control in barley
FRAC mode of action code: 7 + 9

See also cyprodinil

Products

Bontima	Syngenta	187.5:62.5 g/l	EC	14899

Uses
- Brown rust in *spring barley, winter barley*
- Net blotch in *spring barley, winter barley*
- Powdery mildew in *spring barley, winter barley*
- Ramularia leaf spots in *spring barley, winter barley*
- Rhynchosporium in *spring barley, winter barley*

Approval information
- Isopyrazam has not yet been included in Annex 1 under EC Regulation 1107/2009
- Accepted by BBPA for use on malting barley

Restrictions
- Isopyrazam is an SDH respiration inhibitor; Do not apply more than two foliar applications of products containing an SDH inhibitor to any cereal crop

Crop-specific information
- There are no restrictions on succeeding crops in a normal rotation

Environmental safety
- LERAP Category B

Hazard classification and safety precautions
> **Hazard** Harmful, Dangerous for the environment
> **Transport code** 9
> **Packaging group** III
> **UN Number** 3082
> **Risk phrases** R20, R38, R40, R50, R53a, R63
> **Operator protection** A, C, H; U05a, U09a, U11, U20b
> **Environmental protection** E15b, E16a, E16b, E38
> **Storage and disposal** D01, D02, D05, D09a, D10c, D11a, D12a

135   cyprodinil + picoxystrobin

A broad spectrum fungicide mixture for cereals
FRAC mode of action code: 9 + 11

See also picoxystrobin

Products

Acanto Prima	DuPont	30.8% w/w	WG	14971

Uses
- Brown rust in *spring barley, spring wheat, winter barley, winter wheat*
- Eyespot in *spring barley, spring wheat* (moderate control), *winter barley, winter wheat* (moderate control)
- Glume blotch in *spring wheat, winter wheat*
- Net blotch in *spring barley, winter barley*
- Rhynchosporium in *spring barley, winter barley*
- Septoria leaf blotch in *spring wheat, winter wheat*
- Yellow rust in *spring wheat, winter wheat*

Approval information

- Cyprodinil and picoxystrobin included in Annex I under EC Regulation 1107/2009
- Accepted by BBPA for use on malting barley
- Approval expiry 31 Dec 2013 [1]

Efficacy guidance

- Best results obtained from use as a protectant treatment or in the earliest stages of disease development. Further applications may be needed if disease attack is prolonged
- Picoxystrobin is a member of the QoI cross resistance group. Product should be used preventatively and not relied on for its curative potential
- Use product as part of an Integrated Crop Management strategy incorporating other methods of control, including where appropriate other fungicides with a different mode of action. Do not apply more than two foliar applications of QoI containing products to any cereal crop
- There is a significant risk of widespread resistance occurring in *Septoria tritici* populations in UK. Failure to follow resistance management action may result in reduced levels of disease control
- In wheat product must always be used in mixture with another product, recommended for control of the same target disease, that contains a fungicide from a different cross resistance group and is applied at a dose that will give robust control
- Strains of barley powdery mildew resistant to QoIs are common in the UK

Restrictions

- Maximum number of treatments 2 per crop

Crop-specific information

- Latest use: before early milk stage (GS 73) for wheat; before ear just visible (GS 51) for barley

Environmental safety

- Dangerous for the environment
- Very toxic to aquatic organisms
- LERAP Category B

Hazard classification and safety precautions

Hazard Dangerous for the environment
Transport code 9
Packaging group III
UN Number 3077
Risk phrases R50, R53a
Operator protection A, C, H; U05a, U09a, U11, U14, U15, U20a
Environmental protection E15a, E16a, E34, E38
Storage and disposal D01, D02, D09a, D10c, D12a

---

136   2,4-D

A translocated phenoxycarboxylic acid herbicide for cereals, grass and amenity use
HRAC mode of action code: O

See also amitrole + 2,4-D + diuron
*   clopyralid + 2,4-D + MCPA*

Products

1 Depitox	Nufarm UK	500 g/l	SL	13258
2 Dioweed 50	United Phosphorus	500 g/l	SL	13197
3 Headland Staff 500	Headland	500 g/l	SL	13196
4 Herboxone	Headland	500 g/l	SL	13958
5 HY-D Super	Agrichem	500 g/l	SL	13198
6 Zip	AgriChem BV	500 g/l	SL	14806

Uses

- Annual dicotyledons in *amenity grassland, managed amenity turf, spring barley, spring wheat, winter barley, winter rye, winter wheat* [1-6]; *apple orchards, enclosed waters, land immediately adjacent to aquatic areas, pear orchards, spring rye* [1]; *grassland* [1, 2, 6];

FOR FULL CONDITIONS OF USE ALWAYS READ THE PRODUCT LABEL

permanent grassland [3-5]; *spring oats* [6]; *undersown barley, undersown rye, undersown wheat* [1, 5, 6]; *undersown oats* [5, 6]; *undersown spring cereals, undersown winter cereals* [2]; *winter oats* [1-5]
* Perennial dicotyledons in *amenity grassland, managed amenity turf, spring barley, spring wheat, winter barley, winter rye, winter wheat* [1, 3-6]; *apple orchards, enclosed waters, land immediately adjacent to aquatic areas, pear orchards, spring rye* [1]; *grassland* [1, 6]; *permanent grassland* [3-5]; *spring oats* [6]; *undersown barley, undersown rye, undersown wheat* [1, 5, 6]; *undersown oats* [5, 6]; *winter oats* [1, 3-5]

Approval information
* 2,4-D included in Annex I under EC Regulation 1107/2009
* Approved for aquatic weed control [1]. See Section 5 for information on use of herbicides in or near water
* Accepted by BBPA for use on malting barley

Efficacy guidance
* Best results achieved by spraying weeds in seedling to young plant stage when growing actively in a strongly competing crop
* Most effective stage for spraying perennials varies with species. See label for details
* Spray aquatic weeds when in active growth between May and Sep

Restrictions
* Maximum number of treatments normally 1 per crop and in forestry, 1 per yr in grassland and 2 or 3 per yr in amenity turf. Check individual labels
* Do not use on newly sown leys containing clover
* Do not spray grass seed crops after ear emergence
* Do not spray within 6 mth of laying turf or sowing fine grass
* Do not dump surplus herbicide in water or ditch bottoms
* Do not plant conifers until at least 1 mth after treatment
* Do not spray crops stressed by cold weather or drought or if frost expected
* Do not roll or harrow within 7 d before or after spraying
* Do not spray if rain falling or imminent
* Do not mow or roll turf or amenity grass 4 d before or after spraying. The first 4 mowings after treatment must be composted for at least 6 mth before use
* Do not mow grassland or graze for at least 10 d after spraying

Crop-specific information
* Latest use: before 1st node detectable in cereals; end Aug for conifer plantations; before established grassland 25 cm high
* Spray winter cereals in spring when leaf-sheath erect but before first node detectable (GS 31), spring cereals from 5-leaf stage to before first node detectable (GS 15-31)
* Cereals undersown with grass and/or clover, but not with lucerne, may be treated
* Selective treatment of resistant conifers can be made in Aug when growth ceased and plants hardened off, spray must be directed if applied earlier. See label for details

Following crops guidance
* Do not use shortly before or after sowing any crop
* Do not plant succeeding crops within 3 mth [3]
* Do not direct drill brassicas or grass/clover mixtures within 3 wk of application

Environmental safety
* Dangerous for the environment
* Very toxic to aquatic organisms [3]
* Dangerous to aquatic higher plants. Do not contaminate surface waters or ditches with chemical or used container
* 2,4-D is active at low concentrations. Take extreme care to avoid drift onto neighbouring crops, especially beet crops, brassicas, most market garden crops including lettuce and tomatoes under glass, pears and vines
* May be used to control aquatic weeds in presence of fish if used in strict accordance with directions for waterweed control and precautions needed for aquatic use [1]

SEE SECTION 3 FOR PRODUCTS ALSO REGISTERED

- Keep livestock out of treated areas for at least 2 wk following treatment and until poisonous weeds, such as ragwort, have died down and become unpalatable
- Water containing the herbicide must not be used for irrigation purposes within 3 wk of treatment or until the concentration in water is below 0.05 ppm

Hazard classification and safety precautions

Hazard Harmful, Dangerous for the environment [2-6]; Irritant [1]

Transport code 9 [2-4, 6]

Packaging group III [2-4, 6]

UN Number 3082 [2-4, 6]; N/C [1, 5]

Risk phrases R22a [2-6]; R37 [6]; R38, R52 [1]; R41, R53a [1-6]; R43 [2, 5, 6]; R50 [3, 4, 6]; R51 [2, 5]

Operator protection A, C, H [1-6]; M [1]; U05a, U11 [1-6]; U08, U20b [1, 3-6]; U09a, U20a [2]; U14, U15 [5, 6]

Environmental protection E07a, E34 [1-6]; E15a [1, 2]; E15b, E38 [2-6]

Storage and disposal D01, D02, D05 [1-6]; D09a [1, 3-6]; D10a [3-6]; D10b [1]; D10c [2]; D12a [2-6]

Medical advice M03 [1-6]; M05a [2, 5, 6]

137 2,4-D + dicamba

A translocated herbicide for use on turf
HRAC mode of action code: O + O

See also dicamba

Products

1	Landscaper Pro Weed Control + Fertilizer	Everris Ltd	0.8:0.12% w/w	GR	15820
2	New Estermone	Vitax	200:35 g/l	EC	13792
3	Thrust	Nufarm UK	344:120 g/l	SL	15408

Uses

- Annual and perennial weeds in *managed amenity turf* [2]
- Annual dicotyledons in *amenity grassland*, *permanent grassland* [3]; *managed amenity turf* [1]
- Buttercups in *managed amenity turf* [1]
- Perennial dicotyledons in *amenity grassland*, *permanent grassland* [3]

Approval information

- 2,4-D and dicamba included in Annex I under EC Regulation 1107/2009
- Accepted by BBPA for use on malting barley

Efficacy guidance

- Best results achieved by application when weeds growing actively in spring or early summer (later with irrigation and feeding)
- More resistant weeds may need repeat treatment after 3 wk
- Improved control of some weeds can be obtained by use of specifed oil adjuvant [3]
- Do not use during drought conditions

Restrictions

- Maximum number of treatments 2 per yr on amenity grass and established grassland [3]
- Do not treat newly sown or turfed areas or grass less than 1 yr old
- Do not treat grass crops intended for seed production [3]
- Do not treat grass suffering from drought, disease or other adverse factors [3]
- Do not roll or harrow for 7 d before or after treatment [3]
- Do not apply when grassland is flowering [3]
- Avoid spray drift onto cultivated crops or ornamentals
- Do not graze grass for at least seven days after spraying [3]
- Do not mow or roll four days before or after application. The first four mowings after treatment must be composted for at least six months before use [3]

FOR FULL CONDITIONS OF USE ALWAYS READ THE PRODUCT LABEL

Crop-specific information
- Latest use: before grass 25 cm high for amenity grass, established grassland [3]
- The first four mowings after treatment must be composted for at least 6 mth before use

Environmental safety
- Dangerous for the environment [3]
- Toxic to aquatic organisms
- Keep livestock out of treated areas for at least two weeks following treatment and until poisonous weeds, such as ragwort, have died down and become unpalatable [3]

Hazard classification and safety precautions

Hazard Harmful, Dangerous for the environment
UN Number N/C
Risk phrases R22a, R43, R51, R53a [1-3]; R22b, R36, R66 [1, 2]; R37, R41 [3]
Operator protection A, C, H, M [2, 3]; U05a [3]; U08, U14, U15, U20b [1, 2]; U11 [1-3]
Environmental protection E07a, E15a, E38 [1-3]; E34 [3]
Storage and disposal D01, D02 [3]; D05, D09a, D10a, D12a [1-3]
Medical advice M03 [3]; M05b [1, 2]

138   2,4-D + dicamba + dichlorprop-P

A translocated hormone mixture for the control of docks in grassland
HRAC mode of action code: O + O + O

See also dicamba
*   dichlorprop-P*

Products

| Legolas | Nufarm UK | 300:120:400 g/l | SL | 14439 |

Uses
- Creeping thistle in *amenity grassland, grassland*
- Dandelions in *amenity grassland, grassland*
- Docks in *amenity grassland, grassland*
- Sowthistle in *amenity grassland, grassland*

Approval information
- 2,4-D, dicamba and dichlorprop-P included in Annex I under EC Regulation 1107/2009
- Accepted by BBPA for use on malting barley

Efficacy guidance
- Make applications when weed growth is vigorous due to warm, moist conditions

Restrictions
- Treatment will severely damage clover. Do not use where clover is an important component of the sward
- Do not mow or roll four days before or after application. The first four mowings after treatment must be composted for at least six months before use.
- Do not treat newly established grass less than one year old.
- Do not treat grass crops intended for seed production

Crop-specific information
- One application per year on established agricultural grassland, two applications per year on amenity grassland

Following crops guidance
- Do not re-seed within 3 months of treatment
- Avoid drift on to suceptible crops such as tomatoes, vegetables, sugar beet, glasshouse crops, fruit and ornamentals

Environmental safety
- Very toxic to aquatic organisms

SEE SECTION 3 FOR PRODUCTS ALSO REGISTERED

- Keep livestock out of treated areas for at least 2 weeks following treatment and until poisonous weeds such as ragwort have died and become unpaletable

Hazard classification and safety precautions
    Hazard Harmful, Dangerous for the environment
    UN Number N/C
    Risk phrases R20, R21, R22a, R41, R43, R50, R53a
    Operator protection A, C, H; U05a, U11, U14
    Environmental protection E07a, E15a, E34, E38
    Storage and disposal D01, D02, D05, D09a, D10a, D12a
    Medical advice M03

139   2,4-D + dicamba + fluroxypyr

A translocated and contact herbicide mixture for amenity turf
HRAC mode of action code: O + O + O

See also dicamba
*   fluroxypyr*

Products

1	Barclay Holster XL	Barclay	285:52.5:105 g/l	EC	13596
2	Mascot Crossbar	Rigby Taylor	285:52.5:105 g/l	EC	14453
3	Swiftsure	Sherriff Amenity	285:52.5:105 g/l	EC	14468

Uses
- Annual dicotyledons in *managed amenity turf* [1-3]
- Buttercups in *managed amenity turf* [1]
- Clover in *managed amenity turf* [1]
- Daisies in *managed amenity turf* [1]
- Dandelions in *managed amenity turf* [1]
- Yarrow in *managed amenity turf* [1]

Approval information
- 2,4-D, dicamba and fluroxypyr included in Annex I under EC Regulation 1107/2009

Efficacy guidance
- Apply when weeds actively growing (normally between Apr and Sep) and when soil is moist
- Best results obtained from treatment in spring or early summer before weeds begin to flower
- Do not apply if turf is wet or if rainfall expected within 4 h of treatment. Both circumstances will reduce weed control

Restrictions
- Maximum number of treatments 2 per yr
- Do not mow for 3 d before or after treatment
- Avoid overlapping or overdosing, especially on newly sown turf
- Do not spray in drought conditions or if turf under stress from frost, waterlogging, trace element deficiency, pest or disease attack

Crop-specific information
- Latest use: normally Sep for managed amenity turf
- New turf may be treated in spring provided at least 2 mth have elapsed since sowing

Environmental safety
- Dangerous for the environment
- Toxic to aquatic organisms
- Keep livestock out of treated areas for at least two weeks following treatment and until poisonous weeds, such as ragwort, have died down and become unpalatable

Hazard classification and safety precautions
    Hazard Harmful
    Transport code 9

FOR FULL CONDITIONS OF USE ALWAYS READ THE PRODUCT LABEL

Packaging group III
UN Number 3082
Risk phrases R22a, R22b, R36, R38
Operator protection A, C, H, M; U05a, U08, U19a, U20b
Environmental protection E07a, E15a, E34
Storage and disposal D01, D02, D09a, D10a
Medical advice M03

140   2,4-D + dicamba + MCPA + mecoprop-P

A translocated herbicide mixture for grassland
HRAC mode of action code: O + O + O + O

See also dicamba
    MCPA
    mecoprop-P

Products

Longbow	Bayer Environ.	70:20:70:42 g/l	SL	14316

Uses

- Buttercups in *amenity grassland, managed amenity turf*
- Clovers in *amenity grassland, managed amenity turf*
- Daisies in *amenity grassland, managed amenity turf*
- Dandelions in *amenity grassland, managed amenity turf*
- Self-heal in *amenity grassland, managed amenity turf*

Approval information

- 2,4-D, dicamba, MCPA and mecoprop-P included in Annex I under EC Regulation 1107/2009

Environmental safety

- Some pesticides pose a greater threat of contamination of water than others and mecoprop-P is one of these pesticides. Take special care when applying mecoprop-P near water and do not apply if heavy rain is forecast

Hazard classification and safety precautions
    UN Number N/C
    Operator protection A, C, H, M
    Environmental protection E07a, E15a
    Storage and disposal D05, D09a, D10a, D12a

141   2,4-D + dicamba + triclopyr

A translocated herbicide for perennial and woody weed control
HRAC mode of action code: O + O + O

See also dicamba
    triclopyr

Products

Broadsword	United Phosphorus	200:85:65 g/l	EC	09140

Uses

- Annual dicotyledons in *land not intended to bear vegetation, permanent grassland, rotational grass*
- Brambles in *farm forestry, forest, natural surfaces not intended to bear vegetation*
- Docks in *land not intended to bear vegetation, permanent grassland, rotational grass*
- Gorse in *farm forestry, forest, natural surfaces not intended to bear vegetation*
- Green cover in *land temporarily removed from production*
- Japanese knotweed in *farm forestry, forest, natural surfaces not intended to bear vegetation*

SEE SECTION 3 FOR PRODUCTS ALSO REGISTERED

- Perennial dicotyledons in **farm forestry, forest, land not intended to bear vegetation, natural surfaces not intended to bear vegetation, permanent grassland, rotational grass**
- Plantains in **land not intended to bear vegetation, permanent grassland, rotational grass**
- Rhododendrons in **farm forestry, forest, natural surfaces not intended to bear vegetation**
- Stinging nettle in **permanent grassland, rotational grass**
- Thistles in **land not intended to bear vegetation, permanent grassland, rotational grass**
- Woody weeds in **farm forestry, forest, natural surfaces not intended to bear vegetation**

Approval information
- 2,4-D, dicamba and triclopyr included in Annex I under EC Regulation 1107/2009

Efficacy guidance
- Apply as foliar spray to herbaceous or woody weeds. Timing and growth stage for best results vary with species. See label for details
- Dilute with water for stump treatment and apply after felling up to the start of regrowth. Treat any regrowth with a spray to the growing foliage
- May be applied at 1/3 dilution in weed wipers or 1/8 dilution with ropewick applicators

Restrictions
- Maximum total dose per yr equivalent to two full dose treatments
- Do not use on pasture established less than 1 yr or on grass grown for seed
- Where clover a valued constituent of sward only use as a spot treatment
- Do not graze for 7 d or mow for 14 d after treatment
- Do not direct drill grass, clover or brassicas for at least 6 wk after grassland treatment
- Do not plant trees for 1-3 mth after spraying depending on dose applied. See label
- Avoid spray drift into greenhouses or onto crops or ornamentals. Vapour drift may occur in hot conditions

Crop-specific information
- Latest use: before weed flower buds open in grassland, non-cropped land, land temporarily removed from production; end of summer (spray); after felling up to the start of regrowth (cut stump spray) for farm forestry
- Sprays may be applied in pines, spruce and fir providing drift is avoided. Optimum time is mid-autumn when tree growth ceased but weeds not yet senescent

Environmental safety
- Dangerous for the environment
- Very toxic to aquatic organisms
- Flammable
- Docks and other weeds may become increasingly palatable after treatment and may be preferentially grazed
- Keep livestock out of treated areas for at least two weeks following treatment and until poisonous weeds, such as ragwort, have died down and become unpalatable
- Take extreme care to avoid drift onto neighbouring crops, especially beet crops, brassicas, most market garden crops including lettuce and tomatoes under glass, pears and vines

Hazard classification and safety precautions
Hazard Harmful, Flammable, Dangerous for the environment
Transport code 3
Packaging group III
UN Number 3295
Risk phrases R22a, R22b, R36, R37, R38, R50, R53a, R67
Operator protection A, C, H, M; U02a, U04a, U05a, U08, U11, U13, U19a, U20a
Environmental protection E07a, E13b, E34
Consumer protection C01
Storage and disposal D01, D02, D05, D09a, D10b, D12a
Medical advice M03, M05b

142   2,4-D + florasulam

A post-emergence herbicide mixture for control of broad leaved weeds in managed amenity turf
HRAC mode of action code: O + B

See also florasulam

Products

Junction	Rigby Taylor	452:6.25 g/l	ME	12493

Uses
- Clover in *managed amenity turf*
- Daisies in *managed amenity turf*
- Dandelions in *managed amenity turf*
- Plantains in *managed amenity turf*
- Sticky mouse-ear in *managed amenity turf*

Approval information
- 2,4-D and florasulam included in Annex I under EC Regulation 1107/2009

Efficacy guidance
- Best results obtained from treatment of actively growing weeds between Mar and Oct when soil is moist
- Do not apply when rain is imminent or during periods of drought unless irrigation is applied

Restrictions
- Maximum number of treatments on managed amenity turf: 1 per yr

Crop-specific information
- Avoid mowing turf 3 d before and after spraying
- Ensure newly sown turf has established before treating. Turf sown in late summer or autumn should not be sprayed until growth is resumed in the following spring

Following crops guidance
- An interval of 4 wk must elapse between application and re-seeding turf

Environmental safety
- Dangerous for the environment
- Toxic to aquatic organisms

Hazard classification and safety precautions
Hazard Harmful, Dangerous for the environment
Transport code 9
Packaging group III
UN Number 3082
Risk phrases R22a, R43, R51, R53a
Operator protection A, H; U02a, U05a, U08, U14, U20a
Environmental protection E15a, E34, E38
Storage and disposal D01, D02, D09a, D10b, D12a

143   2,4-D + MCPA

A translocated herbicide mixture for cereals and grass
HRAC mode of action code: O + O

See also MCPA

Products

1	Headland Polo	Headland	360:315 g/l	SL	14933
2	Lupo	Nufarm UK	360:315 g/l	SL	14931

SEE SECTION 3 FOR PRODUCTS ALSO REGISTERED

Uses

- Annual dicotyledons in *grassland*, *spring barley*, *spring wheat*, *winter barley*, *winter oats*, *winter wheat* [1, 2]
- Perennial dicotyledons in *grassland* [2]; *spring barley*, *spring wheat*, *winter barley*, *winter oats*, *winter wheat* [1, 2]

Approval information

- 2,4-D and MCPA included in Annex I under EC Regulation 1107/2009
- Accepted by BBPA for use on malting barley

Efficacy guidance

- Best results achieved by spraying weeds in seedling to young plant stage when growing actively in a strongly competing crop
- Most effective stage for spraying perennials varies with species. See label for details

Restrictions

- Maximum number of treatments 1 per crop in cereals, 2 per year in grassland and 3 per year in managed amenity turf [1]; 1 per crop in cereals and 1 per year in grassland [2]
- Do not spray if rain falling or imminent
- Do not cut grass or graze for at least 10 d after spraying
- Do not use on newly sown leys containing clover or other legumes
- Do not spray crops stressed by cold weather or drought or if frost expected
- Do not use shortly before or after sowing any crop
- Do not roll or harrow within 7 d before or after spraying

Crop-specific information

- Spray winter cereals in spring when leaf-sheath erect but before first node detectable (GS 31), spring cereals from 5-leaf stage to before first node detectable (GS 15-31)
- Latest use: before first node detectable (GS 31) in cereals;

Environmental safety

- Dangerous for the environment
- Toxic to aquatic organisms
- Keep livestock out of treated areas for at least two weeks following treatment and until poisonous weeds, such as ragwort, have died down and become unpalatable
- 2,4-D and MCPA are active at low concentrations. Take extreme care to avoid drift onto neighbouring crops, especially beet crops, brassicas, most market garden crops including lettuce and tomatoes under glass, pears and vines

Hazard classification and safety precautions

Hazard Harmful [1, 2]; Dangerous for the environment [1]
Transport code 9 [1]
Packaging group III [1]
UN Number 3082 [1]; N/C [2]
Risk phrases R20, R21, R43, R51 [1]; R22a, R41, R53a [1, 2]; R37, R52 [2]
Operator protection A, C, H, M; U05a, U08, U11, U14, U20b [1, 2]; U15 [1]
Environmental protection E07a, E15a, E34 [1, 2]; E38 [2]
Storage and disposal D01, D02, D05, D09a, D10a [1, 2]; D12a [2]
Medical advice M03 [1, 2]; M05a [1]

144   daminozide

A hydrazide plant growth regulator for use in certain ornamentals

Products

1	B-Nine SG	Certis	85% w/w	SG	14435
2	Dazide Enhance	Fargro	85% w/w	SG	14433

Uses

- Growth regulation in *azaleas*, *bedding plants*, *calibrachoa*, *chrysanthemums*, *hortensia*, *kalanchoes*, *ornamental specimens*, *petunia*, *sunflowers* [2]; *protected ornamentals* [1]

FOR FULL CONDITIONS OF USE ALWAYS READ THE PRODUCT LABEL

SECTION 2

Approval information
- Daminozide included in Annex I under EC Regulation 1107/2009

Efficacy guidance
- Best results obtained by application in late afternoon when glasshouse has cooled down
- Apply a fine spray using compressed air or power sprayers to give good coverage of dry foliage without run off
- Response to treatment differs widely depending on variety, stage of growth and physiological condition of the plant. It is recommended that any new variety is first tested on a small scale to observe if adverse effects occur
- For plants grown in modules or for seedlings prior to pricking out apply at reduced dose on a 7 d regime, starting at cotyledon stage

Restrictions
- Apply only to turgid, well watered plants. Do not water for 24 h after spraying
- Do not mix with other spray chemicals unless specifically recommended
- Do not store product in metal containers

Crop-specific information
- Do not use on chrysanthemum variety Fandango
- Evidence of effectiveness on Azaleas and Hydrangeas is limited
- See label for guidance on a range of bedding plant species

Hazard classification and safety precautions
UN Number N/C
Operator protection A, H; U02a, U08, U19a, U20c [2]; U05a [1, 2]
Environmental protection E15b
Storage and disposal D01, D02, D09a [1, 2]; D10b, D12a [2]
Medical advice M05a [2]

145 dazomet

A methyl isothiocyanate releasing soil fumigant
HRAC mode of action code: Z

Products
1 Basamid	Certis	97% w/w	GR	11324
2 Sterisoil	ChemSource	97% w/w	GR	15841

Uses
- Soil pests in **soil and compost**
- Soil-borne diseases in **soil and compost**

Approval information
- Dazomet included in Annex I under EC Regulation 1107/2009

Efficacy guidance
- Dazomet acts by releasing methyl isothiocyanate in contact with moist soil
- Soil sterilization is carried out after harvesting one crop and before planting the next
- The soil must be of fine tilth, free of clods and evenly moist to the depth of sterilization
- Soil moisture must not be less than 50% of water-holding capacity or oversaturated. If too dry, water at least 7-14 d before treatment
- In order to obtain short treatment times it is recommended to treat soils when soil temperature is above 7°C. Treatment should be used outdoors before winter rains make soil too wet to cultivate - usually early Nov
- For club root control treat only in summer when soil temperature above 10°C
- Where onion white rot is a problem it is unlikely to give effective control where inoculum level high or crop under stress
- Apply granules with suitable applicators, mix into soil immediately to desired depth and seal surface with polythene sheeting, by flooding or heavy rolling. See label for suitable application and incorporation machinery

SEE SECTION 3 FOR PRODUCTS ALSO REGISTERED

- With 'planting through' technique polythene seal is left in place to form mulch into which new crop can be planted

Restrictions
- Maximum number of treatments 1 per batch of soil for protected crops or field crop situation
- Use is limited to one application every third year on the same area
- Crops must not be planted until the cress safety test has been carried out and germination found to be satisfactory
- Do not treat ground where water table may rise into treated layer

Crop-specific information
- Latest use: pre-planting

Following crops guidance
- With 'planting through' technique no gas release cultivations are made and safety test with cress is particularly important. Conduct cress test on soil samples from centre as well as edges of bed. Observe minimum of 30 d from application to cress test in soils at 10°C or above, at least 50 d in soils below 10°C. See label for details
- In all other situations 14-28 d after treatment cultivate lightly to allow gas to disperse and conduct cress test after a further 14-28 d (timing depends on soil type and temperature). Do not treat structures containing live plants or any ground within 1 m of live plants

Environmental safety
- Dangerous for the environment
- Very toxic to aquatic organisms

Hazard classification and safety precautions
  Hazard Harmful, Dangerous for the environment
  Transport code 9
  Packaging group III
  UN Number 3077
  Risk phrases R22a, R50, R53a
  Operator protection A, M; U04a, U05a, U09a, U19a, U20a
  Environmental protection E15a, E34, E38
  Storage and disposal D01, D02, D07, D09a, D11a, D12a
  Medical advice M03, M05a

146 2,4-DB

A translocated phenoxycarboxylic acid herbicide
HRAC mode of action code: O

Products

1 DB Straight	United Phosphorus	400 g/l	SL	13736
2 Headland Spruce	Headland	400 g/l	SL	15089

Uses
- Annual dicotyledons in **grassland, lucerne, red clover, spring barley, spring oats, spring wheat, white clover, winter barley, winter oats, winter wheat**
- Buttercups in **grassland**
- Fat hen in **grassland**
- Penny cress in **grassland**
- Shepherd's purse in **grassland**

Approval information
- 2,4-DB included in Annex I under EC Regulation 1107/2009
- Accepted by BBPA for use on malting barley
- Approval expiry 31 Dec 2013 [1, 2]

Efficacy guidance
- Best results achieved on young seedling weeds under good growing conditions. Treatment less effective in cold weather and dry soil conditions
- Rain within 12 h may reduce effectiveness

Restrictions
- Do not allow spray drift onto neighbouring crops

Crop-specific information
- Latest use: before first node detectable (GS 31) for undersown cereals; fourth trifoliate leaf for lucerne
- In direct sown lucerne spray when seedlings have reached first trifoliate leaf stage. Optimum time 3-4 trifoliate leaves
- Do not treat any lucerne after fourth trifoliate leaf
- In spring barley and spring oats undersown with lucerne spray from when cereal has 1 leaf unfolded and lucerne has first trifoliate leaf
- In spring wheat undersown with lucerne spray from when cereal has 3 leaves unfolded and lucerne has first trifoliate leaf

Following crops guidance
- Do not sow any crop into soil treated with 2,4-DB for at least 3 months after application.

Environmental safety
- Dangerous for the environment
- Toxic to aquatic organisms
- Keep livestock out of treated areas for at least two weeks following treatment and until poisonous weeds, such as ragwort, have died down and become unpalatable

Hazard classification and safety precautions
Hazard Harmful, Dangerous for the environment
Transport code 9
Packaging group III
UN Number 3082
Risk phrases R22a, R41, R43, R51, R53a
Operator protection A, C, H; U05a, U09a, U11, U20a
Environmental protection E07a, E15a, E15b, E34, E38
Storage and disposal D01, D02, D05, D10c, D12a
Medical advice M03, M05a

147   2,4-DB + MCPA

A translocated herbicide for cereals, clovers and leys
HRAC mode of action code: O + O

See also MCPA

Products
Clovermax	Nufarm UK	240:40 g/l	SL	15780

Uses
- Annual dicotyledons in *grassland, oats, spring barley, spring wheat, undersown barley, undersown wheat, winter barley, winter wheat*
- Buttercups in *grassland, oats, spring barley, spring wheat, undersown barley, undersown wheat, winter barley, winter wheat*
- Fat hen in *grassland, oats, spring barley, spring wheat, undersown barley, undersown wheat, winter barley, winter wheat*
- Penny cress in *grassland, oats, spring barley, spring wheat, undersown barley, undersown wheat, winter barley, winter wheat*
- Plantains in *grassland*
- Shepherd's purse in *grassland, oats, spring barley, spring wheat, undersown barley, undersown wheat, winter barley, winter wheat*

SEE SECTION 3 FOR PRODUCTS ALSO REGISTERED

Approval information
- 2,4-DB and MCPA included in Annex I under EC Regulation 1107/2009
- Accepted by BBPA for use on malting barley

Efficacy guidance
- Best results achieved on young seedling weeds under good growing conditions
- Spray thistles and other perennials when 10-20 cm high provided clover at correct stage
- Effectiveness may be reduced by rain within 12 h, by very cold conditions or drought

Restrictions
- Maximum number of treatments 1 per crop
- Do not spray established clover crops or lucerne
- Do not roll or harrow within 7 d before or after spraying
- Do not spray immediately before or after sowing any crop
- DO NOT graze crops within a week before or 2 weeks after applying
- DO NOT apply during drought, rain or if rain is expected
- DO NOT apply in very cold conditions as effectiveness may be reduced
- DO NOT use immediately before or after sowing any crop
- DO NOT use on cereals undersown with Lucerne or in seed mixtures containing Lucerne
- DO NOT spray in windy conditions as the spray drift may cause damage to neighbouring crops. The following crops are particularly susceptible: Beet, Brassicae (e.g. turnips, swedes, oilseed rape), and most market garden crops including lettuce and tomatoes under glass, pears and vines

Crop-specific information
- Latest use: before first node detectable (GS 31) for cereals; 4th trifoliate leaf stage of clover for grass re-seeds
- Apply in spring to winter cereals from leaf sheath erect stage, to spring barley or oats from 2-leaf stage (GS 12), to spring wheat from 5-leaf stage (GS 15)
- Spray clovers as soon as possible after first trifoliate leaf, grasses after 2-3 leaf stage
- Red clover may suffer temporary distortion after treatment

Environmental safety
- Harmful to aquatic organisms
- Harmful to fish or other aquatic life. Do not contaminate surface waters or ditches with chemical or used container
- Keep livestock out of treated areas for at least two weeks following treatment and until poisonous weeds, such as ragwort, have died down and become unpalatable

Hazard classification and safety precautions
Hazard Harmful, Dangerous for the environment
UN Number N/C
Risk phrases R22a, R41, R51, R53a
Operator protection A, C; U05a, U12, U19a, U20b
Environmental protection E07a, E15b, E34, E38
Storage and disposal D01, D02, D05, D09a, D10b, D12a
Medical advice M03

148   deltamethrin

A pyrethroid insecticide with contact and residual activity
IRAC mode of action code: 3

Products

1	Bandu	Headland	25 g/l	EC	10994
2	Decis	Bayer CropScience	25 g/l	EC	07172
3	Decis Protech	Bayer CropScience	15 g/l	EW	11502
4	Delta-M 2.5 EC	AgChem Access	25 g/l	EC	12604
5	Grainstore	Pan Agriculture	25 g/l	EC	14932
6	K-Obiol EC 25	Bayer Environ.	25 g/l	EC	13573
7	K-Obiol ULV 6	Bayer Environ.	0.69% w/v	UL	13572

FOR FULL CONDITIONS OF USE ALWAYS READ THE PRODUCT LABEL

SECTION 2

Products – continued

8 MiteStop   ChemSource   25 g/l   EC   15775

Uses

- Aphids in *amenity vegetation, apples, ornamental plant production, protected cucumbers, protected peppers, protected pot plants, protected tomatoes, spring barley, spring oats, spring wheat, winter barley, winter oats, winter wheat* [1-4]; *blackberries* (off-label), *fodder beet* (off-label), *frise* (off-label), *herbs (see appendix 6)* (off-label), *lamb's lettuce* (off-label), *protected aubergines* (off-label), *scarole* (off-label), *spinach* (off-label) [1, 2]; *bulb onions* (off-label), *garlic* (off-label), *leaf brassicas* (off-label), *leeks* (off-label), *protected choi sum* (off-label), *protected cress* (off-label), *protected frise* (off-label), *protected herbs (see appendix 6)* (off-label), *protected lamb's lettuce* (off-label), *protected lettuce* (off-label), *protected ornamentals, protected pak choi* (off-label), *protected radicchio* (off-label), *protected scarole* (off-label), *protected spinach* (off-label), *rocket* (off-label), *shallots* (off-label) [3]; *carrots* (off-label), *celery (outdoor)* (off-label), *horseradish* (off-label), *lettuce, mallow (althaea spp.)* (off-label), *parsnips* (off-label), *protected celery* (off-label), *protected chinese cabbage* (off-label), *protected rhubarb* (off-label), *protected spring onions* (off-label), *rhubarb* (off-label), *spring onions* (off-label) [1, 3]; *chinese cabbage* (off-label), *durum wheat* (off-label), *spring rye* (off-label), *triticale* (off-label), *winter rye* (off-label) [1-3]; *choi sum* (off-label), *collards* (off-label), *lettuce* (off-label), *pak choi* (off-label), *spring greens* (off-label) [2]; *evening primrose* (off-label), *grass seed crops* (off-label), *marjoram* (off-label), *radicchio* (off-label), *sorrel* (off-label) [2, 3]; *lupins* (off-label), *parsley root* (off-label), *protected cayenne peppers* (off-label) [1]; *nursery stock* [2, 4]
- Apple sucker in *apples* [1-4]
- Barley yellow dwarf virus vectors in *winter barley, winter wheat* [1-4]
- Beet virus yellows vectors in *winter oilseed rape* [1-4]
- Cabbage seed weevil in *mustard, spring oilseed rape, winter oilseed rape* [1-4]
- Cabbage stem flea beetle in *winter oilseed rape* [1-4]
- Cabbage stem weevil in *mustard, spring oilseed rape, winter oilseed rape* [1-4]
- Capsids in *amenity vegetation, apples, ornamental plant production* [1-4]; *nursery stock* [2, 4]
- Caterpillars in *amenity vegetation, apples, broccoli, brussels sprouts, cabbages, cauliflowers, kale, ornamental plant production, plums, protected cucumbers, protected peppers, protected pot plants, protected tomatoes, swedes, turnips* [1-4]; *bulb onions* (off-label), *carrots* (off-label), *celery (outdoor)* (off-label), *chinese cabbage* (off-label), *garlic* (off-label), *horseradish* (off-label), *leeks* (off-label), *lettuce, mallow (althaea spp.)* (off-label), *parsnips* (off-label), *protected celery* (off-label), *protected chinese cabbage* (off-label), *protected choi sum* (off-label), *protected cress* (off-label), *protected frise* (off-label), *protected herbs (see appendix 6)* (off-label), *protected lamb's lettuce* (off-label), *protected lettuce* (off-label), *protected ornamentals, protected pak choi* (off-label), *protected radicchio* (off-label), *protected rhubarb* (off-label), *protected scarole* (off-label), *protected spinach* (off-label), *protected spring onions* (off-label), *rhubarb* (off-label), *rocket* (off-label), *shallots* (off-label), *spring onions* (off-label) [3]; *evening primrose* (off-label), *grass seed crops* (off-label), *marjoram* (off-label), *radicchio* (off-label), *sorrel* (off-label) [2, 3]; *leaf brassicas* (off-label) [1, 3]; *leaf brassicas* (off-label - baby leaf production), *protected brassica seedlings* (off-label), *protected rocket* (off-label), *rocket* (off-label - baby leaf production) [2]; *mustard, protected leaf brassicas* (off-label) [1]; *nursery stock* [2, 4]
- Codling moth in *apples* [1-4]
- Cutworms in *lettuce* [1, 3]
- Damson-hop aphid in *hops, plums* [1-4]
- Flea beetle in *broccoli, brussels sprouts, cabbages, cauliflowers, kale, swedes, turnips* [2, 4]; *bulb onions* (off-label), *carrots* (off-label), *celery (outdoor)* (off-label), *chinese cabbage* (off-label), *garlic* (off-label), *horseradish* (off-label), *leeks* (off-label), *lettuce, mallow (althaea spp.)* (off-label), *parsnips* (off-label), *protected celery* (off-label), *protected chinese cabbage* (off-label), *protected choi sum* (off-label), *protected pak choi* (off-label), *protected rhubarb* (off-label), *protected spring onions* (off-label), *rhubarb* (off-label), *rocket* (off-label), *shallots* (off-label), *spring onions* (off-label) [3]; *evening primrose* (off-label), *grass seed crops* (off-label), *marjoram* (off-label), *protected frise* (off-label), *radicchio* (off-label), *sorrel* (off-label) [2, 3]; *leaf brassicas* (off-label), *protected scarole* (off-label) [1, 3]; *leaf brassicas* (off-label - baby leaf

SEE SECTION 3 FOR PRODUCTS ALSO REGISTERED

production), **protected brassica seedlings** (off-label), **protected rocket** (off-label), **rocket** (off-label - baby leaf production) [2]; **mustard, protected leaf brassicas** (off-label) [1]; **protected cress** (off-label), **protected herbs (see appendix 6)** (off-label), **protected lamb's lettuce** (off-label), **protected lettuce** (off-label), **protected radicchio** (off-label), **protected spinach** (off-label) [1-3]; **sugar beet** [1-4]

- Fruit tree tortrix moth in **plums** [1-4]
- Insect pests in **blackberries** (off-label), **frise** (off-label), **herbs (see appendix 6)** (off-label), **lamb's lettuce** (off-label), **lupins** (off-label), **protected aubergines** (off-label), **scarole** (off-label), **spinach** (off-label) [1-3]; **bulb onions** (off-label), **choi sum** (off-label), **collards** (off-label), **garlic** (off-label), **pak choi** (off-label), **shallots** (off-label), **spring greens** (off-label) [2]; **carrots** (off-label), **celery (outdoor)** (off-label), **fodder beet** (off-label), **horseradish** (off-label), **mallow (althaea spp.)** (off-label), **parsley root** (off-label), **parsnips** (off-label), **protected celery** (off-label), **protected chinese cabbage** (off-label), **protected rhubarb** (off-label), **protected spring onions** (off-label), **rhubarb** (off-label), **spring onions** (off-label) [1, 2]; **chinese cabbage** (off-label) [1]; **durum wheat** (off-label), **protected cayenne peppers** (off-label), **spring rye** (off-label), **triticale** (off-label), **winter rye** (off-label) [1, 3]; **grain** (off-label), **stored grain** (off-label) [7]; **grain stores** [5, 6, 8]; **protected herbs (see appendix 6)** (off-label) [3]; **stored grain, stored pulses** [5-8]
- Leafhoppers in **bulb onions** (off-label), **carrots** (off-label), **celery (outdoor)** (off-label), **chinese cabbage** (off-label), **garlic** (off-label), **horseradish** (off-label), **leaf brassicas** (off-label), **leeks** (off-label), **lettuce, mallow (althaea spp.)** (off-label), **parsnips** (off-label), **protected celery** (off-label), **protected chinese cabbage** (off-label), **protected choi sum** (off-label), **protected pak choi** (off-label), **protected rhubarb** (off-label), **protected spring onions** (off-label), **rhubarb** (off-label), **rocket** (off-label), **shallots** (off-label), **spring onions** (off-label) [3]; **evening primrose** (off-label), **grass seed crops** (off-label), **marjoram** (off-label), **protected frise** (off-label), **radicchio** (off-label), **sorrel** (off-label) [2, 3]; **protected cress** (off-label), **protected herbs (see appendix 6)** (off-label), **protected lamb's lettuce** (off-label), **protected lettuce** (off-label), **protected radicchio** (off-label), **protected spinach** (off-label) [1-3]; **protected scarole** (off-label) [1, 3]
- Mealybugs in **amenity vegetation, ornamental plant production, protected cucumbers, protected peppers, protected pot plants, protected tomatoes** [1-4]; **nursery stock** [2, 4]; **protected ornamentals** [3]
- Pea and bean weevil in **broad beans, peas, spring field beans, winter field beans** [1-4]
- Pea midge in **peas** [2-4]
- Pea moth in **peas** [1-4]
- Pear sucker in **pears** [1-4]
- Phorid flies in **protected edible fungi** (off-label) [1, 3]; **protected mushrooms** (off-label) [3]
- Pollen beetle in **bulb onions** (off-label), **carrots** (off-label), **celery (outdoor)** (off-label), **chinese cabbage** (off-label), **garlic** (off-label), **horseradish** (off-label), **leaf brassicas** (off-label), **leeks** (off-label), **lettuce, mallow (althaea spp.)** (off-label), **parsnips** (off-label), **protected celery** (off-label), **protected chinese cabbage** (off-label), **protected choi sum** (off-label), **protected cress** (off-label), **protected frise** (off-label), **protected herbs (see appendix 6)** (off-label), **protected lamb's lettuce** (off-label), **protected lettuce** (off-label), **protected pak choi** (off-label), **protected radicchio** (off-label), **protected rhubarb** (off-label), **protected scarole** (off-label), **protected spinach** (off-label), **protected spring onions** (off-label), **rhubarb** (off-label), **rocket** (off-label), **shallots** (off-label), **spring onions** (off-label) [3]; **evening primrose** (off-label), **grass seed crops** (off-label), **marjoram** (off-label), **radicchio** (off-label), **sorrel** (off-label) [2, 3]; **mustard, spring oilseed rape, winter oilseed rape** [1-4]
- Raspberry beetle in **raspberries** [1-4]
- Sawflies in **apples, plums** [1-4]
- Scale insects in **amenity vegetation, ornamental plant production, protected cucumbers, protected peppers, protected pot plants, protected tomatoes** [1-4]; **nursery stock** [2, 4]; **protected ornamentals** [3]
- Sciarid flies in **protected edible fungi** (off-label) [1-3]; **protected mushrooms** (off-label) [2, 3]
- Thrips in **amenity vegetation, ornamental plant production** [1-4]; **bulb onions** (off-label), **garlic** (off-label), **shallots** (off-label) [1, 3]; **carrots** (off-label), **celery (outdoor)** (off-label), **chinese cabbage** (off-label), **evening primrose** (off-label), **grass seed crops** (off-label), **horseradish** (off-label), **leaf brassicas** (off-label), **lettuce, mallow (althaea spp.)** (off-label), **marjoram** (off-label), **parsnips** (off-label), **protected celery** (off-label), **protected chinese cabbage** (off-label), **protected choi sum** (off-label), **protected cress** (off-label), **protected frise**

(off-label), **protected herbs (see appendix 6)** *(off-label)*, **protected lamb's lettuce** *(off-label)*, **protected lettuce** *(off-label)*, **protected ornamentals**, **protected pak choi** *(off-label)*, **protected radicchio** *(off-label)*, **protected rhubarb** *(off-label)*, **protected scarole** *(off-label)*, **protected spinach** *(off-label)*, **protected spring onions** *(off-label)*, **radicchio** *(off-label)*, **rhubarb** *(off-label)*, **rocket** *(off-label)*, **sorrel** *(off-label)*, **spring onions** *(off-label)* [3]; **leeks** *(off-label)*, **protected courgettes** *(off-label)*, **protected cucumbers**, **protected gherkins** *(off-label)*, **protected peppers**, **protected tomatoes** [1-3]; **nursery stock** [2, 4]; **protected aubergines** *(off-label)*, **protected cayenne peppers** *(off-label)*, **protected ornamentals** *(off-label)* [2]; **protected non-edible flowers** *(off-label)* [1]

- Tortrix moths in **apples** [1-4]
- Western flower thrips in **protected aubergines** *(off-label)*, **protected cayenne peppers** *(off-label)*, **protected ornamentals** *(off-label)* [2]; **protected courgettes** *(off-label)*, **protected cucumbers**, **protected gherkins** *(off-label)*, **protected peppers**, **protected tomatoes** [1-3]; **protected non-edible flowers** *(off-label)* [1]; **protected ornamentals** [3]
- Whitefly in **amenity vegetation**, **ornamental plant production**, **protected cucumbers**, **protected peppers**, **protected pot plants**, **protected tomatoes** [1-4]; **nursery stock** [2, 4]; **protected ornamentals** [3]

Extension of Authorisation for Minor Use (EAMUs)

- **blackberries** *20071605* [1], *20071705* [2], *20071654* [3]
- **bulb onions** *20071613* [1], *20071698* [2], *20071158* [3]
- **carrots** *20071610* [1], *20071696* [2], *20071158* [3]
- **celery (outdoor)** *20071579* [1], *20071697* [2], *20071158* [3]
- **chinese cabbage** *20071606* [1], *20071612* [1], *20071695* [2], *20071158* [3]
- **choi sum** *20071704* [2]
- **collards** *20082458* [2]
- **durum wheat** *20071603* [1], *20071701* [2], *20071656* [3]
- **evening primrose** *20071694* [2], *20071158* [3]
- **fodder beet** *20071609* [1], *20071702* [2]
- **frise** *20071604* [1], *20071703* [2], *20071655* [3]
- **garlic** *20071613* [1], *20071698* [2], *20071158* [3]
- **grain** *20112491* [7]
- **grass seed crops** *20071694* [2], *20071158* [3]
- **herbs (see appendix 6)** *20071604* [1], *20071703* [2], *20071655* [3]
- **horseradish** *20071610* [1], *20071696* [2], *20071158* [3]
- **lamb's lettuce** *20071604* [1], *20071703* [2], *20071655* [3]
- **leaf brassicas** *20110427* [1], *(baby leaf production) 20071708* [2], *20071158* [3]
- **leeks** *20071614* [1], *20071699* [2], *20071158* [3]
- **lettuce** *20071695* [2]
- **lupins** *20071583* [1], *20071700* [2], *20071653* [3]
- **mallow (althaea spp.)** *20071610* [1], *20071696* [2], *20071158* [3]
- **marjoram** *20071694* [2], *20071158* [3]
- **pak choi** *20082458* [2]
- **parsley root** *20071610* [1], *20071696* [2]
- **parsnips** *20071610* [1], *20071696* [2], *20071158* [3]
- **protected aubergines** *20071608* [1], *20071693* [2], *20071707* [2], *20071649* [3]
- **protected brassica seedlings** *20071708* [2]
- **protected cayenne peppers** *20071607* [1], *20071693* [2], *20071657* [3]
- **protected celery** *20071579* [1], *20071697* [2], *20071158* [3]
- **protected chinese cabbage** *20071579* [1], *20071697* [2], *20071158* [3]
- **protected choi sum** *20071158* [3]
- **protected courgettes** *20071611* [1], *20071693* [2], *20071650* [3]
- **protected cress** *20080175* [1], *20071691* [2], *20071709* [2], *20071158* [3]
- **protected edible fungi** *20071581* [1], *20071692* [2], *20071648* [3]
- **protected frise** *20071691* [2], *20071158* [3]
- **protected gherkins** *20071611* [1], *20071693* [2], *20071650* [3]
- **protected herbs (see appendix 6)** *20080175* [1], *20071691* [2], *20071709* [2], *20071158* [3], *20071652* [3]
- **protected lamb's lettuce** *20080175* [1], *20071691* [2], *20071158* [3]

SEE SECTION 3 FOR PRODUCTS ALSO REGISTERED

- **protected leaf brassicas** *20110427* [1]
- **protected lettuce** *20080175* [1], *20071691* [2], *20071158* [3]
- **protected mushrooms** *20071692* [2], *20071648* [3]
- **protected non-edible flowers** *20071611* [1]
- **protected ornamentals** *20071693* [2]
- **protected pak choi** *20071158* [3]
- **protected radicchio** *20080175* [1], *20071691* [2], *20071158* [3]
- **protected rhubarb** *20071579* [1], *20071697* [2], *20071158* [3]
- **protected rocket** *20071708* [2]
- **protected scarole** *20080175* [1], *20071158* [3]
- **protected spinach** *20080175* [1], *20071691* [2], *20071158* [3]
- **protected spring onions** *20071579* [1], *20071697* [2], *20071158* [3]
- **radicchio** *20071694* [2], *20071158* [3]
- **rhubarb** *20071579* [1], *20071697* [2], *20071158* [3]
- **rocket** *(baby leaf production)* *20071708* [2], *20071158* [3]
- **scarole** *20071604* [1], *20071703* [2], *20071655* [3]
- **shallots** *20071613* [1], *20071698* [2], *20071158* [3]
- **sorrel** *20071694* [2], *20071158* [3]
- **spinach** *20071604* [1], *20071703* [2], *20071655* [3]
- **spring greens** *20082458* [2]
- **spring onions** *20071579* [1], *20071697* [2], *20071158* [3]
- **spring rye** *20071603* [1], *20071701* [2], *20071656* [3]
- **stored grain** *20091011 expires 31 Oct 2013* [7]
- **triticale** *20071603* [1], *20071701* [2], *20071656* [3]
- **winter rye** *20071603* [1], *20071701* [2], *20071656* [3]

Approval information
- Deltamethrin included in Annex I under EC Regulation 1107/2009
- Accepted by BBPA for use on malting barley and hops
- Approval expiry 31 Oct 2013 [5]

Efficacy guidance
- A contact and stomach poison with 3-4 wk persistence, particularly effective on caterpillars and sucking insects
- Normally applied at first signs of damage with follow-up treatments where necessary at 10-14 d intervals. Rates, timing and recommended combinations with other pesticides vary with crop and pest. See label for details
- Spray is rainfast within 1 h
- May be applied in frosty weather provided foliage not covered in ice
- Temperatures above 35°C may reduce effectiveness or persistence

Restrictions
- Maximum number of treatments varies with crop and pest, 4 per crop for wheat and barley, only 1 application between 1 Apr and 31 Aug. See label for other crops
- Do not apply more than 1 aphicide treatment to cereals in summer
- Do not spray crops suffering from drought or other physical stress
- Consult processer before treating crops for processing
- Do not apply to a cereal crop if any product containing a pyrethroid insecticide or dimethoate has been applied to that crop after the start of ear emergence (GS 51)
- Do not spray cereals after 31 Mar in the year of harvest within 6 m of the outside edge of the crop
- Reduced volume spraying must not be used on cereals after 31 Mar in yr of harvest

Crop-specific information
- Latest use: early dough (GS 83) for barley, oats, wheat; before flowering for mustard, oilseed rape; before 31 Mar for grass seed crops [1-3]

Environmental safety
- Dangerous for the environment
- Toxic to aquatic organisms
- Flammable [1, 2]

FOR FULL CONDITIONS OF USE ALWAYS READ THE PRODUCT LABEL

- Extremely dangerous to fish or other aquatic life. Do not contaminate surface waters or ditches with chemical or used container
- High risk to bees. Do not apply to crops in flower or to those in which bees are actively foraging. Do not apply when flowering weeds are present
- Do not apply in tank-mixture with a triazole-containing fungicide when bees are likely to be actively foraging in the crop
- High risk to non-target insects or other arthropods. Do not spray within 6 m of the field boundary
- Broadcast air-assisted LERAP [2-5] (18 m); LERAP Category A [1-5]

Hazard classification and safety precautions
Hazard Harmful, Flammable [1, 2, 4, 5]; Dangerous for the environment [1-5]
Transport code 3 [1-5]; 9 [6-8]
Packaging group III
UN Number 1993 [1-5]; 3082 [6-8]
Risk phrases R20, R22a, R22b, R38, R41, R51 [1, 2, 4, 5]; R50 [3]; R53a [1-5]
Operator protection A, C, H [1-8]; D, M [5-8]; U04a, U05a, U19a [1, 2, 4-8]; U08 [1, 2, 4, 5]; U09a, U20c [6-8]; U10 [1]; U11, U14, U15 [3]; U20b [1-5]
Environmental protection E12a [2-5]; E12c [1]; E12f [1] (cereals, oilseed rape, peas, beans); E12f [2-5] (cereals, oilseed rape, peas, beans - see label for guidance); E15a, E16c, E16d, E38 [1-5]; E15b [6-8]; E17b [2-5] (18 m); E22a [3]; E34 [1, 2, 4-8]
Storage and disposal D01, D02, D09a, D10c [1-8]; D05 [1, 2, 4, 5]; D12a [2-5]
Medical advice M03 [1, 2, 4-8]; M05b [1, 2, 4, 5]

149   desmedipham

A contact phenyl carbamate herbicide available only in mixtures
HRAC mode of action code: C1

150   desmedipham + ethofumesate + lenacil + phenmedipham

A selective contact and residual herbicide mixture for weed control in beet
HRAC mode of action code: C1 + N + C1 + C1

Products
Betanal MaxxPro	Bayer CropScience	47:75:27:60 g/l	OD	15086

Uses
- Annual dicotyledons in *fodder beet*, *mangels*, *sugar beet*
- Annual meadow grass in *fodder beet*, *mangels*, *sugar beet*

Approval information
- Desmedipham, ethofumesate, lenacil and phenmedipham included in Annex I under EC Regulation 1107/2009

Efficacy guidance
- On soils with more than 5% organic matter content, residual activity may be reduced.

Restrictions
- The maximum total dose must not exceed 1.0 kg/ha ethofumesate in any 3 year period.

Following crops guidance
- Beet crops may be sown at any time after the use of Betanal MaxxPro. Any other crop may be sown 3 months after using Betanal MaxxPro. Ploughing (mould board) to a minimum depth of 15 cm should precede preparation of a new seed bed.
- If a crop is suffering from manganese deficiency it may be checked. To avoid crop check, manganese should ideally be applied to the crop first.
- Crops suffering from lime deficiency may also be checked. Growers should ensure that the lime status of the soil is satisfactory before drilling.
- When the temperature is, or is likely to be, above 21˚C (70˚F) on the day of spraying, application should be made after 5 pm otherwise crop check may occur. If crops are subjected to substantial

day to night temperature changes shortly before or after spraying, a check may occur from which the crop may not fully recover.

Hazard classification and safety precautions

Hazard Irritant, Dangerous for the environment
Transport code 9
Packaging group III
UN Number 3082
Risk phrases R41, R43, R50, R53a
Operator protection A, C, H; U05a, U11, U14, U20b
Environmental protection E15b, E38, E39
Storage and disposal D09a, D10b, D12a
Medical advice M03

151 desmedipham + ethofumesate + phenmedipham

A selective contact and residual herbicide for beet
HRAC mode of action code: C1 + N + C1

See also ethofumesate
        phenmedipham

Products

1	Betasana Trio SC	United Phosphorus	15:115:75 g/l	SC	15551
2	Beta-Team	AgriChem BV	25:150:75 g/l	SE	15423
3	Trilogy	United Phosphorus	15:115:75 g/l	SC	15644

Uses

- Annual dicotyledons in **fodder beet**, **mangels**, **sugar beet** [1-3]
- Annual meadow grass in **fodder beet**, **mangels**, **sugar beet** [2]

Approval information

- Desmedipham, ethofumesate and phenmedipham included in Annex I under EC Regulation 1107/2009

Efficacy guidance

- Product recommended for low-volume overall application in a planned spray programme
- Product acts mainly by contact action. A full programme also gives some residual control but this may be reduced on soils with more than 5% organic matter
- Best results achieved from treatments applied at fully expanded cotyledon stage of largest weeds present. Occasional larger weeds will usually be controlled from a full programme of sprays
- Where a pre-emergence band spray has been applied treatment must be timed according to size of the untreated weeds between the rows
- Susceptible weeds may not all be killed by the first spray. Repeat applications as each flush of weeds reaches cotyledon size normally necessary for season long control
- Sequential treatments should be applied when the previous one is still showing an effect on the weeds
- Various mixtures with other beet herbicides are recommended. See label for details

Restrictions

- Maximum total dose 4.5 l/ha [2], 7.0 l/ha [1, 3]
- Do not spray crops stressed by nutrient deficiency, wind damage, pest or disease attack, or previous herbicide treatments. Stressed crops treated under conditions of high light intensity may be checked and not recover fully
- If temperature likely to exceed 21°C spray after 5 pm
- Crystallisation may occur if spray volume exceeds that recommended or spray mixture not used within 2 h, especially if the water temperature is below 5°C
- Before use, wash out sprayer to remove all traces of previous products, especially hormone and sulfonyl urea weedkillers.

FOR FULL CONDITIONS OF USE ALWAYS READ THE PRODUCT LABEL

- The maximum total dose must not exceed 1.0 kg ethofumesate per hectare in any three year period

Crop-specific information
- Latest use: before crop meets between rows
- Apply first treatment when majority of crop plants have reached the fully expanded cotyledon stage
- Frost within 7 d of treatment may cause check from which the crop may not recover

Following crops guidance
- Beet crops may be sown at any time after treatment. Any other crop may be sown 3 mth after treatment following mouldboard ploughing to 15 cm minimum

Environmental safety
- Dangerous for the environment
- Toxic to aquatic organisms

Hazard classification and safety precautions
Hazard Irritant [1, 3]; Dangerous for the environment [1-3]
Transport code 9
Packaging group III
UN Number 3082
Risk phrases R36, R50 [1, 3]; R51 [2]; R53a [1-3]
Operator protection A, H [2]; U08 [1-3]; U20a [2]
Environmental protection E15a [2]; E15b [1, 3]; E38 [1-3]
Storage and disposal D05, D09a, D12a [1-3]; D10b [2]; D10c [1, 3]
Treated seed S06a [1, 3]
Medical advice M03

152   desmedipham + phenmedipham

A mixture of contact herbicides for use in sugar beet
HRAC mode of action code: C1 + C1

See also phenmedipham

Products

1	Beetup Compact SC	United Phosphorus	80:80 g/l	SC	15566
2	Betanal Maxxim	Bayer CropScience	160;160 g/l	EC	14186
3	Betanal Turbo	Bayer CropScience	160;160 g/l	EC	15505
4	Rifle	Makhteshim	160:160 g/l	SE	15705

Uses
- Amaranthus in *sugar beet* [1]
- Annual dicotyledons in *fodder beet* [4]; *sugar beet* [1-4]
- Black nightshade in *fodder beet* [4]; *sugar beet* [1, 4]
- Chickweed in *fodder beet* [4]; *sugar beet* [1, 4]
- Fat hen in *sugar beet* [1]
- Field pansy in *fodder beet* [4]; *sugar beet* [1, 4]
- Field speedwell in *fodder beet*, *sugar beet* [4]
- Groundsel in *fodder beet*, *sugar beet* [4]
- Ivy-leaved speedwell in *fodder beet*, *sugar beet* [4]
- Mayweeds in *fodder beet*, *sugar beet* [4]
- Penny cress in *fodder beet* [4]; *sugar beet* [1, 4]
- Red dead-nettle in *sugar beet* [1]
- Shepherd's purse in *fodder beet* [4]; *sugar beet* [1, 4]

Approval information
- Desmedipham and phenmedipham included in Annex I under EC Regulation 1107/2009

Efficacy guidance

- Product is recommended for low volume overall spraying in a planned programme involving pre- and/or post-emergence treatments at doses recommended for low volume programmes
- Best results obtained from treatment when earliest germinating weeds have reached cotyledon stage
- Further treatments must be applied as each flush of weeds reaches cotyledon stage but allowing a minimum of 7 d between each spray
- Where a pre-emergence band spray has been applied, the first treatment should be timed according to the size of the weeds in the untreated area between the rows
- Product is absorbed by leaves of emerged weeds which are killed by scorching action in 2-10 d. Apply overall as a fine spray to optimise weed cover and spray retention
- Various tank mixtures and sequences recommended to widen weed spectrum and add residual activity - see label for details

Restrictions

- Maximum total dose equivalent to three full dose treatments
- Do not spray crops stressed by nutrient deficiency, frost, wind damage, pest or disease attack, or previous herbicide treatments. Stressed crops may be checked and not recover fully
- If temperature likely to exceed 21°C spray after 5 pm
- Before use, wash out sprayer to remove all traces of previous products, especially hormone and sulfonyl urea weedkillers
- Crystallisation may occur if spray volume exceeds that recommended or spray mixture not used within 2 h, especially if the water temperature is below 5°C
- Product may cause non-reinforced PVC pipes and hoses to soften and swell. Wherever possible, use reinforced PVC or synthetic rubber hoses

Crop-specific information

- Latest use: before crop leaves meet between rows
- Product safe to use on all soil types

Environmental safety

- Dangerous for the environment
- Toxic to aquatic organisms
- Risk to certain non-target insects or other arthropods. Avoid spraying within 6 m of field boundary
- LERAP Category B

Hazard classification and safety precautions

Hazard Irritant [1]; Dangerous for the environment [1-4]

Transport code 9

Packaging group III

UN Number 3082

Risk phrases R36, R38, R43 [1]; R50, R53a [1-3]

Operator protection A [1-3]; U05a, U19a [1-4]; U08 [2, 3]; U09a [1, 4]; U11, U14, U20b [1]; U20a [2-4]

Environmental protection E15b, E16a [1-4]; E16b, E39 [2, 3]; E38 [2-4]

Storage and disposal D01, D02, D09a [1-4]; D05 [1, 4]; D10b, D12a [2-4]; D10c [1]

153   dicamba

A translocated benzoic herbicide available only in mixtures
HRAC mode of action code: O

See also 2,4-D + dicamba
2,4-D + dicamba + dichlorprop-P
2,4-D + dicamba + fluroxypyr
2,4-D + dicamba + MCPA + mecoprop-P
2,4-D + dicamba + triclopyr

FOR FULL CONDITIONS OF USE ALWAYS READ THE PRODUCT LABEL

154   dicamba + dichlorprop-P + MCPA

A translocated herbicide mixture for turf
HRAC mode of action code: O + O + O

See also dichlorprop-P
       MCPA

Products

Intrepid 2	Everris Ltd	20.8:167:167 g/l	SL	11594

Uses
- Annual dicotyledons in **managed amenity turf**
- Perennial dicotyledons in **managed amenity turf**

Approval information
- Dicamba, dichlorprop-P and MCPA included in Annex I under EC Regulation 1107/2009
- Accepted by BBPA for use on malting barley

Efficacy guidance
- Apply as directed on label between Apr and Sep when weeds growing actively
- Do not mow for at least 3 d before and 3-4 d after treatment so that there is sufficient leaf growth for spray uptake and sufficient time for translocation. See label for details
- If re-growth occurs or new weeds germinate re-treatment recommended

Restrictions
- Do not use during drought unless irrigation is carried out before and after treatment
- Do not apply during freezing conditions or when heavy rain is imminent
- Do not treat new turf until established for about 6 mth after seeding or turfing
- The first 4 mowings after treatment should not be used to mulch cultivated plants unless composted for at least 6 mth
- Do not re-seed turf within 8 wk of last treatment

Crop-specific information
- Latest use: end Sep for managed amenity turf

Environmental safety
- Prevent spray running or drifting onto cultivated plants, including shrubs and trees

Hazard classification and safety precautions
    Hazard Harmful
    UN Number N/C
    Risk phrases R22a, R38, R41, R43
    Operator protection A, C; U05a, U08, U11, U19a, U20a
    Environmental protection E15a, E34
    Storage and disposal D01, D02, D05, D09a, D10b
    Medical advice M03, M05a

155   dicamba + MCPA + mecoprop-P

A translocated herbicide for cereals, grassland, amenity grass and orchards
HRAC mode of action code: O + O + O

See also MCPA
      mecoprop-P

Products

1	Field Marshal	United Phosphorus	18:360:80 g/l	SL	08956
2	Grassland Herbicide	United Phosphorus	25:200:200 g/l	SL	14623
3	Greencrop Triathlon	Greencrop	25:200:200 g/l	SL	10956
4	Headland Relay P	Headland	25:200:200 g/l	SL	08580
5	Headland Relay Turf	Headland Amenity	25:200:200 g/l	SL	08935

SEE SECTION 3 FOR PRODUCTS ALSO REGISTERED

Products – continued

6	Headland Transfer	Headland	18:315:50 g/l	SL	11010
7	Headland Trinity	Headland	18:315:50 g/l	SL	10842
8	Hycamba Plus	Agrichem	20.4:125.9:240.2 g/l	SL	10180
9	Hyprone-P	Agrichem	16:101:92 g/l	SL	09125
10	Hysward-P	Agrichem	16:101:92 g/l	SL	09052
11	Impede	Headland	18:315:50 g/l	SL	15601
12	Mircam Plus	Nufarm UK	19.5:245:43.3 g/l	SL	11525
13	Outrun	Barclay	20.4:125.9:240.2 g/l	SL	12838
14	Pasturol Plus	Headland	25:200:200 g/l	SL	10278
15	Pierce	Nufarm UK	31.2:256.2:237.5 g/l	SL	11924
16	Re-act	Everris Ltd	31:256:237 g/l	SL	15718
17	Super Selective Plus	Rigby Taylor	31.25:256.25:238.5 g/l	SL	11928
18	T2 Green	Nufarm UK	31.25: 256.25: 237.5 g/l	SL	11925
19	Tribute	Nomix Enviro	18:252:42 g/l	SL	13864
20	Tribute Plus	Nomix Enviro	18:252:42 g/l	SL	13865
21	UPL Grassland Herbicide	United Phosphorus	25:200:200 g/l	SL	08934

Uses

* Annual dicotyledons in *almonds* (off-label), *chestnuts* (off-label), *hazel nuts* (off-label), *walnuts* (off-label) [6, 7, 12]; *amenity grassland* [2, 3, 7, 8, 10, 13, 14, 16, 18, 20, 21]; *apple orchards*, *durum wheat* (off-label), *pear orchards*, *triticale* (off-label) [7, 12]; *durum wheat*, *rye* (off-label), *triticale* [12]; *grass seed crops* [6, 7, 9, 11, 12, 15]; *grassland*, *green cover on land temporarily removed from production* [20]; *managed amenity turf* [2, 3, 5, 7, 8, 10, 12-14, 16-20]; *permanent grassland* [2-4, 6, 7, 10-12, 14, 15, 21]; *rotational grass* [1-4, 6, 7, 9-12, 14, 15, 21]; *spring barley*, *spring oats*, *spring wheat*, *winter barley*, *winter oats*, *winter wheat* [1, 6, 7, 9, 11, 12]; *spring rye*, *undersown spring cereals*, *undersown winter cereals*, *winter rye* [7]; *undersown barley*, *undersown oats*, *undersown wheat* [9]; *undersown spring cereals* (grass only), *undersown winter cereals* (grass only) [1, 6, 11]
* Docks in *amenity grassland* [10, 21]; *apple orchards*, *pear orchards* [7]; *grass seed crops* [7, 9, 15]; *managed amenity turf* [10]; *permanent grassland*, *rotational grass* [7, 10, 15, 21]
* Perennial dicotyledons in *almonds* (off-label), *chestnuts* (off-label), *hazel nuts* (off-label), *walnuts* (off-label) [6, 7, 12]; *amenity grassland* [2, 3, 7, 8, 10, 13, 14, 16, 18, 20, 21]; *apple orchards*, *pear orchards* [7, 12]; *durum wheat*, *durum wheat* (off-label), *rye* (off-label), *triticale*, *triticale* (off-label) [12]; *grass seed crops* [6, 7, 9, 11, 12, 15]; *grassland*, *green cover on land temporarily removed from production* [20]; *managed amenity turf* [2, 3, 5, 7, 8, 10, 12-14, 16-20]; *permanent grassland* [2-4, 6, 7, 10-12, 14, 15, 21]; *rotational grass* [1-4, 6, 7, 9-12, 14, 15, 21]; *spring barley*, *spring oats*, *spring wheat*, *winter barley*, *winter oats*, *winter wheat* [1, 6, 7, 9, 11, 12]; *spring rye*, *undersown spring cereals*, *undersown winter cereals*, *winter rye* [7]; *undersown barley*, *undersown oats*, *undersown wheat* [9]; *undersown spring cereals* (grass only), *undersown winter cereals* (grass only) [1, 6, 11]

Extension of Authorisation for Minor Use (EAMUs)

* *almonds* 20060514 [6], 20060513 [7], 20121531 [12]
* *chestnuts* 20060514 [6], 20060513 [7], 20121531 [12]
* *durum wheat* 20060526 [7], 20121532 [12]
* *hazel nuts* 20060514 [6], 20060513 [7], 20121531 [12]
* *rye* 20121532 [12]
* *triticale* 20060526 [7], 20121532 [12]
* *walnuts* 20060514 [6], 20060513 [7], 20121531 [12]

Approval information

* Dicamba, MCPA and mecoprop-P included in Annex I under EC Regulation 1107/2009
* Accepted by BBPA for use on malting barley
* Approval expiry 30 Sep 2013 [21]

Efficacy guidance

* Treatment should be made when weeds growing actively. Weeds hardened by winter weather may be less susceptible

FOR FULL CONDITIONS OF USE ALWAYS READ THE PRODUCT LABEL

- For best results apply in fine warm weather, preferably when soil is moist. Do not spray if rain expected within 6 h or in drought
- Application of fertilizer 1-2 wk before spraying aids weed control in turf
- Where a second treatment later in the season is needed in amenity situations and on grass allow 4-6 wk between applications to permit sufficient foliage regrowth for uptake

Restrictions
- Maximum number of treatments (including other mecoprop-P products) or maximum total dose varies with crop and product. See label for details. The total amount of mecoprop-P applied in a single yr must not exceed the maximum total dose approved for any single product for the crop/situation
- Do not apply to cereals after the first node is detectable (GS 31), or to grass under stress from drought or cold weather
- Do not spray cereals undersown with clovers or legumes, to be undersown with grass or legumes or grassland where clovers or other legumes are important
- Do not spray leys established less than 18 mth or orchards established less than 3 yr
- Do not roll or harrow within 7 d before or after treatment, or graze for at least 7 d afterwards (longer if poisonous weeds present)
- Do not use on turf or grass in year of establishment. Allow 6-8 wk after treatment before seeding bare patches
- The first mowings after use should not be used for mulching unless composted for 6 mth
- Turf should not be mown for 24 h before or after treatment (3-4 d for closely mown turf)
- Avoid drift onto all broad-leaved plants outside the target area

Crop-specific information
- Latest use: before first node detectable (GS 31) for cereals; 5-6 wk before head emergence for grass seed crops; mid-Oct for established grass
- HI 7-14 d before cutting or grazing for leys, permanent pasture
- Apply to winter cereals from the leaf sheath erect stage (GS 30), and to spring cereals from the 5 expanded leaf stage (GS 15)
- Spray grass seed crops 4-6 wk before flower heads begin to emerge (timothy 6 wk)
- Turf containing bulbs may be treated once the foliage has died down completely [4, 10]

Environmental safety
- Harmful to aquatic organisms
- Keep livestock out of treated areas for at least 2 wk following treatment and until poisonous weeds, such as ragwort, have died down and become unpalatable
- Harmful to fish or other aquatic life. Do not contaminate surface waters or ditches with chemical or used container
- Some pesticides pose a greater threat of contamination of water than others and mecoprop-P is one of these pesticides. Take special care when applying mecoprop-P near water and do not apply if heavy rain is forecast

Hazard classification and safety precautions
Hazard Harmful [1-8, 11-20]; Irritant [9, 10, 21]; Dangerous for the environment [12, 15-18]
Transport code 9 [19, 20]
Packaging group III [19, 20]
UN Number 3082 [19, 20]; N/C [1-7, 9-18, 21]
Risk phrases R20 [2, 4, 6, 7, 11, 14]; R21 [2-7, 11, 12, 14, 19, 20]; R22a [1-8, 11-18]; R36 [2-4, 9, 10, 12, 14, 19-21]; R38 [6-8, 11-13, 21]; R41 [1, 2, 4-8, 11, 13-18]; R43 [21]; R51 [12, 15-18]; R52 [1, 2, 4-11, 14, 19-21]; R53a [1, 6-12, 15-18, 21]
Operator protection A, C, H, M; U05a [2-21]; U08, U20b [1-14, 19-21]; U11 [1, 2, 4, 6-18, 21]; U15 [2, 4, 14]; U19a [3, 8, 10, 13]
Environmental protection E07a [1-4, 6-20]; E13c [1-11, 13-18, 21]; E15a [1, 12, 15, 16, 18-20]; E23 [19, 20]; E34 [1-7, 11, 12, 14-20]; E38 [2, 4, 14, 19, 20]
Storage and disposal D01, D02, D05, D09a [1-21]; D10a [6, 7, 11, 17]; D10b [2-5, 12, 14-16, 18-20]; D10c [1, 8-10, 13, 21]; D12a [1, 19-21]
Medical advice M03 [1-7, 11, 12, 14-18, 21]; M05a [1, 7-10, 13, 21]

SEE SECTION 3 FOR PRODUCTS ALSO REGISTERED

156   dicamba + mecoprop-P

A translocated post-emergence herbicide for cereals and grassland
HRAC mode of action code: O + O

See also mecoprop-P

Products

1	Di-Farmon R	Headland	42:319 g/l	SL	08472
2	Dockmaster	Nufarm UK	18.7:150 g/l	SL	11651
3	Foundation	Headland	84:600 g/l	SL	11708
4	Headland Saxon	Headland	84:600 g/l	SL	11947
5	High Load Mircam	Nufarm UK	80:600 g/l	SL	11930
6	Hyban-P	Agrichem	18.7:150 g/l	SL	09129
7	Hygrass-P	Agrichem	18.7:150 g/l	SL	09130
8	Mircam	Nufarm UK	18.7:150 g/l	SL	11707
9	Prompt	Headland	84:600 g/l	SL	11948

Uses

- Annual dicotyledons in **canary seed** *(off-label)* [5]; **durum wheat** *(off-label)*, **spring rye** *(off-label)*, **triticale** *(off-label)*, **winter rye** *(off-label)* [3, 4, 9]; **permanent grassland** [1, 3-5, 7, 9]; **rotational grass** [5, 7]; **spring barley, spring oats, spring wheat, winter barley, winter oats, winter wheat** [1-6, 8, 9]; **spring rye, triticale, undersown rye, winter rye** [6]; **undersown barley, undersown oats, undersown wheat** [5, 6]
- Buttercups in **permanent grassland, rotational grass** [2, 6, 8]
- Chickweed in **permanent grassland** [1, 2, 6, 8]; **rotational grass** [2, 5, 6, 8]; **spring barley, spring oats, spring wheat, winter barley, winter oats, winter wheat** [1-6, 8, 9]; **spring rye, triticale, undersown rye, winter rye** [6]; **undersown barley, undersown oats, undersown wheat** [5, 6]
- Cleavers in **permanent grassland** [1]; **rotational grass** [5]; **spring barley, spring oats, spring wheat, winter barley, winter oats, winter wheat** [1-6, 8, 9]; **spring rye, triticale, undersown rye, winter rye** [6]; **undersown barley, undersown oats, undersown wheat** [5, 6]
- Creeping thistle in **permanent grassland, rotational grass** [2, 6, 8]
- Docks in **permanent grassland, rotational grass** [2, 6-8]
- Mayweeds in **permanent grassland** [1]; **rotational grass** [5]; **spring barley, spring oats, spring wheat, winter barley, winter oats, winter wheat** [1-6, 8, 9]; **spring rye, triticale, undersown rye, winter rye** [6]; **undersown barley, undersown oats, undersown wheat** [5, 6]
- Perennial dicotyledons in **permanent grassland** [3-5, 7, 9]; **rotational grass** [7]; **spring barley, spring oats, spring wheat, winter barley, winter oats, winter wheat** [2, 6, 8]; **spring rye, triticale, undersown barley, undersown oats, undersown rye, undersown wheat, winter rye** [6]
- Plantains in **spring barley, spring oats, spring rye, spring wheat, triticale, undersown barley, undersown oats, undersown rye, undersown wheat, winter barley, winter oats, winter rye, winter wheat** [6]
- Polygonums in **permanent grassland** [1]; **rotational grass** [5]; **spring barley, spring oats, spring wheat, winter barley, winter oats, winter wheat** [1-6, 8, 9]; **spring rye, triticale, undersown rye, winter rye** [6]; **undersown barley, undersown oats, undersown wheat** [5, 6]
- Scentless mayweed in **permanent grassland, rotational grass** [2, 6, 8]
- Stinging nettle in **permanent grassland, rotational grass** [6]
- Thistles in **permanent grassland, rotational grass** [7]

Extension of Authorisation for Minor Use (EAMUs)

- **canary seed** *20121416* [5]
- **durum wheat** *20052953* [3], *20052951* [4], *20052952* [9]
- **spring rye** *20052953* [3], *20052951* [4], *20052952* [9]
- **triticale** *20052953* [3], *20052951* [4], *20052952* [9]
- **winter rye** *20052953* [3], *20052951* [4], *20052952* [9]

FOR FULL CONDITIONS OF USE ALWAYS READ THE PRODUCT LABEL

Approval information
- Dicamba and mecoprop-P included in Annex I under EC Regulation 1107/2009
- Accepted by BBPA for use on malting barley

Efficacy guidance
- Best results by application in warm, moist weather when weeds are actively growing

Restrictions
- Maximum number of treatments 1 per crop for cereals and 1 or 2 per yr on grass depending on label. The total amount of mecoprop-P applied in a single yr must not exceed the maximum total dose approved for any single product for the crop/situation
- Do not spray in cold or frosty conditions
- Do not spray if rain expected within 6 h
- Do not treat undersown grass until tillering begins
- Do not spray cereals undersown with clover or legume mixtures
- Do not roll or harrow within 7 d before or after spraying
- Do not treat crops suffering from stress from any cause
- Use product immediately following dilution; do not allow diluted product to stand before use [4]
- Avoid treatment when drift may damage neighbouring susceptible crops

Crop-specific information
- Latest use: before 1st node detectable for cereals; 7 d before cutting or 14 d before grazing grass
- Apply to winter sown crops from 5 expanded leaf stage (GS 15)
- Apply to spring sown cereals from 5 expanded leaf stage but before first node is detectable (GS 15-31)
- Treat grassland just before perennial weeds flower
- Transient crop prostration may occur after spraying but recovery is rapid

Environmental safety
- Dangerous for the environment [1, 3-5, 9]
- Toxic to aquatic organisms
- Harmful to fish or other aquatic life. Do not contaminate surface waters or ditches with chemical or used container
- Keep livestock out of treated areas for at least 2 wk and until foliage of poisonous weeds such as ragwort has died and become unpalatable
- Some pesticides pose a greater threat of contamination of water than others and mecoprop-P is one of these pesticides. Take special care when applying mecoprop-P near water and do not apply if heavy rain is forecast

Hazard classification and safety precautions
Hazard Harmful [1-5, 8, 9]; Irritant [6, 7]; Dangerous for the environment [1, 3-5, 9]
UN Number N/C
Risk phrases R21, R36 [2, 8]; R22a, R38 [1-5, 8, 9]; R41 [1, 3-7, 9]; R51 [1, 3-5, 9]; R52 [2, 6-8]; R53a [1-9]
Operator protection A, C [1-9]; H, M [1-4, 6-9]; U05a, U11 [1-9]; U08 [1, 3, 4, 6, 9]; U09a [2, 5, 7, 8]; U19a [7]; U20a [1]; U20b [2-9]
Environmental protection E07a [1-4, 6-9]; E13c [2, 5-8]; E15a [1, 3, 4, 9]; E34 [1-9]; E38 [1, 3]
Storage and disposal D01, D02, D05, D09a [1-9]; D10b [1-4, 8, 9]; D10c [6, 7]
Medical advice M03 [1-9]; M05a [6, 7]

157   dicamba + prosulfuron

A herbicide mixture for weed control in forage and grain maize
HRAC mode of action code: O + B

Products

Casper	Syngenta	500:50 g/l	WG	15573

Uses
- Annual dicotyledons in **forage maize**, **grain maize**
- Bindweeds in **forage maize**, **grain maize**

SEE SECTION 3 FOR PRODUCTS ALSO REGISTERED

- Docks in **forage maize** *(seedlings only)*, **grain maize** *(seedlings only)*

Approval information
- Dicamba and prosulfuron included in Annex I under EC Regulation 1107/2009

Efficacy guidance
- Always apply in mixture with a non-ionic adjuvant.
- Do not apply in mixture with organo-phosphate insecticides

Following crops guidance
- Following normal harvest wheat, barley, rye, triticale and perennial ryegrass may be sown in the autumn, wheat, barley, rye, triticale, combining peas, maize, field beans, forage kale, broccoli and cauliflower may be sown the following spring but sugar beet, sunflowers or lucerne are not recommended.

Environmental safety
- LERAP Category B

Hazard classification and safety precautions
   Hazard Dangerous for the environment
   Transport code 9
   Packaging group III
   UN Number 3077
   Risk phrases R50, R53a
   Operator protection A, C, H; U02a, U05a, U14, U20b
   Environmental protection E15b, E16a, E38
   Storage and disposal D01, D02, D09a, D10c, D12a

158   dichlorprop-P + ferrous sulphate + MCPA

A herbicide/fertilizer combination for moss and weed control in turf
HRAC mode of action code: O + O

See also ferrous sulphate
*   MCPA*

Products

1	SHL Granular Feed, Weed & Mosskiller	Sinclair	0.2:10.9:0.3% w/w	GR	10972
2	SHL Turf Feed, Weed & Mosskiller	Sinclair	0.2:10.9:0.3% w/w	DP	10973

Uses
- Annual dicotyledons in **managed amenity turf**
- Buttercups in **managed amenity turf**
- Moss in **managed amenity turf**
- Perennial dicotyledons in **managed amenity turf**

Approval information
- Dichlorprop-P, ferrous sulphate and MCPA included in Annex I under EC Regulation 1107/2009

Efficacy guidance
- Apply between Mar and Sep when grass in active growth and soil moist
- A repeat treatment may be needed after 4-6 wk to control perennial weeds or if moss regrows
- Water in if rainfall does not occur within 48 h [2]

Restrictions
- Do not treat newly sown grass for 6 mth after establishment [1]
- Do not apply during drought or freezing conditions or when rain imminent
- Avoid walking on treated areas until it has rained or turf has been watered
- Do not mow for 3-4 d before or after application

FOR FULL CONDITIONS OF USE ALWAYS READ THE PRODUCT LABEL

- Do not use first 4 mowings after treatment for mulching. Mowings should be composted for 6 mth before use
- Do not treat areas of fine turf such as golf or bowling greens [1]
- Avoid contact with tarmac surfaces as staining may occur

Environmental safety
- Avoid drift onto nearby plants and borders [1]

Hazard classification and safety precautions
UN Number N/C
Operator protection U20c
Storage and disposal D01, D09a, D11a

159   dichlorprop-P + MCPA + mecoprop-P

A translocated herbicide mixture for winter and spring cereals
HRAC mode of action code: O + O + O

See also MCPA
       mecoprop-P

Products

Hymec Triple	Agrichem	310:160:130 g/l	SL	15753

Uses
- Annual dicotyledons in *durum wheat, spring barley, spring oats, spring wheat, winter barley, winter oats, winter wheat*
- Chickweed in *durum wheat, spring barley, spring oats, spring wheat, winter barley, winter oats, winter wheat*
- Cleavers in *durum wheat, spring barley, spring oats, spring wheat, winter barley, winter oats, winter wheat*
- Field pansy in *durum wheat, spring barley, spring oats, spring wheat, winter barley, winter oats, winter wheat*
- Mayweeds in *durum wheat, spring barley, spring oats, spring wheat, winter barley, winter oats, winter wheat*
- Poppies in *durum wheat, spring barley, spring oats, spring wheat, winter barley, winter oats, winter wheat*

Approval information
- Dichlorprop-P, MCPA and mecoprop-P included in Annex I under EC Regulation 1107/2009
- Accepted by BBPA for use on malting barley

Efficacy guidance
- Best results obtained if application is made while majority of weeds are at seedling stage but not if temperatures are too low
- Optimum results achieved by spraying when temperature is above 10°C. If temperatures are lower delay spraying until growth becomes more active

Restrictions
- Maximum number of treatments 1 per crop
- Do not spray in windy conditions where spray drift may cause damage to neighbouring crops, especially sugar beet, oilseed rape, peas, turnips and most horticultural crops including lettuce and tomatoes under glass

Crop-specific information
- Latest use: before second node detectable (GS 32) for all crops

Environmental safety
- Harmful to aquatic organisms
- Harmful to fish or other aquatic life. Do not contaminate surface waters or ditches with chemical or used container

SEE SECTION 3 FOR PRODUCTS ALSO REGISTERED

SECTION 2

- Some pesticides pose a greater threat of contamination of water than others and mecoprop-P is one of these pesticides. Take special care when applying mecoprop-P near water and do not apply if heavy rain is forecast

Hazard classification and safety precautions
    Hazard Harmful
    UN Number N/C
    Risk phrases R22a, R41, R52, R53a
    Operator protection A, C, H, M; U05a, U08, U11, U20b
    Environmental protection E13c, E34
    Storage and disposal D01, D02, D05, D09a, D10c
    Medical advice M03

160   difenacoum

An anticoagulant coumarin rodenticide

Products

1	Difenag	Killgerm	0.005% w/w	RB	H8492
2	Neosorexa Bait Blocks	BASF	0.005% w/w	RB	UK12-0360
3	Neosorexa Gold	BASF	0.005% w/w	RB	UK12-0304
4	Neosorexa Gold Ratpacks	BASF	0.005% w/w	RB	UK12-0306
5	Neosorexa Pasta Bait	BASF	0.005% w/w	RB	UK12-0365
6	Sakarat D Pasta Bait	Killgerm	0.005% w/w	RB	UK12-0371
7	Sakarat D Wax Bait	Killgerm	0.005% w/w	RB	UK12-0370
8	Sakarat D Whole Wheat	Killgerm	0.005% w/w	RB	UK12-0301
9	Sorexa D	BASF	0.005% w/w	RB	UK12-0319
10	Sorexa Gel	BASF	0.005% w/w	RB	UK12-0364

Uses
- Mice in *farm buildings*, *farmyards* [2-10]; *farm buildings/yards* [1]
- Rats in *farm buildings*, *farmyards* [2-5, 8, 9]; *farm buildings/yards* [1]

Approval information
- Difenacoum included in Annex I under EC Regulation 1107/2009

Efficacy guidance
- Difenacoum is a chronic poison and rodents need to feed several times before accumulating a lethal dose. Effective against rodents resistant to other commonly used anticoagulants
- Best results achieved by placing baits at points between nesting and feeding places, at entry points, in holes and where droppings are seen
- A minimum of five baiting points normally required for a small infestation; more than 40 for a large infestation
- Inspect bait sites frequently and top up as long as there is evidence of feeding
- Product formulated for application through a skeleton or caulking gun [UK12-0364]
- Maintain a few baiting points to guard against reinfestation after a successful control campaign

Restrictions
- Only for use by farmers, horticulturists and other professional users
- When working in rodent infested areas wear synthetic rubber/PVC gloves to protect against rodent-borne diseases

Environmental safety
- Harmful to wildlife
- Cover baits by placing in bait boxes, drain pipes or under boards to prevent access by children, animals or birds
- Products contain human taste deterrent

Hazard classification and safety precautions
    UN Number N/C [2-10]

FOR FULL CONDITIONS OF USE ALWAYS READ THE PRODUCT LABEL

Operator protection A [10]; U13, U20b
Environmental protection E10b [2-5, 9]; E15a [10]
Storage and disposal D09a, D11a
Vertebrate/rodent control products V01a, V03a, V04a [10]; V01b, V04b [2-7, 9]; V02 [2-7, 9, 10]; V03b [6, 7]; V04c [2-5, 9]
Medical advice M03

161   difenoconazole

A diphenyl-ether triazole protectant and curative fungicide
FRAC mode of action code: 3

See also azoxystrobin + difenoconazole

Products

1	Difcor 250 EC	Q-Chem	250 g/l	EC	13917
2	EA Difcon	European Ag	250 g/l	EC	14965
3	Plover	Syngenta	250 g/l	EC	11763
4	Sandpiper	ChemSource	250 g/l	EC	15004
5	Turnstone	AgChem Access	250 g/l	EC	14019

Uses

- Alternaria in **broccoli, brussels sprouts, cabbages, calabrese, cauliflowers, spring oilseed rape, winter oilseed rape** [1-5]
- Blight in **celeriac** *(off-label)* [1]; **celery (outdoor)** *(off-label)*, **rhubarb** *(off-label)* [1, 3]
- Brown rust in **winter wheat** [1-5]
- Celery leaf spot in **celeriac** *(off-label)* [3]
- Disease control in **borage for oilseed production** *(off-label)*, **canary flower (echium spp.)** *(off-label)*, **durum wheat** *(off-label)*, **evening primrose** *(off-label)*, **grass seed crops** *(off-label)*, **honesty** *(off-label)*, **kale** *(off-label)*, **linseed** *(off-label)*, **mustard** *(off-label)*, **spring rye** *(off-label)*, **triticale** *(off-label)*, **winter rye** *(off-label)* [1]; **protected pinks** *(off-label)*, **protected sweet williams** *(off-label)* [1, 3]
- Foliar disease control in **borage** *(off-label)*, **canary flower (echium spp.)** *(off-label)*, **durum wheat** *(off-label)*, **evening primrose** *(off-label)*, **grass seed crops** *(off-label)*, **honesty** *(off-label)*, **linseed** *(off-label)*, **mustard** *(off-label)*, **spring rye** *(off-label)*, **triticale** *(off-label)*, **winter rye** *(off-label)* [3]
- Late blight in **celeriac** *(off-label)* [1]; **celery (outdoor)** *(off-label)*, **rhubarb** *(off-label)* [1, 3]
- Light leaf spot in **spring oilseed rape, winter oilseed rape** [1-5]
- Ring spot in **broccoli, brussels sprouts, cabbages, calabrese, cauliflowers** [1-5]; **chinese cabbage** *(off-label)*, **choi sum** *(off-label)*, **collards** *(off-label)*, **pak choi** *(off-label)* [1, 3]; **kale** *(off-label)*, **protected chinese cabbage** *(off-label)*, **protected choi sum** *(off-label)*, **protected pak choi** *(off-label)*, **protected pinks** *(off-label - ground grown (or pot grown - SOLA 20050159))*, **protected sweet williams** *(off-label - ground grown (or pot grown - 20050159))*, **protected tat soi** *(off-label)* [3]
- Rust in **asparagus** *(off-label)* [1, 3]; **protected hybrid pinks** *(off-label)*, **protected sweet williams** *(off-label)* [1]; **protected pinks** *(off-label - ground grown (or pot grown - SOLA 20050159))*, **protected sweet williams** *(off-label - ground grown (or pot grown - 20050159))* [3]
- Septoria leaf blotch in **winter wheat** [1-5]
- Stem canker in **spring oilseed rape, winter oilseed rape** [1-5]

Extension of Authorisation for Minor Use (EAMUs)

- **asparagus** *20080617* [1], *20050158* [3]
- **borage** *20060559* [3]
- **borage for oilseed production** *20101167* [1]
- **canary flower (echium spp.)** *20101167* [1], *20060559* [3]
- **celeriac** *20101168* [1], *20060261* [3]
- **celery (outdoor)** *20101169* [1], *20091939* [3]
- **chinese cabbage** *20101165* [1], *20050558* [3]
- **choi sum** *20101165* [1], *20050558* [3]

SEE SECTION 3 FOR PRODUCTS ALSO REGISTERED

- **collards** *20101165* [1], *20050558* [3]
- **durum wheat** *20101164* [1], *20060558* [3]
- **evening primrose** *20101167* [1], *20060559* [3]
- **grass seed crops** *20101164* [1], *20060558* [3]
- **honesty** *20101167* [1], *20060559* [3]
- **kale** *20080616* [1], *20050559* [3]
- **linseed** *20101167* [1], *20060559* [3]
- **mustard** *20101167* [1], *20060559* [3]
- **pak choi** *20101165* [1], *20050558* [3]
- **protected chinese cabbage** *20101124* [3]
- **protected choi sum** *20101124* [3]
- **protected hybrid pinks** *20080613* [1]
- **protected pak choi** *20101124* [3]
- **protected pinks** *20080614* [1], *(ground grown (or pot grown - SOLA 20050159))* *20050156* [3], *20050159* [3]
- **protected sweet williams** *20080613* [1], *20080614* [1], *(ground grown (or pot grown - 20050159))* *20050156* [3], *20050159* [3]
- **protected tat soi** *20101124* [3]
- **rhubarb** *20101169* [1], *20091939* [3]
- **spring rye** *20101164* [1], *20060558* [3]
- **triticale** *20101164* [1], *20060558* [3]
- **winter rye** *20101164* [1], *20060558* [3]

Approval information
- Difenoconazole included in Annex I under EC Regulation 1107/2009

Efficacy guidance
- Improved control of established infections on oilseed rape achieved by mixture with carbendazim. See label
- In brassicas a 3-spray programme should be used starting at the first sign of disease and repeated at 14-21 d intervals
- Product is fully rainfast 2 h after application
- For most effective control of Septoria, apply as part of a programme of sprays which includes a suitable flag leaf treatment
- Difenoconazole is a DMI fungicide. Resistance to some DMI fungicides has been identified in Septoria leaf blotch which may seriously affect performance of some products. For further advice contact a specialist advisor and visit the Fungicide Resistance Action Group (FRAG)-UK website

Restrictions
- Maximum number of treatments 3 per crop for brassicas; 2 per crop for oilseed rape; 1 per crop for wheat [1, 2, 5]
- Maximum total dose equivalent to 3 full dose treatments on brassicas; 2 full dose treatments on oilseed rape; 1 full dose treatment on wheat [3]
- Apply to wheat any time from ear fully emerged stage but before early milk-ripe stage (GS 59-73)

Crop-specific information
- Latest use: before grain early milk-ripe stage (GS 73) for cereals; end of flowering for oilseed rape
- HI brassicas 21 d
- Treat oilseed rape in autumn from 4 expanded true leaf stage (GS 1,4). A repeat spray may be made in spring at the beginning of stem extension (GS 2,0) if visible symptoms develop

Environmental safety
- Dangerous for the environment
- Very toxic to aquatic organisms

Hazard classification and safety precautions
Hazard Irritant [1, 2, 4]; Dangerous for the environment [1-5]
Transport code 9
Packaging group III
UN Number 3082

FOR FULL CONDITIONS OF USE ALWAYS READ THE PRODUCT LABEL

Risk phrases R38, R41, R51 [1, 2, 4]; R50 [3, 5]; R53a [1-5]
Operator protection A, C, H; U05a, U09a, U20b [1-5]; U11 [1, 2, 4]
Environmental protection E15a, E38
Storage and disposal D01, D02, D09a, D12a [1-5]; D05, D07, D10c [3, 5]; D10b [1, 2, 4]

162   difenoconazole + fenpropidin

A triazole morpholine fungicide mixture for beet crops
FRAC mode of action code: 3 + 5

See also fenpropidin

Products

Spyrale	Syngenta	375:100 g/l	EC	12566

Uses
- Powdery mildew in **fodder beet**, **sugar beet**
- Ramularia leaf spots in **fodder beet**, **sugar beet**
- Rust in **fodder beet**, **sugar beet**

Approval information
- Difenoconazole and fenpropidin included in Annex I under EC Regulation 1107/2009

Efficacy guidance
- For best results apply as a preventative treatment or as soon as first symptoms of disease are seen
- Product gives prolonged protection against re-infection but a second application may be needed where crops remain under heavy disease pressure

Restrictions
- Maximum number of treatments 2 per crop of sugar beet or fodder beet

Crop-specific information
- HI: 28 d for fodder beet, sugar beet

Environmental safety
- Dangerous for the environment
- Very toxic to aquatic organisms

Hazard classification and safety precautions
Hazard Harmful, Dangerous for the environment
Transport code 9
Packaging group III
UN Number 3082
Risk phrases R22a, R36, R37, R38, R50, R53a
Operator protection A, C, H; U02a, U04a, U05a, U09a, U10, U11, U14, U15, U20a
Environmental protection E15a, E34, E38
Storage and disposal D01, D02, D05, D07, D09a, D10c, D12a
Medical advice M03

163   difenoconazole + fludioxonil

A triazole + phenylpyrrole seed treatment for use in cereals
FRAC mode of action code: 3 + 12

See also fludioxonil

Products

Celest Extra	Syngenta	25:25 g/l	FS	14050

Uses
- Bunt in **winter wheat**

SEE SECTION 3 FOR PRODUCTS ALSO REGISTERED

- Fusarium foot rot and seedling blight in **winter oats**, **winter wheat**
- Microdochium nivale in **winter oats**, **winter wheat**
- Pyrenophora leaf spot in **winter oats**
- Seed-borne diseases in **winter wheat**
- Seedling blight and foot rot in **winter wheat**
- Septoria seedling blight in **winter wheat**
- Snow mould in **winter wheat**
- Stripe smut in **winter rye**

Approval information
- Difenoconazole and fludioxonil included in Annex I under EC Regulation 1107/2009

Efficacy guidance
- [1] is effective against benzimidazole-resistant and benzimidazole-sensitive strains of *Microdochium nivale*

Crop-specific information
- Under adverse environmental or soil conditions, seed rates should be increased to compensate for a slight drop in germination capacity. Flow rates of seed treated with [1] should be checked before drilling commences.

Hazard classification and safety precautions
Transport code 9
Packaging group III
UN Number 3082
Risk phrases R52, R53a
Operator protection A, H; U05a, U20c
Environmental protection E15a, E38
Storage and disposal D01, D02, D09a, D11a, D12a
Treated seed S01, S02, S03, S04a, S04b, S05, S06a, S07, S08, S09

164   diflubenzuron

A selective, persistent, contact and stomach acting insecticide
IRAC mode of action code: 15

Products

Dimilin Flo	Certis	480 g/l	SC	11056

Uses
- Browntail moth in **amenity vegetation**, **hedges**, **nursery stock**, **ornamental plant production**
- Bud moth in **apples**, **pears**
- Carnation tortrix moth in **amenity vegetation**, **hedges**, **nursery stock**, **ornamental plant production**
- Caterpillars in **broccoli**, **brussels sprouts**, **cabbages**, **calabrese**, **cauliflowers**, **chives** *(off-label)*, **endives** *(off-label)*, **frise** *(off-label)*, **herbs (see appendix 6)** *(off-label)*, **leaf spinach** *(off-label)*, **lettuce** *(off-label)*, **parsley** *(off-label)*, **radicchio** *(off-label)*, **salad brassicas** *(off-label - for baby leaf production)*, **scarole** *(off-label)*
- Clouded drab moth in **apples**, **pears**
- Codling moth in **apples**, **pears**
- Fruit tree tortrix moth in **apples**, **pears**
- Houseflies in **livestock houses**, **manure heaps**, **refuse tips**
- Lackey moth in **amenity vegetation**, **hedges**, **nursery stock**, **ornamental plant production**
- Moths in **almonds** *(off-label)*, **bilberries** *(off-label)*, **blackberries** *(off-label)*, **chestnuts** *(off-label)*, **crab apples** *(off-label)*, **cranberries** *(off-label)*, **gooseberries** *(off-label)*, **hazel nuts** *(off-label)*, **quinces** *(off-label)*, **redcurrants** *(off-label)*, **walnuts** *(off-label)*, **whitecurrants** *(off-label)*
- Oak leaf roller moth in **forest**
- Pear sucker in **pears**
- Phorid flies in **edible fungi** *(off-label - other than mushrooms)*
- Pine beauty moth in **forest**

FOR FULL CONDITIONS OF USE ALWAYS READ THE PRODUCT LABEL

- Pine looper in *forest*
- Plum fruit moth in *plums*
- Rust mite in *apples*, *pears*, *plums*
- Sciarid flies in *edible fungi* (off-label - other than mushrooms)
- Small ermine moth in *amenity vegetation*, *hedges*, *nursery stock*, *ornamental plant production*
- Tortrix moths in *plums*
- Winter moth in *amenity vegetation*, *apples*, *blackcurrants*, *forest*, *hedges*, *nursery stock*, *ornamental plant production*, *pears*, *plums*

Extension of Authorisation for Minor Use (EAMUs)
- *almonds* 20060571
- *bilberries* 20060573
- *blackberries* 20060573
- *chestnuts* 20060571
- *chives* 20051321
- *crab apples* 20060572
- *cranberries* 20060573
- *edible fungi* (other than mushrooms) 20060574
- *endives* 20051321
- *frise* 20051321
- *gooseberries* 20060573
- *hazel nuts* 20060571
- *herbs (see appendix 6)* 20051321
- *leaf spinach* 20051321
- *lettuce* 20051321
- *parsley* 20051321
- *quinces* 20060572
- *radicchio* 20051321
- *redcurrants* 20060573
- *salad brassicas* (for baby leaf production) 20051321
- *scarole* 20051321
- *walnuts* 20060571
- *whitecurrants* 20060573

Approval information
- Approved for aerial application in forestry when average wind velocity does not exceed 18 knots and gusts do not exceed 20 knots. See Section 5 for more information
- Diflubenzuron included in Annex I under EC Regulation 1107/2009

Efficacy guidance
- Most active on young caterpillars and most effective control achieved by spraying as eggs start to hatch
- Dose and timing of spray treatments vary with pest and crop. See label for details
- Addition of wetter recommended for use on brassicas and for pear sucker control in pears

Restrictions
- Maximum number of treatments 3 per yr for apples, pears; 2 per yr for plums, blackcurrants; 2 per crop for brassicas; 1 per yr for forest
- Before treating ornamentals check varietal tolerance on a small sample
- Do not use as a compost drench or incorporated treatment on ornamental crops
- Do not spray protected plants in flower or with flower buds showing colour
- For use only on the food crops specified on the label
- Do not apply directly to livestock/poultry

Crop-specific information
- HI apples, pears, plums, blackcurrants, brassicas 14 d

Environmental safety
- Dangerous for the environment
- Very toxic to aquatic organisms

SEE SECTION 3 FOR PRODUCTS ALSO REGISTERED

- Broadcast air-assisted LERAP [1] (20 m when used in orchards; 10 m when used in blackcurrants, forestry or ornamentals); LERAP Category B [1]

Hazard classification and safety precautions
Hazard Dangerous for the environment
Transport code 9
Packaging group III
UN Number 3082
Risk phrases R50
Operator protection U20c
Environmental protection E05a, E15a, E16a, E16b, E18, E38 [1]; E17b [1] (20 m when used in orchards; 10 m when used in blackcurrants, forestry or ornamentals)
Storage and disposal D05, D09a, D11a, D12b

165   diflufenican

A shoot absorbed pyridinecarboxamide herbicide for winter cereals
HRAC mode of action code: F1

See also bromoxynil + diflufenican + ioxynil
chlorotoluron + diflufenican
clodinafop-propargyl + diflufenican
clopyralid + diflufenican + MCPA

Products

1	Difenikan 500	Goldengrass	500 g/l	SC	14600
2	Diflanil 500 SC	Q-Chem	500 g/l	SC	12489
3	Hurricane SC	Makhteshim	500 g/l	SC	12424
4	Sempra	AgriChem BV	500 g/l	SC	13525
5	Solo D	Sipcam	500 g/l	SC	14667
6	Twister	ChemSource	500 g/l	SC	14885

Uses
- Annual dicotyledons in **durum wheat** [2]; **forest nurseries** *(off-label)*, **game cover** *(off-label)*, **ornamental plant production** *(off-label)* [3]; **grass seed crops** *(off-label)* [2-5]; **oats** *(off-label)* [1-5]; **spring barley** [4, 5]; **triticale, winter barley, winter rye, winter wheat** [2, 4, 5]
- Annual grasses in **grass seed crops** *(off-label)* [2, 4]; **oats** *(off-label)*, **spring barley, triticale, winter barley, winter rye, winter wheat** [4]
- Annual meadow grass in **grass seed crops** *(off-label)*, **oats** *(off-label)* [3]
- Chickweed in **durum wheat** [2]; **spring barley** [5]; **triticale, winter barley, winter rye, winter wheat** [2, 5]
- Cleavers in **durum wheat** [1-3, 6]; **spring barley** [1, 3, 5, 6]; **spring wheat** [1, 3, 6]; **triticale, winter barley, winter rye, winter wheat** [1-3, 5, 6]
- Field pansy in **durum wheat, spring barley, spring wheat, triticale, winter barley, winter rye, winter wheat** [1, 3, 6]
- Field speedwell in **durum wheat** [1-3, 6]; **spring barley** [1, 3, 5, 6]; **spring wheat** [1, 3, 6]; **triticale, winter barley, winter rye, winter wheat** [1-3, 5, 6]
- Ivy-leaved speedwell in **durum wheat** [1-3, 6]; **spring barley** [1, 3, 5, 6]; **spring wheat** [1, 3, 6]; **triticale, winter barley, winter rye, winter wheat** [1-3, 5, 6]
- Mayweeds in **durum wheat, spring barley, spring wheat, triticale, winter barley, winter rye, winter wheat** [1, 3, 6]
- Poppies in **durum wheat, spring barley, spring wheat, triticale, winter barley, winter rye, winter wheat** [1, 3, 6]
- Red dead-nettle in **durum wheat, spring barley, spring wheat, triticale, winter barley, winter rye, winter wheat** [1, 3, 6]
- Volunteer oilseed rape in **durum wheat** [2]; **spring barley** [5]; **triticale, winter barley, winter rye, winter wheat** [2, 5]

Extension of Authorisation for Minor Use (EAMUs)
- **forest nurseries** *20122010* [3]

FOR FULL CONDITIONS OF USE ALWAYS READ THE PRODUCT LABEL

- **game cover** *20122010* [3]
- **grass seed crops** *20121458* [2], *20121243* [3], *20122131* [3], *20121500* [4], *20121501* [5]
- **oats** *20121456* [1], *20121457* [2], *20121242* [3], *20121500* [4], *20121501* [5]
- **ornamental plant production** *20122010* [3]

Approval information
- Diflufenican included in Annex I under EC Regulation 1107/2009
- Accepted by BBPA for use on malting barley

Efficacy guidance
- Best results achieved from treatment of small actively growing weeds in early autumn or spring
- Good weed control depends on efficient burial of trash or straw before or during seedbed preparation
- Loose or fluffy seedbeds should be rolled before application
- The final seedbed should be moist, fine and firm with clods no bigger than fist size
- Ensure good even spray coverage and increase spray volume for post-emergence treatments where the crop or weed foliage is dense
- Activity may be slow under cool conditions and final level of weed control may take some time to appear
- Where cleavers are a particular problem a separate specific herbicide treatment may be required
- Efficacy may be impaired on soils with a Kd factor greater than 6
- Always follow WRAG guidelines for preventing and managing herbicide resistant weeds. See Section 5 for more information

Restrictions
- Maximum number of treatments 1 per crop
- Do not treat broadcast crops [2, 3]
- Do not roll treated crops or harrow at any time after treatment
- Do not apply to soils with more than 10% organic matter or on Sands, or very stony or gravelly soils
- Do not treat after a period of cold frosty weather

Crop-specific information
- Latest use: before end of tillering (GS 29) for wheat and barley [2, 3], pre crop emergence for triticale and winter rye [2, 3]
- Treat only named varieties of rye or triticale [2, 3]

Following crops guidance
- Labels vary slightly but in general ploughing to 150 mm and thoroughly mixing the soil is recommended before drilling or planting any succeeding crops either after crop failure or after normal harvest
- In the event of crop failure only winter wheat or winter barley may be re-drilled immediately after ploughing. Spring crops of wheat, barley, oilseed rape, peas, field beans, sugar beet [2, 3], potatoes, carrots, edible brassicas or onions may be sown provided an interval of 12 wk has elapsed after ploughing
- After normal harvest of a treated crop winter cereals, oilseed rape, field beans, leaf brassicas, sugar beet seed crops and winter onions may be drilled in the following autumn. Other crops listed above for crop failure may be sown in the spring after normal harvest
- Successive treatments with any products containing diflufenican can lead to soil build-up and inversion ploughing to 150 mm must precede sowing any following non-cereal crop. Even where ploughing occurs some crops may be damaged

Environmental safety
- Dangerous for the environment
- Very toxic to aquatic organisms
- LERAP Category B [1, 2, 4-6]

Hazard classification and safety precautions
Hazard Dangerous for the environment
Transport code 9
Packaging group III
UN Number 3082

SEE SECTION 3 FOR PRODUCTS ALSO REGISTERED

Risk phrases R50, R53a
Operator protection A [1-6]; C [1-3, 6]; U05a [1, 3-6]; U08, U13, U19a [1, 3, 6]; U20a [1, 3, 4, 6]
Environmental protection E13a [4]; E13c, E34, E38 [1, 3, 6]; E15a [2, 5]; E16a [1, 2, 4-6]; E16i [3]
Storage and disposal D01, D02 [1, 3-6]; D05 [1-3, 5, 6]; D09a, D10b [1-6]; D12a [2, 4, 5]

166   diflufenican + flufenacet

A contact and residual herbicide mixture for cereals
HRAC mode of action code: F1 + K3

See also flufenacet

Products

1	Firebird	Bayer CropScience	200:400 g/l	SC	12421
2	Liberator	Bayer CropScience	100:400 g/l	SC	12032
3	Liberator	Bayer CropScience	100:400 g/l	SC	15206
4	Regatta	Bayer CropScience	100:400 g/l	SC	15353

Uses

- Annual dicotyledons in *durum wheat* (off-label), *grass seed crops* (off-label), *rye* (off-label), *triticale* (off-label) [2, 3]
- Annual grasses in *winter oats* (off-label) [3]
- Annual meadow grass in *durum wheat* (off-label), *grass seed crops* (off-label), *rye* (off-label), *triticale* (off-label) [2, 3]; *spring barley* (off-label) [3]; *winter barley*, *winter wheat* [1-4]
- Blackgrass in *durum wheat* (off-label), *grass seed crops* (off-label), *rye* (off-label), *triticale* (off-label), *winter oats* (off-label) [3]; *winter barley*, *winter wheat* [1-4]
- Chickweed in *winter barley*, *winter wheat* [1-4]
- Field pansy in *winter barley*, *winter wheat* [1-4]
- Field speedwell in *winter barley*, *winter wheat* [1-4]
- Ivy-leaved speedwell in *winter barley*, *winter wheat* [1-4]
- Mayweeds in *winter barley*, *winter wheat* [1-4]
- Red dead-nettle in *winter barley*, *winter wheat* [2-4]
- Shepherd's purse in *winter barley*, *winter wheat* [1]
- Volunteer oilseed rape in *winter barley*, *winter wheat* [1]
- Wild oats in *spring barley* (off-label), *winter oats* (off-label) [3]

Extension of Authorisation for Minor Use (EAMUs)

- *durum wheat* 20110297 [3], *20121495* [2]
- *grass seed crops* 20110297 [3], *20121495* [2]
- *rye* 20110297 [3], *20121495* [2]
- *spring barley* 20121010 [3]
- *triticale* 20110297 [3], *20121495* [2]
- *winter oats* 20111550 [3]

Approval information

- Diflufenican and flufenacet included in Annex I under EC Regulation 1107/2009
- Accepted by BBPA for use on malting barley

Efficacy guidance

- Best results obtained when there is moist soil at and after application and rain falls within 7 d
- Residual control may be reduced under prolonged dry conditions
- Activity may be slow under cool conditions and final level of weed control may take some time to appear
- Good weed control depends on burying any trash or straw before or during seedbed preparation
- Established perennial grasses and broad-leaved weeds will not be controlled
- Do not use as a stand-alone treatment for blackgrass control. Always follow WRAG guidelines for preventing and managing herbicide resistant weeds. Section 5 for more information

Restrictions

- Maximum number of treatments 1 per crop

FOR FULL CONDITIONS OF USE ALWAYS READ THE PRODUCT LABEL

- Do not treat undersown cereals or those to be undersown
- Do not use on waterlogged soils or soils prone to waterlogging
- Do not use on Sands or Very Light soils, or very stony or gravelly soils, or on soils containing more than 10% organic matter
- Do not treat broadcast crops and treat shallow-drilled crops post-emergence only
- Do not incorporate into the soil or disturb the soil after application by rolling or harrowing
- Avoid treating crops under stress from whatever cause and avoid treating during periods of prolonged or severe frosts

Crop-specific information
- Latest use: before 31 Dec in yr of sowing and before 3rd tiller stage (GS 23) for wheat or 4th tiller stage (GS 24) for barley
- For pre-emergence treatments the seed should be covered with a minimum of 32 mm settled soil

Following crops guidance
- In the event of crop failure wheat, barley or potatoes may be sown provided the soil is ploughed to 15 cm, and a minimum of 12 weeks elapse between treatment and sowing spring wheat or spring barley
- After normal harvest wheat, barley or potatoes my be sown without special cultivations. Soil must be ploughed or cultivated to 15 cm before sowing oilseed rape, field beans, peas, sugar beet, carrots, onions or edible brassicae
- Successive treatments with any products containing diflufenican can lead to soil build-up and inversion ploughing to 150 mm must precede sowing any following non-cereal crop. Even where ploughing occurs some crops may be damaged

Environmental safety
- Dangerous for the environment
- Very toxic to aquatic organisms
- Risk to non-target insects or other arthropods. Avoid spraying within 6 m of the field boundary to reduce the effects on non-target insects or other arthropods
- LERAP Category B

Hazard classification and safety precautions
Hazard Harmful, Dangerous for the environment
Transport code 9
Packaging group III
UN Number 3082
Risk phrases R22a, R43, R48, R50, R53a
Operator protection A, H; U05a, U14
Environmental protection E15a [1, 4]; E16a, E34, E38 [1-4]; E22c [1-3]
Storage and disposal D01, D02, D09a, D10b, D12a
Medical advice M03, M05a

167   diflufenican + flufenacet + flurtamone

A herbicide mixture for use in wheat and barley
HRAC mode of action code: F1 + K3

Products

1	Movon	Bayer CropScience	90:240:120 g/l	SC	14784
2	Vigon	Bayer CropScience	60:240:120 g/l	SC	14785

Uses
- Annual meadow grass in **winter barley, winter wheat** [1, 2]
- Blackgrass in **winter wheat** *(moderately susceptible)* [1, 2]
- Chickweed in **winter barley, winter wheat** [1, 2]
- Field pansy in **winter barley, winter wheat** [1, 2]
- Field speedwell in **winter barley, winter wheat** [1, 2]
- Forget-me-not in **winter wheat** [1, 2]

- Groundsel in **winter wheat** [1, 2]
- Italian ryegrass in **winter wheat** *(moderately susceptible)* [1, 2]
- Ivy-leaved speedwell in **winter wheat** [1, 2]
- Mayweeds in **winter barley** [1]; **winter wheat** [1, 2]

Approval information
- Diflufenican + flufenacet + flurtamone included in Annex 1 under EC Regulation 1107/2009
- Accepted by BBPA for use on malting barley

Environmental safety
- LERAP Category B

Hazard classification and safety precautions
    Hazard Harmful, Dangerous for the environment
    Transport code 9
    Packaging group III
    UN Number 3082
    Risk phrases R22a, R50, R53a, R69
    Operator protection A; U05a
    Environmental protection E15b, E16a, E22c, E34, E38, E39, E40b
    Storage and disposal D01, D02, D09a, D10b, D12a
    Medical advice M03, M05a

168   diflufenican + flupyrsulfuron-methyl

A pyridinecarboxamide and sulfonylurea herbicide mixture for wheat
HRAC mode of action code: F1 + B

See also flupyrsulfuron-methyl

Products

Absolute	DuPont	41.7:8.3% w/w	WG	12558

Uses
- Annual dicotyledons in **winter barley, winter wheat**
- Annual grasses in **winter oats** *(off-label)*
- Blackgrass in **winter barley, winter wheat**
- Fumitory in **winter oats** *(off-label)*

Extension of Authorisation for Minor Use (EAMUs)
- **winter oats** *20120038*

Approval information
- Diflufenican and flupyrsulfuron-methyl included in Annex I under EC Regulation 1107/2009

Efficacy guidance
- Best results achieved from applications made in good growing conditions
- Good spray cover of weeds must be obtained
- Best control of blackgrass obtained from application from the 2-leaf stage but tank mixing or sequencing with a partner product is recommended as a resistance management strategy (see below)
- Growth of weeds is inhibited within hours of treatment but visible symptoms may not be apparent for up to 4 wk
- Product has moderate residual life in soil. Under normal moisture conditions susceptible weeds germinating soon after treatment will be controlled
- Product may be used on all soil types except those with more than 10% organic matter, but residual activity and weed control is reduced by high soil temperatures and when soil is very dry and on highly alkaline soils
- Flupyrsulfuron-methyl is a member of the ALS-inhibitor group of herbicides. To avoid the build up of resistance do not use any product containing an ALS-inhibitor herbicide with claims for control of grass weeds more than once on any crop

FOR FULL CONDITIONS OF USE ALWAYS READ THE PRODUCT LABEL

- Use these products as part of a resistance management strategy that includes cultural methods of control and does not use ALS inhibitors as the sole chemical method of grass weed control

Restrictions
- Maximum number of treatments 1 per crop
- Do not use on wheat undersown with grasses, clover or other legumes, or any other broad-leaved crop
- Do not apply within 7 d of rolling
- Do not treat broadcast crops. Seed should be covered by at least 25 mm of settled soil
- Do not treat any crop suffering from drought, waterlogging, pest or disease attack, nutrient deficiency, or any other stress factors
- Do not use on soils with more than 10% organic matter
- Specific restrictions apply to use in sequence or tank mixture with other sulfonylurea or ALS-inhibiting herbicides. See label for details

Crop-specific information
- Latest use: before 1st node detectable stage (GS 31) for winter wheat
- Under certain conditions slight chlorosis and stunting may occur from which recovery is rapid

Following crops guidance
- Only cereals, oilseed rape or field beans may be sown in the yr of harvest of a treated crop. Ploughing must be carried out before drilling any non-cereal crop
- An interval of at least 20 wk must elapse after treatment before planting wheat, barley, oilseed rape, peas, field beans, sugar beet, potatoes, carrots, brassicas or onions in the following spring, and only after ploughing
- In the event of failure of a treated crop only winter wheat or spring wheat may be sown within 3 mth of treatment, and only after ploughing and cultivating to at least 15 cm. After 3 mth the normal guidance for following crops may be observed
- Successive treatments with any products containing diflufenican can lead to soil build-up and inversion ploughing to 150 mm must precede sowing any following non-cereal crop. Even where ploughing occurs some crops may be damaged

Environmental safety
- Dangerous for the environment
- Very toxic to aquatic organisms
- Take extreme care to avoid damage by drift onto broad-leaved plants outside the target area
- Do not drain or flush spraying equipment onto land planted, or intended for planting, with trees or crops other than cereals
- Poor sprayer cleanout practices may result in inadequate removal of chemical deposits which may lead to damage of subsequently treated non-cereal crops. See label for guidance on cleaning
- LERAP Category B

Hazard classification and safety precautions
Hazard Dangerous for the environment
Transport code 9
Packaging group III
UN Number 3077
Risk phrases R50, R53a
Operator protection U08, U19a, U20b
Environmental protection E15a, E16a, E38
Storage and disposal D09a, D12a

169 diflufenican + flurtamone

A contact and residual herbicide mixture for cereals
HRAC mode of action code: F1 + F1

See also flurtamone

Products

Bacara	Bayer CropScience	100:250 g/l	SC	10744

Uses

- Annual dicotyledons in **grass seed crops** *(off-label)*, **rye** *(off-label)*, **spring barley**, **triticale** *(off-label)*, **winter barley**, **winter oats** *(off-label)*, **winter wheat**
- Annual grasses in **grass seed crops** *(off-label)*, **rye** *(off-label)*, **triticale** *(off-label)*, **winter oats** *(off-label)*
- Annual meadow grass in **spring barley**, **winter barley**, **winter wheat**
- Blackgrass in **spring barley**, **winter barley**, **winter wheat**
- Loose silky bent in **spring barley**, **winter barley**, **winter wheat**
- Volunteer oilseed rape in **spring barley**, **winter barley**, **winter wheat**

Extension of Authorisation for Minor Use (EAMUs)

- **grass seed crops** *20121441*
- **rye** *20121441*
- **triticale** *20121441*
- **winter oats** *20121440*

Approval information

- Diflufenican and flurtamone included in Annex I under EC Regulation 1107/2009
- Accepted by BBPA for use on malting barley

Efficacy guidance

- Apply pre-emergence or from when crop has first leaf unfolded before susceptible weeds pass recommended size
- Best results obtained on firm, fine seedbeds with adequate soil moisture present at and after application. Increase water volume where crop or weed foliage is dense
- Good weed control requires ash, trash and burnt straw to be buried during seed bed preparation
- Loose fluffy seedbeds should be rolled before application and the final seed bed should be fine and firm without large clods
- Speed of control depends on weather conditions and activity can be slow under cool conditions
- Always follow WRAG guidelines for preventing and managing herbicide resistant weeds. See Section 5 for more information

Restrictions

- Maximum number of treatments 1 per crop
- Crops should be drilled to a normal depth of 25 mm and the seed well covered. Do not treat broadcast crops as uncovered seed may be damaged
- Do not treat spring sown cereals, durum wheat, oats, undersown cereals or those to be undersown
- Do not treat frosted crops or when frost is imminent. Severe frost after application, or any other stress, may lead to transient discoloration or scorch
- Do not use on Sands or Very Light soils or those that are very stony or gravelly
- Do not use on waterlogged soils, or on crops subject to temporary waterlogging by heavy rainfall, as there is risk of persistent crop damage which may result in yield loss
- Do not use on soils with more than 10% organic matter
- Do not harrow at any time after application and do not roll autumn treated crops until spring

Crop-specific information

- Latest use: before 2nd node detectable (GS32)
- Take particular care to match spray swaths otherwise crop discoloration and biomass reduction may occur which may lead to yield reduction

FOR FULL CONDITIONS OF USE ALWAYS READ THE PRODUCT LABEL

Following crops guidance

- In the event of crop failure winter wheat may be redrilled immediately after normal cultivation, and winter barley may be sown after ploughing. Fields must be ploughed to a depth of 15 cm and 20 wk must elapse before sowing spring crops of wheat, barley, oilseed rape, peas, field beans or potatoes
- After normal harvest autumn cereals can be drilled after ploughing. Thorough mixing of the soil must take place before drilling field beans, leaf brassicae or winter oilseed rape. For sugar beet seed crops and winter onions complete inversion of the furrow slice is essential
- Do not broadcast or direct drill oilseed rape or other brassica crops as a following crop on treated land. See label for detailed advice on preparing land for subsequent autumn cropping in the normal rotation
- Successive treatments with any products containing diflufenican can lead to soil build-up and inversion ploughing to 150 mm must precede sowing any following non-cereal crop. Even where ploughing occurs some crops may be damaged

Environmental safety

- Dangerous for the environment
- Very toxic to aquatic organisms
- Extremely dangerous to fish or other aquatic life. Do not contaminate surface waters or ditches with chemical or used container
- LERAP Category B

Hazard classification and safety precautions

Hazard Dangerous for the environment
Transport code 9
Packaging group III
UN Number 3082
Risk phrases R50, R53a
Operator protection A, H; U20c
Environmental protection E13a, E16a, E16b, E38
Storage and disposal D05, D09a, D10b, D12a

---

170   diflufenican + glyphosate

A foliar non-selective herbicide mixture
HRAC mode of action code: F1 + G

See also glyphosate

Products

1 Pistol	Bayer Environ.	40:250 g/l	SC	12173
2 Proshield	Everris Ltd	40:250 g/l	SC	14767

Uses

- Annual and perennial weeds in **hard surfaces** [2]; **natural surfaces not intended to bear vegetation, permeable surfaces overlying soil** [1, 2]

Approval information

- Diflufenican and glyphosate included in Annex I under EC Regulation 1107/2009

Efficacy guidance

- Treat when weeds actively growing from Mar to end Sep and before they begin to senesce
- Performance may be reduced if application is made to plants growing under stress, such as drought or water-logging
- Pre-emergence activity is reduced on soils containing more than 10% organic matter or where organic debris has collected
- Perennial weeds such as docks, perennial sowthistle and willowherb are best treated before flowering or setting seed
- Perennial weeds emerging from established rootstocks after treatment will not be controlled
- A rainfree period of at least 6 h (preferably 24 h) should follow spraying for optimum control
- For optimum control do not cultivate or rake after treatment

SEE SECTION 3 FOR PRODUCTS ALSO REGISTERED

Restrictions
- Maximum number of treatments 1 per yr
- May only be used on porous surfaces overlying soil. Must not be used if an impermeable membrane lies between the porous surface and the soil, and must not be used on any non-porous man-made surfaces
- Do not add any wetting agent or adjuvant oil
- Do not spray in windy weather

Following crops guidance
- A period of at least 6 mth must be allowed after treatment of sites that are to be cleared and grubbed before sowing and planting. Soil should be ploughed or dug first to ensure thorough mixing and dilution of any herbicide residues

Environmental safety
- Dangerous for the environment
- Very toxic to aquatic organisms
- Avoid drift onto non-target plants
- Heavy rain after application may wash product onto sensitive areas such as newly sown grass or areas about to be planted
- LERAP Category B

Hazard classification and safety precautions
Hazard Dangerous for the environment
Transport code 9
Packaging group III
UN Number 3082
Risk phrases R50, R53a
Operator protection A, C; U20b
Environmental protection E15a, E16a, E16b, E38
Storage and disposal D01, D02, D09a, D11a, D12a

171 diflufenican + iodosulfuron-methyl-sodium + mesosulfuron-methyl

A contact and residual herbicide mixture containing sulfonyl ureas for winter wheat
HRAC mode of action code: F1 + B + B

See also iodosulfuron-methyl-sodium
mesosulfuron-methyl

Products
Othello	Bayer CropScience	50:2.5:7.5 g/l	OD	12695

Uses
- Annual dicotyledons in **winter wheat**
- Annual meadow grass in **winter wheat**
- Cleavers in **winter wheat**
- Mayweeds in **winter wheat**
- Rough meadow grass in **winter wheat**
- Volunteer oilseed rape in **winter wheat**

Approval information
- Diflufenican, iodosulfuron-methyl-sodium and mesosulfuron-methyl included in Annex I under EC Regulation 1107/2009

Efficacy guidance
- Optimum control obtained when all weeds are emerged at spraying. Activity is primarily via foliar uptake and good spray coverage of the target weeds is essential
- Translocation occurs readily within the target weeds and growth is inhibited within hours of treatment but symptoms may not be apparent for up to 4 wk, depending on weed species, timing of treatment and weather conditions

FOR FULL CONDITIONS OF USE ALWAYS READ THE PRODUCT LABEL

- Iodosulfuron-methyl and mesosulfuron-methyl are both members of the ALS-inhibitor group of herbicides. To avoid the build up of resistance do not use any product containing an ALS-inhibitor herbicide with claims for control of grass weeds more than once on any crop
- Use this product as part of a Resistance Management Strategy that includes cultural methods of control and does not use ALS inhibitors as the sole chemical method of weed control in successive crops. See Section 5 for more information

Restrictions

- Maximum number of treatments 1 per crop
- Do not use on crops undersown with grasses, clover or other legumes or any other broad-leaved crop
- Do not use where annual grass weeds other than annual meadow grass and rough meadow grass are present
- Do not use as a stand-alone treatment for control of common chickweed or common poppy. Only use mixtures with non ALS-inhibitor herbicides for these weeds
- Do not use as the sole means of weed control in successive crops
- Specific restrictions apply to use in sequence or tank mixture with other sulfonylurea or ALS-inhibiting herbicides. See label for details
- Do not apply to crops under stress from any cause
- Do not apply when rain is imminent or during periods of frosty weather
- Specified adjuvant must be used. See label

Crop-specific information

- Latest use: before 2nd node detectable for winter wheat
- Transitory crop effects may occur, particularly on overlaps and after late season/spring applications. Recovery is normally complete and yield not affected

Following crops guidance

- In the event of crop failure winter or spring wheat may be drilled after normal cultivation and an interval of 6 wk
- Winter wheat, winter barley or winter oilseed rape may be drilled in the autumn following normal harvest of a treated crop. Spring wheat, spring barley, spring oilseed rape or sugar beet may be drilled in the following spring
- Where the product has been applied in sequence with a permitted ALS-inhibitor herbicide (see label) only winter or spring wheat or barley, or sugar beet may be sown as following crops
- Successive treatments with any products containing diflufenican can lead to soil build-up and inversion ploughing to 150 mm must precede sowing any following non-cereal crop. Even where ploughing occurs some crops may be damaged

Environmental safety

- Dangerous for the environment
- Very toxic to aquatic organisms
- Take extreme care to avoid drift outside the target area
- LERAP Category B

Hazard classification and safety precautions

Hazard Irritant, Dangerous for the environment
Transport code 9
Packaging group III
UN Number 3082
Risk phrases R36, R50, R53a
Operator protection A, C; U05a, U11, U20b
Environmental protection E15b, E16a, E38
Storage and disposal D01, D02, D05, D09a, D10a, D12a

172   diflufenican + metsulfuron-methyl

A contact and residual herbicide mixture containing sulfonyl ureas for cereals
HRAC mode of action code: F1 + B

Products

Pelican Delta	Headland	60:6% w/w	WG	14312

Uses

- Annual dicotyledons in **spring barley**, **winter barley**, **winter wheat**

Approval information

- Diflufenican and metsulfuron-methyl included in Annex I under EC Regulation 1107/2009

Environmental safety

- LERAP Category B

Hazard classification and safety precautions

Hazard Irritant, Dangerous for the environment
Transport code 9
Packaging group III
UN Number 3077
Risk phrases R36, R50, R53a
Operator protection A, C; U05a, U08, U11, U19a, U20b
Environmental protection E15a, E16a, E34, E38
Storage and disposal D01, D02, D05, D09a, D10b, D12a

173   diflufenican + pendimethalin

A mixture of pyridinecarboxamide and dinitroanaline herbicides, see Section 3
HRAC mode of action code: F1 + K1

See also pendimethalin

Products

1	Churchill	Makhteshim	18.75:300 g/l	EC	14139
2	Omaha	Makhteshim	30:300 g/l	EC	14234

Uses

- Annual dicotyledons in **triticale**, **winter barley**, **winter rye**, **winter wheat** [1, 2]
- Annual grasses in **triticale**, **winter barley**, **winter rye**, **winter wheat** [1]
- Annual meadow grass in **triticale**, **winter barley**, **winter rye**, **winter wheat** [2]

Approval information

- Diflufenican and pendimethalin included in Annex I under EC Regulation 1107/2009

Efficacy guidance

- Do not apply pre-emergence to winter crops drilled after 20th November
- Do not treat broadcast crops

Following crops guidance

- Do not drill autumn-sown broad leaved crops following application to the previous crop [1]
- Plough to 150 mm and thoroughly mix the soil before planting any following crop

Environmental safety

- LERAP Category B

Hazard classification and safety precautions

Hazard Toxic, Flammable, Dangerous for the environment
Transport code 3
Packaging group III
UN Number 1993
Risk phrases R20 [1]; R36, R37, R38 [2]; R43, R50, R53a, R61 [1, 2]

FOR FULL CONDITIONS OF USE ALWAYS READ THE PRODUCT LABEL

Operator protection A, C, H; U08, U13, U19a, U20b [1, 2]; U14, U15, U19c [2]; U19b [1]
Environmental protection E15a, E16a, E34, E38 [1, 2]; E23 [1]
Storage and disposal D09a, D10b, D12a
Medical advice M04a

174   dimethenamid-p

A chloroacetamide herbicide available only in mixtures
HRAC mode of action code: K3

175   dimethenamid-p + metazachlor

A soil acting herbicide mixture for oilseed rape
HRAC mode of action code: K3 + K3

See also metazachlor

Products

1	Muntjac	BASF	200:200 g/l	EC	14540
2	Springbok	BASF	200:200 g/l	EC	12983

Uses
- Annual dicotyledons in *leeks* *(off-label)* [2]; *winter oilseed rape* [1, 2]
- Annual meadow grass in *leeks* *(off-label)* [2]
- Chickweed in *winter oilseed rape* [1, 2]
- Cleavers in *winter oilseed rape* [1, 2]
- Common storksbill in *winter oilseed rape* [1, 2]
- Crane's-bill in *swedes* *(off-label)*, *turnips* *(off-label)* [2]
- Field speedwell in *winter oilseed rape* [1, 2]
- Poppies in *winter oilseed rape* [1, 2]
- Scented mayweed in *winter oilseed rape* [1, 2]
- Shepherd's purse in *swedes* *(off-label)*, *turnips* *(off-label)* [2]; *winter oilseed rape* [1, 2]

Extension of Authorisation for Minor Use (EAMUs)
- *leeks* 20110918 [2]
- *swedes* 20102133 [2]
- *turnips* 20102133 [2]

Approval information
- Dimethenamid-p and metazachlor included in Annex I under EC Regulation 1107/2009

Efficacy guidance
- Best results obtained from treatments to fine, firm and moist seedbeds
- Apply pre- or post-emergence of the crop and ideally before weed emergence
- Residual weed control may be reduced under prolonged dry conditions
- Weeds germinating from depth may not be controlled

Restrictions
- Maximum total dose on winter oilseed rape equivalent to one full dose treatment
- Do not disturb soil after application
- Do not treat broadcast crops until they have attained two fully expanded cotyledons
- Do not use on Sands, Very Light soils, or soils containing 10% organic matter
- Do not apply when heavy rain forecast or on soils waterlogged or prone to waterlogging
- Do not treat crops suffering from stress from any cause
- Do not use pre-emergence when crop seed has started to germinate or if not covered with 15 mm of soil
- Applications shall be limited to a total dose of not more than 1.0 kg metazachlor/ha in a three year period on the same field [1, 2]

SEE SECTION 3 FOR PRODUCTS ALSO REGISTERED

Crop-specific information
- Latest use: before 7th true leaf for winter oilseed rape
- All varieties of winter oilseed rape may be treated

Following crops guidance
- Any crop may follow a normally harvested treated winter oilseed rape crop. Ploughing is not essential before a following cereal crop but is required for all other crops
- In the event of failure of a treated crop winter wheat (excluding durum) or winter barley may be drilled in the same autumn, and any cereal (excluding durum wheat), spring oilseed rape, peas or field beans may be sown in the following spring. Ploughing to at least 150 mm should precede in all cases

Environmental safety
- Dangerous for the environment
- Very toxic to aquatic organisms
- Take extreme care to avoid spray drift onto non-crop plants outside the target area
- Some pesticides pose a greater threat of contamination of water than others and metazachlor is one of these pesticides. Take special care when applying metazachlor near water and do not apply if heavy rain is forecast
- LERAP Category B

Hazard classification and safety precautions
Hazard Harmful, Dangerous for the environment
Transport code 9
Packaging group III
UN Number 3082
Risk phrases R22a, R36, R38, R43, R50, R53a
Operator protection A, C, H; U05a, U08, U20c
Environmental protection E15a, E16a, E34, E38
Storage and disposal D01, D02, D09a, D10c, D12a
Medical advice M03, M05a

176 dimethenamid-p + metazachlor + quinmerac

A soil-acting herbicide mixture for use in oilseed rape
HRAC mode of action code: K3 + K3 + O

See also metazachlor
        quinmerac

Products

1	Elk	BASF	200:200:100	SE	13974
2	Katamaran Turbo	BASF	200:200:100	SE	13999
3	Pontoon	Agroquimicos	200:200:100	SE	15821
4	Shadow	BASF	200:200:100	SE	14000

Uses
- Annual dicotyledons in *winter oilseed rape*
- Annual grasses in *winter oilseed rape*
- Cleavers in *winter oilseed rape*

Approval information
- Dimethenamid-p, metazachlor and quinmerac included in Annex I under EC Regulation 1107/2009

Restrictions
- Applications shall be limited to a total dose of not more than 1.0 kg metazachlor/ha in a three year period on the same field

Environmental safety
- Some pesticides pose a greater threat of contamination of water than others and metazachlor is one of these pesticides. Take special care when applying metazachlor near water and do not apply if heavy rain is forecast
- LERAP Category B

Hazard classification and safety precautions
    Hazard Irritant, Dangerous for the environment
    Transport code 9
    Packaging group III
    UN Number 3082
    Risk phrases R43, R50, R53a
    Operator protection A, H; U05a, U08, U14, U20c
    Environmental protection E07a, E15b, E16a, E34, E38
    Storage and disposal D01, D02, D09a, D10c, D12a
    Medical advice M03, M05a

177   dimethenamid-p + pendimethalin

A chloracetamide and dinitroaniline herbicide mixture for weed control in maize crops
HRAC mode of action code: K1 + K3

Products

Wing - P	BASF	212.5:250 g/l	EC	15425

Uses
- Annual dicotyledons in **bulb onions** *(off-label)*, **bullb onion sets** *(off-label)*, **cabbages** *(off-label)*, **chives** *(off-label)*, **forage maize**, **forage maize (under plastic mulches)** *(off-label)*, **garlic** *(off-label)*, **grain maize**, **grain maize under plastic mulches** *(off-label)*, **herbs (see appendix 6)** *(off-label)*, **leeks** *(off-label)*, **lettuce** *(off-label)*, **salad onions** *(off-label)*, **shallots** *(off-label)*
- Annual grasses in **bulb onions** *(off-label)*, **bullb onion sets** *(off-label)*, **cabbages** *(off-label)*, **chives** *(off-label)*, **forage maize (under plastic mulches)** *(off-label)*, **garlic** *(off-label)*, **grain maize under plastic mulches** *(off-label)*, **herbs (see appendix 6)** *(off-label)*, **leeks** *(off-label)*, **lettuce** *(off-label)*, **salad onions** *(off-label)*, **shallots** *(off-label)*
- Annual meadow grass in **forage maize**, **grain maize**

Extension of Authorisation for Minor Use (EAMUs)
- **bulb onions** *20122250*
- **bullb onion sets** *20122250*
- **cabbages** *20122251*
- **chives** *20122249*
- **forage maize (under plastic mulches)** *20111841*
- **garlic** *20122250*
- **grain maize under plastic mulches** *20111841*
- **herbs (see appendix 6)** *20121052*
- **leeks** *20122248*
- **lettuce** *20121052*
- **salad onions** *20122249*
- **shallots** *20122250*

Approval information
- Dimethenamid-p and pendimethalin are included in Annex 1 under EC Regulation 1107/2009

Efficacy guidance
- Do not use on soil types with more than 10% organic matter
- Best results obtained when rain falls within 7 days of application

Crop-specific information
- Use on maize crops grown under plastic mulches is by EAMU only; do not use in greenhouses, covers or other forms of protection
- Risk of crop damage if heavy rain falls soon after application to stoney or gravelly soils

SEE SECTION 3 FOR PRODUCTS ALSO REGISTERED

- Seed should be covered with at least 5 cm of settled soil

Environmental safety
- LERAP Category B

Hazard classification and safety precautions
    Hazard Harmful, Dangerous for the environment
    Transport code 9
    Packaging group III
    UN Number 3082
    Risk phrases R22a, R22b, R38, R43, R50, R53a
    Operator protection A, H; U05a, U14
    Environmental protection E15b, E16a, E34, E38
    Storage and disposal D01, D02, D09a, D10c, D12a
    Medical advice M03, M05a

178   dimethoate

A contact and systemic organophosphorus insecticide and acaricide
IRAC mode of action code: 1B

Products

1	Danadim Progress	Headland	400 g/l	EC	12208
2	Danadim Progress	Headland	400 g/l	EC	15890

Uses
- Aphids in *broccoli, brussels sprouts, calabrese, calabrese* (off-label), *cauliflowers, celeriac* (off-label), *chinese cabbage* (off-label), *collards* (off-label), *fodder beet* (excluding Myzus persicae), *kale* (off-label), *kohlrabi* (off-label), *lettuce, mangels* (excluding Myzus persicae), *protected broccoli* (off-label), *protected brussels sprouts* (off-label), *protected cabbages* (off-label), *protected calabrese* (off-label), *protected cauliflowers* (off-label), *protected celeriac* (off-label), *protected chinese cabbage* (off-label), *protected collards* (off-label), *protected kale* (off-label), *protected kohlrabi* (off-label), *protected roscoff cauliflowers* (off-label), *protected savoy cabbage* (off-label), *protected spinach* (off-label), *protected spring greens* (off-label), *red beet* (excluding Myzus persicae), *roscoff cauliflowers* (off-label), *savoy cabbage* (off-label), *spring greens* (off-label), *sugar beet* (excluding Myzus persicae), *winter rye* [1]; *durum wheat, rye, spring wheat* [2]; *ornamental plant production, triticale, winter wheat* [1, 2]
- Insect pests in *broccoli, brussels sprouts, calabrese, calabrese* (off-label), *cauliflowers, celeriac* (off-label), *chinese cabbage* (off-label), *collards* (off-label), *compost* (off-label), *kale* (off-label), *kohlrabi* (off-label), *protected broccoli* (off-label), *protected brussels sprouts* (off-label), *protected cabbages* (off-label), *protected calabrese* (off-label), *protected cauliflowers* (off-label), *protected celeriac* (off-label), *protected chinese cabbage* (off-label), *protected collards* (off-label), *protected kale* (off-label), *protected kohlrabi* (off-label), *protected roscoff cauliflowers* (off-label), *protected savoy cabbage* (off-label), *protected spinach* (off-label), *protected spring greens* (off-label), *roscoff cauliflowers* (off-label), *savoy cabbage* (off-label), *spring greens* (off-label) [1]; *protected forest nurseries* (off-label) [2]
- Leaf miner in *fodder beet, mangels, red beet, sugar beet* [1]; *ornamental plant production* [1, 2]
- Red spider mites in *ornamental plant production* [1, 2]
- Wheat bulb fly in *winter wheat* [1]

Extension of Authorisation for Minor Use (EAMUs)
- *calabrese 20122242* [1]
- *celeriac 20122242* [1]
- *chinese cabbage 20122242* [1]
- *collards 20122242* [1]
- *compost 20122243* [1]
- *kale 20122242* [1]
- *kohlrabi 20122242* [1]
- *protected broccoli 20122242* [1]

FOR FULL CONDITIONS OF USE ALWAYS READ THE PRODUCT LABEL

- **protected brussels sprouts** *20122242* [1]
- **protected cabbages** *20122242* [1]
- **protected calabrese** *20122242* [1]
- **protected cauliflowers** *20122242* [1]
- **protected celeriac** *20122242* [1]
- **protected chinese cabbage** *20122242* [1]
- **protected collards** *20122242* [1]
- **protected forest nurseries** *20122130* [2]
- **protected kale** *20122242* [1]
- **protected kohlrabi** *20122242* [1]
- **protected roscoff cauliflowers** *20122242* [1]
- **protected savoy cabbage** *20122242* [1]
- **protected spinach** *20122242* [1]
- **protected spring greens** *20122242* [1]
- **roscoff cauliflowers** *20122242* [1]
- **savoy cabbage** *20122242* [1]
- **spring greens** *20122242* [1]

Approval information
- Dimethoate included in Annex I under EC Regulation 1107/2009
- In 2006 CRD required that all products containing this active ingredient should carry the following warning in the main area of the container label: "Dimethoate is an anticholinesterase organophosphate. Handle with care"

Efficacy guidance
- Chemical has quick knock-down effect and systemic activity lasts for up to 14 d
- With some crops, products differ in range of pests listed as controlled. Uses section above provides summary. See labels for details
- For most pests apply when pest first seen and repeat 2-3 wk later or as necessary. Timing and number of sprays varies with crop and pest. See labels for details
- Best results achieved when crop growing vigorously. Systemic activity reduced when crops suffering from drought or other stress
- In hot weather apply in early morning or late evening
- Where aphids or spider mites resistant to organophosphorus compounds occur control is unlikely to be satisfactory and repeat treatments may result in lower levels of control

Restrictions
- Contains an anticholinesterase organophosphorus compound. Do not use if under medical advice not to work with such compounds
- Maximum number of treatments 6 per crop for Brussels sprouts, broccoli, cauliflower, calabrese, lettuce; 4 per crop for grass seed crops, cereals; 2 per crop for beet crops, triticale, rye; 1 per crop for mangel and sugar beet seed crops
- On beet crops only one treatment per crop may be made for control of leaf miners (max 84 g a.i./ha) and black bean aphid (max 400 g a.i./ha)
- In beet crops resistant strains of peach-potato aphid (*Myzus persicae*) are common and dimethoate products must not be used to control this pest
- Test for varietal susceptibility on all unusual plants or new cultivars
- Do not tank mix with alkaline materials. See label for recommended tank-mixes
- Consult processor before spraying crops grown for processing
- Must not be applied to cereals if any product containing a pyrethroid insecticide or dimethoate has been sprayed after the start of ear emergence (GS 51)

Crop-specific information
- Latest use: before 30 Jun in yr of harvest for beet crops; flowering just complete stage (GS 30) for wheat, rye, triticale; before 31 Mar in yr of harvest for aerial application to cereals
- HI brassicas 21 d; grass seed crops, lettuce, cereals 14 d; ornamental plant production 7 d

Environmental safety
- Dangerous for the environment
- Harmful to aquatic organisms
- Flammable

SEE SECTION 3 FOR PRODUCTS ALSO REGISTERED

- Harmful to game, wild birds and animals
- Harmful to livestock. Keep all livestock out of treated areas for at least 7 d
- Likely to cause adverse effects on beneficial arthropods
- High risk to non-target insects or other arthropods. Do not treat cereals after 1 Apr within 6 m of edge of crop
- High risk to bees. Do not apply to crops in flower or to those in which bees are actively foraging. Do not apply when flowering weeds are present
- Surface residues may also cause bee mortality following spraying
- Dangerous to fish or other aquatic life. Do not contaminate surface waters or ditches with chemical or used container
- Keep in original container, tightly closed, in a safe place, under lock and key
- LERAP Category A

Hazard classification and safety precautions

Hazard Harmful, Flammable
Transport code 6.1
Packaging group III
UN Number 3017
Risk phrases R22a, R43, R52
Operator protection A, H, M; U02a, U04a, U08, U14, U15, U19a, U20b
Environmental protection E06c (7 d); E10b, E12a, E12e, E13b, E16c, E16d, E22a, E34
Storage and disposal D01, D02, D05, D09b, D12b
Medical advice M01, M03

179   dimethomorph

A cinnamic acid fungicide with translaminar activity
FRAC mode of action code: 40

See also ametoctradin + dimethomorph

Products

1 Morph	Makhteshim	500 g/l	SC	15121
2 Paraat	BASF	50% w/w	WP	15445

Uses

- Blight in **potatoes** [1]
- Crown rot in **protected strawberries** *(moderate control)*, **strawberries** *(moderate control)* [2]
- Downy mildew in **brassica leaves and sprouts** *(off-label)*, **chard** *(off-label)*, **herbs (see appendix 6)** *(off-label)*, **lamb's lettuce** *(off-label)*, **lettuce** *(off-label)*, **protected herbs (see appendix 6)** *(off-label)*, **protected lamb's lettuce** *(off-label)*, **protected lettuce** *(off-label)*, **protected ornamentals** *(off-label)*, **rocket** *(off-label)*, **spinach** *(off-label)* [2]
- Root rot in **blackberries**, **protected blackberries** *(moderate control)*, **protected raspberries** *(moderate control)*, **raspberries** *(moderate control)* [2]

Extension of Authorisation for Minor Use (EAMUs)

- **brassica leaves and sprouts** 20122244 *expires 25 Sep 2013* [2]
- **chard** 20122244 *expires 25 Sep 2013* [2]
- **herbs (see appendix 6)** 20122244 *expires 25 Sep 2013* [2]
- **lamb's lettuce** 20122244 *expires 25 Sep 2013* [2]
- **lettuce** 20122244 *expires 25 Sep 2013* [2]
- **protected herbs (see appendix 6)** 20112584 *expires 25 Sep 2013* [2]
- **protected lamb's lettuce** 20112584 *expires 25 Sep 2013* [2]
- **protected lettuce** 20112584 *expires 25 Sep 2013* [2]
- **protected ornamentals** 20112585 *expires 25 Sep 2013* [2]
- **rocket** 20122244 *expires 25 Sep 2013* [2]
- **spinach** 20122244 *expires 25 Sep 2013* [2]

Approval information

- Dimethomorph included in Annex I under EC Regulation 1107/2009

FOR FULL CONDITIONS OF USE ALWAYS READ THE PRODUCT LABEL

Restrictions
- Do not sow clover until the spring following application [1].
- Dimethomorph is a carboxylic acid amide fungicide. No more than three consecutive applications of CAA fungicides should be made.

Crop-specific information
- After application, allow 10 months before sowing clover [2]

Environmental safety
- LERAP Category B [2]

Hazard classification and safety precautions
 Transport code 9
 Packaging group III
 UN Number 3077 [2]; 3082 [1]
 Risk phrases R52 [1]
 Operator protection A, D, H [2]; U05a [1, 2]; U09a, U20b [1]
 Environmental protection E15b, E38 [1, 2]; E16a [2]
 Storage and disposal D01, D02, D09a, D12a [1, 2]; D10a [2]; D10c [1]

180 dimethomorph + mancozeb

A systemic and protectant fungicide for potato blight control
FRAC mode of action code: 40 + M3

See also mancozeb

Products

1	Dunkirke	AgChem Access	7.5:66.7% w/w	WG	13796
2	Invader	BASF	7.5:66.7% w/w	WG	15223

Uses
- Blight in **potatoes** [1, 2]
- Disease control in **chives** *(off-label)*, **cress** *(off-label)*, **forest nurseries** *(off-label)*, **frise, frise** *(off-label)*, **garlic, herbs (see appendix 6)** *(off-label)*, **lamb's lettuce** *(off-label)*, **leaf brassicas** *(off-label)*, **leeks, lettuce** *(off-label)*, **onions, ornamental plant production** *(off-label)*, **poppies for morphine production** *(off-label)*, **radicchio** *(off-label)*, **salad onions** *(off-label)*, **scarole** *(off-label)*, **shallots** [2]
- Downy mildew in **bulb onions, chives** *(off-label)*, **cress** *(off-label)*, **forest nurseries** *(off-label)*, **frise** *(off-label)*, **garlic, herbs (see appendix 6)** *(off-label)*, **lamb's lettuce** *(off-label)*, **leaf brassicas** *(off-label)*, **lettuce** *(off-label)*, **ornamental plant production** *(off-label)*, **poppies for morphine production** *(off-label)*, **radicchio** *(off-label)*, **salad onions** *(off-label)*, **scarole** *(off-label)* [2]
- White tip in **leeks** *(reduction)* [2]

Extension of Authorisation for Minor Use (EAMUs)
- **chives** *20120111 expires 25 Sep 2013* [2]
- **cress** *20120110 expires 25 Sep 2013* [2]
- **forest nurseries** *20120109 expires 25 Sep 2013* [2]
- **frise** *20120110 expires 25 Sep 2013* [2]
- **herbs (see appendix 6)** *20120110 expires 25 Sep 2013* [2]
- **lamb's lettuce** *20120110 expires 25 Sep 2013* [2]
- **leaf brassicas** *20120110 expires 25 Sep 2013* [2]
- **lettuce** *20120110 expires 25 Sep 2013* [2]
- **ornamental plant production** *20120109 expires 25 Sep 2013* [2]
- **poppies for morphine production** *20120112 expires 25 Sep 2013* [2]
- **radicchio** *20120110 expires 25 Sep 2013* [2]
- **salad onions** *20120111 expires 25 Sep 2013* [2]
- **scarole** *20120110 expires 25 Sep 2013* [2]

Approval information
- Dimethomorph and mancozeb included in Annex I under EC Regulation 1107/2009
- Approval expiry 31 Mar 2013 [1]

Efficacy guidance
- Commence treatment as soon as there is a risk of blight infection
- In the absence of a warning treatment should start before the crop meets along the rows
- Repeat treatments every 7-14 d depending in the degree of infection risk
- Irrigated crops should be regarded as at high risk and treated every 7 d
- For best results good spray coverage of the foliage is essential
- To minimise the likelihood of development of resistance these products should be used in a planned Resistance Management strategy. See Section 5 for more information

Restrictions
- Maximum total dose equivalent to eight full dose treatments on potatoes

Crop-specific information
- HI 7 d for all crops

Environmental safety
- Dangerous for the environment
- Very toxic to aquatic organisms
- LERAP Category B

Hazard classification and safety precautions
> **Hazard** Irritant, Dangerous for the environment
> **Transport code** 9
> **Packaging group** III
> **UN Number** 3077
> **Risk phrases** R36, R37, R38, R43, R50, R53a
> **Operator protection** A; U02a, U04a, U05a, U08, U13, U19a, U20b
> **Environmental protection** E16a, E16b, E38
> **Storage and disposal** D01, D02, D05, D09a, D12a
> **Medical advice** M05a

181   dimoxystrobin

A protectant strobilurin fungicide for cereals available only in mixtures
FRAC mode of action code: 11

182   dimoxystrobin + epoxiconazole

A protectant and curative fungicide mixture for cereals
FRAC mode of action code: 11 + 3

See also epoxiconazole

Products

Swing Gold	BASF	133:50 g/l	SC	11658

Uses
- Brown rust in *winter wheat*
- Foliar disease control in *durum wheat* (off-label), *grass seed crops* (off-label), *spring rye* (off-label), *triticale* (off-label), *winter rye* (off-label)
- Fusarium ear blight in *winter wheat*
- Glume blotch in *winter wheat*
- Septoria leaf blotch in *winter wheat*
- Tan spot in *winter wheat*

Extension of Authorisation for Minor Use (EAMUs)
- *durum wheat* 20062642

FOR FULL CONDITIONS OF USE ALWAYS READ THE PRODUCT LABEL

- **grass seed crops** *20062642*
- **spring rye** *20062642*
- **triticale** *20062642*
- **winter rye** *20062642*

Approval information
- Dimoxystrobin and epoxiconazole included in Annex I under EC Regulation 1107/2009

Efficacy guidance
- For best results apply from the start of ear emergence
- Dimoxystrobin is a member of the QoI cross resistance group. Product should be used preventatively and not relied on for its curative potential
- Use product as part of an Integrated Crop Management strategy incorporating other methods of control, including where appropriate other fungicides with a different mode of action. Do not apply more than two foliar applications of QoI containing products to any cereal crop
- There is a significant risk of widespread resistance occurring in *Septoria tritici* populations in UK. Failure to follow resistance management action may result in reduced levels of disease control
- Epoxiconazole is a DMI fungicide. Resistance to some DMI fungicides has been identified in Septoria leaf blotch which may seriously affect performance of some products. For further advice contact a specialist advisor and visit the Fungicide Resistance Action Group (FRAG)-UK website

Restrictions
- Maximum number of treatments 1 per crop
- Do not apply before the start of ear emergence

Crop-specific information
- Latest use: up to and including flowering (GS 69)

Environmental safety
- Dangerous for the environment
- Very toxic to aquatic organisms
- Avoid drift onto neighbouring crops
- LERAP Category B

Hazard classification and safety precautions
Hazard Harmful, Dangerous for the environment
Transport code 9
Packaging group III
UN Number 3082
Risk phrases R20, R22a, R40, R50, R53a, R63
Operator protection A; U05a, U20b
Environmental protection E15a, E16a, E34, E38
Storage and disposal D01, D02, D05, D09a, D10c, D12a
Medical advice M05a

183   diquat

A non-residual bipyridyl contact herbicide and crop desiccant
HRAC mode of action code: D

Products

1	Barclay D-Quat	Barclay	200 g/l	SL	14833
2	Brogue	AgriGuard	200 g/l	SL	15638
3	ChemSource Diquat	ChemSource	200 g/l	SL	14533
4	Clayton Diquat	Clayton	200 g/l	SL	12739
5	Clayton Diquat 200	Clayton	200 g/l	SL	13942
6	Clayton IQ	Clayton	200 g/l	SL	14519
7	Dessicash 200	Sharda	200 g/l	SL	14848
8	Di Novia	Agroquimicos	200 g/l	SL	15840
9	Diqua	Sharda	200 g/l	SL	14032

SEE SECTION 3 FOR PRODUCTS ALSO REGISTERED

Products – continued

10	Diquash	Agrichem	200 g/l	SL	14849
11	Dragoon Gold	Belcrop	200 g/l	SL	14973
12	Knoxdoon	AgChem Access	200 g/l	SL	13987
13	Life Scientific Diquat	Life Scientific	200 g/l	SL	15398
14	Mission	AgriChem BV	200 g/l	SL	13411
15	Quad-Glob 200 SL	Q-Chem	200 g/l	SL	14758
16	Quazar	Belchim	200 g/l	SL	15357
17	Quit	Belchim	200 g/l	SL	14874
18	Reglone	Syngenta	200 g/l	SL	10534
19	Retro	Syngenta	200 g/L	SL	13841
20	Roquat 20	Q-Chem	200 g/l	SL	15703
21	Standon Googly	Standon	200 g/l	SL	13281
22	Standon Yorker	Standon	200 g/l	SL	14751
23	UPL Diquat	United Phosphorus	200 g/l	SL	14944
24	Woomera	Belcrop	200 g/l	SL	14972

Uses

- Annual and perennial weeds in *all edible crops (outdoor)* *(around only)*, *all non-edible crops (outdoor)* *(around only)*, *combining peas* *(dry harvested only)*, *hops*, *linseed*, *ornamental plant production*, *potatoes* *(weed control or dessication)*, *red clover* *(seed crop)*, *spring barley* *(animal feed only)*, *spring field beans*, *spring oats* *(animal feed only)*, *spring oilseed rape*, *sugar beet*, *white clover* *(seed crop)*, *winter barley* *(animal feed only)*, *winter field beans*, *winter oats* *(animal feed only)*, *winter oilseed rape* [14]; *barley for animal feed*, *oats for animal feed* [23]; *forest nurseries* *(off-label)*, *poppies for morphine production* *(off-label)* [15]
- Annual dicotyledons in *all edible crops (outdoor)*, *all non-edible crops (outdoor)*, *potatoes* [16, 17]; *all edible crops (outdoor)* *(around)*, *all non-edible crops (outdoor)* *(around)*, *ornamental plant production*, *sugar beet* [3-13, 15, 18, 20-24]; *fodder beet* [11]; *lucerne* *(off-label)* [18]
- Annual grasses in *all edible crops (outdoor)*, *all non-edible crops (outdoor)*, *potatoes* [16, 17]
- Basal defoliation in *hops* [6]
- Black bindweed in *all edible crops (outdoor)* [19]
- Chemical stripping in *hops* [15]
- Chickweed in *all edible crops (outdoor)*, *all non-edible crops (outdoor)*, *ornamental plant production*, *potatoes*, *sugar beet* [19]
- Crane's-bill in *all edible crops (outdoor)* [19]
- Desiccation in *barley for animal feed*, *oats for animal feed* [16, 17, 23]; *barley seed crops* *(off-label)*, *crambe* *(off-label)*, *hops* *(off-label)*, *oat seed crops* *(off-label)*, *potato dumps* *(off-label)* [18]; *borage for oilseed production* *(off-label)*, *french beans* *(off-label)*, *hemp for oilseed production* *(off-label)*, *herbs (see appendix 6)* *(off-label)*, *poppies for morphine production* *(off-label)*, *poppies for oilseed production* *(off-label)*, *sesame* *(off-label)*, *sunflowers* *(off-label)* [15]; *combining peas*, *linseed*, *red clover*, *spring field beans*, *white clover*, *winter field beans* [2, 16, 17]; *lupins* *(off-label)*, *sweet potato* *(off-label)* [15, 18]; *oilseed rape* [16, 17]; *potatoes* [1, 2, 16, 17]; *spring barley*, *spring oats*, *spring oilseed rape*, *winter barley*, *winter oats*, *winter oilseed rape* [2]
- Fat hen in *all edible crops (outdoor)* [19]
- General weed control in *chives* *(off-label)*, *herbs (see appendix 6)* *(off-label)*, *parsley* *(off-label)*, *poppies for morphine production* *(off-label)* [18]; *potatoes* [3-13, 15, 18, 20-24]
- Grass weeds in *all edible crops (outdoor)* *(around only)*, *all non-edible crops (outdoor)* *(around only)*, *combining peas* *(dry harvested)*, *hops*, *linseed*, *ornamental plant production*, *potatoes* *(weed control and dessication)*, *red clover* *(seed crop)*, *spring barley* *(animal feed)*, *spring field beans*, *spring oats* *(animal feed)*, *spring oilseed rape*, *sugar beet*, *white clover* *(seed crop)*, *winter barley* *(animal feed)*, *winter field beans*, *winter oats* *(animal feed)*, *winter oilseed rape* [14]
- Growth regulation in *hops* [5]

- Harvest management/desiccation in **combining peas**, **linseed**, **red clover**, **spring barley**, **spring field beans**, **spring oats**, **spring oilseed rape**, **white clover**, **winter barley**, **winter field beans**, **winter oats**, **winter oilseed rape** [2]
- Haulm destruction in **potatoes** [3-13, 15, 18, 20-24]
- Perennial dicotyledons in **lucerne** *(off-label)* [18]
- Poppies in **all edible crops (outdoor)** [19]
- Pre-harvest desiccation in **borage for oilseed production** *(off-label)*, **brown mustard** *(off-label)*, **french beans** *(off-label)*, **hemp for oilseed production** *(off-label)*, **poppies for morphine production** *(off-label)*, **poppies for oilseed production** *(off-label)*, **sesame** *(off-label)*, **sunflowers** *(off-label)*, **white mustard** *(off-label)* [18]; **broad beans** [3-5, 7-10, 12, 13, 18, 21, 23]; **clover seed crops**, **combining peas**, **linseed**, **spring field beans**, **spring oilseed rape**, **winter field beans**, **winter oilseed rape** [3-13, 15, 18, 20, 21, 23]; **laid barley** *(stockfeed only)*, **laid oats** *(stockfeed only)* [3-13, 15, 18, 20, 21]
- Sowthistle in **all edible crops (outdoor)** [19]

Extension of Authorisation for Minor Use (EAMUs)
- **barley seed crops** *20111978* [18]
- **borage for oilseed production** *20093348* [15], *20062390* [18]
- **brown mustard** *20062387* [18]
- **chives** *20062388* [18]
- **crambe** *20082092* [18]
- **forest nurseries** *20093350* [15]
- **french beans** *20093346* [15], *20062389* [18]
- **hemp for oilseed production** *20093351* [15], *20062386* [18]
- **herbs (see appendix 6)** *20093345* [15], *20062388* [18]
- **hops** *20102019* [18]
- **lucerne** *20082842* [18]
- **lupins** *20093353* [15], *20073232* [18]
- **oat seed crops** *20111978* [18]
- **parsley** *20062388* [18]
- **poppies for morphine production** *20093349* [15], *20062385* [18]
- **poppies for oilseed production** *20093347* [15], *20062387* [18]
- **potato dumps** *20111882* [18]
- **sesame** *20093347* [15], *20062387* [18]
- **sunflowers** *20093347* [15], *20062387* [18]
- **sweet potato** *20093352* [15], *20070734* [18]
- **white mustard** *20062387* [18]

Approval information
- Diquat included in Annex I under EC Regulation 1107/2009
- All approvals for aquatic weed control expired in 2004
- NOT accepted by BBPA for use on malting barley
- Approval expiry 31 Aug 2013 [12]

Efficacy guidance
- Acts rapidly on green parts of plants and rainfast in 15 min
- Best results for potato desiccation achieved by spraying in bright light and low humidity conditions
- For weed control in row crops apply as overall spray before crop emergence or before transplanting or as an inter-row treatment using a sprayer designed to prevent contamination of crop foliage by spray
- Use of authorised non-ionic wetter for improved weed control essential for most uses except potato desiccation

Restrictions
- Maximum number of treatments 1 per crop in most situations - see labels for details
- Do not apply potato haulm destruction treatment when soil dry. Tubers may be damaged if spray applied during or shortly after dry periods. See label for details of maximum allowable soil-moisture deficit and varietal drought resistance scores
- Do not add wetters to desiccant sprays for potatoes

SEE SECTION 3 FOR PRODUCTS ALSO REGISTERED

- Consult processor before adding non-ionic or other wetter on peas. Staining may be increased
- Do not use treated straw or haulm as animal feed or bedding within 4 d of spraying
- Do not apply through ULV or mistblower equipment

Crop-specific information
- Latest use: before crop emergence for sugar beet and before transplanting ornamentals. Not specified for other crops where use varies according to situation
- HI zero when used as a crop desiccant but see label for advisory intervals
- For pre-harvest desiccation, apply to potatoes when tubers the desired size and to other crops when mature or approaching maturity. See label for details of timing and of period to be left before harvesting potatoes
- If potato tubers are to be stored leave 14 d after treatment before lifting

Environmental safety
- Dangerous for the environment
- Very toxic to aquatic organisms
- Keep all livestock out of treated areas and away from treated water for at least 24 h
- Do not feed treated straw or haulm to livestock within 4 days of spraying
- Do not use on crops if the straw is to be used as animal feed/bedding
- Do not dump surplus herbicide in water or ditch bottoms

Hazard classification and safety precautions
Hazard Very toxic [4]; Toxic [1-3, 5-13, 15-24]; Dangerous for the environment [1-24]
Transport code 6.1 [1]; 8 [2-24]
Packaging group II [1]; III [2-24]
UN Number 1760 [2-24]; 2927 [1]
Risk phrases R22a [1-13, 15-24]; R23 [1-3, 5-24]; R25 [16, 17]; R25b [23]; R26 [4]; R36 [14-17, 20]; R37, R38, R50, R53a [1-24]; R43 [5, 14-17, 20]; R48 [1-13, 15-18, 20-24]; R69 [14, 19]
Operator protection A [1-24]; C [14, 15, 20]; D [21]; H [1-14, 16-19, 21-24]; M [3-20, 22-24]; P [19]; U02a, U05a, U08, U20b [1-24]; U04a [1-15, 18-24]; U04c [14]; U10 [1-3, 6-13, 15, 18-24]; U12, U12b [16, 17]; U15 [15-17, 20]; U19a [1, 3-24]; U19c [2]
Environmental protection E06c [1-15] (24 h); E06c [16] (24 hrs); E06c [17] (24 hours); E06c [18-24] (24 h); E08a [2, 15, 20, 22] (4 d); E09 [1, 3]; E09 [4, 5] (4 d); E09 [6-14]; E09 [16] (4 d); E09 [17-19, 21]; E09 [23] (4 days); E09 [24]; E15a [14]; E15b [1-13, 15-18, 20-24]; E34 [1-24]; E38 [1-3, 6-15, 18-24]
Storage and disposal D01, D02, D09a, D12a [1-24]; D05 [1-13, 15-18, 20-24]; D10a [14]; D10b [4, 5]; D10c [1-3, 6-13, 15-24]
Medical advice M04a

184   dithianon

A protectant and eradicant dicarbonitrile fungicide for scab control
FRAC mode of action code: M9

Products
1	Dartagnan Flo	ChemSource	750 g/l	SC	14979
2	Dartagnan WG	ChemSource	70% w/w	WG	14963
3	Dithianon Flowable	BASF	750 g/l	SC	10219
4	Dithianon WG	BASF	70% w/w	WG	12538

Uses
- Scab in *apples*, *pears* [1-4]; *quinces* (off-label) [3]

Extension of Authorisation for Minor Use (EAMUs)
- *quinces 20061038* [3]

Approval information
- Dithianon included in Annex I under EC Regulation 1107/2009

Efficacy guidance
- Apply at bud-burst and repeat every 7-14 d until danger of scab infection ceases

- Application at high rate within 48 h of a Mills period prevents new infection
- Spray programme also reduces summer infection with apple canker

Restrictions
- Maximum number of treatments on apples and pears 8 per crop [3]; 12 per crop [4]
- Do not use on Golden Delicious apples after green cluster
- Do not mix with lime sulphur or highly alkaline products

Crop-specific information
- HI 4 wk for apples, pears

Environmental safety
- Dangerous for the environment
- Very toxic to aquatic organisms

Hazard classification and safety precautions
Hazard Harmful, Dangerous for the environment
Transport code 9
Packaging group III
UN Number 3077 [2, 4]; 3082 [1, 3]
Risk phrases R20 [1, 3]; R22a, R50, R53a [1-4]; R40 [3, 4]; R41 [2, 4]
Operator protection A [1-4]; D [2, 4]; H [1, 3]; U05a [1-4]; U08, U19a, U20a [1, 3]; U11 [2, 4]
Environmental protection E15a, E34, E38
Storage and disposal D01, D02, D05 [1-4]; D09a, D10b [1, 3]; D12a [2, 4]
Medical advice M03 [1, 3]; M05a [1-4]

185   dithianon + pyraclostrobin

A protectant fungicide mixture for apples and pears
FRAC mode of action code: M9 + 11

See also pyraclostrobin

Products

Maccani	BASF	12:4% w/w	WG	13227

Uses
- Powdery mildew in *apples*
- Scab in *apples*, *pears*

Approval information
- Pyraclostrobin and dithianon included in Annex I under EC Regulation 1107/2009

Efficacy guidance
- Best results obtained from treatments applied from bud burst before disease development and repeated at 10 d intervals
- Water volume should be adjusted according to size of tree and leaf area to ensure good coverage
- Pyraclostrobin is a member of the QoI cross resistance group. Product should be used preventatively at the full recommended dose and not relied on for its curative potential
- Where field performance may be adversely affected apply QoI containing fungicides in mixtures in strict alternation with fungicides from a different cross-resistance group

Restrictions
- Maximum number of treatments 4 per crop. However where the total number of fungicide treatments is less than 12, the maximum number of applications of any QoI containing products should be three
- Consult processors before use on crops destined for fruit preservation

Crop-specific information
- HI 35 d for apples, pears
- Some apple varieties may be considered to be susceptible to russetting caused by the dithianon component during fruit development

SEE SECTION 3 FOR PRODUCTS ALSO REGISTERED

Environmental safety
- Dangerous for the environment
- Very toxic to aquatic organisms
- Broadcast air-assisted LERAP (40 m)

Hazard classification and safety precautions
> **Hazard** Harmful, Dangerous for the environment
> **Transport code** 9
> **Packaging group** III
> **UN Number** 3077
> **Risk phrases** R20, R22a, R36, R50, R53a
> **Operator protection** A, C, D; U05a, U11, U20b
> **Environmental protection** E15b, E34, E38; E17b (40 m)
> **Storage and disposal** D01, D02, D05, D08, D11a, D12a
> **Medical advice** M03, M05a

186 dodeca-8,10-dienyl acetate

An insect sex pheromone
IRAC mode of action code: Not classified

Products

Exosex CM	Exosect	0.1% w/w	DP	12103

Uses
- Codling moth in **apple orchards**, **pear orchards**

Approval information
- Dodeca-8,10-dienyl acetate is not included in Annex 1 under EC Regulation 1107/2009

Efficacy guidance
- Store unopened sachets below 4°C for up to 24 months or store at ambient temperatures and use within 5 months.
- Allow at least 20 m between dispensers and use a minimum of 25 dispensers per heactare.
- Dispenser activity lasts 70 - 90 days depending upon temperatures.

Hazard classification and safety precautions
> **UN Number** N/C
> **Operator protection** A
> **Environmental protection** E15a
> **Storage and disposal** D01, D02
> **Medical advice** M03

187 dodine

A protectant and eradicant guanidine fungicide
FRAC mode of action code: M7

Products

1	Clayton Scabius	Clayton	450 g/l	SC	13339
2	Greencrop Budburst	Greencrop	450 g/l	SC	11042
3	Radspor FL	Truchem	450 g/l	SC	01685

Uses
- Currant leaf spot in **bilberries** *(off-label)*, **blueberries** *(off-label)*, **cranberries** *(off-label)*, **gooseberries** *(off-label)*, **redcurrants** *(off-label)*, **whitecurrants** *(off-label)* [3]; **blackcurrants** [1-3]
- Scab in **apples**, **pears** [1-3]; **quinces** *(off-label)* [3]

Extension of Authorisation for Minor Use (EAMUs)
- *bilberries 20060617* [3]

- **blueberries** *20060617* [3]
- **cranberries** *20060617* [3]
- **gooseberries** *20060617* [3]
- **quinces** *20060616* [3]
- **redcurrants** *20060617* [3]
- **whitecurrants** *20060617* [3]

Approval information
- Dodine included in Annex I under EC Regulation 1107/2009

Efficacy guidance
- Apply protective spray on apples and pears at bud-burst and at 10-14 d intervals until late Jun to early Jul
- Apply post-infection spray within 36 h of rain responsible for initiating infection. Where scab already present spray prevents production of spores
- On blackcurrants commence spraying at early grape stage and repeat at 2-3 wk intervals, and at least once after picking

Restrictions
- Do not apply in very cold weather (under 5°C) or under slow drying conditions to pears or dessert apples during bloom or immediately after petal fall
- Do not mix with lime sulphur or tetradifon
- Consult processors before use on crops grown for processing

Crop-specific information
- Latest use: early Jul for culinary apples; pre-blossom for dessert apples and pears

Environmental safety
- Dangerous for the environment
- Very toxic to aquatic organisms

Hazard classification and safety precautions
Hazard Harmful, Dangerous for the environment
Transport code 9
Packaging group III
UN Number 3082
Risk phrases R21 [2]; R22a, R36, R38, R43, R50, R53a [1-3]; R37, R41 [1, 3]
Operator protection A, C; U05a, U08, U19a, U20b [1-3]; U11 [1, 3]
Environmental protection E15a, E34 [1-3]; E38 [3]
Storage and disposal D01, D02, D09a, D10b [1-3]; D05 [1, 2]; D12a [1]
Medical advice M03 [1, 3]; M04a [2]; M05a [1]

188 epoxiconazole

A systemic, protectant and curative triazole fungicide for use in cereals
FRAC mode of action code: 3

See also boscalid + epoxiconazole
        boscalid + epoxiconazole + pyraclostrobin
        chlorothalonil + epoxiconazole
        dimoxystrobin + epoxiconazole
        epoxiconazole + fluxapyroxad

Products

1 Amber	Makhteshim	125 g/l	SC	15126
2 Bassoon EC	BASF	83 g/l	EC	15219
3 Bowman	Makhteshim	125 g/l	SC	15125
4 Clayton Oust	Clayton	125 g/l	SC	11588
5 Clayton Oust SC	Clayton	125 g/l	SC	14871
6 Cortez	Makhteshim	125 g/l	SC	13932
7 EA Epoxi	European Ag	125 g/l	SC	14832

SEE SECTION 3 FOR PRODUCTS ALSO REGISTERED

Products – continued

8	Epic	BASF	125 g/l	SC	12136
9	Epoxi	Goldengrass	125 g/l	SC	14292
10	Epoxi 125	Goldengrass	125 g/l	SC	14291
11	Epoxi 125 SC	Goldengrass	125 g/l	SC	14442
12	Ignite	BASF	83 g/l	EC	15205
13	Opus	BASF	125 g/l	SC	12057
14	Oracle	ChemSource	125 g/l	SC	14470
15	Oropa	AgChem Access	125 g/l	SC	13435
16	Rubric	Headland	125 g/l	SC	14118
17	Scholar	Becesane	125 g/l	SC	15380
18	Standon Epoxiconazole	Standon	125 g/l	SC	09517
19	Zenith	Becesane	125 g/l	SC	15649

Uses

- Brown rust in **durum wheat**, **rye** [2, 12]; **spring barley, winter barley, winter wheat** [1-19]; **spring rye, winter rye** [1, 3-11, 13-17, 19]; **spring wheat, triticale** [1-17, 19]
- Crown rust in **oats** [2, 12]
- Disease control in **all edible seed crops grown outdoors** (off-label), **all non-edible seed crops grown outdoors** (off-label), **grass seed crops** (off-label) [13]
- Eyespot in **winter barley** (reduction), **winter wheat** (reduction) [1, 3-11, 13-19]
- Foliar disease control in **durum wheat** (off-label) [6, 8]; **grass seed crops** (off-label) [4, 6, 18]
- Fusarium ear blight in **durum wheat** (good reduction), **spring wheat** (good reduction), **winter wheat** (good reduction) [2, 12]; **spring wheat, winter wheat** [1, 3-6, 8-11, 13-15, 19]; **winter wheat** (reduction) [18]
- Glume blotch in **durum wheat** [2, 12]; **spring wheat, triticale** [1-17, 19]; **winter wheat** [1-19]
- Net blotch in **spring barley, winter barley** [1, 3-11, 13-19]; **spring barley** (moderate control), **winter barley** (moderate control) [2, 12]
- Powdery mildew in **oats, rye** [2, 12]; **spring barley, winter barley** [1-19]; **spring oats, spring rye, spring wheat, triticale, winter oats, winter rye** [1, 3-11, 13-17, 19]; **winter wheat** [1, 3-11, 13-19]
- Ramularia leaf spots in **spring barley** (reduction), **winter barley** (reduction) [2, 12]
- Rhynchosporium in **rye** [2, 12]; **spring barley, winter barley** [1-19]; **spring rye, winter rye** [1, 3-11, 13-17, 19]
- Septoria leaf blotch in **durum wheat** [2, 12]; **spring wheat, triticale** [1-17, 19]; **winter wheat** [1-19]
- Sooty moulds in **durum wheat** (reduction) [12]; **spring wheat** (reduction) [1, 3-17, 19]; **winter wheat** (reduction) [1, 3-19]
- Yellow rust in **durum wheat**, **rye** [2, 12]; **spring barley, winter barley, winter wheat** [1-19]; **spring rye, winter rye** [1, 3-11, 13-17, 19]; **spring wheat, triticale** [1-17, 19]

Extension of Authorisation for Minor Use (EAMUs)

- **all edible seed crops grown outdoors** 20112607 [13]
- **all non-edible seed crops grown outdoors** 20112607 [13]
- **durum wheat** 20080981 [6], 20061053 [8]
- **grass seed crops** 20061055 [4], 20080981 [6], 20120113 [13], 20061050 [18]

Approval information

- Accepted by BBPA for use on malting barley (before ear emergence only)
- Epoxiconazole included in Annex I under EC Regulation 1107/2009

Efficacy guidance

- Apply at the start of foliar disease attack
- Optimum effect against eyespot achieved by spraying between leaf-sheath erect and second node detectable stages (GS 30-32)
- Best control of ear diseases of wheat obtained by treatment during ear emergence
- Mildew control improved by use of tank mixtures. See label for details
- For Septoria spray after third node detectable stage (GS 33) when weather favouring disease development has occurred
- Epoxiconazole is a DMI fungicide. Resistance to some DMI fungicides has been identified in Septoria leaf blotch which may seriously affect performance of some products. For further

FOR FULL CONDITIONS OF USE ALWAYS READ THE PRODUCT LABEL

advice contact a specialist advisor and visit the Fungicide Resistance Action Group (FRAG)-UK website

Restrictions

- Maximum total dose equivalent to two full dose treatments
- Product may cause damage to broad-leaved plant species
- Avoid spray drift onto neighbouring crops

Crop-specific information

- Latest use: up to and including flowering just complete (GS 69) in wheat, rye, triticale; up to and including emergence of ear just complete (GS 59) in barley, oats

Environmental safety

- Dangerous for the environment
- Very toxic to aquatic organisms
- LERAP Category B

Hazard classification and safety precautions

Hazard Harmful, Dangerous for the environment

Transport code 9

Packaging group III

UN Number 3082

Risk phrases R20 [2, 12]; R38 [2, 4, 7-19]; R40, R50, R53a, R62, R63 [1-19]

Operator protection A [1-19]; H [2, 12]; U05a, U20b

Environmental protection E15a, E16a [1-19]; E16b [18]; E38 [1-3, 6-17, 19]

Storage and disposal D01, D02, D05, D09a [1-19]; D10b [18]; D10c, D12a [1-17, 19]

Medical advice M05a [1-17, 19]; M05b [18]

189　epoxiconazole + fenpropimorph

A systemic, protectant and curative fungicide mixture for cereals
FRAC mode of action code: 3 + 5

See also fenpropimorph

Products

1	Eclipse	BASF	84:250 g/l	SE	11731
2	Opus Team	BASF	84:250 g/l	SE	11759

Uses

- Brown rust in **spring barley, spring rye, spring wheat, triticale, winter barley, winter rye, winter wheat**
- Eyespot in **winter barley** *(reduction)*, **winter wheat** *(reduction)*
- Foliar disease control in **durum wheat** *(off-label)*, **grass seed crops** *(off-label)*
- Fusarium ear blight in **spring wheat, winter wheat**
- Glume blotch in **spring wheat, triticale, winter wheat**
- Net blotch in **spring barley, winter barley**
- Powdery mildew in **spring barley, spring oats, spring rye, spring wheat, triticale, winter barley, winter oats, winter rye, winter wheat**
- Rhynchosporium in **spring barley, spring rye, winter barley, winter rye**
- Septoria leaf blotch in **spring wheat, triticale, winter wheat**
- Sooty moulds in **spring wheat** *(reduction)*, **winter wheat** *(reduction)*
- Yellow rust in **spring barley, spring rye, spring wheat, triticale, winter barley, winter rye, winter wheat**

Extension of Authorisation for Minor Use (EAMUs)

- **durum wheat** *20062644* [1], *20062645* [2]
- **grass seed crops** *20062644* [1], *20062645* [2]

Approval information

- Epoxiconazole and fenpropimorph included in Annex I under EC Regulation 1107/2009

SEE SECTION 3 FOR PRODUCTS ALSO REGISTERED

- Accepted by BBPA for use on malting barley

Efficacy guidance
- Apply at the start of foliar disease attack
- Optimum effect against eyespot achieved by spraying between leaf-sheath erect and second node detectable stages (GS 30-32)
- Best control of ear diseases obtained by treatment during ear emergence
- For Septoria spray after third node detectable stage (GS 33) when weather favouring disease development has occurred
- Epoxiconazole is a DMI fungicide. Resistance to some DMI fungicides has been identified in Septoria leaf blotch which may seriously affect performance of some products. For further advice contact a specialist advisor and visit the Fungicide Resistance Action Group (FRAG)-UK website

Restrictions
- Maximum total dose equivalent to two full dose treatments
- Product may cause damage to broad-leaved plant species

Crop-specific information
- Latest use: up to and including flowering just complete (GS 69) in wheat, rye, triticale; up to and including emergence of ear just complete (GS 59) in barley, oats

Environmental safety
- Dangerous for the environment
- Toxic to aquatic organisms
- Avoid spray drift onto neighbouring crops
- LERAP Category B

Hazard classification and safety precautions
Hazard Harmful, Dangerous for the environment
Transport code 9
Packaging group III
UN Number 3082
Risk phrases R20, R40, R43, R51, R53a, R62, R63
Operator protection A, H; U05a, U20b
Environmental protection E15a, E16a, E38
Storage and disposal D01, D02, D05, D09a, D10c, D12a
Medical advice M05a

190   epoxiconazole + fenpropimorph + kresoxim-methyl

A protectant, systemic and curative fungicide mixture for cereals
FRAC mode of action code: 3 + 5 + 11

See also fenpropimorph
*   kresoxim-methyl*

Products
Mantra	BASF	125:150:125 g/l	SE	11728

Uses
- Brown rust in **spring barley**, **spring rye**, **spring wheat**, **triticale**, **winter barley**, **winter rye**, **winter wheat**
- Eyespot in **spring oats** (reduction), **spring rye** (reduction), **triticale** (reduction), **winter barley** (reduction), **winter oats** (reduction), **winter rye** (reduction), **winter wheat** (reduction)
- Foliar disease control in **durum wheat** (off-label), **grass seed crops** (off-label)
- Fusarium ear blight in **spring wheat** (reduction), **winter wheat** (reduction)
- Glume blotch in **spring wheat**, **triticale**, **winter wheat**
- Net blotch in **spring barley**, **winter barley**
- Powdery mildew in **spring barley**, **spring oats**, **spring rye**, **triticale**, **winter barley**, **winter oats**, **winter rye**

FOR FULL CONDITIONS OF USE ALWAYS READ THE PRODUCT LABEL

- Rhynchosporium in **spring barley**, **spring rye**, **winter barley**, **winter rye**
- Septoria leaf blotch in **spring wheat**, **triticale**, **winter wheat**
- Sooty moulds in **spring wheat** *(reduction)*, **winter wheat** *(reduction)*
- Yellow rust in **spring barley**, **spring rye**, **spring wheat**, **triticale**, **winter barley**, **winter rye**, **winter wheat**

Extension of Authorisation for Minor Use (EAMUs)
- **durum wheat** *20061101*
- **grass seed crops** *20061101*

Approval information
- Kresoxim-methyl, epoxiconazole and fenpropimorph included in Annex I under EC Regulation 1107/2009
- Accepted by BBPA for use on malting barley (before ear emergence only)

Efficacy guidance
- For best results spray at the start of foliar disease attack and repeat if infection conditions persist
- Optimum effect against eyespot obtained by treatment between leaf sheaths erect and first node detectable stages (GS 30-32)
- For protection against ear diseases apply during ear emergence
- Epoxiconazole is a DMI fungicide. Resistance to some DMI fungicides has been identified in Septoria leaf blotch which may seriously affect performance of some products. For further advice contact a specialist advisor and visit the Fungicide Resistance Action Group (FRAG)-UK website
- Kresoxim-methyl is a member of the QoI cross resistance group. Product should be used preventatively and not relied on for its curative potential
- Use product as part of an Integrated Crop Management strategy incorporating other methods of control, including where appropriate other fungicides with a different mode of action. Do not apply more than two foliar applications of QoI containing products to any cereal crop
- There is a significant risk of widespread resistance occurring in *Septoria tritici* populations in UK. Failure to follow resistance management action may result in reduced levels of disease control
- Strains of barley powdery mildew resistant to QoI's are common in the UK

Restrictions
- Maximum total dose equivalent to two full dose treatments

Crop-specific information
- Latest use: mid flowering (GS 65) for wheat, rye, triticale; completion of ear emergence (GS 59) for barley, oats

Environmental safety
- Dangerous for the environment
- Very toxic to aquatic organisms
- Avoid spray drift onto neighbouring crops. Product may damage broad-leaved species
- LERAP Category B

Hazard classification and safety precautions
Hazard Harmful, Dangerous for the environment
Transport code 9
Packaging group III
UN Number 3082
Risk phrases R40, R50, R53a, R62, R63
Operator protection A, H; U05a, U14, U20b
Environmental protection E15a, E16a, E38
Storage and disposal D01, D02, D05, D09a, D10c, D12a
Medical advice M05a

SEE SECTION 3 FOR PRODUCTS ALSO REGISTERED

191   epoxiconazole + fenpropimorph + metrafenone

A broad spectrum fungicide mixture for cereals
FRAC mode of action code: 3 + 5 + U8

See also fenpropimorph
metrafenone

Products

1 Capalo	BASF	62.5:200:75 g/l	SE	13170
2 Stiletto	BASF	62.5:200:75 g/l	SE	14729

Uses

- Brown rust in **spring barley**, **spring rye**, **spring wheat**, **triticale**, **winter barley**, **winter rye**, **winter wheat**
- Crown rust in **spring oats**, **winter oats**
- Fusarium ear blight in **spring wheat** (reduction), **winter wheat** (reduction)
- Glume blotch in **spring wheat**, **triticale**, **winter wheat**
- Net blotch in **spring barley**, **winter barley**
- Powdery mildew in **spring barley**, **spring oats**, **spring rye**, **spring wheat**, **triticale**, **winter barley**, **winter oats**, **winter rye**, **winter wheat**
- Rhynchosporium in **spring barley**, **spring rye**, **winter barley**, **winter rye**
- Septoria leaf blotch in **spring wheat**, **triticale**, **winter wheat**
- Sooty moulds in **spring wheat** (reduction), **winter wheat** (reduction)
- Yellow rust in **spring barley**, **spring rye**, **spring wheat**, **triticale**, **winter barley**, **winter rye**, **winter wheat**

Approval information

- Epoxiconazole, fenpropimorph and metrafenone included in Annex I under EC Regulation 1107/2009

Efficacy guidance

- For best results spray at the start of foliar disease attack and repeat if infection conditions persist
- Optimum effect against eyespot obtained by treatment between leaf sheaths erect and first node detectable stages (GS 30-32)
- For protection against ear diseases apply during ear emergence
- Treatment for control of Septoria diseases should normally be made after third node detectable stage if the disease is present and favourable weather for disease development has occurred
- Epoxiconazole is a DMI fungicide. Resistance to some DMI fungicides has been identified in Septoria leaf blotch which may seriously affect performance of some products. For further advice contact a specialist advisor and visit the Fungicide Resistance Action Group (FRAG)-UK website
- Should be used as part of a resistance management strategy that includes mixtures or sequences effective against mildew and non-chemical methods

Restrictions

- Maximum number of treatments 2 per crop
- Do not apply more than two applications of any products containing metrafenone to any one crop

Crop-specific information

- Latest use: up to and including beginning of flowering (GS 61) for rye, wheat, triticale; up to and including ear emergence complete (GS 59) for barley, oats

Following crops guidance

- Cereals. oilseed rape, sugar beet, linseed, maize, clover, field beans, peas, turnips, carrots, cauliflowers, onions, lettuce or potatoes may be sown following normal harvest of a treated crop

Environmental safety

- Dangerous for the environment
- Very toxic to aquatic organisms
- Avoid spray drift onto neighbouring crops. Broad-leaved species may be damaged
- LERAP Category B

FOR FULL CONDITIONS OF USE ALWAYS READ THE PRODUCT LABEL

Hazard classification and safety precautions

Hazard Harmful, Dangerous for the environment
Transport code 9
Packaging group III
UN Number 3082
Risk phrases R38, R40, R43, R50, R53a, R62, R63
Operator protection A, H; U05a, U08, U14, U15, U19a, U20b
Environmental protection E16a, E34, E38
Storage and disposal D01, D02, D05, D08, D09a, D10c, D12a
Medical advice M03

192 epoxiconazole + fenpropimorph + pyraclostrobin

A protectant and systemic fungicide mixture for cereals
FRAC mode of action code: 3 + 5 + 11

See also fenpropimorph
pyraclostrobin

Products

1	Demeo	Dow	43:214:114 g/l	SE	15595
2	Diamant	BASF	43:214:114 g/l	SE	14149

Uses

- Brown rust in **spring barley**, **spring wheat**, **winter barley**, **winter wheat**
- Crown rust in **spring oats**, **winter oats**
- Fusarium ear blight in **spring wheat** (reduction), **winter wheat** (reduction)
- Glume blotch in **spring wheat**, **winter wheat**
- Net blotch in **spring barley**, **winter barley**
- Powdery mildew in **spring barley**, **spring oats**, **winter barley**, **winter oats**
- Rhynchosporium in **spring barley**, **winter barley**
- Septoria leaf blotch in **spring wheat**, **winter wheat**
- Yellow rust in **spring barley**, **spring wheat**, **winter barley**, **winter wheat**

Approval information

- Epoxiconazole, fenpropimorph and pyraclostrobin included in Annex I under EC Regulation 1107/2009
- Accepted by BBPA for use on malting barley (before ear emergence only)

Efficacy guidance

- For best results spray at the start of foliar disease attack and repeat if infection conditions persist
- For protection against ear diseases apply during ear emergence
- Epoxiconazole is a DMI fungicide. Resistance to some DMI fungicides has been identified in Septoria leaf blotch which may seriously affect performance of some products. For further advice contact a specialist advisor and visit the Fungicide Resistance Action Group (FRAG)-UK website
- Pyraclostrobin is a member of the QoI cross resistance group. Product should be used preventatively and not relied on for its curative potential
- Use product as part of an Integrated Crop Management strategy incorporating other methods of control, including where appropriate other fungicides with a different mode of action. Do not apply more than two foliar applications of QoI containing products to any cereal crop
- There is a significant risk of widespread resistance occurring in *Septoria tritici* populations in UK. Failure to follow resistance management action may result in reduced levels of disease control
- Strains of barley powdery mildew resistant to QoI's are common in the UK

Restrictions

- Maximum number of treatments 2 per crop

Crop-specific information

- Latest use: up to and including ear emergence just complete (GS 59) for barley, oats; up to and including flowering just complete (GS 69) for wheat

SEE SECTION 3 FOR PRODUCTS ALSO REGISTERED

SECTION 2

Environmental safety
- Dangerous for the environment
- Very toxic to aquatic organisms
- Dangerous to fish or other aquatic life. Do not contaminate surface waters or ditches with chemical or used container
- Avoid spray drift onto neighbouring crops. Product may damage broad-leaved species
- LERAP Category B

Hazard classification and safety precautions
    Hazard Harmful, Dangerous for the environment
    Transport code 9
    Packaging group III
    UN Number 3082
    Risk phrases R20, R22a, R38, R40, R50, R53a, R63
    Operator protection A; U05a, U14, U20b
    Environmental protection E16a, E34, E38
    Storage and disposal D01, D02, D05, D09a, D10c, D12a
    Medical advice M03, M05a

193   epoxiconazole + fluxapyroxad

An SDHI and triazole fungicide mixture for disease control in cereals
FRAC mode of action code: 3 + 7

See also epoxiconazole
       fluxapyroxad

Products

1	Adexar	BASF	62.5:62.5 g/l	EC	15474
2	Pan Epoflux	Pan Agriculture	62.5:62.5 g/l	EC	15889
3	Pexan	BASF	62.5:59.4 g/l	EC	15584

Uses
- Brown rust in *durum wheat*, *rye*, *spring barley*, *spring wheat*, *triticale*, *winter barley*, *winter wheat*
- Crown rust in *oats*
- Eyespot in *durum wheat* (moderate control), *oats* (moderate control), *rye* (moderate control), *triticale* (moderate control), *winter wheat* (moderate control)
- Fusarium ear blight in *durum wheat* (good reduction), *spring wheat* (good reduction), *winter wheat* (good reduction)
- Glume blotch in *durum wheat*, *spring wheat*, *triticale*, *winter wheat*
- Net blotch in *spring barley*, *winter barley*
- Powdery mildew in *durum wheat* (moderate control), *oats* (moderate control), *rye* (moderate control), *spring barley* (moderate control), *spring wheat* (moderate control), *triticale* (moderate control), *winter barley* (moderate control), *winter wheat* (moderate control)
- Ramularia leaf spots in *spring barley*, *winter barley*
- Rhynchosporium in *spring barley*, *winter barley*
- Septoria leaf blotch in *durum wheat*, *spring wheat*, *triticale*, *winter wheat*
- Sooty moulds in *durum wheat* (reduction), *spring wheat* (reduction), *winter wheat* (reduction)
- Tan spot in *durum wheat* (moderate control), *spring wheat* (moderate control), *winter wheat* (moderate control)
- Yellow rust in *durum wheat*, *rye*, *spring barley*, *spring wheat*, *triticale*, *winter barley*, *winter wheat*

Approval information
- Epoxiconazole and fluxapyroxad included in Annex I under EC Regulation 1107/2009

Restrictions
- Avoid drift on to neighbouring crops since it may damage broad-leaved species.

FOR FULL CONDITIONS OF USE ALWAYS READ THE PRODUCT LABEL

Crop-specific information
- After treating a cereal crop with Adexar, cabbage, carrot, clover, dwarf beans, field beans, lettuce, maize, oats, oilseed rape, onions, peas, potatoes, ryegrass, sugar beet, sunflower, winter barley and winter wheat may be sown as the following crop.

Environmental safety
- LERAP Category B

Hazard classification and safety precautions
Hazard Harmful, Dangerous for the environment
Transport code 9
Packaging group III
UN Number 3082
Risk phrases R22a, R40, R43, R50, R53a, R62, R63
Operator protection A [1-3]; H [3]; U05a, U14, U20b
Environmental protection E15b, E16a, E38
Storage and disposal D01, D02, D05, D10c, D12a, D19
Medical advice M05a

194 epoxiconazole + fluxapyroxad + pyraclostrobin

A triazole + SDHI + strobilurin fungicide mixture for disease control in cereals
FRAC mode of action code: 3 + 7 + 11

Products

1	Ceriax	BASF	41.6:41.6:66.6 g/l	EC	15639
2	Vortex	BASF	41.6:41.6:61 g/l	EC	15663

Uses
- Brown rust in **durum wheat**, **rye**, **spring barley**, **spring wheat**, **triticale**, **winter barley**, **winter wheat**
- Crown rust in **oats**
- Eyespot in **durum wheat** *(moderate control)*, **oats** *(moderate control)*, **rye** *(moderate control)*, **triticale** *(moderate control)*, **winter wheat** *(moderate control)*
- Fusarium ear blight in **durum wheat** *(good reduction)*, **spring wheat** *(good reduction)*, **winter wheat** *(good reduction)*
- Glume blotch in **durum wheat**, **spring wheat**, **triticale**, **winter wheat**
- Net blotch in **spring barley**, **winter barley**
- Powdery mildew in **durum wheat** *(moderate control)*, **oats** *(moderate control)*, **rye** *(moderate control)*, **spring barley** *(moderate control)*, **spring wheat** *(moderate control)*, **triticale** *(moderate control)*, **winter barley** *(moderate control)*, **winter wheat** *(moderate control)*
- Ramularia leaf spots in **spring barley**, **winter barley**
- Rhynchosporium in **rye**, **spring barley**, **winter barley**
- Septoria leaf blotch in **durum wheat**, **spring wheat**, **triticale**, **winter wheat**
- Sooty moulds in **durum wheat** *(reduction)*, **spring wheat** *(reduction)*, **winter wheat** *(reduction)*
- Tan spot in **durum wheat** *(moderate control)*, **spring wheat** *(moderate control)*, **winter wheat** *(moderate control)*
- Yellow rust in **durum wheat**, **rye**, **spring barley**, **spring wheat**, **triticale**, **winter barley**, **winter wheat**

Approval information
- Epoxiconazole, fluxapyroxad and pyraclostrobin included in Annex I under EC Regulation 1107/2009

Restrictions
- No more than two applications of any QoI fungicide may be applied to wheat, barley, oats, rye or triticale.
- May cause damage to broad-leaved plant species.

SEE SECTION 3 FOR PRODUCTS ALSO REGISTERED

Following crops guidance
- Cabbage, carrot, clover, dwarf French beans, lettuce, maize, oats, oilseed rape, onions, peas, potatoes, ryegrass, sugarbeet, sunflower, winter barley and winter wheat may be sown as a following crop but the effect on other crops has not been assessed.

Environmental safety
- LERAP Category B

Hazard classification and safety precautions
Hazard Harmful, Dangerous for the environment
Transport code 9
Packaging group III
UN Number 3082
Risk phrases R20, R22a, R40, R43, R50, R53a
Operator protection A, H; U05a, U14, U20b
Environmental protection E15b, E16a, E38
Storage and disposal D01, D02, D05, D07, D10c, D12a
Medical advice M05a

195   epoxiconazole + isopyrazam

A protectant and eradicant fungicide mixture for use in cereals
FRAC mode of action code: 3 + 7

Products

1 Keystone	Syngenta	99:125 g/l	SC	15493
2 Seguris	Syngenta	90:125 g/l	SC	15246

Uses
- Brown rust in *spring barley, triticale, winter barley, winter rye, winter wheat*
- Glume blotch in *winter wheat*
- Net blotch in *spring barley, winter barley*
- Ramularia leaf spots in *spring barley, winter barley*
- Rhynchosporium in *spring barley, winter barley, winter rye*
- Septoria leaf blotch in *triticale, winter wheat*
- Yellow rust in *winter wheat*

Approval information
- Epoxiconazole included in Annex I under EC Regulation 1107/2009 but isopyrazam not yet included
- Approved by BBPA for use on malting barley

Environmental safety
- LERAP Category B

Hazard classification and safety precautions
Hazard Harmful, Dangerous for the environment
Transport code 9
Packaging group III
UN Number 3082
Risk phrases R20, R40, R43, R50, R53a, R62, R63
Operator protection A, H; U09a, U14, U20b
Environmental protection E15b, E16a, E16b, E38
Storage and disposal D01, D02, D05, D09a, D10c, D11a, D12a
Medical advice M05a

196   epoxiconazole + kresoxim-methyl

A protectant, systemic and curative fungicide mixture for cereals
FRAC mode of action code: 3 + 11

See also kresoxim-methyl

Products

Landmark	BASF	125:125 g/l	SC	11730

Uses

- Brown rust in **spring barley**, **spring rye**, **spring wheat**, **triticale**, **winter barley**, **winter rye**, **winter wheat**
- Eyespot in **spring oats** *(reduction)*, **spring rye** *(reduction)*, **triticale** *(reduction)*, **winter barley** *(reduction)*, **winter oats** *(reduction)*, **winter rye** *(reduction)*, **winter wheat** *(reduction)*
- Foliar disease control in **durum wheat** *(off-label)*, **grass seed crops** *(off-label)*
- Fusarium ear blight in **spring wheat** *(reduction)*, **winter wheat** *(reduction)*
- Glume blotch in **spring wheat**, **triticale**, **winter wheat**
- Net blotch in **spring barley**, **winter barley**
- Powdery mildew in **spring barley**, **spring oats**, **spring rye**, **spring wheat**, **triticale**, **winter barley**, **winter oats**, **winter rye**, **winter wheat**
- Rhynchosporium in **spring barley**, **spring rye**, **winter barley**, **winter rye**
- Septoria leaf blotch in **spring wheat**, **triticale**, **winter wheat**
- Sooty moulds in **spring wheat** *(reduction)*, **winter wheat** *(reduction)*
- Yellow rust in **spring barley**, **spring rye**, **spring wheat**, **triticale**, **winter barley**, **winter rye**, **winter wheat**

Extension of Authorisation for Minor Use (EAMUs)

- **durum wheat** *20061084*
- **grass seed crops** *20061084*

Approval information

- Epoxiconazole and kresoxim-methyl included in Annex I under EC Regulation 1107/2009
- Accepted by BBPA for use on malting barley (before ear emergence only)

Efficacy guidance

- For best results spray at the start of foliar disease attack and repeat if infection conditions persist
- Optimum effect against eyespot obtained by treatment between leaf sheaths erect and first node detectable stages (GS 30-32)
- For protection against ear diseases apply during ear emergence
- Epoxiconazole is a DMI fungicide. Resistance to some DMI fungicides has been identified in Septoria leaf blotch which may seriously affect performance of some products. For further advice contact a specialist advisor and visit the Fungicide Resistance Action Group (FRAG)-UK website
- Kresoxim-methyl is a member of the QoI cross resistance group. Product should be used preventatively and not relied on for its curative potential
- Use product as part of an Integrated Crop Management strategy incorporating other methods of control, including where appropriate other fungicides with a different mode of action. Do not apply more than two foliar applications of QoI containing products to any cereal crop
- There is a significant risk of widespread resistance occurring in *Septoria tritici* populations in UK. Failure to follow resistance management action may result in reduced levels of disease control
- Strains of barley powdery mildew resistant to QoI's are common in the UK

Restrictions

- Maximum total dose equivalent to two full dose treatments

Crop-specific information

- Latest use: mid flowering (GS 65) for wheat, rye, triticale; completion of ear emergence (GS 59) for barley, oats

Environmental safety

- Dangerous for the environment

SEE SECTION 3 FOR PRODUCTS ALSO REGISTERED

- Very toxic to aquatic organisms
- Avoid spray drift onto neighbouring crops. Product may damage broad-leaved species
- LERAP Category B

Hazard classification and safety precautions

Hazard Harmful, Dangerous for the environment
Transport code 9
Packaging group III
UN Number 3082
Risk phrases R40, R50, R53a, R62, R63
Operator protection A; U05a, U14, U20b
Environmental protection E15a, E16a, E16b, E38
Storage and disposal D01, D02, D08, D09a, D10c, D12a
Medical advice M05a

197   epoxiconazole + kresoxim-methyl + pyraclostrobin

A protectant, systemic and curative fungicide mixture for cereals
FRAC mode of action code: 3 + 11 + 11

See also kresoxim-methyl
pyraclostrobin

Products

1	Covershield	BASF	50:67:133 g/l	SE	10900
2	Opponent	BASF	50:67:133 g/l	SE	10877

Uses

- Brown rust in *spring barley*, *spring wheat*, *winter barley*, *winter wheat*
- Crown rust in *spring oats*, *winter oats*
- Foliar disease control in *durum wheat* (off-label), *grass seed crops* (off-label), *spring rye* (off-label), *triticale* (off-label), *winter rye* (off-label)
- Glume blotch in *spring wheat*, *winter wheat*
- Net blotch in *spring barley*, *winter barley*
- Rhynchosporium in *spring barley*, *winter barley*
- Septoria leaf blotch in *spring wheat*, *winter wheat*
- Yellow rust in *spring barley*, *spring wheat*, *winter barley*, *winter wheat*

Extension of Authorisation for Minor Use (EAMUs)

- *durum wheat* 20063644 [1], *20060539* [2]
- *grass seed crops* 20063644 [1], *20060539* [2]
- *spring rye* 20063644 [1], *20060539* [2]
- *triticale* 20063644 [1], *20060539* [2]
- *winter rye* 20063644 [1], *20060539* [2]

Approval information

- Epoxiconazole, kresoxim-methyl and pyraclostrobin included in Annex I under EC Regulation 1107/2009
- Accepted by BBPA for use on malting barley (before ear emergence only)

Efficacy guidance

- For best results apply at the start of foliar disease attack
- Good reduction of Fusarium ear blight can be obtained from treatment during flowering
- Yield response may be obtained in the absence of visual disease symptoms
- Best results on Septoria leaf blotch when treated in the latent phase
- Epoxiconazole is a DMI fungicide. Resistance to some DMI fungicides has been identified in Septoria leaf blotch which may seriously affect performance of some products. For further advice contact a specialist advisor and visit the Fungicide Resistance Action Group (FRAG)-UK website
- Kresoxim-methyl and pyraclostrobin are members of the QoI cross resistance group. Product should be used preventatively and not relied on for its curative potential

FOR FULL CONDITIONS OF USE ALWAYS READ THE PRODUCT LABEL

- Use product as part of an Integrated Crop Management strategy incorporating other methods of control, including where appropriate other fungicides with a different mode of action. Do not apply more than two foliar applications of QoI containing products to any cereal crop
- There is a significant risk of widespread resistance occurring in *Septoria tritici* populations in UK. Failure to follow resistance management action may result in reduced levels of disease control

Restrictions
- Maximum total dose equivalent to two full dose treatments

Crop-specific information
- Latest use: before grain watery ripe (GS 71) for wheat; up to and including emergence of ear just complete (GS 59) for barley and oats

Environmental safety
- Dangerous for the environment
- Very toxic to aquatic organisms
- Dangerous to fish or other aquatic life. Do not contaminate surface waters or ditches with chemical or used container
- LERAP Category B

Hazard classification and safety precautions
Hazard Harmful, Dangerous for the environment
Transport code 9
Packaging group III
UN Number 3082
Risk phrases R20, R22a, R40, R50, R53a
Operator protection A; U05a, U14, U20b
Environmental protection E13b [2]; E15a [1]; E16a, E16b, E34, E38 [1, 2]
Storage and disposal D01, D02, D09a, D10c, D12a [1, 2]; D05 [2]
Medical advice M05a

198　epoxiconazole + metconazole

A triazole mixture for use in cereals
FRAC mode of action code: 3 + 3

See also metconazole

Products

1	Brutus	BASF	37.5;27.5 g/l	EC	14353
2	Icarus	BASF	37.5:25.0 g/l	EC	14471
3	Osiris P	BASF	56.25:41.25 g/l	EC	15627

Uses
- Brown rust in *durum wheat, spring wheat, triticale, winter wheat* [1-3]; *rye* [3]; *spring barley, spring rye, winter barley, winter rye* [1, 2]
- Eyespot in *durum wheat* (reduction), *triticale* (reduction), *winter wheat* (reduction) [1-3]; *rye* (reduction) [3]; *spring rye* (reduction), *winter barley* (reduction), *winter rye* (reduction) [1, 2]
- Fusarium ear blight in *durum wheat* (good reduction), *spring wheat* (good reduction), *triticale* (good reduction), *winter wheat* (good reduction) [1, 2]; *durum wheat* (moderate control), *spring wheat* (moderate control), *triticale* (moderate control), *winter wheat* (moderate control) [3]
- Glume blotch in *durum wheat, spring wheat, triticale, winter wheat* [1-3]
- Net blotch in *spring barley, winter barley* [1, 2]
- Powdery mildew in *durum wheat* (moderate control), *spring wheat* (moderate control), *triticale* (moderate control), *winter wheat* (moderate control) [1-3]; *rye* (moderate control) [3]; *spring barley* (moderate control), *winter barley* (moderate control) [1, 2]; *spring rye* (moderate control), *winter rye* (moderate control) [1]; *spring rye* (moderate reduction), *winter rye* (moderate reduction) [2]
- Ramularia leaf spots in *spring barley* [2]; *spring barley* (moderate control) [1]; *winter barley* (moderate control) [1, 2]

SEE SECTION 3 FOR PRODUCTS ALSO REGISTERED

- Rhynchosporium in *rye* (moderate control) [3]; *spring barley, winter barley* [1, 2]
- Septoria leaf blotch in *durum wheat, spring wheat, triticale, winter wheat* [1-3]
- Sooty moulds in *durum wheat* (good reduction), *spring wheat* (good reduction), *winter wheat* (good reduction) [1, 2]; *durum wheat* (moderate control), *spring wheat* (moderate control), *winter wheat* (moderate control) [3]
- Tan spot in *durum wheat* (moderate control), *spring wheat* (moderate control), *winter wheat* (moderate control) [1-3]
- Yellow rust in *durum wheat, spring wheat, triticale, winter wheat* [1-3]; *rye* [3]; *spring barley, spring rye, winter barley, winter rye* [1, 2]

Approval information
- Epoxiconazole and metconazole included in Annex I under EC Regulation 1107/2009

Following crops guidance
- After trreating a cereal crop, oilseed rape, cereals, sugar beet, linseed, maize, clover, beans, peas, carrots, potatoes, lettuce, cabbage, sunflower, ryegrass or onions may be sown as a following crop. The effect on other crops has not been evaluated.

Environmental safety
- LERAP Category B

Hazard classification and safety precautions
Hazard Harmful, Dangerous for the environment
Transport code 9
Packaging group III
UN Number 3082
Risk phrases R40, R43, R53a [1-3]; R50, R62 [3]; R51, R63 [1, 2]
Operator protection A, H; U05a, U14 [1, 2]; U20b [1-3]
Environmental protection E15b, E16a, E34 [1-3]; E38 [3]
Storage and disposal D01, D02, D09a, D10c [1-3]; D05, D12a, D19 [3]
Medical advice M05a

199   epoxiconazole + metrafenone

A broad spectrum fungicide mixture for cereals
FRAC mode of action code: 3 + U8

See also metrafenone

Products

1	Ceando	BASF	83:100 g/l	SC	13271
2	Cloister	BASF	83:100 g/l	SC	14358

Uses
- Brown rust in *spring barley, spring rye, spring wheat, triticale, winter barley, winter rye, winter wheat*
- Crown rust in *spring oats, winter oats*
- Fusarium ear blight in *spring wheat* (reduction), *winter wheat* (reduction)
- Glume blotch in *spring wheat, triticale, winter wheat*
- Net blotch in *spring barley, winter barley*
- Powdery mildew in *spring barley, spring oats, spring rye, spring wheat, triticale, winter barley, winter oats, winter rye, winter wheat*
- Rhynchosporium in *spring barley, spring rye, winter barley, winter rye*
- Septoria leaf blotch in *spring wheat, triticale, winter wheat*
- Sooty moulds in *spring wheat* (reduction), *winter wheat* (reduction)
- Yellow rust in *spring barley, spring rye, spring wheat, triticale, winter barley, winter rye, winter wheat*

Approval information
- Epoxiconazole and metrafenone included in Annex I under EC Regulation 1107/2009

FOR FULL CONDITIONS OF USE ALWAYS READ THE PRODUCT LABEL

Efficacy guidance
- For best results spray at the start of foliar disease attack and repeat if infection conditions persist
- Optimum effect against eyespot obtained by treatment between leaf sheaths erect and first node detectable stages (GS 30-32)
- For protection against ear diseases apply during ear emergence
- Treatment for control of Septoria diseases should normally be made after third node detectable stage if the disease is present and favourable weather for disease development has occurred
- Epoxiconazole is a DMI fungicide. Resistance to some DMI fungicides has been identified in Septoria leaf blotch which may seriously affect performance of some products. For further advice contact a specialist advisor and visit the Fungicide Resistance Action Group (FRAG)-UK website
- Should be used as part of a resistance management strategy that includes mixtures or sequences effective against mildew and non-chemical methods
- Where mildew well established at time of treatment product should be mixed with fenpropimorph. See label

Restrictions
- Maximum number of treatments 2 per crop
- Do not apply more than two applications of any products containing metrafenone to any one crop

Crop-specific information
- Latest use: up to and including beginning of flowering (GS 61) for rye, wheat, triticale; up to and including ear emergence complete (GS 59) for barley, oats

Following crops guidance
- Cereals. oilseed rape, sugar beet, linseed, maize, clover, field beans, peas, turnips, carrots, cauliflowers, onions, lettuce or potatoes may be sown following normal harvest of a treated crop

Environmental safety
- Dangerous for the environment
- Very toxic to aquatic organisms
- Avoid spray drift onto neighbouring crops. Broad-leaved species may be damaged
- LERAP Category B

Hazard classification and safety precautions
Hazard Harmful, Dangerous for the environment
UN Number N/C
Risk phrases R40, R50, R53a, R62, R63
Operator protection A, H; U05a, U20b
Environmental protection E15a, E16a, E34, E38
Storage and disposal D01, D02, D05, D08, D09a, D10c
Medical advice M03, M05a

200   epoxiconazole + prochloraz

A systemic, protectant and curative triazole fungicide mixture for use in cereals
FRAC mode of action code: 3 + 3

See also prochloraz

Products
Ennobe	BASF	62.5 + 225 g/l	DC	14340

Uses
- Brown rust in **spring barley**, **spring rye**, **spring wheat**, **triticale**, **winter barley**, **winter rye**, **winter wheat**
- Eyespot in **spring rye**, **triticale**, **winter barley**, **winter rye**, **winter wheat**
- Fusarium ear blight in **spring wheat** (good reduction), **triticale** (good reduction), **winter wheat** (good reduction)
- Glume blotch in **spring wheat**, **triticale**, **winter wheat**
- Net blotch in **spring barley**, **winter barley**

SEE SECTION 3 FOR PRODUCTS ALSO REGISTERED

SECTION 2

- Powdery mildew in **spring barley** *(moderate control)*, **spring rye** *(moderate control)*, **spring wheat** *(moderate control)*, **triticale** *(moderate control)*, **winter barley**, **winter rye** *(moderate control)*, **winter wheat** *(moderate control)*
- Rhynchosporium in **spring barley**, **spring rye**, **winter barley**, **winter rye**
- Septoria leaf blotch in **spring wheat**, **triticale**, **winter wheat**
- Sooty moulds in **spring wheat** *(good reduction)*, **winter wheat** *(good reduction)*
- Yellow rust in **spring barley**, **spring rye**, **spring wheat**, **triticale**, **winter barley**, **winter rye**, **winter wheat**

Approval information
- Epoxiconazole and prochloraz included in Annex 1 under EC Regulation 1107/2009

Efficacy guidance
- Use in mixture with Compatibility Agent CA 259 when tank-mixing with other products. Order of mixing is not important.

Environmental safety
- LERAP Category B

Hazard classification and safety precautions
Hazard Harmful, Dangerous for the environment
Transport code 9
Packaging group III
UN Number 3082
Risk phrases R20, R22a, R36, R38, R40, R50, R53a, R62, R63
Operator protection A, H; U05a
Environmental protection E15b, E16a, E38
Storage and disposal D01, D02, D08, D10c, D12a
Medical advice M05a

201 epoxiconazole + pyraclostrobin

A protectant, systemic and curative fungicide mixture for cereals
FRAC mode of action code: 3 + 11

See also pyraclostrobin

Products
1	Envoy	BASF	62.5:85 g/l	SE	12750
2	Gemstone	BASF	62.5:80	SE	13918
3	Opera	BASF	50:133 g/l	SE	12167

Uses
- Blight in **forage maize** *(off-label)*, **grain maize** *(off-label)* [3]
- Brown rust in **spring barley**, **spring wheat**, **winter barley**, **winter wheat** [1-3]
- Cercospora leaf spot in **sugar beet** [3]
- Crown rust in **spring oats**, **winter oats** [2, 3]; **spring oats** *(qualified minor use)*, **winter oats** *(qualified minor use)* [1]
- Disease control in **millet** *(off-label)* [3]
- Eyespot in **forage maize** *(off-label)*, **grain maize** *(off-label)* [3]
- Foliar disease control in **durum wheat** *(off-label)*, **fodder beet** *(off-label)*, **grass seed crops** *(off-label)*, **spring rye** *(off-label)*, **triticale** *(off-label)*, **winter rye** *(off-label)* [3]
- Fusarium ear blight in **spring wheat** *(apply during flowering)*, **winter wheat** *(apply during flowering)* [2]; **spring wheat** *(reduction)*, **winter wheat** *(reduction)* [1]
- Glume blotch in **spring wheat**, **winter wheat** [1-3]
- Net blotch in **spring barley**, **winter barley** [1-3]
- Powdery mildew in **sugar beet** [3]
- Ramularia leaf spots in **sugar beet** [3]
- Rhynchosporium in **spring barley**, **winter barley** [1-3]
- Rust in **forage maize** *(off-label)*, **grain maize** *(off-label)*, **sugar beet** [3]
- Septoria leaf blotch in **spring wheat**, **winter wheat** [1-3]

FOR FULL CONDITIONS OF USE ALWAYS READ THE PRODUCT LABEL

- Tan spot in **spring wheat**, **winter wheat** [1, 2]
- Yellow rust in **spring barley**, **spring wheat**, **winter barley**, **winter wheat** [1-3]

Extension of Authorisation for Minor Use (EAMUs)
- **durum wheat** *20062653* [3]
- **fodder beet** *20062654* [3]
- **forage maize** *20120869* [3]
- **grain maize** *20120869* [3]
- **grass seed crops** *20062653* [3]
- **millet** *20120678* [3]
- **spring rye** *20062653* [3]
- **triticale** *20062653* [3]
- **winter rye** *20062653* [3]

Approval information
- Epoxiconazole and pyraclostrobin included in Annex I under EC Regulation 1107/2009
- Accepted by BBPA for use on malting barley (before ear emergence only)

Efficacy guidance
- For best results apply at the start of foliar disease attack
- Good reduction of Fusarium ear blight can be obtained from treatment during flowering
- Yield response may be obtained in the absence of visual disease symptoms
- Epoxiconazole is a DMI fungicide. Resistance to some DMI fungicides has been identified in Septoria leaf blotch which may seriously affect performance of some products. For further advice contact a specialist advisor and visit the Fungicide Resistance Action Group (FRAG)-UK website
- Pyraclostrobin is a member of the QoI cross resistance group. Product should be used preventatively and not relied on for its curative potential
- Use product as part of an Integrated Crop Management strategy incorporating other methods of control, including where appropriate other fungicides with a different mode of action. Do not apply more than two foliar applications of QoI containing products to any cereal crop
- There is a significant risk of widespread resistance occurring in *Septoria tritici* populations in UK. Failure to follow resistance management action may result in reduced levels of disease control

Restrictions
- Maximum number of treatments 2 per crop
- Do not use on sugar beet crops grown for seed [3]

Crop-specific information
- Latest use: before grain watery ripe (GS 71) for wheat; up to and including emergence of ear just complete (GS 59) for barley and oats
- HI 6 wk for sugar beet [3]

Environmental safety
- Dangerous for the environment
- Very toxic to aquatic organisms
- LERAP Category B

Hazard classification and safety precautions
Hazard Harmful, Dangerous for the environment
Transport code 9
Packaging group III
UN Number 3082
Risk phrases R20, R22a, R38, R40, R43, R50, R53a [1-3]; R62, R63 [1, 2]
Operator protection A [1-3]; H [2]; U05a, U14 [1-3]; U20b [1, 3]; U20c [2]
Environmental protection E15a, E16a, E34, E38 [1-3]; E16b [3]
Storage and disposal D01, D02, D09a, D10c, D12a [1-3]; D05 [3]; D08 [2]
Medical advice M03, M05a

SEE SECTION 3 FOR PRODUCTS ALSO REGISTERED

202   esfenvalerate

A contact and ingested pyrethroid insecticide
IRAC mode of action code: 3

Products

1	Clayton Cajole	Clayton	25 g/l	EC	14995
2	Clayton Slalom	Clayton	25 g/l	EC	15054
3	Clayton Vindicate	Clayton	25 g/l	EC	15055
4	Greencrop Cajole Ultra	Greencrop	25 g/l	EC	12967
5	Standon Hounddog	Standon	25 g/l	EC	13201
6	Sumi-Alpha	Interfarm	25 g/l	EC	14023
7	Sven	Interfarm	25 g/l	EC	14859

Uses

- Aphids in **broccoli, brussels sprouts, cabbages, calabrese, cauliflowers, chinese cabbage, combining peas, edible podded peas, kale, kohlrabi, potatoes, spring field beans, vining peas, winter field beans** [1-3, 7]; **grass seed crops** *(off-label)* [6]; **grassland, managed amenity turf, ornamental plant production** [1-3]; **spring barley, spring wheat, winter barley, winter wheat** [1-7]
- Bibionids in **grass seed crops, grassland, managed amenity turf** [7]
- Caterpillars in **broccoli, brussels sprouts, cabbages, calabrese, cauliflowers, chinese cabbage, kale, kohlrabi** [7]
- Insect pests in **ornamental plant production, protected ornamentals** [7]
- Pea and bean weevil in **combining peas, edible podded peas, spring field beans, vining peas, winter field beans** [7]

Extension of Authorisation for Minor Use (EAMUs)

- *grass seed crops 20081266* [6]

Approval information

- Esfenvalerate included in Annex I under EC Regulation 1107/2009
- Accepted by BBPA for use on malting barley

Efficacy guidance

- For best reduction of spread of barley yellow dwarf virus winter sown crops at high risk (e.g. after grass or in areas with history of BYDV) should be treated when aphids first seen or by mid-Oct. Otherwise treat in late Oct-early Nov
- High risk winter sown crops will need a second treatment
- Spring sown crops should be treated from the 2-3 leaf stage if aphids are found colonising in the crop and a further application may be needed before the first node stage if they reinfest
- Product also recommended between onset of flowering and milky ripe stages (GS 61-73) for control of summer cereal aphids

Restrictions

- Maximum number of treatments 3 per crop of which 2 may be in autumn
- Do not use if another pyrethroid or dimethoate has been applied to crop after start of ear emergence (GS 51)
- Do not leave spray solution standing in spray tank

Crop-specific information

- Latest use: 31 Mar in yr of harvest (winter use dose); early milk stage for barley (summer use); late milk stage for wheat (summer use)

Environmental safety

- Dangerous for the environment
- Very toxic to aquatic organisms
- High risk to non-target insects or other arthropods. Do not spray within 6 m of the field boundary
- Flammable
- Store product in dark away from direct sunlight
- LERAP Category A

FOR FULL CONDITIONS OF USE ALWAYS READ THE PRODUCT LABEL

Hazard classification and safety precautions

Hazard Harmful, Flammable, Dangerous for the environment
Transport code 3 [1-3, 6, 7]; 6.1 [4, 5]
Packaging group II [1-3, 6, 7]; III [4, 5]
UN Number 1993 [1-3, 6, 7]; 3351 [4, 5]
Risk phrases R20, R22a, R41, R43, R50, R53a
Operator protection A, C, H; U04a, U05a, U08, U11, U14, U19a, U20b
Environmental protection E15a, E22a [1-4, 6, 7]; E15b, E22b [5]; E16c, E16d, E34 [1-7]; E38 [1-3, 6, 7]
Storage and disposal D01, D02, D09a [1-7]; D05 [4, 5]; D07, D10b [4]; D10c, D12a [1-3, 5-7]
Medical advice M03 [1-7]; M05b [5]

SECTION 2

203   ethofumesate

A benzofuran herbicide for grass weed control in various crops
HRAC mode of action code: N

See also chloridazon + ethofumesate
 desmedipham + ethofumesate + lenacil + phenmedipham
 desmedipham + ethofumesate + metamitron + phenmedipham
 desmedipham + ethofumesate + phenmedipham

Products

1	Barclay Keeper 500 SC	Barclay	500 g/l	SC	14016
2	Etho 500 SC	Goldengrass	500 g/l	SC	15256
3	Ethofol 500 SC	United Phosphorus	500 g/l	SC	15179
4	Ethosat 500	Makhteshim	500 g/l	SC	13050
5	Fumesate 500 SC	Hermoo	500 g/l	SC	14321
6	Oblix 500	AgriChem BV	500 g/l	SC	12349
7	Palamares	Agroquimicos	500 g/l	SC	15653

Uses
- Annual dicotyledons in *amenity grassland, grass seed crops* [1, 2, 4, 7]; *fodder beet, red beet, sugar beet* [1, 2, 4-7]; *managed amenity turf* (off-label) [4]; *mangels* [1, 2, 4, 6, 7]
- Annual grasses in *amenity grassland, grass seed crops* [1, 2]; *fodder beet, red beet, sugar beet* [1, 2, 5, 6]; *forest nurseries* (off-label) [4]; *mangels* [1, 2, 6]
- Annual meadow grass in *amenity grassland, fodder beet, grass seed crops, mangels, red beet, sugar beet* [4, 7]; *managed amenity turf* (off-label) [4]
- Blackgrass in *amenity grassland, fodder beet, grass seed crops, mangels, red beet, sugar beet* [4, 7]; *managed amenity turf* (off-label) [4]
- Cleavers in *fodder beet, sugar beet* [3]

Extension of Authorisation for Minor Use (EAMUs)
- *forest nurseries* 20082846 [4]
- *managed amenity turf* 20081026 [4]

Approval information
- Ethofumesate included in Annex I under EC Regulation 1107/2009

Efficacy guidance
- Most products may be applied pre- or post-emergence of crop or weeds but some restricted to pre-emergence or post-emergence use only. Check label
- Some products recommended for use only in mixtures. Check label
- Volunteer cereals not well controlled pre-emergence, weed grasses should be sprayed before fully tillered
- Grass crops may be sprayed during rain or when wet. Not recommended in very dry conditions or prolonged frost

Restrictions
- The maximum total dose must not exceed 1.0 kg ethofumesate per hectare in any three year period.

SEE SECTION 3 FOR PRODUCTS ALSO REGISTERED

- Do not use on Sands or Heavy soils, Very Light soils containing a high percentage of stones, or soils with more than 5-10% organic matter (percentage varies according to label)
- Do not use on swards reseeded without ploughing
- Clovers will be killed or severely checked
- Do not graze or cut grass for 14 d after, or roll less than 7 d before or after spraying

Crop-specific information
- Latest use: before crops meet across rows for beet crops and mangels; not specified for other crops
- Apply in beet crops in tank mixes with other pre- or post-emergence herbicides. Recommendations vary for different mixtures. See label for details
- Safe timing on beet crops varies with other ingredient of tank mix. See label for details
- In grass crops apply to moist soil as soon as possible after sowing or post-emergence when crop in active growth, normally mid-Oct to mid-Dec. See label for details
- May be used in Italian, hybrid and perennial ryegrass, timothy, cocksfoot, meadow fescue and tall fescue. Apply pre-emergence to autumn-sown leys, post-emergence after 2-3 leaf stage. See label for details

Following crops guidance
- In the event of failure of a treated crop only sugar beet, red beet, fodder beet and mangels may be redrilled within 3 mth of application
- Any crop may be sown 3 mth after application of mixtures in beet crops following ploughing, 5 mth after application in grass crops

Environmental safety
- Dangerous for the environment
- Toxic to aquatic organisms
- Do not empty into drains

Hazard classification and safety precautions
>**Hazard** Irritant [3]; Dangerous for the environment [1, 2, 4, 7]
>**Transport code** 9 [1-4, 7]
>**Packaging group** III [1-4, 7]
>**UN Number** 3082 [1-4, 7]; N/C [5, 6]
>**Risk phrases** R43 [3]; R51 [1, 2, 4, 7]; R52 [5, 6]; R53a [1, 2, 4-7]
>**Operator protection** A [3]; H [3, 6]; U05a [1, 2, 6]; U08 [1, 2, 4, 7]; U09a [3, 5]; U14 [3]; U19a [1-5, 7]; U20a [4, 7]; U20b [1-3, 5]
>**Environmental protection** E15a, E19b [4, 7]; E15b [1-3, 5, 6]; E23 [6]; E34 [1, 2, 4, 6, 7]; E38 [3, 5]
>**Storage and disposal** D01, D02 [3, 6]; D05 [3-5, 7]; D09a [1-7]; D10a [1, 2, 5, 6]; D10b [4, 7]; D10c [3]; D12a [1, 2, 4-7]
>**Medical advice** M05a [3]

204   ethofumesate + metamitron

A contact and residual herbicide mixture for beet crops
HRAC mode of action code: N + C1

See also metamitron

Products

1	Goltix Super	Makhteshim	150:350 g/l	SC	12216
2	Goltix Uno	Makhteshim	150:350 g/l	SC	12602
3	Oblix MT	AgriChem BV	150:350 g/l	SC	14510
4	Torero	Makhteshim	150:350 g/l	SC	11158

Uses
- Annual dicotyledons in **fodder beet**, **mangels**, **sugar beet**
- Annual meadow grass in **fodder beet**, **mangels**, **sugar beet**

FOR FULL CONDITIONS OF USE ALWAYS READ THE PRODUCT LABEL

Approval information
- Ethofumesate and metamitron included in Annex I under EC Regulation 1107/2009

Efficacy guidance
- Best results obtained from a series of treatments applied as an overall fine spray commencing when earliest germinating weeds are no larger than fully expanded cotyledon and the majority of the crop at fully expanded cotyledon
- Apply subsequent sprays as each new flush of weeds reaches early cotyledon and continue until weed emergence ceases
- Product may be used on all soil types but residual activity may be reduced on those with more than 5% organic matter

Restrictions
- Maximum total dose equivalent to three full dose treatments
- The maximum total dose must not exceed 1.0 kg ethofumesate per hectare in any three year period

Crop-specific information
- Latest use: before crop leaves meet between rows
- Crop tolerance may be reduced by stress caused by growing conditions, effects of pests, disease or other pesticides, nutrient deficiency etc

Following crops guidance
- Beet crops may be sown at any time after treatment. Any other crop may be sown after mouldboard ploughing to 15 cm and a minimum interval of 3 mth after treatment

Environmental safety
- Dangerous for the environment
- Very toxic to aquatic organisms
- Do not empty into drains

Hazard classification and safety precautions
Hazard Harmful [3]; Dangerous for the environment [1-4]
Transport code 9
Packaging group III
UN Number 3082
Risk phrases R22a [3]; R51, R53a [1-4]
Operator protection U05a, U19a, U20a [1-4]; U08, U14, U15 [1, 2, 4]
Environmental protection E13c, E34 [1, 2, 4]; E15a [3]; E19b [1-4]
Storage and disposal D01, D02, D09a, D10b, D12a [1-4]; D05 [1, 2, 4]
Medical advice M05a [3]

205   ethofumesate + metamitron + phenmedipham

A contact and residual herbicide mixture for sugar beet
HRAC mode of action code: N + C1 + C1

See also metamitron
phenmedipham

Products

Phemo	AgriChem BV	51:153:51 g/l	SE	14193

Uses
- Annual dicotyledons in *fodder beet*, *mangels*, *red beet*, *sugar beet*
- Annual meadow grass in *fodder beet*, *mangels*, *red beet*, *sugar beet*

Approval information
- Ethofumesate, metamitron and phenmedipham included in Annex I under EC Regulation 1107/2009

Efficacy guidance
- Best results obtained from a series of treatments applied as an overall fine spray commencing when earliest germinating weeds are no larger than fully expanded cotyledon
- Apply subsequent sprays as each new flush of weeds reaches early cotyledon and continue until weed emergence ceases (maximum 3 sprays)
- Product must be applied with Actipron
- Product may follow certain pre-emergence treatments and be used in conjunction with other post-emergence sprays - see label for details
- Where a pre-emergence band spray has been applied, the first treatment should be timed according to the size of the weeds in the untreated area between the rows

Restrictions
- Maximum number of treatments 3 per crop; maximum total dose 6 kg product per ha
- Product may be used on all soil types but residual activity may be reduced on those with more than 5% organic matter
- Crop tolerance may be reduced by stress caused by growing conditions, effects of pests, disease or other pesticides, nutrient deficiency etc
- Only beet crops should be sown within 4 mth of last treatment; winter cereals may be sown after this interval
- Any spring crop may be sown in the year following use
- Mould-board ploughing to 150 mm followed by thorough cultivation recommended before planting any crop.
- The maximum total dose must not exceed 1.0 kg ethofumesate per hectare in any three year period

Crop-specific information
- Latest use: before crop leaves meet between rows

Environmental safety
- Irritating to eyes
- Harmful to fish or other aquatic life. Do not contaminate surface waters or ditches with chemical or used container

Hazard classification and safety precautions
Hazard Dangerous for the environment
Transport code 9
Packaging group III
UN Number 3082
Risk phrases R51, R53a
Operator protection A; U05a, U20c
Environmental protection E15a, E34, E38
Storage and disposal D01, D02, D09a, D11a, D11b, D12a

206 ethofumesate + phenmedipham

A contact and residual herbicide for use in beet crops
HRAC mode of action code: N + C1

See also phenmedipham

Products

1 Brigida	Agroquimicos	200:200 g/l	SC	15658
2 Powertwin	Makhteshim	200:200 g/l	SC	14004
3 Teamforce	AgriChem BV	100:80 g/l	EC	14923
4 Teamforce SE	AgriChem BV	100:80 g/l	SE	15403
5 Twin	Makhteshim	94:97 g/l	EC	13998

Uses
- Annual dicotyledons in *fodder beet*, *mangels*, *sugar beet* [1-5]; *red beet* [4]
- Annual meadow grass in *fodder beet*, *mangels*, *sugar beet* [1-5]; *red beet* [4]
- Blackgrass in *fodder beet*, *mangels*, *sugar beet* [1-5]; *red beet* [4]

FOR FULL CONDITIONS OF USE ALWAYS READ THE PRODUCT LABEL

Approval information
- Ethofumesate and phenmedipham included in Annex I under EC Regulation 1107/2009

Efficacy guidance
- Best results achieved by repeat applications to cotyledon stage weeds. Larger susceptible weeds not killed by first treatment usually checked and controlled by second application
- Apply on all soil types at 5-10 d intervals
- On soils with more than 5-10% organic matter residual activity may be reduced

Restrictions
- Maximum number of treatments normally 3 per crop or maximum total dose equivalent to three full dose treatments, or less - see labels for details
- Must be used in conjunction with specified adjuvant. See label [2]
- Do not spray wet foliage or if rain imminent
- Spray in evening if daytime temperatures above 21°C expected
- Avoid or delay treatment if frost expected within 7 days
- Avoid or delay treating crops under stress from wind damage, manganese or lime deficiency, pest or disease attack etc.
- The maximum total dose must not exceed 1.0 kg ethofumesate per hectare in any three year period

Crop-specific information
- Latest use: before crop foliage meets in the rows
- Check from which recovery may not be complete may occur if treatment made during conditions of sharp diurnal temperature fluctuation

Following crops guidance
- Beet crops may be sown at any time after treatment. Any other crop may be sown after mouldboard ploughing to 15 cm and a minimum interval of 3 mth after treatment

Environmental safety
- Dangerous for the environment [2, 5]
- Toxic to aquatic organisms
- Do not empty into drains
- Extra care necessary to avoid drift because product is recommended for use as a fine spray

Hazard classification and safety precautions
Hazard Harmful [3-5]; Irritant [1, 2]; Dangerous for the environment [1-5]
Transport code 9
Packaging group III
UN Number 3082
Risk phrases R37, R40 [3-5]; R43, R51, R53a [1-5]
Operator protection A [1-5]; H [1-4]; U05a, U08, U19a, U20a [1-5]; U14 [1, 2]
Environmental protection E15a, E19b, E34
Storage and disposal D01, D02, D09a, D10b, D12a [1-5]; D05 [1, 2]
Medical advice M05a

207   ethylene

A gas used for fruit ripening and potato storage

Products
| Biofresh Safestore | Biofresh | 99.9% w/w | GA | 14579 |

Uses
- Sprout suppression in *potatoes*

Approval information
- Ethylene is included in Annex 1 under EC Regulation 1107/2009

Efficacy guidance
- For use in stored fruit or potatoes after harvest

SEE SECTION 3 FOR PRODUCTS ALSO REGISTERED

SECTION 2

Restrictions
- Handling and release of ethylene must only be undertaken by operators suitably trained and competent to carry out the work
- Operators must vacate treated areas immediately after ethylene introduction
- Unprotected persons must be excluded from the treated areas until atmospheres have been thoroughly ventilated for 15 minutes minimum before re-entry
- Ambient atmospheric ethylene concentration must not exceed 1000 ppm. Suitable self-contained breathing apparatus must be worn in atmospheres containing ethylene in excess of 1000 ppm
- A minimum 3 d post treatment period is required before removal of treated crop from storage
- Ethylene treatment must only be undertaken in fully enclosed storage areas that are air tight with appropriate air circulation and venting facilities

208   etoxazole

A mite growth inhibitor
IRAC mode of action code: 10B

Products

1 Borneo	Interfarm	110 g/l	SC	13919
2 Clayton Java	Clayton	110 g/l	SC	15155

Uses
- Mites in **protected aubergines**, **protected tomatoes**
- Spider mites in **protected ornamentals** *(off-label)*, **protected strawberries** *(off-label)*
- Two-spotted spider mite in **protected ornamentals** *(off-label)*, **protected strawberries** *(off-label)*

Extension of Authorisation for Minor Use (EAMUs)
- **protected ornamentals** *20081216* [1], *20111544* [2]
- **protected strawberries** *20092400* [1], *20111545* [2]

Approval information
- Etoxazole included in Annex I under EC Regulation 1107/2009

Hazard classification and safety precautions
Hazard Dangerous for the environment
Transport code 9
Packaging group III
UN Number 3082
Risk phrases R50, R53a
Operator protection A; U05a, U20b
Environmental protection E15a, E34
Storage and disposal D01, D02, D09b, D11a, D12b

209   famoxadone

A strobilurin fungicide available only in mixtures
FRAC mode of action code: 11

See also cymoxanil + famoxadone

210   famoxadone + flusilazole

A contact, preventative and curative fungicide mixture for cereals and oilseed rape
FRAC mode of action code: 11 + 3

See also flusilazole

Products

1 Caynil	DuPont	100:106.7 g/l	EC	14641
2 Charisma	DuPont	100:106.7 g/l	EC	10415
3 Medley	DuPont	100:106.7 g/l	EC	10933

Uses

- Brown rust in **spring barley, winter barley, winter wheat** [2, 3]
- Foliar disease control in **durum wheat** *(off-label)*, **grass seed crops** *(off-label)*, **triticale** *(off-label)* [2, 3]; **spring rye** *(off-label)*, **winter rye** *(off-label)* [2]
- Glume blotch in **winter wheat** [2, 3]
- Light leaf spot in **spring oilseed rape, winter oilseed rape** [1-3]
- Net blotch in **spring barley, winter barley** [2, 3]
- Phoma in **spring oilseed rape** *(suppression)*, **winter oilseed rape** *(suppression)* [1-3]
- Rhynchosporium in **spring barley, winter barley** [2, 3]
- Septoria leaf blotch in **winter wheat** [2, 3]
- Yellow rust in **spring barley, winter barley, winter wheat** [2, 3]

Extension of Authorisation for Minor Use (EAMUs)

- **durum wheat** *20060505* [2], *20060506* [3]
- **grass seed crops** *20060505* [2], *20060506* [3]
- **spring rye** *20060505* [2]
- **triticale** *20060505* [2], *20060506* [3]
- **winter rye** *20060505* [2]

Approval information

- Famoxadone included in Annex I under EC Regulation 1107/2009
- Flusilazole has been reinstated in Annex 1 under Directive 91/414 following an appeal
- Accepted by BBPA for use on malting barley

Efficacy guidance

- Best results obtained from application at early stage of disease development before infection spreads to new growth
- Famoxadone is a member of the QoI cross resistance group. Product should be used preventatively and not relied on for its curative potential
- Use product as part of an Integrated Crop Management strategy incorporating other methods of control, including where appropriate other fungicides with a different mode of action. Do not apply more than two foliar applications of QoI containing products to any cereal crop
- There is a significant risk of widespread resistance occurring in *Septoria tritici* populations in UK. Failure to follow resistance management action may result in reduced levels of disease control
- Flusilazole is a DMI fungicide. Resistance to some DMI fungicides has been identified in Septoria leaf blotch which may seriously affect performance of some products. For further advice contact a specialist advisor and visit the Fungicide Resistance Action Group (FRAG)-UK website

Restrictions

- Maximum number of treatments 2 per crop for wheat and barley; 1 per crop for oilseed rape
- Do not apply to crops under stress
- Do not apply during frosty weather

Crop-specific information

- Latest use: before flowering (GS 60) for winter wheat; before quarter ear emergence (GS 53) for barley; up to decline of flowering (GS 67) for oilseed rape

Environmental safety

- Dangerous for the environment

SEE SECTION 3 FOR PRODUCTS ALSO REGISTERED

- Very toxic to aquatic organisms
- Dangerous to fish or other aquatic life. Do not contaminate surface waters or ditches with chemical or used container
- LERAP Category B

Hazard classification and safety precautions
Hazard Toxic, Dangerous for the environment
Transport code 9
Packaging group III
UN Number 3082
Risk phrases R36, R40, R53a, R61 [1-3]; R50 [1]; R51 [2, 3]
Operator protection A, C [1-3]; D, H, M [1]; U05a, U19a, U20b
Environmental protection E13b, E16a, E16b, E38
Storage and disposal D01, D02, D05, D09a, D10b, D12a
Medical advice M03, M04a

211   fatty acids

A soap concentrate insecticide and acaricide

Products

Savona	Koppert	49% w/w	SL	06057

Uses
- Aphanomyces cochlioides in *choi sum (off-label)*
- Aphids in *asparagus (off-label)*, *aubergines (off-label)*, *blackberries (off-label)*, *blackcurrants (off-label)*, *blueberries (off-label)*, *broad beans*, *broccoli (off-label)*, *brussels sprouts*, *brussels sprouts (off-label)*, *bulb onions (off-label)*, *cabbages*, *cabbages (off-label)*, *calabrese (off-label)*, *carrots (off-label)*, *cauliflowers (off-label)*, *cayenne pepper (off-label)*, *celeriac (off-label)*, *celery (outdoor) (off-label)*, *chicory (off-label)*, *chicory root (off-label)*, *chinese cabbage (off-label)*, *collards (off-label)*, *courgettes (off-label)*, *cress (off-label)*, *cucumbers*, *endives (off-label)*, *fennel (off-label)*, *fodder rape (off-label)*, *frise (off-label)*, *fruit trees*, *garlic (off-label)*, *globe artichoke (off-label)*, *gooseberries (off-label)*, *herbs (see appendix 6) (off-label)*, *horseradish (off-label)*, *jerusalem artichokes (off-label)*, *kale (off-label)*, *kohlrabi (off-label)*, *komatsuna (off-label)*, *lamb's lettuce (off-label)*, *leaf brassicas (off-label)*, *leeks (off-label)*, *lentils (off-label)*, *lettuce*, *loganberries (off-label)*, *lupins (off-label)*, *marrows (off-label)*, *melons (off-label)*, *mooli (off-label)*, *orache (off-label)*, *pak choi (off-label)*, *parsley root (off-label)*, *parsnips (off-label)*, *peas*, *peppers*, *protected herbs (see appendix 6) (off-label)*, *protected tomatoes*, *pumpkins*, *radicchio (off-label)*, *radishes (off-label)*, *raspberries (off-label)*, *red beet (off-label)*, *redcurrants (off-label)*, *rhubarb (off-label)*, *ribes hybrids (off-label)*, *rubus hybrids (off-label)*, *runner beans*, *salad onions (off-label)*, *salsify (off-label)*, *scarole (off-label)*, *seakale (off-label)*, *shallots (off-label)*, *spinach (off-label)*, *spinach beet (off-label)*, *squashes (off-label)*, *strawberries (off-label)*, *swedes (off-label)*, *tomatoes (outdoor)*, *turnips (off-label)*, *watermelons (off-label)*, *whitecurrants (off-label)*, *woody ornamentals*
- Mealybugs in *broad beans*, *brussels sprouts*, *cabbages*, *cucumbers*, *fruit trees*, *lettuce*, *peas*, *peppers*, *protected tomatoes*, *pumpkins*, *runner beans*, *tomatoes (outdoor)*, *woody ornamentals*
- Scale insects in *broad beans*, *brussels sprouts*, *cabbages*, *cucumbers*, *fruit trees*, *lettuce*, *peas*, *peppers*, *protected tomatoes*, *pumpkins*, *runner beans*, *tomatoes (outdoor)*, *woody ornamentals*
- Spider mites in *broad beans*, *brussels sprouts*, *cabbages*, *cucumbers*, *fruit trees*, *lettuce*, *peas*, *peppers*, *protected tomatoes*, *pumpkins*, *runner beans*, *tomatoes (outdoor)*, *woody ornamentals*
- Thrips in *asparagus (off-label)*, *aubergines (off-label)*, *blackberries (off-label)*, *blackcurrants (off-label)*, *broccoli (off-label)*, *brussels sprouts (off-label)*, *bulb onions (off-label)*, *cabbages (off-label)*, *calabrese (off-label)*, *carrots (off-label)*, *cauliflowers (off-label)*, *cayenne pepper (off-label)*, *celeriac (off-label)*, *celery (outdoor) (off-label)*, *chicory (off-label)*, *chicory root (off-label)*, *chinese cabbage (off-label)*, *choi sum (off-label)*, *collards (off-label)*, *courgettes (off-label)*, *cress (off-label)*, *endives (off-label)*, *fennel (off-label)*, *fodder rape (off-label)*, *frise (off-label)*,

garlic (off-label), **globe artichoke** (off-label), **gooseberries** (off-label), **herbs (see appendix 6)** (off-label), **horseradish** (off-label), **jerusalem artichokes** (off-label), **kale** (off-label), **kohlrabi** (off-label), **komatsuna** (off-label), **lamb's lettuce** (off-label), **leaf brassicas** (off-label), **leeks** (off-label), **lentils** (off-label), **loganberries** (off-label), **lupins** (off-label), **marrows** (off-label), **melons** (off-label), **mooli** (off-label), **orache** (off-label), **pak choi** (off-label), **parsley root** (off-label), **parsnips** (off-label), **protected herbs (see appendix 6)** (off-label), **radicchio** (off-label), **radishes** (off-label), **raspberries** (off-label), **red beet** (off-label), **redcurrants** (off-label), **rhubarb** (off-label), **ribes hybrids** (off-label), **rubus hybrids** (off-label), **salad onions** (off-label), **salsify** (off-label), **scarole** (off-label), **seakale** (off-label), **shallots** (off-label), **spinach** (off-label), **spinach beet** (off-label), **squashes** (off-label), **swedes** (off-label), **turnips** (off-label), **watermelons** (off-label)
- Whitefly in **broad beans, brussels sprouts, cabbages, cucumbers, fruit trees, lettuce, peas, peppers, protected tomatoes, pumpkins, runner beans, tomatoes (outdoor), woody ornamentals**

Extension of Authorisation for Minor Use (EAMUs)
- **asparagus** *20100433*
- **aubergines** *20100433*
- **blackberries** *20100433*
- **blackcurrants** *20100433*
- **blueberries** *20102044*
- **broccoli** *20100433*
- **brussels sprouts** *20100433*
- **bulb onions** *20100433*
- **cabbages** *20100433*
- **calabrese** *20100433*
- **carrots** *20100433*
- **cauliflowers** *20100433*
- **cayenne pepper** *20100433*
- **celeriac** *20100433*
- **celery (outdoor)** *20100433*
- **chicory** *20100433*
- **chicory root** *20100433*
- **chinese cabbage** *20100433*
- **choi sum** *20100433*
- **collards** *20100433*
- **courgettes** *20100433*
- **cress** *20100433*
- **endives** *20100433*
- **fennel** *20100433*
- **fodder rape** *20100433*
- **frise** *20100433*
- **garlic** *20100433*
- **globe artichoke** *20100433*
- **gooseberries** *20100433*
- **herbs (see appendix 6)** *20100433*
- **horseradish** *20100433*
- **jerusalem artichokes** *20100433*
- **kale** *20100433*
- **kohlrabi** *20100433*
- **komatsuna** *20100433*
- **lamb's lettuce** *20100433*
- **leaf brassicas** *20100433*
- **leeks** *20100433*
- **lentils** *20100433*
- **loganberries** *20100433*
- **lupins** *20100433*
- **marrows** *20100433*
- **melons** *20100433*
- **mooli** *20100433*

- *orache* 20100433
- *pak choi* 20100433
- *parsley root* 20100433
- *parsnips* 20100433
- *protected herbs (see appendix 6)* 20100433
- *radicchio* 20100433
- *radishes* 20100433
- *raspberries* 20100433
- *red beet* 20100433
- *redcurrants* 20100433
- *rhubarb* 20100433
- *ribes hybrids* 20100433
- *rubus hybrids* 20100433
- *salad onions* 20100433
- *salsify* 20100433
- *scarole* 20100433
- *seakale* 20100433
- *shallots* 20100433
- *spinach* 20100433
- *spinach beet* 20100433
- *squashes* 20100433
- *strawberries* 20102044
- *swedes* 20100433
- *turnips* 20100433
- *watermelons* 20100433
- *whitecurrants* 20102044

Approval information
- Fatty acids C7 - C20 are included in Annex 1 under EC Regulation 1107/2009

Efficacy guidance
- Use only soft or rain water for diluting spray
- Pests must be sprayed directly to achieve any control. Spray all plant parts thoroughly to run off
- For glasshouse use apply when insects first seen and repeat as necessary. For scale insects apply several applications at weekly intervals after egg hatch
- To control whitefly spray when required and use biological control after 12 h

Restrictions
- Do not use on new transplants, newly rooted cuttings or plants under stress
- Do not use on specified susceptible shrubs. See label for details

Crop-specific information
- HI zero

Environmental safety
- Harmful to fish or other aquatic life. Do not contaminate surface waters or ditches with chemical or used container

Hazard classification and safety precautions
Operator protection U20c
Environmental protection E13c
Storage and disposal D05, D09a, D10a

212   fenamidone

A strobilurin fungicide available only in mixtures
FRAC mode of action code: 11

See also fenamidone + fosetyl-aluminium
fenamidone + mancozeb
fenamidone + propamocarb hydrochloride

FOR FULL CONDITIONS OF USE ALWAYS READ THE PRODUCT LABEL

213　fenamidone + fosetyl-aluminium

A strobilurin fungicide mixture
FRAC mode of action code: 11 + 33

See also fenamidone
*　　　　fosetyl-aluminium*

Products

Fenomenal	Bayer CropScience	60:600 g/l	WG	15494

Uses
- Collar rot in **strawberries** *(moderate control from foliar spray)*
- Crown rot in **strawberries** *(moderate control from foliar spray)*
- Phytophthora in **ornamental plant production**, **strawberries**
- Pythium in **ornamental plant production**, **strawberries**
- Red core in **strawberries**

Approval information
- Fenamidone and fosetyl-aluminium included in Annex I under EC Regulation 1107/2009

Efficacy guidance
- Best used preventatively. Do not rely on curative activity
- Hose down dirty plants before dipping
- Due to variations in growing conditions and the range of varieties when treating ornamentals, it is advisable to check the crop tolerance on a small sample before treating the whole crop
- Fenamidone is a member of the QoI cross resistance group. To minimise the likelihood of development of resistance these products should be used in a planned Resistance Management strategy. In addition before application consult and adhere to the latest FRAG-UK resistance guidance on application of QoI fungicides to potatoes. See Section 5 for more information

Crop-specific information
- There is only limited crop safety data on tulip bulbs
- Container-grown Viburnum, Pieris, Gaultheria, Chamaecyparis and Erica have been successfully treated in trials

Following crops guidance
- No limitations on choice of following crops

Environmental safety
- LERAP Category B

Hazard classification and safety precautions
　Hazard Irritant, Dangerous for the environment
　Transport code 9
　Packaging group III
　UN Number 3077
　Risk phrases R36, R50, R53a
　Operator protection M; U05a, U11, U20b
　Environmental protection E15b, E16a, E22c, E34, E38
　Storage and disposal D01, D02, D09a, D11a, D12a

214　fenamidone + propamocarb hydrochloride

A systemic fungicide mixture for potatoes
FRAC mode of action code: 11 + 28

See also propamocarb hydrochloride

Products

1	Consento	Bayer CropScience	75:375 g/l	SC	15150
2	Prompto	Bayer CropScience	75:375 g/l	SC	15156

SEE SECTION 3 FOR PRODUCTS ALSO REGISTERED

SECTION 2

Uses
- Blight in **potatoes**
- Early blight in **potatoes**

Approval information
- Fenamidone and propamocarb hydrochloride included in Annex I under EC Regulation 1107/2009
- Approval expiry 30 Sep 2013 [1, 2]

Efficacy guidance
- Apply as soon as there is risk of blight infection or immediately after a blight warning
- Fenamidone is a member of the QoI cross resistance group. To minimise the likelihood of development of resistance these products should be used in a planned Resistance Management strategy. In addition before application consult and adhere to the latest FRAG-UK resistance guidance on application of QoI fungicides to potatoes. See Section 5 for more information

Restrictions
- Maximum number of treatments (including any other QoI containing fungicide) 6 per yr but no more than three should be applied consecutively. Use in alternation with fungicides from a different cross-resistance group
- Observe a 7 d interval between treatments
- Use only as a protective treatment. Do not use when blight has become readily visible (1% leaf area destroyed)

Crop-specific information
- HI 7 d for potatoes
- All varieties of early, main and seed crop potatoes may be sprayed [2]

Environmental safety
- Dangerous for the environment
- Very toxic to aquatic organisms
- Dangerous to fish or other aquatic life. Do not contaminate surface waters or ditches with chemical or used container
- LERAP Category B

Hazard classification and safety precautions
Hazard Irritant, Dangerous for the environment
Transport code 9
Packaging group III
UN Number 3082
Risk phrases R36, R50, R53a
Operator protection A, C; U05a, U08, U11 [1, 2]; U19a, U20b, U20c [2]; U20a [1]
Environmental protection E13b [1]; E15a [2]; E16a, E38 [1, 2]
Storage and disposal D01, D02, D09a, D10b, D12a [1, 2]; D05 [1]

215   fenbuconazole

A systemic protectant and curative triazole fungicide for top fruit and grapevines
FRAC mode of action code: 3

Products
1 Agrovista Radni	Agrovista	50 g/l	EW	14404
2 Indar 5 EW	Landseer	50 g/l	EW	09518

Uses
- Blossom wilt in **cherries** *(off-label)*, **mirabelles** *(off-label)*, **plums** *(off-label)* [2]
- Brown rot in **cherries** *(off-label)*, **mirabelles** *(off-label)*, **plums** *(off-label)* [2]
- Powdery mildew in **apples** *(reduction)*, **pears** *(reduction)* [1, 2]; **wine grapes** *(off-label)* [2]
- Scab in **apples**, **pears** [1, 2]

Extension of Authorisation for Minor Use (EAMUs)
- **cherries** *20031372* [2]
- **mirabelles** *20031372* [2]

FOR FULL CONDITIONS OF USE ALWAYS READ THE PRODUCT LABEL

- **plums** *20031372* [2]
- **wine grapes** *20032081* [2]

Approval information
- Fenbuconazole included in Annex I under EC Regulation 1107/2009

Efficacy guidance
- Most effective when used as part of a routine preventative programme from bud burst to onset of petal fall
- After petal fall, tank mix with other protectant fungicides to enhance scab control
- See label for recommended spray intervals. In periods of rapid growth or high disease pressure, a 7 d interval should be used

Restrictions
- Maximum total dose equivalent to ten full doses per yr on apples, pears
- Consult processors before using on pears for processing
- Do not harvest for human or animal consumption for at least 4 wk after last application

Crop-specific information
- HI 28 d for apples, pears; 21 d for grapevines; 3 d for cherries, mirabelles, plums
- Safe to use on all main commercial varieties of apples and pears in UK

Environmental safety
- Dangerous for the environment
- Toxic to aquatic organisms

Hazard classification and safety precautions
Hazard Irritant, Dangerous for the environment
Transport code 3 [2]; 9 [1]
Packaging group III
UN Number 1993 [2]; 3082 [1]
Risk phrases R36, R51, R53a [1, 2]; R41 [2]
Operator protection A, C; U05a, U08, U11 [1, 2]; U20a [2]; U20b [1]
Environmental protection E15a, E38
Consumer protection C02a [1] (4 w); C02a [2] (4 wk)
Storage and disposal D01, D02, D11a, D12a [1, 2]; D05 [2]

216   fenhexamid

A protectant hydroxyanilide fungicide for soft fruit and a range of horticultural crops
FRAC mode of action code: 17

Products
1	Agrovista Fenamid	Agrovista	50% w/w	WG	13733
2	Teldor	Bayer CropScience	50% w/w	WG	11229

Uses
- Botrytis in **bilberries** *(off-label)*, **blackberries**, **blackcurrants**, **blueberries** *(off-label)*, **cherries** *(off-label)*, **gooseberries**, **herbs (see appendix 6)** *(off-label)*, **loganberries**, **plums** *(off-label)*, **protected aubergines** *(off-label)*, **protected chilli peppers** *(off-label)*, **protected courgettes** *(off-label)*, **protected cucumbers** *(off-label)*, **protected gherkins** *(off-label)*, **protected peppers** *(off-label)*, **protected squashes** *(off-label)*, **protected tomatoes** *(off-label)*, **raspberries**, **redcurrants**, **rubus hybrids**, **strawberries**, **tomatoes (outdoor)** *(off-label)*, **whitecurrants** [1, 2]; **frise** *(off-label)*, **hops** *(off-label)*, **lamb's lettuce** *(off-label)*, **leaf brassicas** *(off-label - baby leaf production)*, **protected forest nurseries** *(off-label)*, **rocket** *(off-label)*, **scarole** *(off-label)*, **soft fruit** *(off-label)*, **wine grapes** [2]; **outdoor grapes** [1]
- Grey mould in **protected aubergines** *(off-label)*, **protected chilli peppers** *(off-label)*, **protected courgettes** *(off-label)*, **protected cucumbers** *(off-label)*, **protected gherkins** *(off-label)*, **protected peppers** *(off-label)*, **protected squashes** *(off-label)*, **protected tomatoes** *(off-label)* [2]

SEE SECTION 3 FOR PRODUCTS ALSO REGISTERED

Extension of Authorisation for Minor Use (EAMUs)

- *bilberries* *20080485* [1], *20061290* [2]
- *blueberries* *20080485* [1], *20061290* [2]
- *cherries* *20080479* [1], *20031866* [2]
- *frise* *20082062* [2]
- *herbs (see appendix 6)* *20082056* [1], *20082062* [2]
- *hops* *20082926* [2]
- *lamb's lettuce* *20082062* [2]
- *leaf brassicas* *(baby leaf production)* *20082062* [2]
- *plums* *20080479* [1], *20031866* [2]
- *protected aubergines* *20080482* [1], *20042087* [2]
- *protected chilli peppers* *20080481* [1], *20042086* [2]
- *protected courgettes* *20080480* [1], *20042085* [2]
- *protected cucumbers* *20080480* [1], *20042085* [2]
- *protected forest nurseries* *20082926* [2]
- *protected gherkins* *20080480* [1], *20042085* [2]
- *protected peppers* *20080481* [1], *20042086* [2]
- *protected squashes* *20080480* [1], *20042085* [2]
- *protected tomatoes* *20080482* [1], *20042087* [2]
- *rocket* *20082062* [2]
- *scarole* *20082062* [2]
- *soft fruit* *20082926* [2]
- *tomatoes (outdoor)* *20080483* [1], *20042399* [2]

Approval information

- Fenhexamid included in Annex I under EC Regulation 1107/2009

Efficacy guidance

- Use as part of a programme of sprays throughout the flowering period to achieve effective control of Botrytis
- To minimise possibility of development of resistance, no more than two sprays of the product may be applied consecutively. Other fungicides from a different chemical group should then be used for at least two consecutive sprays. If only two applications are made on grapevines, only one may include fenhexamid
- Complete spray cover of all flowers and fruitlets throughout the blossom period is essential for successful control of Botrytis
- Spray programmes should normally start at the start of flowering

Restrictions

- Maximum number of treatments 2 per yr on grapevines; 4 per yr on other listed crops but no more than 2 sprays may be applied consecutively

Crop-specific information

- HI 1 d for strawberries, raspberries, loganberries, blackberries, Rubus hybrids; 3 d for cherries; 7 d for blackcurrants, redcurrants, whitecurrants, gooseberries; 21d for outdoor grapes

Environmental safety

- Dangerous for the environment
- Harmful to fish or other aquatic life. Do not contaminate surface waters or ditches with chemical or used container

Hazard classification and safety precautions

Hazard Dangerous for the environment
Transport code 9
Packaging group III
UN Number 3077
Risk phrases R53a [2]
Operator protection A, H [1]; U08, U19a, U20b
Environmental protection E13c, E38 [2]; E15a [1]
Storage and disposal D05, D09a, D11a [1, 2]; D12a [2]

217   fenoxaprop-P-ethyl

An aryloxyphenoxypropionate herbicide for use in wheat
HRAC mode of action code: A

See also diclofop-methyl + fenoxaprop-P-ethyl

Products

1	Foxtrot EW	Headland	69 g/l	EW	13243
2	Oskar	Headland	69 g/l	EW	13344
3	Polecat	Interfarm	83 g/l	EC	15212
4	Warrant	Interfarm	83 g/l	EC	15211

Uses

- Annual grasses in **grass seed crops** *(off-label)* [1, 2, 4]
- Awned canary grass in **spring wheat**, **winter wheat** [3, 4]
- Blackgrass in **spring barley**, **winter barley** [1, 2]; **spring wheat**, **winter wheat** [1-4]
- Canary grass in **spring barley**, **spring wheat**, **winter barley**, **winter wheat** [1, 2]
- Perennial grasses in **grass seed crops** *(off-label)* [1]
- Rough meadow grass in **spring barley**, **winter barley** [1, 2]; **spring wheat**, **winter wheat** [1-4]
- Wild oats in **spring barley**, **winter barley** [1, 2]; **spring wheat**, **winter wheat** [1-4]

Extension of Authorisation for Minor Use (EAMUs)

- **grass seed crops** *20120446* [1], *20120817* [2], *20113051* [4]

Approval information

- Fenoxaprop-P-ethyl included in Annex I under EC Regulation 1107/2009

Efficacy guidance

- Treat weeds from 2 fully expanded leaves up to flag leaf ligule just visible; for awned canary-grass and rough meadow-grass from 2 leaves to the end of tillering
- A second application may be made in spring where susceptible weeds emerge after an autumn application
- Spray is rainfast 1 h after application
- Dry conditions resulting in moisture stress may reduce effectiveness
- Fenoxaprop-P-ethyl is an ACCase inhibitor herbicide. To avoid the build up of resistance do not apply products containing an ACCase inhibitor herbicide more than twice to any crop. In addition do not use any product containing fenoxaprop-P-ethyl in mixture or sequence with any other product containing the same ingredient
- Use these products as part of a resistance management strategy that includes cultural methods of control and does not use ACCase inhibitors as the sole chemical method of grass weed control
- Applying a second product containing an ACCase inhibitor to a crop will increase the risk of resistance development; only use a second ACCase inhibitor to control different weeds at a different timing
- Always follow WRAG guidelines for preventing and managing herbicide resistant weeds. See Section 5 for more information

Restrictions

- Maximum total dose equivalent to one or two full dose treatments depending on product used
- Do not apply to barley, durum wheat, undersown crops or crops to be undersown
- Do not roll or harrow within 1 wk of spraying
- Do not spray crops under stress, suffering from drought, waterlogging or nutrient deficiency or those grazed or if soil compacted
- Avoid spraying immediately before or after a sudden drop in temperature or a period of warm days/cold nights
- Do not mix with hormone weedkillers

Crop-specific information

- Latest use: before flag leaf sheath extending (GS 41)
- Treat from crop emergence to flag leaf fully emerged (GS 41).

SEE SECTION 3 FOR PRODUCTS ALSO REGISTERED

- Product may be sprayed in frosty weather provided crop hardened off but do not spray wet foliage or leaves covered with ice
- Broadcast crops should be sprayed post-emergence after plants have developed well-established root system

Environmental safety
- Dangerous for the environment
- Very toxic to aquatic organisms

Hazard classification and safety precautions
    Hazard Irritant, Dangerous for the environment
    Transport code 9
    Packaging group III
    UN Number 3082
    Risk phrases R38, R51, R53a [1-4]; R43 [1, 2]
    Operator protection A, C, H; U05a, U08, U14, U20b [1-4]; U11 [3, 4]; U19a [1, 2]
    Environmental protection E15a, E38 [1-4]; E34 [1, 2]
    Storage and disposal D01, D02, D05, D09a, D10b [1-4]; D12a [3, 4]

218   fenoxycarb

An insect specific growth regulator for top fruit
IRAC mode of action code: 7B

Products

Insegar WG	Syngenta	25% w/w	WG	09789

Uses
- Summer-fruit tortrix moth in **apples**, **pears**

Approval information
- Fenoxycarb included in Annex I under EC Regulation 1107/2009

Efficacy guidance
- Best results from application at 5th instar stage before pupation. Product prevents transformation from larva to pupa
- Correct timing best identified from pest warnings
- Because of mode of action rapid knock-down of pest is not achieved and larvae continue to feed for a period after treatment
- Adequate water volume necessary to ensure complete coverage of leaves

Restrictions
- Maximum number of treatments 2 per crop
- Consult processors before use

Crop-specific information
- HI 42 d for apples, pears
- Use on all varieties of apples and pears

Environmental safety
- Dangerous for the environment
- Toxic to aquatic organisms
- High risk to bees. Do not apply to crops in flower or to those in which bees are actively foraging. Do not apply when flowering weeds are present
- Risk to certain non-target insects or other arthropods. See directions for use
- Broadcast air-assisted LERAP (8 m)
- Apply to minimise off-target drift to reduce effects on non-target organisms. Some margin of safety to beneficial arthropods is indicated.
- Broadcast air-assisted LERAP (8 m)

Hazard classification and safety precautions
    Hazard Dangerous for the environment

FOR FULL CONDITIONS OF USE ALWAYS READ THE PRODUCT LABEL

Transport code 9
Packaging group III
UN Number 3077
Risk phrases R51, R53a
Operator protection A, C, H; U05a, U20a, U23a
Environmental protection E12a, E12e, E15a, E22b, E38; E17b (8 m)
Storage and disposal D01, D02, D07, D09a, D11a, D12a

219   fenpropidin

A systemic, curative and protective piperidine (morpholine) fungicide
FRAC mode of action code: 5

See also difenoconazole + fenpropidin

Products

1	Instinct	Headland	750 g/l	EC	14512
2	Tern	Syngenta	750 g/l	EC	08660

Uses
- Brown rust in **spring barley**, **winter barley** [1, 2]
- Powdery mildew in **durum wheat** *(off-label)*, **rye** *(off-label)*, **triticale** *(off-label)* [2]; **spring barley**, **spring wheat**, **winter barley**, **winter wheat** [1, 2]
- Rhynchosporium in **spring barley** *(moderate control)*, **winter barley** *(moderate control)* [1, 2]
- Rust in **durum wheat** *(off-label)*, **rye** *(off-label)*, **triticale** *(off-label)* [2]
- Yellow rust in **spring barley** *(moderate control)*, **spring wheat** *(moderate control)*, **winter barley** *(moderate control)*, **winter wheat** *(moderate control)* [1, 2]

Extension of Authorisation for Minor Use (EAMUs)
- **durum wheat** *20121529* [2]
- **rye** *20121529* [2]
- **triticale** *20121529* [2]

Approval information
- Accepted by BBPA for use on malting barley
- Fenpropidin included in Annex I under EC Regulation 1107/2009

Efficacy guidance
- Best results obtained when applied at early stage of disease development. See label for details of recommended timing alone and in mixtures
- Disease control enhanced by vapour-phase activity. Control can persist for 4-6 wk
- Alternate with triazole fungicides to discourage build-up of resistance

Restrictions
- Maximum number of treatments 3 per crop (up to 2 in yr of harvest) for winter crops; 2 per crop for spring crops
- Treated crops must not be harvested for human or animal consumption for at least 5 wk after the last application

Crop-specific information
- Latest use: up to and including ear emergence complete (GS 59).
- HI 5 wk

Environmental safety
- Dangerous for the environment
- Very toxic to aquatic organisms

Hazard classification and safety precautions
Hazard Harmful, Dangerous for the environment
Transport code 9
Packaging group III
UN Number 3082

SEE SECTION 3 FOR PRODUCTS ALSO REGISTERED

Risk phrases R22a, R37 [2]; R38, R41 [1]; R50, R53a [1, 2]
Operator protection A, C, H; U02a, U05a [1, 2]; U04a, U10, U20a [2]; U09a, U11, U19a, U20b [1]
Environmental protection E15a, E34, E38
Consumer protection C02a (5 wk)
Storage and disposal D01, D02, D05, D09a, D12a [1, 2]; D10c [1]
Medical advice M03

220 fenpropidin + prochloraz + tebuconazole

A contact and systemic fungicide mixture for cereals
FRAC mode of action code: 5 + 3 + 3

See also fenpropidin
        fenpropidin + prochloraz
        fenpropidin + tebuconazole
        prochloraz
        prochloraz + tebuconazole
        tebuconazole

Products

Artemis	Makhteshim	150:200:150 g/l	EC	14178

Uses
- Brown rust in *spring wheat*, *winter wheat*
- Ear diseases in *spring wheat*, *winter wheat*
- Eyespot in *spring wheat*, *winter wheat*
- Glume blotch in *spring wheat*, *winter wheat*
- Powdery mildew in *spring wheat*, *winter wheat*
- Septoria leaf blotch in *spring wheat*, *winter wheat*
- Yellow rust in *spring wheat*, *winter wheat*

Approval information
- Fenpropidin, prochloraz and tebuconazole included in Annex I under EC Regulation 1107/2009

Efficacy guidance
- Best disease control and yield benefit obtained when applied at early stage of disease development before infection spreads to new growth
- To protect the flag leaf and ear from Septoria diseases apply from flag leaf emergence to ear fully emerged (GS 37-59)
- Applications once foliar symptoms of *Septoria tritici* are already present on upper leaves will be less effective
- Resistance to some DMI fungicides has been identified in Septoria leaf blotch which may seriously affect the performance of some products.

Crop-specific information
- Do not apply when temperatures are high.

Hazard classification and safety precautions
Hazard Harmful, Dangerous for the environment
Transport code 9
Packaging group III
UN Number 3082
Risk phrases R22a, R38, R41, R48, R50, R53a, R63
Operator protection A, C, H; U02a, U04a, U05a, U09a, U10, U11, U15, U19a, U20b
Environmental protection E15a, E34, E38
Storage and disposal D01, D02, D05, D12a
Medical advice M03, M05a

221   fenpropimorph

A contact and systemic morpholine fungicide
FRAC mode of action code: 5

See also azoxystrobin + fenpropimorph
       epoxiconazole + fenpropimorph
       epoxiconazole + fenpropimorph + kresoxim-methyl
       epoxiconazole + fenpropimorph + metrafenone
       epoxiconazole + fenpropimorph + pyraclostrobin

Products

1	Clayton Spigot	Clayton	750 g/l	EC	11560
2	Corbel	BASF	750 g/l	EC	00578
3	Crebol	AgChem Access	750 g/l	EC	13922
4	Fenprop 750	Goldengrass	750 g/l	EC	15388
5	Propimorf 750	Goldengrass	750 g/l	EC	14410
6	Standon Fenpropimorph 750	Standon	750 g/l	EC	08965

Uses

* Alternaria in **carrots** *(off-label)*, **horseradish** *(off-label)*, **parsley root** *(off-label)*, **parsnips** *(off-label)*, **salsify** *(off-label)* [2]
* Brown rust in **spring barley**, **spring wheat**, **triticale**, **winter barley**, **winter wheat** [1-6]; **undersown barley**, **undersown wheat** [5]
* Crown rot in **carrots** *(off-label)*, **horseradish** *(off-label)*, **parsley root** *(off-label)*, **parsnips** *(off-label)*, **salsify** *(off-label)* [1, 2]; **mallow (althaea spp.)** *(off-label)* [1]
* Foliar disease control in **durum wheat** *(off-label)* [2]; **grass seed crops** *(off-label)* [1, 6]
* Powdery mildew in **bilberries** *(off-label)*, **blackcurrants** *(off-label)*, **blueberries** *(off-label)*, **carrots** *(off-label)*, **cranberries** *(off-label)*, **dewberries** *(off-label)*, **gooseberries** *(off-label)*, **hops** *(off-label)*, **horseradish** *(off-label)*, **loganberries** *(off-label)*, **parsley root** *(off-label)*, **parsnips** *(off-label)*, **raspberries** *(off-label)*, **redcurrants** *(off-label)*, **salsify** *(off-label)*, **strawberries** *(off-label)*, **tayberries** *(off-label)*, **whitecurrants** *(off-label)* [1, 2]; **durum wheat** *(off-label)*, **fodder beet** *(off-label)*, **mallow (althaea spp.)** *(off-label)*, **red beet** *(off-label)*, **sugar beet seed crops** *(off-label)* [1]; **spring barley**, **spring oats**, **spring wheat**, **winter barley**, **winter oats**, **winter rye**, **winter wheat** [1-6]; **undersown barley**, **undersown wheat** [5]
* Rhynchosporium in **spring barley**, **winter barley** [1-6]; **undersown barley** [5]
* Rust in **fodder beet** *(off-label)*, **sugar beet seed crops** *(off-label)* [2]; **leeks** [1]; **red beet** *(off-label)* [1, 2]
* Yellow rust in **spring barley**, **spring wheat**, **triticale**, **winter barley**, **winter wheat** [1-6]; **undersown barley**, **undersown wheat** [5]

Extension of Authorisation for Minor Use (EAMUs)

* **bilberries** *20111798* [1], *20040804* [2]
* **blackcurrants** *20111798* [1], *20040804* [2]
* **blueberries** *20111798* [1], *20040804* [2]
* **carrots** *20111796* [1], *20023753* [2]
* **cranberries** *20111798* [1], *20040804* [2]
* **dewberries** *20111798* [1], *20040804* [2]
* **durum wheat** *20111799* [1], *20061460* [2]
* **fodder beet** *20111797* [1], *20023757* [2]
* **gooseberries** *20111798* [1], *20040804* [2]
* **grass seed crops** *20061060* [1], *20061057* [6]
* **hops** *20111794* [1], *20023759* [2]
* **horseradish** *20111796* [1], *20023753* [2]
* **loganberries** *20111798* [1], *20040804* [2]
* **mallow (althaea spp.)** *20111796* [1]
* **parsley root** *20111796* [1], *20023753* [2]
* **parsnips** *20111796* [1], *20023753* [2]
* **raspberries** *20111798* [1], *20040804* [2]

- **red beet** *20111795* [1], *20023751* [2]
- **redcurrants** *20111798* [1], *20040804* [2]
- **salsify** *20111796* [1], *20023753* [2]
- **strawberries** *20111798* [1], *20040804* [2]
- **sugar beet seed crops** *20111797* [1], *20023757* [2]
- **tayberries** *20111798* [1], *20040804* [2]
- **whitecurrants** *20111798* [1], *20040804* [2]

Approval information
- Accepted by BBPA for use on malting barley and hops
- Fenpropimorph included in Annex I under EC Regulation 1107/2009

Efficacy guidance
- On all crops spray at start of disease attack. See labels for recommended tank mixes. Follow-up treatments may be needed if disease pressure remains high
- Product rainfast after 2 h

Restrictions
- Maximum number of treatments 2 per crop for spring cereals; 3 or 4 per crop for winter cereals; 6 per crop for leeks
- Consult processors before using on crops for processing

Crop-specific information
- HI cereals 5 wk
- Scorch may occur on cereals if applied during frosty weather or in high temperatures
- Some leaf spotting may occur on undersown clovers
- Leeks should be treated every 2-3 wk as required, up to 6 applications

Environmental safety
- Dangerous for the environment
- Very toxic to aquatic organisms

Hazard classification and safety precautions
Hazard Harmful [1-5]; Irritant [6]; Dangerous for the environment [1-6]
Transport code 9
Packaging group III
UN Number 3082
Risk phrases R22a [1]; R38, R53a [1-6]; R50 [2-5]; R51 [1, 6]; R63 [1-5]
Operator protection A [1-6]; C [1, 6]; U05a, U08, U14, U15, U19a, U20b
Environmental protection E15a [1-6]; E38 [2-5]
Storage and disposal D01, D02, D09a [1-6]; D05 [1, 5, 6]; D08 [2-5]; D10b [6]; D10c, D12a [1-5]
Medical advice M05a

222   fenpropimorph + flusilazole

A broad-spectrum eradicant and protectant fungicide mixture for cereals
FRAC mode of action code: 5 + 3

See also flusilazole

Products

1	Colstar	DuPont	375:160 g/l	EC	12175
2	Pluton	DuPont	375:160 g/l	EC	12200

Uses
- Brown rust in **spring barley**, **winter barley**, **winter wheat** [1, 2]
- Foliar disease control in **durum wheat** *(off-label)*, **grass seed crops** *(off-label)* [1, 2]; **spring rye** *(off-label)*, **triticale** *(off-label)*, **winter rye** *(off-label)* [1]
- Glume blotch in **winter wheat** [1, 2]
- Net blotch in **spring barley**, **winter barley** [1, 2]
- Powdery mildew in **spring barley**, **winter barley**, **winter wheat** [1, 2]

FOR FULL CONDITIONS OF USE ALWAYS READ THE PRODUCT LABEL

- Rhynchosporium in *spring barley*, *winter barley* [1, 2]
- Septoria leaf blotch in *winter wheat* [1, 2]
- Yellow rust in *spring barley*, *winter barley*, *winter wheat* [1, 2]

Extension of Authorisation for Minor Use (EAMUs)
- *durum wheat* 20060977 [1], 20060985 [2]
- *grass seed crops* 20060977 [1], 20060985 [2]
- *spring rye* 20060977 [1]
- *triticale* 20060977 [1]
- *winter rye* 20060977 [1]

Approval information
- Fenpropimorph and flusilazole included in Annex I under EC Regulation 1107/2009
- Accepted by BBPA for use on malting barley

Efficacy guidance
- Disease control is more effective if treatment made at an early stage of disease development
- Treat winter cereals in spring or early summer before diseases spread to new growth
- Spring barley should be treated when diseases are first evident
- Treatment may be repeated after 3-4 wk if necessary
- Flusilazole is a DMI fungicide. Resistance to some DMI fungicides has been identified in Septoria leaf blotch which may seriously affect performance of some products. For further advice contact a specialist advisor and visit the Fungicide Resistance Action Group (FRAG)-UK website

Restrictions
- Maximum number of treatments 2 per crop of barley or wheat
- Do not apply to crops under stress
- Do not apply during frosty weather

Crop-specific information
- Latest use: before beginning of anthesis (GS 60) for winter wheat; up to and including completion of ear emergence (GS 59) for barley

Environmental safety
- Dangerous for the environment
- Toxic to aquatic organisms
- Dangerous to fish or other aquatic life. Do not contaminate surface waters or ditches with chemical or used container

Hazard classification and safety precautions
Hazard Toxic, Dangerous for the environment
Transport code 9
Packaging group III
UN Number 3082
Risk phrases R36, R40, R51, R53a, R61
Operator protection A, C; U05a, U11, U19a, U20b
Environmental protection E13b, E15a, E38
Storage and disposal D01, D02, D05, D09a, D10b, D12a
Medical advice M04a

223 fenpropimorph + kresoxim-methyl

A protectant and systemic fungicide mixture for cereals
FRAC mode of action code: 5 + 11

See also kresoxim-methyl

Products

Ensign	BASF	300:150 g/l	SE	11729

SEE SECTION 3 FOR PRODUCTS ALSO REGISTERED

SECTION 2

Uses

- Glume blotch in **spring wheat** *(reduction)*, **triticale** *(reduction)*, **winter wheat** *(reduction)*
- Powdery mildew in **spring barley**, **spring oats**, **spring rye**, **triticale**, **winter barley**, **winter oats**, **winter rye**
- Rhynchosporium in **spring barley**, **spring rye**, **triticale**, **winter barley**, **winter rye**
- Septoria leaf blotch in **spring wheat** *(reduction)*, **triticale** *(reduction)*, **winter wheat** *(reduction)*

Extension of Authorisation for Minor Use (EAMUs)

- **durum wheat** *20122660*

Approval information

- Fenpropimorph and kresoxim-methyl included in Annex I under EC Regulation 1107/2009
- Accepted by BBPA for use on malting barley

Efficacy guidance

- For best results spray at the start of foliar disease attack and repeat if infection conditions persist
- Kresoxim-methyl is a member of the QoI cross resistance group. Product should be used preventatively and not relied on for its curative potential
- Use product as part of an Integrated Crop Management strategy incorporating other methods of control, including where appropriate other fungicides with a different mode of action. Do not apply more than two foliar applications of QoI containing products to any cereal crop
- There is a significant risk of widespread resistance occurring in *Septoria tritici* populations in UK. Strains of barley powdery mildew resistant to QoI's are common in UK. Failure to follow resistance management action may result in reduced levels of disease control

Restrictions

- Maximum total dose equivalent to two full dose treatments for all crops

Crop-specific information

- Latest use: completion of ear emergence (GS 59) for barley and oats; completion of flowering (GS 69) for wheat, rye and triticale

Environmental safety

- Dangerous for the environment
- Very toxic to aquatic organisms

Hazard classification and safety precautions

Hazard Harmful, Dangerous for the environment
Transport code 9
Packaging group III
UN Number 3082
Risk phrases R40, R50, R53a, R63
Operator protection A, H; U05a, U14, U20b
Environmental protection E15a, E38
Storage and disposal D01, D02, D05, D09a, D10c, D12a
Medical advice M05a

224   fenpropimorph + pyraclostrobin

A protectant and curative fungicide mixture for cereals
FRAC mode of action code: 5 +11

See also pyraclostrobin

Products

Jenton	BASF	375:100 g/l	EC	11898

Uses

- Brown rust in **spring barley**, **spring wheat**, **winter barley**, **winter wheat**
- Crown rust in **spring oats**, **winter oats**
- Drechslera leaf spot in **spring wheat**, **winter wheat**
- Glume blotch in **spring wheat**, **winter wheat**

FOR FULL CONDITIONS OF USE ALWAYS READ THE PRODUCT LABEL

- Net blotch in **spring barley, winter barley**
- Powdery mildew in **spring barley, spring oats, winter barley, winter oats**
- Rhynchosporium in **spring barley, winter barley**
- Septoria leaf blotch in **spring wheat, winter wheat**
- Yellow rust in **spring barley, spring wheat, winter barley, winter wheat**

Extension of Authorisation for Minor Use (EAMUs)
- **grass seed crops** *20122661*
- **triticale** *20122661*

Approval information
- Fenpropimorph and pyraclostrobin included in Annex I under EC Regulation 1107/2009
- Accepted by BBPA for use on malting barley

Efficacy guidance
- Best results obtained from treatment at the start of foliar disease attack
- Yield response may be obtained in the absence of visual disease
- Pyraclostrobin is a member of the QoI cross resistance group. Product should be used preventatively and not relied on for its curative potential
- Use product as part of an Integrated Crop Management strategy incorporating other methods of control, including where appropriate other fungicides with a different mode of action. Do not apply more than two foliar applications of QoI containing products to any cereal crop
- There is a significant risk of widespread resistance occurring in *Septoria tritici* populations in UK. Strains of barley powdery mildew resistant to QoI's are common in UK. Failure to follow resistance management action may result in reduced levels of disease control

Restrictions
- Maximum number of treatments 2 per crop

Crop-specific information
- Latest use: up to and including emergence of ear just complete (GS 59) for barley

Environmental safety
- Dangerous for the environment
- Very toxic to aquatic organisms
- LERAP Category B

Hazard classification and safety precautions
Hazard Harmful, Dangerous for the environment
Transport code 9
Packaging group III
UN Number 3082
Risk phrases R20, R22a, R36, R38, R50, R53a, R63
Operator protection A, C; U05a, U14, U20b
Environmental protection E15a, E16a, E34, E38
Storage and disposal D01, D02, D05, D09a, D10c, D12a
Medical advice M03, M05a

225   fenpropimorph + quinoxyfen

A systemic fungicide mixture for cereals
FRAC mode of action code: 5 + 13

See also quinoxyfen

Products
Orka	Dow	250:66.7 g/l	EW	08879

Uses
- Powdery mildew in **durum wheat, spring barley, spring oats, spring rye, spring wheat, triticale, winter barley, winter oats, winter rye, winter wheat**

Extension of Authorisation for Minor Use (EAMUs)
- *grass seed crops* 20122662

Approval information
- Fenpropimorph and quinoxyfen included in Annex I under Regulation 1107/2009
- Accepted by BBPA for use on malting barley

Efficacy guidance
- For best results treat at early stage of disease development before infection spreads to new crop growth. Further treatment may be necessary if disease pressure remains high
- For control of established infections and broad spectrum disease control use in tank mixtures. See label
- Product rainfast after 1 h
- Systemic activity may be reduced in severe drought

Restrictions
- Maximum total dose 3.0 l product per ha
- Apply only in the spring from mid-tillering stage (GS 25)

Crop-specific information
- Latest use: when first awns visible (GS 49)
- Crop scorch may occur when treatment made in high temperatures

Environmental safety
- Dangerous for the environment
- Very toxic to aquatic organisms
- LERAP Category B

Hazard classification and safety precautions
Hazard Harmful, Dangerous for the environment
Transport code 9
Packaging group III
UN Number 3082
Risk phrases R43, R50, R53a, R61
Operator protection A, H; U05a, U14
Environmental protection E15a, E16a, E34, E38
Storage and disposal D01, D02, D12a

226   fenpyroximate

A mitochondrial electron transport inhibitor (METI) acaricide for apples
IRAC mode of action code: 21

Products
Sequel	Certis	51.3 g/l	SC	12657

Uses
- Fruit tree red spider mite in *apples*
- Red spider mites in *plums* (off-label), *protected strawberries* (off-label), *strawberries* (off-label)

Extension of Authorisation for Minor Use (EAMUs)
- *plums* 20093060
- *protected strawberries* 20093061
- *strawberries* 20093061

Approval information
- Fenpyroximate included in Annex I under EC Regulation 1107/2009

Efficacy guidance
- Kills motile stages of fruit tree red spider mite. Best results achieved if applied in warm weather
- Total spray cover of trees essential. Use higher volumes for large trees
- Apply when majority of winter eggs have hatched

FOR FULL CONDITIONS OF USE ALWAYS READ THE PRODUCT LABEL

Restrictions
- Maximum number of treatments 1 per yr or total dose equivalent to one full dose treatment
- Other mitochondrial electron transport inhibitor (METI) acaricides should not be applied to the same crop in the same calendar yr either separately or in mixture
- Do not apply when apple crops or pollinator are in flower
- Consult processor before use on crops for processing

Crop-specific information
- HI 2 wk

Environmental safety
- Dangerous for the environment
- Very toxic to aquatic organisms
- Risk to non-target insects or other arthropods
- Broadcast air-assisted LERAP (40 m)

Hazard classification and safety precautions
> **Hazard** Harmful, Dangerous for the environment
> **Transport code** 9
> **Packaging group** III
> **UN Number** 3082
> **Risk phrases** R20, R36, R43, R50, R53a
> **Operator protection** A, C, H; U05a, U08, U14, U15, U20b
> **Environmental protection** E13b, E22c; E17b (40 m)
> **Storage and disposal** D01, D02, D05, D09a, D10c, D12a

227   ferric phosphate

A molluscicide bait for controlling slugs and snails

Products

1	Derrex	Certis	2.97% w/w	GB	15351
2	Ferramol Max	Certis	2.97% w/w	GB	14463
3	Sluggo	Omex	1% w/w	GB	12529
4	Sluggo	Omex	1% w/w	GB	14788
5	Sluxx	Certis	2.97% w/w	GB	14462

Uses
- Slugs in *all edible crops (outdoor)*, *all non-edible crops (outdoor)*, *protected crops* [3, 4]
- Slugs and snails in *all edible crops (outdoor and protected)*, *all non-edible crops (outdoor)*, *amenity vegetation* [1, 2, 5]
- Snails in *all edible crops (outdoor)*, *all non-edible crops (outdoor)*, *protected crops* [3, 4]

Approval information
- Ferric phosphate included in Annex 1 under EC Regulation 1107/2009

Efficacy guidance
- Treat as soon as damage first seen preferably in early evening. Repeat as necessary to maintain control
- Best results obtained from moist soaked granules. This will occur naturally on moist soils or in humid conditions
- Ferric phosphate does not cause excessive slime secretion and has no requirement to collect moribund slugs from the soil surface
- Active ingredient is degraded by micro-organisms to beneficial plant nutrients

Restrictions
- Maximum total dose equivalent to four full dose treatments

Crop-specific information
- Latest use not specified for any crop

SEE SECTION 3 FOR PRODUCTS ALSO REGISTERED

Hazard classification and safety precautions
UN Number N/C
Operator protection U05a [1, 3-5]; U20a [1, 5]; U20b [3, 4]
Environmental protection E15a [1, 3, 4]; E34 [1, 3-5]
Storage and disposal D01, D09a [1, 3-5]

228 ferrous sulphate

A herbicide/fertilizer combination for moss control in turf

See also dicamba + dichlorprop-P + ferrous sulphate + MCPA
dichlorprop-P + ferrous sulphate + MCPA

Products

1	Elliott's Lawn Sand	Elliott	9.3% w/w	SA	04860
2	Elliott's Mosskiller	Elliott	24.6% w/w	GR	04909
3	Ferromex Mosskiller Concentrate	Omex	16.4% w/w	SL	13180
4	Greenmaster Autumn	Everris Ltd	18.2% w/w	GR	15808
5	Greenmaster Mosskiller	Everris Ltd	27% w/w	GR	15796
6	Greentec Mosskiller Pro	Headland Amenity	16.78% w/w	GR	14645
7	Landscaper Pro Moss Control + Fertilizer	Everris Ltd	21.7% w/w	GR	15795
8	Mascot Micronised Lawn Sand	Rigby Taylor	8.2% w/w	SA	14783
9	SHL Lawn Sand	Sinclair	5.4% w/w	SA	05254
10	Taylors Lawn Sand	Rigby Taylor	4.1% w/w	SA	04451

Uses

- Moss in **amenity grassland** [3, 5, 6]; **lawns** [8]; **managed amenity turf** [1-7, 9, 10]

Approval information

- Ferrous sulphate included in Annex 1 under EC Regulation 1107/2009

Efficacy guidance

- For best results apply when turf is actively growing and the soil is moist
- Fertilizer component of most products encourages strong root growth and tillering
- Mow 3 d before treatment and do not mow for 3-4 d afterwards
- Water after 2 d if no rain
- Rake out dead moss thoroughly 7-14 d after treatment. Re-treatment may be necessary for heavy infestations

Restrictions

- Maximum number of treatments - see labels
- Do not apply during drought or when heavy rain expected
- Do not apply in frosty weather or when the ground is frozen
- Do not walk on treated areas until well watered
- Do not apply product undiluted [3]

Crop-specific information

- If spilt on paving, concrete, clothes etc brush off immediately to avoid discolouration
- Observe label restrictions for interval before cutting after treatment

Environmental safety

- Harmful to fish or other aquatic life. Do not contaminate surface waters or ditches with chemical or used container

Hazard classification and safety precautions

Hazard Harmful [6]
UN Number N/C [1-7, 9, 10]
Risk phrases R22a, R36, R38 [6]
Operator protection A [3]; U20a [1, 2, 8]; U20b [3, 5, 6]; U20c [4, 7, 9, 10]

FOR FULL CONDITIONS OF USE ALWAYS READ THE PRODUCT LABEL

Environmental protection E13c [4, 7]; E15a [1, 2, 5, 6, 8, 10]; E15b [3]
Storage and disposal D01 [3, 5, 6, 9]; D09a [1-10]; D11a [1, 2, 4-10]; D12a [3]

229   ferrous sulphate + MCPA + mecoprop-P

A translocated herbicide and moss killer mixture
HRAC mode of action code: O + O

See also MCPA
mecoprop-P

Products

Renovator Pro	Everris Ltd	0.49:0.29:16.3% w/w	GR	15815

Uses
- Annual dicotyledons in ***managed amenity turf***
- Moss in ***managed amenity turf***
- Perennial dicotyledons in ***managed amenity turf***

Approval information
- Ferrous sulphate, MCPA and mecoprop-P included in Annex I under EC Regulation 1107/2009

Efficacy guidance
- Apply from Apr to Sep when weeds are growing
- For best results apply when light showers or heavy dews are expected
- Apply with a suitable calibrated fertilizer distributor
- Retreatment may be necessary after 6 wk if weeds or moss persist
- For best control of moss scarify vigorously after 2 wk to remove dead moss
- Where regrowth of moss or weeds occurs a repeat treatment may be made after 6 wk
- Avoid treatment of wet grass or during drought. If no rain falls within 48 h water in thoroughly

Restrictions
- Maximum number of treatments 3 per yr
- Do not treat new turf until established for 6 mth
- The first 4 mowings after treatment should not be used to mulch cultivated plants unless composted at least 6 mth
- Avoid walking on treated areas until it has rained or they have been watered
- Do not re-seed or turf within 8 wk of last treatment
- Do not cut grass for at least 3 d before and at least 4 d after treatment
- Do not apply during freezing conditions or when rain imminent

Environmental safety
- Keep livestock out of treated areas for up to two weeks following treatment and until poisonous weeds, such as ragwort, have died down and become unpalatable
- Harmful to fish or other aquatic life. Do not contaminate surface waters or ditches with chemical or used container
- Do not empty into drains
- Some pesticides pose a greater threat of contamination of water than others and mecoprop-P is one of these pesticides. Take special care when applying mecoprop-P near water and do not apply if heavy rain is forecast

Hazard classification and safety precautions
UN Number N/C
Operator protection U20b
Environmental protection E07a, E13c, E19b
Storage and disposal D01, D09a, D12a
Medical advice M05a

230　flazasulfuron

A sulfonylurea herbicide for non-crop use
HRAC mode of action code: B

Products

1	Chikara Weed Control	Belchim	25% w/w	WG	14189
2	Chikita	ChemSource	25% w/w	WG	14675
3	Clayton Apt	Clayton	25% w/w	WG	15157
4	Paradise	Pan Agriculture	25% w/w	WG	14504
5	PureFlazasul	Pure Amenity	25% w/w	WG	15135

Uses

- Annual and perennial weeds in *natural surfaces not intended to bear vegetation, permeable surfaces overlying soil, railway tracks*

Approval information

- Flazasulfuron included in Annex 1 under EC Regulation 1107/2009

Environmental safety

- LERAP Category B

Hazard classification and safety precautions

Hazard Dangerous for the environment
Transport code 9
Packaging group III
UN Number 3077
Risk phrases R50, R53a
Operator protection A, H, M; U02a, U05a, U20b [1-5]; U08 [1, 2, 5]; U09a [3, 4]
Environmental protection E15a, E16a, E16b, E38
Storage and disposal D01, D02, D09a, D11a, D12a

231　flonicamid

A selective feeding blocker aphicide
IRAC mode of action code: 9C

Products

1	Mainman	Belchim	50% w/w	WG	13123
2	Mutiny	ChemSource	50% w/w	WG	14805
3	Teppeki	Belchim	50% w/w	WG	12402

Uses

- Aphids in *apples, hops* (off-label), *pears* [1]; *ornamental plant production* (off-label) [1, 3]; *potatoes, winter wheat* [2, 3]
- Tobacco whitefly in *protected ornamentals* (off-label) [1]
- Whitefly in *protected ornamentals* (off-label) [1]

Extension of Authorisation for Minor Use (EAMUs)

- *hops* 20120619 *expires 31 Aug 2013* [1]
- *ornamental plant production* 20120618 *expires 31 Aug 2013* [1], 20120621 *expires 31 Aug 2013* [3]
- *protected ornamentals* 20120620 *expires 31 Aug 2013* [1]

Approval information

- Flonicamid included in Annex 1 under EC Regulation 1107/2009

Efficacy guidance

- Apply when warning systems forecast significant aphid infestations
- Persistence of action is 21 d

- Do not apply more than two consecutive treatments of flonicamid. If further treatment is needed use an insecticide with a different mode of action

Restrictions
- Maximum number of treatments 2 per crop for potatoes, winter wheat
- Must not be applied to winter wheat before 50% ear emerged stage (GS 53)
- Use a maximum of two consecutive applications of [1] in apples and pears. If further treatments are required, use an insecticide with a different mode of action before applying the final application of [1]

Crop-specific information
- HI: potatoes 14 d; winter wheat 28 d

Environmental safety
- Dangerous for the environment
- Harmful to aquatic organisms

Hazard classification and safety precautions
  Hazard Dangerous for the environment [2, 3]
  UN Number N/C
  Risk phrases R52, R53a
  Operator protection A, H, M [1-3]; C [2, 3]; E [1]; U05a [1-3]; U13, U14, U15 [1]
  Environmental protection E15a
  Storage and disposal D01, D02, D12b [1-3]; D03, D08 [1]; D07 [2, 3]

232 florasulam

A triazolopyrimidine herbicide for cereals
HRAC mode of action code: B

See also 2,4-D + florasulam
*   clopyralid + florasulam + fluroxypyr*

Products

1	Barton WG	Dow	25% w/w	WG	13284
2	Boxer	Dow	50 g/l	SC	09819

Uses
- Annual dicotyledons in *farm forestry* (off-label), *forest nurseries* (off-label), *game cover* (off-label), *grass seed crops* (off-label), *ornamental plant production* (off-label), *spring rye* (off-label), *triticale* (off-label), *winter rye* (off-label) [2]; *spring barley, spring oats, spring wheat, winter barley, winter oats, winter wheat* [1, 2]
- Chickweed in *spring barley, spring oats, spring wheat, winter barley, winter oats, winter wheat* [1, 2]
- Cleavers in *spring barley, spring oats, spring wheat, winter barley, winter oats, winter wheat* [1, 2]
- Mayweeds in *spring barley, spring oats, spring wheat, winter barley, winter oats, winter wheat* [1, 2]
- Volunteer oilseed rape in *spring barley, spring oats, spring wheat, winter barley, winter oats, winter wheat* [1, 2]

Extension of Authorisation for Minor Use (EAMUs)
- *farm forestry* 20082826 [2]
- *forest nurseries* 20082826 [2]
- *game cover* 20082826 [2]
- *grass seed crops* 20060997 [2]
- *ornamental plant production* 20082826 [2]
- *spring rye* 20060997 [2]
- *triticale* 20060997 [2]
- *winter rye* 20060997 [2]

SECTION 2

SEE SECTION 3 FOR PRODUCTS ALSO REGISTERED

Approval information
- Florasulam included in Annex I under EC Regulation 1107/2009
- Accepted by BBPA for use on malting barley

Efficacy guidance
- Best results obtained from treatment of small actively growing weeds in good conditions
- Apply in autumn or spring once crop has 3 leaves
- Product is mainly absorbed by leaves of weeds and is effective on all soil types
- Florasulam is a member of the ALS-inhibitor group of herbicides

Restrictions
- Maximum total dose on any crop equivalent to one full dose treatment
- Do not roll or harrow within 7 d before or after application
- Do not spray when crops under stress from cold, drought, pest damage, nutrient deficiency or any other cause
- Specific restrictions apply to use in sequence or tank mixture with other sulfonylurea or ALS-inhibiting herbicides. See label for details

Crop-specific information
- Latest use: up to and including flag leaf just visible stage for all crops

Following crops guidance
- Where the product has been used in mixture with certain named products (see label) only cereals or grass may be sown in the autumn following harvest [2]
- Unless otherwise restricted cereals, oilseed rape, field beans, grass or vegetable brassicas as transplants may be sown as a following crop in the same calendar yr as treatment. Oilseed rape may show some temporary reduction of vigour after a dry summer, but yields are not affected
- In addition to the above, linseed, peas, sugar beet, potatoes, maize, clover (for use in grass/clover mixtures) or carrots may be sown in the calendar yr following treatment
- In the event of failure of a treated crop in spring only spring wheat, spring barley, spring oats, maize or ryegrass may be sown

Environmental safety
- Dangerous for the environment
- Very toxic to aquatic organisms
- See label for detailed instructions on tank cleaning

Hazard classification and safety precautions
Hazard Dangerous for the environment
Transport code 9
Packaging group III
UN Number 3077 [1]; 3082 [2]
Risk phrases R50, R53a
Operator protection U05a
Environmental protection E15a [2]; E15b [1]; E34, E38 [1, 2]
Storage and disposal D01, D02, D05, D09a, D10b, D12a

233   florasulam + fluroxypyr

A post-emergence herbicide mixture for cereals
HRAC mode of action code: B + O

See also fluroxypyr

Products

1	Cabadex	Headland Amenity	2.5:100 g/l	SE	13948
2	GF 184	Dow	2.5:100 g/l	SE	10878
3	Headland Sure-fire Gold	Headland	1.05:100 g/l	SE	15130
4	Hiker	Dow	1.0:100 g/l	SE	11451
5	Hunter	Dow	2.5:100 g/l	SE	12836
6	Slalom	Dow	2.5:100 g/l	SE	13772

FOR FULL CONDITIONS OF USE ALWAYS READ THE PRODUCT LABEL

Products – continued

7	Spitfire	Dow	5:100 g/l	SE	15101
8	Starane Gold	Dow	1.0:100 g/l	SE	10879
9	Starane Vantage	Dow	1.0:100 g/l	SE	10922
10	Starane XL	Dow	2.5:100 g/l	SE	10921
11	Trafalgar	Pan Amenity	2.5:100 g/l	SE	14888

Uses

- Annual and perennial weeds in *amenity grassland, lawns, managed amenity turf* [1]
- Annual dicotyledons in *amenity grassland, lawns, managed amenity turf* [11]; *durum wheat* (off-label), *spring rye* (off-label), *triticale* (off-label), *winter rye* (off-label) [4, 8-10]; *game cover* (off-label) [2, 8, 10]; *grass seed crops* (off-label) [2, 4, 8-10]; *oats, rye, triticale* [7]; *ornamental plant production* (off-label) [10]; *spring barley, spring wheat, winter barley, winter wheat* [2, 5-7, 10]; *spring oats, winter oats* [2, 5, 6, 10]
- Black bindweed in *oats, rye, spring barley, spring wheat, triticale, winter barley, winter wheat* [7]
- Black nightshade in *oats, rye, spring barley, spring wheat, triticale, winter barley, winter wheat* [7]
- Buttercups in *amenity grassland, lawns, managed amenity turf* [1]
- Charlock in *oats, rye, spring barley, spring wheat, triticale, winter barley, winter wheat* [7]
- Chickweed in *amenity grassland, lawns, managed amenity turf* [11]; *oats, rye, triticale* [7]; *spring barley, spring wheat, winter barley, winter wheat* [2-10]; *spring oats, winter oats* [2-6, 8-10]
- Cleavers in *oats, rye, triticale* [7]; *spring barley, spring wheat, winter barley, winter wheat* [2-10]; *spring oats, winter oats* [2-6, 8-10]
- Clover in *amenity grassland, lawns, managed amenity turf* [1]
- Daisies in *amenity grassland, lawns, managed amenity turf* [1]
- Dandelions in *amenity grassland, lawns, managed amenity turf* [1]
- Forget-me-not in *oats, rye, spring barley, spring wheat, triticale, winter barley, winter wheat* [7]
- Knotgrass in *oats, rye, spring barley, spring wheat, triticale, winter barley, winter wheat* [7]
- Mayweeds in *oats, rye, triticale* [7]; *spring barley, spring wheat, winter barley, winter wheat* [2, 5-7, 10]; *spring oats, winter oats* [2, 5, 6, 10]
- Plantains in *amenity grassland, lawns, managed amenity turf* [1]
- Poppies in *oats, rye, spring barley, spring wheat, triticale, winter barley, winter wheat* [7]
- Redshank in *oats, rye, spring barley, spring wheat, triticale, winter barley, winter wheat* [7]
- Volunteer beans in *oats, rye, spring barley, spring wheat, triticale, winter barley, winter wheat* [7]
- Volunteer oilseed rape in *oats, rye, spring barley, spring wheat, triticale, winter barley, winter wheat* [7]

Extension of Authorisation for Minor Use (EAMUs)

- *durum wheat* 20072810 [4], 20072811 [8], 20072812 [9], 20072815 [10]
- *game cover* 20082858 [2], 20082878 [8], 20082904 [10]
- *grass seed crops* 20072809 [2], 20072810 [4], 20072811 [8], 20072812 [9], 20072815 [10]
- *ornamental plant production* 20082904 [10]
- *spring rye* 20072810 [4], 20072811 [8], 20072812 [9], 20072815 [10]
- *triticale* 20072810 [4], 20072811 [8], 20072812 [9], 20072815 [10]
- *winter rye* 20072810 [4], 20072811 [8], 20072812 [9], 20072815 [10]

Approval information

- Florasulam and fluroxypyr included in Annex I under EC Regulation 1107/2009
- Accepted by BBPA for use on malting barley

Efficacy guidance

- Best results obtained when weeds are small and growing actively
- Products are mainly absorbed through weed foliage. Cleavers emerging after application will not be controlled
- Florasulam is a member of the ALS-inhibitor group of herbicides

SEE SECTION 3 FOR PRODUCTS ALSO REGISTERED

Restrictions
- Maximum total dose equivalent to one full dose treatment
- Do not roll or harrow 7 d before or after application
- Do not spray when crops are under stress from cold, drought, pest damage or nutrient deficiency
- Do not apply through CDA applicators
- Specific restrictions apply to use in sequence or tank mixture with other sulfonylurea or ALS-inhibiting herbicides. See label for details

Crop-specific information
- Latest use: before flag leaf sheath extended (before GS 41) for spring barley and spring wheat; before flag leaf sheath opening (before GS 47) for winter barley and winter wheat; before second node detectable (before GS 32) for winter oats

Following crops guidance
- Cereals, oilseed rape, field beans or grass may follow treated crops in the same yr. Oilseed rape may suffer temporary vigour reduction after a dry summer
- In addition to the above, linseed, peas, sugar beet, potatoes, maize or clover may be sown in the calendar yr following treatment
- In the event of failure of a treated crop in the spring, only spring cereals, maize or ryegrass may be planted

Environmental safety
- Dangerous for the environment
- Toxic to aquatic organisms
- Take extreme care to avoid drift onto non-target crops or plants
- LERAP Category B [7]

Hazard classification and safety precautions
Hazard Irritant, Dangerous for the environment [1-11]; Flammable [7]
Transport code 3 [7]; 9 [1-6, 8-11]
Packaging group III
UN Number 1993 [7]; 3082 [1-6, 8-11]
Risk phrases R36, R38, R53a, R67 [1-11]; R43 [7]; R50 [1]; R51 [2-11]
Operator protection A [1-11]; C [1-10]; H [7]; U05a, U11, U19a [1-11]; U08 [1]; U09a [7]; U14, U20b [1, 7]
Environmental protection E15a [2-11]; E15b [1]; E16a [7]; E34, E38 [1-11]
Storage and disposal D01, D02, D09a, D10c, D12a [1-11]; D05 [11]
Medical advice M05a [7, 11]

234   florasulam + pyroxsulam

A mixture of two triazolopirimidine sulfonamides for winter wheat
HRAC mode of action code: B

See also pyroxsulam

Products

Broadway Star	Dow	1.42:7.08% w/w	WG	14319

Uses
- Charlock in *triticale, winter rye, winter wheat*
- Chickweed in *triticale, winter rye, winter wheat*
- Cleavers in *triticale, winter rye, winter wheat*
- Field pansy in *triticale, winter rye, winter wheat*
- Field speedwell in *triticale, winter rye, winter wheat*
- Geranium species in *triticale, winter rye, winter wheat*
- Ivy-leaved speedwell in *triticale, winter rye, winter wheat*
- Mayweeds in *triticale, winter rye, winter wheat*
- Poppies in *triticale, winter rye, winter wheat*
- Ryegrass in *triticale, winter rye, winter wheat*
- Sterile brome in *triticale, winter rye, winter wheat*

FOR FULL CONDITIONS OF USE ALWAYS READ THE PRODUCT LABEL

- Volunteer oilseed rape in *triticale, winter rye, winter wheat*
- Wild oats in *triticale, winter rye, winter wheat*

Approval information
- Florasulam included in Annex I under EC Regulation 1107/2009 while pyroxsulam is awaiting inclusion.

Efficacy guidance
- Rainfast within 1 hour of application
- Requires an authorised adjuvant at application and recommended for use in a programme with herbicides employing a different mode of action.

Crop-specific information
- Crop injury may occur if applied in tank mixture with plant growth regulators - allow a minimum interval of 7 days.
- Crop injury may occur if applied in tank mixture with OP insecticides, MCPB or dicamba - allow a minimum interval of 14 days

Following crops guidance
- Crop failure before 1st Feb - plough and allow 6 weeks to elapse and then drill spring wheat, spring barley, grass or maize.
- Crop failure after 1st Feb - plough and allow 6 weeks to elapse before drilling grass or maize.

Environmental safety
- Take extreme care to avoid drift on to susceptible crops, non-target plants or waterways
- LERAP Category B

Hazard classification and safety precautions
Hazard Dangerous for the environment
Transport code 9
Packaging group III
UN Number 3077
Risk phrases R50, R53a
Operator protection U05a, U14, U15, U20a
Environmental protection E15b, E16a, E34, E38, E39
Storage and disposal D01, D02, D09a, D12a

235   fluazifop-P-butyl

A phenoxypropionic acid grass herbicide for broadleaved crops
HRAC mode of action code: A

Products

1	Clayton Maximus	Clayton	125 g/l	EC	12543
2	Fusilade Max	Syngenta	125 g/l	EC	11519
3	Greencrop Bantry	Greencrop	125 g/l	EC	12737
4	Howitzer	ChemSource	125 g/l	EC	14942

Uses
- Annual grasses in *almonds* (off-label), *apples* (off-label), *apricots* (off-label), *asparagus* (off-label), *bilberries* (off-label), *blackberries* (off-label), *blueberries* (off-label), *borage* (off-label), *broad beans (dry-harvested)* (off-label), *cabbages* (off-label), *canary flower (echium spp.)* (off-label), *cherries* (off-label), *chicory* (off-label), *choi sum* (off-label), *cob nuts* (off-label), *collards* (off-label), *cranberries* (off-label), *evening primrose* (off-label), *figs* (off-label), *filberts* (off-label), *fodder beet* (off-label), *garlic* (off-label), *globe artichoke* (off-label), *grapevines* (off-label), *haricot beans* (off-label), *hazel nuts* (off-label), *herbs (see appendix 6)* (off-label), *honesty* (off-label), *hybridberry* (off-label), *linseed* (off-label), *lucerne* (off-label), *lupins* (off-label), *mallow (althaea spp.)* (off-label), *mustard* (off-label), *navy beans* (off-label), *non-edible flowers* (off-label), *ornamental plant production* (off-label), *pak choi* (off-label), *parsley root* (off-label), *parsnips* (off-label), *peaches* (off-label), *pears* (off-label), *plums* (off-label), *protected non-edible flowers* (off-label), *protected ornamentals* (off-label), *quinces* (off-label), *red beet*

(off-label), **redcurrants** *(off-label)*, **shallots** *(off-label)*, **spring field beans** *(off-label)*, **spring greens** *(off-label)*, **swedes** *(off-label)*, **tic beans** *(off-label)*, **turnips** *(off-label)*, **walnuts** *(off-label)*, **whitecurrants** *(off-label)*, **winter field beans** *(off-label)* [2]; **blackcurrants, bulb onions, carrots, combining peas, farm forestry, fodder beet, gooseberries, hops, linseed, raspberries, spring field beans, spring oilseed rape, spring oilseed rape for industrial use, strawberries, sugar beet, swedes** *(stockfeed only)*, **turnips** *(stockfeed only)*, **vining peas, winter field beans, winter oilseed rape, winter oilseed rape for industrial use** [1-4]; **field margins, flax, flax for industrial use, linseed for industrial use** [2, 4]; **kale** *(stockfeed only)* [1, 3]
- Barren brome in **field margins** [2, 4]
- Blackgrass in **hops, spring field beans, winter field beans** [1-4]
- Green cover in **land temporarily removed from production** [1-4]
- Perennial grasses in **almonds** *(off-label)*, **apples** *(off-label)*, **apricots** *(off-label)*, **asparagus** *(off-label)*, **bilberries** *(off-label)*, **blackberries** *(off-label)*, **blueberries** *(off-label)*, **borage** *(off-label)*, **broad beans (dry-harvested)** *(off-label)*, **cabbages** *(off-label)*, **canary flower (echium spp.)** *(off-label)*, **cherries** *(off-label)*, **chicory** *(off-label)*, **choi sum** *(off-label)*, **cob nuts** *(off-label)*, **collards** *(off-label)*, **cranberries** *(off-label)*, **evening primrose** *(off-label)*, **figs** *(off-label)*, **filberts** *(off-label)*, **fodder beet** *(off-label)*, **garlic** *(off-label)*, **globe artichoke** *(off-label)*, **grapevines** *(off-label)*, **haricot beans** *(off-label)*, **hazel nuts** *(off-label)*, **herbs (see appendix 6)** *(off-label)*, **honesty** *(off-label)*, **hybridberry** *(off-label)*, **linseed** *(off-label)*, **lucerne** *(off-label)*, **lupins** *(off-label)*, **mallow (althaea spp.)** *(off-label)*, **mustard** *(off-label)*, **navy beans** *(off-label)*, **non-edible flowers** *(off-label)*, **ornamental plant production** *(off-label)*, **pak choi** *(off-label)*, **parsley root** *(off-label)*, **parsnips** *(off-label)*, **peaches** *(off-label)*, **pears** *(off-label)*, **plums** *(off-label)*, **protected non-edible flowers** *(off-label)*, **protected ornamentals** *(off-label)*, **quinces** *(off-label)*, **red beet** *(off-label)*, **redcurrants** *(off-label)*, **shallots** *(off-label)*, **spring field beans** *(off-label)*, **spring greens** *(off-label)*, **swedes** *(off-label)*, **tic beans** *(off-label)*, **turnips** *(off-label)*, **walnuts** *(off-label)*, **whitecurrants** *(off-label)*, **winter field beans** *(off-label)* [2]; **blackcurrants, bulb onions, carrots, combining peas, farm forestry, fodder beet, gooseberries, hops, linseed, raspberries, spring field beans, spring oilseed rape, spring oilseed rape for industrial use, strawberries, sugar beet, swedes** *(stockfeed only)*, **turnips** *(stockfeed only)*, **vining peas, winter field beans, winter oilseed rape, winter oilseed rape for industrial use** [1-4]; **flax, flax for industrial use, linseed for industrial use** [2, 4]; **kale** *(stockfeed only)* [1, 3]
- Volunteer cereals in **blackcurrants, bulb onions, carrots, combining peas, farm forestry, fodder beet, gooseberries, hops, linseed, raspberries, spring field beans, spring oilseed rape, spring oilseed rape for industrial use, strawberries, sugar beet, swedes** *(stockfeed only)*, **turnips** *(stockfeed only)*, **vining peas, winter field beans, winter oilseed rape, winter oilseed rape for industrial use** [1-4]; **field margins, flax, flax for industrial use, linseed for industrial use** [2, 4]; **kale** *(stockfeed only)* [1, 3]
- Wild oats in **blackcurrants, bulb onions, carrots, combining peas, farm forestry, fodder beet, gooseberries, hops, linseed, raspberries, spring field beans, spring oilseed rape, spring oilseed rape for industrial use, strawberries, sugar beet, swedes** *(stockfeed only)*, **turnips** *(stockfeed only)*, **vining peas, winter field beans, winter oilseed rape, winter oilseed rape for industrial use** [1-4]; **field margins, flax, flax for industrial use, linseed for industrial use** [2, 4]; **kale** *(stockfeed only)* [1, 3]

Extension of Authorisation for Minor Use (EAMUs)
- **almonds** *20121261 expires 31 Dec 2013* [2]
- **apples** *20121261 expires 31 Dec 2013* [2]
- **apricots** *20121261 expires 31 Dec 2013* [2]
- **asparagus** *20121323 expires 31 Dec 2013* [2]
- **bilberries** *20121258 expires 31 Dec 2013* [2]
- **blackberries** *20121315 expires 31 Dec 2013* [2]
- **blueberries** *20121258 expires 31 Dec 2013* [2]
- **borage** *20121320 expires 31 Dec 2013* [2]
- **broad beans (dry-harvested)** *20121321 expires 31 Dec 2013* [2]
- **cabbages** *20121317 expires 31 Dec 2013* [2]
- **canary flower (echium spp.)** *20121320 expires 31 Dec 2013* [2]
- **cherries** *20121261 expires 31 Dec 2013* [2]
- **chicory** *20121322 expires 31 Dec 2013* [2]

FOR FULL CONDITIONS OF USE ALWAYS READ THE PRODUCT LABEL

- **choi sum** *20121321 expires 31 Dec 2013* [2]
- **cob nuts** *20121261 expires 31 Dec 2013* [2]
- **collards** *20121321 expires 31 Dec 2013* [2]
- **cranberries** *20121258 expires 31 Dec 2013* [2]
- **evening primrose** *20121320 expires 31 Dec 2013* [2]
- **figs** *20121261 expires 31 Dec 2013* [2]
- **filberts** *20121261 expires 31 Dec 2013* [2]
- **fodder beet** *20121256 expires 31 Dec 2013* [2]
- **garlic** *20121321 expires 31 Dec 2013* [2]
- **globe artichoke** *20121260 expires 31 Dec 2013* [2]
- **grapevines** *20121316 expires 31 Dec 2013* [2]
- **haricot beans** *20121321 expires 31 Dec 2013* [2]
- **hazel nuts** *20121261 expires 31 Dec 2013* [2]
- **herbs (see appendix 6)** *20121319 expires 31 Dec 2013* [2]
- **honesty** *20121320 expires 31 Dec 2013* [2]
- **hybridberry** *20121315 expires 31 Dec 2013* [2]
- **linseed** *20121320 expires 31 Dec 2013* [2]
- **lucerne** *20121318 expires 31 Dec 2013* [2]
- **lupins** *20121255 expires 31 Dec 2013* [2]
- **mallow (althaea spp.)** *20121259 expires 31 Dec 2013* [2]
- **mustard** *20121320 expires 31 Dec 2013* [2]
- **navy beans** *20121321 expires 31 Dec 2013* [2]
- **non-edible flowers** *20121321 expires 31 Dec 2013* [2]
- **ornamental plant production** *20121321 expires 31 Dec 2013* [2]
- **pak choi** *20121321 expires 31 Dec 2013* [2]
- **parsley root** *20121259 expires 31 Dec 2013* [2]
- **parsnips** *20121321 expires 31 Dec 2013* [2]
- **peaches** *20121261 expires 31 Dec 2013* [2]
- **pears** *20121261 expires 31 Dec 2013* [2]
- **plums** *20121261 expires 31 Dec 2013* [2]
- **protected non-edible flowers** *20121321 expires 31 Dec 2013* [2]
- **protected ornamentals** *20121321 expires 31 Dec 2013* [2]
- **quinces** *20121261 expires 31 Dec 2013* [2]
- **red beet** *20121321 expires 31 Dec 2013* [2]
- **redcurrants** *20121258 expires 31 Dec 2013* [2]
- **shallots** *20121257 expires 31 Dec 2013* [2]
- **spring field beans** *20121321 expires 31 Dec 2013* [2]
- **spring greens** *20121321 expires 31 Dec 2013* [2]
- **swedes** *20121321 expires 31 Dec 2013* [2]
- **tic beans** *20121321 expires 31 Dec 2013* [2]
- **turnips** *20121321 expires 31 Dec 2013* [2]
- **walnuts** *20121261 expires 31 Dec 2013* [2]
- **whitecurrants** *20121258 expires 31 Dec 2013* [2]
- **winter field beans** *20121321 expires 31 Dec 2013* [2]

Approval information
- Fluazifop-P-butyl included in Annex I under EC Regulation 1107/2009
- Accepted by BBPA for use on hops
- Approval expiry 31 Dec 2013 [1-4]

Efficacy guidance
- Best results achieved by application when weed growth active under warm conditions with adequate soil moisture.
- Spray weeds from 2-expanded leaf stage to fully tillered, couch from 4 leaves when majority of shoots have emerged, with a second application if necessary
- Control may be reduced under dry conditions. Do not cultivate for 2 wk after spraying couch
- Annual meadow grass is not controlled
- May also be used to remove grass cover crops
- Fluazifop-P-butyl is an ACCase inhibitor herbicide. To avoid the build up of resistance do not apply products containing an ACCase inhibitor herbicide more than twice to any crop. In addition do not

use any product containing fluazifop-P-butyl in mixture or sequence with any other product containing the same ingredient
- Use these products as part of a resistance management strategy that includes cultural methods of control and does not use ACCase inhibitors as the sole chemical method of grass weed control

Restrictions
- Maximum number of treatments 1 per crop or yr for all crops
- Do not sow cereals or grass crops for at least 8 wk after application of high rate or 2 wk after low rate
- Do not apply through CDA sprayer, with hand-held equipment or from air
- Avoid treatment before spring growth has hardened or when buds opening
- Do not treat bush and cane fruit or hops between flowering and harvest
- Consult processors before treating crops intended for processing
- Oilseed rape, linseed and flax for industrial use must not be harvested for human or animal consumption nor grazed
- Do not use for forestry establishment on land not previously under arable cultivation or improved grassland
- Treated vegetation in field margins, land temporarily removed from production etc, must not be grazed or harvested for human or animal consumption and unprotected persons must be kept out of treated areas for at least 24 h

Crop-specific information
- Latest use: before 50% ground cover for swedes, turnips; before 5 leaf stage for spring oilseed rape; before flowering for blackcurrants, gooseberries, hops, raspberries, strawberries; before flower buds visible for field beans, peas, linseed, flax, winter oilseed rape; 2 wk before sowing cereals or grass for field margins, land temporarily removed from production
- HI beet crops, kale, carrots 8 wk; onions 4 wk; oilseed rape for industrial use 2 wk
- Apply to sugar and fodder beet from 1-true leaf to 50% ground cover
- Apply to winter oilseed rape from 1-true leaf to established plant stage
- Apply to spring oilseed rape from 1-true leaf but before 5-true leaves
- Apply in fruit crops after harvest. See label for timing details on other crops
- Before using on onions or peas use crystal violet test to check that leaf wax is sufficient

Environmental safety
- Dangerous for the environment
- Very toxic to aquatic organisms

Hazard classification and safety precautions
Hazard Harmful, Dangerous for the environment
Transport code 9
Packaging group III
UN Number 3082
Risk phrases R38, R50, R53a, R63
Operator protection A, C, H, M; U05a, U08, U20b
Environmental protection E15a [1-4]; E38 [2, 4]
Storage and disposal D01, D02, D05, D09a, D12a [1-4]; D10b [1]; D10c [2-4]
Medical advice M05b [1]

236   fluazinam

A dinitroaniline fungicide for use in potatoes
FRAC mode of action code: 29

See also cymoxanil + fluazinam

Products

1 ChemSource Fluazinam	ChemSource	500 g/l	SC	15045
2 Clayton Solstice	Clayton	500 g/l	SC	13943
3 Floozee	AgChem Access	500 g/l	SC	13856

FOR FULL CONDITIONS OF USE ALWAYS READ THE PRODUCT LABEL

Products – continued

4	Greencrop Solanum	Greencrop	500 g/l	SC	11052
5	Nando 500 SC	Nufarm UK	500 g/l	SC	14372
6	Shirlan	Syngenta	500 g/l	SC	10573
7	Tizca	Headland	500 g/l	SC	13877
8	Volley	Makhteshim	500 g/l	SC	13591

SECTION 2

Uses
- Blight in **potatoes** [1-8]
- Phytophthora root rot in **blackberries** *(off-label)*, **protected blackberries** *(off-label)*, **protected raspberries** *(off-label)*, **protected rubus hybrids** *(off-label)*, **raspberries** *(off-label)*, **rubus hybrids** *(off-label)* [5, 8]; **container-grown blackberry** *(off-label)*, **container-grown raspberry** *(off-label)*, **container-grown rubus hybrids** *(off-label)* [8]
- Powdery scab in **seed potatoes** *(off-label)* [6]
- Root rot in **blackberries** *(off-label)*, **protected blackberries** *(off-label)*, **protected raspberries** *(off-label)*, **protected rubus hybrids** *(off-label)*, **raspberries** *(off-label)*, **rubus hybrids** *(off-label)* [6, 7]
- Scab in **seed potatoes** *(off-label)* [3, 5, 7, 8]

Extension of Authorisation for Minor Use (EAMUs)
- **blackberries** *20092990* [5], *20032168* [6], *20080597* [7], *20073759* [8]
- **container-grown blackberry** *20073759* [8]
- **container-grown raspberry** *20073759* [8]
- **container-grown rubus hybrids** *20073759* [8]
- **protected blackberries** *20092990* [5], *20032168* [6], *20080597* [7], *20073759* [8]
- **protected raspberries** *20092990* [5], *20032168* [6], *20080597* [7], *20073759* [8]
- **protected rubus hybrids** *20092990* [5], *20032168* [6], *20080597* [7], *20073759* [8]
- **raspberries** *20092990* [5], *20032168* [6], *20080597* [7], *20073759* [8]
- **rubus hybrids** *20092990* [5], *20032168* [6], *20080597* [7], *20073759* [8]
- **seed potatoes** *20100375* [3], *20092989* [5], *20060893* [6], *20080598* [7], *20073757* [8]

Approval information
- Fluazinam included in Annex I under EC Regulation 1107/2009

Efficacy guidance
- Commence treatment at the first blight risk warning (before blight enters the crop). Products are rainfast within 1 h
- In the absence of a warning, treatment should start before foliage of adjacent plants meets in the rows
- Spray at 7-14 d intervals (5-10 d intervals [6]) depending on severity of risk (see label)
- Ensure complete coverage of the foliage and stems, increasing volume as haulm growth progresses, in dense crops and if blight risk increases

Restrictions
- Maximum number of treatments 10 per crop on potatoes [4]
- Maximum total dose equivalent to 7.5 full dose treatments [6, 8]
- Do not use with hand-held sprayers

Crop-specific information
- HI 0 - 10 d for potatoes. Check label
- Ensure complete kill of potato haulm before lifting and do not lift crops for storage while there is any green tissue left on the leaves or stem bases

Environmental safety
- Dangerous for the environment
- Very toxic to aquatic organisms
- LERAP Category B

Hazard classification and safety precautions
> **Hazard** Harmful [1, 5]; Irritant [2-4, 6, 7]; Dangerous for the environment [1-8]
> **Transport code** 9
> **Packaging group** III

SEE SECTION 3 FOR PRODUCTS ALSO REGISTERED

UN Number 3082
Risk phrases R36 [1-6]; R41 [7]; R43 [1-7]; R50, R53a [1-8]; R63 [1, 5]
Operator protection A, C, H; U02a, U04a, U08 [1-6, 8]; U05a [1-7]; U11 [7]; U14 [1-3, 5-7]; U15,
U20a [1-3, 5, 6]; U20b [4, 7, 8]
Environmental protection E15a, E16a [1-8]; E16b, E34 [1-3, 5, 6, 8]; E38 [1-3, 5-8]
Storage and disposal D01, D02, D05 [1-7]; D09a [1-8]; D10a [8]; D10b [4]; D10c [1-3, 5-7]; D12a
[1-3, 5-8]
Medical advice M03 [1-6, 8]; M05a [1-3, 5, 6]

237   fluazinam + metalaxyl-M

A mixture of contact and systemic fungicides for potatoes
FRAC mode of action code: 29 + 4

See also metalaxyl-M

Products

Epok	Belchim	400:200 g/l	EC	11997

Uses
• Blight in **potatoes**

Approval information
• Fluazinam and metalaxyl-M included in Annex I under EC Regulation 1107/2009

Efficacy guidance
• Commence treatment at the first blight risk warning (before blight enters the crop). Products are rainfast within 1 h
• In the absence of a warning, treatment should start before foliage of adjacent plants meets in the rows
• Spray at 7-14 d intervals depending on severity of risk (see label)
• Ensure complete coverage of the foliage and stems, increasing volume as haulm growth progresses, in dense crops and if blight risk increases
• Metalaxyl-M works best on young actively growing foliage and efficacy declines with the onset of senescence. Therefore it is recommended that the product is only used for the first part of the blight control program, which should be completed with a reliable protectant fungicide

Restrictions
• Maximum number of treatments 3 per crop

Crop-specific information
• HI 7d for potatoes

Environmental safety
• Dangerous for the environment
• Very toxic to aquatic organisms
• LERAP Category B

Hazard classification and safety precautions
Hazard Harmful, Dangerous for the environment
Transport code 9
Packaging group III
UN Number 3082
Risk phrases R20, R36, R38, R40, R43, R50, R53a
Operator protection A, C, H; U05a, U07, U11, U14, U15, U19a, U20a
Environmental protection E15a, E16a, E16b, E34, E38
Storage and disposal D01, D02, D09a, D10b, D12a
Medical advice M05a

FOR FULL CONDITIONS OF USE ALWAYS READ THE PRODUCT LABEL

238   fludioxonil

A phenylpyrrole fungicide seed treatment for wheat and barley
FRAC mode of action code: 12

See also chlorothalonil + fludioxonil + propiconazole
cymoxanil + fludioxonil + metalaxyl-M
cyprodinil + fludioxonil
difenoconazole + fludioxonil
difenoconazole + fludioxonil + tebuconazole

Products

1	Beret Gold	Syngenta	25 g/l	FS	12625
2	Maxim 100FS	Syngenta	100 g/l	FS	15683
3	Medallion	Syngenta	125 g/l	SC	15287
4	Pure Gold	Pure Amenity	125 g/l	SC	15559

Uses

- Anthracnose in **amenity grassland** *(reduction)*, **managed amenity turf** *(reduction)* [3, 4]
- Black dot in **potatoes** *(some reduction)* [2]
- Black scurf in **potatoes** [2]
- Bunt in **durum wheat** *(off-label)*, **rye** *(off-label)*, **spring wheat** *(seed treatment)*, **triticale** *(off-label)*, **winter wheat** *(seed treatment)* [1]
- Covered smut in **spring barley** *(seed treatment)*, **winter barley** *(seed treatment)* [1]
- Drechslera leaf spot in **amenity grassland** *(useful levels of control)*, **managed amenity turf** *(useful levels of control)* [3, 4]
- Fusarium foot rot and seedling blight in **durum wheat** *(off-label)*, **rye** *(off-label)*, **spring barley** *(seed treatment)*, **spring oats** *(seed treatment)*, **spring wheat** *(seed treatment)*, **triticale** *(off-label)*, **winter barley** *(seed treatment)*, **winter oats** *(seed treatment)*, **winter wheat** *(seed treatment)* [1]
- Fusarium patch in **amenity grassland**, **managed amenity turf** [3, 4]
- Leaf stripe in **spring barley** *(seed treatment - reduction)*, **winter barley** *(seed treatment - reduction)* [1]
- Microdochium nivale in **amenity grassland**, **managed amenity turf** [3, 4]
- Pyrenophora leaf spot in **spring oats** *(seed treatment)*, **winter oats** *(seed treatment)* [1]
- Septoria seedling blight in **durum wheat** *(off-label)*, **rye** *(off-label)*, **spring wheat** *(seed treatment)*, **triticale** *(off-label)*, **winter wheat** *(seed treatment)* [1]
- Silver scurf in **potatoes** *(reduction)* [2]
- Snow mould in **durum wheat** *(off-label)*, **rye** *(off-label)*, **spring barley** *(seed treatment)*, **spring oats** *(seed treatment)*, **spring wheat** *(seed treatment)*, **triticale** *(off-label)*, **winter barley** *(seed treatment)*, **winter oats** *(seed treatment)*, **winter wheat** *(seed treatment)* [1]

Extension of Authorisation for Minor Use (EAMUs)

- **durum wheat** *20121081* [1]
- **rye** *20121081* [1]
- **triticale** *20121081* [1]

Approval information

- Fludioxonil included in Annex I under EC Regulation 1107/2009
- Accepted by BBPA for use on malting barley

Efficacy guidance

- Apply direct to seed using conventional seed treatment equipment. Continuous flow treaters should be calibrated using product before use
- Effective against benzimidazole-resistant strains of *Microdochium nivale*

Restrictions

- Maximum number of treatments 1 per seed batch
- Do not apply to cracked, split or sprouted seed
- Sow treated seed within 6 mth

SEE SECTION 3 FOR PRODUCTS ALSO REGISTERED

Crop-specific information
- Latest use: before drilling
- Product may reduce flow rate of seed through drill. Recalibrate with treated seed before drilling

Environmental safety
- Dangerous for the environment
- Toxic to aquatic organisms
- Do not use treated seed as food or feed
- Treated seed harmful to game and wildlife
- LERAP Category B [3, 4]

Hazard classification and safety precautions
Hazard Dangerous for the environment
Transport code 9
Packaging group III
UN Number 3082
Risk phrases R51, R53a
Operator protection A [1, 2]; H [1]; U05a, U20b
Environmental protection E03, E15a [1]; E15b, E34 [2-4]; E16a, E16b [3, 4]; E38 [1-4]
Storage and disposal D01, D02, D09a, D12a [1-4]; D05, D11a [1, 2]; D10c [2-4]
Treated seed S01, S03 [2]; S02, S05 [1, 2]; S07 [1]

239   fludioxonil + flutriafol

A fungicide seed treatment mixture for cereals
FRAC mode of action code: 12 + 3

See also flutriafol

Products

Beret Multi	Syngenta	25:25 g/l	FS	13017

Uses
- Bunt in **spring wheat** (seed treatment), **winter wheat** (seed treatment)
- Covered smut in **spring barley** (seed treatment), **winter barley** (seed treatment)
- Fusarium foot rot and seedling blight in **spring barley** (seed treatment), **spring oats** (seed treatment), **spring wheat** (seed treatment), **winter barley** (seed treatment), **winter oats** (seed treatment), **winter wheat** (seed treatment)
- Leaf stripe in **spring barley** (seed treatment), **winter barley** (seed treatment)
- Loose smut in **spring barley** (seed treatment), **spring wheat** (seed treatment), **winter barley** (seed treatment), **winter wheat** (seed treatment)
- Pyrenophora leaf spot in **spring oats** (seed treatment), **winter oats** (seed treatment)
- Septoria seedling blight in **spring wheat** (seed treatment), **winter wheat** (seed treatment)
- Snow mould in **spring barley** (seed treatment), **spring wheat** (seed treatment), **winter barley** (seed treatment), **winter wheat** (seed treatment)

Approval information
- Fludioxonil and flutriafol included in Annex I under EC Regulation 1107/2009

Efficacy guidance
- Apply directly to the seed using conventional seed treatment equipment
- Calibrate continuous flow treaters before use
- Effective against benzimidazole-resistant strains of *Microdochium nivale*

Restrictions
- Maximum number of treatments 1 per seed batch
- Do not apply to cracked, split or sprouted seed

Crop-specific information
- Latest use: before drilling for all cereals
- Treated seed may affect the flow rate through drills. Recalibrate with treated seed before drilling

FOR FULL CONDITIONS OF USE ALWAYS READ THE PRODUCT LABEL

Environmental safety
- Dangerous for the environment
- Toxic to aquatic organisms

Hazard classification and safety precautions
    Hazard Dangerous for the environment
    Transport code 9
    Packaging group III
    UN Number 3082
    Risk phrases R51, R53a
    Operator protection A, H; U05a, U20c
    Environmental protection E03, E15a, E38
    Storage and disposal D01, D02, D05, D09a, D10c, D11a, D12a
    Treated seed S02, S05, S07

240   fludioxonil + metalaxyl M

A seed dressing for use in forage maize
FRAC mode of action code: 12 + 4

Products

Maxim XL	Syngenta	25:9.69 g/l	FS	14136

Uses
- Damping off in **forage maize**
- Fusarium in **forage maize**
- Pythium in **forage maize**

Approval information
- Fludioxonil and metalaxyl-M included in Annex I under EC Regulation 1107/2009

Restrictions
- For advice on resistance management, refer to the latest Fungicide Resistance Action Group (FRAG) guidelines

Hazard classification and safety precautions
    UN Number N/C
    Risk phrases R52, R53a
    Operator protection A, H; U05a
    Environmental protection E34, E38
    Storage and disposal D01, D02, D05, D09a, D10c, D11a, D12a
    Treated seed S01, S02, S03, S04a, S05, S07

241   fludioxonil + metalaxyl-M + thiamethoxam

A seed dressing for use in oilseed rape, fodder rape and mustard
FRAC mode of action code: 12 + 4 + 4A

See also metalaxyl-M
    thiamethoxam

Products

Cruiser OSR	Syngenta	8:32.3:280 g/l	FS	14496

Uses
- Alternaria in **fodder rape** *(moderate control)*, **mustard** *(moderate control)*, **spring oilseed rape** *(moderate control)*, **winter oilseed rape** *(moderate control)*
- Cabbage stem flea beetle in **fodder rape** *(reduction of damage)*, **mustard** *(reduction of damage)*, **spring oilseed rape** *(reduction of damage)*, **winter oilseed rape** *(reduction of damage)*
- Downy mildew in **fodder rape**, **linseed** *(off-label)*, **mustard**, **spring oilseed rape**, **winter oilseed rape**

SEE SECTION 3 FOR PRODUCTS ALSO REGISTERED

- Flea beetle in **fodder rape** *(reduction of damage)*, **mustard** *(reduction of damage)*, **spring oilseed rape** *(reduction of damage)*, **winter oilseed rape** *(reduction of damage)*
- Peach-potato aphid in **fodder rape** *(early season only)*, **mustard** *(early season only)*, **spring oilseed rape** *(early season only)*, **winter oilseed rape** *(early season only)*
- Phoma lingam in **fodder rape**, **mustard**, **spring oilseed rape**, **winter oilseed rape**
- Pythium in **fodder rape** *(reduction)*, **linseed** *(off-label)*, **mustard** *(reduction)*, **poppies for morphine production** *(off-label)*, **spring oilseed rape** *(reduction)*, **winter oilseed rape** *(reduction)*

Extension of Authorisation for Minor Use (EAMUs)
- **linseed** 20102863
- **poppies for morphine production** 20093080

Approval information
- Fludioxonil, metalaxyl-M and thiamethoxam included in Annex I under EC Regulation 1107/2009

Efficacy guidance
- Contains a neonicotinoid insecticide. The first subsequent foliar insecticide spray should contain an insecticide with a different mode of action to combat resistance - consult the IRAC website for details of modes of action.
- Contains metalaxyl-M, a phenylamide fungicide. Do not use fungicide sprays containing metalaxyl-M on crops grown from treated seed

Crop-specific information
- For use on all varieties of oilseed rape, fodder rape and mustard

Hazard classification and safety precautions
Hazard Dangerous for the environment
UN Number N/C
Risk phrases R50, R53a
Operator protection A, D, H; U05a, U07, U20b, U20e
Environmental protection E15b, E34, E38
Storage and disposal D01, D02, D09a, D10a, D10d, D11a, D12a, D14
Treated seed S01, S02, S03, S04a, S04b, S05, S06b, S07, S08, S09

242   fludioxonil + tefluthrin

A fungicide and insecticide seed treatment mixture for cereals
IRAC mode of action code: Not classified

See also tefluthrin

Products
Austral Plus	Syngenta	10:40 g/l	FS	13314

Uses
- Bunt in **spring wheat**, **winter wheat**
- Covered smut in **spring barley**, **winter barley**
- Fusarium foot rot and seedling blight in **spring barley**, **spring oats**, **spring wheat**, **winter barley**, **winter oats**, **winter wheat**
- Leaf stripe in **spring barley** *(partial control)*, **winter barley** *(partial control)*
- Pyrenophora leaf spot in **spring oats**, **winter oats**
- Seed-borne diseases in **triticale seed crop** *(off-label)*
- Septoria seedling blight in **spring wheat**, **winter wheat**
- Snow mould in **spring barley**, **spring wheat**, **winter barley**, **winter wheat**
- Wheat bulb fly in **spring barley**, **spring wheat**, **winter barley**, **winter wheat**
- Wireworm in **spring barley**, **spring oats**, **spring wheat**, **winter barley**, **winter oats**, **winter wheat**

Extension of Authorisation for Minor Use (EAMUs)
- **triticale seed crop** 20080982

Approval information
- Fludioxonil and tefluthrin included in Annex 1 under EC Regulation 1107/2009
- Accepted by BBPA for use on malting barley

Efficacy guidance
- Apply direct to seed using conventional seed treatment equipment. Continuous flow treaters should be calibrated using product before use
- Best results obtained from seed drilled into a firm even seedbed
- Tefluthrin is released into soil after drilling and repels or kills larvae of wheat bulb fly and wireworm attacking below ground. Pest control may be reduced by deep or shallow drilling
- Where egg counts of wheat bulb fly or population counts of wireworm indicate a high risk of severe attack follow up spray treatments may be needed
- Control of leaf stripe in barley may not be sufficient in crops grown for seed certification
- Effective against benzimidazole-resistant strains of *Microdochium nivale*
- Under adverse soil or environmental conditions seed rates should be increased to compensate for possible reduced germination capacity

Restrictions
- Maximum number of treatments 1 per batch
- Store treated seed in cool dry conditions and drill within 3 mth
- Do not use on seed above 16% moisture content or on sprouted, cracked or damaged seed

Crop-specific information
- Latest use: before drilling

Environmental safety
- Dangerous for the environment
- Very toxic to aquatic organisms

Hazard classification and safety precautions
    Hazard Dangerous for the environment
    Transport code 9
    Packaging group III
    UN Number 3082
    Risk phrases R50, R53a
    Operator protection A, C, D, H; U02a, U04a, U05a, U07, U08, U14, U20b
    Environmental protection E03, E15b, E34, E38
    Storage and disposal D01, D02, D05, D09b, D10c, D11a, D12a
    Treated seed S02, S04b, S05, S07, S08

---

243   flufenacet

A broad spectrum oxyacetamide herbicide available only in mixtures
HRAC mode of action code: K3

See also diflufenican + flufenacet
diflufenican + flufenacet + flurtamone

---

244   flufenacet + isoxaflutole

A residual herbicide mixture for maize
HRAC mode of action code: K3 + F2

See also isoxaflutole

Products

1	Amethyst	AgChem Access	48:10% w/w	WG	14079
2	Cadou Star	Bayer CropScience	48:10% w/w	WG	13242

SEE SECTION 3 FOR PRODUCTS ALSO REGISTERED

Uses

- Annual dicotyledons in **forage maize**, **grain maize** [1, 2]; **forage maize (under plastic mulches)** *(off-label)*, **game cover** *(off-label)*, **hops** *(off-label)*, **miscanthus** *(off-label)*, **ornamental plant production** *(off-label)*, **sweetcorn** *(off-label)*, **sweetcorn under plastic mulches** *(off-label)*, **top fruit** *(off-label)* [2]
- Annual grasses in **forage maize (under plastic mulches)** *(off-label)*, **game cover** *(off-label)*, **hops** *(off-label)*, **miscanthus** *(off-label)*, **ornamental plant production** *(off-label)*, **sweetcorn** *(off-label)*, **sweetcorn under plastic mulches** *(off-label)*, **top fruit** *(off-label)* [2]
- Grass weeds in **forage maize**, **grain maize** [1, 2]

Extension of Authorisation for Minor Use (EAMUs)

- **forage maize (under plastic mulches)** *20080960 expires 30 Sep 2013* [2]
- **game cover** *20082829 expires 30 Sep 2013* [2]
- **hops** *20082829 expires 30 Sep 2013* [2]
- **miscanthus** *20082829 expires 30 Sep 2013* [2]
- **ornamental plant production** *20082829 expires 30 Sep 2013* [2]
- **sweetcorn** *20111301 expires 30 Sep 2013* [2]
- **sweetcorn under plastic mulches** *20111301 expires 30 Sep 2013* [2]
- **top fruit** *20082829 expires 30 Sep 2013* [2]

Approval information

- Flufenacet and isoxaflutole included in Annex I under EC Regulation 1107/2009
- Approval expiry 30 Sep 2013 [1, 2]

Efficacy guidance

- Best results obtained from applications to a fine, firm seedbed in the presence of some soil moisture
- Efficacy may be reduced on cloddy seedbeds or under prolonged dry conditions
- Established perennial grasses and broad-leaved weeds growing from rootstocks will not be controlled
- Always follow WRAG guidelines for preventing and managing herbicide resistant weeds. See Section 5 for more information

Restrictions

- Maximum number of treatments 1 per crop
- Do not use on soils both above 70% sand and less than 2% organic matter
- Do not use on maize crops intended for seed production

Crop-specific information

- Latest use: before crop emergence
- Ideally treatment should be made within 4 d of sowing, before the maize seeds have germinated

Following crops guidance

- In the event of failure of a treated crop maize, sweet corn or potatoes may be grown after surface cultivation or ploughing
- After normal harvest of a treated crop oats or barley may be sown after ploughing and wheat, rye, triticale, sugar beet, field beans, peas, soya, sorghum or sunflowers may be sown after surface cultivation or ploughing.

Environmental safety

- Dangerous for the environment
- Very toxic to aquatic organisms
- Take care to avoid drift over other crops. Beet crops and sunflowers are particularly sensitive
- LERAP Category B

Hazard classification and safety precautions

Hazard Harmful, Dangerous for the environment
Transport code 9
Packaging group III
UN Number 3077
Risk phrases R22a, R36, R43, R48, R50, R53a, R63
Operator protection A, C, H; U05a, U11, U20b

FOR FULL CONDITIONS OF USE ALWAYS READ THE PRODUCT LABEL

Environmental protection E15b, E16a, E34, E38
Storage and disposal D01, D02, D05, D09a, D10a, D12a
Medical advice M03

245   flufenacet + metribuzin

A herbicide mixture for potatoes
HRAC mode of action code: K3 + C1

See also metribuzin

Products

Artist	Bayer CropScience	24:17.5% w/w	WP	11239

Uses
- Annual dicotyledons in **bilberries** *(off-label)*, **blackcurrants** *(off-label)*, **blueberries** *(off-label)*, **cranberries** *(off-label)*, **early potatoes**, **gooseberries** *(off-label)*, **maincrop potatoes**, **redcurrants** *(off-label)*, **ribes hybrids** *(off-label)*
- Annual meadow grass in **bilberries** *(off-label)*, **blackcurrants** *(off-label)*, **blueberries** *(off-label)*, **cranberries** *(off-label)*, **early potatoes**, **gooseberries** *(off-label)*, **maincrop potatoes**, **redcurrants** *(off-label)*, **ribes hybrids** *(off-label)*

Extension of Authorisation for Minor Use (EAMUs)
- *bilberries 20111355*
- *blackcurrants 20111355*
- *blueberries 20111355*
- *cranberries 20111355*
- *gooseberries 20111355*
- *redcurrants 20111355*
- *ribes hybrids 20111355*

Approval information
- Flufenacet and metribuzin included in Annex I under EC Regulation 1107/2009

Efficacy guidance
- Product acts through root uptake and needs sufficient soil moisture at and shortly after application
- Effectiveness is reduced under dry soil conditions
- Residual activity is reduced on mineral soils with a high organic matter content and on peaty or organic soils
- Ensure application is made evenly to both sides of potato ridges
- Perennial weeds are not controlled

Restrictions
- Maximum total dose equivalent to one full dose treatment
- Potatoes must be sprayed before emergence of crop and weeds
- See label for list of tolerant varieties. Do not treat Maris Piper grown on Sands or Very Light soils
- Do not use on Sands
- On stony or gravelly soils there is risk of crop damage especially if heavy rain falls soon after application

Crop-specific information
- Latest use: before potato crop emergence
- Consult processor before use on crops for processing

Following crops guidance
- Before drilling or planting any succeeding crop soil must be mouldboard ploughed to at least 15 cm as soon as possible after lifting and no later than end Dec
- In W Cornwall on soils with more than 5% organic matter treated early potatoes may be followed by summer planted brassica crops 14 wk after treatment and after mouldboard ploughing.

SEE SECTION 3 FOR PRODUCTS ALSO REGISTERED

Elsewhere cereals or winter beans may be grown in the same year if at least 16 wk have elapsed since treatment
- In the yr following treatment any crop may be grown except lettuce or radish, or vegetable brassica crops on silt soils in Lincs

Environmental safety
- Dangerous for the environment
- Very toxic to aquatic organisms
- Take care to avoid spray drift onto neighbouring crops, especially lettuce or brassicas

Hazard classification and safety precautions
Hazard Harmful, Dangerous for the environment
Transport code 9
Packaging group III
UN Number 3077
Risk phrases R22a, R43, R48, R50, R53a
Operator protection A, D, H; U05a, U08, U13, U14, U19a, U20b
Environmental protection E15a, E38
Storage and disposal D01, D02, D09a, D11a, D12a
Medical advice M03

246 flufenacet + pendimethalin

A broad spectrum residual and contact herbicide mixture for winter cereals
HRAC mode of action code: K3 + K1

See also pendimethalin

Products

1	Crystal	BASF	60:300 g/l	EC	13914
2	Rock	Goldengrass	60:300 g/l	EC	14928
3	Shooter	BASF	60:300 g/l	EC	14106
4	Trooper	BASF	60:300 g/l	EC	13924

Uses
- Annual dicotyledons in *game cover* (off-label) [1]; *winter barley*, *winter wheat* [1-4]
- Annual grasses in *winter barley*, *winter wheat* [1-4]
- Annual meadow grass in *game cover* (off-label) [1]; *winter barley*, *winter wheat* [1-4]
- Blackgrass in *game cover* (off-label) [1]; *winter barley*, *winter wheat* [1-4]
- Chickweed in *winter barley*, *winter wheat* [1-4]
- Corn marigold in *winter barley*, *winter wheat* [1-4]
- Field speedwell in *winter barley*, *winter wheat* [1-4]
- Ivy-leaved speedwell in *winter barley*, *winter wheat* [1-4]

Extension of Authorisation for Minor Use (EAMUs)
- *game cover* 20090450 expires 31 Dec 2013 [1]

Approval information
- Flufenacet and pendimethalin included in Annex I under EC Regulation 1107/2009
- Accepted by BBPA for use on malting barley
- Approval expiry 31 Dec 2013 [1-4]

Efficacy guidance
- Best results achieved when applied from pre-emergence of weeds up to 2 leaf stage but post emergence treatment is not recommended on clay soils
- Product requires some soil moisture to be activated ideally from rain within 7 d of application. Prolonged dry conditions may reduce residual control
- Product is slow acting and final level of weed control may take some time to appear
- For effective weed control seed bed preparations should ensure even incorporation of any trash, straw and ash to 15 cm
- Efficacy may be reduced on soils with more than 6% organic matter

FOR FULL CONDITIONS OF USE ALWAYS READ THE PRODUCT LABEL

- Always follow WRAG guidelines for preventing and managing herbicide resistant weeds. See Section 5 for more information

Restrictions
- Maximum total dose equivalent to one full dose treatment
- For pre-emergence treatments seed should be covered with at least 32 mm settled soil. Shallow drilled crops should be treated post-emergence only
- Do not treat undersown crops
- Avoid spraying during periods of prolonged or severe frosts
- Do not use on stony or gravelly soils or those with more than 10% organic matter
- Pre-emergence treatment may only be used on crops drilled before 30 Nov. All crops must be treated before 31 Dec in yr of planting
- Concentrated or diluted product may stain clothing or skin

Crop-specific information
- Latest use: before third tiller stage (GS 23) and before 31 Dec in yr of planting
- Very wet weather before and after treatment may result in loss of crop vigour and reduced yield, particularly where soils become waterlogged

Following crops guidance
- Any crop may follow a failed or normally harvested treated crop provided ploughing to at least 15 cm is carried out beforehand

Environmental safety
- Dangerous for the environment
- Very toxic to aquatic organisms
- Risk to certain non-target insects or other arthropods
- Some products supplied in small volume returnable packs. Follow instructions for use
- LERAP Category B

Hazard classification and safety precautions
> **Hazard** Harmful, Dangerous for the environment
> **Transport code** 3
> **Packaging group** III
> **UN Number** 1993
> **Risk phrases** R22a, R38, R50, R53a [1-4]; R22b, R40 [1, 2, 4]
> **Operator protection** A, C, H; U02a, U05a [1-4]; U14, U20b [3]; U20c [1, 2, 4]
> **Environmental protection** E15a [1, 2, 4]; E15b [3]; E16a, E16b, E22b, E34, E38 [1-4]
> **Storage and disposal** D01, D02, D09a, D12a [1-4]; D05 [2]; D08, D10c [3]; D10a [1, 2, 4]
> **Medical advice** M03 [1-4]; M05b [1, 2, 4]

247 flumioxazine

A phenyphthalimide herbicide for winter wheat
HRAC mode of action code: E

Products

1	Digital	Interfarm	300 g/l	SC	13561
2	Guillotine	Interfarm	300 g/l	SC	13562
3	Sumimax	Interfarm	300 g/l	SC	13548

Uses
- Annual dicotyledons in *bulb onions* (off-label), *carrots* (off-label), *ornamental plant production* (off-label), *parsnips* (off-label), *vining peas* (off-label), *winter wheat* [1-3]; *farm forestry* (off-label), *forest nurseries* (off-label) [1]; *hops* (off-label), *soft fruit* (off-label), *top fruit* (off-label) [2, 3]
- Annual grasses in *winter oats* (off-label) [1-3]
- Annual meadow grass in *winter wheat* [1-3]
- Blackgrass in *winter oats* (off-label) [1-3]
- Chickweed in *winter wheat* [1-3]
- Cleavers in *winter wheat* [1-3]

SEE SECTION 3 FOR PRODUCTS ALSO REGISTERED

- Groundsel in **bulb onions** *(off-label)*, **carrots** *(off-label)*, **ornamental plant production** *(off-label)*, **parsnips** *(off-label)*, **vining peas** *(off-label)* [1-3]; **farm forestry** *(off-label)*, **forest nurseries** *(off-label)* [1]; **hops** *(off-label)*, **soft fruit** *(off-label)*, **top fruit** *(off-label)* [2, 3]
- Loose silky bent in **winter wheat** [1-3]
- Mayweeds in **winter wheat** [1-3]
- Ryegrass in **winter oats** *(off-label)* [1-3]
- Speedwells in **winter wheat** [1-3]
- Volunteer oilseed rape in **bulb onions** *(off-label)*, **carrots** *(off-label)*, **ornamental plant production** *(off-label)*, **parsnips** *(off-label)*, **vining peas** *(off-label)*, **winter wheat** [1-3]; **farm forestry** *(off-label)*, **forest nurseries** *(off-label)* [1]; **hops** *(off-label)*, **soft fruit** *(off-label)*, **top fruit** *(off-label)* [2, 3]
- Volunteer potatoes in **bulb onions** *(off-label)*, **carrots** *(off-label)*, **ornamental plant production** *(off-label)*, **parsnips** *(off-label)*, **vining peas** *(off-label)* [1-3]; **farm forestry** *(off-label)*, **forest nurseries** *(off-label)* [1]; **hops** *(off-label)*, **soft fruit** *(off-label)*, **top fruit** *(off-label)* [2, 3]

Extension of Authorisation for Minor Use (EAMUs)
- **bulb onions** *20091114* [1], *20091116* [2], *20091108* [3]
- **carrots** *20091112* [1], *20091117* [2], *20091111* [3]
- **farm forestry** *20082844* [1]
- **forest nurseries** *20082844* [1]
- **hops** *20082897* [2], *20082881* [3]
- **ornamental plant production** *20082844* [1], *20082897* [2], *20082881* [3]
- **parsnips** *20091112* [1], *20091117* [2], *20091111* [3]
- **soft fruit** *20082897* [2], *20082881* [3]
- **top fruit** *20082897* [2], *20082881* [3]
- **vining peas** *20091113* [1], *20091115* [2], *20091109* [3]
- **winter oats** *20093121* [1], *20092506* [2], *20093122* [3]

Approval information
- Flumioxazine is not listed in Annex 1 under EC Regulation 1107/2009

Efficacy guidance
- Best results obtained from applications to moist soil early post-emergence of the crop when weeds are germinating up to 1 leaf stage
- Flumioxazin is contact acting and relies on weeds germinating in moist soil and coming into sufficient contact with the herbicide
- Flumioxazin remains in the surface layer of the soil and does not migrate to lower layers
- Contact activity is long lasting during the autumn
- Seed beds should be firm, fine and free from clods. Crops should be drilled to 25 mm and well covered by soil
- Efficacy is reduced on soils with more than 10% organic matter
- Flumioxazin is a member of the protoporphyrin oxidase (PPO) inhibitor herbicides. To avoid the build up of resistance do not use PPO-inhibitor herbicides more than once on any crop
- Use as part of a resistance management strategy that includes cultural methods of control and does not use PPO-inhibitors as the sole chemical method of grass weed control

Restrictions
- Maximum number of treatments 1 per crop for winter wheat
- Only apply to crops that have been hardened by cool weather. Do not treat crops with lush or soft growth
- Do not apply to waterlogged soils or to soils with more than 10% organic matter
- Do not follow an application with another pesticide for 14 d
- Do not treat undersown crops
- Do not treat broadcast crops until they are past the 3 leaf stage
- Do not treat crops under stress for any reason, or during prolonged frosty weather
- Do not roll or harrow for 2 wk before treatment or at any time afterwards

Crop-specific information
- Latest use: before 5th true leaf stage (GS 15) for winter wheat
- Light discolouration of leaf margins can occur after treatment and treatment of soft crops will cause transient leaf bleaching

FOR FULL CONDITIONS OF USE ALWAYS READ THE PRODUCT LABEL

Following crops guidance
- Any crop may be sown following normal harvest of a treated crop
- In the event of failure of a treated crop spring cereals, spring oilseed rape, sugar beet, maize or potatoes may be redrilled after ploughing

Environmental safety
- Dangerous for the environment
- Very toxic to aquatic organisms
- Take extreme care to avoid drift onto plants outside the target area
- LERAP Category B

Hazard classification and safety precautions
Hazard Toxic, Dangerous for the environment
Transport code 9
Packaging group III
UN Number 3082
Risk phrases R50, R53a, R61
Operator protection A, H; U05a, U19a
Environmental protection E15b, E16a, E34, E38
Storage and disposal D01, D02, D05, D07, D09a, D12b
Medical advice M04a

248   fluopicolide

An benzamide fungicide available only in mixtures
FRAC mode of action code: 43

249   fluopicolide + propamocarb hydrochloride

A protectant and systemic fungicide mixture for potato blight
FRAC mode of action code: 43 + 28

See also propamocarb hydrochloride

Products

1	Beyond	ChemSource	62.5:625 g/l	SC	15023
2	Infinito	Bayer CropScience	62.5:625 g/l	SC	12644

Uses
- Blight in **potatoes** [1, 2]
- Disease control in **cabbages** *(off-label)*, **cauliflowers** *(off-label)* [2]

Extension of Authorisation for Minor Use (EAMUs)
- *cabbages 20101096* [2]
- *cauliflowers 20101096* [2]

Approval information
- Fluopicolide and propamocarb hydrochloride included in Annex I under EC Regulation 1107/2009

Efficacy guidance
- Commence spray programme before infection appears as soon as weather conditions favourable for disease development occur. At latest the first treatment should be made as the foliage meets along the rows
- Repeat treatments at 7-10 day intervals according to disease incidence and weather conditions
- Reduce spray interval if conditions are conducive to the spread of blight
- Spray as soon as possible after irrigation
- Increase water volume in dense crops
- When used from full canopy development to haulm desiccation as part of a full blight protection programme tubers will be protected from late blight after harvest and tuber blight incidence will be reduced

SEE SECTION 3 FOR PRODUCTS ALSO REGISTERED

- To reduce the development of resistance product should be used in single or block applications with fungicides from a different cross-resistance group

Restrictions
- Maximum total dose on potatoes equivalent to four full dose treatments
- Do not apply if rainfall or irrigation is imminent. Product is rainfast in 1 h provided spray has dried on leaf
- Do not apply as a curative treatment when blight is present in the crop
- Do not apply more than 3 consecutive treatments of the product

Crop-specific information
- HI 7 d for potatoes
- All varieties of potatoes, including seed crops, may be treated

Environmental safety
- Dangerous for the environment
- Toxic to aquatic organisms

Hazard classification and safety precautions
> **Hazard** Dangerous for the environment
> **Transport code** 9
> **Packaging group** III
> **UN Number** 3082
> **Risk phrases** R51, R53a
> **Operator protection** A, H; U05a, U08, U11, U19a, U20a
> **Environmental protection** E15a, E38
> **Storage and disposal** D01, D02, D09a, D10b, D12a

250 fluopyram

An SDHI fungicide only available in mixtures
FRAC mode of action code: 7

251 fluopyram + prothioconazole + tebuconazole

An SDHI and triazole mixture for seed treatment in winter barley
FRAC mode of action code: 3 + 7

Products

Raxil Star	Bayer CropScience	20.8:103.1:63.2 g/l	FS	15197

Uses
- Covered smut in *spring barley, winter barley*
- Fusarium in *spring barley, winter barley*
- Leaf stripe in *spring barley, winter barley*
- Loose smut in *spring barley, winter barley*
- Microdochium nivale in *spring barley, winter barley*
- Net blotch in *spring barley* (seed borne), *winter barley* (seed borne)

Approval information
- Prothioconazole and tebuconazole included in Annex 1 under EC Regulation 1107/2009 but fluopyram not yet included.
- Accepted by BBPA for use on cereals

Efficacy guidance
- Seed should be drilled to a depth of 40 mm into a well prepared and firm seedbed. If seed is present on the soil surface, or spills have occurred, then, if conditions are appropriate, the field should be harrowed then rolled to ensure good incorporation
- Treated seed must not be sown between 1 February and 31 August

FOR FULL CONDITIONS OF USE ALWAYS READ THE PRODUCT LABEL

- Do not use on seed with more than 16% moisture content or on sprouted, cracked, skinned or otherwise damaged seed

Hazard classification and safety precautions

Hazard Harmful, Dangerous for the environment
Transport code 9
Packaging group III
UN Number 3082
Risk phrases R40, R52, R53a, R63
Operator protection A, C, D, H; U05a, U12b, U20b, U20e
Environmental protection E15b, E34, E38
Storage and disposal D01, D02, D05, D09a, D10d, D12a, D14
Treated seed S01, S02, S04b, S05, S07, S08

252 fluoxastrobin

A protectant stobilurin fungicide available in mixtures
FRAC mode of action code: 11

253 fluoxastrobin + prothioconazole

A strobilurin and triazole fungicide mixture for cereals
FRAC mode of action code: 11 + 3

See also prothioconazole

Products

1	Fandango	Bayer CropScience	100:100 g/l	EC	12276
2	Firefly 155	Bayer CropScience	45:110 g/l	EC	14818
3	Kurdi	AgChem Access	100+100 g/l	EC	13531
4	Maestro	Bayer CropScience	100:100 g/l	EC	12307
5	Unicur	Bayer CropScience	100:100 g/l	EC	14776

Uses

- Botrytis in **bulb onions** *(useful reduction)* [5]
- Botrytis squamosa in **bulb onions** *(control)* [5]
- Brown rust in **spring barley**, **winter barley** [1, 3, 4]; **spring wheat**, **winter rye**, **winter wheat** [1-4]
- Cladosporium in **bulb onions** *(useful reduction)* [5]
- Crown rust in **spring oats**, **winter oats** [1, 3, 4]
- Downy mildew in **bulb onions** *(control)* [5]
- Eyespot in **spring barley** *(reduction)*, **spring oats**, **winter barley** *(reduction)*, **winter oats** [1, 3, 4]; **spring oats** *(reduction of incidence and severity)*, **winter oats** *(reduction of incidence and severity)* [2]; **spring wheat** [4]; **spring wheat** *(reduction)* [1-3]; **winter rye** *(reduction)*, **winter wheat** *(reduction)* [1-4]
- Foliar disease control in **forest nurseries** *(off-label)* [1]
- Fusarium root rot in **spring wheat** *(reduction)*, **winter wheat** *(reduction)* [1, 3]
- Glume blotch in **spring wheat** [1-3]; **spring wheat** *(reduction)* [4]; **winter wheat** [1-4]
- Late ear diseases in **spring barley**, **spring wheat**, **winter barley**, **winter wheat** [1, 3, 4]
- Net blotch in **spring barley**, **winter barley** [1, 3, 4]
- Powdery mildew in **spring barley**, **spring oats**, **winter barley**, **winter oats** [1, 3, 4]; **spring wheat**, **winter rye**, **winter wheat** [1-4]
- Rhynchosporium in **spring barley**, **winter barley** [1, 3, 4]; **winter rye** [1-4]
- Rust in **bulb onions** *(useful reduction)* [5]
- Septoria leaf blotch in **spring wheat**, **winter wheat** [1-4]
- Sharp eyespot in **spring wheat** *(reduction)*, **winter wheat** *(reduction)* [1, 3]
- Sooty moulds in **spring wheat** *(reduction)*, **winter wheat** *(reduction)* [1, 3]
- Take-all in **spring wheat** *(reduction)*, **winter barley** *(reduction)*, **winter wheat** *(reduction)* [1, 3, 4]

SEE SECTION 3 FOR PRODUCTS ALSO REGISTERED

- Tan spot in **spring wheat** [1, 3]; **winter wheat** [1, 3, 4]
- Yellow rust in **spring wheat**, **winter wheat** [1-4]

Extension of Authorisation for Minor Use (EAMUs)
- **forest nurseries** *20090226* [1]

Approval information
- Fluoxastrobin and prothioconazole included in Annex I under EC Regulation 1107/2009
- Accepted by BBPA for use on malting barley
- Approval expiry 30 Apr 2013 [3]

Efficacy guidance
- Seed treatment products must be applied by manufacturer's recommended treatment application equipment
- Treated cereal seed should preferably be drilled in the same season
- Follow-up treatments will be needed later in the season to give protection against air-borne and splash-borne diseases
- Best results on foliar diseases obtained from treatment at early stages of disease development. Further treatment may be needed if disease attack is prolonged [1, 4]
- Foliar applications to established infections of any disease are likely to be less effective [1, 4]
- Best control of cereal ear diseases obtained by treatment during ear emergence [1, 4]
- Fluoxastrobin is a member of the QoI cross resistance group. Foliar product should be used preventatively and not relied on for its curative potential
- Use product as part of an Integrated Crop Management strategy incorporating other methods of control, including where appropriate other fungicides with a different mode of action. Do not apply more than two foliar applications of QoI containing products to any cereal crop
- There is a significant risk of widespread resistance occurring in *Septoria tritici* populations in UK. Failure to follow resistance management action may result in reduced levels of disease control
- Strains of wheat and barley powdery mildew resistant to QoI's are common in the UK. Control of wheat mildew can only be relied on from the triazole component
- Prothioconazole is a DMI fungicide. Resistance to some DMI fungicides has been identified in Septoria leaf blotch which may seriously affect performance of some products. For further advice contact a specialist advisor and visit the Fungicide Resistance Action Group (FRAG)-UK website
- Where specific control of wheat mildew is required this should be achieved through a programme of measures including products recommended for the control of mildew that contain a fungicide from a different cross-resistance group and applied at a dose that will give robust control

Restrictions
- Maximum total dose of foliar sprays equivalent to two full dose treatments for the crop [1, 4]

Crop-specific information
- Latest use: before grain milky ripe for spray treatments on wheat and rye; beginning of flowering for barley, oats [1, 4]
- Some transient leaf chlorosis may occur after treatment of wheat or barley but this has not been found to affect yield [1, 4]

Environmental safety
- Dangerous for the environment
- Toxic to aquatic organisms
- Risk to non-target insects or other arthropods. Avoid spraying within 6 m of the field boundary to reduce the effects on non-target insects or other arthropods [1, 4]
- LERAP Category B

Hazard classification and safety precautions
Hazard Dangerous for the environment
Transport code 9
Packaging group III
UN Number 3082
Risk phrases R51, R53a
Operator protection A [1-5]; H [2, 5]; U05a [1-5]; U09a, U20b [2]; U09b, U20a [1, 3-5]

FOR FULL CONDITIONS OF USE ALWAYS READ THE PRODUCT LABEL

Environmental protection E15a, E16a, E34, E38 [1-5]; E22c [1, 3-5]
Storage and disposal D01, D05, D09a, D10b, D12a
Medical advice M03

254   fluoxastrobin + prothioconazole + trifloxystrobin

A triazole and strobilurin fungicide mixture for cereals
FRAC mode of action code: 11 + 3 + 11

See also prothioconazole
trifloxystrobin

Products

Jaunt	Bayer CropScience	75:150:75 g/l	EC	12350

Uses
- Brown rust in *durum wheat, rye, spring barley, spring wheat, triticale, winter barley, winter wheat*
- Eyespot in *durum wheat* (reduction), *rye* (reduction), *spring barley* (reduction), *spring wheat* (reduction), *triticale* (reduction), *winter barley* (reduction), *winter wheat* (reduction)
- Glume blotch in *durum wheat, rye, spring wheat, triticale, winter wheat*
- Late ear diseases in *durum wheat, rye, spring barley, spring wheat, triticale, winter barley, winter wheat*
- Net blotch in *spring barley, winter barley*
- Powdery mildew in *durum wheat, rye, spring barley, spring wheat, triticale, winter barley, winter wheat*
- Rhynchosporium in *spring barley, winter barley*
- Septoria leaf blotch in *durum wheat, rye, spring wheat, triticale, winter wheat*
- Tan spot in *durum wheat, rye, spring wheat, triticale, winter wheat*
- Yellow rust in *durum wheat, rye, spring wheat, triticale, winter wheat*

Approval information
- Fluoxastrobin, prothioconazole and trifloxystrobin included in Annex I under EC Regulation 1107/2009
- Accepted by BBPA for use on malting barley

Efficacy guidance
- Best results obtained from treatment at early stages of disease development. Further treatment may be needed if disease attack is prolonged
- Applications to established infections of any disease are likely to be less effective
- Best control of cereal ear diseases obtained by treatment during ear emergence
- Fluoxastrobin and trifloxystrobin are members of the QoI cross resistance group. Product should be used preventatively and not relied on for its curative potential
- Use product as part of an Integrated Crop Management strategy incorporating other methods of control, including where appropriate other fungicides with a different mode of action. Do not apply more than two foliar applications of QoI containing products to any cereal crop
- There is a significant risk of widespread resistance occurring in *Septoria tritici* populations in UK. Failure to follow resistance management action may result in reduced levels of disease control
- Strains of wheat and barley powdery mildew resistant to QoIs are common in the UK. Control of wheat mildew can only be relied on from the triazole component
- Where specific control of wheat mildew is required this should be achieved through a programme of measures including products recommended for the control of mildew that contain a fungicide from a different cross-resistance group and applied at a dose that will give robust control
- Prothioconazole is a DMI fungicide. Resistance to some DMI fungicides has been identified in Septoria leaf blotch which may seriously affect performance of some products. For further advice contact a specialist advisor and visit the Fungicide Resistance Action Group (FRAG)-UK website

---

SEE SECTION 3 FOR PRODUCTS ALSO REGISTERED

Restrictions
- Maximum total dose equivalent to two full dose treatments

Crop-specific information
- Latest use: before grain milky ripe for winter wheat; up to beginning of anthesis (GS 61) for barley

Environmental safety
- Dangerous for the environment
- Very toxic to aquatic organisms
- Risk to non-target insects or other arthropods. Avoid spraying within 6 m of the field boundary to reduce the effects on non-target insects or other arthropods
- LERAP Category B

Hazard classification and safety precautions
Hazard Irritant, Dangerous for the environment
Transport code 9
Packaging group III
UN Number 3082
Risk phrases R37, R50, R53a
Operator protection A, C, H; U05a, U09b, U19a, U20b
Environmental protection E15a, E16a, E22c, E34, E38
Storage and disposal D01, D02, D05, D09a, D10b, D12a
Medical advice M03

255   flupyrsulfuron-methyl

A sulfonylurea herbicide for winter wheat and winter barley
HRAC mode of action code: B

See also carfentrazone-ethyl + flupyrsulfuron-methyl
        diflufenican + flupyrsulfuron-methyl

Products

1	Bullion	DuPont	50% w/w	WG	14058
2	Ductis SX	DuPont	50% w/w	WG	15426
3	Exceed SX	DuPont	50% w/w	WG	15427
4	Lexus SX	DuPont	50% w/w	WG	12979
5	Oklar SX	DuPont	50% w/w	WG	15037
6	Oriel 50SX	DuPont	50% w/w	WG	14640
7	Staka SX	DuPont	50% w/w	WG	15428

Uses
- Annual dicotyledons in *triticale, winter oats, winter rye* [1, 3-7]; *winter barley* [1-4, 7]; *winter linseed* (off-label) [4]; *winter wheat* [1-7]
- Blackgrass in *triticale, winter oats, winter rye* [1, 3, 4, 7]; *winter barley, winter wheat* [1-4, 7]; *winter linseed* (off-label) [4]

Extension of Authorisation for Minor Use (EAMUs)
- *winter linseed* 20063714 [4]

Approval information
- Flupyrsulfuron-methyl included in Annex I under EC Regulation 1107/2009
- Accepted by BBPA for use on malting barley

Efficacy guidance
- Best results achieved from applications made in good growing conditions
- Good spray cover of weeds must be obtained
- For pre-emergence control of blackgrass application must be made in tank mixture with specified products (see label). Best control of blackgrass post-emergence in wheat obtained from application from the 1-leaf stage
- Winter barley must be treated pre-emergence and used in mixture. See label

FOR FULL CONDITIONS OF USE ALWAYS READ THE PRODUCT LABEL

- Growth of weeds is inhibited within hours of treatment but visible symptoms may not be apparent for up to 4 wk
- Flupyrsulfuron-methyl has moderate residual life in soil. Under normal moisture conditions susceptible weeds germinating soon after treatment will be controlled
- May be used on all soil types but residual activity and weed control is reduced on highly alkaline soils
- Flupyrsulfuron-methyl is a member of the ALS-inhibitor group of herbicides. To avoid the build up of resistance do not use any product containing an ALS-inhibitor herbicide with claims for control of grass weeds more than once on any crop. However, approval had now been granted for application of flupyrsulfuron-methyl in mixture with a non-ALS herbicide pre-emergence of blackgrass in sequence with post-emergence applications of iodosulfuron + mesosulfuron products.
- Use these products as part of a resistance management strategy that includes cultural methods of control and does not use ALS inhibitors as the sole chemical method of grass weed control

Restrictions
- Maximum number of treatments 1 per crop
- Do not use on wheat undersown with grasses, clover or other legumes, or any other broad-leaved crop
- Do not apply within 7 d of rolling
- Do not treat any crop suffering from drought, waterlogging, pest or disease attack, nutrient deficiency, or any other stress factors
- Specific restrictions apply to use in sequence or tank mixture with other sulfonylurea or ALS-inhibiting herbicides. See label for details

Crop-specific information
- Latest use: pre-emergence or before 1st node detectable (GS 31) for wheat; pre-emergence for barley
- Latest use for winter oats, rye & triticale is Dec 31st in year before harvest
- Under certain conditions chlorosis and stunting may occur, from which recovery is rapid

Following crops guidance
- Only cereals, oilseed rape, field beans, clover or grass may be sown in the yr of harvest of a treated crop
- In the event of crop failure only winter or spring wheat may be sown within 3 mth of treatment. Land should be ploughed and cultivated to 15 cm minimum before resowing. After 3 mth crops may be sown as shown above

Environmental safety
- Dangerous for the environment
- Very toxic to aquatic organisms
- Take extreme care to avoid drift onto broad-leaved plants outside the target area or onto ponds, waterways or ditches, or onto land intended for cropping
- Spraying equipment should not be drained or flushed onto land planted, or to be planted, with trees or crops other than cereals and should be thoroughly cleansed after use - see label for instructions

Hazard classification and safety precautions
Hazard Dangerous for the environment
Transport code 9
Packaging group III
UN Number 3077
Risk phrases R50, R53a [1-4, 7]
Operator protection U08, U19a, U20b
Environmental protection E15a [1-7]; E38 [1-4, 7]
Storage and disposal D09a, D11a [1-7]; D12a [1-4, 7]

SEE SECTION 3 FOR PRODUCTS ALSO REGISTERED

256   flupyrsulfuron-methyl + pyroxsulam

A sulfonyl urea and triazolinone herbicide mixture for weed control in winter wheat
HRAC mode of action code: B

Products

1	GF-2070	Dow	3.71:6.94% w/w	WG	15414
2	Unite	Dow	3.71:6.94% w/w	WG	15758

Uses

- Annual dicotyledons in *winter wheat* [1, 2]
- Blackgrass in *winter wheat* (moderately susceptible) [1, 2]
- Brome grasses in *winter wheat* [2]
- Ryegrass in *winter wheat* (from seed) [1, 2]
- Wild oats in *winter wheat* [1, 2]

Approval information

- Flupyrsulfuron-methyl included in Annex I under EC Regulation 1107/2009 while pyroxsulam is awaiting inclusion.

Efficacy guidance

- Residual activity in soil will be reduced by high temperatures due to increased degradation.
- Apply in mixture with an authorised adjuvant.
- Do not apply more than one ALS product per crop for grass weed control.
- In some situations slight chlorosis mat be observed after treatment. Crops quickly grow out of these symptoms and there is no effect on final grain yield.

Restrictions

- Do not apply in mixture with plant growth regulators - maintain an interval of at least 7 days.
- Do not apply in mixture with an organophosphate insecticide - maintain an interval of at least 14 days.

Following crops guidance

- Only cereals, oilseed rape, field beans, red and white clover or grass may be sown in the year of harvest.
- In the event of crop failure, only sow winter or spring wheat within 3 months of application and, before sowing, plough or thoroughly cultivate to a minimum depth of 15 cm.
- No cultivation restrictions prior to planting following crops.

Environmental safety

- LERAP Category B

Hazard classification and safety precautions

Hazard Dangerous for the environment
Transport code 9
Packaging group III
UN Number 3077
Risk phrases R50, R53a
Operator protection U05a, U14, U15, U20a
Environmental protection E15b, E16a, E34, E38
Storage and disposal D01, D02, D09a, D12a

257   flupyrsulfuron-methyl + thifensulfuron-methyl

A sulfonylurea herbicide mixture for winter wheat and winter oats
HRAC mode of action code: B + B

See also thifensulfuron-methyl

Products

1	Lancer	Headland	10:40% w/w	WG	13031

Products – continued
2 Lexus Millenium DuPont 10:40% w/w WG 09206

Uses
- Annual dicotyledons in *winter oats*, *winter wheat*
- Blackgrass in *winter oats*, *winter wheat*

Approval information
- Flupyrsulfuron-methyl and thifensulfuron-methyl included in Annex I under EC Regulation 1107/2009

Efficacy guidance
- Best results obtained when applied to small actively growing weeds
- Good spray cover of weeds must be obtained
- Increased degradation of active ingredient in high soil temperatures reduces residual activity
- Product has moderate residual life in soil. Under normal moisture conditions susceptible weeds germinating soon after treatment will be controlled
- Product may be used on all soil types but residual activity and weed control are reduced on highly alkaline soils
- Blackgrass should be treated from 1 leaf up to first node stage
- Flupyrsulfuron-methyl and thifensulfuron-methyl are members of the ALS-inhibitor group of herbicides. To avoid the build up of resistance do not use any product containing an ALS-inhibitor herbicide with claims for control of grass weeds more than once on any crop
- Use these products as part of a resistance management strategy that includes cultural methods of control and does not use ALS inhibitors as the sole chemical method of grass weed control

Restrictions
- Maximum number of treatments 1 per crop
- Do not use on wheat undersown with grasses or legumes, or any other broad-leaved crop
- Do not apply within 7 d of rolling
- Do not treat any crop suffering from drought, waterlogging, pest or disease attack, nutrient deficiency, or any other stress factors
- Contact contract agents before use on crops grown for seed
- Specific restrictions apply to use in sequence or tank mixture with other sulfonylurea or ALS-inhibiting herbicides. See label for details

Crop-specific information
- Latest use: before 1st node detectable (GS 31) on wheat; before 31 Dec in yr of sowing on oats [2]
- Slight chlorosis, speckling and stunting may occur in certain conditions. Recovery is rapid and yield not affected

Following crops guidance
- Only cereals, oilseed rape, field beans or grass may be sown in the yr of harvest of a treated crop. Any crop may be sown the following spring
- In the event of crop failure only winter wheat may be sown before normal harvest date. Land should be ploughed and cultivated to 15 cm minimum before resowing

Environmental safety
- Dangerous for the environment
- Very toxic to aquatic organisms
- LERAP Category B

Hazard classification and safety precautions
Hazard Irritant, Dangerous for the environment
Transport code 9
Packaging group III
UN Number 3077
Risk phrases R43, R50, R53a
Operator protection A, H; U05a, U08, U14, U19a, U20b
Environmental protection E15a, E16a, E38
Storage and disposal D01, D02, D09a, D11a, D12a

SEE SECTION 3 FOR PRODUCTS ALSO REGISTERED

258   fluquinconazole

A protectant, eradicant and systematic triazole fungicide for winter cereals
FRAC mode of action code: 3

Products

1	Flamenco	BASF	100 g/l	SC	11699
2	Galmano	Bayer CropScience	167 g/l	FS	11650
3	Jockey Solo	BASF	167 g/l	FS	14281

Uses

- Brown rust in **winter wheat** [1]
- Bunt in **winter wheat** (seed treatment) [2]
- Foliar disease control in **grass seed crops** (off-label), **triticale** (off-label) [1]
- Glume blotch in **winter wheat** [1]
- Powdery mildew in **winter wheat** [1]
- Seed-borne diseases in **durum wheat** (off-label), **triticale** (off-label) [2]; **winter barley, winter wheat** [3]
- Septoria leaf blotch in **winter wheat** [1]; **winter wheat** (seed treatment) [2]
- Take-all in **winter wheat** (seed treatment - reduction) [2]
- Yellow rust in **winter wheat** [1]; **winter wheat** (seed treatment) [2]

Extension of Authorisation for Minor Use (EAMUs)

- **durum wheat** 20063065 [2]
- **grass seed crops** 20063472 [1]
- **triticale** 20063472 [1], 20063065 [2]

Approval information

- Fluquinconazole included in Annex 1 under EC Regulation 1107/2009
- Accepted by BBPA on malting barley before ear emergence only

Efficacy guidance

- Best results obtained from spray treatments when disease first becomes active in crop but before infection spreads to younger leaves [1]
- Adequate disease protection throughout the season will usually require a programme of at least two fungicide treatments [1]
- May also be applied as a protectant at end of ear emergence (but no later) if crop still disease free [1]
- With seed treatments ensure good even coverage of seed to obtain reliable disease control [2]
- Seed treatment provides early control of Septoria leaf blotch and yellow rust but may require foliar treatment for later infection [2]
- Fluquinconazole is a DMI fungicide. Resistance to some DMI fungicides has been identified in Septoria leaf blotch which may seriously affect performance of some products. For further advice contact a specialist advisor and visit the Fungicide Resistance Action Group (FRAG)-UK website

Restrictions

- Maximum total spray dose equivalent to two full doses [1]
- Maximum number of seed treatments 1 per batch [2]
- Do not apply spray treatments after the start of anthesis (GS 59) [1]
- Do not mix with any other products [2]
- Do not treat cracked, split or sprouted seed [2]
- Do not use treated seed as food or feed [2]

Crop-specific information

- Latest use: before beginning of anthesis (GS 59) for spray treatment; before drilling for seed treatment
- Ensure good spray coverage and increase volume in dense crops [1]
- Delayed emergence may result when treated seed is drilled into heavy or poorly drained soils which then become wet or waterlogged [2]

FOR FULL CONDITIONS OF USE ALWAYS READ THE PRODUCT LABEL

Environmental safety
- Dangerous for the environment
- Toxic to aquatic organisms
- Dangerous to fish or other aquatic life. Do not contaminate surface waters or ditches with chemical or used container
- Special PPE requirements and precautions apply where product supplied in returnable packs. Check label

Hazard classification and safety precautions
Hazard Toxic, Dangerous for the environment
Transport code 9
Packaging group III
UN Number 3082
Risk phrases R22a, R51, R53a [1-3]; R25b [3]; R36, R43 [1]; R48 [1, 2]
Operator protection A, H [1-3]; C, M [1]; D [2, 3]; U05a [1-3]; U07 [2]; U11, U14, U15 [1]; U20b [2, 3]; U25 [3]
Environmental protection E13b, E36a [2]; E15a [1, 3]; E34, E38 [1-3]
Storage and disposal D01, D02, D12a [1-3]; D05 [1, 2]; D08 [2]; D09a, D14 [2, 3]; D10c [1]; D10d [3]
Treated seed S01, S03, S04b, S06a [2]; S02, S05, S07, S08 [2, 3]
Medical advice M03 [1, 2]; M04a [1, 3]

259   fluquinconazole + prochloraz

A broad-spectrum triazole mixture for use as a spray and seed treatment in winter cereals
FRAC mode of action code: 3 + 3

See also prochloraz

Products

1	Epona	BASF	167:31 g/l	FS	12444
2	Foil	BASF	54:174 g/l	SE	11700
3	Jockey	BASF	167:31 g/l	FS	11689

Uses
- Brown rust in **winter barley**, **winter wheat** [2]; **winter wheat** *(seed treatment - moderate control)* [1]
- Bunt in **winter wheat** *(seed treatment)* [1, 3]
- Covered smut in **winter barley** *(seed treatment)* [3]
- Foliar disease control in **durum wheat** *(off-label)*, **grass seed crops** *(off-label)*, **spring rye** *(off-label)*, **triticale** *(off-label)*, **winter rye** *(off-label)* [2]
- Fusarium foot rot and seedling blight in **winter barley** *(seed treatment)* [3]
- Fusarium root rot in **winter wheat** *(seed treatment)* [1, 3]
- Loose smut in **winter barley** *(seed treatment)* [3]
- Powdery mildew in **winter barley**, **winter wheat** *(moderate control)* [2]
- Rhynchosporium in **winter barley** [2]
- Septoria leaf blotch in **winter wheat** [2]; **winter wheat** *(seed treatment)* [3]; **winter wheat** *(seed treatment - moderate control)* [1]
- Take-all in **winter barley** *(seed treatment - reduction)* [3]; **winter wheat** *(seed treatment - reduction)* [1, 3]
- Yellow rust in **winter barley** *(seed treatment)*, **winter wheat** *(seed treatment)* [3]; **winter wheat** [2]; **winter wheat** *(seed treatment - moderate control)* [1]

Extension of Authorisation for Minor Use (EAMUs)
- *durum wheat* 20060982 [2]
- *grass seed crops* 20060982 [2]
- *spring rye* 20060982 [2]
- *triticale* 20060982 [2]
- *winter rye* 20060982 [2]

SEE SECTION 3 FOR PRODUCTS ALSO REGISTERED

Approval information
- Fluquinconazole and prochloraz included under EC Regulation 1107/2009
- Accepted by BBPA for use on malting barley (before ear emergence only)

Efficacy guidance
- Best results obtained from spray treatment when disease first becomes active in crop but before infection spreads to younger leaves [2]
- When treating seed ensure good even coverage of seed to obtain reliable disease control [1, 3]
- Seed treatment provides early control of *Septoria* leaf spot and rust but may require foliar treatment for later infection [1, 3]
- Fluquinconazole and prochloraz are DMI fungicides. Resistance to some DMI fungicides has been identified in Septoria leaf blotch which may seriously affect performance of some products. For further advice contact a specialist advisor and visit the Fungicide Resistance Action Group (FRAG)-UK website

Restrictions
- Maximum number of treatments 1 per seed batch (seed dressing)
- Maximum total dose equivalent to two full doses (spray)
- Do not treat cracked, split or sprouted seed [1, 3]
- Do not use treated seed as food or feed [1, 3]

Crop-specific information
- Latest use: before 1st awns visible (GS 47) for winter barley; before beginning of anthesis (GS 59) for winter wheat; before drilling (seed treatment)
- Ensure good spray coverage and increase volume in dense crops [2]
- Delayed emergence may result when treated seed is drilled into heavy or poorly drained soils which then become wet or waterlogged [1, 3]

Environmental safety
- Dangerous for the environment
- Toxic to aquatic organisms
- Dangerous to fish or other aquatic life. Do not contaminate surface waters or ditches with chemical or used container
- Product supplied in returnable packs for which special PPE requirements and precautions apply. Check label

Hazard classification and safety precautions
Hazard Toxic [1, 3]; Harmful, Irritant [2]; Dangerous for the environment [1-3]
Transport code 9
Packaging group III
UN Number 3082
Risk phrases R22a, R48, R51, R53a [1-3]; R36, R43, R66 [2]
Operator protection A, H [1-3]; C [2]; D [3]; G [1]; U05a [1-3]; U07 [3]; U20b [1, 3]
Environmental protection E15a, E34, E38 [1-3]; E36a [3]
Storage and disposal D01, D02, D09a, D12a [1-3]; D05, D10c [2]; D08, D11a [1, 3]; D14 [3]
Treated seed S01, S03, S04b, S06a [3]; S02, S05, S07, S08 [1, 3]
Medical advice M03, M05b [2]; M04a [1, 3]

FOR FULL CONDITIONS OF USE ALWAYS READ THE PRODUCT LABEL

260   fluroxypyr

A post-emergence pyridinecarboxylic acid herbicide
HRAC mode of action code: O

See also 2,4-D + dicamba + fluroxypyr
aminopyralid + fluroxypyr
bromoxynil + fluroxypyr
bromoxynil + fluroxypyr + ioxynil
clopyralid + florasulam + fluroxypyr
clopyralid + fluroxypyr + MCPA
clopyralid + fluroxypyr + triclopyr
florasulam + fluroxypyr

Products

1	Agriguard Fluroxypyr	AgriGuard	200 g/l	EC	13335
2	Barclay Hurler	Barclay	200 g/l	EC	13458
3	Casino	AgriGuard	25% w/w	WG	14352
4	Crescent	AgriGuard	200 g/l	EC	13330
5	Flurostar 200	Globachem	200 g/l	EC	12903
6	Fluroxy 200	Goldengrass	200 g/l	EC	14421
7	Fluxyr 200 EC	AgriChem BV	200 g/l	EC	14086
8	Gala	Dow	200 g/l	EC	12019
9	Greencrop Reaper	Greencrop	200 g/l	EC	12261
10	Hatchet Xtra	AgriGuard	200 g/l	EC	13213
11	Headland Sure-fire	Headland	200 g/l	EC	15088
12	Klever	Globachem	200 g/l	EC	13295
13	Minstrel	United Phosphorus	200 g/l	EC	13745
14	Skora	AgChem Access	200 g/l	EC	13422
15	Standon Homerun 2	Standon	200 g/l	EC	13890
16	Starane 2	Dow	200 g/l	EC	12018
17	Tandus	Nufarm UK	200 g/l	EC	13432
18	Tomahawk	Makhteshim	200 g/l	EC	09249

Uses

- Annual and perennial weeds in **durum wheat**, **forage maize**, **spring barley**, **spring oats**, **spring wheat**, **triticale**, **winter barley**, **winter oats**, **winter rye**, **winter wheat** [8, 15]; **grassland** [8]; **permanent grassland**, **rotational grass**, **spring rye** [15]
- Annual dicotyledons in **almonds** (off-label), **apple orchards** (off-label), **chestnuts** (off-label), **crab apples** (off-label), **hazel nuts** (off-label), **pears** (off-label), **poppies for morphine production** (off-label), **sweetcorn** (off-label), **walnuts** (off-label) [18]; **bulb onions** (off-label), **garlic** (off-label), **leeks** (off-label), **shallots** (off-label) [8, 16, 18]; **durum wheat**, **forage maize**, **spring barley**, **spring oats**, **spring wheat**, **triticale**, **winter barley**, **winter oats**, **winter rye**, **winter wheat** [1-7, 9-14, 16-18]; **farm forestry** (off-label), **forest nurseries** (off-label), **game cover** (off-label), **miscanthus** (off-label), **ornamental plant production** (off-label) [16, 18]; **grassland** [7, 13]; **newly sown grass leys** [6, 15, 17]; **permanent grassland**, **rotational grass** [1-6, 9-12, 14, 16-18]; **spring rye** [1-6, 9-14, 16, 18]
- Black bindweed in **bulb onions** (off-label), **garlic** (off-label), **shallots** (off-label) [16]; **durum wheat**, **spring barley**, **spring oats**, **spring wheat**, **triticale**, **winter barley**, **winter oats**, **winter rye**, **winter wheat** [1-6, 8-18]; **forage maize** [1-6, 8-16, 18]; **grassland** [8, 13]; **newly sown grass leys** [6, 15]; **permanent grassland**, **rotational grass** [1-6, 9-12, 14-16, 18]; **spring rye** [1-6, 9-16, 18]
- Chickweed in **bulb onions** (off-label), **garlic** (off-label), **leeks** (off-label), **shallots** (off-label) [16]; **durum wheat**, **forage maize**, **spring barley**, **spring oats**, **spring wheat**, **triticale**, **winter barley**, **winter oats**, **winter rye**, **winter wheat** [1-6, 8-18]; **grassland** [8, 13]; **newly sown grass leys** [6, 15, 17]; **permanent grassland**, **rotational grass** [1-6, 9-12, 14-18]; **spring rye** [1-6, 9-16, 18]
- Cleavers in **apple orchards** (off-label), **bulb onions** (off-label), **garlic** (off-label), **leeks** (off-label), **pear orchards** (off-label), **poppies for morphine production** (off-label), **shallots** (off-label) [16]; **durum wheat**, **spring barley**, **spring oats**, **spring wheat**, **triticale**, **winter barley**, **winter**

oats, *winter rye*, *winter wheat* [1-6, 8-18]; *forage maize* [1-6, 8-16, 18]; *grassland* [8, 13]; *millet* *(off-label - grown for wild bird seed production)* [18]; *newly sown grass leys* [6, 15]; *permanent grassland*, *rotational grass* [1-6, 9-12, 14-16, 18]; *spring rye* [1-6, 9-16, 18]

- Docks in *apple orchards* *(off-label)*, *pear orchards* *(off-label)* [16]; *durum wheat*, *spring barley*, *spring oats*, *spring wheat*, *triticale*, *winter barley*, *winter oats*, *winter rye*, *winter wheat* [1-6, 8-18]; *forage maize* [1-6, 8-16, 18]; *grassland* [8]; *newly sown grass leys* [6]; *permanent grassland*, *rotational grass* [1-6, 9-12, 14-16, 18]; *spring rye* [1-6, 9-16, 18]
- Forget-me-not in *bulb onions* *(off-label)*, *garlic* *(off-label)*, *shallots* *(off-label)* [16]; *durum wheat*, *spring barley*, *spring oats*, *spring wheat*, *triticale*, *winter barley*, *winter oats*, *winter rye*, *winter wheat* [1-6, 8-18]; *forage maize* [1-6, 8-16, 18]; *grassland* [8, 13]; *newly sown grass leys* [6, 15]; *permanent grassland*, *rotational grass* [1-6, 9-12, 14-16, 18]; *spring rye* [1-6, 9-16, 18]
- Hemp-nettle in *durum wheat*, *spring barley*, *spring oats*, *spring wheat*, *triticale*, *winter barley*, *winter oats*, *winter rye*, *winter wheat* [1-6, 8-18]; *forage maize* [1-6, 8-16, 18]; *grassland* [8, 13]; *newly sown grass leys* [6, 15]; *permanent grassland*, *rotational grass* [1-6, 9-12, 14-16, 18]; *spring rye* [1-6, 9-16, 18]
- Poppies in *poppies for morphine production* *(off-label)* [16]
- Stinging nettle in *apple orchards* *(off-label)*, *pear orchards* *(off-label)* [16]
- Volunteer potatoes in *bulb onions* *(off-label)* [8, 16, 18]; *durum wheat*, *forage maize*, *triticale*, *winter oats*, *winter rye* [8, 15]; *garlic* *(off-label)*, *leeks* *(off-label)*, *shallots* *(off-label)* [8, 18]; *grassland* [8]; *permanent grassland*, *rotational grass*, *spring rye* [15]; *poppies for morphine production* *(off-label)*, *sweetcorn* *(off-label)* [16, 18]; *spring barley*, *spring oats*, *spring wheat* [8, 15, 17]; *winter barley*, *winter wheat* [1-6, 8-18]

Extension of Authorisation for Minor Use (EAMUs)
- *almonds* *20101920* [18]
- *apple orchards* *20040988* [16], *20101920* [18]
- *bulb onions* *20111769* [8], *20081404* [16], *20101919* [18]
- *chestnuts* *20101920* [18]
- *crab apples* *20101920* [18]
- *farm forestry* *20082925* [16], *20101910* [18]
- *forest nurseries* *20082925* [16], *20101910* [18]
- *game cover* *20082925* [16], *20082907* [18], *20101910* [18]
- *garlic* *20111769* [8], *20081404* [16], *20101919* [18]
- *hazel nuts* *20101920* [18]
- *leeks* *20111770* [8], *20100390* [16], *20101916* [18]
- *millet* *(grown for wild bird seed production)* *20080936* [18]
- *miscanthus* *20082925* [16], *20101910* [18]
- *ornamental plant production* *20082925* [16], *20101910* [18]
- *pear orchards* *20040988* [16]
- *pears* *20101920* [18]
- *poppies for morphine production* *20052059* [16], *20101908* [18]
- *shallots* *20111769* [8], *20081404* [16], *20101919* [18]
- *sweetcorn* *20051696* [16], *20101907* [18]
- *walnuts* *20101920* [18]

Approval information
- Fluroxypyr included in Annex I under EC Regulation 1107/2009
- Accepted by BBPA for use on malting barley
- Approval expiry 31 Dec 2013 [15]

Efficacy guidance
- Best results achieved under good growing conditions in a strongly competing crop
- A number of tank mixtures with HBN and other herbicides are recommended for use in autumn and spring to extend range of species controlled. See label for details
- Spray is rainfast in 1 h

Restrictions
- Maximum number of treatments 1 per crop or yr or maximum total dose equivalent to one full dose treatment

FOR FULL CONDITIONS OF USE ALWAYS READ THE PRODUCT LABEL

- Do not apply in any tank-mix on triticale or forage maize
- Do not use on crops undersown with clovers or other legumes
- Do not treat crops suffering stress caused by any factor
- Do not roll or harrow for 7 d before or after treatment
- Do not spray if frost imminent
- Do not apply before 1 Mar in yr of harvest [17]
- Do not use on crops grown for seed production [17]
- Straw from treated crops must not be returned directly to the soil but must be removed and used only for livestock bedding

Crop-specific information

- Latest use: before flag leaf sheath opening (GS 47) for winter wheat and barley; before flag leaf sheath extending (GS 41) for spring wheat and barley; before second node detectable (GS 32) for oats, rye, triticale and durum wheat; before 7 leaves unfolded and before buttress roots appear for maize
- Apply to new leys from 3 expanded leaf stage
- Timing varies in tank mixtures. See label for details
- Crops undersown with grass may be sprayed provided grasses are tillering

Following crops guidance

- Clovers, peas, beans and other legumes must not be sown for 12 mth following treatment at the highest dose

Environmental safety

- Dangerous for the environment
- Very toxic to aquatic organisms
- Flammable
- Keep livestock out of treated areas for at least 3 d following treatment and until poisonous weeds, such as ragwort, have died down and become unpalatable
- Wash spray equipment thoroughly with water and detergent immediately after use. Traces of product can damage susceptible plants sprayed later

Hazard classification and safety precautions

Hazard Harmful [1, 2, 4-18]; Irritant [3]; Flammable [1, 2, 4-16, 18]; Dangerous for the environment [1-18]

Transport code 3 [1, 2, 4-6, 8-16, 18]; 9 [3, 7, 17]

Packaging group III

UN Number 1933 [11, 16]; 1993 [1, 2, 4-6, 8-10, 12-15, 18]; 3077 [3]; 3082 [7, 17]

Risk phrases R22b, R67 [1, 2, 4-18]; R36 [1, 3-5, 8, 10, 12]; R37 [1, 2, 4-8, 10-16, 18]; R38 [1, 2, 4, 5, 8, 10, 12]; R43 [1, 2, 4, 5, 8, 10, 12, 17]; R50 [1, 3-5, 10, 12]; R51 [2, 6-9, 11, 13-18]; R53a [1-18]

Operator protection A [1-6, 8, 10, 12, 13, 15, 17]; C [1, 3-6, 8, 10, 12, 13, 15]; H [1, 2, 4-6, 8, 10, 12, 13, 15, 17]; U05a [1-5, 8, 10, 12]; U08, U19a [1-18]; U11 [1, 3-5, 8, 10, 12]; U14 [1, 2, 4, 5, 8, 10, 12, 17]; U20a [8]; U20b [1-7, 9-16, 18]

Environmental protection E06a [3, 4] (3 d); E07a [1, 5-7, 9-18]; E07c [2] (14 d); E07c [8] (3 days); E15a [6, 7, 9, 11, 13-16, 18]; E15b [1-5, 8, 10, 12, 17]; E34 [1-5, 8, 10, 12, 18]; E38 [1, 2, 4-8, 10-18]

Storage and disposal D01 [1-5, 8, 10, 12]; D02 [1-6, 8, 10, 12]; D05 [1, 2, 4-6, 8-10, 12, 18]; D09a [1-18]; D10a [2]; D10b [1, 3-7, 9-18]; D10c [8]; D12a [1, 2, 4-7, 10-18]

Treated seed S06a [17]

Medical advice M03 [1-5, 8, 10, 12]; M05b [1, 2, 4-18]

261   fluroxypyr + metsulfuron-methyl

A foliar-applied herbicide mixture for use in wheat and barley
HRAC mode of action code: O + B

Products

Croupier	AgriGuard	25:1% w/w	WG	15643

SEE SECTION 3 FOR PRODUCTS ALSO REGISTERED

Uses
- Annual dicotyledons in *spring barley*, *spring wheat*, *winter barley*, *winter wheat*

Approval information
- Fluroxypyr and metsulfuron-methyl included in Annex I under EC Regulation 1107/2009

Efficacy guidance
- See label for ALS herbicides allowed in mixture with [1]
- Weed control may be reduced in dry soil conditions

Environmental safety
- LERAP Category B

Hazard classification and safety precautions
    Hazard Irritant, Dangerous for the environment
    Transport code 9
    Packaging group III
    UN Number 3077
    Risk phrases R38, R41, R50, R53a
    Operator protection A, C, H; U05a, U09a, U19a, U20b
    Environmental protection E15b, E16a, E16b, E34, E38
    Storage and disposal D01, D02, D09a, D11a, D12a
    Medical advice M05a

262   fluroxypyr + triclopyr

A foliar acting herbicide for docks in grassland
HRAC mode of action code: O + O

See also triclopyr

Products

1 Dovekie	AgChem Access	100:100 g/l	EC	14121
2 Doxstar	Dow	100:100 g/l	EC	11063
3 Doxstar Pro	Dow	150:150 g/l	EC	15664

Uses
- Chickweed in *rotational grass* [1, 2]
- Docks in *amenity grassland*, *grassland* [3]; *permanent grassland*, *rotational grass* [1, 2]

Approval information
- Fluroxypyr and triclopyr included in Annex I under EC Regulation 1107/2009

Efficacy guidance
- Seedling docks in new leys are controlled up to 50 mm diameter. In established grass apply in spring or autumn or, at lower dose, in spring and autumn on docks up to 200 mm. A second application in the subsequent yr may be needed
- Allow 2-3 wk after cutting or grazing to allow sufficient regrowth of docks to occur before spraying
- Control may be reduced if rain falls within 2 h of application
- To allow maximum translocation to the roots of docks do not cut grass for 28 d after spraying

Restrictions
- Maximum total dose equivalent to one full dose treatment
- Do not roll or harrow for 10 d before or 7 d after spraying
- Do not spray in drought, very hot or very cold weather

Crop-specific information
- Latest use: 7 d before grazing or harvest of grass
- Grass less than one yr old may be treated at half dose from the third leaf visible stage
- Clover will be killed or severely checked by treatment

FOR FULL CONDITIONS OF USE ALWAYS READ THE PRODUCT LABEL

Following crops guidance

- Do not sow kale, turnips, swedes or grass mixtures containing clover by direct drilling or minimum cultivation techniques within 6 wk of application

Environmental safety

- Dangerous for the environment
- Toxic to aquatic organisms
- Keep livestock out of treated areas for at least 7 d following treatment and until poisonous weeds, such as ragwort, have died down and become unpalatable
- Do not allow drift to come into contact with crops, amenity plantings, gardens, ponds, lakes or watercourses
- Wash spray equipment thoroughly with water and detergent immediately after use. Traces of product can damage susceptible plants sprayed later

Hazard classification and safety precautions

Hazard Harmful, Flammable [1, 2]; Irritant [3]; Dangerous for the environment [1-3]
Transport code 3
Packaging group III
UN Number 1993
Risk phrases R22a, R22b, R37, R38, R51, R67 [1, 2]; R43, R53a [1-3]; R50 [3]
Operator protection A [1-3]; C [1, 2]; U02a, U05a, U08, U19a, U20b [1, 2]; U14 [1-3]
Environmental protection E07a, E15a, E34 [1, 2]; E38 [1-3]
Consumer protection C01 [1, 2]
Storage and disposal D01, D02, D05, D12a [1-3]; D09a, D10b [1, 2]
Medical advice M05a [3]; M05b [1, 2]

263　flurtamone

A carotenoid synthesis inhibitor available only in mixtures
HRAC mode of action code: F1

See also diflufenican + flufenacet + flurtamone
*　　　　diflufenican + flurtamone*

264　flusilazole

A systemic, protective and curative triazole fungicide for cereals, oilseed rape and sugar beet
FRAC mode of action code: 3

See also carbendazim + flusilazole
*　　　　chlorothalonil + flusilazole*
*　　　　famoxadone + flusilazole*
*　　　　fenpropimorph + flusilazole*

Products

1	Capitan 25	DuPont	250 g/l	EW	13067
2	Genie 25	DuPont	250 g/l	EW	13065
3	Lyric	DuPont	250 g/l	EW	13022
4	Nustar 25	DuPont	250 g/l	EW	13921
5	Sanction 25	DuPont	250 g/l	EW	13066

Uses

- Brown rust in **spring barley**, **winter barley**, **winter wheat** [1-5]
- Canker in **brussels sprouts** *(off-label)*, **swedes** *(off-label)*, **turnips** *(off-label)* [1-3, 5]
- Disease control in **fodder beet** *(off-label)* [1]
- Eyespot in **forage maize** *(off-label)* [1-3, 5]; **grain maize** *(off-label)* [1]; **spring barley**, **winter barley**, **winter wheat** [1-5]
- Foliar disease control in **brussels sprouts** *(off-label)*, **fodder beet** *(off-label)*, **forage maize** *(off-label)*, **swedes** *(off-label)*, **turnips** *(off-label)* [4]; **grain maize** *(off-label)* [3, 4]
- Glume blotch in **spring barley**, **winter barley** [1-3, 5]; **winter wheat** [1-5]

SEE SECTION 3 FOR PRODUCTS ALSO REGISTERED

- Light leaf spot in **spring oilseed rape, winter oilseed rape** [1-5]
- Net blotch in **spring barley, winter barley** [1-5]
- Powdery mildew in **spring barley, winter barley** [1-3, 5]; **spring barley** *(moderate control)*, **winter barley** *(moderate control)* [4]; **sugar beet, winter wheat** [1-5]
- Rhynchosporium in **spring barley, winter barley** [1-3, 5]; **spring barley** *(moderate control)*, **winter barley** *(moderate control)* [4]
- Rust in **sugar beet** [1-5]
- Septoria leaf blotch in **winter wheat** [1-5]
- Yellow rust in **spring barley, winter barley, winter wheat** [1-5]

Extension of Authorisation for Minor Use (EAMUs)
- **brussels sprouts** *20070102* [1], *20070101* [2], *20063998* [3], *20081104* [4], *20070100* [5]
- **fodder beet** *20101760* [1], *20091618* [4]
- **forage maize** *20080197* [1], *20101759* [1], *20080198* [2], *20080178* [3], *20081102* [4], *20080196* [5]
- **grain maize** *20101759* [1], *20081044* [3], *20081102* [4]
- **swedes** *20073284* [1], *20073283* [2], *20073270* [3], *20081103* [4], *20073285* [5]
- **turnips** *20073284* [1], *20073283* [2], *20073270* [3], *20081103* [4], *20073285* [5]

Approval information
- Flusilazole included in Annex 1 under EC Regulation 1107/2009
- Accepted by BBPA for use on malting barley

Efficacy guidance
- Best results obtained from treatment at onset of disease development
- Best control of eyespot in cereals achieved by spraying between leaf-sheath erect and second node detectable stages (GS 30-32)
- Product active against both MBC-sensitive and MBC-resistant strains of eyespot
- Rain occurring within 2 h after application may reduce effectiveness
- See label for recommended tank-mixes to give broader spectrum control
- Flusilazole is a DMI fungicide. Resistance to some DMI fungicides has been identified in Septoria leaf blotch which may seriously affect performance of some products. For further advice contact a specialist advisor and visit the Fungicide Resistance Action Group (FRAG)-UK website

Restrictions
- Maximum number of treatments (including other products containing flusilazole) on cereals and oilseed rape depends on dose and timing - see labels for details. High rate must not be used more than once in any crop
- Maximum total dose on sugar beet equivalent to one full dose
- Do not apply to crops under stress or during frosty weather

Crop-specific information
- Latest use: high dose before 3rd node detectable (GS 33) plus reduced dose before early milk stage (GS 72) on winter wheat and barley; before first flowers open (GS 4,0) on oilseed rape
- HI 7 wk for sugar beet
- Treat oilseed rape in autumn and/or spring at stem extension stage
- On sugar beet apply at an early stage of disease development, usually in early Aug

Environmental safety
- Dangerous for the environment
- Toxic to aquatic organisms
- Harmful to fish or other aquatic life. Do not contaminate surface waters or ditches with chemical or used container

Hazard classification and safety precautions
Hazard Toxic, Dangerous for the environment
Transport code 9
Packaging group III
UN Number 3082
Risk phrases R22a, R38, R40, R51, R53a, R61

FOR FULL CONDITIONS OF USE ALWAYS READ THE PRODUCT LABEL

Operator protection A, C [1-5]; H, M [4]; U05a, U11, U19a [1-5]; U16a, U20b, U23a [4]; U20a [1-3, 5]
Environmental protection E13c, E34, E38
Storage and disposal D01, D02, D10b, D12a [1-5]; D05 [1-3, 5]; D09a [4]
Medical advice M04a

265   flutolanil

An carboxamide fungicide for treatment of potato seed tubers
FRAC mode of action code: 7

Products

1 Rhino	Certis	460 g/l	SC	14311
2 Rhino DS	Certis	6% w/w	DS	12763

Uses
- Black scurf in **potatoes** *(tuber treatment)*
- Stem canker in **potatoes** *(tuber treatment)*

Approval information
- Flutolanil included in Annex I under EC Regulation 1107/2009

Efficacy guidance
- Apply to clean tubers before chitting, prior to planting, or at planting
- Apply flowable concentrate through canopied, hydraulic or spinning disc equipment (with or without electrostatics) mounted on a rolling conveyor or table [1]
- Flowable concentrate may be diluted with water up to 2.0 l per tonne to improve tuber coverage. Disease in areas not covered by spray will not be controlled [1]
- Dry powder may be applied via an on-planter applicator [2]

Restrictions
- Maximum number of treatments 1 per batch of seed tubers
- Check with processor before use on crops for processing

Crop-specific information
- Latest use: at planting
- Seed tubers should be of good quality and free from bacterial rots, physical damage or virus infection, and should not be sprouted to such an extent that mechanical damage to the shoots will occur during treatment or planting

Environmental safety
- Harmful to aquatic organisms

Hazard classification and safety precautions
Hazard Irritant
Transport code 9 [1]
Packaging group III [1]
UN Number 3082 [1]; N/C [2]
Risk phrases R36, R52, R53a [2]; R43, R51 [1]
Operator protection A, H [1, 2]; C [2]; U04a, U05a, U19a, U20a [1, 2]; U14 [1]
Environmental protection E15a, E34, E38
Storage and disposal D01, D02, D09a, D12a [1, 2]; D05, D10a [1]; D11a [2]
Treated seed S01, S03, S04a, S05 [1, 2]; S02 [2]
Medical advice M03

SEE SECTION 3 FOR PRODUCTS ALSO REGISTERED

266   flutriafol

A broad-spectrum triazole fungicide for cereals
FRAC mode of action code: 3

See also chlorothalonil + flutriafol
       fludioxonil + flutriafol

Products

1 Consul	Headland	125 g/l	SC	12976
2 Pointer	Headland	125 g/l	SC	12975

Uses

- Brown rust in **spring barley, winter barley, winter wheat**
- Powdery mildew in **spring barley, winter barley, winter wheat**
- Rhynchosporium in **spring barley, winter barley**
- Septoria leaf blotch in **winter wheat**
- Yellow rust in **spring barley, winter barley, winter wheat**

Approval information

- Flutriafol included in Annex I under EC Regulation 1107/2009
- Accepted by BBPA for use on malting barley

Efficacy guidance

- Best results obtained from treatment in early stages of disease development. See label for detailed guidance on spray timing for specific diseases and the need for repeat treatments
- Tank mix options available on the label to broaden activity spectrum
- Good spray coverage essential for optimum performance
- Flutriafol is a DMI fungicide. Resistance to some DMI fungicides has been identified in Septoria leaf blotch which may seriously affect performance of some products. For further advice contact a specialist advisor and visit the Fungicide Resistance Action Group (FRAG)-UK website

Restrictions

- Maximum number of treatments 2 per crop (including other products containing flutriafol)

Crop-specific information

- Latest use: before early grain milky ripe stage (GS 73)
- Flag leaf tip scorch on wheat caused by stress may be increased by fungicide treatment

Environmental safety

- Harmful to aquatic organisms
- Harmful to fish or other aquatic life. Do not contaminate surface waters or ditches with chemical or used container

Hazard classification and safety precautions

Hazard Harmful
UN Number N/C
Risk phrases R43 [1]; R48, R52 [1, 2]
Operator protection A, H; U05a, U09a, U19a, U20b
Environmental protection E13c, E38
Storage and disposal D01, D02, D09a, D10c

267   fluxapyroxad

Also known as Xemium, it is an SDHI fungicide for use in cereals.
FRAC mode of action code: 7

See also epoxiconazole + fluxapyroxad
       epoxiconazole + fluxapyroxad + pyraclostrobin
       fluxapyroxad + metconazole

268   folpet

A multi-site protectant fungicide for disease control in wheat and barley
FRAC mode of action code: M4

Products

1 Arizona	Makhteshim	500 g/l	SC	15318
2 Phoenix	Makhteshim	500 g/l	SC	15259

Uses
- Leaf spot in **spring wheat** *(reduction)*, **winter wheat** *(reduction)*
- Rhynchosporium in **spring barley** *(reduction)*, **winter barley** *(reduction)*

Approval information
- Folpet included in Annex I under EC Regulation 1107/2009
- Accepted by BBPA for use on cereals

Efficacy guidance
- Folpet is a protectant fungicide and so the first application must be made before the disease becomes established in the crop. A second application should be timed to protect new growth.
- Folpet only has contact activity and so good coverage of the target foliage is essential for good activity.

Environmental safety
- LERAP Category B

Hazard classification and safety precautions
 Hazard Harmful, Dangerous for the environment
 Transport code 9
 Packaging group III
 UN Number 3082
 Risk phrases R40, R50, R53a
 Operator protection A, H; U05a, U08, U11, U14, U15, U20a
 Environmental protection E15b, E16a, E34, E38
 Storage and disposal D01, D02, D09a, D12a
 Medical advice M03

269   fosetyl-aluminium

A systemic phosphonic acid fungicide for various horticultural crops
FRAC mode of action code: 33

See also fenamidone + fosetyl-aluminium
        fosetyl-aluminium + propamocarb hydrochloride

Products

Plant Trust	Everris Ltd	16.5% w/w	CG	15779

Uses
- Disease control in **ornamental plant production**

Approval information
- Fosetyl-aluminium included in Annex I under EC Regulation 1107/2009

Restrictions
- Check tolerance of ornamental species before large-scale treatment

Crop-specific information
- Use by wet incorporation in blocking compost for protected lettuce only from Sep to Apr and follow all directions carefully to avoid severe crop injury. Crop maturity may be delayed by a few days

SEE SECTION 3 FOR PRODUCTS ALSO REGISTERED

- Apply as drench to rooted cuttings of hardy nursery stock after first potting and repeat mthly. Up to 6 applications may be needed
- See label for details of application to capillary benches and trays of young plants.

Environmental safety
- Harmful to aquatic organisms

Hazard classification and safety precautions
Operator protection U20c
Environmental protection E15a
Storage and disposal D09a, D11a

270 fosetyl-aluminium + propamocarb hydrochloride

A systemic and protectant fungicide mixture for use in horticulture
FRAC mode of action code: 33 + 28

See also propamocarb hydrochloride

Products

1	Pan Cradle	Pan Agriculture	310:530 g/l	SL	15923
2	Previcur Energy	Bayer CropScience	310:530 g/l	SL	15367

Uses
- Damping off in **protected aubergines** *(off-label)*, **protected chilli peppers** *(off-label)*, **protected courgettes** *(off-label)*, **protected forest nurseries** *(off-label)*, **protected gherkins** *(off-label)*, **protected herbs (see appendix 6)** *(off-label)*, **protected hops** *(off-label)*, **protected marrows** *(off-label)*, **protected pumpkins** *(off-label)*, **protected soft fruit** *(off-label)*, **protected squashes** *(off-label)*, **protected sweet peppers** *(off-label)*, **protected top fruit** *(off-label)*, **protected watermelon** *(off-label)* [2]; **protected broccoli, protected brussels sprouts, protected cabbages, protected calabrese, protected cauliflowers, protected chinese cabbage, protected collards, protected kale, protected melons, protected radishes** [1, 2]
- Downy mildew in **lettuce, protected broccoli, protected brussels sprouts, protected cabbages, protected calabrese, protected cauliflowers, protected chinese cabbage, protected collards, protected kale, protected lettuce, protected melons, protected radishes** [1, 2]; **protected forest nurseries** *(off-label)*, **protected herbs (see appendix 6)** *(off-label)*, **protected hops** *(off-label)*, **protected soft fruit** *(off-label)*, **protected top fruit** *(off-label)*, **spinach** *(off-label)* [2]
- Pythium in **lettuce, protected cucumbers, protected lettuce, protected tomatoes** [1, 2]; **ornamental plant production** *(off-label)* [2]

Extension of Authorisation for Minor Use (EAMUs)
- *ornamental plant production* 20111557 [2]
- *protected aubergines* 20111553 [2]
- *protected chilli peppers* 20111553 [2]
- *protected courgettes* 20111556 [2]
- *protected forest nurseries* 20122045 [2]
- *protected gherkins* 20111556 [2]
- *protected herbs (see appendix 6)* 20111555 [2], 20122227 [2]
- *protected hops* 20122045 [2]
- *protected marrows* 20111556 [2]
- *protected pumpkins* 20111556 [2]
- *protected soft fruit* 20122045 [2]
- *protected squashes* 20111556 [2]
- *protected sweet peppers* 20111553 [2]
- *protected top fruit* 20122045 [2]
- *protected watermelon* 20111556 [2]
- *spinach* 20112452 [2]

Approval information
- Fosetyl-aluminium and propamocarb hydrochloride included in Annex I under EC Regulation 1107/2009

Hazard classification and safety precautions
Hazard Irritant
UN Number N/C
Risk phrases R43
Operator protection A, H, M; U04a, U05a, U08, U14, U19a, U20b
Environmental protection E15a
Storage and disposal D09a, D11a
Medical advice M03

271 fosthiazate

An organophosphorus contact nematicide for potatoes
IRAC mode of action code: 1B

Products

Nemathorin 10G	Syngenta	10% w/w	FG	11003

Uses
- Nematodes in *hops (off-label)*, *ornamental plant production (off-label)*, *soft fruit (off-label)*
- Potato cyst nematode in *potatoes*
- Spraing vectors in *potatoes (reduction)*
- Wireworm in *potatoes (reduction)*

Extension of Authorisation for Minor Use (EAMUs)
- *hops* 20082912 *expires 31 Dec 2013*
- *ornamental plant production* 20082912 *expires 31 Dec 2013*
- *soft fruit* 20082912 *expires 31 Dec 2013*

Approval information
- Fosthiazate included in Annex I under EC Regulation 1107/2009
- In 2006 CRD required that all products containing this active ingredient should carry the following warning in the main area of the container label: "Fosthiazate is an anticholinesterase organophosphate. Handle with care"

Efficacy guidance
- Application best achieved using equipment such as Horstine Farmery Microband Applicator, Matco or Stocks Micrometer applicators together with a rear mounted powered rotary cultivator
- Granules must not become wet or damp before use

Restrictions
- Contains an anticholinesterase organophosphorus compound. Do not use if under medical advice not to work with such compounds
- Maximum number of treatments 1 per crop
- Product must only be applied using tractor-mounted/drawn direct placement machinery. Do not use air assisted broadcast machinery other than that referenced on the product label
- Do not allow granules to stand overnight in the application hopper
- Do not apply more than once every four years on the same area of land
- Do not use on crops to be harvested less than 17 wk after treatment
- Consult before using on crops intended for processing

Crop-specific information
- Latest use: at planting
- HI 17 wk for potatoes

Environmental safety
- Dangerous for the environment
- Toxic to aquatic organisms
- Dangerous to game, wild birds and animals

SEE SECTION 3 FOR PRODUCTS ALSO REGISTERED

- Dangerous to livestock. Keep all livestock out of treated areas for at least 13 wk
- Incorporation to 10-15 cm and ridging up of treated soil must be carried out immediately after application. Powered rotary cultivators are preferred implements for incorporation but discs, power, spring tine or Dutch harrows may be used provided two passes are made at right angles
- To protect birds and wild mammals remove spillages
- Failure completely to bury granules immediately after application is hazardous to wildlife
- To protect groundwater do not apply any product containing fosthiazate more than once every four yr

Hazard classification and safety precautions
Hazard Harmful, Dangerous for the environment
Transport code 9
Packaging group III
UN Number 3077
Risk phrases R22a, R43, R51, R53a
Operator protection A, E, G, H, K, M; U02a, U04a, U05a, U09a, U13, U14, U19a, U20a
Environmental protection E06b (13 wk); E15b, E34, E38
Storage and disposal D01, D02, D05, D09a, D11a, D14
Medical advice M01, M03, M05a

272　fuberidazole

A benzimidazole (MBC) fungicide available only in mixtures
FRAC mode of action code: 1

See also bitertanol + fuberidazole
*　　　　bitertanol + fuberidazole + imidacloprid*

273　fuberidazole + imidacloprid + triadimenol

A broad spectrum systemic fungicide and insecticide seed treatment for winter cereals
FRAC mode of action code: 1 + IRAC 4A + FRAC 3

See also imidacloprid
*　　　　triadimenol*

Products

Tripod Plus	Makhteshim	15:117:125 g/l	LS	13168

Uses
- Blue mould in **winter wheat** *(seed treatment)*
- Brown foot rot in **winter barley** *(seed treatment)*
- Brown rust in **winter barley** *(seed treatment)*, **winter wheat** *(seed treatment)*
- Bunt in **winter wheat** *(seed treatment)*
- Covered smut in **winter barley** *(seed treatment)*
- Fusarium foot rot and seedling blight in **winter barley** *(seed treatment - reduction)*, **winter oats** *(seed treatment - reduction)*, **winter wheat** *(seed treatment - reduction)*
- Loose smut in **winter barley** *(seed treatment)*, **winter oats** *(seed treatment)*, **winter wheat** *(seed treatment)*
- Net blotch in **winter barley** *(seed treatment - seed-borne only)*
- Powdery mildew in **winter barley** *(seed treatment)*, **winter oats** *(seed treatment)*, **winter wheat** *(seed treatment)*
- Pyrenophora leaf spot in **winter oats** *(seed treatment)*
- Septoria leaf blotch in **winter wheat** *(seed treatment)*
- Septoria seedling blight in **winter wheat** *(seed treatment - reduction)*
- Snow rot in **winter barley** *(seed treatment - reduction)*
- Virus vectors in **winter barley** *(seed treatment)*, **winter oats** *(seed treatment)*, **winter wheat** *(seed treatment)*

FOR FULL CONDITIONS OF USE ALWAYS READ THE PRODUCT LABEL

- Wireworm in **winter barley** *(reduction of damage)*, **winter oats** *(reduction of damage)*, **winter wheat** *(reduction of damage)*
- Yellow rust in **winter barley** *(seed treatment)*, **winter wheat** *(seed treatment)*

Approval information
- Fuberidazole, imidacloprid and triadimenol included in Annex I under EC Regulation 1107/2009
- Accepted by BBPA for use on malting barley

Efficacy guidance
- Apply through recommended seed treatment machinery
- Evenness of seed cover improved by simultaneous application of equal volumes of product and water
- Calibrate drill for treated seed and drill at 4 cm into firm, well prepared seedbed
- Use minimum 125 kg treated seed per ha and increase seed rate as drilling season progresses. Lower drilling rates and/or early drilling affect duration of BYDV protection needed and may require follow-up aphicide treatment
- When aphid activity unusually late, or is heavy and prolonged in areas of high risk, and mild weather predominates, follow-up treatment may be required
- In addition to seed-borne diseases early attacks of various foliar, air-borne diseases are controlled or suppressed. See label for details

Restrictions
- Maximum number of treatments 1 per batch of seed
- Do not use on naked oats
- Do not drill treated winter wheat seed after end of Nov
- Do not handle seed unnecessarily
- Do not use treated seed as food or feed
- Treated seed should not be broadcast, but drilled to a depth of 4 cm in a well prepared seedbed
- Germination tests should be done on all batches of seed to be treated to ensure seed viability and suitability for treatment
- Do not use on seed with moisture content above 16%, on sprouted, cracked or skinned seed or on seed already treated with another seed treatment

Crop-specific information
- Latest use: before drilling
- Store treated seed in cool, dry, well-ventilated store and drill as soon as possible, preferably in season of purchase
- Treatment may accentuate effects of adverse seedbed conditions on crop emergence

Environmental safety
- Dangerous to fish or other aquatic life. Do not contaminate surface waters or ditches with chemical or used container
- Dangerous to birds, game and other wildlife. Treated seed should not be left on the soil surface. Bury spillages
- Seed should be drilled to a depth of 4 cm into a well-prepared seed bed
- If seed is present on the soil surface, or if spills have occurred, the field should be harrowed and rolled if conditions are appropriate to ensure good incorporation

Hazard classification and safety precautions
Transport code 9
Packaging group III
UN Number 3082
Risk phrases R52, R53a
Operator protection A, H; U04a, U05a, U07, U13, U14, U20b
Environmental protection E03, E36a, E38
Storage and disposal D01, D02, D05, D09a, D12a
Treated seed S01, S02, S03, S04c, S05, S06a, S07, S08

SEE SECTION 3 FOR PRODUCTS ALSO REGISTERED

274 fuberidazole + triadimenol

A broad spectrum systemic fungicide seed treatment for cereals
FRAC mode of action code: 1 + 3

See also triadimenol

Products

Tripod	Makhteshim	22.5:187.5 g/l	FS	13014

Uses

- Blue mould in **spring wheat** *(seed treatment)*, **triticale** *(seed treatment)*, **winter wheat** *(seed treatment)*
- Brown foot rot in **spring barley** *(seed treatment)*, **spring oats** *(seed treatment)*, **spring rye** *(seed treatment)*, **spring wheat** *(seed treatment)*, **triticale** *(seed treatment)*, **winter barley** *(seed treatment)*, **winter oats** *(seed treatment)*, **winter rye** *(seed treatment)*, **winter wheat** *(seed treatment)*
- Bunt in **spring wheat** *(seed treatment)*, **winter wheat** *(seed treatment)*
- Covered smut in **spring barley** *(seed treatment)*, **winter barley** *(seed treatment)*
- Leaf stripe in **spring barley** *(seed treatment)*
- Loose smut in **spring barley** *(seed treatment)*, **spring oats** *(seed treatment)*, **spring wheat** *(seed treatment)*, **winter barley** *(seed treatment)*, **winter oats** *(seed treatment)*, **winter wheat** *(seed treatment)*
- Pyrenophora leaf spot in **spring oats** *(seed treatment)*, **winter oats** *(seed treatment)*

Approval information

- Fuberidazole and triadimenol included in Annex I under EC Regulation 1107/2009
- Accepted by BBPA for use on malting barley

Efficacy guidance

- Apply through recommended seed treatment machinery
- Calibrate drill for treated seed and drill at 2.5-4 cm into firm, well prepared seedbed
- Product will give control of early attacks of foliar air-borne diseases such as powdery mildew, yellow rust, brown rust, crown rust, *Septoria*. On autumn-sown crops control of rust is prolonged and usually extends over winter, leading to a delayed build-up of the disease in the following spring, often resulting in fewer fungicide sprays being required to maintain control during spring/summer
- Product will often delay and reduce the likelihood and extent of lodging in autumn sown wheat.

Restrictions

- Maximum number of treatments 1 per batch of seed
- Do not use on naked oats
- Do not drill treated winter wheat or rye seed after end of Nov. Seed rate should be increased as drilling season progresses
- Germination tests should be done on all batches of seed to be treated to ensure seed viability and suitability for treatment
- Do not use on seed with moisture content above 16%, on sprouted, cracked or skinned seed or on seed already treated with another seed treatment
- Do not use treated seed as food or feed

Crop-specific information

- Latest use: before drilling
- Treated spring wheat may be drilled in autumn up to end of Nov or from Feb onwards
- Store treated seed in cool, dry, well-ventilated store and drill as soon as possible, preferably in season of purchase
- Treatment may accentuate effects of adverse seedbed conditions on crop emergence

Environmental safety

- Dangerous for the environment
- Harmful to aquatic organisms

FOR FULL CONDITIONS OF USE ALWAYS READ THE PRODUCT LABEL

Hazard classification and safety precautions
    Hazard Dangerous for the environment
    Transport code 9
    Packaging group III
    UN Number 3082
    Risk phrases R52, R53a
    Operator protection A; U20b
    Environmental protection E03, E15a, E34
    Storage and disposal D05, D09a, D11a
    Treated seed S01, S02, S03, S04a, S05, S06a, S07

275   gibberellins

A plant growth regulator for use in top fruit and grassland

Products

1	Berelex	Interfarm	1 g per tablet	TB	14123
2	GIBB 3	Globachem	10% w/w	TB	12290
3	GIBB Plus	Globachem	10 g/l	SL	11875
4	Novagib	Fine	10 g/l	SL	08954
5	Regulex 10 SG	Interfarm	10% w/w	SG	14116
6	Smartgrass	Interfarm	40% w/w	SG	15628

Uses
- Growth regulation in **crab apples** *(off-label)*, **ornamental crop production (seed)** *(off-label)*, **quinces** *(off-label)* [5]; **farm forestry** *(off-label)*, **forest** *(off-label)* [3, 5]; **grassland** [6]; **pears** *(off-label)* [4, 5]
- Improved germination in **nothofagus** *(do not exceed 2.5 g per 5 litres when used as a seed soak)* [5]
- Increasing fruit set in **pears** [1, 2]
- Increasing germination in **nothofagus** *(qualified minor use)* [3]
- Increasing yield in **celery (outdoor)** [1, 2]; **cherries** *(off-label)*, **protected cherries** *(off-label)*, **rhubarb** [1]; **protected rhubarb** [2]
- Reducing fruit russeting in **apples** [3, 4]; **pears** *(off-label)*, **quinces** *(off-label)* [3]
- Russet reduction in **apples** [5]

Extension of Authorisation for Minor Use (EAMUs)
- **cherries** *20091072* [1]
- **crab apples** *20091078* [5]
- **farm forestry** *20071675* [3], *20091076* [5]
- **forest** *20071675* [3], *20091076* [5]
- **ornamental crop production (seed)** *20091077* [5]
- **pears** *20061627* [3], *20012756* [4], *20091078* [5]
- **protected cherries** *20091072* [1]
- **quinces** *20061627* [3], *20091078* [5]

Approval information
- Gibberellins included in Annex 1 under EC Regulation 1107/2009

Efficacy guidance
- For optimum results on apples spray under humid, slow drying conditions and ensure good spray cover [3, 4]
- Treat apples and pears immediately after completion of petal fall and repeat as directed on the label [3, 4]
- Nothofagus seeds should be stirred into the solution, soaked for 24 hr and then sown immediately [3]
- Fruit set in pears can be improved when blossom is spare, setting is poor or where frost has killed many flowers [2]

Restrictions
- Maximum total dose varies with crop and product. Check label

SEE SECTION 3 FOR PRODUCTS ALSO REGISTERED

- Prepared spray solutions are unstable. Do not leave in the sprayer during meal breaks or overnight
- Return bloom may be reduced in yr following treatment
- Consult before treating apple crops grown for processing [3, 4]
- Avoid storage at temperatures above 32˚C [2]

Crop-specific information
- HI zero for apples [3-5]
- Apply to apples at completion of petal fall and repeat 3 or 4 times at 7-10 d intervals. Number of sprays and spray interval depend on weather conditions and dose (see labels) [3, 4]
- Good results achieved on apple varieties Cox's Orange Pippin, Discovery, Golden Delicious and Karmijn. For other cultivars test on a small number of trees [3, 4]
- Split treatment on pears allows second spray to be omitted if pollinating conditions become very favourable [2]
- Pear variety Conference usually responds well to treatment; Doyenne du Comice can be variable [2]
- On celery the interval between application and harvest should never exceed 4 wk to avoid risk of premature seeding and root splitting [2]
- Treat washed rhubarb crowns on transfer to forcing shed [2]

Hazard classification and safety precautions
UN Number N/C [1-5]
Operator protection U05a [5]; U08 [2]; U09a [1]; U20b [4]; U20c [1-3, 5]
Environmental protection E15a [1-5]
Storage and disposal D01 [4, 5]; D02 [5]; D03, D05, D07 [4]; D09a [1-5]; D10b [3-5]

276   Gliocladium catenulatum

A fungus that acts as an antagonist against other fungi

Products

Prestop	Fargro	32% w/w	WP	15103

Uses
- Botrytis in *all protected edible crops* (moderate control), *all protected non-edible crops* (moderate control), *bilberries* (off-label - moderate control), *blackcurrants* (off-label - moderate control), *blueberries* (off-label - moderate control), *brassica root crops* (off-label - moderate control), *bulb vegetables* (off-label - moderate control), *cane fruit* (off-label - moderate control), *carrots* (off-label - moderate control), *celeriac* (off-label - moderate control), *chicory root* (off-label - moderate control), *cranberries* (off-label - moderate control), *cress* (off-label - moderate control), *endives* (off-label - moderate control), *fruiting vegetables* (off-label - moderate control), *gooseberries* (off-label - moderate control), *herbs (see appendix 6)* (off-label - moderate control), *hops* (off-label - moderate control), *jerusalem artichokes* (off-label - moderate control), *lamb's lettuce* (off-label - moderate control), *land cress* (off-label - moderate control), *legumes* (off-label - moderate control), *lettuce* (off-label - moderate control), *orache* (off-label - moderate control), *ornamental plant production* (off-label - moderate control), *parsley root* (off-label - moderate control), *parsnips* (off-label - moderate control), *red beet* (off-label - moderate control), *redcurrants* (off-label - moderate control), *ribes hybrids* (off-label - moderate control), *rocket* (off-label - moderate control), *salad burnet* (off-label - moderate control), *salsify* (off-label - moderate control), *spinach* (off-label - moderate control), *spinach beet* (off-label - moderate control), *stem vegetables* (off-label - moderate control), *strawberries* (moderate control), *table grapes* (off-label - moderate control), *top fruit* (off-label - moderate control), *vegetable brassicas* (off-label - moderate control), *wine grapes* (off-label - moderate control)
- Didymella in *all protected edible crops* (moderate control), *all protected non-edible crops* (moderate control), *strawberries* (moderate control)
- Fusarium in *all protected edible crops* (moderate control), *all protected non-edible crops* (moderate control), *bilberries* (off-label - moderate control), *blackcurrants* (off-label - moderate control), *blueberries* (off-label - moderate control), *brassica root crops* (off-label - moderate

control), **bulb vegetables** (off-label - moderate control), **cane fruit** (off-label - moderate control), **carrots** (off-label - moderate control), **celeriac** (off-label - moderate control), **chicory root** (off-label - moderate control), **cranberries** (off-label - moderate control), **cress** (off-label - moderate control), **endives** (off-label - moderate control), **fruiting vegetables** (off-label - moderate control), **gooseberries** (off-label - moderate control), **herbs (see appendix 6)** (off-label - moderate control), **hops** (off-label - moderate control), **jerusalem artichokes** (off-label - moderate control), **lamb's lettuce** (off-label - moderate control), **land cress** (off-label - moderate control), **legumes** (off-label - moderate control), **lettuce** (off-label - moderate control), **orache** (off-label - moderate control), **ornamental plant production** (off-label - moderate control), **parsley root** (off-label - moderate control), **parsnips** (off-label - moderate control), **red beet** (off-label - moderate control), **redcurrants** (off-label - moderate control), **ribes hybrids** (off-label - moderate control), **rocket** (off-label - moderate control), **salad burnet** (off-label - moderate control), **salsify** (off-label - moderate control), **spinach** (off-label - moderate control), **spinach beet** (off-label - moderate control), **stem vegetables** (off-label - moderate control), **strawberries** (moderate control), **table grapes** (off-label - moderate control), **top fruit** (off-label - moderate control), **vegetable brassicas** (off-label - moderate control), **wine grapes** (off-label - moderate control)

- Phytophthora in **all protected edible crops** (moderate control), **all protected non-edible crops** (moderate control), **bilberries** (off-label - moderate control), **blackcurrants** (off-label - moderate control), **blueberries** (off-label - moderate control), **brassica root crops** (off-label - moderate control), **bulb vegetables** (off-label - moderate control), **cane fruit** (off-label - moderate control), **carrots** (off-label - moderate control), **celeriac** (off-label - moderate control), **chicory root** (off-label - moderate control), **cranberries** (off-label - moderate control), **cress** (off-label - moderate control), **endives** (off-label - moderate control), **fruiting vegetables** (off-label - moderate control), **gooseberries** (off-label - moderate control), **herbs (see appendix 6)** (off-label - moderate control), **hops** (off-label - moderate control), **jerusalem artichokes** (off-label - moderate control), **lamb's lettuce** (off-label - moderate control), **land cress** (off-label - moderate control), **legumes** (off-label - moderate control), **lettuce** (off-label - moderate control), **orache** (off-label - moderate control), **ornamental plant production** (off-label - moderate control), **parsley root** (off-label - moderate control), **parsnips** (off-label - moderate control), **red beet** (off-label - moderate control), **redcurrants** (off-label - moderate control), **ribes hybrids** (off-label - moderate control), **rocket** (off-label - moderate control), **salad burnet** (off-label - moderate control), **salsify** (off-label - moderate control), **spinach** (off-label - moderate control), **spinach beet** (off-label - moderate control), **stem vegetables** (off-label - moderate control), **strawberries** (moderate control), **table grapes** (off-label - moderate control), **top fruit** (off-label - moderate control), **vegetable brassicas** (off-label - moderate control), **wine grapes** (off-label - moderate control)

- Pythium in **all protected edible crops** (moderate control), **all protected non-edible crops** (moderate control), **bilberries** (off-label - moderate control), **blackcurrants** (off-label - moderate control), **blueberries** (off-label - moderate control), **brassica root crops** (off-label - moderate control), **bulb vegetables** (off-label - moderate control), **cane fruit** (off-label - moderate control), **carrots** (off-label - moderate control), **celeriac** (off-label - moderate control), **chicory root** (off-label - moderate control), **cranberries** (off-label - moderate control), **cress** (off-label - moderate control), **endives** (off-label - moderate control), **fruiting vegetables** (off-label - moderate control), **gooseberries** (off-label - moderate control), **herbs (see appendix 6)** (off-label - moderate control), **hops** (off-label - moderate control), **jerusalem artichokes** (off-label - moderate control), **lamb's lettuce** (off-label - moderate control), **land cress** (off-label - moderate control), **legumes** (off-label - moderate control), **lettuce** (off-label - moderate control), **orache** (off-label - moderate control), **ornamental plant production** (off-label - moderate control), **parsley root** (off-label - moderate control), **parsnips** (off-label - moderate control), **red beet** (off-label - moderate control), **redcurrants** (off-label - moderate control), **ribes hybrids** (off-label - moderate control), **rocket** (off-label - moderate control), **salad burnet** (off-label - moderate control), **salsify** (off-label - moderate control), **spinach** (off-label - moderate control), **spinach beet** (off-label - moderate control), **stem vegetables** (off-label - moderate control), **strawberries** (moderate control), **table grapes** (off-label - moderate control), **top fruit** (off-label - moderate control), **vegetable brassicas** (off-label - moderate control), **wine grapes** (off-label - moderate control)

- Rhizoctonia in **all protected edible crops** (moderate control), **all protected non-edible crops** (moderate control), **bilberries** (off-label - moderate control), **blackcurrants** (off-label - moderate

SEE SECTION 3 FOR PRODUCTS ALSO REGISTERED

control), **blueberries** *(off-label - moderate control)*, **brassica root crops** *(off-label - moderate control)*, **bulb vegetables** *(off-label - moderate control)*, **cane fruit** *(off-label - moderate control)*, **carrots** *(off-label - moderate control)*, **celeriac** *(off-label - moderate control)*, **chicory root** *(off-label - moderate control)*, **cranberries** *(off-label - moderate control)*, **cress** *(off-label - moderate control)*, **endives** *(off-label - moderate control)*, **fruiting vegetables** *(off-label - moderate control)*, **gooseberries** *(off-label - moderate control)*, **herbs (see appendix 6)** *(off-label - moderate control)*, **hops** *(off-label - moderate control)*, **jerusalem artichokes** *(off-label - moderate control)*, **lamb's lettuce** *(off-label - moderate control)*, **land cress** *(off-label - moderate control)*, **legumes** *(off-label - moderate control)*, **lettuce** *(off-label - moderate control)*, **orache** *(off-label - moderate control)*, **ornamental plant production** *(off-label - moderate control)*, **parsley root** *(off-label - moderate control)*, **parsnips** *(off-label - moderate control)*, **red beet** *(off-label - moderate control)*, **redcurrants** *(off-label - moderate control)*, **ribes hybrids** *(off-label - moderate control)*, **rocket** *(off-label - moderate control)*, **salad burnet** *(off-label - moderate control)*, **salsify** *(off-label - moderate control)*, **spinach** *(off-label - moderate control)*, **spinach beet** *(off-label - moderate control)*, **stem vegetables** *(off-label - moderate control)*, **strawberries** *(moderate control)*, **table grapes** *(off-label - moderate control)*, **top fruit** *(off-label - moderate control)*, **vegetable brassicas** *(off-label - moderate control)*, **wine grapes** *(off-label - moderate control)*

Extension of Authorisation for Minor Use (EAMUs)

- **bilberries** *(moderate control)* 20120564
- **blackcurrants** *(moderate control)* 20120564
- **blueberries** *(moderate control)* 20120564
- **brassica root crops** *(moderate control)* 20120564
- **bulb vegetables** *(moderate control)* 20120564
- **cane fruit** *(moderate control)* 20120564
- **carrots** *(moderate control)* 20120564
- **celeriac** *(moderate control)* 20120564
- **chicory root** *(moderate control)* 20120564
- **cranberries** *(moderate control)* 20120564
- **cress** *(moderate control)* 20120564
- **endives** *(moderate control)* 20120564
- **fruiting vegetables** *(moderate control)* 20120564
- **gooseberries** *(moderate control)* 20120564
- **herbs (see appendix 6)** *(moderate control)* 20120564
- **hops** *(moderate control)* 20120564
- **jerusalem artichokes** *(moderate control)* 20120564
- **lamb's lettuce** *(moderate control)* 20120564
- **land cress** *(moderate control)* 20120564
- **legumes** *(moderate control)* 20120564
- **lettuce** *(moderate control)* 20120564
- **orache** *(moderate control)* 20120564
- **ornamental plant production** *(moderate control)* 20120564
- **parsley root** *(moderate control)* 20120564
- **parsnips** *(moderate control)* 20120564
- **red beet** *(moderate control)* 20120564
- **redcurrants** *(moderate control)* 20120564
- **ribes hybrids** *(moderate control)* 20120564
- **rocket** *(moderate control)* 20120564
- **salad burnet** *(moderate control)* 20120564
- **salsify** *(moderate control)* 20120564
- **spinach** *(moderate control)* 20120564
- **spinach beet** *(moderate control)* 20120564
- **stem vegetables** *(moderate control)* 20120564
- **table grapes** *(moderate control)* 20120564
- **top fruit** *(moderate control)* 20120564
- **vegetable brassicas** *(moderate control)* 20120564
- **wine grapes** *(moderate control)* 20120564

FOR FULL CONDITIONS OF USE ALWAYS READ THE PRODUCT LABEL

Approval information
- Gliocladium catenulatum included in Annex 1 under EC Regulation 1107/2009

Hazard classification and safety precautions
UN Number N/C
Operator protection A, D, H; U05a, U10, U14, U15, U20a
Environmental protection E15b, E34
Storage and disposal D01, D02, D09a, D10a
Medical advice M03

277 glufosinate-ammonium

A non-selective, non-residual phosphinic acid contact herbicide
HRAC mode of action code: H

Products

1	Finale	Bayer Environ.	120 g/l	SL	10092
2	Harvest	Bayer CropScience	150 g/l	SL	07321
3	Kaspar	Certis	150 g/l	SL	11214
4	Kibosh	AgChem Access	150 g/l	SL	13976
5	Kurtail	Progreen	150 g/l	SL	15329
6	PureReap	Pure Amenity	150 g/l	SL	15132

Uses
- Annual and perennial weeds in *all tree nuts, apple orchards, bilberries, blueberries, cane fruit, cherries, cranberries, cultivated land/soil, currants, damsons, fallows, forest, non-crop farm areas, pear orchards, plums, potatoes* (not for seed), *strawberries, stubbles, sugar beet, vegetables, wine grapes* [2]; *managed amenity turf* [1]
- Annual dicotyledons in *all tree nuts, apple orchards, cane fruit, cherries, cultivated land/soil, currants, damsons, fallows, forest, headlands* (uncropped), *non-crop farm areas, pear orchards, plums, strawberries, sugar beet, vegetables, ware potatoes, wine grapes* [3-6]; *nursery stock, woody ornamentals* [1]
- Annual grasses in *all tree nuts, apple orchards, cane fruit, cherries, cultivated land/soil, currants, damsons, fallows, forest, headlands* (uncropped), *non-crop farm areas, pear orchards, plums, strawberries, sugar beet, vegetables, ware potatoes, wine grapes* [3-6]; *nursery stock, woody ornamentals* [1]
- Green cover in *land temporarily removed from production* [2-6]
- Harvest management/desiccation in *combining peas, ware potatoes* [3-6]; *combining peas* (not for seed), *potatoes* [2]; *linseed, spring field beans, spring oilseed rape, winter field beans, winter oilseed rape* [2-6]
- Line marking preparation in *managed amenity turf* [1]
- Perennial dicotyledons in *all tree nuts, apple orchards, cane fruit, cherries, currants, damsons, fallows, forest, headlands* (uncropped), *land temporarily removed from production, non-crop farm areas, pear orchards, plums, strawberries, wine grapes* [3-6]; *nursery stock, woody ornamentals* [1]
- Perennial grasses in *all tree nuts, apple orchards, cane fruit, cherries, currants, damsons, fallows, forest, headlands* (uncropped), *land temporarily removed from production, non-crop farm areas, pear orchards, plums, strawberries, wine grapes* [3-6]; *nursery stock, woody ornamentals* [1]
- Sward destruction in *permanent grassland* [2-6]

Approval information
- Glufosinate-ammonium included in Annex I under EC Regulation 1107/2009
- Accepted by BBPA for use on malting barley and hops
- Approval expiry 31 Mar 2013 [4, 6]

Efficacy guidance
- Activity quickest under warm, moist conditions. Light rainfall 3-4 h after application will not affect activity. Do not spray wet foliage or if rain likely within 6 h

SECTION 2

- For weed control uses treat when weeds growing actively. Deep rooted weeds may require second treatment
- On uncropped headlands apply in May/Jun to prevent weeds invading field
- Glufosinate kills all green tissue but does not harm mature bark

Restrictions

- Maximum number of treatments 4 per crop for potatoes (including 2 desiccant uses); 2 per crop (including 1 desiccant use) for oilseed rape, dried peas, field beans, linseed, wheat and barley; 3 per yr for fruit and forestry; 2 per yr for strawberries, non-crop land and land temporarily removed from production, round fruit trees, canes or bushes; 1 per crop for sugar beet, vegetables and other crops; 1 per yr for grassland destruction, on cultivated land prior to planting edible crops
- Pre-harvest desiccation sprays should not be used on seed crops of wheat, barley, peas or potatoes but may be used on seed crops of oilseed rape, field beans and linseed [2, 3]
- Do not desiccate potatoes in exceptionally wet weather or in saturated soil. See label for details
- Do not spray potatoes after emergence if grown from small or diseased seed or under very dry conditions
- Application for weed control and line-marking uses in sports turf must be between 1 Mar and 30 Sep [1]
- Do not spray hedge bottoms or on broadcast crops [2]

Crop-specific information

- Latest use: 30 Sep for use on non-crop land, top fruit, soft fruit, cane fruit, bush fruit, forestry; pre-drilling, pre-planting or pre-emergence in sugar beet, vegetables and other crops; before winter dormancy for grassland destruction.
- HI potatoes, oilseed rape, combining peas, field beans, linseed 7 d; wheat, barley 14 d
- Ploughing or other cultivations can follow 4 h after spraying
- Crops can normally be sown/planted immediately after spraying or sprayed post-drilling. On sand, very light or immature peat soils allow at least 3 d before sowing/planting or expected emergence
- For weed control in potatoes apply pre-emergence or up to 10% emergence on earlies and seed crops, up to 40% on maincrop, on plants up to 15 cm high
- In sugar beet and vegetables apply just before crop emergence, using stale seedbed technique
- In top and soft fruit, grapevines and forestry apply between 1 Mar and 30 Sep as directed sprays
- For grass destruction apply before winter dormancy occurs. Heavily grazed fields should show active regrowth. Plough from the day after spraying
- Apply pre-harvest desiccation treatments 10-21 d before harvest (14-21 d for oilseed rape). See label for timing details on individual crops
- For potato haulm desiccation apply to listed varieties (not seed crops) at onset of senescence, 14-21 d before harvest

Environmental safety

- Harmful to fish or other aquatic life. Do not contaminate surface waters or ditches with chemical or used container
- Keep livestock out of treated areas until foliage of any poisonous weeds such as ragwort has died and become unpalatable
- Risk to certain non-target insects or other arthropods [2, 3]
- Treated pea haulm may be fed to livestock from 7 d after spraying, treated grain from 14 d [2]
- Product supplied in small volume returnable container - see label for filling and mixing instructions [2]

Hazard classification and safety precautions

Hazard Harmful [2-6]; Irritant [1]
Transport code 6.1
Packaging group III
UN Number 2902
Risk phrases R21, R22a, R41 [2-6]; R36 [1]
Operator protection A [2-6]; B [1]; C, H, M [1-6]; U04a, U11, U13, U14, U15, U19a, U20a [2-6]; U05a, U08 [1-6]; U20b [1]
Environmental protection E07a [1-6]; E13c [1, 3-6]; E15a, E22c [2]; E22b [3-6]; E34 [2-6]
Storage and disposal D01, D02, D09a, D10b [1-6]; D05 [1]; D12a [2-6]

FOR FULL CONDITIONS OF USE ALWAYS READ THE PRODUCT LABEL

Medical advice M03 [2-6]

278   glyphosate

A translocated non-residual glycine derivative herbicide
HRAC mode of action code: G

See also diflufenican + glyphosate
diuron + glyphosate
flufenacet + glyphosate + metosulam
glyphosate + pyraflufen-ethyl
glyphosate + sulfosulfuron

Products

1	Amega Duo	Nufarm UK	540 g/l	SL	13358
2	Amega Duo	Nufarm UK	540 g/l	SL	14131
3	Asteroid	Headland	360 g/l	SL	11118
4	Azural	Monsanto	360 g/l	SL	14361
5	Barclay Barbarian	Barclay	360 g/l	SL	12714
6	Barclay Gallup 360	Barclay	360 g/l	SL	14988
7	Barclay Gallup Amenity	Barclay	360 g/l	SL	13250
8	Barclay Gallup Biograde 360	Barclay	360 g/l	SL	15188
9	Barclay Gallup Biograde Amenity	Barclay	360 g/l	SL	12716
10	Barclay Gallup Hi-Aktiv	Barclay	490 g/l	SL	14987
11	Barclay Glyde 144	Barclay	144 g/l	SL	15362
12	Buggy XTG	Sipcam	36% w/w	SG	12951
13	CDA Vanquish Biactive	Bayer Environ.	120 g/l	UL	12586
14	Charger C	AgChem Access	360 g/l	SL	14216
15	Clinic Ace	Nufarm UK	360 g/l	SL	12980
16	Clinic Ace	Nufarm UK	360 g/l	SL	14040
17	Credit DST	Nufarm UK	540 g/l	SL	13822
18	Credit DST	Nufarm UK	540 g/l	SL	14066
19	Discman Biograde	Barclay	216 g/l	SL	12856
20	Dow Agrosciences Glyphosate 360	Dow	360 g/l	SL	12720
21	Ecoplug Max	Monsanto	68 % w/w	GR	14741
22	Envision	Headland	450 g/l	SL	10569
23	Etna	AgriChem BV	360 g/l	SL	14674
24	Frontsweep	Nomix Enviro	360 g/l	SL	14700
25	Gallup Hi- Aktiv Amenity	Barclay	360 g/l	SL	12898
26	Glyfo_TDI	Q-Chem	360 g/l	SL	14743
27	Glyfos	Headland	360 g/l	SL	10995
28	Glyfos Dakar	Headland	68% w/w	SG	13054
29	Glyfos Dakar Pro	Headland Amenity	68% w/w	SG	13147
30	Glyfos Proactive	Nomix Enviro	360 g/l	SL	11976
31	Glyfos Supreme	Headland	450 g/l	SL	12371
32	Glyfosat 36	Goldengrass	360 g/l	SL	14297
33	Glyfo-TDI	Q-Chem	360 g/l	SL	13940
34	Glyphogan	Makhteshim	360 g/l	SL	12668
35	Glypho-Rapid 450	Barclay	450 g/l	SL	13882
36	Hilite	Nomix Enviro	144 g/l	RH	13871
37	Kernel	Headland	480 g/l	SL	10993
38	Manifest	Headland	360 g/l	SL	11041
39	Mascot Hi-Aktiv Amenity	Rigby Taylor	490 g/l	SL	15178
40	Monsanto Amenity Glyphosate	Monsanto	360 g/l	SL	15227
41	Nomix Conqueror	Nomix Enviro	144 g/l	UL	12370

SEE SECTION 3 FOR PRODUCTS ALSO REGISTERED

Products – continued

42	Nufosate Ace	Nufarm UK	360 g/l	SL	13794
43	Nufosate Ace	Nufarm UK	360 g/l	SL	14959
44	Pitch	Nufarm UK	540 g/l	SL	15485
45	Pitch	Nufarm UK	540 g/l	SL	15693
46	Proliance Quattro	Syngenta	360 g/l	SL	13670
47	Pure Glyphosate 360	Pure Amenity	360 g/l	SL	14834
48	Rattler	Nufarm UK	540 g/l	SL	15522
49	Rattler	Nufarm UK	540 g/l	SL	15692
50	Reaper	AgChem Access	360 g/l	SL	14924
51	Rhizeup	Clayton	360 g/l	SL	15043
52	Roundup Biactive	Monsanto	360 g/l	SL	10320
53	Roundup Energy	Monsanto	450 g/l	SL	12945
54	Roundup Flex	Monsanto	480 g/l	SL	15541
55	Roundup Max	Monsanto	68% w/w	SG	12952
56	Roundup Pro Biactive	Monsanto	360 g/l	SL	10330
57	Roundup ProBiactive 450	Monsanto	450 g/l	SL	12778
58	Roundup Ultimate	Monsanto	540 g/l	SL	13081
59	Rustler	ChemSource	360 g/l	SL	14951
60	Rustler Pro-Green	ChemSource	360 g/l	SL	15454
61	Samurai	Monsanto	360 g/l	SL	12674
62	Shyfo	Sharda	360 g/l	SL	15040
63	Slingshot	ChemSource	360 g/l	SL	14505
64	Snapper	Nufarm UK	550 g/l	SL	15489
65	Snapper	Nufarm UK	550 g/l	SL	15695
66	Stacato	Sipcam	360 g/l	SL	12675
67	Stirrup	Nomix Enviro	144 g/l	RH	13875
68	Symbol	United Phosphorus	360 g/l	SL	14769
69	Tanker	Nufarm UK	540 g/l	SL	15016
70	Tanker	Nufarm UK	540 g/l	SL	15694
71	Touchdown Quattro	Syngenta	360 g/l	SL	10608
72	Trustee Amenity	Barclay	450 g/l	SL	12897

Uses

- Annual and perennial weeds in *all edible crops (outdoor and protected)* [1, 2, 15-18, 35, 42, 43, 66, 68-70]; *all edible crops (outdoor)* [15-17, 26, 33, 34, 42-45, 48, 49, 54, 59, 64, 65]; *all edible crops (outdoor)* (before planting), *all non-edible crops (outdoor)* (before planting) [5, 6, 8, 23, 28, 31, 32, 47, 50, 51, 60, 62, 63, 71]; *all non-edible crops (outdoor)* [1, 2, 15-18, 26, 33-35, 42-45, 48, 49, 54, 59, 64-66, 68-70]; *all non-edible seed crops grown outdoors* [15-17, 42, 43]; *almonds* (off-label), *apples* (off-label), *apricots* (off-label), *asparagus* (off-label), *chestnuts* (off-label), *cob nuts* (off-label), *crab apples* (off-label), *grapevines* (off-label), *hazel nuts* (off-label), *peaches* (off-label), *pears* (off-label), *quinces* (off-label), *rhubarb* (off-label), *walnuts* (off-label) [3, 52]; *amenity grassland*, *amenity grassland* (wiper application), *amenity vegetation* (wiper application), *broad-leaved trees*, *conifers*, *farm buildings/yards*, *fencelines*, *land clearance*, *non-crop areas*, *paths and drives*, *walls*, *woody ornamentals* [56]; *amenity vegetation* [1, 2, 4, 5, 8, 11, 13, 15-18, 24, 32, 36, 40-45, 48-51, 54, 57, 59, 60, 63-67, 69, 70]; *apple orchards*, *pear orchards* [1-3, 5, 6, 8, 10, 12, 14-18, 20, 22, 23, 27, 28, 31, 32, 34, 35, 37, 38, 42-45, 47-51, 53, 54, 58-66, 68-70]; *apples*, *pears* [26, 33, 41]; *asparagus* [4-6, 8, 15, 16, 23, 24, 32, 34, 47, 50, 51, 54, 59, 60, 62]; *bilberries* (off-label), *blackcurrants* (off-label), *blueberries* (off-label), *cranberries* (off-label), *gooseberries* (off-label), *redcurrants* (off-label), *vaccinium spp.* (off-label), *whitecurrants* (off-label) [3, 41, 52]; *bog myrtle* (off-label) [57]; *buckwheat* (off-label), *canary seed* (off-label), *lupins* (off-label), *millet* (off-label), *quinoa* (off-label), *sorghum* (off-label) [53]; *bulb onions* [1, 2, 15-18, 42-45, 48, 49, 54, 59, 62, 64-66, 69, 70]; *cherries*, *plums* [1-6, 8, 10, 12, 14-18, 20, 22-24, 26-28, 31-34, 37, 38, 41-45, 47-54, 58-66, 68-70]; *cherries* (off-label), *fruit trees* (off-label), *plums* (off-label) [3]; *combining peas*, *spring barley*, *spring field beans*, *spring wheat*, *winter barley*, *winter field beans*, *winter wheat* [1-3, 15-18, 26, 33, 35, 42-45, 48, 49, 54, 59, 64-66, 68-71]; *cultivated land/soil* [3, 20, 61]; *damsons* [1-3, 5, 6, 8, 10, 12, 14, 18, 20, 22, 26-28, 31-34, 37, 38, 41, 47, 50-53, 58, 60, 61, 63, 66, 68-70]; *durum wheat* [1-3, 15-18, 35, 42-45, 48, 49, 54, 59, 64-66, 68-70]; *enclosed waters*, *open*

SECTION 2

waters [7, 9, 25, 26, 33, 35, 39, 41, 51, 72]; ***farm forestry*** [26, 33, 62]; ***farm forestry*** *(off-label)*, ***game cover*** *(off-label)*, ***hemp*** *(off-label)*, ***hops*** *(off-label)*, ***miscanthus*** *(off-label)*, ***soft fruit*** *(off-label)*, ***top fruit*** *(off-label)*, ***woad*** *(off-label)* [52, 53, 55, 58]; ***field crops*** *(wiper application)*, ***non-crop farm areas*** [20, 61]; ***forest*** [1, 2, 4-12, 14-20, 22-27, 30-38, 40-51, 54, 56, 57, 59, 60, 62-67, 69, 70, 72]; ***forest nurseries*** [4, 15-17, 23, 24, 42, 43, 46, 59, 62]; ***forestry plantations*** [1, 2, 17, 18, 44, 45, 48, 49, 64-66, 69, 70]; ***grassland*** [1, 2, 4, 15-18, 24, 26, 33, 35, 42-45, 48, 49, 59, 64-66, 68-70]; ***green cover on land temporarily removed from production*** [1, 2, 4, 15-18, 24, 35, 36, 42-45, 48, 49, 54, 59, 64-66, 68-70]; ***hard surfaces*** [4-11, 15-17, 19, 23-26, 31-36, 39-51, 54, 56, 57, 59, 60, 63-65, 67, 72]; ***industrial sites***, ***road verges*** [30, 56]; ***kentish cobnuts*** *(off-label)*, ***wine grapes*** *(off-label)* [52]; ***land immediately adjacent to aquatic areas*** [7, 9-11, 25, 26, 33, 35, 39, 41, 51, 52, 56, 57, 72]; ***land not intended to bear vegetation*** [10, 12-14, 20, 22, 27, 30, 37, 38, 52, 55, 56, 61]; ***land prior to cultivation*** [31]; ***leeks***, ***sugar beet***, ***swedes***, ***turnips*** [1, 2, 4, 15-18, 24, 26, 33, 35, 42-45, 48, 49, 54, 59, 62, 64-66, 69, 70]; ***linseed*** [1-4, 15-18, 24, 26, 33, 35, 42-45, 48, 49, 54, 64-66, 68-71]; ***managed amenity turf*** *(pre-establishment)* [36, 67]; ***managed amenity turf*** *(pre-establishment only)* [41]; ***mustard*** [1, 2, 15-18, 26, 33, 35, 42-45, 48, 49, 54, 59, 64-66, 68-71]; ***natural surfaces not intended to bear vegetation*** [1, 2, 4-11, 15-19, 23-26, 31-36, 39-51, 54, 57, 59, 60, 63-70, 72]; ***oilseed rape*** [4, 24]; ***onions*** [26, 33, 35]; ***ornamental plant production*** [41, 67]; ***ornamental plant production*** *(off-label)* [15, 56, 57]; ***permanent grassland*** [12, 20, 34, 52, 53, 61]; ***permeable surfaces overlying soil*** [1, 2, 4-11, 15-19, 23-26, 31-36, 39, 40, 42-51, 54, 57, 59, 60, 63-70, 72]; ***poppies for morphine production*** *(off-label)* [15]; ***potatoes*** [54]; ***rye*** *(off-label)* [53, 55]; ***spring oats***, ***winter oats*** [1-3, 15-18, 26, 33, 35, 42-45, 48, 49, 54, 59, 64-66, 68-70]; ***spring oilseed rape***, ***winter oilseed rape*** [1-3, 15-18, 26, 33, 35, 42-45, 48, 49, 54, 64-66, 68-71]; ***stubbles*** [3-6, 8, 10, 12, 20, 23, 24, 32, 34, 47, 50-53, 55, 58, 60-63, 71]; ***vining peas*** [1, 2, 4, 15-18, 24, 42-45, 48, 49, 54, 59, 62, 64-66, 69, 70]; ***water or waterside areas*** [19]

- Annual dicotyledons in ***bulb onions***, ***durum wheat*** [4-6, 8, 23, 24, 28, 47, 50-53, 55, 58, 60]; ***combining peas*** [5, 6, 8, 23, 28, 47, 50, 51, 55, 60]; ***cultivated land/soil*** [10]; ***leeks***, ***linseed***, ***mustard***, ***spring field beans***, ***spring oilseed rape***, ***sugar beet***, ***swedes***, ***winter field beans*** [5, 6, 8, 23, 28, 47, 50-53, 55, 58, 60]; ***peas*** [52]; ***permanent grassland*** [52, 53, 58]; ***spring barley***, ***spring oats***, ***spring wheat***, ***winter barley***, ***winter oats***, ***winter wheat*** [5, 6, 8, 14, 22, 23, 27, 28, 31, 37, 38, 47, 50-53, 55, 58, 60]; ***turnips*** [28, 52, 53, 55, 58]; ***vining peas*** [5, 6, 8, 23, 47, 50, 51, 53, 55, 58, 60]; ***winter oilseed rape*** [5, 6, 8, 23, 28, 32, 47, 50-53, 55, 58, 60, 62, 63]
- Annual grasses in ***bulb onions***, ***durum wheat*** [4, 24, 28, 52, 53, 55, 58]; ***combining peas*** [28, 55]; ***cultivated land/soil*** [10, 14, 22, 27, 37, 38]; ***leeks***, ***linseed***, ***mustard***, ***spring barley***, ***spring field beans***, ***spring oats***, ***spring oilseed rape***, ***spring wheat***, ***sugar beet***, ***swedes***, ***turnips***, ***winter barley***, ***winter field beans***, ***winter oats***, ***winter oilseed rape***, ***winter wheat*** [28, 52, 53, 55, 58]; ***peas*** [52]; ***permanent grassland*** [52, 53, 58]; ***vining peas*** [53, 55, 58]
- Aquatic weeds in ***enclosed waters***, ***land immediately adjacent to aquatic areas*** [25, 52, 54, 56, 72]; ***open waters*** [10, 25, 52, 54, 57, 72]
- Black bent in ***combining peas***, ***spring field beans***, ***spring oilseed rape***, ***stubbles***, ***winter field beans***, ***winter oilseed rape*** [10, 14, 22, 27, 31, 37, 38]; ***durum wheat***, ***spring barley***, ***spring oats***, ***spring wheat***, ***winter barley***, ***winter oats***, ***winter wheat*** [10]; ***linseed*** [14, 22, 27, 31, 37, 38]
- Bolters in ***sugar beet*** *(wiper application)* [10, 20, 28, 52, 53, 58, 61]
- Bracken in ***amenity vegetation***, ***fencelines***, ***land clearance***, ***non-crop areas***, ***paths and drives*** [56]; ***farm forestry*** [62]; ***forest*** [5, 6, 8, 10, 12, 14, 20, 22, 23, 27, 29-32, 34, 37, 38, 40, 47, 50, 51, 54, 56, 57, 60, 62, 63]; ***forest nurseries*** [23, 62]; ***tolerant conifers*** [30, 56]
- Brambles in ***tolerant conifers*** [30, 56]
- Canary grass in ***aquatic areas*** [28]; ***enclosed waters***, ***land immediately adjacent to aquatic areas*** [29]
- Chemical thinning in ***farm forestry*** [62]; ***forest*** [4-6, 8, 10, 14, 22-24, 27, 29, 31, 32, 34, 37, 38, 40, 46, 47, 50, 51, 54, 56, 57, 60, 62, 63]; ***forest*** *(stump treatment)* [12]; ***forest nurseries*** [4, 23, 24, 46, 62]
- Couch in ***combining peas***, ***spring field beans***, ***winter field beans*** [3, 4, 10, 12, 14, 20, 22, 24, 27, 28, 31, 34, 37, 38, 52, 53, 58, 61, 71]; ***durum wheat*** [3, 4, 10, 24, 28, 52, 53, 58]; ***forest*** [10, 29, 34, 56, 57]; ***linseed*** [3, 12, 14, 22, 27, 28, 31, 34, 37, 38, 52, 53, 58, 71]; ***mustard*** [4, 12, 24, 34, 52, 53, 58, 71]; ***oats*** [4, 24]; ***permanent grassland*** [12]; ***spring barley***, ***spring wheat***, ***winter barley***, ***winter wheat*** [3, 4, 10, 12, 20, 24, 28, 34, 52, 53, 58, 61, 71]; ***spring oats***,

winter oats [3, 10, 12, 20, 28, 34, 52, 53, 58, 61]; *spring oilseed rape, stubbles, winter oilseed rape* [3, 10, 12, 14, 20, 22, 27, 28, 31, 34, 37, 38, 52, 53, 58, 61, 71]

- Creeping bent in *aquatic areas* [28]; *combining peas, spring field beans, spring oilseed rape, stubbles, winter field beans, winter oilseed rape* [10, 14, 22, 27, 31, 37, 38]; *durum wheat, spring barley, spring oats, spring wheat, winter barley, winter oats, winter wheat* [10]; *enclosed waters, land immediately adjacent to aquatic areas* [29]; *linseed* [14, 22, 27, 31, 37, 38]

- Desiccation in *buckwheat* (off-label), *canary seed* (off-label), *hemp* (off-label), *lupins* (off-label), *millet* (off-label), *quinoa* (off-label), *sorghum* (off-label) [53]; *buckwheat* (off-label - for wild bird seed production), *canary seed* (off-label - for wild bird seed production), *hemp* (off-label - for wild bird seed production), *millet* (off-label - for wild bird seed production), *quinoa* (off-label - for wild bird seed production), *sorghum* (off-label - for wild bird seed production) [55]; *combining peas, durum wheat, linseed, mustard, spring barley, spring field beans, spring oats, spring oilseed rape, spring wheat, winter barley, winter field beans, winter oats, winter oilseed rape, winter wheat* [54]; *poppies for morphine production* (off-label) [15, 53]; *rye* (off-label) [53, 55]

- Destruction of crops in *all edible crops (outdoor and protected)* [1, 2, 69, 70]; *all edible crops (outdoor)* [4, 15-17, 24, 34, 42-45, 48, 49, 64, 65]; *all non-edible crops (outdoor)* [4, 24, 34, 44, 45, 48, 49, 64, 65]; *all non-edible seed crops grown outdoors* [15-17, 42, 43]; *grassland* [1, 2, 44, 45, 48, 49, 54, 64, 65, 69, 70]; *green cover on land temporarily removed from production* [4, 24]

- Destruction of short term leys in *grassland* [36]

- Grass weeds in *all edible crops (outdoor and protected), all non-edible crops (outdoor), apple orchards, combining peas, durum wheat, enclosed waters, grassland, green cover on land temporarily removed from production, land immediately adjacent to aquatic areas, leeks, linseed, mustard, onions, open waters, pear orchards, spring barley, spring field beans, spring oats, spring oilseed rape, spring wheat, sugar beet, swedes, turnips, winter barley, winter field beans, winter oats, winter oilseed rape, winter wheat* [35]; *forest, hard surfaces, natural surfaces not intended to bear vegetation, permeable surfaces overlying soil* [19, 35]; *water or waterside areas* [19]

- Green cover in *grassland* [23]; *land temporarily removed from production* [3, 5, 6, 8, 10, 12, 14, 22, 23, 27, 28, 31, 32, 37, 38, 47, 50-53, 55, 58, 60, 62, 63, 71]

- Growth suppression in *forestry plantations* [15, 16, 42, 43]; *tree stumps* [2, 15-17, 42-45, 48, 49, 64, 65]

- Harvest management/desiccation in *combining peas* [1, 2, 4-6, 8, 12, 15-17, 20, 23, 24, 32, 34, 42-45, 47-53, 58, 60-65, 69, 70]; *durum wheat* [1, 2, 4-6, 8, 15-17, 20, 23, 24, 28, 32, 42-45, 47-53, 58, 60-65, 69, 70]; *grassland* [4, 24, 62]; *linseed* [1, 2, 4-6, 8, 10, 12, 15-17, 20, 23, 24, 32, 34, 42-45, 47-53, 58, 60-65, 69, 70]; *mustard* [1, 2, 4-6, 8, 12, 15-17, 20, 23, 24, 32, 34, 42-45, 47-53, 58, 60-65, 69-71]; *oats, oilseed rape* [4, 24]; *spring barley, winter barley* [1, 2, 4-6, 8, 12, 14-17, 22-24, 27, 28, 31, 32, 34, 37, 38, 42-45, 47-53, 58, 60, 62-65, 69, 70]; *spring field beans, winter field beans* [4-6, 8, 12, 20, 23, 24, 32, 34, 44, 45, 47-53, 58, 60-65]; *spring oats, winter oats* [1, 2, 5, 6, 8, 12, 14-17, 20, 22, 23, 27, 28, 31, 32, 34, 37, 38, 42-45, 47-53, 58, 60-65, 69, 70]; *spring oilseed rape, winter oilseed rape* [1-3, 5, 6, 8, 10, 12, 14-17, 20, 22, 23, 27, 28, 31, 32, 34, 37, 38, 42-45, 47-53, 58, 60-65, 69-71]; *spring wheat, winter wheat* [1, 2, 4-6, 8, 12, 14-17, 20, 22-24, 27, 28, 31, 32, 34, 37, 38, 42-45, 47-53, 58, 60-65, 69, 70]

- Heather in *farm forestry* [62]; *forest* [5, 6, 8, 10, 12, 14, 20, 22, 23, 27, 29-32, 37, 38, 40, 47, 50, 51, 57, 60, 62, 63]; *forest nurseries* [23, 62]

- Perennial dicotyledons in *combining peas, linseed, spring field beans, spring oilseed rape, winter field beans, winter oilseed rape* [14, 22, 27, 28, 31, 37, 38]; *forest* [40]; *forest* (wiper application) [56]; *permanent grassland* (wiper application) [52, 53, 58]

- Perennial grasses in *aquatic areas* [3, 10, 12, 20, 22, 27, 30, 37, 38, 56]; *combining peas, linseed, spring field beans, stubbles, winter field beans* [28]; *tolerant conifers* [30, 56]

- Pre-harvest desiccation in *combining peas, linseed, spring barley, spring field beans, spring wheat, winter barley, winter field beans, winter wheat* [71]; *poppies for morphine production* (off-label) [52]

- Reeds in *aquatic areas* [3, 10, 12, 20, 22, 27, 28, 30, 37, 38, 52, 56]; *enclosed waters, land immediately adjacent to aquatic areas* [10, 29, 54, 57]; *open waters* [54]

- Rhododendrons in *farm forestry* [62]; *forest* [5, 6, 8, 10, 12, 14, 20, 22, 23, 27, 29-32, 37, 38, 40, 47, 50, 51, 56, 57, 60, 62, 63]; *forest nurseries* [23, 62]

FOR FULL CONDITIONS OF USE ALWAYS READ THE PRODUCT LABEL

- Rushes in **aquatic areas** [3, 10, 12, 20, 22, 27, 28, 30, 37, 38, 52, 56]; **enclosed waters**, **land immediately adjacent to aquatic areas** [10, 29, 54, 57]; **forest** [34]; **open waters** [54]
- Sedges in **aquatic areas** [3, 10, 12, 20, 22, 27, 28, 30, 37, 38, 52, 56]; **enclosed waters**, **land immediately adjacent to aquatic areas** [10, 29, 54, 57]; **open waters** [54]
- Sprout suppression in **tree stumps** [54]
- Sucker control in **apple orchards**, **cherries**, **pear orchards**, **plums** [4, 24, 52, 53, 61]; **damsons** [52, 53, 61]
- Sucker inhibition in **tree stumps** [1, 4, 21, 24, 69, 70]
- Sward destruction in **amenity grassland** [56]; **permanent grassland** [3, 5, 6, 8, 10, 12, 14, 20, 22, 27, 28, 31, 32, 34, 37, 38, 47, 50-53, 58, 60, 61, 63, 71]; **rotational grass** [5, 6, 8, 32, 47, 50, 51, 60, 63, 71]
- Total vegetation control in **all edible crops (outdoor)** *(pre-sowing/planting)*, **all non-edible crops (outdoor)** *(pre-sowing/planting)* [53, 57]; **amenity vegetation** [13, 30, 56]; **fencelines**, **industrial sites**, **road verges** [30, 56]; **hard surfaces** [29, 30, 56]; **land not intended to bear vegetation** [13, 30, 56, 71]; **natural surfaces not intended to bear vegetation**, **permeable surfaces overlying soil** [29]
- Volunteer cereals in **bulb onions** [4, 24, 28, 52, 53, 55, 58]; **combining peas** [28, 55]; **cultivated land/soil** [10, 14, 22, 27, 37, 38]; **durum wheat** [4, 24, 52, 53, 55, 58]; **leeks**, **linseed**, **mustard**, **spring field beans**, **sugar beet**, **swedes**, **turnips**, **winter field beans** [28, 52, 53, 55, 58]; **peas** [52]; **spring barley**, **spring oats**, **spring oilseed rape**, **spring wheat**, **winter barley**, **winter oats**, **winter oilseed rape**, **winter wheat** [52, 53, 55, 58]; **stubbles** [3, 10, 12, 14, 20, 22, 27, 28, 31, 34, 37, 38, 52, 53, 58, 61, 71]; **vining peas** [53, 55, 58]
- Volunteer potatoes in **stubbles** [3, 10, 12, 20, 28, 34, 52, 53, 58, 61, 71]
- Waterlilies in **aquatic areas** [3, 10, 12, 20, 22, 27, 28, 30, 37, 38, 56]; **enclosed waters** [10, 29, 54, 57]; **land immediately adjacent to aquatic areas**, **open waters** [54]
- Weed beet in **sugar beet** *(wiper application)* [28, 61]
- Wild oats in **durum wheat**, **spring barley**, **spring oats**, **spring wheat**, **winter barley**, **winter oats**, **winter wheat** [20, 61]
- Woody weeds in **farm forestry** [62]; **forest** [5, 6, 8, 10, 12, 14, 20, 22, 23, 27, 29-32, 34, 37, 38, 40, 47, 50, 51, 56, 57, 60, 62, 63]; **forest nurseries** [23, 62]; **tolerant conifers** [30, 56]

Extension of Authorisation for Minor Use (EAMUs)
- **almonds** *20102476* [3], *20072045* [52]
- **apples** *20102476* [3], *20072045* [52]
- **apricots** *20102476* [3], *20072045* [52]
- **asparagus** *20071476* [3], *20062036* [52]
- **bilberries** *20071475* [3], *20120715* [41], *20062035* [52]
- **blackcurrants** *20071475* [3], *20120715* [41], *20062035* [52]
- **blueberries** *20071475* [3], *20120715* [41], *20062035* [52]
- **bog myrtle** *20082727* [57]
- **buckwheat** *20071839* [53], *(for wild bird seed production) 20070582* [55]
- **canary seed** *20071839* [53], *(for wild bird seed production) 20070582* [55]
- **cherries** *20102476* [3]
- **chestnuts** *20071477* [3], *20062028* [52], *20072045* [52]
- **cob nuts** *20071477* [3], *20072045* [52]
- **crab apples** *20102476* [3], *20072045* [52]
- **cranberries** *20071475* [3], *20120715* [41], *20062035* [52]
- **farm forestry** *20082869* [52], *20082888* [53], *20082891* [55], *20082892* [58]
- **fruit trees** *20102476* [3]
- **game cover** *20082869* [52], *20082888* [53], *20082891* [55], *20082892* [58]
- **gooseberries** *20071475* [3], *20120715* [41], *20062035* [52]
- **grapevines** *20071479* [3], *20062031* [52]
- **hazel nuts** *20071477* [3], *20062028* [52], *20072045* [52]
- **hemp** *20082869* [52], *20071840* [53], *(for wild bird seed production) 20070581* [55], *20070581* [55], *20082892* [58]
- **hops** *20082869* [52], *20082888* [53], *20082891* [55], *20082892* [58]
- **kentish cobnuts** *20062028* [52]
- **lupins** *20071279* [53]
- **millet** *20071839* [53], *(for wild bird seed production) 20070582* [55]

SECTION 2

- **miscanthus** *20082869* [52], *20082888* [53], *20082891* [55], *20082892* [58]
- **ornamental plant production** *20122277* [15], *20082877* [56], *20082890* [57]
- **peaches** *20102476* [3], *20072045* [52]
- **pears** *20102476* [3], *20072045* [52]
- **plums** *20102476* [3]
- **poppies for morphine production** *20122278* [15], *20062030* [52], *20081058* [53]
- **quinces** *20102476* [3], *20072045* [52]
- **quinoa** *20071840* [53], *(for wild bird seed production) 20070581* [55]
- **redcurrants** *20071475* [3], *20120715* [41], *20062035* [52]
- **rhubarb** *20071478* [3], *20062029* [52]
- **rye** *20111701* [53], *20111702* [55]
- **soft fruit** *20082869* [52], *20082888* [53], *20082891* [55], *20082892* [58]
- **sorghum** *20071839* [53], *(for wild bird seed production) 20070582* [55]
- **top fruit** *20082869* [52], *20082888* [53], *20082891* [55], *20082892* [58]
- **vaccinium spp.** *20071475* [3], *20120715* [41], *20062035* [52]
- **walnuts** *20071477* [3], *20062028* [52], *20072045* [52]
- **whitecurrants** *20071475* [3], *20120715* [41], *20062035* [52]
- **wine grapes** *20062031* [52]
- **woad** *20082869* [52], *20082888* [53], *20082891* [55], *20082892* [58]

Approval information
- Glyphosate included in Annex I under EC Regulation 1107/2009
- Accepted by BBPA for use on malting barley and hops

Efficacy guidance
- For best results apply to actively growing weeds with enough leaf to absorb chemical
- For most products a rainfree period of at least 6 h (preferably 24 h) should follow spraying
- Adjuvants are obligatory for some products and recommended for some uses with others. See labels
- Mixtures with other pesticides or fertilizers may lead to reduced control.
- Products are formulated as isopropylamine, ammonium, potassium, or trimesium salts of glyphosate and may vary in the details of efficacy claims. See individual product labels
- Some products in ready-to-use formulations for use through hand-held applicators. See label for instructions [13, 36, 67].
- If using to treat hard surfaces, ensure spraying takes place only when weeds are actively growing (normally March to October) and is confined only to visible weeds including those in the 30cm swath covering the kerb edge and road gulley – do not overspray drains.
- When applying products through rotary atomisers the spray droplet spectra must have a minimum Volume Median Diameter (VMD) of 200 microns
- With wiper application weeds should be at least 10 cm taller than crop
- Annual weed grasses should have at least 5 cm of leaf and annual broad-leaved weeds at least 2 expanded true leaves
- Perennial grass weeds should have 4-5 new leaves and be at least 10 cm long when treated. Perennial broad-leaved weeds should be treated at or near flowering but before onset of senescence
- Volunteer potatoes and polygonums are not controlled by harvest-aid rates
- Bracken must be treated at full frond expansion
- Fruit tree suckers best treated in late spring
- Chemical thinning treatment can be applied as stump spray or stem injection
- In order to allow translocation, do not cultivate before spraying and do not apply other pesticides, lime, fertilizer or farmyard manure within 5 d of treatment
- Recommended intervals after treatment and before cultivation vary. See labels

Restrictions
- Maximum total dose per crop or season normally equivalent to one full dose treatment on field and edible crops and no restriction for non-crop uses. However some older labels indicate a maximum number of treatments. Check for details
- Do not treat cereals grown for seed or undersown crops
- Consult grain merchant before treating crops grown on contract or intended for malting
- Do not use treated straw as a mulch or growing medium for horticultural crops

FOR FULL CONDITIONS OF USE ALWAYS READ THE PRODUCT LABEL

- For use in nursery stock, shrubberies, orchards, grapevines and tree nuts care must be taken to avoid contact with the trees. Do not use in orchards established less than 2 yr and keep off low-lying branches
- Certain conifers may be sprayed overall in dormant season. See label for details
- Use a tree guard when spraying in established forestry plantations
- Do not spray root suckers in orchards in late summer or autumn
- Do not use under glass or polythene as damage to crops may result
- Do not mix, store or apply in galvanised or unlined mild steel containers or spray tanks
- Do not leave diluted chemical in spray tanks for long periods and make sure that tanks are well vented

Crop-specific information
- Harvest intervals: 4 wk for blackcurrants, blueberries; 14 d for linseed, oilseed rape; 8 d for mustard; 5-7 d for all other edible crops. Check label for exact details
- Latest use: for most products 2-14 d before cultivating, drilling or planting a crop in treated land; after harvest (post-leaf fall) but before bud formation in the following season for nuts and most fruit and vegetable crops; before fruit set for grapevines. See labels for details

Following crops guidance
- Decaying remains of plants killed by spraying must be dispersed before direct drilling
- Crops may be drilled 48 h after application. Trees and shrubs may be planted 7 d after application.

Environmental safety
- Products differ in their hazard and environmental safety classification. See labels
- Do not dump surplus herbicide in water or ditch bottoms or empty into drains
- Check label for maximum permitted concentration in treated water
- The Environment Agency or Local River Purification Authority must be consulted before use in or near water
- Take extreme care to avoid drift and possible damage to neighbouring crops or plants
- Treated poisonous plants must be removed before grazing or conserving
- Do not use in covered areas such as greenhouses or under polythene
- For field edge treatment direct spray away from hedge bottoms
- Some products require livestock to be excluded from treated areas and do not permit treated forage to be used for hay, silage or bedding. Check label for details
- LERAP Category B [58]

Hazard classification and safety precautions
Hazard Harmful [20, 32, 40, 51, 61-63, 66, 68]; Irritant [4-7, 12, 14-16, 21, 23-26, 33, 34, 42-45, 47-50, 55, 58, 64, 65, 67]; Dangerous for the environment [4-7, 14-16, 20, 21, 23-26, 32-34, 36-38, 40-45, 47-51, 54, 55, 58, 61-68, 71]

Transport code 9 [4-7, 15, 16, 20, 21, 23, 24, 26, 27, 30-34, 36-38, 42-45, 47-49, 55, 58, 61, 64-68]

Packaging group III [4-7, 15, 16, 20, 21, 23, 26, 27, 30-34, 36-38, 42-45, 47-49, 55, 58, 61, 64-68]

UN Number 3077 [21, 55]; 3082 [4-7, 15, 16, 20, 23, 24, 26, 27, 30-34, 36-38, 42-45, 47-49, 58, 61, 64-68]; N/C [1-3, 8-14, 17-19, 22, 25, 28, 29, 35, 39, 41, 46, 50-54, 56, 57, 59, 60, 62, 63, 69-72]

Risk phrases R20 [5, 6, 20, 23, 40, 47, 50, 51, 61, 66, 68]; R36 [20, 34, 67]; R38 [58]; R41 [4-6, 12, 14-16, 21, 23, 24, 26, 32, 33, 40, 42-45, 47-51, 55, 61-66, 68]; R50 [36, 41, 58]; R51 [4-6, 14-16, 20, 21, 23, 24, 26, 32-34, 37-40, 42-45, 47-51, 55, 61-68]; R52 [28, 29, 31, 53, 57, 71]; R53a [1, 2, 4-6, 14-18, 20, 21, 23, 24, 26, 28, 29, 31-34, 36-45, 47-51, 53-55, 57, 58, 61-71]

Operator protection A [1-20, 22-72]; C [3-10, 12, 14-16, 20-24, 26, 27, 30-35, 37-40, 42-45, 47-53, 55-57, 59-66, 68]; D [3, 22, 27, 30, 37, 38, 59]; F [24, 34, 61]; H [1-3, 5-14, 17-19, 22-31, 33-41, 44, 45, 47-50, 52, 54-61, 64, 65, 67-72]; M [1-3, 5-14, 17-20, 22-31, 33-38, 40, 41, 44, 45, 47-61, 64, 65, 67-70, 72]; N [14, 20, 24, 26, 33, 34, 61]; U02a [1, 2, 4, 7-11, 15-20, 24-29, 31-38, 41-45, 48, 49, 55, 60, 62-70, 72]; U05a [1, 2, 5, 6, 14, 17, 18, 20, 21, 23, 26, 33, 34, 36, 39-41, 44, 45, 47-51, 55, 58, 61, 64, 65, 67-71]; U08 [7-10, 20, 21, 25-29, 31, 33, 34, 37, 38, 55, 60, 72]; U09a [11, 14, 19, 35, 36, 41, 67, 68]; U11 [4-6, 12, 14-16, 20, 21, 23, 24, 26, 28, 29, 32-34, 39, 40, 42, 43, 47, 50, 51, 55, 61-63, 66-68]; U12 [44, 45, 48, 49, 64, 65]; U14 [26, 33]; U15 [34, 44, 45, 48, 49,

64, 65, 68]; U16b [67]; U19a [7-11, 13, 14, 19, 20, 25-29, 31, 33-38, 41, 55, 60, 67, 68, 72]; U20a [5, 6, 19, 21, 23, 35, 39, 40, 47, 50-52, 58, 61, 71]; U20b [1-4, 7-11, 13-18, 20, 22, 24-34, 36-38, 41-46, 48, 49, 53-57, 59, 60, 62-70, 72]

Environmental protection E06a [46] (2 weeks); E06d, E08b [71]; E13c [10, 27, 34, 36, 41, 72]; E15a [3, 13, 20, 22, 30, 31, 34, 37, 38, 52, 55, 56, 58, 59, 68]; E15b [1, 2, 4-9, 11, 12, 15-19, 21, 23-26, 28, 29, 32, 33, 35, 36, 40-51, 53, 54, 57, 60-67, 69-71]; E16a, E16b [58]; E19a [3, 10, 20, 22, 27, 37, 38, 59, 72]; E19b [14]; E34 [13, 17, 18, 31, 34, 69, 70]; E36a [17, 18, 69, 70]; E38 [4-6, 12, 15, 16, 20, 21, 23, 24, 26, 28, 29, 31-34, 40, 42-45, 47-51, 54, 55, 58, 61-68, 71]

Storage and disposal D01 [1-3, 5-9, 11-14, 17-21, 23, 25-31, 33-41, 44, 45, 47-61, 64, 65, 67-71]; D02 [1, 2, 5-9, 11-14, 17-21, 23, 25, 26, 28-30, 33-36, 39-41, 44, 45, 47-58, 60, 61, 64, 65, 67-71]; D05 [1, 2, 4-11, 13, 15-19, 21, 23-25, 30-32, 34-36, 40-45, 47-51, 53-58, 60-72]; D07 [55]; D09a [3-16, 19-38, 40-68, 71, 72]; D10a [7-9, 13, 25, 60]; D10b [3, 10, 14, 20, 22, 26, 33-35, 37, 59, 72]; D10c [1, 2, 4-6, 15-18, 23, 24, 27, 30-32, 38, 40, 42-54, 56-58, 61-66, 68-71]; D11a [11, 12, 19, 28, 29, 36, 41, 55, 67]; D12a [4-6, 14-16, 20, 21, 23, 24, 26, 28, 29, 31-34, 36, 40-43, 47, 50, 51, 54, 55, 58, 61-63, 66-68, 71]; D12b [37, 38]; D14 [17, 18, 69, 70]

Medical advice M03 [71]; M05a [4, 14-16, 24, 32, 39, 42, 43, 62, 63, 66]

279 glyphosate + pyraflufen-ethyl

A herbicide mixture for non-selective weed control around amenity plants and on areas not intended to bear vegetation
HRAC mode of action code: G + E

See also glyphosate
  pyraflufen-ethyl

Products

Hammer	Everris Ltd	240:0.67 g/l	SE	15060

Uses
• Annual and perennial weeds in *amenity vegetation, land not intended to bear vegetation, natural surfaces not intended to bear vegetation, ornamental plant production, permeable surfaces overlying soil*

Approval information
• Glyphosate and pyraflufen-ethyl included in Annex I under EC Regulation 1107/2009

Hazard classification and safety precautions
  Transport code 9
  Packaging group III
  UN Number 3082
  Operator protection A, H, M; U02a, U08, U20b
  Environmental protection E15b
  Storage and disposal D01, D02, D05, D09a, D10c

280 glyphosate + sulfosulfuron

A translocated non-residual glycine derivative herbicide + a sulfonyl urea herbicide
HRAC mode of action code: G + B

See also sulfosulfuron

Products

Nomix Dual	Nomix Enviro	120:2.22 g/l	RH	13420

Uses
• Annual and perennial weeds in *amenity vegetation, hard surfaces, natural surfaces not intended to bear vegetation, permeable surfaces overlying soil*

Approval information
• Glyphosate and sulfosulfuron included in Annex I under EC Regulation 1107/2009

FOR FULL CONDITIONS OF USE ALWAYS READ THE PRODUCT LABEL

Hazard classification and safety precautions
  Hazard Dangerous for the environment
  Transport code 9
  Packaging group III
  UN Number 3082
  Risk phrases R50, R53a
  Operator protection A, H, M; U02a, U09a, U19a, U20b
  Environmental protection E15b
  Storage and disposal D01, D05, D09a, D11a

281   hymexazol

A systemic heteroaromatic fungicide for pelleting sugar beet seed
FRAC mode of action code: 32

Products

Tachigaren 70 WP	Sumi Agro	70% w/w	WP	12568

Uses
  • Black leg in **sugar beet** *(seed treatment)*

Approval information
  • Hymexazol included in Annex 1 under EC Regulation 1107/2009

Efficacy guidance
  • Incorporate into pelleted seed using suitable seed pelleting machinery

Restrictions
  • Maximum number of treatments 1 per batch of seed
  • Do not use treated seed as food or feed

Crop-specific information
  • Latest use: before planting sugar beet seed

Environmental safety
  • Harmful to aquatic organisms
  • Harmful to fish or other aquatic life. Do not contaminate surface waters or ditches with chemical or used container
  • Treated seed harmful to game and wildlife

Hazard classification and safety precautions
  Hazard Irritant, Highly flammable
  Transport code 3
  Packaging group III
  UN Number 1325
  Risk phrases R41, R52, R53a
  Operator protection A, C, F; U05a, U11, U20b
  Environmental protection E03, E13c
  Storage and disposal D01, D02, D09a, D11a
  Treated seed S01, S02, S03, S04a, S05

282   imazalil

A systemic and protectant imidazole fungicide
FRAC mode of action code: 3

See also guazatine + imazalil

Products

1	Fungazil 100 SL	Certis	100 g/l	SL	14999
2	Fungazil 50LS	Certis	50 g/l	LS	14069

SEE SECTION 3 FOR PRODUCTS ALSO REGISTERED

Uses

- Dry rot in **seed potatoes, ware potatoes** [1]
- Gangrene in **seed potatoes, ware potatoes** [1]
- Leaf stripe in **spring barley, winter barley** [2]
- Net blotch in **spring barley, winter barley** [2]
- Silver scurf in **seed potatoes, ware potatoes** [1]
- Skin spot in **seed potatoes, ware potatoes** [1]

Approval information

- Imazalil included in Annex I under EC Regulation 1107/2009
- Approved by BBPA for use on malting barley

Efficacy guidance

- For best control of skin and wound diseases of ware potatoes treat as soon as possible after harvest, preferably within 7-10 d, before any wounds have healed [1]

Restrictions

- Maximum number of treatments 1 per batch of ware tubers [1]; 2 per batch of seed tubers [1];
- Consult processor before treating potatoes for processing [1]
- Do not treat protected cucumbers in bright sunny conditions [1]

Crop-specific information

- Latest use: during storage and before chitting for seed potatoes [1]
- Apply to clean soil-free potatoes post-harvest before putting into store, or at first grading. A further treatment may be applied in early spring before planting [1]
- Apply through canopied hydraulic or spinning disc equipment preferably diluted with up to two litres water per tonne of potatoes to obtain maximum skin cover and penetration [1]
- Use on ware potatoes subject to discharges of imazalil from potato washing plants being within emission limits set by the UK monitoring authority [1]

Environmental safety

- Dangerous for the environment
- Toxic to aquatic organisms
- Do not empty into drains [1]
- Personal protective equipment requirements may vary for each pack size. Check label

Hazard classification and safety precautions

Hazard Irritant [1, 2]; Flammable [2]
Transport code 9 [2]
Packaging group III [2]
UN Number 3082 [2]; N/C [1]
Risk phrases R36 [2]; R41, R52, R53a [1]
Operator protection A, H [1, 2]; C [1]; U04a, U11, U14, U19a, U20a [1]; U05a [1, 2]; U09a, U20b [2]
Environmental protection E13c [2]; E15a, E19b, E34 [1]
Storage and disposal D01, D02, D05, D09a [1, 2]; D10a [2]; D10c, D12a [1]
Treated seed S01, S02, S03, S04a, S05 [1, 2]; S06a [1]; S07 [2]
Medical advice M03 [2]

283   imazalil + ipconazole

An imidazole + triazole seed treatment for barley
FRAC mode of action code: 3

Products

Racona i-MIX	Chemtura	50:20 g/l	MS	15574

Uses

- Fusarium foot rot and seedling blight in **spring barley** *(useful protection)*, **winter barley** *(useful protection)*
- Leaf stripe in **spring barley**

FOR FULL CONDITIONS OF USE ALWAYS READ THE PRODUCT LABEL

- Loose smut in **spring barley**, **winter barley**
- Microdochium nivale in **spring barley** *(useful protection)*, **winter barley** *(useful protection)*

Approval information
- Imazalil included in Annex I under EC Regulation 1107/2009 while ipconazole is awaiting inclusion

Hazard classification and safety precautions
Hazard Harmful, Dangerous for the environment
Transport code 9
Packaging group III
UN Number 3082
Risk phrases R51, R53a
Operator protection A, H; U05a, U09b, U20c
Environmental protection E15b, E34, E38
Storage and disposal D01, D02, D09a, D10a
Treated seed S01, S02, S03, S04d, S05, S07
Medical advice M03

284 imazalil + pencycuron

A fungicide mixture for treatment of seed potatoes
FRAC mode of action code: 3 + 20

See also pencycuron

Products
Monceren IM	Bayer CropScience	0.6:12.5% w/w	DS	11426

Uses
- Black scurf in **potatoes** *(tuber treatment)*
- Silver scurf in **potatoes** *(tuber treatment - reduction)*
- Stem canker in **potatoes** *(tuber treatment - reduction)*

Approval information
- Imazalil and pencycuron included in Annex I under EC Regulation 1107/2009

Efficacy guidance
- Apply to clean seed tubers during planting (see label for suitable method) or sprinkle over tubers in chitting trays before loading into planter. It is essential to obtain an even distribution over tubers
- If seed tubers become damp from light rain distribution of product should not be affected. Tubers in the hopper should be covered if a shower interrupts planting

Restrictions
- Maximum number of treatments 1 per batch of tubers
- Operators must wear suitable respiratory equipment and gloves when handling product and when riding on planter. Wear gloves when handling treated tubers
- Do not use on tubers which have previously been treated with a dry powder seed treatment or hot water
- Do not use treated tubers for human or animal consumption

Crop-specific information
- Latest use: immediately before planting
- May be used on seed tubers previously treated with a liquid fungicide but not before 8 wk have elapsed if this contained imazalil

Environmental safety
- Harmful to aquatic organisms
- Harmful to fish or other aquatic life. Do not contaminate surface waters or ditches with chemical or used container
- Treated seed harmful to game and wildlife

SEE SECTION 3 FOR PRODUCTS ALSO REGISTERED

Hazard classification and safety precautions
 UN Number N/C
 Risk phrases R52, R53a
 Operator protection A, D; U20b
 Environmental protection E03, E13c
 Storage and disposal D09a, D11a
 Treated seed S01, S02, S03, S04a, S05, S06a

285   imazamox

An imidazolinone contact and residual herbicide available only in mixtures
HRAC mode of action code: B

See also imazamox + metazachlor
        imazamox + pendimethalin

286   imazamox + metazachlor

A contact and residual herbicide mixture for weed control in oilseed rape
HRAC mode of action code: B + K3

Products

Cleranda	BASF	17.5:375 g/l	SC	15036

Uses
 • Annual dicotyledons in **winter oilseed rape**
 • Annual grasses in **winter oilseed rape**

Approval information
 • Imazamox and metazachlor are included in Annex I under EC Regulation 1107/2009

Efficacy guidance
 • Must only be used for weed control in CLEARFIELD oilseed rape hybrids. Treatment of an oilseed rape variety that is not a 'CLEARFIELD HYBRID' will result in complete crop loss.

Restrictions
 • Applications shall be limited to a total dose of not more than 1.0 kg metazachlor/ha in a three year period on the same field
 • To avoid the buildup of resistance, do not apply this or any other product containing an ALS inhibitor herbicide with claims for grass weed control more than once to any crop.

Following crops guidance
 • Wheat, barley, oats, oilseed rape, field beans, combining peas and sugar beet can follow normally harvested oilseed rape treated with [1].
 • In the event of crop failure Clearfield oilseed rape may be drilled 4 weeks after application after thorough mixing of the soil to distribute residues. If winter field beans are to be planted, allow 10 weeks after application and plough before planting. Wheat and barley also require ploughing before planting but may be drilled 8 weeks after application.

Environmental safety
 • LERAP Category B

Hazard classification and safety precautions
 Transport code 9
 Packaging group III
 UN Number 3082
 Operator protection A, H; U05a
 Environmental protection E07d, E15b, E16a, E40b
 Storage and disposal D01, D02, D09a, D10c

FOR FULL CONDITIONS OF USE ALWAYS READ THE PRODUCT LABEL

287 imazamox + pendimethalin

A pre-emergence broad-spectrum herbicide mixture for legumes
HRAC mode of action code: B + K1

See also pendimethalin

Products

Nirvana	BASF	16.7:250 g/l	EC	14256

Uses
- Annual dicotyledons in **broad beans** *(off-label)*, **combining peas**, **hops** *(off-label)*, **ornamental plant production** *(off-label)*, **soft fruit** *(off-label)*, **spring field beans**, **top fruit** *(off-label)*, **vining peas**, **winter field beans**

Extension of Authorisation for Minor Use (EAMUs)
- **broad beans** 20092891 expires 30 Jun 2013
- **hops** 20092894 expires 30 Jun 2013
- **ornamental plant production** 20092894 expires 30 Jun 2013
- **soft fruit** 20092894 expires 30 Jun 2013
- **top fruit** 20092894 expires 30 Jun 2013

Approval information
- Imazamox and pendimethalin included in Annex I under EC Regulation 1107/2009
- Approval expiry 30 Jun 2013 [1]

Efficacy guidance
- Best results obtained from applications to fine firm seedbeds in the presence of adequate moisture
- Weed control may be reduced on cloddy seedbeds and on soils with over 6% organic matter
- Residual control may be reduced under prolonged dry conditions

Restrictions
- Maximum number of treatments 1 per crop
- Seed must be drilled to at least 2.5 cm of settled soil
- Do not use on soils containing more than 10% organic matter
- Do not apply to soils that are waterlogged or are prone to waterlogging
- Do not apply if heavy rain is forecast
- Do not soil incorporate the product or disturb the soil after application
- Consult processors before use on crops destined for processing

Crop-specific information
- Latest use: pre-crop emergence for all crops
- Inadequately covered seed may result in cupping of the leaves after application from which recovery is normally complete
- Crop damage may occur on stony or gravelly soils especially if heavy rain follows treatment
- Winter oilseed rape and other brassica crops should not be drilled as the following crop.

Following crops guidance
- Winter wheat or winter barley may be drilled as a following crop provided 3 mth have elapsed since treatment and the land has been at least cultivated by a non-inversion technique such as discing
- Winter oilseed rape or other brassica crops should not be drilled as following crops
- A minimum of 12 mth must elapse between treatment and sowing red beet, sugar beet or spinach

Environmental safety
- Dangerous for the environment
- Very toxic to aquatic organisms
- LERAP Category B

Hazard classification and safety precautions
   Hazard Irritant, Dangerous for the environment

SEE SECTION 3 FOR PRODUCTS ALSO REGISTERED

SECTION 2

Transport code 9
Packaging group III
UN Number 3082
Risk phrases R38, R43, R50, R53a
Operator protection A, H; U05a, U14, U20b
Environmental protection E15b, E16a, E38
Storage and disposal D01, D02, D09a, D10c, D12a
Medical advice M03

288   imazaquin

An imidazolinone herbicide and plant growth regulator available only in mixtures
HRAC mode of action code: B

See also chlormequat + 2-chloroethylphosphonic acid + imazaquin
chlormequat + imazaquin

289   imidacloprid

A neonicotinoid insecticide for seed, soil, peat or foliar treatment
IRAC mode of action code: 4A

See also beta-cyfluthrin + imidacloprid
bitertanol + fuberidazole + imidacloprid
fuberidazole + imidacloprid + triadimenol

Products

1	Clayton Divot	Clayton	0.5% w/w	GR	15451
2	Couraze	Solufeed	5% w/w	GR	13862
3	Imidasect 5GR	Fargro	5 % w/w	GR	14574
4	Intercept 5GR	Everris Ltd	5% w/w	GR	14091
5	Intercept 70WG	Everris Ltd	70% w/w	WG	08585
6	Merit Turf	Bayer Environ.	0.5% w/w	GR	12415
7	Nuprid 600FS	Nufarm UK	600 g/l	FS	13855
8	Valiant	Sherriff Amenity	5% w/w	WG	14456

Uses

- Aphids in *bedding plants*, *herbaceous perennials*, *ornamental plant production*, *pot plants* [5]; *fodder beet*, *sugar beet* [7]; *ornamental plant production* (container grown) [3, 4]; *protected ornamentals* [2-4]
- Chafer grubs in *managed amenity turf* [1, 6]
- Flea beetle in *fodder beet*, *sugar beet* [7]
- Glasshouse whitefly in *bedding plants*, *herbaceous perennials*, *ornamental plant production*, *pot plants* [5]; *ornamental plant production* (container grown), *protected ornamentals* [3, 4]
- Insect pests in *lawns* [8]
- Leaf miner in *fodder beet*, *sugar beet* [7]
- Leatherjackets in *managed amenity turf* [1, 6]
- Millipedes in *fodder beet*, *sugar beet* [7]
- Pygmy beetle in *fodder beet*, *sugar beet* [7]
- Sciarid flies in *bedding plants*, *herbaceous perennials*, *ornamental plant production*, *pot plants* [5]; *ornamental plant production* (container grown), *protected ornamentals* [3, 4]
- Springtails in *fodder beet*, *sugar beet* [7]
- Symphylids in *fodder beet*, *sugar beet* [7]
- Tobacco whitefly in *bedding plants*, *herbaceous perennials*, *ornamental plant production*, *pot plants* [5]; *ornamental plant production* (container grown), *protected ornamentals* [3, 4]
- Vine weevil in *bedding plants*, *herbaceous perennials*, *ornamental plant production*, *pot plants* [5]; *ornamental plant production* (container grown), *protected ornamentals* [3, 4]

FOR FULL CONDITIONS OF USE ALWAYS READ THE PRODUCT LABEL

- Whitefly in **protected ornamentals** [2]

Approval information
- Imidacloprid is included in Annex I under EC Regulation 1107/2009
- Accepted by BBPA for use on hops
- Approval expiry 31 Jul 2013 [4, 5]

Efficacy guidance
- Treated seed should be drilled within 18 months of purchase [7]
- Best results on managed amenity turf achieved when treatment is followed by sufficient irrigation to move chemical through the thatch to wet the top 2.5 cm of soil [6]
- To avoid disturbing uniformity of treatment on turf do not mow until irrigation or rainfall has occurred [6]
- To minimise likelihood of resistance when using in compost adopt a planned programme to alternate with pesticides of different types or use other measures [5]
- When applied as drench or incorporated as granules in moist compost imidacloprid is readily absorbed and translocated to aerial parts of plant [5]

Restrictions
- Maximum number of treatments 1 per batch of seed [7] or growing medium [5]; 1 per yr for managed amenity turf [6]
- Must not be used in compost that has already been treated with an imidacloprid-containing product, nor should compost be subsequently re-treated with imidacloprid [5]
- For use only on container grown ornamentals [5]
- Product must not be used on crops for human or animal consumption and treated compost must not be re-used for this purpose [5]
- Do not use treated seed as food or feed
- Do not treat saturated or waterlogged turf, or when raining or in windy conditions [6]
- Avoid puddling or run-off of irrigation water after treatment [6]
- Do not use grass clippings from treated turf in situ, or use them for mulch [6]

Crop-specific information
- Latest use: before drilling for sugar beet; before sowing or planting for brassicas, bedding plants, hardy ornamental nursery stock, pot plants
- Apply to sugar beet seed as part of the normal commercial pelleting process using special treatment machinery [7]
- On managed amenity turf application should be made prior to egg-laying of the target pests [6]
- Low levels of leaf scorch on some *Agrostis* turf species if application is made in mixture with fertilisers [6]

Environmental safety
- High risk to bees. Do not apply to crops in flower or to those in which bees are actively foraging. Do not apply when flowering weeds are present [5]
- Treated seed harmful to game and wildlife [7]

Hazard classification and safety precautions
    Hazard Harmful [5, 7]; Irritant [1, 6]
    Transport code 9 [5, 7]
    Packaging group III [5, 7]
    UN Number 3077 [5]; 3082 [7]; N.C [3]; N/C [1, 2, 4, 6, 8]
    Risk phrases R22a [5, 7]; R36 [1, 6]; R43 [7]
    Operator protection A [1-4, 6-8]; C, H [7]; U04a, U13, U14 [7]; U05a [1, 3-8]; U20b [2-5]; U20c [1, 6-8]
    Environmental protection E12a, E12e [5]; E15a [1-8]; E34 [2, 5, 7]
    Storage and disposal D01 [1-7]; D02 [1, 3-7]; D09a, D11a [1-8]; D12a [8]
    Medical advice M03 [7]

SEE SECTION 3 FOR PRODUCTS ALSO REGISTERED

290   4-indol-3-ylbutyric acid

A plant growth regulator promoting the rooting of cuttings

Products

1	Chryzoplus Grey 0.8%	Fargro	0.8% w/w	DP	09094
2	Chryzopon Rose 0.1%	Fargro	0.1% w/w	DP	09092
3	Chryzosan White 0.6%	Fargro	0.6% w/w	DP	09093
4	Chryzotek Beige	Fargro	0.4% w/w	DP	09081
5	Chryzotop Green	Fargro	0.25% w/w	DP	09085
6	Rhizopon AA Powder (0.5%)	Fargro	0.5% w/w	DP	09082
7	Rhizopon AA Powder (1%)	Fargro	1% w/w	DP	09084
8	Rhizopon AA Powder (2%)	Fargro	2% w/w	DP	09083
9	Rhizopon AA Tablets	Fargro	50 mg a.i.	WT	09086

Uses
- Rooting of cuttings in *ornamental plant production*

Approval information
- 4-indol-3-ylbutyric acid not included in Annex 1 under EC Regulation 1107/2009

Efficacy guidance
- Dip base of cuttings into powder immediately before planting
- Powders or solutions of different concentration are required for different types of cutting. Lowest concentration for softwood, intermediate for semi-ripe, highest for hardwood
- See label for details of concentration and timing recommended for different species
- Use of planting holes recommended for powder formulations to ensure product is not removed on insertion of cutting. Cuttings should be watered in if necessary

Restrictions
- Maximum number of treatments 1 per situation
- Use of too strong a powder or solution may cause injury to cuttings
- No unused moistened powder should be returned to container

Crop-specific information
- Latest use: before cutting insertion for ornamental specimens

Hazard classification and safety precautions
  UN Number N/C
  Operator protection U19a, U20a
  Environmental protection E15a
  Storage and disposal D09a, D11a

291   indoxacarb

An oxadiazine insecticide for caterpillar control in a range of crops
IRAC mode of action code: 22A

Products

1	Explicit	DuPont	30% w/w	WG	15359
2	Rumo	DuPont	30% w/w	WG	14883
3	Steward	DuPont	30% w/w	WG	13149

Uses
- Caterpillars in *all edible seed crops grown outdoors* (off-label), *all non-edible seed crops grown outdoors* (off-label) [3]; *apples*, *cherries* (off-label), *forest nurseries* (off-label), *grapevines* (off-label), *hops* (off-label), *ornamental plant production* (off-label), *pears*, *protected forest nurseries* (off-label), *protected hops* (off-label), *protected soft fruit* (off-

label), **protected top fruit** *(off-label)*, **soft fruit** *(off-label)*, **top fruit** *(off-label)* [1, 3]; **broccoli, brussels sprouts** *(off-label)*, **cabbages, cauliflowers, protected aubergines, protected courgettes, protected cucumbers, protected marrows, protected melons, protected ornamentals, protected peppers, protected pumpkins, protected squashes, protected tomatoes, sweetcorn** *(off-label)* [1-3]; **protected all edible seed crops** *(off-label)*, **protected all non-edible seed crops** *(off-label)* [1]

- Diamond-back moth in **sweetcorn** *(off-label)* [3]
- Insect pests in **all edible seed crops grown outdoors** *(off-label)*, **all non-edible seed crops grown outdoors** *(off-label)*, **forest nurseries** *(off-label)*, **hops** *(off-label)*, **ornamental plant production** *(off-label)*, **protected forest nurseries** *(off-label)*, **protected hops** *(off-label)* [2]
- Pollen beetle in **spring oilseed rape, winter oilseed rape** [1-3]

Extension of Authorisation for Minor Use (EAMUs)

- **all edible seed crops grown outdoors** *20101537* [2], *20082905* [3]
- **all non-edible seed crops grown outdoors** *20101537* [2], *20082905* [3]
- **brussels sprouts** *20112093* [1], *20101537* [2], *20081485* [3]
- **cherries** *20112092* [1], *20092300* [3]
- **forest nurseries** *20112090* [1], *20101538* [2], *20082905* [3]
- **grapevines** *20112091* [1], *20072289* [3]
- **hops** *20112090* [1], *20101537* [2], *20082905* [3]
- **ornamental plant production** *20112090* [1], *20101537* [2], *20082905* [3]
- **protected all edible seed crops** *20112090* [1]
- **protected all non-edible seed crops** *20112090* [1]
- **protected forest nurseries** *20112090* [1], *20101537* [2], *20082905* [3]
- **protected hops** *20112090* [1], *20101537* [2], *20082905* [3]
- **protected soft fruit** *20112090* [1], *20082905* [3]
- **protected top fruit** *20112090* [1], *20082905* [3]
- **soft fruit** *20112090* [1], *20082905* [3]
- **sweetcorn** *20112089* [1], *20101536* [2], *20072288* [3]
- **top fruit** *20112090* [1], *20082905* [3]

Approval information

- Indoxacarb included in Annex I under EC Regulation 1107/2009

Efficacy guidance

- Best results in brassica and protected crops obtained from treatment when first caterpillars are detected, or when damage first seen, or 7-10 d after trapping first adults in pheromone traps
- In apples and pears apply at egg-hatch
- Subsequent treatments in all crops may be applied at 8-14 d intervals
- Indoxacarb acts by ingestion and contact. Only larval stages are controlled but there is some ovicidal action against some species

Restrictions

- Maximum number of treatments 3 per crop or yr for apples, pears, brassica crops; 6 per yr for protected crops and 1 per crop for oilseed rape
- Do not apply to any crop suffering from stress from any cause

Crop-specific information

- HI: brassica crops, protected crops 1 d; apples, pears 7 d

Following crops guidance

- When treating protected crops, observe the maximum concentration permitted.

Environmental safety

- Dangerous for the environment
- Toxic to aquatic organisms
- In accordance with good agricultural practice apply in early morning or late evening when bees are less active
- Broadcast air-assisted LERAP (15)

Hazard classification and safety precautions

Hazard Harmful, Dangerous for the environment

SEE SECTION 3 FOR PRODUCTS ALSO REGISTERED

Transport code 9
Packaging group III
UN Number 3077
Risk phrases R22a, R51, R53a
Operator protection A, D; U04a, U05a
Environmental protection E15b, E34, E38; E17b (15)
Storage and disposal D01, D02, D09a, D12a
Medical advice M03

292   iodosulfuron-methyl-sodium

A post-emergence sulfonylurea herbicide for winter sown cereals
HRAC mode of action code: B

See also amidosulfuron + iodosulfuron-methyl-sodium
       diflufenican + iodosulfuron-methyl-sodium + mesosulfuron-methyl

Products

1 Hussar	Bayer CropScience	5% w/w	WG	12364
2 Klaxon	Bayer CropScience	5% w/w	WG	15174

Uses

- Annual dicotyledons in **durum wheat** *(off-label)*, **grass seed crops** *(off-label)* [1]
- Chickweed in **spring barley**, **triticale**, **winter rye**, **winter wheat** [1, 2]; **winter barley** [2]
- Cleavers in **triticale**, **winter rye**, **winter wheat** [1, 2]
- Fat hen in **spring barley** [1, 2]; **winter barley** [2]
- Field pansy in **spring barley** [1, 2]; **winter barley** [2]
- Field speedwell in **spring barley**, **triticale**, **winter rye**, **winter wheat** [1, 2]; **winter barley** [2]
- Hemp-nettle in **spring barley** [1, 2]; **winter barley** [2]
- Italian ryegrass in **triticale** *(from seed)*, **winter rye** *(from seed)*, **winter wheat** *(from seed)* [1, 2]
- Ivy-leaved speedwell in **triticale**, **winter rye**, **winter wheat** [1, 2]
- Mayweeds in **spring barley**, **triticale**, **winter rye**, **winter wheat** [1, 2]; **winter barley** [2]
- Perennial ryegrass in **triticale** *(from seed)*, **winter rye** *(from seed)*, **winter wheat** *(from seed)* [1, 2]
- Red dead-nettle in **spring barley**, **triticale**, **winter rye**, **winter wheat** [1, 2]; **winter barley** [2]
- Ryegrass in **durum wheat** *(off-label - from seed)*, **grass seed crops** *(off-label - from seed)* [1]
- Volunteer oilseed rape in **spring barley** [1, 2]; **winter barley** [2]

Extension of Authorisation for Minor Use (EAMUs)

- **durum wheat** *(from seed)* 20080912 expires 31 Dec 2013 [1], 20080912 expires 31 Dec 2013 [1]
- **grass seed crops** *(from seed)* 20080912 expires 31 Dec 2013 [1], 20080912 expires 31 Dec 2013 [1]

Approval information

- Iodosulfuron-methyl-sodium included in Annex I under EC Regulation 1107/2009
- Accepted by BBPA for use on malting barley
- Approval expiry 31 Dec 2013 [1, 2]

Efficacy guidance

- Best results obtained from treatment in warm weather when soil is moist and the weeds are growing actively
- Weeds must be present at application to be controlled
- Weed control is slow especially under cool dry conditions
- Dry conditions resulting in moisture stress may reduce effectiveness
- Occasionally weeds may only be stunted but they will normally have little or no competitive effect on the crop
- Iodosulfuron is a member of the ALS-inhibitor group of herbicides. To avoid the build up of resistance do not use any product containing an ALS-inhibitor herbicide with claims for control of grass weeds more than once on any crop

FOR FULL CONDITIONS OF USE ALWAYS READ THE PRODUCT LABEL

- Use these products as part of a resistance management strategy that includes cultural methods of control and does not use ALS inhibitors as the sole chemical method of grass weed control

Restrictions
- Maximum number of treatments 1 per crop
- Must only be applied between 1 Feb in yr of harvest and specified latest time of application
- Do not apply to undersown crops, or crops to be undersown
- Do not roll or harrow within 1 wk of spraying
- Do not spray crops under stress from any cause or if the soil is compacted
- Treat broadcast crops after the plants have a well-established root system
- Do not spray if rain imminent or frost expected
- Do not apply in mixture or in sequence with any other ALS inhibitor

Crop-specific information
- Latest use: before third node detectable (GS 33)

Following crops guidance
- No restrictions apply to the sowing of cereal crops or sugar beet in the spring of the yr following treatment

Environmental safety
- Dangerous for the environment
- Very toxic to aquatic organisms
- Dangerous to fish or other aquatic life. Do not contaminate surface waters or ditches with chemical or used container
- Take extreme care to avoid damage by drift onto broad-leaved plants outside the target area or onto ponds, waterways and ditches
- Observe carefully label instructions for sprayer cleaning
- LERAP Category B

Hazard classification and safety precautions
Hazard Irritant, Dangerous for the environment
Transport code 9
Packaging group III
UN Number 3077
Risk phrases R41, R50, R53a
Operator protection A, C, H; U05a, U08, U11, U14, U15, U20b
Environmental protection E15b, E16a, E16b, E38
Storage and disposal D01, D02, D10a, D12a

293   iodosulfuron-methyl-sodium + mesosulfuron-methyl

A sulfonyl urea herbicide mixture for winter wheat
HRAC mode of action code: B + B

See also mesosulfuron-methyl

Products

1	Atlantis WG	Bayer CropScience	0.6:3.0% w/w	WG	12478
2	Hatra	Bayer CropScience	2:10 g/l	OD	14524
3	Horus	Bayer CropScience	2:10 g/l	OD	14541
4	Nautilus	ChemSource	2:10 g/l	OD	14961
5	Nemo	AgChem Access	0.6:3.0% w/w	WG	14140
6	Pacifica	Bayer CropScience	1.0:3.0% w/w	WG	12049

Uses
- Annual dicotyledons in *durum wheat* (off-label), *spring rye* (off-label), *triticale* (off-label), *winter rye* (off-label) [1, 6]; *spring wheat* (off-label) [1]
- Annual grasses in *durum wheat* (off-label), *spring rye* (off-label), *triticale* (off-label), *winter rye* (off-label) [1, 6]; *spring wheat* (off-label) [1]
- Annual meadow grass in *winter wheat* [1-6]

- Blackgrass in **winter wheat** [1-6]
- Chickweed in **winter wheat** [1-6]
- Italian ryegrass in **winter wheat** [1-6]
- Mayweeds in **winter wheat** [1-6]
- Perennial ryegrass in **winter wheat** [1-6]
- Rough meadow grass in **winter wheat** [1-6]
- Wild oats in **winter wheat** [1-6]

Extension of Authorisation for Minor Use (EAMUs)
- **durum wheat** *20063529 expires 31 Dec 2013* [1], *20081264 expires 31 Dec 2013* [6]
- **spring rye** *20063529 expires 31 Dec 2013* [1], *20081264 expires 31 Dec 2013* [6]
- **spring wheat** *20082946 expires 31 Dec 2013* [1]
- **triticale** *20063529 expires 31 Dec 2013* [1], *20081264 expires 31 Dec 2013* [6]
- **winter rye** *20063529 expires 31 Dec 2013* [1], *20081264 expires 31 Dec 2013* [6]

Approval information
- Iodosulfuron-methyl-sodium and mesosulfuron-methyl included in Annex I under EC Regulation 1107/2009
- Approval expiry 31 Dec 2013 [1-6]

Efficacy guidance
- Optimum grass weed control obtained when all grass weeds are emerged at spraying. Activity is primarily via foliar uptake and good spray coverage of the target weeds is essential
- Translocation occurs readily within the target weeds and growth is inhibited within hours of treatment but symptoms may not be apparent for up to 4 wk, depending on weed species, timing of treatment and weather conditions
- Residual activity is important for best results and is optimised by treatment on fine moist seedbeds. Avoid application under very dry conditions
- Residual efficacy may be reduced by high soil temperatures and cloddy seedbeds
- Iodosulfuron-methyl and mesosulfuron-methyl are both members of the ALS-inhibitor group of herbicides. To avoid the build up of resistance do not use any product containing an ALS-inhibitor herbicide with claims for control of grass weeds more than once on any crop
- Use these products as part of a Resistance Management Strategy that includes cultural methods of control and does not use ALS inhibitors as the sole chemical method of grass weed control. See Section 5 for more information

Restrictions
- Maximum number of treatments 1 per crop with a maximum total dose equivalent to one full dose treatment
- Do not use on crops undersown with grasses, clover or other legumes or any other broad-leaved crop
- Do not use as a stand-alone treatment for blackgrass, ryegrass or chickweed control
- Do not use as the sole means of weed control in successive crops
- Do not use in mixture or in sequence with any other ALS-inhibitor herbicide except those (if any) specified on the label
- Do not apply earlier than 1 Feb in the yr of harvest [6]
- Do not apply to crops under stress from any cause
- Do not apply when rain is imminent or during periods of frosty weather
- Specified adjuvant must be used. See label

Crop-specific information
- Latest use: flag leaf just visible (GS 39)
- Winter wheat may be treated from the two-leaf stage of the crop
- Safety to crops grown for seed not established
- Aviod application before a severe frost or transient yellowing of the crop may occur.

Following crops guidance
- In the event of crop failure sow only winter wheat in the same cropping season
- Only winter wheat or winter barley (or winter oilseed rape [1]) may be sown in the year of harvest of a treated crop

FOR FULL CONDITIONS OF USE ALWAYS READ THE PRODUCT LABEL

- Spring wheat, spring barley, sugar beet or spring oilseed rape may be drilled in the following spring. Plough before drilling oilseed rape

Environmental safety
- Dangerous for the environment
- Very toxic to aquatic organisms
- Dangerous to fish or other aquatic life. Do not contaminate surface waters or ditches with chemical or used container
- Take extreme care to avoid drift onto plants outside the target area or on to ponds, waterways or ditches
- LERAP Category B

Hazard classification and safety precautions
    Hazard Irritant, Dangerous for the environment
    Transport code 9
    Packaging group III
    UN Number 3077 [1, 5, 6]; 3082 [2-4]
    Risk phrases R36, R38 [2-4]; R40 [6]; R41 [1, 5, 6]; R50, R53a [1-6]
    Operator protection A, C, H; U05a, U11, U20b
    Environmental protection E15a, E16a, E38
    Storage and disposal D01, D02, D05, D09a, D10a, D12a

294   ioxynil

A contact acting HBN herbicide for use in turf and onions
HRAC mode of action code: C3

See also bromoxynil + diflufenican + ioxynil
bromoxynil + fluroxypyr + ioxynil
bromoxynil + ioxynil
bromoxynil + ioxynil + mecoprop-P
bromoxynil + ioxynil + triasulfuron
dichlorprop-P + ioxynil

Products
    Totril   Bayer CropScience   225 g/l   EC   14633

Uses
- Annual dicotyledons in **bulb onions**, **chives** (off-label), **garlic**, **leeks**, **salad onions**, **shallots**, **welsh onions** (off-label)

Extension of Authorisation for Minor Use (EAMUs)
- **chives** 20110428
- **welsh onions** 20110428

Approval information
- Ioxynil included in Annex I under EC Regulation 1107/2009
- Ioxynil was reviewed in 1995 and approvals for home garden use, and most hand held applications revoked
- Accepted by BBPA for use on malting barley

Efficacy guidance
- Best results on seedling to 4-leaf stage weeds in active growth during mild weather

Restrictions
- Maximum number of treatments up to 4 per crop at split doses on onions; 1 per crop on carrots, garlic, leeks, shallots
- Do not apply by hand-held equipment or at concentrations higher than those recommended

Crop-specific information
- Latest use: pre-emergence for carrots, parsnips
- HI bulb onions, salad onions, shallots, garlic, leeks 14 d

SEE SECTION 3 FOR PRODUCTS ALSO REGISTERED

- Apply to sown onion crops as soon as possible after plants have 3 true leaves or to transplanted crops when established

Environmental safety
- Dangerous for the environment
- Very toxic to aquatic organisms
- Flammable
- Harmful to bees. Do not apply to crops in flower or to those in which bees are actively foraging. Do not apply when flowering weeds are present
- Keep livestock out of treated areas for at least 6 wk after treatment and until foliage of any poisonous weeds such as ragwort has died and become unpalatable

Hazard classification and safety precautions
Hazard Harmful, Flammable, Dangerous for the environment
Transport code 3
Packaging group III
UN Number 1993
Risk phrases R22a, R22b, R36, R37, R43, R50, R53a, R63, R66, R67
Operator protection A, C; U05a, U08, U19a, U20b, U23a
Environmental protection E06a (6 wk); E12d, E12e, E15a, E34, E38
Storage and disposal D01, D02, D05, D09a, D10b, D12a
Medical advice M03, M05b

295　ipconazole

A triazole fungicide for disease control in cereals
FRAC mode of action code: 3

See also imazalil + ipconazole

Products

Rancona 15 ME	Chemtura	15 g/l	MS	14407

Uses
- Seed-borne diseases in *spring barley*, *spring wheat*, *winter barley*, *winter wheat*
- Soil-borne diseases in *spring barley*, *spring wheat*, *winter barley*, *winter wheat*

Approval information
- Annex 1 approval is pending under EC Regulation 1107/2009
- Accepted by BBPA for use on malting barley

Hazard classification and safety precautions
UN Number N/C
Operator protection A, H; U05a, U09a, U20b
Environmental protection E03, E15b, E34
Storage and disposal D01, D02, D09a, D10a
Treated seed S01, S02, S03, S04d, S05, S07
Medical advice M03

296　iprodione

A protectant dicarboximide fungicide with some eradicant activity
FRAC mode of action code: 2

See also carbendazim + iprodione

Products

1	Cavron Green	Belcrop	250 g/l	SC	15340
2	Chipco Green	Bayer Environ.	255 g/l	SC	13843
3	Green Turf King	Standon	255 g/l	SC	14348
4	Grisu	Sipcam	500 g/l	SC	15375

FOR FULL CONDITIONS OF USE ALWAYS READ THE PRODUCT LABEL

Products – continued

5	Mascot Rayzor	Rigby Taylor	255 g/l	SC	14903
6	Panto	Pan Amenity	255 g/l	SC	14307
7	Prime Turf	ChemSource	255 g/l	SC	14617
8	Pure Turf	Pure Amenity	255 g/l	SC	14851
9	Rovral AquaFlo	BASF	500 g/l	SC	14206
10	Rovral WG	BASF	75% w/w	WG	13811
11	Surpass Pro	Headland Amenity	250 g/l	SC	15368

Uses

- Alternaria in **brassica seed crops**, **lettuce**, **oilseed rape**, **ornamental plant production**, **protected lettuce**, **protected strawberries**, **protected tomatoes**, **raspberries**, **salad onions**, **strawberries** [4]; **brassica seed crops** (pre-storage only), **dwarf beans** (off-label), **french beans** (off-label), **garlic** (off-label), **grapevines** (off-label), **mange-tout peas** (off-label), **protected herbs (see appendix 6)** (off-label), **runner beans** (off-label), **shallots** (off-label), **spring oilseed rape**, **winter oilseed rape** [10]; **broccoli** (off-label), **brussels sprouts** (off-label), **cabbages** (off-label), **calabrese** (off-label), **cauliflowers** (off-label), **chinese cabbage** (off-label), **collards** (off-label), **kale** (off-label) [9]; **brussels sprouts**, **cauliflowers** [4, 10]
- Anthracnose in **managed amenity turf** [2]
- Botrytis in **brassica seed crops**, **brussels sprouts**, **cauliflowers**, **oilseed rape** [4]; **bulb onions**, **cabbages** (off-label), **dwarf beans** (off-label), **forest nurseries** (off-label), **french beans** (off-label), **garlic** (off-label), **grapevines** (off-label), **hops** (off-label), **mange-tout peas** (off-label), **ornamental plant production** (off-label), **pears** (off-label), **protected aubergines** (off-label), **protected forest nurseries** (off-label), **protected herbs (see appendix 6)** (off-label), **protected hops** (off-label), **protected ornamentals** (off-label), **protected rhubarb** (off-label), **protected soft fruit** (off-label), **protected top fruit** (off-label), **red beet** (off-label), **runner beans** (off-label), **shallots** (off-label), **soft fruit** (off-label), **spring oilseed rape**, **top fruit** (off-label), **winter oilseed rape** [10]; **lettuce**, **ornamental plant production**, **protected lettuce**, **protected strawberries**, **protected tomatoes**, **raspberries**, **salad onions**, **strawberries** [4, 10]
- Brown patch in **amenity grassland** [7, 8]; **managed amenity turf** [1, 7, 8, 11]
- Collar rot in **bulb onions**, **garlic** (off-label), **salad onions**, **shallots** (off-label) [10]
- Disease control in **managed amenity turf** [5, 6]; **ornamental plant production** (off-label) [10]
- Dollar spot in **amenity grassland** [2, 7, 8]; **managed amenity turf** [1, 2, 7, 8, 11]
- Fusarium patch in **amenity grassland** [2, 7, 8]; **managed amenity turf** [1-3, 7, 8, 11]
- Melting out in **amenity grassland** [2, 7, 8]; **managed amenity turf** [1, 2, 7, 8, 11]
- Red thread in **amenity grassland** [2, 7, 8]; **managed amenity turf** [1-3, 7, 8, 11]
- Rust in **managed amenity turf** [2]
- Sclerotinia in **oilseed rape** (moderate control) [4]
- Seed-borne diseases in **ornamental crop production (seed)** [9]
- Snow mould in **amenity grassland** [2, 7, 8]; **managed amenity turf** [1, 2, 7, 8, 11]

Extension of Authorisation for Minor Use (EAMUs)

- **broccoli** 20090708 [9]
- **brussels sprouts** 20090708 [9]
- **cabbages** 20090708 [9], 20091845 expires 31 Dec 2013 [10]
- **calabrese** 20090708 [9]
- **cauliflowers** 20090708 [9]
- **chinese cabbage** 20090708 [9]
- **collards** 20090708 [9]
- **dwarf beans** 20081760 expires 31 Dec 2013 [10]
- **forest nurseries** 20082902 expires 31 Dec 2013 [10]
- **french beans** 20081760 expires 31 Dec 2013 [10]
- **garlic** 20081768 expires 31 Dec 2013 [10]
- **grapevines** 20081758 expires 31 Dec 2013 [10]
- **hops** 20082902 expires 31 Dec 2013 [10]
- **kale** 20090708 [9]
- **mange-tout peas** 20081760 expires 31 Dec 2013 [10]
- **ornamental plant production** 20113200 expires 31 Dec 2013 [10]
- **pears** 20081761 expires 31 Dec 2013 [10]
- **protected aubergines** 20081767 expires 31 Dec 2013 [10]

SEE SECTION 3 FOR PRODUCTS ALSO REGISTERED

- **protected forest nurseries** *20082902 expires 31 Dec 2013* [10]
- **protected herbs (see appendix 6)** *20081757 expires 31 Dec 2013* [10]
- **protected hops** *20082902 expires 31 Dec 2013* [10]
- **protected ornamentals** *20113200 expires 31 Dec 2013* [10]
- **protected rhubarb** *20081763 expires 31 Dec 2013* [10]
- **protected soft fruit** *20082902 expires 31 Dec 2013* [10]
- **protected top fruit** *20082902 expires 31 Dec 2013* [10]
- **red beet** *20081762 expires 31 Dec 2013* [10]
- **runner beans** *20081760 expires 31 Dec 2013* [10]
- **shallots** *20081768 expires 31 Dec 2013* [10]
- **soft fruit** *20082902 expires 31 Dec 2013* [10]
- **top fruit** *20082902 expires 31 Dec 2013* [10]

Approval information
- Iprodione included in Annex I under EC Regulation 1107/2009
- Accepted by BBPA for use on malting barley
- Approval expiry 31 Dec 2013 [3, 6, 8, 10]

Efficacy guidance
- Many diseases require a programme of 2 or more sprays at intervals of 2-4 wk. Recommendations vary with disease and crop - see label for details
- Use as a drench to control cabbage storage diseases. Spray ornamental pot plants and cucumbers to run-off
- Apply turf and amenity grass treatments after mowing to dry grass, free of dew. Do not mow again for at least 24 h [2]

Restrictions
- Maximum number of treatments or maximum total dose equivalent to 1 per batch for seed treatments; 1 per crop on cereals, cabbage (as drench), 2 per crop on field beans and stubble turnips; 3 per crop on brassicas (including seed crops), oilseed rape, protected winter lettuce (Oct-Feb); 4 per crop on strawberries, grapevines, salad onions, cucumbers; 5 per crop on raspberries; 6 per crop on bulb onions, tomatoes, turf; 7 per crop on lettuce (Mar-Sep)
- A minimum of 3 wk must elapse between treatments on leaf brassicas
- Treated brassica seed crops not to be used for human or animal consumption
- See label for pot plants showing good tolerance. Check other species before applying on a large scale
- Do not treat oilseed rape seed that is cracked or broken or of low viability
- Do not treat oats

Crop-specific information
- Latest use: pre-planting for seed treatments; at planting for potatoes; before grain watery ripe (GS 69) for wheat and barley; 21 d before consumption by livestock for stubble turnips
- HI strawberries, protected tomatoes 1 d; outdoor tomatoes 2 d; bulb onions, salad onions, raspberries, lettuce 7 d; brassicas, brassica seed crops, oilseed rape, mustard, field beans, stubble turnips 21 d
- Turf and amenity grass may become temporarily yellowed if frost or hot weather follows treatment
- Personal protective equipment requirements may vary for each pack size. Check label

Environmental safety
- Dangerous for the environment
- Very toxic to aquatic organisms
- See label for guidance on disposal of spent drench liquor
- Treatment harmless to *Encarsia* or *Phytoseiulus* being used for integrated pest control
- Broadcast air-assisted LERAP [10] (18 m); LERAP Category B [3, 10]

Hazard classification and safety precautions
Hazard Harmful, Dangerous for the environment
Transport code 9
Packaging group III
UN Number 3077 [10]; 3082 [1-9, 11]

FOR FULL CONDITIONS OF USE ALWAYS READ THE PRODUCT LABEL

Risk phrases R36 [1, 2, 5-8, 10, 11]; R38 [1, 2, 5-8, 11]; R40 [1, 2, 4-11]; R50, R53a [1-11]
Operator protection A [1-11]; C [1, 2, 5-8, 10, 11]; H [4, 9, 10]; P [3]; U04a, U08 [1, 2, 5-8, 11]; U05a [1-11]; U09a [3]; U11 [4, 9, 10]; U16b, U19a, U20b [10]; U20c [1-9, 11]
Environmental protection E15a [1, 2, 4-9, 11]; E15b, E16a [3, 10]; E16b [3]; E17b [10] (18 m); E34 [1-11]; E38 [1, 2, 4-11]
Storage and disposal D01, D02, D09a [1-11]; D05, D10b [1, 2, 5-8, 11]; D10a [3]; D11a [10]; D12a [1, 2, 4-11]
Treated seed S01, S02, S03, S04a, S05 [4, 9]
Medical advice M03 [1, 2, 4-9, 11]; M04a [3]; M05a [3, 4, 9, 10]

297   iprodione + thiophanate-methyl

A protectant and systemic fungicide for oilseed rape
FRAC mode of action code: 2 + 1

See also thiophanate-methyl

Products

Compass	BASF	167:167 g/l	SC	11740

Uses

- Alternaria in **carrots** *(off-label)*, **horseradish** *(off-label)*, **parsnips** *(off-label)*, **spring oilseed rape**, **winter oilseed rape**
- Botrytis in **meadowfoam (limnanthes alba)** *(off-label)*
- Chocolate spot in **spring field beans, winter field beans**
- Crown rot in **carrots** *(off-label)*, **horseradish** *(off-label)*, **parsnips** *(off-label)*
- Grey mould in **spring oilseed rape, winter oilseed rape**
- Light leaf spot in **spring oilseed rape, winter oilseed rape**
- Sclerotinia stem rot in **spring oilseed rape, winter oilseed rape**
- Stem canker in **spring oilseed rape, winter oilseed rape**

Extension of Authorisation for Minor Use (EAMUs)

- **carrots** *20040525*
- **horseradish** *20040525*
- **meadowfoam (limnanthes alba)** *20040507*
- **parsnips** *20040525*

Approval information

- Iprodione and thiophanate-methyl included in Annex I under EC Regulation 1107/2009

Efficacy guidance

- Timing of sprays on oilseed rape varies with disease, see label for details
- For season-long control product should be applied as part of a disease control programme

Restrictions

- Maximum number of treatments (refers to total sprays containing benomyl, carbendazim or thiophanate methyl) 2 per crop for oilseed rape, field beans, meadowfoam; 3 per crop for carrots, horseradish, parsnips

Crop-specific information

- Latest use: before end of flowering for oilseed rape, field beans
- HI meadowfoam, field beans, oilseed rape 3 wk; carrots, horseradish, parsnips 28 d
- Treatment may extend duration of green leaf in winter oilseed rape

Environmental safety

- Dangerous for the environment
- Very toxic to aquatic organisms

Hazard classification and safety precautions

Hazard Harmful, Dangerous for the environment
Transport code 9
Packaging group III

SEE SECTION 3 FOR PRODUCTS ALSO REGISTERED

UN Number 3082
Risk phrases R40, R43, R50, R53a, R68
Operator protection A, C, H, M; U05a, U20c
Environmental protection E15a, E38
Storage and disposal D01, D02, D08, D09a, D10b, D12a
Medical advice M05a

298   isopyrazam

An SDHI fungicide available only in mixtures
FRAC mode of action code: 7

See also cyprodinil + isopyrazam
epoxiconazole + isopyrazam

299   isoxaben

A soil-acting benzamide herbicide for use in grass and fruit
HRAC mode of action code: L

Products

Flexidor 125	Landseer	125 g/l	SC	10946

Uses
- Annual dicotyledons in *almonds* (off-label), *amenity vegetation*, *apple orchards*, *bilberries* (off-label), *blackberries*, *blackcurrants*, *blueberries* (off-label), *canary grass* (off-label - grown for game cover), *carrots* (off-label), *cherries*, *chestnuts* (off-label), *clovers* (off-label - grown for game cover), *courgettes* (off-label), *cranberries* (off-label), *forage maize* (off-label - grown for game cover), *forestry transplants*, *gooseberries*, *hazel nuts* (off-label), *hops*, *marrows* (off-label), *millet* (off-label - grown for game cover), *ornamental plant production*, *pear orchards*, *plums*, *protected carrots* (off-label - temporary protection), *protected courgettes* (off-label), *protected marrows* (off-label), *protected ornamentals* (off-label), *protected parsnips* (off-label - temporary protection), *protected pumpkins* (off-label), *protected squashes* (off-label), *pumpkins* (off-label), *quinoa* (off-label - grown for game cover), *raspberries*, *redcurrants* (off-label), *rubus hybrids* (off-label), *squashes* (off-label), *strawberries*, *sunflowers* (off-label - grown for game cover), *walnuts* (off-label), *whitecurrants* (off-label), *wine grapes*
- Cleavers in *asparagus* (off-label)
- Fat hen in *asparagus* (off-label)
- Groundsel in *asparagus* (off-label)
- Knotgrass in *asparagus* (off-label)

Extension of Authorisation for Minor Use (EAMUs)
- *almonds* 20061111
- *asparagus* 20050893
- *bilberries* 20061113
- *blueberries* 20061113
- *canary grass* (grown for game cover) 20050896
- *carrots* 20050895
- *chestnuts* 20061111
- *clovers* (grown for game cover) 20050896
- *courgettes* 20050894
- *cranberries* 20061113
- *forage maize* (grown for game cover) 20050896
- *hazel nuts* 20061111
- *marrows* 20050894
- *millet* (grown for game cover) 20050896
- *protected carrots* (temporary protection) 20050892
- *protected courgettes* 20050894
- *protected marrows* 20050894

FOR FULL CONDITIONS OF USE ALWAYS READ THE PRODUCT LABEL

- **protected ornamentals** *20050891*
- **protected parsnips** *(temporary protection) 20050892*
- **protected pumpkins** *20050894*
- **protected squashes** *20050894*
- **pumpkins** *20050894*
- **quinoa** *(grown for game cover) 20050896*
- **redcurrants** *20061113*
- **rubus hybrids** *20061112*
- **squashes** *20050894*
- **sunflowers** *(grown for game cover) 20050896*
- **walnuts** *20061111*
- **whitecurrants** *20061113*

Approval information
- Isoxaben included in Annex 1 under EC Regulation 1107/2009
- Accepted by BBPA for use on hops

Efficacy guidance
- When used alone apply pre-weed emergence
- Effectiveness is reduced in dry conditions. Weed seeds germinating at depth are not controlled
- Activity reduced on soils with more than 10% organic matter. Do not use on peaty soils
- Various tank mixtures are recommended for early post-weed emergence treatment (especially for grass weeds). See label for details

Restrictions
- Maximum number of treatments 1 per crop for all edible crops; 2 per yr on amenity vegetation and non-edible crops

Crop-specific information
- Latest use: before 1 Apr in yr of harvest for edible crops

Following crops guidance
- See label for details of crops which may be sown in the event of failure of a treated crop

Environmental safety
- Keep all livestock out of treated areas for at least 50 d

Hazard classification and safety precautions
UN Number N/C
Operator protection A, C; U05a, U20b
Environmental protection E06a (50 d); E15a
Storage and disposal D01, D05, D09a, D11a

300   isoxaben + terbuthylazine

A contact and residual herbicide for use in peas and spring field beans
HRAC mode of action code: L + C1

See also terbuthylazine

Products

Skirmish	Syngenta	75:420 g/l	SC	08444

Uses
- Annual dicotyledons in **broad beans** *(off-label)*, **combining peas**, **lupins** *(off-label)*, **spring field beans**, **vining peas**
- Annual grasses in **broad beans** *(off-label)*, **lupins** *(off-label)*

Extension of Authorisation for Minor Use (EAMUs)
- **broad beans** *20121685 expires 31 Dec 2013*
- **lupins** *20121686 expires 31 Dec 2013*

Approval information

- Isoxaben and terbuthylazine included in Annex 1 under EC Regulation 1107/2009.
- Approval expiry 31 Dec 2013
- Approval expiry 31 Dec 2013 [1]

Efficacy guidance

- May be applied pre- or post-emergence of crop but before second node stage (GS 102)
- Product will give residual control of germinating weeds on mineral soils for up to 8 wk
- Product is slow acting and control may not be evident for 7-10 d or more after spraying
- Best results achieved when soil surface damp and with a fine, firm tilth. Do not use on very cloddy or stony soil
- Rain after spraying will normally improve weed control but excessive rainfall, very dry conditions or unusually low soil temperatures may lead to unsatisfactory control

Restrictions

- Maximum total dose equivalent to one full dose treatment on peas and spring field beans
- Do not use on forage peas
- Do not use on soils lighter than Coarse Sandy Loam, on very stony soils or soils with more than 10% organic matter
- Pea seed should be covered by at least 25 mm of soil
- Heavy rain after application may cause some crop damage, especially on light soils. Do not use on soils where surface water is likely to accumulate
- For post-emergence treatment a crystal violet test for cuticle wax is advised

Crop-specific information

- Latest use: before second node stage (GS 102) for peas; pre-emergence for spring field beans; beans that are emerging at application will be severely damaged or killed
- Product may be used on all varieties of spring sown vining and combining peas and spring field beans but pea varieties Vedette and Printana may be damaged from which recovery may not be complete

Environmental safety

- Dangerous for the environment
- Very toxic to aquatic organisms

Hazard classification and safety precautions

Hazard Harmful, Dangerous for the environment
Transport code 9
Packaging group III
UN Number 3082
Risk phrases R22a, R50, R53a
Operator protection A; U05a, U20a
Environmental protection E15a, E38
Storage and disposal D01, D02, D05, D07, D09a, D10c, D12a

301   isoxaflutole

An isoxazole herbicide available only in mixtures
HRAC mode of action code: F2

See also flufenacet + isoxaflutole

302   kresoxim-methyl

A protectant strobilurin fungicide for apples
FRAC mode of action code: 11

See also epoxiconazole + fenpropimorph + kresoxim-methyl
*   epoxiconazole + kresoxim-methyl*
*   epoxiconazole + kresoxim-methyl + pyraclostrobin*
*   fenpropimorph + kresoxim-methyl*

Products

1	Beem WG	AgChem Access	50% w/w	WG	13807
2	Stroby WG	BASF	50% w/w	WG	08653

Uses

- Black spot in **protected roses, roses** [1, 2]
- Powdery mildew in **all edible seed crops grown outdoors** (off-label), **all non-edible seed crops grown outdoors** (off-label), **gooseberries** (off-label), **hops** (off-label), **protected soft fruit** (off-label), **quinces** (off-label), **redcurrants** (off-label), **soft fruit** (off-label), **table grapes** (off-label), **whitecurrants** (off-label), **wine grapes** (off-label) [2]; **apples** (reduction), **blackcurrants, protected roses, protected strawberries, roses, strawberries** [1, 2]
- Scab in **apples** [1, 2]

Extension of Authorisation for Minor Use (EAMUs)

- **all edible seed crops grown outdoors** *20082917* [2]
- **all non-edible seed crops grown outdoors** *20082917* [2]
- **gooseberries** *20060656* [2]
- **hops** *20082917* [2]
- **protected soft fruit** *20082917* [2]
- **quinces** *20060657* [2]
- **redcurrants** *20060656* [2]
- **soft fruit** *20082917* [2]
- **table grapes** *20092711* [2]
- **whitecurrants** *20060656* [2]
- **wine grapes** *20092711* [2]

Approval information

- Kresoxim-methyl included in Annex I under EC Regulation 1107/2009
- Accepted by BBPA on malting barley
- Approved for use in ULV systems

Efficacy guidance

- Activity is protectant. Best results achieved from treatments prior to disease development. See label for timing details on each crop. Treatments should be repeated at 10-14 d intervals but note limitations below
- To minimise the likelihood of development of resistance to strobilurin fungicides these products should be used in a planned Resistance Management strategy. See Section 5 for more information
- Product may be applied in ultra low volumes (ULV) but disease control may be reduced

Restrictions

- Maximum number of treatments 4 per yr on apples; 3 per yr on other crops. See notes in Efficacy about limitations on consecutive treatments
- Consult before using on crops intended for processing

Crop-specific information

- HI 14 d for blackcurrants, protected strawberries, strawberries; 35 d for apples
- On apples do not spray product more than twice consecutively and separate each block of two consecutive treatments with at least two applications from a different cross-resistance group. For all other crops do not apply consecutively and use a maximum of once in every three fungicide sprays
- Product should not be used as final spray of the season on apples

SEE SECTION 3 FOR PRODUCTS ALSO REGISTERED

Environmental safety
- Dangerous for the environment
- Very toxic to aquatic organisms
- Harmless to ladybirds and predatory mites
- Harmless to honey bees and may be applied during flowering. Nevertheless local beekeepers should be notified when treatment of orchards in flower is to occur
- Broadcast air-assisted LERAP (5 m)

Hazard classification and safety precautions
Hazard Harmful, Dangerous for the environment
Transport code 9
Packaging group III
UN Number 3077
Risk phrases R40, R50, R53a
Operator protection U05a, U20b
Environmental protection E15a, E38; E17b (5 m)
Storage and disposal D01, D02, D08, D09a, D10c, D12a
Medical advice M05a

303 lambda-cyhalothrin

A quick-acting contact and ingested pyrethroid insecticide
IRAC mode of action code: 3

Products

1	Clayton Lanark	Clayton	100 g/l	CS	12942
2	Clayton Sparta	Clayton	50 g/l	EC	13457
3	EA Lambda 50	European Ag	50 g/l	EC	15041
4	EA Lambda C	European Ag	100 g/l	CS	15093
5	Hallmark with Zeon Technology	Syngenta	100 g/l	CS	12629
6	Karate 2.5WG	Syngenta	2.5% w/w	GR	14060
7	Karis 10 CS	Headland	100 g/l	CS	15312
8	Life Scientific Lambda-Cyhalothrin	Life Scientific	100 g/l	CS	15342
9	Major	Makhteshim	50 g/l	CS	14527
10	Markate 50	Agrovista	50 g/l	EC	13529
11	Seal Z	AgChem Access	100 g/l	CS	14201

Uses
- Aphids in *all edible seed crops grown outdoors* (off-label), *all non-edible seed crops grown outdoors* (off-label), *farm forestry* (off-label), *miscanthus* (off-label) [5, 9]; *borage for oilseed production* (off-label), *bulb onions* (off-label), *canary flower (echium spp.)* (off-label), *evening primrose* (off-label), *frise* (off-label), *garlic* (off-label), *grass seed crops* (off-label), *honesty* (off-label), *leeks* (off-label), *mustard* (off-label), *outdoor herbs* (off-label), *salad onions* (off-label), *shallots* (off-label), *short rotation coppice willow* (off-label), *spring linseed* (off-label), *spring rye* (off-label), *triticale* (off-label), *winter linseed* (off-label), *winter rye* (off-label) [9]; *chestnuts* (off-label), *hazel nuts* (off-label), *walnuts* (off-label) [5, 8-10]; *cob nuts* (off-label) [8, 10]; *combining peas, vining peas* [2, 3, 10]; *crambe* (off-label), *forest nurseries* (off-label), *hops* (off-label), *ornamental plant production* (off-label), *protected aubergines* (off-label), *protected forest nurseries* (off-label), *protected ornamentals* (off-label), *protected pak choi* (off-label), *protected peppers* (off-label), *protected tat soi* (off-label), *protected tomatoes* (off-label), *soft fruit* (off-label), *top fruit* (off-label), *willow (short rotation coppice)* (off-label) [5]; *durum wheat* [1, 3-7, 10, 11]; *edible podded peas, spring field beans, spring oilseed rape, sugar beet, winter field beans, winter oilseed rape* [3, 10]; *filberts* (off-label) [8]; *fodder beet* (off-label) [1, 9]; *lamb's lettuce* (off-label), *leaf brassicas* (off-label) [8, 9]; *lupins* (off-label) [9, 10]; *potatoes* [1-5, 8-11]; *spring barley, spring oats* [1, 4, 5, 7, 8, 11]; *spring wheat* [1-5, 7, 8, 10, 11]; *winter barley, winter oats* [1, 4-8, 11]; *winter wheat* [1-8, 10, 11]

FOR FULL CONDITIONS OF USE ALWAYS READ THE PRODUCT LABEL

- Barley yellow dwarf virus vectors in **durum wheat** [1, 4-7, 11]; **winter barley, winter oats, winter wheat** [1, 4-8, 11]
- Beet leaf miner in **sugar beet** [1, 4, 5, 8, 9, 11]
- Beet virus yellows vectors in **spring oilseed rape** [1, 4, 5, 7, 8, 11]; **winter oilseed rape** [1, 4-8, 11]
- Beetles in **combining peas, durum wheat, edible podded peas, potatoes, spring field beans, spring oilseed rape, spring wheat, sugar beet, vining peas, winter field beans, winter oilseed rape, winter wheat** [3, 10]
- Brassica pod midge in **spring oilseed rape, winter oilseed rape** [2]
- Cabbage seed weevil in **celeriac** (off-label), **radishes** (off-label) [8]; **spring oilseed rape, winter oilseed rape** [1, 2, 4, 5, 7, 8, 11]
- Cabbage stem flea beetle in **spring oilseed rape** [1, 4, 5, 7-9, 11]; **winter oilseed rape** [1, 4-9, 11]
- Cabbage stem weevil in **celeriac** (off-label), **radishes** (off-label) [1, 9, 10]
- Capsids in **blackberries** (off-label), **dewberries** (off-label), **raspberries** (off-label), **rubus hybrids** (off-label) [1, 5, 8-10]; **grapevines** (off-label) [1, 5, 8, 10]; **outdoor grapes** (off-label) [9]
- Carrot fly in **carrots** (off-label), **parsnips** (off-label) [10]; **celeriac** (off-label), **radishes** (off-label) [1, 8-10]; **celery (outdoor)** (off-label), **horseradish** (off-label), **mallow (althaea spp.)** (off-label), **parsley root** (off-label) [1, 5, 8-10]; **fennel** (off-label) [5, 8, 9]
- Caterpillars in **beetroot** (off-label) [1, 5, 8, 9]; **broccoli, brussels sprouts, cabbages, calabrese, cauliflowers** [1, 4, 5, 8, 11]; **celery (outdoor)** (off-label), **navy beans** (off-label) [1, 5, 8-10]; **combining peas, durum wheat, edible podded peas, potatoes, spring field beans, spring oilseed rape, spring wheat, sugar beet, vining peas, winter field beans, winter oilseed rape, winter wheat** [3, 10]; **dwarf pak choi** (off-label), **protected pak choi** (off-label), **protected tat soi** (off-label) [8-10]; **french beans** (off-label) [1, 5, 10]; **red beet** (off-label) [10]; **runner beans** (off-label) [1, 8-10]; **swedes** (off-label), **turnips** (off-label) [5, 8, 10]
- Clay-coloured weevil in **blackberries** (off-label), **dewberries** (off-label), **raspberries** (off-label), **rubus hybrids** (off-label) [1, 5, 8-10]
- Cutworms in **beetroot** (off-label) [1, 5, 8, 9]; **carrots** [1, 4, 5, 8, 9, 11]; **celeriac** (off-label), **radishes** (off-label) [1, 8-10]; **chicory** (off-label), **fennel** (off-label) [5, 8, 9]; **fodder beet** (off-label), **red beet** (off-label) [10]; **lettuce** [1, 4, 5, 8, 11]; **outdoor lettuce, potatoes** (if present at time of application) [9]; **parsnips** [4, 5, 8, 9, 11]; **sugar beet** [1, 2, 4, 5, 8, 9, 11]
- Flea beetle in **corn gromwell** (off-label) [5]; **crambe** (off-label), **protected pak choi** (off-label), **protected tat soi** (off-label) [8-10]; **fodder beet** (off-label) [10]; **poppies for morphine production** (off-label) [1, 5, 8-10]; **spring oilseed rape** [1, 2, 4, 5, 7-9, 11]; **sugar beet** [1, 2, 4, 5, 8, 9, 11]; **winter oilseed rape** [1, 2, 4-9, 11]
- Frit fly in **sweetcorn** (off-label) [1, 5, 8-10]
- Grain aphid in **durum wheat, spring barley, spring oats, spring wheat, winter barley, winter oats, winter wheat** [9]
- Insect pests in **all edible seed crops grown outdoors** (off-label), **all non-edible seed crops grown outdoors** (off-label), **celery (outdoor)** (off-label), **farm forestry** (off-label), **forest nurseries** (off-label), **hops** (off-label), **lupins** (off-label), **miscanthus** (off-label), **ornamental plant production** (off-label), **protected forest nurseries** (off-label), **protected ornamentals** (off-label), **soft fruit** (off-label), **top fruit** (off-label) [5, 8, 10]; **beetroot** (off-label), **poppies for morphine production** (off-label) [5, 8]; **blackberries** (off-label), **blackcurrants** (off-label), **bulb onions** (off-label), **dewberries** (off-label), **garlic** (off-label), **gooseberries** (off-label), **leeks** (off-label), **raspberries** (off-label), **redcurrants** (off-label), **rubus hybrids** (off-label), **salad onions** (off-label), **shallots** (off-label), **whitecurrants** (off-label) [1, 5, 8-10]; **borage for oilseed production** (off-label), **canary flower (echium spp.)** (off-label), **evening primrose** (off-label), **fodder beet** (off-label), **grass seed crops** (off-label), **herbs (see appendix 6)** (off-label), **honesty** (off-label), **horseradish** (off-label), **mallow (althaea spp.)** (off-label), **mustard** (off-label), **navy beans** (off-label), **parsley root** (off-label), **runner beans** (off-label), **spring linseed** (off-label), **triticale** (off-label), **winter linseed** (off-label) [1, 5, 8, 10]; **broad beans** (off-label), **cob nuts** (off-label), **dwarf beans** (off-label), **protected all edible seed crops** (off-label), **short rotation coppice willow** (off-label) [8, 10]; **carrots** (off-label), **land cress** (off-label), **parsnips** (off-label), **protected all non-edible seed crops** (off-label), **red beet** (off-label) [10]; **celeriac** (off-label), **radishes** (off-label), **spring rye** (off-label), **winter rye** (off-label) [5, 10]; **chestnuts** (off-label), **hazel nuts** (off-label), **walnuts** (off-label) [5, 8-10]; **chicory** (off-label), **fennel** (off-label), **filberts** (off-label) [8]; **crambe** (off-label), **protected aubergines** (off-label), **protected pak choi** (off-

label), **protected peppers** *(off-label)*, **protected strawberries** *(off-label)*, **protected tat soi** *(off-label)*, **protected tomatoes** *(off-label)*, **strawberries** *(off-label)*, **willow (short rotation coppice)** *(off-label)* [5]; **french beans** *(off-label)* [1, 10]; **frise** *(off-label)*, **lamb's lettuce** *(off-label)*, **leaf brassicas** *(off-label)*, **scarole** *(off-label)* [1, 5, 8]; **grapevines** *(off-label)* [1]; **rye** *(off-label)* [1, 8]; **sweetcorn** *(off-label)* [1, 8, 10]

- Leaf midge in **blackcurrants** *(off-label)*, **gooseberries** *(off-label)*, **redcurrants** *(off-label)*, **whitecurrants** *(off-label)* [1, 5, 9, 10]
- Leaf miner in **fodder beet** *(off-label)* [10]; **sugar beet** [2]
- Midges in **blackcurrants** *(off-label)*, **gooseberries** *(off-label)*, **redcurrants** *(off-label)*, **whitecurrants** *(off-label)* [8]
- Pea and bean weevil in **combining peas**, **spring field beans**, **vining peas**, **winter field beans** [1, 2, 4, 5, 8, 9, 11]; **edible podded peas** [1, 4, 5, 8, 9, 11]
- Pea aphid in **combining peas**, **edible podded peas**, **vining peas** [1, 4, 5, 8, 9, 11]
- Pea midge in **combining peas**, **edible podded peas**, **vining peas** [1, 4, 5, 8, 9, 11]
- Pea moth in **combining peas**, **vining peas** [1, 2, 4, 5, 8, 9, 11]; **edible podded peas** [1, 4, 5, 8, 9, 11]
- Pear sucker in **pears** [1, 2, 4, 5, 8, 11]; **winter wheat** [2]
- Pod midge in **spring oilseed rape**, **winter oilseed rape** [1, 4, 5, 7-9, 11]
- Pollen beetle in **corn gromwell** *(off-label)* [5]; **crambe** *(off-label)* [8-10]; **poppies for morphine production** *(off-label)* [1, 5, 8-10]; **spring oilseed rape**, **winter oilseed rape** [1, 2, 4, 5, 7-9, 11]
- Rose-grain aphid in **durum wheat**, **spring barley**, **spring oats**, **spring wheat**, **winter barley**, **winter oats**, **winter wheat** [9]
- Sawflies in **blackcurrants** *(off-label)*, **gooseberries** *(off-label)*, **redcurrants** *(off-label)*, **whitecurrants** *(off-label)* [1, 5, 8-10]; **short rotation coppice willow** *(off-label)* [8-10]
- Seed beetle in **broad beans** *(off-label)* [1, 5, 8-10]
- Seed weevil in **spring oilseed rape**, **winter oilseed rape** [9]
- Silver Y moth in **beetroot** *(off-label)* [8, 9]; **celery (outdoor)** *(off-label)*, **dwarf beans** *(off-label)*, **navy beans** *(off-label)* [8-10]; **french beans** *(off-label)* [5, 10]; **red beet** *(off-label)* [10]; **runner beans** *(off-label)* [5, 8-10]
- Thrips in **all edible seed crops grown outdoors** *(off-label)*, **all non-edible seed crops grown outdoors** *(off-label)*, **farm forestry** *(off-label)*, **miscanthus** *(off-label)* [9]; **broad beans** *(off-label)* [1, 5, 9]; **bulb onions** *(off-label)*, **garlic** *(off-label)*, **leeks** *(off-label)*, **salad onions** *(off-label)*, **shallots** *(off-label)* [1, 5, 8, 9]
- Tobacco whitefly in **all edible seed crops grown outdoors** *(off-label)*, **all non-edible seed crops grown outdoors** *(off-label)*, **borage for oilseed production** *(off-label)*, **canary flower (echium spp.)** *(off-label)*, **evening primrose** *(off-label)*, **farm forestry** *(off-label)*, **honesty** *(off-label)*, **miscanthus** *(off-label)*, **mustard** *(off-label)*, **spring linseed** *(off-label)*, **winter linseed** *(off-label)* [9]
- Wasps in **grapevines** *(off-label)* [1, 5, 8, 10]; **outdoor grapes** *(off-label)* [9]
- Weevils in **combining peas**, **durum wheat**, **edible podded peas**, **potatoes**, **spring field beans**, **spring oilseed rape**, **spring wheat**, **sugar beet**, **vining peas**, **winter field beans**, **winter oilseed rape**, **winter wheat** [3, 10]
- Wheat-blossom midge in **winter wheat** [1, 4, 5, 7, 8, 11]
- Whitefly in **all edible seed crops grown outdoors** *(off-label)*, **all non-edible seed crops grown outdoors** *(off-label)*, **borage for oilseed production** *(off-label)*, **canary flower (echium spp.)** *(off-label)*, **evening primrose** *(off-label)*, **farm forestry** *(off-label)*, **honesty** *(off-label)*, **miscanthus** *(off-label)*, **mustard** *(off-label)*, **spring linseed** *(off-label)*, **winter linseed** *(off-label)* [9]; **broccoli**, **brussels sprouts**, **cabbages**, **calabrese**, **cauliflowers** [1, 4, 5, 8, 11]
- Willow aphid in **short rotation coppice willow** *(off-label)* [8-10]
- Willow beetle in **short rotation coppice willow** *(off-label)* [8-10]
- Willow sawfly in **short rotation coppice willow** *(off-label)* [8-10]
- Yellow cereal fly in **winter wheat** [1, 2, 4, 5, 7-9, 11]

Extension of Authorisation for Minor Use (EAMUs)

- *all edible seed crops grown outdoors* 20082944 [5], *20111433* [8], *20093498* [9], *20102880* [10]
- *all non-edible seed crops grown outdoors* 20082944 [5], *20111429* [8], *20111433* [8], *20093498* [9], *20102880* [10]
- *beetroot* 20063761 [1], *20060743* [5], *20111440* [8], *20093488* [9]

FOR FULL CONDITIONS OF USE ALWAYS READ THE PRODUCT LABEL

- **blackberries** *20063755* [1], *20060728* [5], *20111426* [8], *20093478* [9], *20073266* [10]
- **blackcurrants** *20063752* [1], *20060727* [5], *20111425* [8], *20093477* [9], *20073269* [10]
- **borage for oilseed production** *20063748* [1], *20060634* [5], *20111422* [8], *20093473* [9], *20073258* [10]
- **broad beans** *20063764* [1], *20060753* [5], *20111444* [8], *20093493* [9], *20073234* [10]
- **bulb onions** *20063756* [1], *20060730* [5], *20111427* [8], *20093479* [9], *20073256* [10]
- **canary flower (echium spp.)** *20063748* [1], *20060634* [5], *20111422* [8], *20093473* [9], *20073258* [10]
- **carrots** *20080201* [10]
- **celeriac** *20063757* [1], *20060731* [5], *20111428* [8], *20093480* [9], *20080204* [10]
- **celery (outdoor)** *20063762* [1], *20060744* [5], *20111441* [8], *20093490* [9], *20073257* [10]
- **chestnuts** *20060742* [5], *20111439* [8], *20093487* [9], *20080206* [10]
- **chicory** *20060740* [5], *20111438* [8], *20093486* [9]
- **cob nuts** *20111439* [8], *20080206* [10]
- **corn gromwell** *20121047* [5]
- **crambe** *20081046* [5], *20111431* [8], *20093496* [9], *20102876* [10]
- **dewberries** *20063755* [1], *20060728* [5], *20111426* [8], *20093478* [9], *20073266* [10]
- **dwarf beans** *20111437* [8], *20093485* [9], *20080202* [10]
- **evening primrose** *20063664* [1], *20060634* [5], *20111422* [8], *20093473* [9], *20073258* [10]
- **farm forestry** *20082944* [5], *20111433* [8], *20093498* [9], *20102880* [10]
- **fennel** *20060733* [5], *20111436* [8], *20093484* [9]
- **filberts** *20111439* [8]
- **fodder beet** *20063665* [1], *20060637* [5], *20111424* [8], *20093476* [9], *20073233* [10]
- **forest nurseries** *20082944* [5], *20111433* [8], *20102880* [10]
- **french beans** *20063758* [1], *20060739* [5], *20080202* [10]
- **frise** *20063751* [1], *20060636* [5], *20111445* [8], *20093475* [9]
- **garlic** *20063756* [1], *20060730* [5], *20111427* [8], *20093479* [9], *20073256* [10]
- **gooseberries** *20063752* [1], *20060727* [5], *20111425* [8], *20093477* [9], *20073269* [10]
- **grapevines** *20063747* [1], *20060266* [5], *20111420* [8], *20080205* [10]
- **grass seed crops** *20063749* [1], *20060624* [5], *20111421* [8], *20093472* [9], *20073260* [10]
- **hazel nuts** *20060742* [5], *20111439* [8], *20093487* [9], *20080206* [10]
- **herbs (see appendix 6)** *20063751* [1], *20060636* [5], *20111445* [8], *20073259* [10]
- **honesty** *20063748* [1], *20060634* [5], *20111422* [8], *20093473* [9], *20073258* [10]
- **hops** *20082944* [5], *20111433* [8], *20102880* [10]
- **horseradish** *20071307* [1], *20071301* [5], *20111429* [8], *20093495* [9], *20080201* [10]
- **lamb's lettuce** *20063751* [1], *20060636* [5], *20111439* [8], *20093475* [9]
- **land cress** *20102878* [10]
- **leaf brassicas** *20063751* [1], *20060636* [5], *20111439* [8], *20093475* [9]
- **leeks** *20063756* [1], *20060730* [5], *20111427* [8], *20093479* [9], *20073256* [10]
- **lupins** *20060635* [5], *20111423* [8], *20093474* [9], *20102877* [10]
- **mallow (althaea spp.)** *20071307* [1], *20071301* [5], *20111429* [8], *20093495* [9], *20080201* [10]
- **miscanthus** *20082944* [5], *20111433* [8], *20093498* [9], *20102880* [10]
- **mustard** *20063664* [1], *20060634* [5], *20111422* [8], *20093473* [9], *20073258* [10]
- **navy beans** *20063758* [1], *20060739* [5], *20111437* [8], *20093485* [9], *20080202* [10]
- **ornamental plant production** *20082944* [5], *20111433* [8], *20102880* [10]
- **outdoor grapes** *20093471* [9]
- **outdoor herbs** *20093475* [9]
- **parsley root** *20071307* [1], *20071301* [5], *20111429* [8], *20093495* [9], *20080201* [10]
- **parsnips** *20080201* [10]
- **poppies for morphine production** *20063763* [1], *20060749* [5], *20111443* [8], *20093492* [9], *20102875* [10]
- **protected all edible seed crops** *20111433* [8], *20102880* [10]
- **protected all non-edible seed crops** *20102880* [10]
- **protected aubergines** *20121994* [5]
- **protected forest nurseries** *20082944* [5], *20111433* [8], *20102880* [10]
- **protected ornamentals** *20082944* [5], *20111433* [8], *20102880* [10]
- **protected pak choi** *20081263* [5], *20111432* [8], *20093497* [9], *20102879* [10]
- **protected peppers** *20121994* [5]
- **protected strawberries** *20111705* [5]

SEE SECTION 3 FOR PRODUCTS ALSO REGISTERED

- **protected tat soi** *20081263* [5], *20111432* [8], *20093497* [9], *20102879* [10]
- **protected tomatoes** *20121994* [5]
- **radishes** *20063757* [1], *20060731* [5], *20111428* [8], *20093480* [9], *20080204* [10]
- **raspberries** *20063755* [1], *20060728* [5], *20111426* [8], *20093478* [9], *20073266* [10]
- **red beet** *20073254* [10]
- **redcurrants** *20063752* [1], *20060727* [5], *20111425* [8], *20093477* [9], *20073269* [10]
- **rubus hybrids** *20063755* [1], *20060728* [5], *20111426* [8], *20093478* [9], *20073266* [10]
- **runner beans** *20063758* [1], *20060739* [5], *20111437* [8], *20093485* [9], *20080202* [10]
- **rye** *20063749* [1], *20111421* [8]
- **salad onions** *20063756* [1], *20060730* [5], *20111427* [8], *20093479* [9], *20073256* [10]
- **scarole** *20063751* [1], *20060636* [5], *20111445* [8]
- **shallots** *20063756* [1], *20060730* [5], *20111427* [8], *20093479* [9], *20073256* [10]
- **short rotation coppice willow** *20111442* [8], *20093491* [9], *20073268* [10]
- **soft fruit** *20082944* [5], *20111433* [8], *20102880* [10]
- **spring linseed** *20063748* [1], *20060634* [5], *20111422* [8], *20093473* [9], *20073258* [10]
- **spring rye** *20060624* [5], *20093472* [9], *20073260* [10]
- **strawberries** *20111705* [5]
- **swedes** *20101856* [5], *20111434* [8], *20102911* [10]
- **sweetcorn** *20063760* [1], *20060732* [5], *20111435* [8], *20093481* [9], *20080203* [10]
- **top fruit** *20082944* [5], *20111433* [8], *20102880* [10]
- **triticale** *20063749* [1], *20060624* [5], *20111421* [8], *20093472* [9], *20073260* [10]
- **turnips** *20101856* [5], *20111434* [8], *20102911* [10]
- **walnuts** *20060742* [5], *20111439* [8], *20093487* [9], *20080206* [10]
- **whitecurrants** *20063752* [1], *20060727* [5], *20111425* [8], *20093477* [9], *20073269* [10]
- **willow (short rotation coppice)** *20060748* [5]
- **winter linseed** *20063748* [1], *20060634* [5], *20111422* [8], *20093473* [9], *20073258* [10]
- **winter rye** *20060624* [5], *20093472* [9], *20073260* [10]

Approval information
- Lambda-cyhalothrin included in Annex I under EC Regulation 1107/2009
- Accepted by BBPA for use on malting barley

Efficacy guidance
- Best results normally obtained from treatment when pest attack first seen. See label for detailed recommendations on each crop
- Timing for control of barley yellow dwarf virus vectors depends on specialist assessment of the level of risk in the area
- Repeat applications recommended in some crops where prolonged attack occurs, up to maximum total dose. See label for details
- Where strains of aphids resistant to lambda-cyhalothrin occur control is unlikely to be satisfactory
- Addition of wetter recommended for control of certain pests in brassicas and oilseed rape
- Use of sufficient water volume to ensure thorough crop penetration recommended for optimum results
- Use of drop-legged sprayer gives improved results in crops such as Brussels sprouts

Restrictions
- Maximum number of applications or maximum total dose per crop varies - see labels
- Do not apply to a cereal crop if any product containing a pyrethroid insecticide or dimethoate has been applied to the crop after the start of ear emergence (GS 51)
- Do not spray cereals in the spring/summer (ie after 1 Apr) within 6 m of edge of crop

Crop-specific information
- Latest use before late milk stage on cereals; before end of flowering for winter oilseed rape
- HI 3 d for radishes, red beet; 7 d for lettuce [1, 5]; 14 d for carrots and parsnips; 25 d for peas, field beans; 6 wk for spring oilseed rape; 8 wk for sugar beet

Environmental safety
- Dangerous for the environment
- Very toxic to aquatic organisms
- Flammable [2]
- Risk to certain non-target insects or other arthropods [5]

FOR FULL CONDITIONS OF USE ALWAYS READ THE PRODUCT LABEL

- To protect non-target arthropods respect an untreated buffer zone of 5 m to non-crop land
- Broadcast air-assisted LERAP [1, 4, 5, 7, 8, 11] (25 m); Broadcast air-assisted LERAP [2] (38 m); LERAP Category A [2, 3, 10]; LERAP Category B [1, 4-9, 11]

Hazard classification and safety precautions

Hazard Harmful, Dangerous for the environment [1-11]; Corrosive [3, 10]; Flammable [2]

Transport code 3 [2]; 8 [3, 10]; 9 [1, 4-9, 11]

Packaging group III

UN Number 1760 [3, 10]; 1993 [2]; 3077 [6]; 3082 [1, 4, 5, 7-9, 11]

Risk phrases R20, R22a, R53a [1-11]; R21, R37, R67 [2]; R22b [2, 3, 9, 10]; R34 [3, 10]; R36, R38 [2, 6]; R43 [1, 4-9, 11]; R50 [1, 2, 4-8, 11]; R51 [3, 9, 10]

Operator protection A, H [1-11]; C [2, 3, 6, 9, 10]; J, K, M [6]; U02a [1, 3-11]; U04a [3, 10]; U05a [1-8, 10, 11]; U08 [1-5, 7, 8, 10, 11]; U09a [6, 9]; U11 [3, 6, 10]; U14 [1, 4-8, 11]; U19a [2, 3, 9, 10]; U20a [2, 3, 10]; U20b [1, 4-9, 11]

Environmental protection E12a, E16d [3, 10]; E15b [2, 3, 6, 9, 10]; E16a [1, 4-9, 11]; E16b [1, 4, 5, 7, 8, 11]; E16c [2, 3, 10]; E17b [1] (25 m); E17b [2] (38 m); E17b [4, 5, 7, 8, 11] (25 m); E22a [3, 6, 10]; E22b [3-5, 7-11]; E34 [1-5, 7-11]; E38 [1, 3-11]

Storage and disposal D01, D02 [1-8, 10, 11]; D05 [2, 9]; D07 [2]; D09a [1-5, 7-11]; D10b [2, 3, 10]; D10c [1, 4, 5, 7-9, 11]; D12a [1, 2, 4-9, 11]

Medical advice M03 [1, 4-8, 11]; M04a [3, 10]; M05b [2, 3, 9, 10]

304   Lecanicillium lecanii

A fungal parasite of aphids and whitefly

Products

Mycotal	Koppert	16.1% w/w	WP	04782

Uses

- Whitefly in *aubergines*, *protected beans*, *protected bilberries* (off-label), *protected blackberries* (off-label), *protected blackcurrants* (off-label), *protected blueberry* (off-label), *protected cayenne peppers* (off-label), *protected chives* (off-label), *protected cranberries* (off-label), *protected cucumbers*, *protected edible podded peas* (off-label), *protected gooseberries* (off-label), *protected herbs (see appendix 6)* (off-label), *protected lettuce*, *protected loganberries* (off-label), *protected ornamentals*, *protected parsley* (off-label), *protected peppers*, *protected raspberries* (off-label), *protected redcurrants* (off-label), *protected ribes hybrids* (off-label), *protected rubus hybrids* (off-label), *protected salad brassicas* (off-label), *protected strawberries* (off-label), *protected table grapes* (off-label), *protected tomatoes*, *protected wine grapes* (off-label)

Extension of Authorisation for Minor Use (EAMUs)

- *protected bilberries* 20101247
- *protected blackberries* 20101248
- *protected blackcurrants* 20101247
- *protected blueberry* 20101247
- *protected cayenne peppers* 20070862
- *protected chives* 20070861
- *protected cranberries* 20101247
- *protected edible podded peas* 20070863
- *protected gooseberries* 20101247
- *protected herbs (see appendix 6)* 20070861
- *protected loganberries* 20101248
- *protected parsley* 20070861
- *protected raspberries* 20101248
- *protected redcurrants* 20101247
- *protected ribes hybrids* 20101247
- *protected rubus hybrids* 20101248
- *protected salad brassicas* 20070861
- *protected strawberries* 20101249

- **protected table grapes** *20101247*
- **protected wine grapes** *20101247*

Approval information
- *Lecanicillium* included in Annex I under EC Regulation 1107/2009

Efficacy guidance
- *Lecanicillium lecanii* is a pathogenic fungus that infects the target pests and destroys them
- Apply spore powder as spray as part of biological control programme keeping the spray liquid well agitated
- Pre-soak the product for 2-4 h before application to rehydrate the spores and assist in dispersion
- Treat before infestations build to high levels and repeat as directed on the label
- Spray during late afternoon and early evening directing spray onto underside of leaves and to growing points
- Best results require minimum 80% relative humidity and 18°C within the crop canopy
- Product highly infective to many aphid species except the chrysanthemum aphid. Follow specific label directions for this pest

Restrictions
- Never use in tank mixture
- Do not use a fungicide within 3 d of treatment. Pesticides containing captan, chlorothalonil, fenarimol, dichlofluanid, imazalil, maneb, prochloraz, quinomethionate, thiram or tolylfluanid may not be used on the same crop
- Keep in a refrigerated store at 2-6°C but do not freeze

Environmental safety
- Products have negligible effects on commercially available natural predators or parasites but consult manufacturer before using with a particular biological control agent for the first time

Hazard classification and safety precautions
UN Number N/C
Operator protection U19a, U20b
Environmental protection E15a
Storage and disposal D09a, D11a

305   lenacil

A residual, soil-acting uracil herbicide for beet crops
HRAC mode of action code: C1

See also desmedipham + ethofumesate + lenacil + phenmedipham

Products

1 Lenazar Flo	Hermoo	440 g/l	SC	14791
2 Pirapima	Agroquimicos	440 g/l	SC	15777
3 Venzar Flowable	DuPont	440 g/l	SC	06907

Uses
- Annual dicotyledons in **blackberries** *(off-label)*, **blackcurrants** *(off-label)*, **blueberries** *(off-label)*, **cranberries** *(off-label)*, **farm woodland** *(off-label)*, **gooseberries** *(off-label)*, **herbs (see appendix 6)** *(off-label)*, **raspberries** *(off-label)*, **redcurrants** *(off-label)*, **ribes hybrids** *(off-label)*, **rubus hybrids** *(off-label)*, **spinach** *(off-label)*, **spinach beet** *(off-label)*, **strawberries** *(off-label)*, **whitecurrants** *(off-label)* [3]; **fodder beet, mangels, red beet, sugar beet** [1-3]
- Annual grasses in **blackcurrants** *(off-label)*, **blueberries** *(off-label)*, **cranberries** *(off-label)*, **gooseberries** *(off-label)*, **herbs (see appendix 6)** *(off-label)*, **redcurrants** *(off-label)*, **ribes hybrids** *(off-label)*, **spinach** *(off-label)*, **spinach beet** *(off-label)*, **whitecurrants** *(off-label)* [3]
- Annual meadow grass in **farm woodland** *(off-label)* [3]; **fodder beet, mangels, red beet, sugar beet** [1-3]
- Black bindweed in **fodder beet** *(off-label)* [3]

Extension of Authorisation for Minor Use (EAMUs)
- **blackberries** *20093246* [3]
- **blackcurrants** *970704* [3]
- **blueberries** *970704* [3]
- **cranberries** *970704* [3]
- **farm woodland** *971282* [3]
- **fodder beet** *20111765* [3]
- **gooseberries** *970704* [3]
- **herbs (see appendix 6)** *970703* [3]
- **raspberries** *20093246* [3]
- **redcurrants** *970704* [3]
- **ribes hybrids** *970704* [3]
- **rubus hybrids** *20093246* [3]
- **spinach** *970703* [3]
- **spinach beet** *970703* [3]
- **strawberries** *20093246* [3]
- **whitecurrants** *970704* [3]

Approval information
- Lenacil included in Annex I under EC Regulation 1107/2009

Efficacy guidance
- Best results, especially from pre-emergence treatments, achieved on fine, even, firm and moist soils free from clods. Continuing presence of moisture from rain or irrigation gives improved residual control of later germinating weeds. Effectiveness may be reduced by dry conditions
- On beet crops may be used pre- or post-emergence, alone or in mixture to broaden weed spectrum
- Apply overall or as band spray to beet crops pre-drilling incorporated, pre- or post-emergence
- All labels have limitations on soil types that may be treated. Residual activity reduced on soils with high OM content

Restrictions
- Maximum number of treatments in beet crops 1 pre-emergence + 3 post-emergence per crop; 2 per yr on established woody ornamentals and roses
- See label for soil type restrictions
- Do not use any other residual herbicide within 3 mth of the initial application to fruit or ornamental crops
- Do not treat crops under stress from drought, low temperatures, nutrient deficiency, pest or disease attack, or waterlogging

Crop-specific information
- Latest use: pre-emergence for red beet, fodder beet, spinach, spinach beet, mangels; before leaves meet over rows when used on these crops post-emergence; 24 h after planting new strawberry runners or before flowering for established strawberry crops, blackcurrants, gooseberries, raspberries
- Heavy rain after application to beet crops may cause damage especially if followed by very hot weather
- Reduction in beet stand may occur where crop emergence or vigour is impaired by soil capping or pest attack
- Strawberry runner beds to be treated should be level without depressions around the roots
- New soft fruit cuttings should be planted at least 15 cm deep and firmed before treatment
- Check varietal tolerance of ornamentals before large scale treatment

Following crops guidance
- Succeeding crops should not be planted or sown for at least 4 mth (6 mth on organic soils) after treatment following ploughing to at least 150 mm.
- Only beet crops, mangels or strawberries may be sown within 4 months of treatment and no further applications of lenacil should be made for at least 4 months [3]

Environmental safety
- Dangerous for the environment
- Very toxic to aquatic organisms

SEE SECTION 3 FOR PRODUCTS ALSO REGISTERED

Hazard classification and safety precautions

Hazard Dangerous for the environment
Transport code 9
Packaging group III
UN Number 3082
Risk phrases R50, R53a
Operator protection A [1-3]; C [1]; U05a, U08, U19a, U20a
Environmental protection E13c, E38
Storage and disposal D01, D02, D05, D09a, D10a, D12a

306   lenacil + triflusulfuron-methyl

A foliar and residual herbicide mixture for sugar beet
HRAC mode of action code: C1 + B

See also triflusulfuron-methyl

Products

Safari Lite WSB	DuPont	71.4:5.4% w/w	WG	12169

Uses
- Annual dicotyledons in *fodder beet* *(off-label)*, *sugar beet*
- Volunteer oilseed rape in *sugar beet*

Extension of Authorisation for Minor Use (EAMUs)
- *fodder beet* 20101934

Approval information
- Lenacil and triflusulfuron-methyl included in Annex I under EC Regulation 1107/2009

Efficacy guidance
- Best results obtained when weeds are small and growing actively
- Product recommended for use in a programme of treatments in tank mixture with a suitable herbicide partner to broaden the weed spectrum
- Ensure good spray cover of weeds. Apply when first weeds have emerged provided crop has reached cotyledon stage
- Susceptible plants cease to grow almost immediately after treatment and symptoms can be seen 5-10 d later
- Weed control may be reduced in very dry soil conditions
- Triflusulfuron-methyl is a member of the ALS-inhibitor group of herbicides

Restrictions
- Maximum number of treatments on sugar beet 3 per crop and do not apply more than 4 applications of any product containing triflusulfuron-methyl
- A maximum total dose of 60 g/ha triflusulfuron may only be applied every third year on the same field.
- Do not apply to crops suffering from stress caused by drought, water-logging, low temperatures, pest or disease attack, nutrient deficiency or any other factors affecting crop growth
- Do not use on Sands, stony or gravelly soils or on soils with more than 10% organic matter
- Do not apply when temperature above or likely to exceed 21°C on day of spraying or under conditions of high light intensity

Crop-specific information
- Latest use: before crop leaves meet between rows for sugar beet

Following crops guidance
- Only cereals may be sown in the same calendar yr as a treated sugar beet crop. Any crop may be sown in the following spring
- In the event of crop failure sow only sugar beet within 4 mth of treatment

Environmental safety
- Dangerous for the environment

FOR FULL CONDITIONS OF USE ALWAYS READ THE PRODUCT LABEL

- Very toxic to aquatic organisms
- Take extreme care to avoid drift onto broad-leaved plants outside the target area or onto surface waters or ditches, or land intended for cropping
- Spraying equipment should not be drained or flushed onto land planted, or to be planted, with trees or crops other than cereals and should be thoroughly cleansed after use - see label for instructions
- LERAP Category B

Hazard classification and safety precautions

Hazard Irritant, Dangerous for the environment
Transport code 9
Packaging group III
UN Number 3077
Risk phrases R36, R37, R38, R43, R50, R53a
Operator protection A, C; U05a, U08, U11, U19a, U20b, U22a
Environmental protection E15a, E16a, E38
Storage and disposal D01, D02, D09a, D12a

307   linuron

A contact and residual urea herbicide for various field crops
HRAC mode of action code: C2

See also 2,4-DB + linuron + MCPA
clomazone + linuron

Products

1	Afalon	Makhteshim	450 g/l	SC	14187
2	Datura	AgriChem BV	500 g/l	SC	14915
3	Nightjar	AgChem Access	450 g/l	SC	14656

Uses

- Annual dicotyledons in **asparagus, celeriac, chervil, dill, forest nurseries, lovage, ornamental plant production** [2]; **bulb onions** *(off-label)*, **combining peas, garlic** *(off-label)*, **leeks** *(off-label)*, **shallots** *(off-label)*, **spring field beans** [1, 3]; **carrots, parsley, parsnips, potatoes** [1-3]; **celeriac** *(off-label)*, **dandelions** *(off-label)*, **dwarf beans** *(off-label)*, **french beans** *(off-label)*, **game cover** *(off-label)*, **ginseng root** *(off-label)*, **herbs (see appendix 6)** *(off-label)*, **liquorice** *(off-label)*, **mallow (althaea spp.)** *(off-label)*, **nettle** *(off-label)*, **ornamental plant production** *(off-label)*, **parsley root** *(off-label)*, **runner beans** *(off-label)*, **valerian root** *(off-label)* [1]
- Annual grasses in **bulb onions** *(off-label)*, **celeriac** *(off-label)*, **dandelions** *(off-label)*, **dwarf beans** *(off-label)*, **french beans** *(off-label)*, **game cover** *(off-label)*, **garlic** *(off-label)*, **ginseng root** *(off-label)*, **herbs (see appendix 6)** *(off-label)*, **leeks** *(off-label)*, **liquorice** *(off-label)*, **mallow (althaea spp.)** *(off-label)*, **nettle** *(off-label)*, **ornamental plant production** *(off-label)*, **parsley root** *(off-label)*, **runner beans** *(off-label)*, **shallots** *(off-label)*, **valerian root** *(off-label)* [1]
- Annual meadow grass in **carrots, combining peas, parsley, parsnips, potatoes, spring field beans** [1]

Extension of Authorisation for Minor Use (EAMUs)

- **bulb onions** *20090875 expires 31 Dec 2013* [1], *20100649 expires 31 Dec 2013* [3]
- **celeriac** *20090885 expires 31 Dec 2013* [1]
- **dandelions** *20090861 expires 31 Dec 2013* [1]
- **dwarf beans** *20090859 expires 31 Dec 2013* [1]
- **french beans** *20090859 expires 31 Dec 2013* [1]
- **game cover** *20082802 expires 31 Dec 2013* [1]
- **garlic** *20090875 expires 31 Dec 2013* [1], *20100649 expires 31 Dec 2013* [3]
- **ginseng root** *20090861 expires 31 Dec 2013* [1]
- **herbs (see appendix 6)** *20090860 expires 31 Dec 2013* [1]
- **leeks** *20090864 expires 31 Dec 2013* [1], *20101135 expires 31 Dec 2013* [3]
- **liquorice** *20090861 expires 31 Dec 2013* [1]
- **mallow (althaea spp.)** *20090876 expires 31 Dec 2013* [1]

SEE SECTION 3 FOR PRODUCTS ALSO REGISTERED

- **nettle** *20090861 expires 31 Dec 2013* [1]
- **ornamental plant production** *20090877 expires 31 Dec 2013* [1]
- **parsley root** *20090876 expires 31 Dec 2013* [1]
- **runner beans** *20090859 expires 31 Dec 2013* [1]
- **shallots** *20090875 expires 31 Dec 2013* [1], *20100649 expires 31 Dec 2013* [3]
- **valerian root** *20090861 expires 31 Dec 2013* [1]

Approval information
- Linuron included in Annex I under EC Regulation 1107/2009
- Accepted by BBPA for use on malting barley
- Approval expiry 31 Dec 2013 [1, 3]

Efficacy guidance
- Many weeds controlled pre-emergence or post-emergence to 2-3 leaf stage, some (annual meadow grass, mayweed) only susceptible pre-emergence. See label for details
- Best results achieved by application to firm, moist soil of fine tilth
- Little residual effect on soil with more than 10% organic matter

Restrictions
- Maximum total dose equivalent to one full dose treatment
- Do not use on undersown cereals or crops grown on Sands or Very Light soils or soils heavier than Sandy Clay Loam or with more than 10% organic matter
- Do not apply to emerged crops of carrots, parsnips or parsley under stress
- Do not apply by hand-held sprayers

Crop-specific information
- Latest use: pre-emergence for most crops (products differ, see label for details).
- HI onions, garlic 8 wk; celeriac 12 wk; leeks 16 wk
- Apply to potatoes well earthed up to a rounded ridge pre-crop emergence and do not cultivate after spraying
- Apply to carrots at any time after drilling on organic soils, and within 4 d of drilling on other soils. Apply post-emergence as soon as weeds appear but after first rough leaf stage.
- Recommendations for parsnips, parsley and celery vary. See label for details

Following crops guidance
- Potatoes, carrots and parsnips may be planted at any time after application. Lettuce should not be grown within 12 mth of treatment. Transplanted brassicas may be grown from 3 mth after treatment

Environmental safety
- Dangerous for the environment
- Very toxic to aquatic organisms
- LERAP Category B

Hazard classification and safety precautions
Hazard Toxic, Dangerous for the environment
Transport code 9 [1, 2]
Packaging group III [1, 2]
UN Number 3082 [1, 2]
Risk phrases R22a, R48, R61 [1, 3]; R25b, R63 [2]; R40, R50, R53a, R62 [1-3]
Operator protection A, H [1-3]; C [3]; U05a, U13, U14, U15, U19a [1-3]; U08, U19c, U20b [2]
Environmental protection E15b, E16a, E38
Storage and disposal D02, D05, D09a, D11a, D12a [1-3]; D08 [1, 3]
Medical advice M04a

308   magnesium phosphide

A phosphine generating compound used to control insect pests in stored commodities

Products

Degesch Plates	Rentokil	56% w/w	GE	07603

Uses
- Insect pests in **stored grain**

Approval information
- Magnesium phosphide included in Annex I under EC Regulation 1107/2009
- Accepted by BBPA for use in stores for malting barley

Efficacy guidance
- Product acts as fumigant by releasing poisonous hydrogen phosphide gas on contact with moisture in the air
- Place plates on the floor or wall of the building or on the surface of the commodity. Exposure time varies depending on temperature and pest. See label

Restrictions
- Magnesium phosphide is subject to the Poisons Rules 1982 and the Poisons Act 1972. See Section 5 for more information
- Only to be used by professional operators trained in the use of magnesium phosphide and familiar with the precautionary measures to be observed. See label for full precautions

Environmental safety
- Highly flammable
- Prevent access to buildings under fumigation by livestock, pets and other non-target mammals and birds
- Dangerous to fish or other aquatic life. Do not contaminate surface waters or ditches with chemical or used container
- Keep in original container, tightly closed, in a safe place, under lock and key
- Do not allow plates or their spent residues to come into contact with food other than raw cereal grains
- Remove used plates after treatment. Do not bulk spent plates and residues: spontaneous ignition could result
- Keep livestock out of treated areas

Hazard classification and safety precautions
Hazard Very toxic, Highly flammable, Dangerous for the environment
Transport code 4.3
Packaging group I
UN Number 2011
Risk phrases R21, R26, R28, R50
Operator protection A, D, H; U01, U05b, U07, U13, U19a, U20a
Environmental protection E02a (4 h min); E02b, E13b, E34
Storage and disposal D01, D02, D05, D07, D09b, D11b
Medical advice M04a

309   maleic hydrazide

A pyridazine plant growth regulator suppressing sprout and bud growth

See also fatty acids + maleic hydrazide

Products
1 Fazor	Dow	60% w/w	GR	13679
2 Shaver	ChemSource	60% w/w	SG	15335
3 Source II	Chiltern	60% w/w	GR	13618

Uses
- Annual grasses in **amenity grassland** *(off-label)*, **hops** *(off-label)*, **ornamental specimens** *(off-label)*, **soft fruit** *(off-label)* [1]
- Growth regulation in **amenity vegetation** *(off-label)*, **carrots** *(off-label)*, **garlic** *(off-label)*, **parsnips** *(off-label)*, **shallots** *(off-label)* [1]; **bulb onions**, **potatoes** [1-3]

Extension of Authorisation for Minor Use (EAMUs)
- **amenity grassland** 20082790 expires 31 Dec 2013 [1]

SEE SECTION 3 FOR PRODUCTS ALSO REGISTERED

- **amenity vegetation** *20111726 expires 31 Dec 2013* [1]
- **carrots** *20081326 expires 31 Dec 2013* [1]
- **garlic** *20072795 expires 31 Dec 2013* [1]
- **hops** *20082848 expires 31 Dec 2013* [1]
- **ornamental specimens** *20082848 expires 31 Dec 2013* [1]
- **parsnips** *20081326 expires 31 Dec 2013* [1]
- **shallots** *20072795 expires 31 Dec 2013* [1]
- **soft fruit** *20082848 expires 31 Dec 2013* [1]

Approval information
- Maleic hydrazide included in Annex I under EC Regulation 1107/2009
- Approval expiry 31 Dec 2013 [1-3]

Efficacy guidance
- Apply to grass at any time of yr when growth active, best when growth starting in Apr-May and repeated when growth recommences
- Uniform coverage and dry weather necessary for effective results
- Accurate timing essential for good results on potatoes but rain or irrigation within 24 h may reduce effectiveness on onions and potatoes
- Mow 2-3 d before and 5-10 d after spraying for best results. Need for mowing reduced for up to 6 wk
- When used for suppression of volunteer potatoes treatment may also give some suppression of sprouting in store but separate treatment will be necessary if sprouting occurs

Restrictions
- Maximum number of treatments 2 per yr on amenity grass, land not intended for cropping and land adjacent to aquatic areas; 1 per crop on onions, potatoes and on or around tree trunks
- Do not apply in drought or when crops are suffering from pest, disease or herbicide damage. Do not treat fine turf or grass seeded less than 8 mth previously
- Do not treat potatoes within 3 wk of applying a haulm desiccant or if temperatures above 26°C
- Consult processor before use on potato crops for processing

Crop-specific information
- Latest use: 3 wk before haulm destruction for potatoes; before 50% necking for onions
- HI onions 1 wk; potatoes 3 wk
- Apply to onions at 10% necking and not later than 50% necking stage when the tops are still green
- Only treat onions in good condition and properly cured, and do not treat more than 2 wk before maturing. Treated onions may be stored until Mar but must then be removed to avoid browning
- Apply to second early or maincrop potatoes at least 3 wk before haulm destruction
- Only treat potatoes of good keeping quality; not on seed, first earlies or crops grown under polythene

Environmental safety
- Only apply to grass not to be used for grazing
- Do not use treated water for irrigation purposes within 3 wk of treatment or until concentration in water falls below 0.02 ppm
- Maximum permitted concentration in water 2 ppm
- Do not dump surplus product in water or ditch bottoms
- Avoid drift onto nearby vegetables, flowers or other garden plants

Hazard classification and safety precautions
Transport code 9
Packaging group III
UN Number 3077
Operator protection A, H; U08, U20b
Environmental protection E15a
Storage and disposal D09a, D11a

FOR FULL CONDITIONS OF USE ALWAYS READ THE PRODUCT LABEL

310   maleic hydrazide + pelargonic acid

A pyridazine plant growth regulator suppressing sprout and bud growth + a naturally occuring 9 carbon chain acid
HRAC mode of action code: Not classified

Products

Finalsan Plus   Certis   30:186.7 g/l   SL   15147

Uses
- Algae in *amenity vegetation, natural surfaces not intended to bear vegetation, ornamental plant production, permeable surfaces overlying soil*
- Annual and perennial weeds in *amenity vegetation, natural surfaces not intended to bear vegetation, ornamental plant production, permeable surfaces overlying soil*
- Moss in *amenity vegetation, natural surfaces not intended to bear vegetation, ornamental plant production, permeable surfaces overlying soil*

Approval information
- Maleic hydrazide and pelargonic acid included in Annex I under EC Regulation 1107/2009

Hazard classification and safety precautions
UN Number N/C
Operator protection A, C, H; U05a, U12, U15
Environmental protection E15a
Storage and disposal D01, D02
Medical advice M05a

311   maltodextrin

A polysaccharide used as a food additive and with activity against red spider mites
IRAC mode of action code: Not classified

Products

1 Eradicoat	Certis	598 g/l	SL	13724
2 Majestik	Certis	598 g/l	SL	14831

Uses
- Aphids in *all protected edible crops, all protected non-edible crops*
- Spider mites in *all protected edible crops, all protected non-edible crops*
- Whitefly in *all protected edible crops, all protected non-edible crops*

Approval information
- Maltodextrin is awaiting inclusion in Annex 1 under EC Regulation 1107/2009

Hazard classification and safety precautions
Hazard Irritant
UN Number N/C [2]
Risk phrases R43
Operator protection A, H; U05a, U09a, U14, U19a, U20b
Storage and disposal D01, D02, D09a, D10b, D12a

312   mancozeb

A protective dithiocarbamate fungicide for potatoes and other crops
FRAC mode of action code: M3

See also ametoctradin + mancozeb
benalaxyl + mancozeb
benthiavalicarb-isopropyl + mancozeb
chlorothalonil + mancozeb
cymoxanil + mancozeb
dimethomorph + mancozeb
fenamidone + mancozeb

Products

1	Dithane 945	Interfarm	80% w/w	WP	15715
2	Karamate Dry Flo Neotec	Landseer	75% w/w	WG	14632
3	Laminator 75 WG	Interfarm	75% w/w	WG	15667
4	Laminator Flo	Interfarm	455 g/l	SC	15345
5	Malvi	AgChem Access	75% w/w	WG	14981
6	Manzate 75 WG	United Phosphorus	75% w/w	WG	15052
7	Penncozeb 80 WP	United Phosphorus	80% w/w	WP	14718
8	Penncozeb WDG	United Phosphorus	75% w/w	WG	14719
9	Quell Flo	Interfarm	455 g/l	SC	15237

Uses

* Blight in **potatoes** [1, 3-9]; **tomatoes (outdoor)** [3, 5]
* Botrytis in **flower bulbs** [6]
* Brown rust in **durum wheat** *(useful control)*, **spring wheat** *(useful control)*, **winter wheat** *(useful control)* [1]; **spring wheat**, **winter wheat** [4, 9]
* Disease control in **amenity vegetation**, **bulb onions**, **carrots**, **courgettes**, **forest nurseries** *(off-label)*, **hops** *(off-label)*, **ornamental plant production**, **parsnips**, **shallots**, **soft fruit** *(off-label)*, **top fruit** *(off-label)* [2]; **apples**, **table grapes** [3, 5]; **farm forestry** *(off-label)*, **ornamental plant production** *(off-label)* [1]; **wine grapes** [2, 3, 5]
* Downy mildew in **bulb onions** [6-8]; **lettuce** *(off-label)* [2]
* Early blight in **potatoes** [1]
* Scab in **apples** [2, 6-8]; **pears** [2]
* Septoria leaf blotch in **durum wheat** *(reduction)* [1]; **spring wheat**, **winter wheat** [4, 7-9]; **spring wheat** *(reduction)*, **winter wheat** *(reduction)* [1, 6]
* Sooty moulds in **spring wheat**, **winter wheat** [4, 9]
* Yellow rust in **winter wheat** [4, 9]

Extension of Authorisation for Minor Use (EAMUs)

* **farm forestry** *20122057* [1]
* **forest nurseries** *20122047 expires 14 Oct 2013* [2]
* **hops** *20122047 expires 14 Oct 2013* [2]
* **lettuce** *20101082 expires 14 Oct 2013* [2]
* **ornamental plant production** *20122057* [1]
* **soft fruit** *20122047 expires 14 Oct 2013* [2]
* **top fruit** *20122047 expires 14 Oct 2013* [2]

Approval information

* Mancozeb included in Annex I under EC Regulation 1107/2009
* Accepted by BBPA for use on malting barley
* Approval expiry 18 Oct 2013 [6, 8]
* Approval expiry 14 Oct 2013 [7]

Efficacy guidance

* Mancozeb is a protectant fungicide and will give moderate control, suppression or reduction of the cereal diseases listed if treated before they are established but in many cases mixture with carbendazim is essential to achieve satisfactory results. See labels for details

FOR FULL CONDITIONS OF USE ALWAYS READ THE PRODUCT LABEL

- May be recommended for suppression or control of mildew in cereals depending on product and tank mix. See label for details

Restrictions
- Maximum number of treatments varies with crop and product used - check labels for details
- Check labels for minimum interval that must elapse between treatments
- On protected lettuce only 2 post-planting applications of mancozeb or of any combination of products containing EBDC fungicide (mancozeb, maneb, thiram, zineb) either as a spray or a dust are permitted within 2 wk of planting out and none thereafter.
- Avoid treating wet cereal crops or those suffering from drought or other stress
- Keep dry formulations away from fire and sparks
- Use dry formulations immediately. Do not store

Crop-specific information
- Latest use: before early milk stage (GS 73) for cereals; before 6 true leaf stage and before 31 Dec for winter oilseed rape.
- HI potatoes 7 d; outdoor lettuce 14 d; protected lettuce 21 d; apples, blackcurrants 28 d
- Apply to potatoes before haulm meets across rows (usually mid-Jun) or at earlier blight warning, and repeat every 7-14 d depending on conditions and product used (see label)
- May be used on potatoes up to desiccation of haulm
- On oilseed rape apply as soon as disease develops between cotyledon and 5-leaf stage (GS 1,0-1,5)
- Apply to cereals from 4-leaf stage to before early milk stage (GS 71). Recommendations vary, see labels for details
- Treat winter oilseed rape before 6 true leaf stage (GS 1,6) and before 31 Dec

Environmental safety
- Dangerous for the environment
- Very toxic to aquatic organisms
- Harmful to fish or other aquatic life. Do not contaminate surface waters or ditches with chemical or used container
- Do not empty into drains
- LERAP Category B [2]

Hazard classification and safety precautions
Hazard Harmful [3, 7, 8]; Irritant [1, 2, 4-6, 9]; Dangerous for the environment [1-3, 5-9]
Transport code 9
Packaging group III
UN Number 3077 [1-3, 5-8]; 3082 [4, 9]
Risk phrases R36 [5]; R37 [1-3, 6, 9]; R43 [1-4, 6-9]; R50, R53a [1-3, 5-9]; R63 [3, 7, 8]
Operator protection A [1-6, 8, 9]; C [5]; D [1-3, 6, 8]; H [1, 8]; U05a [1-6, 8, 9]; U08 [4, 5, 9]; U11 [5]; U14 [1-3, 6-9]; U19a [5, 7, 8]; U20b [1-6, 9]
Environmental protection E13c [4]; E15a [1-3, 5, 6, 9]; E15b [7, 8]; E16a [2]; E19b [8]; E34 [4, 5, 9]; E38 [1-3, 5-9]
Storage and disposal D01, D12a [1-3, 5-9]; D02 [1-9]; D05 [1-6, 9]; D07 [4, 5]; D09a [1-6, 8, 9]; D10a [5]; D11a [1-4, 6, 9]
Medical advice M05a [7, 8]

313 mancozeb + metalaxyl-M

A systemic and protectant fungicide mixture
FRAC mode of action code: M3 + 4

See also metalaxyl-M

Products
| Fubol Gold WG | Syngenta | 64:4% w/w | WG | 14605 |

Uses
- Blight in **potatoes**

SEE SECTION 3 FOR PRODUCTS ALSO REGISTERED

- Disease control in **forest nurseries** *(off-label)*, **herbs (see appendix 6)** *(off-label)*, **hops** *(off-label)*, **lettuce** *(off-label)*, **ornamental plant production** *(off-label)*, **protected forest nurseries** *(off-label)*, **protected hops** *(off-label)*, **protected ornamentals** *(off-label)*, **protected soft fruit** *(off-label)*, **soft fruit** *(off-label)*
- Downy mildew in **bulb onions** *(useful control)*, **poppies for morphine production** *(off-label)*, **protected herbs (see appendix 6)** *(off-label)*, **rhubarb** *(off-label)*, **salad onions** *(off-label)*, **shallots** *(useful control)*
- Phytophthora fruit rot in **apple orchards** *(off-label)*
- White blister in **cabbages** *(off-label)*

Extension of Authorisation for Minor Use (EAMUs)
- **apple orchards** *20101734*
- **cabbages** *20101733*
- **forest nurseries** *20120217*
- **herbs (see appendix 6)** *20101779*
- **hops** *20120217*
- **lettuce** *20101779*
- **ornamental plant production** *20120217*
- **poppies for morphine production** *20102751*
- **protected forest nurseries** *20120217*
- **protected herbs (see appendix 6)** *20111498*
- **protected hops** *20120217*
- **protected ornamentals** *20120217*
- **protected soft fruit** *20120217*
- **rhubarb** *20101735*
- **salad onions** *20101736*
- **soft fruit** *20120217*

Approval information
- Mancozeb and metalaxyl-M included in Annex I under EC Regulation 1107/2009

Efficacy guidance
- Commence potato blight programme before risk of infection occurs as crops begin to meet along the rows and repeat every 7-14 d according to blight risk. Do not exceed a 14 d interval between sprays
- If infection risk conditions occur earlier than the above growth stage commence spraying potatoes immediately
- Complete the potato blight programme using a protectant fungicide starting no later than 10 d after the last phenylamide spray. At least 2 such sprays should be applied
- To minimise the likelihood of development of resistance these products should be used in a planned Resistance Management strategy. See Section 5 for more information

Crop-specific information
- Latest use: before end of active potato haulm growth or before end Aug, whichever is earlier
- HI 7 d for potatoes
- After treating early potatoes destroy and remove any remaining haulm after harvest to minimise blight pressure on neighbouring maincrop potatoes

Environmental safety
- Dangerous for the environment
- Very toxic to aquatic organisms
- Do not harvest crops for human consumption for at least 7 d after final application
- LERAP Category B

Hazard classification and safety precautions
    Hazard Harmful, Dangerous for the environment
    Transport code 9
    Packaging group III
    UN Number 3077
    Risk phrases R37, R43, R50, R53a, R63
    Operator protection A; U05a, U08, U20b

FOR FULL CONDITIONS OF USE ALWAYS READ THE PRODUCT LABEL

Environmental protection E15a, E16a, E34, E38
Consumer protection C02a (7 d)
Storage and disposal D01, D02, D05, D09a, D11a, D12a

314   mancozeb + zoxamide

A protectant fungicide mixture for potatoes
FRAC mode of action code: M3 + 22

See also zoxamide

Products

1 Electis 75WG	Gowan	66.7:8.3% w/w	WG	14195
2 Roxam 75WG	Gowan	66.7:8.3% w/w	WG	14191

Uses
• Blight in **potatoes**

Approval information
• Mancozeb and zoxamide included in Annex I under EC Regulation 1107/2009

Efficacy guidance
• Apply as protectant spray on potatoes immediately risk of blight in district or as crops begin to meet along the rows and repeat every 7-14 d according to blight risk
• Do not use if potato blight present in crop. Products are not curative
• Spray irrigated potato crops as soon as possible after irrigation once the crop leaves are dry

Restrictions
• Maximum number of treatments 10 per crop for potatoes

Crop-specific information
• HI 7 d for potatoes

Environmental safety
• Dangerous for the environment
• Very toxic to aquatic organisms
• Keep away from fire and sparks
• LERAP Category B

Hazard classification and safety precautions
Hazard Irritant, Dangerous for the environment
Transport code 9
Packaging group III
UN Number 3077
Risk phrases R37, R43, R50, R53a
Operator protection A, H; U05a, U14, U20b
Environmental protection E15a, E16a, E16b, E38
Storage and disposal D01, D02, D05, D09a, D11a, D12a

315   mandipropamid

A mandelamide fungicide for the control of potato blight
FRAC mode of action code: 40

Products

1 Pergardo Uni	Fargro	250 g/l	SC	15734
2 Revus	Syngenta	250 g/l	SC	13484

Uses
• Blight in **potatoes** [2]

SECTION 2

- Disease control in **chard, chicory, cress, endives, frise, herbs (see appendix 6), lamb's lettuce, land cress, lettuce, protected chard, protected chicory, protected cress, protected endives, protected frise, protected herbs (see appendix 6), protected lamb's lettuce, protected land cress, protected lettuce, protected purslane, protected red mustard, protected rocket, protected salad greens, protected spinach, protected spinach beet, purslane, radicchio, red mustard, rocket, salad greens, spinach, spinach beet** [1]; **ornamental plant production** *(off-label)* [2]
- Downy mildew in **chard, chicory, cress, endives, frise, herbs (see appendix 6), lamb's lettuce, land cress, lettuce, protected chard, protected chicory, protected cress, protected endives, protected frise, protected herbs (see appendix 6), protected lamb's lettuce, protected land cress, protected lettuce, protected ornamentals** *(off-label)*, **protected purslane, protected red mustard, protected rocket, protected salad greens, protected spinach, protected spinach beet, purslane, radicchio, red mustard, rocket, salad greens, spinach, spinach beet** [1]; **ornamental plant production** *(off-label)* [1, 2]

Extension of Authorisation for Minor Use (EAMUs)
- **ornamental plant production** *20121605* [1], *20082867* [2], *20120487* [2]
- **protected ornamentals** *20121605* [1]

Approval information
- Mandipropamid is awaiting inclusion in Annex 1 under EC Regulation 1107/2009

Efficacy guidance
- Mandipropamid acts preventatively by preventing spore germination and inhibiting mycelial growth during incubation. Apply immediately after blight warning or as soon as local conditions favour disease development but before blight enters the crop
- Spray at 7-10 d intervals reducing the interval as blight risk increases
- Spray programme should include a complete haulm desiccant to prevent tuber infection at harvest
- Eliminate other potential infection sources
- To minimise the likelihood of development of resistance this product should be used in a planned Resistance Management strategy. See Section 5 for more information
- See label for details of tank mixtures that may be used as part of a resistance management strategy

Restrictions
- Maximum number of treatments 4 per crop. Do not apply more than 3 treatments of this, or any other fungicide in the same resistance category, consecutively

Crop-specific information
- HI 3 d for potatoes

Hazard classification and safety precautions
UN Number N/C
Risk phrases R52, R53a
Operator protection A, H; U05a, U20b
Environmental protection E15b, E34, E38
Storage and disposal D01, D02, D05, D09a, D10c, D12a
Medical advice M03 [1, 2]; M05a [2]

316   maneb

A protectant dithiocarbamate fungicide
FRAC mode of action code: M3

See also carbendazim + maneb

Products

1	Trimangol 80	United Phosphorus	80% w/w	WP	06871
2	Trimangol WDG	United Phosphorus	75	WG	06992
3	Trimazone	United Phosphorus	80% w/w	WP	14726

FOR FULL CONDITIONS OF USE ALWAYS READ THE PRODUCT LABEL

Uses

- Blight in **potatoes** [1-3]
- Brown rust in **spring barley**, **winter barley** [1, 2]; **spring wheat**, **winter wheat** [1-3]
- Disease control in **grass seed crops** (off-label) [2]; **protected tomatoes** [1, 2]
- Downy mildew in **bulb onions** [3]
- Glume blotch in **spring wheat**, **winter wheat** [1-3]
- Net blotch in **spring barley**, **winter barley** [1, 2]
- Rhynchosporium in **spring barley**, **winter barley** [1, 2]
- Scab in **apples** [1-3]
- Septoria leaf blotch in **spring wheat**, **winter wheat** [1-3]
- Sooty moulds in **spring barley**, **winter barley** [1, 2]; **spring wheat**, **winter wheat** [1-3]
- Yellow rust in **spring barley**, **winter barley** [1, 2]; **spring wheat**, **winter wheat** [1-3]

Extension of Authorisation for Minor Use (EAMUs)

- **grass seed crops** 20100031 [2]

Approval information

- Maneb included in Annex I under EC Regulation 1107/2009
- Accepted by BBPA on malting barley
- Approved for aerial application [1, 2]. See Section 5 for more information
- Approval expiry 14 Oct 2013 [3]

Efficacy guidance

- On potatoes first application should be made before blight infection occurs, and further applications made to protect new growth
- Maneb is a protectant fungicide and will give moderate control, suppression or reduction of the cereal diseases listed but mixture with carbendazim is essential to achieve satisfactory results
- Best results on cereals obtained from a programme of a preventative treatment before disease established, an early application at about first node stage (GS 31), a late application after flag leaf emergence (GS 37) and during ear emergence before watery ripe stage (GS 71)

Restrictions

- Maximum number of treatments 2 per crop on cereals; not specified on potatoes
- Do not apply if frost or rain expected, if crop wet or suffering from drought or physical or chemical stress
- A minimum of 7 d must elapse between applications to wheat and barley, and 10 d for potatoes

Crop-specific information

- Latest use: before grain milky-ripe (GS 73) for wheat, barley; before flag leaf sheath opening (GS 45) for barley; HI potatoes 7 d
- Apply to potatoes before blight infection occurs, at blight warning or before haulms meet in row and repeat every 10-14 d

Hazard classification and safety precautions

Hazard Harmful [3]; Irritant [1]; Dangerous for the environment [1, 3]
Transport code 9
Packaging group III
UN Number 3077
Risk phrases R37, R43, R50, R53a [1, 3]; R63 [3]
Operator protection A, D [1, 3]; H [3]; U05a, U14, U20b [1, 3]
Environmental protection E15a [1]; E15b [3]; E38 [1, 3]
Storage and disposal D01, D02, D09a, D11a, D12a [1, 3]
Medical advice M05a [3]

SEE SECTION 3 FOR PRODUCTS ALSO REGISTERED

317   MCPA

A translocated phenoxycarboxylic acid herbicide for cereals and grassland
HRAC mode of action code: O

See also 2,4-D + dicamba + MCPA + mecoprop-P
2,4-D + dichlorprop-P + MCPA + mecoprop-P
2,4-D + MCPA
2,4-DB + linuron + MCPA
2,4-DB + MCPA
bentazone + MCPA + MCPB
bifenox + MCPA + mecoprop-P
clopyralid + 2,4-D + MCPA
clopyralid + diflufenican + MCPA
clopyralid + fluroxypyr + MCPA
dicamba + dichlorprop-P + ferrous sulphate + MCPA
dicamba + dichlorprop-P + MCPA
dicamba + MCPA + mecoprop-P
dichlorprop-P + ferrous sulphate + MCPA
dichlorprop-P + MCPA
dichlorprop-P + MCPA + mecoprop-P
ferrous sulphate + MCPA + mecoprop-P
ferrous sulphate monohydrate + MCPA + mecoprop-P

Products

1	Agrichem MCPA 500	AgriChem BV	500 g/l	SL	15836
2	Agritox	Nufarm UK	500 g/l	SL	14894
3	Agroxone	Headland	500 g/l	SL	14909
4	Easel	Nufarm UK	750 g/l	SL	15548
5	Headland Spear	Headland	500 g/l	SL	14910
6	HY-MCPA	Agrichem	500 g/l	SL	14927
7	Larke	Nufarm UK	750 g/l	SL	14914
8	MCPA 25%	Nufarm UK	250 g/l	SL	14893
9	MCPA 50	United Phosphorus	500 g/l	SL	14908
10	Nufarm MCPA 750	Nufarm UK	750 g/l	SL	14892

Uses

- Annual and perennial weeds in **grass seed crops**, **grassland** [1, 2, 4, 7, 8, 10]
- Annual dicotyledons in **farm forestry** *(off-label)*, **game cover** *(off-label)* [3]; **grass seed crops**, **grassland** [3, 5, 6, 9]; **listed cereals u/sown with grass only**, **listed cereals u/sown with red clover** [9]; **miscanthus** *(off-label)* [5]; **spring barley**, **spring oats**, **spring wheat**, **winter barley**, **winter oats**, **winter wheat** [1-10]; **spring rye**, **winter rye** [1, 2, 4, 7-10]; **undersown barley**, **undersown oats**, **undersown rye**, **undersown wheat** [1, 2, 4, 6-8, 10]; **undersown barley** *(red clover or grass)*, **undersown wheat** *(red clover or grass)* [3, 5]
- Charlock in **spring barley**, **spring oats**, **spring wheat**, **winter barley**, **winter oats**, **winter wheat** [1-10]; **spring rye** [1, 2, 4, 7, 8, 10]; **undersown barley** *(red clover or grass)*, **undersown wheat** *(red clover or grass)* [3, 5]; **winter rye** [1, 2, 4, 7-10]
- Fat hen in **spring barley**, **spring oats**, **spring wheat**, **winter barley**, **winter oats**, **winter wheat** [1-10]; **spring rye** [1, 2, 4, 7, 8, 10]; **undersown barley** *(red clover or grass)*, **undersown wheat** *(red clover or grass)* [3, 5]; **winter rye** [1, 2, 4, 7-10]
- Hemp-nettle in **spring barley**, **spring oats**, **spring wheat**, **winter barley**, **winter oats**, **winter wheat** [1-10]; **spring rye** [1, 2, 4, 7, 8, 10]; **undersown barley** *(red clover or grass)*, **undersown wheat** *(red clover or grass)* [3, 5]; **winter rye** [1, 2, 4, 7-10]
- Perennial dicotyledons in **farm forestry** *(off-label)*, **game cover** *(off-label)* [3]; **grass seed crops**, **undersown barley** *(red clover or grass)*, **undersown wheat** *(red clover or grass)* [3, 5]; **grassland** [3, 5, 9]; **listed cereals u/sown with grass only**, **listed cereals u/sown with red clover** [9]; **miscanthus** *(off-label)* [5]; **spring barley**, **spring oats**, **spring wheat**, **winter barley**, **winter oats**, **winter wheat** [1-10]; **spring rye**, **winter rye** [1, 2, 4, 7-10]

- Wild radish in **spring barley**, **spring oats**, **spring wheat**, **winter barley**, **winter oats**, **winter wheat** [1-10]; **spring rye** [1, 2, 4, 7, 8, 10]; **undersown barley** *(red clover or grass)*, **undersown wheat** *(red clover or grass)* [3, 5]; **winter rye** [1, 2, 4, 7-10]

Extension of Authorisation for Minor Use (EAMUs)
- **farm forestry** *20122061* [3]
- **game cover** *20122061* [3]
- **miscanthus** *20122050* [5]

Approval information
- MCPA included in Annex I under EC Regulation 1107/2009
- Accepted by BBPA for use on malting barley

Efficacy guidance
- Best results achieved by application to weeds in seedling to young plant stage under good growing conditions when crop growing actively
- Spray perennial weeds in grassland before flowering. Most susceptible growth stage varies between species. See label for details
- Do not spray during cold weather, drought, if rain or frost expected or if crop wet

Restrictions
- Maximum number of treatments normally 1 per crop or yr except grass (2 per yr) for some products. See label
- Do not treat grass within 3 mth of germination and preferably not in the first yr of a direct sown ley or after reseeding
- Do not use on cereals before undersowing
- Do not roll, harrow or graze for a few days before or after spraying; see label
- Do not use on grassland where clovers are an important part of the sward
- Do not use on any crop suffering from stress or herbicide damage
- Avoid spray drift onto nearby susceptible crops

Crop-specific information
- Latest use: before 1st node detectable (GS 31) for cereals; 4-6 wk before heading for grass seed crops; before crop 15-25 cm high for linseed
- Apply to winter cereals in spring from fully tillered, leaf sheath erect stage to before first node detectable (GS 31)
- Apply to spring barley and wheat from 5-leaves unfolded (GS 15), to oats from 1-leaf unfolded (GS 11) to before first node detectable (GS 31)
- Apply to cereals undersown with grass after grass has 2-3 leaves unfolded
- Recommendations for crops undersown with legumes vary. Red clover may withstand low doses after 2-trifoliate leaf stage, especially if shielded by taller weeds but white clover is more sensitive. See label for details
- Apply to grass seed crops from 2-3 leaf stage to 5 wk before head emergence
- Temporary wilting may occur on linseed but without long term effects

Following crops guidance
- Do not direct drill brassicas or legumes within 6 wk of spraying grassland

Environmental safety
- Harmful to aquatic organisms
- MCPA is active at low concentrations. Take extreme care to avoid drift onto neighbouring crops, especially beet crops, brassicas, most market garden crops including lettuce and tomatoes under glass, pears and vines
- Keep livestock out of treated areas for at least 2 wk and until foliage of poisonous weeds such as ragwort has died and become unpalatable
- LERAP Category B

Hazard classification and safety precautions
    Hazard Harmful [1-10]; Dangerous for the environment [3-6]
    Transport code 9 [9]
    Packaging group III [9]
    UN Number 3082 [9]; N/C [1-8, 10]

SECTION 2

SEE SECTION 3 FOR PRODUCTS ALSO REGISTERED

Risk phrases R20, R21 [3, 9]; R22a [3-10]; R37 [1, 2, 4-8, 10]; R41 [1, 2, 4-10]; R50 [3-7, 10]; R51 [8]; R53a [3-8, 10]

Operator protection A, C; U05a, U08, U11 [1-10]; U09a [3]; U12 [4]; U15 [5]; U20a [9]; U20b [1-8, 10]

Environmental protection E06a [6] (2 wk); E07a [1-5, 7-10]; E15a [1-3, 6-10]; E15b, E38 [4, 5]; E16a, E34 [1-10]; E40b [5]

Storage and disposal D01, D02, D05, D09a [1-10]; D10a [1-4, 6-8, 10]; D10b [5]; D10c [9]; D12a [4]

Medical advice M03 [1-10]; M05a [6, 9]

318   MCPA + mecoprop-P

A translocated selective herbicide for amenity grass
HRAC mode of action code: O + O

See also mecoprop-P

Products

1	Cleanrun Pro	Everris Ltd	0.49:0.29% w/w	GR	15828
2	Greenmaster Extra	Everris Ltd	0.49:0.29 % w/w	GR	15817

Uses
- Annual dicotyledons in *managed amenity turf*
- Perennial dicotyledons in *managed amenity turf*

Approval information
- MCPA and mecoprop-P included in Annex I under EC Regulation 1107/2009

Efficacy guidance
- Apply from Apr to Sep, when weeds growing actively and have large leaf area available for chemical absorption

Restrictions
- The total amount of mecoprop-P applied in a single year must not exceed the maximum total dose approved for any single product for use on turf
- Avoid contact with cultivated plants
- Do not use first 4 mowings as compost or mulch unless composted for 6 mth
- Do not treat newly sown or turfed areas for at least 6 mth
- Do not reseed bare patches for 8 wk after treatment
- Do not apply when heavy rain expected or during prolonged drought. Irrigate after 1-2 d unless rain has fallen
- Do not mow within 2-3 d of treatment
- Treat areas planted with bulbs only after the foliage has died down
- Avoid walking on treated areas until it has rained or irrigation has been applied

Crop-specific information
- Granules contain NPK fertilizer to encourage grass growth

Environmental safety
- Take extreme care to avoid drift onto neighbouring crops, especially beet crops, brassicas, most market garden crops including lettuce and tomatoes under glass, pears and vines
- Harmful to fish or other aquatic life. Do not contaminate surface waters or ditches with chemical or used container
- Keep livestock out of treated areas for at least 2 wk and until foliage of any poisonous weeds such as ragwort has died and become unpalatable
- Some pesticides pose a greater threat of contamination of water than others and mecoprop-P is one of these pesticides. Take special care when applying mecoprop-P near water and do not apply if heavy rain is forecast

Hazard classification and safety precautions
   UN Number N/C

FOR FULL CONDITIONS OF USE ALWAYS READ THE PRODUCT LABEL

Operator protection A, C, H, M; U20b
Environmental protection E07a, E13c, E19b
Storage and disposal D01, D09a, D12a
Medical advice M05a

319   MCPB

A translocated phenoxycarboxylic acid herbicide
HRAC mode of action code: O

See also bentazone + MCPA + MCPB
bentazone + MCPB
MCPA + MCPB

Products

1	Bellmac Straight	United Phosphorus	400 g/l	SL	14448
2	Butoxone	Headland	400 g/l	SL	14406
3	Tropotox	Nufarm UK	400 g/l	SL	14450

Uses

- Annual dicotyledons in **combining peas**, **vining peas** [1-3]; **game cover** *(off-label)* [1, 2]
- Docks in **combining peas**, **vining peas** [3]
- Perennial dicotyledons in **combining peas**, **game cover** *(off-label)*, **vining peas** [1, 2]
- Thistles in **combining peas**, **vining peas** [3]

Extension of Authorisation for Minor Use (EAMUs)

- **game cover** *20122052* [1], *20122053* [2]

Approval information

- MCPB included in Annex I under EC Regulation 1107/2009

Efficacy guidance

- Best results achieved by spraying young seedling weeds in good growing conditions
- Best results on perennials by spraying before flowering
- Effectiveness may be reduced by rain within 12 h, by very cold or dry conditions

Restrictions

- Maximum number of treatments 1 per crop or yr.
- Do not roll or harrow for 7-10 d before or after treatment (check label)

Crop-specific information

- Latest use: first node detectable stage (GS 31) for cereals; before flower buds appear in terminal leaf (GS 201) for peas; before flower buds form for clover; before weeds damaged by frost for cane and bush fruit
- Apply to undersown cereals from 2-leaves unfolded to first node detectable (GS 12-31), and after first trifoliate leaf stage of clover
- Red clover seedlings may be temporarily damaged but later growth is normal
- Apply to white clover seed crops in Mar to early Apr, not after mid-May, and allow 3 wk before cutting and closing up for seed
- Apply to peas from 3-6 leaf stage but before flower bud detectable (GS 103-201). Consult PGRO (see Appendix 2) or label for information on susceptibility of cultivars.
- Do not use on leguminous crops not mentioned on the label
- Apply to cane and bush fruit after harvest and after shoot growth ceased but before weeds are damaged by frost, usually in late Aug or Sep; direct spray onto weeds as far as possible

Environmental safety

- Harmful to aquatic organisms
- Harmful to fish or other aquatic life. Do not contaminate surface waters or ditches with chemical or used container
- Keep livestock out of treated areas until foliage of any poisonous weeds such as ragwort has died and become unpalatable

SEE SECTION 3 FOR PRODUCTS ALSO REGISTERED

- Take extreme care to avoid drift onto neighbouring sensitive crops

Hazard classification and safety precautions
   Hazard Harmful [1-3]; Dangerous for the environment [2, 3]
   Transport code 9 [3]
   Packaging group III [3]
   UN Number 3082 [3]; N/C [1, 2]
   Risk phrases R22a, R38, R41 [1-3]; R51 [3]; R52 [1]; R53a [1, 3]
   Operator protection A, C; U05a, U08, U20b [1-3]; U11, U14, U15 [1, 2]; U19a [3]
   Environmental protection E07a, E34 [1-3]; E13c [1, 2]; E15a, E38 [3]
   Storage and disposal D01, D02, D09a [1-3]; D05 [1, 2]; D10b [2, 3]; D10c [1]
   Medical advice M03, M05a

320   mecoprop-P

A translocated phenoxycarboxylic acid herbicide for cereals and grassland
HRAC mode of action code: O

See also 2,4-D + dicamba + MCPA + mecoprop-P
   2,4-D + dichlorprop-P + MCPA + mecoprop-P
   2,4-D + mecoprop-P
   bifenox + MCPA + mecoprop-P
   bromoxynil + ioxynil + mecoprop-P
   carfentrazone-ethyl + mecoprop-P
   dicamba + MCPA + mecoprop-P
   dicamba + mecoprop-P
   dichlorprop-P + MCPA + mecoprop-P
   diflufenican + mecoprop-P
   ferrous sulphate + MCPA + mecoprop-P
   ferrous sulphate monohydrate + MCPA + mecoprop-P
   fluroxypyr + mecoprop-P
   MCPA + mecoprop-P

Products

1	Clenecorn Super	Nufarm UK	600 g/l	SL	14628
2	Compitox Plus	Nufarm UK	600 g/l	SL	14390
3	Duplosan KV	Nufarm UK	600 g/l	SL	13971
4	Headland Charge	Headland	600 g/l	SL	14394
5	Isomec	Nufarm UK	600 g/l	SL	14385
6	Optica	Headland	600 g/l	SL	14373

Uses
- Annual dicotyledons in **amenity grassland**, **durum wheat** *(off-label)*, **spring durum wheat** *(off-label)*, **spring rye** *(off-label)*, **spring triticale** *(off-label)*, **triticale** *(off-label)*, **winter rye** *(off-label)* [1-3, 5]; **game cover** *(off-label)*, **miscanthus** *(off-label)* [3, 6]; **grass seed crops**, **managed amenity turf**, **spring barley**, **spring oats**, **spring wheat**, **winter barley**, **winter oats**, **winter wheat** [1-6]; **permanent grassland**, **rotational grass** [4, 6]
- Chickweed in **amenity grassland** [1-3, 5]; **grass seed crops**, **managed amenity turf**, **spring barley**, **spring oats**, **spring wheat**, **winter barley**, **winter oats**, **winter wheat** [1-6]; **permanent grassland**, **rotational grass** [4, 6]
- Cleavers in **amenity grassland** [1-3, 5]; **grass seed crops**, **managed amenity turf**, **spring barley**, **spring oats**, **spring wheat**, **winter barley**, **winter oats**, **winter wheat** [1-6]; **permanent grassland**, **rotational grass** [4, 6]
- Perennial dicotyledons in **amenity grassland**, **grass seed crops**, **managed amenity turf**, **spring barley**, **spring oats**, **spring wheat**, **winter barley**, **winter oats**, **winter wheat** [1-3, 5]; **game cover** *(off-label)*, **miscanthus** *(off-label)* [3]

Extension of Authorisation for Minor Use (EAMUs)
- **durum wheat** *20101042* [1], *20093214* [2], *20093211* [3], *20101043* [5]
- **game cover** *20111129* [3], *20111125* [6]

FOR FULL CONDITIONS OF USE ALWAYS READ THE PRODUCT LABEL

- *miscanthus* *20111129* [3], *20111125* [6]
- *spring durum wheat* *20101042* [1], *20093214* [2], *20093211* [3], *20101043* [5]
- *spring rye* *20101042* [1], *20093214* [2], *20093211* [3], *20101043* [5]
- *spring triticale* *20101042* [1], *20093214* [2], *20093211* [3], *20101043* [5]
- *triticale* *20101042* [1], *20093214* [2], *20093211* [3], *20101043* [5]
- *winter rye* *20101042* [1], *20093214* [2], *20093211* [3], *20101043* [5]

Approval information
- Mecoprop-P included in Annex I under EC Regulation 1107/2009
- Accepted by BBPA for use on malting barley

Efficacy guidance
- Best results achieved by application to seedling weeds which have not been frost hardened, when soil warm and moist and expected to remain so for several days

Restrictions
- Maximum number of treatments normally 1 per crop for spring cereals and 1 per yr for newly sown grass; 2 per crop or yr for winter cereals and grass crops. Check labels for details
- The total amount of mecoprop-P applied in a single yr must not exceed the maximum total dose approved for any single product for the crop/situation
- Do not spray cereals undersown with clovers or legumes or to be undersown with legumes or grasses
- Do not spray grass seed crops within 5 wk of seed head emergence
- Do not spray crops suffering from herbicide damage or physical stress
- Do not spray during cold weather, periods of drought, if rain or frost expected or if crop wet
- Do not roll or harrow for 7 d before or after treatment

Crop-specific information
- Latest use: generally before 1st node detectable (GS 31) for spring cereals and before 3rd node detectable (GS 33) for winter cereals, but individual labels vary; 5 wk before emergence of seed head for grass seed crops
- Spray winter cereals from 1 leaf stage in autumn up to and including first node detectable in spring (GS 10-31) or up to second node detectable (GS 32) if necessary. Apply to spring cereals from first fully expanded leaf stage (GS 11) but before first node detectable (GS 31)
- Spray cereals undersown with grass after grass starts to tiller
- Spray newly sown grass leys when grasses have at least 3 fully expanded leaves and have begun to tiller. Any clovers will be damaged

Environmental safety
- Harmful to aquatic organisms
- Harmful to fish or other aquatic life. Do not contaminate surface waters or ditches with chemical or used container
- Keep livestock out of treated areas for at least 2 wk and until foliage of any poisonous weeds, such as ragwort, has died and become unpalatable
- Take extreme care to avoid drift onto neighbouring crops, especially beet crops, brassicas, most market garden crops including lettuce and tomatoes under glass, pears and vines
- Some pesticides pose a greater threat of contamination of water than others and mecoprop-P is one of these pesticides. Take special care when applying mecoprop-P near water and do not apply if heavy rain is forecast

Hazard classification and safety precautions
Hazard Harmful, Dangerous for the environment
Transport code 9 [4, 6]
Packaging group III [4, 6]
UN Number 3082 [4, 6]; N/C [1-3, 5]
Risk phrases R22a, R38, R41, R51, R53a [1-6]; R52 [4, 6]
Operator protection A, C [1-6]; H [1-3, 5]; P [4, 6]; U05a, U08, U11, U20b [1-6]; U15 [4, 6]
Environmental protection E07a, E13c [4, 6]; E15b [1-3, 5]; E34, E38 [1-6]
Storage and disposal D01, D02 [1-6]; D05, D10b [4, 6]; D09a, D10c [1-3, 5]
Medical advice M03

SEE SECTION 3 FOR PRODUCTS ALSO REGISTERED

321   mepanipyrim

An anilinopyrimidine fungicide for use in horticulture
FRAC mode of action code: 9

Products
Frupica SC	Certis	450 g/l	SC	12067

Uses
- Botrytis in **courgettes** *(off-label)*, **forest nurseries** *(off-label)*, **protected strawberries**, **strawberries**

Extension of Authorisation for Minor Use (EAMUs)
- **courgettes** *20093235*
- **forest nurseries** *20082853*

Approval information
- Mepanipyrim included in Annex I under EC Regulation 1107/2009

Efficacy guidance
- Product is protectant and should be applied as a preventative spray when conditions favourable for Botrytis development occur
- To maintain Botrytis control use as part of a programme with other fungicides that control the disease
- To minimise the possibility of development of resistance adopt resistance management procedures by using products from different chemical groups as part of a mixed spray programme

Restrictions
- Maximum number of treatments 2 per crop (including other anilinopyrimidine products)
- Consult processor before use on crops for processing
- Use spray mixture immediately after preparation

Crop-specific information
- HI 3 d

Environmental safety
- Dangerous for the environment
- Very toxic to aquatic organisms
- LERAP Category B

Hazard classification and safety precautions
> **Hazard** Dangerous for the environment
> **Transport code** 9
> **Packaging group** III
> **UN Number** 3082
> **Risk phrases** R50, R53a
> **Operator protection** A, H; U20c
> **Environmental protection** E16a, E16b, E34
> **Storage and disposal** D01, D02, D05, D10c, D11a, D12b

322   mepiquat chloride

A quaternary ammonium plant growth regulator available only in mixtures

See also 2-chloroethylphosphonic acid + mepiquat chloride
chlormequat + 2-chloroethylphosphonic acid + mepiquat chloride
chlormequat + mepiquat chloride

323 mepiquat chloride + prohexadione-calcium

A growth regulator mixture for cereals

See also prohexadione-calcium

Products

1 Canopy	BASF	300:50 g/l	SC	13181
2 Standon Midget	Standon	300:50 g/l	SC	13803

Uses

- Increasing yield in **durum wheat** *(off-label)*, **triticale** [1]; **winter barley, winter wheat** [1, 2]
- Lodging control in **durum wheat** *(off-label)*, **triticale** [1]; **winter barley, winter wheat** [1, 2]

Extension of Authorisation for Minor Use (EAMUs)

- **durum wheat** *20072403* [1]

Approval information

- Mepiquat chloride and prohexadione-calcium included in Annex I under EC Regulation 1107/2009
- Accepted by BBPA for use on malting barley

Efficacy guidance

- Best results obtained from treatments applied to healthy crops from the beginning of stem extension

Restrictions

- Maximum total dose equivalent to one full dose treatment on all crops
- Do not apply to any crop suffering from physical stress caused by waterlogging, drought or other conditions
- Do not treat on soils with a substantial moisture deficit
- Consult grain merchant or processor before use on crops for bread making or brewing. Effects on these processes have not been tested

Crop-specific information

- Latest use: before flag leaf fully emerged on wheat and barley

Following crops guidance

- Any crop may follow a normally harvested treated crop. Ploughing is not essential.

Environmental safety

- Harmful to aquatic organisms
- Avoid spray drift onto neighbouring crops

Hazard classification and safety precautions

Hazard Harmful
UN Number N/C
Risk phrases R22a, R52, R53a
Operator protection A; U05a, U20b
Environmental protection E15a, E34, E38
Storage and disposal D01, D02, D05, D09a, D10c
Medical advice M05a

324 meptyldinocap

A protectant dinitrophenyl fungicide for powdery mildew control
FRAC mode of action code: 29

Products

Kindred	Landseer	350 g/l	EC	13891

Uses

- Powdery mildew in **apples** *(off-label)*, **crab apples** *(off-label)*, **pears** *(off-label)*, **protected strawberries**, **quinces** *(off-label)*, **table grapes**, **wine grapes**

SEE SECTION 3 FOR PRODUCTS ALSO REGISTERED

Extension of Authorisation for Minor Use (EAMUs)
- *apples* 20092664
- *crab apples* 20092664
- *pears* 20092664
- *quinces* 20092664

Approval information
- Meptyldinocap is awaiting inclusion in Annex 1 under EC Regulation 1107/2009

Environmental safety
- LERAP Category B

Hazard classification and safety precautions
Hazard Harmful, Flammable, Dangerous for the environment
Transport code 3
Packaging group III
UN Number 1993
Risk phrases R22a, R36, R38, R43, R50, R53a, R67
Operator protection A, C, H; U04a, U05a, U11, U14, U19a, U20a
Environmental protection E15b, E16a, E34, E38
Storage and disposal D01, D02, D09a, D10a, D12a
Medical advice M05a

325   mesosulfuron-methyl

A sulfonyl urea herbicide for cereals available only in mixtures
HRAC mode of action code: B

See also diflufenican + iodosulfuron-methyl-sodium + mesosulfuron-methyl
iodosulfuron-methyl-sodium + mesosulfuron-methyl

326   mesotrione

A foliar applied triketone herbicide for maize
HRAC mode of action code: F2

Products

1	Callisto	Syngenta	100 g/l	SC	12323
2	Greencrop Goldcob	Greencrop	100 g/l	SC	12863
3	Kalypstowe	AgChem Access	100 g/l	SC	13844
4	Marvel	ChemSource	100 g/l	SC	14508

Uses
- Annual dicotyledons in *asparagus* (off-label), *game cover* (off-label), *linseed* (off-label), *poppies for morphine production* (off-label), *sweetcorn* (off-label) [1]; *forage maize* [1-4]
- Annual grasses in *game cover* (off-label), *sweetcorn* (off-label) [1]
- Annual meadow grass in *grain maize* [1, 3, 4]; *linseed* (off-label) [1]
- Volunteer oilseed rape in *forage maize* [1-4]; *grain maize* [1, 3, 4]; *linseed* (off-label) [1]

Extension of Authorisation for Minor Use (EAMUs)
- *asparagus* 20111113 expires 30 Sep 2013 [1]
- *game cover* 20082830 expires 30 Sep 2013 [1]
- *linseed* 20071706 expires 30 Sep 2013 [1]
- *poppies for morphine production* 20113148 expires 30 Sep 2013 [1]
- *sweetcorn* 20051893 expires 30 Sep 2013 [1]

Approval information
- Mesotrione included in Annex I under EC Regulation 1107/2009
- Approval expiry 30 Sep 2013 [2, 4]
- Approval expiry 30 Jun 2013 [3]

FOR FULL CONDITIONS OF USE ALWAYS READ THE PRODUCT LABEL

SECTION 2

Efficacy guidance
- Best results obtained from treatment of young actively growing weed seedlings in the presence of adequate soil moisture
- Treatment in poor growing conditions or in dry soil may give less reliable control
- Activity is mostly by foliar uptake with some soil uptake
- To minimise the possible development of resistance where continuous maize is grown the product should not be used for more than two consecutive seasons

Restrictions
- Maximum number of treatments 1 per crop of forage maize
- Do not use on seed crops or on sweetcorn varieties
- Do not spray when crop foliage wet or when excessive rainfall is expected to follow application
- Do not treat crops suffering from stress from cold or drought conditions, or when wide temperature fluctuations are anticipated

Crop-specific information
- Latest use: 8 leaves unfolded stage (GS 18) for forage maize
- Treatment under adverse conditions may cause mild to moderate chlorosis. The effect is transient and does not affect yield

Following crops guidance
- Winter wheat, durum wheat, winter barley or ryegrass may follow a normally harvested treated crop of maize. Oilseed rape may be sown provided it is preceded by deep ploughing to more than 15 cm
- In the spring following application only forage maize, ryegrass, spring wheat or spring barley may be sown
- In the event of crop failure maize may be re-seeded immediately. Some slight crop effects may be seen soon after emergence but these are normally transient

Environmental safety
- Dangerous for the environment
- Very toxic to aquatic organisms
- Take extreme care to avoid drift onto all plants outside the target area
- LERAP Category B

Hazard classification and safety precautions
Hazard Irritant, Dangerous for the environment
UN Number N/C
Risk phrases R36, R50, R53a
Operator protection A, C; U05a, U20b [1-4]; U08, U19a [2]; U09a [1, 3, 4]
Environmental protection E15b, E16a, E16b [1-4]; E38 [1, 3, 4]
Storage and disposal D01, D02, D05, D09a, D12a [1-4]; D10b [2]; D10c, D11a [1, 3, 4]
Medical advice M05a [2]

327 mesotrione + terbuthylazine

A foliar and soil acting herbicide mixture for maize
HRAC mode of action code: F2 + C1

See also terbuthylazine

Products

1 Calaris	Syngenta	70:330 g/l	SC	12405
2 Clayton Faize	Clayton	70:330 g/l	SC	13810
3 Destiny	AgChem Access	70:330 g/l	SC	14159
4 Pan Theta	Pan Agriculture	70:330 g/l	SC	14501

Uses
- Annual dicotyledons in *forage maize*, *grain maize* [1-4]; *sweetcorn* (off-label) [1]
- Annual meadow grass in *forage maize*, *grain maize* [1-4]; *sweetcorn* (off-label) [1]

SEE SECTION 3 FOR PRODUCTS ALSO REGISTERED

Extension of Authorisation for Minor Use (EAMUs)
- *sweetcorn* *20051892* [1]

Approval information
- Mesotrione and terbuthylazine included in Annex 1 under EC Regulation 1107/2009

Efficacy guidance
- Best results obtained from treatment of young actively growing weed seedlings in the presence of adequate soil moisture
- Treatment in poor growing conditions or in dry soil may give less reliable control
- Residual weed control is reduced on soils with more than 10% organic matter
- To minimise the possible development of resistance where continuous maize is grown the product should not be used for more than two consecutive seasons

Restrictions
- Maximum number of treatments 1 per crop of forage maize
- Do not use on seed crops or on sweetcorn varieties
- Do not spray when crop foliage wet or when excessive rainfall is expected to follow application
- Do not treat crops suffering from stress from cold or drought conditions, or when wide temperature fluctuations are anticipated
- Do not apply on Sands or Very Light soils

Crop-specific information
- Latest use: 8 leaves unfolded stage (GS 18) for forage maize
- Treatment under adverse conditions may cause mild to moderate chlorosis. The effect is transient and does not affect yield

Following crops guidance
- Winter wheat, durum wheat, winter barley or ryegrass may follow a normally harvested treated crop of maize. Oilseed rape may be sown provided it is preceded by deep ploughing to more than 15 cm
- In the spring following application forage maize, ryegrass, spring wheat or spring barley may be sown
- Spinach, beet crops, peas, beans, lettuce and cabbages must not be sown in the yr following application
- In the event of crop failure maize may be re-seeded immediately. Some slight crop effects may be seen soon after emergence but these are normally transient

Environmental safety
- Dangerous for the environment
- Very toxic to aquatic organisms
- Take extreme care to avoid drift onto all plants outside the target area
- LERAP Category B

Hazard classification and safety precautions
Hazard Harmful, Dangerous for the environment
Transport code 9
Packaging group III
UN Number 3082
Risk phrases R22a, R50, R53a
Operator protection A; U05a
Environmental protection E15a, E16a, E16b, E34, E38
Storage and disposal D01, D02, D05, D09a, D10c, D12a

328   metalaxyl-M

A phenylamide systemic fungicide
FRAC mode of action code: 4

See also chlorothalonil + metalaxyl-M
cymoxanil + fludioxonil + metalaxyl-M
fluazinam + metalaxyl-M
fludioxonil + metalaxyl-M + thiamethoxam
mancozeb + metalaxyl-M

Products

1	Apron XL	Syngenta	339.2 g/l	ES	14654
2	Clayton Tine	Clayton	465 g/l	SL	14072
3	Divot	ChemSource	465 g/l	SL	15413
4	SL 567A	Syngenta	465.2 g/l	SL	12380
5	Subdue	Fargro	465 g/l	SL	12503

Uses

* Cavity spot in **carrots, parsnips** *(off-label)* [4]; **carrots** *(reduction only)* [2, 3]; **parsnips** *(off-label - reduction)* [2]
* Crown rot in **protected water lilies** *(off-label)*, **water lilies** *(off-label)* [4]
* Damping off in **watercress** *(off-label)* [4]
* Downy mildew in **asparagus** *(off-label)*, **blackberries** *(off-label)*, **broccoli** *(off-label)*, **calabrese** *(off-label)*, **collards** *(off-label)*, **grapevines** *(off-label)*, **herbs (see appendix 6)** *(off-label)*, **hops** *(off-label)*, **horseradish** *(off-label)*, **kale** *(off-label)*, **protected cucumbers** *(off-label)*, **protected herbs (see appendix 6)** *(off-label)*, **protected soft fruit** *(off-label)*, **protected spinach** *(off-label)*, **protected spinach beet** *(off-label)*, **protected water lilies** *(off-label)*, **raspberries** *(off-label)*, **rubus hybrids** *(off-label)*, **salad onions** *(off-label)*, **soft fruit** *(off-label)*, **spinach** *(off-label)*, **spinach beet** *(off-label)*, **water lilies** *(off-label)*, **watercress** *(off-label)* [4]
* Phytophthora in **amenity vegetation** *(off-label)*, **forest nurseries** *(off-label)* [5]
* Phytophthora root rot in **ornamental plant production**, **protected ornamentals** [5]
* Pythium in **beetroot** *(off-label)*, **broccoli**, **brussels sprouts**, **bulb onions**, **cabbages**, **calabrese**, **cauliflowers**, **chard** *(off-label)*, **chinese cabbage**, **herbs (see appendix 6)** *(off-label)*, **kohlrabi**, **ornamental plant production** *(off-label)*, **radishes** *(off-label)*, **shallots** *(off-label)*, **spinach** [1]; **ornamental plant production**, **protected ornamentals** [5]
* Root malformation disorder in **red beet** *(off-label)* [4]
* Storage rots in **cabbages** *(off-label)* [4]
* White blister in **horseradish** *(off-label)* [4]

Extension of Authorisation for Minor Use (EAMUs)

* **amenity vegetation** *20120383* [5]
* **asparagus** *20051502* [4]
* **beetroot** *20102539* [1]
* **blackberries** *20072195* [4]
* **broccoli** *20112502* [4]
* **cabbages** *20062117* [4]
* **calabrese** *20112502* [4]
* **chard** *20120526* [1]
* **collards** *20112048* [4]
* **forest nurseries** *20120383* [5]
* **grapevines** *20051504* [4]
* **herbs (see appendix 6)** *20102539* [1], *20051507* [4]
* **hops** *20051500* [4]
* **horseradish** *20051499* [4], *20061040* [4]
* **kale** *20112048* [4]
* **ornamental plant production** *20102539* [1]
* **parsnips** *(reduction)* *20111033* [2], *20051508* [4]
* **protected cucumbers** *20051503* [4]
* **protected herbs (see appendix 6)** *20051507* [4]

SEE SECTION 3 FOR PRODUCTS ALSO REGISTERED

- ***protected soft fruit*** *20082937* [4]
- ***protected spinach*** *20051507* [4]
- ***protected spinach beet*** *20051507* [4]
- ***protected water lilies*** *20051501* [4]
- ***radishes*** *20102539* [1]
- ***raspberries*** *20072195* [4]
- ***red beet*** *20051307* [4]
- ***rubus hybrids*** *20072195* [4]
- ***salad onions*** *20072194* [4]
- ***shallots*** *20121635* [1]
- ***soft fruit*** *20082937* [4]
- ***spinach*** *20051507* [4]
- ***spinach beet*** *20051507* [4]
- ***water lilies*** *20051501* [4]
- ***watercress*** *20072193* [4]

Approval information
- Metalaxyl-M included in Annex I under EC Regulation 1107/2009
- Accepted by BBPA for use on hops

Efficacy guidance
- Best results achieved when applied to damp soil or potting media
- Treatments to ornamentals should be followed immediately by irrigation to wash any residues from the leaves and allow penetration to the rooting zone [5]
- Efficacy may be reduced in prolonged dry weather [4]
- Results may not be satisfactory on soils with high organic matter content [4]
- Control of cavity spot on carrots overwintered in the ground or lifted in winter may be lower than expected [4]
- Use in an integrated pest management strategy and, where appropriate, alternate with products from different chemical groups
- Product should ideally be used preventatively and the number of phenylamide applications should be limited to 1-2 consecutive treatments
- Always follow FRAG guidelines for preventing and managing fungicide resistance. See Section 5 for more information

Restrictions
- Maximum number of treatments on ornamentals 1 per situation for media treatment. See label for details of drench treatment of protected ornamentals [5]
- Maximum total dose on carrots equivalent to one full dose treatment [4]
- Do not use where carrots have been grown on the same site within the previous eight yrs [4]
- Consult before use on crops intended for processing [4]
- Do not re-use potting media from treated plants for subsequent crops [5]
- Disinfect pots thoroughly prior to re-use [5]

Crop-specific information
- Latest use: 6 wk after drilling for carrots [4]
- Because of the large number of species and ornamental cultivars susceptibility should be checked before large scale treatment [5]
- Treatment of *Viburnum* and *Prunus* species not recommended [5]

Environmental safety
- Harmful to aquatic organisms
- Limited evidence suggests that metalaxyl-M is not harmful to soil dwelling predatory mites

Hazard classification and safety precautions
Hazard Harmful
UN Number N/C
Risk phrases R22a, R52, R53a [1-5]; R37 [2-5]
Operator protection A [1-5]; C [5]; D [1]; H [1, 5]; U02a, U05a, U20b [1-5]; U04a, U10, U19a [2-5]
Environmental protection E15b, E34, E38
Storage and disposal D01, D02, D05, D09a, D12a [1-5]; D07, D10c [2-5]; D11a [1]

FOR FULL CONDITIONS OF USE ALWAYS READ THE PRODUCT LABEL

Treated seed S02, S04d, S05, S07, S08, S09 [1]
Medical advice M03 [2-5]; M05a [1]

329   metaldehyde

A molluscicide bait for controlling slugs and snails

Products

1	Allure	Chiltern	1.5% w/w	PT	12651
2	Appeal	Chiltern	1.5% w/w	PT	12022
3	Arresto 3WX	Sipcam	3% w/w	RB	15492
4	Attract	Chiltern	1.5% w/w	PT	12023
5	Carakol 3	Makhteshim	3% w/w	PT	14309
6	Certis Metaldehyde 3	Certis	3% w/w	PT	14337
7	Certis Red 3	Certis	3% w/w	PT	14061
8	Condor 3	Doff Portland	3% w/w	PT	14324
9	Corso 3SSP	Sipcam	3% w/w	RB	15491
10	Desire	Chiltern	1.5% w/w	PT	14048
11	Doff Horticultural Slug Killer Blue Mini Pellets	Doff Portland	3% w/w	RB	11463
12	Enzo	Makhteshim	3% w/w	PT	14306
13	Escar-go 3	Chiltern	3% w/w	PT	14044
14	ESP	De Sangosse	5% w/w	PT	12999
15	Gusto 3	Makhteshim	3% w/w	PT	14308
16	Helimax S	De Sangosse	5% w/w	RB	13109
17	Lynx H	De Sangosse	3% w/w	RB	14426
18	Osarex W	De Sangosse	3% w/w	RB	14428
19	Pesta H	De Sangosse	3% w/w	RB	14427
20	Regel	De Sangosse	5% w/w	RB	13115
21	Slugdown 3	Doff Portland	3% w/w	PT	14630
22	Steadfast 3	Doff Portland	3% w/w	PT	14367
23	Super 3	Unicrop	3% w/w	RB	14370
24	TDS Major	De Sangosse	4% w/w	CB	13462
25	TDS Metarex Amba	De Sangosse	4% w/w	CB	13461
26	Tempt	Chiltern	3% w/w	PT	14227
27	Trigger 3	Certis	3% w/w	GB	14304
28	Trounce	Chiltern	3% w/w	PT	14222
29	Verve	ChemSource	3% w/w	PT	15333

Uses

* Slugs in *all edible crops (outdoor)*, *all non-edible crops (outdoor)* [2, 4, 11, 14, 16, 24, 25]; *cultivated land/soil* [2, 4, 11]; *natural surfaces not intended to bear vegetation* [16, 24, 25]; *protected crops* [2, 4, 24, 25]
* Slugs and snails in *all edible crops (outdoor and protected)* [1, 10, 13, 26, 28]; *all edible crops (outdoor and protected)* (do not apply to cauliflowers), *amenity grassland*, *managed amenity turf* [6, 7, 27]; *all edible crops (outdoor)* [8, 17-19, 21-23]; *all edible crops (outdoor)* (excluding potato and cauliflower) [3, 9, 29]; *all edible crops (outdoor)* (excluding potatoes) [5, 12, 15]; *all edible crops except potatoes and cauliflowers* [20]; *all non-edible crops (outdoor)* [1, 3, 5, 8-10, 12, 13, 15, 17-23, 26, 28, 29]; *cauliflowers* [1, 17-20, 26, 28]; *cultivated land/soil*, *protected crops* [10, 13]; *natural surfaces not intended to bear vegetation* [6-8, 17-23, 27]; *potatoes* [1, 3, 5-7, 9, 12, 15, 17-20, 23, 26-29]
* Snails in *all edible crops (outdoor)*, *all non-edible crops (outdoor)* [2, 4, 11, 14, 16, 24, 25]; *cultivated land/soil* [2, 4, 11]; *natural surfaces not intended to bear vegetation* [16, 24, 25]; *protected crops* [2, 4, 24, 25]

Approval information

* Metaldehyde included in Annex I under EC Regulation 1107/2009
* Accepted by BBPA for use on malting barley and hops

SEE SECTION 3 FOR PRODUCTS ALSO REGISTERED

Efficacy guidance
- Apply pellets by hand, fiddle drill, fertilizer distributor, by air (check label) or in admixture with seed. See labels for rates and timing.
- Best results achieved from an even spread of granules applied during mild, damp weather when slugs and snails most active. May be applied in standing crops
- To establish the need for pellet application on winter wheat or winter oilseed rape, monitor for slug activity. Where bait traps are used, use a foodstuff attractive to slugs e.g. chicken layer's mash
- Varieties of oilseed rape low in glucosinolates can be more acceptable to slugs than "single low" varieties and control may not be as good
- To prevent slug build up apply at end of season to brassicas and other leafy crops
- To reduce tuber damage in potatoes apply twice in Jul and Aug
- For information on slug trapping and damage risk assessment refer to HGCA Topic Sheets No. 84 (winter wheat) and 85 (winter oilseed rape), available from the HGCA website (www.hgca.com)

Restrictions
- Do not apply when rain imminent or water glasshouse crops within 4 d of application
- Take care to avoid lodging of pellets in the foliage when making late applications to edible crops.
- The maximum total dose of metaldehyde must not exceed 700 g active substance/ha/year.

Crop-specific information
- Put slug traps out before cultivation, when the soil surface is visibly moist and the weather mild (5-25°C) (see label for guidance)
- For winter wheat, a catch of 4 or more slugs/trap indicates a possible risk, where soil and weather conditions favour slug activity
- For winter oilseed rape a catch of 4 or more slugs in standing cereals, or 1 or more in cereal stubble, if other conditions were met, would indicate possible risk of damage

Environmental safety
- Dangerous to game, wild birds and animals
- Some products contain proprietary cat and dog deterrent
- Keep poultry out of treated areas for at least 7 d
- Do not use slug pellets in traps in winter wheat or winter oilseed rape since they are a potential hazard to wildlife and pets
- Some pesticides pose a greater threat of contamination of water than others and metaldehyde is one of these pesticides. Take special care when applying metaldehyde near water and do not apply if heavy rain is forecast

Hazard classification and safety precautions
    UN Number N/C
    Operator protection A, H [1-10, 12-29]; J [14]; U05a [1-10, 12, 13, 15, 17-19, 22, 24-29]; U15 [20]; U20a [3, 5, 9, 12, 15, 29]; U20b [6-8, 17-19, 22, 27]; U20c [1, 2, 4, 10, 11, 13, 14, 16, 20, 21, 23-26, 28]
    Environmental protection E05b [1-7] (7 d); E05b [8] (7 days); E05b [9] (7 d); E05b [10] (7 days); E05b [11, 12] (7 d); E05b [13] (7 days); E05b [14-16] (7 d); E05b [17-20] (7 days); E05b [21] (7 d); E05b [22] (7 days); E05b [23-29] (7 d); E06b [10, 13] (7 d); E07a, E34 [8, 17-19, 22]; E10a [1-5, 8-26, 28, 29]; E10c [6, 7, 27]; E15a [1-29]
    Storage and disposal D01, D09a [1-10, 12-29]; D02 [1-10, 12, 13, 15-20, 22, 24-29]; D05 [6, 7, 27]; D07 [1, 2, 4, 6-8, 10, 11, 13, 16-28]; D11a [1-15, 17-23, 26-29]; D11b [16, 24, 25]; D12a [16]; D12b [20]
    Treated seed S04a [8, 14, 16-20, 22, 24, 25]
    Medical advice M04a [6, 7, 27]; M05a [8, 16-20, 22, 24, 25]

FOR FULL CONDITIONS OF USE ALWAYS READ THE PRODUCT LABEL

330   metamitron

A contact and residual triazinone herbicide for use in beet crops
HRAC mode of action code: C1

See also chloridazon + chlorpropham + metamitron
chloridazon + metamitron
chlorpropham + metamitron
desmedipham + ethofumesate + metamitron + phenmedipham
ethofumesate + metamitron
ethofumesate + metamitron + phenmedipham

Products

1	Bettix 70 WG	United Phosphorus	70% w/w	WG	11154
2	Bettix Flo	United Phosphorus	700 g/l	SC	11959
3	Celmitron 70% WDG	AgriChem BV	70% w/w	WG	14408
4	Defiant SC	United Phosphorus	700 g/l	SC	12302
5	Goltix Flowable	Makhteshim	700 g/l	SC	12851
6	Goltix WG	Makhteshim	70% w/w	WG	11539
7	Meta Flo	ChemSource	700 g/l	SC	15486
8	Meta WDG	ChemSource	70% w/w	WG	15010
9	Mitron 70 WG	Hermoo	70% w/w	WG	11516
10	Mitron 700 SC	Belcrop	700 g/l	SC	15621
11	Mitron 90 WG	Hermoo	90% w/w	WG	10888
12	Mitron SC	Hermoo	700 g/l	SC	11643
13	Skater	Makhteshim	700 g/l	SC	12857
14	Target SC	AgriChem BV	700 g/l	SC	13307

Uses

- Annual dicotyledons in **asparagus** *(off-label)*, **forest** *(off-label)*, **herbs (see appendix 6)** *(off-label)*, **horseradish** *(off-label)*, **parsnips** *(off-label)* [5, 13]; **cardoons** *(off-label)*, **celery (outdoor)** *(off-label)*, **rhubarb** *(off-label)*, **strawberries** *(off-label)* [5, 6, 13]; **fodder beet**, **sugar beet** [1-14]; **mangels** [1-8, 10-14]; **red beet** [1-8, 10-12, 14]
- Annual grasses in **fodder beet**, **sugar beet** [1-6, 8-13]; **horseradish** *(off-label)*, **parsnips** *(off-label)*, **strawberries** *(off-label)* [5]; **mangels** [1-6, 8, 10-13]; **red beet** [1-6, 8, 10-12]
- Annual meadow grass in **asparagus** *(off-label)*, **forest** *(off-label)*, **herbs (see appendix 6)** *(off-label)* [5, 13]; **cardoons** *(off-label)*, **celery (outdoor)** *(off-label)*, **rhubarb** *(off-label)* [5, 6]; **fodder beet**, **sugar beet** [1-14]; **horseradish** *(off-label)*, **parsnips** *(off-label)* [13]; **mangels** [1-8, 10-14]; **red beet** [1-8, 10-12, 14]; **strawberries** *(off-label)* [6, 13]
- Fat hen in **fodder beet**, **sugar beet** [1-6, 8-13]; **mangels** [1-6, 8, 10-13]; **red beet** [1-6, 8, 10-12]
- General weed control in **asparagus** *(off-label)*, **forest** *(off-label)*, **horseradish** *(off-label)*, **parsnips** *(off-label)* [6]
- Groundsel in **chives** *(off-label)*, **parsley** *(off-label)* [6]; **herbs (see appendix 6)** *(off-label)* [5, 6]
- Polygonums in **cardoons** *(off-label)*, **celery (outdoor)** *(off-label)*, **rhubarb** *(off-label)* [13]

Extension of Authorisation for Minor Use (EAMUs)

- **asparagus** *20100065* [5], *20041758* [6], *20100066* [13]
- **cardoons** *20093520* [5], *20093518* [6], *20070844* [13]
- **celery (outdoor)** *20093520* [5], *20093518* [6], *20070844* [13]
- **chives** *20041756* [6]
- **forest** *20093519* [5], *20041757* [6], *20093525* [13]
- **herbs (see appendix 6)** *20093521* [5], *20041756* [6], *20093522* [13]
- **horseradish** *20070513* [5], *20061637* [6], *20093524* [13]
- **parsley** *20041756* [6]
- **parsnips** *20070513* [5], *20061637* [6], *20093524* [13]
- **rhubarb** *20093520* [5], *20093518* [6], *20070844* [13]
- **strawberries** *20070856* [5], *20093517* [6], *20093523* [13]

Approval information

- Metamitron included in Annex I under EC Regulation 1107/2009

SEE SECTION 3 FOR PRODUCTS ALSO REGISTERED

Efficacy guidance
- May be used pre-emergence alone or post-emergence in tank mixture or with an authorised adjuvant oil
- Low dose programme (LDP). Apply a series of low-dose post-weed emergence sprays, including adjuvant oil, timing each treatment according to weed emergence and size. See label for details and for recommended tank mixes and sequential treatments. On mineral soils the LDP should be preceded by pre-drilling or pre-emergence treatment
- Traditional application. Apply either pre-drilling before final cultivation with incorporation to 8-10 cm, or pre-crop emergence at or soon after drilling into firm, moist seedbed to emerged weeds from cotyledon to first true leaf stage
- On emerged weeds at or beyond 2-leaf stage addition of adjuvant oil advised
- For control of wild oats and certain other weeds, tank mixes with other herbicides or sequential treatments are recommended. See label for details

Restrictions
- Maximum total dose equivalent to three full dose treatments for most products. Check label
- Using traditional method post-crop emergence on mineral soils do not apply before first true leaves have reached 1 cm long

Crop-specific information
- Latest use: before crop foliage meets across rows for beet crops
- HI herbs 6 wk
- Crop tolerance may be reduced by stress caused by growing conditions, effects of pests, disease or other pesticides, nutrient deficiency etc

Following crops guidance
- Only sugar beet, fodder beet or mangels may be drilled within 4 mth after treatment. Winter cereals may be sown in same season after ploughing, provided 16 wk passed since last treatment

Environmental safety
- Dangerous for the environment
- Very toxic to aquatic organisms
- Dangerous to fish or other aquatic life. Do not contaminate surface waters or ditches with chemical or used container
- Do not empty into drains

Hazard classification and safety precautions
Hazard Harmful [1-8, 11, 12, 14]; Dangerous for the environment [1-14]
Transport code 9
Packaging group III
UN Number 3077 [1, 3, 6, 8, 9, 11]; 3082 [2, 4, 5, 7, 10, 12-14]
Risk phrases R22a [1-8, 11, 12, 14]; R43 [2, 4]; R50 [1, 2, 4, 5, 9-12]; R51 [3, 6-8, 13, 14]; R53a [1-14]
Operator protection A [2, 4-6, 10, 12]; U05a [1, 7, 9, 10, 14]; U08 [7, 9, 10, 14]; U09a [2, 4]; U14 [2, 4, 7, 9, 10, 14]; U19a [1, 3, 6-10, 14]; U20a [3, 6, 8-11]; U20b [1, 5, 7, 12, 14]; U20c [2, 4, 13]
Environmental protection E13b [1, 2, 4]; E15a [3, 5-14]; E19b [5, 11-13]; E34 [5, 7, 12, 14]; E38 [3, 6-10, 14]
Storage and disposal D01 [1, 2, 4, 5, 7, 9-12, 14]; D02 [1, 2, 4, 5, 7, 9, 10, 12, 14]; D05 [2, 4]; D09a [1-14]; D10c [2, 4, 13]; D11a [1, 3, 5-12, 14]; D12a [1-8, 11-14]
Medical advice M03 [1, 2, 4, 5, 7, 12, 14]; M04a [11]; M05a [1, 2, 4, 5, 12]

331   metam-sodium

A methyl isothiocyanate producing sterilant for glasshouse, nursery and outdoor soils
HRAC mode of action code: Z

Products

Metam 510	Taminco	510 g/l	LI	09796

Uses

- Brown rot in *potting soils*, *soils*
- Millipedes in *potting soils*, *soils*
- Potato cyst nematode in *potting soils*, *soils*
- Root rot in *potting soils*, *soils*
- Root-knot nematodes in *potting soils*, *soils*
- Symphylids in *potting soils*, *soils*
- Weed seeds in *potting soils*, *soils*
- Wireworm in *potting soils*, *soils*

Approval information

- Metam-sodium is not included in Annex 1 under EC Regulation 1107/2009

Efficacy guidance

- Metam-sodium is a partial soil sterilant and acts by breaking down in contact with soil to release methyl isothiocyanate (MIT)
- Apply to glasshouse soils as a drench, or inject undiluted to 20 cm at 30 cm intervals and seal immediately, or apply to surface and rotavate
- May also be used by mixing into potting soils
- Apply when soil temperatures exceed 7°C, preferably above 10°C, between 1 Apr and 31 Oct. Soil must be of fine tilth, free from debris and with 'potting moisture' content. If soil is too dry postpone treatment and water soil

Restrictions

- No plants must be present during treatment
- Crops must not be planted until a cress germination test has been completed satisfactorily
- Do not treat glasshouses within 2 m of growing crops. Fumes are damaging to all plants
- Avoid using in equipment incorporating natural rubber parts

Crop-specific information

- Latest use: pre-planting of crop

Following crops guidance

- Do not plant until soil is entirely free of fumes

Environmental safety

- Dangerous for the environment
- Very toxic to aquatic organisms
- Keep unprotected persons, livestock and pets out of treated areas for at least 24 h following treatment
- When diluted breakdown commences almost immediately. Only quantities for immediate use should be made up
- Divert or block drains which could carry solution under untreated glasshouses
- After treatment allow sufficient time (several weeks) for residues to dissipate and aerate soil by forking. Time varies with soil and season. Soils with high clay or organic matter content will retain gas longer than lighter soils

Hazard classification and safety precautions

Hazard Corrosive, Dangerous for the environment
Transport code 8
Packaging group III
UN Number 3267
Risk phrases R22a, R34, R43, R50, R53a
Operator protection A, C, H, M; U02a, U04a, U05a, U08, U11, U16a, U19a, U20c
Environmental protection E02a (24 hours); E15a, E34, E38
Storage and disposal D01, D02, D09a, D10a, D12a
Medical advice M03, M04a

SEE SECTION 3 FOR PRODUCTS ALSO REGISTERED

332   Metarhizium anisopliae

A naturally occurring insect parasitic fungus, Metarhizium anisopliae var. anisopliae strain F52
IRAC mode of action code: Not classified

Products

Met52 granular bioinsecticide	Fargro	2% w/w	GR	15168

Uses

* Cabbage root fly in **vegetable brassicas** *(off-label)*
* Leatherjackets in **bilberries** *(off-label)*, **cranberries** *(off-label)*, **loganberries** *(off-label)*, **ribes hybrids** *(off-label)*, **rubus hybrids** *(off-label)*, **table grapes** *(off-label)*, **wine grapes** *(off-label)*
* Lettuce root aphid in **herbs (see appendix 6)** *(off-label)*, **leafy vegetables** *(off-label)*
* Midges in **amenity vegetation** *(off-label)*, **bilberries** *(off-label)*, **blueberries** *(off-label)*, **container-grown ornamentals** *(off-label)*, **cranberries** *(off-label)*, **herbs (see appendix 6)** *(off-label)*, **leaf brassicas** *(off-label)*, **loganberries** *(off-label)*, **ornamental plant production** *(off-label)*, **ribes hybrids** *(off-label)*, **rubus hybrids** *(off-label)*, **table grapes** *(off-label)*, **top fruit** *(off-label)*, **wine grapes** *(off-label)*
* Sciarid flies in **amenity vegetation** *(off-label)*, **bilberries** *(off-label)*, **blueberries** *(off-label)*, **container-grown ornamentals** *(off-label)*, **cranberries** *(off-label)*, **edible fungi** *(off-label)*, **herbs (see appendix 6)** *(off-label)*, **leaf brassicas** *(off-label)*, **leafy vegetables** *(off-label)*, **loganberries** *(off-label)*, **ornamental plant production** *(off-label)*, **ribes hybrids** *(off-label)*, **rubus hybrids** *(off-label)*, **table grapes** *(off-label)*, **top fruit** *(off-label)*, **wine grapes** *(off-label)*
* Thrips in **amenity vegetation** *(off-label)*, **bilberries** *(off-label)*, **blueberries** *(off-label)*, **container-grown ornamentals** *(off-label)*, **cranberries** *(off-label)*, **herbs (see appendix 6)** *(off-label)*, **leaf brassicas** *(off-label)*, **leafy vegetables** *(off-label)*, **loganberries** *(off-label)*, **ornamental plant production** *(off-label)*, **ribes hybrids** *(off-label)*, **rubus hybrids** *(off-label)*, **table grapes** *(off-label)*, **top fruit** *(off-label)*, **wine grapes** *(off-label)*
* Vine weevil in **amenity vegetation** *(off-label)*, **blackberries**, **blackcurrants**, **blueberries**, **gooseberries**, **ornamental plant production**, **raspberries**, **redcurrants**, **strawberries**

Extension of Authorisation for Minor Use (EAMUs)

* **amenity vegetation** *20111568, 20111997*
* **bilberries** *20111568, 20111997*
* **blueberries** *20111997*
* **container-grown ornamentals** *20111997*
* **cranberries** *20111568, 20111997*
* **edible fungi** *20111568*
* **herbs (see appendix 6)** *20111568, 20111997*
* **leaf brassicas** *20111997*
* **leafy vegetables** *20111568*
* **loganberries** *20111568, 20111997*
* **ornamental plant production** *20111997*
* **ribes hybrids** *20111568, 20111997*
* **rubus hybrids** *20111568, 20111997*
* **table grapes** *20111568, 20111997*
* **top fruit** *20111997*
* **vegetable brassicas** *20111568*
* **wine grapes** *20111568, 20111997*

Approval information

* Metarhizium anisopliae included in Annex 1 under EC Regulation 1107/2009

Efficacy guidance

* Incorporate into growing media at any growth stage

Hazard classification and safety precautions
    UN Number N/C
    Risk phrases R70
    Operator protection A, H; U14

Storage and disposal D01

333   metazachlor

A residual anilide herbicide for use in brassicas, nurseries and forestry
HRAC mode of action code: K3

See also clomazone + metazachlor
dimethenamid-p + metazachlor
dimethenamid-p + metazachlor + quinmerac
imazamox + metazachlor

Products

1	Butisan S	BASF	500 g/l	SC	11733
2	Clayton Buzz	Clayton	500 g/l	SC	12509
3	Clayton Metazachlor 50 SC	Clayton	500 g/l	SC	11719
4	EA Metazachlor	European Ag	500 g/l	SC	15039
5	Fuego 50	Makhteshim	500 g/l	SC	11473
6	Greencrop Monogram	Greencrop	500 g/l	SC	12048
7	Mashona	AgChem Access	500 g/l	SC	13006
8	Metaz 500 SC	Goldengrass	500 g/l	SC	14413
9	Rapsan 500 SC	Globachem	500 g/l	SC	13916
10	Segundo	Agroquimicos	500 g/l	SC	15710
11	Standon Metazachlor 500	Standon	500 g/l	SC	12012
12	Sultan 50 SC	Makhteshim	500 g/l	SC	10418

Uses

* Annual dicotyledons in *almonds, apples, apricots, cherries, chestnuts, crab apples, forest nurseries, hazel nuts, nectarines, ornamental plant production, peaches, pears, plums, quinces, walnuts* [10, 12]; *amenity vegetation* [2]; *borage for oilseed production (off-label), canary flower (echium spp.) (off-label)* [1, 3, 6, 11, 12]; *broccoli* [1-4, 6-8, 10-12]; *brussels sprouts, cabbages, cauliflowers, swedes, turnips* [1-4, 6-12]; *calabrese* [1-4, 7-12]; *chinese cabbage (off-label), choi sum (off-label), collards (off-label), kale (off-label), leeks (off-label), pak choi (off-label)* [1, 12]; *evening primrose (off-label)* [1, 3, 8, 11, 12]; *farm forestry* [3, 10, 12]; *farm woodland* [1, 9, 11]; *forest* [1-3, 6, 9-12]; *honesty (off-label), mustard (off-label)* [1, 3, 6, 8, 11, 12]; *kohlrabi (off-label), linseed (off-label), spring corn gromwell (off-label), winter corn gromwell (off-label)* [1]; *nursery fruit trees and bushes, ornamental trees* [1-3, 6, 9, 11]; *shrubs* [1-4, 6, 9-12]; *spring linseed (off-label), winter linseed (off-label)* [12]; *spring oilseed rape, winter oilseed rape* [1-12]
* Annual grasses in *amenity vegetation* [2]; *broccoli, brussels sprouts, cabbages, calabrese, cauliflowers, swedes, turnips* [7]; *farm woodland* [1, 9, 11]; *forest, ornamental trees* [1-3, 6, 9, 11]; *shrubs* [1-4, 6, 9-12]; *spring oilseed rape, winter oilseed rape* [5, 7]
* Annual meadow grass in *almonds, apples, apricots, cherries, chestnuts, crab apples, forest, forest nurseries, hazel nuts, nectarines, ornamental plant production, peaches, pears, plums, quinces, walnuts* [10, 12]; *amenity vegetation* [2]; *borage for oilseed production (off-label), canary flower (echium spp.) (off-label)* [1, 3, 6, 11, 12]; *broccoli, brussels sprouts, cabbages, cauliflowers, spring oilseed rape, swedes, turnips, winter oilseed rape* [1-4, 6, 8-12]; *calabrese* [1-4, 8-12]; *chinese cabbage (off-label), choi sum (off-label), collards (off-label), kale (off-label), leeks (off-label), pak choi (off-label)* [1, 12]; *evening primrose (off-label)* [1, 3, 8, 11, 12]; *farm forestry* [3, 10, 12]; *honesty (off-label), mustard (off-label)* [1, 3, 6, 8, 11, 12]; *kohlrabi (off-label), linseed (off-label), spring corn gromwell (off-label), winter corn gromwell (off-label)* [1]; *nursery fruit trees and bushes, ornamental trees* [1-3, 6, 9, 11]; *shrubs* [1-4, 6, 9-12]; *spring linseed (off-label), winter linseed (off-label)* [12]
* Blackgrass in *almonds, apples, apricots, cherries, chestnuts, crab apples, forest, forest nurseries, hazel nuts, nectarines, ornamental plant production, peaches, pears, plums, quinces, walnuts* [10, 12]; *amenity vegetation* [2]; *broccoli* [1-4, 6, 8, 10-12]; *brussels sprouts, cabbages, cauliflowers, spring oilseed rape, swedes, turnips, winter oilseed rape*

SEE SECTION 3 FOR PRODUCTS ALSO REGISTERED

[1-4, 6, 8-12]; *calabrese* [1-4, 8-12]; *farm forestry* [3, 10, 12]; *nursery fruit trees and bushes*, *ornamental trees* [1-3, 6, 9, 11]; *shrubs* [1-4, 6, 9-12]

Extension of Authorisation for Minor Use (EAMUs)
- *borage for oilseed production* *20112440* [1], *20112442* [3], *20112444* [6], *20112447* [11], *20112448* [12]
- *canary flower (echium spp.)* *20112440* [1], *20112442* [3], *20112444* [6], *20112447* [11], *20112448* [12]
- *chinese cabbage* *20112439* [1], *20112449* [12]
- *choi sum* *20112439* [1], *20112449* [12]
- *collards* *20112439* [1], *20112449* [12]
- *evening primrose* *20112440* [1], *20112442* [3], *20112445* [8], *20112447* [11], *20112448* [12]
- *honesty* *20112440* [1], *20112442* [3], *20112444* [6], *20112445* [8], *20112447* [11], *20112448* [12]
- *kale* *20112439* [1], *20112449* [12]
- *kohlrabi* *20120518* [1]
- *leeks* *20112441* [1], *20112450* [12]
- *linseed* *20112440* [1]
- *mustard* *20112440* [1], *20112442* [3], *20112444* [6], *20112445* [8], *20112447* [11], *20112448* [12]
- *pak choi* *20112439* [1], *20112449* [12]
- *spring corn gromwell* *20110182* [1]
- *spring linseed* *20112448* [12]
- *winter corn gromwell* *20110182* [1]
- *winter linseed* *20112448* [12]

Approval information
- Metazachlor is included in Annex I under EC Regulation 1107/2009

Efficacy guidance
- Activity is dependent on root uptake. For pre-emergence use apply to firm, moist, clod-free seedbed
- Some weeds (chickweed, mayweed, blackgrass etc) susceptible up to 2- or 4-leaf stage. Moderate control of cleavers achieved provided weeds not emerged and adequate soil moisture present
- Split pre- and post-emergence treatments recommended for certain weeds in winter oilseed rape on light and/or stony soils
- Effectiveness is reduced on soils with more than 10% organic matter
- Always follow WRAG guidelines for preventing and managing herbicide resistant weeds. See Section 5 for more information

Restrictions
- Maximum number of treatments 1 per crop for spring oilseed rape, swedes, turnips and brassicas; 2 per crop for winter oilseed rape (split dose treatment); 3 per yr for ornamentals, nursery stock, nursery fruit trees, forestry and farm forestry
- Do not use on sand, very light or poorly drained soils
- Do not treat protected crops or spray overall on ornamentals with soft foliage
- Do not spray crops suffering from wilting, pest or disease
- Do not spray broadcast crops or if a period of heavy rain forecast
- When used on nursery fruit trees any fruit harvested within 1 yr of treatment must be destroyed
- Applications shall be limited to a total dose of not more than 1.0 kg metazachlor/ha in a three year period on the same field

Crop-specific information
- Latest use: pre-emergence for swedes and turnips; before 10 leaf stage for spring oilseed rape; before end of Jan for winter oilseed rape
- HI brassicas 6 wk
- On winter oilseed rape may be applied pre-emergence from drilling until seed chits, post-emergence after fully expanded cotyledon stage (GS 1,0) or by split dose technique depending on soil and weeds. See label for details
- On spring oilseed rape may also be used pre-weed-emergence from cotyledon to 10-leaf stage of crop (GS 1,0-1,10)

FOR FULL CONDITIONS OF USE ALWAYS READ THE PRODUCT LABEL

- With pre-emergence treatment ensure seed covered by 15 mm of well consolidated soil. Harrow across slits of direct-drilled crops
- Ensure brassica transplants have roots well covered and are well established. Direct drilled brassicas should not be treated before 3 leaf stage
- In ornamentals and hardy nursery stock apply after plants established and hardened off as a directed spray or, on some subjects, as an overall spray. See label for list of tolerant subjects. Do not treat plants in containers

Following crops guidance
- Any crop can follow normally harvested treated winter oilseed rape. See label for details of crops which may be planted after spring treatment and in event of crop failure

Environmental safety
- Dangerous for the environment
- Very toxic to aquatic organisms
- Keep livestock out of treated areas until foliage of any poisonous weeds such as ragwort has died and become unpalatable
- Keep livestock out of treated areas of swede and turnip for at least 5 wk following treatment
- Some pesticides pose a greater threat of contamination of water than others and metazachlor is one of these pesticides. Take special care when applying metazachlor near water and do not apply if heavy rain is forecast

Hazard classification and safety precautions
Hazard Harmful, Dangerous for the environment
Transport code 9
Packaging group III
UN Number 3082
Risk phrases R22a, R43, R50 [1-12]; R38 [1-8, 10, 12]; R41 [9]; R53a [1-3, 7-9]
Operator protection A, C, H [1-12]; M [1-7, 9-12]; U05a, U19a [1-12]; U08, U20b [1-8, 10-12]; U09a, U20a [9]; U14 [1, 4, 5, 7, 8, 10, 12]; U15 [4, 5, 10, 12]
Environmental protection E06a [1-4, 6-12] (5 wk for swedes, turnips); E07a, E15a, E34 [1-12]; E38 [1, 2, 7, 8]
Consumer protection C02a [9] (5 weeks for swedes and turnips)
Storage and disposal D01, D02, D09a [1-12]; D05 [2-6, 8-12]; D10b [3, 6, 11]; D10c [1, 2, 4, 5, 7-10, 12]; D12a [1-3, 7, 8]; D12b [4, 5, 10, 12]
Medical advice M03 [1-12]; M05a [1-3, 7, 8]

334   metazachlor + quinmerac

A residual herbicide mixture for oilseed rape
HRAC mode of action code: K3 + O

See also quinmerac

Products

1	Arenales	Agroquimicos	333:83 g/l	SC	15814
2	Boomerang	BASF	400:100 g/l	SC	14043
3	Naspar TDI	Q-Chem	400:100 g/l	SC	14584
4	Noah	ChemSource	400:100 g/l	SC	15855
5	Novall	BASF	400:100 g/l	SC	12031
6	Oryx	BASF	333:83 g/l	SC	13527
7	Palometa	Agroquimicos	375:125 g/l	SC	15833

Uses
- Annual dicotyledons in **borage** *(off-label)*, **canary flower (echium spp.)** *(off-label)*, **evening primrose** *(off-label)*, **honesty** *(off-label)*, **linseed** *(off-label)*, **mustard** *(off-label)* [5]; **winter oilseed rape** [1-7]
- Annual grasses in **borage** *(off-label)*, **canary flower (echium spp.)** *(off-label)*, **evening primrose** *(off-label)*, **honesty** *(off-label)*, **linseed** *(off-label)*, **mustard** *(off-label)* [5]; **winter oilseed rape** [2]
- Annual meadow grass in **winter oilseed rape** [1, 3-7]

SEE SECTION 3 FOR PRODUCTS ALSO REGISTERED

- Blackgrass in **winter oilseed rape** [1, 3-7]
- Cleavers in **winter oilseed rape** [1-7]
- Poppies in **winter oilseed rape** [1, 3-7]

Extension of Authorisation for Minor Use (EAMUs)
- **borage** *20060548* [5]
- **canary flower (echium spp.)** *20060548* [5]
- **evening primrose** *20060548* [5]
- **honesty** *20060548* [5]
- **linseed** *20060548* [5]
- **mustard** *20060548* [5]

Approval information
- Metazachlor and quinmerac included in Annex I under EC Regulation 1107/2009

Efficacy guidance
- Activity is dependent on root uptake. Pre-emergence treatments should be applied to firm moist seedbeds. Applications to dry soil do not become effective until after rain has fallen
- Maximum activity achieved from treatment before weed emergence for some species
- Weed control may be reduced if excessive rain falls shortly after application especially on light soils
- May be used on all soil types except Sands, Very Light Soils, and soils containing more than 10% organic matter. Crop vigour and/or plant stand may be reduced on brashy and stony soils

Restrictions
- Maximum total dose equivalent to one full dose treatment
- Damage may occur in waterlogged conditions. Do not use on poorly drained soils
- Do not treat stressed crops. In frosty conditions transient scorch may occur
- Applications shall be limited to a total dose of not more than 1.0 kg metazachlor/ha in a three year period on the same field

Crop-specific information
- Latest use: end Jan in yr of harvest
- To ensure crop safety it is essential that crop seed is well covered with soil to 15 mm. Loose or puffy seedbeds must be consolidated before treatment. Do not use on broadcast crops
- Crop vigour and possibly plant stand may be reduced if excessive rain falls shortly after treatment especially on light soils

Following crops guidance
- In the event of crop failure after use, wheat or barley may be sown in the autumn after ploughing to 15 cm. Spring cereals or brassicas may be planted after ploughing in the spring

Environmental safety
- Dangerous for the environment
- Very toxic to aquatic organisms
- Keep livestock out of treated areas until foliage of any poisonous weeds such as ragwort has died and become unpalatable
- To reduce risk of movement to water do not apply to dry soil or if heavy rain is forecast. On clay soils create a fine consolidated seedbed
- Some pesticides pose a greater threat of contamination of water than others and metazachlor is one of these pesticides. Take special care when applying metazachlor near water and do not apply if heavy rain is forecast
- LERAP Category B

Hazard classification and safety precautions
Hazard Irritant, Dangerous for the environment
Transport code 9 [1-6]
Packaging group III [1-6]
UN Number 3082 [1-6]
Risk phrases R43, R50, R53a
Operator protection A [1-7]; C, H [1-6]; U04a [1-6]; U05a, U08, U14, U19a, U20b [1-7]
Environmental protection E07a, E15a, E16a, E38

FOR FULL CONDITIONS OF USE ALWAYS READ THE PRODUCT LABEL

Storage and disposal D01, D02, D08, D09a, D10c, D12a
Medical advice M05a

335   metconazole

A triazole fungicide for cereals and oilseed rape
FRAC mode of action code: 3

See also boscalid + metconazole
       epoxiconazole + metconazole
       fluxapyroxad + metconazole

Products

1	Caramba 90	BASF	90 g/l	SL	15524
2	Juventus	BASF	90 g/l	SL	13340
3	Juventus	BASF	90 g/l	EC	15528
4	Sunorg Pro	BASF	90 g/l	SL	15433

Uses

- Alternaria in *spring oilseed rape*, *winter oilseed rape* [1-4]
- Ascochyta in *combining peas* (reduction), *lupins* (qualified minor use), *vining peas* (reduction) [1-4]
- Botrytis in *combining peas* (reduction), *lupins* (qualified minor use), *vining peas* (reduction) [1-4]
- Brown rust in *durum wheat*, *rye*, *spring barley*, *spring wheat*, *triticale*, *winter barley*, *winter wheat* [1-4]
- Disease control in *broad beans* (off-label), *spring linseed* (off-label), *winter linseed* (off-label) [4]
- Fusarium ear blight in *durum wheat* (reduction) [2-4]; *rye* (reduction) [1, 2, 4]; *spring wheat* (reduction), *triticale* (reduction), *winter wheat* (reduction) [1-4]
- Growth regulation in *winter oilseed rape* [4]
- Light leaf spot in *spring oilseed rape*, *winter oilseed rape* [1, 2, 4]; *spring oilseed rape* (reduction), *winter oilseed rape* (reduction) [3]
- Mycosphaerella in *combining peas* (reduction), *vining peas* (reduction) [1-4]
- Net blotch in *durum wheat* (reduction) [1]; *spring barley* (reduction), *winter barley* (reduction) [1-4]
- Phoma in *spring oilseed rape* (reduction), *winter oilseed rape* (reduction) [1-4]
- Powdery mildew in *durum wheat* (moderate control), *rye* (moderate control), *spring barley* (moderate control), *spring wheat* (moderate control), *triticale* (moderate control), *winter wheat* (moderate control) [1-4]; *winter barley* [1, 2, 4]; *winter barley* (moderate control) [3]
- Rhynchosporium in *rye* (moderate control), *spring barley* (reduction), *winter barley* (reduction) [3]; *spring barley*, *winter barley* [1, 2, 4]
- Rust in *combining peas*, *lupins* (qualified minor use), *spring field beans*, *vining peas*, *winter field beans* [1-4]
- Septoria leaf blotch in *durum wheat* [2-4]; *rye* [1, 2, 4]; *spring wheat*, *triticale*, *winter wheat* [1-4]
- Yellow rust in *durum wheat* [2-4]; *rye*, *spring wheat*, *triticale*, *winter wheat* [1-4]; *spring barley*, *winter barley* [3]

Extension of Authorisation for Minor Use (EAMUs)

- *broad beans* 20111928 [4]
- *spring linseed* 20111929 [4]
- *winter linseed* 20111929 [4]

Approval information

- Metconazole included in Annex I under EC Regulation 1107/2009
- Accepted by BBPA for use on malting barley
- Approval expiry 31 Jul 2013 [2]

Efficacy guidance
- Best results from application to healthy, vigorous crops when disease starts to develop
- Good spray cover of the target is essential for best results. Spray volume should be increased to improve spray penetration into dense crops
- Metconazole is a DMI fungicide. Resistance to some DMI fungicides has been identified in Septoria leaf blotch which may seriously affect performance of some products. For further advice contact a specialist advisor and visit the Fungicide Resistance Action Group (FRAG)-UK website

Restrictions
- Maximum total dose equivalent to two full dose treatments on all crops
- Do not apply to oilseed rape crops that are damaged or stressed from previous treatments, adverse weather, nutrient deficiency or pest attack. Spring application may lead to reduction of crop height
- The addition of adjuvants is neither advised nor necessary and can lead to enhanced growth regulatory effects on stressed crops of oilseed rape
- Ensure sprayer is free from residues of previous treatments that may harm the crop, especially oilseed rape. Use of a detergent cleaner is advised before and after use
- Do not apply with pyrethroid insecticides on oilseed rape at flowering [1, 2]
- Maintain an interval of at least 14 days between applications to oilseed rape, peas, beans and lupins, 21 days between applications to cereals [3]
- To protect birds, only one application is allowed on cereals before end of tillering (GS29) [3]

Crop-specific information
- Latest use: up to and including milky ripe stage (GS 71) for cereals; 10% pods at final size for oilseed rape
- HI: 14 d for peas, field beans, lupins
- Spring treatments on oilseed rape can reduce the height of the crop
- Treatment for Septoria leaf spot in wheat should be made before second node detectable stage and when weather favouring development of the disease has occurred. If conditions continue to favour disease development a follow-up treatment may be needed
- Treat mildew infections in cereals before 3% infection on any green leaf. A specific mildewicide will improve control of established infections
- Treat yellow rust on wheat before 1% infection on any leaf or as preventive treatment after GS 39
- For brown rust in cereals spray susceptible varieties before any of top 3 leaves has more than 2% infection
- Peas should be treated at the start of flowering and repeat 3-4 wk later if required
- Field beans and lupins should be treated at petal fall and repeat 3-4 wk later if required

Following crops guidance
- Only cereals, oilseed rape, sugar beet, linseed, maize, clover, beans, peas, carrots, potatoes or onions may be sown as following crops after treatment

Environmental safety
- Dangerous for the environment
- Very toxic to aquatic organisms
- LERAP Category B but avoid treatment close to field boundary, even if permitted by LERAP assessment, to reduce effects on non-target insects or other arthropods
- LERAP Category B

Hazard classification and safety precautions
   Hazard Harmful [3]; Irritant [1, 2, 4]; Dangerous for the environment [1-4]
   Transport code 9
   Packaging group III
   UN Number 3082
   Risk phrases R36, R51, R53a, R63
   Operator protection A, C [1-4]; H [3]; U02a, U05a, U20c
   Environmental protection E15a, E16a, E34, E38 [1-4]; E16b [3]
   Storage and disposal D01, D02, D05, D06c, D09a, D12a [1-4]; D10a [1, 2, 4]; D10c [3]
   Medical advice M03, M05a [3]

FOR FULL CONDITIONS OF USE ALWAYS READ THE PRODUCT LABEL

336   methiocarb

A stomach acting carbamate molluscicide and insecticide
IRAC mode of action code: 1A

Products

1 Cobra	Interfarm	4% w/w	PT	14561
2 Decoy Wetex	Bayer CropScience	2% w/w	PT	11266
3 Draza Forte	Bayer CropScience	4% w/w	PT	13306
4 Huron	Bayer CropScience	3% w/w	PT	11288
5 Karan	Bayer CropScience	3% w/w	PT	11289
6 Mesurol	Bayer CropScience	500 g/l	FS	15311
7 Rivet	Bayer CropScience	3% w/w	PT	11300
8 Zeal Plus	ChemSource	4% w/w	PT	15850

Uses

- Cutworms in *sugar beet* (reduction) [1-5, 7, 8]
- Frit fly in *fodder maize, grain maize, sweetcorn* [6]
- Leatherjackets in *all cereals* (reduction), *potatoes* (reduction), *sugar beet* (reduction) [1-5, 7, 8]; *rotational grass* (seed admixture) [2, 4, 5, 7]
- Millipedes in *sugar beet* (reduction) [1-5, 7, 8]
- Slugs in *all cereals, all non-edible crops (outdoor)* (outdoor only), *brussels sprouts, cabbages, cauliflowers, forage maize, leaf spinach, lettuce, potatoes, spring oilseed rape, sugar beet, sunflowers, winter oilseed rape* [1-5, 7, 8]; *rotational grass* (seed admixture) [2, 4, 5, 7]; *strawberries* [1, 2, 4, 5, 7]
- Slugs and snails in *strawberries* [3, 8]
- Strawberry seed beetle in *strawberries* [1, 2, 4, 5, 7]

Approval information

- Methiocarb included in Annex I under EC Regulation 1107/2009
- In 2006 CRD required that all products containing this active ingredient should carry the following warning in the main area of the container label: "Methiocarb is an anticholinesterase carbamate. Handle with care"
- Accepted by BBPA for use on malting barley

Efficacy guidance

- Use as a surface, overall application when pests active (normally mild, damp weather), pre-drilling or post-emergence. May also be used on cereals or ryegrass in admixture with seed at time of drilling
- Also reduces populations of cutworms and millipedes
- Best on potatoes in late Jul to Aug
- Apply to strawberries before strawing down to prevent seed beetles contaminating crop
- See label for details of suitable application equipment

Restrictions

- This product contains an anticholinesterase carbamate compound. Do not use if under medical advice not to work with such compounds
- Maximum number of treatments 3 per crop for potatoes; 2 per crop for cereals, cauliflowers, maize, oilseed rape, sunflowers; 1 per crop for cabbages, lettuce, leaf spinach, sugar beet; 1 per yr for strawberries. Other crops vary according to product - see labels
- Do not treat any protected crops
- Do not allow pellets to lodge in edible crops

Crop-specific information

- Latest use: before first node detectable for cereals, maize; before three visibly extended internodes for oilseed rape, sunflowers
- HI 7 d for spinach, strawberries; 14 d for Brussels sprouts, cabbages, cauliflowers, lettuce; 18 d for potatoes; 6 mth for sugar beet

Environmental safety

- Dangerous for the environment [3]

SEE SECTION 3 FOR PRODUCTS ALSO REGISTERED

- Very toxic to aquatic organisms [3]
- Harmful to aquatic organisms [2, 4, 5, 7]
- Dangerous to game, wild birds and animals
- Risk to certain non-target insects or other arthropods
- Avoid surface broadcasting application within 6 m of field boundary to reduce effects on non-target species
- Admixed seed should be drilled and not broadcast, and may not be applied from the air

Hazard classification and safety precautions
Hazard Toxic [6]; Harmful [1-5, 7, 8]; Dangerous for the environment [1, 3, 6, 8]
Transport code 6.1 [6]; 9 [1-5, 7, 8]
Packaging group III
UN Number 2992 [6]; 3077 [1-5, 7, 8]
Risk phrases R22a [1-5, 7, 8]; R25 [6]; R50 [1, 3, 6, 8]; R52 [2, 4, 5, 7]; R53a [1-8]
Operator protection A, H [1-8]; D [6]; J [2, 4, 5, 7]; U05a [1-8]; U14, U19d [6]; U20b [1-5, 7, 8]
Environmental protection E05b [1-5, 7, 8] (7 d); E10a, E22b [1-8]; E13b [2, 4, 5, 7]; E15a [6]; E34 [1-5, 7, 8]
Storage and disposal D01, D02, D09a [1-8]; D10d, D12b [6]; D11a [1-5, 7, 8]
Treated seed S01, S02, S03, S04a, S04b, S05, S06a, S07, S08 [6]
Medical advice M02, M03 [1-5, 7, 8]; M04a [6]

337   methoxyfenozide

A moulting accelerating diacylhydrazine insecticide
IRAC mode of action code: 18A

Products

1 Agrovista Trotter	Agrovista	240 g/l	SC	14236
2 Runner	Landseer	240 g/l	SC	15629

Uses
- Codling moth in *apples*, *pears* [1, 2]
- Insect pests in *hops* (off-label), *ornamental plant production* (off-label), *soft fruit* (off-label) [2]
- Tortrix moths in *apples*, *pears* [1, 2]
- Winter moth in *apples*, *pears* [1, 2]

Extension of Authorisation for Minor Use (EAMUs)
- *hops* 20120448 [2]
- *ornamental plant production* 20120448 [2]
- *soft fruit* 20120448 [2]

Approval information
- Methoxyfenozide included in Annex 1 under EC Regulation 1107/2009

Efficacy guidance
- To achieve best results uniform coverage of the foliage and full spray penetration of the leaf canopy is important, particularly when spraying post-blossom
- For maximum effectiveness on winter moth and tortrix spray pre-blossom when first signs of active larvae are seen, followed by a further spray in June if larvae of the summer generation are present
- For codling moth spray post-blossom to coincide with early to peak egg deposition. Follow-up treatments will normally be needed
- Methoxyfenozide is a moulting accelerating compound (MAC) and may be used in an anti-resistance strategy with other top fruit insecticides (including chitin biosynthesis inhibitors and juvenile hormones) which have a different mode of action
- To reduce further the likelihood of resistance development use at full recommended dose in sufficient water volume to achieve required spray penetration

Restrictions
- Maximum number of treatments 3 per yr but no more than two should be sprayed consecutively

FOR FULL CONDITIONS OF USE ALWAYS READ THE PRODUCT LABEL

SECTION 2

Crop-specific information
- HI 14 d for apples, pears

Environmental safety
- Risk to non-target insects or other arthropods
- Broadcast air-assisted LERAP (5 m)

Hazard classification and safety precautions
UN Number N/C
Operator protection U20b
Environmental protection E15b, E16b, E22c; E17b (5 m)
Storage and disposal D05, D09a, D11a

338   1-methylcyclopropene

An inhibitor of ethylene production for use in stored apples

Products
1	SmartFresh	Landseer	3.3% w/w	SP	11799
2	Stay Fresh	ChemSource	3.3% w/w	SP	15760

Uses
- Ethylene inhibition in *apples* (post-harvest use) [1, 2]; *pears* (off-label - post harvest only) [1]
- Scald in *apples* (post-harvest use) [1, 2]; *pears* (off-label - post harvest only) [1]
- Storage rots in *broccoli* (off-label), *brussels sprouts* (off-label), *cabbages* (off-label), *calabrese* (off-label), *cauliflowers* (off-label) [1]

Extension of Authorisation for Minor Use (EAMUs)
- *broccoli* 20111502 [1]
- *brussels sprouts* 20111503 [1]
- *cabbages* 20111503 [1]
- *calabrese* 20111502 [1]
- *cauliflowers* 20111502 [1]
- *pears* (post harvest only) 20102424 [1]

Approval information
- Methylcyclopropene included in Annex 1 under EC Regulation 1107/2009

Efficacy guidance
- Best results obtained from treatment of fruit in good condition and of proper quality for long-term storage
- Effects may be reduced in fruit that is in poor condition or ripe prior to storage or harvested late
- Product acts by releasing vapour into store when mixed with water
- Apply as soon as possible after harvest
- Treatment controls superficial scald and maintains fruit firmness and acid content for 3-6 mth in normal air, and 6-9 mth in controlled atmosphere storage
- Ethylene production recommences after removal from storage

Restrictions
- Maximum number of treatments 1 per batch of apples
- Must only be used by suitably trained and competent persons in fumigation operations
- Consult processors before treatment of fruit destined for processing or cider making
- Do not apply in mixture with other products
- Ventilate all areas thoroughly with all refrigeration fans operating at maximum power for at least 15 min before re-entry

Crop-specific information
- Latest use: 7d after harvest of apples
- Product tested on Granny Smith, Gala, Jonagold, Bramley and Cox. Consult distributor or supplier before treating other varieties

SEE SECTION 3 FOR PRODUCTS ALSO REGISTERED

Environmental safety
- Unprotected persons must be kept out of treated stores during the 24 h treatment period
- Prior to application ensure that the store can be properly and promptly sealed

Hazard classification and safety precautions
    UN Number N/C
    Operator protection U05a
    Environmental protection E02a (24 h); E15a, E34
    Consumer protection C12
    Storage and disposal D01, D02, D07, D14

339   metrafenone

A benzophenone protectant and curative fungicide for cereals
FRAC mode of action code: U8

See also epoxiconazole + fenpropimorph + metrafenone
                epoxiconazole + metrafenone
                fenpropimorph + metrafenone

Products

1	Attenzo	BASF	300 g/l	SC	11917
2	Flexity	BASF	300 g/l	SC	11775
3	Lexi	AgChem Access	300 g/l	SC	14890

Uses
- Disease control/foliar feed in **ornamental plant production** *(off-label)* [2]
- Eyespot in **spring wheat** *(reduction)*, **winter wheat** *(reduction)* [1-3]
- Powdery mildew in **forest nurseries** *(off-label)* [1]; **ornamental plant production** [2]; **spring barley**, **spring oats** *(evidence of mildew control on oats is limited)*, **spring wheat**, **winter barley**, **winter oats** *(evidence of mildew control on oats is limited)*, **winter wheat** [1-3]

Extension of Authorisation for Minor Use (EAMUs)
- **forest nurseries** *20082814* [1]
- **ornamental plant production** *20082850* [2]

Approval information
- Metrafenone included in Annex I under EC Regulation 1107/2009
- Accepted by BBPA for use on malting barley

Efficacy guidance
- Best results obtained from treatment at the start of foliar disease attack
- Activity against mildew in wheat is mainly protectant with moderate curative control in the latent phase; activity is entirely protectant in barley
- Useful reduction of eyespot in wheat is obtained if treatment applied at GS 30-32
- Should be used as part of a resistance management strategy that includes mixtures or sequences effective against mildew and non-chemical methods

Restrictions
- Maximum number of treatments 2 per crop
- Avoid the use of sequential applications of Flexity unless it is used in tank mixture with other products active against powdery mildew employing a different mode of action

Crop-specific information
- Latest use: beginning of flowering (GS 31) for wheat, barley, oats

Following crops guidance
- Cereals, oilseed rape, sugar beet, linseed, maize, clover, field beans, peas, turnips, carrots, cauliflowers, onions, lettuce or potatoes may follow a treated cereal crop

Environmental safety
- Dangerous for the environment
- Toxic to aquatic organisms

FOR FULL CONDITIONS OF USE ALWAYS READ THE PRODUCT LABEL

Hazard classification and safety precautions

    Hazard Irritant, Dangerous for the environment
    UN Number N/C
    Risk phrases R43, R51, R53a
    Operator protection A, H; U05a, U20b
    Environmental protection E15a, E34
    Storage and disposal D01, D02, D05, D09a, D10c
    Medical advice M03

340   metribuzin

A contact and residual triazinone herbicide for use in potatoes
HRAC mode of action code: C1

See also clomazone + metribuzin
       flufenacet + metribuzin
       mecoprop-P + metribuzin

Products

1	Sarabi	AgChem Access	70% w/w	WG	13453
2	Sencorex WG	Interfarm	70% w/w	WG	14747
3	Shotput	Makhteshim	70% w/w	WG	13788
4	Shotput	Makhteshim	70% w/w	WG	15968

Uses

- Annual dicotyledons in **asparagus** *(off-label)*, **carrots** *(off-label)*, **mallow (althaea spp.)** *(off-label)*, **parsnips** *(off-label)*, **sweet potato** *(off-label)* [2, 4]; **early potatoes, maincrop potatoes** [1-4]
- Annual grasses in **asparagus** *(off-label)*, **carrots** *(off-label)*, **mallow (althaea spp.)** *(off-label)*, **parsnips** *(off-label)*, **sweet potato** *(off-label)* [2, 4]; **early potatoes, maincrop potatoes** [1-4]
- Volunteer oilseed rape in **asparagus** *(off-label)*, **carrots** *(off-label)*, **mallow (althaea spp.)** *(off-label)*, **parsnips** *(off-label)*, **sweet potato** *(off-label)* [2, 4]; **early potatoes, maincrop potatoes** [1-4]

Extension of Authorisation for Minor Use (EAMUs)

- **asparagus** *20111358* [2], *20122425* [4]
- **carrots** *20111357* [2], *20122424* [4]
- **mallow (althaea spp.)** *20111357* [2], *20122424* [4]
- **parsnips** *20111357* [2], *20122424* [4]
- **sweet potato** *20111356* [2], *20122426* [4]

Approval information

- Metribuzin included in Annex I under EC Regulation 1107/2009
- Approval expiry 31 Mar 2013 [1]

Efficacy guidance

- Best results achieved on weeds at cotyledon to 1-leaf stage
- May be applied pre- or post-emergence of crop on named maincrop varieties and cv. Marfona; pre-emergence only on named early varieties
- Apply to moist soil with well-rounded ridges and few clods
- Activity reduced by dry conditions and on soils with high organic matter content
- On fen and moss soils pre-planting incorporation to 10-15 cm gives increased activity. Incorporate thoroughly and evenly
- With named maincrop and second early potato varieties on soils with more than 10% organic matter shallow pre- or post-planting incorporation may be used. See label for details
- Effective control using a programme of reduced doses is made possible by using a spray of smaller droplets, thus improving retention

Restrictions

- Maximum total dose equivalent to one full dose treatment on early potato varieties and one and a third full doses on maincrop varieties

SEE SECTION 3 FOR PRODUCTS ALSO REGISTERED

- Only certain varieties may be treated. Apply pre-emergence only on named first earlies, pre- or post-emergence on named second earlies. On named maincrop varieties apply pre-emergence (except for certain varieties on Sands or Very Light soils) or post-emergence. See label
- All post-emergence treatments must be carried out before longest shoots reach 15 cm
- Do not cultivate after treatment
- Some recommended varieties may be sensitive to post-emergence treatment if crop under stress

Crop-specific information
- Latest use: pre-crop emergence for named early potato varieties; before most advanced shoots have reached 15 cm for post-emergence treatment of potatoes
- On stony or gravelly soils there is risk of crop damage, especially if heavy rain falls soon after application
- When days are hot and sunny delay spraying until evening

Following crops guidance
- Ryegrass, cereals or winter beans may be sown in same season provided at least 16 wk elapsed after treatment and ground ploughed to 15 cm and thoroughly cultivated as soon as possible after harvest and no later than end Dec
- In W Cornwall on soil with more than 5% organic matter early potatoes treated as recommended may be followed by summer planted brassica crops provided the soil has been ploughed, spring rainfall has been normal and at least 14 wk have elapsed since treatment
- Do not grow any vegetable brassicas, lettuces or radishes on land treated the previous yr. Other crops may be sown normally in spring of next yr

Environmental safety
- Dangerous for the environment
- Very toxic to aquatic organisms
- Do not empty into drains
- LERAP Category B

Hazard classification and safety precautions
Hazard Harmful, Dangerous for the environment
Transport code 9
Packaging group III
UN Number 3077
Risk phrases R22a, R50, R53a [1-4]; R43 [3, 4]
Operator protection A, H [3, 4]; U05a, U14, U20b [3, 4]; U08, U13, U19a [1-4]; U20a [1, 2]
Environmental protection E15a, E16a, E16b [1-4]; E19b [3, 4]; E38 [1, 2]
Storage and disposal D01, D02 [3, 4]; D09a, D11a, D12a [1-4]
Medical advice M05a [3, 4]

341   metsulfuron-methyl

A contact and residual sulfonylurea herbicide used in cereals, linseed and set-aside
HRAC mode of action code: B

See also carfentrazone-ethyl + metsulfuron-methyl
diflufenican + metsulfuron-methyl
flupyrsulfuron-methyl + metsulfuron-methyl
fluroxypyr + metsulfuron-methyl
mecoprop-P + metsulfuron-methyl

Products
1	Alias SX	DuPont	20% w/w	SG	13398
2	Answer SX	DuPont	20% w/w	SG	15632
3	Cimarron	Headland	50% w/w	TB	13408
4	Finy	AgriChem BV	20% w/w	WG	12855
5	Firecrest	European Ag	20% w/w	SG	14486
6	Forge	Interfarm	20% w/w	SG	12848

FOR FULL CONDITIONS OF USE ALWAYS READ THE PRODUCT LABEL

Products – continued

7	Goldron-M	Goldengrass	20% w/w	SG	13466
8	Jubilee SX	DuPont	20% w/w	SG	12203
9	Lorate	DuPont	20% w/w	SG	12743
10	Metro 20	AgriGuard	20% w/w	WG	15633
11	Metro Clear	AgriGuard	20% w/w	SG	15624
12	Reprisal	ChemSource	20% w/w	SG	15056
13	Simba SX	DuPont	20% w/w	SG	15503
14	Snicket	AgChem Access	20% w/w	SG	15681
15	Standon Metso XT	Standon	20% w/w	SG	13880
16	Standon Mexxon Xtra	Standon	20% w/w	SG	13931

Uses

- Annual dicotyledons in **durum wheat** [6]; **farm forestry** *(off-label)*, **forest nurseries** *(off-label)*, **ornamental plant production** *(off-label)* [8]; **game cover** *(off-label)*, **miscanthus** *(off-label)* [8, 9]; **green cover on land temporarily removed from production** [4, 6, 15, 16]; **linseed**, **spring barley**, **spring oats**, **spring wheat**, **triticale**, **winter barley**, **winter oats**, **winter wheat** [1, 2, 4-16]
- Chickweed in **durum wheat** [6]; **linseed**, **spring barley**, **spring oats**, **spring wheat**, **triticale**, **winter barley**, **winter oats**, **winter wheat** [1, 2, 5-14]
- Docks in **grassland** [3]
- Green cover in **land temporarily removed from production** [1, 2, 5, 7-9, 12-14]
- Mayweeds in **durum wheat** [6]; **linseed**, **spring barley**, **spring oats**, **spring wheat**, **triticale**, **winter barley**, **winter oats**, **winter wheat** [1, 2, 5-14]

Extension of Authorisation for Minor Use (EAMUs)

- **farm forestry** *20082859* [8]
- **forest nurseries** *20082859* [8]
- **game cover** *20082859* [8], *20082860* [9]
- **miscanthus** *20082859* [8], *20082860* [9]
- **ornamental plant production** *20082859* [8]

Approval information

- Metsulfuron-methyl included in Annex I under EC Regulation 1107/2009
- Accepted by BBPA for use on malting barley

Efficacy guidance

- Best results achieved on small, actively growing weeds up to 6-true leaf stage. Good spray cover is important
- Weed control may be reduced in dry soil conditions
- Commonly used in tank-mixture on wheat and barley with other cereal herbicides to improve control of resistant dicotyledons (cleavers, fumitory, ivy-leaved speedwell), larger weeds and grasses. See label for recommended mixtures
- Metsulfuron-methyl is a member of the ALS-inhibitor group of herbicides and products should be used in a planned Resistance Management strategy. See Section 5 for more information

Restrictions

- Maximum number of treatments 1 per crop or per yr (for set-aside)
- Product must only be used after 1 Feb
- Do not apply within 7 d of rolling
- Do not use on cereal crops undersown with grass or legumes
- Do not use in tank mixture on oats, triticale or linseed. On linseed allow at least 7 d before or after other treatments
- Do not tank mix with chlorpyrifos. Allow at least 14 d before or after chlorpyrifos treatments
- Consult contract agents before use on a cereal crop for seed
- Do not use on any crop suffering stress from drought, waterlogging, frost, deficiency, pest or disease attack or apply within 7 d of rolling
- Specific restrictions apply to use in sequence or tank mixture with other sulfonylurea or ALS-inhibiting herbicides. See label for details

SEE SECTION 3 FOR PRODUCTS ALSO REGISTERED

- Spraying equipment should not be drained or flushed onto land planted, or to be planted, with trees or crops other than cereals and should be thoroughly cleansed after use - see label for instructions

Crop-specific information
- Latest use: before flag leaf sheath extending stage for cereals (GS 41); before flower buds visible or crop 30 cm tall for linseed [1, 8]; before 1 Aug in yr of treatment for land not being used for crop production [1, 8]
- Apply after 1 Feb to wheat, oats and triticale from 2-leaf (GS 12), and to barley from 3-leaf (GS 13) until flag-leaf sheath extending (GS 41)
- On linseed allow at least 7 d (10 d if crop is growing poorly or under stress) before or after other treatments
- Use in set-aside when a full green cover is established and made up predominantly of grassland, wheat, barley, oats or triticale. Do not use on seedling grasses

Following crops guidance
- Only cereals, oilseed rape, field beans or grass may be sown in same calendar year after treating cereals with the product alone. Other restrictions apply to tank mixtures. See label for details
- Only cereals should be planted within 16 mth of applying to a linseed crop or set-aside
- In the event of failure of a treated crop sow only wheat within 3 mth after treatment

Environmental safety
- Dangerous for the environment
- Very toxic to aquatic organisms
- Take extreme care to avoid damage by drift onto broad-leaved plants outside the target area, onto surface waters or ditches or onto land intended for cropping
- A range of broad leaved species will be fully or partially controlled when used in land temporarily removed from production, hence product may not be suitable where wild flower borders or other forms of conservation headland are being developed
- Before use on land temporarily removed from production as part of grant-aided scheme, ensure compliance with the management rules
- Green cover on land temporarily removed from production must not be grazed by livestock or harvested for human or animal consumption or used for animal bedding [8]
- LERAP Category B [1, 2, 4-8, 10-16]

Hazard classification and safety precautions
Hazard Dangerous for the environment [1, 2, 4-16]
Transport code 9
Packaging group III
UN Number 3077
Risk phrases R50, R53a [1, 2, 4-16]
Operator protection U05a [4]; U08 [10, 11]; U09a [15, 16]; U19a [10, 11, 15, 16]; U20b [3, 10, 11, 15, 16]
Environmental protection E07a [3]; E14a [10, 11]; E15b [1-16]; E16a [1, 2, 4-8, 10-16]; E16b [2, 10, 11]; E23 [4]; E34 [10, 11, 15, 16]; E38 [1, 2, 5-16]
Storage and disposal D01 [4, 10, 11, 15, 16]; D02, D12b [4]; D09a [2-4, 7, 10, 11, 15, 16]; D11a [2, 4, 7, 10, 11]; D12a [1-3, 5-16]; D17, D18 [3]
Medical advice M05a [10, 11, 15, 16]

342   metsulfuron-methyl + thifensulfuron-methyl

A contact residual and translocated sulfonylurea herbicide mixture for use in cereals
HRAC mode of action code: B + B

See also thifensulfuron-methyl

Products

1	Avro SX	DuPont	2.9:42.9 % w/w	SG	14313
2	Chimera SX	DuPont	2.9:42.9% w/w	SG	13171
3	Concert SX	DuPont	4:40% w/w	SG	12288

FOR FULL CONDITIONS OF USE ALWAYS READ THE PRODUCT LABEL

Products – continued

4	Finish SX	DuPont	6.7:33.3% w/w	SG	12259
5	Harmony M SX	DuPont	4:40% w/w	SG	12258
6	Pennant	Headland	4:40% w/w	SG	13350
7	Presite SX	DuPont	6.7:33.3% w/w	SG	12291
8	Refine Max SX	DuPont	6.7:33.3% w/w	SG	15622

Uses

- Annual dicotyledons in *durum wheat* (off-label) [7]; *miscanthus* (off-label) [2, 5]; *spring barley*, *spring wheat*, *winter wheat* [1-8]; *winter barley* [1, 2, 4, 7, 8]; *winter oats* [4, 7, 8]
- Black bindweed in *spring barley*, *spring wheat*, *winter barley*, *winter wheat* [1]
- Charlock in *spring barley*, *spring wheat*, *winter barley*, *winter wheat* [1]
- Chickweed in *durum wheat* (off-label) [7]; *spring barley*, *spring wheat*, *winter wheat* [1-8]; *winter barley* [1, 2, 4, 7, 8]; *winter oats* [4, 7, 8]
- Creeping thistle in *spring barley*, *spring wheat*, *winter barley*, *winter wheat* [1, 2]
- Fat hen in *spring barley*, *spring wheat*, *winter barley*, *winter wheat* [1]
- Field pansy in *durum wheat* (off-label) [7]; *spring barley*, *spring wheat*, *winter barley*, *winter wheat* [1, 4, 7, 8]; *winter oats* [4, 7, 8]
- Field speedwell in *durum wheat* (off-label) [7]; *spring barley*, *spring wheat*, *winter barley*, *winter wheat* [1, 4, 7, 8]; *winter oats* [4, 7, 8]
- Forget-me-not in *spring barley*, *spring wheat*, *winter barley*, *winter wheat* [1]
- Hemp-nettle in *spring barley*, *spring wheat*, *winter barley*, *winter wheat* [1]
- Ivy-leaved speedwell in *durum wheat* (off-label) [7]; *spring barley*, *spring wheat*, *winter barley*, *winter oats*, *winter wheat* [4, 7, 8]
- Knotgrass in *durum wheat* (off-label) [7]; *spring barley*, *spring wheat*, *winter barley*, *winter wheat* [1, 4, 7, 8]; *winter oats* [4, 7, 8]
- Mayweeds in *durum wheat* (off-label) [7]; *spring barley*, *spring wheat*, *winter wheat* [1-8]; *winter barley* [1, 2, 4, 7, 8]; *winter oats* [4, 7, 8]
- Polygonums in *spring barley*, *spring wheat*, *winter wheat* [2, 3, 5, 6]; *winter barley* [2]
- Poppies in *spring barley*, *spring wheat*, *winter barley*, *winter wheat* [1]
- Red dead-nettle in *spring barley*, *spring wheat*, *winter wheat* [1-3, 5, 6]; *winter barley* [1, 2]
- Redshank in *spring barley*, *spring wheat*, *winter barley*, *winter wheat* [1]
- Shepherd's purse in *spring barley*, *spring wheat*, *winter wheat* [1-3, 5, 6]; *winter barley* [1, 2]

Extension of Authorisation for Minor Use (EAMUs)

- *durum wheat* 20070452 [7]
- *miscanthus* 20082835 [2], 20082916 [5]

Approval information

- Metsulfuron-methyl and thifensulfuron-methyl included in Annex I under EC Regulation 1107/2009
- Accepted by BBPA for use on malting barley

Efficacy guidance

- Best results by application to small, actively growing weeds up to 6-true leaf stage
- Ensure good spray cover
- Susceptible weeds stop growing almost immediately but symptoms may not be visible for about 2 wk
- Effectiveness may be reduced by heavy rain or if soil conditions very dry
- Metsulfuron-methyl and thifensulfuron-methyl are members of the ALS-inhibitor group of herbicides and products should be used in a planned Resistance Management strategy. See Section 5 for more information

Restrictions

- Maximum number of treatments 1 per crop
- Products may only be used after 1 Feb
- Do not use on any crop suffering stress from drought, waterlogging, frost, deficiency, pest or disease attack or any other cause
- Do not use on crops undersown with grasses, clover or legumes, or any other broad leaved crop
- Specific restrictions apply to use in sequence or tank mixture with other sulfonylurea or ALS-inhibiting herbicides. See label for details

SEE SECTION 3 FOR PRODUCTS ALSO REGISTERED

- Do not apply within 7 d of rolling
- Consult contract agents before use on a cereal crop grown for seed

Crop-specific information
- Latest use: before flag leaf sheath extending (GS 39) for barley and wheat, before 2nd node (GS32) for winter oats

Following crops guidance
- Only cereals, oilseed rape, field beans or grass may be sown in same calendar year after treatment
- Additional constraints apply after use of certain tank mixtures. See label
- In the event of crop failure sow only winter wheat within 3 mth after treatment and after ploughing and cultivating to a depth of at least 15 cm

Environmental safety
- Dangerous for the environment
- Very toxic to aquatic organisms
- Take extreme care to avoid damage by drift onto broad-leaved plants outside the target area, or onto ponds, waterways or ditches
- Spraying equipment should not be drained or flushed onto land planted, or to be planted, with trees or crops other than cereals and should be thoroughly cleansed after use - see label for instructions
- LERAP Category B

Hazard classification and safety precautions
Hazard Dangerous for the environment
Transport code 9
Packaging group III
UN Number 3077
Risk phrases R50, R53a
Operator protection U05a [4, 7, 8]; U08, U19a, U20b [1-8]
Environmental protection E15a [4, 7, 8]; E15b [3, 5, 6]; E16a, E38 [1-8]
Storage and disposal D01, D02 [4, 7, 8]; D09a, D12a [1-8]; D11a [3-8]

343   metsulfuron-methyl + tribenuron-methyl

A sulfonylurea herbicide mixture for cereals
HRAC mode of action code: B + B

See also tribenuron-methyl

Products

1	Ally Max SX	DuPont	14.3:14.3% w/w	SG	14835
2	BiPlay SX	DuPont	11.1:22.2% w/w	SG	14836
3	Traton SX	DuPont	11.1:22.2% w/w	SG	14837

Uses
- Annual dicotyledons in **durum wheat** *(off-label)*, **rye** *(off-label)*, **spring oats**, **spring rye** *(off-label)* [1]; **spring barley**, **spring wheat**, **triticale**, **winter barley**, **winter oats**, **winter wheat** [1-3]
- Chickweed in **spring barley**, **spring wheat**, **triticale**, **winter barley**, **winter oats**, **winter wheat** [1-3]; **spring oats** [1]
- Field pansy in **spring barley**, **spring wheat**, **triticale**, **winter barley**, **winter oats**, **winter wheat** [1-3]
- Field speedwell in **spring oats** [1]
- Hemp-nettle in **spring barley**, **spring wheat**, **triticale**, **winter barley**, **winter oats**, **winter wheat** [1-3]; **spring oats** [1]
- Mayweeds in **spring barley**, **spring wheat**, **triticale**, **winter barley**, **winter oats**, **winter wheat** [1-3]; **spring oats** [1]
- Red dead-nettle in **spring barley**, **spring wheat**, **triticale**, **winter barley**, **winter oats**, **winter wheat** [1-3]; **spring oats** [1]

FOR FULL CONDITIONS OF USE ALWAYS READ THE PRODUCT LABEL

- Volunteer oilseed rape in **spring barley**, **spring oats**, **spring wheat**, **triticale**, **winter barley**, **winter oats**, **winter wheat** [1]
- Volunteer sugar beet in **spring barley**, **spring oats**, **spring wheat**, **triticale**, **winter barley**, **winter oats**, **winter wheat** [1]

Extension of Authorisation for Minor Use (EAMUs)
- **durum wheat** *20101080* [1]
- **rye** *20101080* [1]
- **spring rye** *20101080* [1]

Approval information
- Metsulfuron-methyl and tribenuron-methyl included in Annex I under EC Regulation 1107/2009
- Accepted by BBPA for use on malting barley

Efficacy guidance
- Best results obtained when applied to small actively growing weeds
- Product acts by foliar and root uptake. Good spray cover essential but performance may be reduced when soil conditions are very dry and residual effects may be reduced by heavy rain
- Weed growth inhibited within hours of treatment and many show marked colour changes as they die back. Full effects may not be apparent for up to 4 wk
- Metsulfuron-methyl and tribenuron-methyl are members of the ALS-inhibitor group of herbicides and products should be used in a planned Resistance Management strategy. See Section 5 for more information

Restrictions
- Maximum number of treatments 1 per crop
- Product must only be used after 1 Feb and after crop has three leaves
- Do not apply to a crop suffering from drought, water-logging, low temperatures, pest or disease attack, nutrient deficiency, soil compaction or any other stress
- Do not use on crops undersown with grasses, clover or other legumes
- Do not apply within 7 d of rolling
- Specific restrictions apply to use in sequence or tank mixture with other sulfonylurea or ALS-inhibiting herbicides. See label for details

Crop-specific information
- Latest use: before flag leaf sheath extending

Following crops guidance
- Only cereals, field beans, grass or oilseed rape may be sown in the same calendar yr as harvest of a treated crop [1-3]
- In the event of failure of a treated crop only winter wheat may be sown within 3 mth of treatment, and only after ploughing and cultivation to 15 cm min

Environmental safety
- Dangerous for the environment
- Very toxic to aquatic organisms
- Some non-target crops are highly sensitive. Take extreme care to avoid drift outside the target area, or onto ponds, waterways or ditches
- Spraying equipment should be thoroughly cleaned in accordance with manufacturer's instructions
- LERAP Category B [2, 3]

Hazard classification and safety precautions
Hazard Irritant, Dangerous for the environment
Transport code 9
Packaging group III
UN Number 3077
Risk phrases R43, R50, R53a
Operator protection A, H; U08, U19a, U20b
Environmental protection E15a, E38 [1-3]; E16a [2, 3]
Storage and disposal D01, D02, D09a, D11a, D12a

SEE SECTION 3 FOR PRODUCTS ALSO REGISTERED

344   myclobutanil

A systemic, protectant and curative triazole fungicide
FRAC mode of action code: 3

Products

1	Clayton Lithium	Clayton	200 g/l	EW	15803
2	CS Myclobutanil	ChemSource	200 g/l	EW	15434
3	Masalon	Rigby Taylor	45 g/l	EW	12385
4	PureMyclobutanil	Pure Amenity	200 g/l	EW	15370
5	Systhane 20EW	Landseer	200 g/l	EW	09396

Uses

- American gooseberry mildew in **blackcurrants**, **gooseberries** [1, 2, 4, 5]
- Black spot in **ornamental plant production**, **roses** [1, 2, 4, 5]
- Blossom wilt in **cherries** *(off-label)*, **mirabelles** *(off-label)* [5]
- Brown rot in **apricots** *(off-label)*, **peaches** *(off-label)* [5]
- Fusarium patch in **managed amenity turf** [3]
- Plum rust in **plums** *(off-label)* [5]
- Powdery mildew in **apples**, **ornamental plant production**, **pears**, **roses**, **strawberries** [1, 2, 4, 5]; **apricots** *(off-label)*, **artichokes** *(off-label)*, **courgettes** *(off-label)*, **crab apples** *(off-label)*, **gherkins** *(off-label)*, **hops** *(off-label)*, **peaches** *(off-label)*, **protected aubergines** *(off-label)*, **protected blackberries** *(off-label)*, **protected courgettes** *(off-label)*, **protected cucumbers** *(off-label)*, **protected gherkins** *(off-label)*, **protected raspberries** *(off-label)*, **protected rubus hybrids** *(off-label)*, **protected tomatoes** *(off-label)*, **quinces** *(off-label)*, **redcurrants** *(off-label)*, **whitecurrants** *(off-label)*, **wine grapes** *(off-label)* [5]
- Rust in **ornamental plant production**, **roses** [1, 2, 4, 5]
- Scab in **apples**, **pears** [1, 2, 4, 5]

Extension of Authorisation for Minor Use (EAMUs)

- **apricots** *20070626* [5]
- **artichokes** *20070624* [5]
- **cherries** *991535* [5]
- **courgettes** *20070627* [5]
- **crab apples** *20080504* [5]
- **gherkins** *20070627* [5]
- **hops** *20021412* [5]
- **mirabelles** *991535* [5]
- **peaches** *20070626* [5]
- **plums** *20012459* [5]
- **protected aubergines** *20070625* [5]
- **protected blackberries** *20051189* [5]
- **protected courgettes** *20070627* [5]
- **protected cucumbers** *20070627* [5]
- **protected gherkins** *20070627* [5]
- **protected raspberries** *20051189* [5]
- **protected rubus hybrids** *20051189* [5]
- **protected tomatoes** *20070625* [5]
- **quinces** *20080504* [5]
- **redcurrants** *20080505* [5]
- **whitecurrants** *20080505* [5]
- **wine grapes** *20040116* [5]

Approval information

- Accepted by BBPA for use on hops
- Myclobutanil included in Annex 1 under EC Regulation 1107/2009

FOR FULL CONDITIONS OF USE ALWAYS READ THE PRODUCT LABEL

Efficacy guidance
- Best results achieved when used as part of routine preventive spray programme from bud burst to end of flowering in apples and pears and from just before the signs of mildew infection in blackcurrants and gooseberries [5]
- In strawberries commence spraying at, or just prior to, first flower. Post-harvest sprays may be required on mildew-susceptible varieties where mildew is present and likely to be damaging [5]
- Spray at 7-14 d intervals depending on disease pressure and dose applied [5]
- For improved scab control on apples in post-blossom period tank-mix with mancozeb or captan [5]
- Apply alone from mid-Jun for control of secondary mildew on apples and pears
- On roses spray at first signs of disease and repeat every 2 wk. In high risk areas spray when leaves emerge in spring, repeat 1 wk later and then continue normal programme [5]
- Treatment of managed amenity turf may be carried out at any time of year before, or at, the first sign of disease [3]
- Myclobutanil products should be used in conjunction with other fungicides with a different mode of action to reduce the possibility of resistance developing

Restrictions
- Maximum total dose equivalent to ten full dose treatments in apples and pears; six full dose treatments in blackcurrants, gooseberries, strawberries
- Maximum number of treatments on managed amenity turf: two per yr [3]
- Do not mow turf within 24 hr after treatment [3]

Crop-specific information
- HI 28 d (grapevines); 21 d (cherries, mirabelles); 14 d (apples, pears, blackcurrants, gooseberries, hops); 3 d (plums, protected blackberries, protected raspberries, protected Rubus hybrids, strawberries)
- Product may be applied to newly sown turf after the two-leaf stage but in view of the large number of turf grass cultivars a safety test on a small area is recommended before large scale treatment [3]

Environmental safety
- Dangerous for the environment
- Toxic (harmful [3]) to aquatic organisms
- LERAP Category B [3]

Hazard classification and safety precautions
  Hazard Harmful, Dangerous for the environment [1, 2, 4, 5]
  Transport code 9 [1, 2, 4, 5]
  Packaging group III [1, 2, 4, 5]
  UN Number 3082 [1, 2, 4, 5]; N/C [3]
  Risk phrases R22b, R51, R63 [1, 2, 4, 5]; R52 [3]; R53a [1-5]
  Operator protection A, H [1-5]; J [1, 2, 4, 5]; U08, U20a [1-5]; U15 [1, 2, 4, 5]
  Environmental protection E15a [1-5]; E16a [3]
  Storage and disposal D01, D02 [1, 2, 4, 5]; D05, D10c, D12a [1-5]
  Medical advice M05b [1, 2, 4, 5]

345   1-naphthylacetic acid

A plant growth regulator to promote rooting of cuttings

See also 4-indol-3-ylbutyric acid + 1-naphthylacetic acid

Products

1	Rhizopon B Powder (0.1%)	Fargro	0.1% w/w	DP	09089
2	Rhizopon B Powder (0.2%)	Fargro	0.2% w/w	DP	09090
3	Rhizopon B Tablets	Fargro	25 mg a.i.	WT	09091
4	Tipoff	Globachem	57 g/l	EC	15728

SEE SECTION 3 FOR PRODUCTS ALSO REGISTERED

Uses
- Growth regulation in *apples*, *blackberries* (off-label), *cherries*, *hibridberry* (off-label), *pears*, *plums*, *raspberries* [4]
- Rooting of cuttings in *ornamental plant production* [1-3]

Extension of Authorisation for Minor Use (EAMUs)
- *blackberries* 20120980 [4]
- *hibridberry* 20120980 [4]

Approval information
- 1-naphthylacetic acid included in Annex I under EC Regulation 1107/2009

Efficacy guidance
- Dip moistened base of cuttings into powder immediately before planting [1, 2]

Restrictions
- Maximum number of treatments 1 per cutting

Crop-specific information
- Latest use: before cutting insertion for ornamental specimens

Hazard classification and safety precautions
    UN Number N/C
    Operator protection U08, U19a, U20a [1-3]; U20c [4]
    Environmental protection E15a [1-4]; E34 [1-3]
    Storage and disposal D05 [4]; D09a, D11a [1-4]

346   napropamide

A soil applied alkanamide herbicide for oilseed rape, fruit and woody ornamentals
HRAC mode of action code: K3

Products

1	AC 650	United Phosphorus	450 g/l	SC	11102
2	Associate	Belchim	450 g/l	SC	13933
3	Devrinol	United Phosphorus	450 g/l	SC	09374
4	MAC-Napropamide 450 SC	AgChem Access	450 g/l	SC	13972
5	Nappa	ChemSource	450 g/l	SC	14881

Uses
- Annual dicotyledons in *bilberries* (off-label), *blackberries* (off-label), *blueberries* (off-label), *chinese cabbage* (off-label), *choi sum* (off-label), *collards* (off-label), *evening primrose* (off-label), *farm forestry* (off-label), *forest* (off-label), *honesty* (off-label), *lamb's lettuce* (off-label), *mustard* (off-label), *pak choi* (off-label), *redcurrants* (off-label), *rocket* (off-label), *rubus hybrids* (off-label), *whitecurrants* (off-label) [3]; *blackcurrants*, *broccoli*, *brussels sprouts*, *cabbages*, *calabrese*, *cauliflowers*, *forest nurseries*, *gooseberries*, *kale*, *ornamental plant production*, *raspberries*, *strawberries*, *woody ornamentals* [3-5]; *linseed* (off-label) [1, 3]; *winter oilseed rape* [1, 5]
- Annual grasses in *bilberries* (off-label), *blackberries* (off-label), *blueberries* (off-label), *chinese cabbage* (off-label), *choi sum* (off-label), *collards* (off-label), *evening primrose* (off-label), *farm forestry* (off-label), *forest* (off-label), *honesty* (off-label), *lamb's lettuce* (off-label), *mustard* (off-label), *pak choi* (off-label), *redcurrants* (off-label), *rocket* (off-label), *rubus hybrids* (off-label), *whitecurrants* (off-label) [3]; *blackcurrants*, *broccoli*, *brussels sprouts*, *cabbages*, *calabrese*, *cauliflowers*, *forest nurseries*, *gooseberries*, *kale*, *ornamental plant production*, *raspberries*, *strawberries*, *woody ornamentals* [3-5]; *linseed* (off-label) [1, 3]; *winter oilseed rape* [1, 5]
- Annual meadow grass in *broccoli*, *brussels sprouts*, *cabbages*, *calabrese*, *cauliflowers*, *kale*, *winter oilseed rape* [2]
- Chickweed in *broccoli*, *brussels sprouts*, *cabbages*, *calabrese*, *cauliflowers*, *kale*, *winter oilseed rape* [2]

- Cleavers in **blackcurrants, broccoli, brussels sprouts, cabbages, calabrese, cauliflowers, forest nurseries, gooseberries, kale, ornamental plant production, raspberries, strawberries, woody ornamentals** [3-5]; **winter oilseed rape** [5]
- Groundsel in **blackcurrants, broccoli, brussels sprouts, cabbages, calabrese, cauliflowers, forest nurseries, gooseberries, kale, ornamental plant production, raspberries, strawberries, woody ornamentals** [3-5]; **winter oilseed rape** [5]
- Knotgrass in **broccoli, brussels sprouts, cabbages, calabrese, cauliflowers, kale, winter oilseed rape** [2]
- Mayweeds in **broccoli, brussels sprouts, cabbages, calabrese, cauliflowers, kale, winter oilseed rape** [2]

Extension of Authorisation for Minor Use (EAMUs)
- **bilberries** *20090403* [3]
- **blackberries** *20090399* [3]
- **blueberries** *20090403* [3]
- **chinese cabbage** *20111741* [3]
- **choi sum** *20111741* [3]
- **collards** *20111741* [3]
- **evening primrose** *20090400* [3]
- **farm forestry** *20090409* [3]
- **forest** *20090409* [3]
- **honesty** *20090400* [3]
- **lamb's lettuce** *20111742* [3]
- **linseed** *20090397* [1], *20090400* [3]
- **mustard** *20090400* [3]
- **pak choi** *20111741* [3]
- **redcurrants** *20090403* [3]
- **rocket** *20111742* [3]
- **rubus hybrids** *20090399* [3]
- **whitecurrants** *20090403* [3]

Approval information
- Napropamide included in Annex I under EC Regulation 1107/2009

Efficacy guidance
- Best results obtained from treatment pre-emergence of weeds but product may be used in conjunction with contact herbicides for control of emerged weeds. Otherwise remove existing weeds before application
- Weed control may be reduced where spray is mixed too deeply in the soil
- Manure, crop debris or other organic matter may reduce weed control
- Napropamide broken down by sunlight, so application during such conditions not recommended. Most crops recommended for treatment between Nov and end-Feb
- Apply to winter oilseed rape pre-emergence or as pre-drilling treatment in tank-mixture with trifluralin and incorporate within 30 min [1]
- Post-emergence use of specific grass weedkilller recommended where volunteer cereals are a serious problem
- Increase water volume to ensure adequate dampening of compost when treating containerised nursery stock with dense leaf canopy [3]

Restrictions
- Maximum number of treatments 1 per crop or yr
- Do not use on Sands
- Do not use on soils with more than 10% organic matter
- Strawberries, blackcurrants, gooseberries and raspberries must be treated between 1 Nov and end Feb [3]
- Applications to strawberries must only be made where the mature foliage has been previously removed [3]
- Applications to ornamentals and forest nurseries must be made after 1 Nov and before end Apr
- Do not treat ornamentals in containers of less than 1 litre [3]

SEE SECTION 3 FOR PRODUCTS ALSO REGISTERED

SECTION 2

- Some phytotoxicity seen on yellow and golden varieties of conifers and container grown alpines. On any ornamental variety treat only a small number of plants in first season [3]
- Consult processors before use on any crop for processing

Crop-specific information

- Latest use: before transplanting for brassicas; pre-emergence for winter oilseed rape; before end Feb for strawberries, bush and cane fruit; before end of Apr for field and container grown ornamental trees and shrubs
- Where minimal cultivation used to establish oilseed rape, tank-mixture may be applied directly to stubble and mixed into top 25 mm as part of surface cultivations [1]
- Apply up to 14 d prior to drilling winter oilseed rape [1]
- Apply to strawberries established for at least one season or to maiden crops as long as planted carefully and no roots exposed, between 1 Nov and end Feb. Do not treat runners of poor vigour or with shallow roots, or runner beds [3]
- Newly planted ornamentals should have no roots exposed. Do not treat stock of poor vigour or with shallow roots. Treatments made in Mar and Apr must be followed by 25 mm irrigation within 24 h [3]
- Bush and cane fruit must be established for at least 10 mth before spraying and treated between 1 Nov and end Feb [3]

Following crops guidance

- After use in fruit or ornamentals no crop can be drilled within 7 mth of treatment. Leaf, flowerhead, root and fodder brassica crops may be drilled after 7 mth; potatoes, maize, peas or dwarf beans after 9 mth; autumn sown wheat or grass after 18 mth; any crop after 2 yr
- After use in oilseed rape only oilseed rape, swedes, fodder turnips, brassicas or potatoes should be sown within 12 mth of application
- Soil should be mould-board ploughed to a depth of at least 200 mm before drilling or planting any following crop

Environmental safety

- Dangerous for the environment
- Toxic to aquatic organisms

Hazard classification and safety precautions

Hazard Irritant, Dangerous for the environment
Transport code 9
Packaging group III
UN Number 3082
Risk phrases R36, R38, R51, R53a
Operator protection A [1-5]; C [2-5]; U05a, U09a, U11, U20b
Environmental protection E13c [2]; E15a [3-5]; E34, E38 [1, 3-5]
Storage and disposal D01, D02, D05, D09a, D10c [1-5]; D12a [1, 2]; D12b [3-5]

347   nicosulfuron

A sulfonylurea herbicide for maize
HRAC mode of action code: B

See also mesotrione + nicosulfuron

Products

1	Accent	DuPont	75% w/w	WG	15392
2	Agrotech-Nicosulfron 40 SC	AgChem Access	40 g/l	SC	12938
3	Bishop	AgChem Access	40 g/l	SC	14055
4	Bishop Extra	ChemSource	60 g/l	SC	15824
5	Clayton Delilah	Clayton	40 g/l	SC	12610
6	Clayton Delilah XL	Clayton	60 g/l	SC	14443
7	Clayton Myth	Clayton	40 g/l	SC	14489
8	Clayton Myth XL	Clayton	60 g/l	SC	14466

FOR FULL CONDITIONS OF USE ALWAYS READ THE PRODUCT LABEL

Products – continued

9	Crew	Nufarm UK	40 g/l	OD	15770
10	Entail	Headland	240 g/l	OD	15671
11	Greencrop Folklore	Greencrop	40 g/l	SC	12867
12	Nico 4	Goldengrass	40 g/l	SC	14293
13	Samson Extra 6%	Syngenta	60 g/l	OD	15662
14	Standon Frontrunner	Standon	40 g/l	SC	13308

Uses

- Amaranthus in *forage maize*, *grain maize* [4, 6, 8, 9]
- Annual dicotyledons in *forage maize* [1-3, 5, 7, 10-14]; *grain maize* [10, 13]; *sweetcorn (off-label)* [6]
- Annual grasses in *sweetcorn (off-label)* [6]
- Annual meadow grass in *forage maize* [1-9, 11, 12, 14]; *grain maize* [4, 6, 8, 9]; *sweetcorn (off-label)* [6]
- Black bindweed in *forage maize*, *grain maize* [4, 6, 8]
- Black nightshade in *forage maize*, *grain maize* [4, 6, 8]
- Blackgrass in *forage maize* [1]
- Chickweed in *forage maize* [2-8, 11, 12, 14]; *grain maize* [4, 6, 8]
- Cockspur grass in *forage maize* [1]
- Couch in *forage maize* [1-3, 5, 7, 11, 12, 14]
- Fat hen in *forage maize*, *grain maize* [4, 6, 8]
- Grass weeds in *forage maize*, *grain maize* [10, 13]
- Groundsel in *forage maize*, *grain maize* [9]
- Italian ryegrass in *forage maize*, *grain maize* [4, 6, 8]
- Knotgrass in *forage maize*, *grain maize* [4, 6, 8]
- Mayweeds in *forage maize* [2-9, 11, 12, 14]; *grain maize* [4, 6, 8, 9]
- Perennial grasses in *sweetcorn (off-label)* [6]
- Perennial ryegrass in *forage maize*, *grain maize* [4, 6, 8]
- Redshank in *forage maize*, *grain maize* [4, 6, 8]
- Ryegrass in *forage maize* [1-3, 5, 7, 9, 11, 12, 14]; *grain maize* [9]
- Shepherd's purse in *forage maize* [2-9, 11, 12, 14]; *grain maize* [4, 6, 8, 9]
- Volunteer cereals in *forage maize* [1]
- Volunteer oilseed rape in *forage maize* [2, 3, 5, 7, 11, 12, 14]
- Wild oats in *forage maize* [1]

Extension of Authorisation for Minor Use (EAMUs)

- *sweetcorn* 20121687 [6]

Approval information

- Nicosulfuron included in Annex I under EC Regulation 1107/2009

Efficacy guidance

- Product should be applied post-emergence between 2 and 8 crop leaf stage and to emerged weeds from the 2-leaf stage
- Product acts mainly by foliar activity. Ensure good spray cover of the weeds
- Nicosulfuron is a member of the ALS-inhibitor group of herbicides. To avoid the build up of resistance do not use any product containing an ALS-inhibitor herbicide with claims for control of grass weeds more than once on any crop
- Use these products as part of a resistance management strategy that includes cultural methods of control and does not use ALS inhibitors as the sole chemical method of grass weed control

Restrictions

- Maximum number of treatments 1 per crop
- Do not use if an organophosphorus soil insecticide has been used on the same crop
- Do not mix with foliar or liquid fertilisers or specified herbicides. See label for details
- Do not apply in mixture, or sequence, with any other sulfonyl-urea containing product
- Do not treat crops under stress
- Do not apply if rainfall is forecast to occur within 6 h of application

SEE SECTION 3 FOR PRODUCTS ALSO REGISTERED

SECTION 2

Crop-specific information
- Latest use: up to and including 8 true leaves for maize
- Some transient yellowing may be seen from 1-2 wk after treatment
- Only healthy maize crops growing in good field conditions should be treated

Following crops guidance
- Winter wheat (not undersown) may be sown 4 mth after treatment; maize may be sown in the spring following treatment; all other crops may be sown from the next autumn

Environmental safety
- Dangerous for the environment [11, 14]
- Very toxic to aquatic organisms [11, 14]
- LERAP Category A [9]; LERAP Category B [1-8, 10-14]

Hazard classification and safety precautions

Hazard Harmful [4, 6, 8, 10, 13]; Irritant [2, 3, 5, 7, 9, 11, 12, 14]; Dangerous for the environment [1, 2, 4, 6-11, 13, 14]

Transport code 9

Packaging group III

UN Number 3077 [1]; 3082 [2-14]

Risk phrases R20, R43 [4, 6, 8-10, 13]; R36 [4, 6, 8, 10, 13]; R38 [2, 3, 5, 7, 9, 11, 12, 14]; R50, R53a [1, 2, 4, 6-11, 13, 14]; R53b [7]

Operator protection A, H [2-14]; C [4, 6, 8, 10, 13]; U02a, U20b [2-8, 10-14]; U05a [1-14]; U10 [2, 4, 6-8, 10, 11, 13, 14]; U11 [4, 6, 8, 10, 13]; U14 [2-14]; U15 [9]; U19a [4, 6, 8-10, 13]

Environmental protection E15a [3, 5, 7, 11, 12]; E15b [1, 2, 9, 10, 13, 14]; E16a [1-8, 10-14]; E16b [2-5, 7, 8, 10-14]; E16c [9]; E34 [1, 9]; E38 [1, 2, 9-11, 13, 14]; E40b [1]

Storage and disposal D01, D02 [1-3, 5, 7, 9-14]; D05 [2, 11, 14]; D09a [1, 3, 5, 7, 9, 12, 14]; D10b [3, 5, 7, 12]; D10c [2, 9-11, 13, 14]; D11a [1]; D12a [1, 2, 10, 11, 13, 14]; D12b [9]

Medical advice M03 [9]; M05a [2-8, 10-14]

348 oxadiazon

A residual and contact oxadiazole herbicide for fruit and ornamentals
HRAC mode of action code: E

Products

1 Clayton Oxen FL	Clayton	250 g/l	EC	12861
2 Festival	Bayer Environ.	250 g/l	EC	14606
3 Ronstar 2G	Certis	2% w/w	GR	12965
4 Ronstar Liquid	Certis	250 g/l	EC	11215
5 Standon Roxx L	Standon	250 g/l	EC	13309

Uses
- Annual dicotyledons in *almonds* (off-label), *bilberries* (off-label), *blackberries* (off-label), *blueberries* (off-label), *chestnuts* (off-label), *cranberries* (off-label), *hazel nuts* (off-label), *redcurrants* (off-label), *rubus hybrids* (off-label), *walnuts* (off-label), *whitecurrants* (off-label) [4]; *amenity vegetation* (off-label), *ornamental plant production* [3]; *apple orchards, blackcurrants, gooseberries, hops, pear orchards, raspberries, wine grapes, woody ornamentals* [1, 4, 5]
- Annual grasses in *almonds* (off-label), *bilberries* (off-label), *blackberries* (off-label), *blueberries* (off-label), *chestnuts* (off-label), *cranberries* (off-label), *hazel nuts* (off-label), *redcurrants* (off-label), *rubus hybrids* (off-label), *walnuts* (off-label), *whitecurrants* (off-label) [4]; *amenity vegetation* (off-label), *ornamental plant production* [3]; *apple orchards, blackcurrants, gooseberries, hops, pear orchards, raspberries, wine grapes, woody ornamentals* [1, 4, 5]
- Annual meadow grass in *ornamental plant production* [2]
- Bindweeds in *almonds* (off-label), *bilberries* (off-label), *blackberries* (off-label), *blueberries* (off-label), *chestnuts* (off-label), *cranberries* (off-label), *hazel nuts* (off-label), *redcurrants* (off-label), *rubus hybrids* (off-label), *walnuts* (off-label), *whitecurrants* (off-label) [4]; *apple orchards, blackcurrants, gooseberries, hops, pear orchards, raspberries, wine grapes, woody ornamentals* [1, 4, 5]

FOR FULL CONDITIONS OF USE ALWAYS READ THE PRODUCT LABEL

- Black bindweed in *ornamental plant production* [2]
- Charlock in *ornamental plant production* [2]
- Cleavers in *almonds* (off-label), *bilberries* (off-label), *blackberries* (off-label), *blueberries* (off-label), *chestnuts* (off-label), *cranberries* (off-label), *hazel nuts* (off-label), *redcurrants* (off-label), *rubus hybrids* (off-label), *walnuts* (off-label), *whitecurrants* (off-label) [4]; *apple orchards, blackcurrants, gooseberries, hops, pear orchards, raspberries, wine grapes, woody ornamentals* [1, 4, 5]
- Corn spurrey in *ornamental plant production* [2]
- Fat hen in *ornamental plant production* [2]
- Groundsel in *ornamental plant production* [2]
- Knotgrass in *almonds* (off-label), *bilberries* (off-label), *blackberries* (off-label), *blueberries* (off-label), *chestnuts* (off-label), *cranberries* (off-label), *hazel nuts* (off-label), *redcurrants* (off-label), *rubus hybrids* (off-label), *walnuts* (off-label), *whitecurrants* (off-label) [4]; *apple orchards, blackcurrants, gooseberries, hops, pear orchards, raspberries, wine grapes, woody ornamentals* [1, 4, 5]; *ornamental plant production* [2]
- Mayweeds in *ornamental plant production* [2]
- Red dead-nettle in *ornamental plant production* [2]
- Redshank in *ornamental plant production* [2]
- Shepherd's purse in *ornamental plant production* [2]
- Small nettle in *ornamental plant production* [2]
- Sowthistle in *ornamental plant production* [2]
- Speedwells in *ornamental plant production* [2]
- Sun spurge in *ornamental plant production* [2]
- Wild radish in *ornamental plant production* [2]

Extension of Authorisation for Minor Use (EAMUs)
- *almonds* 20063295 [4]
- *amenity vegetation* 20102995 [3]
- *bilberries* 20063296 [4]
- *blackberries* 20063297 [4]
- *blueberries* 20063296 [4]
- *chestnuts* 20063295 [4]
- *cranberries* 20063296 [4]
- *hazel nuts* 20063295 [4]
- *redcurrants* 20063296 [4]
- *rubus hybrids* 20063297 [4]
- *walnuts* 20063295 [4]
- *whitecurrants* 20063296 [4]

Approval information
- Oxadiazon included in Annex I under EC Regulation 1107/2009
- Accepted by BBPA for use on hops

Efficacy guidance
- Apply as a directed spray or as a spot treatment as directed. Avoid direct contact with the crop. See label [1, 4, 5]
- Apply granules as soon after potting as possible and before crop plant is making soft growth [3]
- Best results from spray treatments obtained when weeds are dry at application and for a period afterwards. Later rain or overhead watering is needed for effective results [1, 4, 5]
- Residual activity reduced on soils with more than 10% organic matter and, in these conditions, post-emergence treatment is more effective [1, 4, 5]
- Best results on bindweed when first shoots are 10-15 cm long [1, 4, 5]
- Adjust water volume to achieve good coverage when weed density is high [1, 4, 5]

Restrictions
- Maximum number of treatments 2 per yr for ornamental plant production [3]
- Maximum total dose per yr equivalent to one full dose treatment on edible crops [1, 4, 5]
- See label for list of ornamental species which may be treated with granules. Treat small numbers of other species to check safety. Do not treat Hydrangea, Spiraea or Genista [3]
- Do not cultivate after treatment [1, 4, 5]

SEE SECTION 3 FOR PRODUCTS ALSO REGISTERED

- Do not treat container stock under glass or use on plants rooted in media with high sand or non-organic content. Do not apply to plants with wet foliage [3]
- Consult processors before treatment of crops intended for processing [1, 4, 5]

Crop-specific information
- Latest use: Jul for apples, grapevines, hops, pears; Jun for raspberries, woody ornamentals
- Apply spray to apples and pears from Jan to Jul, avoiding young growth [1, 4, 5]
- Treat bush fruit from Jan to bud-break, avoiding bushes, grapevines in Feb/Mar before start of new growth or in Jun/Jul avoiding foliage [1, 4, 5]
- Treat hops cropped for at least 2 yr in Feb or in Jun/Jul after deleafing [1, 4, 5]
- Treat woody ornamentals from Jan to Jun, avoiding young growth. Do not spray container stock overall [1, 4, 5]

Following crops guidance
- A period of at least 6 mth must elapse between treatment at half the recommended maximum dose and planting a following crop. 12 mth must elapse after use of the maximum recommended dose. In either case soil must be ploughed to min 15 cm and cultivated before sowing or planting a following crop [1, 4, 5]

Environmental safety
- Flammable [1, 4, 5]
- Dangerous for the environment [1, 4, 5]
- Very toxic to aquatic organisms [1, 4, 5]
- Harmful to aquatic organisms [3]

Hazard classification and safety precautions
Hazard Harmful, Flammable, Dangerous for the environment [1, 2, 4, 5]
Transport code 3 [1, 2, 4, 5]; 9 [3]
Packaging group III
UN Number 1993 [1, 2, 4, 5]; 3077 [3]
Risk phrases R22b, R36, R38, R50 [1, 2, 4, 5]; R41 [4]; R52 [3]; R53a [1-5]; R67 [1, 4, 5]
Operator protection A, C [1-5]; H [2, 5]; U02a [1, 5]; U05a, U20b [1-5]; U08 [1, 4, 5]; U09a [2, 3]; U14, U15 [2-5]; U19a [3]
Environmental protection E15a [1, 3-5]; E38 [4]
Storage and disposal D01, D02, D09a [1-5]; D05 [2, 5]; D10a [1, 2, 4, 5]; D11a [3]; D12a [1, 3-5]
Medical advice M05b [1, 4, 5]

349   oxamyl

A soil-applied, systemic carbamate nematicide and insecticide
IRAC mode of action code: 1A

Products
1	Tecka	Agroquimicos	10% w/w	GR	15831
2	Vydate 10G	DuPont	10% w/w	GR	02322

Uses
- American serpentine leaf miner in **ornamental plant production** *(off-label)*, **protected aubergines** *(off-label)*, **protected ornamentals** *(off-label)*, **protected tomatoes** *(off-label)* [2]
- Aphids in **fodder beet** *(off-label)* [2]; **potatoes, sugar beet** [1, 2]
- Docking disorder vectors in **fodder beet** *(off-label)* [2]; **sugar beet** [1, 2]
- Eelworm in **bulb onions** *(off-label)*, **garlic** *(off-label)*, **shallots** *(off-label)* [2]
- Free-living nematodes in **potatoes** [1, 2]
- Insect pests in **garlic** *(off-label)* [2]
- Leaf miner in **ornamental plant production** *(off-label)*, **protected aubergines** *(off-label)*, **protected ornamentals** *(off-label)*, **protected tomatoes** *(off-label)* [2]
- Mangold fly in **fodder beet** *(off-label)* [2]; **sugar beet** [1, 2]
- Millipedes in **fodder beet** *(off-label)* [2]; **sugar beet** [1, 2]
- Potato cyst nematode in **potatoes** [1, 2]
- Pygmy beetle in **fodder beet** *(off-label)* [2]; **sugar beet** [1, 2]

FOR FULL CONDITIONS OF USE ALWAYS READ THE PRODUCT LABEL

- South American leaf miner in ***ornamental plant production*** *(off-label)*, ***protected aubergines*** *(off-label)*, ***protected ornamentals*** *(off-label)*, ***protected tomatoes*** *(off-label)* [2]
- Spraing vectors in ***potatoes*** [1, 2]
- Stem nematodes in ***bulb onions*** *(off-label)*, ***garlic*** *(off-label)*, ***shallots*** *(off-label)* [2]; ***carrots***, ***parsnips*** [1, 2]

Extension of Authorisation for Minor Use (EAMUs)
- ***bulb onions*** *20061890* [2]
- ***fodder beet*** *20061107* [2]
- ***garlic*** *20061890* [2]
- ***ornamental plant production*** *20122322* [2]
- ***protected aubergines*** *20122322* [2]
- ***protected ornamentals*** *20122322* [2]
- ***protected tomatoes*** *20122322* [2]
- ***shallots*** *20061890* [2]

Approval information
- Oxamyl included in Annex I under EC Regulation 1107/2009
- In 2006 CRD required that all products containing this active ingredient should carry the following warning in the main area of the container label: "Oxamyl is an anticholinesterase carbamate. Handle with care"

Efficacy guidance
- Apply granules with suitable applicator before drilling or planting. See label for details of recommended machines
- In potatoes incorporate thoroughly to 10 cm and plant within 1-2 d
- In sugar beet apply in seed furrow at drilling

Restrictions
- Oxamyl is subject to the Poisons Rules 1982 and the Poisons Act 1972. See Section 5 for more information
- Contains an anticholinesterase carbamate compound. Do not use if under medical advice not to work with such compounds
- Maximum number of treatments 1 per crop or yr
- Keep in original container, tightly closed, in a safe place, under lock and key
- Wear protective gloves if handling treated compost or soil within 2 wk after treatment
- Allow at least 12 h, followed by at least 1 h ventilation, before entry of unprotected persons into treated glasshouses

Crop-specific information
- Latest use: at drilling/planting for vegetables; before drilling/planting for potatoes and peas.
- HI:potatoes, sugar beet, carrots, parsnips 12 wk

Environmental safety
- Dangerous for the environment
- Toxic to aquatic organisms
- Dangerous to fish or other aquatic life. Do not contaminate surface waters or ditches with chemical or used container
- Dangerous to game, wild birds and animals. Bury spillages

Hazard classification and safety precautions
Hazard Toxic, Dangerous for the environment
Transport code 6.1
Packaging group II
UN Number 2757
Risk phrases R23, R25, R51, R53a
Operator protection A, B, H, K, M; C (or D); U02a, U04a, U05a, U07, U09a, U13, U19a, U20a
Environmental protection E10a, E13b, E34, E36a, E38
Storage and disposal D01, D02, D09a, D11a, D14
Medical advice M04a

350   paclobutrazol

A triazole plant growth regulator for ornamentals and fruit

Products

1 Agrovista Paclo	Agrovista	250 g/l	SC	13424
2 Bonzi	Syngenta Bioline	4 g/l	SC	13623
3 Cultar	Syngenta	250 g/l	SC	10523
4 Pirouette	Fine	4 g/l	SC	13073
5 Pirouette	Fine	4 g/l	SC	13137

Uses

- Control of shoot growth in *apples*, *pears* [1, 3]
- Growth regulation in *cherries* *(off-label)*, *plums* *(off-label)* [1, 3]; *quinces* *(off-label)* [1]
- Improving colour in *poinsettias* [2, 4, 5]
- Increasing flowering in *azaleas*, *bedding plants*, *begonias*, *kalanchoes*, *lilies*, *roses*, *tulips* [2, 4, 5]
- Increasing fruit set in *apples*, *pears* [1, 3]; *quinces* *(off-label)* [3]
- Stem shortening in *azaleas*, *bedding plants*, *begonias*, *kalanchoes*, *lilies*, *poinsettias*, *roses*, *tulips* [2, 4, 5]

Extension of Authorisation for Minor Use (EAMUs)

- *cherries* 20080461 [1], 20031235 [3]
- *plums* 20080461 [1], 20031235 [3]
- *quinces* 20080462 [1], 20062993 [3]

Approval information

- Paclobutrazol included in Annex I under EC Regulation 1107/2009
- Approval expiry 31 May 2013 [5]

Efficacy guidance

- Chemical is active via both foliage and root uptake. For best results apply in dull weather when relative humidity not high

Restrictions

- Maximum number of treatments 1 per specimen for some species [4]
- Maximum total dose equivalent to 3 full dose treatments for pears and 4 full dose treatments for apples [1, 3]
- Some varietal restrictions apply in top fruit (see label for details) [1, 3]
- Do not use on trees of low vigour or under stress [1, 3]
- Do not use on trees from green cluster to 2 wk after full petal fall [1, 3]
- Do not use in underplanted orchards or those recently interplanted [1, 3]

Crop-specific information

- HI apples, pears 14 d [1, 3]
- Apply to apple and pear trees under good growing conditions as pre-blossom spray (apples only) and post-blossom at 7-14 d intervals [1, 3]
- Timing and dose of orchard treatments vary with species and cultivar. See label [1, 3]

Following crops guidance

- Chemical has residual soil activity which can affect growth of following crops. Withhold treatment from orchards due for grubbing to allow the following intervals between the last treatment and planting the next crop: apples, pears 1 yr; beans, peas, onions 2 yr; stone fruit 3 yr; cereals, grass, oilseed rape, carrots 4 yr; market brassicas 6 yr; potatoes and other crops 7 yr [1, 3]

Environmental safety

- Harmful to aquatic organisms
- Do not use on food crops [4]
- Keep livestock out of treated areas for at least 2 years after treatment [1, 3]

FOR FULL CONDITIONS OF USE ALWAYS READ THE PRODUCT LABEL

Hazard classification and safety precautions
  Hazard Irritant [1, 3]
  UN Number N/C
  Risk phrases R36, R52, R53a [1, 3]
  Operator protection A, H, M [1, 3]; U05a, U20a [1-5]; U08 [2, 4, 5]
  Environmental protection E06a [1, 3] (2 yr); E13c [4, 5]; E15a [2-5]; E15b [1]; E34, E38 [1, 3]
  Consumer protection C01 [2, 4, 5]; C02a [1, 3] (2 wk)
  Storage and disposal D01, D02, D09a [1-5]; D05 [2, 4, 5]; D10b [4, 5]; D10c [1-3]; D12a [1, 3]
  Medical advice M03 [1, 3]

SECTION 2

351   pelargonic acid

A naturally occuring 9 carbon acid terminating in a carboxylic acid only available in mixtures
HRAC mode of action code: Not classified

See also maleic hydrazide + pelargonic acid

352   penconazole

A protectant triazole fungicide with antisporulant activity
FRAC mode of action code: 3

See also captan + penconazole

Products
1 Agrovista Penco	Agrovista	100 g/l	EC	13425
2 Topas	Syngenta	100 g/l	EC	09717
3 Topenco 100 EC	Globachem	100 g/l	EC	11972

Uses
• Powdery mildew in **apples** [1-3]; **bilberries** (off-label), **blackcurrants**, **blueberries** (off-label), **cranberries** (off-label), **gooseberries** (off-label), **hops**, **ornamental trees**, **protected strawberries** (off-label), **redcurrants** (off-label), **strawberries** (off-label), **whitecurrants** (off-label) [1, 2]
• Rust in **roses** [1, 2]
• Scab in **ornamental trees** [1, 2]

Extension of Authorisation for Minor Use (EAMUs)
• **bilberries** *20080477* [1], *20061106* [2]
• **blueberries** *20080477* [1], *20061106* [2]
• **cranberries** *20080477* [1], *20061106* [2]
• **gooseberries** *20080477* [1], *20061106* [2]
• **protected strawberries** *20080478* [1], *20073651* [2]
• **redcurrants** *20080477* [1], *20061106* [2]
• **strawberries** *20080478* [1], *20073651* [2]
• **whitecurrants** *20080477* [1], *20061106* [2]

Approval information
• Accepted by BBPA for use on hops
• Penconazole included in Annex I under EC Regulation 1107/2009

Efficacy guidance
• Use as a protectant fungicide by treating at the earliest signs of disease
• Treat crops every 10-14 d (every 7-10 d in warm, humid weather) at first sign of infection or as a protective spray ensuring complete coverage. See label for details of timing
• Increase dose and volume with growth of hops but do not exceed 2000 l/ha. Little or no activity will be seen on established powdery mildew
• Antisporulant activity reduces development of secondary mildew in apples

SEE SECTION 3 FOR PRODUCTS ALSO REGISTERED

Restrictions
- Maximum number of treatments 10 per yr for apples, 6 per yr for hops, 4 per yr for blackcurrants
- Check for varietal susceptibility in roses. Some defoliation may occur after repeat applications on Dearest

Crop-specific information
- HI apples, hops 14 d; currants, bilberries, blueberries, cranberries, gooseberries 4 wk

Environmental safety
- Dangerous for the environment
- Toxic to aquatic organisms

Hazard classification and safety precautions
Hazard Irritant, Dangerous for the environment
Transport code 3
Packaging group III
UN Number 1915
Risk phrases R36, R51, R53a [1-3]; R38 [3]
Operator protection A, C; U02a, U05a, U20b [1-3]; U08 [1, 2]; U11 [3]
Environmental protection E15a [1-3]; E22c [3]; E38 [1, 2]
Storage and disposal D01, D02, D05, D09a [1-3]; D07 [1, 3]; D10b [3]; D10c, D12a [1, 2]
Medical advice M05b

353 pencycuron

A non-systemic urea fungicide for use on seed potatoes
FRAC mode of action code: 20

See also imazalil + pencycuron

Products

1	CS Pencycuron	ChemSource	12.5% w/w	DS	15349
2	Monceren DS	Bayer CropScience	12.5% w/w	DS	11292

Uses
- Black scurf in **potatoes** *(tuber treatment)*
- Stem canker in **potatoes** *(tuber treatment)*

Approval information
- Pencycuron included in Annex I under EC Regulation 1107/2009

Efficacy guidance
- Provides control of tuber-borne disease and gives some reduction of stem canker
- Seed tubers should be of good quality and vigour, and free from soil deposits when treated
- Dust formulations should be applied immediately before, or during, the planting process by treating seed tubers in chitting trays, in bulk bins immediately before planting or in hopper at planting. Liquid formulations may be applied by misting equipment at any time, into or out of store but treatment over a roller table at the end of grading out, or at planting, is usually most convenient
- Whichever method is chosen, complete and even distribution on the tubers is essential for optimum efficacy
- Apply in accordance with detailed guidelines in manufacturer's literature
- If rain interrupts planting, cover dust treated tubers in hopper

Restrictions
- Maximum number of treatments 1 per batch of seed potatoes
- Treated tubers must be used only as seed and not for human or animal consumption
- Do not use on tubers previously treated with a dry powder seed treatment or hot water
- To prevent rapid absorption through damaged 'eyes' seed tubers should only be treated before the 'eyes' begin to open or immediately before or during planting
- Tubers removed from cold storage must be allowed to attain a temperature of at least 8°C before treatment

FOR FULL CONDITIONS OF USE ALWAYS READ THE PRODUCT LABEL

- Use of suitable dust mask is mandatory when applying dust, filling the hopper or riding on planter [2]
- Treated tubers in bulk bins must be allowed to dry before being stored, especially when intended for cold storage
- Some internal staining of boxes used to store treated tubers may occur. Ware tubers should not be stored or supplied to packers in such stained boxes

Crop-specific information
- Latest use: at planting

Environmental safety
- Do not use treated seed as food or feed
- Treated seed harmful to game and wildlife

Hazard classification and safety precautions
  UN Number N/C
  Operator protection A, F; U20b
  Environmental protection E03, E15a
  Storage and disposal D09a, D11a
  Treated seed S01, S02, S03, S04a, S05, S06a

354   pendimethalin

A residual dinitroaniline herbicide for cereals and other crops
HRAC mode of action code: K1

See also bentazone + pendimethalin
chlorotoluron + pendimethalin
diflufenican + pendimethalin
dimethenamid-p + pendimethalin
flufenacet + pendimethalin
imazamox + pendimethalin
isoproturon + pendimethalin

Products

1	Anthem	Makhteshim	400 g/l	SC	15761
2	Blazer M	Headland	330 g/l	EC	15084
3	Clinker	ChemSource	400 g/l	CS	15453
4	Fastnet	Sipcam	330 g/l	EC	14068
5	Nighthawk 330 EC	AgChem Access	330 g/l	EC	14083
6	PDM 330 EC	BASF	330 g/l	EC	13406
7	Pendragon	AgriGuard	400 g/l	SC	15512
8	Sherman	Makhteshim	330 g/l	EC	13859
9	Stamp 330 EC	AgChem Access	330 g/l	EC	13995
10	Stomp Aqua	BASF	455 g/l	CS	14664
11	Trample	AgChem Access	330 g/l	EC	14173
12	Yomp	Becesane	330 g/l	EC	15609

Uses
- Annual dicotyledons in *almonds* (off-label), *asparagus* (off-label), *bilberries* (off-label), *blueberries* (off-label), *chestnuts* (off-label), *cranberries* (off-label), *cress* (off-label), *evening primrose* (off-label), *farm forestry* (off-label), *fennel* (off-label), *frise* (off-label), *garlic* (off-label), *grass seed crops* (off-label), *hazel nuts* (off-label), *herbs (see appendix 6)* (off-label), *lamb's lettuce* (off-label), *lettuce* (off-label), *lupins* (off-label), *mallow (althaea spp.)* (off-label), *parsley root* (off-label), *quinces* (off-label), *radicchio* (off-label), *redcurrants* (off-label), *runner beans* (off-label), *salad onions* (off-label), *scarole* (off-label), *shallots* (off-label), *walnuts* (off-label), *whitecurrants* (off-label) [8, 10]; *apple orchards, pear orchards* [5, 8]; *apples, pears* [1-3, 6, 7, 10, 12]; *blackberries, blackcurrants, broccoli, brussels sprouts, cabbages, calabrese, carrots, cauliflowers, cherries, gooseberries, leeks, loganberries, parsnips, plums, raspberries, rubus hybrids* [1-3, 5-8, 10, 12]; *bog myrtle* (off-label), *broad beans* (off-label),

broccoli *(off-label)*, **brussels sprouts** *(off-label)*, **bulb onions** *(off-label)*, **cabbages** *(off-label)*, **calabrese** *(off-label)*, **carrots** *(off-label)*, **cauliflowers** *(off-label)*, **celery (outdoor)** *(off-label)*, **chervil** *(off-label)*, **collards** *(off-label)*, **coriander** *(off-label)*, **dill** *(off-label)*, **forage soya bean** *(off-label)*, **forest** *(off-label)*, **french beans**, **game cover** *(off-label)*, **hops** *(off-label)*, **horseradish** *(off-label)*, **kale** *(off-label)*, **leaf brassicas** *(off-label)*, **leeks** *(off-label)*, **miscanthus** *(off-label)*, **navy beans**, **parsley** *(off-label)*, **parsnips** *(off-label)*, **protected forest nurseries** *(off-label)*, **rhubarb** *(off-label)*, **spring wheat** *(off-label)*, **sweetcorn** *(off-label)*, **sweetcorn under plastic mulches** *(off-label)*, **top fruit** *(off-label)* [10]; **bulb onions** [1-3, 6-8, 10, 12]; **combining peas, durum wheat, forage maize, potatoes, spring barley, sunflowers, triticale, winter barley, winter rye, winter wheat** [1-12]; **daffodils grown for galanthamine production** *(off-label)*, **onion sets** *(off-label)*, **salad brassicas** *(off-label)* [8]; **edible podded peas** *(off-label)*, **vining peas** *(off-label)* [5, 8, 10]; **forage maize (under plastic mulches), grain maize, grain maize under plastic mulches** [1, 2, 6, 8, 10, 12]; **forest nurseries** *(off-label)*, **ornamental plant production** *(off-label)* [6, 10]; **spring field beans** *(off-label)*, **winter field beans** *(off-label)* [2, 4, 6, 8, 10]; **strawberries** [2, 3, 5-8, 10, 12]

- Annual grasses in **almonds** *(off-label)*, **asparagus** *(off-label)*, **bilberries** *(off-label)*, **blueberries** *(off-label)*, **chestnuts** *(off-label)*, **cranberries** *(off-label)*, **cress** *(off-label)*, **daffodils grown for galanthamine production** *(off-label)*, **evening primrose** *(off-label)*, **farm forestry** *(off-label)*, **frise** *(off-label)*, **garlic** *(off-label)*, **hazel nuts** *(off-label)*, **herbs (see appendix 6)** *(off-label)*, **lamb's lettuce** *(off-label)*, **lettuce** *(off-label)*, **lupins** *(off-label)*, **mallow (althaea spp.)** *(off-label)*, **onion sets** *(off-label)*, **parsley root** *(off-label)*, **quinces** *(off-label)*, **radicchio** *(off-label)*, **redcurrants** *(off-label)*, **runner beans** *(off-label)*, **salad brassicas** *(off-label)*, **salad onions** *(off-label)*, **scarole** *(off-label)*, **shallots** *(off-label)*, **walnuts** *(off-label)*, **whitecurrants** *(off-label)* [8]; **apples, blackberries, blackcurrants, broccoli, brussels sprouts, cabbages, calabrese, carrots, cauliflowers, cherries, gooseberries, leeks, loganberries, parsnips, pears, plums, raspberries** [1-3, 6, 10, 12]; **bog myrtle** *(off-label)*, **broccoli** *(off-label)*, **brussels sprouts** *(off-label)*, **calabrese** *(off-label)*, **cauliflowers** *(off-label)*, **celery (outdoor)** *(off-label)*, **collards** *(off-label)*, **kale** *(off-label)*, **rhubarb** *(off-label)*, **spring wheat** *(off-label)* [10]; **combining peas, durum wheat, forage maize, potatoes, spring barley, sunflowers, triticale, winter barley, winter rye, winter wheat** [1-4, 6, 9-12]; **edible podded peas** *(off-label)*, **vining peas** *(off-label)* [5, 8]; **fennel** *(off-label)*, **grass seed crops** *(off-label)* [8, 10]; **forage maize (under plastic mulches), grain maize, grain maize under plastic mulches** [1, 2, 6, 8, 10, 12]; **forest nurseries** *(off-label)*, **ornamental plant production** *(off-label)* [6]; **rubus hybrids** [1-3, 5, 6, 8, 10, 12]; **spring field beans** *(off-label)*, **winter field beans** *(off-label)* [2, 4-6, 8]; **strawberries** [2, 3, 6, 10, 12]
- Annual meadow grass in **apple orchards, pear orchards** [5, 8]; **apples, pears, rubus hybrids** [7]; **blackberries, blackcurrants, broccoli, brussels sprouts, cabbages, calabrese, carrots, cauliflowers, cherries, combining peas, durum wheat, forage maize, gooseberries, leeks, loganberries, parsnips, plums, potatoes, raspberries, spring barley, strawberries, sunflowers, triticale, winter barley, winter rye, winter wheat** [5, 7, 8]; **bulb onions** [7, 8]; **cabbages** *(off-label)*, **forage soya bean** *(off-label)* [10]
- Blackgrass in **apple orchards, blackberries, blackcurrants, broccoli, brussels sprouts, cabbages, calabrese, carrots, cauliflowers, cherries, combining peas, durum wheat, gooseberries, leeks, loganberries, parsnips, pear orchards, plums, raspberries, spring barley, strawberries, sunflowers, triticale, winter barley, winter rye, winter wheat** [5, 8]; **bulb onions** [8]; **forage soya bean** *(off-label)* [10]
- Cleavers in **winter field beans** *(off-label)* [5]
- Knotgrass in **edible podded peas** *(off-label)*, **vining peas** *(off-label)* [5]
- Rough meadow grass in **apple orchards, blackberries, blackcurrants, broccoli, brussels sprouts, cabbages, calabrese, carrots, cauliflowers, cherries, combining peas, durum wheat, forage maize, gooseberries, leeks, loganberries, parsnips, pear orchards, plums, potatoes, raspberries, spring barley, strawberries, sunflowers, triticale, winter barley, winter rye, winter wheat** [5, 8]; **bulb onions** [8]
- Volunteer oilseed rape in **apple orchards, pear orchards, spring barley, sunflowers** [5, 8]; **apples, pears** [1, 2, 12]; **blackberries, blackcurrants, broccoli, brussels sprouts, cabbages, calabrese, carrots, cauliflowers, cherries, gooseberries, leeks, loganberries, parsnips, plums, raspberries, rubus hybrids** [1, 2, 5, 8, 12]; **bulb onions** [1, 2, 8, 12]; **combining peas, durum wheat, forage maize, potatoes, triticale, winter barley, winter rye, winter wheat** [1, 2, 5, 6, 8, 9, 11, 12]; **forage maize (under plastic mulches), grain maize, grain maize under**

SECTION 2

plastic mulches [1, 2, 6, 12]; *spring field beans* (off-label), *winter field beans* (off-label) [2, 6]; *strawberries* [2, 5, 8, 12]
- Wild oats in *durum wheat*, *triticale*, *winter barley*, *winter rye*, *winter wheat* [5, 8]

Extension of Authorisation for Minor Use (EAMUs)

- *almonds* 20090812 expires 31 Dec 2013 [8], 20092916 expires 03 Sep 2013 [10]
- *asparagus* 20090809 expires 31 Dec 2013 [8], 20092920 expires 03 Sep 2013 [10]
- *bilberries* 20090817 expires 31 Dec 2013 [8], 20092907 expires 03 Sep 2013 [10]
- *blueberries* 20090817 expires 31 Dec 2013 [8], 20092907 expires 03 Sep 2013 [10]
- *bog myrtle* 20120323 expires 03 Sep 2013 [10]
- *broad beans* 20092922 expires 03 Sep 2013 [10]
- *broccoli* 20112451 expires 03 Sep 2013 [10]
- *brussels sprouts* 20112451 expires 03 Sep 2013 [10]
- *bulb onions* 20092914 expires 03 Sep 2013 [10]
- *cabbages* 20100650 expires 03 Sep 2013 [10]
- *calabrese* 20112451 expires 03 Sep 2013 [10]
- *carrots* 20093526 expires 03 Sep 2013 [10]
- *cauliflowers* 20112451 expires 03 Sep 2013 [10]
- *celery (outdoor)* 20092924 expires 03 Sep 2013 [10], 20111287 expires 03 Sep 2013 [10]
- *chervil* 20092913 expires 03 Sep 2013 [10]
- *chestnuts* 20090812 expires 31 Dec 2013 [8], 20092916 expires 03 Sep 2013 [10]
- *collards* 20112451 expires 03 Sep 2013 [10]
- *coriander* 20092913 expires 03 Sep 2013 [10]
- *cranberries* 20090817 expires 31 Dec 2013 [8], 20092907 expires 03 Sep 2013 [10]
- *cress* 20090819 expires 31 Dec 2013 [8], 20092921 expires 03 Sep 2013 [10]
- *daffodils grown for galanthamine production* 20090807 expires 31 Dec 2013 [8]
- *dill* 20092913 expires 03 Sep 2013 [10]
- *edible podded peas* 20111007 expires 31 Dec 2013 [5], 20090813 expires 31 Dec 2013 [8], 20092909 expires 03 Sep 2013 [10]
- *evening primrose* 20090804 expires 31 Dec 2013 [8], 20092915 expires 03 Sep 2013 [10]
- *farm forestry* 20090806 expires 31 Dec 2013 [8], 20092918 expires 03 Sep 2013 [10]
- *fennel* 20090802 expires 31 Dec 2013 [8], 20092924 expires 03 Sep 2013 [10], 20111287 expires 03 Sep 2013 [10]
- *forage soya bean* 20121634 [10]
- *forest* 20092918 expires 03 Sep 2013 [10]
- *forest nurseries* 20082862 expires 31 Dec 2013 [6], 20092919 expires 03 Sep 2013 [10]
- *french beans* 20092908 expires 03 Sep 2013 [10]
- *frise* 20090819 expires 31 Dec 2013 [8], 20092921 expires 03 Sep 2013 [10]
- *game cover* 20092919 expires 03 Sep 2013 [10]
- *garlic* 20090810 expires 31 Dec 2013 [8], 20092914 expires 03 Sep 2013 [10]
- *grass seed crops* 20090811 expires 31 Dec 2013 [8], 20092923 expires 03 Sep 2013 [10]
- *hazel nuts* 20090812 expires 31 Dec 2013 [8], 20092916 expires 03 Sep 2013 [10]
- *herbs (see appendix 6)* 20090819 expires 31 Dec 2013 [8], 20092921 expires 03 Sep 2013 [10]
- *hops* 20092919 expires 03 Sep 2013 [10]
- *horseradish* 20093526 expires 03 Sep 2013 [10]
- *kale* 20112451 expires 03 Sep 2013 [10]
- *lamb's lettuce* 20090819 expires 31 Dec 2013 [8], 20092921 expires 03 Sep 2013 [10]
- *leaf brassicas* 20092921 expires 03 Sep 2013 [10]
- *leeks* 20092914 expires 03 Sep 2013 [10]
- *lettuce* 20090819 expires 31 Dec 2013 [8], 20092921 expires 03 Sep 2013 [10]
- *lupins* 20090818 expires 31 Dec 2013 [8], 20092909 expires 03 Sep 2013 [10]
- *mallow (althaea spp.)* 20090819 expires 31 Dec 2013 [8], 20092921 expires 03 Sep 2013 [10]
- *miscanthus* 20092919 expires 03 Sep 2013 [10]
- *navy beans* 20092908 expires 03 Sep 2013 [10]
- *onion sets* 20090810 expires 31 Dec 2013 [8]
- *ornamental plant production* 20082862 expires 31 Dec 2013 [6], 20092919 expires 03 Sep 2013 [10]
- *parsley* 20092913 expires 03 Sep 2013 [10]

SEE SECTION 3 FOR PRODUCTS ALSO REGISTERED

- **parsley root** *20090803 expires 31 Dec 2013* [8], *20092911 expires 03 Sep 2013* [10], *20093526 expires 03 Sep 2013* [10]
- **parsnips** *20093526 expires 03 Sep 2013* [10]
- **protected forest nurseries** *20092919 expires 03 Sep 2013* [10]
- **quinces** *20090814 expires 31 Dec 2013* [8], *20092917 expires 03 Sep 2013* [10]
- **radicchio** *20090819 expires 31 Dec 2013* [8], *20092921 expires 03 Sep 2013* [10]
- **redcurrants** *20090817 expires 31 Dec 2013* [8], *20092907 expires 03 Sep 2013* [10]
- **rhubarb** *20092924 expires 03 Sep 2013* [10], *20111287 expires 03 Sep 2013* [10]
- **runner beans** *20090805 expires 31 Dec 2013* [8], *20092909 expires 03 Sep 2013* [10]
- **salad brassicas** *20090819 expires 31 Dec 2013* [8]
- **salad onions** *20090810 expires 31 Dec 2013* [8], *20092914 expires 03 Sep 2013* [10]
- **scarole** *20090819 expires 31 Dec 2013* [8], *20092921 expires 03 Sep 2013* [10]
- **shallots** *20090810 expires 31 Dec 2013* [8], *20092914 expires 03 Sep 2013* [10]
- **spring field beans** *20102697 expires 31 Dec 2013* [2], *20092685 expires 31 Dec 2013* [4], *20092551 expires 31 Dec 2013* [5], *20070876 expires 31 Dec 2013* [6], *20090404 expires 31 Dec 2013* [8], *20092909 expires 03 Sep 2013* [10]
- **spring wheat** *20092910 expires 03 Sep 2013* [10]
- **sweetcorn** *20092912 expires 03 Sep 2013* [10]
- **sweetcorn under plastic mulches** *20092912 expires 03 Sep 2013* [10]
- **top fruit** *20092919 expires 03 Sep 2013* [10]
- **vining peas** *20111007 expires 31 Dec 2013* [5], *20090813 expires 31 Dec 2013* [8], *20092909 expires 03 Sep 2013* [10]
- **walnuts** *20090812 expires 31 Dec 2013* [8], *20092916 expires 03 Sep 2013* [10]
- **whitecurrants** *20090817 expires 31 Dec 2013* [8], *20092907 expires 03 Sep 2013* [10]
- **winter field beans** *20102697 expires 31 Dec 2013* [2], *20092685 expires 31 Dec 2013* [4], *20092552 expires 31 Dec 2013* [5], *20070876 expires 31 Dec 2013* [6], *20090808 expires 31 Dec 2013* [8], *20092909 expires 03 Sep 2013* [10]

Approval information
- Pendimethalin included in Annex I under EC Regulation 1107/2009
- Accepted by BBPA for use on malting barley and hops
- Approval expiry 14 Jun 2013 [3]
- Approval expiry 31 Dec 2013 [5, 6, 8, 11, 12]
- Approval expiry 31 Mar 2013 [9]

Efficacy guidance
- Apply as soon as possible after drilling. Weeds are controlled as they germinate and emerged weeds will not be controlled by use of the product alone
- For effective blackgrass control apply not more than 2 d after final cultivation and before weed seeds germinate
- Best results by application to fine firm, moist, clod-free seedbeds when rain follows treatment. Effectiveness reduced by prolonged dry weather after treatment
- Effectiveness reduced on soils with more than 6% organic matter. Do not use where organic matter exceeds 10%
- Any trash, ash or straw should be incorporated evenly during seedbed preparation
- Do not disturb soil after treatment
- Apply to potatoes as soon as possible after planting and ridging in tank-mix with metribuzin but note that this will restrict the varieties that may be treated
- Always follow WRAG guidelines for preventing and managing herbicide resistant weeds. See Section 5 for more information

Restrictions
- Maximum number of treatments 1 per crop or yr
- Maximum total dose equivalent to one full dose treatment on most crops; 2 full dose treatments on leeks
- May be applied pre-emergence of cereal crops sown before 30 Nov provided seed covered by at least 32 mm soil, or post-emergence to early tillering stage (GS 23)
- Do not undersow treated crops
- Do not use on crops suffering stress due to disease, drought, waterlogging, poor seedbed conditions or chemical treatment or on soils where water may accumulate

FOR FULL CONDITIONS OF USE ALWAYS READ THE PRODUCT LABEL

Crop-specific information

- Latest use: pre-emergence for spring barley, carrots, lettuce, fodder maize, parsnips, parsley, combining peas, potatoes, onions and leeks; before transplanting for brassicas; before leaf sheaths erect for winter cereals; before bud burst for blackcurrants, gooseberries, cane fruit, hops; before flower trusses emerge for strawberries; 14 d after transplanting for leaf herbs
- Do not use on spring barley after end Mar (mid-Apr in Scotland on some labels) because dry conditions likely. Do not apply to dry seedbeds in spring unless rain imminent
- Apply to combining peas as soon as possible after sowing. Do not spray if plumule less than 13 mm below soil surface
- Apply to potatoes up to 7 d before first shoot emerges
- Apply to drilled crops as soon as possible after drilling but before crop and weed emergence
- Apply in top fruit, bush fruit and hops from autumn to early spring when crop dormant
- In cane fruit apply to weed free soil from autumn to early spring, immediately after planting new crops and after cutting out canes in established crops
- Apply in strawberries from autumn to early spring (not before Oct on newly planted bed). Do not apply pre-planting or during flower initiation period (post-harvest to mid-Sep)
- Apply pre-emergence in drilled onions or leeks, not on Sands, Very Light, organic or peaty soils or when heavy rain forecast
- Apply to brassicas after final plant-bed cultivation but before transplanting. Avoid unnecessary soil disturbance after application and take care not to introduce treated soil into the root zone when transplanting. Follow transplanting with specified post-planting treatments - see label
- Do not use on protected crops or in greenhouses

Following crops guidance

- Before ryegrass is drilled after a very dry season plough or cultivate to at least 15 cm. If treated spring crops are to be followed by crops other than cereals, plough or cultivate to at least 15 cm
- In the event of crop failure land must be ploughed or thoroughly cultivated to at least 15 cm. See label for minimum intervals that should elapse between treatment and sowing a range of replacement crops

Environmental safety

- Dangerous for the environment
- Very toxic to aquatic organisms
- Dangerous to fish or other aquatic life. Do not contaminate surface waters or ditches with chemical or used container
- LERAP Category B [1-3, 5-12]

Hazard classification and safety precautions

Hazard Harmful [2, 4-6, 8, 9, 11, 12]; Flammable [5, 8]; Dangerous for the environment [1-12]
Transport code 9
Packaging group III
UN Number 3082
Risk phrases R22a, R36, R38 [2, 4, 6, 9, 11, 12]; R22b [5, 8]; R50 [1-4, 6, 7, 9, 11, 12]; R51 [5, 8, 10]; R53a [1-12]; R70 [10]
Operator protection A [1-7, 9-12]; C [2, 5, 6, 9, 11, 12]; H [1, 7, 10]; U02a [1, 7, 10]; U05a [1, 2, 4-12]; U08, U19a [1-3, 5-12]; U11 [2, 4, 6, 9, 11, 12]; U13 [1-3, 6, 7, 9-12]; U14 [10]; U20b [1, 2, 6, 7, 9, 11, 12]; U20c [3]
Environmental protection E15a [5, 8]; E15b [1, 2, 6, 7, 9-12]; E16a [1-3, 5-12]; E16b [1, 7]; E34 [2, 5, 6, 8, 9, 11, 12]; E38 [1-12]; E39 [10]
Storage and disposal D01, D12a [1-12]; D02 [1, 2, 4-12]; D05 [1, 5, 7, 8, 10]; D08 [2, 6, 9, 11, 12]; D09a, D10b [1, 2, 5-12]
Medical advice M03, M05b [2, 5, 6, 8, 9, 11, 12]; M05a [1, 3, 4, 7]

SEE SECTION 3 FOR PRODUCTS ALSO REGISTERED

355   pendimethalin + picolinafen

A post-emergence broad-spectrum herbicide mixture for winter cereals
HRAC mode of action code: K1 + F1

See also picolinafen

Products

1	Chronicle	BASF	320:16 g/l	SC	15394
2	Galivor	BASF	320:14.5 g/l	SC	14986
3	Orient	BASF	330:7.5 g/l	SC	12541
4	PicoMax	BASF	320:16 g/l	SC	13456
5	Picona	BASF	320:16 g/l	SC	13428
6	PicoPro	BASF	320:16 g/l	SC	13454
7	Sienna	BASF	300:7.5 g/l	SC	14991

Uses

- Annual dicotyledons in *winter barley, winter wheat*
- Annual meadow grass in *winter barley, winter wheat*
- Chickweed in *winter barley, winter wheat*
- Cleavers in *winter barley, winter wheat*
- Field speedwell in *winter barley, winter wheat*
- Ivy-leaved speedwell in *winter barley, winter wheat*
- Rough meadow grass in *winter barley, winter wheat*

Approval information

- Pendimethalin and picolinafen included in Annex I under EC Regulation 1107/2009

Efficacy guidance

- Best results obtained on crops growing in a fine, firm tilth and when rain falls within 7 d of application
- Loose or cloddy seed beds must be consolidated prior to application otherwise reduced weed control may occur
- Residual weed control may be reduced on soils with more than 6% organic matter or under prolonged dry conditions
- Always follow WRAG guidelines for preventing and managing herbicide resistant weeds. See Section 5 for more information

Restrictions

- Maximum number of treatments 1 per crop
- Do not treat undersown cereals or those to be undersown
- Do not roll emerged crops before treatment nor autumn treated crops until the following spring
- Do not use on stony or gravelly soils, on soils that are waterlogged or prone to waterlogging, or on soils with more than 10% organic matter
- Do not treat crops under stress from any cause
- Do not apply pre-emergence to crops drilled after 30 Nov [3]
- Consult processor before treating crops for processing [3]

Crop-specific information

- Latest use: before pseudo-stem erect stage (GS 30)
- Transient bleaching may occur after treatment but it does not lead to yield loss
- Do not use pre-emergence on crops drilled after 30th November [4]
- For pre-emergence applications, seed should be covered with a minimum of 3.2 cm settled soil [5, 6]

Following crops guidance

- In the event of failure of a treated crop plough to at least 15 cm to ensure any residues are evenly dispersed
- In the event of failure of a treated crop allow at least 8 wk from the time of treatment before re-drilling either spring wheat or spring barley. After a normally harvested crop there are no restrictions on following crops apart from rye-grass, as indicated below [4-6]
- In a very dry season, plough or cultivate to at least 15 cm before drilling rye-grass

FOR FULL CONDITIONS OF USE ALWAYS READ THE PRODUCT LABEL

- A minimum of 5 mth must follow autumn applications before sowing spring barley, spring wheat, maize, peas, potatoes, spring field beans, dwarf beans, turnips, Brussels sprouts, cabbage, calabrese, linseed, carrots, parsnips, parsley, cauliflower [3]
- After spring or summer applications a minimum of 2 mth must follow before sowing spring field beans, dwarf beans, broad beans, peas, turnips, Brussels sprouts, cabbage, calabrese, cauliflower, carrots, parsnips, parsley, linseed and any crop may be sown after 5 mth except red beet, sugar beet and spinach, for which an interval of 12 mth must be allowed [3]

Environmental safety
- Dangerous for the environment
- Very toxic to aquatic organisms
- Product binds strongly to soil minimising likelihood of movement into groundwater
- LERAP Category B

Hazard classification and safety precautions
Hazard Dangerous for the environment
Transport code 9
Packaging group III
UN Number 3082
Risk phrases R50, R53a
Operator protection A, H [1, 3, 4, 7]; U02a, U05a, U20c
Environmental protection E15a [3]; E15b [1, 2, 4-7]; E16a, E34, E38 [1-7]
Storage and disposal D01, D02, D05, D09a, D10c, D12a
Medical advice M03

356 pendimethalin + pyroxsulam

A dinitroaniline + triazolopyrimidine herbicide mixture for use in winter wheat, rye and triticale
HRAC mode of action code: K1 + B

Products

Broadway Sunrise	Dow	314:5.4 g/l	OD	14960

Uses
- Annual dicotyledons in *triticale*, *winter rye*, *winter wheat*
- Blackgrass in *triticale* (MS only), *winter rye* (MS only), *winter wheat* (MS only)
- Ryegrass in *triticale* (from seed), *winter rye* (from seed), *winter wheat* (from seed)
- Sterile brome in *triticale*, *winter rye*, *winter wheat*
- Wild oats in *triticale*, *winter rye*, *winter wheat*

Approval information
- Pendimethalin included in Annex I under EC Regulation 1107/2009 while pyroxsulam is awaiting inclusion.

Efficacy guidance
- To avoid the build-up of resistance, do not apply this or any other product containing an ALS inhibitor with claims for control of grass weeds more than once to any crop.

Following crops guidance
- In the event of crop failure, plough and sow only spring barley, spring wheat or maize at least 5 months after application and cultivate soil to at least 15 cm.

Environmental safety
- LERAP Category B

Hazard classification and safety precautions
Hazard Irritant, Dangerous for the environment
Transport code 9
Packaging group III
UN Number 3082
Risk phrases R38, R43, R50, R53a

SEE SECTION 3 FOR PRODUCTS ALSO REGISTERED

Operator protection A, H; U05a, U14, U15, U20b
Environmental protection E15b, E16a, E34, E38
Storage and disposal D01, D02, D05, D10c, D12a

357   pendimethalin + terbuthylazine

A residual dinitroaniline and triazine herbicide mixture for weed control in peas, beans, potatoes and forage maize
HRAC mode of action code: K1 + C1

Products

Bullet XL	Makhteshim	64:270 g/l	SE	14469

Uses
- Annual dicotyledons in **forage maize**
- Annual grasses in **forage maize**

Approval information
- Pendimethalin and terbuthylazine included in Annex 1 under EC Regulation 1107/2009

Environmental safety
- LERAP Category B

Hazard classification and safety precautions
Hazard Irritant, Dangerous for the environment
Transport code 9
Packaging group III
UN Number 3082
Risk phrases R41, R50, R53a
Operator protection A, C, H; U05a, U11, U19a, U20b
Environmental protection E15b, E16a, E38
Storage and disposal D01, D02, D05, D09a, D11a, D12a
Medical advice M05a

358   petroleum oil

An insecticidal and acaricidal hydrocarbon oil

Products

Croptex Spraying Oil	Certis	710 g/l	EC	-

Uses
- Mealybugs in **bush fruit, cane fruit, fruit trees, hops, nursery stock, protected cucumbers, protected pot plants, protected tomatoes, wine grapes**
- Red spider mites in **bush fruit, cane fruit, fruit trees, hops, nursery stock, protected cucumbers, protected pot plants, protected tomatoes, wine grapes**
- Scale insects in **bush fruit, cane fruit, fruit trees, hops, nursery stock, protected cucumbers, protected pot plants, protected tomatoes, wine grapes**

Approval information
- Product not controlled by Control of Pesticides Regulations because it acts by physical means only

Efficacy guidance
- Spray at 1% (0.5% on tender foliage) to wet plants thoroughly, particularly the underside of leaves, and repeat as necessary
- Petroleum oil acts by blocking pest breathing pores and making leaf surfaces inhospitable to pests seeking to attack or attach to the leaf surface
- On outdoor crops apply when dormant or on plants with a known tolerance
- On plants of unknown sensitivity test first on a small scale

FOR FULL CONDITIONS OF USE ALWAYS READ THE PRODUCT LABEL

- Mixtures with certain pesticides may damage crop plants. If mixing, spray a few plants to test for tolerance before treating larger areas

Restrictions
- Do not mix with sulphur or iprodione products or use such mixtures within 28 d of treatment
- Do not treat protected crops in bright sunshine unless the glass is well shaded
- Test species tolerance before large scale treatment

Crop-specific information
- HI zero
- Treat grapevines before flowering

Hazard classification and safety precautions
  Hazard Harmful
  Risk phrases R22a, R36, R37, R38
  Operator protection A, C; U05a, U08, U11, U19a
  Environmental protection E34
  Storage and disposal D01, D02, D09a, D10a, D12a
  Medical advice M03

359 phenmedipham

A contact phenyl carbamate herbicide for beet crops and strawberries
HRAC mode of action code: C1

See also desmedipham + ethofumesate + lenacil + phenmedipham
desmedipham + ethofumesate + metamitron + phenmedipham
desmedipham + ethofumesate + phenmedipham
desmedipham + phenmedipham
ethofumesate + metamitron + phenmedipham
ethofumesate + phenmedipham

Products

1	Agrichem PMP EC	AgriChem BV	160 g/l	SE	15230
2	Betasana SC	United Phosphorus	160 g/l	SC	14209
3	Corzal	AgriChem BV	157 g/l	SE	14192
4	Shrapnel	Makhteshim	320 g/l	SC	14088

Uses
- Annual dicotyledons in *chard (off-label)*, *herbs (see appendix 6) (off-label)*, *seakale (off-label)*, *spinach (off-label)*, *spinach beet (off-label)* [3]; *fodder beet*, *mangels*, *red beet*, *sugar beet* [1-4]; *strawberries* [4]

Extension of Authorisation for Minor Use (EAMUs)
- *chard* 20111766 [3]
- *herbs (see appendix 6)* 20111766 [3]
- *seakale* 20111766 [3]
- *spinach* 20111766 [3]
- *spinach beet* 20111766 [3]

Approval information
- Phenmedipham included in Annex I under EC Regulation 1107/2009
- Approval expiry 31 Dec 2013 [2]

Efficacy guidance
- Best results achieved by application to young seedling weeds, preferably cotyledon stage, under good growing conditions when low doses are effective
- If using a low-dose programme 2-3 repeat applications at 7-10 d intervals are recommended on mineral soils, 3-5 applications may be needed on organic soils
- Addition of adjuvant oil may improve effectiveness on some weeds
- Various tank-mixtures with other beet herbicides recommended. See label for details

SEE SECTION 3 FOR PRODUCTS ALSO REGISTERED

- Use of certain pre-emergence herbicides is recommended in combination with post-emergence treatment. See label for details

Restrictions

- Maximum number of treatments and maximum total dose varies with crop and product used. See label for details
- At high temperatures (above 21°C) reduce rate and spray after 5 pm
- Do not apply immediately after frost or if frost expected
- Do not spray wet foliage or if rain imminent
- Do not spray crops stressed by wind damage, nutrient deficiency, pest or disease attack etc. Do not roll or harrow for 7 d before or after treatment
- Do not use on strawberries under cloches or polythene tunnels
- Consult processor before use on crops for processing

Crop-specific information

- Latest use: before crop leaves meet between rows for beet crops; before flowering for strawberries
- Apply to beet crops at any stage as low dose/low volume spray or from fully developed cotyledon stage with full rate. Apply to red beet after fully developed cotyledon stage
- Apply to strawberries at any time when weeds in susceptible stage, except in period from start of flowering to picking

Following crops guidance

- Beet crops may follow at any time after a treated crop. 3 mth must elapse from treatment before any other crop is sown and must be preceded by mould-board ploughing to 15 cm

Environmental safety

- Dangerous for the environment
- Very toxic to aquatic organisms
- Harmful to fish or other aquatic life. Do not contaminate surface waters or ditches with chemical or used container
- LERAP Category B [1, 3]

Hazard classification and safety precautions

Hazard Irritant [2, 4]; Dangerous for the environment [1-4]

Transport code 9

Packaging group III

UN Number 3082

Risk phrases R36, R51 [2]; R38 [4]; R43 [2, 4]; R50 [1, 3, 4]; R53a [1-4]

Operator protection A [1-4]; C [2]; H [2, 4]; U05a [3, 4]; U08 [2-4]; U11, U20c [2]; U14, U19a [2, 4]; U20a [4]; U20b [1]

Environmental protection E15a, E19b [4]; E15b [1-3]; E16a [1, 3]; E16b [3]; E23 [1]; E34 [3, 4]; E38 [1, 3, 4]

Storage and disposal D01 [2-4]; D02 [2, 3]; D05 [1, 2, 4]; D09a, D12a [1-4]; D10b [1, 3, 4]; D10c [2]

Medical advice M05a [4]

360 d-phenothrin

A non-systemic pyrethroid insecticide available only in mixtures
IRAC mode of action code: 3A

361   d-phenothrin + tetramethrin

A pyrethroid insecticide mixture for control of flying insects
IRAC mode of action code: 3 + 3

See also tetramethrin

Products

Killgerm ULV 500	Killgerm	36.8:18.4 g/l	UL	H4647

Uses
- Flies in **agricultural premises**
- Grain storage mite in **grain stores**
- Mosquitoes in **agricultural premises**
- Wasps in **agricultural premises**

Approval information
- Product approved for ULV application. See label for details
- D-phenothrin and tetramethrin are not included in Annex 1 under EC Regulation 1107/2009

Efficacy guidance
- Close doors and windows and spray in all directions for 3-5 sec. Keep room closed for at least 10 min

Restrictions
- For use only by professional operators
- Do not use space sprays containing pyrethrins or pyrethroid more than once per week in intensive or controlled environment animal houses in order to avoid development of resistance. If necessary, use a different control method or product

Crop-specific information
- May be used in the presence of poultry and livestock

Environmental safety
- Dangerous for the environment
- Toxic to aquatic organisms
- Do not apply directly to livestock/poultry
- Remove exposed milk and collect eggs before application. Protect milk machinery and containers from contamination

Hazard classification and safety precautions
Hazard Harmful, Dangerous for the environment
Transport code 9
Packaging group III
UN Number 3082
Risk phrases R22b, R51, R53a
Operator protection A, C, D, E, H; U02b, U09b, U14, U19a, U20a, U20b
Environmental protection E05a, E15a, E38
Consumer protection C06, C07, C08, C09, C11, C12
Storage and disposal D01, D05, D06a, D09a, D10a, D12a
Medical advice M05b

362   Phlebiopsis gigantea

A fungal protectant against root and butt rot in conifers

Products

PG Suspension	Forest Research	0.5% w/w	SC	11772

Uses
- Fomes root and butt rot in **farm forestry**, **forest**, **forest nurseries**

SEE SECTION 3 FOR PRODUCTS ALSO REGISTERED

Approval information
- Phlebiopsis gigantea included in Annex 1 under EC Regulation 1107/2009

Efficacy guidance
- Store sachets in the original box in a cool dry place between 2°C and 15°C out of the sun.
- Use the diutes solution within 24 hours of mixing. Ensure complete coversge of stumps for optimum efficacy.

Hazard classification and safety precautions
  UN Number N/C
  Operator protection A; U20c
  Environmental protection E15a
  Storage and disposal D11a

363   physical pest control

Products that work by physical action only

Products

1	SB Plant Invigorator	Fargro	-	SL	-
2	Silico-Sec	Interfarm	>90% w/w	DS	-

Uses
- Aphids in *all edible crops (outdoor and protected)*, *ornamental plant production*, *protected ornamentals* [1]
- Mealybugs in *all edible crops (outdoor and protected)*, *ornamental plant production*, *protected ornamentals* [1]
- Mites in *stored grain* [2]
- Powdery mildew in *all edible crops (outdoor and protected)*, *ornamental plant production*, *protected ornamentals* [1]
- Scale insects in *all edible crops (outdoor and protected)*, *ornamental plant production*, *protected ornamentals* [1]
- Spider mites in *all edible crops (outdoor and protected)*, *ornamental plant production*, *protected ornamentals* [1]
- Suckers in *all edible crops (outdoor and protected)*, *ornamental plant production*, *protected ornamentals* [1]
- Whitefly in *all edible crops (outdoor and protected)*, *ornamental plant production*, *protected ornamentals* [1]

Approval information
- Products included in this profile are not subject to the Control of Pesticides Regulations/Plant Protection Products Regulations because they act by physical means only

Efficacy guidance
- Products act by physical means following direct contact with spray. They may therefore be used at any time of year on all pest growth stages
- Ensure thorough spray coverage of plant, paying special attention to growing points and the underside of leaves
- Treat as soon as target pests are seen and repeat as often as necessary

Restrictions
- No limit on the number of treatments and no minimum interval between applications
- Before large scale use on a new crop, treat a few plants to check for crop safety
- Do not treat ornamental crops when in flower

Crop-specific information
- HI zero

Environmental safety
- May be used in conjunction with biological control agents. Spray 24 h before they are introduced

Hazard classification and safety precautions
> **UN Number** N/C
> **Operator protection** F [2]; U10, U14, U15 [1]
> **Storage and disposal** D01, D02, D05, D06d [1]

364   picloram

A persistent, translocated pyridine carboxylic acid herbicide for non-crop areas
HRAC mode of action code: O

See also 2,4-D + picloram
clopyralid + picloram

Products

1	Cupari	Agroquimicos	240 g/l	SC	15883
2	Pantheon 2	Pan Agriculture	240 g/l	SC	14052
3	Tordon 22K	Dow	240 g/l	SC	05083

Uses
- Annual and perennial weeds in **green cover on land temporarily removed from production, land not intended to bear vegetation, natural surfaces not intended to bear vegetation** [1]; **land not intended for cropping** [1, 2]
- Annual dicotyledons in **land not intended for cropping** [3]
- Bracken in **green cover on land temporarily removed from production, land not intended to bear vegetation, natural surfaces not intended to bear vegetation** [1]; **land not intended for cropping** [1-3]
- Brambles in **green cover on land temporarily removed from production, land not intended to bear vegetation, natural surfaces not intended to bear vegetation** [1]; **land not intended for cropping** [1, 2]
- Japanese knotweed in **green cover on land temporarily removed from production, land not intended to bear vegetation, natural surfaces not intended to bear vegetation** [1]; **land not intended for cropping** [1, 3]
- Perennial dicotyledons in **land not intended for cropping** [3]
- Woody weeds in **green cover on land temporarily removed from production, land not intended to bear vegetation, natural surfaces not intended to bear vegetation** [1]; **land not intended for cropping** [1, 3]

Approval information
- Picloram included in Annex I under EC Regulation 1107/2009

Efficacy guidance
- May be applied at any time of year. Best results achieved by application as foliage spray in late winter to early spring
- For bracken control apply 2-4 wk before frond emergence
- Clovers are highly sensitive and eliminated at very low doses
- Persists in soil for up to 2 yr

Restrictions
- Maximum number of treatments 1 per yr on land not intended for cropping
- Do not apply around desirable trees or shrubs where roots may absorb chemical
- Do not apply on slopes where chemical may be leached onto areas of desirable plants

Environmental safety
- Harmful to fish or other aquatic life. Do not contaminate surface waters or ditches with chemical or used container
- Keep livestock out of treated areas for at least 2 wk and until foliage of any poisonous weeds such as ragwort has died and become unpalatable

Hazard classification and safety precautions
> **Hazard** Irritant
> **Transport code** 9

SEE SECTION 3 FOR PRODUCTS ALSO REGISTERED

Packaging group III
UN Number 3082
Risk phrases R43
Operator protection A, H; U05a, U08, U14, U20b
Environmental protection E07a, E13c
Storage and disposal D01, D02, D09a [1-3]; D10a [2]; D10b [1, 3]

365   picolinafen

A pyridinecarboxamide herbicide for cereals;
HRAC mode of action code: F1

See also flupyrsulfuron-methyl + picolinafen
pendimethalin + picolinafen

Products

Vixen	BASF	75% w/w	WG	13621

Uses
* Chickweed in **winter barley, winter wheat**
* Cleavers in **winter barley, winter wheat**
* Field pansy in **winter barley, winter wheat**
* Field speedwell in **winter barley, winter wheat**
* Ivy-leaved speedwell in **winter barley, winter wheat**
* Shepherd's purse in **winter barley, winter wheat**

Approval information
* Picolinafen included in Annex I under EC Regulation 1107/2009

Efficacy guidance
* Picolinafen is contact acting. Ensure good spray cover of the target weeds
* Best results obtained from treatment of small actively growing weeds
* Weeds emerging after application will not be controlled
* Always follow WRAG guidelines for preventing and managing herbicide resistant weeds. See Section 5 for more information

Restrictions
* Maximum number of treatments 1 per crop
* Do not disturb soil surface after application
* Do not roll emerged crops prior to application
* Do not treat undersown crops or those to be undersown
* Do not spray during periods of prolonged or severe frost
* Do not apply to crops under stress from any cause

Crop-specific information
* Latest use: before pseudo-stem erect stage (GS 30) for wheat and barley
* Treatment may cause transient bleaching in some crops

Following crops guidance
* No restrictions on following crops after normal harvest of a treated crop
* In the event of failure of a treated crop further crops of wheat or barley may be drilled after a minimum interval of 8 wk from treatment and after ploughing to at least 15 cm

Environmental safety
* Dangerous for the environment
* Very toxic to aquatic organisms
* Take extreme care to avoid drift onto plants outside the target area
* LERAP Category B

Hazard classification and safety precautions
Hazard Dangerous for the environment
Transport code 9

FOR FULL CONDITIONS OF USE ALWAYS READ THE PRODUCT LABEL

Packaging group III
UN Number 3077
Risk phrases R50, R53a
Operator protection A, D; U05a, U08, U13, U19a, U20b
Environmental protection E15a, E16a, E38
Storage and disposal D01, D02, D09a, D10b, D12a

366 picoxystrobin

A broad spectrum strobilurin fungicide for cereals
FRAC mode of action code: 11

See also chlorothalonil + picoxystrobin
cyproconazole + picoxystrobin
cyprodinil + picoxystrobin

Products

1	Flanker	DuPont	250 g/l	SC	14760
2	Galileo	DuPont	250 g/l	SC	13252
3	Oranis	DuPont	250 g/l	SC	15185

Uses

- Brown rust in **spring barley, spring wheat, winter barley, winter wheat** [1-3]
- Crown rust in **spring oats, winter oats** [1-3]
- Eyespot in **spring wheat** *(reduction)*, **winter wheat** *(reduction)* [1-3]
- Foliar disease control in **ornamental plant production** *(off-label)* [2]
- Glume blotch in **spring wheat, winter wheat** [1-3]
- Late ear diseases in **spring wheat, winter wheat** [1-3]
- Net blotch in **spring barley, winter barley** [1-3]
- Powdery mildew in **spring barley, spring oats, winter barley, winter oats** [1-3]
- Rhynchosporium in **spring barley, winter barley** [1-3]
- Sclerotinia in **spring oilseed rape, winter oilseed rape** [1-3]
- Septoria leaf blotch in **spring wheat, winter wheat** [1-3]
- Yellow rust in **spring wheat, winter wheat** [1-3]

Extension of Authorisation for Minor Use (EAMUs)

- **ornamental plant production** *20082855 expires 31 Dec 2013* [2]

Approval information

- Picoxystrobin included in Annex I under EC Regulation 1107/2009
- Accepted by BBPA for use on malting barley
- Approval expiry 31 Dec 2013 [1-3]

Efficacy guidance

- Best results achieved from protectant treatments made before disease establishes in the crop
- Use of tank mixtures is recommended where disease has become established at the time of treatment
- Persistence of yellow rust control is less than for other foliar diseases but can be improved by use of an appropriate tank mixture
- Eyespot control is not reliable and an appropriate tank mix should be used where crops are at risk
- Picoxystrobin is a member of the QoI cross resistance group. Product should be used preventatively and not relied on for its curative potential
- Use product as part of an Integrated Crop Management strategy incorporating other methods of control, including where appropriate other fungicides with a different mode of action. Do not apply more than two foliar applications of QoI containing products to any cereal crop
- There is a significant risk of widespread resistance occurring in *Septoria tritici* populations in UK. Failure to follow resistance management action may result in reduced levels of disease control
- On cereal crops product must always be used in mixture with another product, recommended for control of the same target disease, that contains a fungicide from a different cross resistance group and is applied at a dose that will give robust control

SEE SECTION 3 FOR PRODUCTS ALSO REGISTERED

- Strains of barley powdery mildew resistant to QoIs are common in the UK

Restrictions
- Maximum number of treatments 2 per crop per year for ornamental plant production, 2 per year for cereals and 1 per year for oilseed rape

Crop-specific information
- Latest use: grain watery ripe (GS71) for wheat, barley and oats, flowers declining (majority of petals fallen) for oilseed rape.

Environmental safety
- Dangerous for the environment
- Very toxic to aquatic organisms

Hazard classification and safety precautions
Hazard Harmful, Dangerous for the environment
Transport code 9
Packaging group III
UN Number 3082
Risk phrases R50, R53a
Operator protection A; U05a, U09a, U14, U15, U20a
Environmental protection E15a, E38
Storage and disposal D01, D02, D05, D09a, D10c, D12a

367   pinoxaden

A phenylpyrazoline grass weed herbicide for cereals
HRAC mode of action code: A

See also clodinafop-propargyl + pinoxaden
florasulam + pinoxaden

Products

1 Axial	Syngenta	100 g/l	EC	12521
2 Clayton Tonto	Clayton	100 g/l	EC	14436
3 Greencrop Helvick	Greencrop	100 g/l	EC	13052
4 Liberate	ChemSource	45 g/l	EC	14872
5 PureRecovery	Pure Amenity	45 g/l	EC	15137
6 Quail	ChemSource	100 g/l	EC	14506
7 Rescue	Syngenta	45 g/l	EC	14518

Uses
- Annual grasses in *game cover* (off-label) [1]
- Blackgrass in *spring barley*, *winter barley* [1, 2, 6]
- Italian ryegrass in *spring barley*, *winter barley*, *winter wheat* [1-3, 6]; *spring wheat* [1, 6]
- Perennial grasses in *game cover* (off-label) [1]
- Perennial ryegrass in *spring barley*, *winter barley*, *winter wheat* [1-3, 6]; *spring wheat* [1, 6]
- Ryegrass in *amenity grassland* (off-label - reduction), *managed amenity turf* (off-label - reduction) [7]; *amenity grassland* (reduction only), *managed amenity turf* (reduction only) [4, 5, 7]
- Wild oats in *spring barley*, *winter barley*, *winter wheat* [1-3, 6]; *spring wheat* [1, 6]

Extension of Authorisation for Minor Use (EAMUs)
- *amenity grassland* (reduction) 20111290 [7]
- *game cover* 20082815 [1]
- *managed amenity turf* (reduction) 20111290 [7]

Approval information
- Pinoxaden has not yet been included in Annex 1 under EC Regulation 1107/2009

Efficacy guidance

- Best results obtained from treatment when all grass weeds have emerged. There is no residual activity
- Broad-leaved weeds are not controlled
- Treat before emerged weed competition reduces yield
- Blackgrass in winter and spring barley is also controlled when used as part of an integrated control strategy. Product is not recommended for blackgrass control in winter wheat
- Grass weed control may be reduced if rain falls within 1 hr of application
- Pinoxaden is an ACCase inhibitor herbicide. To avoid the build up of resistance do not apply products containing an ACCase inhibitor herbicide more than twice to any crop. In addition do not use any product containing pinoxaden in mixture or sequence with any other product containing the same ingredient
- Use these products as part of a resistance management strategy that includes cultural methods of control and does not use ACCase inhibitors as the sole chemical method of grass weed control
- Applying a second product containing an ACCase inhibitor to a crop will increase the risk of resistance development; only use a second ACCase inhibitor to control different weeds at a different timing
- Always follow WRAG guidelines for preventing and managing herbicide resistant weeds. See Section 5 for more information

Restrictions

- Maximum number of treatments 1 per crop
- Do not spray crops under stress or suffering from waterlogging, pest attack, disease or frost damage
- Do not spray crops undersown with grass mixtures
- Avoid the use of hormone-containing herbicides in mixture or in sequence. Allow 21 d after, or 7 d before, hormone application
- Product must be used with prescribed adjuvant. See label

Crop-specific information

- Latest use: before flag leaf extending stage (GS 41)
- Spray cereals in autumn, winter or spring from the 2-leaf stage (GS 12)

Following crops guidance

- There are no restrictions on succeeding crops in a normal rotation
- In the event of failure of a treated crop ryegrass, maize, oats or any broad-leaved crop may be planted after a minimum interval of 4 wk from application

Environmental safety

- Dangerous for the environment
- Toxic to aquatic organisms
- Do not allow spray to drift onto neighbouring crops of oats, ryegrass or maize

Hazard classification and safety precautions

Hazard Harmful [4, 5, 7]; Irritant [1-3, 6]; Dangerous for the environment [1-7]
Transport code 9
Packaging group III
UN Number 3082
Risk phrases R22b, R43 [4, 5, 7]; R36 [1-3, 6]; R38, R51, R53a [1-7]
Operator protection A, H [1-7]; C [1-3, 6]; U05a, U20b [1-7]; U09a [1-3, 6]
Environmental protection E15b, E38
Storage and disposal D01, D02, D05, D09a, D10c, D12a [1-7]; D11a [1-3, 6]
Medical advice M05b [4, 5, 7]

SEE SECTION 3 FOR PRODUCTS ALSO REGISTERED

368   pirimicarb

A carbamate insecticide for aphid control
IRAC mode of action code: 1A

Products

1 Aphox	Syngenta	50% w/w	WG	10515
2 Clayton Pirimicarb 50	Clayton	50% w/w	WG	12910
3 Phantom	Syngenta	50% w/w	WG	11954
4 Pirimate	European Ag	50% w/w	WG	14539
5 Pirimicarb 50	Goldengrass	50% w/w	WG	14295
6 Reynard	ChemSource	50% w/w	WG	14523
7 Standon Pirimicarb 50	Standon	50% w/w	WG	13290

Uses

*   Aphids in **apples, blackcurrants, cherries, chinese cabbage, collards, cucumbers, durum wheat, forest nurseries, gooseberries, ornamental plant production, pears, peppers, protected carnations, protected chrysanthemums, protected cinerarias, protected cyclamen, protected lettuce, protected roses, protected tomatoes, raspberries, redcurrants, spring rye, strawberries, tomatoes (outdoor), triticale, winter rye** [1-3, 5-7]; **bilberries** *(off-label)*, **blueberries** *(off-label)*, **cranberries** *(off-label)*, **evening primrose** *(off-label)*, **linseed** *(off-label)*, **mallow (althaea spp.)** *(off-label)*, **mustard** *(off-label)*, **protected lamb's lettuce** *(off-label)*, **protected leaf brassicas** *(off-label)*, **salad brassicas** *(for baby leaf production)* [1]; **blackberries** *(off-label)*, **celeriac** *(off-label)*, **celery (outdoor)** *(off-label)*, **chicory root** *(off-label)*, **choi sum** *(off-label)*, **courgettes** *(off-label)*, **endives** *(off-label)*, **fennel** *(off-label)*, **fodder beet** *(off-label)*, **frise** *(off-label)*, **gherkins** *(off-label)*, **herbs (see appendix 6)** *(off-label)*, **honesty** *(off-label)*, **horseradish** *(off-label)*, **leaf brassicas** *(off-label)*, **marrows** *(off-label)*, **pak choi** *(off-label)*, **parsley** *(off-label)*, **parsley root** *(off-label)*, **plums** *(off-label)*, **poppies for morphine production** *(off-label)*, **protected aubergines** *(off-label)*, **protected cayenne peppers** *(off-label)*, **protected courgettes** *(off-label)*, **protected endives** *(off-label)*, **protected frise** *(off-label)*, **protected gherkins** *(off-label)*, **protected herbs (see appendix 6)** *(off-label)*, **protected horseradish** *(off-label)*, **protected parsley** *(off-label)*, **protected radicchio** *(off-label)*, **protected radishes** *(off-label)*, **protected scarole** *(off-label)*, **protected spinach** *(off-label)*, **protected spinach beet** *(off-label)*, **radicchio** *(off-label)*, **radishes** *(off-label)*, **red beet** *(off-label)*, **rhubarb** *(off-label)*, **rubus hybrids** *(off-label)*, **scarole** *(off-label)*, **spinach** *(off-label)*, **spinach beet** *(off-label)*, **sweetcorn** *(off-label)* [1, 3]; **broad beans, broccoli, brussels sprouts, cabbages, calabrese, carrots, cauliflowers, dwarf beans, forage maize, kale, parsnips, peas, potatoes, spring barley, spring field beans, spring oats, spring oilseed rape, spring wheat, sugar beet, swedes, sweetcorn, turnips, winter barley, winter field beans, winter oats, winter oilseed rape, winter wheat** [1-7]; **lamb's lettuce** *(off-label)*, **protected parsley root** *(off-label)*, **salsify** *(off-label)* [3]; **lettuce** [1, 5-7]; **lettuce** *(outdoor crops)* [2, 3]; **runner beans** [1, 2, 4-7]
*   Leaf curling plum aphid in **plums** *(off-label)* [1]
*   Mealy plum aphid in **plums** *(off-label)* [1]
*   Thrips in **endives** *(off-label)*, **parsley** *(off-label)*, **protected endives** *(off-label)*, **protected parsley** *(off-label)* [3]

Extension of Authorisation for Minor Use (EAMUs)

*   **bilberries** *20102319* [1]
*   **blackberries** *20102316* [1], *20102354* [3]
*   **blueberries** *20102319* [1]
*   **celeriac** *20102314* [1], *20102351* [3]
*   **celery (outdoor)** *20102320* [1], *20102347* [3]
*   **chicory root** *20102313* [1], *20102349* [3]
*   **choi sum** *20102317* [1], *20102361* [3]
*   **courgettes** *20102328* [1], *20102350* [3]
*   **cranberries** *20102319* [1]
*   **endives** *20102312* [1], *20102343* [3]
*   **evening primrose** *20102325* [1]
*   **fennel** *20102314* [1], *20102351* [3]

FOR FULL CONDITIONS OF USE ALWAYS READ THE PRODUCT LABEL

- **fodder beet** *20102323* [1], *20102356* [3]
- **frise** *20102321* [1], *20102357* [3]
- **gherkins** *20102328* [1], *20102350* [3]
- **herbs (see appendix 6)** *20102321* [1], *20102350* [3]
- **honesty** *20102325* [1], *20102326* [1], *20102345* [3]
- **horseradish** *20102324* [1], *20102328* [1], *20102350* [3], *20102358* [3]
- **lamb's lettuce** *20102357* [3]
- **leaf brassicas** *20102311* [1], *20102346* [3], *20102357* [3]
- **linseed** *20102325* [1]
- **mallow (althaea spp.)** *20102324* [1]
- **marrows** *20102328* [1], *20102350* [3]
- **mustard** *20102325* [1]
- **pak choi** *20102317* [1], *20102361* [3]
- **parsley** *20102312* [1], *20102343* [3]
- **parsley root** *20102328* [1], *20102350* [3]
- **plums** *20102318* [1], *20102355* [3]
- **poppies for morphine production** *20102322* [1], *20102353* [3]
- **protected aubergines** *20102315* [1], *20102360* [3]
- **protected cayenne peppers** *20102315* [1], *20102360* [3]
- **protected courgettes** *20102327* [1], *20102348* [3]
- **protected endives** *20102312* [1], *20102343* [3]
- **protected frise** *20102321* [1], *20102357* [3]
- **protected gherkins** *20102327* [1], *20102348* [3]
- **protected herbs (see appendix 6)** *20102321* [1], *20102350* [3]
- **protected horseradish** *20102328* [1], *20102350* [3]
- **protected lamb's lettuce** *20102321* [1]
- **protected leaf brassicas** *20102321* [1]
- **protected parsley** *20102328* [1], *20102343* [3]
- **protected parsley root** *20102350* [3]
- **protected radicchio** *20102321* [1], *20102357* [3]
- **protected radishes** *20102328* [1], *20102350* [3]
- **protected scarole** *20102321* [1], *20102357* [3]
- **protected spinach** *20102328* [1], *20102350* [3]
- **protected spinach beet** *20102328* [1], *20102350* [3]
- **radicchio** *20102321* [1], *20102357* [3]
- **radishes** *20102328* [1], *20102350* [3]
- **red beet** *20102314* [1], *20102351* [3]
- **rhubarb** *20102320* [1], *20102347* [3]
- **rubus hybrids** *20102316* [1], *20102354* [3]
- **salsify** *20102351* [3]
- **scarole** *20102321* [1], *20102357* [3]
- **spinach** *20102328* [1], *20102350* [3]
- **spinach beet** *20102328* [1], *20102350* [3]
- **sweetcorn** *20102328* [1], *20102350* [3]

Approval information
- Pirimicarb included in Annex I under EC Regulation 1107/2009
- Approved for aerial application on cereals [1, 3]. See Section 5 for more information
- Accepted by BBPA for use on malting barley
- In 2006 CRD required that all products containing this active ingredient should carry the following warning in the main area of the container label: "Pirimicarb is an anticholinesterase carbamate. Handle with care"

Efficacy guidance
- Chemical has contact, fumigant and translaminar activity
- Best results achieved under warm, calm conditions when plants not wilting and spray does not dry too rapidly. Little vapour activity at temperatures below 15˚C
- Apply as soon as aphids seen or warning issued and repeat as necessary
- Addition of non-ionic wetter recommended for use on brassicas

SEE SECTION 3 FOR PRODUCTS ALSO REGISTERED

- On cucumbers and tomatoes a root drench is preferable to spraying when using predators in an integrated control programme
- Where aphids resistant to pirimicarb occur control is unlikely to be satisfactory

Restrictions
- Contains an anticholinesterase carbamate compound. Do not use if under medical advice not to work with such compounds
- Maximum number of treatments not specified in some cases but normally 2-6 depending on crop
- When treating ornamentals check safety by treating a small number of plants first
- Spray equipment must only be used where the operator's normal working position is within a closed cab on a tractor or self-propelled sprayer when making air-assisted applications to apples or other top fruit

Crop-specific information
- Latest use in accordance with harvest intervals below
- HI oilseed rape, cereals, maize, sweetcorn 14 d (sweetcorn off-label 3 d); grassland, chicory, plums, swedes and turnips 7 d; lettuce 3 d; cucumbers, tomatoes and peppers under glass 2 d; protected courgettes and gherkins 24 h; other edible crops 3 d; flowers and ornamentals zero

Environmental safety
- Dangerous for the environment
- Very toxic to aquatic organisms
- Dangerous to fish or other aquatic life. Do not contaminate surface waters or ditches with chemical or used container
- Chemical has little effect on bees, ladybirds and other insects and is suitable for use in integrated control programmes on apples and pears
- Keep all livestock out of treated areas for at least 7 d. Bury or remove spillages

Hazard classification and safety precautions
Hazard Toxic, Dangerous for the environment
Transport code 6.1
Packaging group III
UN Number 2757
Risk phrases R20, R25, R36, R50, R53a
Operator protection A, C, D, H, J, M; U05a, U08, U19a [1-7]; U20a [2, 3]; U20b [1, 4-7]
Environmental protection E06c [1-7] (7 d); E15a [2]; E15b, E38 [1, 3-7]; E34 [1-7]
Storage and disposal D01, D02, D09a, D11a, D12a [1-7]; D07 [2]
Medical advice M02, M04a [1-7]; M03 [1, 3-7]; M04b [2]

---

369   pirimiphos-methyl

A contact, fumigant and translaminar organophosphorus insecticide
IRAC mode of action code: 1B

Products

1 Actellic 50 EC	Syngenta	500 g/l	EC	12726
2 Actellic Smoke Generator No. 20	Syngenta	22.5% w/w	FU	15739
3 Clayton Galic	Clayton	500 g/l	EC	15374
4 Flycatcher	AgChem Access	500 g/l	EC	14135
5 Pan PMT	Pan Agriculture	500 g/l	EC	13996

Uses
- Flour beetle in **stored grain** [1, 3-5]
- Flour moth in **stored grain** [1, 3-5]
- Grain beetle in **stored grain** [1, 3-5]
- Grain storage mite in **stored grain** [1, 3-5]
- Grain storage pests in **grain stores** [2]
- Grain weevil in **stored grain** [1, 3-5]
- Insect pests in **grain stores** [2]

FOR FULL CONDITIONS OF USE ALWAYS READ THE PRODUCT LABEL

- Mites in **grain stores** [2]
- Warehouse moth in **stored grain** [1, 3-5]

Approval information
- Pirimiphos-methyl included in Annex I under EC Regulation 1107/2009
- In 2006 PSD required that all products containing this active ingredient should carry the following warning in the main area of the container label: "Pirimiphos-methyl is an anticholinesterase organophosphate. Handle with care"
- Accepted by BBPA for use in stores for malting barley

Efficacy guidance
- Chemical acts rapidly and has short persistence in plants, but persists for long periods on inert surfaces
- Best results for protection of stored grain achieved by cleaning store thoroughly before use and employing a combination of pre-harvest and grain/seed treatments [1, 2]
- Best results for admixture treatment obtained when grain stored at 15% moisture or less. Dry and cool moist grain coming into store but then treat as soon as possible, ideally as it is loaded [1]
- Surface admixture can be highly effective on localised surface infestations but should not be relied on for long-term control unless application can be made to the full depth of the infestation [1]
- Where insect pests resistant to pirimiphos-methyl occur control is unlikely to be satisfactory

Restrictions
- Contains an organophosphorus anticholinesterase compound. Do not use if under medical advice not to work with such compounds
- Maximum number of treatments 2 per grain store [1, 2]; 1 per batch of stored grain [1]
- Do not apply surface admixture or complete admixture using hand held equipment [1]
- Surface admixture recommended only where the conditions for store preparation, treatment and storage detailed in the label can be met [1]

Crop-specific information
- Latest use: before storing grain [1, 2]
- Disinfect empty grain stores by spraying surfaces and/or fumigation and treat grain by full or surface admixture. Treat well before harvest in late spring or early summer and repeat 6 wk later or just before harvest if heavily infested. See label for details of treatment and suitable application machinery
- Treatment volumes on structural surfaces should be adjusted according to surface porosity. See label for guidelines
- Treat inaccessible areas with smoke generating product used in conjunction with spray treatment of remainder of store
- Predatory mites (*Cheyletus*) found in stored grain feeding on infestations of grain mites may survive treatment and can lead to rejection by buyers. They do not remain for long but grain should be inspected carefully before selling [1, 2]

Environmental safety
- Dangerous for the environment
- Very toxic to aquatic organisms [1]
- Toxic to aquatic organisms [2]
- Highly flammable [2]
- Flammable [1]
- Ventilate fumigated or fogged spaces thoroughly before re-entry
- Unprotected persons must be kept out of fumigated areas within 3 h of ignition and for 4 h after treatment
- Wildlife must be excluded from buildings during treatment [1]
- Keep away from combustible materials [2]

Hazard classification and safety precautions
Hazard Harmful, Dangerous for the environment [1-5]; Highly flammable [2]; Flammable [1, 3-5]
Transport code 3 [1, 3-5]; 9 [2]
Packaging group III
UN Number 1993 [1, 3-5]; 3077 [2]

SEE SECTION 3 FOR PRODUCTS ALSO REGISTERED

Risk phrases R20, R50, R53a [1-5]; R22a, R22b, R36, R37 [1, 3-5]
Operator protection A, C, J, M [1, 3-5]; D, H [1-5]; U04a, U09a, U16a, U24 [1, 3-5]; U05a, U19a, U20b [1-5]; U14 [2]
Environmental protection E02a [2] (4 h); E15a, E34, E38 [1-5]
Consumer protection C09 [1, 3-5]; C12 [1-5]
Storage and disposal D01, D02, D09a, D12a [1-5]; D05, D10c [1, 3-5]; D07, D11a [2]
Treated seed S06a [1, 3-5]
Medical advice M01 [1-5]; M03 [2]; M05b [1, 3-5]

370   potassium bicarbonate (commodity substance)

An inorganic fungicide for use in horticulture. Approval valid until 31/8/2019

Products

potassium bicarbonate	various	100% w/w	AP	00000

Uses
- Disease control in *all edible crops (outdoor and protected)*, *all non-edible crops (outdoor)*, *all protected non-edible crops*

Approval information
- Approval for the use of potassium bicarbonate as a commodity substance was granted on 26 July 2005 by Ministers under regulation 5 of the Control of Pesticides Regulations 1986
- Potassium bicarbonate is being supported for review in the fourth stage of the EC Review Programme under EC Regulation 1107/2009. Whether or not it is included in Annex I, the existing commodity chemical approval will have to be revoked because substances listed in Annex I must be approved for marketing and use under the EC regime. If potassium bicarbonate is included in Annex I, products containing it will need to gain approval in the normal way if they carry label claims for pesticidal activity

Efficacy guidance
- Adjuvants authorised by CRD may be used in conjunction with potassium bicarbonate
- Crop phytotoxicity can occur

Restrictions
- Food grade potassium bicarbonate must only be used

371   prochloraz

A broad-spectrum protectant and eradicant imidazole fungicide
FRAC mode of action code: 3

See also carbendazim + prochloraz
epoxiconazole + prochloraz
fenpropidin + prochloraz
fenpropidin + prochloraz + tebuconazole
fluquinconazole + prochloraz

Products

1	Cloraz 400	Goldengrass	400 g/l	EC	14657
2	Mirage 40 EC	Makhteshim	400 g/l	EC	06770
3	Octave	Everris Ltd	46% w/w	WP	09275
4	Poraz	BASF	450 g/l	EC	11701
5	Prelude 20LF	Agrichem	200 g/l	LS	04371

Uses
- Alternaria in *spring oilseed rape* [1, 2]; *winter oilseed rape* [1, 2, 4]
- Disease control in *herbs (see appendix 6)* (off-label), *lamb's lettuce* (off-label), *leaf brassicas* (off-label), *lettuce* (off-label), *parsley* (off-label), *protected herbs (see appendix 6)* (off-label), *protected leaf brassicas* (off-label), *protected lettuce* (off-label), *protected parsley* (off-label),

SECTION 2

protected radicchio (off-label), *protected rocket* (off-label), *protected savory* (off-label), *protected scarole* (off-label), *radicchio* (off-label), *rocket* (off-label), *savory* (off-label), *scarole* (off-label) [3]; *winter linseed* (off-label) [2]

- Eyespot in *spring wheat* [1, 2]; *winter barley*, *winter rye*, *winter wheat* [1, 2, 4]
- Foliar disease control in *triticale* (off-label) [2]
- Fungus diseases in *hardy ornamentals*, *ornamental plant production*, *woody ornamentals* [3]
- Glume blotch in *spring wheat* [1, 2]; *winter wheat* [1, 2, 4]
- Grey mould in *spring oilseed rape* [1, 2]; *winter oilseed rape* [1, 2, 4]
- Light leaf spot in *spring oilseed rape* [1, 2]; *winter oilseed rape* [1, 2, 4]
- Net blotch in *spring barley*, *winter barley* [1, 2, 4]
- Phoma in *spring oilseed rape* [1, 2]; *winter oilseed rape* [1, 2, 4]
- Powdery mildew in *spring barley*, *spring wheat*, *winter barley*, *winter wheat* [4]; *winter rye* [1, 2, 4]
- Rhynchosporium in *spring barley*, *winter barley*, *winter rye* [1, 2, 4]
- Sclerotinia stem rot in *spring oilseed rape* [1, 2]; *winter oilseed rape* [1, 2, 4]
- Seed-borne diseases in *flax* (seed treatment), *hemp for oilseed production* (off-label - seed treatment), *linseed* (seed treatment) [5]
- Septoria leaf blotch in *spring wheat* [1, 2]; *winter rye* [4]; *winter wheat* [1, 2, 4]
- White leaf spot in *spring oilseed rape* [1, 2]; *winter oilseed rape* [1, 2, 4]

Extension of Authorisation for Minor Use (EAMUs)
- *hemp for oilseed production* (seed treatment) 20052562 [5]
- *herbs (see appendix 6)* 20121415 [3]
- *lamb's lettuce* 20121415 [3]
- *leaf brassicas* 20121415 [3]
- *lettuce* 20121415 [3]
- *parsley* 20121415 [3]
- *protected herbs (see appendix 6)* 20121415 [3]
- *protected leaf brassicas* 20121415 [3]
- *protected lettuce* 20121415 [3]
- *protected parsley* 20121415 [3]
- *protected radicchio* 20121415 [3]
- *protected rocket* 20121415 [3]
- *protected savory* 20121415 [3]
- *protected scarole* 20121415 [3]
- *radicchio* 20121415 [3]
- *rocket* 20121415 [3]
- *savory* 20121415 [3]
- *scarole* 20121415 [3]
- *triticale* 20062686 [2]
- *winter linseed* 20062686 [2], 20062687 [2]

Approval information
- Prochloraz included in Annex 1 under EC Regulation 1107/2009
- Accepted by BBPA for use on malting barley

Efficacy guidance
- Spray cereals at first signs of disease. Protection of winter crops through season usually requires at least 2 treatments. See label for details of rates and timing. Treatment active against strains of eyespot resistant to benzimidazole fungicides [2, 4]
- Tank mixes with other fungicides recommended to improve control of rusts in wheat and barley. See label for details [2, 4]
- A period of at least 3 h without rain should follow spraying [2, 4]
- Can be used through most seed treatment machines if good even seed coverage is obtained. Check drill calibration before drilling treated seed. Use inert seed flow agent supplied by manufacturer [5]
- Use methylated spirits, rather than water, to clean residual material from seed treatment machinery, then carry out final rinse with water and detergent [5]

SEE SECTION 3 FOR PRODUCTS ALSO REGISTERED

- Apply as drench against soil diseases, as a spray against aerial diseases, as a dip for cuttings, or as a drench at propagation. Spray applications may be repeated at 10-14 d intervals. Under mist propagation use 7 d intervals [3]
- Prochloraz is a DMI fungicide. Resistance to some DMI fungicides has been identified in Septoria leaf blotch which may seriously affect performance of some products. For further advice contact a specialist advisor and visit the Fungicide Resistance Action Group (FRAG)-UK website

Restrictions
- Maximum number of treatments 1 per batch of seed for flax, linseed [5]; 2 per crop for foliar spray uses
- Do not treat linseed varieties Linda, Bolas, Karen, Laura, Mikael, Norlin, Moonraker, Abbey, Agriace [5]
- Do not use treated seed as food or feed [5]

Crop-specific information
- Latest use: before drilling for seed treatments on flax or linseed [5]; watery ripe to early milk stage (GS 71-73) for cereals
- HI 6 wk for cereals
- Application with other fungicides may cause cereal crop scorch [2, 4]

Environmental safety
- Dangerous for the environment
- Very toxic to aquatic organisms
- Treated seed harmful to game and wildlife. Product contains red dye for easy identification of treated seed [5]

Hazard classification and safety precautions
Hazard Harmful [4, 5]; Irritant [1-3]; Dangerous for the environment [1-5]
Transport code 3 [1]; 9 [2-5]
Packaging group III
UN Number 1993 [1]; 3077 [3]; 3082 [2, 4, 5]
Risk phrases R22a [4, 5]; R36, R53a [1-5]; R38 [1-4]; R50 [3, 4]; R51 [1, 2, 5]
Operator protection A, C; U05a, U20b [1-5]; U08 [1, 2, 4, 5]; U09a, U19a [3]; U11 [1, 2, 5]; U14 [1, 2, 4]; U15 [1, 2]
Environmental protection E15a [1-5]; E34 [4]; E38 [4, 5]
Storage and disposal D01, D09a [1-5]; D02, D10b [1-4]; D05 [1, 2, 5]; D10a [5]; D12a [3-5]
Treated seed S02, S03, S04b, S05, S06a, S07 [5]
Medical advice M03, M05a [4]; M05b [1, 2]

372   prochloraz + propiconazole

A broad spectrum fungicide mixture for wheat, barley and oilseed rape
FRAC mode of action code: 3 + 3

See also propiconazole

Products

1	Bumper Excell	Makhteshim	400:90 g/l	EC	11015
2	Bumper P	Makhteshim	400:90 g/l	EC	08548
3	Greencrop Twinstar	Greencrop	400:90 g/l	EC	09516
4	MAC-Prochloraz Plus 490 EC	AgChem Access	400:90 g/l	EC	13616

Uses
- Disease control in **grass seed crops** *(off-label)* [1]
- Eyespot in **spring barley**, **spring wheat** [1, 2, 4]; **winter barley**, **winter wheat** [1-4]
- Light leaf spot in **spring oilseed rape**, **winter oilseed rape** [1, 2, 4]
- Phoma in **spring oilseed rape**, **winter oilseed rape** [1, 2, 4]
- Rhynchosporium in **spring barley** [1, 2, 4]; **winter barley** [1-4]
- Septoria leaf blotch in **spring wheat** [1, 2, 4]; **winter wheat** [1-4]

FOR FULL CONDITIONS OF USE ALWAYS READ THE PRODUCT LABEL

Extension of Authorisation for Minor Use (EAMUs)
- **grass seed crops** 20120366 expires 31 Dec 2013 [1]

Approval information
- Propiconazole and prochloraz included under EC Regulation 1107/2009
- Accepted by BBPA for use on malting barley
- Approval expiry 31 Dec 2013 [1, 4]

Efficacy guidance
- Best results obtained from treatment when disease is active but not well established
- Treat wheat normally from flag leaf ligule just visible stage (GS 39) but earlier if there is a high risk of Septoria
- Treat barley from the first node detectable stage (GS 31)
- If disease pressure persists a second application may be necessary
- Prochloraz and propiconazole are DMI fungicides. Resistance to some DMI fungicides has been identified in Septoria leaf blotch which may seriously affect performance of some products. For further advice contact a specialist advisor and visit the Fungicide Resistance Action Group (FRAG)-UK website

Restrictions
- Maximum number of treatments 2 per crop

Crop-specific information
- Latest use: before grain watery ripe (GS 71) for cereals; before most seeds green for winter oilseed rape

Environmental safety
- Dangerous for the environment
- Very toxic to aquatic organisms

Hazard classification and safety precautions
Hazard Irritant, Dangerous for the environment
Transport code 9
Packaging group III
UN Number 3082
Risk phrases R36, R53a [1-4]; R50 [3]; R51 [1, 2, 4]
Operator protection A, C; U05a, U08, U20b [1-4]; U11, U15 [1, 2, 4]
Environmental protection E15a
Storage and disposal D01, D02, D05, D09a, D10b
Medical advice M05a [1, 2, 4]

373   prochloraz + proquinazid + tebuconazole

A broad spectrum systemic fungicide mixture for cereals
FRAC mode of action code: 3 + U7 + 3

See also proquinazid
tebuconazole

Products

Vareon	DuPont	320:40:160 g/l	EC	14376

Uses
- Brown rust in **spring barley, spring wheat, winter barley, winter wheat**
- Crown rust in **spring oats** (qualified minor use recommendation), **winter oats** (qualified minor use recommendation)
- Ear diseases in **spring wheat** (reduction only), **winter wheat** (reduction only)
- Eyespot in **spring barley, spring wheat, winter barley, winter wheat**
- Glume blotch in **spring wheat, winter wheat**
- Net blotch in **spring barley** (moderate control), **winter barley** (moderate control)
- Powdery mildew in **spring barley, spring oats, spring wheat, winter barley, winter oats, winter wheat**

SEE SECTION 3 FOR PRODUCTS ALSO REGISTERED

SECTION 2

- Rhynchosporium in **spring barley** *(moderate control)*, **winter barley** *(moderate control)*
- Septoria leaf blotch in **spring wheat** *(moderate control)*, **winter wheat** *(moderate control)*
- Yellow rust in **spring barley**, **spring wheat**, **winter barley**, **winter wheat**

Approval information
- Prochloraz, proquinazid and tebuconazole included in Annex 1 under EC Regulation 1107/2009
- Accepted by BBPA for use on malting barley

Efficacy guidance
- Resistance to DMI fungicides has been identified in Septoria leaf blotch and may seriously affect the performance of some products. Advice on resistance management is available from the Fungicide Action Group (FRAG) web site

Restrictions
- Maximum number of applications is 2 per crop

Crop-specific information
- Latest application in wheat is before full flowering (GS65)
- Latest application in barley is before beginning of heading (GS49)

Environmental safety
- Avoid spraying within 5 m of the field boundary to reduce the effect on non-target insects or other arthropods
- LERAP Category B

Hazard classification and safety precautions
Hazard Harmful, Dangerous for the environment
Transport code 9
Packaging group III
UN Number 3082
Risk phrases R22a, R36, R40, R43, R50, R53a, R63
Operator protection A, C, H
Environmental protection E15a, E16a, E22c, E34, E38
Storage and disposal D01, D02, D09a, D12a

374 prochloraz + tebuconazole

A broad spectrum systemic fungicide mixture for cereals, oilseed rape and amenity use
FRAC mode of action code: 3 + 3

See also tebuconazole

Products

1 Agate	Makhteshim	267:133 g/l	EC	14482
2 Astute	Sherriff Amenity	267:133 g/l	EC	14897
3 Monkey	Makhteshim	267:133 g/l	EW	12906
4 Orius P	Makhteshim	267:133 g/l	EW	12880
5 Pan Proteb	Pan Amenity	267:132 g/l	EC	14896
6 PureProteb	Pure Amenity	267:132 g/l	EC	15469
7 Throttle	Headland Amenity	267:133 g/l	EC	13900

Uses
- Brown rust in **spring barley**, **spring rye**, **spring wheat**, **winter barley**, **winter rye**, **winter wheat** [1, 3, 4]
- Eyespot in **spring barley**, **spring rye**, **spring wheat**, **winter barley**, **winter rye**, **winter wheat** [1, 3, 4]
- Fusarium patch in **managed amenity turf** [2, 5-7]
- Glume blotch in **spring wheat**, **winter wheat** [1, 3, 4]
- Late ear diseases in **spring wheat**, **winter wheat** [1, 3, 4]
- Net blotch in **spring barley**, **winter barley** [1, 3, 4]

FOR FULL CONDITIONS OF USE ALWAYS READ THE PRODUCT LABEL

- Powdery mildew in *spring barley, spring rye, spring wheat, winter barley, winter rye, winter wheat* [1, 3, 4]
- Rhynchosporium in *spring barley, spring rye, winter barley, winter rye* [1, 3, 4]
- Sclerotinia stem rot in *spring oilseed rape, winter oilseed rape* [1, 3, 4]
- Septoria leaf blotch in *spring wheat, winter wheat* [1, 3, 4]
- Yellow rust in *spring barley, spring rye, spring wheat, winter barley, winter rye, winter wheat* [1, 3, 4]

Approval information
- Prochloraz and tebuconazole included in Annex I under EC Regulation 1107/2009
- Accepted by BBPA for use on malting barley

Efficacy guidance
- Complete protection of winter cereals will usually require a programme of at least two treatments [1, 3]
- Protection of flag leaf and ear from Septoria diseases requires treatment at flag leaf emergence and repeated when ear fully emerged [1, 3]
- Two treatments separated by a 2 wk gap may be needed for difficult Fusarium patch attacks
- To avoid resistance, do not apply repeated applications of the single product on the same crop against the same disease [7]
- Optimum application timing is normally when disease first seen but varies with main target disease - see labels
- Prochloraz and tebuconazole are DMI fungicides. Resistance to some DMI fungicides has been identified in Septoria leaf blotch which may seriously affect performance of some products. For further advice contact a specialist advisor and visit the Fungicide Resistance Action Group (FRAG)-UK website

Restrictions
- Maximum total dose on cereals equivalent to two full dose treatments
- Maximum number of treatments on managed amenity turf 2 per yr

Crop-specific information
- Latest use: before grain milky ripe (GS 73) for cereals [3]; most seeds green for oilseed rape [3]
- HI 6 wk for cereals [3]
- Occasionally transient leaf speckling may occur after treating wheat. Yield responses should not be affected

Environmental safety
- Dangerous for the environment
- Very toxic to aquatic organisms

Hazard classification and safety precautions
Hazard Harmful, Dangerous for the environment
Transport code 9
Packaging group III
UN Number 3082
Risk phrases R22a, R36, R43, R50, R53a [1-7]; R63 [5-7]
Operator protection A, C, H [1-7]; M [5-7]; U05a [1-7]; U08, U20b [7]; U09b, U20a [1-4]; U11 [5-7]; U14 [1-6]
Environmental protection E13b [1-4]; E15a [5-7]; E34, E38 [1-7]
Storage and disposal D01, D02, D09a, D12a [1-7]; D05 [1-6]; D10b [5-7]; D10c [1-4]
Medical advice M03 [1-7]; M05a [5, 6]

SEE SECTION 3 FOR PRODUCTS ALSO REGISTERED

375   prochloraz + thiram

A seed treatment fungicide mixture for oilseed rape
FRAC mode of action code: 3 + M3

See also thiram

Products

Agrichem Hy-Pro Duet	Agrichem	150:333 g/l	FS	14018

Uses
- Phoma in **winter oilseed rape** *(seed treatment)*

Approval information
- Prochloraz and thiram included in Annex I under EC Regulation 1107/2009

Efficacy guidance
- Can be applied through most types of seed treatment machinery
- Good coverage of the seed surface essential for best results
- Product is active against seed-borne disease. It will not control leaf spotting from leaf spores on trash on the seed bed

Restrictions
- Maximum number of treatments 1 per crop
- Seed must be of satisfactory quality and moisture content
- Sow seed as soon as possible after treatment
- Do not store treated seed from one season to the next
- Do not co-apply product with water or mix with any other seed treatment product

Crop-specific information
- Latest use: pre-sowing
- Product will also control soil-borne fungi that cause damping-off of rape seedlings

Following crops guidance
- Any crop may be grown following a treated winter oilseed rape crop

Hazard classification and safety precautions
   Hazard Harmful, Dangerous for the environment
   Transport code 9
   Packaging group III
   UN Number 3082
   Risk phrases R20, R22a, R36, R38, R43, R50, R53a
   Operator protection A, C, H; U02a, U08, U14, U15, U20a
   Environmental protection E15a, E36b
   Storage and disposal D01, D02, D05, D08, D09a, D12b, D13, D15
   Treated seed S01, S02, S03, S04d, S05, S08
   Medical advice M04a

376   prochloraz + triticonazole

A broad spectrum fungicide seed treatment mixture for winter cereals
FRAC mode of action code: 3 + 3

See also triticonazole

Products

Kinto	BASF	60:20 g/l	FS	12038

Uses
- Bunt in **durum wheat** *(off-label - seed treatment)*, **winter wheat** *(seed treatment)*
- Covered smut in **winter barley** *(seed treatment)*, **winter oats** *(seed treatment)*

- Fusarium foot rot and seedling blight in **durum wheat** *(off-label - seed treatment)*, **triticale** *(seed treatment)*, **winter barley** *(seed treatment)*, **winter oats** *(seed treatment)*, **winter rye** *(seed treatment)*, **winter wheat** *(seed treatment)*
- Leaf stripe in **winter barley** *(seed treatment)*
- Loose smut in **durum wheat** *(off-label - seed treatment)*, **winter barley** *(seed treatment)*, **winter oats** *(seed treatment)*, **winter wheat** *(seed treatment)*
- Septoria seedling blight in **durum wheat** *(off-label - seed treatment)*, **winter wheat** *(seed treatment)*

Extension of Authorisation for Minor Use (EAMUs)
- **durum wheat** *(seed treatment)* 20061094

Approval information
- Prochloraz and triticonazole included in Annex I under EC Regulation 1107/2009
- Accepted by BBPA for use on malting barley

Efficacy guidance
- Apply undiluted through conventional seed treatment machine
- Calibrate seed drill with treated seed before sowing
- Bag treated seed immediately and keep in a dry, draught free store
- Ensure good even coverage of seed to achieve best results
- Treatment may cause some slight delay in crop emergence

Restrictions
- Maximum number of treatments 1 per batch of seed
- Do not treat seed with moisture content above 16% and do not allow moisture content of treated seed to exceed 16%
- Do not treat cracked, split or sprouted seed
- Do not handle treated seed unnecessarily
- Do not use treated seed as food or feed

Crop-specific information
- Latest use: before drilling wheat, barley, oats, rye or triticale

Environmental safety
- Dangerous for the environment
- Toxic to aquatic organisms

Hazard classification and safety precautions
Hazard Dangerous for the environment
Transport code 9
Packaging group III
UN Number 3082
Risk phrases R51, R53a
Operator protection A, H; U05a, U07, U20b
Environmental protection E15a, E34, E38
Storage and disposal D01, D02, D08, D09a, D11a, D12a
Treated seed S01, S02, S03, S04a, S04b, S05, S07

377   prohexadione-calcium

A cyclohexanecarboxylate growth regulator for use in apples

See also mepiquat chloride + prohexadione-calcium

Products

Regalis	BASF	10% w/w	WG	12414

Uses
- Control of shoot growth in **apples**, **pears** *(off-label)*
- Growth regulation in **grapevines** *(off-label)*, **hops** *(off-label)*, **ornamental plant production** *(off-label)*, **soft fruit** *(off-label)*

SEE SECTION 3 FOR PRODUCTS ALSO REGISTERED

Extension of Authorisation for Minor Use (EAMUs)
- *grapevines* 20101467
- *hops* 20082866
- *ornamental plant production* 20082866
- *pears* 20102622
- *soft fruit* 20082866

Approval information
- Prohexadione-calcium included in Annex I under EC Regulation 1107/2009

Efficacy guidance
- For a standard orchard treat at the start of active growth and again after 3-5 wk depending on growth conditions
- Alternatively follow the first application with four reduced rate applications at two wk intervals

Restrictions
- Maximum total dose equivalent to two full dose treatments
- Do not apply in conjunction with calcium based foliar fertilisers

Crop-specific information
- HI 55 d for apples

Environmental safety
- Harmful to aquatic organisms
- Avoid spray drift onto neighbouring crops and other non-target plants
- Product does not harm natural insect predators when used as directed

Hazard classification and safety precautions
  UN Number N/C
  Risk phrases R52, R53a
  Operator protection U05a
  Environmental protection E38
  Storage and disposal D01, D02, D09a, D10c, D12a

378 propamocarb hydrochloride

A translocated protectant carbamate fungicide
FRAC mode of action code: 28

See also chlorothalonil + propamocarb hydrochloride
fenamidone + propamocarb hydrochloride
fluopicolide + propamocarb hydrochloride
fosetyl-aluminium + propamocarb hydrochloride
mancozeb + propamocarb hydrochloride

Products

Proplant	Fargro	722 g/l	SL	15422

Uses
- Damping off in **protected broccoli, protected brussels sprouts, protected calabrese, protected cauliflowers**
- Downy mildew in **protected broccoli, protected brussels sprouts, protected calabrese, protected cauliflowers**
- Peronospora spp in **protected broccoli, protected brussels sprouts, protected calabrese, protected cauliflowers**
- Phytophthora in **protected broccoli, protected brussels sprouts, protected calabrese, protected cauliflowers**
- Pythium in **protected broccoli, protected brussels sprouts, protected calabrese, protected cauliflowers**
- Root rot in **protected broccoli, protected brussels sprouts, protected calabrese, protected cauliflowers**

Approval information

- Propamocarb hydrochloride included in Annex I under EC Regulation 1107/2009

Efficacy guidance

- Chemical is absorbed through roots and translocated throughout plant
- Incorporate in compost before use or drench moist compost or soil before sowing, pricking out, striking cuttings or potting up
- Drench treatment can be repeated at 3-6 wk intervals
- Concentrated solution is corrosive to all metals other than stainless steel
- May also be applied in trickle irrigation systems
- To prevent root rot in tulip bulbs apply as dip for 20 min but full protection only achieved by soil drench treatment as well

Restrictions

- Maximum number of treatments 4 per crop for cucumbers, tomatoes, peppers, aubergines; 3 per crop for outdoor and protected lettuce; 1 per crop for listed brassicas; 1 compost incorporation and/or 1 drench treatment for leeks, onions; 1 dip and 1 soil drench for flower bulbs
- Do not apply to lettuce crops grown in non-soil systems
- When applied over established seedlings rinse off foliage with water (except lettuce) and do not apply under hot, dry conditions
- On plants of unknown tolerance test first on a small scale
- Do not apply in a recirculating irrigation/drip system
- Store away from seeds and fertilizers
- Consult processor before use on crops grown for processing

Crop-specific information

- Latest use: before transplanting for brassicas
- HI 14 d for cucumbers, tomatoes, peppers, aubergines; 4 wk for brassicas; 19 wk for leeks, onions

Hazard classification and safety precautions

Hazard Irritant
UN Number N/C
Risk phrases R43, R52
Operator protection A, H, M; U02a, U05a, U08, U19a, U20b
Environmental protection E15a, E34
Storage and disposal D01, D02, D05, D09a, D10c

379 propaquizafop

A phenoxy alkanoic acid foliar acting herbicide for grass weeds in a range of crops
HRAC mode of action code: A

Products

1	Clayton Orleans	Clayton	100 g/l	EC	14452
2	Clayton Orleans	Clayton	100 g/l	EC	14827
3	Cleancrop GYR 2	Makhteshim	100 g/l	EC	14578
4	Falcon	Makhteshim	100 g/l	EC	14552
5	Greencrop Satchmo	Greencrop	100 g/l	EC	10748
6	Longhorn	AgChem Access	100 g/l	EC	12512
7	MAC-Propaquizafop 100 EC	AgChem Access	100 g/l	EC	13978
8	PQF 100	Goldengrass	100 g/l	EC	14414
9	Shogun	Makhteshim	100 g/l	EC	14567
10	Standon Propaquizafop	Standon	100 g/l	EC	11536
11	Standon Zing PQF	Standon	100 g/l	EC	13898
12	Standon Zing PQF	Standon	100 g/l	EC	14949

Uses

- Annual grasses in ***beet leaves*** *(off-label)*, ***bog myrtle*** *(off-label)*, ***celery (outdoor)*** *(off-label)*, ***herbs (see appendix 6)*** *(off-label)*, ***honesty*** *(off-label)*, ***horseradish*** *(off-label)*, ***leeks*** *(off-label)*,

SEE SECTION 3 FOR PRODUCTS ALSO REGISTERED

mallow (althaea spp.) *(off-label)*, **poppies for morphine production** *(off-label)*, **red beet** *(off-label)*, **rhubarb** *(off-label)*, **spinach** *(off-label)*, **spring linseed** *(off-label)* [3, 4, 9]; **borage** *(off-label)* [4]; **borage for oilseed production** *(off-label)* [3, 9, 10]; **bulb onions, carrots, combining peas, early potatoes, farm forestry, fodder beet, linseed, parsnips, spring oilseed rape, sugar beet, swedes, turnips, winter field beans, winter oilseed rape** [1-12]; **canary flower (echium spp.)** *(off-label)* [3, 4, 9, 10]; **cut logs, forest, forest nurseries** [2-4, 7, 9]; **evening primrose** *(off-label)*, **garlic** *(off-label)*, **lupins** *(off-label)*, **mustard** *(off-label)*, **shallots** *(off-label)* [3-5, 9, 10]; **maincrop potatoes** [1, 2, 4-12]; **potatoes** [3]; **spring field beans** [1-5, 7-12]

- Annual meadow grass in **beet leaves** *(off-label)*, **bog myrtle** *(off-label)*, **borage** *(off-label)*, **celery (outdoor)** *(off-label)*, **herbs (see appendix 6)** *(off-label)*, **honesty** *(off-label)*, **horseradish** *(off-label)*, **leeks** *(off-label)*, **mallow (althaea spp.)** *(off-label)*, **poppies for morphine production** *(off-label)*, **red beet** *(off-label)*, **rhubarb** *(off-label)*, **spinach** *(off-label)*, **spring linseed** *(off-label)* [4]; **borage for oilseed production** *(off-label)* [10]; **bulb onions, carrots, combining peas, cut logs, early potatoes, farm forestry, fodder beet, forest, forest nurseries, linseed, parsnips, potatoes, spring field beans, spring oilseed rape, sugar beet, swedes, turnips, winter field beans, winter oilseed rape** [3]; **canary flower (echium spp.)** *(off-label)* [4, 10]; **evening primrose** *(off-label)*, **garlic** *(off-label)*, **lupins** *(off-label)*, **mustard** *(off-label)*, **shallots** *(off-label)* [4, 5, 10]
- Couch in **beet leaves** *(off-label)*, **bog myrtle** *(off-label)*, **borage** *(off-label)*, **canary flower (echium spp.)** *(off-label)*, **celery (outdoor)** *(off-label)*, **evening primrose** *(off-label)*, **garlic** *(off-label)*, **herbs (see appendix 6)** *(off-label)*, **honesty** *(off-label)*, **horseradish** *(off-label)*, **leeks** *(off-label)*, **lupins** *(off-label)*, **mallow (althaea spp.)** *(off-label)*, **mustard** *(off-label)*, **poppies for morphine production** *(off-label)*, **red beet** *(off-label)*, **rhubarb** *(off-label)*, **shallots** *(off-label)*, **spinach** *(off-label)*, **spring linseed** *(off-label)* [4]
- Grass weeds in **corn gromwell** *(off-label)* [4]
- Perennial grasses in **borage for oilseed production** *(off-label)*, **canary flower (echium spp.)** *(off-label)* [10]; **bulb onions, carrots, combining peas, early potatoes, farm forestry, fodder beet, linseed, parsnips, spring oilseed rape, sugar beet, swedes, turnips, winter field beans, winter oilseed rape** [1-12]; **cut logs, forest, forest nurseries** [2-4, 7, 9]; **evening primrose** *(off-label)*, **garlic** *(off-label)*, **lupins** *(off-label)*, **mustard** *(off-label)*, **shallots** *(off-label)* [5, 10]; **maincrop potatoes** [1, 2, 4-12]; **potatoes** [3]; **red beet** *(off-label)* [3, 9]; **spring field beans** [1-5, 7-12]
- Volunteer cereals in **beet leaves** *(off-label)*, **bog myrtle** *(off-label)*, **borage** *(off-label)*, **celery (outdoor)** *(off-label)*, **herbs (see appendix 6)** *(off-label)*, **honesty** *(off-label)*, **horseradish** *(off-label)*, **leeks** *(off-label)*, **mallow (althaea spp.)** *(off-label)*, **poppies for morphine production** *(off-label)*, **red beet** *(off-label)*, **rhubarb** *(off-label)*, **spinach** *(off-label)*, **spring linseed** *(off-label)* [4]; **borage for oilseed production** *(off-label)* [10]; **canary flower (echium spp.)** *(off-label)* [4, 10]; **evening primrose** *(off-label)*, **garlic** *(off-label)*, **lupins** *(off-label)*, **mustard** *(off-label)*, **shallots** *(off-label)* [4, 5, 10]

Extension of Authorisation for Minor Use (EAMUs)

- **beet leaves** *20092740* [3], *20092447* [4], *20092784* [9]
- **bog myrtle** *20092739* [3], *20092454* [4], *20092790* [9]
- **borage** *20092450* [4]
- **borage for oilseed production** *20092737* [3], *20092789* [9], *20080893* [10]
- **canary flower (echium spp.)** *20092737* [3], *20092450* [4], *20092789* [9], *20080893* [10]
- **celery (outdoor)** *20092735* [3], *20092448* [4], *20092788* [9]
- **corn gromwell** *20111308* [4]
- **evening primrose** *20092737* [3], *20092450* [4], *20080879* [5], *20092789* [9], *20080893* [10]
- **garlic** *20092732* [3], *20092451* [4], *20080880* [5], *20092785* [9], *20080894* [10]
- **herbs (see appendix 6)** *20092740* [3], *20092447* [4], *20092784* [9]
- **honesty** *20092737* [3], *20092450* [4], *20092789* [9]
- **horseradish** *20092738* [3], *20092455* [4], *20092787* [9]
- **leeks** *20092740* [3], *20092447* [4], *20092784* [9]
- **lupins** *20092736* [3], *20092452* [4], *20080878* [5], *20092783* [9], *20080892* [10]
- **mallow (althaea spp.)** *20092733* [3], *20092449* [4], *20092786* [9]
- **mustard** *20092737* [3], *20092450* [4], *20080879* [5], *20092789* [9], *20080893* [10]
- **poppies for morphine production** *20092734* [3], *20092453* [4], *20092791* [9]
- **red beet** *20092740* [3], *20092447* [4], *20092784* [9]

FOR FULL CONDITIONS OF USE ALWAYS READ THE PRODUCT LABEL

- **rhubarb** *20092735* [3], *20092448* [4], *20092788* [9]
- **shallots** *20092732* [3], *20092451* [4], *20080880* [5], *20092785* [9], *20080894* [10]
- **spinach** *20092740* [3], *20092447* [4], *20092784* [9]
- **spring linseed** *20092737* [3], *20092450* [4], *20092789* [9]

Approval information
- Propaquizafop included in Annex 1 under EC Regulation 1107/2009

Efficacy guidance
- Apply to emerged weeds when they are growing actively with adequate soil moisture
- Activity is slower under cool conditions
- Broad-leaved weeds and any weeds germinating after treatment are not controlled
- Annual meadow grass up to 3 leaves checked at low doses and severely checked at highest dose
- Spray barley cover crops when risk of wind blow has passed and before there is serious competition with the crop
- Various tank mixtures and sequences recommended for broader spectrum weed control in oilseed rape, peas and sugar beet. See label for details
- Severe couch infestations may require a second application at reduced dose when regrowth has 3-4 leaves unfolded
- Products contain surfactants. Tank mixing with adjuvants not required or recommended
- Propaquizafop is an ACCase inhibitor herbicide. To avoid the build up of resistance do not apply products containing an ACCase inhibitor herbicide more than twice to any crop. In addition do not use any product containing propaquizafop in mixture or sequence with any other product containing the same ingredient
- Use these products as part of a resistance management strategy that includes cultural methods of control and does not use ACCase inhibitors as the sole chemical method of grass weed control
- Applying a second product containing an ACCase inhibitor to a crop will increase the risk of resistance development; only use a second ACCase inhibitor to control different weeds at a different timing
- Always follow WRAG guidelines for preventing and managing herbicide resistant weeds. See Section 5 for more information

Restrictions
- Maximum number of treatments 1 or 2 per crop or yr. See label for details
- See label for list of tolerant tree species
- Do not treat seed potatoes

Crop-specific information
- Latest use: before crop flower buds visible for winter oilseed rape, linseed, field beans; before 8 fully expanded leaf stage for spring oilseed rape; before weeds are covered by the crop for potatoes, sugar beet, fodder beet, swedes, turnips; when flower buds visible for peas
- HI onions, carrots, early potatoes, parsnips 4 wk; combining peas, early potatoes 7 wk; sugar beet, fodder beet, maincrop potatoes, swedes, turnips 8 wk; field beans 14 wk
- Application in high temperatures and/or low soil moisture content may cause chlorotic spotting especially on combining peas and field beans
- Overlaps at the highest dose can cause damage from early applications to carrots and parsnips

Following crops guidance
- In the event of a failed treated crop an interval of 2 wk must elapse between the last application and redrilling with winter wheat or winter barley. 4 wk must elapse before sowing oilseed rape, peas or field beans, and 16 wk before sowing ryegrass or oats

Environmental safety
- Dangerous for the environment
- Toxic to aquatic organisms
- Risk to certain non-target insects or other arthropods. See directions for use

Hazard classification and safety precautions
Hazard Harmful [3, 4, 9]; Irritant [1, 2, 5-8, 10-12]; Dangerous for the environment [1-12]
Transport code 9

SEE SECTION 3 FOR PRODUCTS ALSO REGISTERED

Packaging group III
UN Number 3082
Risk phrases R20, R41 [3]; R36, R38 [1, 2, 4-12]; R51, R53a [1-12]
Operator protection A, C [1-12]; H [3, 6, 9]; U02a, U05a [1, 2, 4-12]; U08, U19a, U20b [1-12]; U11 [4, 7, 10-12]; U12 [3]; U13 [1-4, 6-9]; U14, U15 [1-4, 6-12]; U23a [5]
Environmental protection E15a, E22b [1-12]; E34 [1, 2, 4, 6-9]; E38 [3]
Storage and disposal D01, D02, D05, D09a [1-12]; D10a, D12a [3]; D10b [1, 2, 4-12]; D12b, D14 [1, 2, 4, 6-9]
Medical advice M05b

380 propiconazole

A systemic, curative and protectant triazole fungicide
FRAC mode of action code: 3

See also azoxystrobin + propiconazole
chlorothalonil + cyproconazole + propiconazole
chlorothalonil + fludioxonil + propiconazole
chlorothalonil + propiconazole
cyproconazole + propiconazole
fenbuconazole + propiconazole
fenpropidin + propiconazole
fenpropidin + propiconazole + tebuconazole
prochloraz + propiconazole
propiconazole + tebuconazole
propiconazole + trifloxystrobin

Products

1	Banner Maxx	Syngenta	156 g/l	EC	13167
2	Bounty	AgChem Access	250 g/l	EC	14616
3	Bumper 250 EC	Makhteshim	250 g/l	EC	14399
4	Mascot Exocet	Rigby Taylor	156 g/l	EC	15598
5	Propi 25 EC	Sharda	250 g/l	EC	14939
6	Spaniel	Pan Amenity	156 g/l	EC	15530

Uses

- Anthracnose in **amenity grassland** *(qualified minor use)*, **managed amenity turf** *(qualified minor use)* [1, 4, 6]
- Brown patch in **amenity grassland** *(qualified minor use)*, **managed amenity turf** *(qualified minor use)* [1, 4, 6]
- Brown rust in **spring barley**, **spring wheat**, **winter barley**, **winter rye**, **winter wheat** [2, 3, 5]
- Crown rust in **grass seed crops**, **permanent grassland**, **spring oats**, **winter oats** [2, 3, 5]
- Disease control in **forest nurseries** *(off-label)*, **honesty** *(off-label)*, **hops** *(off-label)*, **protected forest nurseries** *(off-label)*, **soft fruit** *(off-label)* [3]
- Dollar spot in **amenity grassland**, **managed amenity turf** [1, 4, 6]
- Downy mildew in **garlic** *(off-label)*, **onions** *(off-label)*, **shallots** *(off-label)* [3]
- Drechslera leaf spot in **grass seed crops**, **permanent grassland** [2, 3, 5]
- Fusarium patch in **amenity grassland**, **managed amenity turf** [1, 4, 6]
- Glume blotch in **spring wheat**, **winter rye**, **winter wheat** [2, 3, 5]
- Light leaf spot in **spring oilseed rape** *(reduction)*, **winter oilseed rape** *(reduction)* [2, 3, 5]
- Mildew in **grass seed crops**, **permanent grassland** [2, 3, 5]
- Powdery mildew in **spring barley**, **spring oats**, **spring wheat**, **winter barley**, **winter oats**, **winter rye**, **winter wheat** [2, 3, 5]
- Ramularia leaf spots in **sugar beet** *(reduction)* [2, 3, 5]
- Rhynchosporium in **grass seed crops**, **permanent grassland**, **spring barley**, **winter barley**, **winter rye** [2, 3, 5]
- Rust in **ornamental plant production** *(off-label)*, **protected ornamentals** *(off-label)* [3]; **sugar beet** [2, 3, 5]
- Septoria leaf blotch in **spring wheat**, **winter rye**, **winter wheat** [2, 3, 5]

FOR FULL CONDITIONS OF USE ALWAYS READ THE PRODUCT LABEL

- Sooty moulds in **spring wheat**, **winter wheat** [2, 3, 5]
- Yellow rust in **spring barley**, **spring wheat**, **winter barley**, **winter wheat** [2, 3, 5]

Extension of Authorisation for Minor Use (EAMUs)

- **forest nurseries** *20111133* [3]
- **garlic** *20090705* [3]
- **honesty** *20090706* [3]
- **hops** *20111133* [3]
- **onions** *20090705* [3]
- **ornamental plant production** *20090707* [3]
- **protected forest nurseries** *20111133* [3]
- **protected ornamentals** *20090707* [3]
- **shallots** *20090705* [3]
- **soft fruit** *20111133* [3]

Approval information

- Propiconazole included in Annex I under EC Regulation 1107/2009
- Accepted by BBPA for use on malting barley

Efficacy guidance

- Best results achieved by applying at early stage of disease development. Recommended spray programmes vary with crop, disease, season, soil type and product.
- For optimum turf quality and control use in conjunction with turf management practices that promote good plant health [1]
- Propiconazole is a DMI fungicide. Resistance to some DMI fungicides has been identified in Septoria leaf blotch which may seriously affect performance of some products. For further advice contact a specialist advisor and visit the Fungicide Resistance Action Group (FRAG)-UK website

Restrictions

- On oilseed rape do not apply during flowering.
- Grass seed crops must be treated in yr of harvest
- A minimum interval of 14 d must elapse between treatments on leeks and 21 d on sugar beet
- Avoid spraying crops or turf under stress, eg during cold weather or periods of frost

Crop-specific information

- Latest use: before 4 pairs of true leaves for honesty
- HI cereals, grass seed crops 35 d; oilseed rape, sugar beet, garlic, onions, shallots, grass for ensiling 28 d
- May be used on all common turf grass species [1]

Environmental safety

- Dangerous for the environment
- Toxic to aquatic organisms
- Risk to non-target insects or other arthropods [1]
- When treating turf with vehicle mounted or drawn hydraulic boom sprayers avoid spraying within 5 m of unmanaged land to reduce effects on non-target arthropods [1]

Hazard classification and safety precautions

Hazard Harmful [1, 4, 6]; Dangerous for the environment [1-6]
Transport code 9
Packaging group III
UN Number 3082
Risk phrases R20, R43, R52 [1, 4, 6]; R51 [2, 3, 5]; R53a [1-6]
Operator protection A [1-4, 6]; C [2, 3]; H [1, 4, 6]; U02a, U05a, U19a, U20b [1-6]; U08 [2, 3, 5]; U09a, U14 [1, 4, 6]
Environmental protection E15a, E34 [2, 3, 5]; E15b, E22c, E38 [1, 4, 6]
Storage and disposal D01, D02, D05, D09a [1-6]; D10b [2, 3, 5]; D10c, D12a [1, 4, 6]
Treated seed S06a [2, 3, 5]
Medical advice M05a [2, 3, 5]

SEE SECTION 3 FOR PRODUCTS ALSO REGISTERED

SECTION 2

381   propoxycarbazone-sodium

A sulfonylaminocarbonyltriazolinone residual grass weed herbicide for winter wheat
HRAC mode of action code: B

See also iodosulfuron-methyl-sodium + propoxycarbazone-sodium

Products

Attribut	Interfarm	70% w/w	SG	14749

Uses
- Annual grasses in **miscanthus** *(off-label)*
- Blackgrass in **winter wheat**
- Couch in **miscanthus** *(off-label)*, **winter wheat**

Extension of Authorisation for Minor Use (EAMUs)
- **miscanthus** *20093179*

Approval information
- Propoxycarbazone-sodium included in Annex I under EC Regulation 1107/2009

Efficacy guidance
- Activity by root and foliar absorption but depends on presence of sufficient soil moisture to ensure root uptake by weeds. Control is enhanced by use of an adjuvant to encourage foliar uptake
- Best results obtained from treatments applied when weed grasses are growing actively. Symptoms may not become apparent for 3-4 wk after application
- Ensure good even spray coverage
- To achieve best control of couch and reduction of infestation in subsequent crop treat between 2 true leaves and first node stage of the weed
- Blackgrass should be treated after using specialist blackgrass herbicides with a different mode of action
- Propoxycarbazone-sodium is a member of the ALS-inhibitor group of herbicides. To avoid the build up of resistance do not use any product containing an ALS-inhibitor herbicide with claims for control of grass weeds more than once on any crop
- Use these products as part of a resistance management strategy that includes cultural methods of control and does not use ALS inhibitors as the sole chemical method of grass weed control

Restrictions
- Maximum total dose equivalent to one full dose treatment
- Do not use in a programme with other aceto-lactase synthesis (ALS) inhibitors
- Do not apply when temperature near or below freezing
- Avoid treatment under dry soil conditions

Crop-specific information
- Latest use: third node detectable in wheat (GS 33)

Following crops guidance
- Only winter wheat, field beans or winter barley may be sown in the autumn following spring treatment
- Any crop may be grown in the spring on land treated during the previous calendar yr

Environmental safety
- Dangerous for the environment
- Very toxic to aquatic organisms
- Take extreme care to avoid drift onto adjacent plants or land as this could result in severe damage
- LERAP Category B

Hazard classification and safety precautions
Hazard Dangerous for the environment
Transport code 9
Packaging group III
UN Number 3077

FOR FULL CONDITIONS OF USE ALWAYS READ THE PRODUCT LABEL

Risk phrases R50, R53a
Operator protection A; U05a, U20b
Environmental protection E15b, E16a, E16b, E38
Storage and disposal D01, D02, D09a, D11a, D12a

382   propyzamide

A residual benzamide herbicide for use in a wide range of crops
HRAC mode of action code: K1

See also clopyralid + propyzamide

Products

1	Careca	AgriChem BV	500 g/l	SC	14948
2	Cohort	Makhteshim	400 g/l	SC	15035
3	Conform	AgChem Access	400 g/l	SC	14602
4	Dennis	Interfarm	80% w/w	WG	15081
5	Engage	Interfarm	50% w/w	WP	14233
6	Flomide	Interfarm	400 g/l	SC	14223
7	Flomide 2	Interfarm	400 g/l	SC	15080
8	Hazard	ChemSource	80% w/w	WG	15022
9	Judo	Headland	400 g/l	SC	15346
10	Kerb 50 W	Dow	50% w/w	WP	13715
11	Kerb Flo	Dow	400 g/l	SC	13716
12	Kerb Flo 500	Dow	500 g/l	SC	15586
13	Kerb Granules	Barclay	4% w/w	GR	14213
14	Menace 80 EDF	Dow	80% w/w	WG	13714
15	Pizamide	ChemSource	400 g/l	SC	14684
16	Pizza 400 SC	Goldengrass	400 g/l	SC	14430
17	Ponder	ChemSource	400 g/l	SC	14693
18	Ponder WP	ChemSource	50% w/w	WP	15293
19	Prop 400	ChemSource	400 g/l	SC	15832
20	Propel Flo	Clayton	400 g/l	SC	15044
21	Proper Flo	Globachem	400 g/l	SC	15102
22	Propyz	Barclay	400 g/l	SC	15083
23	Propyzamide 400	European Ag	400 g/l	SC	14685
24	Prova	Belcrop	400 g/l	SC	15641
25	PureFlo	Pure Amenity	400 g/l	SC	15138
26	Quaver Flo	Dow	400 g/l	SC	14203
27	Relva	Belcrop	400 g/l	SC	14873
28	Setanta 50 WP	AgriGuard	50% w/w	WP	15642
29	Setanta Flo	AgriGuard	400 g/l	SC	15791
30	Solitaire	AgriGuard	400 g/l	SC	15637
31	Solitaire 50 WP	AgriGuard	50% w/w	WP	15636
32	Standon Santa Fe 50 WP	Standon	50% w/w	WP	14966
33	Standon Santa Fe Flo	Standon	400 g/l	SC	14727
34	Stroller	AgChem Access	400 g/l	SC	14603

Uses

- Annual and perennial weeds in *amenity vegetation* [29, 30]
- Annual dicotyledons in *all edible seed crops grown outdoors* (off-label), *all non-edible seed crops grown outdoors* (off-label), *game cover* (off-label), *hops* (off-label) [10, 11]; *almonds* (off-label), *bilberries* (off-label), *blueberries* (off-label), *bog myrtle* (off-label), *broccoli* (off-label), *calabrese* (off-label), *cauliflowers* (off-label), *cherries* (off-label), *chestnuts* (off-label), *chicory root* (off-label), *cob nuts* (off-label), *courgettes* (off-label), *cranberries* (off-label), *cress* (off-label), *endives* (off-label), *frise* (off-label), *hazel nuts* (off-label), *herbs (see appendix 6)* (off-label), *lamb's lettuce* (off-label), *leaf brassicas* (off-label), *marrows* (off-label), *mirabelles* (off-label), *protected endives* (off-label), *protected forest nurseries* (off-label), *protected herbs (see appendix 6)* (off-label), *protected lettuce* (off-label), *pumpkins* (off-label), *quinces* (off-

SEE SECTION 3 FOR PRODUCTS ALSO REGISTERED

label), **radicchio** *(off-label)*, **salad brassicas** *(off-label)*, **scarole** *(off-label)*, **squashes** *(off-label)*, **table grapes** *(off-label)*, **walnuts** *(off-label)*, **wine grapes** *(off-label)* [11]; **amenity vegetation** [1-3, 8, 10, 11, 14-19, 22-26, 32-34]; **apple orchards**, **clover seed crops**, **pear orchards**, **plums** [1-3, 5, 6, 8, 10, 11, 14-19, 22, 23, 25, 26, 29, 30, 32-34]; **blackberries**, **blackcurrants**, **gooseberries**, **loganberries**, **raspberries** *(England only)*, **redcurrants**, **strawberries** [2, 3, 5, 6, 8, 10, 11, 14-19, 22, 23, 25, 26, 29, 30, 32-34]; **brassica seed crops**, **lettuce** *(outdoor crops)*, **rhubarb** *(outdoor)* [8, 14]; **farm forestry**, **forest nurseries** [1-4, 8, 10, 11, 14-20, 22-25, 28-34]; **fodder rape seed crops**, **kale seed crops**, **turnip seed crops** [1-3, 5, 6, 10, 11, 15-19, 21-23, 25, 26, 29, 30, 32-34]; **forest** [1-5, 10, 11, 13, 15-25, 28-34]; **hedges** [1-3, 8, 10, 11, 14-19, 22-25, 28-34]; **lettuce** [2-6, 10, 11, 15-21, 23, 26, 28-34]; **lucerne** [2, 3, 5, 6, 8, 10, 11, 14-19, 23, 25, 26, 29, 30, 32-34]; **red clover, white clover** [21]; **rhubarb** [1-3, 5, 6, 10, 11, 15-19, 22, 23, 25, 26, 29, 30, 32-34]; **roses** [5]; **sugar beet seed crops** [1-3, 5, 6, 8, 10, 11, 14-19, 21-23, 25, 26, 29, 30, 32-34]; **top fruit** *(off-label)* [10]; **trees and shrubs** [5, 13]; **winter field beans** [2-6, 8, 10-12, 14-21, 23, 25, 26, 28-34]; **winter oilseed rape** [1-6, 8-12, 14-23, 25-34]; **woody ornamentals** [5, 8, 10, 13, 14]

- Annual grasses in **all edible seed crops grown outdoors** *(off-label)*, **all non-edible seed crops grown outdoors** *(off-label)*, **game cover** *(off-label)*, **hops** *(off-label)* [10, 11]; **almonds** *(off-label)*, **bilberries** *(off-label)*, **blueberries** *(off-label)*, **bog myrtle** *(off-label)*, **broccoli** *(off-label)*, **calabrese** *(off-label)*, **cauliflowers** *(off-label)*, **cherries** *(off-label)*, **chestnuts** *(off-label)*, **chicory root** *(off-label)*, **cob nuts** *(off-label)*, **courgettes** *(off-label)*, **cranberries** *(off-label)*, **cress** *(off-label)*, **endives** *(off-label)*, **frise** *(off-label)*, **hazel nuts** *(off-label)*, **herbs (see appendix 6)** *(off-label)*, **lamb's lettuce** *(off-label)*, **leaf brassicas** *(off-label)*, **marrows** *(off-label)*, **mirabelles** *(off-label)*, **protected endives** *(off-label)*, **protected forest nurseries** *(off-label)*, **protected herbs (see appendix 6)** *(off-label)*, **protected lettuce** *(off-label)*, **pumpkins** *(off-label)*, **quinces** *(off-label)*, **radicchio** *(off-label)*, **salad brassicas** *(off-label)*, **scarole** *(off-label)*, **squashes** *(off-label)*, **table grapes** *(off-label)*, **walnuts** *(off-label)*, **wine grapes** *(off-label)* [11]; **amenity vegetation** [1-3, 8, 10, 11, 14-19, 22-26, 32-34]; **apple orchards**, **clover seed crops**, **pear orchards**, **plums** [1-3, 5, 6, 8, 10, 11, 14-19, 22, 23, 25, 26, 29, 30, 32-34]; **blackberries**, **blackcurrants**, **gooseberries**, **loganberries**, **raspberries** *(England only)*, **redcurrants**, **strawberries** [2, 3, 5, 6, 8, 10, 11, 14-19, 22, 23, 25, 26, 29, 30, 32-34]; **brassica seed crops**, **lettuce** *(outdoor crops)*, **rhubarb** *(outdoor)* [8, 14]; **farm forestry**, **forest nurseries** [1-4, 7, 8, 10, 11, 14-20, 22-25, 28-34]; **fodder rape seed crops**, **kale seed crops**, **turnip seed crops** [1-3, 5, 6, 10, 11, 15-19, 21-23, 25, 26, 29, 30, 32-34]; **forest** [1-5, 7, 10, 11, 13, 15-25, 28-34]; **hedges** [1-3, 8, 10, 11, 14-19, 22-25, 28-34]; **lettuce** [2-7, 10, 11, 15-21, 23, 26, 28-34]; **lucerne** [2, 3, 5, 6, 8, 10, 11, 14-19, 23, 25, 26, 29, 30, 32-34]; **red clover, white clover** [21]; **rhubarb** [1-3, 5, 6, 10, 11, 15-19, 22, 23, 25, 26, 29, 30, 32-34]; **roses** [5]; **sugar beet seed crops** [1-3, 5, 6, 8, 10, 11, 14-19, 21-23, 25, 26, 29, 30, 32-34]; **top fruit** *(off-label)* [10]; **trees and shrubs** [5, 13]; **winter field beans** [2-8, 10, 11, 14-21, 23, 25, 26, 28-34]; **winter oilseed rape** [1-11, 14-23, 25-34]; **woody ornamentals** [5, 8, 10, 13, 14]
- Annual meadow grass in **winter field beans**, **winter oilseed rape** [12]
- Blackgrass in **winter field beans**, **winter oilseed rape** [12]
- Chickweed in **farm forestry**, **forest**, **forest nurseries**, **lettuce**, **winter field beans**, **winter oilseed rape** [7]
- Field speedwell in **farm forestry**, **forest**, **forest nurseries**, **lettuce**, **winter field beans**, **winter oilseed rape** [7]
- Horsetails in **forest** [8, 14]
- Perennial grasses in **all edible seed crops grown outdoors** *(off-label)*, **all non-edible seed crops grown outdoors** *(off-label)*, **game cover** *(off-label)*, **hops** *(off-label)* [10, 11]; **almonds** *(off-label)*, **bilberries** *(off-label)*, **blueberries** *(off-label)*, **bog myrtle** *(off-label)*, **broccoli** *(off-label)*, **calabrese** *(off-label)*, **cauliflowers** *(off-label)*, **cherries** *(off-label)*, **chestnuts** *(off-label)*, **chicory root** *(off-label)*, **cob nuts** *(off-label)*, **courgettes** *(off-label)*, **cranberries** *(off-label)*, **cress** *(off-label)*, **endives** *(off-label)*, **frise** *(off-label)*, **hazel nuts** *(off-label)*, **herbs (see appendix 6)** *(off-label)*, **lamb's lettuce** *(off-label)*, **leaf brassicas** *(off-label)*, **marrows** *(off-label)*, **mirabelles** *(off-label)*, **protected endives** *(off-label)*, **protected forest nurseries** *(off-label)*, **protected herbs (see appendix 6)** *(off-label)*, **protected lettuce** *(off-label)*, **pumpkins** *(off-label)*, **quinces** *(off-label)*, **radicchio** *(off-label)*, **salad brassicas** *(off-label)*, **scarole** *(off-label)*, **squashes** *(off-label)*, **table grapes** *(off-label)*, **walnuts** *(off-label)*, **wine grapes** *(off-label)* [11]; **amenity vegetation** [1-3, 8, 10, 11, 14-19, 22, 23, 25, 26, 32-34]; **apple orchards**, **pear orchards**, **plums** [1-3, 5, 6, 8, 10, 11, 14-19, 22, 23, 25, 26, 29, 30, 32-34]; **blackberries**, **blackcurrants**, **gooseberries**, **loganberries**, **raspberries** *(England only)*, **redcurrants** [2, 3, 5, 6, 8, 10, 11, 14-19, 22, 23, 25,

26, 29, 30, 32-34]; *clover seed crops*, *rhubarb* [1-3, 5, 6, 10, 11, 15-19, 22, 23, 25, 26, 29, 30, 32-34]; *farm forestry*, *forest nurseries* [1-3, 8, 10, 11, 14-20, 22, 23, 25, 29, 30, 32-34]; *fodder rape seed crops*, *kale seed crops*, *sugar beet seed crops*, *turnip seed crops* [1-3, 5, 6, 10, 11, 15-19, 21-23, 25, 26, 29, 30, 32-34]; *forest* [1-3, 5, 8, 10, 11, 13-23, 25, 29, 30, 32-34]; *hedges* [1-3, 8, 10, 11, 14-19, 22, 23, 25, 29, 30, 32-34]; *lettuce* [2, 3, 5, 6, 10, 11, 15-21, 23, 26, 29, 30, 32-34]; *lucerne* [2, 3, 5, 6, 10, 11, 15-19, 23, 25, 26, 29, 30, 32-34]; *red clover*, *white clover* [21]; *rhubarb* (outdoor) [8, 14]; *roses* [5]; *strawberries* [2, 3, 5, 6, 10, 11, 15-19, 22, 23, 25, 26, 29, 30, 32-34]; *top fruit* (off-label) [10]; *trees and shrubs* [5, 13]; *winter field beans* [2, 3, 5, 6, 10, 11, 15-21, 23, 25, 26, 29, 30, 32-34]; *winter oilseed rape* [1-3, 5, 6, 9-11, 15-23, 25-27, 29, 30, 32-34]; *woody ornamentals* [5, 8, 10, 13, 14]
- Sedges in *forest* [8, 14]
- Volunteer cereals in *sugar beet seed crops* [8, 14]; *winter field beans*, *winter oilseed rape* [8, 12, 14]
- Wild oats in *sugar beet seed crops* [8, 14]; *winter field beans*, *winter oilseed rape* [8, 12, 14]

Extension of Authorisation for Minor Use (EAMUs)
- *all edible seed crops grown outdoors* 20082943 [10], 20082942 [11]
- *all non-edible seed crops grown outdoors* 20082943 [10], 20082942 [11]
- *almonds* 20082420 [11]
- *bilberries* 20082419 [11]
- *blueberries* 20082419 [11]
- *bog myrtle* 20120509 [11]
- *broccoli* 20091902 [11]
- *calabrese* 20091902 [11]
- *cauliflowers* 20091902 [11]
- *cherries* 20082418 [11]
- *chestnuts* 20082420 [11]
- *chicory root* 20091530 [11]
- *cob nuts* 20082420 [11]
- *courgettes* 20082416 [11]
- *cranberries* 20082419 [11]
- *cress* 20082410 [11]
- *endives* 20082410 [11]
- *frise* 20082411 [11]
- *game cover* 20082943 [10], 20082942 [11]
- *hazel nuts* 20082420 [11]
- *herbs (see appendix 6)* 20082412 [11]
- *hops* 20082943 [10], 20082414 [11]
- *lamb's lettuce* 20082410 [11]
- *leaf brassicas* 20082410 [11]
- *marrows* 20082416 [11]
- *mirabelles* 20082418 [11]
- *protected endives* 20082415 [11]
- *protected forest nurseries* 20082942 [11]
- *protected herbs (see appendix 6)* 20082412 [11]
- *protected lettuce* 20082415 [11]
- *pumpkins* 20082416 [11]
- *quinces* 20082413 [11]
- *radicchio* 20082411 [11]
- *salad brassicas* 20082410 [11]
- *scarole* 20082411 [11]
- *squashes* 20082416 [11]
- *table grapes* 20082417 [11]
- *top fruit* 20082943 [10]
- *walnuts* 20082420 [11]
- *wine grapes* 20082417 [11]

Approval information
- Propyzamide included in Annex I under EC Regulation 1107/2009
- Some products may be applied through CDA equipment. See labels for details

SEE SECTION 3 FOR PRODUCTS ALSO REGISTERED

- Accepted by BBPA for use on hops
- Approval expiry 13 Sep 2013 [33]

Efficacy guidance
- Active via root uptake. Weeds controlled from germination to young seedling stage, some species (including many grasses) also when established
- Best results achieved by winter application to fine, firm, moist soil. Rain is required after application if soil dry
- Uptake is slow and and may take up to 12 wk
- Excessive organic debris or ploughed-up turf may reduce efficacy
- For heavy couch infestations a repeat application may be needed in following winter
- Always follow WRAG guidelines for preventing and managing herbicide resistant weeds. See Section 5 for more information

Restrictions
- Maximum number of treatments 1 per crop or yr
- Maximum total dose equivalent to one full dose treatment for all crops
- Do not treat protected crops
- Apply to listed edible crops only between 1 Oct and the date specified as the latest time of application (except lettuce)
- Do not apply in windy weather and avoid drift onto non-target crops
- Do not use on soils with more than 10% organic matter except in forestry

Crop-specific information
- Latest use: labels vary but normally before 31 Dec in year before harvest for rhubarb, lucerne, strawberries and winter field beans; before 31 Jan for other crops
- HI: 6 wk for edible crops
- Apply as soon as possible after 3-true leaf stage of oilseed rape (GS 1,3) and seed brassicas, after 4-leaf stage of sugar beet for seed, within 7 d after sowing but before emergence for field beans, after perennial crops established for at least 1 season, strawberries after 1 yr
- Only apply to strawberries on heavy soils. Do not use on matted row crops
- Only apply to field beans on medium and heavy soils
- Only apply to established lucerne not less than 7 d after last cut
- In lettuce lightly incorporate in top 25 mm pre-drilling or irrigate on dry soil
- See label for lists of ornamental and forest species which may be treated

Following crops guidance
- Following an application between 1 Apr and 31 Jul at any dose the following minimum intervals must be observed before sowing the next crop: lettuce 0 wk; broad beans, chicory, clover, field beans, lucerne, radishes, peas 5 wk; brassicas, celery, leeks, oilseed rape, onions, parsley, parsnips 10 wk
- Following an application between 1 Aug and 31 Mar at any dose the following minimum intervals must be observed before sowing the next crop: lettuce 0 wk; broad beans, chicory, clover, field beans, lucerne, radishes, peas 10 wk; brassicas, celery, leeks, oilseed rape, onions, parsley, parsnips 25 wk or after 15 Jun, whichever occurs sooner
- Cereals or grasses or other crops not listed may be sown 30 wk after treatment up to 840 g ai/ha between 1 Aug and 31 Mar or 40 wk after treatment at higher doses at any time and after mouldboard ploughing to at least 15 cm
- A period of at least 9 mth must elapse between applications of propyzamide to the same land

Environmental safety
- Dangerous for the environment
- Very toxic to aquatic organisms
- Some pesticides pose a greater threat of contamination of water than others and propyzamide is one of these pesticides. Take special care when applying propyzamide near water and do not apply if heavy rain is forecast

Hazard classification and safety precautions
Hazard Harmful, Dangerous for the environment
Transport code 9 [1-21, 23-34]
Packaging group III [1-21, 23-34]

FOR FULL CONDITIONS OF USE ALWAYS READ THE PRODUCT LABEL

UN Number 3077 [4, 5, 8, 10, 13, 14, 18, 28, 31, 32]; 3082 [1-3, 6, 7, 9, 11, 12, 15-17, 19-21, 23-27, 29, 30, 33, 34]

Risk phrases R40, R53a [1-34]; R50 [2-4, 6-11, 14-21, 23-31, 33, 34]; R51 [1, 5, 12, 13, 22, 32]

Operator protection A, H [1-5, 7-34]; D [2-4, 8-11, 14-19, 22, 23, 25, 28, 31, 32, 34]; M [2-4, 7-11, 14-19, 22-34]; U05a [13]; U20c [1-12, 14-34]

Environmental protection E15a [1-9, 11-17, 19-23, 25, 33, 34]; E15b [10, 18, 24, 26-32]; E34 [1, 20, 21, 28, 31-33]; E38 [1-27, 29, 30, 33, 34]; E39 [24, 26, 27, 29, 30]

Consumer protection C02a [1-11, 14-23, 25, 28, 31-34] (6 wk)

Storage and disposal D01, D02 [1, 13]; D05, D11a [1-12, 14-23, 25, 28, 31-34]; D07 [1, 20, 21, 33]; D09a [1-23, 25, 28, 31-34]; D12a [1-23, 25, 33, 34]

Medical advice M03 [13]

383   proquinazid

A quinazolinone fungicide for powdery mildew control in cereals
FRAC mode of action code: U7

See also chlorothalonil + proquinazid
    prochloraz + proquinazid + tebuconazole

Products

1	Justice	DuPont	200 g/l	EC	12835
2	Talius	DuPont	200 g/l	EC	12752

Uses

- Powdery mildew in **durum wheat** *(off-label)*, **spring barley**, **spring oats**, **spring rye**, **spring wheat**, **triticale**, **winter barley**, **winter oats**, **winter rye**, **winter wheat** [1, 2]; **forest nurseries** *(off-label)*, **soft fruit** *(off-label)*, **top fruit** *(off-label)* [2]; **grapevines** *(off-label)* [1]

Extension of Authorisation for Minor Use (EAMUs)

- **durum wheat** *20090418* [1], *20090419* [2]
- **forest nurseries** *20090420* [2]
- **grapevines** *20111763* [1]
- **soft fruit** *20090420* [2]
- **top fruit** *20090420* [2]

Approval information

- Proquinazid included in Annex 1 under EC Regulation 1107/2009
- Accepted by BBPA on malting barley

Efficacy guidance

- Best results obtained from preventive treatment before disease is established in the crop
- Where mildew has already spread to new growth a tank mix with a curative fungicide with an alternative mode of action should be used
- Use as part of an integrated crop management (ICM) strategy incorporating other methods of control or fungicides with different modes of action

Restrictions

- Maximum number of treatments 2 per crop
- Do not apply to any crop suffering from stress from any cause
- Avoid application in either frosty or hot, sunny conditions

Crop-specific information

- Latest use: before beginning of heading (GS 49) for barley, oats, rye, triticale; before full flowering (GS 65) for wheat

Environmental safety

- Dangerous for the environment
- Toxic to aquatic organisms
- Dangerous to fish or other aquatic life. Do not contaminate surface waters or ditches with chemical or used container

SEE SECTION 3 FOR PRODUCTS ALSO REGISTERED

- LERAP Category B

Hazard classification and safety precautions
Hazard Harmful, Dangerous for the environment
Transport code 9
Packaging group III
UN Number 3082
Risk phrases R38, R40, R41, R51, R53a
Operator protection A, C, H; U05a, U11
Environmental protection E13b, E16a, E16b, E34, E38
Storage and disposal D01, D02, D05, D09a, D10b, D11a, D12a
Medical advice M03, M05a

384   prosulfocarb

A thiocarbamate herbicide for grass and broad-leaved weed control in cereals and potatoes
HRAC mode of action code: N

See also clodinafop-propargyl + prosulfocarb

Products

1 Auros	Syngenta	800 g/l	EC	15049
2 Defy	Syngenta	800 g/l	EC	12606
3 Dian	AgChem Access	800 g/l	EC	13567

Uses
- Annual dicotyledons in *durum wheat* (off-label), *leeks* (off-label), *spring field beans* (off-label), *spring onions* (off-label), *spring wheat* (off-label), *winter field beans* (off-label), *winter linseed* (off-label) [2]; *potatoes, winter barley, winter wheat* [1-3]
- Annual grasses in *bulb onions* (off-label), *bullb onion sets* (off-label), *carrots* (off-label), *celeriac* (off-label), *celery (outdoor)* (off-label), *durum wheat* (off-label), *garlic* (off-label), *herbs (see appendix 6)* (off-label), *horseradish* (off-label), *leeks* (off-label), *parsley root* (off-label), *parsnips* (off-label), *poppies for morphine production* (off-label), *rye* (off-label), *salsify* (off-label), *shallots* (off-label), *spring barley* (off-label), *spring field beans* (off-label), *spring onions* (off-label), *spring wheat* (off-label), *triticale* (off-label), *winter field beans* (off-label), *winter linseed* (off-label) [2]
- Annual meadow grass in *potatoes, winter barley, winter wheat* [1-3]; *spring barley* (off-label), *winter linseed* (off-label) [2]
- Black nightshade in *herbs (see appendix 6)* (off-label) [2]
- Chickweed in *celeriac* (off-label) [2]; *potatoes, winter barley, winter wheat* [1-3]
- Cleavers in *carrots* (off-label), *herbs (see appendix 6)* (off-label), *horseradish* (off-label), *parsley root* (off-label), *parsnips* (off-label), *salsify* (off-label), *spring field beans* (off-label), *spring wheat* (off-label), *winter field beans* (off-label) [2]; *potatoes, winter barley, winter wheat* [1-3]
- Fat hen in *carrots* (off-label), *celeriac* (off-label), *horseradish* (off-label), *parsley root* (off-label), *parsnips* (off-label), *salsify* (off-label) [2]
- Field speedwell in *potatoes, winter barley, winter wheat* [1-3]
- Fumitory in *bulb onions* (off-label), *bullb onion sets* (off-label), *garlic* (off-label), *herbs (see appendix 6)* (off-label), *leeks* (off-label), *shallots* (off-label), *spring onions* (off-label), *spring wheat* (off-label) [2]
- Ivy-leaved speedwell in *potatoes, winter barley, winter wheat* [1-3]
- Knotgrass in *carrots* (off-label), *horseradish* (off-label), *parsley root* (off-label), *parsnips* (off-label), *salsify* (off-label) [2]
- Loose silky bent in *potatoes, winter barley, winter wheat* [1-3]
- Mayweeds in *carrots* (off-label), *celeriac* (off-label), *horseradish* (off-label), *parsley root* (off-label), *parsnips* (off-label), *salsify* (off-label) [2]
- Polygonums in *bulb onions* (off-label), *bullb onion sets* (off-label), *celeriac* (off-label), *garlic* (off-label), *leeks* (off-label), *shallots* (off-label) [2]
- Rough meadow grass in *potatoes, winter barley, winter wheat* [1-3]

FOR FULL CONDITIONS OF USE ALWAYS READ THE PRODUCT LABEL

- Small nettle in *celeriac* *(off-label)* [2]
- Thistles in *celeriac* *(off-label)* [2]
- Volunteer oilseed rape in *spring onions* *(off-label)* [2]

Extension of Authorisation for Minor Use (EAMUs)
- *bulb onions 20121115* [2]
- *bullb onion sets 20121115* [2]
- *carrots 20121103* [2]
- *celeriac 20121100* [2]
- *celery (outdoor) 20121101* [2]
- *durum wheat 20121109* [2]
- *garlic 20121115* [2]
- *herbs (see appendix 6) 20121102* [2]
- *horseradish 20121103* [2]
- *leeks 20121114* [2]
- *parsley root 20121103* [2]
- *parsnips 20121103* [2]
- *poppies for morphine production 20121111* [2]
- *rye 20121107* [2]
- *salsify 20121103* [2]
- *shallots 20121115* [2]
- *spring barley 20121106* [2]
- *spring field beans 20121112* [2]
- *spring onions 20121110* [2]
- *spring wheat 20121113* [2]
- *triticale 20121107* [2]
- *winter field beans 20121108* [2]
- *winter linseed 20092012* [2]

Approval information
- Prosulfocarb included in Annex I under EC Regulation 1107/2009
- Accepted by BBPA for use on malting barley
- Approval expiry 30 Jun 2013 [1]

Efficacy guidance
- Best results obtained in cereals from treatment of crops in a firm moist seedbed, free from clods
- Pre-emergence use will reduce blackgrass populations but the product should only be used against this weed as part of a management strategy involving sequences with products of alternative modes of action
- Always follow WRAG guidelines for preventing and managing herbicide resistant weeds. See Section 5 for more information

Restrictions
- Maximum number of treatments 1 per crop for winter barley, winter wheat and potatoes
- Do not apply to crops under stress from any cause. Transient yellowing can occur from which recovery is complete
- Winter cereals must be covered by 3 cm of settled soil

Crop-specific information
- Latest use: at emergence (soil rising over emerging potato shoots) for potatoes, up to and including early tillering (GS 21) for winter barley, winter wheat
- When applied pre-emergence to cereals crop emergence may occasionally be slowed down but yield is not affected
- For potatoes, complete ridge formation before application and do not disturb treated soil afterwards
- Peas are severely damaged or killed.

Following crops guidance
- Do not sow field or broad beans within 12 mth of treatment
- In the event of failure of a treated cereal crop, winter wheat, winter barley or transplanted brassicas may be re-sown immediately. Seed brassicas can be sown 14 weeks after application

SEE SECTION 3 FOR PRODUCTS ALSO REGISTERED

provided that the land is ploughed first. In the following spring sunflowers, maize, flax, spring cereals, peas, oilseed rape or soya beans may be sown without ploughing, and carrots, lettuce, onions, sugar beet or potatoes may be sown or planted after ploughing

Environmental safety
- Dangerous for the environment
- Very toxic to aquatic organisms
- LERAP Category B

Hazard classification and safety precautions
Hazard Irritant, Dangerous for the environment
Transport code 9
Packaging group III
UN Number 3082
Risk phrases R38, R43, R50, R53a
Operator protection A, C, H; U02a, U05a, U08, U14, U15, U20b
Environmental protection E15b, E16a, E16b, E38
Storage and disposal D01, D02, D05, D09a, D10c, D12a
Medical advice M05a

385 prosulfuron

A contact and residual sulfonyl urea herbicide for use in maize crops
HRAC mode of action code: B

See also bromoxynil + prosulfuron
dicamba + prosulfuron

Products

1	Clayton Kibo	Clayton	75% w/w	WG	15822
2	Peak	Syngenta	75% w/w	WG	15521

Uses
- Annual dicotyledons in *forage maize*, *grain maize* [1, 2]; *game cover* *(off-label)* [2]
- Black bindweed in *forage maize*, *grain maize* [1, 2]
- Chickweed in *forage maize*, *grain maize* [1, 2]
- Fumitory in *forage maize*, *grain maize* [1, 2]
- Groundsel in *forage maize*, *grain maize* [1, 2]
- Knotgrass in *forage maize*, *grain maize* [1, 2]
- Mayweeds in *forage maize*, *grain maize* [1, 2]
- Redshank in *forage maize*, *grain maize* [1, 2]
- Scarlet pimpernel in *forage maize*, *grain maize* [1, 2]
- Shepherd's purse in *forage maize*, *grain maize* [1, 2]
- Sowthistle in *forage maize*, *grain maize* [1, 2]

Extension of Authorisation for Minor Use (EAMUs)
- *game cover* 20120906 [2]

Approval information
- Prosulfuron included in Annex I under EC Regulation 1107/2009

Efficacy guidance
- For optimum efficacy, apply when weeds are at the 2 - 4 leaf stage

Restrictions
- Do not apply to forage maize or grain maize grown for seed production
- Do not apply with orgaono-phosphate insecticides

Following crops guidance
- In the event of crop failure, wait 4 weeks after treatment and then re-sow

FOR FULL CONDITIONS OF USE ALWAYS READ THE PRODUCT LABEL

- After normal harvest wheat, barley and winter beans may be sown as a following crop in the autumn once the soil has been ploughed to 15 cms. In spring, wheat, barley, peas or beans may be sown but do not sow any other crop at this time.

Environmental safety
- LERAP Category B

Hazard classification and safety precautions
Hazard Harmful, Dangerous for the environment
Transport code 9
Packaging group III
UN Number 3077
Risk phrases R22a, R50, R53a
Operator protection A, H; U05a, U14, U20c
Environmental protection E15b, E16a, E38
Storage and disposal D01, D02, D05, D09b, D10c, D12a

386   prothioconazole

A systemic, protectant and curative triazole fungicide
FRAC mode of action code: 3

See also bixafen + fluoxystrobin + prothioconazole
bixafen + prothioconazole
bixafen + prothioconazole + spiroxamine
bixafen + prothioconazole + tebuconazole
clothianidin + prothioconazole
clothianidin + prothioconazole + tebuconazole + triazoxide
fluopyram + prothioconazole
fluopyram + prothioconazole + tebuconazole
fluoxastrobin + prothioconazole
fluoxastrobin + prothioconazole + tebuconazole
fluoxastrobin + prothioconazole + trifloxystrobin

Products

1	Proline 275	Bayer CropScience	275 g/l	EC	14790
2	Prothio	ChemSource	250 g/l	EC	14945
3	Redigo	Bayer CropScience	100 g/l	FS	12085
4	Rudis	Bayer CropScience	480 g/l	SC	14122

Uses

- Alternaria in **broccoli, brussels sprouts, cabbages, calabrese, cauliflowers** [4]
- Alternaria blight in **carrots, parsnips, swedes, turnips** [4]
- Brown rust in **durum wheat** [2]; **spring barley, winter barley, winter rye, winter wheat** [1, 2]; **spring wheat** [1]
- Bunt in **durum wheat** *(seed treatment)*, **spring rye** *(seed treatment)*, **spring wheat** *(seed treatment)*, **triticale** *(seed treatment)*, **winter rye** *(seed treatment)*, **winter wheat** *(seed treatment)* [3]
- Covered smut in **spring barley** *(seed treatment)*, **winter barley** *(seed treatment)* [3]
- Crown rust in **spring oats, winter oats** [1]
- Disease control in **mustard** *(off-label)* [1]
- Eyespot in **durum wheat** [2]; **spring barley, winter barley, winter rye, winter wheat** [1, 2]; **spring oats, spring wheat, winter oats** [1]
- Fusarium foot rot and seedling blight in **durum wheat** *(seed treatment)*, **spring barley** *(seed treatment)*, **spring oats** *(seed treatment)*, **spring rye** *(seed treatment)*, **spring wheat** *(seed treatment)*, **triticale** *(seed treatment)*, **winter barley** *(seed treatment)*, **winter oats** *(seed treatment)*, **winter rye** *(seed treatment)*, **winter wheat** *(seed treatment)* [3]
- Glume blotch in **durum wheat** [2]; **spring wheat** [1]; **winter wheat** [1, 2]
- Late ear diseases in **durum wheat** [2]; **spring barley, winter barley, winter wheat** [1, 2]; **spring wheat** [1]

SEE SECTION 3 FOR PRODUCTS ALSO REGISTERED

- Leaf blotch in **leeks** *(useful reduction)* [4]
- Leaf stripe in **spring barley** *(seed treatment)*, **winter barley** *(seed treatment)* [3]
- Light leaf spot in **broccoli**, **brussels sprouts**, **cabbages**, **calabrese**, **cauliflowers** [4]
- Loose smut in **durum wheat** *(seed treatment)*, **spring barley** *(seed treatment)*, **spring oats** *(seed treatment)*, **spring wheat** *(seed treatment)*, **winter barley** *(seed treatment)*, **winter oats** *(seed treatment)*, **winter wheat** *(seed treatment)* [3]
- Net blotch in **spring barley**, **winter barley** [1, 2]
- Phoma in **broccoli**, **brussels sprouts**, **cabbages**, **calabrese**, **cauliflowers** [4]
- Powdery mildew in **broccoli**, **brussels sprouts**, **cabbages**, **calabrese**, **carrots**, **cauliflowers**, **parsnips**, **swedes**, **turnips** [4]; **durum wheat** [2]; **spring barley**, **winter barley**, **winter rye**, **winter wheat** [1, 2]; **spring oats**, **spring wheat**, **winter oats** [1]
- Purple blotch in **leeks** [4]
- Rhynchosporium in **spring barley**, **winter barley**, **winter rye** [1, 2]
- Ring spot in **broccoli**, **brussels sprouts**, **cabbages**, **calabrese**, **cauliflowers** [4]
- Rust in **leeks** [4]
- Sclerotinia rot in **carrots**, **parsnips** [4]
- Sclerotinia stem rot in **winter oilseed rape** [1, 2]
- Septoria leaf blotch in **durum wheat** [2]; **spring wheat** [1]; **winter wheat** [1, 2]
- Stemphylium in **leeks** *(useful reduction)* [4]
- Tan spot in **durum wheat** [2]; **spring wheat** [1]; **winter wheat** [1, 2]
- Yellow rust in **durum wheat** [2]; **spring wheat** [1]; **winter barley**, **winter wheat** [1, 2]

Extension of Authorisation for Minor Use (EAMUs)
- **mustard** *20111003* [1]

Approval information
- Accepted by BBPA for use on malting barley
- Prothioconazole included in Annex I under EC Regulation 1107/2009

Efficacy guidance
- Seed treatments must be applied by manufacturer's recommended treatment application equipment [3]
- Treated cereal seed should preferably be drilled in the same season [3]
- Follow-up treatments will be needed later in the season to give protection against air-borne and splash-borne diseases [3]
- Best results on cereal foliar diseases obtained from treatment at early stages of disease development. Further treatment may be needed if disease attack is prolonged [1]
- Foliar applications to established infections of any disease are likely to be less effective [1]
- Best control of cereal ear diseases obtained by treatment during ear emergence [1]
- Treat oilseed rape at early to full flower [1]
- Prothioconazole is a DMI fungicide. Resistance to some DMI fungicides has been identified in Septoria leaf blotch which may seriously affect performance of some products. For further advice contact a specialist advisor and visit the Fungicide Resistance Action Group (FRAG)-UK website

Restrictions
- Maximum number of seed treatments one per batch [3]
- Maximum total dose equivalent to one full dose treatment on oilseed rape, two full dose treatments on barley, oats; three full dose treatments on wheat, rye [1]
- Seed treatment must be fully re-dispersed and homogeneous before use [3]
- Do not use on seed with more than 16% moisture content, or on sprouted, cracked or skinned seed [3]
- All seed batches should be tested to ensure they are suitable for treatment [3]
- Treated winter barley seed must be used within the season of treatment; treated winter wheat seed should preferably be drilled in the same season [3]
- Do not make repeated treatments of the product alone to the same crop against pathogens such as powdery mildew. Use tank mixtures or alternate with fungicides having a different mode of action [3]

Crop-specific information
- Latest use: pre-drilling for seed treatments [3]; before grain milky ripe for foliar sprays on winter rye, winter wheat; beginning of flowering for barley, oats [1]
- HI 56 d for winter oilseed rape [1]

Environmental safety
- Dangerous for the environment [1]
- Toxic to aquatic organisms [1]
- Harmful to aquatic organisms [3]
- LERAP Category B [1, 2, 4]

Hazard classification and safety precautions
Hazard Irritant [1-3]; Dangerous for the environment [1, 2, 4]
Transport code 9 [1, 2, 4]
Packaging group III [1, 2, 4]
UN Number 3082 [1, 2, 4]; N/C [3]
Risk phrases R36 [1, 2]; R43, R52 [3]; R51 [1, 2, 4]; R53a [1-4]
Operator protection A, H [1-4]; C [1, 2]; U05a, U20b [1, 2, 4]; U07, U14, U20a [3]; U09b [1, 2]
Environmental protection E15a [1-3]; E15b [4]; E16a, E38 [1, 2, 4]; E34 [1-4]; E36a [3]
Storage and disposal D01, D02, D09a, D12a [1-4]; D05 [1-3]; D10b [1, 2, 4]; D14 [3]
Treated seed S01, S02, S03, S04a, S04b, S05, S06b, S07, S08 [3]
Medical advice M03 [1, 2, 4]

387   prothioconazole + spiroxamine

A broad spectrum fungicide mixture for cereals
FRAC mode of action code: 3 + 5

See also spiroxamine

Products

1	Helix	Bayer CropScience	160:300 g/l	EC	12264
2	Pan Hale	Pan Agriculture	160:300 g/l	EC	15924
3	Spiral	AgChem Access	160:300 g/l	EC	15437

Uses
- Brown rust in *spring barley, winter barley, winter rye, winter wheat*
- Crown rust in *spring oats, winter oats*
- Eyespot in *spring barley, spring oats, winter barley, winter oats, winter rye, winter wheat*
- Glume blotch in *winter wheat*
- Late ear diseases in *winter rye, winter wheat*
- Net blotch in *spring barley, winter barley*
- Powdery mildew in *spring barley, spring oats, winter barley, winter oats, winter rye, winter wheat*
- Rhynchosporium in *spring barley, winter barley, winter rye*
- Septoria leaf blotch in *winter wheat*
- Tan spot in *winter wheat*
- Yellow rust in *spring barley, winter barley, winter wheat*

Approval information
- Prothioconazole and spiroxamine included in Annex I under EC Regulation 1107/2009
- Accepted by BBPA for use on malting barley

Efficacy guidance
- Best results obtained from treatment at early stages of disease development. Further treatment may be needed if disease attack is prolonged
- Applications to established infections of any disease are likely to be less effective
- Best control of cereal ear diseases obtained by treatment during ear emergence
- Prothioconazole is a DMI fungicide. Resistance to some DMI fungicides has been identified in Septoria leaf blotch which may seriously affect performance of some products. For further

SEE SECTION 3 FOR PRODUCTS ALSO REGISTERED

SECTION 2

advice contact a specialist advisor and visit the Fungicide Resistance Action Group (FRAG)-UK website

Restrictions
- Maximum total dose equivalent to two full dose treatments on barley and oats and three full dose treatments on wheat and rye

Crop-specific information
- Latest use: before grain watery ripe for rye, oats, winter wheat; up to beginning of anthesis for barley

Environmental safety
- Dangerous for the environment
- Very toxic to aquatic organisms
- LERAP Category B

Hazard classification and safety precautions
Hazard Harmful, Dangerous for the environment
Transport code 9
Packaging group III
UN Number 3082
Risk phrases R20, R22a, R36, R38, R50, R53a
Operator protection A, C, H; U05a, U09b, U11, U19a, U20b
Environmental protection E15a, E16a, E34, E38
Storage and disposal D01, D02, D05, D09a, D10b, D12a
Medical advice M03

388   prothioconazole + spiroxamine + tebuconazole

A broad spectrum fungicide mixture for cereals
FRAC mode of action code: 3 + 5 + 3

See also spiroxamine
tebuconazole

Products
Cello	Bayer CropScience	100:250:100 g/l	EC	13178

Uses
- Brown rust in **spring barley, winter barley, winter rye, winter wheat**
- Crown rust in **spring oats, winter oats**
- Eyespot in **spring barley** *(reduction)*, **spring oats, winter barley** *(reduction)*, **winter oats, winter rye** *(reduction)*, **winter wheat** *(reduction)*
- Glume blotch in **winter wheat**
- Late ear diseases in **spring barley, winter barley, winter wheat**
- Net blotch in **spring barley, winter barley**
- Powdery mildew in **spring barley, spring oats, winter barley, winter oats, winter rye, winter wheat**
- Rhynchosporium in **spring barley, winter barley, winter rye**
- Septoria leaf blotch in **winter wheat**
- Yellow rust in **spring barley, winter barley, winter wheat**

Approval information
- Prothioconazole, spiroxamine and tebuconazole included in Annex I under EC Regulation 1107/2009

Efficacy guidance
- Best results obtained from treatment at early stages of disease development. Further treatment may be needed if disease attack is prolonged
- Applications to established infections of any disease are likely to be less effective
- Best control of cereal ear diseases obtained by treatment during ear emergence

- Prothioconazole and tebuconazole are DMI fungicides. Resistance to some DMI fungicides has been identified in Septoria leaf blotch which may seriously affect performance of some products. For further advice contact a specialist advisor and visit the Fungicide Resistance Action Group (FRAG)-UK website

Restrictions
- Maximum total dose equivalent to two full dose treatments

Crop-specific information
- Latest use: before grain milky ripe stage for rye, wheat; up to beginning of anthesis for barley and oats

Environmental safety
- Dangerous for the environment
- Very toxic to aquatic organisms
- LERAP Category B

Hazard classification and safety precautions
> **Hazard** Harmful, Dangerous for the environment
> **Transport code** 9
> **Packaging group** III
> **UN Number** 3082
> **Risk phrases** R20, R36, R37, R38, R43, R50, R53a, R63
> **Operator protection** A, C, H; U05a, U09b, U11, U14, U19a, U20b
> **Environmental protection** E15a, E16a, E34, E38
> **Storage and disposal** D01, D02, D05, D09a, D10b, D12a
> **Medical advice** M03

389   prothioconazole + tebuconazole

A triazole fungicide mixture for cereals
FRAC mode of action code: 3 + 3

See also tebuconazole

Products

1 Corinth	Bayer CropScience	80;160 g/l	EC	14199
2 Kestrel	Bayer CropScience	160+80 g/L	EC	13809
3 Prosaro	Bayer CropScience	125:125 g/l	EC	12263
4 Proteb	AgChem Access	125:125 g/l	EC	13879
5 Standon Mastana	Standon	125:125 g/l	EC	13795

Uses
- Brown rust in *spring barley*, *winter barley*, *winter rye*, *winter wheat* [2-5]; *spring wheat* [2, 3]
- Crown rust in *spring oats*, *winter oats* [2-4]
- Eyespot in *spring barley*, *winter barley*, *winter rye*, *winter wheat* [2, 5]; *spring barley (reduction)*, *winter barley (reduction)*, *winter rye (reduction)*, *winter wheat (reduction)* [3, 4]; *spring oats*, *winter oats* [2-4]; *spring wheat* [2, 3]
- Foliar disease control in *grass seed crops (off-label)* [3]
- Fusarium ear blight in *spring barley*, *spring wheat*, *winter barley*, *winter wheat* [2]; *winter wheat (moderate control)* [5]
- Glume blotch in *spring wheat* [2, 3]; *winter wheat* [2-5]
- Late ear diseases in *spring barley*, *winter barley*, *winter wheat* [3, 4]; *spring wheat* [3]
- Light leaf spot in *oilseed rape* [2]; *oilseed rape (moderate control only)* [1]; *spring oilseed rape*, *winter oilseed rape* [3, 4]; *winter oilseed rape (moderate control)* [5]
- Net blotch in *spring barley*, *winter barley* [2-5]; *winter rye* [5]
- Phoma in *spring oilseed rape* [3, 4]; *winter oilseed rape* [3-5]
- Phoma leaf spot in *oilseed rape* [1, 2]
- Powdery mildew in *spring barley*, *spring oats*, *winter barley*, *winter oats*, *winter rye*, *winter wheat* [2-5]; *spring wheat* [2, 3]

SEE SECTION 3 FOR PRODUCTS ALSO REGISTERED

- Rhynchosporium in *spring barley, winter barley, winter rye* [2-5]
- Sclerotinia stem rot in *oilseed rape* [1, 2]; *spring oilseed rape* [3, 4]; *winter oilseed rape* [3-5]
- Septoria leaf blotch in *spring wheat* [2, 3]; *winter wheat* [2-5]
- Sooty moulds in *spring barley, spring wheat, winter barley, winter wheat* [2]; *winter wheat (reduction)* [5]
- Stem canker in *oilseed rape* [1]
- Tan spot in *spring wheat* [2, 3]; *winter wheat* [2-4]
- Yellow rust in *spring barley, winter barley, winter wheat* [2-5]; *spring wheat* [2, 3]; *winter rye* [5]

Extension of Authorisation for Minor Use (EAMUs)
- *grass seed crops* 20081806 [3]

Approval information
- Prothioconazole and tebuconazole included in Annex I under EC Regulation 1107/2009
- Accepted by BBPA for use on malting barley

Efficacy guidance
- Best results on cereal foliar diseases obtained from treatment at early stages of disease development. Further treatment may be needed if disease attack is prolonged
- On oilseed rape apply a protective treatment in autumn/winter for Phoma followed by a further spray in early spring from the onset of stem elongation, if necessary. For control of Sclerotinia apply at early to full flower
- Applications to established infections of any disease are likely to be less effective
- Best control of cereal ear diseases obtained by treatment during ear emergence
- Prothioconazole and tebuconazole are DMI fungicides. Resistance to some DMI fungicides has been identified in Septoria leaf blotch which may seriously affect performance of some products. For further advice contact a specialist advisor and visit the Fungicide Resistance Action Group (FRAG)-UK website

Restrictions
- Maximum total dose equivalent to two full dose treatments on barley, oats, oilseed rape; three full dose treatments on wheat and rye

Crop-specific information
- Latest use: before grain milky ripe for rye, wheat; beginning of flowering for barley, oats
- HI 56 d for oilseed rape

Environmental safety
- Dangerous for the environment
- Toxic to aquatic organisms
- Risk to non-target insects or other arthropods. Avoid spraying within 6 m of the field boundary to reduce the effects on non-target insects or other arthropods
- LERAP Category B

Hazard classification and safety precautions
Hazard Harmful [1, 2, 5]; Irritant [3, 4]; Dangerous for the environment [1-5]
Transport code 9
Packaging group III
UN Number 3082
Risk phrases R36 [1, 2]; R38, R43, R51, R53a, R63 [1-5]
Operator protection A, H [1-5]; C [1, 2]; U05a [1-5]; U09b, U20b [2-5]; U11, U20c [1]; U19a [1, 3-5]
Environmental protection E15a, E38 [1-4]; E15b [5]; E16a, E34 [1-5]; E22b [1]; E22c [2-5]
Storage and disposal D01, D02, D09a, D12a [1-5]; D05, D10b [1-4]; D10a [5]
Medical advice M03

390   prothioconazole + trifloxystrobin

A triazole and strobilurin fungicide mixture for cereals
FRAC mode of action code: 3 + 11

See also trifloxystrobin

Products

1 Mobius	Bayer CropScience	175;150 g/l	SC	13395
2 Zephyr	Bayer CropScience	175:88 g/l	SC	13174

Uses

* Brown rust in **durum wheat**, **rye**, **spring barley**, **triticale**, **winter barley**, **winter wheat** [1, 2]; **spring wheat** [2]
* Ear diseases in **durum wheat**, **rye**, **triticale**, **winter wheat** [1]
* Eyespot in **durum wheat**, **rye**, **triticale**, **winter wheat** [1, 2]; **spring barley** *(reduction)*, **spring wheat**, **winter barley** *(reduction)* [2]; **spring barley** *(reduction in severity)*, **winter barley** *(reduction in severity)* [1]
* Glume blotch in **durum wheat**, **rye**, **triticale**, **winter wheat** [1, 2]; **spring wheat** [2]
* Net blotch in **spring barley**, **winter barley** [1, 2]
* Powdery mildew in **durum wheat**, **rye**, **spring barley**, **triticale**, **winter barley**, **winter wheat** [1, 2]; **spring wheat** [2]
* Rhynchosporium in **spring barley**, **winter barley** [1, 2]
* Septoria leaf blotch in **durum wheat**, **rye**, **triticale**, **winter wheat** [1, 2]; **spring wheat** [2]
* Yellow rust in **durum wheat**, **rye**, **spring barley**, **triticale**, **winter barley**, **winter wheat** [1, 2]; **spring wheat** [2]

Approval information

* Prothioconazole and trifloxystrobin included in Annex I under EC Regulation 1107/2009
* Accepted by BBPA for use on malting barley

Efficacy guidance

* Best results obtained from treatment at early stages of disease development. Further treatment may be needed if disease attack is prolonged
* Applications to established infections of any disease are likely to be less effective
* Best control of cereal ear diseases obtained by treatment during ear emergence
* Prothioconazole is a DMI fungicide. Resistance to some DMI fungicides has been identified in Septoria leaf blotch which may seriously affect performance of some products. For further advice contact a specialist advisor and visit the Fungicide Resistance Action Group (FRAG)-UK website
* Trifloxystrobin is a member of the QoI cross resistance group. Product should be used preventatively and not relied on for its curative potential
* Use product as part of an Integrated Crop Management strategy incorporating other methods of control, including where appropriate other fungicides with a different mode of action. Do not apply more than two foliar applications of QoI containing products to any cereal crop
* There is a significant risk of widespread resistance occurring in *Septoria tritici* populations in UK. Failure to follow resistance management action may result in reduced levels of disease control
* Strains of wheat and barley powdery mildew resistant to QoIs are common in the UK. Control of wheat mildew can only be relied on from the triazole component
* Where specific control of wheat mildew is required this should be achieved through a programme of measures including products recommended for the control of mildew that contain a fungicide from a different cross-resistance group and applied at a dose that will give robust control

Restrictions

* Maximum total dose equivalent to two full dose treatments

Crop-specific information

* Latest use: before grain milky ripe for wheat; beginning of flowering for barley

Environmental safety

* Dangerous for the environment

SEE SECTION 3 FOR PRODUCTS ALSO REGISTERED

SECTION 2

- Very toxic to aquatic organisms
- LERAP Category B

Hazard classification and safety precautions
Hazard Irritant, Dangerous for the environment
Transport code 9
Packaging group III
UN Number 3082
Risk phrases R37, R50, R53a [1, 2]; R70 [1]
Operator protection A, C, H; U05a, U09b, U19a [1, 2]; U20b [2]; U20c [1]
Environmental protection E15a, E16a, E38 [1, 2]; E34 [2]
Storage and disposal D01, D02, D09a, D10c, D12a
Medical advice M03

391   pymetrozine

A novel azomethine insecticide
IRAC mode of action code: 9B

Products

1	Chess WG	Syngenta Bioline	50% w/w	WG	13310
2	MyZus	ChemSource	50% w/w	WG	15903
3	Plenum WG	Syngenta	50% w/w	WG	10652
4	Shogi	ChemSource	50% w/w	WG	15736

Uses

- Aphids in **bilberries** *(off-label)*, **blackberries** *(off-label)*, **blackcurrants** *(off-label)*, **blueberries** *(off-label)*, **celeriac** *(off-label)*, **celery (outdoor)** *(off-label)*, **chard** *(off-label)*, **chinese cabbage** *(off-label)*, **choi sum** *(off-label)*, **collards** *(off-label)*, **cranberries** *(off-label)*, **forest nurseries** *(off-label)*, **frise** *(off-label)*, **gooseberries** *(off-label)*, **kale** *(off-label)*, **lamb's lettuce** *(off-label)*, **lettuce** *(off-label)*, **loganberries** *(off-label)*, **pak choi** *(off-label)*, **radicchio** *(off-label)*, **raspberries** *(off-label)*, **redcurrants** *(off-label)*, **rubus hybrids** *(off-label)*, **salad brassicas** *(off-label - for baby leaf production)*, **soft fruit** *(off-label)*, **spinach** *(off-label)*, **spinach beet** *(off-label)*, **strawberries** *(off-label)*, **sweetcorn** *(off-label)*, **vaccinium spp.** *(off-label)*, **whitecurrants** *(off-label)* [3]; **brassica seed beds** *(off-label)*, **protected aubergines** *(off-label)*, **protected blackberries** *(off-label)*, **protected blackcurrants** *(off-label)*, **protected celery** *(off-label)*, **protected chilli peppers** *(off-label)*, **protected chives** *(off-label)*, **protected courgettes** *(off-label)*, **protected cucumbers** *(off-label)*, **protected endives** *(off-label)*, **protected forest nurseries** *(off-label)*, **protected gherkins** *(off-label)*, **protected gooseberries** *(off-label)*, **protected herbs (see appendix 6)** *(off-label)*, **protected hops** *(off-label)*, **protected lettuce** *(off-label)*, **protected loganberries** *(off-label)*, **protected marrows** *(off-label)*, **protected melons** *(off-label)*, **protected okra** *(off-label)*, **protected ornamentals** *(off-label)*, **protected parsley** *(off-label)*, **protected peppers** *(off-label)*, **protected pumpkins** *(off-label)*, **protected raspberries** *(off-label)*, **protected redcurrants** *(off-label)*, **protected ribes hybrids** *(off-label)*, **protected rubus hybrids** *(off-label)*, **protected salad brassicas** *(off-label - for baby leaf production)*, **protected soft fruit** *(off-label)*, **protected squashes** *(off-label)*, **protected strawberries** *(off-label)*, **protected tomatoes** *(off-label)*, **protected top fruit** *(off-label)* [1]; **ornamental plant production, protected cucumbers, protected ornamentals** [1, 4]; **seed potatoes, ware potatoes** [2, 3]
- Cabbage aphid in **chinese cabbage** *(off-label)*, **choi sum** *(off-label)*, **collards** *(off-label)*, **kale** *(off-label)*, **pak choi** *(off-label)* [3]; **protected chinese cabbage** *(off-label)*, **protected choi sum** *(off-label)*, **protected pak choi** *(off-label)* [1]
- Damson-hop aphid in **hops** *(off-label)* [3]
- Glasshouse whitefly in **protected tomatoes** *(off-label)* [1]
- Mealy aphid in **broccoli** *(useful levels of control)*, **brussels sprouts** *(useful levels of control)*, **cabbages** *(useful levels of control)*, **calabrese** *(useful levels of control)*, **cauliflowers** *(useful levels of control)* [2, 3]; **chinese cabbage** *(off-label)*, **choi sum** *(off-label)*, **collards** *(off-label)*, **kale** *(off-label)*, **pak choi** *(off-label)* [3]
- Peach-potato aphid in **broccoli, brussels sprouts, cabbages, calabrese, cauliflowers** [2, 3]; **chinese cabbage** *(off-label)*, **choi sum** *(off-label)*, **collards** *(off-label)*, **kale** *(off-label)*, **pak choi**

(off-label) [3]; **protected aubergines** *(off-label)*, **protected endives** *(off-label)*, **protected peppers** *(off-label)* [1]

- Pollen beetle in **oilseed rape** [2, 3]
- Tobacco whitefly in **protected tomatoes** *(off-label)* [1]
- Whitefly in **protected aubergines** *(off-label)*, **protected chilli peppers** *(off-label)*, **protected courgettes** *(off-label)*, **protected cucumbers** *(off-label)*, **protected forest nurseries** *(off-label)*, **protected gherkins** *(off-label)*, **protected hops** *(off-label)*, **protected melons** *(off-label)*, **protected okra** *(off-label)*, **protected ornamentals** *(off-label)*, **protected peppers** *(off-label)*, **protected pumpkins** *(off-label)*, **protected soft fruit** *(off-label)*, **protected squashes** *(off-label)*, **protected tomatoes** *(off-label)*, **protected top fruit** *(off-label)* [1]

Extension of Authorisation for Minor Use (EAMUs)

- **bilberries** *20061702* [3]
- **blackberries** *20061633* [3]
- **blackcurrants** *20060946* [3]
- **blueberries** *20061702* [3]
- **brassica seed beds** *20070788* [1]
- **celeriac** *20051062* [3]
- **celery (outdoor)** *20051062* [3]
- **chard** *20111664* [3]
- **chinese cabbage** *20082246* [3]
- **choi sum** *20082246* [3]
- **collards** *20082081* [3]
- **cranberries** *20061702* [3]
- **forest nurseries** *20082920* [3]
- **frise** *20070060* [3]
- **gooseberries** *20061702* [3]
- **hops** *20031423* [3]
- **kale** *20082081* [3]
- **lamb's lettuce** *20070060* [3]
- **lettuce** *20070060* [3]
- **loganberries** *20061702* [3]
- **pak choi** *20082246* [3]
- **protected aubergines** *20070501* [1], *20092024* [1]
- **protected blackberries** *20070498* [1]
- **protected blackcurrants** *20070504* [1]
- **protected celery** *20070503* [1]
- **protected chilli peppers** *20092024* [1]
- **protected chinese cabbage** *20090649* [1]
- **protected chives** *20070502* [1]
- **protected choi sum** *20090649* [1]
- **protected courgettes** *20081012* [1], *20092024* [1]
- **protected cucumbers** *20092024* [1]
- **protected endives** *20092025* [1]
- **protected forest nurseries** *20082834* [1]
- **protected gherkins** *20081012* [1], *20092024* [1]
- **protected gooseberries** *20070504* [1]
- **protected herbs (see appendix 6)** *20070502* [1]
- **protected hops** *20082834* [1]
- **protected lettuce** *20070502* [1]
- **protected loganberries** *20070504* [1]
- **protected marrows** *20081012* [1]
- **protected melons** *20081012* [1], *20092024* [1]
- **protected okra** *20092024* [1]
- **protected ornamentals** *20082834* [1]
- **protected pak choi** *20090649* [1]
- **protected parsley** *20070502* [1]
- **protected peppers** *20070501* [1], *20092024* [1]
- **protected pumpkins** *20081012* [1], *20092024* [1]

SECTION 2

SEE SECTION 3 FOR PRODUCTS ALSO REGISTERED

- *protected raspberries* *20070498* [1]
- *protected redcurrants* *20070504* [1]
- *protected ribes hybrids* *20070504* [1]
- *protected rubus hybrids* *20070504* [1]
- *protected salad brassicas* *(for baby leaf production)* *20070502* [1]
- *protected soft fruit* *20082834* [1]
- *protected squashes* *20081012* [1], *20092024* [1]
- *protected strawberries* *20070499* [1]
- *protected tomatoes* *20070501* [1], *20092024* [1]
- *protected top fruit* *20082834* [1]
- *radicchio* *20070060* [3]
- *raspberries* *20061633* [3]
- *redcurrants* *20061702* [3]
- *rubus hybrids* *20061702* [3]
- *salad brassicas* *(for baby leaf production)* *20070060* [3]
- *soft fruit* *20082920* [3]
- *spinach* *20111664* [3]
- *spinach beet* *20111664* [3]
- *strawberries* *20060461* [3]
- *sweetcorn* *20041318* [3]
- *vaccinium spp.* *20061702* [3]
- *whitecurrants* *20061702* [3]

Approval information
- Pymetrozine included in Annex I under EC Regulation 1107/2009
- Accepted by BBPA for use on hops

Efficacy guidance
- Pymetrozine moves systemically in the plant and acts by preventing feeding leading to death by starvation in 1-4 d. There is no immediate knockdown
- Aphids controlled include those resistant to organophosphorus and carbamate insecticides
- To prevent development of resistance do not use continuously or as the sole method of control
- Best results achieved by starting spraying as soon as aphids seen in crop and repeating as necessary.
- To limit spread of persistent viruses such as potato leaf roll virus, apply from 90% crop emergence

Restrictions
- Maximum total dose equivalent to two full dose treatments on ware potatoes; three full dose treatments on seed potatoes, leaf spinach [3]; four full dose treatments on cucumbers, ornamentals [1]
- Consult processors before use on potatoes for processing
- Check tolerance of ornamental species before large scale use. See label for list of species known to have been treated without damage. Visible spray deposits may be seen on leaves of some species [1]

Crop-specific information
- HI cucumbers 3 d; potatoes, leaf spinach 7 d

Environmental safety
- High risk to bees (outdoor use only). Do not apply to crops in flower or to those in which bees are actively foraging. Do not apply when flowering weeds are present
- Avoid spraying within 6 m of field boundaries to reduce effects on non-target insects or arthropods. Risk to certain non-target insects and arthropods.

Hazard classification and safety precautions
Hazard Harmful
UN Number N/C
Risk phrases R40
Operator protection A [1-4]; H [2, 3]; U05a, U20c
Environmental protection E12a, E12e, E15a

FOR FULL CONDITIONS OF USE ALWAYS READ THE PRODUCT LABEL

Storage and disposal D01, D02, D09a, D11a

392   pyraclostrobin

A protectant and curative strobilurin fungicide for cereals
FRAC mode of action code: 11

See also boscalid + epoxiconazole + pyraclostrobin
boscalid + pyraclostrobin
dithianon + pyraclostrobin
epoxiconazole + fenpropimorph + pyraclostrobin
epoxiconazole + fluxapyroxad + pyraclostrobin
epoxiconazole + kresoxim-methyl + pyraclostrobin
epoxiconazole + pyraclostrobin
fenpropimorph + pyraclostrobin

Products

1	Comet 200	BASF	200 g/l	EC	12639
2	Flyer	BASF	250 g/l	EC	12654
3	Mascot Eland	Rigby Taylor	20% w/w	WG	14549
4	Platoon 250	BASF	250 g/l	EC	12640
5	Tucana	BASF	250 g/l	EC	10899
6	Vanguard	Sherriff Amenity	20% w/w	WG	13838
7	Vivid	BASF	250 g/l	EC	10898

Uses

- Brown rust in **spring barley**, **spring wheat**, **winter barley**, **winter wheat** [1, 2, 4, 5, 7]
- Crown rust in **spring oats**, **winter oats** [1, 2, 4, 5, 7]
- Dollar spot in **managed amenity turf** *(useful reduction)* [3, 6]
- Foliar disease control in **durum wheat** *(off-label)*, **grass seed crops** *(off-label)*, **spring rye** *(off-label)*, **triticale** *(off-label)*, **winter rye** *(off-label)* [5, 7]; **forage maize** *(off-label)* [1]; **ornamental plant production** *(off-label)* [7]
- Fusarium patch in **managed amenity turf** *(moderate control)* [3]; **managed amenity turf** *(moderate control only)* [6]
- Glume blotch in **spring wheat**, **winter wheat** [1, 2, 4, 5, 7]
- Net blotch in **spring barley**, **winter barley** [1, 2, 4, 5, 7]
- Red thread in **managed amenity turf** [3, 6]
- Rhynchosporium in **spring barley** *(moderate)*, **winter barley** *(moderate)* [1, 2, 4, 5, 7]
- Septoria leaf blotch in **spring wheat**, **winter wheat** [1, 2, 4, 5, 7]
- Yellow rust in **spring barley**, **spring wheat**, **winter barley**, **winter wheat** [1, 2, 4, 5, 7]

Extension of Authorisation for Minor Use (EAMUs)

- **durum wheat** *20062652* [5], *20062649* [7]
- **forage maize** *20090928* [1]
- **grass seed crops** *20062652* [5], *20062649* [7]
- **ornamental plant production** *20082884* [7]
- **spring rye** *20062652* [5], *20062649* [7]
- **triticale** *20062652* [5], *20062649* [7]
- **winter rye** *20062652* [5], *20062649* [7]

Approval information

- Pyraclostrobin included in Annex I under EC Regulation 1107/2009
- Accepted by BBPA for use on malting barley (before ear emergence only)

Efficacy guidance

- For best results apply at the start of disease attack on cereals [1, 2, 4, 5, 7]
- For Fusarium Patch treat early as severe damage to turf can occur once the disease is established [3, 6]
- Regular turf aeration, appropriate scarification and judicious use of nitrogenous fertiliser will assist the control of Fusarium Patch [3, 6]

SEE SECTION 3 FOR PRODUCTS ALSO REGISTERED

SECTION 2

- Best results on Septoria glume blotch achieved when used as a protective treatment and against Septoria leaf blotch when treated in the latent phase [1, 2, 4, 5, 7]
- Yield response may be obtained in the absence of visual disease symptoms [1, 2, 4, 5, 7]
- Pyraclostrobin is a member of the QoI cross resistance group. Product should be used preventatively and not relied on for its curative potential
- Use product as part of an Integrated Crop Management strategy incorporating other methods of control, including where appropriate other fungicides with a different mode of action. Do not apply more than two foliar applications of QoI containing products to any cereal crop or to grass
- There is a significant risk of widespread resistance occurring in *Septoria tritici* populations in UK. Failure to follow resistance management action may result in reduced levels of disease control [1, 2, 4, 5, 7]
- On cereal crops product must always be used in mixture with another product, recommended for control of the same target disease, that contains a fungicide from a different cross resistance group and is applied at a dose that will give robust control [1, 2, 4, 5, 7]
- Late application to maize crops can give a worthwhile increase in biomass

Restrictions
- Maximum number of treatments 2 per crop on cereals [1, 2, 4, 5, 7]
- Maximum total dose on turf equivalent to two full dose treatments [3, 6]
- Do not apply during drought conditions or to frozen turf [3, 6]

Crop-specific information
- Latest use: before grain watery ripe (GS 71) for wheat; up to and including emergence of ear just complete (GS 59) for barley and oats [1, 2, 4, 5, 7]
- Avoid applying to turf immediately after cutting or 48 h before mowing [3, 6]

Environmental safety
- Dangerous for the environment
- Very toxic to aquatic organisms
- LERAP Category A [6]; LERAP Category B [1-7]

Hazard classification and safety precautions
Hazard Harmful, Dangerous for the environment
Transport code 6.1 [1, 2, 4, 5, 7]; 9 [3, 6]
Packaging group III
UN Number 2902 [1, 2, 4, 5, 7]; 3077 [3, 6]
Risk phrases R20, R50 [1-7]; R22a, R38, R53a [1, 2, 4, 5, 7]
Operator protection A [1-7]; D [3, 6]; U05a, U14, U20b [1, 2, 4, 5, 7]
Environmental protection E15a, E16d [6]; E15b, E16a, E38 [1-7]; E16b [1-5, 7]; E34 [1, 2, 4, 5, 7]
Storage and disposal D01, D02 [1, 2, 4, 5, 7]; D08, D09a, D10c, D12a [1-7]
Medical advice M03, M05a [1, 2, 4, 5, 7]

393   pyraflufen-ethyl

A phenylpyrazole herbicide for potatoes
HRAC mode of action code: E

See also glyphosate + pyraflufen-ethyl

Products
Quickdown	Certis	26.5 g/l	EC	15396

Uses
- Annual dicotyledons in **potatoes**
- Desiccation in **potatoes**

Approval information
- Pyraflufen-ethyl included in Annex I under EC Regulation 1107/2009

FOR FULL CONDITIONS OF USE ALWAYS READ THE PRODUCT LABEL

Efficacy guidance
- Use with a methylated vegetable oil adjuvant [1]

Following crops guidance
- After cultivation to at least 20 cms winter cereals can be sown as a following crop but the safety to following broad-leaved crops has not yet been established.

Environmental safety
- LERAP Category B

Hazard classification and safety precautions
    Hazard Harmful, Dangerous for the environment
    Transport code 9
    Packaging group III
    UN Number 3082
    Risk phrases R20, R38, R41, R50
    Operator protection A, C, H; U05a, U09a, U11, U14, U15
    Environmental protection E15b, E16a, E34
    Storage and disposal D01, D02, D05, D09a, D10a, D12a, D12b
    Medical advice M03, M05b

394   pyrethrins

A non-persistent, contact acting insecticide extracted from Pyrethrum
IRAC mode of action code: 3

Products

1	Dairy Fly Spray	B H & B	0.75 g/l	AL	H5579
2	Killgerm ULV 400	Killgerm	30 g/l	UL	H4838
3	Pyblast	Agropharm	30 g/l	UL	H7485
4	Pyrethrum 5 EC	Agropharm	50 g/l	EC	12685
5	Pyrethrum Spray	Killgerm	3.3 g/l	UL	H4636
6	Spruzit	Certis	4.59 g/l	EC	13438

Uses
- Aphids in *all edible crops (outdoor)*, *all non-edible crops (outdoor)*, *protected crops* [6]; *broccoli, brussels sprouts, bush fruit, cabbages, calabrese, cane fruit, cauliflowers, lettuce, ornamental plant production, protected broccoli, protected brussels sprouts, protected bush fruit, protected cabbages, protected calabrese, protected cane fruit, protected cauliflowers, protected cayenne peppers* (off-label), *protected chilli peppers* (off-label), *protected lettuce, protected ornamentals, protected peppers* (off-label), *protected tomatoes, tomatoes (outdoor)* [4]
- Caterpillars in *all edible crops (outdoor)*, *all non-edible crops (outdoor)*, *protected crops* [6]; *broccoli, brussels sprouts, bush fruit, cabbages, calabrese, cane fruit, cauliflowers, lettuce, ornamental plant production, protected broccoli, protected brussels sprouts, protected bush fruit, protected cabbages, protected calabrese, protected cane fruit, protected cauliflowers, protected lettuce, protected ornamentals, protected tomatoes, tomatoes (outdoor)* [4]
- Flea beetle in *broccoli, brussels sprouts, cabbages, calabrese, cauliflowers, protected broccoli, protected brussels sprouts, protected cabbages, protected calabrese, protected cauliflowers, protected tomatoes, tomatoes (outdoor)* [4]; *protected crops* [6]
- Insect pests in *dairies, farm buildings, livestock houses, poultry houses* [1-3, 5]; *glasshouses* [5]
- Macrolophus caliginosus in *protected tomatoes, protected tomatoes* (off-label) [4]
- Mealybugs in *protected crops* [6]; *protected tomatoes, protected tomatoes* (off-label) [4]
- Scale insects in *protected crops* [6]
- Spider mites in *all edible crops (outdoor)*, *all non-edible crops (outdoor)*, *protected crops* [6]
- Thrips in *all edible crops (outdoor)*, *all non-edible crops (outdoor)*, *protected crops* [6]

SEE SECTION 3 FOR PRODUCTS ALSO REGISTERED

- Whitefly in *broccoli, brussels sprouts, cabbages, calabrese, cauliflowers, protected broccoli, protected brussels sprouts, protected cabbages, protected calabrese, protected cauliflowers, protected tomatoes, tomatoes (outdoor)* [4]; *protected crops* [6]

Extension of Authorisation for Minor Use (EAMUs)
- *protected cayenne peppers* 20091005 [4]
- *protected chilli peppers* 20091005 [4]
- *protected peppers* 20091005 [4]
- *protected tomatoes* 20063026 [4]

Approval information
- Pyrethrins have been included in Annex 1 under EC Regulation 1107/2009
- Products formulated for ULV application [1-3]. May be applied through fogging machine or sprayer [1, 3, 5]. See label for details
- Accepted by BBPA for use in empty grain stores

Efficacy guidance
- For indoor fly control close doors and windows and spray or apply fog as appropriate [1-3]
- For best fly control outdoors spray during early morning or late afternoon and evening when conditions are still [1-3, 5]
- For all uses ensure good spray coverage of the target area or plants by increasing spray volume where necessary
- Best results on outdoor or protected crops achieved from treatment at first signs of pest attack in early morning or evening [6]
- To avoid possibility of development of resistance do not spray more frequently than once per week

Restrictions
- Maximum number of treatments on edible crops 3 per crop or season [4]
- For use only by professional operators [1-3, 5]
- Do not allow spray to contact open food products or food preparing equipment or utensils [1-3, 5]
- Remove exposed milk and collect eggs before application [1-3, 5]
- Do not treat plants [1-3, 5]
- Avoid direct application to open flowers [6]
- Do not mix with, or apply closely before or after, products containing dithianon or tolylfluanid [6]
- Do not use space sprays containing pyrethrins or pyrethroid more than once per week in intensive or controlled environment animal houses in order to avoid development of resistance. If necessary, use a different control method or product [1-3, 5]
- Store away from strong sunlight [4]

Crop-specific information
- HI: 24 h for edible crops [4]
- Some plant species including *Ageratum*, ferns, *Ficus, Lantana, Poinsettia, Petroselium crispum*, and some strawberries may be sensitive especially where more than one application is made. Test before large scale treatment [6]

Environmental safety
- Dangerous for the environment [2-6]
- Very toxic to aquatic organisms [1, 4, 5]; toxic to aquatic organisms [2, 3, 6]
- High risk to bees. Do not apply to crops in flower or to those in which bees are actively foraging. Do not apply when flowering weeds are present [4]
- Risk to non-target insects or other arthropods [4]
- Do not apply directly to livestock and exclude all persons and animals during treatment [1-3, 5]
- Wash spray equipment thoroughly after use to avoid traces of pyrethrum causing damage to susceptible crops sprayed later [4]
- LERAP Category A [4]

Hazard classification and safety precautions
Hazard Harmful [1, 2, 5]; Irritant [1, 3, 4]; Dangerous for the environment [2-6]
Transport code 9 [1, 3-6]
Packaging group III [1, 3-6]
UN Number 3082 [1, 3-6]; N/C [2]

FOR FULL CONDITIONS OF USE ALWAYS READ THE PRODUCT LABEL

Risk phrases R22a, R36, R38 [1]; R22b [2, 5]; R41 [3, 4]; R50 [4]; R51 [2, 3, 5, 6]; R53a [2-6]
Operator protection A [1-6]; B [1]; C [2, 3, 5]; D [2, 3]; E [1-3]; H [2-6]; U02b, U09b, U14, U20a [2]; U05a [1, 4, 5]; U09a [1, 5]; U11 [3, 4]; U15 [3]; U19a [1-4]; U20b [1-3, 5]; U20c [4]; U20d [6]
Environmental protection E02c [3]; E02d, E05c [5]; E05a [1-3, 5]; E12a, E12e, E16c, E16d, E22c [4]; E13c [1]; E15a [2-5]; E15b [6]; E19b [3, 6]; E38 [2-4, 6]
Consumer protection C04 [1]; C05 [5]; C06 [1, 2, 5]; C07, C09, C11 [1-3, 5]; C08 [1-3]; C10 [1, 3]; C12 [2, 3, 5]
Storage and disposal D01, D09a [1-6]; D02 [1, 3-6]; D05 [2, 4]; D06a, D10a [2]; D11a [1, 4, 6]; D12a [2-6]
Medical advice M05a [3]; M05b [2, 5]

395   pyridate

A contact phenylpyridazine herbicide for cereals, maize and brassicas but approval expired 31/12/2011
HRAC mode of action code: C3

Products

Lentagran WP	Belchim	45% w/w	WP	14162

Uses

- Annual dicotyledons in *asparagus* (off-label), *broccoli* (off-label), *calabrese* (off-label), *cauliflowers* (off-label), *chives* (off-label), *collards* (off-label), *fodder rape* (off-label), *game cover* (off-label), *garlic* (off-label), *kale* (off-label), *leeks* (off-label), *lupins* (off-label), *oilseed rape* (off-label), *salad onions* (off-label), *shallots* (off-label), *spring greens* (off-label)
- Black nightshade in *asparagus* (off-label), *broccoli* (off-label), *brussels sprouts*, *bulb onions*, *cabbages*, *calabrese* (off-label), *cauliflowers* (off-label), *chives* (off-label), *collards* (off-label), *fodder rape* (off-label), *game cover* (off-label), *garlic* (off-label), *kale* (off-label), *leeks* (off-label), *lupins* (off-label), *oilseed rape* (off-label), *salad onions* (off-label), *shallots* (off-label), *spring greens* (off-label)
- Cleavers in *asparagus* (off-label), *broccoli* (off-label), *brussels sprouts*, *bulb onions*, *cabbages*, *calabrese* (off-label), *cauliflowers* (off-label), *chives* (off-label), *collards* (off-label), *fodder rape* (off-label), *game cover* (off-label), *garlic* (off-label), *kale* (off-label), *leeks* (off-label), *lupins* (off-label), *oilseed rape* (off-label), *salad onions* (off-label), *shallots* (off-label), *spring greens* (off-label)
- Fat hen in *asparagus* (off-label), *broccoli* (off-label), *brussels sprouts*, *bulb onions*, *cabbages*, *calabrese* (off-label), *cauliflowers* (off-label), *chives* (off-label), *collards* (off-label), *fodder rape* (off-label), *game cover* (off-label), *garlic* (off-label), *kale* (off-label), *leeks* (off-label), *lupins* (off-label), *oilseed rape* (off-label), *salad onions* (off-label), *shallots* (off-label), *spring greens* (off-label)
- Fumitory in *asparagus* (off-label), *broccoli* (off-label), *calabrese* (off-label), *cauliflowers* (off-label), *chives* (off-label), *collards* (off-label), *fodder rape* (off-label), *game cover* (off-label), *garlic* (off-label), *kale* (off-label), *leeks* (off-label), *lupins* (off-label), *oilseed rape* (off-label), *salad onions* (off-label), *shallots* (off-label), *spring greens* (off-label)
- Groundsel in *asparagus* (off-label), *broccoli* (off-label), *calabrese* (off-label), *cauliflowers* (off-label), *chives* (off-label), *collards* (off-label), *fodder rape* (off-label), *game cover* (off-label), *garlic* (off-label), *kale* (off-label), *leeks* (off-label), *lupins* (off-label), *oilseed rape* (off-label), *salad onions* (off-label), *shallots* (off-label), *spring greens* (off-label)

Extension of Authorisation for Minor Use (EAMUs)

- *asparagus* 20091039
- *broccoli* 20090786
- *calabrese* 20090786
- *cauliflowers* 20090786
- *chives* 20120267
- *collards* 20090785
- *fodder rape* 20093230
- *game cover* 20090788
- *garlic* 20092862

SEE SECTION 3 FOR PRODUCTS ALSO REGISTERED

- **kale** *20090785*
- **leeks** *20090784*
- **lupins** *20090787*
- **oilseed rape** *20093230*
- **salad onions** *20120267*
- **shallots** *20092862*
- **spring greens** *20090785*

Approval information
- Pyridate included in Annex I under EC Regulation 1107/2009

Efficacy guidance
- Best results achieved by application to actively growing weeds at 6-8 leaf stage when temperatures are above 8°C before crop foliage forms canopy

Restrictions
- Maximum number of treatments 1 per crop
- Do not apply in mixture with or within 14 d of any other product which may result in dewaxing of crop foliage
- Do not use on crops suffering stress from frost, drought, disease or pest attack

Crop-specific information
- Latest use: before 7 leaf stage for maize; before 5 true leaves for onions; before flower buds visible for oilseed rape
- HI 6 wk for Brussels sprouts, cabbages
- Apply to cabbages and Brussels sprouts after 4 fully expanded leaf stage. Allow 2 wk after transplanting before treating

Environmental safety
- Dangerous for the environment
- Toxic to aquatic organisms

Hazard classification and safety precautions
Hazard Irritant, Dangerous for the environment
Transport code 9
Packaging group III
UN Number 3077
Risk phrases R43, R51, R53a
Operator protection A, C; U05a, U14, U22a
Environmental protection E15a, E38
Storage and disposal D01, D02, D09a, D10c, D12a

396   pyrimethanil

An anilinopyrimidine fungicide for apples and strawberries
FRAC mode of action code: 9

See also chlorothalonil + pyrimethanil

Products

1	Scala	BASF	400 g/l	SC	15222
2	Walia	ChemSource	400 g/l	SC	15787

Uses
- Botrytis in **bilberries** *(off-label)*, **blackberries** *(off-label)*, **blackcurrants** *(off-label)*, **blueberries** *(off-label)*, **cranberries** *(off-label)*, **forest nurseries** *(off-label)*, **gooseberries** *(off-label)*, **grapevines** *(off-label)*, **ornamental plant production** *(off-label)*, **pears** *(off-label)*, **protected aubergines** *(off-label)*, **protected blackberries** *(off-label)*, **protected herbs (see appendix 6)** *(off-label)*, **protected lettuce** *(off-label)*, **protected raspberries** *(off-label)*, **protected strawberries**, **protected tomatoes** *(off-label)*, **quinces** *(off-label)*, **raspberries** *(off-label)*,

redcurrants *(off-label)*, **top fruit** *(off-label)*, **vaccinium spp.** *(off-label)*, **whitecurrants** *(off-label)* [1]; **strawberries** [1, 2]
- Disease control in **forest nurseries** *(off-label)*, **top fruit** *(off-label)* [1]
- Scab in **apples** [1, 2]; **pears** *(off-label)*, **quinces** *(off-label)* [1]

Extension of Authorisation for Minor Use (EAMUs)
- **bilberries** *20110291* [1]
- **blackberries** *20110293* [1]
- **blackcurrants** *20110291* [1]
- **blueberries** *20110291* [1]
- **cranberries** *20110291* [1]
- **forest nurseries** *20122054* [1]
- **gooseberries** *20110291* [1]
- **grapevines** *20110283* [1]
- **ornamental plant production** *20111315* [1]
- **pears** *20110295* [1]
- **protected aubergines** *20110282* [1]
- **protected blackberries** *20110292* [1]
- **protected herbs (see appendix 6)** *20110287* [1]
- **protected lettuce** *20110287* [1]
- **protected raspberries** *20110292* [1]
- **protected tomatoes** *20110282* [1]
- **quinces** *20110295* [1]
- **raspberries** *20110293* [1]
- **redcurrants** *20110291* [1]
- **top fruit** *20122054* [1]
- **vaccinium spp.** *20110291* [1]
- **whitecurrants** *20110291* [1]

Approval information
- Pyrimethanil included in Annex I under EC Regulation 1107/2009. However a Standing Committee decision in July 2007 to reduce to the limit of determination the MRL set for pyrimenthanil in cane fruit (other than blackberries and raspberries) and the MRL for brassicas grown for baby leaf production led to immediate revocation of SOLAs for these crops

Efficacy guidance
- On apples a programme of sprays will give early season control of scab. Season long control can be achieved by continuing programme with other approved fungicides
- In strawberries product should be used as part of a programme of disease control treatments which should alternate with other materials to prevent or limit development of less sensitive strains of grey mould

Restrictions
- Maximum number of treatments 5 per yr for apples; 4 per yr for protected crops; 3 per yr for grapevines; 2 per yr for cane fruit, bush fruit, strawberries, tomatoes
- Product does not taint apples. Processors should be consulted before use on strawberries

Crop-specific information
- Latest use: before end of flowering for apples
- HI protected cane fruit, strawberries 1 d; aubergines, tomatoes 3 d; protected lettuce 14 d; bush fruit, grapevines 21 d
- All varieties of apples and strawberries may be treated
- Treat apples from bud burst at 10-14 d intervals
- In strawberries start treatments at white bud to give maximum protection of flowers against grey mould and treat every 7-10 d. Product should not be used more than once in a 3 or 4 spray programme

Environmental safety
- Harmful to aquatic organisms
- Product has negligible effect on hoverflies and lacewings. Limited evidence indicates some margin of safety to *Typhlodromus pyri*

SEE SECTION 3 FOR PRODUCTS ALSO REGISTERED

- Broadcast air-assisted LERAP [1] (20); Broadcast air-assisted LERAP [2] (20 m); LERAP Category B [1]

Hazard classification and safety precautions
    UN Number N/C
    Risk phrases R52, R53a
    Operator protection A, H, M [1]; U08, U20b
    Environmental protection E15a [1, 2]; E16a [1]; E17b [1] (20); E17b [2] (20 m)
    Storage and disposal D01, D02, D05, D08, D09a, D10b, D12a

397   pyriofenone

A protectant fungicide with some curative activity against mildew in the latent phase
FRAC mode of action code: U8

Products

Property 180 SC	Belchim	180 g/l	SC	15661

Uses
- Powdery mildew in **spring wheat**, **winter wheat**

Approval information
- Pyrifenone is not yet included in Annex 1 under EC Regulation 1107/2009

Restrictions
- Only two applications of benzoylpyridine fungicides are permitted per crop per season. Where a second application is required it should be applied in mixture with fungicides using a different mode of action such as fenpropidin or fenpropimorph.

Following crops guidance
- Ollseed rape, maize, potato, sugar beet, kidney bean, soybean, pea, onion, turnip, lettuce, flax, oat, barley and wheat may be sown as a following crop.

Hazard classification and safety precautions
    Hazard Harmful, Dangerous for the environment
    Transport code 9
    Packaging group III
    UN Number 3082
    Risk phrases R40, R51, R53a
    Operator protection A, H; U05a, U20b
    Environmental protection E15b, E34, E38
    Storage and disposal D01, D02, D05, D09a, D12a, D20
    Medical advice M05a

398   quinmerac

A residual herbicide available only in mixtures
HRAC mode of action code: O

See also chloridazon + quinmerac
        dimethenamid-p + metazachlor + quinmerac
        metazachlor + quinmerac

399   quinoclamine

A selective moss killer for moderate control of moss in managed amenity turf

Products

Mogeton	Certis	25% w/w	WP	15837

FOR FULL CONDITIONS OF USE ALWAYS READ THE PRODUCT LABEL

Uses
* Moss in **managed amenity turf** *(moderate control)*

Approval information
* Quinoclamine included in Annex 1 under EC Regulation 1107/2009

Environmental safety
* LERAP Category A

Hazard classification and safety precautions
Hazard Harmful, Dangerous for the environment
Transport code 9
Packaging group III
UN Number 3077
Risk phrases R22a, R36, R43, R48, R50, R53a, R63
Operator protection A, C, G, H; U05a, U11, U19a, U19e
Environmental protection E15a, E16c
Storage and disposal D01, D02, D12a
Medical advice M03, M05a

400　quinoxyfen

A systemic protectant quinoline fungicide for cereals
FRAC mode of action code: 13

See also cyproconazole + quinoxyfen
　　　fenpropimorph + quinoxyfen

Products
Fortress	Dow	500 g/l	SC	08279

Uses
* Powdery mildew in **blackcurrants** *(off-label)*, **blueberries** *(off-label)*, **cranberries** *(off-label)*, **durum wheat**, **forest nurseries** *(off-label)*, **gooseberries** *(off-label)*, **hops** *(off-label)*, **ornamental plant production** *(off-label)*, **protected forest nurseries** *(off-label)*, **protected ornamentals** *(off-label)*, **protected strawberries** *(off-label)*, **redcurrants** *(off-label)*, **spring barley**, **spring oats**, **spring rye**, **spring wheat**, **strawberries** *(off-label)*, **sugar beet**, **triticale**, **whitecurrants** *(off-label)*, **winter barley**, **winter oats**, **winter rye**, **winter wheat**

Extension of Authorisation for Minor Use (EAMUs)
* **blackcurrants** *20111459*
* **blueberries** *20111459*
* **cranberries** *20111459*
* **forest nurseries** *20082852*
* **gooseberries** *20111459*
* **hops** *20061579*
* **ornamental plant production** *20082852*
* **protected forest nurseries** *20082852*
* **protected ornamentals** *20082852*
* **protected strawberries** *20041923*
* **redcurrants** *20111459*
* **strawberries** *20041923*
* **whitecurrants** *20111459*

Approval information
* Quinoxyfen included in Annex I under Regulation 1107/2009
* Accepted by BBPA for use on malting barley (before ear emergence only)

Efficacy guidance
* For best results treat at early stage of disease development before infection spreads to new crop growth. Further treatment may be necessary if disease pressure remains high
* Product not curative and will not control latent or established disease infections

SEE SECTION 3 FOR PRODUCTS ALSO REGISTERED

- For broad spectrum control in cereals use in tank mixtures - see label
- Product rainfast after 1 h
- Systemic activity may be reduced in severe drought
- Use as part of an integrated crop management (ICM) strategy incorporating other methods of control or fungicides with different modes of action

Restrictions
- Maximum total dose equivalent to two full dose treatments in cereals; see labels for split dose recommendation in sugar beet
- On cereals apply only in the spring from mid-tillering stage (GS 25)

Crop-specific information
- Latest use: first awns visible stage (GS 49) for cereals
- HI sugar beet 28 d

Environmental safety
- Dangerous for the environment
- Very toxic to aquatic organisms
- LERAP Category B

Hazard classification and safety precautions
Hazard Irritant, Dangerous for the environment
Transport code 9
Packaging group III
UN Number 3082
Risk phrases R43, R50, R53a
Operator protection A, C, H; U05a, U14
Environmental protection E15a, E16a, E16b, E34, E38
Storage and disposal D01, D02, D12a

401   quizalofop-P-ethyl

An aryl phenoxypropionic acid post-emergence herbicide for grass weed control
HRAC mode of action code: A

Products
1 Leopard 5 EC	Makhteshim	50 g/l	EC	13720
2 Pilot Ultra	Nissan	50 g/l	SC	12929

Uses
- Annual grasses in *combining peas, fodder beet, linseed, mangels, red beet, spring field beans, spring oilseed rape, sugar beet, vining peas, winter field beans, winter oilseed rape*
- Couch in *combining peas, fodder beet, linseed, mangels, red beet, spring field beans, spring oilseed rape, sugar beet, vining peas, winter field beans, winter oilseed rape*
- Perennial grasses in *combining peas, fodder beet, linseed, mangels, red beet, spring field beans, spring oilseed rape, sugar beet, vining peas, winter field beans, winter oilseed rape*
- Volunteer cereals in *combining peas, fodder beet, linseed, mangels, red beet, spring field beans, spring oilseed rape, sugar beet, vining peas, winter field beans, winter oilseed rape*

Approval information
- Quizalofop-P has been included in Annex 1 under EC Regulation 1107/2009

Efficacy guidance
- Best results achieved by application to emerged weeds growing actively in warm conditions with adequate soil moisture
- Weed control may be reduced under conditions such as drought that limit uptake and translocation
- Annual meadow-grass is not controlled
- For effective couch control do not hoe beet crops within 21 d after spraying
- At least 2 h without rain should follow application otherwise results may be reduced

FOR FULL CONDITIONS OF USE ALWAYS READ THE PRODUCT LABEL

- Quizalofop-P-ethyl is an ACCase inhibitor herbicide. To avoid the build up of resistance do not apply products containing an ACCase inhibitor herbicide more than twice to any crop. In addition do not use any product containing quizalofop-P-ethyl in mixture or sequence with any other product containing the same ingredient
- Use these products as part of a resistance management strategy that includes cultural methods of control and does not use ACCase inhibitors as the sole chemical method of grass weed control
- Applying a second product containing an ACCase inhibitor to a crop will increase the risk of resistance development; only use a second ACCase inhibitor to control different weeds at a different timing
- Always follow WRAG guidelines for preventing and managing herbicide resistant weeds. See Section 5 for more information

Restrictions
- Maximum number of treatments 1 on all recommended crops
- Do not spray crops under stress from any cause or in frosty weather
- Consult processor before use on crops recommended for processing
- An interval of at least 3 d must elapse between treatment and use of another herbicide on beet crops, 14 d on oilseed rape, 21 d on linseed and other recommended crops
- Avoid drift onto neighbouring crops

Crop-specific information
- HI 16 wk for beet crops; 11 wk for oilseed rape, linseed; 8 wk for field beans; 5 wk for peas
- In some situations treatment can cause yellow patches on foliage of peas, especially vining varieties. Symptoms usually rapidly and completely outgrown
- May cause taint in peas

Following crops guidance
- In the event of failure of a treated crop broad-leaved crops may be resown after a minimum interval of 2 wk, and cereals after 2-6 wk depending on dose applied
- Onions, leeks and maize are not recommended to follow a failed treated crop

Environmental safety
- Dangerous for the environment
- Toxic to aquatic organisms
- Flammable

Hazard classification and safety precautions
    Hazard Harmful, Flammable [1]; Dangerous for the environment [1, 2]
    Transport code 3
    Packaging group III
    UN Number 1993
    Risk phrases R22b, R37, R38, R41, R67 [1]; R51, R53a [1, 2]
    Operator protection A, C [1, 2]; H [2]; U05a, U11, U19a [1]; U09a [2]; U20b [1, 2]
    Environmental protection E13b [2]; E15a [1]; E34, E38 [1, 2]
    Storage and disposal D01, D02, D05, D10b [1]; D09a, D12a [1, 2]; D10c [2]
    Medical advice M03, M05b [1]

---

402   quizalofop-P-tefuryl

An aryloxyphenoxypropionate herbicide for grass weed control
HRAC mode of action code: A

Products

Panarex	Certis	40 g/l	EC	12532

Uses
- Blackgrass in *amenity vegetation* (off-label), *cereal cover crops*, *combining peas*, *fodder beet*, *natural surfaces not intended to bear vegetation* (off-label), *ornamental plant production* (off-label), *permeable surfaces overlying soil* (off-label), *potatoes*, *spring field*

beans, **spring linseed**, **spring oilseed rape**, **sugar beet**, **winter field beans**, **winter linseed**, **winter oilseed rape**
- Couch in **amenity vegetation** *(off-label)*, **cereal cover crops**, **combining peas**, **fodder beet**, **natural surfaces not intended to bear vegetation** *(off-label)*, **ornamental plant production** *(off-label)*, **permeable surfaces overlying soil** *(off-label)*, **potatoes**, **spring field beans**, **spring linseed**, **spring oilseed rape**, **sugar beet**, **winter field beans**, **winter linseed**, **winter oilseed rape**
- Italian ryegrass in **cereal cover crops**, **combining peas**, **fodder beet**, **potatoes**, **spring field beans**, **spring linseed**, **spring oilseed rape**, **sugar beet**, **winter field beans**, **winter linseed**, **winter oilseed rape**
- Perennial ryegrass in **amenity vegetation** *(off-label)*, **cereal cover crops**, **combining peas**, **fodder beet**, **natural surfaces not intended to bear vegetation** *(off-label)*, **ornamental plant production** *(off-label)*, **permeable surfaces overlying soil** *(off-label)*, **potatoes**, **spring field beans**, **spring linseed**, **spring oilseed rape**, **sugar beet**, **winter field beans**, **winter linseed**, **winter oilseed rape**
- Volunteer cereals in **cereal cover crops**, **combining peas**, **fodder beet**, **potatoes**, **spring field beans**, **spring linseed**, **spring oilseed rape**, **sugar beet**, **winter field beans**, **winter linseed**, **winter oilseed rape**
- Wild oats in **cereal cover crops**, **combining peas**, **fodder beet**, **potatoes**, **spring field beans**, **spring linseed**, **spring oilseed rape**, **sugar beet**, **winter field beans**, **winter linseed**, **winter oilseed rape**

Extension of Authorisation for Minor Use (EAMUs)
- **amenity vegetation** *20060364*
- **natural surfaces not intended to bear vegetation** *20060364*
- **ornamental plant production** *20060364*
- **permeable surfaces overlying soil** *20060364*

Approval information
- Quizalofop-P has been included in Annex 1 under EC Regulation 1107/2009

Efficacy guidance
- Best results on annual grass weeds when growing actively and treated from 2 leaves to the start of tillering
- Treat cover crops when they have served their purpose and the threat of wind blow has passed
- Best results on couch achieved when the weed is growing actively and commencing new rhizome growth
- Grass weeds germinating after treatment will not be controlled
- Treatment quickly stops growth and visible colour changes to the leaf tips appear after about 7 d. Complete kill takes 3-4 wk under good growing conditions
- Quizalofop-P-tefuryl is an ACCase inhibitor herbicide. To avoid the build up of resistance do not apply products containing an ACCase inhibitor herbicide more than twice to any crop. In addition do not use any product containing quizalofop-P-tefuryl in mixture or sequence with any other product containing the same ingredient
- Use these products as part of a resistance management strategy that includes cultural methods of control and does not use ACCase inhibitors as the sole chemical method of grass weed control
- Applying a second product containing an ACCase inhibitor to a crop will increase the risk of resistance development; only use a second ACCase inhibitor to control different weeds at a different timing
- Always follow WRAG guidelines for preventing and managing herbicide resistant weeds. See Section 5 for more information

Restrictions
- Maximum number of treatments 1 per crop
- Consult processors before use on peas or potatoes for processing
- Do not treat crops and weeds growing under stress from any cause

Crop-specific information
- HI 60 d for all crops
- Treat oilseed rape from the fully expanded cotyledon stage

FOR FULL CONDITIONS OF USE ALWAYS READ THE PRODUCT LABEL

- Treat linseed, peas and field beans from 2-3 unfolded leaves
- Treat sugar beet from 2 unfolded leaves

Following crops guidance
- In the event of failure of a treated crop any broad-leaved crop may be planted at any time. Cereals may be drilled from 4 wk after treatment

Environmental safety
- Dangerous for the environment
- Very toxic to aquatic organisms
- Risk to certain non-target insects or other arthropods. For advice on risk management and use in Integrated Pest Management (IPM) see directions for use
- Avoid spraying within 6 m of the field boundary to reduce effects on non-target insects and other arthropods

Hazard classification and safety precautions
Hazard Irritant, Dangerous for the environment
Transport code 9
Packaging group III
UN Number 3082
Risk phrases R41, R43, R50, R53a
Operator protection A, C, H; U02a, U04a, U05a, U10, U11, U14, U15, U19a, U20b
Environmental protection E15a, E22b, E38
Storage and disposal D01, D02, D09a, D10b, D12a

403 rimsulfuron

A selective systemic sulfonylurea herbicide
HRAC mode of action code: B

Products

1	Caesar	AgChem Access	25% w/w	SG	15272
2	Clayton Bramble	Clayton	25% w/w	SG	15281
3	Clayton Nero	Clayton	25% w/w	SG	15511
4	Emperor	ChemSource	25% w/w	SG	15331
5	Titus	Makhteshim	25% w/w	SG	15050

Uses
- Annual dicotyledons in *forage maize*, *potatoes* [1-5]; *forest nurseries* (off-label), *ornamental plant production* (off-label) [5]
- Charlock in *forage maize*, *potatoes* [5]
- Chickweed in *forage maize*, *potatoes* [5]
- Cleavers in *forage maize*, *potatoes* [5]
- Hemp-nettle in *forage maize*, *potatoes* [5]
- Red dead-nettle in *forage maize*, *potatoes* [5]
- Redshank in *forage maize*, *potatoes* [5]
- Scentless mayweed in *forage maize*, *potatoes* [5]
- Small nettle in *forage maize*, *potatoes* [5]
- Volunteer oilseed rape in *forage maize*, *potatoes* [1-5]

Extension of Authorisation for Minor Use (EAMUs)
- *forest nurseries* 20122062 [5]
- *ornamental plant production* 20122062 [5]

Approval information
- Rimsulfuron included in Annex I under EC Regulation 1107/2009

Efficacy guidance
- Product should be used with a suitable adjuvant or a suitable herbicide tank-mix partner. See label for details

SEE SECTION 3 FOR PRODUCTS ALSO REGISTERED

- Product acts by foliar action. Best results obtained from good spray cover of small actively growing weeds. Effectiveness is reduced in very dry conditions
- Weed spectrum can be broadened by tank mixture with other herbicides. See label for details
- Susceptible weeds cease growth immediately and symptoms can be seen 10 d later
- Rimsulfuron is a member of the ALS-inhibitor group of herbicides

Restrictions
- Maximum number of treatments 1 per crop
- Do not treat maize previously treated with organophosphorus insecticides
- Do not apply to potatoes grown for certified seed
- Consult processor before use on crops grown for processing
- Avoid high light intensity (full sunlight) and high temperatures on the day of spraying
- Do not treat during periods of substantial diurnal temperature fluctuation or when frost anticipated
- Do not apply to any crop stressed by drought, water-logging, low temperatures, pest or disease attack, nutrient or lime deficiency
- Do not apply to forage maize treated with organophosphate insecticides
- Do not apply to forage maize undersown with grass or clover

Crop-specific information
- Latest use: before most advanced potato plants are 25 cm high; before 4-collar stage of fodder maize
- All varieties of ware potatoes may be treated, but variety restrictions of any tank-mix partner must be observed
- Only certain named varieties of forage maize may be treated. See label

Following crops guidance
- Only winter wheat should follow a treated crop in the same calendar yr
- Only barley, wheat or maize should be sown in the spring of the yr following treatment
- In the second autumn after treatment any crop except brassicas or oilseed rape may be drilled

Environmental safety
- Dangerous for the environment
- Toxic to aquatic organisms
- Extremely dangerous to fish or other aquatic life. Do not contaminate surface waters or ditches with chemical or used container
- Herbicide is very active. Take particular care to avoid drift onto plants outside the target area
- Spraying equipment should not be drained or flushed onto land planted, or to be planted, with trees or crops other than potatoes or forage maize and should be thoroughly cleansed after use - see label for instructions

Hazard classification and safety precautions
Hazard Dangerous for the environment
Transport code 9
Packaging group III
UN Number 3077
Risk phrases R51, R53a
Operator protection A [3, 5]; U08, U19a, U20b
Environmental protection E13a, E34, E38
Storage and disposal D01, D09a, D11a, D12a

404 silthiofam

A thiophene carboxamide fungicide seed dressing for cereals
FRAC mode of action code: 38

Products

1	Latitude	Monsanto	125 g/l	FS	10695
2	Meridian	AgChem Access	125 g/l	FS	14676

FOR FULL CONDITIONS OF USE ALWAYS READ THE PRODUCT LABEL

Uses

- Seed-borne diseases in **durum wheat** *(off-label)*, **spring rye** *(off-label)*, **triticale** *(off-label)*, **winter rye** *(off-label)* [1]
- Take-all in **spring wheat** *(seed treatment)*, **winter barley** *(seed treatment)*, **winter wheat** *(seed treatment)* [1, 2]

Extension of Authorisation for Minor Use (EAMUs)

- **durum wheat** 20060973 expires 31 Dec 2013 [1]
- **spring rye** 20060973 expires 31 Dec 2013 [1]
- **triticale** 20060973 expires 31 Dec 2013 [1]
- **winter rye** 20060973 expires 31 Dec 2013 [1]

Approval information

- Silthiofam included in Annex I under EC Regulation 1107/2009
- Accepted by BBPA for use on malting barley
- Approval expiry 31 Dec 2013 [1, 2]

Efficacy guidance

- Apply using approved seed treatment equipment which has been accurately calibrated
- Apply simultaneously (i.e. not in mixture) with a standard seed treatment
- Drill treated seed in the season of purchase. The viability of treated seed and fungicide activity may be reduced by physical storage
- Drill treated seed at 2.5-4 cm into a well prepared firm seedbed
- As precaution against possible development of disease resistance do not treat more than three consecutive susceptible cereal crops in any one rotation

Restrictions

- Maximum number of treatments one per batch of seed
- Do not use on seed with more than 16% moisture content, or on sprouted, cracked or skinned seed
- Test germination of all seed batches before treatment

Crop-specific information

- Latest use: immediately prior to drilling

Hazard classification and safety precautions

UN Number N/C
Operator protection A, H; U20b
Environmental protection E03, E15a, E34
Storage and disposal D05, D09a, D10a
Treated seed S01, S02, S04b, S05, S06a, S07

405 S-metolachlor

A chloroacetamide residual herbicide for use in forage maize and grain maize
HRAC mode of action code: K3

Products

1	Clayton Smelter	Clayton	960 g/l	EC	14937
2	Dual Gold	Syngenta	960 g/l	EC	14649
3	Duel Star	ChemSource	960 g/l	EC	15072

Uses

- Annual dicotyledons in **forest nurseries** *(off-label)*, **french beans** *(off-label)*, **garlic** *(off-label)*, **herbs (see appendix 6)** *(off-label)*, **lettuce** *(off-label)*, **onions** *(off-label)*, **ornamental plant production** *(off-label)*, **red beet** *(off-label)*, **runner beans** *(off-label)*, **shallots** *(off-label)*, **swedes** *(off-label)*, **turnips** *(off-label)* [2]
- Annual grasses in **begonias** *(off-label)*, **broccoli** *(off-label)*, **brussels sprouts** *(off-label)*, **cabbages** *(off-label)*, **calabrese** *(off-label)*, **cauliflowers** *(off-label)*, **chinese cabbage** *(off-label)*, **collards** *(off-label)*, **kale** *(off-label)* [2]; **chicory** *(off-label)*, **chicory root** *(off-label)*, **dwarf beans**

SECTION 2

(off-label), **endives** *(off-label),* **runner beans** *(off-label),* **strawberries** *(off-label)* [1, 2]; **french beans** *(off-label)* [1]

- Annual meadow grass in **forage maize**, **grain maize** [1-3]; **forest nurseries** *(off-label),* **french beans** *(off-label),* **garlic** *(off-label),* **herbs (see appendix 6)** *(off-label),* **lettuce** *(off-label),* **onions** *(off-label),* **ornamental plant production** *(off-label),* **red beet** *(off-label),* **runner beans** *(off-label),* **shallots** *(off-label),* **swedes** *(off-label),* **turnips** *(off-label)* [2]
- Black nightshade in **chicory** *(off-label),* **chicory root** *(off-label),* **endives** *(off-label)* [1, 2]
- Chickweed in **forage maize** *(moderately susceptible),* **grain maize** *(moderately susceptible)* [1-3]
- Fat hen in **forage maize** *(moderately susceptible),* **grain maize** *(moderately susceptible)* [1-3]
- Field speedwell in **forage maize**, **grain maize** [1-3]
- Mayweeds in **forage maize**, **grain maize** [1-3]
- Red dead-nettle in **forage maize**, **grain maize** [1-3]
- Sowthistle in **forage maize**, **grain maize** [1-3]

Extension of Authorisation for Minor Use (EAMUs)
- **begonias** *20101390* [2], *20111256* [2]
- **broccoli** *20103142* [2], *20111258* [2]
- **brussels sprouts** *20103143* [2], *20111259* [2]
- **cabbages** *20103143* [2], *20111259* [2]
- **calabrese** *20103142* [2], *20111258* [2]
- **cauliflowers** *20103142* [2], *20111258* [2]
- **chicory** *20111262* [1], *20111257* [2]
- **chicory root** *20103107* [1], *20101391* [2]
- **chinese cabbage** *20103144* [2], *20111260* [2]
- **collards** *20103144* [2], *20111260* [2]
- **dwarf beans** *20103103* [1], *20111254* [2]
- **endives** *20103107* [1], *20111262* [1], *20101391* [2], *20111257* [2]
- **forest nurseries** *20120501* [2]
- **french beans** *20103103* [1], *20101388* [2]
- **garlic** *20110840* [2]
- **herbs (see appendix 6)** *20120594* [2]
- **kale** *20103144* [2], *20111260* [2]
- **lettuce** *20120594* [2]
- **onions** *20110840* [2]
- **ornamental plant production** *20120501* [2]
- **red beet** *20111006* [2]
- **runner beans** *20103103* [1], *20101388* [2], *20111254* [2]
- **shallots** *20110840* [2]
- **strawberries** *20103104* [1], *20101389* [2], *20111255* [2]
- **swedes** *20111006* [2]
- **turnips** *20111006* [2]

Approval information
- S-metolachlor included in Annex I under EC Regulation 1107/2009

Environmental safety
- LERAP Category B

Hazard classification and safety precautions
Hazard Irritant, Dangerous for the environment
Transport code 9
Packaging group III
UN Number 3082
Risk phrases R43, R50, R53a
Operator protection A [1-3]; C [2, 3]; H [1]; U05a, U09a, U20b
Environmental protection E15b, E16a, E16b, E38
Storage and disposal D01, D02, D09a, D10c, D11a, D12a

FOR FULL CONDITIONS OF USE ALWAYS READ THE PRODUCT LABEL

406 sodium hypochlorite (commodity substance)

An inorganic horticultural bactericide for use in mushrooms

Products

sodium hypochlorite	various	100% w/v	ZZ	-

Uses
- Bacterial blotch in **mushrooms**

Approval information
- Sodium hypochlorite included in Annex 1 under EC Regulation 1107/2009
- Approval as a comodity substance valid until 31 Aug 2019

Restrictions
- Maximum concentration 315 mg/litre of water
- Mixing and loading must only take place in a ventilated area
- Must only be used by suitably trained and competent operators

Crop-specific information
- HI 1 d

Environmental safety
- Harmful to fish or other aquatic life. Do not contaminate surface waters or ditches with chemical or used container

Hazard classification and safety precautions
    Operator protection A, C, H
    Environmental protection E13c

407 spinosad

A selective insecticide derived from naturally occurring soil fungi (naturalyte)
IRAC mode of action code: 5

Products

1	Conserve	Fargro	120 g/l	SC	12058
2	Tracer	Landseer	480 g/l	SC	12438

Uses
- Cabbage moth in **brussels sprouts, cabbages, cauliflowers** [2]
- Cabbage white butterfly in **brussels sprouts, cabbages, cauliflowers** [2]
- Codling moth in **apples, pears** [2]
- Diamond-back moth in **brussels sprouts, cabbages, cauliflowers** [2]
- Insect pests in **blackberries** *(off-label)*, **blackcurrants** *(off-label)*, **celery leaves, cress, frise, grapevines** *(off-label)*, **herbs (see appendix 6), hops** *(off-label)*, **lamb's lettuce, lettuce, ornamental specimens** *(off-label)*, **protected herbs (see appendix 6)** *(off-label)*, **radicchio, raspberries** *(off-label)*, **redcurrants** *(off-label)*, **scarole, soft fruit** *(off-label)*, **top fruit** *(off-label)*, **whitecurrants** *(off-label)* [2]
- Moths in **hops** *(off-label)*, **ornamental specimens** *(off-label)*, **soft fruit** *(off-label)*, **top fruit** *(off-label)* [2]
- Sawflies in **blackcurrants** *(off-label)*, **redcurrants** *(off-label)*, **whitecurrants** *(off-label)* [2]
- Small white butterfly in **brussels sprouts, cabbages, cauliflowers** [2]
- Thrips in **blackberries** *(off-label)*, **bulb onions, leeks, raspberries** *(off-label)*, **salad onions** [2]
- Tortrix moths in **apples, pears** [2]
- Western flower thrips in **protected aubergines, protected cucumbers, protected ornamentals, protected peppers, protected tomatoes** [1]
- Winter moth in **blackcurrants** *(off-label)*, **redcurrants** *(off-label)*, **whitecurrants** *(off-label)* [2]

Extension of Authorisation for Minor Use (EAMUs)
- **blackberries** *20092118* [2]

SEE SECTION 3 FOR PRODUCTS ALSO REGISTERED

SECTION 2

- **blackcurrants** *20113223* [2]
- **grapevines** *20122222* [2]
- **hops** *20082908* [2]
- **ornamental specimens** *20082908* [2]
- **protected herbs (see appendix 6)** *20081290* [2]
- **raspberries** *20092118* [2]
- **redcurrants** *20113223* [2]
- **soft fruit** *20082908* [2]
- **top fruit** *20082908* [2]
- **whitecurrants** *20113223* [2]

Approval information
- Spinosad included in Annex I under EC Regulation 1107/2009

Efficacy guidance
- Product enters insects by contact from a treated surface or ingestion of treated plant material therefore good spray coverage is essential
- Some plants, for example Fuchsia flowers, can provide effective refuges from spray deposits and control of western flower thrips may be reduced [1]
- Apply to protected crops when western flower thrip nymphs or adults are first seen [1]
- Monitor western flower thrip development carefully to see whether further applications are necessary. A 2 spray programme at 5-7 d intervals may be needed when conditions favour rapid pest development [1]
- Treat top fruit and field crops when pests are first seen or at very first signs of crop damage [2]
- Ensure a rain-free period of 12 h after treatment before applying irrigation [2]
- To reduce possibility of development of resistance, adopt resistance management measures. See label and Section 5 for more information

Restrictions
- Maximum number of treatments 4 per crop for brassicas, onions, leeks [2]; 1 pre-blossom and 3 post blossom for apples, pears [2]
- Apply no more than 2 consecutive sprays. Rotate with another insecticide with a different mode of action or use no further treatment after applying the maximum number of treatments
- Establish whether any incoming plants have been treated and apply no more than 3 consecutive sprays. Maximum number of treatments 6 per structure per yr
- Avoid application in bright sunlight or into open flowers [1]

Crop-specific information
- HI 3 d for protected cucumbers, brassicas; 7 d for apples, pears, onions, leeks
- Test for tolerance on a small number of ornamentals or cucumbers before large scale treatment
- Some spotting of african violet flowers may occur

Environmental safety
- Dangerous for the environment
- Very toxic (toxic [1]) to aquatic organisms
- Whenever possible use an Integrated Pest Management system. Spinosad presents low risk to beneficial arthropods
- Product has low impact on many insect and mite predators but is harmful to adults of most parasitic wasps. Most beneficials may be introduced to treated plants when spray deposits are dry but an interval of 2 wk should elapse before introduction of parasitic wasps. See label for details
- Treatment may cause temporary reduction in abundance of insect and mite predators if present at application
- Exposure to direct spray is harmful to bees but dry spray deposits are harmless. Treatment of field crops should not be made in the heat of the day when bees are actively foraging
- Broadcast air-assisted LERAP [2] (40 m); LERAP Category B [2]

Hazard classification and safety precautions
Hazard Dangerous for the environment
Transport code 9
Packaging group III

FOR FULL CONDITIONS OF USE ALWAYS READ THE PRODUCT LABEL

UN Number 3082
Risk phrases R50 [2]; R51 [1]; R53a [1, 2]
Operator protection A, H [2]; U05a [1]; U08, U20b [2]
Environmental protection E15a, E16a [2]; E15b [1]; E17b [2] (40 m); E34, E38 [1, 2]
Storage and disposal D01, D05 [1]; D02, D10b, D12a [1, 2]

408   spirodiclofen

A tetronic acid which inhibits lipid biosynthesis
IRAC mode of action code: 23

Products

Envidor	Bayer CropScience	240 g/l	SC	13947

Uses
- Mussel scale in *apples, pears*
- Pear sucker in *pears*
- Red spider mites in *apples, pears*
- Rust mite in *apples, pears*
- Spider mites in *hops* (off-label), *ornamental plant production* (off-label), *protected ornamentals* (off-label), *protected strawberries* (off-label), *soft fruit* (off-label), *strawberries* (off-label)
- Two-spotted spider mite in *apples, pears, protected strawberries* (off-label), *strawberries* (off-label)

Extension of Authorisation for Minor Use (EAMUs)
- *hops* 20093372 expires 31 Jul 2013, 20120253
- *ornamental plant production* 20093366 expires 31 Jul 2013, 20120254
- *protected ornamentals* 20093366 expires 31 Jul 2013, 20120254
- *protected strawberries* 20093371 expires 31 Jul 2013, 20120255
- *soft fruit* 20093372 expires 31 Jul 2013, 20120253
- *strawberries* 20093371 expires 31 Jul 2013, 20120255

Approval information
- Spirodiclofen included in Annex 1 under EC Regulation 1107/2009

Restrictions
- Consult processor on crops grown for processing or for cider production

Environmental safety
- Broadcast air-assisted LERAP (18 m)

Hazard classification and safety precautions
Hazard Harmful
UN Number N/C
Risk phrases R40, R43, R52, R53a
Operator protection A, H; U05a, U20c
Environmental protection E12c, E12f, E15b, E22c, E34, E38; E17b (18 m)
Storage and disposal D01, D02, D09a, D10c, D12a
Medical advice M05a

409   spiromesifen

A tetronic acid insecticide for protected tomatoes
IRAC mode of action code: 23

Products

Oberon	Certis	240 g/l	SC	11819

Uses
- Red spider mites in *protected ornamentals* (off-label)

SEE SECTION 3 FOR PRODUCTS ALSO REGISTERED

- Spider mites in **protected aubergines** *(off-label)*, **protected cayenne peppers** *(off-label)*, **protected cucumbers** *(off-label)*, **protected gherkins** *(off-label)*, **protected peppers** *(off-label)*, **protected strawberries** *(off-label)*
- Whitefly in **inert substrate tomatoes**, **nft tomatoes**, **protected aubergines** *(off-label)*, **protected cayenne peppers** *(off-label)*, **protected cucumbers** *(off-label)*, **protected gherkins** *(off-label)*, **protected ornamentals** *(off-label)*, **protected peppers** *(off-label)*, **protected strawberries** *(off-label)*

Extension of Authorisation for Minor Use (EAMUs)
- **protected aubergines** *20050959, 20063645*
- **protected cayenne peppers** *20062149*
- **protected cucumbers** *20050958*
- **protected gherkins** *20050958*
- **protected ornamentals** *20041718*
- **protected peppers** *20062149*
- **protected strawberries** *20050957*

Approval information
- Spiromesifen is awaiting inclusion in Annex 1 under EC Regulation 1107/2009

Efficacy guidance
- Apply to run off and ensure that all sides of the crop are covered
- Apply when infestation first observed and repeat 7-10 d later if necessary
- Always follow guidelines for preventing and managing insect resistance. See Section 5 for more information

Restrictions
- Maximum number of treatments 2 per cropping cycle
- Do not use on any outdoor crops
- Product should be used in rotation with compounds with different modes of action

Crop-specific information
- HI 3 d for tomatoes

Environmental safety
- Dangerous for the environment
- Very toxic to aquatic organisms

Hazard classification and safety precautions
    Hazard Irritant, Dangerous for the environment
    Transport code 9
    Packaging group III
    UN Number 3082
    Risk phrases R43, R50, R53a
    Operator protection A, H; U05a, U14, U20b
    Environmental protection E15a, E34, E38
    Storage and disposal D01, D02, D05, D09a, D11a, D12a

410   spirotetramat

Tetramic acid derivative that acts on lipid synthesis
IRAC mode of action code: 23

Products

Movento	Bayer CropScience	150 g/l	OD	14446

Uses
- Aphids in **blackcurrants** *(off-label)*, **blueberries** *(off-label)*, **broccoli**, **brussels sprouts**, **cabbages**, **calabrese**, **cauliflowers**, **chard** *(off-label)*, **chinese cabbage** *(off-label)*, **choi sum** *(off-label)*, **cress** *(off-label)*, **endives** *(off-label)*, **frise** *(off-label)*, **gooseberries** *(off-label)*, **hops** *(off-label)*, **lamb's lettuce** *(off-label)*, **leaf brassicas** *(off-label)*, **lettuce**, **mustard** *(off-label)*, **pak**

choi *(off-label)*, **protected chard** *(off-label)*, **protected cress** *(off-label)*, **protected endives** *(off-label)*, **protected frise** *(off-label)*, **protected lamb's lettuce** *(off-label)*, **protected leaf brassicas** *(off-label)*, **protected rocket** *(off-label)*, **protected scarole** *(off-label)*, **protected spinach** *(off-label)*, **redcurrants** *(off-label)*, **rocket** *(off-label)*, **scarole** *(off-label)*, **spinach** *(off-label)*, **tatsoi** *(off-label)*, **whitecurrants** *(off-label)*, **wine grapes** *(off-label)*

- Damson-hop aphid in **hops** *(off-label)*
- Insect pests in **forest nurseries** *(off-label)*, **ornamental plant production** *(off-label)*
- White tip in **chard** *(off-label)*
- Whitefly in **broccoli**, **brussels sprouts**, **cabbages**, **calabrese**, **cauliflowers**, **chinese cabbage** *(off-label)*, **choi sum** *(off-label)*, **cress** *(off-label)*, **endives** *(off-label)*, **frise** *(off-label)*, **lamb's lettuce** *(off-label)*, **leaf brassicas** *(off-label)*, **mustard** *(off-label)*, **pak choi** *(off-label)*, **protected chard** *(off-label)*, **protected cress** *(off-label)*, **protected endives** *(off-label)*, **protected frise** *(off-label)*, **protected lamb's lettuce** *(off-label)*, **protected leaf brassicas** *(off-label)*, **protected rocket** *(off-label)*, **protected scarole** *(off-label)*, **protected spinach** *(off-label)*, **rocket** *(off-label)*, **scarole** *(off-label)*, **spinach** *(off-label)*, **tatsoi** *(off-label)*

Extension of Authorisation for Minor Use (EAMUs)

- **blackcurrants** *20121401*
- **blueberries** *20121401*
- **chard** *20102410*
- **chinese cabbage** *20101095*
- **choi sum** *20101095*
- **cress** *20102410*
- **endives** *20102410*
- **forest nurseries** *20111987*
- **frise** *20102410*
- **gooseberries** *20121401*
- **hops** *20102117*
- **lamb's lettuce** *20102410*
- **leaf brassicas** *20102410*
- **mustard** *20102410*
- **ornamental plant production** *20111987*
- **pak choi** *20101095*
- **protected chard** *20102410*
- **protected cress** *20102410*
- **protected endives** *20102410*
- **protected frise** *20102410*
- **protected lamb's lettuce** *20102410*
- **protected leaf brassicas** *20102410*
- **protected rocket** *20102410*
- **protected scarole** *20102410*
- **protected spinach** *20102410*
- **redcurrants** *20121401*
- **rocket** *20102410*
- **scarole** *20102410*
- **spinach** *20102410*
- **tatsoi** *20101095*
- **whitecurrants** *20121401*
- **wine grapes** *20113085*

Approval information

- Spirotetramat has not yet been added to Annex 1 under EC Regulation 1107/2009

Hazard classification and safety precautions

Hazard Irritant, Dangerous for the environment
Transport code 9
Packaging group III
UN Number 3082
Risk phrases R43, R51, R53a
Operator protection A; U05a, U14, U20b

SEE SECTION 3 FOR PRODUCTS ALSO REGISTERED

Environmental protection E15b, E22c, E34, E38
Storage and disposal D02, D09a, D10c, D12a
Medical advice M03

411 spiroxamine

A spiroketal amine fungicide for cereals
FRAC mode of action code: 5

See also bixafen + prothioconazole + spiroxamine
*   prothioconazole + spiroxamine*
*   prothioconazole + spiroxamine + tebuconazole*

Products

Torch	Bayer CropScience	500 g/l	EW	11258

Uses
- Brown rust in **spring barley**, **spring rye**, **spring wheat**, **winter barley**, **winter rye**, **winter wheat**
- Powdery mildew in **spring barley**, **spring rye**, **spring wheat**, **winter barley**, **winter rye**, **winter wheat**
- Rhynchosporium in **spring barley** *(reduction)*, **spring rye** *(reduction)*, **winter barley** *(reduction)*, **winter rye** *(reduction)*
- Yellow rust in **spring barley**, **spring rye**, **spring wheat**, **winter barley**, **winter rye**, **winter wheat**

Approval information
- Spiroxamine included in Annex I under EC Regulation 1107/2009
- Accepted by BBPA for use on malting barley

Efficacy guidance
- For best results treat at an early stage of disease development before infection spreads to new growth
- To reduce the risk of development of resistance avoid repeat treatments on diseases such as powdery mildew. If necessary tank mix or alternate with other non-morpholine fungicides

Restrictions
- Maximum total dose equivalent to two full dose treatments

Crop-specific information
- Latest use: before caryopsis watery ripe (GS 71) for wheat, rye; ear emergence complete (GS 59) for barley

Environmental safety
- Dangerous for the environment
- Very toxic to aquatic organisms
- LERAP Category B

Hazard classification and safety precautions
Hazard Harmful, Dangerous for the environment
Risk phrases R20, R22a, R38, R41, R43, R50, R53a
Operator protection A, C, H; U05a, U11, U13, U14, U15, U20a
Environmental protection E13b, E16a, E34, E38
Storage and disposal D01, D02, D05, D09a, D10b, D12a
Medical advice M03

412   spiroxamine + tebuconazole

A broad spectrum systemic fungicide mixture for cereals
FRAC mode of action code: 5 + 3

See also tebuconazole

Products

Thyme	Interfarm	250:133 g/l	EW	14745

Uses

- Brown rust in **spring barley, spring rye, spring wheat, winter barley, winter rye, winter wheat**
- Disease control in **grass seed crops** *(off-label)*
- Fusarium ear blight in **spring wheat, winter wheat**
- Glume blotch in **spring wheat, winter wheat**
- Net blotch in **spring barley, winter barley**
- Powdery mildew in **spring barley, spring rye, spring wheat, winter barley, winter rye, winter wheat**
- Rhynchosporium in **spring barley, winter barley**
- Septoria leaf blotch in **spring wheat, winter wheat**
- Sooty moulds in **spring wheat, winter wheat**
- Yellow rust in **spring barley, spring rye, spring wheat, winter barley, winter rye, winter wheat**

Extension of Authorisation for Minor Use (EAMUs)

- **grass seed crops** *20093172*

Approval information

- Spiroxamine and tebuconazole included in Annex I under EC Regulation 1107/2009
- Accepted by BBPA for use on malting barley

Efficacy guidance

- Best results achieved from treatment at early stage of disease development before infection spreads to new growth
- To protect flag leaf and ear from Septoria apply from flag leaf emergence to ear fully emerged (GS 37-59). Earlier treatment may be necessary where there is high disease risk
- Control of rusts, powdery mildew, leaf blotch and net blotch may require second treatment 2-3 wk later
- Tebuconazole is a DMI fungicide. Resistance to some DMI fungicides has been identified in Septoria leaf blotch which may seriously affect performance of some products. For further advice contact a specialist advisor and visit the Fungicide Resistance Action Group (FRAG)-UK website

Restrictions

- Maximum total dose equivalent to two full dose treatments
- Do not use on durum wheat

Crop-specific information

- Latest use: before caryopsis watery ripe (GS 71) for rye, wheat; ear emergence complete (GS 59) for barley
- Some transient leaf speckling may occur on wheat but this has not been shown to reduce yield reponse or disease control

Environmental safety

- Dangerous for the environment
- Very toxic to aquatic organisms
- Dangerous to fish or other aquatic life. Do not contaminate surface waters or ditches with chemical or used container
- LERAP Category B

SEE SECTION 3 FOR PRODUCTS ALSO REGISTERED

Hazard classification and safety precautions

Hazard Harmful, Dangerous for the environment
Transport code 9
Packaging group III
UN Number 3082
Risk phrases R20, R22a, R38, R41, R43, R50, R53a
Operator protection A, C, H; U05a, U11, U13, U14, U15, U20b
Environmental protection E13b, E16a, E16b, E34, E38
Storage and disposal D01, D02, D05, D09a, D10b, D12a
Medical advice M03

413   sulfosulfuron

A sulfonylurea herbicide for grass and broad-leaved weed control in winter wheat
HRAC mode of action code: B

See also glyphosate + sulfosulfuron

Products

Monitor	Monsanto	80% w/w	WG	12236

Uses

- Annual dicotyledons in *game cover* (off-label), *grass seed crops* (off-label), *spring rye* (off-label), *triticale* (off-label), *winter rye* (off-label)
- Brome grasses in *winter wheat* (moderate control of barren brome)
- Chickweed in *winter wheat*
- Cleavers in *winter wheat*
- Couch in *winter wheat* (moderate control only)
- Grass weeds in *game cover* (off-label), *grass seed crops* (off-label), *spring rye* (off-label), *triticale* (off-label), *winter rye* (off-label)
- Loose silky bent in *winter wheat*
- Mayweeds in *winter wheat*
- Onion couch in *winter wheat* (moderate control only)

Extension of Authorisation for Minor Use (EAMUs)

- *game cover* 20082861
- *grass seed crops* 20081279
- *spring rye* 20081279
- *triticale* 20081279
- *winter rye* 20081279

Approval information

- Sulfosulfuron included in Annex I under EC Regulation 1107/2009

Efficacy guidance

- For best results treat in early spring when annual weeds are small and growing actively. Avoid treatment when weeds are dormant for any reason
- Best control of onion couch is achieved when the weed has more than two leaves. Effects on bulbils or on growth in the following yr have not been examined
- An extended period of dry weather before or after treatment may result in reduced control
- The addition of a recommended surfactant is essential for full activity
- Specific follow-up treatments may be needed for complete control of some weeds
- Use only where competitively damaging weed populations have emerged otherwise yield may be reduced
- Sulfosulfuron is a member of the ALS-inhibitor group of herbicides. To avoid the build up of resistance do not use any product containing an ALS-inhibitor herbicide with claims for control of grass weeds more than once on any crop
- Use these products as part of a resistance management strategy that includes cultural methods of control and does not use ALS inhibitors as the sole chemical method of grass weed control

FOR FULL CONDITIONS OF USE ALWAYS READ THE PRODUCT LABEL

Restrictions
- Maximum total dose equivalent to one treatment at full dose
- Do not treat crops under stress
- Apply only after 1 Feb and from 3 expanded leaf stage
- Do not treat durum wheat or any undersown wheat crop
- Do not use in mixture or in sequence with any other sulfonyl urea herbicide on the same crop

Crop-specific information
- Latest use: flag leaf ligule just visible for winter wheat (GS 39)

Following crops guidance
- In the autumn following treatment winter wheat, winter rye, winter oats, triticale, winter oilseed rape, winter peas or winter field beans may be sown on any soil, and winter barley on soils with less than 60% sand
- In the next spring following application crops of wheat, barley, oats, maize, peas, beans, linseed, oilseed rape, potatoes or grass may be sown
- In the second autumn following application winter linseed may be sown, and winter barley on soils with more than 60% sand
- Sugar beet or any other crop not mentioned above must not be drilled until the second spring following application
- Where winter oilseed rape is to be sown in the autumn following treatment soil cultivation to a minimum of 10 cm is recommended

Environmental safety
- Dangerous for the environment
- Very toxic to aquatic organisms
- Extremely dangerous to fish or other aquatic life. Do not contaminate surface waters or ditches with chemical or used container
- Take extreme care to avoid drift onto broad leaved plants or other crops, or onto ponds, waterways or ditches.
- Follow detailed label instructions for cleaning the sprayer to avoid damage to sensitive crops during subsequent use
- LERAP Category B

Hazard classification and safety precautions
Hazard Dangerous for the environment
UN Number N/C
Risk phrases R50, R53a
Operator protection U20b
Environmental protection E15a, E16a, E16b, E34, E38
Storage and disposal D09a, D11a, D12a

414   sulphur

A broad-spectrum inorganic protectant fungicide, foliar feed and acaricide
FRAC mode of action code: M2

Products

1	Cosavet DF	Nufarm UK	80% w/w	WG	11477
2	Headland Sulphur	Headland	800 g/l	SC	12879
3	Kumulus DF	BASF	80% w/w	WG	04707
4	Microthiol Special	United Phosphorus	80% w/w	MG	06268
5	Sulphur Flowable	United Phosphorus	800 g/l	SC	07526
6	Thiovit Jet	Syngenta	80% w/w	WG	10928

Uses
- Disease control/foliar feed in *bilberries* (off-label), *blueberries* (off-label), *cranberries* (off-label), *redcurrants* (off-label), *whitecurrants* (off-label) [3]
- Foliar feed in *all top fruit*, *bush fruit*, *cane fruit*, *edible brassicas*, *managed amenity turf*, *potatoes*, *spring oats*, *sugar beet*, *swedes*, *turnips*, *vegetables*, *winter oats* [4]; *permanent*

SEE SECTION 3 FOR PRODUCTS ALSO REGISTERED

grassland, **spring oilseed rape** [4, 6]; **spring barley**, **spring wheat**, **winter barley**, **winter oilseed rape**, **winter wheat** [3, 4, 6]
- Gall mite in **blackcurrants** [1, 3]
- Powdery mildew in **apples**, **strawberries** [1-3, 5]; **aubergines** *(off-label)*, **combining peas** *(off-label)*, **protected chilli peppers** *(off-label)*, **protected chives** *(off-label)*, **protected parsley** *(off-label)*, **vining peas** *(off-label)* [6]; **borage for oilseed production** *(off-label)*, **parsnips** *(off-label)*, **protected aubergines** *(off-label)*, **protected cayenne peppers** *(off-label)*, **protected cucumbers** *(off-label)*, **protected herbs (see appendix 6)** *(off-label)*, **protected peppers** *(off-label)*, **protected tomatoes** *(off-label)* [4, 6]; **fodder beet** *(off-label)* [2, 4, 6]; **gooseberries**, **wine grapes** [1, 3]; **grapevines**, **turnips** [4]; **grass seed crops** *(off-label)*, **spring barley**, **spring wheat**, **winter barley**, **winter wheat** [2, 4, 5]; **hops** [1-3, 5, 6]; **pears** [2, 5]; **quinces** *(off-label)* [5]; **spring oats**, **winter oats** [2, 4]; **spring oilseed rape**, **spring rye**, **triticale** *(off-label)*, **winter oilseed rape**, **winter rye** [2]; **sugar beet** [1-6]; **swedes** [2, 4-6]; **table grapes** [1]; **turnips** *(off-label)* [2, 5, 6]
- Scab in **apples** [1-3, 5]; **pears** [2, 5]; **quinces** *(off-label)* [2]

Extension of Authorisation for Minor Use (EAMUs)
- **aubergines** *20023652* [6]
- **bilberries** *20061042* [3]
- **blueberries** *20061042* [3]
- **borage for oilseed production** *20072027* [4], *20061719* [6]
- **combining peas** *20081361* [6]
- **cranberries** *20061042* [3]
- **fodder beet** *20062690* [2], *20072025* [4], *20061383* [6]
- **grass seed crops** *20062691* [2], *20061043* [4], *20061375* [5]
- **parsnips** *20072026* [4], *20023654* [6]
- **protected aubergines** *20072024* [4], *20111032* [6]
- **protected cayenne peppers** *20072024* [4], *20111032* [6]
- **protected chilli peppers** *20081360* [6]
- **protected chives** *20023652* [6]
- **protected cucumbers** *20072024* [4], *20023652* [6], *20111032* [6]
- **protected herbs (see appendix 6)** *20072024* [4], *20023652* [6], *20111032* [6]
- **protected parsley** *20023652* [6]
- **protected peppers** *20072024* [4], *20023652* [6], *20111032* [6]
- **protected tomatoes** *20072024* [4], *20023652* [6], *20111032* [6]
- **quinces** *20062688* [2], *20061374* [5]
- **redcurrants** *20061042* [3]
- **triticale** *20062691* [2]
- **turnips** *20062689* [2], *20061373* [5], *20061384* [6]
- **vining peas** *20081361* [6]
- **whitecurrants** *20061042* [3]

Approval information
- Sulphur is included in Annex 1 under EC Regulation 1107/2009
- Accepted by BBPA for use on malting barley and hops (before burr)

Efficacy guidance
- Apply when disease first appears and repeat 2-3 wk later. Details of application rates and timing vary with crop, disease and product. See label for information
- Sulphur acts as foliar feed as well as fungicide and with some crops product labels vary in whether treatment recommended for disease control or growth promotion
- In grassland best results obtained at least 2 wk before cutting for hay or silage, 3 wk before grazing
- Treatment unlikely to be effective if disease already established in crop

Restrictions
- Maximum number of treatments normally 2 per crop for grassland, sugar beet, parsnips, swedes, hops, protected herbs; 3 per yr for blackcurrants, gooseberries; 4 per crop on apples, pears but labels vary

FOR FULL CONDITIONS OF USE ALWAYS READ THE PRODUCT LABEL

- Do not use on sulphur-shy apples (Beauty of Bath, Belle de Boskoop, Cox's Orange Pippin, Lanes Prince Albert, Lord Derby, Newton Wonder, Rival, Stirling Castle) or pears (Doyenne du Comice)
- Do not use on gooseberry cultivars Careless, Early Sulphur, Golden Drop, Leveller, Lord Derby, Roaring Lion, or Yellow Rough
- Do not use on apples or gooseberries when young, under stress or if frost imminent
- Do not use on fruit for processing, on grapevines during flowering or near harvest on grapes for wine-making
- Do not use on hops at or after burr stage
- Do not spray top or soft fruit with oil or within 30 d of an oil-containing spray

Crop-specific information
- Latest use: before burr stage for hops; before end Sep for parsnips, swedes, sugar beet; fruit swell for gooseberries; milky ripe stage for cereals
- HI cutting grass for hay or silage 2 wk; grazing grassland 3 wk

Environmental safety
- Sulphur products are attractive to livestock and must be kept out of their reach
- Do not empty into drains

Hazard classification and safety precautions
   UN Number N/C
   Operator protection A, H [6]; U02a, U08, U14 [1, 6]; U05a [6]; U20a [2]; U20b [4, 5]; U20c [1, 3, 6]
   Environmental protection E15a [1-6]; E34 [2]
   Storage and disposal D01 [4, 6]; D02 [6]; D05 [5, 6]; D09a [1-6]; D10b [2]; D10c [5]; D11a [1, 3, 4, 6]

415   tau-fluvalinate

A contact pyrethroid insecticide for cereals and oilseed rape
IRAC mode of action code: 3

Products

1	Greencrop Malin	Greencrop	240 g/l	EW	11787
2	Klartan	Makhteshim	240 g/l	EW	11074
3	Mavrik	Makhteshim	240 g/l	EW	10612
4	Revolt	Makhteshim	240 g/l	EW	13383

Uses
- Aphids in **borage for oilseed production** *(off-label)* [3]; **durum wheat** *(off-label)*, **evening primrose** *(off-label)*, **grass seed crops** *(off-label)*, **honesty** *(off-label)*, **mustard** *(off-label)*, **spring linseed** *(off-label)*, **spring rye** *(off-label)*, **triticale** *(off-label)*, **winter rye** *(off-label)* [2, 3]; **spring barley, spring oilseed rape, spring wheat, winter barley, winter oilseed rape, winter wheat** [1-4]
- Barley yellow dwarf virus vectors in **winter barley, winter wheat** [1-4]
- Cabbage stem flea beetle in **winter oilseed rape** [2-4]
- Insect pests in **grass seed crops** *(off-label)*, **mustard** *(off-label)* [1]
- Pollen beetle in **spring oilseed rape, winter oilseed rape** [1-4]

Extension of Authorisation for Minor Use (EAMUs)
- **borage for oilseed production** *20061366* [3]
- **durum wheat** *20061369* [2], *20061365* [3]
- **evening primrose** *20061370* [2], *20061366* [3]
- **grass seed crops** *20061368* [1], *20061369* [2], *20061365* [3]
- **honesty** *20061370* [2], *20061366* [3]
- **mustard** *20061367* [1], *20061370* [2], *20061366* [3]
- **spring linseed** *20061370* [2], *20061366* [3]
- **spring rye** *20061369* [2], *20061365* [3]
- **triticale** *20061369* [2], *20061365* [3]
- **winter rye** *20061369* [2], *20061365* [3]

SEE SECTION 3 FOR PRODUCTS ALSO REGISTERED

Approval information
- Tau-fluvalinate included in Annex 1 under EC Regulation 1107/2009
- Accepted by BBPA for use on malting barley

Efficacy guidance
- For BYDV control on winter cereals follow local warnings or spray high risk crops in mid-Oct and make repeat application in late autumn/early winter if aphid activity persists
- For summer aphid control on cereals spray once when aphids present on two thirds of ears and increasing
- On oilseed rape treat peach potato aphids in autumn in response to local warning and repeat if necessary
- Best control of pollen beetle in oilseed rape obtained from treatment at green to yellow bud stage and repeat if necessary
- Good spray cover of target essential for best results
- Do not mix with boron or products containing boron [3]

Restrictions
- Maximum total dose equivalent to two full dose treatments on oilseed rape. See label for dose rates on cereals
- A minimum of 14 d must elapse between applications to cereals

Crop-specific information
- Latest use: before caryopsis watery ripe (GS 71) for barley; before flowering for oilseed rape; before kernel medium milk (GS 75) for wheat

Environmental safety
- Dangerous for the environment
- Very toxic to aquatic organisms
- High risk to non-target insects or other arthropods. Do not spray within 6 m of the field boundary [1-3]
- Avoid spraying oilseed rape within 6 m of field boundary to reduce effects on certain non-target species or other arthropods
- Must not be applied to cereals if any product containing a pyrethroid insecticide or dimethoate has been sprayed after the start of ear emergence (GS 51)
- LERAP Category A

Hazard classification and safety precautions
Hazard Irritant [1]; Dangerous for the environment [1-4]
Transport code 9
Packaging group III
UN Number 3082
Risk phrases R36, R38, R51 [1]; R50 [2-4]; R53a [1-4]
Operator protection A, C, H; U05a, U10, U11, U19a, U20a
Environmental protection E15a, E16c, E16d, E22a
Storage and disposal D01, D02, D05, D10c [1-4]; D09a [2-4]; D09b [1]

416   tebuconazole

A systemic triazole fungicide for cereals and other field crops
FRAC mode of action code: 3

See also bixafen + prothioconazole + tebuconazole
carbendazim + tebuconazole
chlorothalonil + tebuconazole
clothianidin + prothioconazole + tebuconazole + triazoxide
difenoconazole + fludioxonil + tebuconazole
fenpropidin + prochloraz + tebuconazole
fenpropidin + propiconazole + tebuconazole
fenpropidin + tebuconazole
fluopyram + prothioconazole + tebuconazole
fluoxastrobin + prothioconazole + tebuconazole
imidacloprid + tebuconazole + triazoxide
prochloraz + proquinazid + tebuconazole
prochloraz + tebuconazole
propiconazole + tebuconazole
prothioconazole + spiroxamine + tebuconazole
prothioconazole + tebuconazole
prothioconazole + tebuconazole + triazoxide
spiroxamine + tebuconazole

Products

1	Chani	AgChem Access	250 g/l	EW	13325
2	Clayton Tebucon EW	Clayton	250 g/l	EW	14824
3	Deacon	Makhteshim	200 g/l	EW	14270
4	EA Tebuzole	European Ag	250 g/l	EW	15000
5	Eiger	ChemSource	250 g/l	EW	15058
6	Fathom	Headland	200 g/l	EC	15328
7	Fezan	Sipcam	250 g/l	EW	13681
8	Folicur	Bayer CropScience	250 g/l	EW	11278
9	Gizmo	Nufarm UK	60 g/l	FS	14150
10	Greencrop Tabloid	Greencrop	250 g/l	EW	11969
11	Mitre	Makhteshim	200 g/l	EW	12901
12	Orian	ChemSource	200 g/l	EW	14798
13	Orius 20 EW	Makhteshim	200 g/l	EW	12311
14	Riza	Headland	250 g/l	EW	12696
15	Tebcon	Goldengrass	250 g/l	EW	14300
16	Tebucon 250	Goldengrass	250 g/l	EW	14789
17	Tebucur 250	Globachem	250 g/l	EW	13975

Uses

- Alternaria in **brussels sprouts** [12, 13]; **cabbages, spring oilseed rape, winter oilseed rape** [1-6, 8, 10-17]; **carrots** [1-6, 8, 10, 14-17]; **horseradish** [1-3, 5, 6, 8, 10, 15, 16]; **horseradish** *(off-label)* [14]; **parsnips** [4, 5, 14, 17]
- Blight in **blackberries** *(off-label)*, **raspberries** *(off-label)*, **rubus hybrids** *(off-label)* [8, 11, 13]
- Botrytis in **daffodils** *(off-label)* [13]; **daffodils** *(off-label - for galanthamine production)* [8, 11]; **daffodils grown for galanthamine production** *(off-label)* [3]; **linseed** *(reduction)* [1-3, 6, 8, 10, 15, 16]
- Brown rust in **spring barley, spring rye, spring wheat, winter barley, winter rye, winter wheat** [1-8, 10-17]; **spring oats, winter oats** [6]; **triticale** *(off-label)* [13]
- Cane blight in **blackberries** *(off-label)*, **raspberries** *(off-label)*, **rubus hybrids** *(off-label)* [3, 8, 11, 13]
- Canker in **apples** *(off-label)*, **crab apples** *(off-label)*, **pears** *(off-label)*, **quinces** *(off-label)* [3, 6, 8, 11, 13, 14]; **chestnuts** *(off-label)*, **cob nuts** *(off-label)*, **walnuts** *(off-label)* [3, 8, 11, 13, 14]
- Chocolate spot in **spring field beans, winter field beans** [1-6, 8, 10-17]
- Crown rust in **spring oats, winter oats** [4, 5, 14, 17]; **spring oats** *(reduction)*, **winter oats** *(reduction)* [1, 2, 8, 10, 15, 16]

SEE SECTION 3 FOR PRODUCTS ALSO REGISTERED

- Disease control in **borage for oilseed production** *(off-label)*, **canary flower (echium spp.)** *(off-label)*, **evening primrose** *(off-label)*, **honesty** *(off-label)*, **mustard** *(off-label)* [3, 6, 13, 17]; **grass seed crops** *(off-label)*, **mallow (althaea spp.)** *(off-label)*, **parsley root** *(off-label)* [3, 13, 17]; **kohlrabi** *(off-label)* [3, 13]; **triticale** *(off-label)* [3, 6, 17]
- Foliar disease control in **borage** *(off-label)*, **grass seed crops** *(off-label)* [8, 11]; **borage for oilseed production** *(off-label)*, **horseradish** *(off-label)* [14]; **canary flower (echium spp.)** *(off-label)*, **evening primrose** *(off-label)*, **honesty** *(off-label)*, **mallow (althaea spp.)** *(off-label)*, **mustard** *(off-label)*, **parsley root** *(off-label)*, **triticale** *(off-label)* [8, 11, 14]
- Fungus diseases in **kohlrabi** *(off-label)* [8, 11]
- Fusarium ear blight in **spring oats**, **winter oats** [6]; **spring wheat**, **winter wheat** [1-6, 8, 10-17]; **triticale** *(off-label)* [13]
- Glume blotch in **spring oats**, **winter oats** [6]; **spring wheat**, **winter wheat** [1-8, 10-17]; **triticale** *(off-label)* [13]
- Leaf stripe in **spring barley** [9]
- Light leaf spot in **brussels sprouts** [12, 13]; **cabbages**, **spring oilseed rape**, **winter oilseed rape** [1-6, 8, 10-17]
- Lodging control in **spring oilseed rape**, **winter oilseed rape** [1-3, 6, 8, 10-13, 15, 16]
- Loose smut in **spring barley**, **winter barley** [9]
- Net blotch in **spring barley**, **winter barley** [1-8, 10-17]
- Phoma in **spring oilseed rape**, **winter oilseed rape** [1-6, 8, 10-17]
- Powdery mildew in **brussels sprouts** [12, 13]; **cabbages**, **swedes**, **turnips** [1-6, 8, 10-17]; **carrots**, **linseed** [1-6, 8, 10, 14-17]; **chives** *(off-label)* [3, 8, 11, 13, 14]; **horseradish** [5]; **horseradish** *(off-label)* [14]; **parsnips** [1-6, 8, 14-17]; **spring barley**, **spring rye**, **spring wheat**, **winter barley**, **winter rye**, **winter wheat** [1-8, 10-17]; **spring oats**, **winter oats** [1, 2, 4-8, 10, 14-17]; **triticale** *(off-label)* [13]
- Rhynchosporium in **spring barley**, **spring rye**, **winter barley**, **winter rye** [1-8, 10-17]
- Ring spot in **broccoli** *(off-label)*, **cauliflowers** *(off-label)*, **chinese cabbage** *(off-label)* [3, 6, 8, 11, 13, 14]; **brussels sprouts** [12, 13]; **cabbages** [1-6, 8, 10-17]; **calabrese** *(off-label)* [6, 8, 11, 13, 14]; **choi sum** *(off-label)*, **pak choi** *(off-label)* [3, 8, 11, 13, 14]; **kohlrabi** *(off-label)* [14]; **spring oilseed rape** *(reduction)*, **winter oilseed rape** *(reduction)* [1-3, 6, 8, 10-13, 15, 16]
- Rust in **borage for oilseed production** *(off-label)*, **canary flower (echium spp.)** *(off-label)*, **evening primrose** *(off-label)*, **grass seed crops** *(off-label)*, **honesty** *(off-label)*, **mallow (althaea spp.)** *(off-label)*, **mustard** *(off-label)*, **parsley root** *(off-label)* [17]; **broad beans** *(off-label)*, **chives** *(off-label)* [3, 8, 11, 13, 14]; **dwarf beans** *(off-label)*, **french beans** *(off-label)*, **runner beans** *(off-label)* [3, 6, 8, 9, 11, 13, 14]; **leeks**, **spring field beans**, **winter field beans** [1-6, 8, 10-17]; **triticale** *(off-label)* [6, 17]
- Sclerotinia in **carrots** [1-6, 8, 10, 14-17]; **horseradish** [5]; **horseradish** *(off-label)* [14]; **parsnips** [4, 5, 14, 17]
- Sclerotinia stem rot in **spring oilseed rape**, **winter oilseed rape** [1-6, 8, 10-17]
- Septoria leaf blotch in **spring oats**, **winter oats** [6]; **spring wheat**, **winter wheat** [1-8, 10-17]; **triticale** *(off-label)* [13]
- Sooty moulds in **spring oats**, **winter oats** [6]; **spring wheat**, **winter wheat** [1-6, 8, 10-17]
- Stem canker in **spring oilseed rape**, **winter oilseed rape** [1-6, 8, 10-17]
- White rot in **bulb onions** *(off-label)*, **garlic** *(off-label)*, **salad onions** *(off-label)*, **shallots** *(off-label)* [3, 6, 8, 9, 11, 13, 14]; **onion sets** *(off-label)* [3, 13, 14]
- Yellow rust in **spring barley**, **spring rye**, **spring wheat**, **winter barley**, **winter rye**, **winter wheat** [1-8, 10-17]; **spring oats**, **winter oats** [6]; **triticale** *(off-label)* [13]

Extension of Authorisation for Minor Use (EAMUs)

- **apples** *20090147* [3], *20112097* [6], *20082159* [8], *20071368* [11], *20071329* [13], *20070554* [14]
- **blackberries** *20091407* [3], *20082160* [8], *20091406* [11], *20091399* [13]
- **borage** *20062694* [8], *20071359* [11]
- **borage for oilseed production** *20090113* [3], *20122123* [6], *20071337* [13], *20070558* [14], *20122270* [17]
- **broad beans** *20090089* [3], *20031878* [8], *20071364* [11], *20071338* [13], *20070549* [14]
- **broccoli** *20090091* [3], *20122122* [6], *20031874* [8], *20071371* [11], *20071326* [13], *20070546* [14]
- **bulb onions** *20090090* [3], *20122121* [6], *20031877 expires 31 Dec 2013* [8], *20031879* [8], *20101121* [9], *20071365* [11], *20071369* [11], *20071327* [13], *20070550* [14]

FOR FULL CONDITIONS OF USE ALWAYS READ THE PRODUCT LABEL

- **calabrese** *20122122* [6], *20031874* [8], *20071371* [11], *20071326* [13], *20070546* [14]
- **canary flower (echium spp.)** *20090113* [3], *20122123* [6], *20062694* [8], *20071359* [11], *20071337* [13], *20070558* [14], *20122270* [17]
- **cauliflowers** *20090091* [3], *20122122* [6], *20031874* [8], *20071371* [11], *20071326* [13], *20070546* [14]
- **chestnuts** *20090116* [3], *20082158* [8], *20071372* [11], *20071321* [13], *20070555* [14]
- **chinese cabbage** *20090091* [3], *20122122* [6], *20031874* [8], *20071371* [11], *20071326* [13], *20070546* [14]
- **chives** *20090088* [3], *20082084* [8], *20082085* [11], *20082087* [13], *20082088* [14]
- **choi sum** *20090145* [3], *20082249* [8], *20082250* [11], *20082251* [13], *20082252* [14]
- **cob nuts** *20090116* [3], *20082158* [8], *20071372* [11], *20071321* [13], *20070555* [14]
- **crab apples** *20090147* [3], *20112097* [6], *20082159* [8], *20071368* [11], *20071329* [13], *20070554* [14]
- **daffodils** *(for galanthamine production)* *20041516* [8], *(for galanthamine production)* *20071370* [11], *20071328* [13]
- **daffodils grown for galanthamine production** *20090095* [3]
- **dwarf beans** *20090094* [3], *20122120* [6], *20031880* [8], *20071357* [11], *20071320* [13], *20070551* [14]
- **evening primrose** *20090113* [3], *20122123* [6], *20062694* [8], *20071359* [11], *20071337* [13], *20070558* [14], *20122270* [17]
- **french beans** *20090094* [3], *20122120* [6], *20031880* [8], *20071357* [11], *20071320* [13], *20070551* [14]
- **garlic** *20090090* [3], *20122121* [6], *20031879* [8], *20101121* [9], *20071369* [11], *20071327* [13], *20070550* [14]
- **grass seed crops** *20090144* [3], *20081764* [8], *20071362* [11], *20071340* [13], *20122273* [17]
- **honesty** *20090113* [3], *20122123* [6], *20062694* [8], *20071359* [11], *20071337* [13], *20070558* [14], *20122270* [17]
- **horseradish** *20070556* [14]
- **kohlrabi** *20090093* [3], *20031881* [8], *20071360* [11], *20071334* [13], *20070552* [14]
- **mallow (althaea spp.)** *20090143* [3], *20062692* [8], *20071363* [11], *20071339* [13], *20070556* [14], *20122272* [17]
- **mustard** *20090113* [3], *20122123* [6], *20062694* [8], *20071359* [11], *20071337* [13], *20070558* [14], *20122270* [17]
- **onion sets** *20090096* [3], *20071336* [13], *20070548* [14]
- **pak choi** *20090145* [3], *20082249* [8], *20082250* [11], *20082251* [13], *20082252* [14]
- **parsley root** *20090143* [3], *20062692* [8], *20071363* [11], *20071339* [13], *20070556* [14], *20122272* [17]
- **pears** *20090147* [3], *20112097* [6], *20082159* [8], *20071368* [11], *20071329* [13], *20070554* [14]
- **quinces** *20090147* [3], *20112097* [6], *20082159* [8], *20071368* [11], *20071329* [13], *20070554* [14]
- **raspberries** *20091407* [3], *20082160* [8], *20091406* [11], *20091399* [13]
- **rubus hybrids** *20091407* [3], *20082160* [8], *20091406* [11], *20091399* [13]
- **runner beans** *20090094* [3], *20122120* [6], *20031880* [8], *20071357* [11], *20071320* [13], *20070551* [14]
- **salad onions** *20090092* [3], *20122119* [6], *20042408* [8], *20101121* [9], *20071361* [11], *20071322* [13], *20070553* [14]
- **shallots** *20090090* [3], *20122121* [6], *20031879* [8], *20101121* [9], *20071369* [11], *20071327* [13], *20070550* [14]
- **triticale** *20090148* [3], *20122118* [6], *20062693* [8], *20071366* [11], *20071335* [13], *20070557* [14], *20122271* [17]
- **walnuts** *20090116* [3], *20082158* [8], *20071372* [11], *20071321* [13], *20070555* [14]

Approval information
- Tebuconazole is included in Annex 1 under EC Regulation 1107/2009
- Accepted by BBPA for use on malting barley
- Approval expiry 31 Dec 2013 [16]

Efficacy guidance
- For best results apply at an early stage of disease development before infection spreads to new crop growth

SEE SECTION 3 FOR PRODUCTS ALSO REGISTERED

- To protect flag leaf and ear from Septoria diseases apply from flag leaf emergence to ear fully emerged (GS 37-59). Earlier application may be necessary where there is a high risk of infection
- Improved control of established cereal mildew can be obtained by tank mixture with fenpropimorph
- For light leaf spot control in oilseed rape apply in autumn/winter with a follow-up spray in spring/summer if required
- For control of most other diseases spray at first signs of infection with a follow-up spray 2-4 wk later if necessary. See label for details
- For disease control in cabbages a 3-spray programme at 21-28 d intervals will give good control
- Tebuconazole is a DMI fungicide. Resistance to some DMI fungicides has been identified in Septoria leaf blotch which may seriously affect performance of some products. For further advice contact a specialist advisor and visit the Fungicide Resistance Action Group (FRAG)-UK website

Restrictions
- Maximum total dose equivalent to 1 full dose treatment on linseed; 2 full dose treatments on wheat, barley, rye, oats, field beans, swedes, turnips, onions; 2.25 full dose treatments on cabbages; 2.5 full dose treatments on oilseed rape; 3 full dose treatments on leeks, parsnips, carrots
- Do not treat durum wheat
- Apply only to listed oat varieties (see label) and do not apply to oats in tank mixture
- Do not apply before swedes and turnips have a root diameter of 2.5 cm, or before heart formation in cabbages, or before button formation in Brussels sprouts
- Consult processor before use on crops for processing

Crop-specific information
- Latest use: before grain milky-ripe for cereals (GS 71); when most seed green-brown mottled for oilseed rape (GS 6,3); before brown capsule for linseed
- HI field beans, swedes, turnips, linseed, winter oats 35 d; market brassicas, carrots, horseradish, parsnips 21 d; leeks 14 d
- Some transient leaf speckling on wheat or leaf reddening/scorch on oats may occur but this has not been shown to reduce yield response to disease control

Environmental safety
- Dangerous for the environment
- Toxic to aquatic organisms
- LERAP Category B [6]

Hazard classification and safety precautions
Hazard Harmful [1, 2, 4, 5, 8-11, 14-17]; Irritant [3, 6, 7, 12, 13]; Dangerous for the environment [1, 2, 4, 5, 8, 10, 11, 14-17]
Transport code 9
Packaging group III
UN Number 3082
Risk phrases R20 [1, 2, 8, 11, 15, 16]; R22a, R41, R51 [1, 2, 4, 5, 8, 10, 11, 14-17]; R36 [3, 6, 7, 12, 13]; R38 [3-7, 10, 12-14, 17]; R52 [3, 6, 7, 9, 12, 13]; R53a [1-17]; R63 [9, 17]
Operator protection A, H [1-17]; C [1-8, 10-17]; U05a, U11, U20a [1-8, 10-17]; U14, U15 [3, 6, 7, 12, 13]; U19a [4, 5, 10, 14, 17]; U20b [9]
Environmental protection E03, E15b [9]; E15a, E34 [1-8, 10-17]; E16a [6]; E38 [1, 2, 8, 9, 11, 15, 16]
Storage and disposal D01, D02 [1-8, 10-17]; D05, D10b [1, 2, 4, 5, 8, 10, 11, 14-17]; D09a [1-17]; D11a [9]; D12a [1, 2, 8, 11, 15, 16]
Treated seed S01, S02, S03, S04a, S05, S07 [9]
Medical advice M03 [1, 2, 4, 5, 8, 10, 11, 14-17]

FOR FULL CONDITIONS OF USE ALWAYS READ THE PRODUCT LABEL

417   tebuconazole + triadimenol

A broad spectrum systemic fungicide for cereals
FRAC mode of action code: 3 + 3

See also triadimenol

Products

Veto F	Interfarm	225:75 g/l	EC	14748

Uses

- Brown rust in **spring barley, spring rye, spring wheat, winter barley, winter rye, winter wheat**
- Crown rust in **spring oats, winter oats**
- Fusarium ear blight in **spring wheat, winter wheat**
- Glume blotch in **spring wheat, winter wheat**
- Net blotch in **spring barley, winter barley**
- Powdery mildew in **spring barley, spring oats, spring rye, spring wheat, winter barley, winter oats, winter rye, winter wheat**
- Rhynchosporium in **spring barley, winter barley**
- Septoria leaf blotch in **spring wheat, winter wheat**
- Sooty moulds in **spring wheat, winter wheat**
- Yellow rust in **spring barley, spring rye, spring wheat, winter barley, winter rye, winter wheat**

Approval information

- Tebuconazole and triadimenol included in Annex I under EC Regulation 1107/2009
- Accepted by BBPA for use on malting barley

Efficacy guidance

- For best results apply at an early stage of disease development before infection spreads to new crop growth
- To protect flag leaf and ear from Septoria diseases apply from flag leaf emergence to ear fully emerged (GS 37-59). Earlier application may be necessary where there is a high risk of infection
- For control of rust, powdery mildew, leaf and net blotch apply at first signs of disease with a second application 2-3 wk later if necessary
- Tebuconazole is a DMI fungicide. Resistance to some DMI fungicides has been identified in Septoria leaf blotch which may seriously affect performance of some products. For further advice contact a specialist advisor and visit the Fungicide Resistance Action Group (FRAG)-UK website

Restrictions

- Maximum total dose equivalent to two full dose treatments
- Do not use on durum wheat

Crop-specific information

- Latest use: before grain milky-ripe (GS 71)
- Some transient leaf speckling may occur on wheat or leaf reddening/scorch on oats but this has not been shown to reduce yield response or disease control

Environmental safety

- Harmful to aquatic organisms
- Harmful to fish or other aquatic life. Do not contaminate surface waters or ditches with chemical or used container

Hazard classification and safety precautions

UN Number N/C
Risk phrases R52, R53a
Operator protection A, C, H; U05a, U09b, U20a
Environmental protection E13c
Storage and disposal D01, D02, D05, D09a, D10c
Medical advice M03

SEE SECTION 3 FOR PRODUCTS ALSO REGISTERED

418   tebuconazole + trifloxystrobin

A conazole and stobilurin fungicide mixture for wheat and vegetable crops
FRAC mode of action code: 3 + 11

See also trifloxystrobin

Products

1 Clayton Bestow	Clayton	200:100 g/l	SC	14996
2 Dedicate	Bayer Environ.	200:100 g/l	SC	13612
3 Mascot Fusion	Rigby Taylor	200:100 g/l	EC	14113
4 Nativo 75 WG	Bayer CropScience	50:25% w/w	WG	13057

Uses

- Alternaria in **broccoli, brussels sprouts, cabbages, calabrese, carrots, cauliflowers** [4]
- Anthracnose in **amenity grassland, managed amenity turf** [3]
- Dollar spot in **amenity grassland, managed amenity turf** [3]
- Foliar disease control in **horseradish** *(off-label)*, **parsley root** *(off-label)*, **parsnips** *(off-label)*, **salsify** *(off-label)* [4]
- Fusarium in **amenity grassland, managed amenity turf** [1, 2]
- Fusarium patch in **amenity grassland, managed amenity turf** [3]
- Light leaf spot in **broccoli, brussels sprouts, cabbages, calabrese, cauliflowers** [4]
- Melting out in **amenity grassland, managed amenity turf** [2]
- Phoma leaf spot in **broccoli, brussels sprouts, cabbages, calabrese, cauliflowers** [4]
- Powdery mildew in **broccoli, brussels sprouts, cabbages, calabrese, carrots, cauliflowers, wine grapes** *(off-label)* [4]
- Purple blotch in **leeks** [4]
- Red thread in **amenity grassland, managed amenity turf** [1-3]
- Ring spot in **broccoli, brussels sprouts, cabbages, calabrese, cauliflowers** [4]
- Rust in **amenity grassland, managed amenity turf** [2]; **leeks** [4]
- Sclerotinia rot in **carrots** [4]
- White blister in **broccoli, brussels sprouts, cabbages, calabrese, cauliflowers** [4]
- White tip in **leeks** [4]

Extension of Authorisation for Minor Use (EAMUs)

- **horseradish** *20064163* [4]
- **parsley root** *20064163* [4]
- **parsnips** *20064163* [4]
- **salsify** *20064163* [4]
- **wine grapes** *20090730* [4]

Approval information

- Tebuconazole and trifloxystrobin included in Annex I under EC Regulation 1107/2009

Efficacy guidance

- Best results obtained from treatment at early stages of disease development. Further treatment may be needed if disease attack is prolonged
- Applications to established infections of any disease are likely to be less effective
- Treatment may give some control of *Stemphyllium botryosum* and leaf blotch on leeks [4]
- Tebuconazole is a DMI fungicide. Resistance to some DMI fungicides has been identified in Septoria leaf blotch which may seriously affect performance of some products. For further advice contact a specialist advisor and visit the Fungicide Resistance Action Group (FRAG)-UK website
- Trifloxystrobin is a member of the QoI cross resistance group. Product should be used preventatively and not relied on for its curative potential
- Use product as part of an Integrated Crop Management strategy incorporating other methods of control, including where appropriate other fungicides with a different mode of action. Do not apply more than two foliar applications of QoI containing products to any cereal crop, broccoli, calabrese or cauliflower. Do not apply more than three applications to Brussels sprouts, cabbage, carrots or leeks

FOR FULL CONDITIONS OF USE ALWAYS READ THE PRODUCT LABEL

- There is a significant risk of widespread resistance occurring in *Septoria tritici* populations in UK. Failure to follow resistance management action may result in reduced levels of disease control
- Strains of wheat powdery mildew resistant to QoIs are common in the UK. Control of wheat mildew can only be relied on from the triazole component
- Where specific control of wheat mildew is required this should be achieved through a programme of measures including products recommended for the control of mildew that contain a fungicide from a different cross-resistance group and applied at a dose that will give robust control
- Do not apply to grass during drought conditions nor to frozen turf [2]

Restrictions
- Maximum number of treatments 2 per crop for broccoli, calabrese, cauliflowers; 3 per crop for Brussels sprouts, cabbages, carrots, leeks [4]
- In addition to the maximum number of treatments per crop a maximum of 3 applications may be applied to the same ground in one calendar year [4]
- Consult processor before use on vegetable crops for processing [4]

Crop-specific information
- Latest use: before milky ripe stage on winter wheat
- HI 35 d for wheat; 21 d for all other crops
- Performance against leaf spot diseases of brassicas, *Alternaria* leaf blight of carrots, and rust in leeks may be improved by mixing with an approved sticker/wetter [4]

Environmental safety
- Dangerous for the environment
- Very toxic to aquatic organisms

Hazard classification and safety precautions
Hazard Harmful, Dangerous for the environment
Transport code 9
Packaging group III
UN Number 3077 [4]; 3082 [1-3]
Risk phrases R37 [1-3]; R50, R53a, R63 [1-4]; R70 [1, 2]
Operator protection A [1-4]; H [1, 3, 4]; P [2]; U05a [1, 2, 4]; U09a [1, 2]; U09b [3, 4]; U19a [1-3]; U20b [1-4]
Environmental protection E15a, E34, E38
Storage and disposal D01, D02, D09a, D10b, D12a [1-4]; D05 [3]
Medical advice M03

419   tebufenpyrad

A pyrazole mitochondrial electron transport inhibitor (METI) aphicide and acaricide
IRAC mode of action code: 21

Products

| Masai | BASF | 20% w/w | WB | 13082 |

Uses
- Bud mite in **almonds** *(off-label)*, **chestnuts** *(off-label)*, **hazel nuts** *(off-label)*, **walnuts** *(off-label)*
- Damson-hop aphid in **hops**, **plums** *(off-label)*
- Gall mite in **bilberries** *(off-label)*, **blackberries** *(off-label)*, **blackcurrants** *(off-label)*, **blueberries** *(off-label)*, **cranberries** *(off-label)*, **gooseberries** *(off-label)*, **protected rubus hybrids** *(off-label)*, **raspberries** *(off-label)*, **redcurrants** *(off-label)*, **vaccinium spp.** *(off-label)*, **whitecurrants** *(off-label)*
- Red spider mites in **apples**, **blackberries** *(off-label)*, **pears**, **protected rubus hybrids** *(off-label)*, **raspberries** *(off-label)*
- Two-spotted spider mite in **hops**, **protected roses**, **strawberries**

Extension of Authorisation for Minor Use (EAMUs)
- *almonds* 20080133
- *bilberries* 20080131

SEE SECTION 3 FOR PRODUCTS ALSO REGISTERED

- *blackberries* 20100204
- *blackcurrants* 20080131
- *blueberries* 20080131
- *chestnuts* 20080133
- *cranberries* 20080131
- *gooseberries* 20080131
- *hazel nuts* 20080133
- *plums* 20080132
- *protected rubus hybrids* 20100204
- *raspberries* 20100204
- *redcurrants* 20080131
- *vaccinium spp.* 20080131
- *walnuts* 20080133
- *whitecurrants* 20080131

Approval information
- Tebufenpyrad included in Annex 1 under EC Regulation 1107/2009
- Accepted by BBPA for use on hops

Efficacy guidance
- Acts on eggs (except winter eggs) and all motile stages of spider mites up to adults
- Treat spider mites from 80% egg hatch but before mites become established
- For effective control total spray cover of the crop is required
- Product can be used in a programme to give season-long control of damson-hop aphids coupled with mite control
- Where aphids resistant to tebufenpyrad occur in hops control is unlikely to be satisfactory and repeat treatments may result in lower levels of control. Where possible use different active ingredients in a programme

Restrictions
- Maximum total dose equivalent to one full dose treatment on apples, pears, strawberries; 3 full dose treatments on hops
- Other mitochondrial electron transport inhibitor (METI) acaricides should not be applied to the same crop in the same calendar yr either separately or in mixture
- Do not treat apples before 90% petal fall
- Small-scale testing of rose varieties to establish tolerance recommended before use
- Inner liner of container must not be removed

Crop-specific information
- Latest use: end of burr stage for hops
- HI strawberries 3 d; apples, bilberries, blackcurrants, blueberries, cranberries, gooseberries, pears, redcurrants, whitecurrants 7d; blackberries, plums, raspberries 21 d
- Product has no effect on fruit quality or finish

Environmental safety
- Dangerous for the environment
- Very toxic to aquatic organisms
- High risk to bees. Do not apply to crops in flower or to those in which bees are actively foraging. Do not apply when flowering weeds are present
- Broadcast air-assisted LERAP (18 m); LERAP Category B

Hazard classification and safety precautions
    Hazard Harmful, Dangerous for the environment
    Transport code 9
    Packaging group III
    UN Number 3077
    Risk phrases R20, R22a, R50, R53a
    Operator protection A; U02a, U05a, U09a, U13, U14, U20b
    Environmental protection E12a, E12e, E15a, E16a, E16b, E38; E17b (18 m)
    Storage and disposal D01, D02, D09a, D11a, D12a
    Medical advice M05a

FOR FULL CONDITIONS OF USE ALWAYS READ THE PRODUCT LABEL

420   tefluthrin

A soil acting pyrethroid insecticide seed treatment
IRAC mode of action code: 3

See also fludioxonil + tefluthrin

Products

Force ST	Syngenta	200 g/l	CF	11752

Uses
- Bean seed fly in **bulb onions** *(off-label)*, **chard** *(off-label - seed treatment)*, **chives** *(off-label - seed treatment)*, **green mustard** *(off-label - seed treatment)*, **herbs (see appendix 6)** *(off-label - seed treatment)*, **leaf spinach** *(off-label - seed treatment)*, **leeks** *(off-label)*, **miduna** *(off-label - seed treatment)*, **mizuna** *(off-label - seed treatment)*, **pak choi** *(off-label - seed treatment)*, **parsley** *(off-label - seed treatment)*, **red mustard** *(off-label - seed treatment)*, **salad onions** *(off-label)*, **shallots** *(off-label)*, **spinach beet** *(off-label - seed treatment)*, **tatsoi** *(off-label - seed treatment)*
- Carrot fly in **carrots** *(off-label - seed treatment)*, **parsnips** *(off-label - seed treatment)*
- Millipedes in **fodder beet** *(seed treatment)*, **sugar beet** *(seed treatment)*
- Onion fly in **bulb onions** *(off-label)*, **leeks** *(off-label)*, **salad onions** *(off-label)*, **shallots** *(off-label)*
- Pygmy beetle in **fodder beet** *(seed treatment)*, **sugar beet** *(seed treatment)*
- Springtails in **fodder beet** *(seed treatment)*, **sugar beet** *(seed treatment)*
- Symphylids in **fodder beet** *(seed treatment)*, **sugar beet** *(seed treatment)*

Extension of Authorisation for Minor Use (EAMUs)
- **bulb onions** *20120604*
- **carrots** *(seed treatment) 20050547*
- **chard** *(seed treatment) 20050545*
- **chives** *(seed treatment) 20050545*
- **green mustard** *(seed treatment) 20050545*
- **herbs (see appendix 6)** *(seed treatment) 20050545*
- **leaf spinach** *(seed treatment) 20050545*
- **leeks** *20120604*
- **miduna** *(seed treatment) 20050545*
- **mizuna** *(seed treatment) 20050545*
- **pak choi** *(seed treatment) 20050545*
- **parsley** *(seed treatment) 20050545*
- **parsnips** *(seed treatment) 20050547*
- **red mustard** *(seed treatment) 20050545*
- **salad onions** *20120604*
- **shallots** *20120604*
- **spinach beet** *(seed treatment) 20050545*
- **tatsoi** *(seed treatment) 20050545*

Approval information
- Tefluthrin included in Annex 1 under EC Regulation 1107/2009
- Accepted by BBPA for use on malting barley

Efficacy guidance
- Apply during process of pelleting beet seed. Consult manufacturer for details of specialist equipment required
- Micro-capsule formulation allows slow release to provide a protection zone around treated seed during establishment

Restrictions
- Maximum number of treatments 1 per batch of seed
- Sow treated seed as soon as possible. Do not store treated seed from one drilling season to next
- If used in areas where soil erosion by wind or water likely, measures must be taken to prevent this happening

SEE SECTION 3 FOR PRODUCTS ALSO REGISTERED

- Can cause a transient tingling or numbing sensation to exposed skin. Avoid skin contact with product, treated seed and dust throughout all operations in the seed treatment plant and at drilling

Crop-specific information
- Latest use: before drilling seed
- Treated seed must be drilled within the season of treatment

Environmental safety
- Dangerous for the environment
- Very toxic to aquatic organisms
- Keep treated seed secure from people, domestic stock/pets and wildlife at all times during storage and use
- Treated seed harmful to game and wild life. Bury spillages
- In the event of seed spillage clean up as much as possible into the related seed sack and bury the remainder completely
- Do not apply treated seed from the air
- Keep livestock out of areas drilled with treated seed for at least 80 d

Hazard classification and safety precautions
Hazard Harmful, Dangerous for the environment
Transport code 9
Packaging group III
UN Number 3082
Risk phrases R20, R22b, R43, R50, R53a
Operator protection A, D, E, H; U02a, U04a, U05a, U08, U20b
Environmental protection E06a (80 d); E15a, E34, E38
Storage and disposal D01, D02, D05, D09a, D11a, D12a
Treated seed S02, S03, S04b, S05, S07
Medical advice M05b

421 tepraloxydim

A systemic post-emergence herbicide for control of annual grass weeds
HRAC mode of action code: A

Products

1	Aramo	BASF	50 g/l	EC	10280
2	Clayton Rally	Clayton	50 g/l	EC	13415
3	Omarra	AgChem Access	50 g/l	EC	13693
4	Standon Tonga	Standon	50 g/l	EC	13394
5	Tepra	Goldengrass	50 g/l	EC	14160

Uses
- Annual grasses in *all edible seed crops grown outdoors* (off-label), *all non-edible seed crops grown outdoors* (off-label), *borage* (off-label), *broad beans* (off-label), *broccoli* (off-label), *calabrese* (off-label), *canary flower (echium spp.)* (off-label), *chickpeas* (off-label), *edamame beans* (off-label), *evening primrose* (off-label), *forest nurseries* (off-label), *game cover* (off-label), *golf courses* (off-label), *honesty* (off-label), *hops* (off-label), *lentils* (off-label), *lupins* (off-label), *mustard* (off-label), *ornamental plant production* (off-label), *salad onions* (off-label), *swedes* (off-label), *top fruit* (off-label), *turnips* (off-label) [1]; *bulb onions, cabbages, carrots, cauliflowers, combining peas, fodder beet, land temporarily removed from production, leeks, linseed, spring field beans, sugar beet, vining peas, winter field beans, winter oilseed rape* [1-3, 5]; *flax, linseed for industrial use, winter oilseed rape for industrial use* [1, 3]; *garlic* (off-label), *mallow (althaea spp.)* (off-label), *parsnips* (off-label), *red beet* (off-label), *shallots* (off-label) [1, 5]
- Annual meadow grass in *all edible seed crops grown outdoors* (off-label), *all non-edible seed crops grown outdoors* (off-label), *forest nurseries* (off-label), *game cover* (off-label), *golf courses* (off-label), *hops* (off-label), *ornamental plant production* (off-label), *poppies for morphine production* (off-label), *salad onions* (off-label), *swedes* (off-label), *top fruit* (off-

FOR FULL CONDITIONS OF USE ALWAYS READ THE PRODUCT LABEL

label), **turnips** *(off-label)* [1]; **bulb onions, cabbages, carrots, cauliflowers, combining peas, fodder beet, leeks, linseed, spring field beans, sugar beet, vining peas, winter field beans, winter oilseed rape** [4]; **garlic** *(off-label)*, **mallow (althaea spp.)** *(off-label)*, **parsnips** *(off-label)*, **red beet** *(off-label)*, **shallots** *(off-label)* [5]

- Blackgrass in **bulb onions, cabbages, carrots, cauliflowers, combining peas, fodder beet, leeks, linseed, spring field beans, sugar beet, vining peas, winter field beans, winter oilseed rape** [4]
- Green cover in **land temporarily removed from production** [4]
- Perennial grasses in **all edible seed crops grown outdoors** *(off-label)*, **all non-edible seed crops grown outdoors** *(off-label)*, **borage** *(off-label)*, **broccoli** *(off-label)*, **calabrese** *(off-label)*, **canary flower (echium spp.)** *(off-label)*, **evening primrose** *(off-label)*, **forest nurseries** *(off-label)*, **game cover** *(off-label)*, **golf courses** *(off-label)*, **honesty** *(off-label)*, **hops** *(off-label)*, **lupins** *(off-label)*, **mustard** *(off-label)*, **ornamental plant production** *(off-label)*, **swedes** *(off-label)*, **top fruit** *(off-label)*, **turnips** *(off-label)* [1]; **bulb onions, cabbages, carrots, cauliflowers, combining peas, fodder beet, land temporarily removed from production, leeks, linseed, spring field beans, sugar beet, vining peas, winter field beans, winter oilseed rape** [1-3, 5]; **flax, linseed for industrial use, winter oilseed rape for industrial use** [1, 3]; **garlic** *(off-label)*, **mallow (althaea spp.)** *(off-label)*, **parsnips** *(off-label)*, **red beet** *(off-label)*, **shallots** *(off-label)* [1, 5]
- Volunteer barley in **bulb onions, cabbages, carrots, cauliflowers, combining peas, fodder beet, leeks, linseed, spring field beans, sugar beet, vining peas, winter field beans, winter oilseed rape** [4]
- Volunteer cereals in **bulb onions, cabbages, carrots, cauliflowers, combining peas, fodder beet, land temporarily removed from production, leeks, linseed, spring field beans, sugar beet, vining peas, winter field beans, winter oilseed rape** [1-3, 5]; **flax, linseed for industrial use, winter oilseed rape for industrial use** [1, 3]
- Volunteer wheat in **bulb onions, cabbages, carrots, cauliflowers, combining peas, fodder beet, leeks, linseed, spring field beans, sugar beet, vining peas, winter field beans, winter oilseed rape** [4]
- Wild oats in **red beet** *(off-label)* [1]

Extension of Authorisation for Minor Use (EAMUs)
- **all edible seed crops grown outdoors** *20082813* [1]
- **all non-edible seed crops grown outdoors** *20082813* [1]
- **borage** *20064019* [1]
- **broad beans** *20101553* [1]
- **broccoli** *20103121* [1]
- **calabrese** *20103121* [1]
- **canary flower (echium spp.)** *20064019* [1]
- **chickpeas** *20101553* [1]
- **edamame beans** *20101553* [1]
- **evening primrose** *20064019* [1]
- **forest nurseries** *20082813* [1]
- **game cover** *20082813* [1]
- **garlic** *20064021* [1], *20092128* [5]
- **golf courses** *20091527* [1]
- **honesty** *20064019* [1]
- **hops** *20082813* [1]
- **lentils** *20101553* [1]
- **lupins** *20064022* [1]
- **mallow (althaea spp.)** *20064023* [1], *20092129* [5]
- **mustard** *20064019* [1]
- **ornamental plant production** *20082813* [1]
- **parsnips** *20064023* [1], *20092129* [5]
- **poppies for morphine production** *20064026* [1]
- **red beet** *20071844* [1], *20092127* [5]
- **salad onions** *20101146* [1]
- **shallots** *20064021* [1], *20092128* [5]
- **swedes** *20090082* [1]

SECTION 2

SEE SECTION 3 FOR PRODUCTS ALSO REGISTERED

- **top fruit** *20082813* [1]
- **turnips** *20090082* [1]

Approval information
- Tepraloxydim included in Annex I under EC Regulation 1107/2009

Efficacy guidance
- Best results obtained from applications when weeds small and have not begun to compete with crop
- Only emerged weeds are controlled
- Cool conditions slow down activity and very dry conditions reduce activity by interfering with uptake and translocation
- Foliar death of susceptible weeds evident after 3-4 wks in warm conditions
- Reduced doses must not be used on resistant grass weed populations
- Tepraloxydim is an ACCase inhibitor herbicide. To avoid the build up of resistance do not apply products containing an ACCase inhibitor herbicide more than twice to any crop. In addition do not use any product containing tepraloxydim in mixture or sequence with any other product containing the same ingredient
- Use these products as part of a resistance management strategy that includes cultural methods of control and does not use ACCase inhibitors as the sole chemical method of grass weed control
- Applying a second product containing an ACCase inhibitor to a crop will increase the risk of resistance development; only use a second ACCase inhibitor to control different weeds at a different timing
- Always follow WRAG guidelines for preventing and managing herbicide resistant weeds. See Section 5 for more information

Restrictions
- Maximum number of treatments 1 per crop or yr
- Consult processors or contract agents before treatment of crops intended for processing or for seed
- Do not apply to crops damaged or stressed by factors such as previous herbicide treatments, pest or disease attack
- Do not spray if rain or frost expected, or if foliage is wet
- Do not treat oilseed rape with very low vigour and poor yield potential. Overlapping on oilseed rape may cause damage and reduce yields
- On peas a satisfactory crystal violet leaf wax test must be carried out before treatment. Winter varieties may be treated only in the spring
- For sugar beet, fodder beet, linseed and green cover on land temporarily removed from production, applications are prohibited between 1 Nov and 31 Mar
- For field beans, peas, leeks, bulb onions, carrots, cabbages and cauliflowers applications are prohibited between 1 Nov and 1 Mar

Crop-specific information
- Latest use : before end Nov or before crop has 9 true leaves, whichever is first for winter oilseed rape; before flower buds visible for flax [1], linseed; before head formation for cabbages, cauliflowers
- HI 3 wk for carrots; 4 wk for bulb onions, leeks; 5 wk for peas; 8 wk for fodder beet, field beans, sugar beet
- Treatment prohibited between 1 Nov and 31 Mar on golf courses (off-label use), sugar beet, fodder beet, linseed and green cover on land temporarily removed from production
- Green cover must be fully established on land temporarily removed from production before treatment, and treated plants must not be grazed by livestock or harvested for human or animal consumption

Following crops guidance
- In the event of failure of a treated crop, wheat or barley may be drilled after 2 wk following normal seedbed cultivations, maize or Italian rye-grass may be planted after 8 wk following cultivation to 20 cm
- Graminaceous crops other than those mentioned above should not follow a treated crop in the rotation.

FOR FULL CONDITIONS OF USE ALWAYS READ THE PRODUCT LABEL

- Broad leaved crops may be planted at any time following a failed or normally harvested treated crop

Environmental safety
- Dangerous for the environment
- Toxic to aquatic organisms

Hazard classification and safety precautions
> **Hazard** Harmful, Dangerous for the environment
> **Transport code** 9
> **Packaging group** III
> **UN Number** 3082
> **Risk phrases** R22b, R38, R51, R53a, R63, R66 [1-5]; R40, R62 [2, 4]
> **Operator protection** A; U05a, U08, U20b, U23a [1-5]; U19a [4]
> **Environmental protection** E15a [1-3, 5]; E15b [4]; E38 [1, 3, 5]
> **Storage and disposal** D01, D02, D09a, D10c, D12a [1-5]; D05 [2, 4]; D07 [2]; D08 [1, 3-5]
> **Medical advice** M04a [4]; M05b [1-5]

422 terbuthylazine

A triazine herbicide available only in mixtures
HRAC mode of action code: C1

See also bromoxynil + terbuthylazine
isoxaben + terbuthylazine
mesotrione + terbuthylazine
pendimethalin + terbuthylazine

423 tetramethrin

A contact acting pyrethroid insecticide
IRAC mode of action code: 3

See also d-phenothrin + tetramethrin

Products

Killgerm Py-Kill W	Killgerm	10.2% w/v	EC	H4632

Uses
- Flies in **agricultural premises**, **livestock houses**

Approval information
- Tetramethrin not included in Annex 1 under EC Regulation 1107/2009

Efficacy guidance
- Dilute in accordance with directions and apply as space or surface spray

Restrictions
- For use only by professional operators
- Maximum number of treatments 1 per wk in agricultural premises and livestock houses
- Do not apply directly on food or livestock
- Remove exposed milk before application. Protect milk machinery and containers from contamination
- Do not use space sprays containing pyrethrins or pyrethroid more than once per wk in intensive or controlled environment animal houses in order to avoid development of resistance. If necessary, use a different control method or product

Environmental safety
- Dangerous for the environment
- Toxic to aquatic organisms

SEE SECTION 3 FOR PRODUCTS ALSO REGISTERED

Hazard classification and safety precautions

> **Hazard** Harmful, Flammable, Dangerous for the environment
> **UN Number** N/C
> **Risk phrases** R22b, R41, R51, R53a
> **Operator protection** A, H; U02b, U09a, U19a, U20b
> **Environmental protection** E15a
> **Consumer protection** C05, C06, C07, C08, C09, C10, C11
> **Storage and disposal** D01, D09a, D12a
> **Medical advice** M05b

424   thiabendazole

A systemic, curative and protectant benzimidazole (MBC) fungicide
FRAC mode of action code: 1

See also imazalil + thiabendazole
metalaxyl + thiabendazole

Products

1 Hykeep	Agrichem	2% w/w	DP	12704
2 Storite Clear Liquid	Frontier	220 g/l	SL	12706
3 Storite Excel	Frontier	500 g/l	SC	12705
4 Tezate 220 SL	AgriChem BV	220 g/l	SL	14007

Uses

- Basal stem rot in **narcissi** *(off-label)* [2, 4]
- Dry rot in **potatoes**, **seed potatoes** [4]; **seed potatoes** *(tuber treatment - post-harvest)* [3]; **ware potatoes** *(tuber treatment - post-harvest)* [1, 3]
- Fusarium basal neck rot in **narcissi** [4]; **narcissi** *(post-lifting or pre-planting dip)*, **narcissi** *(post-lifting spray)* [2]
- Gangrene in **potatoes**, **seed potatoes** [4]; **seed potatoes** *(tuber treatment - post-harvest)* [3]; **ware potatoes** *(tuber treatment - post-harvest)* [1, 3]
- Neck rot in **narcissi** *(off-label)* [2, 4]
- Silver scurf in **potatoes**, **seed potatoes** [4]; **seed potatoes** *(tuber treatment - post-harvest)* [3]; **ware potatoes** *(tuber treatment - post-harvest)* [1, 3]
- Skin spot in **potatoes**, **seed potatoes** [4]; **seed potatoes** *(tuber treatment - post-harvest)* [3]; **ware potatoes** *(tuber treatment - post-harvest)* [1, 3]

Extension of Authorisation for Minor Use (EAMUs)

- **narcissi** *20070924* [2], *20091180* [4]

Approval information

- Thiabendazole included in Annex I under EC Regulation 1107/2009
- Use of thiabendazole products on ware potatoes requires that the discharge of thiabendazole to receiving water from washing plants is kept within emission limits set by the UK monitoring authority

Efficacy guidance

- For best results tuber treatments should be applied as soon as possible after lifting and always within 24 hr
- Dust treatments should be applied evenly over the whole tuber surface
- Thiabendazole should only be used on ware potatoes where there is a likely risk of disease during the storage period and in combination with good storage hygiene and maintenance [1, 3]
- Benzimidazole tolerant strains of silver scurf and skin spot are common in UK and tolerant strains of dry rot have been reported. To reduce the chance of such strains increasing benzimidazole based products should not be used more than once in the cropping cycle [1, 3]
- On narcissi treat as the crop goes into store and only if fungicidal treatment is essential. Adopt a resistance management strategy [2]

FOR FULL CONDITIONS OF USE ALWAYS READ THE PRODUCT LABEL

Restrictions
- Maximum number of treatments 1 per batch for ware or seed potato tuber treatments [1, 3]; 1 per yr for narcissi bulbs [2]
- Treated seed potatoes must not be used for food or feed [3]
- Do not remove treated potatoes from store for sale, processing or consumption for at least 21 d after application [1, 3]
- Do not mix with any other product
- Off-label use as post-lifting cold water dip or pre-planting hot water dip not to be used on narcissus bulbs grown in the Isles of Scilly [2]

Crop-specific information
- Latest use: 21 d before removal from store for sale, processing or consumption for ware potatoes [1, 3]
- Apply to potatoes as soon as possible after harvest using suitable equipment and always within 2 wk of lifting provided the skins are set. See label for details [1, 3]
- Potatoes should only be treated by systems that provide an accurate dose to tubers not carrying excessive quantities of soil [1, 3]
- Use as a post-lifting treatment to reduce basal and neck rot in narcissus bulbs. Ensure bulbs are clean [2]

Environmental safety
- Dangerous for the environment
- Toxic to aquatic organisms

Hazard classification and safety precautions
> **Hazard** Irritant [3]; Dangerous for the environment [1-4]
> **Transport code** 9
> **Packaging group** III
> **UN Number** 3077 [1]; 3082 [2-4]
> **Risk phrases** R43 [3]; R51 [2-4]; R52 [1]; R53a [1-4]
> **Operator protection** A, H [1-4]; C, D [2-4]; U05a [2-4]; U14 [3]; U19a [1-4]; U20b [4]; U20c [1-3]
> **Environmental protection** E15b [1-4]; E38 [2-4]
> **Storage and disposal** D01, D09a, D12a [1-4]; D02, D10c [2-4]; D05 [2, 3]; D07, D11a [1]
> **Treated seed** S03, S05 [4]; S04a [2, 4]

425   thiacloprid

A chloronicotinyl insecticide for use in agriculture and horticulture
IRAC mode of action code: 4A

Products

1	Agrovista Reggae	Agrovista	480 g/l	SC	13706
2	Biscaya	Bayer CropScience	240 g/l	OD	15014
3	Calypso	Bayer CropScience	480 g/l	SC	11257
4	Exemptor	Everris Ltd	10% w/w	GR	15615
5	Standon Zero Tolerance	Standon	240 g/l	OD	13546
6	Zubarone	AgChem Access	240 g/l	OD	15798

Uses
- Aphids in *all edible seed crops grown outdoors* (off-label), *all non-edible seed crops grown outdoors* (off-label), *ornamental plant production* (off-label) [2]; *bedding plants, hardy ornamental nursery stock, ornamental plant production, pot plants, protected ornamentals* [4]; *carrots, parsnips, peas, potatoes* [2, 6]; *leaf brassicas* (off-label - baby leaf production), *protected herbs (see appendix 6)* (off-label), *protected lamb's lettuce* (off-label), *protected lettuce* (off-label) [1]; *seed potatoes, ware potatoes* [2, 5, 6]
- Asparagus beetle in *asparagus* (off-label) [3]
- Beetles in *bedding plants, hardy ornamental nursery stock, ornamental plant production, pot plants, protected ornamentals* [4]
- Bud mite in *almonds* (off-label), *chestnuts* (off-label), *cob nuts* (off-label), *hazel nuts* (off-label), *walnuts* (off-label) [1, 3]; *filberts* (off-label) [3]

SEE SECTION 3 FOR PRODUCTS ALSO REGISTERED

- Cabbage aphid in *broccoli, brussels sprouts, cabbages, calabrese, cauliflowers* [2, 6]; *chinese cabbage* (off-label), *choi sum* (off-label), *pak choi* (off-label), *swedes* (off-label), *tatsoi* (off-label), *turnips* (off-label) [2]
- Capsids in *protected strawberries* (off-label) [3]; *strawberries* (off-label) [1, 3]
- Common green capsid in *protected blackberries* (off-label), *protected raspberries* (off-label) [3]
- Damson-hop aphid in *plums* (off-label) [1, 3]
- Gall midge in *protected blueberry* (off-label) [1]
- Insect pests in *bilberries* (off-label), *blackberries* (off-label), *blackcurrants* (off-label), *blueberries* (off-label), *courgettes* (off-label), *cranberries* (off-label), *gherkins* (off-label), *gooseberries* (off-label), *hops* (off-label), *marrows* (off-label), *ornamental plant production* (off-label), *pears* (off-label), *protected forest nurseries* (off-label), *raspberries* (off-label), *redcurrants* (off-label), *rubus hybrids* (off-label), *soft fruit* (off-label), *top fruit* (off-label), *whitecurrants* (off-label) [1, 3]; *cherries* (off-label), *mirabelles* (off-label), *protected blueberry* (off-label) [3]; *cherries* (off-label - under temporary protective rain covers), *mirabelles* (off-label - under temporary protective rain covers), *protected blackberries* (off-label), *protected raspberries* (off-label), *protected strawberries* (off-label) [1]
- Leaf miner in *protected aubergines* (off-label), *protected courgettes* (off-label), *protected cucumbers* (off-label), *protected ornamentals* (off-label), *protected peppers* (off-label), *protected tomatoes* (off-label) [1, 3]
- Mealy aphid in *broccoli, brussels sprouts, cabbages, calabrese, cauliflowers* [2, 6]; *chinese cabbage* (off-label), *choi sum* (off-label), *pak choi* (off-label), *swedes* (off-label), *tatsoi* (off-label), *turnips* (off-label) [2]
- Moths in *protected blueberry* (off-label) [1]
- Palm thrips in *protected aubergines* (off-label), *protected courgettes* (off-label), *protected cucumbers* (off-label), *protected ornamentals* (off-label), *protected tomatoes* (off-label) [3]
- Pea midge in *peas* [2, 6]
- Peach-potato aphid in *leaf brassicas* (off-label) [3]; *leaf brassicas* (off-label - baby leaf production) [1]; *protected herbs (see appendix 6)* (off-label), *protected lamb's lettuce* (off-label), *protected lettuce* (off-label) [1, 3]
- Pear midge in *pears* (off-label) [1, 3]
- Pollen beetle in *mustard* [2, 6]; *spring oilseed rape, winter oilseed rape* [2, 5, 6]
- Raspberry beetle in *protected blackberries* (off-label), *protected raspberries* (off-label) [3]
- Rosy apple aphid in *apples* [1, 3]
- Tarnished plant bug in *protected blackberries* (off-label), *protected raspberries* (off-label) [3]
- Thrips in *protected aubergines* (off-label), *protected courgettes* (off-label), *protected cucumbers* (off-label), *protected ornamentals* (off-label), *protected peppers* (off-label), *protected tomatoes* (off-label) [1, 3]
- Vine weevil in *bedding plants, hardy ornamental nursery stock, ornamental plant production, pot plants, protected ornamentals* [4]
- Western flower thrips in *protected aubergines* (off-label), *protected courgettes* (off-label), *protected cucumbers* (off-label), *protected ornamentals* (off-label), *protected peppers* (off-label), *protected tomatoes* (off-label) [1, 3]
- Wheat-blossom midge in *spring wheat, winter wheat* [2, 6]
- Whitefly in *bedding plants, hardy ornamental nursery stock, ornamental plant production, pot plants, protected ornamentals* [4]; *protected aubergines* (off-label), *protected courgettes* (off-label), *protected cucumbers* (off-label), *protected ornamentals* (off-label), *protected tomatoes* (off-label) [1, 3]; *protected peppers* (off-label) [1]
- Willow aphid in *horseradish* (off-label), *parsley root* (off-label), *red beet* (off-label), *salsify* (off-label) [2]

Extension of Authorisation for Minor Use (EAMUs)
- *all edible seed crops grown outdoors* 20110392 [2]
- *all non-edible seed crops grown outdoors* 20110392 [2]
- *almonds* 20080471 [1], 20060341 [3]
- *asparagus* 20101802 [3]
- *bilberries* 20080466 [1], 20060335 [3]
- *blackberries* 20080475 [1], 20060336 [3]
- *blackcurrants* 20080466 [1], 20060335 [3]

FOR FULL CONDITIONS OF USE ALWAYS READ THE PRODUCT LABEL

- **blueberries** *20080466* [1], *20060335* [3]
- **cherries** *(under temporary protective rain covers) 20080469* [1], *20060338* [3]
- **chestnuts** *20080471* [1], *20060341* [3]
- **chinese cabbage** *20120671* [2]
- **choi sum** *20120671* [2]
- **cob nuts** *20080471* [1], *20060341* [3]
- **courgettes** *20102032* [1], *20081006* [3]
- **cranberries** *20080466* [1], *20060335* [3]
- **filberts** *20060341* [3]
- **gherkins** *20102032* [1], *20081006* [3]
- **gooseberries** *20080466* [1], *20060335* [3]
- **hazel nuts** *20080471* [1], *20060341* [3]
- **hops** *20102034* [1], *20082831* [3]
- **horseradish** *20110805* [2]
- **leaf brassicas** *(baby leaf production) 20080470* [1], *20060340* [3]
- **marrows** *20102032* [1], *20081006* [3]
- **mirabelles** *(under temporary protective rain covers) 20080469* [1], *20060338* [3]
- **ornamental plant production** *20102034* [1], *20110392* [2], *20082831* [3]
- **pak choi** *20120671* [2]
- **parsley root** *20110805* [2]
- **pears** *20080464* [1], *20060324* [3]
- **plums** *20080468* [1], *20060337* [3]
- **protected aubergines** *20080474* [1], *20063728* [3], *20121997* [3]
- **protected blackberries** *20080467* [1], *20070534* [3]
- **protected blueberry** *20102035* [1], *20081332* [3]
- **protected courgettes** *20080474* [1], *20063728* [3], *20121997* [3]
- **protected cucumbers** *20080474* [1], *20063728* [3], *20121997* [3]
- **protected forest nurseries** *20102034* [1], *20082831* [3]
- **protected herbs (see appendix 6)** *20080472* [1], *20060453* [3]
- **protected lamb's lettuce** *20080472* [1], *20060453* [3]
- **protected lettuce** *20080472* [1], *20060453* [3]
- **protected ornamentals** *20080474* [1], *20063728* [3], *20121997* [3]
- **protected peppers** *20080474* [1], *20063728* [3]
- **protected raspberries** *20080467* [1], *20070534* [3]
- **protected strawberries** *20102033* [1], *20060334* [3]
- **protected tomatoes** *20080474* [1], *20063728* [3], *20121997* [3]
- **raspberries** *20080475* [1], *20060336* [3]
- **red beet** *20110805* [2]
- **redcurrants** *20080466* [1], *20060335* [3]
- **rubus hybrids** *20080475* [1], *20060336* [3]
- **salsify** *20110805* [2]
- **soft fruit** *20102034* [1], *20082831* [3]
- **strawberries** *20080465* [1], *20060333* [3]
- **swedes** *20110393* [2]
- **tatsoi** *20120671* [2]
- **top fruit** *20102034* [1], *20082831* [3]
- **turnips** *20110393* [2]
- **walnuts** *20080471* [1], *20060341* [3]
- **whitecurrants** *20080466* [1], *20060335* [3]

Approval information
- Thiacloprid included in Annex I under EC Regulation 1107/2009

Efficacy guidance
- Best results in apples obtained by a programme of sprays commencing pre-blossom at the first sign of aphids [3]
- For best pear midge control treat under warm conditions (mid-day) when the adults are flying (temperatures higher than 12˚C). Applications on cool damp days are unlikely to be successful [3]

SEE SECTION 3 FOR PRODUCTS ALSO REGISTERED

- Best results in field crops obtained from treatments when target pests reach threshold levels. For wheat blossom midge use pheromone or sticky yellow traps to monitor adult activity in crops at risk [2, 5]
- Treatment of ornamentals and protected ornamentals is by incorporation into peat-based growing media prior to sowing or planting using suitable automated equipment [4]
- Ensure thorough mixing in the compost to achieve maximum control. Top dressing is ineffective [4]
- Unless transplants have been treated with a suitable pesticide prior to planting in treated compost, full protection against target pests is not guaranteed [4]
- Minimise the possibility of the development of resistance by altenating insecticides with different modes of action in the programme
- In dense canopies and on larger trees increase water volume to ensure full coverage

Restrictions
- Maximum number of treatments 2 per yr on apples, seed potatoes; 1 per yr on other field crops and for compost incorporation
- Consult processor before use on crops for processing [2, 5]
- Do not mix treated compost with any other bulky materials such as perlite or bark [4]
- Use treated compost as soon as possible, and within 4 wk of mixing [4]

Crop-specific information
- Latest use: up to and including flowering just complete (GS 69) for wheat; before sowing or planting for treated compost
- HI apples and potatoes 14 d; oilseed rape and mustard 30 d
- Test tolerance of ornamental species before large scale use [4]

Environmental safety
- Dangerous for the environment [2, 4, 5]
- Very toxic or toxic to aquatic organisms
- Risk to certain non-target insects or other arthropods. See directions for use [3]
- Broadcast air-assisted LERAP [1] (18 m); Broadcast air-assisted LERAP [3] (30 m)

Hazard classification and safety precautions
> **Hazard** Harmful [1-6]; Dangerous for the environment [2, 4-6]
> **Transport code** 6.1 [1, 3]; 9 [4]
> **Packaging group** III [1, 3, 4]
> **UN Number** 2902 [1, 3]; 3077 [4]; N/C [2, 5, 6]
> **Risk phrases** R20, R52 [1, 3]; R22a [1-3, 5, 6]; R36, R38, R51 [2, 5, 6]; R40, R53a [1-6]; R43 [1, 3, 4]; R50 [4]
> **Operator protection** A [1-6]; C [2, 5, 6]; D [4]; H [1, 3]; U05a [1-6]; U11, U19a [2, 5, 6]; U14 [1, 3]; U16a [2, 6]; U20b [1-3, 5, 6]; U20c [4]
> **Environmental protection** E13c, E22b [3]; E15a, E38 [4]; E15b [1, 2, 5, 6]; E17b [1] (18 m); E17b [3] (30 m); E22c [1]; E34 [1-6]
> **Storage and disposal** D01, D02, D09a [1-6]; D10b [1, 3]; D10c [2, 5, 6]; D11a [4]; D12a [1, 2, 5, 6]
> **Treated seed** S04a [4]
> **Medical advice** M03 [1-6]; M05a [1]; M05b [3]

426   thiamethoxam

A neonicotinoid insecticide for apples, pears, potatoes and beet crops
IRAC mode of action code: 4A

See also fludioxonil + metalaxyl-M + thiamethoxam

Products

1	Actara	Syngenta	25% w/w	WG	13728
2	Cruiser SB	Syngenta	600 g/l	FS	15012
3	Tara	ChemSource	25% w/w	WG	15003

FOR FULL CONDITIONS OF USE ALWAYS READ THE PRODUCT LABEL

Uses

- Aphids in **potatoes, potatoes grown for seed** [1, 3]
- Beet leaf miner in **fodder beet** *(seed treatment)*, **sugar beet** *(seed treatment)* [2]
- Flea beetle in **fodder beet** *(seed treatment)*, **sugar beet** *(seed treatment)*, **swedes** *(off-label)* [2]
- Foliar disease control in **ornamental plant production** *(off-label)* [1]
- Millipedes in **fodder beet** *(seed treatment)*, **sugar beet** *(seed treatment)* [2]
- Pygmy beetle in **fodder beet** *(seed treatment)*, **sugar beet** *(seed treatment)* [2]
- Springtails in **fodder beet** *(seed treatment)*, **sugar beet** *(seed treatment)* [2]
- Symphylids in **fodder beet** *(seed treatment)*, **sugar beet** *(seed treatment)* [2]
- Wireworm in **fodder beet** *(seed treatment - reduction)*, **sugar beet** *(seed treatment - reduction)* [2]

Extension of Authorisation for Minor Use (EAMUs)

- **ornamental plant production** *20082801* [1]
- **swedes** *20111292* [2]

Approval information

- Thiamethoxam included in Annex I under EC Regulation 1107/2009

Efficacy guidance

- Apply to sugar beet or fodder beet seed as part of the normal commercial pelleting process using special treatment machinery [2]
- Seed drills must be suitable for use with polymer-coated seeds. Standard drill settings should not need to be changed [2]
- Drill treated seed into a firm even seedbed. Poor seedbed quality or seedbed conditions may results in delayed emergence and poor establishment [2]
- Where very high populations of soil pests are present protection may be inadequate to achieve an optimum plant stand [2]
- Use in line with latest IRAG guidelines

Restrictions

- Maximum number of treatments 1 per seed batch [2]
- Avoid deep or shallow drilling which may adversely affect establishment and reduce the level of pest control [2]
- Do not use herbicides containing lenacil pre-emergence on treated crops [2]

Crop-specific information

- Latest use: before drilling for fodder beet, sugar beet [2]

Environmental safety

- Dangerous for the environment
- Very toxic to aquatic organisms

Hazard classification and safety precautions

Hazard Dangerous for the environment
UN Number N/C
Risk phrases R50, R53a
Operator protection A, D, H [2]; U05a, U07 [2]; U05b, U20c [1, 3]
Environmental protection E03, E34, E36a [2]; E12c, E12e [1, 3]; E15b, E38 [1-3]
Storage and disposal D01, D02, D09a, D12a [1-3]; D05, D11a, D14 [2]; D10c [1, 3]
Treated seed S02, S04b, S05, S06a, S06b, S08 [2]

SEE SECTION 3 FOR PRODUCTS ALSO REGISTERED

427   thifensulfuron-methyl

A translocated sulfonylurea herbicide
HRAC mode of action code: B

See also carfentrazone-ethyl + thifensulfuron-methyl
               flupyrsulfuron-methyl + thifensulfuron-methyl
               fluroxypyr + thifensulfuron-methyl + tribenuron-methyl
               metsulfuron-methyl + thifensulfuron-methyl

Products

Pinnacle	Headland	50% w/w	SG	12285

Uses

- Annual dicotyledons in **soya beans** *(off-label)*
- Docks in **permanent grassland, rotational grass**
- Green cover in **land temporarily removed from production**

Extension of Authorisation for Minor Use (EAMUs)

- **soya beans** *20112063*

Approval information

- Thifensulfuron-methyl included in Annex I under EC Regulation 1107/2009
- Accepted by BBPA for use on malting barley

Efficacy guidance

- Best results achieved from application to small emerged weeds when growing actively. Broad-leaved docks are susceptible during the rosette stage up to onset of stem extension
- Ensure good spray coverage and apply to dry foliage
- Susceptible weeds stop growing almost immediately but symptoms may not be visible for about 2 wk
- Product sold in twin pack with mecoprop-P to provide option for improved control of cleavers
- Only broad-leaved docks (*Rumex obtusifolius*) are controlled; curled docks (*Rumex crispus*) are resistant
- Docks with developing or mature seed heads should be topped and the regrowth treated later
- Established docks with large tap roots may require follow-up treatment
- High populations of docks in grassland will require further treatment in following yr
- Thifensulfuron-methyl is a member of the ALS-inhibitor group of herbicides and products should be used in a planned Resistance Management strategy. See Section 5 for more information

Restrictions

- Maximum number of treatments 1 per crop for cereals or one per yr for grassland and cereals. Must only be applied from 1 Feb in yr of harvest
- Do not treat new leys in year of sowing
- Do not treat where nutrient imbalances, drought, waterlogging, low temperatures, lime deficiency, pest or disease attack have reduced crop or sward vigour
- Do not roll or harrow within 7 d of spraying
- Do not graze grass crops within 7 d of spraying
- Specific restrictions apply to use in sequence or tank mixture with other sulfonylurea or ALS-inhibiting herbicides. See label for details
- Only one application of a sulfonylurea product may be applied per calendar yr to grassland and green cover on land temporarily removed from production

Crop-specific information

- Latest use: before 1 Aug on grass; flag leaf fully emerged stage (GS 39) on cereals
- On grass apply 7-10 d before grazing and do not graze for 7 d afterwards
- Product may cause a check to both sward and clover which is usually outgrown

Following crops guidance

- Only grass or cereals may be sown within 4 wk of application to grassland or setaside, or in the event of failure of any treated crop
- No restrictions apply after normal harvest of a treated cereal crop

---

FOR FULL CONDITIONS OF USE ALWAYS READ THE PRODUCT LABEL

Environmental safety
- Dangerous for the environment
- Very toxic to aquatic organisms
- Keep livestock out of treated areas for at least 7 d following treatment
- Take extreme care to avoid drift onto broad-leaved plants outside the target area or onto surface waters or ditches, or land intended for cropping
- Spraying equipment should not be drained or flushed onto land planted, or to be planted, with trees or crops other than cereals and should be thoroughly cleansed after use - see label for instructions

Hazard classification and safety precautions
Hazard Dangerous for the environment
Transport code 9
Packaging group III
UN Number 3077
Risk phrases R50, R53a
Operator protection U19a, U20b
Environmental protection E06a (7 d); E15a, E38
Storage and disposal D09a, D11a, D12a

428   thifensulfuron-methyl + tribenuron-methyl

A mixture of two sulfonylurea herbicides for cereals
HRAC mode of action code: B + B

See also tribenuron-methyl

Products

1	Calibre SX	DuPont	33.3:16.7% w/w	SG	15032
2	Inka SX	DuPont	25:25% w/w	SG	13601
3	Ratio SX	DuPont	40:10% w/w	SG	12601

Uses
- Annual dicotyledons in *game cover (off-label)*, *spring oats*, *spring rye*, *triticale*, *winter oats*, *winter rye* [2]; *spring barley*, *spring wheat*, *winter barley*, *winter wheat* [1-3]
- Charlock in *spring barley*, *spring wheat*, *winter barley*, *winter wheat* [1-3]; *spring oats*, *spring rye*, *triticale*, *winter oats*, *winter rye* [2]
- Chickweed in *spring barley*, *spring wheat*, *winter barley*, *winter wheat* [1-3]; *spring oats*, *spring rye*, *triticale*, *winter oats*, *winter rye* [2]
- Docks in *spring barley*, *spring oats*, *spring rye*, *spring wheat*, *triticale*, *winter barley*, *winter oats*, *winter rye*, *winter wheat* [2]
- Mayweeds in *spring barley*, *spring wheat*, *winter barley*, *winter wheat* [1-3]; *spring oats*, *spring rye*, *triticale*, *winter oats*, *winter rye* [2]

Extension of Authorisation for Minor Use (EAMUs)
- *game cover* 20101408 [2]

Approval information
- Thifensulfuron-methyl and tribenuron-methyl included in Annex I under EC Regulation 1107/2009
- Accepted by BBPA for use on malting barley

Efficacy guidance
- Apply when weeds are small and actively growing
- Apply after 1 Feb [1-3]
- Apply after end of Feb in year of harvest [2]
- Ensure good spray cover of the weeds
- Susceptible weeds cease growth almost immediately after application and symptoms become evident 2 wk later
- Effectiveness reduced by rain within 4 h of treatment and in very dry conditions
- Various tank mixtures recommended to broaden weed control spectrum [1]

SEE SECTION 3 FOR PRODUCTS ALSO REGISTERED

- Thifensulfuron-methyl and tribenuron-methyl are members of the ALS-inhibitor group of herbicides and products should be used in a planned Resistance Management strategy. See Section 5 for more information

Restrictions
- Maximum number of treatments 1 per crop
- Do not apply to cereals undersown with grass, clover or other legumes, or any other broad-leaved crop
- Do not apply within 7 d of rolling
- Specific restrictions apply to use in sequence or tank mixture with other sulfonylurea or ALS-inhibiting herbicides. See label for details
- Do not apply to any crop suffering from stress
- Consult contract agents before use on crops grown for seed

Crop-specific information
- Latest use: before flag leaf ligule first visible (GS 39) for all crops

Following crops guidance
- Only cereals, field beans, grass or oilseed rape may be sown in the same calendar year as harvest of a treated crop
- In the event of failure of a treated crop sow only a cereal crop within 3 mth of product application and after ploughing and cultivating to at least 15 cm. After 3 mth field beans or oilseed rape may also be sown [2]

Environmental safety
- Dangerous for the environment
- Very toxic to aquatic organisms
- Spraying equipment should not be drained or flushed onto land planted, or to be planted, with trees or crops other than cereals and should be thoroughly cleansed after use - see label for instructions
- Take particular care to avoid damage by drift onto broad-leaved plants outside the target area or onto surface waters or ditches
- LERAP Category B

Hazard classification and safety precautions
Hazard Irritant [1, 3]; Dangerous for the environment [1-3]
Transport code 9
Packaging group III
UN Number 3077
Risk phrases R43 [1, 3]; R50, R53a [1-3]
Operator protection A, H [1, 3]; U05a, U08, U20a [1, 3]; U14 [3]; U20c [2]
Environmental protection E15a [1, 3]; E15b [2]; E16a, E38 [1-3]; E16b [1]
Storage and disposal D01, D02 [1, 3]; D09a, D11a, D12a [1-3]

429   thiophanate-methyl

A thiophanate fungicide with protectant and curative activity
FRAC mode of action code: 1

See also iprodione + thiophanate-methyl

Products
1 Cercobin WG	Certis	70% w/w	WG	13854
2 Taurus	Certis	70% w/w	WG	15508
3 Topsin WG	Certis	70% w/w	WG	13988

Uses
- Disease control in **hops** (off-label - do not harvest for human or animal consumption (including idling) within 12 months of treatment.), **soft fruit** (off-label) [1]

FOR FULL CONDITIONS OF USE ALWAYS READ THE PRODUCT LABEL

- Fusarium in **durum wheat**, **spring oilseed rape** [3]; **durum wheat** *(reduction)*, **spring wheat** *(reduction)*, **triticale** *(reduction)*, **winter wheat** *(reduction)* [2]; **spring wheat**, **triticale**, **winter wheat** [1, 3]; **winter oilseed rape** [2, 3]
- Fusarium diseases in **protected ornamentals** *(off-label)* [1]
- Mycotoxins in **durum wheat** *(reduction)*, **spring wheat** *(reduction)*, **triticale** *(reduction)*, **winter wheat** *(reduction)* [2]; **durum wheat** *(reduction only)*, **spring oilseed rape** *(reduction only)* [3]; **spring wheat** *(reduction only)*, **triticale** *(reduction only)*, **winter wheat** *(reduction only)* [1, 3]; **winter oilseed rape** *(reduction only)* [2, 3]
- Root diseases in **container-grown ornamentals** *(off-label)*, **protected ornamentals** *(off-label)* [1]
- Sclerotinia stem rot in **spring oilseed rape**, **winter oilseed rape** [2]
- Verticillium wilt in **protected tomatoes** *(off-label)* [1]

Extension of Authorisation for Minor Use (EAMUs)
- **container-grown ornamentals** *20111655* [1]
- **hops** *(do not harvest for human or animal consumption (including idling) within 12 months of treatment.) 20082833* [1]
- **protected ornamentals** *20111827* [1], *20111887* [1]
- **protected tomatoes** *20091969* [1]
- **soft fruit** *20082833* [1]

Approval information
- Thiophanate-methyl included in Annex I under EC Regulation 1107/2009

Efficacy guidance
- To avoid the development of resistance, a maximum of 2 applications of any MBC product (thiophanate-methyl or carbendazim) are allowed in any one crop. Avoid using MBC fungicides alone.

Environmental safety
- Dangerous for the environment
- Very toxic to aquatic organisms
- LERAP Category B

Hazard classification and safety precautions
Hazard Harmful, Dangerous for the environment [2]
Transport code 9
Packaging group III
UN Number 3077
Risk phrases R20, R22a, R43, R51, R53a, R68 [2]
Operator protection A; U19a [2, 3]
Environmental protection E15b, E16a, E16b [1-3]; E38 [2, 3]
Storage and disposal D01, D02, D12a [2, 3]; D10b [1-3]
Medical advice M05a [2, 3]

430   thiram

A protectant dithiocarbamate fungicide
FRAC mode of action code: M3

See also carboxin + thiram
*   prochloraz + thiram*

Products

1	Agrichem Flowable Thiram	Agrichem	600 g/l	FS	10784
2	Tercero	Agroquimicos	80% w/w	WG	15819
3	Thianosan DG	Unicrop	80% w/w	WG	13404
4	Thiraflo	Chemtura	480 g/l	FS	13338
5	Thyram Plus	Agrichem	600 g/l	FS	10785
6	Triptam	Certis	80% w/w	WG	14014

SEE SECTION 3 FOR PRODUCTS ALSO REGISTERED

Uses

- Botrytis in *apples, ornamental plant production, pears, protected strawberries, strawberries* [2, 6]; *ornamental plant production* (except Hydrangea), *tulips* [3]
- Botrytis fruit rot in *apples* [3]
- Carnation rust in *carnations* [3]
- Chrysanthemum brown rust in *chrysanthemums* [3]
- Damping off in *broad beans* (seed treatment), *combining peas* (seed treatment), *dwarf beans* (seed treatment), *forage maize* (seed treatment), *grass seed* (seed treatment), *runner beans* (seed treatment), *spring field beans* (seed treatment), *spring oilseed rape* (seed treatment), *vining peas* (seed treatment), *winter field beans* (seed treatment), *winter oilseed rape* (seed treatment) [1, 4, 5]; *bulb onions* (seed treatment), *cabbages* (seed treatment), *cauliflowers* (seed treatment), *edible podded peas* (seed treatment), *kale* (off-label), *leeks* (seed treatment), *millet* (off-label), *radishes* (seed treatment), *salad onions* (seed treatment), *sorghum* (off-label), *soya beans* (seed treatment - qualified minor use), *turnip seed crops* (off-label), *turnips* (seed treatment) [1, 5]; *cress* (off-label), *endives* (off-label), *frise* (off-label), *garlic* (off-label - seed treatment), *lamb's lettuce* (off-label), *lettuce* (off-label), *lupins* (off-label - seed treatment), *poppies for morphine production* (off-label - seed treatment), *radicchio* (off-label), *rhubarb* (off-label - seed treatment), *scarole* (off-label), *shallots* (off-label - seed treatment), *swedes* (off-label - seed treatment) [1]; *kale seed crops* (off-label), *turnips* (off-label) [5]
- Gloeosporium in *apples* [3]
- Grey mould in *chrysanthemums, freesias, protected strawberries, strawberries* [3]
- Pythium in *poppies for morphine production* (off-label - seed treatment) [1]
- Scab in *apples, pears* [3]
- Seed-borne diseases in *carrots* (seed soak), *celery (outdoor)* (seed soak), *fodder beet* (seed soak), *mangels* (seed soak), *parsley* (seed soak), *red beet* (seed soak), *sugar beet* (seed soak) [1]; *garlic* (off-label - seed treatment), *lupins* (off-label - seed treatment), *shallots* (off-label - seed treatment), *swedes* (off-label - seed treatment) [5]

Extension of Authorisation for Minor Use (EAMUs)

- *cress* 20070792 [1]
- *endives* 20070792 [1]
- *frise* 20070792 [1]
- *garlic* (seed treatment) 20052395 [1], (seed treatment) 20052391 [5]
- *kale* 20120996 [1], 20120724 [5], 20121006 [5]
- *kale seed crops* 20121006 [5]
- *lamb's lettuce* 20070792 [1]
- *lettuce* 20070792 [1]
- *lupins* (seed treatment) 20052394 [1], (seed treatment) 20052389 [5]
- *millet* 20120996 [1], 20120724 [5], 20121006 [5]
- *poppies for morphine production* (seed treatment) 20040681 [1]
- *radicchio* 20070792 [1]
- *rhubarb* (seed treatment) 20052396 [1]
- *scarole* 20070792 [1]
- *shallots* (seed treatment) 20052395 [1], (seed treatment) 20052391 [5]
- *sorghum* 20120996 [1], 20120724 [5], 20121006 [5]
- *swedes* (seed treatment) 20052397 [1], (seed treatment) 20052390 [5]
- *turnip seed crops* 20120996 [1], 20121006 [5]
- *turnips* 20120724 [5]

Approval information

- Thiram included in Annex I under EC Regulation 1107/2009

Efficacy guidance

- Spray before onset of disease and repeat every 7-14 d. Spray interval varies with crop and disease. See label for details [3]
- Seed treatments may be applied through most types of seed treatment machinery, from automated continuous flow machines to smaller batch treating apparatus [1, 4, 5]
- Co-application of 175 ml water per 100 kg of seed likely to improve evenness of seed coverage [1, 4, 5]

Restrictions
- Maximum number of treatments 3 per crop for protected winter lettuce (thiram based products only; 2 per crop if sequence of thiram and other EBDC fungicides used) [3]; 2 per crop for protected summer lettuce [3]; 1 per batch of seed for seed treatments [1, 4, 5]
- Do not apply to hydrangeas [3]
- Notify processor before dusting or spraying crops for processing [3]
- Do not dip roots of forestry transplants [3]
- Do not spray when rain imminent [3]
- Do not treat seed of tomatoes, peppers or aubergines [1, 4, 5]
- Soya bean seed treatments restricted to crops that are to be harvested as a mature pulse crop only [1, 5]
- Treated seed should not be stored from one season to the next [1, 4, 5]

Crop-specific information
- Latest use: pre-drilling for seed treatments and seed soaks [1, 4, 5]; 21 d after planting out or 21 d before harvest, whichever is earlier, for protected winter lettuce; 14 d after planting out or 21 d before harvest, whichever is earlier, for spraying protected summer lettuce [3]
- Seed to be treated should be of satisfactory quality and moisture content [1, 4, 5]
- Follow label instructions for treating small quantities of seed [1, 4, 5]
- For use on tulips, chrysanthemums and carnations add non-ionic wetter [3]

Environmental safety
- Dangerous for the environment [1, 4, 5]
- Very toxic to aquatic organisms [1, 4, 5]
- Do not use treated seed as food or feed [1, 4, 5]
- Treated seed harmful to game and wildlife [1, 4, 5]
- A red dye is available from manufacturer to colour treated seed, but not recommended as part of the seed steep. See label for details [1]

Hazard classification and safety precautions
Hazard Harmful, Dangerous for the environment
Transport code 9
Packaging group III
UN Number 3077 [2, 3, 6]; 3082 [1, 4, 5]
Risk phrases R20, R36, R38 [1, 5]; R22a, R48, R50, R53a [1-6]; R43 [1-3, 5, 6]
Operator protection A [1-6]; P [2, 6]; U02a, U11 [1, 5]; U05a, U20b [1-6]; U08 [1, 3-5]; U09a [2, 6]; U14, U19a [2, 3, 6]
Environmental protection E13b [2, 3, 6]; E15a, E34 [1, 5]; E38 [4]
Storage and disposal D01, D09a [1-6]; D02, D05 [1, 4, 5]; D10a [4]; D10c, D12b [1, 5]; D11a, D12a [2, 3, 6]
Treated seed S01, S02, S03, S04b, S05, S06a, S07 [1, 4, 5]; S04a [1, 5]
Medical advice M03 [4]; M04a [1, 5]

431   tolclofos-methyl

A protectant nitroaniline fungicide for soil-borne diseases
FRAC mode of action code: 14

Products

1	Basilex	Everris Ltd	50% w/w	WP	07494
2	Rizolex 10D	Interfarm	10% w/w	DP	14204
3	Rizolex Flowable	Interfarm	500 g/l	FS	14207

Uses
- Black scurf and stem canker in *potatoes* (tuber treatment) [2]; *potatoes* (tuber treatment only to be used with automatic planters) [3]
- Bottom rot in *protected lettuce* [1]
- Damping off and foot rot in *radishes* (off-label) [1]
- Damping off and wirestem in *brussels sprouts*, *cabbages*, *calabrese*, *cauliflowers*, *chinese cabbage* [1]

SEE SECTION 3 FOR PRODUCTS ALSO REGISTERED

- Foot rot in **ornamental plant production** [1]
- Rhizoctonia in **broccoli** *(off-label)*, **leaf brassicas** *(off-label)*, **protected celery** *(off-label)*, **protected radishes** *(off-label)*, **swedes** *(off-label)*, **turnips** *(off-label)* [1]; **potatoes** *(chitted seed treatment)* [3]
- Root rot in **ornamental plant production** [1]

Extension of Authorisation for Minor Use (EAMUs)
- **broccoli** *20102302* [1]
- **leaf brassicas** *20102301* [1]
- **protected celery** *20102291* [1]
- **protected radishes** *20102290* [1]
- **radishes** *20102292* [1]
- **swedes** *20102293* [1]
- **turnips** *20102293* [1]

Approval information
- Tolclofos-methyl included in Annex I under EC Regulation 1107/2009
- In 2006 CRD required that all products containing this active ingredient should carry the following warning in the main area of the container label: "Tolclofos-methyl is an anticholinesterase organophosphate. Handle with care"
- May be applied by misting equipment mounted over roller table. See label for details [2]

Efficacy guidance
- Apply dust to seed potatoes during hopper loading [2]
- Apply flowable formulation to clean tubers with suitable misting equipment over a roller table. Spray as potatoes taken into store (first earlies) or as taken out of store (second earlies, maincrop, crops for seed) pre-chitting [3]
- Do not mix flowable formulation with any other product [3]
- To control Rhizoctonia in vegetables and ornamentals apply as drench before sowing, pricking out or planting [1]
- On established seedlings and pot plants apply as drench and rinse off foliage [1]

Restrictions
- Tolclofos-methyl is an atypical organophosphorus compound which has weak anticholinesterase activity. Do not use if under medical advice not to work with such compounds
- Maximum number of treatments 1 per batch for seed potatoes; 1 per crop for lettuce, brassicas, swedes, turnips; 1 at each stage of growth (ie sowing, pricking out, potting) to a maximum of 3, for ornamentals
- Only to be used with automatic planters [2]
- Not recommended for use on seed potatoes where hot water treatment used or to be used [2]
- Do not apply as overhead drench to vegetables or ornamentals when hot and sunny [1]
- Do not use on heathers [1]
- Must not be used via hand-held equipment
- Application to protected crops must only be made where the operator is outside the structure at the time of treatment

Crop-specific information
- Latest use: at planting of seed potatoes [2]; before transplanting for lettuce, brassicas, ornamental specimens [1]; before 2 true lvs for swedes, turnips [1]

Environmental safety
- Dangerous for the environment [1-3]
- Very toxic to aquatic organisms
- Treated tubers to be used as seed only, not for food or feed [2]
- LERAP Category B [1]

Hazard classification and safety precautions
Hazard Irritant, Dangerous for the environment
Transport code 9
Packaging group III
UN Number 3077 [1, 2]; 3082 [3]
Risk phrases R36, R37, R38, R51 [1-3]; R53a [1, 3]

FOR FULL CONDITIONS OF USE ALWAYS READ THE PRODUCT LABEL

Operator protection A, D, H [1-3]; C [3]; E [2, 3]; J, M [2]; U05a, U11 [2]; U16a [3]; U19a, U20b [1-3]
Environmental protection E15a [2, 3]; E16a [1]
Storage and disposal D01, D02, D12a [2]; D08 [2, 3]; D09a [1-3]; D10a [3]; D11a [1, 2]
Medical advice M01 [1]; M04a [2, 3]

432   tralkoxydim

SECTION 2

A foliar applied oxime herbicide for grass weed control in cereals.
HRAC mode of action code: A

Products

Grasp 400 SC	Nufarm UK	400 g/l	SC	13594

Uses
- Awned canary grass in *durum wheat, spring barley, spring wheat, triticale, winter barley, winter rye, winter wheat*
- Blackgrass in *durum wheat, spring barley, spring wheat, triticale, winter barley, winter rye, winter wheat*
- Italian ryegrass in *durum wheat, spring barley, spring wheat, triticale, winter barley, winter rye, winter wheat*
- Perennial ryegrass in *durum wheat, spring barley, spring wheat, triticale, winter barley, winter rye, winter wheat*
- Wild oats in *durum wheat, spring barley, spring wheat, triticale, winter barley, winter rye, winter wheat*
- Yorkshire fog in *durum wheat, spring barley, spring wheat, triticale, winter barley, winter rye, winter wheat*

Approval information
- Accepted by BBPA for use on malting barley
- Tralkoxydim included in Annex I under EC Regulation 1107/2009

Efficacy guidance
- Product leaf-absorbed and translocated rapidly to growing points. Best results achieved after completion of emergence of grass weeds when growing actively in competitive crops under warm humid conditions with adequate soil moisture
- Activity not dependent on soil type or condition. Weeds germinating after application will not be controlled
- Best control of wild oats obtained from 2 leaf to 1st node detectable stage of weeds, and of blackgrass up to 3 tillers
- Tralkoxydim is an ACCase inhibitor herbicide. To avoid the build up of resistance do not apply products containing an ACCase inhibitor herbicide more than twice to any crop. In addition do not use any product containing tralkoxydim in mixture or sequence with any other product containing the same ingredient
- Use these products as part of a resistance management strategy that includes cultural methods of control and does not use ACCase inhibitors as the sole chemical method of grass weed control
- Applying a second product containing an ACCase inhibitor to a crop will increase the risk of resistance development; only use a second ACCase inhibitor to control different weeds at a different timing
- Always follow WRAG guidelines for preventing and managing herbicide resistant weeds. See Section 5 for more information

Restrictions
- Maximum number of treatments 1 per crop. See label
- Authorised adjuvant must always be added. See label
- Do not use on oats
- Do not spray undersown crops or crops to be undersown
- Do not spray when foliage wet or covered in ice or crop otherwise under stress
- Do not spray if a protracted period of cold weather forecast

SEE SECTION 3 FOR PRODUCTS ALSO REGISTERED

- Do not spray crops under stress from chemical treatment, grazing, pest attack, mineral deficiency or low fertility. Treatment of stressed crops may be followed by transient foliar discolouration
- Do not roll or harrow within 1 wk of spraying
- Restrictions apply to mixtures or sequences with phenoxy hormone or sulfonylurea herbicides. See label

Crop-specific information
- Latest use: before booting (GS 41)
- Apply to winter cereals from 2 leaves unfolded. If necessary winter cereals may be sprayed twice: once in autumn and once in spring
- Apply to spring cereals from end of tillering

Environmental safety
- Take care to avoid drift onto neighbouring crops, especially oats

Hazard classification and safety precautions
UN Number N/C
Operator protection A, H; U02a, U04a, U05a, U08, U20b
Environmental protection E15b
Storage and disposal D01, D02, D05, D09a, D10c

433   tri-allate

A soil-acting thiocarbamate herbicide for grass weed control
HRAC mode of action code: N

Products

Avadex Excel 15G	Gowan	15% w/w	GR	12109

Uses
- Annual grasses in **durum wheat** *(off-label)*, **linseed** *(off-label)*, **spring oilseed rape** *(off-label)*, **winter oilseed rape** *(off-label)*
- Blackgrass in **broad beans** *(off-label)*, **canary seed** *(off-label - for wild bird seed production)*, **combining peas, durum wheat, fodder beet, lucerne, lupins** *(off-label)*, **mangels, red beet, red clover, sainfoin, spring barley, spring field beans, sugar beet, triticale, vetches, vining peas, white clover, winter barley, winter field beans, winter wheat**
- Meadow grasses in **broad beans** *(off-label)*, **canary seed** *(off-label - for wild bird seed production)*, **combining peas, durum wheat, fodder beet, lucerne, lupins** *(off-label)*, **mangels, red beet, red clover, sainfoin, spring barley, spring field beans, sugar beet, triticale, vetches, vining peas, white clover, winter barley, winter field beans, winter wheat**
- Wild oats in **broad beans** *(off-label)*, **canary seed** *(off-label - for wild bird seed production)*, **combining peas, durum wheat, fodder beet, lucerne, lupins** *(off-label)*, **mangels, red beet, red clover, sainfoin, spring barley, spring field beans, sugar beet, triticale, vetches, vining peas, white clover, winter barley, winter field beans, winter wheat**

Extension of Authorisation for Minor Use (EAMUs)
- **broad beans** 20063531
- **canary seed** *(for wild bird seed production)* 20070867
- **durum wheat** 20063532
- **linseed** 20101945
- **lupins** 20063530
- **spring oilseed rape** 20100135
- **winter oilseed rape** 20100135

Approval information
- Tri-allate included in Annex 1 under EC Regulation 1107/2009
- Approved for aerial application on wheat, barley, rye, triticale, field beans, peas, fodder beet, sugar beet, red beet, lucerne, sainfoin, vetches, clover. See Section 5 for more information
- Accepted by BBPA for use on malting barley

FOR FULL CONDITIONS OF USE ALWAYS READ THE PRODUCT LABEL

Efficacy guidance
- Incorporate or apply to surface pre-emergence (post-emergence application possible in winter cereals up to 2-leaf stage of wild oats)
- Do not use on soils with more than 10% organic matter
- Wild oats controlled up to 2-leaf stage
- If applied to dry soil, rainfall needed for full effectiveness, especially with granules on surface. Do not use if top 5-8 cm bone dry
- Do not apply with spinning disc granule applicator; see label for suitable types
- Do not apply to cloddy seedbeds
- Use sequential treatments to improve control of barren brome and annual dicotyledons (see label for details)

Restrictions
- Maximum number of treatments 1 per crop
- Consolidate loose, puffy seedbeds before drilling to avoid chemical contact with seed
- Drill cereals well below treated layer of soil (see label for safe drilling depths)
- Do not use on direct-drilled crops or undersow grasses into treated crops
- Do not sow oats or grasses within 1 yr of treatment

Crop-specific information
- Latest use: pre-drilling for beet crops; before crop emergence for field beans, spring barley, peas, forage legumes; before first node detectable stage (GS 31) for winter wheat, winter barley, durum wheat, triticale, winter rye

Environmental safety
- Irritating to eyes and skin
- May cause sensitization by skin contact
- Harmful to fish or other aquatic life. Do not contaminate surface waters or ditches with chemical or used container

Hazard classification and safety precautions
> **Hazard** Irritant
> **Transport code** 9
> **Packaging group** III
> **UN Number** 3077
> **Risk phrases** R36, R38, R43
> **Operator protection** A, C, H; U02a, U05a, U20a
> **Environmental protection** E13c
> **Storage and disposal** D01, D02, D09a, D11a

434   triazoxide

A benzotriazine fungicide available only in mixtures
FRAC mode of action code: 35

See also clothianidin + prothioconazole + tebuconazole + triazoxide
imidacloprid + tebuconazole + triazoxide
prothioconazole + tebuconazole + triazoxide
tebuconazole + triazoxide

SEE SECTION 3 FOR PRODUCTS ALSO REGISTERED

435   tribenuron-methyl

A foliar acting sulfonylurea herbicide with some root activity for use in cereals
HRAC mode of action code: B

*See also fluroxypyr + thifensulfuron-methyl + tribenuron-methyl
        metsulfuron-methyl + tribenuron-methyl
        thifensulfuron-methyl + tribenuron-methyl*

Products

1	Helmstar	Rotam Agrochemical	75% w/w	WG	15070
2	Thor	Nufarm UK	50% w/w	WG	15239
3	Titanium	ChemSource	50% w/w	SG	15007
4	Triad	Headland	50% w/w	TB	12751

Uses

- Annual dicotyledons in **durum wheat**, **farm forestry** *(off-label)*, **forest nurseries** *(off-label)*, **green cover on land temporarily removed from production**, **miscanthus** *(off-label)* [2]; **spring barley** [2-4]; **spring oats**, **spring wheat**, **triticale**, **winter barley**, **winter oats**, **winter rye** [2, 3]; **undersown spring barley** [4]; **winter wheat** [1-3]
- Charlock in **spring barley**, **spring oats**, **spring wheat**, **triticale**, **winter barley**, **winter oats**, **winter rye**, **winter wheat** [3]
- Chickweed in **spring barley** [3, 4]; **spring oats**, **spring wheat**, **triticale**, **winter barley**, **winter oats**, **winter rye**, **winter wheat** [3]; **undersown spring barley** [4]
- Mayweeds in **spring barley**, **spring oats**, **spring wheat**, **triticale**, **winter barley**, **winter oats**, **winter rye**, **winter wheat** [3]

Extension of Authorisation for Minor Use (EAMUs)

- **farm forestry** *20122064* [2]
- **forest nurseries** *20122064* [2]
- **miscanthus** *20122064* [2]

Approval information

- Tribenuron-methyl included in Annex I under EC Regulation 1107/2009Accepted by BBPA for use on malting barley

Efficacy guidance

- Best control achieved when weeds small and actively growing
- Good spray cover must be achieved since larger weeds often become less susceptible
- Susceptible weeds cease growth almost immediately after treatment and symptoms can be seen in about 2 wk
- Weed control may be reduced when conditions very dry
- Tribenuron-methyl is a member of the ALS-inhibitor group of herbicides and products should be used in a planned Resistance Management strategy. See Section 5 for more information

Restrictions

- Maximum number of treatments 1 per crop
- Specific restrictions apply to use in sequence or tank mixture with other sulfonylurea or ALS-inhibiting herbicides. See label for details
- Do not apply to crops undersown with grass, clover or other broad-leaved crops
- Do not apply to any crop suffering stress from any cause or not actively growing
- Do not apply within 7 d of rolling

Crop-specific information

- Latest use: up to and including flag leaf ligule/collar just visible (GS 39)
- Apply in autumn or in spring from 3 leaf stage of crop

Following crops guidance

- Only cereals, field beans or oilseed rape may be sown in the same calendar yr as harvest of a treated crop
- In the event of crop failure sow only a cereal within 3 mth of application. After 3 mth field beans or oilseed rape may also be sown

FOR FULL CONDITIONS OF USE ALWAYS READ THE PRODUCT LABEL

Environmental safety
- Dangerous for the environment
- Very toxic to aquatic organisms
- Take extreme care to avoid drift onto broad-leaved plants outside the target area or onto surface waters or ditches, or land intended for cropping
- Spraying equipment should not be drained or flushed onto land planted, or to be planted, with trees or crops other than cereals and should be thoroughly cleansed after use - see label for instructions

Hazard classification and safety precautions
>**Hazard** Irritant [2-4]; Dangerous for the environment [1-4]
>**Transport code** 9
>**Packaging group** III
>**UN Number** 3077
>**Risk phrases** R43, R50, R53a [1-4]; R70 [1, 2]
>**Operator protection** A [1-4]; H [1-3]; U05a, U08, U20b
>**Environmental protection** E07b [4] (3 weeks); E07e [4]; E15a [2-4]; E15b [1]; E38 [1-3]
>**Storage and disposal** D01, D02, D09a, D11a [1-4]; D12a [1-3]

436 Trichoderma asperellum (Strain T34)

>Biological control agent

Products

T34 Biocontrol	Fargro	10.832% w/w	WP	15603

Uses
- Fusarium foot rot and seedling blight in **container-grown crops**, **ornamental plant production**
- Fusarium root rot in **container-grown crops** *(off-label)*, **forest nurseries** *(off-label)*, **protected forest nurseries** *(off-label)*, **protected ornamentals** *(off-label)*

Extension of Authorisation for Minor Use (EAMUs)
- **container-grown crops** *20121118*
- **forest nurseries** *20121118*
- **protected forest nurseries** *20121118*
- **protected ornamentals** *20121118*

Approval information
- Trichoderma asperellum awaiting inclusion in Annex 1 under EC Regulation 1107/2009

Efficacy guidance
- May be applied by spraying, through an irrigation system or by root dipping.

Restrictions
- Efficacy has been demonstrated on peat and coir composts but should be checked on the more unusual composts before large scale use.

Crop-specific information
- Crop safety has been demonstrated on a range of carnation species. It is advisable to check the safety to other species on a sample of the population before large-scale use.

Hazard classification and safety precautions
>**Hazard** Harmful
>**Risk phrases** R42, R43
>**Operator protection** A, D, H; U05a, U09a, U19a, U20b
>**Environmental protection** E15a
>**Storage and disposal** D01, D02, D05, D09a, D09b, D10c
>**Medical advice** M03, M04a

SEE SECTION 3 FOR PRODUCTS ALSO REGISTERED

437   triclopyr

A pyridinecarboxylic acid herbicide for perennial and woody weed control
HRAC mode of action code: O

See also 2,4-D + dicamba + triclopyr
2,4-D + triclopyr
aminopyralid + triclopyr
clopyralid + fluroxypyr + triclopyr
clopyralid + triclopyr
florasulam + triclopyr
fluroxypyr + triclopyr

Products

1	Garlon 4	Dow	480 g/l	EC	05090
2	Nomix Garlon 4	Nomix Enviro	480 g/l	EC	13868
3	PureWoody	Pure Amenity	480 g/l	EC	15133
4	Thrash	Pan Agriculture	480 g/l	EC	15067
5	Timbrel	Dow	480 g/l	EC	05815
6	Woody	AgChem Access	480 g/l	EC	14163

Uses
- Annual dicotyledons in *green cover on land temporarily removed from production* [4]
- Brambles in *forest, land not intended to bear vegetation* [1-6]
- Broom in *forest, land not intended to bear vegetation* [1-6]
- Brush clearance in *industrial sites* [1-6]
- Docks in *forest, land not intended to bear vegetation* [1-6]
- Gorse in *forest, land not intended to bear vegetation* [1-6]
- Perennial dicotyledons in *asparagus (off-label)* [5]; *forest, industrial sites, land not intended to bear vegetation* [1-6]; *green cover on land temporarily removed from production* [4]
- Rhododendrons in *forest, land not intended to bear vegetation* [1-6]
- Scrub clearance in *industrial sites* [1-6]
- Stinging nettle in *forest, land not intended to bear vegetation* [1-6]
- Woody weeds in *forest, industrial sites, land not intended to bear vegetation* [1-6]

Extension of Authorisation for Minor Use (EAMUs)
- *asparagus 20102938* [5]

Approval information
- Triclopyr included in Annex I under EC Regulation 1107/2009
- Approval expiry 31 Aug 2013 [6]

Efficacy guidance
- Apply in grassland as spot treatment or overall foliage spray when weeds in active growth in spring or summer. Details of dose and timing vary with species. See label
- Apply to woody weeds as summer foliage, winter shoot, basal bark, cut stump or tree injection treatment
- Apply foliage spray in water when leaves fully expanded but not senescent
- Do not spray in drought, in very hot or cold conditions
- Control may be reduced if rain falls within 2 h of application
- Control of rhododendron can be variable. If higher than 1.8 m cut stump treatment recommended. A follow-up shoot treatment may be required
- See label for maximum concentrations when applying in oil, water or via watering can

Restrictions
- Maximum number of treatments 2 per yr
- Do not apply to grass leys less than 1 yr old
- Do not drill kale, swedes, turnips, grass or mixtures containing clover within 6 wk of treatment. Allow at least 6 wk before planting trees
- Not to be used on food crops
- Do not apply through hand held rotary atomisers

FOR FULL CONDITIONS OF USE ALWAYS READ THE PRODUCT LABEL

- Not to be applied in or near water

Crop-specific information
- Latest use: 6 wk before replanting; 7 d before grazing
- Uses on land not intended for cropping include grassland of no agricultural interest such as roadside verges, railway and motorway embankments
- Apply winter shoot, basal bark or cut stump sprays in paraffin or diesel oil. Dose and timing vary with species. See label for details
- Inject undiluted or 1:1 dilution into cuts spaced every 7.5 cm round trunk
- Clover will be killed or severely checked by application in grassland

Environmental safety
- Dangerous for the environment
- Very toxic to aquatic organisms
- Do not allow spray to drift onto agricultural or horticultural crops, amenity plantings, gardens, ponds, lakes or water courses. Vapour drift may occur under hot conditions
- Keep livestock out of treated areas for at least 7 d and until foliage of any poisonous weeds such as buttercups or ragwort has died and become unpalatable
- LERAP Category B

Hazard classification and safety precautions
  Hazard Harmful, Dangerous for the environment
  Transport code 9
  Packaging group III
  UN Number 3082
  Risk phrases R22a, R22b, R38, R43, R50, R53a
  Operator protection A, C, H, M; U02a, U05a, U08, U14, U20b
  Environmental protection E07a, E15a, E16a, E16b, E23, E34, E38
  Consumer protection C01
  Storage and disposal D01, D02, D09a, D10b, D12a
  Medical advice M05b

438   trifloxystrobin

A protectant strobilurin fungicide for cereals and managed amenity turf
FRAC mode of action code: 11

See also cyproconazole + trifloxystrobin
  fluoxastrobin + prothioconazole + trifloxystrobin
  propiconazole + trifloxystrobin
  prothioconazole + trifloxystrobin
  tebuconazole + trifloxystrobin

Products

1	Martinet	AgChem Access	500 g/l	SC	14614
2	Pan Aquarius	Pan Agriculture	50% w/w	WG	13018
3	PureFloxy	Pure Amenity	50% w/w	WG	15473
4	Scorpio	Bayer Environ.	50% w/w	WG	12293
5	Swift SC	Bayer CropScience	500 g/l	SC	11227

Uses
- Brown rust in *spring barley*, *winter barley*, *winter wheat* [1, 5]
- Foliar disease control in *durum wheat* (off-label), *grass seed crops* (off-label), *ornamental specimens* (off-label), *spring rye* (off-label), *triticale* (off-label), *winter rye* (off-label) [5]
- Fusarium patch in *amenity grassland*, *managed amenity turf* [2-4]
- Glume blotch in *winter wheat* [1, 5]
- Net blotch in *spring barley*, *winter barley* [1, 5]
- Red thread in *amenity grassland*, *managed amenity turf* [2-4]
- Rhynchosporium in *spring barley*, *winter barley* [1, 5]
- Septoria leaf blotch in *winter wheat* [1, 5]

SEE SECTION 3 FOR PRODUCTS ALSO REGISTERED

Extension of Authorisation for Minor Use (EAMUs)
- ***durum wheat*** *20061287 expires 30 Sep 2013* [5]
- ***grass seed crops*** *20061287 expires 30 Sep 2013* [5]
- ***ornamental specimens*** *20082882 expires 30 Sep 2013* [5]
- ***spring rye*** *20061287 expires 30 Sep 2013* [5]
- ***triticale*** *20061287 expires 30 Sep 2013* [5]
- ***winter rye*** *20061287 expires 30 Sep 2013* [5]

Approval information
- Trifloxystrobin included in Annex I under EC Regulation 1107/2009
- Accepted by BBPA for use on malting barley
- Approval expiry 30 Sep 2013 [1, 2, 5]

Efficacy guidance
- Should be used protectively before disease is established in crop. Further treatment may be necessary if disease attack prolonged
- Treat grass after cutting and do not mow for at least 48 h afterwards to allow adequate systemic movement [2, 4]
- Trifloxystrobin is a member of the QoI cross resistance group. Product should be used preventatively and not relied on for its curative potential
- Use product as part of an Integrated Crop Management strategy incorporating other methods of control, including where appropriate other fungicides with a different mode of action. Do not apply more than two foliar applications of QoI containing products to any cereal crop
- There is a significant risk of widespread resistance occurring in *Septoria tritici* populations in UK. Failure to follow resistance management action may result in reduced levels of disease control
- On cereal crops product must always be used in mixture with another product, recommended for control of the same target disease, that contains a fungicide from a different cross resistance group and is applied at a dose that will give robust control
- Strains of barley powdery mildew resistant to QoIs are common in the UK

Restrictions
- Maximum number of treatments 2 per crop per yr
- Do not apply to turf during dought conditions or to frozen turf [2, 4]

Crop-specific information
- HI barley, wheat 35 d [5]

Environmental safety
- Dangerous for the environment
- Very toxic to aquatic organisms
- LERAP Category B [2-4]

Hazard classification and safety precautions
Hazard Irritant [2-4]; Dangerous for the environment [1-5]
Transport code 9
Packaging group III
UN Number 3077 [2-4]; 3082 [1, 5]
Risk phrases R43 [2-4]; R50, R53a [1-5]
Operator protection A, H; U02a, U09a, U19a [1, 5]; U05a, U20b [1-5]; U11, U13, U14, U15 [2-4]
Environmental protection E13b [1, 5]; E15b, E16a, E16b, E34 [2-4]; E38 [1-5]
Consumer protection C02a [1, 5] (35 d)
Storage and disposal D01, D02, D09a, D10c, D12a
Medical advice M03 [2-4]

439   triflusulfuron-methyl

A sulfonyl urea herbicide for beet crops
HRAC mode of action code: B

See also lenacil + triflusulfuron-methyl

Products

Debut	DuPont	50% w/w	WG	07804

Uses
- Annual dicotyledons in *chicory* (off-label), *fodder beet*, *red beet* (off-label), *sugar beet*
- Charlock in *red beet* (off-label)
- Cleavers in *red beet* (off-label)
- Flixweed in *red beet* (off-label)
- Fool's parsley in *red beet* (off-label)
- Nipplewort in *red beet* (off-label)
- Wild chrysanthemum in *red beet* (off-label)

Extension of Authorisation for Minor Use (EAMUs)
- *chicory* 20120743
- *red beet* 20120742

Approval information
- Triflusulfuron-methyl included in Annex I under EC Regulation 1107/2009

Efficacy guidance
- Product should be used with a recommended adjuvant or a suitable herbicide tank-mix partner - see label for details
- Product acts by foliar action. Best results obtained from good spray cover of small actively growing weeds
- Susceptible weeds cease growth immediately and symptoms can be seen 5-10 d later
- Best results achieved from a programme of up to 4 treatments starting when first weeds have emerged with subsequent applications every 5-14 d when new weed flushes at cotyledon stage
- Weed spectrum can be broadened by tank mixture with other herbicides. See label for details
- Product may be applied overall or via band sprayer
- Triflusulfuron-methyl is a member of the ALS-inhibitor group of herbicides

Restrictions
- Maximum number of treatments 4 per crop
- Do not apply to any crop stressed by drought, water-logging, low temperatures, pest or disease attack, nutrient or lime deficiency
- A maximum total dose of 60 g/ha triflusulfuron may only be applied every third year on the same field.

Crop-specific information
- Latest use: before crop leaves meet between rows
- HI 4 wk for red beet
- All varieties of sugar beet and fodder beet may be treated from early cotyledon stage until the leaves begin to meet between the rows

Following crops guidance
- Only winter cereals should follow a treated crop in the same calendar yr. Any crop may be sown in the next calendar yr
- After failure of a treated crop, sow only spring barley, linseed or sugar beet within 4 mth of spraying unless prohibited by tank-mix partner

Environmental safety
- Dangerous for the environment
- Very toxic to aquatic organisms
- Extremely dangerous to fish or other aquatic life. Do not contaminate surface waters or ditches with chemical or used container

SEE SECTION 3 FOR PRODUCTS ALSO REGISTERED

- Take extreme care to avoid drift onto broad-leaved plants outside the target area or onto surface waters or ditches, or land intended for cropping
- Spraying equipment should not be drained or flushed onto land planted, or to be planted, with trees or crops other than sugar beet and should be thoroughly cleansed after use - see label for instructions
- LERAP Category B

Hazard classification and safety precautions
Hazard Dangerous for the environment
Transport code 9
Packaging group III
UN Number 3077
Risk phrases R50, R53a
Operator protection A; U05a, U08, U19a, U20a
Environmental protection E13a, E15a, E16a, E16b, E38
Storage and disposal D01, D02, D05, D09a, D11a, D12a

440　trinexapac-ethyl

A novel cyclohexanecarboxylate plant growth regulator for cereals, turf and amenity grassland

Products

1	Clipless	Headland Amenity	120 g/l	ME	15435
2	Cutaway	Syngenta	121 g/l	SL	14445
3	Moddus	Syngenta	250 g/l	EC	15151
4	Optimus	Makhteshim	175 g/l	EC	15249
5	Pan Tepee	Pan Agriculture	250 g/l	EC	15867
6	Primo Maxx	Syngenta	121 g/l	SL	14780
7	Pure Max	Pure Amenity	121 g/l	SL	14952
8	Shorten	Becesane	250 g/l	EC	15377
9	Shrink	AgChem Access	250 g/l	EC	15660
10	Standon Pygmy	Standon	250 g/l	EC	15169
11	Tacet	Goldengrass	250 g/l	EC	15258
12	Tempo	Syngenta	250 g/l	EC	15170

Uses
- Growth regulation in *canary seed (off-label)*, *spring wheat (off-label)* [4]; *durum wheat, ryegrass seed crops, spring barley, spring oats, spring rye, triticale, winter barley, winter oats, winter rye, winter wheat* [8-11]; *forest nurseries (off-label)*, *ornamental plant production (off-label)* [3]
- Growth retardation in *amenity grassland* [1, 2, 6, 7]; *managed amenity turf* [1, 6, 7]; *spring red wheat (off-label)* [12]
- Lodging control in *durum wheat, ryegrass seed crops, spring barley, spring oats, spring rye, triticale, winter barley, winter oats, winter rye, winter wheat* [3-5, 12]; *spring red wheat (off-label)* [3, 12]

Extension of Authorisation for Minor Use (EAMUs)
- *canary seed* 20111263 [4]
- *forest nurseries* 20122055 [3]
- *ornamental plant production* 20103062 [3]
- *spring red wheat* 20121385 [3], 20113201 [12]
- *spring wheat* 20111264 [4]

Approval information
- Trinexapac-ethyl included in Annex I under EC Regulation 1107/2009
- Accepted by BBPA for use on malting barley

Efficacy guidance
- Best results on cereals and ryegrass seed crops obtained from treatment from the leaf sheath erect stage [3, 5]

FOR FULL CONDITIONS OF USE ALWAYS READ THE PRODUCT LABEL

- Best results on turf achieved from application to actively growing weed free turf grass that is adequately fertilized and watered and is not under stress. Adequate soil moisture is essential [7]
- Turf should be dry and weed free before application [7]
- Environmental conditions, management and cultural practices that affect turf growth and vigour will influence effectiveness of treatment [7]
- Repeat treatments on turf up to the maximum approved dose may be made as soon as growth resumes [7]

Restrictions
- Maximum total dose equivalent to one full dose on cereals, ryegrass seed crops [3, 5]
- Maximum number of treatments on turf equivalent to five full dose treatments [7]
- Do not apply if rain or frost expected or if crop wet. Products are rainfast after 12 h
- Only use on crops at risk of lodging [3, 5]
- Do not apply within 12 h of mowing turf [7]
- Do not treat newly sown turf [7]
- Not to be used on food crops [7]
- Do not compost or mulch grass clippings [7]

Crop-specific information
- Latest use: before 2nd node detectable (GS 32) for oats, ryegrass seed crops; before 3rd node detectable (GS 33) for durum wheat, spring barley, triticale, rye; before flag leaf sheath extending (GS 41) for winter barley, winter wheat
- On wheat apply as single treatment between leaf sheath erect stage (GS 30) and flag leaf fully emerged (GS 39) [3, 5]
- On barley, rye, triticale and durum wheat apply as single treatment between leaf sheath erect stage (GS 30) and second node detectable (GS 32), or on winter barley at higher dose between flag leaf just visible (GS 37) and flag leaf fully emerged (GS 39) [3, 5]
- On oats and ryegrass seed crops apply between leaf sheath erect stage (GS 30) and first node detectable stage (GS 32) [3, 5]
- Treatment may cause ears of cereals to remain erect through to harvest [3, 5]
- Turf under stress when treated may show signs of damage [7]
- Any weed control in turf must be carried out before application of the growth regulator [7]

Environmental safety
- Dangerous for the environment [3, 5]
- Toxic to aquatic organisms [3, 5]

Hazard classification and safety precautions
Hazard Irritant, Dangerous for the environment [3-5, 8-12]
Transport code 9 [1, 3-5, 8-12]
Packaging group III [1, 3-5, 8-12]
UN Number 3082 [1, 3-5, 8-12]; N/C [2, 6, 7]
Risk phrases R43, R51, R53a [3-5, 8-12]
Operator protection A, C [1-12]; H, K [1, 2, 6, 7]; U05a [1-12]; U15, U20c [3-5, 8-12]; U20b [1, 2, 6, 7]
Environmental protection E15a, E38 [3-5, 8-12]
Consumer protection C01 [1, 2, 6, 7]
Storage and disposal D01, D02, D05, D10c [1-12]; D09a, D12a [3-5, 8-12]

441   triticonazole

A triazole fungicide available only in mixtures
FRAC mode of action code: 3

See also guazatine + triticonazole
imazalil + triticonazole
prochloraz + triticonazole

SEE SECTION 3 FOR PRODUCTS ALSO REGISTERED

442   urea

Commodity substance for fungicide treatment of cut tree stumps

443   warfarin

A coumarin anti-coagulant rodenticide

Products

1	Grey Squirrel Bait	Killgerm	0.02% w/w	RB	14807
2	Sewarin Extra	Killgerm	0.05% w/w	RB	H6810
3	Warfarin 0.5% Concentrate	B H & B	0.5% w/w	CB	H6815
4	Warfarin Ready Mixed Bait	B H & B	0.025% w/w	RB	H6816

Uses

- Grey squirrels in *agricultural premises*, *forest*, *trees* *(nuts)* [1]
- Rats in *agricultural premises* [2-4]

Approval information

- Warfarin included in Annex I under EC Regulation 1107/2009

Efficacy guidance

- For rodent control place ready-to-use or prepared baits at many points wherever rats active. Out of doors shelter bait from weather [2-4]
- Inspect baits frequently and replace or top up as long as evidence of feeding. Do not underbait [2-4]
- Use grey squirrel baits when bark stripping is evident in farm forestry or forestry or where there is risk of damage to trees grown for their nuts [1]
- Use squirrel bait in specially constructed hoppers and inspect every 2-3 d. Place bait hoppers so as to prevent rainwater entry and replace as necessary [1]

Restrictions

- For use only by local authorities, professional operators providing a pest control service and persons occupying industrial, agricultural or horticultural premises
- For use only between 15 Mar and 15 Aug for tree protection [1]
- Squirrel bait hoppers must be clearly labelled "Poison" or carry the instruction "Tree protection, do not disturb" [1]
- Warfarin baits for grey squirrel control must only be used outdoors in specially constructed hoppers which comply with the Grey Squirrel Order 1973, in Scotland and specified counties of England and Wales. See label [1]
- The use of warfarin to control grey squirrels is illegal unless the provisions of the Grey Squirrels Order 1973 are observed [1]

Environmental safety

- Harmful to wildlife
- Prevent access to baits by children and animals, especially cats, dogs and pigs
- Rodent bodies must be searched for and burned or buried, not placed in refuse bins or rubbish tips. Remains of bait and containers must be removed after treatment and burned or buried
- Bait must not be used where food, feed or water could become contaminated
- If squirrel bait hoppers have obviously been disturbed by badgers or other animals, change the site or lift onto tables or platforms [1]
- Warfarin baits must nor be used outdoors where pine martens are known to occur naturally [1]
- Under the terms of the Wildlife and Countryside Act 1981 warfarin squirrel baits must not be used outdoors in England, Scotland or Wales where red squrrels are known to occur [1]

Hazard classification and safety precautions
  UN Number N/C [1, 2, 4]
  Operator protection A [1, 2]; C, D, E, H [2]; U13 [1-4]; U20a [2]; U20b [1, 3, 4]

FOR FULL CONDITIONS OF USE ALWAYS READ THE PRODUCT LABEL

Environmental protection E10b, E15a [1]
Storage and disposal D09a [1-4]; D10a [3, 4]; D11a [1, 2]
Vertebrate/rodent control products V01a, V03a [3, 4]; V01b [1, 2]; V02 [1-4]; V03b, V04b [2]; V04a [1, 3, 4]
Medical advice M03 [2]

444   zeta-cypermethrin

A contact and stomach acting pyrethroid insecticide
IRAC mode of action code: 3

Products

1 Angri	AgChem Access	100 g/l	EW	13730
2 Fury 10 EW	Belchim	100 g/l	EW	12248
3 Minuet EW	Belchim	100 g/l	EW	12304

Uses

- Aphids in **broad beans** *(off-label)*, **durum wheat** *(off-label)*, **fodder beet** *(off-label)*, **grass seed crops** *(off-label)*, **lupins** *(off-label)*, **spring rye** *(off-label)*, **triticale** *(off-label)*, **winter rye** *(off-label)* [2]; **spring barley, spring oats, spring wheat, winter barley, winter oats, winter wheat** [1-3]
- Barley yellow dwarf virus vectors in **spring barley, spring wheat, winter barley, winter wheat** [1-3]
- Cabbage seed weevil in **spring oilseed rape, winter oilseed rape** [1-3]
- Cabbage stem flea beetle in **spring oilseed rape, winter oilseed rape** [1-3]
- Cutworms in **potatoes, sugar beet** [1-3]
- Flax flea beetle in **linseed** [1-3]
- Flea beetle in **spring oilseed rape, winter oilseed rape** [1-3]
- Insect pests in **broad beans** *(off-label)*, **durum wheat** *(off-label)*, **fodder beet** *(off-label)*, **grass seed crops** *(off-label)*, **lupins** *(off-label)*, **spring rye** *(off-label)*, **triticale** *(off-label)*, **winter rye** *(off-label)* [2]
- Large flax flea beetle in **linseed** [1-3]
- Pea and bean weevil in **combining peas, spring field beans, vining peas, winter field beans** [1-3]
- Pea aphid in **combining peas, vining peas** [1-3]
- Pea moth in **combining peas, vining peas** [1-3]
- Pod midge in **spring oilseed rape, winter oilseed rape** [1-3]
- Pollen beetle in **spring oilseed rape, winter oilseed rape** [1-3]
- Rape winter stem weevil in **spring oilseed rape, winter oilseed rape** [1-3]

Extension of Authorisation for Minor Use (EAMUs)

- **broad beans** *20063573* [2]
- **durum wheat** *20063570* [2]
- **fodder beet** *20063572* [2]
- **grass seed crops** *20063570* [2]
- **lupins** *20063571* [2]
- **spring rye** *20063570* [2]
- **triticale** *20063570* [2]
- **winter rye** *20063570* [2]

Approval information

- Zeta-cypermethrin included in Annex I under EC Regulation 1107/2009
- Accepted by BBPA for use on malting barley

Efficacy guidance

- On winter cereals spray when aphids first found in the autumn for BYDV control. A second spray may be required on late drilled crops or in mild conditions
- For summer aphids on cereals spray when treatment threshold reached
- For listed pests in other crops spray when feeding damage first seen or when treatment threshold reached. Under high infestation pressure a second treatment may be necessary

SEE SECTION 3 FOR PRODUCTS ALSO REGISTERED

- Best results for pod midge and seed weevil control in oilseed rape obtained from treatment after pod set but before 80% petal fall
- Pea moth treatments should be applied according to ADAS/PGRO warnings or when economic thresholds reached as indicated by pheromone traps
- Treatments for cutworms should be made at egg hatch and repeated no sooner than 10 d later

Restrictions
- Maximum total dose equivalent to two full dose treatments on all crops [2, 3]
- Consult processors before use on crops for processing

Crop-specific information
- Latest use: before 4 true leaves for linseed, before end of flowering for oilseed rape, cereals
- HI potatoes, field beans, peas 14 d; sugar beet 60 d

Environmental safety
- Dangerous for the environment
- Very toxic to aquatic organisms
- High risk to non-target insects or other arthropods. Do not spray within 6 m of the field boundary
- LERAP Category A

Hazard classification and safety precautions
    Hazard Harmful, Dangerous for the environment
    Transport code 6.1
    Packaging group III
    UN Number 3352
    Risk phrases R20, R22a, R43, R50, R53a
    Operator protection A, C, H; U05a, U08, U14, U15, U19a, U20b
    Environmental protection E15b, E16c, E16d, E22a, E34, E38
    Storage and disposal D01, D02, D09a, D10b, D12a
    Medical advice M05a

445   zoxamide

A substituted benzamide fungicide available only in mixtures
FRAC mode of action code: 22

See also mancozeb + zoxamide

SECTION 3
PRODUCTS ALSO REGISTERED

Products also Registered

Products listed in the table below have not been notified for inclusion in Section 2 of this edition of the *Guide*. These products may legally be stored and used in accordance with their label until their approval expires, but they may not still be available for purchase.

Product	Approval holder	MAPP No.	Expiry Date
1 abamectin			
Abam	CMI	15535	
Abamex 18 EC	MAC	15806	
Acaramik	Rotam Agrochemical	14344	
Hi-Mectin	Hockley	15420	
Mectinide	AgriGuard	13953	
2 acetamiprid			
Acetamex 20 SP	MAC	15888	31 Dec 2014
Gazelle	Certis	12909	31 Dec 2014
3 acetic acid			
Natural Weed Spray NO. 1	Punya	12328	
New-Way Weed Spray	Punya	15319	31 Aug 2019
OWK	UK Organic	15363	28 Feb 2022
OWK	UK Organic	15161	30 Apr 2013
4 acibenzolar-S-methyl			
Bion	Syngenta	09803	31 Dec 2015
5 alpha-cypermethrin			
Alert	BASF	13632	
Alpha C 6 ED	Techneat	13611	28 Feb 2015
Alphamex 100 EC	MAC	14071	
Alphax 100	Agrifarm	15647	
Fastac	BASF	13604	
Fastac Dry	BASF	10221	
6 aluminium ammonium sulphate			
Curb S	Sphere	13564	
Guardsman	Chiltern	05494	
Rezist Liquid	Barrett	14643	
Sphere ASBO	Amenity Land	14884	31 Aug 2014
7 aluminium phosphide			
Degesch Fumigation Pellets	Rentokil	11436	
Detia Gas Ex-P	Igrox	09802	
Detia Gas-Ex-B	Detia Degesch	06927	
Detia Gas-Ex-T	Igrox	03792	
Phostoxin I	Rentokil	05694	
Talunex	Certis	14608	
10 ametoctradin + dimethomorph			
Percos	BASF	15248	30 Sep 2014
Resplend	BASF	14975	30 Sep 2014
11 ametoctradin + mancozeb			
Decabane	BASF	14985	30 Sep 2014

EXPIRY DATE IS 31/12/2021 UNLESS OTHERWISE SHOWN

Product	Approval holder	MAPP No.	Expiry Date
12 amidosulfuron			
Amidosulf	Goldengrass	14409	
Barclay Cleave	Barclay	11340	
Eagle	Interfarm	15945	
Landgold Amidosulfuron	Teliton	12110	31 Mar 2013
Pursuit	Bayer CropScience	07333	
Squire	Bayer CropScience	08715	30 Jun 2014
Squire Ultra	Bayer CropScience	13125	30 Jun 2014
13 amidosulfuron + iodosulfuron-methyl-sodium			
Sekator	Bayer CropScience	12634	30 Nov 2013
14 aminopyralid			
Pro-Banish	Dow	14730	31 Jan 2015
15 aminopyralid + fluroxypyr			
Halcyon	Dow	14709	31 Jan 2015
18 amitrole			
Aminotriazole Technical	Nufarm UK	11137	
Weedazol Pro	Nufarm UK	11995	
22 azoxystrobin			
5504	Syngenta	12351	31 Jul 2021
Alpha Azoxystrobin	Makhteshim	14400	31 Dec 2013
Amicron	AgriGuard	13336	31 Jul 2021
Astrobin 250	Euro	15336	31 Jul 2021
Astrobin 250	Euro	15354	31 Jul 2021
Barclay ZX	Barclay	11336	31 Dec 2013
Chamane	United Phosphorus	15922	20 Dec 2015
Clayton Putter	Syngenta	15827	31 Jul 2021
Clayton Stobik	Clayton	09440	31 Dec 2013
Cleancrop Celeb	United Agri	13294	31 Dec 2013
Euclid	Indigrow	15926	31 Jul 2021
Hi Strobin 25	Hockley	15468	31 Jul 2021
Jumbo 250 SC	Aako	14900	24 Feb 2013
Kingdom Turf	AgriGuard	13833	31 Jul 2021
Landgold Strobilurin 250	Teliton	12128	31 Dec 2013
Me2 Azoxystrobin	Me2	09654	31 Dec 2013
Ortiva	Syngenta	10542	31 Jul 2021
Priori	Syngenta	10543	31 Jul 2021
Quake	AA Tech Inc	15294	31 Dec 2013
Reconcile	Becesane	15379	31 Jul 2021
RouteOne Roxybin 25	Albaugh UK	15289	31 Jul 2021
RouteOne Roxybin 25	Albaugh UK	13653	30 Apr 2015
Strobiplus 250 EC	Agform	15094	31 Dec 2013
UPL Azoxystrobin	United Phosphorus	15572	30 Sep 2016
Zoxis	Agriphar	15483	31 Dec 2016
Zoxy	Agriphar	15247	31 Mar 2013
24 azoxystrobin + cyproconazole			
Cecure	AgriGuard	14889	
26 azoxystrobin + fenpropimorph			
RouteOne Roxypro FP	Albaugh UK	13634	30 Apr 2015
RouteOne Roxypro FP	Albaugh UK	15291	

EXPIRY DATE IS 31/12/2021 UNLESS OTHERWISE SHOWN

Product	Approval holder	MAPP No.	Expiry Date
28 Bacillus subtilis			
Serenade ASO	Fargro	14318	31 Aug 2013
32 benalaxyl + mancozeb			
Galben M	Belchim	14661	30 Sep 2013
Galben M	Belchim	15707	29 Mar 2016
Intro Plus	Belchim	15725	29 Mar 2016
Intro Plus	Belchim	14666	31 Oct 2013
Tairel	Belchim	15724	29 Mar 2016
Tairel	Belchim	14659	31 Oct 2013
34 bentazone			
Basagran	BASF	00188	31 Dec 2015
Bently	Chem-Wise	14210	31 Dec 2015
Buddy	EZCrop	15727	31 Dec 2015
Euro Benta 480	Euro	14707	31 Dec 2015
Hockley Bentazone 48	Hockley	15543	31 Dec 2015
IT Bentazone	I T Agro	13132	31 Dec 2015
RouteOne Benta 48	Albaugh UK	13662	30 Apr 2015
RouteOne Benta 48	Albaugh UK	15266	31 Dec 2015
RouteOne Bentazone 48	Albaugh UK	15269	31 Dec 2015
RouteOne Bentazone 48	Albaugh UK	13664	30 Apr 2015
Zone 48	AgriGuard	13165	31 Dec 2015
38 bentazone + pendimethalin			
Impuls	BASF	13372	31 Dec 2013
40 benthiavalicarb-isopropyl + mancozeb			
En-Garde	Certis	14901	31 Dec 2018
41 benzoic acid			
MENNO Florades	Brinkman	15091	31 May 2014
42 benzyladenine			
MaxCel	Interfarm	15708	29 Mar 2016
45 beta-cyfluthrin + imidacloprid			
Chinook	Bayer CropScience	13696	
46 bifenazate			
Inter Bifenazate 240 SC	Iticon	14543	30 Nov 2015
47 bifenox			
Aspire	AgriGuard	14303	30 Jun 2014
Cleancrop Diode	United Agri	14620	30 Jun 2014
Oram	Makhteshim	13410	30 Jun 2014
Sabine	Makhteshim	13439	30 Jun 2014
48 bifenox + MCPA + mecoprop-P			
Quickfire	Headland Amenity	13053	31 May 2014
Quickfire	Makhteshim	14882	30 Jun 2014
Sirocco	Makhteshim	13051	30 Jun 2014
52 bixafen			
Bixafen EC125	Bayer CropScience	15951	20 Sep 2016
59 boscalid + epoxiconazole			
BAR	BASF	14775	
Maitre	Agro Trade	13878	

SECTION 3

EXPIRY DATE IS 31/12/2021 UNLESS OTHERWISE SHOWN

Product	Approval holder	MAPP No.	Expiry Date
RouteOne Rocker	Albaugh UK	15275	
RouteOne Rocker	Albaugh UK	13642	30 Apr 2015
Splice	BASF	12315	
Totem	BASF	12692	
Venture	BASF	12316	
60 boscalid + epoxiconazole + pyraclostrobin			
Nebula	BASF	15229	02 Feb 2015
65 bromoxynil			
Akocynil 225 EC	Aako	15688	31 Aug 2017
Alpha Bromolin 225 EC	Makhteshim	14864	28 Feb 2015
67 bromoxynil + diflufenican + ioxynil			
Capture	Bayer CropScience	09982	30 Jun 2014
Capture	Nufarm UK	12514	30 Jun 2014
70 bromoxynil + ioxynil			
Shamseer-2	DAPT	15173	28 Feb 2015
74 bromoxynil + terbuthylazine			
Alpha Bromotril PT	Makhteshim	09435	
Cleancrop Amaize	United Agri	11990	
78 Candida Oleophila Strain 0			
NEXY 1	BioNext	13609	31 Jul 2015
79 captan			
Akotan 80 WG	Aako	14490	31 Mar 2013
Alpha Captan 50 WP	Makhteshim	04797	31 Mar 2013
Alpha Captan 83 WP	Makhteshim	04806	31 Mar 2013
PP Captan 80 WG	Calliope SAS	12435	
PP Captan 83	Calliope SAS	12330	31 Mar 2013
81 carbendazim			
Clayton Am-Carb	Clayton	11906	
Cleancrop Curve	United Agri	11774	
Mascot Systemic	Scotts	09132	
Turf Systemic Fungicide	Barclay	13291	
Turfclear	Scotts	07506	
89 carbetamide			
Carbetamex	Makhteshim	13045	
Kartouch 60 WG	Aako	15417	
Riot	Makhteshim	13380	
Scrum	Makhteshim	13386	
93 carfentrazone-ethyl			
Aurora 40 WG	Belchim	11614	30 Sep 2013
Carfen	AgriGuard	13005	30 Sep 2013
Harrier	Belchim	14164	30 Sep 2013
Platform	Belchim	11615	30 Sep 2013
96 carfentrazone-ethyl + metsulfuron-methyl			
Ally Express	DuPont	08640	30 Sep 2013
Asset Express	AgriGuard	14378	30 Sep 2013
99 chloridazon			
Pyramin FL	BASF	11628	30 Jun 2014

EXPIRY DATE IS 31/12/2021 UNLESS OTHERWISE SHOWN

Product	Approval holder	MAPP No.	Expiry Date
Tempest	AgriGuard	12688	30 Jun 2014

101 chloridazon + ethofumesate

Gremlin	Sipcam	09468	30 Jun 2014

102 chloridazon + metamitron

Volcan Combi FL	Sipcam	11442	

104 chlormequat

3C Chlormequat 750	BASF	13984	
AgriGuard Chlormequat 720	AgriGuard	14120	
AgriGuard Chlormequat 720	AgriGuard	14125	
AgriGuard Chlormequat 750	AgriGuard	14144	
AgriGuard Chlormequat 750	AgriGuard	14180	
Alpha Chlormequat 460	Makhteshim	04804	31 Jul 2014
Alpha Pentagan	Makhteshim	04794	
Alpha Pentagan Extra	Makhteshim	04796	
Atlas 5C Quintacel	Nufarm UK	11130	
Barclay Holdup	Barclay	14804	30 Nov 2015
Barclay Holdup	Barclay	11365	31 Mar 2014
Barclay Take 5	Barclay	11368	31 Mar 2014
Barclay Take 5	Barclay	14799	30 Nov 2015
Barleyquat B	Taminco	13965	
BASF 3C Chlormequat 600	BASF	04077	
BASF 3C Chlormequat 720	BASF	06514	
BASF 3C Chlormequat 750	BASF	06878	
Chlormequat 46	Nufarm UK	11504	31 Jan 2014
C-Kwat 720	Goldengrass	15608	30 Sep 2013
Clayton CCC 750	Clayton	07952	
CleanCrop Transist	United Agri	15124	
C-Quat	Enigma	15668	
Hockley CMQ 720	Hockley	15527	
Mandops Chlormequat 460	Taminco	13969	30 Nov 2015
Midget	Nufarm UK	12283	
Terbine	Nufarm UK	11407	31 Jan 2014
Tricol	Nufarm UK	12682	
Turpin	Makhteshim	15057	

105 chlormequat + 2-chloroethylphosphonic acid

Barclay Banshee XL	Barclay	11339	
Sypex	BASF	04650	
Terpal C	BASF	07062	

106 chlormequat + 2-chloroethylphosphonic acid + imazaquin

Satellite	BASF	10395	

110 2-chloroethylphosphonic acid

Cerone	Interfarm	15944	31 Jan 2020
Hi-Phone 48	Hockley	15506	31 Jan 2020
Padawan	SFP Europe	15577	31 Jan 2020
Padawan	Drax	14853	30 Jun 2013
Ripe-On	AgriGuard	15020	08 Feb 2014

111 2-chloroethylphosphonic acid + mepiquat chloride

Barclay Banshee	Barclay	11343	31 Aug 2014
CleanCrop Fonic M	United Agri	09553	31 Aug 2014
RouteOne Mepiquat Plus	Albaugh UK	13636	31 Aug 2014
RouteOne Mepiquat Plus	Albaugh UK	15300	31 Aug 2014

SECTION 3

EXPIRY DATE IS 31/12/2021 UNLESS OTHERWISE SHOWN

Product	Approval holder	MAPP No.	Expiry Date
Terpitz	Me2	09634	31 Aug 2014

113 chloropicrin

Chloropicrin Fumigant	Dewco-Lloyd	04216	23 Jun 2013
Custo-Fume	Custodian	13272	23 Jun 2013
K & S Chlorofume	K & S Fumigation	08722	23 Jun 2013

114 chlorothalonil

Agriguard Chlorothalonil	AgriGuard	14839	28 Feb 2016
Balear 720 SC	Agriphar	15545	08 Nov 2015
Chlorostar	Biologic HC	15376	03 Mar 2015
CleanCrop Wanderer	United Agri	15447	07 Jul 2015
Hockley Lonil 50	Hockley	15430	03 Mar 2015
IT Chlorothalonil	I T Agro	15845	31 Aug 2018
Jupital	Syngenta	14558	03 Mar 2015
L S Chlorothalonil	Kilcullen	15326	31 May 2015
Life Scientific Chlorothalonil 500	Life Scientific	15813	31 Aug 2018
LS Chlorothalonil XL	Agrovista	15443	31 Aug 2018
Rover 500	Sipcam	15496	31 Aug 2018
Supreme	AgriGuard	14838	28 Feb 2016
Thalonil 500	Euro	14926	28 Feb 2016
UPL Chlorothalonil	United Phosphorus	15472	31 Aug 2018
X-SEPT	Albaugh UK	15431	07 Jul 2015

115 chlorothalonil + cymoxanil

Mixanil	Sipcam	15675	18 Apr 2016

116 chlorothalonil + cyproconazole

Bravo Xtra	Syngenta	11824	
Cleancrop Cyprothal	United Agri	09580	

117 chlorothalonil + cyproconazole + propiconazole

Apache	Syngenta	14255	

120 chlorothalonil + flusilazole

Midas	DuPont	13285	
Scout	AgriGuard	14779	

121 chlorothalonil + flutriafol

Argon	Headland	12312	
Halo	Headland	11546	

126 chlorothalonil + propamocarb hydrochloride

Merlin	Bayer CropScience	07943	31 Mar 2013
Pan Wizard	Pan Agriculture	11953	31 Mar 2013

127 chlorothalonil + propiconazole

Oxana	Arysta	15177	31 May 2014
SIP 313	Sipcam	14755	31 May 2014

130 chlorothalonil + tebuconazole

Crafter	Nufarm UK	15895	

131 chlorothalonil + tetraconazole

Eminent Star	Isagro	10447	

133 chlorotoluron + diflufenican

Agena	Nufarm UK	14051	30 Jun 2014
Buckler	Nufarm UK	13476	30 Jun 2014

EXPIRY DATE IS 31/12/2021 UNLESS OTHERWISE SHOWN

Product	Approval holder	MAPP No.	Expiry Date
Dicurane XL	Makhteshim	14569	30 Jun 2014
Gloster	Makhteshim	14037	30 Jun 2014
Steel	Nufarm UK	14111	30 Jun 2014
Tremor	Makhteshim	15001	30 Jun 2014

135 chlorpropham

Aceto Chlorpropham 50M	Aceto	14134	31 Jul 2017
Aceto Sprout Nip	Aceto	14156	31 Jul 2017
Aceto Sprout Nip Pellet	Aceto	15397	12 Jun 2015
Aceto Sprout Nip Ultra	Aceto	15456	11 Aug 2015
CIPC Gold	Dormfresh	15674	31 Jul 2017
Cleancrop Amigo	United Agri	15292	31 Jan 2015
Gro-Stop HN	Certis	14146	31 Jan 2015
Gro-Stop Innovator	Certis	14147	31 Jan 2015
Intruder	AgriChem BV	15076	31 Jul 2017
LS Chlorpropham 300HN	Life Scientific	14957	31 Aug 2014
MSS CIPC 50LF	United Phosphorus	14635	31 Jan 2015
MSS Sprout Nip	United Phosphorus	15676	31 Jul 2017
Neo-Stop 300 HN	Agriphar	15261	31 Jan 2015
Neo-Stop Starter	Agriphar	15651	19 Feb 2016
ProStore HN	Belchim	15002	31 Jan 2015
Tuberprop Easy	AgriChem BV	15701	31 Jul 2017

138 chlorpyrifos

Agriguard Chlorpyrifos	AgriGuard	13298	
Akofos 480 EC	Aako	14211	
Alpha Chlorpyrifos 48 EC	Makhteshim	13532	
Clayton Pontoon	Clayton	13219	
Crossfire 480	Dow	12516	
Pirisect	AgriGuard	13371	
Tipulex	Indigrow	15757	

141 cinidon-ethyl

Lotus	Nufarm UK	13744	31 Mar 2014

147 clofentezine

Acaristop 500 SC	Aako	14405	

148 clomazone

Blanco	Makhteshim	15592	
Calico	EZCrop	15878	
Clomate	Albaugh UK	15565	09 Jan 2016
Clone	AgriGuard	14320	
Echo	AgriGuard	14982	
IT Clomazone	Inter-Trade	14695	
Mazone 360	Euro	14898	
Mazone 360-II	Euro	15704	

150 clomazone + metazachlor

Centium Plus	Belchim	14634	

152 clopyralid

Cliophar 400	Agriphar	15008	17 Jun 2014
Glopyr 400	Globachem	15009	17 Jun 2014
Lontrel 72SG	Dow	15235	14 Feb 2015
Shield SG	Dow	15389	14 Feb 2015

EXPIRY DATE IS 31/12/2021 UNLESS OTHERWISE SHOWN

Product	Approval holder	MAPP No.	Expiry Date
155 clopyralid + florasulam + fluroxypyr			
Bofix FFC	Dow	14179	30 Jun 2018
156 clopyralid + fluroxypyr + MCPA			
Interfix	Iticon	15762	31 Oct 2018
158 clopyralid + picloram			
Euro Pyralid Extra	Euro	14681	
Legara	AgChem Access	13888	30 Jun 2013
Nugget	AgriGuard	13121	
Prevail	Dow	13205	
161 clothianidin			
Deter	Bayer CropScience	12411	31 Jul 2016
NipsIT Inside	Interfarm	14744	31 Jan 2019
170 copper sulphate			
Cuproxat FL	Nufarm UK	13241	
171 cyazofamid			
Rithfir	DuPont	14816	30 Jun 2013
RouteOne Roazafod	Albaugh UK	15271	30 Jun 2013
172 cyazofamid + cymoxanil			
Ranman Super	Belchim	14024	
173 cyazofamid + polyalkyleneoxide modheptamethylsiloxane			
RouteOne Roazafod	Albaugh UK	13666	30 Apr 2015
175 Cydia pomonella GV			
Cyd-X Xtra	Certis	14397	
176 cyflufenamid			
Cosine	Certis	15347	31 Mar 2015
NF-149 SC	Certis	15835	21 Aug 2016
Vegas	Certis	15238	30 Apr 2014
177 cymoxanil			
Cymostraight 45	Belchim	14615	
Cymoxanil 45% WG	Globachem	13727	
Sipcam C 50	Sipcam	10610	
179 cymoxanil + fluazinam			
Kunshi	Syngenta	15631	24 Nov 2015
Shirlan Forte	Syngenta	15542	24 Nov 2015
181 cymoxanil + mancozeb			
Agriguard Cymoxanil Plus	AgriGuard	10893	
Besiege	DuPont	08086	
Clayton Krypton WDG	United Phosphorus	15497	
Cleancrop Xanilite	United Agri	10050	
Curzate M68	DuPont	08072	
Hockley Cymozeb Combi	Hockley	15504	
Me2 Cymoxeb	Me2	09486	
Nautile DG	Nufarm UK	13416	31 Oct 2013
Rhapsody	DuPont	11958	
Rhythm	Interfarm	09636	
Solace	Nufarm UK	11936	
Standon Cymoxanil Extra	Standon	09442	

EXPIRY DATE IS 31/12/2021 UNLESS OTHERWISE SHOWN

Product	Approval holder	MAPP No.	Expiry Date
Zetanil	Sipcam	11993	
Zetanil WG	Sipcam	15488	20 Sep 2015

183 cypermethrin

Afrisect 10	Agriphar	13159	
Curfew	AgriGuard	14289	
Cymbaz	CMI	H7475	
Cyperkill 10	Agriphar	13157	
Landgold Cyper 100	Goldengrass	14105	
MAC-Cypermethrin 100 EC	MAC	14934	
Sherpa 100 EC	SBM	13570	31 Aug 2013
Signal 300 ES	Chemtura	15949	19 Sep 2016
Supasect	Agriphar	15536	
Toppel 100 EC	United Phosphorus	13704	31 Aug 2013

185 cyproconazole + cyprodinil

Radius	Syngenta	09387	31 May 2013

187 cyproconazole + propiconazole

Alto Xtra	Syngenta	13058	31 Mar 2013
Menara	Syngenta	13035	28 Feb 2013

189 cyproconazole + trifloxystrobin

Sphere	Bayer CropScience	11429	31 May 2013

191 cyprodinil + fludioxonil

Switch	Syngenta	13185	30 Apr 2014

194 2,4-D

2,4-D Amine 500	Nufarm UK	14360	30 Jun 2018
Damine	Agriphar	13366	30 Jun 2018
Growell 2,4-D Amine	Nufarm UK	13146	30 Jun 2018
Herboxone 60	Headland	14080	30 Jun 2018
Maton	Headland	13234	30 Jun 2018

195 2,4-D + dicamba

Compo Floranid + Herbicide	Compo	14503	30 Jun 2018
Landscaper Pro Weed Control + Fertilizer	Scotts	14128	30 Jun 2015
Lawn Builder Plus Weed Control	Scotts	08499	
Lawn Spot Weeder	Bayer CropScience	13116	
Magneto	Nufarm UK	12339	
Scotts Feed and Weed	Scotts	13076	
Thrust	Nufarm UK	12230	

196 2,4-D + dicamba + dichlorprop-P

NUB-041	Nufarm UK	14287	31 Mar 2013

198 2,4-D + dicamba + MCPA + mecoprop-P

Dicophar	Agriphar	13256	

199 2,4-D + dicamba + triclopyr

Broadshot	Agriphar	15475	
Cleancrop Broadshot	United Agri	11664	31 Mar 2013
Cleancrop Broadshot	Agriphar	15594	
Greengard	SumiAgro Amenity	11715	31 Jul 2013
Kaskara	Agriphar	15593	
Scimitar	Agriphar	15567	

SECTION 3

EXPIRY DATE IS 31/12/2021 UNLESS OTHERWISE SHOWN

Product	Approval holder	MAPP No.	Expiry Date
202 2,4-D + MCPA			
Agroxone Combi	Nufarm UK	14907	30 Jun 2018
204 2,4-D + picloram			
Atladox Hi	Nomix Enviro	13867	31 Mar 2013
205 2,4-D + triclopyr			
Genoxone ZX EC	Agriphar	13131	
206 daminozide			
B-Nine SG	Chemtura	14434	28 Feb 2016
207 dazomet			
Assassin	AgriGuard	13737	
Basamid	Kanesho	12895	
208 2,4-DB			
Butoxone DB	Nufarm UK	14100	28 Feb 2014
Butoxone DB	Nufarm UK	14840	31 Dec 2013
210 2,4-DB + MCPA			
Butoxone DB Extra	Nufarm UK	15741	30 Jun 2016
Redlegor	United Phosphorus	15783	30 Jun 2016
211 deltamethrin			
Agriguard Deltamethrin	AgriGuard	10770	
Agrotech Deltamethrin	Agrotech-Trading	12165	
Cleancrop Decathlon	United Agri	12834	
Delta-M 2.5 EC	MAC	14212	
Deltason-D	DAPT	15531	
Euro Delta 25	Euro	15823	
GAT Decline 2.5 EC	GAT Micro	15769	15 Aug 2016
Grain-Tect ULV	Limagrain	15253	31 Oct 2013
Landgold Deltaland	Teliton	12114	
Milentus Deltamethrin	Milentus	12219	
RouteOne Deltam 10	Albaugh UK	15274	
RouteOne Deltam 10	Albaugh UK	13615	30 Apr 2015
214 desmedipham + ethofumesate + metamitron + phenmedipham			
Betanal Quattro	Bayer CropScience	12779	
215 desmedipham + ethofumesate + phenmedipham			
Betanal Expert	Bayer CropScience	14034	31 Aug 2018
D.E.P. 251	Goldengrass	14922	31 Aug 2018
217 dicamba			
Cadence	Barclay	09578	
Cadence	Syngenta	08796	
Oceal	Rotam Agrochemical	15618	23 Jan 2016
218 dicamba + dichlorprop-P + ferrous sulphate + MCPA			
Renovator 2	Scotts	11411	
219 dicamba + dichlorprop-P + MCPA			
Intrepid 2	Everris Ltd	15738	
220 dicamba + MCPA + mecoprop-P			
ALS Premier Selective Plus	Amenity Land	08940	
Banlene Super	Certis	14420	

EXPIRY DATE IS 31/12/2021 UNLESS OTHERWISE SHOWN

Product	Approval holder	MAPP No.	Expiry Date
Broadband	Indigrow	15063	
Mascot Super Selective Plus	Barclay	12839	
Mircam Plus	Nufarm UK	15868	20 Jun 2016
Mircam Super	Nufarm UK	11836	
Nomix Turf	Nomix Enviro	14375	
Nomix Turf Plus	Nomix Enviro	14377	
Premier Amenity Selective	Maxwell	15421	
Re-Act	Scotts	12231	30 Jun 2015
Trireme	Nufarm UK	15908	30 Jun 2015
Trireme	Nufarm UK	11524	
Tritox	Everris Ltd	15737	
Tritox	Scotts	07764	30 Jun 2015

221 dicamba + mecoprop-P

Camber	Headland	09901	
Foundation	Syngenta	08475	
Headland Swift	Headland	11945	24 Jun 2016
Optica Forte	Nufarm UK	14334	28 Feb 2014
Optica Forte	Nufarm UK	14845	24 Jun 2016

228 dichlorprop-P + MCPA + mecoprop-P

Hymec Triple	Agrichem	09949	30 Nov 2013
Optica Trio	Nufarm UK	15717	30 Nov 2016
Optica Trio	Nufarm UK	14099	31 Oct 2013

229 diclofop-methyl + fenoxaprop-P-ethyl

Corniche	Bayer CropScience	08947	
Tigress Ultra	Bayer CropScience	08946	

231 difenacoum

Sakarat D (Whole Wheat)	Killgerm	H7109	
Sakarat D Wax Bait	Killgerm	H7489	

232 difenoconazole

COM 302 09 F AL	Westland Horticulture	14021	31 Mar 2013
COM 302 8 F EC	Westland Horticulture	14020	31 Mar 2013
Fungus Attack 2 RTU	Westland Horticulture	14454	30 Sep 2013
MAC-Difenoconazole 250 EC	MAC	14935	
Plant Rescue Fungus Ready to Use	Westland Horticulture	14646	

235 difenoconazole + fludioxonil + tebuconazole

Celest Trio	Syngenta	15510	11 Oct 2015

236 diflubenzuron

Diflox Flow	Hockley	15640	
Dimilin 25-WP	Chemtura	08902	
Dimilin Flo	Chemtura	08769	

237 diflufenican

Alpha Diflufenican 500 SC	Makhteshim	13182	30 Jun 2014
Beeper 500 SC	Aako	15358	30 Jun 2021
Champion	AgriGuard	14599	30 Jun 2014
Crater	Makhteshim	13084	30 Jun 2014
Diflufen 500	Euro	15262	30 Jun 2014
Diflufenican GL 500	Globachem	12531	30 Jun 2014
Dina 50	Globachem	13627	30 Jun 2014
Hi-Flufenican 50	Hockley	15499	30 Jun 2014
MAC-Diflufenican 500 SC	MAC	14109	30 Jun 2014

SECTION 3

EXPIRY DATE IS 31/12/2021 UNLESS OTHERWISE SHOWN

Product	Approval holder	MAPP No.	Expiry Date
Overlord	Makhteshim	13521	30 Jun 2021
RouteOne Flucan 50 UK	Albaugh UK	14130	30 Jun 2014
RouteOne Flucan 50 UK	Albaugh UK	15295	30 Jun 2014
Solo 500	Sipcam	15526	30 Jun 2014
Viola	Bayer CropScience	13321	30 Jun 2014

238 diflufenican + flufenacet

Bomber	Goldengrass	15098	30 Jun 2014
Cleancrop Marauder	United Agri	15933	30 Jun 2016
Firebird	Bayer CropScience	14826	30 Jun 2016
Flufenican	Euro	15386	31 Dec 2013
Plumage	Goldengrass	15085	30 Jun 2014
Regatta	Bayer CropScience	12054	30 Jun 2014

241 diflufenican + flurtamone

Graduate	Bayer CropScience	10776	30 Jun 2014

242 diflufenican + glyphosate

Proshield	Scotts	14767	

244 diflufenican + mecoprop-P

Atom	Nufarm UK	15038	30 Jun 2014
Conga	Nufarm UK	15011	30 Jun 2014
Pixie	Nufarm UK	13884	30 Jun 2014

247 dimethachlor

Terridox	Syngenta	15876	30 Jun 2022

252 dimethoate

BASF Dimethoate 40	BASF	00199	31 Mar 2013
Danadim	Headland	11550	
Danadim	Headland	15960	03 Sep 2016
Dimethoate 40	BASF	13949	31 Mar 2013
Rogor L40	Headland	15071	
Rogor L40	Sipcam	07611	31 Mar 2013
Rogor L40	Headland	15918	20 Sep 2016
Sector	Headland	15959	03 Sep 2016
Sector	Cheminova	10492	

253 dimethomorph

Murphy 500 SC	Aako	15203	30 Sep 2017

254 dimethomorph + mancozeb

Invader	BASF	11978	31 Mar 2013
Saracen	BASF	12005	31 Mar 2013
Saracen	BASF	15250	26 Jan 2015

258 diquat

A1412A2	Syngenta	13440	31 Dec 2015
Balista	Novastar	15186	30 Jun 2018
Barclay DQ-200	Barclay	15260	01 Feb 2014
BD-200	Barclay	14918	01 Feb 2014
Brogue	AgriGuard	14465	31 Aug 2013
Brogue	AgriGuard	13328	31 Dec 2015
CleanCrop Flail	United Agri	14870	31 Dec 2015
Di-Quattro	Agriphar	15712	30 Jun 2018
Dragoon	Hermoo	13927	31 Dec 2015
Dragoon Gold	Belcrop	14053	30 Jun 2014

EXPIRY DATE IS 31/12/2021 UNLESS OTHERWISE SHOWN

Product	Approval holder	MAPP No.	Expiry Date
Hockley Diquat 20	Hockley	15585	31 Dec 2015
I.T. Diquat	I T Agro	13557	31 Dec 2015
Inter Diquat	I T Agro	14196	31 Dec 2015
Kalahari	Barclay	14917	01 Feb 2014
Life Scientific Diquat 200	Life Scientific	15811	30 Jun 2018
Quad	Q-Chem	13396	31 Dec 2013
Quad S	Q-Chem	13578	31 Dec 2015
Quat 200	Euro	15042	31 Dec 2015
Standon Googly	Standon	12995	31 Dec 2015
Tuber Mission	AgriChem BV	14542	31 Dec 2015
Woomera	Hermoo	14009	30 Jun 2014

259 dithianon

Barclay Cluster	Barclay	11347	31 May 2013
RouteOne Dith-WG 70	Albaugh UK	13851	30 Apr 2015
RouteOne Dith-WG 70	Albaugh UK	15276	

265 dodine

Barclay Dodex	Barclay	11351	
Dodifun 400 SC	Belcrop	11657	31 May 2013
Syllit 400 SC	Agriphar	13363	
Syllit 400 SC	Chimac-Agriphar	11079	31 May 2013

266 epoxiconazole

Agriguard Epoxiconazole	AgriGuard	09407	
Agrotech-Epoxiconazole 125 SC	Agrotech-Trading	12382	
Ajax 125 SC	Robbin Aalbers	15352	
Bassoon	BASF	14402	31 Oct 2021
Blitz	EZCrop	15654	
Crown	AgriGuard	14830	
Epox 125	Euro	15387	
Hi-Poxi 125	Hockley	15450	
Landgold Epoxiconazole	Teliton	12117	
Milentus Epoxiconazole	Milentus	12363	
Overture	AgriGuard	12587	
RouteOne Epro 125	Albaugh UK	13640	30 Apr 2015
RouteOne Epro 125	Albaugh UK	15278	
Starburst	Chem-Wise	13981	
Strand	Headland	14261	
Supo	Makhteshim	13903	
Torrent	AgriGuard	14374	

267 epoxiconazole + fenpropimorph

Optiorb	AgriGuard	13426	
Tango Super	BASF	13283	

268 epoxiconazole + fenpropimorph + kresoxim-methyl

Allegro Plus	BASF	12218	
Asana	BASF	11934	
BAS 493F	BASF	11748	
Bullseye	AgriGuard	12278	
Cleancrop Chant	United Agri	11746	
Mastiff	BASF	11747	

270 epoxiconazole + fenpropimorph + pyraclostrobin

Demeo	Dow	15595	
Diamant	BASF	11557	
RouteOne Super 3	Albaugh UK	13721	30 Apr 2015

SECTION 3

Product	Approval holder	MAPP No.	Expiry Date
RouteOne Super 3	Albaugh UK	15299	
Tourmaline	Agro Trade	13934	

271 epoxiconazole + fluxapyroxad
Morex	BASF	15553	05 Oct 2015

274 epoxiconazole + kresoxim-methyl
Allegro	BASF	12220	
Barclay Avalon	Barclay	11337	31 Dec 2013
Clayton Gantry	Clayton	09482	31 Dec 2013
Cleancrop Kresoxazole	United Agri	09698	31 Dec 2013
Serial Duo	AgriGuard	12252	
Standon Kresoxim-Epoxiconazole	Standon	09281	31 Dec 2013

279 epoxiconazole + pyraclostrobin
Euro	Me2	11511	
Ibex	BASF	12168	

282 ethanol
Ethy-Gen II	Banana-Rite	13412	
Ethy-Gen II	Ripe Rite	15839	
Restrain Fuel	Restrain	14520	

283 ethofumesate
Alpha Ethofumesate	Makhteshim	13055	31 Jan 2019
Barclay Keeper 500 SC	Barclay	13430	31 Jan 2019
Catalyst	AgriGuard	13266	31 Jan 2019
Kubist Flo	Nufarm UK	12987	31 Jan 2019
Nortron Flo	Bayer CropScience	12986	31 Jan 2019
Stelga 500 SC	Novastar	14737	31 Jan 2019

284 ethofumesate + metamitron
Galahad	Bayer CropScience	10727	
Goltix Plus	Aako	14461	

285 ethofumesate + metamitron + phenmedipham
Cross 41 WG	Aako	15742	
MAUK 540	Makhteshim	11545	

286 ethofumesate + phenmedipham
Betosip Combi FL	Sipcam	14403	31 Aug 2018
Duo 400 SC	AgriGuard	14232	
Duo 400 SC	AgriGuard	14323	31 Aug 2018
Fenlander 2	Makhteshim	14030	31 Aug 2018
Gemini	EZCrop	15670	31 Aug 2018
Thunder	Makhteshim	14031	31 Aug 2018

287 ethoprophos
Mocap 10G	Certis	15207	

288 ethylene
Biofresh Safestore	Freshpallet	15729	28 Feb 2022
Ethylene	Biofresh	14177	31 Jul 2013

291 famoxadone + flusilazole
Jenga	DuPont	14357	
Mandolin	DuPont	14521	

EXPIRY DATE IS 31/12/2021 UNLESS OTHERWISE SHOWN

Product	Approval holder	MAPP No.	Expiry Date
292 fatty acids			
Finalsan	Certis	13102	
Finalsan Moss Control for Lawns Concentrate	Vitax	13141	
Finalsan Weed Killer RTU	Vitax	13143	
NEU 1170 H	Sinclair	15754	
NEU1170 H	Growing Success	12971	31 May 2016
Organic Moss Control for Lawns Concentrate	Growing Success	13140	31 May 2016
Organic Weed Killer RTU	Growing Success	13142	
Safers Insecticidal Soap	Woodstream	07197	
297 fenamidone + propamocarb hydrochloride			
Consento	Bayer CropScience	11889	30 Jun 2013
Prompto	Bayer CropScience	14391	30 Jun 2013
298 fenazaquin			
Matador 200 SC	Margarita	11058	
299 fenbuconazole			
Indar 5 EW	Whelehan	09644	
300 fenbuconazole + propiconazole			
Graphic	Interfarm	10987	30 Apr 2013
302 fenhexamid			
Druid	AgriGuard	13901	31 Dec 2015
RouteOne Fenhex 50	Albaugh UK	15282	31 Dec 2015
RouteOne Fenhex 50	Albaugh UK	13665	30 Apr 2015
304 fenoxaprop-P-ethyl			
Cheetah Energy	Bayer CropScience	13599	
Cheetah Super	Bayer CropScience	08723	
Triumph	Bayer CropScience	10902	
Warrant	Bayer CropScience	13806	31 Jan 2015
306 fenpropidin			
Alpha Fenpropidin 750 EC	Makhteshim	12843	30 Jun 2014
Cleancrop Fulmar	United Agri	12033	30 Jun 2014
Instinct	Headland	12317	30 Jun 2014
Mallard	Syngenta	08662	30 Jun 2014
Patrol	Syngenta	10531	30 Jun 2014
309 fenpropidin + propiconazole			
Prophet	Syngenta	08433	30 Jun 2014
311 fenpropidin + tebuconazole			
Eros	Makhteshim	14112	
312 fenpropimorph			
Cleancrop Fenpro	United Agri	09885	
Cleancrop Fenpropimorph	United Agri	09445	
Keetak	BASF	06950	
Landgold Fenpropimorph 750	Teliton	12118	31 Mar 2013
Marnoch Phorm	Me2	11087	
Raven	AgriGuard	13188	
316 fenpropimorph + pyraclostrobin			
BAS 528 00f	BASF	11444	

EXPIRY DATE IS 31/12/2021 UNLESS OTHERWISE SHOWN

Product	Approval holder	MAPP No.	Expiry Date
Jemker	BASF	12225	
317 fenpropimorph + quinoxyfen			
Jackdraw	BASF	13769	
318 fenpyroximate			
NNI-850 5 SC	Certis	12658	
319 ferric phosphate			
Ferramol	Certis	12274	31 Dec 2015
Ferrox	Neudorff	14356	31 Dec 2015
NEU 1181 M	Neudorff	14355	31 Dec 2015
NEU 1185	Neudorff	14736	31 Dec 2015
320 ferrous sulphate			
Aitken's Lawn Sand	Aitken	05253	
Greenmaster Autumn	Scotts	12196	31 Jul 2016
Greenmaster Mosskiller	Scotts	12197	31 Jul 2016
Greentec Mosskiller	Headland Amenity	12518	
Landscaper Pro Moss Control + Fertilizer	Scotts	14317	31 Jul 2016
Maxicrop Moss Killer & Conditioner	Maxicrop	04635	
Moss Control Plus Lawn Fertilizer	Miracle	07912	
No More Moss Lawn Feed	Wolf	10754	
Scotts Moss Control	Scotts	12928	
321 ferrous sulphate + MCPA + mecoprop-P			
Renovator Pro	Scotts	12204	31 Jul 2016
324 flazasulfuron			
Chikara	Nomix Enviro	13775	31 May 2014
Flazasulf 25	Euro	15723	31 May 2014
Railtrax	PSI	15139	31 May 2014
325 flonicamid			
RouteOne Ski	Albaugh UK	13777	30 Apr 2015
RouteOne Ski	Albaugh UK	15296	31 May 2014
326 florasulam			
Flora 50	Euro	15700	30 Jun 2018
Primus 25 SC	Dow	10175	30 Jun 2018
RouteOne Florasul 50	Albaugh UK	13668	30 Apr 2015
RouteOne Florasul 50	Albaugh UK	15285	30 Jun 2018
Spitfire Solo	Dow	14756	30 Jun 2018
327 florasulam + fluroxypyr			
Flurosulam XL	Euro	15236	30 Jun 2018
Image	Dow	13771	30 Jun 2018
328 florasulam + pinoxaden			
Axial One	Syngenta	15213	31 May 2014
331 fluazifop-P-butyl			
Clayton Crowe	Clayton	12572	31 Dec 2013
Clean Crop Clifford	United Agri	13597	31 Dec 2013
Fusilade 250 EW	Syngenta	10525	31 Dec 2013
PP 007	Syngenta	10533	31 Dec 2013
RouteOne Fluazifop +	Albaugh UK	13641	31 Dec 2013
RouteOne Fluazifop +	Albaugh UK	15288	30 Sep 2013

EXPIRY DATE IS 31/12/2021 UNLESS OTHERWISE SHOWN

Product	Approval holder	MAPP No.	Expiry Date
RouteOne Fluazifop +	Albaugh UK	15679	31 Dec 2013
Wizzard	Syngenta	10539	31 Dec 2013

332 fluazinam

Alpha Fluazinam 50SC	Makhteshim	13622	31 Aug 2014
Barclay Cobbler	Barclay	11348	31 Aug 2014
Blizzard	AgriGuard	12854	
Blizzard	AgriGuard	13831	31 Aug 2014
Boyano	Belchim	14571	
Comandanti	Sipcam	15411	
Euro Zinam 500	Euro	14858	
Float	Headland	14950	
Fluazinam 500	Goldengrass	14411	
FOLY 500 SC	Aako	14920	31 Aug 2014
Hockley Fluazinam 50	Hockley	15520	
Ibiza 500	Belchim	14393	
Ibiza 500	Q-Chem	14002	31 Aug 2014
Landgold Fluazinam	Teliton	12119	31 Aug 2014
Legacy	Syngenta	13735	
Ohayo	Syngenta	13734	
Shirlan Programme	Syngenta	10574	
Standon Fluazinam 500	Standon	08670	31 Aug 2014

333 fluazinam + metalaxyl-M

RouteOne Fluazmet M	Albaugh UK	13650	31 Aug 2014
RouteOne Fluazmet M	Albaugh UK	15290	

334 fludioxonil

Beret Gold	BASF	12656	30 Apr 2014

338 fludioxonil + tefluthrin

Austral Plus	Bayer CropScience	13376	

341 flufenacet + isoxaflutole

RouteOne Oxanet 481	Albaugh UK	13828	30 Apr 2015
RouteOne Oxanet 481	Albaugh UK	15309	30 Sep 2013

343 flufenacet + pendimethalin

Clayton Glacier	Clayton	15920	31 Dec 2013
Cleancrop Hector	Agrii	15932	31 Dec 2013
Ice	BASF	13930	31 Dec 2013
Kruos	Chem-Wise	15880	31 Dec 2013

348 fluopyram + prothioconazole

Propulse	Bayer CropScience	15735	31 Jul 2015

350 fluoxastrobin

Bayer UK 831	Bayer CropScience	12091	31 Jul 2018

351 fluoxastrobin + prothioconazole

Clayton Edge	Clayton	12911	30 Apr 2013
Firefly	Bayer CropScience	13692	31 Jan 2021
Prostrob 20	Chem-Wise	15698	31 Jan 2021
Redigo Twin TXC	Bayer CropScience	14259	04 Feb 2013

354 flupyrsulfuron-methyl

Ductis SX	DuPont	15426	30 Jun 2018
Spelio SX	DuPont	15424	30 Jun 2018

SECTION 3

EXPIRY DATE IS 31/12/2021 UNLESS OTHERWISE SHOWN

Product	Approval holder	MAPP No.	Expiry Date
359 fluquinconazole			
Sahara	Bayer CropScience	11905	31 Dec 2013
Triplex	BASF	12443	
360 fluquinconazole + prochloraz			
Baron	BASF	11686	
Galmano Plus	Bayer CropScience	11645	31 Dec 2013
361 fluroxypyr			
Agriguard Fluroxypyr	AgriGuard	12525	30 Jun 2024
Agrostar	Agro Trade	14925	30 Jun 2024
Agrotech Fluroxypyr	Agrotech-Trading	12294	30 Jun 2024
Awac	Globachem	13296	30 Jun 2024
Barclay Hudson	Barclay	13606	30 Jun 2024
Cascade	Chem-Wise	13939	30 Jun 2024
Cerastar	Cera Chem	13183	30 Jun 2024
Cerfix EC	Novastar	15117	31 Dec 2013
Chopper	AgriGuard	15165	30 Jun 2024
Clean Crop Gallifrey 200	Globachem	13813	30 Jun 2024
Crescent	AgriGuard	12559	30 Jun 2024
Flixfit 20 EC	GAT Micro	15684	26 May 2015
Floxy	Agriphar	13367	31 Dec 2013
Floxy	Chimac-Agriphar	12984	31 Dec 2013
Flurox 180	Stockton	13959	30 Jun 2024
Flurox 200	Stockton	13938	30 Jun 2024
Forban EC	Novastar	14862	31 Dec 2013
GAL-GONE	Globachem	13821	30 Jun 2024
GAT Stake 20 EC	GAT Micro	15399	26 May 2015
Hatchet	AgriGuard	12524	30 Jun 2024
Hatchet 200 EC	AgriGuard	15097	30 Jun 2024
Hockley Fluroxypyr 20	Hockley	15448	30 Jun 2024
Hy-Flox 200	Agrichem	15215	31 Dec 2013
Kildock Extra	Agrichem	15210	31 Dec 2013
NUB-010	Nufarm UK	13829	31 Dec 2013
Standon Homerun	Standon	13177	30 Jun 2024
Taipan	Cera Chem	14155	07 Dec 2015
Tomahawk 2	Makhteshim	15231	07 Mar 2015
Triton	EZCrop	15673	30 Jun 2024
364 fluroxypyr + metsulfuron-methyl			
Croupier	AgriGuard	15217	31 Aug 2013
365 fluroxypyr + thifensulfuron-methyl + tribenuron-methyl			
GEX 353	DuPont	15114	31 Dec 2013
368 flusilazole			
Punch 25	DuPont	14497	
369 flutolanil			
NNF-136	Nihon	14302	31 Aug 2014
Rhino	Certis	13101	30 Aug 2014
370 flutriafol			
Impact	Headland	12776	
371 fluxapyroxad			
BAS 700	BASF	15552	05 Oct 2015
Imtrex	BASF	15509	05 Oct 2015

EXPIRY DATE IS 31/12/2021 UNLESS OTHERWISE SHOWN

Product	Approval holder	MAPP No.	Expiry Date
372 fluxapyroxad + metconazole			
Librax	BASF	15887	04 Sep 2016
374 fosetyl-aluminium			
Plant Trust	Scotts	15208	31 Jul 2016
375 fosetyl-aluminium + propamocarb hydrochloride			
Previcur Energy	Bayer CropScience	13342	31 Mar 2013
380 garlic extract			
Eagle Green Care	ECOspray	14989	22 Jun 2014
NEMguard granules	ECOspray	15254	17 Feb 2015
382 gibberellins			
Gibb Plus 10 SL	Q-Chem	14968	
383 Gliocladium catenulatum			
Prestop Mix	Fargro	15104	31 Mar 2015
384 glufosinate-ammonium			
Basta	Bayer CropScience	13820	31 Mar 2013
Challenge	Bayer CropScience	07306	
Challenge 60	Fargro	08236	31 Mar 2013
Finale 150	Bayer CropScience	15233	
Weedex	Novastar	15115	
385 glyphosate			
Accelerate	Headland	13390	30 Jun 2018
Acomac	CP AGRO	14254	30 Jun 2018
Acrion	Bayer Environ.	12677	30 Jun 2018
Amenity Glyphosate 360	Monsanto	13000	28 Feb 2015
Asteroid Pro	Headland	15373	30 Jun 2018
Barclay Gallup 360	Barclay	12659	31 May 2014
Barclay Gallup Biograde 360	Barclay	12660	31 Dec 2014
Barclay Gallup Hi-Aktiv	Barclay	12661	31 May 2014
Barclay Garryowen	Barclay	12715	30 Jun 2018
Boom efekt	Pinus	15606	30 Jun 2018
Cleancrop Corral	United Agri	15220	30 Jun 2018
CleanCrop Hoedown	United Agri	15404	30 Jun 2018
Cleancrop Hoedown	United Agri	12913	30 Jun 2018
CleanCrop Hoedown Green	United Agri	15402	21 Oct 2014
Cleancrop Tungsten	United Agri	13049	30 Jun 2018
Clinic	Nufarm UK	12678	30 Jun 2018
Clipper	Makhteshim	14820	30 Jun 2018
Cosmic NG	Fargro	15646	30 Jun 2018
Emrald Gly 360	Emrald	15078	30 Jun 2018
Emrald Green	Emrald	15146	21 Oct 2014
Euro Glyfo 360	Euro	14691	30 Jun 2018
Euro Glyfo 450	Euro	14936	30 Jun 2018
Frontierland Glyphosate	Soyl	14638	31 Oct 2013
Glisor 360	Menora	14974	07 Jul 2014
Glister	Sinon EU	12990	30 Jun 2018
Glister Ultra	Sinon EU	15209	30 Jun 2018
Glycel	Excel	15068	30 Jun 2018
Glydate	Nufarm UK	12679	30 Jun 2018
Glyder	Agform	15442	25 Jul 2015
Glyder 360	Agform	14006	30 Jun 2018
Glyfos Gold	Nomix Enviro	10570	30 Jun 2018

SECTION 3

EXPIRY DATE IS 31/12/2021 UNLESS OTHERWISE SHOWN

Product	Approval holder	MAPP No.	Expiry Date
Glyfos Monte	Headland	15069	30 Jun 2018
Glymark	Nomix Enviro	13870	30 Jun 2018
Glyper	SBM	14383	30 Jun 2018
GlyphoBUZZ 36 SL	GAT Micro	15172	30 Jun 2018
GlyphoFIT 36 SL	GAT Micro	14777	30 Jun 2018
Glyphosate 360	Monsanto	12669	30 Jun 2018
Glyweed	Sabero	14444	30 Jun 2018
Habitat	Barclay	12717	31 Jan 2015
Habitat	Barclay	15199	30 Jun 2018
HH-001	CP AGRO	14772	30 Jun 2018
Hi-Fosate	Hockley	15062	30 Jun 2018
HY-GLO 360	Agrichem	15194	30 Jun 2018
HY-Gly 360	Agrichem	14919	30 Jun 2018
Jali 450	DAPT	15930	30 Jun 2018
KN 540	Monsanto	12009	30 Jun 2018
Kraken	Bayer CropScience	15175	30 Jun 2018
Landmaster	Albaugh UK	15759	30 Jun 2018
Master Gly 36T	Albaugh UK	15731	30 Jun 2018
Mentor	Monsanto	15224	30 Jun 2018
MON 76473	Monsanto	15348	28 Mar 2015
MON 79351	Monsanto	12860	30 Jun 2018
MON 79376	Monsanto	12664	30 Jun 2018
MON 79545	Monsanto	12663	30 Jun 2018
MON 79632	Monsanto	14841	30 Jun 2018
MON 79632	Monsanto	12582	28 Feb 2014
Montana	Sapec	14843	30 Jun 2018
Motif	Monsanto	15241	30 Jun 2018
Nomix Frontclear	Nomix Enviro	15180	01 Dec 2014
Nomix G	Nomix Enviro	13872	30 Jun 2018
Nomix Nova	Nomix Enviro	13873	30 Jun 2018
Nomix Revenge	Nomix Enviro	13874	30 Jun 2018
Nufosate	Nufarm UK	12699	30 Jun 2018
Onslaught	Procam	14564	30 Jun 2018
Oxalis NG	Fargro	15612	30 Jun 2018
Pitch	Nufarm UK	15485	28 Sep 2015
Pontil 360	Barclay	14322	30 Jun 2018
Rhizeup	Clayton	15043	30 Jun 2018
Rodeo	Monsanto	15242	30 Jun 2018
Romany	Greencrop	12681	30 Jun 2018
Rosate 36	Albaugh UK	14459	30 Jun 2018
Rosate 36 SL	Albaugh UK	14866	13 May 2014
Rosate 36 SL	Albaugh UK	15263	13 May 2014
Rosate Green	Albaugh UK	15122	21 Oct 2014
Roundup	Monsanto	12645	30 Jun 2018
Roundup Ace	Monsanto	12772	30 Jun 2018
Roundup Advance	Monsanto	15540	30 Jun 2018
Roundup Amenity	Monsanto	12672	30 Jun 2018
Roundup Assure	Monsanto	15537	28 Mar 2015
Roundup Biactive 3G	Monsanto	13409	30 Jun 2018
Roundup Biactive Dry	Monsanto	12646	30 Jun 2018
Roundup Bio	Monsanto	15538	28 Mar 2015
Roundup Express	Monsanto	12526	30 Jun 2018
Roundup Gold	Monsanto	10975	30 Jun 2018
Roundup Greenscape	Monsanto	12731	30 Jun 2018
Roundup Klik	Monsanto	12866	30 Jun 2018
Roundup Metro	Monsanto	13989	28 Feb 2014
Roundup Metro	Monsanto	14842	30 Jun 2018

EXPIRY DATE IS 31/12/2021 UNLESS OTHERWISE SHOWN

Product	Approval holder	MAPP No.	Expiry Date
Roundup ProBio	Monsanto	15539	28 Mar 2015
Roundup Pro-Green	Monsanto	11907	30 Jun 2018
Roundup ProVantage	Monsanto	15534	30 Jun 2018
Roundup Provide	Monsanto	12953	30 Jun 2018
Roundup Rail	Monsanto	12671	30 Jun 2018
Roundup Star	Monsanto	15470	30 Jun 2018
Roundup Ultimate	Monsanto	12774	30 Jun 2018
Roundup Ultra ST	Monsanto	12732	30 Jun 2018
RouteOne Glyphosate 360	Albaugh UK	14458	30 Jun 2018
RouteOne Rosate 36	Albaugh UK	13889	30 Apr 2015
RouteOne Rosate 36	Albaugh UK	15280	30 Jun 2018
RouteOne Rosate 360	Albaugh UK	15283	30 Jun 2018
RouteOne Rosate 360	Albaugh UK	15284	30 Jun 2018
RouteOne Rosate 360	Albaugh UK	13659	30 Apr 2015
RouteOne Rosate 360	Albaugh UK	13661	30 Apr 2015
Scorpion	Cardel	12673	30 Jun 2018
Silvio	Rotam Agrochemical	14626	30 Jun 2018
Snapper	Nufarm UK	15489	28 Sep 2015
Stamen	AgriGuard	14895	30 Jun 2018
Statis	AgriGuard	13079	30 Jun 2018
Statis 360	AgriGuard	14568	30 Jun 2018
Tangent	Headland Amenity	11872	30 Jun 2018
Task 360	Albaugh UK	14570	30 Jun 2018
Toro 360	Sipcam	15652	30 Jun 2018
Total 360 WDC	Extreme Green	14958	30 Jun 2018
Tumbleweed Pro-Active	Scotts	12729	30 Jun 2018
Typhoon 360	Makhteshim	14817	30 Jun 2018
Ultramax	PSI	15931	30 Jun 2018
Vival	Belchim	14550	30 Jun 2018
Wither 36	Albaugh UK	15768	30 Jun 2018
Wither 36	Amenity Tech	14867	30 Jun 2018
Wopro Glyphosate	B.V. Industrie	15462	30 Jun 2018

386 glyphosate + pyraflufen-ethyl

Thunderbolt	Nichino	14102	25 May 2015

387 glyphosate + sulfosulfuron

Nomix Blade	Nomix Enviro	13907	30 Jun 2018
Nomix Duplex	Nomix Enviro	14953	30 Jun 2018

392 imazalil

Fungazil 100 SL	BASF	11762	31 Dec 2013
Imaz 100 SL	AgriChem BV	13322	31 Dec 2013
Imaz 200 EC	AgriChem BV	12991	31 Dec 2013
Magnate 100 SL	Makhteshim	11705	30 Sep 2013
Magnate 100 SL	Makhteshim	14660	31 Jul 2021
Neozil 10 SL	Laboratorios Agrochem	13034	31 Jul 2021
Sphinx	BASF	11764	31 Jul 2021

395 imazalil + thiabendazole

Storite Super	Syngenta	13135	31 Dec 2015

399 imazamox + pendimethalin

Nirvana	BASF	13220	30 Jun 2013

401 imidacloprid

Admire	Bayer CropScience	11234	
Gaucho	Bayer CropScience	11281	

SECTION 3

EXPIRY DATE IS 31/12/2021 UNLESS OTHERWISE SHOWN

Product	Approval holder	MAPP No.	Expiry Date
Gaucho FS	Bayer CropScience	11282	
Imidachem	AgriChem BV	14232	
Imidamex 700 WG	MAC	15879	
Imidasect 60 FS	Bioscientific	14573	31 Mar 2013
Imidasect 70 WS	Biologic HC	14572	
Intercept 5GR	Everris Ltd	15620	
Intercept 70 WG	Everris Ltd	15617	
Jive	Makhteshim	14514	
Merit Forest	Bayer CropScience	15583	
Mido 70% WDG	Sharda	14174	
Neptune	Makhteshim	12992	
Neutron Turf	Indigrow	15467	
Picus 600 FS	Headland	14598	30 Sep 2014
Picus 600 FS	Agrichem	15065	31 Jan 2014

404 4-indol-3-ylbutyric acid

Product	Approval holder	MAPP No.	Expiry Date
Seradix 1	Hortichem	10422	31 May 2013
Seradix 1	Certis	11330	31 May 2013
Seradix 2	Certis	11331	31 May 2013
Seradix 2	Hortichem	10423	31 May 2013
Seradix 3	Hortichem	10424	31 May 2013
Seradix 3	Certis	11332	31 May 2013

405 4-indol-3-ylbutyric acid + 1-naphthylacetic acid

Product	Approval holder	MAPP No.	Expiry Date
Synergol	Certis	07386	31 May 2013

408 iodosulfuron-methyl-sodium + mesosulfuron-methyl

Product	Approval holder	MAPP No.	Expiry Date
Greencrop Doonbeg 2	Greencrop	13025	31 Dec 2013
Idosem 36	Chem-Wise	15913	31 Dec 2013
Mesiodo 35	Euro	15305	31 Dec 2013
Ocean	Goldengrass	14422	31 Dec 2013
Octavian	Bayer CropScience	14604	31 Dec 2013
RouteOne Seafarer	Albaugh UK	15317	31 Dec 2013
RouteOne Seafarer	Albaugh UK	13654	30 Apr 2015
Teliton Ocean	Teliton	13163	31 Mar 2013

409 iodosulfuron-methyl-sodium + propoxycarbazone-sodium

Product	Approval holder	MAPP No.	Expiry Date
Caliban Duo	Headland	14283	31 Mar 2013

412 iprodione

Product	Approval holder	MAPP No.	Expiry Date
Hi-Prodione 50	BASF	15498	30 Jun 2016
MAC-Iprodione 255 SC	MAC	14781	31 Dec 2013
Mascot Rayzor	Rigby Taylor	14329	31 Dec 2013
Prime-Turf	ChemSource	14617	31 Jul 2014
Superturf	AgriGuard	14495	31 Dec 2013

417 isoxaben

Product	Approval holder	MAPP No.	Expiry Date
Agriguard Isoxaben	AgriGuard	11652	
Flexidor	Dow	05121	
Flexidor 125	Dow	05104	
Gallery 125	Rigby Taylor	06889	
Knot Out	Vitax	05163	31 May 2013
Pan Isoxaben	Pan Amenity	15955	

420 kresoxim-methyl

Product	Approval holder	MAPP No.	Expiry Date
Kresoxy 50 WG	AgriGuard	12275	30 Jun 2024

EXPIRY DATE IS 31/12/2021 UNLESS OTHERWISE SHOWN

Product	Approval holder	MAPP No.	Expiry Date
421 lambda-cyhalothrin			
CleanCrop Corsair	United Agri	14124	31 Dec 2015
Dalda 5	Globachem	13688	31 Dec 2015
Eminentos 10 CS	GAT Micro	15854	10 Mar 2015
Euro Lambda 100 CS	Euro	14794	31 Dec 2015
Hockley Lambda 5EC	Hockley	15516	31 Dec 2015
Hockley Lambda CS	Hockley	15476	10 Mar 2015
IT Lambda	Inter-Trade	15119	31 Dec 2015
Jackpot 2	AgriGuard	12888	31 Dec 2015
Judo 5CS	Syngenta	15106	30 Jun 2013
Katana	AgriGuard	15825	12 Sep 2016
Kendo	Syngenta	15562	30 Jun 2018
Komodo 10 EC	United Phosphorus	14703	09 Oct 2013
Laidir 10 CS	GAT Micro	15461	10 Mar 2015
Lambd'Africa	Astec	15697	30 Jun 2018
Life Scientific Lambda-cyhalothrin 100	Life Scientific	15812	31 Aug 2014
MAC-Lambda-Cyhalothrin 50 EC	MAC	14941	31 Dec 2015
Markate 50	Agrovista	13529	31 Dec 2015
Reparto	Sipcam	15452	31 Dec 2015
Sparviero	Sipcam	15687	26 Apr 2016
Stealth	Syngenta	14551	31 Dec 2015
Warrior	Syngenta	13857	30 Jun 2018
423 laminarin			
Vacciplant	Goemar	13260	31 Mar 2015
425 lenacil			
Agriguard Lenacil	AgriGuard	10488	
Fernpath Lenzo Flo	AgriGuard	11919	
Venzar 80 WP	DuPont	09981	
427 linuron			
Alpha Linuron 50SC	Makhteshim	14215	31 Aug 2013
Aredios 45 SC	Novastar	14586	31 Dec 2013
Kaedyn 500	DAPT	14739	31 Dec 2013
Linurex 50 SC	Makhteshim	14652	31 Dec 2013
Messidor 45 SC	Novastar	14710	31 Dec 2013
428 magnesium phosphide			
Detia Gas-Ex-B Forte	Detia Degesch	10661	
429 maleic hydrazide			
Cleancrop Malahide	United Agri	13629	31 Dec 2013
Fazor	Chemtura	13617	31 Dec 2013
Gro-Slo	Agriphar	15851	30 Jun 2016
Himalaya	Agriphar	15214	30 Jun 2016
430 maleic hydrazide + pelargonic acid			
Ultima	Certis	13347	30 Nov 2014
432 mancozeb			
Agrizeb	Agriphar	14770	30 Nov 2013
Cleancrop Mancozeb	United Agri	15226	26 Nov 2013
Cleancrop Mandrake	United Agri	14792	30 Jun 2016
Dithane 945	Indofil	14547	31 May 2016
Dithane 945	Interfarm	14621	31 Dec 2018
Dithane Dry Flowable Neotec	Interfarm	15720	31 Mar 2014

SECTION 3

EXPIRY DATE IS 31/12/2021 UNLESS OTHERWISE SHOWN

Product	Approval holder	MAPP No.	Expiry Date
Dithane Dry Flowable Neotec	Indofil	14631	31 May 2016
Dithane NT Dry Flowable	Interfarm	14704	31 Mar 2014
Laminator 80 WP	Interfarm	15666	26 Nov 2013
Manfil 75 WG	Interfarm	15958	31 Dec 2018
Manfil 75 WG	Indofil	14768	30 Oct 2016
Manfil 80 WP	Indofil	14766	31 Oct 2016
Manfil 80 WP	Interfarm	15956	26 Nov 2013
Micene 80	Sipcam	14622	02 Jul 2013
Micene DF	Sipcam	14705	13 Sep 2013
Zebra WDG	Headland	14875	30 Jun 2016

434 mancozeb + propamocarb hydrochloride

Tattoo	Bayer CropScience	07293	31 Mar 2013

435 mancozeb + zoxamide

Unikat 75WG	Gowan	14200	29 Mar 2016

438 MCPA

Agritox 50	Nufarm UK	14814	30 Apr 2016
Dow MCPA Amine 50	Dow	14911	18 Mar 2014
Go-Low Power	DuPont	13812	30 Apr 2016
POL-MCPA 500 SL	Zaklady	14912	18 Mar 2014
Tasker 75	Headland	14913	30 Apr 2016

440 MCPA + mecoprop-P

Cleanrun Pro	Scotts	15073	31 May 2014
Cleanrun Pro	Scotts	15596	31 Aug 2015
Greenmaster Extra	Scotts	15075	31 Jul 2016

442 mecoprop-P

Clenecorn Super	Nufarm UK	14628	31 May 2014

447 mepiquat chloride + prohexadione-calcium

Medax Top	BASF	13282	

450 mesotrione

Hockley Mesotrione 10	Hockley	15519	30 Sep 2013
Labyrinth	EZCrop	15702	31 Mar 2016
RouteOne Trione 10	Albaugh UK	14765	30 Sep 2013

451 mesotrione + nicosulfuron

Elumis	Syngenta	15800	31 Aug 2016

452 mesotrione + terbuthylazine

RouteOne Mesot	Albaugh UK	13576	30 Apr 2015
RouteOne Mesot	Albaugh UK	15302	

454 metalaxyl-M

Fongarid Gold	Syngenta	12547	30 Jun 2018

455 metaldehyde

Allure	Chiltern	11089	
Antares	De Sangosse	12998	31 May 2013
Appeal	Chiltern	12713	
Attract	Chiltern	12712	
Carakol Plus	Makhteshim	13980	31 Dec 2013
Cargo 3 GB	Aako	15784	
CDP H	De Sangosse	14432	
Certis Deal 5	Certis	13488	31 May 2013

Product	Approval holder	MAPP No.	Expiry Date
Certis Metaldehyde 5	Certis	13490	31 Mar 2013
Certis Metaldehyde 5	Certis	14457	
Certis Red 5	Certis	14479	
Certis Red 5	Certis	13491	30 Apr 2013
Chiltern Hundreds	Chiltern	12710	
Chiltern Hundreds	Chiltern	10072	
Clartex	De Sangosse	12935	31 May 2013
Cleancrop Hyde 2	United Agri	13199	31 May 2013
Cleancrop Hyde 3	Makhteshim	14310	31 Dec 2013
CleanCrop Jekyll 5	United Agri	14117	30 Jun 2013
CleanCrop Jekyll 5	United Agri	14534	31 May 2013
Cleancrop Jekyll D3	United Agri	14369	
Cleancrop Potent	United Agri	14228	
Cleancrop Tremolo	United Agri	14225	
Delicia Slug-lentils	Barclay	13613	31 May 2013
Duro	Makhteshim	14347	31 Dec 2013
ESP W	De Sangosse	14429	
Gusto 4	Makhteshim	14165	31 Dec 2013
Helimax	De Sangosse	13108	31 May 2013
Helimax W	De Sangosse	14424	
Keel Over	De Sangosse	13110	31 May 2013
Keel Over W	De Sangosse	14431	
Limagri GR	SBM	13475	31 May 2013
Metadisque	Certis	14665	31 May 2013
MetaPads	Certis	14658	31 May 2013
Metarex Amba	De Sangosse	13245	31 May 2013
Metarex Green	De Sangosse	13069	31 May 2013
Metarex RG	De Sangosse	13070	
Molotov	Chiltern	08295	
Molotov	Chiltern	12650	
Oreto	Makhteshim	14258	31 Dec 2013
Portland 3	Doff Portland	15610	
Prego	Makhteshim	13582	31 Dec 2013
Regel Star	De Sangosse	13071	31 May 2013
Regel W	De Sangosse	14425	
Regiment	Certis	14338	31 May 2013
Sloggy	SBM	13373	31 May 2013
Slugdown 3	Doff Portland	14366	31 Aug 2013
Slug-Lentils	Frunol Delicia	13613	31 May 2013
Slugoids 3	Doff Portland	15623	
Spinner	Certis	14127	31 May 2013
Steam	Chiltern	12473	
Terminator	Chiltern	14229	
Tinto	Certis	14336	31 May 2013
Top Gun	Certis	14325	31 May 2013
Trigger 5	Certis	14480	
Trigger 5	Certis	13493	30 Apr 2013
Trojan	Chiltern	14224	

456 metamitron

Product	Approval holder	MAPP No.	Expiry Date
Alpha Metamitron	Makhteshim	11081	
Alpha Metamitron 70 SC	Makhteshim	12868	
Defiant WG	United Phosphorus	12300	
Goldbeet	Makhteshim	11538	
Goltix 90	Makhteshim	11578	
Goltix Compact	Aako	14863	
Hockley Metamitron 70	Hockley	15515	

SECTION 3

Product	Approval holder	MAPP No.	Expiry Date
Homer	Makhteshim	11544	
Lektan	Makhteshim	11543	
Marquise	Makhteshim	08738	
Metamitron 700	Euro	15699	
Mitron 700	Euro	15400	31 Aug 2015
Predator WG	AgriGuard	12630	

457 metam-sodium

Discovery	United Phosphorus	10416	31 Dec 2014
Fumetham	Chemical Nutrition	10047	31 Dec 2014
Metham Sodium 400	United Phosphorus	08051	31 Dec 2014
Sistan	Unicrop	01957	31 Dec 2014
Sistan 38	Unicrop	08646	31 Dec 2014
Sistan 51	Unicrop	10046	31 Dec 2014
Vapam	Willmot Pertwee	09194	31 Dec 2014

459 metazachlor

Agrotech Metazachlor 500 SC	Agrotech-Trading	12454	
Alpha Metazachlor 50 SC	Makhteshim	10669	
Butey	Chem-Wise	14176	
Fuego	Makhteshim	11177	
Landgold Metazachlor 50 SC	Teliton	12133	31 Mar 2013
Makila 500 SC	Novastar	14492	
Marksman	AgriGuard	12456	
Medusa	EZCrop	15766	
Metachlor	Euro	15532	
Metazachlore GL 500	Globachem	12528	
Mezzanine	AgriGuard	13719	
RouteOne Metaz 50	Albaugh UK	13648	30 Apr 2015
RouteOne Metaz 50	Albaugh UK	15306	

460 metazachlor + quinmerac

Katamaran	BASF	11732	
Metamerac	Euro	15221	
Rapsan TDI	Q-Chem	13920	31 Jul 2013
Syndicate	AgriGuard	12262	

461 metconazole

Caramba	BASF	15337	30 Nov 2019

464 methoxyfenozide

Runner	Bayer CropScience	12940	31 Aug 2013

467 1-methylcyclopropene

SmartFresh SmartTabs	Landseer	12684	31 Mar 2016

471 metribuzin

AgriGuard Metribuzin	AgriGuard	14084	31 Mar 2013
Buzin WG	Goldengrass	14415	31 Mar 2013
Citation 70	United Phosphorus	09370	31 Mar 2013
Greencrop Majestic	Greencrop	12422	31 Mar 2013
Landgold Metribuzin WG	Teliton	12468	31 Mar 2013
Metribuzin WG	Goldengrass	14315	
Milentus Metribuzin 70 WG	Milentus	12224	31 Mar 2013
Promor	Bayer CropScience	12502	31 Mar 2013
Python	Makhteshim	13802	
Rapture	AgriGuard	14536	31 Mar 2013
Sencorex WG	Bayer CropScience	11304	30 Nov 2013

EXPIRY DATE IS 31/12/2021 UNLESS OTHERWISE SHOWN

Product	Approval holder	MAPP No.	Expiry Date
472 metsulfuron-methyl			
Accurate	Headland	13224	31 Dec 2015
Alias	DuPont	13397	30 Jun 2018
Ally SX	DuPont	12059	30 Jun 2018
Asset	AgriGuard	13902	31 Dec 2015
Avail XT	AgriGuard	13194	31 Dec 2015
Cleancrop Mondial	United Agri	13353	31 Dec 2015
Landgold Metsulfuron	Teliton	13421	31 Dec 2015
Metro 20	AgriGuard	13496	31 Aug 2013
Minx PX	Nufarm UK	13337	31 Dec 2015
Pike	Nufarm UK	12746	30 Jun 2018
Pike PX	Nufarm UK	13249	31 Dec 2015
Quest	EZCrop	15659	31 Dec 2015
Revenge 20	Agform	14449	31 Dec 2015
Revenge 20 SG	Agform	13633	31 Dec 2015
RouteOne Romet 20	Albaugh UK	13657	30 Apr 2015
RouteOne Romet 20	Albaugh UK	15277	31 Dec 2015
Savvy	Rotam Agrochemical	14266	31 Dec 2015
Savvy Clear	Rotam Agrochemical	14474	10 Mar 2013
Savvy Clear	Rotam Agrochemical	15183	10 Mar 2013
Savvy Premium	Rotam Agrochemical	14782	15 Dec 2013
Tuko	Terrachem	15611	30 Jun 2018
Victor	EZCrop	15669	30 Jun 2018
473 metsulfuron-methyl + thifensulfuron-methyl			
Accurate Extra	Headland	14167	31 Dec 2015
Aria	AgriGuard	14771	31 Dec 2015
Choir	Nufarm UK	14081	30 Jun 2018
Ergon	Rotam Agrochemical	14382	31 Dec 2015
474 metsulfuron-methyl + tribenuron-methyl			
Boudha	Rotam Agrochemical	15158	18 Nov 2014
475 myclobutanil			
Aristocrat	AgriGuard	12343	
Crassus	Landseer	13899	
RouteOne Butanil 20	Albaugh UK	14773	
RouteOne Robut 20	Albaugh UK	15273	30 Sep 2015
RouteOne Robut 20	Albaugh UK	15685	
RouteOne Robut 20	Albaugh UK	13646	30 Sep 2015
Systhane 20 EW	Whelehan	09397	31 Mar 2013
Systhane 6 W	Dow	10808	31 Mar 2013
476 1-naphthylacetic acid			
Tipoff	Hatrim	12292	31 Dec 2013
478 napropamide			
Banweed	United Phosphorus	09376	
Devrinol	United Phosphorus	09375	
Jouster	AgriGuard	12741	
Naprop	Globachem	13402	
479 nicosulfuron			
Clayton Myth XL	Clayton	15877	01 Mar 2016
Euro Nico 40	Euro	14778	30 Jun 2014
Fornet 6 OD	Belchim	15891	01 Mar 2016
Landgold Nicosulfuron	Teliton	13464	30 Jun 2014
Maizon	AgriGuard	13222	30 Jun 2014

SECTION 3

EXPIRY DATE IS 31/12/2021 UNLESS OTHERWISE SHOWN

Product	Approval holder	MAPP No.	Expiry Date
Nico Pro 4SC	Syngenta	15645	30 Jun 2014
RouteOne Strong 4	Albaugh UK	13830	30 Jun 2014
RouteOne Strong 4	Albaugh UK	15298	30 Jun 2014
Samson	Syngenta	12141	30 Jun 2014
Samson Extra 6%	Syngenta	13436	30 Sep 2013

480 oxadiazon

Noble Oxadiazon	Barclay	12889	

485 penconazole

Pencomex 100 EC	MAC	15902	
RouteOne Pencon 10	Albaugh UK	13651	30 Apr 2015
RouteOne Pencon 10	Albaugh UK	15310	

486 pencycuron

Agriguard Pencycuron	AgriGuard	10445	31 May 2013
Me2 Penny	Me2	10369	31 May 2013
Monceren Flowable	Bayer CropScience	11425	
Pan Pencycuron P	Pan Agriculture	10494	31 May 2013
Pency Flowable	Globachem	12637	31 May 2013
Standon Pencycuron DP	Standon	08774	31 May 2013
Tubercare 12.5 DS	AgriChem BV	12194	31 May 2013

487 pendimethalin

Akolin 330 EC	Aako	14282	31 Dec 2013
Alpha Pendimethalin 330 EC	Makhteshim	13815	31 Dec 2013
Aquarius	Makhteshim	14712	30 Jun 2016
Bunker	Makhteshim	13816	30 Jun 2016
Campus 400 CS	Aako	15366	14 Jun 2013
Cinder	Makhteshim	14526	30 Jun 2016
Claymore	BASF	13441	31 Dec 2013
CleanCrop Stomp	United Agri	13443	31 Dec 2013
Convey 330EC	AgriGuard	14535	31 Dec 2013
Hockley Pendi 330	Hockley	15513	31 Dec 2013
Pendi 330	Euro	15807	31 Dec 2013
Pendimet 400	Goldengrass	14715	31 Dec 2013
Pendulum	AgriGuard	15234	30 Apr 2013
RouteOne Penthal 330	Albaugh UK	15314	31 Dec 2013
RouteOne Penthal 330	Albaugh UK	13768	30 Apr 2015
Shuttle	AgriGuard	14041	31 Jul 2013
Sovereign	BASF	13442	31 Dec 2013
Stomp 400 SC	BASF	13405	31 Dec 2013

488 pendimethalin + picolinafen

Flight	BASF	12534	30 Jun 2016
PicoStomp	BASF	13455	30 Jun 2016

491 penflufen

Emesto Prime FS	Bayer CropScience	15446	03 Aug 2015

495 phenmedipham

Alpha Phenmedipham 320 SC	Makhteshim	14070	28 Feb 2015
Beetup Flo	United Phosphorus	14328	31 Dec 2013
Betamax	AgriGuard	14371	31 Dec 2013
Betanal Flow	Bayer CropScience	13893	28 Feb 2015
Crotale	Sipcam	13909	28 Feb 2015
Dancer Flow	Sipcam	14395	28 Feb 2015
Galcon SC	Novastar	14607	31 Dec 2013

EXPIRY DATE IS 31/12/2021 UNLESS OTHERWISE SHOWN

Product	Approval holder	MAPP No.	Expiry Date
Herbasan Flow	Nufarm UK	13894	28 Feb 2015
Mandolin Flow	Nufarm UK	13895	28 Feb 2015
Pump	Makhteshim	14087	28 Feb 2015
Rubie	EZCrop	15655	31 Dec 2013

500 picloram

RouteOne Loram	Albaugh UK	15678	
RouteOne Loram 24	Albaugh UK	13951	30 Apr 2015
RouteOne Loram 24	Albaugh UK	15297	30 Sep 2013
Tordon 22K	Frontier	15682	
Tordon 22K	Dow	13869	31 Mar 2013

501 picolinafen

AC 900001	BASF	10714	30 Jun 2018

502 picoxystrobin

Acanto	DuPont	13043	31 Dec 2013

503 pinoxaden

RouteOne Roaxe 100	Albaugh UK	13655	30 Jun 2013
RouteOne Roaxe 100	Albaugh UK	15267	31 Aug 2013
RouteOne Roaxe 100	Chem-Wise	15630	31 May 2014
Symmetry	AgriGuard	13577	31 May 2014

504 pirimicarb

Agrotech-Pirimicarb 50 WG	Agrotech-Trading	12269	
Arena	AgriGuard	14028	
Cleancrop Miricide	United Agri	11776	
Hockley Pirimicarb WG	Hockley	15529	
Milentus Pirimicarb	Milentus	12268	
Piri 50	Euro	14956	
RouteOne Primro 50 WG	Albaugh UK	15315	
RouteOne Primro 50 WG	Albaugh UK	13644	30 Apr 2015

505 pirimiphos-methyl

Actellic Smoke Generator No 20	Syngenta	10540	31 Mar 2013
Actellic Smoke Generator No. 10	Syngenta	10448	31 Mar 2013
Ultrasect 50EC	AgriGuard	15030	

507 prochloraz

Alpha Prochloraz 40 EC	Makhteshim	11002	31 Dec 2013
Barclay Eyetak 40	Barclay	11352	
Mirage 45 SC	Makhteshim	10894	
Octave	BASF	11741	
Panache 40	DAPT	10632	31 Dec 2013
Prelude 20 LF	BASF	11904	
Prospero	Makhteshim	11931	31 Dec 2013
Scarlett	AgriGuard	14271	
Scotts Octave	Everris Ltd	15756	
Sporgon 50 WP	Sylvan	08802	
Sportak 45 EW	BASF	11696	
Sportak 45 HF	BASF	11697	
Sprint	AgriGuard	14272	30 Jun 2013

508 prochloraz + propiconazole

Mambo	Makhteshim	13387	31 Dec 2013

EXPIRY DATE IS 31/12/2021 UNLESS OTHERWISE SHOWN

Product	Approval holder	MAPP No.	Expiry Date
510 prochloraz + tebuconazole			
Agate	Makhteshim	13024	30 Apr 2013
System Turf	AgriGuard	14538	
514 propamocarb hydrochloride			
Edipro	Agriphar	15564	31 Mar 2020
Edito	Agriphar	14869	31 Jul 2013
Filex	Scotts	07631	31 Mar 2013
Flash	AgriGuard	11989	31 Mar 2013
Pan PCH	Pan Agriculture	12842	31 Mar 2013
Previcur N	Bayer CropScience	08575	31 Mar 2013
Propamex-I 604 SL	MAC	15901	31 Mar 2020
Proplant	Agriphar	15422	31 Mar 2020
Proplant	Fargro	13359	31 Mar 2013
515 propaquizafop			
Agil	Makhteshim	11048	31 Jul 2013
Agil	Makhteshim	14562	
Alpha Propaquizafop 100 EC	Makhteshim	13689	31 Jul 2013
Alpha Propaquizafop 100 EC	Makhteshim	14563	
Barclay Rebel	Barclay	11648	
Bulldog	Makhteshim	11723	31 Jul 2013
Bulldog	Makhteshim	14565	
Falcon	Makhteshim	10585	30 Jun 2013
Landgold PQF 100	Teliton	12127	31 Mar 2013
Noble Propaquizafop	Barclay	12890	
Osprey	Chem-Wise	13731	
Raptor	Makhteshim	14566	
Raptor	Makhteshim	11092	31 Jul 2013
RouteOne Quizro 10	Albaugh UK	13643	30 Apr 2015
RouteOne Quizro 10	Albaugh UK	15316	
Shogun	Makhteshim	10584	30 Jun 2013
Zealot	AgriGuard	12548	
516 propiconazole			
Anode	Makhteshim	14447	31 May 2014
Barclay Propizole	Barclay	15113	30 Nov 2016
Hockley Propicon 25	Hockley	15517	31 May 2014
Matsuri	Sumi Agro	15544	30 Nov 2016
Profiol 250	PSI	15393	31 May 2014
Propicon 250	Goldengrass	15255	31 May 2014
Tonik	AgriGuard	14761	31 May 2014
Zolex	ITACA	15713	30 May 2016
517 propiconazole + tebuconazole			
Cogito	Syngenta	13847	
520 propoxycarbazone-sodium			
Attribut	Bayer CropScience	12730	30 Nov 2013
521 propyzamide			
Artax Flo	Bayer CropScience	15051	13 Sep 2013
Edge 400	Albaugh UK	14683	14 Sep 2014
Emrald Prop 400 FL	Emrald	15079	14 Sep 2013
Emrald Prop 800 WG	Emrald	15077	31 Mar 2014
Hentsch	Rotam Agrochemical	15616	31 May 2016
Hockley Propyzamide 40	Hockley	15449	30 Sep 2016
KeMiChem - Propyzamide 400 SC	KeMiChem	14706	31 Mar 2014

EXPIRY DATE IS 31/12/2021 UNLESS OTHERWISE SHOWN

Product	Approval holder	MAPP No.	Expiry Date
Kerb Pro Granules	Barclay	14280	30 Sep 2016
Levada	Certis	15743	30 Sep 2016
MAC-Propyzamide 400 SC	MAC	14786	31 Mar 2014
Master Prop 40	Albaugh UK	15829	30 Sep 2016
Megaflo	Dow	15034	31 Mar 2014
Pentaflo	Dow	15523	17 Oct 2015
Pizza Flo	Goldengrass	15764	13 Sep 2013
Precis	Dow	14202	31 Mar 2014
Propyzamide Flo	Albaugh UK	14732	14 Sep 2014
Prova	Belcrop	15344	31 May 2013
Prova	Nomix Enviro	15560	31 Aug 2013
Prova	Frontier	15641	28 Feb 2014
Pyzamid 400 SC	Euro	15257	31 Mar 2014
RouteOne Zamide Flo	Albaugh UK	14559	31 Mar 2014
Setanta 50 WP	AgriGuard	15636	30 Sep 2016
Setanta 50 WP	AgriGuard	14494	31 Mar 2014
Setanta 50 WP	AgriGuard	14943	31 Aug 2013
Setanta Flo	AgriGuard	14678	31 Aug 2013
Setanta Flo	AgriGuard	15635	31 Jul 2016
Setanta Flo	AgriGuard	14493	31 Mar 2014
Shamal	Rotam Agrochemical	15686	30 Sep 2016
Solitaire	Certis	15792	30 Sep 2016
Solitaire	AgriGuard	14698	31 Aug 2013
Solitaire 50 WP	AgriGuard	14962	31 Aug 2013
Verdah 400	DAPT	14680	31 Mar 2014
Verge 400	Albaugh UK	14682	14 Sep 2014
Zamide 80 WG	Albaugh UK	14723	31 Mar 2014
Zamide Flo	Albaugh UK	14679	14 Sep 2014
Zammo	Headland	15313	03 Mar 2015

523 prosulfocarb

A8545G	Syngenta	14796	
Fidox	Syngenta	15385	
Jade	Syngenta	15558	
RouteOne Rosulfocarb 80	Albaugh UK	15286	
RouteOne Rosulfocarb 80	Albaugh UK	13635	31 Jan 2015
Sincere	AgriGuard	14947	

525 prothioconazole

Banguy	PSI	15140	31 Jul 2018
Proline	Bayer CropScience	12084	31 Jul 2018

526 prothioconazole + spiroxamine

Pikazole 46	Chem-Wise	15691	31 Jan 2021
RouteOne Prothioxamine	Albaugh UK	14904	27 Jan 2014

528 prothioconazole + tebuconazole

Capo	Sipcam	15429	
Redigo Pro	Bayer CropScience	15145	31 Jan 2021

531 Pseudomonas chlororaphis MA 342

Cerall	Chemtura	14546	11 Jun 2013

533 pyraclostrobin

BAS 500 06F	BASF	12338	31 May 2014
BASF Insignia	BASF	11900	31 May 2014
Comet	BASF	10875	31 May 2014
Insignia	Vitax	11865	31 May 2014

SECTION 3

EXPIRY DATE IS 31/12/2021 UNLESS OTHERWISE SHOWN

Product	Approval holder	MAPP No.	Expiry Date
LEY	BASF	14774	31 May 2014
Platoon	BASF	12325	31 May 2014
534 pyraflufen-ethyl			
OS159	Ceres	12723	31 Dec 2015
539 pyroxsulam			
Avocet	Dow	14829	31 Jan 2015
542 quinoxyfen			
Apres	Dow	08881	01 Sep 2014
Clean Crop QFN	United Agri	11966	31 Mar 2013
Erysto	Dow	08697	31 Mar 2013
543 quizalofop-P-ethyl			
Visor	Makhteshim	14513	
544 quizalofop-P-tefuryl			
Rango	Certis	14039	
545 rimsulfuron			
Clayton Rasp	Clayton	15952	31 Jan 2017
Rimsulf 250	Euro	15802	31 Jan 2017
554 spiroxamine			
Spirostar	Bioscientific	15265	31 Dec 2013
Spirostar	Agriphar	15356	28 Feb 2015
Torch Extra	Bayer CropScience	12140	31 Dec 2013
Zenon	Bayer CropScience	11232	31 Dec 2013
555 spiroxamine + tebuconazole			
Sage	Interfarm	11303	30 Nov 2013
556 sulfosulfuron			
Fosulfuron	Euro	15164	31 Dec 2015
557 sulfuryl fluoride			
Profume	Dow	12035	30 Apr 2013
558 sulphur			
Headland Venus	Headland	10611	
Solfa WG	Nufarm UK	11602	
Venus	Headland	11856	
559 tau-fluvalinate			
Alpha Tau-Fluvalinate 240 EW	Makhteshim	13605	31 May 2013
560 tebuconazole			
Agrotech-Tebuconazole 250 EW	Agrotech-Trading	12523	
Alpha Tebuconazole 20 EW	Makhteshim	12893	
Bezel	Bayer CropScience	12753	
CHA 1640	Headland	14583	
Euro Tebu 250	Euro	14754	
Hockley Tebucon 25	Hockley	15500	
Icon	AgriGuard	12202	
Manicure	Chem-Wise	13906	
Odin	Rotam Agrochemical	13468	
RouteOne Tebbro 25	RouteOne	13647	30 Apr 2015
RouteOne Tebbro 25	Albaugh UK	15301	

Product	Approval holder	MAPP No.	Expiry Date
Savannah	Rotam Agrochemical	14821	
Starpro	Rotam Agrochemical	14823	
TEB250	Goldengrass	15755	
Toledo	Rotam Agrochemical	14036	
561 tebuconazole + triadimenol			
Silvacur	Bayer CropScience	11309	
Veto F	Bayer CropScience	14274	30 Nov 2013
563 tebuconazole + trifloxystrobin			
Continuum	Indigrow	15600	
Eradicate	Pan Agriculture	13986	
Inter Tebloxy	Iticon	15412	
Pandora	Pan Agriculture	14115	
Spectrum	Indigrow	15407	31 Jul 2013
564 tebufenpyrad			
Masai G	BASF	10224	
566 tefluthrin			
Force ST	Bayer CropScience	11671	
567 tepraloxydim			
Assimilate	AgriGuard	14860	31 May 2015
Landgold Tepraloxydim	Teliton	13379	31 May 2015
Omara	Chem-Wise	13682	31 May 2015
RouteOne Tepralox 50	Albaugh UK	13667	30 Apr 2015
RouteOne Tepralox 50	Albaugh UK	15303	31 May 2015
569 tetraconazole			
Domark	Isagro	13048	
Juggler	Sipcam	13077	
572 thiacloprid			
Biscaya	Bayer CropScience	12471	31 Jul 2014
Exemptor	Scotts	13122	31 Jul 2013
Thiaclomex 480 SC	MAC	15900	31 Dec 2014
573 thiamethoxam			
Centric	Syngenta	13954	31 Jan 2017
Cruiser SB	Syngenta	12958	31 Jul 2014
574 thifensulfuron-methyl			
Harmony SX	DuPont	12181	30 Jun 2018
Prospect SX	DuPont	12212	31 Dec 2015
576 thiophanate-methyl			
Loredo	AgriGuard	15202	28 Feb 2016
Thiofin WG	Q-Chem	15384	29 Feb 2016
578 tolclofos-methyl			
Rizolex 50 WP	Interfarm	14217	
Rizolex Flowable	Interfarm	14207	
581 triadimenol			
Bayfidan	Bayer CropScience	11417	
582 tri-allate			
Avadex BW Granular	Gowan	12104	

SECTION 3

EXPIRY DATE IS 31/12/2021 UNLESS OTHERWISE SHOWN

Product	Approval holder	MAPP No.	Expiry Date
585 tribenuron-methyl			
Helmstar 75 WG	Helm	15070	28 Feb 2016
Nuance	Headland	14813	28 Feb 2016
Quantum	DuPont	15190	29 Feb 2016
Quantum SX	DuPont	15189	29 Feb 2016
Taxi	Nufarm UK	15733	31 Aug 2018
Thor	Nufarm UK	14240	
587 triclopyr			
Topper	Agriphar	15719	30 Nov 2019
588 trifloxystrobin			
Flint	Bayer CropScience	11259	30 Sep 2013
Mascot Defender	Rigby Taylor	14065	30 Sep 2013
Micro-Turf	AgriGuard	12988	30 Sep 2013
Pan Tees	Pan Agriculture	12894	30 Sep 2013
Twist	Bayer CropScience	11230	30 Sep 2013
Twist 500 SC	Bayer CropScience	11231	30 Sep 2013
Zest SC	Bayer CropScience	11233	30 Sep 2013
590 triflusulfuron-methyl			
Standon Triflusulfuron	Standon	09487	
591 trinexapac-ethyl			
Cleancrop Alatrin	United Agri	15196	30 Apr 2017
Duet Max	ChemSource	14802	30 Jun 2014
Freeze	Headland	15871	03 Oct 2016
Maintain	Headland	15626	10 Aug 2015
Scitec	Syngenta	15588	31 Oct 2019
Trinex 222	Euro	15322	30 Apr 2017
Upkeep	Makhteshim	15225	31 Oct 2019
594 warfarin			
Sakarat Ready-to-Use (Cut Wheat Base)	Killgerm	H6807	
Sewarin P	Killgerm	H6811	
595 zeta-cypermethrin			
RouteOne Zeta 10	Albaugh UK	13608	30 Apr 2015
RouteOne Zeta 10	Albaugh UK	15308	
Symphony	AgriGuard	13021	

SECTION 4
ADJUVANTS

Adjuvants

Adjuvants are not themselves classed as pesticides and there is considerable misunderstanding over the extent to which they are legally controlled under the Food and Environment Protection Act. An adjuvant is a substance other than water which enhances the effectiveness of a pesticide with which it is mixed. Consent C(i)5 under the Control of Pesticides Regulations allows that an adjuvant can be used with a pesticide only if that adjuvant is authorised and on a list published from time to time in *The Pesticide Register*. An authorised adjuvant has an *adjuvant number* and may have specific requirements about the circumstances in which it may be used.

Adjuvant product labels must be consulted for full details of authorised use, but the table below provides a summary of the label information to indicate the area of use of the adjuvant. Label precautions refer to the keys given in Appendix 4, and may include warnings about products harmful or dangerous to fish. The table includes all adjuvants notified by suppliers as available in 2013.

Product	Supplier	Adj. No.	Type
Abacus	De Sangosse	A0543	vegetable oil
Contains	20% w/w alkoxylated alcohols -EAC 1, 9% w/w oil (tall oil fatty acids - EAC 2), 53.43% w/w oil (rapeseed fatty acid esters - EAC 3)		
Use with	All approved pesticides on all edible crops when used at half their recommended dose or less, and on all non-edible crops up their full recommended dose. Also at a maximum concentration of 0.1% with approved pesticides on listed crops up to specified growth stages		
Protective clothing	A, C, H		
Precautions	R36, U05a, U11, U14, U19a, U20b, E15a, E19b, E34, D01, D02, D05, D10a, D12a, H04		
Activator 90	De Sangosse	A0547	non-ionic surfactant/wetter
Contains	375g/kg alkoxylated alcohols (EAC 1), 375g/kg alkoxylated alcohols (EAC 2), 150g/kg oil (tall oil fatty acids - EAC 3)		
Use with	All approved pesticides on all edible crops when used at half their recommended dose or less, and on all non-edible crops up their full recommended dose. Also at a maximum concentration of 0.1% with approved pesticides on listed crops up to specified growth stages		
Protective clothing	A, C, H		
Precautions	R36, R38, R53a, U02a, U04a, U05a, U10, U11, U20b, E15a, E19b, D01, D02, D05, D10a, H04		
Addit	Koppert	A0693	spreader/sticker/wetter
Contains	780.2 g/l oil (rapeseed triglycerides)		
Use with	Mycotal at 0.25% spray solution and all approved pesticides at half or less than half the approved pesticide rate		
Adigor	Syngenta	A0522	wetter
Contains	47% w/w methylated rapeseed oil		
Use with	Topik, Axial, Trazos, Amazon and Viscount on cereals in accordance with recommendations on the respective herbicide labels		
Protective clothing	A, C, H		
Precautions	R43, R51, R53a, U02a, U05a, U09a, U20b, E15b, E38, D01, D02, D05, D09a, D10c, D12a, H04, H11		

SECTION 4

Product	Supplier	Adj. No.	Type
Admix-P	De Sangosse	A0301	wetter
Contains	80% w/w trisiloxane organosilicone copolymers (EAC 1)		
Use with	A wide range of pesticides applied as corm, tuber, onion and other bulb treatments in seed production and in seed potato treatment		
Protective clothing	A, C, H		
Precautions	R20, R21, R22a, R36, R43, R48, R51, R58, U11, U15, U19a, E15a, E19b, D01, D02, D05, D10a, D12a, H03, H11		
Agropen	Intracrop	A0227	vegetable oil
Contains	95% refined rapeseed oil		
Use with	Recommended rates of approved pesticides up to the growth stage indicated for specified crops, and with half or less than the recommended rate on these crops after the stated growth stages. Also for use with pesticides on grassland at half their recommended rate		
Protective clothing	A, C		
Precautions	U08, U20b, E13c, E37, D09a, D10b		
Amber	Interagro	A0367	vegetable oil
Contains	95% w/w methylated rapeseed oil		
Use with	Sugar beet herbicides, oilseed rape herbicides, cereal graminicides and a wide range of other pesticides that have a label recommendation for use with authorised adjuvant oils on specified crops. See label for details		
Protective clothing	A, C		
Precautions	U05a, U20b, E15a, D01, D02, D05, D09a, D10a		
Anoint	Intracrop	A0236	spreader/vegetable oil/wetter
Contains	95% rapeseed oil		
Use with	Pesticides that have a recommendation for the addition of a wetter/spreader		
Protective clothing	A, C		
Precautions	R36, R38, U05a, U08, U19a, U20b, E13c, E34, D01, D02, D09a, D10a, H04		
Arma	Interagro	A0306	penetrant
Contains	500 g/l alkoxylated fatty amine + 500 g/l polyoxyethylene monolaurate		
Use with	Cereal growth regulators, cereal herbicides, cereal fungicides, oilseed rape fungicides and a wide range of other pesticides on specified crops		
Protective clothing	A, C		
Precautions	R51, R58, U05a, E15a, E34, E37, D01, D02, D05, D09a, D10a, H11		
Artisan	De Sangosse	A0518	adjuvant
Contains	70% w/w oil (rapeseed fatty acid esters) (EAC 1)		
Use with	All approved pesticides at 0.75% spray solution on non-edible crops and all approved pesticides at half or less than half the approved pesticide rate at 0.75% spray solution on edible crops		
Protective clothing	A, C, H		
Precautions	U19a, U20b, E15a, E19b, E34, D01, D02, D05, D10a, D12a		
Asu-Flex	Greenaway	A0677	spreader/sticker/wetter
Contains	10.0 % w/w rapeseed oil		
Use with	Asulox, Greencrop Found, I T Asulam, Inter Asulam and Spitfire		
Protective clothing	A, C		
Precautions	U11, U12, U15, U20b, E15a, E34, D02		

Product	Supplier	Adj. No.	Type
BackRow	Interagro	A0472	mineral oil
Contains	60% w/w refined paraffinic petroleum oil		
Use with	Pre-emergence herbicides. Refer to label or contact supplier for further details		
Protective clothing	A		
Precautions	R22b, R38, U02a, U05a, U08, U20b, E15a, D01, D02, D05, D09a, D10a, M05b, H03		
Ballista	Interagro	A0524	spreader/wetter
Contains	10% w/w alkoxylated triglycerides		
Use with	All approved pesticides and plant growth regulators for use in winter and spring cereals applied at full rate up to and including GS 52, and at half the approved rate thereafter. May be applied with approved pesticides for other specified crops at full rate up to specified growth stages and at half rate thereafter		
Protective clothing	A		
Precautions	U05a, U20b, E15a, E34, D01, D02, D09a		
Banka	Interagro	A0245	spreader/wetter
Contains	29.2% w/w alkyl pyrrolidones		
Use with	Potato fungicides and a wide range of other pesticides on specified crops		
Protective clothing	A, C		
Precautions	R38, R41, R52, R58, U02a, U05a, U11, U19a, U20b, E15a, E34, E37, D01, D02, D05, D09a, H04		
Barramundi	Interagro	A0376	mineral oil
Contains	95% w/w mineral oil		
Use with	Sugar beet herbicides, oilseed rape herbicides, cereal graminicides and a wide range of other pesticides that have a label recommendation for use with authorised adjuvant oils on specified crops. Refer to label or contact supplier for further details		
Protective clothing	A		
Precautions	R22b, R38, U02a, U05a, U08, U20b, E15a, D01, D02, D05, D09a, D10a, M05b, H03		
Bazuka	Intracrop	A0241	spreader/wetter
Contains	800 g/l polyalkyleneoxide modified heptamethyltrisiloxane		
Use with	All fungicides applied to spring and winter sown wheat, spring and winter sown barley, spring and winter sown oats, triticale, rye and durum wheat where the pesticide label recommends the addition of wetters and spreaders		
Protective clothing	A, C		
Precautions	R21, R22a, R43, U02a, U05a, U08, U19a, U20b, E13c, E34, D01, D02, D09a, D10a, M03, H03, H04		
Bio Syl	Intracrop	A0385	spreader/sticker/wetter
Contains	1% w/w polyoxyethylen-alpha-methyl-omega-[3-(1,3,3,3-tetramethyl-3-trimethylsiloxy)-disiloxanyl] propylether and 32.67% w/w ethylene oxide condensate		
Use with	Recommended rates of approved pesticides on a range of specified fruit and vegetable crops and on managed amenity turf. See label for details		
Protective clothing	A, C		
Precautions	R36, R38, U05a, U08, U20c, E15a, D01, D02, D10a, H04		

SECTION 4

Product	Supplier	Adj. No.	Type
Bioduo	Intracrop	A0606	wetter
Contains	85% alkyl polyglycol ether and fatty acids		
Use with	A wide range of pesticides used in grassland, agriculture and horticulture and with pesticides used in non-crop situations		
Protective clothing	A, C		
Precautions	R22a, R36, R38, U05a, U08, U20c, E13c, E34, D01, D02, D09a, D10a, M03, H03, H04, H08		
Biofilm	Intracrop	A0634	anti-drift agent/anti-transpirant/sticker/ UV screen/wetter
Contains	96% poly-1-p-menthene		
Use with	All approved fungicides and insecticides on edible crops and all approved formulations of glyphosate used pre-harvest on wheat, barley, oilseed rape, stubble, and in non-crop situations and grassland destruction		
Protective clothing	A, C		
Precautions	U19a, U20c, E13c, D09a, D11a		
BioPower	Bayer CropScience	A0617	wetter
Contains	6.7% w/w 3,6-dioxaeicosylsulphate sodium salt and 20.2% w/w 3,6-dioxaoctadecylsulphate sodium salt		
Use with	Atlantis and all other approved cereal herbicides		
Protective clothing	A, C		
Precautions	R36, R38, U02a, U05a, U08, U13, U19a, U20b, E13c, E34, D01, D02, D05, D09a, D10a, H04		
Biothene	Intracrop	A0633	anti-drift agent/anti-transpirant/sticker/ UV screen/wetter
Contains	96% poly-1-p-menthene		
Use with	All approved fungicides and insecticides on edible crops up to 30 d before harvest, and all approved formulations of glyphosate used pre-harvest on wheat, barley, oilseed rape, stubble, and in non-crop situations and grassland destruction. Must not be used in mixture with adjuvant oils or surfactants		
Precautions	U19a, U20c, E13c, D09a, D11a		
Bond	De Sangosse	A0556	extender/sticker/wetter
Contains	10% w/w alkoxylated alcohols (EAC 1), 45% w/w styrene-butadiene copolymers (EAC 2)		
Use with	All approved potato blight fungicides. Also with all approved pesticides on all edible crops when used at half their recommended dose or less, and on all non-edible crops up their full recommended dose. Also at a maximum concentration of 0.14% with approved pesticides on listed crops up to specified growth stages		
Protective clothing	A, C, H		
Precautions	R36, R38, U11, U14, U16b, U19a, E15a, E19b, D01, D02, D05, D10a, D12a, H04		

Product	Supplier	Adj. No.	Type
Break-Thru S 240	PP Products	A0431	wetter

Contains	85% w/w polyalkylene oxide modified heptamethyl trisiloxane
Use with	All approved pesticides which have a recommendation for use with a wetting agent applied to cereals, permanent grassland and rotational grass. Also with all approved pesticides at full rate on non-edible crops and on listed edible crops up to specified growth stages and on all edible crops when used at half or less of their approved full rate
Protective clothing	A, C
Precautions	R20, R21, R38, R41, R51, R53a, U11, U16b, U19a, E19b, D01, D02, D09a, D10a, M05a, H03, H11

Product	Supplier	Adj. No.	Type
Broad-Flex	Greenaway	A0680	spreader/sticker/wetter

Contains	10.0 % w/w rapeseed oil
Use with	Broad Sword or Green Guard
Protective clothing	A, C
Precautions	U11, U12, U15, U20b, E15a, E34, D02

Product	Supplier	Adj. No.	Type
Buzz	De Sangosse	A0520	wetter

Contains	375g/l alkoxylated alcohols (EAC 1), 250g/l alkoxylated coconut amines (EAC 2) 50g/l oil (tall oil fatty acids) (EAC 3)
Use with	All approved pesticides at half or less than half the approved pesticide rate at 0.1% spray solution on edible crops
Protective clothing	A, C, H
Precautions	R36, R38, R53a, U02a, U04a, U05a, U10, U11, U20b, E15a, E19b, D01, D02, D05, D10a, H04

Product	Supplier	Adj. No.	Type
Byo-Flex	Greenaway	A0545	sticker/wetter

Contains	10.0 % w/w rapeseed oil
Use with	GLY 490 (MAPP 12718)
Protective clothing	A, C
Precautions	U11, U12, U15, U20b, E34

Product	Supplier	Adj. No.	Type
Byte	De Sangosse	A0517	adjuvant

Contains	70% w/w oil (rapeseed fatty acid esters) (EAC 1)
Use with	All approved pesticides at 0.75% spray solution on non-edible crops and all approved pesticides at half or less than half the approved pesticide rate at 0.75% spray solution on edible crops
Protective clothing	A, C, H
Precautions	U19a, U20b, E15a, E19b, E34, D01, D02, D05, D10a, D12a

Product	Supplier	Adj. No.	Type
C-Cure	Interagro	A0467	mineral oil

Contains	60% w/w refined mineral oil
Use with	Pre-emergence herbicides
Protective clothing	A
Precautions	R22b, R38, U02a, U05a, U08, U20b, E15a, D01, D02, D05, D09a, D10a, M05b, H03

Product	Supplier	Adj. No.	Type
Ceres Platinum	Interagro	A0445	penetrant/spreader

Contains	500 g/l alkoxylated fatty amine + 500 g/l polyoxyethylene monolaurate
Use with	Cereal growth regulators, cereal herbicides, cereal fungicides, oilseed rape fungicides and a wide range of other pesticides on specified crops
Precautions	R51, R58, U05a, E15a, E34, D01, D02, D05, D09a, D10a, H11

SECTION 4

Product	Supplier	Adj. No.	Type
Cerround	Helena	A0510	adjuvant/vegetable oil

Contains — 80.0 % w/w oil (soybean fatty acid esters) (EAC 1) and 12.0 % w/w trisiloxane organosilicone copolymers (EAC 2)

Use with — Approved pesticides on non-edible crops, cereals and stubbles of all edible crops, grassland (destruction), and pesticides approved for use on growing edible crops when used at half recommended dose or less. On specified crops product may be used at a maximum spray concentration of 0.5% with approved pesticides at their full approved rate up to the growth stages shown in the label

Protective clothing — A, H

Precautions — R38, R41, U02a, U05a, U08, U11, U19a, U20b, E13c, E34, E37, D01, D02, D05, D09a, M03, H04

Product	Supplier	Adj. No.	Type
Clayton Astra	Clayton	A0724	penetrant/spreader

Contains — 500 g/l alkoxylated sorbitan esters and 500 g/l alkoxylated tallow amines

Use with — All approved pesticides up to a maximum conc of 0.15% of the total spray volume

Precautions — R51, R53a, U05a, E15a, E34, D01, D02, D05, D09a, D10b, D12a, H11

Product	Supplier	Adj. No.	Type
Clayton Dolmen	Clayton	A0725	adjuvant/spreader

Contains — 64% w/w trisiloxane organosilicone copolymers

Use with — All approved pesticides up to a maximum conc of 0.2% of the total spray volume

Protective clothing — A, C, H

Precautions — R20, R21, R36, R38, R51, R53a, U02a, U05a, U09a, U11, U16b, U19a, U20b, D01, D02, D05, D09a, D10a, H03, H11

Product	Supplier	Adj. No.	Type
Clayton Union	Clayton	A0686	adjuvant

Contains — 47% w/w methylated rapeseed oil

Use with — Clayton Tonto as 0.5% of total spray volume

Precautions — R43, R50, R53a, U02a, U05a, U09a, U20b, E15b, D01, D02, D05, D10c, D12a, H04, H11

Product	Supplier	Adj. No.	Type
Codacide Oil	Microcide	A0629	vegetable oil

Contains — 95% w/w oil (rapeseed triglycerides) (EAC 1)

Use with — All approved pesticides and tank mixes. See label for details

Protective clothing — A, C

Precautions — U20b, D09a, D10b

Product	Supplier	Adj. No.	Type
Companion Gold	Agrovista	A0723	acidifier/buffering agent/drift retardant/extender

Contains — 16% w/w ammonium sulphate + 0.95% w/w polyacrylamide

Use with — Diquat or glyphosate on oilseed rape at 0.5% solution; glyphosate on wheat, rye or triticale at 0.5% solution and with all approved pesticide on non-edible crops at 1% solution or all approved pesticides at 50% dose rate or less on edible crops

Protective clothing — A, C, H

Precautions — U05a, U14, U15, U19a, U19c, U20b, E15a, D02, D05, D09a, D10a

Product	Supplier	Adj. No.	Type
Compliment	United Agri	A0705	spreader/vegetable oil/wetter

Contains — 75% mixed fatty acid esters of rapeseed oil

Use with — All approved pesticides in non-edible crops, all approved pesticides on edible crops when used at half or less their recommended rate, morpholine or triazine fungicides on cereals, and with all pesticides on listed crops up to specified growth stages

Protective clothing — A, C

Precautions — R43, R52, R53a, U14, E38, D01, D05, D07, D08, H04

Conjoin	AgriGuard	A0676	adjuvant

Contains — 47% w/w methylated rapeseed oil

Use with — Amazon, Symmetry, Topik, Traxos and Viscount

Protective clothing — A, C, H

Precautions — R43, R50, R53a, U02a, U05a, U08, U20b, E13c, E15b, D01, D02, D09a, D10c, H04, H11

Contact Plus	Interagro	A0418	mineral oil

Contains — 95% w/w mineral oil

Use with — Sugar beet herbicides, oilseed rape herbicides, cereal graminicides and a wide range of other pesticides that have a label recommendation for use with authorised adjuvant oils on specified crops. Refer to label or contact supplier for further details

Protective clothing — A, C

Precautions — R22b, R38, U02a, U05a, U08, U20b, E15a, D01, D02, D09a, D10a, M05b, H03

County Mark	Greenaway	A0689	sticker/wetter

Contains — 10% w/w rapeseed oil

Use with — 'Greenaway Gly-490' (MAPP 12718) and all approved 490 g/l glyphosate products

Protective clothing — A, C

Precautions — U11, U12, U15, U20b, E34

Cropspray 11-E	Petro-Lube	A0537	adjuvant/mineral oil

Contains — 99% w/w oil (petroleum oils) (EAC 1)

Use with — All approved pesticides on all edible crops when used at half their recommended dose or less, and on all non-edible crops up to their full recommended dose. Also with listed herbicides on a range of specified crops and with all pesticides on a range of specified crops up to specified growth stages.

Protective clothing — A, C

Precautions — R22a, U10, U16b, U19a, E15a, E19b, E34, E37, D01, D02, D05, D10a, D12a, M05b, H03

Dash HC	BASF	A0729	non-ionic surfactant/wetter

Contains — 348.75 g/l oil (fatty acid esters) and 209.25 g/l alkoxylated alcohols-phosphate esters

Use with — For use with Cleranda at 1.0 l/ha in a water volume of 100 - 400 l/ha on Clearfield oilseed rape

Protective clothing — A, C, H

Precautions — U05a, U12, U14, D01, D02, D10c, M05b, H03

SECTION 4

Product	Supplier	Adj. No.	Type
Designer	De Sangosse	A0660	drift retardant/extender/sticker/wetter
Contains	25% w/w styrene-butadiene copolymers (EAC 1), 7.1% w/w trisiloxane organosilicone copolymers (EAC 2)		
Use with	A wide range of fungicides, insecticides and trace elements for cereals and specified agricultural and horticultural crops		
Protective clothing	A, C, H		
Precautions	R36, R38, R52, U02a, U11, U14, U15, U19a, E15a, E37, D01, D02, D05, D09a, D10a, H04		
Desikote Max	Taminco	A0666	anti-transpirant/extender/sticker/wetter
Contains	400 g/l di-1-p-menthene		
Use with	Approved pesticides on all edible crops (except herbicides on peas) and as an anti-transpirant on vegetables before transplanting, evergreens, deciduous trees, shrubs, bushes and on turf		
Protective clothing	A		
Precautions	R38, R50, R53a, U05a, U14, U20b, E15a, E34, E37, D01, D09a, D10b, D12b, M03, H04, H11		
Diagor	AgChem Access	A0671	wetter
Contains	47% w/w methylated rapeseed oil		
Use with	Topik, Axial, Trazos, Amazon and Viscount on cereals in accordance with recommendations on the respective herbicide labels		
Protective clothing	A, C, H		
Precautions	R43, R51, R53a, U02a, U05a, U09a, U20b, E15b, E38, D01, D02, D05, D09a, D10c, D12a, H04, H11		
Drill	De Sangosse	A0544	adjuvant
Contains	15% w/w alkoxylated alcohols (EAC 1), 7.5% w/w oil (tall oil fatty acids - EAC 2), 63.34% w/w oil (rapeseed fatty acid esters) (EAC 3)		
Use with	All approved pesticides on all edible crops when used at half their recommended dose or less, and on all non-edible crops up to their full recommended dose. Also at a maximum concentration of 0.1% with approved pesticides on listed crops up to specified growth stages		
Protective clothing	A, C, H		
Precautions	R36, U05a, U11, U14, U19a, U20b, E15a, E19b, E34, D01, D02, D05, D10a, D12a, H04		
Eco-flex	Greenaway	A0696	sticker/wetter
Contains	10% w/w refined rapeseed oil		
Use with	Approved formulations of glyphosate, 2,4-D		
Protective clothing	A, C		
Precautions	U11, U12, U15, U20b, E34		
Elan	Intracrop	A0392	spreader/sticker/wetter
Contains	82% w/w polyoxyethylen-alpha-methyl-omega-disiloxanyl] propylether		
Use with	Recommended rates of approved pesticides up to the growth stage indicated for specified crops, and with half or less than the recommended rate on these crops after the stated growth stages. Also for use with recommended rates of herbicides on managed amenity turf		
Protective clothing	A, C, H		
Precautions	R36, R38, U05a, U08, U20c, E13c, E15a, E37, D01, D02, D09a, D10a, H04		

Product	Supplier	Adj. No.	Type
Emerald	Intracrop	A0636	anti-drift agent/anti-transpirant/extender/ UV screen

Contains	96% di-1-p-menthene
Use with	Recommended rates of approved pesticides up to the growth stage indicated for specified crops, and with half or less than the recommended rate on these crops after the stated growth stages. Also for use alone on transplants, turf, fruit crops, glasshouse crops and Christmas trees. Must not be used in mixture with adjuvant oils or surfactants
Precautions	U19a, U20c, E13c, E37, D09a, D11a

Product	Supplier	Adj. No.	Type
Esterol	Interagro	A0330	vegetable oil

Contains	95% w/w methylated rapeseed oil
Use with	Plant protection products that have a label recommendation for use with adjuvant oils. Refer to label or contact supplier for further details
Precautions	U05a, U20b, E15a, D01, D02, D09a, D10a

Product	Supplier	Adj. No.	Type
Euroagkem Pace	Intracrop	A0562	spreader/wetter

Contains	42% w/w propionic acid
Use with	All approved formulations of chlormequat, all approved formulations of glyphosate, diquat, fenoxaprop-P-ethyl, tralkoxydim, clodinafop-propargyl, fluazifop-P-butyl, cycloxydim and propaquizafop at half or less than half the approved pesticide rate in edible crops, at full rate in non-edible crops.
Protective clothing	A, C
Precautions	U02a, U04a, U05a, U08, U10, U11, U13, U14, U15, U19a, U20b, E15a, D01, D02, D09a, D10a, M05a, H05

Product	Supplier	Adj. No.	Type
Euroagkem Pen-e-trate	Intracrop	A0564	spreader/wetter

Contains	350 g/l propionic acid
Use with	All approved pesticides which have a recommendation for use with a wetting agent; pesticides must be used at half approved dose rate or less on edible crops, up to full dose rate on non-edible crops
Protective clothing	A, C
Precautions	U05a, U08, U11, U14, U15, U19a, U20c, E13e, E15a, D01, D02, D09a, D10b, M03, H05

Product	Supplier	Adj. No.	Type
Felix	Intracrop	A0178	spreader/wetter

Contains	60% ethylene oxide condensate
Use with	Mecoprop, 2,4-D in cereals and amenity turf, and a range of grass weedkillers in agriculture. See label for details
Protective clothing	A, C
Precautions	R36, R38, U05a, U08, U20a, E15a, D01, D02, D09a, D10a, H04, H08

Product	Supplier	Adj. No.	Type
Firebrand	Barclay	-	fertiliser/water conditioner

Contains	500 g/l ammonium sulphate
Use with	Use at 0.5% v/v in the spray solution with glyphosate
Precautions	U08, U20a, E15a, D09a, D10b

SECTION 4

Product	Supplier	Adj. No.	Type
Gateway	United Agri	A0651	extender/sticker/wetter

Contains 73% w/v synthetic latex solution and 8.5% w/v polyether modified trisiloxane

Use with All approved pesticides in crops not destined for human or animal consumption, all approved pesticides on edible crops when used at half or less their recommended rate, and with all pesticides on listed crops up to specified growth stages

Protective clothing A, C, H

Precautions R41, R52, R53a, U11, U15, E37, E38, D01, D05, D07, D08, H04

Gladiator	De Sangosse	A0542	penetrant/wetter

Contains 190g/kg alkoxylated alcohols (EAC 1), 190g/kg alkoxylated alcohols (EAC 2), 75g/kg oil (tall oil fatty acids) (EAC 3), 210 g/kg alkoxylated tallow amines (EAC 4)

Use with All approved pesticides on all edible crops when used at half their recommended dose or less, and on all non-edible crops up their full recommended dose. Also at a maximum concentration of 0.1% with approved pesticides on listed crops up to specified growth stages

Protective clothing A, C, H

Precautions R36, R38, R41, R52, R58, U05a, U10, U11, U19a, E15a, E19b, E34, E37, D01, D02, D05, D10a, D12a, M03, H04

Gly-Flex	Greenaway	A0588	spreader/sticker/wetter

Contains 95% refined rapeseed oil

Use with GLY-490 (MAPP 12718) or any approved formulations of 490 g/l glyphosate

Protective clothing A, C

Precautions R36, U11, U12, U15, U20b, E13c, E34, E37, H04

Gly-Plus A	Greenaway	A0736	spreader/sticker/wetter

Contains 10.0 % w/w oil (rapeseed triglycerides) (EAC 1)

Use with Any approved 360 g/l glyphosate

Protective clothing A, C

Precautions R36, U05a, U08, U11, U12, U15, U20b, E13c, E34, E37, E38, H04

Green Gold	Intracrop	A0250	spreader/wetter

Contains 95% refined rapeseed oil

Use with All pesticides which have a recommendation for the addition of a wetter/spreader

Precautions U08, U20b, E13c, E34, D09a, D10b

Greencrop Astra	Greencrop	A0417	adjuvant/penetrant/spreader

Contains 500 g/l alkoxylated fatty amine and 500 g/l polyoxylene monolaurate

Use with Approved pesticides on a range of specified arable crops, and in forestry, managed amenity turf, stubbles and non-crop areas and situations

Precautions U05a, E13c, E34, D01, D02, D05, D09a, D10b

Greencrop Dolmen	Greencrop	A0419	adjuvant/spreader

Contains 64% w/w polyalkeneoxide modified heptamethyltrisiloxane

Use with Approved pesticides on a range of specified arable crops, and in forestry, managed amenity turf, stubbles and non-crop areas and situations

Protective clothing A, C, H

Precautions R21, R22a, R38, R41, R43, U02a, U05a, U08, U11, E13c, E34, D01, D02, D09a, D10b, M03, H03, H04

Product	Supplier	Adj. No.	Type
Grenadier	Intracrop	A0608	spreader/sticker/wetter
Contains	700 g/l polyoxyethylene (5-8 EO) C10-C15 primary alcohol and 150 g/l fatty acids		
Use with	All approved pesticides on edible and non-edible crops when used at half their approved dose or less.		
Protective clothing	A, C		
Grip	De Sangosse	A0558	extender/sticker/wetter
Contains	10% w/w alkoxylated alcohols (EAC 1), 45% w/w styrene-butadiene copolymers (EAC 2)		
Use with	All approved potato blight fungicides. Also with all approved pesticides on all edible crops when used at half their recommended dose or less, and on all non-edible crops up their full recommended dose. Also at a maximum concentration of 0.14% with approved pesticides on listed crops up to specified growth stages		
Protective clothing	A, C, H		
Precautions	R36, R38, U11, U14, U16b, U19a, E15a, E19b, D01, D02, D05, D10a, D12a, H04		
Grounded	Helena	A0456	mineral oil
Contains	732 g/l petroleum oils		
Use with	All approved pesticides on edible and non-edible crops when used at half their approved dose or less. Also with approved pesticides on specified crops, up to specified growth stages, at up to their full approved dose		
Protective clothing	A, C		
Precautions	R53a, U02a, U05a, U08, U20b, E15a, E34, E37, D01, D02, D05, D09a, D10c, H11		
Guide	De Sangosse	A0528	acidifier/drift retardant/penetrant
Contains	9.39% w/w alkoxylated alcohols (EAC 1), 35.0% w/w propionic acid (EAC 2), 35.0% w/w soybean phospholipids (EAC 3)		
Use with	All approved pesticides on all edible crops when used at half their recommended dose or less, and on all non-edible crops up their full recommended dose. Also with morpholine fungicides on cereals and with pirimicarb on legumes when used at full dose; with iprodione on brassicas when used at 75% dose, and at a maximum concentration of 0.5% with all pesticides on listed crops at full dose up to specified growth stages		
Protective clothing	A, C, H		
Precautions	R36, R38, U11, U14, U15, U19a, E15a, E19b, D01, D02, D05, D10a, D12a, H04		
Headland Fortune	Headland	A0703	penetrant/spreader/vegetable oil/wetter
Contains	75% w/w mixed methylated fatty acid esters of seed oil and N-butanol		
Use with	Herbicides and fungicides in a wide range of crops. See label for details		
Protective clothing	A, C		
Precautions	R43, U02a, U05a, U14, U20a, E15a, E34, E38, D01, D02, D05, D09a, D10b, H04		

SECTION 4

Product	Supplier	Adj. No.	Type
Headland Guard 2000	Headland	A0652	extender/sticker
Contains	10% w/w styrene/butadiene co-polymers		
Use with	All approved pesticides on all edible crops when used at half their recommended dose or less, and on all non-edible crops up their full recommended dose. Also at a maximum concentration of 0.1% with approved pesticides on listed crops up to specified growth stages		
Protective clothing	A, C		
Precautions	E15a, D01, D02		
Headland Guard Pro	Headland Amenity	A0423	extender/sticker
Contains	52% synthetic latex solution		
Use with	All approved pesticides, micronutrients and sea-weed based plant growth stimulants in amenity grass, managed amenity turf and amenity vegetation		
Precautions	U05a, U08, U19a, U20b, E13b, E34, E37, D01, D02, D05, D10b		
Headland Inflo XL	Headland Amenity	A0329	wetter
Contains	85% w/w polyalkylene oxide modified heptamethyl siloxane		
Use with	Any pesticide for use on amenity grassland or managed amenity turf (except any product applied in or near water) where the use of a wetting, spreading and penetrating surfactant is recommended to improve foliar coverage, and with any approved pesticide on crops not intended for human or animal consumption		
Protective clothing	A, C		
Precautions	R21, R22a, R38, R41, R43, R51, R58, U02a, U05a, U08, U14, U19a, U20b, E15a, E34, E38, D01, D02, D05, D09a, D10b, M03, H03, H11		
Headland Intake	Headland	A0074	penetrant
Contains	450 g/l propionic acid		
Use with	All approved pesticides on any crop not intended for human or animal consumption, and with all approved pesticides on beans, peas, edible podded peas, oilseed rape, linseed, sugar beet, cereals (except triazole fungicides), maize, Brussels sprouts, potatoes, cauliflowers. See label for detailed advice on timing on these crops		
Protective clothing	A, C		
Precautions	R34, U02a, U05a, U10, U11, U14, U15, U19a, E34, D01, D02, D05, D09b, D10b, M04a, H05		
Headland Rhino	Headland	A0328	wetter
Contains	85% w/w polyalkylene oxide modified heptamethyl siloxane		
Use with	Any herbicide, systemic fungicide, systemic insecticide or plant growth regulator (except any product applied in or near water) where the use of a wetting, spreading and penetrating surfactant is recommended to improve foliar coverage		
Protective clothing	A, C		
Precautions	R21, R22a, R38, R41, R43, R51, R58, U02a, U05a, U08, U11, U14, U19a, U20b, E13c, E34, D01, D02, D05, D09a, D10b, M03, H03, H11		

Product	Supplier	Adj. No.	Type
Impala	Intracrop	A0607	spreader/sticker/wetter
Contains	700 g/l polyoxyethylene (5-8 EO) C10-C15 primary alcohol and 150 g/l fatty acids		
Use with	All approved pesticides at half or less than half the approved pesticide rate on edible crops and all approved pesticides on non-edible crops		
Intracrop Agwet	Intracrop	A0616	sticker/wetter
Contains	900 g/l polyoxyethylene (5-8 EO) C10-C15 primary alcohol		
Use with	All approved pesticides at half or less than half the approved pesticide rate on edible crops and all approved pesticides on non-edible crops		
Intracrop Agwet Gtx	Intracrop	A0646	sticker/wetter
Contains	500 g/l polyoxyethylene (5-8 EO) C10-C15 primary alcohol		
Use with	All approved pesticides at half or less than half the approved pesticide rate on edible crops and all approved pesticides on non-edible crops		
Intracrop Archer	Intracrop	A0242	spreader/wetter
Contains	800 g/l polyalkyleneoxide modified heptamethyltrisiloxane		
Use with	All fungicides used in cereals where the addition of wetters/spreaders is recommended on the pesticide label		
Protective clothing	A, C		
Precautions	R21, R22a, R43, U02a, U05a, U08, U19a, U20b, E13c, E34, D01, D02, D09a, D10a, M03, H03, H04		
Intracrop BLA	Intracrop	A0453	anti-drift agent/anti-transpirant/extender/ sticker/UV screen
Contains	41.5% w/w styrene butadiene co-polymer solution		
Use with	All potato blight fungicides. Also with recommended rates of approved pesticides in certain non-crop situations and up to the growth stage indicated for specified crops, and with half or less than the recommended rate on these crops after the stated growth stages. Also with pesticides in grassland at half or less their recommended rate		
Precautions	U08, U20b, E13c, E34, E37, D01, D05, D09a, D10a		
Intracrop Bla-Tex	Intracrop	A0656	anti-drift agent/anti-transpirant/sticker/ UV screen
Contains	22.0 % w/w styrene-butadiene co-polymer		
Use with	All approved pesticides at half or less than half the approved pesticide rate on edible crops, all approved pesticides on non edible crops and with all blight fungicides on potatoes		
Intracrop Boost	Intracrop	A0709	spreader/sticker/wetter
Contains	32.67 % w/w polyoxyethylene (5-8 EO) C10-C15 primary alcohol and 1.0 % w/w polyoxyethylen-[-alpha-]-methyl-[-omega-]-[3-(1,3,3,3-tetramethyl-3-trimethylsiloxy)-disiloxanyl] propylether;		
Use with	All approved pesticides at half or less than half the approved pesticide rate on edible crops and all approved pesticides on non-edible crops		
Intracrop Cogent	Intracrop	A0707	spreader/sticker/wetter
Contains	32.67 % w/w polyoxyethylene (5-8 EO) C10-C15 primary alcohol and 1.0 % w/w polyoxyethylen-[-alpha-]-methyl-[-omega-]-[3-(1,3,3,3-tetramethyl-3-trimethylsiloxy)-disiloxanyl] propylether		
Use with	All approved pesticides at half or less than half the approved pesticide rate on edible crops and all approved pesticides on non-edible crops		

SECTION 4

Product	Supplier	Adj. No.	Type
Intracrop Dictate	Intracrop	A0673	adjuvant
Contains	91% w/w methylated rapeseed oil		
Use with	All approved pesticides on non-edible crops; all approved pesticides applied at half or less than half dose on edible crops		
Protective clothing	A, C		
Precautions	U19a, U20b, D05, D09a, D10b		
Intracrop Era	Intracrop	A0664	spreader/sticker/wetter
Contains	27.0 % w/w polyoxyethylen-[-alpha-]-methyl-[-omega-]-[3-(1,3,3,3-tetramethyl-3-trimethylsiloxy)-disiloxanyl] propylether		
Use with	All approved pesticides at half or less than half the approved rate		
Protective clothing	A, C		
Precautions	U05a, U08, U20b, E13e, D01, D02, D09a, D10a		
Intracrop Evoque	Intracrop	A0754	spreader/wetter
Contains	652 g/l oil (rapeseed fatty acid esters) (EAC 1), 112 g/l trisiloxane organosilicone copolymers (EAC 2), 22.5 g/l alkoxylated alcohols (EAC 3) and 19.1 g/l alkoxylated alcohols (EAC 4)		
Use with	All authorised plant protection products on non-edible crops and all authorised plant protection products at half or less than half the authorised plant protection product rate on edible crops. Max conc is 0.2% of spray volume.		
Intracrop F16	Intracrop	A0752	spreader/wetter
Contains	652 g/l oil (rapeseed fatty acid esters) (EAC 1), 112 g/l trisiloxane organosilicone copolymers (EAC 2), 22.5 g/l alkoxylated alcohols (EAC 3) and 19.1 g/l alkoxylated alcohols (EAC 4).		
Use with	All authorised plant protection products on non-edible crops and all authorised plant protection products at half or less than half the authorised plant protection product rate on edible crops. Max conc is 0.2% of spray volume.		
Intracrop Green Oil	Intracrop	A0262	wetter
Contains	95% refined rapeseed oil		
Use with	Recommended rates of approved pesticides up to the growth stage indicated for specified crops, and with half or less than the recommended rate on these crops after the stated growth stages. Also for use with pesticides on grassland at half their recommended rate		
Precautions	U08, U20b, E13c, E37, D09a, D10b		
Intracrop Impetus	Intracrop	A0647	spreader/wetter
Contains	50.0 % w/w polyoxyethylene (5-8 EO) C10-C15 primary alcohol		
Use with			
Intracrop Inca	Intracrop	A0701	adjuvant
Contains	840 g/l methylated rapeseed oil		
Use with	All approved pesticides at half or less than half the approved pesticide rate on edible crops and all approved pesticides on non-edible crops		
Intracrop Incite	Intracrop	A0702	adjuvant
Contains	840 g/l methylated rapeseed oil		
Use with	All approved pesticides at half or less than half the approved pesticide rate on edible crops and all approved pesticides on non-edible crops		

Product	Supplier	Adj. No.	Type
Intracrop Iona	Intracrop	A0583	spreader/wetter
Contains	90.0 % w/w polyoxyethylene (3EO) C12-C15 primary alcohol		
Use with	All approved pesticides at half their recommended dose rate on edible crops and all approved pesticides on non-edible crops. For edible crops see label for latest growth stages at application.		
Intracrop Iona Low Foam	Intracrop	A0582	spreader/wetter
Contains	89.8 % w/w polyoxyethylene (3EO) C12-C15 primary alcohol		
Use with	All approved pesticides at half their recommended dose rate on edible crops and all approved pesticides on non-edible crops. For edible crops see label for latest growth stages at application.		
Intracrop Mica	Intracrop	A0663	spreader/sticker/wetter
Contains	27.0 % w/w polyoxyethylen-[-alpha-]-methyl-[-omega-]-[3-(1,3,3,3-tetramethyl-3-trimethylsiloxy)-disiloxanyl] propylether		
Use with	All approved pesticides at half or less than half the approved rate		
Protective clothing	A, C		
Precautions	U05a, U08, U20c, E13e, D01, D02, D09a, D10a		
Intracrop Micotex	Intracrop	A0688	activator/spreader/wetter
Contains	25.0 % w/w ethylene oxide-propylene oxide copolymers (EAC 1)		
Use with	All approved pesticides on non-edible crops and non crop production uses; All approved pesticides at half or less than half the approved pesticide rate on edible crops		
Protective clothing	A, C		
Precautions	R36, R38, U05b, U08, U20c, E15a, D01, D02, D09a, D10b, H04		
Intracrop Neotex	Intracrop	A0460	anti-drift agent/anti-transpirant/extender/sticker/UV screen
Contains	42.5% w/w styrene butadiene co-polymer solution		
Use with	Recommended rates of approved pesticides up to specified growth stages of a wide range of agricultural arable and horticultural crops, and for non-crop uses. Use on edible crops beyond specified growth stages, and in grass, should only be with half recommended rates of the pesticide or less. See label for details of growth stage restrictions. In addition may be used with all potato blight fungicides at their recommended rates of use up to the latest recommended timing of the fungicide		
Precautions	U08, U20b, E13c, E34, E37, D01, D05, D09a, D10a		
Intracrop Novatex	Intracrop	A0566	anti-drift agent/anti-transpirant/extender/sticker/UV screen
Contains	42.5% w/w styrene butadiene co-polymer solution		
Use with	Recommended rates of approved pesticides up to the growth stage indicated for specified crops, and with half or less than the recommended rate on these crops after the stated growth stages. Also for use with recommended rates of pesticides on grassland and specified non-crop situations		
Precautions	U08, U20b, E13c, E34, E37, D01, D09a, D10a		

SECTION 4

Product	Supplier	Adj. No.	Type
Intracrop Perm-E8	Intracrop	A0565	spreader/wetter

Contains 42% w/w propionic acid

Use with All approved formulations of chlormequat, all approved formulations of glyphosate, diquat, fenoxaprop-P-ethyl, tralkoxydim, clodinafop-propargyl, fluazifop-P-butyl, cycloxydim and propaquizafop at half or less than half the approved pesticide rate in edible crops, at full rate in non-edible crops.

Protective clothing A, C

Precautions U02a, U04a, U05a, U08, U10, U11, U13, U14, U15, U19a, U20b, D01, D02, D09a, D10a, H05

Product	Supplier	Adj. No.	Type
Intracrop Predict	Intracrop	A0503	adjuvant/vegetable oil

Contains 91% w/w methylated rapeseed oil

Use with All approved pesticides on non-edible crops, and all pesticides approved for use on growing edible crops when used at half recommended dose or less. On specified crops product may be used at a maximum spray concentration of 1% with approved pesticides at their full approved rate up to the growth stages shown in the label

Precautions U19a, U20b, D05, D09a, D10b

Product	Supplier	Adj. No.	Type
Intracrop Protocol	Intracrop	A0687	spreader/wetter

Contains 25.0 % w/w ethylene oxide-propylene oxide copolymers (EAC 1)

Use with All approved pesticides on all non-edible crops and non crop production; all approved pesticides at half or less than half the approved pesticide rate on all edible crops

Protective clothing A, C

Precautions R36, R38, U05a, U08, U20c, E15a, D01, D02, D09a, D10b, H04

Product	Supplier	Adj. No.	Type
Intracrop Questor	Intracrop	A0495	activator/non-ionic surfactant/spreader

Contains 75% polyoxyethylene polypropoxypropanol

Use with All approved pesticides on non-edible crops and pesticides used in non-crop production, and all pesticides approved for use on growing edible crops when used at half recommended dose or less. On specified crops product may be used at a maximum spray concentration of 0.3% with approved pesticides at their full approved rate up to the growth stages shown in the label

Protective clothing A, C

Precautions R36, R38, U04a, U05a, U08, U19a, E13b, D01, D02, D09a, D10b, M03, M05a, H04

Product	Supplier	Adj. No.	Type
Intracrop Rapeze	Intracrop	A0643	vegetable oil

Contains 842 g/l methylated rapeseed oil

Use with All approved pesticides at half or less than half the approved pesticide rate on edible crops and all approved pesticides on non-edible crops at 1.78% spray solution.

Protective clothing A, C

Product	Supplier	Adj. No.	Type
Intracrop Rapide	Intracrop	A0563	penetrant/surfactant

Contains 40% propionic acid

Use with A wide range of pesticides and growth regulators, especially chlormequat

Protective clothing A, C

Precautions R34, R36, R38, U02a, U05a, U08, U19a, U20a, D01, D02, D09a, D10a, H04, H05

Product	Supplier	Adj. No.	Type
Intracrop Rapide Beta	Intracrop	A0672	wetter
Contains	350 g/l propionic acid and 100 g/l polyoxyethylene (5-8 EO) C10-C15 primary alcohol		
Use with	All approved pesticides which have a recommendation for use with a wetting agent; pesticides must be used at half approved dose rate or less on edible crops, up to full dose rate on non-edible crops		
Protective clothing	A, C		
Precautions	U02a, U04a, U05a, U08, U10, U11, U13, U14, U15, U19a, U20b, E15a, D01, D02, D09a, D10a, M05a, H05		
Intracrop Retainer	Intracrop	A0508	adjuvant/vegetable oil
Contains	60% w/w methylated rapeseed oil		
Use with	All approved pesticides on non-edible crops, and all pesticides approved for use on growing edible crops when used at half recommended dose or less. On specified crops product may be used at a maximum spray concentration of 1% with approved pesticides at their full approved rate up to the growth stages shown in the label		
Precautions	U19a, U20b, D05, D09a, D10b		
Intracrop Retainer NF	Intracrop	A0711	adjuvant
Contains	91.0 % w/w methylated rapeseed oil		
Use with	All approved pesticides at half or less than half the approved pesticide rate on edible crops and all approved pesticides on non-edible crops		
Intracrop Rigger	Intracrop	A0572	adjuvant/vegetable oil
Contains	91.7% w/w methylated rapeseed oil		
Use with	All approved pesticides on non-edible crops, and all pesticides approved for use on growing edible crops when used at half recommended dose or less. On specified crops product may be used at a maximum spray concentration of 1.78% with approved pesticides at their full approved rate up to the growth stages shown in the label		
Precautions	U19a, U20b, D05, D09a, D10b		
Intracrop Salute	Intracrop	A0683	spreader/sticker/wetter
Contains	27% w/w polyoxyethylen-methyl- [3-(1,3,3,3-tetramethyl-3-trimethylsiloxy) disiloxanyl] propylether		
Use with	All approved pesticides on non-edible crops and non crop production uses; All approved pesticides at half or less than half the approved pesticide rate on edible crops		
Protective clothing	A, C		
Precautions	U05a, U08, U20c, E13e, D01, D02, D09a, D10a		
Intracrop Sapper	Intracrop	A0684	spreader/sticker/wetter
Contains	27% w/w polyoxyethylene-methyl-[3-(1,3,3,3-tetramethyl- 3-trimethylsiloxy) disiloxanyl] propylether		
Use with	All approved pesticides on non-edible crops and non crop production uses; All approved pesticides at half or less than half the approved pesticide rate on edible crops		
Protective clothing	A, C		
Precautions	U05a, U08, U20c, E13c, D01, D02, D09a, D10a		

SECTION 4

677

Product	Supplier	Adj. No.	Type
Intracrop Saturn	Intracrop	A0494	activator/non-ionic surfactant/spreader
Contains	75% polyoxyethylene polypropoxypropanol		
Use with	All approved pesticides on non-edible crops and pesticides used in non-crop production, and all pesticides approved for use on growing edible crops when used at half recommended dose or less. On specified crops product may be used at a maximum spray concentration of 0.3% with approved pesticides at their full approved rate up to the growth stages shown in the label		
Protective clothing	A, C		
Precautions	R36, R38, U04a, U05a, U08, U19a, E13b, D01, D02, D09a, D10b, M03, M05a, H04		
Intracrop Signal XL	Intracrop	A0659	anti-drift agent/anti-transpirant/extender/ sticker/UV screen
Contains	62.5% liquid concentrate formulation containing 22% w/w latex solidsco- formulated with 20% alcohol ethoxylate		
Use with	All approved pesticides on non-edible crops and non crop production uses; All approved pesticides at half or less than half the approved pesticide rate on edible crops		
Protective clothing	A, C		
Precautions	U05a, U19a, U20b, E13c, E34, D01, D02, D09a, D10b, M04a, M05a		
Intracrop Spread-n-wet	Intracrop	A0624	spreader/wetter
Contains	900 g/l polyoxyethylene (5-8 EO) C10-C15 primary alcohol		
Use with			
Intracrop Sprinter	Intracrop	A0513	spreader/wetter
Contains	192 g/l primary alcohol ethoxylate		
Use with	Recommended rates of approved pesticides up to the growth stage indicated for specified crops, and with half or less than the recommended rate on these crops after the stated growth stages. Also with herbicides on managed amenity turf at recommended rates, and on grassland at half or less than recommended rates		
Protective clothing	A, C		
Precautions	R36, R38, R41, U05a, U08, U11, U14, U15, U20c, E15a, D01, D02, D09a, D10b, M03, H04		
Intracrop Status	Intracrop	A0506	adjuvant/vegetable oil
Contains	91% w/w methylated rapeseed oil		
Use with	All approved pesticides on non-edible crops, and all pesticides approved for use on growing edible crops when used at half recommended dose or less. On specified crops product may be used at a maximum spray concentration of 1.0% with approved pesticides at their full approved rate up to the growth stages shown in the label		
Precautions	U19a, U20b, D05, D09a, D10b		
Intracrop Stay-Put	Intracrop	A0507	vegetable oil
Contains	91% w/w methylated rapeseed oil		
Use with	Recommended rates of approved pesticides for non-crop uses and with recommended rates of approved pesticides up to the growth stage indicated for specified crops, and with half or less than the recommended rate on these crops after the stated growth stages		
Precautions	U19a, U20b, D05, D09a, D10b		

Product	Supplier	Adj. No.	Type
Intracrop Super Rapeze MSO	Intracrop	A0571	adjuvant/vegetable oil
Contains	840 g/l oil (rapeseed fatty acid esters) (EAC 1)		
Use with	All approved pesticides on non-edible crops, and all pesticides approved for use on growing edible crops when used at half recommended dose or less. On specified crops product may be used at a maximum spray concentration of 1.78% with approved pesticides at their full approved rate up to the growth stages shown in the label		
Precautions	U19a, U20b, D05, D09a, D10b		
Intracrop Tonto	Intracrop	A0708	spreader/sticker/wetter
Contains	32.67 % w/w polyoxyethylene (5-8 EO) C10-C15 primary alcohol and 1.0 % w/w polyoxyethylen-[-alpha-]-methyl-[-omega-]-[3-(1,3,3,3-tetramethyl-3-trimethylsiloxy)-disiloxanyl] propylether		
Use with	All approved pesticides at half or less than half the approved pesticide rate on edible crops and all approved pesticides on non-edible crops		
Intracrop Warrior	Intracrop	A0514	spreader/wetter
Contains	192 g/l primary alcohol ethylene		
Use with	Recommended rates of approved pesticides up to the growth stage indicated for specified crops, and with half or less than the recommended rate on these crops after the stated growth stages. Also with herbicides on managed amenity turf at recommended rates, and on grassland at half or less than recommended rates		
Protective clothing	A, C		
Precautions	R36, R38, R41, U05a, U08, U11, U14, U15, U20c, E15a, D01, D02, D09a, D10b, M03, H04		
Intracrop Zenith	Intracrop	A0665	spreader/sticker/wetter
Contains	27.0 % w/w polyoxyethylen-[-alpha-]-methyl-[-omega-]-[3-(1,3,3,3-tetramethyl-3-trimethylsiloxy)-disiloxanyl] propylether co-formulated with a betaine latex polymer emulsifier complex		
Use with	All approved pesticides at half or less than half the approved pesticide rate		
Protective clothing	A, C		
Precautions	R36, R38, R41, U05a, U08, U11, U20c, E13e, D01, D02, D09a, D10a, H04		
Jogral	Intracrop	A0226	cationic surfactant
Contains	800 g/l tallow amine ethoxylate		
Use with	Glyphosate		
Protective clothing	A, C		
Precautions	R22a, U08, U19a, U20a, E13b, E34, D01, D02, D06b, D10b, D11a, M03, H03, H08		
Kantor	Interagro	A0623	spreader/wetter
Contains	790 g/l alkoxylated triglycerides		
Use with	All approved pesticides		
Precautions	U05a, U20b, E15b, E19b, E34, D01, D02, D09a, D10a, D12a		

SECTION 4

Product	Supplier	Adj. No.	Type
Katalyst	Interagro	A0450	penetrant/water conditioner
Contains	90% w/w alkoxylated fatty amine		
Use with	Glyphosate and a wide range of other pesticides on specified crops. Refer to label or contact supplier for further details		
Protective clothing	A, C		
Precautions	R22a, R36, R38, R50, R58, U02a, U05a, U08, U19a, U20b, E15a, E34, E37, D01, D02, D05, D09a, D10a, M03, H03, H11		
Kinetic	Helena	A0252	spreader/wetter
Contains	80% w/w ethylene oxide & propylene oxide copolymers + 16% w/w trisiloxane copolymers		
Use with	Approved pesticides on non-edible crops and cereals and stubbles of all edible crops when used at full recommended dose, and with pesticides approved for use on growing edible crops when used at half recommended dose or less. On specified crops product may be used at a maximum spray concentration of 0.2% with approved pesticides at their full approved rate up to the growth stages shown in the label		
Protective clothing	A, H		
Precautions	R38, R41, U02a, U05a, U08, U19a, U20b, E13c, E34, E37, D01, D02, D05, D09a, D10c, M03, H04		
Klipper	Amega	A0260	spreader/wetter
Contains	600 g/l ethylene oxide condensate		
Use with	Approved salt formulations of mecoprop alone or in mixtures with 2,4-D on managed amenity turf		
Protective clothing	A, C		
Precautions	R36, R38, U05a, U08, U20c, E15a, D01, D02, D09a, D10a, H04, H08		
Leaf-Koat	Helena	A0511	spreader/wetter
Contains	80% w/w ethylene oxide & propylene oxide copolymers + 16% w/w trisiloxane copolymers		
Use with	Approved pesticides on non-edible crops and cereals and stubbles of all edible crops when used at full recommended dose, and with pesticides approved for use on growing edible crops when used at half recommended dose or less. On specified crops product may be used at a maximum spray concentration of 0.2% with approved pesticides at their full approved rate up to the growth stages shown in the label		
Protective clothing	A, H		
Precautions	R38, R41, U02a, U05a, U08, U19a, U20b, E13c, E34, E37, D01, D02, D05, D09a, D10c, M03, H04		
Level	United Agri	A0654	extender/sticker
Contains	10% w/w styrene-butadiene copolymers (EAC 1)		
Use with	All approved pesticides in non-edible crops, all approved pesticides on edible crops when used at half or less their recommended rate, potato blight fungicides and with all pesticides on listed crops up to specified growth stages		
Precautions	E37, D01, D02, D05		

Product	Supplier	Adj. No.	Type
Li-700	De Sangosse	A0529	acidifier/drift retardant/penetrant
Contains	9.39% w/w alkoxylated alcohols (EAC 1), 35.0% w/w propionic acid (EAC 2), 35.0% w/w soybean phospholipids (EAC 3)		
Use with	All approved pesticides on all edible crops when used at half their recommended dose or less, and on all non-edible crops up their full recommended dose. Also with morpholine fungicides on cereals and with pirimicarb on legumes when used at full dose; with iprodione on brassicas when used at 75% dose, and at a maximum concentration of 0.5% with all pesticides on listed crops at full dose up to specified growth stages		
Protective clothing	A, C, H		
Precautions	R36, R38, U11, U14, U15, U19a, E15a, E19b, D01, D02, D05, D10a, D12a, H04		
Logic	Microcide	A0288	vegetable oil
Contains	95% w/w oil (rapeseed triglycerides) (EAC 1)		
Use with	All approved pesticides on edible and non-edible crops for ground or aerial application		
Protective clothing	A, C		
Precautions	U19a, U20b, D09a, D10b		
Logic Oil	Microcide	A0630	vegetable oil
Contains	95% w/w oil (rapeseed triglycerides) (EAC 1)		
Use with	All approved pesticides on edible and non-edible crops for ground or aerial application		
Protective clothing	A, C		
Precautions	U19a, U20b, D09a, D10b		
Low Down	Helena	A0459	mineral oil
Contains	732 g/l petroleum oils		
Use with	All approved pesticides on edible and non-edible crops when used at half their approved dose or less. Also with approved pesticides on specified crops, up to specified growth stages, at up to their full approved dose		
Protective clothing	A, C		
Precautions	R53a, U02a, U05a, U08, U20b, E15a, E34, E37, D01, D02, D05, D09a, D10c, H11		
Mangard	Taminco	A0667	adjuvant/anti-transpirant
Contains	96% w/w di-1-p-menthene		
Use with	Glyphosate and a wide range of other herbicides, fungicides and insecticides. See label for details		
Precautions	U20c, E15a, D09a, D10a		
Meco-Flex	Greenaway	A0678	spreader/sticker/wetter
Contains	10.0 % w/w oil (rapeseed triglycerides) (EAC 1)		
Use with	Re-Act, Headland Relay Depitox		
Protective clothing	A, C		
Precautions	U11, U12, U15, U20b, E15a, E34, D02		

SECTION 4

Product	Supplier	Adj. No.	Type
Mixture B NF	Amega	A0570	non-ionic surfactant/spreader/wetter

Contains: 42.5% polyoxyethylene (3EO) C12-C15 primary alcohol and 38.25 % w/w polyoxyethylene (7EO) C12-C15 primary alcohol

Use with: All approved pesticides on non-edible crops at their full recommended rate. Also with Timbrel (MAPP 05815) and all approved formulations of glyphosate in non-crop situations. Also with a range of specified herbicides when used at less than half their recommended rate on forest and grassland, and with all approved pesticides at full rate on a wide range of specified edible crops up to specified growth stages

Protective clothing: A, C

Precautions: R21, R22a, R36, R38, U05a, U08, U20b, E13c, E34, E37, D01, D02, D09a, D10a, M03, H03, H04

Nettle	Interagro	A0466	mineral oil

Contains: 60% w/w mineral oil

Use with: Cereal graminicides and a wide range of other pesticides that have a label recommendation for use with authorised adjuvant oils on specified crops. Refer to label or contact supplier for further details

Protective clothing: A

Precautions: R22b, R38, U02a, U05a, U08, U20b, E15a, D01, D02, D05, D09a, D10a, M05b, H03

Newman Cropspray 11E	De Sangosse	A0530	adjuvant

Contains: 99% w/w oil (petroleum oils) (EAC 1)

Use with: All approved pesticides on all edible crops when used at half their recommended dose or less, and on all non-edible crops up to their full recommended dose. Also with listed herbicides on a range of specified crops and with all pesticides on a range of specified crops up to specified growth stages.

Protective clothing: A, C

Precautions: R22a, U10, U16b, U19a, E15a, E19b, E34, E37, D01, D02, D05, D10a, D12a, M05b, H03

Newman's T-80	De Sangosse	A0192	spreader/wetter

Contains: 800g/l alkoxylated tallow amines (EAC 1)

Use with: Glyphosate

Protective clothing: A, C, H

Precautions: R22a, R37, R38, R41, R50, R58, R67, U05a, U11, U14, U19a, U20a, E15a, E19b, E34, E37, D01, D02, D05, D10a, D12a, M03, M04a, H03, H08, H11

Nion	Amega	A0415	spreader/wetter

Contains: 90% (900 g/l) ethylene oxide condensate

Use with: Recommended rates of approved pesticides up to specified growth stages of a wide range of agricultural arable and horticultural crops. Use on edible crops beyond specified growth stages should only be with half recommended rates of the pesticide or less. See label for details of growth stage restrictions.

Protective clothing: A, C

Precautions: R36, R38, U05a, U08, U19a, U20b, E13c, E37, D01, D02, D09a, D10a, H04

Product	Supplier	Adj. No.	Type
Nu Film P	Intracrop	A0635	anti-drift agent/anti-transpirant/sticker/ UV screen/wetter

Contains	96% poly-1-p-menthene
Use with	Glyphosate and many other pesticides and growth regulators for which a protectant is recommended. Do not use in mixture with adjuvant oils or surfactants
Precautions	U19a, U20b, E13c, D09a, D10a

Product	Supplier	Adj. No.	Type
Nufarm Cropoil	Nufarm UK	A0447	mineral oil

Contains	99% highly refined paraffinic oil
Use with	Approved pesticides for use on certain specified crops (cereals, combinable break crops, field and dwarf beans, peas, oilseed rape, brassicas, potatoes, carrots, parsnips, sugar beet, fodder beet, mangels, red beet, maize, sweetcorn, onions, leeks, horticultural crops, forestry, amenity, grassland)
Protective clothing	A, C
Precautions	U19a, U20b, E13c, D05, D09a, D10b, H04

Product	Supplier	Adj. No.	Type
Output	Nufarm UK	A0644	mineral oil/surfactant

Contains	60% w/w mineral oil and 40% w/w surfactants
Use with	Grasp
Protective clothing	A, C
Precautions	R36, R38, R52, U05a, U08, U10, U19a, U20b, E15a, E38, D01, D02, D09a, D10b, D12a, H04

Product	Supplier	Adj. No.	Type
Pan Oasis	Pan Agriculture	A0411	vegetable oil

Contains	95% w/w methylated rapeseed oil
Use with	A wide range of pesticides that have a label recommendation for use with authorised adjuvant oils. Contact distributor for further details
Protective clothing	A
Precautions	R36, U02a, U05a, U08, U20b, E15a, D01, D02, D05, D09a, D10a, H04

Product	Supplier	Adj. No.	Type
Pan Panorama	Pan Agriculture	A0412	mineral oil

Contains	95% w/w mineral oil
Use with	A wide range of pesticides that have a label recommendation for use with adjuvant oils. Contact distributor for details
Protective clothing	A
Precautions	R22b, R38, U02a, U05a, U08, U20b, E13c, D01, D02, D05, D09a, D10a, M05b, H03

Product	Supplier	Adj. No.	Type
Paramount	United Agri	A0585	wetter

Contains	85% polyalkylene oxide modified heptamethyl siloxane
Use with	Approved pesticides on cereals and grassland for which a wetter is recommended, all approved pesticides in non-edible crops, all approved pesticides on edible crops when used at half or less their recommended rate, and with all pesticides on listed crops up to specified growth stages
Protective clothing	A, C
Precautions	R21, R22a, R38, R41, R51, R53a, U11, U14, U16b, E37, E38, D01, D05, D07, D08, H03, H11

SECTION 4

Product	Supplier	Adj. No.	Type
Phase II	De Sangosse	A0622	vegetable oil
Contains	95.2% w/w oil (rapeseed fatty acid esters) (EAC 1)		
Use with	Pesticides approved for use in sugar beet, oilseed rape, cereals (for grass weed control), and other specified agricultural and horticultural crops		
Protective clothing	A, C, H		
Precautions	U19a, U20b, E15a, E19b, E34, D01, D02, D05, D10a, D12a		
Pik-Flex	Greenaway	A0681	spreader/sticker/wetter
Contains	10.0 % w/w oil (rapeseed triglycerides) (EAC 1)		
Use with	Tordon 22K or Tordon 101		
Protective clothing	A, C		
Precautions	U11, U12, U15, U20b, E15a, E34, D02		
Pin-o-Film	Intracrop	A0637	anti-drift agent/anti-transpirant/sticker/ UV screen/wetter
Contains	96% di-1-p-menthene		
Use with	Recommended rates of approved pesticides up to the growth stage indicated for specified crops, and with half or less than the recommended rate on these crops after the stated growth stages. Also for use with pesticides on grassland at half or less than their recommended rates. Must not be used in mixture with adjuvant oils or surfactants		
Precautions	U19a, U20c, E13c, C02a, D09a, D11a		
Planet	Intracrop	A0605	non-ionic surfactant/spreader/wetter
Contains	85% alkyl polyglycol ether and fatty acid		
Use with	Any spray for which additional wetter is recommended		
Precautions	R22a, R36, U05a, U08, U19a, U20b, E13c, D01, D09a, D10a, D11a, H03, H04, H08		
Prima	De Sangosse	A0531	mineral oil
Contains	99% w/w oil (petroleum oils) (EAC 1)		
Use with	All pesticides on all crops when used up to 50% of maximum approved dose for that use (mixtures with Roundup must only be used for treatment of stubbles). Also with all approved pesticides at full dose on listed crops (see label). Also with listed herbicides on specified crops		
Protective clothing	A, C		
Precautions	R22a, U10, U16b, U19a, E15a, E19b, E34, E37, D01, D02, D05, D10a, D12a, M05b, H03		
Profit Oil	Microcide	A0631	extender/sticker/wetter
Contains	95% w/w oil (rapeseed triglycerides) (EAC 1)		
Use with	All approved pesticides		
Protective clothing	A, C		
Precautions	U19a, U20b, D09a, D10b		
Pryz-Flex	Greenaway	A0679	spreader/sticker/wetter
Contains	10.0 % w/w oil (rapeseed triglycerides) (EAC 1)		
Use with	'Greenaway Gly-490' (MAPP 12718) and all approved 490 g/l glyphosate products on permeable surfaces overlying soil or with 'Kerb Flo' (MAPP 13716) in forest situations.		
Protective clothing	A, C		
Precautions	U11, U12, U15, U20b, E34, E37		

Product	Supplier	Adj. No.	Type
Ranger	De Sangosse	A0532	acidifier/drift retardant/penetrant

Ranger

Contains: 9.39% w/w alkoxylated alcohols (EAC 1), 35.0% w/w propionic acid (EAC 2), 35.0% w/w soybean phospholipids (EAC 3)

Use with: All approved pesticides on all edible crops when used at half their recommended dose or less, and on all non-edible crops up their full recommended dose. Also with morpholine fungicides on cereals and with pirimicarb on legumes when used at full dose; with iprodione on brassicas when used at 75% dose, and at a maximum concentration of 0.5% with all pesticides on listed crops at full dose up to specified growth stages

Protective clothing: A, C, H

Precautions: R36, R38, U11, U14, U15, U19a, E15a, E19b, D01, D02, D05, D10a, D12a, H04

Reward Oil — Microcide — A0632 — extender/sticker/wetter

Contains: 95% w/w oil (rapeseed triglycerides) (EAC 1)

Use with: All approved pesticides

Protective clothing: A, C

Precautions: U19a, U20b, D09a, D10b

Roller — Agrovista — A0694 — spreader/wetter

Contains: 832 g/l ethylene oxide and propylene oxide copolymers + 169.3 g/l trisiloxane organosilicone copolymers.

Use with: All approved pesticides up to their maximum dose at a max conc of 0.2% vol/vol.

Protective clothing: A, C

Precautions: R36, R38, R41, U02a, U08, U20b, E13c, E34, E40c, D01, D02, D05, D09a, D10a, H04

Ryda — Interagro — A0168 — cationic surfactant/wetter

Contains: 800 g/l polyethoxylated tallow amine

Use with: Glyphosate

Protective clothing: A, C

Precautions: R22a, R36, R38, R41, R51, R58, U02a, U05a, U08, U13, U19a, U20a, E15a, E34, E37, D01, D02, D05, D09a, D10b, M03, H03, H08, H11

Saracen — Interagro — A0368 — vegetable oil

Contains: 95% w/w methylated vegetable oil

Use with: Cereal graminicides and a wide range of other pesticides that have a recommendation for use with authorised adjuvant oils on specified crops. Refer to label or contact supplier for further details

Precautions: U05a, U20b, E15a, D01, D02, D09a, D10a

SAS 90 — Intracrop — A0311 — spreader/wetter

Contains: 100% polyoxyethylen-alpha-methyl-omega-[3-(1,3,3,3-tetramethyl-3-trimethylsiloxy)-disiloxanyl] propylether

Use with: A wide range of pesticides on specified crops in agriculture and horticulture, and on amenity vegetation and land not intended for cropping

Precautions: U02a, U08, U19a, U20b, E13c, E34, D09a, D10a

SECTION 4

Product	Supplier	Adj. No.	Type
Scout	De Sangosse	A0533	acidifier/drift retardant/penetrant
Contains	9.39% w/w alkoxylated alcohols (EAC 1), 35.0% w/w propionic acid (EAC 2), 35.0% w/w soybean phospholipids (EAC 3)		
Use with	All approved pesticides on all edible crops when used at half their recommended dose or less, and on all non-edible crops up their full recommended dose. Also with morpholine fungicides on cereals and with pirimicarb on legumes when used at full dose; with iprodione on brassicas when used at 75% dose, and at a maximum concentration of 0.5% with all pesticides on listed crops at full dose up to specified growth stages		
Protective clothing	A, C, H		
Precautions	R36, R38, U11, U14, U15, U19a, E15a, E19b, D01, D02, D05, D10a, D12a, H04		
Siltex	Intracrop	A0398	anti-drift agent/anti-transpirant/spreader/ sticker/UV screen/wetter
Contains	27% w/w polyoxyethylen-[-alpha-]-methyl-[-omega-]-[3-(1,3,3,3-tetramethyl-3-trimethylsiloxy)-disiloxanyl] propylether		
Use with	Approved pesticides used at full dose on a range of specified fruit and vegetable crops, and on managed amenity turf		
Protective clothing	A, C		
Precautions	U05a, U08, U20c, E13c, E37, D01, D02, D09a, D10a		
Silwet L-77	De Sangosse	A0640	drift retardant/spreader/wetter
Contains	80% w/w trisiloxane organosilicone copolymers (EAC 1).		
Use with	All approved fungicides on winter and spring sown cereals; all approved pesticides applied at 50% or less of their full approved dose. A wide range of other uses. See label or contact supplier for details		
Protective clothing	A, C, H		
Precautions	R20, R21, R22a, R36, R43, R48, R51, R58, U11, U15, U19a, E15a, E19b, D01, D02, D05, D10a, D12a, H03, H11		
Slippa	Interagro	A0206	spreader/wetter
Contains	655 g/l polyalkyleneoxide modified heptamethyltrisiloxane		
Use with	Cereal fungicides and a wide range of other pesticides and trace elements on specified crops		
Protective clothing	A, C, H		
Precautions	R20, R38, R41, R43, R48, R51, R58, U02a, U05a, U08, U11, U19a, U20a, E15a, E34, E37, D01, D02, D05, D09a, D10a, M03, H03, H11		
SM-99	De Sangosse	A0534	mineral oil
Contains	99% w/w oil (petroleum oils) (EAC 1)		
Use with	All pesticides on all crops when used up to 50% of maximum approved dose for that use (mixtures with Roundup must only be used for treatment of stubbles). Also with all approved pesticides at full dose on listed crops (see label). Also with listed herbicides on specified crops		
Protective clothing	A, C		
Precautions	R22a, U10, U16b, U19a, E15a, E19b, E34, E37, D01, D02, D05, D10a, D12a, M05b, H03		
Solar	Intracrop	A0225	activator/non-ionic surfactant/spreader
Contains	75% polypropoxypropanol		
Use with	Foliar applied plant growth regulators		
Protective clothing	A, C		
Precautions	R36, R38, U04a, U05a, U08, U20a, E13c, E34, D01, D02, D09a, D10b, D11a, M03, H04		

Product	Supplier	Adj. No.	Type
Spartan	Interagro	A0375	penetrant/water conditioner
Contains	500 g/l alkoxylated fatty amine + 400 g/l polyoxyethylene monolaurate		
Use with	Cereal growth regulators, cereal herbicides, oilseed rape fungicides and a wide range of other pesticides on specified crops. Refer to label or contact supplier for further details		
Protective clothing	A, C		
Precautions	R22a, R36, R38, R51, R58, U02a, U05a, U19a, U20b, E15a, E34, E37, D01, D02, D05, D09a, D10a, M03, H03, H11		
Speed	Headland Amenity	A0710	spreader/surfactant
Contains	18.5% w/w trisiloxane organosilicone copolymers (EAC1).		
Use with	All approved pesticides on amenity grassland and managed amenity turf at 0.25% spray solution		
Protective clothing	A, C, H		
Sprayfast	Taminco	A0668	extender/sticker/wetter
Contains	334 g/l di-1-p-menthene		
Use with	Glyphosate and other pesticides, growth regulators or nutrients for which a coating agent is approved and recommended		
Precautions	U20b, E13c, D09a, D10a		
Spray-fix	De Sangosse	A0559	extender/sticker/wetter
Contains	10% w/w alkoxylated alcohols (EAC 1), 45% w/w styrene-butadiene copolymers (EAC 2)		
Use with	All approved potato blight fungicides. Also with all approved pesticides on all edible crops when used at half their recommended dose or less, and on all non-edible crops up their full recommended dose. Also at a maximum concentration of 0.14% with approved pesticides on listed crops up to specified growth stages		
Protective clothing	A, C, H		
Precautions	R36, R38, U11, U14, U16b, U19a, E15a, E19b, D01, D02, D05, D10a, D12a, H04		
Spraygard	Taminco	A0669	extender/sticker/wetter
Contains	400 g/l di-1-p-menthene		
Use with	Approved pesticides on all edible crops (except herbicides on peas) and as an anti-transpirant on vegetables before transplanting, evergreens, deciduous trees, shrubs, bushes and on turf		
Protective clothing	A		
Precautions	R38, R50, R53a, U05a, U14, U20b, E15a, E34, E37, D01, D09a, D10b, D12b, M03, H04, H11		
Spraymac	De Sangosse	A0549	acidifier/non-ionic surfactant
Contains	100g/l alkoxylated alcohols (EAC 1), 350g/l propionic acid (EAC 2)		
Use with	All approved pesticides on all edible crops when used at half their recommended dose or less, and on all non-edible crops up their full recommended dose. Also at a maximum concentration of 0.5% with approved pesticides on listed crops up to specified growth stages		
Protective clothing	A, C, H		
Precautions	R34, U04a, U10, U11, U14, U19a, U20b, E15a, E19b, E34, D01, D02, D05, D10a, D12a, M04a, H05		

SECTION 4

Product	Supplier	Adj. No.	Type
Stamina	Interagro	A0202	penetrant
Contains	100% w/w alkoxylated fatty amine		
Use with	Glyphosate and a wide range of other pesticides on specified crops		
Protective clothing	A, C		
Precautions	R22a, R38, R50, R58, U02a, U05a, U20a, E15a, E34, D01, D02, D05, D09a, D10a, M03, H03, H11		
Standon Shiva	Standon	A0519	wetter
Contains	20.2 % w/w sodium salt of 3,6-dioxaoctadecylsulphate and 6.7% w/w sodium salt of 3,6-dioxaeicosylsulphate		
Use with	Approved cereal herbicides		
Protective clothing	A, C		
Precautions	R36, R38, R52, U02a, U05a, U08, U13, U19a, E15a, D01, D02, D05, D09a, D10b, H04		
Stika	De Sangosse	A0557	extender/sticker/wetter
Contains	10% w/w alkoxylated alcohols (EAC 1), 22.5% w/w styrene-butadione copolymers (EAC 2)		
Use with	All approved potato blight fungicides. Also with all approved pesticides on all edible crops when used at half their recommended dose or less, and on all non-edible crops up their full recommended dose. Also at a maximum concentration of 0.14% with approved pesticides on listed crops up to specified growth stages		
Protective clothing	A, C, H		
Precautions	R36, R38, U11, U14, U16b, U19a, E15a, E19b, D01, D02, D05, D10a, D12a, H04		
Super Nova	Interagro	A0364	penetrant
Contains	500 g/l alkoxylated fatty amine + 500 g/l polyoxyethylene monolaurate		
Use with	Cereal growth regulators, cereal herbicides, cereal fungicides, oilseed rape fungicides and a wide range of other pesticides on specified crops		
Precautions	R51, R58, U05a, E15a, E34, D01, D02, D05, D09a, D10a, H11		
Sward	Amega	A0215	spreader/wetter
Contains	15.2% w/w polyalkylene oxide modified heptomethyltrisiloxane		
Use with	Specified pesticides in amenity situations		
Protective clothing	A, C		
Precautions	R41, R43, U05a, U08, U20c, E13c, D01, D02, D09a, D10a, H04		
Talzene	Intracrop	A0233	wetter
Contains	800 g/l polyoxyethylene tallow amine		
Use with	Herbicides and desiccants up to their latest recommended time of application in agriculture, horticulture, amenity and forestry		
Protective clothing	A, C		
Precautions	R22a, R36, R38, R41, R48, U05a, U08, U11, U13, U19a, U20b, E13c, E34, D01, D02, D09a, D10a, M03, H03		

Product	Supplier	Adj. No.	Type
TM 1088	Intracrop	A0675	wetter

Contains: 750 g/l polyoxyethylene, polypropoxypropanol

Use with: All approved pesticides on non-edible crops and non crop production uses; All approved pesticides at half or less than half the approved pesticide rate on edible crops

Protective clothing: A, C

Precautions: R36, R38, U04a, U05a, U08, U19a, U20b, E13b, E34, D01, D02, D09a, D10a, M03, M05a, H04

Product	Supplier	Adj. No.	Type
Toil	Interagro	A0248	vegetable oil

Contains: 95% w/w methylated rapeseed oil

Use with: Sugar beet herbicides, oilseed rape herbicides, cereal graminicides and a wide range of other pesticides that have a label recommendation for use with authorised adjuvant oils on specified crops. Refer to label or contact supplier for further details

Precautions: U05a, U20b, E15a, D01, D02, D05, D09a, D10a

Product	Supplier	Adj. No.	Type
Torpedo-II	De Sangosse	A0541	penetrant/wetter

Contains: 190g/kg alkoxylated alcohols (EAC 1), 190g/kg alkoxylated alcohols (EAC 2), 75g/kg oil (tall oil fatty acids) (EAC 3), 210 g/kg alkoxylated tallow amines (EAC 4)

Use with: All approved pesticides on all edible crops when used at half their recommended dose or less, and on all non-edible crops up their full recommended dose. Also at a maximum concentration of 0.1% with approved pesticides on listed crops up to specified growth stages

Protective clothing: A, C, H

Precautions: R36, R38, R41, R52, R58, U05a, U10, U11, U19a, E15a, E19b, E34, E37, D01, D02, D05, D10a, D12a, M03, H04

Product	Supplier	Adj. No.	Type
Transact	United Agri	A0584	penetrant

Contains: 40% w/w propionic acid

Use with: Plant growth regulators in cereals, all approved pesticides in non-edible crops, all approved pesticides on edible crops when used at half or less their recommended rate, and with all pesticides on listed crops up to specified growth stages

Protective clothing: A, C

Precautions: R34, U10, U11, U14, U15, U19a, D01, D05, D09b, M04a, H05

Product	Supplier	Adj. No.	Type
Transcend	Helena	A0333	adjuvant/vegetable oil

Contains: 80% w/w oil (soybean fatty acid esters) + 12% w/w trisiloxane organosilicone copolymers.

Use with: Approved pesticides on non-edible crops, cereals and stubbles of all edible crops, grassland (destruction), and pesticides approved for use on growing edible crops when used at half recommended dose or less. On specified crops product may be used at a maximum spray concentration of 0.5% with approved pesticides at their full approved rate up to the growth stages shown in the label

Protective clothing: A, H

Precautions: R38, R41, U02a, U05a, U08, U11, U19a, U20b, E13c, E34, E37, D01, D02, D05, D09a, M03, H04

SECTION 4

Product	Supplier	Adj. No.	Type
Validate	De Sangosse	A0500	adjuvant/wetter

Contains 25% w/w alkoxylated alcohols (EAC 1), 25% w/w oil (soybean fatty acid esters) (EAC 2), 50% w/w soybean phospholipids (EAC 3)

Use with Approved pesticides on non-edible crops where the addition of a wetter/spreader or adjuvant oil is recommended on the pesticide label, and pesticides approved for use on growing edible crops when used at half recommended dose or less. On specified crops product may be used at a maximum spray concentration of 0.5% with approved pesticides at their full approved rate up to the growth stages shown in the label

Protective clothing A, C, H

Precautions R51, R53a, U05a, U11, U14, U19a, U20b, E15a, E19b, E34, D01, D02, D10a, D12a, H11

Velocity	Agrovista	A0697	adjuvant

Contains 745g/l oil (rapeseed fatty acid esters) + 103g/l trisiloxane organosilicone copolymers.

Use with All approved pesticides up to their full approved rate at a maximum conc of 0.5% vol/vol.

Protective clothing A, C

Precautions U02a, U08, U20b, E13c, E34, E40c, D01, D02, D09a, D10a, H04

Verdant	Agrovista	A0714	wetter

Contains 8% w/w alkoxylated alcohol + 63% w/w alkoxylated alkyl polyglucoside + 10% w/w alkyl polyglucoside

Use with All approved pesticides at 50% or less on edible crops and with all approved pesticides on non-edible crops, non-crop production and stubbles of edible crops at 0.25% spray sloution.

Protective clothing A, C, H

Precautions R38, R41, U02a, U05a, U11, U14, U15, U20b, E15a, E34, D02, D05, D09a, D10a, H04

Wetcit	Plant Solutions	A0586	penetrant/wetter

Contains 8.15% alcohol ethoxylate

Use with All approved pesticides on all edible crops when used at half their recommended dose or less, and on all non-edible crops up their full recommended dose. Also at a maximum concentration of 0.25% with approved pesticides on listed crops up to specified growth stages at full rate, and at half rate thereafter

Precautions R22a, R38, R41, U08, U11, U19a, U20b, E13c, E15a, E34, E37, D05, D08, D09a, D10b, H03

X-Wet	De Sangosse	A0548	non-ionic surfactant/wetter

Contains 375g/kg alkoxylated alcohols (EAC 1), 375g/kg alkoxylated alcohols (EAC 2), 150g/kg oil (tall oil fatty acids) (EAC 3)

Use with All approved pesticides on all edible crops when used at half their recommended dose or less, and on all non-edible crops up their full recommended dose. Also at a maximum concentration of 0.1% with approved pesticides on listed crops up to specified growth stages

Protective clothing A, C, H

Precautions R36, R38, R53a, U02a, U04a, U05a, U10, U11, U20b, E15a, E19b, D01, D02, D05, D10a, H04

Product	Supplier	Adj. No.	Type
Zarado	De Sangosse	A0516	vegetable oil

Contains 70% w/w oil (rapeseed oil fatty acid esters) as emulsifiable concentrate (EAC 1).

Use with All approved pesticides on edible and non-edible crops when used at half their approved dose or less. Also with approved pesticides on specified crops, up to specified growth stages, at up to their full approved dose

Protective clothing A, C, H

Precautions R43, R52, U05a, U13, U19a, U20b, E15a, E19b, E34, D01, D02, D05, D10a, M05a, H04

Product	Supplier	Adj. No.	Type
Zeal	Interagro	A0685	water conditioner/wetter

Contains 60% w/w alkoxylated fatty amine

Use with Trace elements such as calcium, copper, manganese and sulphur and with macronutrients such as phosphates

Protective clothing A, C, H

Precautions R22a, R36, R38, R51, R53a, U05a, U09a, U11, U14, U15, U19a, E15b, E34, D01, D02, D09a, D10a, M03, H03, H11

Product	Supplier	Adj. No.	Type
Zigzag	Headland	A0692	extender/sticker/wetter

Contains 36.9% w/v styrene/butadiene co-polymer and 6.375% w/v polyether-modified trisiloxane

Use with Approved glyphosate formulations on cereal crops; All approved pesticides applied at half-rate or less; All approved pesticides for use on crops not destined for human or animal consumption; With all approved pesticides on edible crops up to the latest growth stage specified on the label.

Protective clothing C, H

Precautions U11, U15, E38, D01

Product	Supplier	Adj. No.	Type
Zinzan	Certis	A0600	extender/spreader/wetter

Contains 70% 1,2 bis (2-ethylhexyloxycarbonyl) ethanesulponate

Use with All approved pesticides on non-edible crops, and all pesticides approved for use on growing edible crops when used at half recommended dose or less. On specified edible crops product may be used at a maximum spray concentration of 0.075% with approved pesticides at their full approved rate up to the growth stages shown in the label

Protective clothing A, C, P

Precautions R38, R41, U11, U14, U20c, D01, D09a, D10a, H04

SECTION 4

SECTION 5
USEFUL INFORMATION

Pesticide Legislation

Anyone who advertises, sells, supplies, stores or uses a pesticide is bound by legislation, including those who use pesticides in their own homes, gardens or allotments. There are numerous UK statutory controls, but the major legal instruments are outlined below.

Regulation 1107/2009 — The Replacement for EU 91/414

Regulation 1107/2009 entered into force within the EU on 14 Dec 2009 and was applied to new approval applications from 14th June 2011. Since that date all pesticides that hold a current approval under 91/414 have been deemed to be approved under 1107/2009 and the new criteria for approval will only be applied when the active substance comes up for review. The new regulation largely mirrors the previous regulation but requires additional approval criteria to be met to weed out the more hazardous chemicals. These include assessing the effect on vulnerable groups, taking into account known synergistic and cumulative effects, considering the effect on coastal and estuarine waters, effects on the behaviour of non-target organisms, biodiversity and the ecosystem. However, these effects will only be evaluated when scientific methods approved by EFSA (European Food Safety Authority) have been developed.

The additional criteria will require that no CMR's (substances which are Carcinogens, Mutagens or toxic to Reproduction), no endocrine disruptors (not to be defined until 2015!) and no POP (Persistent Organic Pollutants), PBT (chemicals that are Persistent, Bioaccumulative or Toxic) or vPVB (very Persistent and very Bioaccumulative) will gain approval unless the exposure is negligible. However, note that for EFSA the term 'negligible' has yet to receive a precise definition.

If an active substance passes these hurdles, the first approval for basic substances will be granted for 10 years (15 years for 'low risk' substances, 7 years for candidates for substitution – see below) and then at each subsequent renewal a further 15 years will be granted. Actives will be divided between 5 categories – low risk substances (e.g. pheromones, semiochemicals, micro-organisms and natural plant extracts), basic substances, candidates for substitution and then safeners/synergists and co-formulants. Candidates for substitution will be those chemicals which have just scraped past the approval thresholds but are deemed to still carry some degree of hazard. They will be withdrawn when significantly safer alternatives are available. This includes physical methods of control but any alternative must not have significant economic or practical disadvantages, must minimise the risk of the development of resistance and the consequences for any 'minor uses' of the original product must also be considered.

The Food and Environment Protection Act 1986 (FEPA)

FEPA introduced statutory powers to control pesticides with the aims of protecting human beings, creatures and plants, safeguarding the environment, ensuring safe, effective and humane methods of controlling pests and making pesticide information available to the public. This was supplemented by Control of Pesticides Regulations 1986 (COPR) which has now been replaced by EC Regulation 1107/2009 – see above. Details are given on the websites of the Chemicals Regulation Directorate (CRD) and the Health and Safety Executive (HSE). Together these regulations mean that:

- Only approved products may be sold, supplied, stored, advertised or used.
- No advertisement may contain any claim for safety beyond that which is permitted in the approved label text.
- Only products specifically approved for the purpose may be applied from the air.
- A recognised Storeman's Certificate of Competence is required by anyone who stores for sale or supply pesticides approved for agricultural use.
- A recognised Certificate of Competence is required by anyone who gives advice when selling or supplying pesticides approved for agricultural use.
- Users of pesticides must comply with the Conditions of Approval relating to use.
- A recognised Certificate of Competence is required for all contractors and persons born after 31 December 1964 applying pesticides approved for agricultural use (unless working under direct supervision of a certificate holder). Proposals are now in place that every user

SECTION 5

of agricultural pesticides will have to hold a Certificate of Competence, regardless of age or supervision. A suitable transition period will allow this requirement to be enacted.

- Only those adjuvants authorised by CRD may be used.
- Regarding tank-mixes, 'no person shall combine or mix for use two or more pesticides which are anti-cholinesterase compounds unless the approved label of at least one of the pesticide products states that the mixture may be made; and no person shall combine or mix for use two or more pesticides if all the conditions of the Approval relating to this use cannot be complied with'.

Dangerous Preparations Directive (1999/45/EC)

The Dangerous Preparations Directive (DPD) came into force in UK for pesticide and biocidal products on 30 July 2004. Its aim is to achieve a uniform approach across all Member States to the classification, packaging and labelling of most dangerous preparations, including crop protection products. The Directive is implemented in the UK under the Chemicals (Hazard Information and Packaging for Supply) Regulations 2002, often referred to under the acronym CHIP3. In most cases the Regulations have led to additional hazard symbols, and associated risk and safety phrases relating to environmental and health hazards, appearing on the label. All products affected by the DPD entering the supply chain from the implementation date above must be so labelled. The new environmental hazard classifications and risk phrases are included.

Under the DPD, the labels of all plant protection products now state 'To avoid risks to man and the environment, comply with the instructions for use'. As the instruction applies to all products, it has not been repeated in each fact sheet.

The Review Programme

The review programme of existing active substances will continue under EC Directive 1107/2009. The programme is designed to ensure that all available plant protection products are supported by up-to-date information on safety and efficacy. As this will now be under 1107/2009, the new standards for approval will be applied at the due date and any that are deemed to be hazardous chemicals will be withdrawn. Any that are just within the safety standards yet to be defined will be called 'candidates for substitution' and granted only a seven-year renewal to allow time for significantly safer alternatives to be developed. If these are available at the time of renewal, the approval will be withdrawn. Chemicals deemed to be low risk, such as natural plant extracts, microorganisms, semiochemicals and pheromones, will be renewed for 15 years, while most active substances will be renewed for a further 10 years.

Control of Substances Hazardous to Health Regulations 1988 (COSHH)

The COSHH regulations, which came into force on 1 October 1989, were made under the Health and Safety at Work Act 1974, and are also important as a means of regulating the use of pesticides. The regulations cover virtually all substances hazardous to health, including those pesticides classed as Very Toxic, Toxic, Harmful, Irritant or Corrosive, other chemicals used in farming or industry, and substances with occupational exposure limits. They also cover harmful micro-organisms, dusts and any other material, mixture, or compound used at work which can harm people's health.

The original Regulations, together with all subsequent amendments, have been consolidated into a single set of regulations: The Control of Substances Hazardous to Health Regulations 1994 (COSHH 1994).

The basic principle underlying the COSHH regulations is that the risks associated with the use of any substance hazardous to health must be assessed before it is used, and the appropriate measures taken to control the risk. The emphasis is changed from that pertaining under the Poisonous Substances in Agriculture Regulations 1984 (now repealed) – whereby the principal method of ensuring safety was the use of protective clothing – to the prevention or control of exposure to hazardous substances by a combination of measures. In order of preference the measures should be:

(a) substitution with a less hazardous chemical or product
(b) technical or engineering controls (e.g. the use of closed handling systems, etc.)
(c) operational controls (e.g. operators located in cabs fitted with air-filtration systems, etc.)
(d) use of personal protective equipment (PPE), which includes protective clothing.

Consideration must be given as to whether it is necessary to use a pesticide at all in a given situation and, if so, the product posing the least risk to humans, animals and the environment must be selected. Where other measures do not provide adequate control of exposure and the use of PPE is necessary, the items stipulated on the product label must be used as a minimum. It is essential that equipment is properly maintained and the correct procedures adopted. Where necessary, the exposure of workers must be monitored, health checks carried out, and employees instructed and trained in precautionary techniques. Adequate records of all operations involving pesticide application must be made and retained for at least 3 years

Biocidal Products Regulations

These regulations concern disinfectants, preservatives and pest control products but only the latter are listed in this book. The regulation aims to harmonise the European market for biocidal products, to provide a high level of safety for humans, animals and the environment and to ensure that the products are effective against the target organisms. Products that have been reviewed and included under these regulations are issued with a new number of the format UK-2012-xxxx and these are summarised in this book as UK12-xxxx beginning with the difenacoum products.

Certificates of Competence – the roles of BASIS and NPTC

COPR, COSHH and other legislation places certain obligations on those who handle and use pesticides. Minimum standards are laid down for the transport, storage and use of pesticides, and the law requires those who act as storekeepers, sellers and advisors to hold recognised Certificates of Competence.

BASIS

BASIS is an independent Registration Scheme for the pesticide industry. It is responsible for organising training courses and examinations to enable such staff to obtain a Certificate of Competence.

In addition, BASIS undertakes an annual assessment of pesticide supply stores, enabling distributors, contractors and seedsmen to meet their obligations under the Code of Practice for Suppliers of Pesticides. Further information can be obtained from BASIS.

Certificates of Competence

Storage

- BASIS Certificate of Competence in the Storage and Handling of Crop Protection Products

Sale and supply

- BASIS Certificate in Crop Protection (Agriculture)
- BASIS Certificate in Crop Protection (Commercial Horticulture)
- BASIS Certificate in Crop Protection (Amenity Horticulture)
- BASIS Certificate in Crop Protection (Forestry)
- BASIS Certificate in Crop Protection (Seed Treatment)
- BASIS Certificate in Crop Protection (Seed Sellers)
- BASIS Certificate in Crop Protection (Field Vegetables)
- BASIS Certificate in Crop Protection (Potatoes)
- BASIS Certificate in Crop Protection (Aquatic)
- BASIS Certificate in Crop Protection (Indoor Landscaping)
- BASIS Certificate in Crop Protection (Grassland and Forage Crops)
- BASIS/LEAF ICM Certificate
- BASIS Advanced Certificate
- FACTS (Fertiliser Advisers Certification and Training Scheme) Certificate

SECTION 5

NPTC

Certain spray operators also require certificates of competence under the Control of Pesticides Regulations. NPTC's Pesticides Award is a recognised Certificate of Competence under COPR and it is aimed principally at those people who use pesticide products approved for use in agriculture, horticulture (including amenity horticulture) and forestry. All contractors, and anyone else born after 31 December 1964, must possess a certificate if they spray such products unless they are working under the direct supervision of a certificate holder.

Because the required spraying skills vary widely among the uses listed above, candidates are assessed under one (or more) modules that are most appropriate for their professional work. Assessments are carried out by an approved NPTC or Scottish Skills Testing Service Assessor. All candidates must first complete a foundation module (PA1), for which a certificate is not issued, before taking one of the specialist modules. Certificate holders who change their work so that a different specialist module becomes more appropriate may need to obtain a new certificate under the new module.

Holders are required to produce on demand their Certificate of Competence for inspection to any authorised person. Further information can be obtained from NPTC.

Certificates of Competence for spray operators

- PA1 Foundation Module
- PA2 Ground Crop Sprayers – mounted or trailed
- PA3 Broadcast Air Blast Sprayer. Variable Geometry Boom Air Assisted Sprayer
- PA4 Granule Applicator – Mounted or Trailed
- PA5 Boat Mounted Applicators
- PA6 Hand Held Applicators
- PA7 Aerial Application
- PA8 Mixer/Loader
- PA9 Fogging, Misting and Smokes
- PA10 Dipping Bulbs, Corms, Plant Material or Containers
- PA11 Seed Treating Equipment
- PA12 Application of Pesticides to Material as a Continuous or Batch Process

Maximum Residue Levels

A small number of pesticides are liable to leave residues in foodstuffs, even when used correctly. Where residues can occur, statutory limits, known as maximum residue levels (MRLs), have been established. MRLs provide a check that products have been used as directed; **they are not safety limits**. However, they do take account of consumer safety because they are set at levels that ensure normal dietary intake of residues presents no risk to health. Wide safety margins are built in, and eating food containing residues above the MRL does not automatically imply a risk to health. Nevertheless, it is an offence to put into circulation any produce where the MRL is exceeded.

The surrounding legislation is complex. MRLs may be specified by several different bodies. The UK has set statutory MRLs since 1988. The European Union intends eventually to introduce MRLs for all pesticide/commodity combinations. These are being introduced initially by a series of priority lists, but will subsequently be covered by the review programme under EC Regulation 1107/2009. However, in cases where no information is available, EU Regulation 2000/42/EC requires many MRLs to be set at the limit of determination (LOD). As a result, certain approvals are being withdrawn where such use would leave residues above the MRL set in the Directive.

MRLs apply to imported as well as home-produced foodstuffs. Details of those that have been set have been published in *The Pesticides (Maximum Residue Levels in Crops, Food and Feeding Stuffs) Regulations 1994*, and successive amendments to these Regulations. These Statutory Instruments are available from The Stationery Office (www.thestationeryoffice.com).

MRLs are set for many chemicals not currently marketed in Britain. Because of this and the ever-changing information on MRLs, direct access is provided to the comprehensive MRL databases on the Chemicals Regulation Directorate website (www.pesticides.gov.uk). This online database sets out in table form the levels specified by UK Regulations, EC Directives, and the Codex Alimentarius for each commodity.

Approval (On-label and Off-label)

Only officially approved pesticides can be marketed and used in the UK. Approvals are granted by UK Government Ministers in response to applications that are supported by satisfactory data on safety, efficacy and, where relevant, humaneness. The Chemicals Regulation Directorate (CRD, www.pesticides.gov.uk) comes under the Health and Safety Executive (HSE, www.hse.gov.uk), and is the UK Government Agency for regulating agricultural pesticides and plant protection products. The HSE currently fulfils the same role for other pesticides, with the two organisations in the process of being merged. The main focus of the regulatory process in both bodies is the protection of human health and the environment.

Statutory Conditions of Use

Approvals are normally granted only in relation to individual products and for specified uses. It is an offence to use non-approved products or to use approved products in a manner that does not comply with the statutory conditions of use, except where the crop or situation is the subject of an off-label extension of use (see below).

Statutory conditions have been laid down for the use of individual products and may include:

- field of use (e.g. agriculture, horticulture etc.)
- crop or situations for which treatment is permitted
- maximum individual dose
- maximum number of treatments or maximum total dose
- maximum area or quantity which may be treated
- latest time of application or harvest interval
- operator protection or training requirements
- environmental protection
- any other specific restrictions relating to particular pesticides.

Products must display these statutory conditions in a boxed area on the label entitled 'Important Information', or words to that effect. At the bottom of the boxed area must be shown a bold text statement: '**Read the label before use. Using this product in a manner that is inconsistent with the label may be an offence. Follow the Code of Practice for Using Plant Protection Products**'. This requirement came into effect in October 2006, and replaced the previous 'Statutory Box'.

Types of Approval

Where there were once three levels of approval (full; provisional; experimental permit), there will now be just two levels (authorisation for use; limited approval for research and development). Provisional approval used to be granted while further data were generated to justify label claims, but under the new regulations, decisions on authorisation will be reached much more quickly, and this 'half-way stage' is deemed to be unnecessary. The official list of approved products, excluding those approved for research and development only, are shown on the websites of CRD and the Health and Safety Executive (HSE).

Withdrawal of Approval

Product approvals may be reviewed, amended, suspended or revoked at any time. Revocation may occur for various reasons, such as commercial withdrawal, or failure by the approval holder to meet data requirements.

From September 2007, where an approval is being revoked for purely administrative, 'housekeeping' reasons, the existing approval will be revoked and in its place will be issued:

- an approval for advertisement, sale and supply by any person for 24 months, and
- an approval for storage and use by any person for 48 months.

Where an approval is replaced by a newer approval, such that there are no safety concerns with the previous approval, but the newer updated approval is more appropriate in terms of reflecting the latest regulatory standard, the existing approval will be revoked and in its place will be issued:

- an approval for advertisement, sale and supply by any person for 12 months, and
- an approval for storage and use by any person for 24 months.

Where there is a need for tighter control of the withdrawal of the product from the supply chain, for example failure to meet data submission deadlines, the current timelines will be retained:

- immediate revocation for advertisement, sale or supply for the approval holder
- approval for 6 months for advertisement, sale and supply by 'others', and
- approval for storage and use by any person for 18 months.

Immediate revocation and product withdrawal remain an option where serious concerns are identified, with immediate revocation of all approvals and approval for storage only by anyone for 3 months, to allow for disposal of product.

The expiry date shown in the product fact sheet is the final date of legal use of the product.

Off-label Extension of Use

Products may legally be used in a manner not covered by the printed label in several ways:

- In accordance with an Extension of Authorisation for Minor Use (EAMU formerly known as SOLA). EAMUs are uses for which individuals or organisations other than the manufacturers have sought approval. The Notices of Approval are published by CRD and are widely available from ADAS or NFU offices. Users of EAMUs must first obtain a copy of the relevant Notice of Approval and comply strictly with the conditions laid down therein. Users of this Guide will find details of extant EAMUs and direct links to the CRD website, where EAMU notices can also be accessed.
- In tank mixture with other approved pesticides in accordance with Consent C(i) made under FEPA. Full details of Consent C(i) are given in Annex A of Guide to Pesticides on the PSD website, but there are two essential requirements for tank mixes. First, all the conditions of approval of all the components of a mixture must be complied with. Second, no person may mix or combine pesticides that are cholinesterase compounds unless allowed by the label of at least one of the pesticides in the mixture.
- In conjunction with authorised adjuvants.
- In reduced spray volume under certain conditions.
- In the use of certain herbicides on specified set-aside areas subject to restrictions, which differ between Scotland and the rest of the UK.
- By mutual recognition of a use fully approved in another Member State of the European Union and authorised by CRD.

Although approved, off-label uses are not endorsed by manufacturers and such treatments are made entirely at the risk of the user.

Using Crop Protection Chemicals

Use of Herbicides In or Near Water

Products in this *Guide* approved for use in or near water are listed in Table 5.1. Before use of any product in or near water, the appropriate water regulatory body (Environment Agency/Local Rivers Purification Authority; or in Scotland the Scottish Environmental Protection Agency) must be consulted. Guidance and definitions of the situation covered by approved labels are given in the Defra publication *Guidelines for the Use of Herbicides on Weeds in or near Watercourses and Lakes.* Always read the label before use.

Table 5.1 Products approved for use in or near water

Chemical	Product
2,4-D	Depitox
glyphosate	Asteroid, Barclay Gallup 360, Barclay Gallup Amenity, Barclay Gallup Biograde Amenity, Barclay Gallup Hi-Aktiv, Barclay Glyde 144, Buggy XTG, Discman Biograde, Dow Agrosciences Glyphosate 360, Envision, Gallup Hi- Aktiv Amenity, Glyfo_TDI, Glyfos Dakar, Glyfos Dakar Pro, Glyfos Proactive, Glyfo-TDI, Glypho-Rapid 450, Kernel, Manifest, Mascot Hi-Aktiv Amenity, Pure Glyphosate 360, Roundup Biactive, Roundup Pro Biactive, Roundup ProBiactive 450, Rustler, Trustee Amenity

Use of Pesticides in Forestry

Table 5.2 Products in this *Guide* approved for use in forestry

Chemical	Product	Use
2,4-D + dicamba + triclopyr	Broadsword	Herbicide
aluminium ammonium sulphate	Curb Crop Spray Powder, Liquid Curb Crop Spray, Sphere ASBO	Animal deterrent/repellent
chlorpyrifos	Dursban WG, EA Pyrifos, Equity, Govern, Parapet	Acaricide, Insecticide
cycloxydim	Laser	Herbicide
cypermethrin	Forester	Insecticide
diflubenzuron	Dimilin Flo	Insecticide
ferric phosphate	Derrex, Ferramol Max, Sluxx	Molluscicide
fluazifop-P-butyl	Clayton Maximus, Fusilade Max, Greencrop Bantry, Howitzer	Herbicide
glufosinate-ammonium	Harvest, Kaspar, Kibosh, Kurtail, PureReap	Herbicide
glyphosate	Amega Duo, Azural, Barclay Gallup 360, Barclay Gallup Amenity, Barclay Gallup Biograde Amenity, Barclay Gallup Hi-Aktiv, Barclay Glyde 144, Buggy XTG, Charger C, Clinic Ace, Credit DST, Discman Biograde, Dow Agrosciences Glyphosate 360, Ecoplug Max, Envision, Etna, Gallup Hi- Aktiv Amenity, Glyfo_TDI, Glyfos, Glyfos Dakar Pro, Glyfos Proactive, Glyfos Supreme, Glyfosat 36, Glyfo-TDI, Glyphogan, Glypho-Rapid 450, Hilite, Kernel, Manifest, Mascot Hi-Aktiv Amenity, Monsanto Amenity Glyphosate, Nomix Conqueror, Nufosate Ace, Pitch, Pure Glyphosate 360, Rattler, Reaper, Rhizeup, Roundup Pro Biactive, Roundup ProBiactive 450, Shyfo, Slingshot, Snapper, Stacato, Stirrup, Tanker, Trustee Amenity	Herbicide
isoxaben	Flexidor 125	Herbicide
metazachlor	Clayton Buzz, Clayton Metazachlor 50 SC, EA Metazachlor, Greencrop Monogram, Rapsan 500 SC, Segundo, Standon Metazachlor 500, Sultan 50 SC	Herbicide
napropamide	MAC-Napropamide 450 SC	Herbicide
Phlebiopsis gigantea	PG Suspension	Biological agent

Chemical	Product	Use
propaquizafop	Clayton Orleans, Cleancrop GYR 2, Falcon, Greencrop Satchmo, MAC-Propaquizafop 100 EC, PQF 100, Shogun, Standon Propaquizafop, Standon Zing PQF	Herbicide
propyzamide	Careca, Cohort, Dennis, Engage, Flomide 2, Hazard, Kerb 50 W, Kerb Granules, Menace 80 EDF, Pizza 400 SC, Ponder WP, Propel Flo, Proper Flo, Propyz, PureFlo, Quaver Flo, Setanta 50 WP, Solitaire 50 WP, Standon Santa Fe 50 WP, Standon Santa Fe Flo	Herbicide
pyrethrins	Spruzit	Insecticide
triclopyr	Garlon 4, Nomix Garlon 4, PureWoody, Thrash, Timbrel, Woody	Herbicide

SECTION 5

Pesticides Used as Seed Treatments

Information on the target for these products can be found in the relevant pesticide profile in Section 2.

Table 5.3 Products used as seed treatments (including treatments on seed potatoes)

Chemical	Product	Formulation	Crop(s)
beta-cyfluthrin + clothianidin	Modesto	FS	Poppies for morphine production, Winter oilseed rape
	Poncho Beta	FS	Fodder beet, Sugar beet
beta-cyfluthrin + imidacloprid	Chinook Blue	LS	Evening primrose, Honesty, Linseed, Mustard, Spring oilseed rape, Winter oilseed rape
	Chinook Colourless	LS	Evening primrose, Honesty, Linseed, Mustard, Winter oilseed rape
carboxin + thiram	Anchor	FS	Spring barley, Spring oats, Spring rye, Spring wheat, Triticale, Winter barley, Winter oats, Winter rye, Winter wheat
clothianidin + prothioconazole	Redigo Deter	FS	Durum wheat, Triticale, Winter barley, Winter oats, Winter rye, Winter wheat
cymoxanil + fludioxonil + metalaxyl-M	Wakil XL	WS	Broad beans, Carrots, Combining peas, Lupins, Parsnips, Poppies for morphine production, Red beet, Spring field beans, Vining peas, Winter field beans
difenoconazole + fludioxonil	Celest Extra	FS	Winter oats, Winter rye, Winter wheat
fludioxonil	Beret Gold	FS	Durum wheat, Rye, Spring barley, Spring oats, Spring wheat, Triticale, Winter barley, Winter oats, Winter wheat
	Maxim 100FS	FS	Potatoes
fludioxonil + flutriafol	Beret Multi	FS	Spring barley, Spring oats, Spring wheat, Winter barley, Winter oats, Winter wheat
fludioxonil + metalaxyl M	Maxim XL	FS	Forage maize
fludioxonil + metalaxyl-M + thiamethoxam	Cruiser OSR	FS	Fodder rape, Linseed, Mustard, Poppies for morphine production, Spring oilseed rape, Winter oilseed rape

Chemical	Product	Formulation	Crop(s)
fludioxonil + tefluthrin	Austral Plus	FS	Spring barley, Spring oats, Spring wheat, Triticale seed crop, Winter barley, Winter oats, Winter wheat
fluopyram + prothioconazole + tebuconazole	Raxil Star	FS	Spring barley, Winter barley
fluquinconazole	Galmano	FS	Durum wheat, Triticale, Winter wheat
	Jockey Solo	FS	Winter barley, Winter wheat
fluquinconazole + prochloraz	Epona	FS	Winter wheat
	Jockey	FS	Winter barley, Winter wheat
flutolanil	Rhino DS	DS	Potatoes
fuberidazole + imidacloprid + triadimenol	Tripod Plus	LS	Winter barley, Winter oats, Winter wheat
fuberidazole + triadimenol	Tripod	FS	Spring barley, Spring oats, Spring rye, Spring wheat, Triticale, Winter barley, Winter oats, Winter rye, Winter wheat
imazalil	Fungazil 50LS	LS	Spring barley, Winter barley
imazalil + pencycuron	Monceren IM	DS	Potatoes
imidacloprid	Nuprid 600FS	FS	Fodder beet, Sugar beet
metalaxyl-M	Apron XL	ES	Beetroot, Broccoli, Brussels sprouts, Bulb onions, Cabbages, Calabrese, Cauliflowers, Chard, Chinese cabbage, Herbs (see appendix 6), Kohlrabi, Ornamental plant production, Radishes, Shallots, Spinach
methiocarb	Mesurol	FS	Fodder maize, Grain maize, Sweetcorn
pencycuron	CS Pencycuron	DS	Potatoes
	Monceren DS	DS	Potatoes
physical pest control	Silico-Sec	DS	Stored grain
prochloraz	Prelude 20LF	LS	Flax, Hemp for oilseed production, Linseed
prochloraz + thiram	Agrichem Hy-Pro Duet	FS	Winter oilseed rape

SECTION 5

Chemical	Product	Formulation	Crop(s)
prochloraz + triticonazole	Kinto	FS	Durum wheat, Triticale, Winter barley, Winter oats, Winter rye, Winter wheat
prothioconazole	Redigo	FS	Durum wheat, Spring barley, Spring oats, Spring rye, Spring wheat, Triticale, Winter barley, Winter oats, Winter rye, Winter wheat
silthiofam	Latitude	FS	Durum wheat, Spring rye, Spring wheat, Triticale, Winter barley, Winter rye, Winter wheat
	Meridian	FS	Spring wheat, Winter barley, Winter wheat
tebuconazole	Gizmo	FS	Bulb onions, Garlic, Salad onions, Shallots, Spring barley, Winter barley
thiamethoxam	Cruiser SB	FS	Fodder beet, Sugar beet, Swedes
thiram	Agrichem Flowable Thiram	FS	Broad beans, Bulb onions, Cabbages, Carrots, Cauliflowers, Celery (outdoor), Combining peas, Cress, Dwarf beans, Edible podded peas, Endives, Fodder beet, Forage maize, Frise, Garlic, Grass seed, Kale, Lamb's lettuce, Leeks, Lettuce, Lupins, Mangels, Millet, Parsley, Poppies for morphine production, Radicchio, Radishes, Red beet, Rhubarb, Runner beans, Salad onions, Scarole, Shallots, Sorghum, Soya beans, Spring field beans, Spring oilseed rape, Sugar beet, Swedes, Turnip seed crops, Turnips, Vining peas, Winter field beans, Winter oilseed rape
	Thiraflo	FS	Broad beans, Combining peas, Dwarf beans, Forage maize, Grass seed, Runner beans, Spring field beans, Spring oilseed rape, Vining peas, Winter field beans, Winter oilseed rape

Chemical	Product	Formulation	Crop(s)
	Thyram Plus	FS	Broad beans, Bulb onions, Cabbages, Cauliflowers, Combining peas, Dwarf beans, Edible podded peas, Forage maize, Garlic, Grass seed, Kale, Kale seed crops, Leeks, Lupins, Millet, Radishes, Runner beans, Salad onions, Shallots, Sorghum, Soya beans, Spring field beans, Spring oilseed rape, Swedes, Turnip seed crops, Turnips, Vining peas, Winter field beans, Winter oilseed rape
tolclofos-methyl	Rizolex Flowable	FS	Potatoes

SECTION 5

Aerial Application of Pesticides

New aerial spraying permit arrangements came into force in June 2012 and those undertaking aerial applications must ensure that the spraying is done in line with an Approved Application Plan that is subject to approval by CRD. Plans will only be approved where there is no viable alternative method of application or where aerial application results in reduced impact on human health and/or the environment compared to land-based application. Thus applications to bracken, forestry and possibly blight sprays to potatoes may meet these requirements but few others are likely to do so. Template application plans are available from the CRD web site and once all the necessary details have been received by CRD they undertake to approve or reject the plan within 10 working days.

Table 5.4 Products approved for aerial application

Chemical	Product	Crop(s)
2-chloroethylphosphonic acid	Becki	Winter barley
	Cerone	Winter barley
chlormequat	3C Chlormequat 720	Spring oats, Spring rye, Spring wheat, Triticale, Winter barley, Winter oats, Winter rye, Winter wheat
	Agrovista 3 See 750	Spring oats, Spring wheat, Winter oats, Winter wheat
	CCC 720	Spring oats, Spring rye, Spring wheat, Triticale, Winter barley, Winter oats, Winter rye, Winter wheat
	Hive	Winter wheat
	Mirquat	Spring oats, Spring wheat, Winter oats, Winter wheat
	New 5C Cycocel	Spring oats, Spring rye, Spring wheat, Triticale, Winter barley, Winter oats, Winter rye, Winter wheat
	New 5C Quintacel	Spring oats, Spring rye, Spring wheat, Triticale, Winter barley, Winter oats, Winter rye, Winter wheat
	Sigma PCT	Spring wheat, Winter barley, Winter wheat
copper oxychloride	Cuprokylt	Potatoes
diflubenzuron	Dimilin Flo	Forest
dimethoate	Danadim Progress	Durum wheat, Fodder beet, Rye, Spring wheat, Sugar beet, Triticale, Winter rye, Winter wheat
mancozeb	Laminator Flo	Potatoes
	Quell Flo	Potatoes

Chemical	Product	Crop(s)
metaldehyde	Doff Horticultural Slug Killer Blue Mini Pellets	All edible crops (outdoor), All non-edible crops (outdoor), Cultivated land/soil
	ESP	All edible crops (outdoor), All non-edible crops (outdoor)
pirimicarb	Aphox	Durum wheat, Spring barley, Spring oats, Spring rye, Spring wheat, Triticale, Winter barley, Winter oats, Winter rye, Winter wheat
	Phantom	Durum wheat, Spring barley, Spring oats, Spring rye, Spring wheat, Triticale, Winter barley, Winter oats, Winter rye, Winter wheat
	Pirimate	Spring barley, Spring oats, Spring wheat, Winter barley, Winter oats, Winter wheat
	Pirimicarb 50	Durum wheat, Spring barley, Spring oats, Spring rye, Spring wheat, Triticale, Winter barley, Winter oats, Winter rye, Winter wheat
	Reynard	Durum wheat, Spring barley, Spring oats, Spring rye, Spring wheat, Triticale, Winter barley, Winter oats, Winter rye, Winter wheat
	Standon Pirimicarb 50	Durum wheat, Spring barley, Spring oats, Spring rye, Spring wheat, Triticale, Winter barley, Winter oats, Winter rye, Winter wheat
tri-allate	Avadex Excel 15G	Combining peas, Fodder beet, Lucerne, Red beet, Red clover, Sainfoin, Spring barley, Spring field beans, Sugar beet, Triticale, Vetches, Vining peas, White clover, Winter barley, Winter field beans, Winter wheat

SECTION 5

Resistance Management

Pest species are, by definition, adaptable organisms. The development of resistance to some crop protection chemicals is just one example of this adaptability. Repeated use of products with the same mode of action will clearly favour those individuals in the pest population able to tolerate the treatment. This leads to a situation where the tolerant (or resistant) individuals can dominate the population and the product becomes ineffective. In general, the more rapidly the pest species reproduces and the more mobile it is, the faster the emergence of resistant populations, although some weeds seem able to evolve resistance more quickly than would be expected. In the UK, key independent research organisations, chemical manufacturers and other organisations have collaborated to share knowledge and expertise on resistance issues through three action groups. Participants include ADAS, the Chemicals Regulation Directorate, universities, colleges and the Home Grown Cereals Authority (HGCA). The groups have a common aim of monitoring resistance in the UK and devising and publishing management strategies designed to combat it where it occurs.

- **Weed Resistance Action Group (WRAG)**, formed in 1989 (Secretary: Dr Stephen Moss, Rothamsted Research, Harpenden, Herts AL5 2JQ *Tel: 01582 7631330 ext. 2521*).
- **Fungicide Resistance Action Group (FRAG)**, formed in 1995 (Secretary: Mr Paul Ashby, HSE, Email: paul.ashby@hse.gsi.gov.uk).
- **Insecticide Resistance Action Group (IRAG)**, formed in 1997 (Secretary: Tom Pope, ADAS Boxworth, Battlegate Road, Cambridge, Cambs CB3 8NN. *Tel: 01954 267666* Email: tom.pope@adas.co.uk).

The above groups publish detailed advice on resistance management relevant for each sector and, in some cases, specific to a pest problem. This information, together with further details about the function of each group, can be obtained from the Chemicals Regulation Directorate website (www.pesticides.gov.uk).

The speed at which resistance appears depends on the mode of action of the crop protection chemicals, as well as the manner in which they are used. Resistance among insects and fungal diseases has been evident for much longer than weed resistance to herbicides, but examples in all three categories are now widespread and increasing. This has created a need for agreement on the advice given for the use of crop protection chemicals in order to reduce the likelihood of the development of resistance and to avoid the loss of potentially valuable products in the chemical armoury. Mixing or alternating modes of action is one of the guiding principles of resistance management. To assist appropriate product choices, mode of action codes, published by the international resistance action committees (see below), are shown in the respective active ingredient fact sheets. The product label and/or a professional advisor should always be consulted before making decisions. The general guidelines for resistance management are similar for all three problem areas.

Preparation in advance

- Be aware of factors that favour the development of resistance, such as repeated annual use of the same product, and assess the risk.
- Plan ahead and aim to integrate all possible means of control.
- Use cultural measures, such as rotations, stubble hygiene, variety selection and, for fungicides, removal of primary inoculum sources, to reduce reliance on chemical control.
- Monitor crops regularly.
- Keep aware of local resistance problems.
- Monitor effectiveness of actions taken and take professional advice, especially in cases of unexplained poor control.

Using crop protection products

- Optimise product efficacy by using them as directed, at the right time, in good condition.
- Treat pest problems early.
- Mix or alternate chemicals with different modes of action.
- Avoid repeated applications of very low doses.
- Keep accurate field records.

Label guidance depends on the appropriate strategy for the product. Most frequently it consists of a warning of the possibility of poor performance due to resistance, and a restriction on the number of treatments that should be applied in order to minimise the development of resistance. This information is summarised in the profiles in the active ingredient fact sheets, but detailed guidance must always be obtained by reading the label itself before use.

International Action Committees

Resistance to crop protection products is an international problem. Agrochemical industry collaboration on a global scale is via three action committees whose aims are to support a coordinated industry approach to the management of resistance worldwide. In particular they produce lists of crop protection chemicals classified according to their mode of action. These lists and other information can be obtained from the respective websites.

Herbicide Resistance Action Committee – www.hracglobal.com

Fungicide Resistance Action Committee – www.frac.info

Insecticide Resistance Action Committee – www.irac-online.org

Poisons and Poisoning

Chemicals Subject to the Poison Law

Certain products are subject to the provisions of the Poisons Act 1972, the Poisons List order 1982 and the Poisons Rules 1982 (copies obtainable from The Stationery Office, www.tso.co.uk). These Rules include general and specific provisions for the storage and sale and supply of listed non-medicine poisons. Full details can be accessed in Annex C of the Guide to Pesticides on the Chemicals Regulation Directorate (CRD) website. The nature of the formulation and the concentration of the active ingredient allow some products to be exempted from the Rules, while others with the same active ingredient are included (see below). The chemicals approved for use in the UK are specified under Parts I and II of the Poisons List as follows.

Part I Poisons (sale restricted to registered retail pharmacists and to registered non-pharmacy businesses provided sales do not take place on retail premises):

- aluminium phosphide
- chloropicrin
- magnesium phosphide.

Part II Poisons (sale restricted to registered retail pharmacists and listed sellers registered with a local authority):

- formaldehyde
- oxamyl (a).

Notes

(a) Granular formulations that do not contain more than 12% w/w of this or a combination of similarly flagged poisons are exempt.

Occupational Exposure Limits

A fundamental requirement of the COSHH Regulations is that exposure of employees to substances hazardous to health should be prevented or adequately controlled. Exposure by inhalation is usually the main hazard, and in order to measure the adequacy of control of exposure by this route various substances have been assigned occupational exposure limits.

There are two types of occupational exposure limits defined under COSHH: Occupational Exposure Standards (OES) and Maximum Exposure Limits (MEL). The key difference is that an OES is set at a level at which there is no indication of risk to health; for a MEL a residual risk may exist and the level takes socio-economic factors into account. In practice, MELs have been most often allocated to carcinogens and to other substances for which no threshold of effect can be identified and for which there is no doubt about the seriousness of the effects of exposure.

OESs and MELs are set on the recommendation of the Advisory Committee on Toxic Substances (ACTS). Full details are published by HSE in EH 40/95 Occupational Exposure Limits 1995, ISBN 0 7176 0876 X.

As far as pesticides are concerned, OESs and MELs have been set for relatively few active ingredients. This is because pesticide products usually contain other substances in their formulation, including solvents, which may have their own OES/MEL. In practice inhalation of solvent may be at least, or more, important than that of the active ingredient. These factors are taken into account by the regulators when approving a pesticide product under the Control of Pesticides Regulations. This indicates one of the reasons why a change of pesticide formulation usually necessitates a new approval assessment under COPR.

First Aid Measures

If pesticides are handled in accordance with the required safety precautions, as given on the container label, poisoning should not occur. It is difficult, however, to guard completely against the occasional accidental exposure. Thus, if a person handling, or exposed to, pesticides becomes ill, it is a wise precaution to apply first aid measures appropriate to pesticide poisoning even though the cause of illness may eventually prove to have been quite different. An employer has a legal duty to make adequate first aid provision for employees. Regular pesticide users should consider appointing a trained first aider even if numbers of employees are not large, since there is a specific hazard.

The first essential in a case of suspected poisoning is for the person involved to stop work, to be moved away from any area of possible contamination and for a doctor to be called at once. If no doctor is available the patient should be taken to hospital as quickly as possible. In either event it is most important that the name of the chemical being used should be recorded and preferably the whole product label or leaflet should be shown to the doctor or hospital concerned.

Some pesticides, which are unlikely to cause poisoning in normal use, are extremely toxic if swallowed accidentally or deliberately. In such cases get the patient to hospital as quickly as possible, with all the information you have. Some labels now include Material Safety Data Sheets and these contain valuable information for both the first aider and medical staff.

General Measures

Measures appropriate in all cases of suspected poisoning include the following:

- Remove any protective or other contaminated clothing (taking care to avoid personal contamination).
- Wash any contaminated areas carefully with water or with soap and water if available.
- In cases of eye contamination, flush with plenty of clean water for at least 15 minutes.
- Lay the patient down, keep at rest and under shelter. Cover with one clean blanket or coat, etc. Avoid overheating.
- Monitor level of consciousness, breathing and pulse rate.
- If consciousness is lost, place the casualty in the recovery position (on his/her side with head down and tongue forward to prevent inhalation of vomit).

Reporting of Pesticide Poisoning

Any cases of poisoning by pesticides must be reported without delay to an HM Agricultural Inspector of the Health and Safety Executive. In addition any cases of poisoning by substances named in schedule 2 of The Reporting of Injuries, Diseases and Dangerous Occurrences Regulations 1985, must also be reported to HM Agricultural Inspectorate (this includes organophosphorus chemicals, mercury and some fumigants).

Cases of pesticide poisoning should also be reported to the manufacturer concerned.

Additional Information

General advice on the safe use of pesticides is given in a range of Health and Safety Executive leaflets available from HSE Books. The major agrochemical companies are able to provide authoritative medical advice about their own pesticide products. Useful information is now available from the Material Safety Data Sheet available from the manufacturer or the manufacturer's website.

A useful booklet, *Guidelines for Emergency Measures in Cases of Pesticide Poisoning*, is available from CropLife International.

New arrangements for the provision of information to doctors about poisons and the management of poisonings have been introduced as part of the modernisation of the National Poisons Information Service (NPIS). The Service provides a year-round, 24-hour-a-day service for healthcare staff on the diagnosis, treatment and management of patients who may have been poisoned. The new arrangements are aimed at moving away from the telephone as the first point of contact for poisons information, to the use by doctors of an online database, supported by a second-tier, consultant-led information service for more complex clinical advice. NPIS no longer provides direct information on poisoning to members of the public. **Anyone suspecting poisoning by pesticides should seek professional medical help immediately via their GP or NHS Direct (www.nhsdirect.nhs.uk) or NHS 24 (www.nhs24.com).**

Environmental Protection

Environmental Land Management: 8 simple steps for arable farmers

Choosing the right measures, putting them in the right place, and managing them in the right way will make all the difference to your farm environment. The general principles given here should be considered in conjunction with local priorities for soil and water protection and wildlife conservation. This approach complements best practice in soil, crop, fertiliser and pesticide management.

An improved farm environment can be achieved by good management of around 4% of the arable area where high quality habitats are maintained or created. However, the actual area required on your farm will depend on factors such as the area of vulnerable soils and length of watercourses.

If you need further advice then consult a competent environmental adviser.

What you can do

It is important to have a balance of environmental measures that contribute to each of the relevant points below to achieve improved environmental benefits.

1. **Look after established wildlife habitats**
   Start by assessing what you already have on the farm! Maintaining, or where necessary, restoring any existing wildlife habitats, such as woodland, ponds, flower-rich grassland or field margins, is critical to the survival of much of the wildlife on the farm, and may count towards some of the following measures without the need to create new habitats. Unproductive land can be used to create new habitats to complement what you already have.

2. **Maximise the environmental value of field boundaries**
   Hedgerow management and ditch management on a 2–3 year rotation boosts flowers, fruit and refuges for wildlife. This is most suited to hedges dominated by hawthorn and blackthorn, and ditches where rotational management will not compromise the drainage function. Establish new hedgerow trees to maintain or restore former numbers within the landscape.

3. **Create a network of grass margins**
   The highest priority is to buffer watercourses, ideally with a minimum of 5m buffer strips. Grass margins can also be used to boost beneficial insects and small mammals, and buffer hedges, ponds and other environmental features. Beetle banks can be used to reduce soil erosion and run-off on slopes greater than 1:20 and boost beneficial insects in fields greater than 20 ha.

4. **Establish flower rich habitats**
   Evidence suggests that a network of flower-rich margins on 1% of arable land will support beneficial insects and a wealth of wildlife that feeds on insects. Assess whether this is best done by allowing arable plants in the seed bank to germinate, establishing perennial margins with a grass and wildflower mix, or using nectar flower mixtures. Improving the linkages between these features on the farm will also help wildlife move across the landscape.

5. **Provide winter food for birds with weedy over-wintered stubbles or wild bird cover**
   Provision of seed for wildlife is best achieved by leaving over-wintered stubbles unsprayed and uncultivated until at least mid-February on at least 5% of arable land, or growing seed-rich crops as wild bird cover on 2% of arable land.

6. **Use of spring cropping or in-field measures** to help ground-nesting birds
   Spring crops provide better habitat for a range of plants and insects, and birds such as lapwings and skylarks. Use rotational fallows, skylark plots in winter cereals or (if breeding lapwings occur) fallow plots to support ground-nesting birds where spring cropping forms less than 25% of the arable area. Fallow plots should not be created on land liable to runoff or erosion. Evidence suggests that at least 20 skylark plots or a 1ha fallow plot per 100 ha would support ground-nesting birds.

7. **Use winter cover crops** to protect water
   You should consider whether a winter cover crop (e.g. mustard) is necessary to capture residual nitrogen on cultivated land left fallow through the winter. This is not necessary if the stubble is retained until at least mid-February and forms a green cover.
8. **Establish in-field grass areas** to reduce soil erosion and run-off
   Land liable to act as channels for soil erosion or run-off (e.g. steep slopes or field corners) should be converted to in-field grass areas.

Remember

- Right measures
- Right place
- Right management

Protection of Bees

Honey bees

Honey bees are a source of income for their owners and important to farmers and growers as pollinators of their crops. It is irresponsible and unnecessary to use pesticides in such a way that may endanger them. Pesticides vary in their toxicity to bees, but those that present a special hazard carry a specific warning in the precautions section of the label. They are indicated in this *Guide* in the hazard classification and safety precautions section of the pesticide profile.

Product labels indicate the necessary environmental precautions to take, but where use of an insecticide on a flowering crop is contemplated the British Beekeepers Association have produced the following guidelines for growers:

- Target insect pests with the most appropriate product.
- Choose a product that will cause minimal harm to beneficial species.
- Follow the manufacturer's instructions carefully.
- Inspect and monitor crops regularly.
- Avoid spraying crops in flower or where bees are actively foraging.
- Keep down flowering weeds.
- Spray late in the day, in still conditions.
- Avoid excessive spray volume and run-off.
- Adjust sprayer pressure to reduce production of fine droplets and drift.
- Give local beekeepers as much warning of your intention as possible.

Wild bees

Wild bees also play an important role. Bumblebees are useful pollinators of spring flowering crops and fruit trees because they forage in cool, dull weather when honey bees are inactive. They play a particularly important part in pollinating field beans, red and white clover, lucerne and borage. Bumblebees nest and overwinter in field margins and woodland edges. Avoidance of direct or indirect spray contamination of these areas, in addition to the creation of hedgerows and field margins, and late cutting or grazing of meadows, all help the survival of these valuable insects.

The Campaign Against Illegal Poisoning of Wildlife

The Campaign Against Illegal Poisoning of Wildlife, aimed at protecting some of Britain's rarest birds of prey and wildlife whilst also safeguarding domestic animals, was launched in March 1991 by the (then) Ministry of Agriculture, Fisheries and Food and the Department of the Environment, Transport and the Regions. The main objective is to deter those who may be considering using pesticides illegally.

The Campaign is supported by a range of organisations associated with animal welfare, nature preservation, field sports and gamekeeping including the RSPB, English Nature, the Countryside Alliance and the Game Conservancy Trust.

The three objectives are:

- To advise farmers, gamekeepers and other land managers on legal ways of controlling pests;
- To advise the public on how to report illegal poisoning incidents and to respect the need for legal alternatives;
- To investigate incidents and prosecute offenders.

A freephone number (0800 321 600) is available to make it easier for the public to report incidents, and numerous leaflets, posters, postcards and stickers have been created to publicise the existence of the Campaign. A video has also been produced to illustrate the many talks, demonstrations and exhibitions.

The Campaign arose from the results of the Wildlife Incident Investigation Scheme for the investigation of possible cases of illegal poisoning. Under this scheme, all reported incidents are considered and thoroughly investigated where appropriate. Enforcement action is taken wherever sufficient evidence of an offence can be obtained and numerous prosecutions have been made since the start of the Campaign.

Further information about the Campaign is available from: Chemicals Regulation Directorate, Research Co-ordination and Environmental Policy Branch, Room 317, Mallard House, 3 Peasholme Green, York YO1 7PX, or from the website: www.caip-uk.info/default.aspx. E-mail: wiis@hse.gsi.gov.uk

Water Quality

Even when diluted, some pesticides are potentially dangerous to fish and other aquatic life. Not only this, but many watercourses and groundwaters are sources of drinking water, and it requires only a tiny amount of contamination to breach the stringent European Union water quality standards. The EEC Drinking Water Directive sets a maximum admissible level in drinking water for any pesticide, regardless of its toxicity, at 1 part in 10,000 million. As little as 250 grams could be enough to cause the daily supply to a city the size of London to exceed the permitted levels (although it would be very unlikely to present a health hazard to consumers).

The protection of groundwater quality is therefore vital. The Food and Environment Protection Act 1985 (FEPA) places a special obligation on users of pesticides to "safeguard the environment and in particular avoid the pollution of water". Under the Water Resources Act 1991 it is an offence to pollute any controlled waters (watercourses or groundwater), either deliberately or accidentally. Protection of controlled waters from pollution is the responsibility of the Environment Agency (in England and Wales) and the Scottish Environmental Protection Agency. Addresses for both organisations can be found under Contacts.

Users of pesticides therefore have a duty to adopt responsible working practices and, unless they are applying herbicides in or near water, to prevent them getting into water. Guidance on how to achieve this is given in the Defra *Code of Good Agricultural Practice for the Protection of Water*.

The duty of care covers not only the way in which a pesticide is sprayed, but also its storage, preparation and disposal of surplus, sprayer washings and the container. Products that are a major hazard to fish, other aquatic life or aquatic higher plants carry one of several specific label precautions in their fact sheet, depending on the assessed hazard level.

Where to get information

For general enquiries telephone 0645 333 111 (Environment Agency) or 01786 457 700 (SEPA). Both Agencies operate a 24-hour emergency hotline for reporting all environmental incidents relating to air, land and water:

0800 80 70 60 (for incidents in England and Wales)

0345 73 72 71 (for incidents in Scotland)

In addition, printed literature concerning the protection of water is available from the Environment Agency and the Crop Protection Association (see Key References).

SECTION 5

Protecting surface waters

Surface waters are particularly vulnerable to contamination. One of the best ways of preventing those pesticides that carry the greatest risk to aquatic wildlife from reaching surface waters is to prohibit their application within a boundary adjacent to the water. Such areas are known as no-spray, or buffer, zones. Certain products are restricted in this way and have a legally binding label precaution to make sure the potential exposure of aquatic organisms to pesticides that might harm them is minimised.

Before 1999 the protected zones were measured from the edge of the water. The distances were 2 metres for hand-held or knapsack sprayers, 6 metres for ground crop sprayers, and a variable distance (but often 18 metres) for broadcast air-assisted applications, such as in orchards. The introduction of LERAPs (see below) has changed the method of measuring buffer zones.

'Surface water' includes lakes, ponds, reservoirs, streams, rivers and watercourses (natural or artificial). It also includes temporarily or seasonally dry ditches, which have the potential to carry water at different times of the year. Buffer zone restrictions do not necessarily apply to all products containing the same active ingredient. Those in formulations that are not likely to contaminate surface water through spray drift do not pose the same risk to aquatic life and are not subject to the restrictions.

Local Environmental Risk Assessment for Pesticides (LERAPs)

Local Environment Risk Assessments for Pesticides (LERAPs) were introduced in March 1999, and revised guidelines were issued in January 2002. They give users of most products currently subject to a buffer zone restriction the option of continuing to comply with the existing buffer zone restriction (using the new method of measurement), or carrying out a LERAP and possibly reducing the size of the buffer zone as a result. In either case, there is a new legal obligation for the user to record his decision, including the results of the LERAP.

The scheme has changed the method of measuring the buffer zone. Previously the zone was measured from the edge of the water, but it is now the distance from the top of the bank of the watercourse to the edge of the spray area.

The LERAP provides a mechanism for taking into account other factors that may reduce the risk, such as dose reduction, the use of low drift nozzles, and whether the watercourse is dry or flowing. The previous arrangements applied the same restriction regardless of whether there was actually water present. Now there is a standard zone of 1 m from the top of a dry ditch bank.

Other factors to include in a LERAP that may allow a reduction in the buffer zone are:

- The size of the watercourse, because the wider it is, the greater the dilution factor, and the lower the risk of serious pollution.
- The dose applied. The lower the dose, the less is the risk.
- The application equipment. Sprayers and nozzles are star-rated according to their ability to reduce spray drift fallout. Equipment offering the greatest reductions achieves the highest rating of three stars. The scheme was originally restricted to ground crop sprayers; new, more flexible rules introduced in February 2002 included broadcast air-assisted orchard and hop sprayers.
- Other changes introduced in 2002 allow the reduction of a buffer zone if there is an appropriate living windbreak between the sprayed area and a watercourse.

Not all products that had a label buffer zone restriction are included in the LERAP scheme. The option to reduce the buffer zone does not to apply to organophosphorus or synthetic pyrethroid insecticides. This group are classified as Category A products. All other products that had a label buffer zone restriction are classified as Category B. In addition some products have a 'Broadcast air-assisted LERAP' classification. The wording of the buffer zone label precautions has been amended for all products to take account of the new method of measurement, and whether or not the particular product qualifies for inclusion in the LERAP scheme.

Products in this *Guide* that are in Category A or B, or have a broadcast air-assisted LERAP, are identified with an appropriate icon on the product fact sheet. Updates to the list are published regularly by CRD and details can be obtained from their website at www.pesticides.gov.uk.

The introduction of LERAPs was an important step forward because it demonstrated a willingness to reduce the impact of regulation on users of pesticides by allowing flexibility where local conditions make it safe to do so. This places a legal responsibility on the user to ensure the risk assessment is done either by himself or by the spray operator or by a professional consultant or advisor. It is compulsory to record the LERAP and make it available for inspection by enforcement authorities.

More details of the LERAP arrangements and guidance on how to carry out assessments are published in the Ministry booklets PB5621 *Local Environmental Risk Assessment for Pesticides – Horizontal Boom Sprayers*, and booklet PB6533 *Local Environment Risk Assessment for Pesticides – Broadcast Air-Assisted Sprayers*. Additional booklets, PB2088 *Keeping Pesticides Out of Water*, and PB3160 *Is Your Sprayer Fit for Work* give general practical guidance.

Groundwater Regulations

Groundwater Regulations were introduced in 1999 to complete the implementation in UK of the EU Groundwater Directive (Protection of Groundwater Against Pollution Caused by Certain Dangerous Substances – 80/68/EEC). These Regulations help prevent the pollution of groundwater by controlling discharges or disposal of certain substances, including all pesticides.

Groundwater is defined under the Regulations as any water contained in the ground below the water table. Pesticides must not enter groundwater unless it is deemed by the appropriate Agency to be permanently unsuitable for other uses. The Agricultural Waste Regulations were introduced in May 2006 and apply to the disposal of all farm wastes, including pesticides. With certain exemptions farmers are required to obtain a waste management licence for most waste disposal activities.

As far as pesticides are concerned the new Regulations made little difference to existing controls, except for the disposal of empty containers and the disposal of sprayer washings and rinsings. It remains an offence to dispose of pesticides onto land without official authorisation. Normal use of a pesticide in accordance with product approval does not require authorisation. This includes spraying the washings and rinsings back on the crop provided that, in so doing, the maximum approved dose for that product on that crop is not exceeded. However, those wishing to use a lined biobed for this purpose need to obtain an exemption from the Agency.

In practice the best advice to farmers and growers is to plan to use all diluted spray within the crop and to dispose of all washings via the same route making sure that they stay within the conditions of approval of the product. The enforcing agencies for these Regulations are the Environment Agency (in England and Wales) and the Scottish Environmental Protection Agency. The Agencies can serve notice at any time to modify the conditions of an authorisation where necessary to prevent pollution of groundwater.

Integrated Farm Management (IFM)

Integrated farm management is a method of farming that balances the requirements of running a profitable farming business with the adoption of responsible and sensitive environmental management. It is a whole-farm, long-term strategy that combines the best of modern technology with some basic principles of good farming practice. It is a realistic way forward that addresses the justifiable concerns of the environmental impact of modern farming practices, at the same time as ensuring that the industry remains viable and continues to provide wholesome, affordable food. IFM embraces arable and livestock management. The phrase 'integrated crop management' is sometimes used where a farm is wholly arable.

Pest control is essential in any management system. Much can be done to minimise the incidence and impact of pests, but their presence is almost inevitable. IFM ensures that, where a pest problem needs to be contained, the action taken is the best combination of all available options. Pesticides are one of these options, and form an essential, but by no means exclusive, part of pest control strategy.

Where chemicals are to be used, the choice of product should be made not only with the pest problem in mind, but with an awareness of the environmental and social risks that might accompany its use. The aim should be to use as much as necessary, but as little as possible. A major part of the approval process aims to safeguard the environment, so that any approved product, when used as directed, will not cause long-term harm to wildlife or the environment. This is achieved by specifying on the label detailed rules for the way in which a product may be used.

The skill in implementing an IFM pest control strategy is the decision on how easily these rules may be complied with in the particular situation where use is contemplated. Although many chemical options may be available, some are likely to be more suitable than others. This *Guide* lists, under the heading **Special precautions/Environmental safety**, the key label precautions that need to be considered in this context, although the actual product label must always be read before a product is used.

Campaign for the Farmed Environment

With the demise of set-aside, it was realised that the environmental benefits achieved by this scheme could be lost. Under the threat of legislation to be brought in during 2012, the Campaign for the Farmed Environment (CFE) scheme seeks to retain or even exceed the environmental benefits achieved by set-aside. It has three main themes:

- farming for cleaner water and a healthier soil
- helping farmland birds thrive
- improving the environment for farm wildlife.

The scheme encourages farmers to sign up for Entry Level Stewardship (ELS) schemes and to target their already extensive knowledge of habitat management to greater effect. If successful, it will help to fend off the proposal that farmers of cultivated land in England will have to adopt environmental management options on up to 6% of their land.

SECTION 6
APPENDICES

Appendix 1
Suppliers of Pesticides and Adjuvants

AgChem Access: AgChemAccess Limited
Cedar House,
41 Thorpe Road
Norwich
Norfolk
NR1 1ES
UK
Tel: 01603 624413
Fax: 01759 371971
Email: martin@agchemaccess.co.uk
Web: www.agchemaccess.com

Agform: Agform Ltd
Maidstone Heath
Blundell Lane
Bursledon
Southampton
SO31 1AA
Tel: 023 8040 7831
Fax: 023 8040 7198
Email: info@agform.com

Agrichem: Agrichem (International) Ltd
Industrial Estate
Station Road
Whittlesey
Cambs.
PE7 2EY
Tel: (01733) 204019
Fax: (01733) 204162
Email:
alison.williamson@agrichem.co.uk
Web: www.agrichem.co.uk

AgriChem BV: AgriChem BV
Koopvaardijweg 9
4906 CV Oosterhout
Postbus 295
4900 AG Oosterhout
Netherlands
Tel: 0031 1624 31931
Fax: 0031 1624 56797
Email: info@agrichem.com
Web: www.agrichem.com

AgriGuard: AgriGuard Ltd
Unit 1
Broomfield Business Park
Malahide
Co. Dublin
Ireland
Tel: (+353) 1 846 2044
Fax: (+353) 1 846 2489
Email: info@agriguard.ie
Web: www.agriguard.ie

Agropharm: Agropharm Limited
Buckingham Place
Church Road
Penn
High Wycombe
Bucks.
HP10 8LN
Tel: (01494) 816575
Fax: (01494) 816578
Email: sales@agropharm.co.uk
Web: www.agropharm.co.uk

Agroquimicos : Agroquimicos Genericos
Calle San Vicente Martir
no 85, 8, 46007
Valencia,
Spain

Agrovista: Agrovista UK Ltd
Cambridge House
Nottingham Road
Stapleford
Nottingham
NG9 8AB
Tel: (0115) 939 0202
Fax: (0115) 921 8498
Email: enquiries@agrovista.co.uk
Web: www.agrovista.co.uk

Amega: Amega Sciences
Lanchester Way
Royal Oak Industrial Estate
Daventry
Northants.
NN11 5PH
Tel: (01327) 704444
Fax: (01327) 71154
Email: admin@amega-sciences.com
Web: www.amega-sciences.com

Antec Biosentry: Antec Biosentry
DuPont Animal Health Solutions
Windham Road
Chilton Industrial Estate
Sudbury
Suffolk
CO10 2XD
Tel: (01787) 377305
Fax: (01787) 310846
Email: biosecurity@antecint.com
Web: www.antecint.com

Arysta: Arysta Life Sciences
  Route d'Artix, BP80
  64150
  Nogueres
  France

B H & B: Battle Hayward & Bower Ltd
  Victoria Chemical Works
  Crofton Drive
  Allenby Road Industrial Estate
  Lincoln
  LN3 4NP
  Tel: (01522) 529206
  Fax: (01522) 538960

B.V. Industrie: B.V. Industrie
  en Handelsonderneming SIMONIS
  PO Box 620
  7000 Ap Doetinchem
  The Netherlands.

Barclay:
Barclay Chemicals Manufacturing Ltd
  Damastown Way
  Damastown Industrial Park
  Mulhuddart
  Dublin 15
  Ireland
  Tel: (+353) 1 811 2900
  Fax: (+353) 1 822 4678
  Email: margaret@barclay.ie
  Web: www.barclay.ie

Barrier: Barrier BioTech Ltd
  36/37 Haverscroft Industrial Estate
  New Road
  Attleborough
  Norfolk
  NR17 1YE
  Tel: (01953) 456363
  Fax: (01953) 455594
  Email: sales@barrier-biotech.com
  Web: www.barrier-biotech.com

BASF: BASF plc
  Agricultural Divison
  PO Box 4, Earl Road
  Cheadle Hulme
  Cheshire
  SK8 6QG
  Tel: (0845) 602 2553
  Fax: (0161) 485 2229
  Web: www.agricentre.co.uk

Bayer CropScience:
Bayer CropScience Limited
  230 Cambridge Science Park
  Milton Road
  Cambridge
  CB4 0WB
  Tel: (01223) 226500
  Fax: (01223) 426240
  Web: www.bayercropscience.co.uk

Bayer Environ.:
Bayer Environmental Science
  230 Cambridge Science Park
  Milton Road
  Cambridge
  CB4 0WB
  Tel: (01223) 226680
  Fax: (01223) 226635
  Email: claire.hazell@bayer.com
  Web: www.bayer-escience.co.uk

Becesane : Becesane s.r.o.
  Rohacova, 188/377
  130 00
  PRAHA 3
  Czech Republic

Belchim: Belchim Crop Protection Ltd
  Unit 1b, Fenice Court
  Phoenix Park
  Eaton Socon
  St Neots
  Cambs.
  PE19 8EP
  Tel: (01480) 403333
  Fax: (01480) 403444
  Email: info@belchim.com
  Web: www.belchim.com

Belcrop: Belcrop (UK)
  21 Victoria Rd
  Wargrave
  Berkshire
  RG10 8AD
  UK
  Tel: (0118) 940 4264
  Fax: (0118) 940 4264
  Email: JOHNHUDSON23@aol.com
  Web: www.BELCROP.com

Certis: Certis
  Suite 5, 1 Riverside
  Granta Park
  Great Abington
  Cambs.
  CB21 6AD
  Tel: 0845 373 0305
  Fax: 01223 891210
  Email: certis@certiseurope.co.uk
  Web: www.certiseurope.co.uk

ChemSource: ChemSource Ltd
  5 Jupiter House
  Calleva Park
  Aldermaston
  Reading
  Berkshire
  RG7 4NN
  UK
  Tel: 0845 459 4039
  Email: james@chemsource.co.uk

Chemtura: Chemtura Europe Ltd
  Kennet House
  4 Langley Quay
  Slough
  Berks.
  SL3 6EH
  Tel: (01753) 603000
  Fax: (01753) 603077
  Web: www.chemtura.com

Chiltern: Chiltern Farm Chemicals Ltd
  East Mellwaters
  Stainmore
  Bowes
  Barnard Castle
  Co. Durham
  DL12 9RH
  Tel: (01833) 628282
  Fax: (01833) 628020
  Web: www.chilternfarm.com

Clayton: Clayton Plant Protection Ltd
  Bracetown Business Park
  Clonee
  Co. Meath
  Ireland
  Tel: (+353) 1 821 0127
  Fax: (+353) 81 841 1084
  Email: info@cppltd.eu
  Web: www.cppltd.eu

DAPT: DAPT Agrochemicals Ltd
  14 Monks Walk
  Southfleet
  Gravesend
  Kent
  DA13 9NZ
  Tel: (01474) 834448
  Fax: (01474) 834449
  Email: rkjltd@supanet.com

De Sangosse: De Sangosse Ltd
  Hillside Mill
  Quarry Lane
  Swaffham Bulbeck
  Cambridge
  CB25 0LU
  Tel: (01223) 811215
  Fax: (01223) 810020
  Email: info@desangosse.co.uk
  Web: www.desangosse.co.uk

Doff Portland: Doff Portland Ltd
  Aerial Way
  Hucknall
  Nottingham
  NG15 6DW
  Tel: (0115) 963 2842
  Fax: (0115) 963 8657
  Email: info@doff.co.uk
  Web: www.doff.co.uk

Dow: Dow AgroSciences Ltd
  Latchmore Court
  Brand Street
  Hitchin
  Herts.
  SG5 1NH
  Tel: (01462) 457272
  Fax: (01462) 426605
  Email: fhihotl@dow.com
  Web: www.dowagro.com/uk

DuPont: DuPont (UK) Ltd
  Crop Protection Products Department
  Wedgwood Way
  Stevenage
  Herts.
  SG1 4QN
  Tel: (01438) 734450
  Fax: (01438) 734452
  Web: www.gbr.ag.dupont.com

Elliott: Thomas Elliott Fertilisers
  Selby Place
  Stanley Industrial Estate
  Skelmersdale
  Lancashire
  WN8 8EF
  Tel: 01522 811981
  Fax: 01522 810238
  Email:
  christinehall@thomaselliot.demon.co.uk

European Ag: European Agrichemicals Ltd
  Unit NR2, Hurst House
  City Road
  Radnage
  Bucks
  HP14 4DW
  UK
  Tel: 01494 483728
  Email: Radnage

Everris Ltd: Everris Limited
  Epsilon House
  West Road
  Ipswich
  Suffolk
  IP3 9FJ
  Tel: 01473 237100
  Fax: 01473 830386
  Email: prof.sales@everris.com

SECTION 6

Exosect: Exosect Ltd
Leylands Business Park
Colden Common
Winchester
SO21 1TH
Tel: 02380 60 395
Email: sue.harris@exosect.com
Web: www.exosect.com

Fargro: Fargro Ltd
Toddington Lane
Littlehampton
Sussex
BN17 7PP
Tel: (01903) 721591
Fax: (01903) 730737
Email: info@fargro.co.uk
Web: www.fargro.co.uk

Fine: Fine Agrochemicals Ltd
Hill End House
Whittington
Worcester
WR5 2RQ
Tel: (01905) 361800
Fax: (01905) 361810
Email: enquire@fine.eu
Web: www.fine.eu

Forest Research: Forest Research
Alice Holt Lodge
Farnham
Surrey
GU10 4LH

Frontier: Frontier Agriculture Ltd
Fleet Road Industrial Estate
Holbeach
Spalding
Lincs.
PE12 8LY
Tel: (01406) 421405
Email: michael.hales@frontierag.co.uk
Web: www.bankscargill.co.uk

Globachem: Globachem NV
Brustem Industriepark
Lichtenberglaan 2019
3800 Sint-Truiden
Belgium
Tel: (+32) 11 69 01 73
Fax: (+32) 11 68 15 65
Email: globachem@globachem.com
Web: www.globachem.com

Goldengrass: Goldengrass Ltd
PO Box 280
Ware
SG12 8XY
Tel: 01920 486253
Fax: 01920 485123
Email: contact@goldengrass.uk.com

Gowan:
Gowan Comercio e Servicos Limitada
2 Leazes Cresent
Hexham
Northumberland
NE46 3JX
Tel: 07584 052323
Email: DLamb@GOWANCO.com

Greenaway: Greenaway Amenity Ltd
7 Browntoft Lane
Donington
Spalding
Lincs.
PE11 4TQ
Tel: (01775) 821031
Fax: (01775) 821034
Email: greenawayamenity@aol.com
Web: www.greenawaycda.com

Greencrop: Greencrop Technology Ltd
See Clayton Plant Protection Ltd

Headland: Headland Agrochemicals Ltd
Rectors Lane
Pentre
Deeside
Flintshire
CH5 2DH
Tel: (01244) 537370
Fax: (01244) 532097
Email: enquiry@headlandgroup.com
Web: www.headland-ag.co.uk

Headland Amenity:
Headland Amenity Limited
1010 Cambourne Business Park
Cambourne
Cambs.
CB3 6DP
Tel: (01223) 597834
Fax: (01223) 598052
Email: info@headlandamenity.com
Web: www.headlandamenity.com

Helena: Helena Chemical Company
Cambridge House
Nottingham Road
Stapleford
Nottingham
NG9 8AB
Tel: (0115) 939 0202
Fax: (0115) 939 8031

Hermoo: Hermoo (UK)
21 Victoria Road
Wargrave
Berks.
RG10 8AD
Tel: (0118) 940 4264
Fax: (0118) 940 4264
Email: johnhudson23@aol.com
Web: www.hermoo.com

Interagro: Interagro (UK) Ltd
230 Avenue West,
Skyline 120,
Great Notley
Braintree
Essex
CM77 7AA
Tel: (01376) 552703
Fax: (01376) 567608
Email: info@interagro.co.uk
Web: www.interagro.co.uk

Interfarm: Interfarm (UK) Ltd
Kinghams's Place
36 Newgate Street
Doddington
Cambs.
PE15 0SR
Tel: (01354) 741414
Fax: (01354) 741004
Email: technical@interfarm.co.uk
Web: www.interfarm.co.uk

Intracrop: Intracrop
Little Hay
Broadwell
Lechlade
Glos.
GL7 3QS
Tel: (01367) 860255
Fax: (01926) 634798
Email: catherine@a-o-c.co.uk

Irish Drugs: Irish Drugs Ltd
Burnfoot
Lifford
Co. Donegal
Ireland
Tel: (+353) 74 9368104
Fax: (+353) 74 9368311
Email: jgmcivor@eircom.net
Web: www.idl-home.com

K & S Fumigation:
K & S Fumigation Services Ltd
Asparagus Farm
Court Lodge Road
Appledore
Ashford
Kent
TN26 2DH
Tel: (01233) 758252
Fax: (01233) 758343
Email: david@kstreatments.co.uk

Killgerm: Killgerm Chemicals Ltd
115 Wakefield Road
Flushdyke
Ossett
W. Yorks.
WF5 9AR
Tel: (01924) 268400
Fax: (01924) 264757
Email: info@killgerm.com
Web: www.killgerm.com

Koppert: Koppert (UK) Ltd
Unit 8, Tudor Rose Court
53 Hollands Road
Haverhill
Suffolk
CB9 8PJ
Tel: (01440) 704488
Fax: (01440) 704487
Email: info@koppert.co.uk
Web: www.koppert.com

Landseer: Landseer Ltd
Lodge Farm
Goat Hall Lane
Galleywood
Chelmsford
Essex
CM2 8PH
Tel: (01245) 357109
Fax: (01245) 494165
Web: www.lanfruit.co.uk

Life Scientific: Life Scientific Limited
Unit 12, NovaUCD
Belfield Innovation Park
University College Dublin
Dublin 4
Ireland
Tel: 00353 1 283 2024

SECTION 6

Makhteshim: Makhteshim-Agan (UK) Ltd
Unit 15
Thatcham Business Village
Colthrop Way
Thatcham
Berks.
RG19 4LW
Tel: (01635) 860555
Fax: (01635) 861555
Email: admin@mauk.co.uk
Web: www.mauk.co.uk

Microcide: Microcide Ltd
Shepherds Grove
Stanton
Bury St. Edmunds
Suffolk
IP31 2AR
Tel: (01359) 251077
Fax: (01359) 251545
Email: microcide@microcide co.uk
Web: www.microcide.co.uk

Monsanto: Monsanto (UK) Ltd
PO Box 663
Cambourne
Cambridge
CB1 0LD
UK
Tel: (01954) 717575
Fax: (01954) 717579
Email:
technical.helpline.uk@monsanto.com
Web: www.monsanto-ag.co.uk

Nissan: Nissan Chemical Europe Sarl
Parc d'Affaires de Crecy
2 Rue Claude Chappe
69371 St Didier au Mont d'Or
France
Tel: (+33) 4 376 44020
Fax: (+33) 4 376 46874

Nomix Enviro:
Nomix Enviro, A Division of Frontier
Agriculture Ltd
The Grain Silos
Weyhill Road
Andover
Hampshire
SP10 3NT
UK
Tel: 01264 388050
Fax: 01522 866176
Email: nomixenviro@frontierag.co.uk
Web: www.nomix.co.uk

Nufarm UK: Nufarm UK Ltd
Wyke Lane
Wyke
West Yorkshire
BD12 9EJ
UK
Tel: 01274 69 1234
Fax: 01274 69 1176
Email: infouk@uk.nufarm.com
Web: www.nufarm.co.uk

Omex: Omex Agriculture Ltd
Bardney Airfield
Tupholme
Lincoln
LN3 5TP
Tel: (01526) 396000
Fax: (01526) 396001
Email: enquire@omex.com
Web: www.omex.co.uk

Pan Agriculture: Pan Agriculture Ltd
8 Cromwell Mews
Station Road
St Ives
Huntingdon
Cambs.
PE27 5HJ
Tel: (01480) 467790
Fax: (01480) 467041
Email: info@panagriculture.co.uk

Pan Amenity: Pan Amenity Ltd
8 Cromwell Mews
Station Road
St Ives
Cambs.
PE27 5HJ
UK
Tel: 01480 467790
Fax: 01480 467041

Petro-Lube: Petro-Lube (UK) Limited
Lonsto House, Suit No: 10
276 Chase Road
Southgate
London
N14 6HA

Plant Solutions: Plant Solutions
Pyports
Downside Bridge Road
Cobham
Surrey
KT11 3EH
Tel: (01932) 576699
Fax: (01932) 868973
Email: sales@plantsolutionsltd.com
Web: www.plantsolutionsltd.com

PP Products: PP Products
Longmynd
Tunstead Road
Hoveton
Norwich
NR12 8QN
Tel: (01603) 784367
Fax: (01603) 784367
Email: mail@ppproducts.co.uk
Web: www.ppproducts.co.uk

Procam: Procam Group Ltd
Industrial Estate
Saxon Way
Melbourne
Royston
Herts
S98 6DN

Progreen:
Progreen Weed Control Solutions Limited
Kellington House
South Fen Business Park
South Fen Road
Bourne
Lincolnshire
PE10 0DN

Pure Amenity : Pure Amenity Limited
Unit 4A, Primrose Hill
Buttercrambe Road
Stamford Bridge
North Yorks
YO41 1AW
Tel: 0845 257 4710
Email: martin@pureamenity.co.uk

Q-Chem: Q-CHEM nv
Leeuwerweg 138
BE-3803
Belgium
Tel: (+32) 11 785 717
Fax: (+32) 11 681 565
Email: q-chem@skynet.be

Rentokil: Rentokil Initial plc
7-8 Foundry Court,
Foundry Lane
Horsham
W. Sussex
RH13 5PY
Tel: (01403) 214122
Fax: (01403) 214101
Email: keira.corrie@rentokil-initial.com
Web: www.rentokil.com

Rigby Taylor: Rigby Taylor Ltd
Rigby Taylor House
Crown Lane
Horwich
Bolton
Lancs.
BL6 5HP
Tel: (01204) 677777
Fax: (01204) 677715
Email: info@rigbytaylor.com
Web: www.rigbytaylor.com

Ripe Rite : Ripe Rite Limited
Meadowside
Prees Green
Whitchurch
Shropshire
SY13 2BL

Rotam Agrochemical: Rotam Europe Ltd
Hamilton House
Mabledon Place
London
WC1H 9BB
Tel: 0207 953 0447

Scotts: The Scotts Company (UK) Ltd
Paper Mill Lane
Bramford
Ipswich
Suffolk
IP8 4BZ
Tel: (01473) 830492
Fax: (01473) 830814
Web: www.scottsprofessional.co.uk

Sharda:
Sharda Worldwide Exports PVT Ltd
Domnic Holm
29th Road
Bandra (West)
Mumbai
400050
India
Tel: 0032 477 707036

Sherriff Amenity:
Sherriff Amenity Services
The Pines
Fordham Road
Newmarket
Cambs
CB8 7LG
Tel: (01638) 721888
Fax: (01638) 721815
Web: www.sherriffamenity.com

SECTION 6

Sinclair: William Sinclair Horticulture Ltd
Firth Road
Lincoln
LN6 7AH
Tel: (01522) 537561
Fax: (01522) 513609
Web: www.william-sinclair.co.uk

Sipcam: Sipcam UK Ltd
3 The Barn
27 Kneesworth Street
Royston
Herts.
SG8 5AB
Tel: (01763) 212100
Fax: (01763) 212101
Email: paul@sipcamuk.co.uk

Solufeed: Solufeed Ltd
Highground Orchards, Highground Lane
Barnham
Bognor Regis
West Sussex
PO22 0BT
Tel: 01243 554090
Fax: 01243 554568
Email: enquiries@solufeed.com

Sphere: Sphere Laboratories (London) Ltd
c/o Mainswood
Putley Common
Ledbury
Herefordshire
HR8 2RF
Tel: (07875) 283542
Fax: (01531) 670517
Email:
Fiona.Homes@Sphere-London.co.uk

Standon: Standon Chemicals Ltd
48 Grosvenor Square
London
W1K 2HT
Tel: (020) 7493 8648
Fax: (020) 7493 4219

Sumi Agro: Sumi Agro Europe Limited
Vintners' Place
68 Upper Thames Street
London
EC4V 3BJ
Tel: (020) 7246 3697
Fax: (020) 7246 3799
Email: summit@summit-agro.com

Syngenta:
Syngenta Crop Protection UK Limited
CPC4
Capital Park
Fulbourn
Cambridge
CB21 5XE
Tel: 01223 883400
Fax: (01223) 493700
Web: www.syngenta-crop.co.uk

Syngenta Bioline: Syngenta Bioline
Telstar Nursery
Holland Road
Little Clacton
Essex
CO16 9QG
Tel: (01255) 863200
Fax: (01255) 863206
Email: syngenta.bioline@syngenta.com
Web: www.syngenta-bioline.co.uk

Taminco: Taminco UK Ltd
15 Rampton Drift
Longstanton
Cambridge
CB24 3EH
Tel: 01954 789941
Email: philip.forster@taminco.com
Web: www.taminco.com

Truchem: Truchem Ltd
The Knoll
The Cross
Horsley
Stroud
Gloucestershire
GL6 0PR
Tel: (01453) 833293
Fax: (01453) 833293
Email: nichollstruchem@aol.com

Unicrop: Universal Crop Protection Ltd
Park House
Maidenhead Road
Cookham
Berks.
SL6 9DS
Tel: (01628) 526083
Fax: (01628) 810457
Email: enquiries@unicrop.com

United Agri: United Agri Products Ltd
The Crossways
Alconbury Hill
Huntingdon
PE28 4JH
Tel: (01480) 418000
Fax: (01480) 418010

United Phosphorus:
United Phosphorus Ltd
  The Centre
  Birchwood Park
  Warrington
  Cheshire
  WA3 6YN
  Tel: (01925) 819999
  Fax: (01925) 856075
  Email: lpinto@uniphos.com
  Web: www.upleurope.com

Vitax: Vitax Ltd
  Owen Street
  Coalville
  Leicester
  LE67 3DE
  Tel: (01530) 510060
  Fax: (01530) 510299
  Email: info@vitax.co.uk
  Web: www.vitax.co.uk

SECTION 6

Appendix 2
Useful Contacts

Agriculture Industries Confederation Ltd
Confederation House
East of England Showground
Peterborough PE2 6XE
Tel: (01733) 385236
Fax: (01733) 385270
Web: www.agindustries.org.uk

BASIS Ltd
Bank Chambers
34 St John Street
Ashbourne
Derbyshire DE6 1GH
Tel: (01335) 343945/346138
Fax: (01335) 346488
Web: www.basis-reg.co.uk

British Beekeepers' Association
National Agricultural Centre
Stoneleigh
Kenilworth
Warwickshire CV8 2LZ
Tel: (024) 7669 6679
Web: www.bbka.org.uk

British Beer & Pub Association
Market Towers
1 Nine Elms Lane
London SW8 5NQ
Tel: (0207) 627 9191
Fax: (0207) 627 9123

British Crop Production Council (BCPC)
7 Omni Business Centre
Omega Park
Alton
Hampshire GU34 2QD
Tel: (01420) 593200
Fax: (01420) 593209
Web: www.bcpc.org

British Pest Control Association
1 Gleneagles House
Vernon Gate, South Street
Derby DE1 1UP
Tel: (01332) 294288
Fax: (01332) 295904
Web: www.bpca.org.uk

Chemicals Regulation Directorate
Mallard House
King's Pool
3 Peasholme Green
York YO1 2PX
Tel: (01904) 640500
Fax: (01904) 455733
Web: www.pesticides.gov.uk

Crop Protection Association Ltd
2 Swan Court
Cygnet Park
Hampton
Peterborough PE7 8GX
Tel: (01733) 355370
Fax: (01733) 355371
Web: www.cropprotection.org.uk

CropLife International
Avenue Louise 143
B-1050 Brussels
Belgium
Tel: (+32) 2 542 0410
Fax: (+32) 2 542 0419
Web: www.croplife.org

Department of Agriculture and Rural Development (Northern Ireland)
Pesticides Section
Dundonald House
Upper Newtownards Road
Belfast BT4 3SB
Tel: (028) 9052 4704
Fax: (028) 9052 4059
Web: www.dardni.gov.uk

Department of Environment, Food and Rural Affairs (Defra)
Nobel House
17 Smith Square
London SW1P 3JR
Tel: (020) 7238 6000
Fax: (020) 7238 6591
Web: www.defra.gov.uk

Environment Agency
Rio House, Waterside Drive
Aztec West
Almondsbury
Bristol BS12 4UD
Tel: (01454) 624400
Fax: (01454) 624409
Web: www.environment-agency.gov.uk

European Crop Protection Association
(ECPA)
Avenue E van Nieuwenhuyse 6
B-1160 Brussels
Belgium
Tel: (+32) 2 663 1550
Fax: (+32) 2 663 1560
Web: www.ecpa.be

Farmers' Union of Wales
Llys Amaeth
Queen's Square
Aberystwyth
Dyfed SY23 2EA
Tel: (01970) 612755
Fax: (01970)612755
Web: www.fuw.org.uk

Forestry Commission
231 Corstorphine Road
Edinburgh EH12 7AT
Tel: (0131) 334 0303
Fax: (0131) 334 3047
Web: www.forestry.gov.uk

Health and Safety Executive
Information Services
Room 318 Daniel House
Stanley Precinct, Bootle
Merseyside L20 3TW
Tel: (0151) 951 3191
Web: www.hse.gov.uk

Health and Safety Executive
Biocides & Pesticides Assessment Unit
Room 123 Magdalen House, Bootle
Merseyside L20 3QZ
Tel: (0151) 951 3535
Fax: (0151) 951 3317

Health and Safety Executive – Books
PO Box 1999
Sudbury
Suffolk CO10 2WA
Tel: (01787) 881165
Fax: (01787) 313995

Horticulture Development Council
Bradbourne House
East Malling
Kent ME19 6DZ
Tel: (01732) 848383
Fax: (01732) 848498
Web: www.hdc.org.uk

Lantra (previously ATB-LandBase)
National Agricultural Centre
Stoneleigh
Kenilworth
Warwickshire CV8 2LG
Tel: (024) 7669 6996
Fax: (024) 7669 6732
Web: www.lantra.co.uk

National Association of Agricultural Contractors (NAAC)
Samuelson House, Paxton Road
Orton Centre, Peterborough
Cambs PE2 5LT
Tel: (01733) 362920
Fax: (01733) 362921
Web: www.naac.co.uk

National Chemicals Emergency Centre
Building 329H, Harwell
Didcot
Oxon, OX11 0QJ
Tel: (0870) 190 6621
Fax: (0870) 190 6620
Web: http://the-ncec.com/

National Farmers' Union
Agriculture House
Stoneleigh Park
Stoneleigh
Warwickshire CV8 2TZ
Tel: (024) 7685 0662
Fax: (024) 7685 8501
Web: www.nfu.org.uk

National Poisons Information Service
(Birmingham Unit)
City Hospital
Birmingham B18 7QH
Tel: (0121) 507 4123
Fax: (0121) 507 5580
Web: www.npis.org

National Turfgrass Council
Hunter's Lodge
Dr Brown's Road
Minchinhampton
Glos. GL6 9BT
Tel: (01453) 883588
Fax: (01453) 731449

NPTC
National Agricultural Centre
Stoneleigh
Kenilworth
Warwickshire CV8 2LG
Tel: (024) 7669 6553
Fax: (020) 7331 7313
Web: www.nptc.org.uk

SECTION 6

**Processors and Growers Research
Organisation**
The Research Station
Great North Road, Thornhaugh
Peterborough, Cambs. PE8 6HJ
Tel: (01780) 782585
Fax: (01780) 783993
Web: www.pgro.co.uk

Scottish Beekeepers' Association
North Trinity House
114 Trinity Road
Edinburgh EH5 3JZ
Tel: (0131) 552 5341
Web: www.scottishbeekeepers.org.uk

**Scottish Environment Protection
Agency (SEPA)**
Erskine Court
The Castle Business Park
Stirling FK9 4TR
Tel: (01786) 457 700
Fax: (01786) 446 885
Web: www.sepa.org.uk

TSO (The Stationery Office)
Publications Centre
PO Box 276
London SW8 5DT
Tel: (020) 7873 9090 (orders)
Tel: (020) 7873 0011 (enquiries)
Fax: (020) 7873 8200
Web: www.thestationeryoffice.com

Ulster Beekeepers' Association
57 Liberty Road
Carrickfergus
Co. Antrim BT38 9DJ
Tel: (01960) 362998
Web: www.ubka.org

Welsh Beekeepers' Association
Trem y Clawdd
Fron Isaf, Chirk
Wrexham
Clwyd LL14 5AH
Tel/Fax: (01691) 773300

Appendix 3
Keys to Crop and Weed Growth Stages

Decimal Code for the Growth Stages of Cereals
Illustrations of these growth stages can be found in the reference indicated below and in some company product manuals.

0 Germination

00	Dryseed
03	Imbibition complete
05	Radicle emerged from caryopsis
07	Coleoptile emerged from caryopsis
09	Leaf at coleoptile tip

1 Seedling growth

10	First leaf through coleoptile
11	First leaf unfolded
12	2 leaves unfolded
13	3 leaves unfolded
14	4 leaves unfolded
15	5 leaves unfolded
16	6 leaves unfolded
17	7 leaves unfolded
18	8 leaves unfolded
19	9 or more leaves unfolded

2 Tillering

20	Main shoot only
21	Main shoot and 1 tiller
22	Main shoot and 2 tillers
23	Main shoot and 3 tillers
24	Main shoot and 4 tillers
25	Main shoot and 5 tillers
26	Main shoot and 6 tillers
27	Main shoot and 7 tillers
28	Main shoot and 8 tillers
29	Main shoot and 9 or more tillers

3 Stem elongation

30	Ear at 1 cm
31	1st node detectable
32	2nd node detectable
33	3rd node detectable
34	4th node detectable
35	5th node detectable
36	6th node detectable
37	Flag leaf just visible
39	Flag leaf ligule/collar just visible

4 Booting

41	Flag leaf sheath extending
43	Boots just visibly swollen
45	Boots swollen
47	Flag leaf sheath opening
49	First awns visible

5 Inflorescence

51	First spikelet of inflorescence just visible
52	1/4 of inflorescence emerged
55	1/2 of inflorescence emerged
57	3/4 of inflorescence emerged
59	Emergence of inflorescence completed

6 Anthesis

60	
61	Beginning of anthesis
64	
65	Anthesis half way
68	
69	Anthesis complete

7 Milk development

71	Caryopsis watery ripe
73	Early milk
75	Medium milk
77	Late milk

8 Dough development

83	Early dough
85	Soft dough
87	Hard dough

9 Ripening

91	Caryopsis hard (difficult to divide by thumb-nail)
92	Caryopsis hard (can no longer be dented by thumb-nail)
93	Caryopsis loosening in daytime

(From Tottman, 1987. *Annals of Applied Biology*, **110**, 441–454)

SECTION 6

Stages in Development of Oilseed Rape

Illustrations of these growth stages can be found in the reference indicated below and in some company product manuals.

0 Germination and emergence

1 Leaf production

1,0 Both cotyledons unfolded and green
1,1 First true leaf
1,2 Second true leaf
1,3 Third true leaf
1,4 Fourth true leaf
1,5 Fifth true leaf
1,10 About tenth true leaf
1,15 About fifteenth true leaf

2 Stem extension

2,0 No internodes ('rosette')
2,5 About five internodes

3 Flower bud development

3,0 Only leaf buds present
3,1 Flower buds present but enclosed
by   leaves
3,3 Flower buds visible from above
     ('green bud')
3,5 Flower buds raised above leaves
3,6 First flower stalks extending
3,7 First flower buds yellow ('yellow
     bud')

4 Flowering

4,0 First flower opened
4,1 10% all buds opened
4,3 30% all buds opened
4,5 50% all buds opened

5 Pod development

5,3 30% potential pods
5,5 50% potential pods
5,7 70% potential pods
5,9 All potential pods

6 Seed development

6,1 Seeds expanding
6,2 Most seeds translucent but full size
6,3 Most seeds green
6,4 Most seeds green-brown mottled
6,5 Most seeds brown
6,6 Most seeds dark brown
6,7 Most seeds black but soft
6,8 Most seeds black and hard
6,9 All seeds black and hard

7 Leaf senescence

8 Stem senescence

8,1 Most stem green
8,5 Half stem green
8,9 Little stem green

9 Pod senescence

9,1 Most pods green
9,5 Half pods green
9.9 Few pods green

(From Sylvester-Bradley, 1985. *Aspects of Applied Biology*, **10**, 395–400)

Stages in Development of Peas

Illustrations of these growth stages can be found in the reference indicated below and in some company product manuals.

0   Germination and emergence

000   Dry seed
001   Imbibed seed
002   Radicle apparent
003   Plumule and radicle apparent
004   Emergence

1   Vegetative stage

101   First node (leaf with one pair leaflets, no tendril)
102   Second node (leaf with one pair leaflets, simple tendril)
103   Third node (leaf with one pair leaflets, complex tendril)
•
l0x   X nodes (leaf with more than one pair leaflets, complex tendril)
•
•
10n   Last recorded node

2   Reproductive stage (main stem)

201   Enclosed buds
202   Visible buds
203   First open flower
204   Pod set (small immature pod)
205   Flat pod
206   Pod swell (seeds small, immature)
207   Podfill
208   Pod green, wrinkled
209   Pod yellow, wrinkled (seeds rubbery)
210   Dry seed

3   Senescence stage

301   Desiccant application stage. Lower pods dry and brown, middle yellow, upper green. Overall moisture content of seed less than 45%
302   Pre-harvest stage. Lower and middle pods dry and brown, upper yellow. Overall moisture content of seed less than 30%
303   Dry harvest stage. All pods dry and brown, seed dry

(From Knott, 1987. *Annals of Applied Biology*, **111**, 233–244)

SECTION 6

Stages in Development of Faba Beans

Illustrations of these growth stages can be found in the reference indicated below and in some company product manuals.

0   Germination and emergence

000   Dry seed
001   Imbibed seed
002   Radicle apparent
003   Plumule and radicle apparent
004   Emergence
005   First leaf unfolding
006   First leaf unfolded

1   Vegetative stage

101   First node
102   Second node
103   Third node
•
•
l0x   X nodes
•
•
10n   N, last recorded node

2   Reproductive stage (main stem)

201   Flower buds visible
203   First open flowers
204   First pod set
205   Pods fully formed, green
207   Pod fill, pods green
209   Seed rubbery, pods pliable, turning black
210   Seed dry and hard, pods dry and black

3   Pod senescence

301   10% pods dry and black
•
•
305   50% pods dry and black
•
•
308   80% pods dry and black, some upper pods green
309   90% pods dry and black, most seed dry. Desiccation stage.
310   All pods dry and black, seed hard. Pre-harvest (glyphosate application stage) 4 Stem senescence

4   Stem senescence

401   10% stem brown/black
•
405   50% stem brown/black
•
•
409   90% stem brown/black
410   All stems brown/black. All pods dry and black, seed hard.

(From Knott, 1990. *Annals of Applied Biology*, **116**, 391–404)

Stages in Development of Potato

Illustrations of these growth stages can be found in the reference indicated below and in some company product manuals.

0   Seed germination and seedling emergence

000   Dry seed
001   Imbibed seed
002   Radicle apparent
003   Elongation of hypocotyl
004   Seedling emergence
005   Cotyledons unfolded

1   Tuber dormancy

100   Innate dormancy (no sprout development under favourable conditions)
150   Enforced dormancy (sprout development inhibited by environmental conditions)

2   Tuber sprouting

200   Dormancy break, sprout development visible
21x   Sprout with 1 node
22x   Sprout with 2 nodes
•
•
29x   Sprout with 9 nodes
21x(2) Second generation sprout with 1 node
22x(2) Second generation sprout with 2 nodes
•
•
29x(2) Second generation sprout with 9 nodes

Where x = 1, sprout >2 mm;
2, 2-5 mm; 3, 5-20 mm;
4, 20-30 mm; 5, 50-100 mm;
6, 100-150 mm long

3   Emergence and shoot expansion

300   Main stem emergence
301   Node 1
302   Node 2
•
•
319   Node 19
Second order branch
321   Node 1
•
•
Nth   order branch
3N1   Node 1
•
•
3N9   Node 9

4   Flowering

Primary flower
400   No flowers
410   Appearance of flower bud
420   Flower unopen
430   Flower open
440   Flower closed
450   Berry swelling
460   Mature berry
Second order flowers
410(2) Appearance of flower bud
420(2) Flower unopen
430(2) Flower open
440(2) Flower closed
450(2) Berry swelling
460(2) Mature berry

5   Tuber development

500   No stolons
510   Stolon initials
520   Stolon elongation
530   Tuber initiation
540   Tuber bulking (>10 mm diam)
550   Skin set
560   Stolon development

6   Senescence

600   Onset of yellowing
650   Half leaves yellow
670   Yellowing of stems
690   Completely dead

(From Jefferies & Lawson, 1991. *Annals of Applied Biology*, **119**, 387–389)

SECTION 6

Stages in Development of Linseed

Illustrations of these growth stages can be found in the reference indicated below and in some company product manuals.

0   Germination and emergence

00   Dry seed
01   Imbibed seed
02   Radicle apparent
04   Hypocotyl extending
05   Emergence
07   Cotyledon unfolding from seed case
09   Cotyledons unfolded and fully expanded

1   Vegetative stage (of main stem)

10   True leaves visible
12   First pair of true leaves fully expanded
13   Third pair of true leaves fully expanded
1n n leaf fully expanded

2   Basal branching

21   One branch
22   Two branches
23   Three branches
2n n branches

3   Flower bud development (on main stem)

31   Enclosed bud visible in leaf axils
33   Bud extending from axil
35   Corymb formed
37   Buds enclosed but petals visible
39   First flower open

4   Flowering (whole plant)

41   10% of flowers open
43   30% of flowers open
45   50% of flowers open
49   End of flowering

5   Capsule formation (whole plant)

51   10% of capsules formed
53   30% 0f capsules formed
55   50% of capsules formed
59   End of capsule formation

6   Capsule senescence (on most advanced plant)

61   Capsules expanding
63   Capsules green and full size
65   Capsules turning yellow
67   Capsules all yellow brown but soft
69   Capsules brown, dry and senesced

7   Stem senescence (whole plant)

71   Stems mostly green below panicle
73   Most stems 30% brown
75   Most stems 50% brown
77   Stems 75% brown
79   Stems completely brown

8   Stems rotting (retting)

81   Outer tissue rotting
85   Vascular tissue easily removed
89   Stems completely collapsed

9   Seed development (whole plant)

91   Seeds expanding
92   Seeds white but full size
93   Most seeds turning ivory yellow
94   Most seeds turning brown
95   All seeds brown and hard
98   Some seeds shed from capsule
99   Most seeds shed from capsule

(From Jefferies & Lawson, 1991. *(From Freer, 1991. Aspects of Applied Biology, **28**, 33–40)*

Stages in Development of Annual Grass Weeds
Illustrations of these growth stages can be found in the reference indicated below and in some company product manuals.

0   Germination and emergence

00	Dry seed
01	Start of imbibition
03	Imbibition complete
05	Radicle emerged from caryopsis
07	Coleoptile emerged from caryopsis
09	Leaf just at coleoptile tip

1   Seedling growth

10	First leaf through coleoptile
11	First leaf unfolded
12	2 leaves unfolded
13	3 leaves unfolded
14	4 leaves unfolded
15	5 leaves unfolded
16	6 leaves unfolded
17	7 leaves unfolded
18	8 leaves unfolded
19	9 or more leaves unfolded

2   Tillering

20	Main shoot only
21	Main shoot and 1 tiller
22	Main shoot and 2 tillers
23	Main shoot and 3 tillers
24	Main shoot and 4 tillers
25	Main shoot and 5 tillers
26	Main shoot and 6 tillers
27	Main shoot and 7 tillers
28	Main shoot and 8 tillers
29	Main shoot and 9 or more tillers

3   Stem elongation

31	First node detectable
32	2nd node detectable
33	3rd node detectable
34	4th node detectable
35	5th node detectable
36	6th node detectable
37	Flag leaf just visible
39	Flag leaf ligule just visible

4   Booting

41	Flag leaf sheath extending
43	Boots just visibly swollen
45	Boots swollen
47	Flag leaf sheath opening
49	First awns visible

5   Inflorescence emergence

51	First spikelet of inflorescence just visible
53	1/4 of inflorescence emerged
55	1/2 of inflorescence emerged
57	3/4 of inflorescence emerged
59	Emergence of inflorescencecompleted

6   Anthesis

61	Beginning of anthesis
65	Anthesis half-way
69	Anthesis complete

(From Lawson & Read, 1992. *Annals of Applied Biology*, **12**, 211–214)

SECTION 6

Growth Stages of Annual Broad-leaved Weeds

Preferred Descriptive Phrases

Illustrations of these growth stages can be found in the reference indicated below and in some company product manuals.

Pre-emergence
Early cotyledons
Expanded cotyledons
One expanded true leaf
Two expanded true leaves
Four expanded true leaves
Six expanded true leaves
Plants up to 25 mm across/high

Plants up to 50 mm across/high
Plants up to 100 mm across/high
Plants up to 150 mm across/high
Plants up to 250 mm across/high
Flower buds visible
Plant flowering
Plant senescent

(From Lutman & Tucker, 1987. *Annals of Applied Biology,* **110**, 683–687)

Appendix 4
Key to Hazard Classifications and
Safety Precautions

Every product label contains information to warn users of the risks from using the product, together with precautions that must be followed in order to minimise the risks. A hazard classification (if any) and the symbol are shown with associated risk phrases, followed by a series of safety numbers under the heading **Hazard classification and safety precautions**.

The codes are defined below, under the same sub-headings as they appear in the pesticide profiles.

Where a product label specifies the use of personal protective equipment (PPE), the requirements are listed under the sub-heading **Operator protection**, using letter codes to denote the protective items, e.g. handling the concentrate, cleaning equipment etc., but it is not possible to list them separately. The lists of PPE are therefore an indication of what the user may need to have available to use the product in different ways. **When making a COSHH assessment it is therefore essential that the product label is consulted for information on the particular use that is being assessed**.

Where the generalised wording includes a phrase such as '... for *xx* days', the specific requirement for each pesticide is shown in brackets after the code.

Hazard

H01	Very toxic
H02	Toxic
H03	Harmful
H04	Irritant
H05	Corrosive
H06	Extremely flammable
H07	Highly flammable
H08	Flammable
H09	Oxidising agent
H10	Explosive
H11	Dangerous for the environment

Risk phrases

R08	Contact with combustible material may cause fire
R09	Explosive when mixed with combustible material
R16	Explosive when mixed with oxidising substances
R20	Harmful by inhalation
R21	Harmful in contact with skin
R22a	Harmful if swallowed
R22b	May cause lung damage if swallowed
R23	Toxic by inhalation
R24	Toxic in contact with skin
R25	Toxic if swallowed
R25b	Toxic; Danger of serious damage to health by prolonged exposure if swallowed
R26	Very toxic by inhalation
R27	Very toxic in contact with skin
R28	Very toxic if swallowed
R34	Causes burns
R35	Causes severe burns
R36	Irritating to eyes
R37	Irritating to respiratory system
R38	Irritating to skin

R39	Danger of very serious irreversible effects
R40	Limited evidence of a carcinogenic effect
R41	Risk of serious damage to eyes
R42	May cause sensitization by inhalation
R43	May cause sensitization by skin contact
R45	May cause cancer
R46	May cause heritable genetic damage
R48	Danger of serious damage to health by prolonged exposure
R50	Very toxic to aquatic organisms
R51	Toxic to aquatic organisms
R52	Harmful to aquatic organisms
R53a	May cause long-term adverse effects in the aquatic environment
R53b	Dangerous to aquatic organisms
R54	Toxic to flora
R55	Toxic to fauna
R56	Toxic to soil organisms
R57	Toxic to bees
R58	May cause long-term adverse effects in the environment
R60	May impair fertility
R60b	May cause heritable genetic damage
R61	May cause harm to the unborn child
R62	Possible risk of impaired fertility
R63	Possible risk of harm to the unborn child
R64	May cause harm to breast-fed babies
R66	Repeated exposure may cause skin dryness or cracking
R67	Vapours may cause drowsiness and dizziness
R68	Possible risk of irreversible effects
R69	Danger of serious damage to health by prolonged oral exposure.
R70	May produce an allergic reaction

Operator protection

A	Suitable protective gloves (the product label should be consulted for any specific requirements about the material of which the gloves should be made)
B	Rubber gauntlet gloves
C	Face-shield
D	Approved respiratory protective equipment
E	Goggles
F	Dust mask
G	Full face-piece respirator
H	Coverall
J	Hood
K	Apron/Rubber apron
L	Waterproof coat
M	Rubber boots
N	Waterproof jacket and trousers
P	Suitable protective clothing
U01	To be used only by operators instructed or trained in the use of chemical/product/type of produce and familiar with the precautionary measures to be observed
U02a	Wash all protective clothing thoroughly after use, especially the inside of gloves
U02b	Avoid excessive contamination of coveralls and launder regularly
U02c	Remove and wash contaminated gloves immediately
U03	Wash splashes off gloves immediately
U04a	Take off immediately all contaminated clothing
U04b	Take off immediately all contaminated clothing and wash underlying skin. Wash clothes before re-use
U04c	Wash clothes before re-use
U05a	When using do not eat, drink or smoke
U05b	When using do not eat, drink, smoke or use naked lights
U06	Handle with care and mix only in a closed container
U07	Open the container only as directed (returnable containers only)
U08	Wash concentrate/dust from skin or eyes immediately

U09a	Wash any contamination/splashes/dust/powder/concentrate from skin or eyes immediately
U09b	Wash any contamination/splashes/dust/powder/concentrate from eyes immediately
U10	After contact with skin or eyes wash immediately with plenty of water
U11	In case of contact with eyes rinse immediately with plenty of water and seek medical advice
U12	In case of contact with skin rinse immediately with plenty of water and seek medical advice
U12b	After contact with skin, take off immediately all contaminated clothing and wash immediately with plenty of water
U13	Avoid contact by mouth
U14	Avoid contact with skin
U15	Avoid contact with eyes
U16a	Ensure adequate ventilation in confined spaces
U16b	Use in a well ventilated area
U18	Extinguish all naked flames, including pilot lights, when applying the fumigant/dust/liquid/product
U19a	Do not breathe dust/fog/fumes/gas/smoke/spray mist/vapour. Avoid working in spray mist
U19b	Do not work in confined spaces or enter spaces in which high concentrations of vapour are present. Where this precaution cannot be observed distance breathing or self-contained breathing apparatus must be worn, and the work should be done by trained operators
U19c	In case of insufficient ventilation, wear suitable respiratory equipment
U19d	Wear suitable respiratory equipment during bagging and stacking of treated seed
U19e	During fumigation/spraying wear suitable respiratory equipment
U20a	Wash hands and exposed skin before eating, drinking or smoking and after work
U20b	Wash hands and exposed skin before eating and drinking and after work
U20c	Wash hands before eating and drinking and after work
U20d	Wash hands after use
U20e	Wash hands and exposed skin after cleaning and re-calibrating equipment
U21	Before entering treated crops, cover exposed skin areas, particularly arms and legs
U22a	Do not touch sachet with wet hands or gloves/Do not touch water soluble bag directly
U22b	Protect sachets from rain or water
U23a	Do not apply by knapsack sprayer/hand-held equipment
U23b	Do not apply through hand held rotary atomisers
U23c	Do not apply via tractor mounted horizontal boom sprayers
U24	Do not handle grain unnecessarily
U25	Open the container only as directed
U26	Keep unprotected workers out of treated areas for at least 5 days after treatment.

Environmental protection

E02a	Keep unprotected persons/animals out of treated/fumigation areas for at least xx hours/days
E02b	Prevent access by livestock, pets and other non-target mammals and birds to buildings under fumigation and ventilation
E02c	Vacate treatment areas before application
E02d	Exclude all persons and animals during treatment
E03	Label treated seed with the appropriate precautions, using the printed sacks, labels or bag tags supplied
E05a	Do not apply directly to livestock/poultry
E05b	Keep poultry out of treated areas for at least xx days/weeks
E05c	Do not apply directly to animals
E06a	Keep livestock out of treated areas for at least xx days/weeks after treatment
E06b	Dangerous to livestock. Keep all livestock out of treated areas/away from treated water for at least xx days/weeks. Bury or remove spillages
E06c	Harmful to livestock. Keep all livestock out of treated areas/away from treated water for at least xx days/weeks. Bury or remove spillages
E06d	Exclude livestock from treated fields. Livestock may not graze or be fed treated forage nor may it be used for hay silage or bedding

E07a Keep livestock out of treated areas for up to two weeks following treatment and until poisonous weeds, such as ragwort, have died down and become unpalatable

E07b Dangerous to livestock. Keep livestock out of treated areas/away from treated water for at least xx weeks and until foliage of any poisonous weeds, such as ragwort, has died and become unpalatable

E07c Harmful to livestock. Keep livestock out of treated areas/away from treated water for at least xx days/weeks and until foliage of any poisonous weeds such as ragwort has died and become unpalatable

E07d Keep livestock out of treated areas for up to 4-6 weeks following treatment and until poisonous weeds, such as ragwort, have died down and become unpalatable

E07e Do not take grass crops for hay or silage for at least 21 days after application

E08a Do not feed treated straw or haulm to livestock within xx days/weeks of spraying

E08b Do not use on grassland if the crop is to be used as animal feed or bedding

E09 Do not use straw or haulm from treated crops as animal feed or bedding for at least xx days after last application

E10a Dangerous to game, wild birds and animals

E10b Harmful to game, wild birds and animals

E10c Dangerous to game, wild birds and animals. All spillages must be buried or removed

E11 Paraquat can be harmful to hares; spray stubbles early in the day

E12a High risk to bees

E12b Extremely dangerous to bees

E12c Dangerous to bees

E12d Harmful to bees

E12e Do not apply to crops in flower or to those in which bees are actively foraging. Do not apply when flowering weeds are present

E12f Do not apply to crops in flower, or to those in which bees are actively foraging, except as directed on [crop]. Do not apply when flowering weeds are present

E12g Apply away from bees

E13a Extremely dangerous to fish or other aquatic life. Do not contaminate surface waters or ditches with chemical or used container

E13b Dangerous to fish or other aquatic life. Do not contaminate surface waters or ditches with chemical or used container

E13c Harmful to fish or other aquatic life. Do not contaminate surface waters or ditches with chemical or used container

E13d Apply away from fish

E13e Harmful to fish or other aquatic life. The maximum concentration of active ingredient in treated water must not exceed XX ppm or such lower concentration as the appropriate water regulatory body may require.

E14a Extremely dangerous to aquatic higher plants. Do not contaminate surface waters or ditches with chemical or used container

E14b Dangerous to aquatic higher plants. Do not contaminate surface waters or ditches with chemical or used container

E15a Do not contaminate surface waters or ditches with chemical or used container

E15b Do not contaminate water with product or its container. Do not clean application equipment near surface water. Avoid contamination via drains from farmyards or roads

E15c To protect groundwater, do not apply to grass leys less than 1 year old

E16a Do not allow direct spray from horizontal boom sprayers to fall within 5 m of the top of the bank of a static or flowing waterbody, unless a Local Environment Risk Assessment for Pesticides (LERAP) permits a narrower buffer zone, or within 1 m of the top of a ditch which is dry at the time of application. Aim spray away from water

E16b Do not allow direct spray from hand-held sprayers to fall within 1 m of the top of the bank of a static or flowing waterbody. Aim spray away from water

E16c Do not allow direct spray from horizontal boom sprayers to fall within 5 m of the top of the bank of a static or flowing waterbody, or within 1m of the top of a ditch which is dry at the time of application. Aim spray away from water. This product is not eligible for buffer zone reduction under the LERAP horizontal boom sprayers scheme.

E16d Do not allow direct spray from hand-held sprayers to fall within 1 m of the top of the bank of a static or flowing waterbody. Aim spray away from water. This product is not eligible for buffer zone reduction under the LERAP horizontal boom sprayers scheme scheme.

E16e Do not allow direct spray from horizontal boom sprayers to fall within 5 m of the top of the bank of a static or flowing water body or within 1 m from the top of any ditch which is dry at the time of application. Spray from hand held sprayers must not in any case be allowed to fall within 1 m of the top of the bank of a static or flowing water body. Always direct spray away from water. The LERAP scheme does not extend to adjuvants. This product is therefore not eligible for a reduced buffer zone under the LERAP scheme.

E16f Do not allow direct spray/granule applications from vehicle mounted/drawn hydraulic sprayers/applicators to fall within 6 m of surface waters or ditches/Do not allow direct spray/granule applications from hand-held sprayers/applicators to fall within 2 m of surface waters or ditches. Direct spray/applications away from water

E16g Do not allow direct spray from train sprayers to fall within 5 m of the top of the bank of any static or flowing waterbody.

E16h Do not spray cereals after 31st March in the year of harvest within 6 metres of the edge of the growing crop

E16i Do not allow direct spray from horizontal boom sprayers to fall within 12 metres of the top of the bank of a static or flowing water body, or within 1

E17a Do not allow direct spray from broadcast air-assisted sprayers to fall within xx m of surface waters or ditches. Direct spray away from water

E17b Do not allow direct spray from broadcast air-assisted sprayers to fall within xx m of the top of the bank of a static or flowing waterbody, unless a Local Environmental Risk Assessment for Pesticides (LERAP) permits a narrower buffer zone, or within 5 m of the top of a ditch which is dry at the time of application. Aim spray away from water

E18 Do not spray from the air within 250 m horizontal distance of surface waters or ditches

E19a Do not dump surplus herbicide in water or ditch bottoms

E19b Do not empty into drains

E20 Prevent any surface run-off from entering storm drains

E21 Do not use treated water for irrigation purposes within xx days/weeks of treatment

E22a High risk to non-target insects or other arthropods. Do not spray within 6 m of the field boundary

E22b Risk to certain non-target insects or other arthropods. For advice on risk management and use in Integrated Pest Management (IPM) see directions for use

E22c Risk to non-target insects or other arthropods

E23 Avoid damage by drift onto susceptible crops or water courses

E34 Do not re-use container for any purpose/Do not re-use container for any other purpose

E35 Do not burn this container

E36a Do not rinse out container (returnable containers only)

E36b Do not open or rinse out container (returnable containers only)

E37 Do not use with any pesticide which is to be applied in or near water

E38 Use appropriate containment to avoid environmental contamination

E39 Extreme care must be taken to avoid spray drift onto non-crop plants outside the target area

E40 To protect groundwater/soil organisms the maximum total dose of this or other products containing ethofumesate MUST NOT exceed 1.0 kg ethofumesate per hectare in any three year period

E40b To protect aquatic organisms respect an unsprayed buffer zone to surface waters in line with LERAP requirements

E40c This product must not be used with any pesticide to be applied in or near water

E40d To protect non-target arthropods respect an untreated buffer zone of 5m to non crop land

E41 Hay for silage must not be cut from treated crops for at least 21 days after treatment

Consumer protection

C01 Do not use on food crops

C02a Do not harvest for human or animal consumption for at least xx days/weeks after last application

C02b Do not remove from store for sale or processing for at least 21 days after application

C02c Do not remove from store for sale or processing for at least 2 days after application

747

SECTION 6

C04	Do not apply to surfaces on which food/feed is stored, prepared or eaten
C05	Remove/cover all foodstuffs before application
C06	Remove exposed milk before application
C07	Collect eggs before application
C08	Protect food preparing equipment and eating utensils from contamination during application
C09	Cover water storage tanks before application
C10	Protect exposed water/feed/milk machinery/milk containers from contamination
C11	Remove all pets/livestock/fish tanks before treatment/spraying
C12	Ventilate treated areas thoroughly when smoke has cleared/Ventilate treated rooms thoroughly before occupying

Storage and disposal

D01	Keep out of reach of children
D02	Keep away from food, drink and animal feeding-stuffs
D03	Store away from seeds, fertilizers, fungicides and insecticides
D04	Store well away from corms, bulbs, tubers and seeds
D05	Protect from frost
D06a	Store away from heat
D06b	Do not store near heat or open flame
D06c	Do not store in direct sunlight
D06d	Do not store above 30/35 °C
D07	Store under cool, dry conditions
D08	Store in a safe, dry, frost-free place designated as an agrochemical store
D09a	Keep in original container, tightly closed, in a safe place
D09b	Keep in original container, tightly closed, in a safe place, under lock and key
D09c	Store unused sachets in a safe place. Do not store half-used sachets
D10a	Wash out container thoroughly and dispose of safely
D10b	Wash out container thoroughly, empty washings into spray tank and dispose of safely
D10c	Rinse container thoroughly by using an integrated pressure rinsing device or manually rinsing three times. Add washings to sprayer at time of filling and dispose of container safely
D10d	Do not rinse out container
D11a	Empty container completely and dispose of safely/Dispose of used generator safely
D11b	Empty container completely and dispose of it in the specified manner
D11c	Ventilate empty containers until no phosphine is detected and then dispose of as hazardous waste via an authorised waste-disposal contractor
D12a	This material (and its container) must be disposed of in a safe way
D12b	This material and its container must be disposed of as hazardous waste
D13	Treat used container as if it contained pesticide
D14	Return empty container as instructed by supplier (returnable containers only)
D15	Store container in purpose built chemical store until returned to supplier for refilling (returnable containers only)
D16	Do not store below 4 degrees Centigrade
D17	Use immediately on removal of foil
D18	Place the tablets whole into the spray tank - do not break or crumble the tablets
D19	Do not empty into drains
D20	Clean all equipment after use

Treated Seed

S01	Do not handle treated seed unnecessarily
S02	Do not use treated seed as food or feed
S03	Keep treated seed secure from people, domestic stock/pets and wildlife at all times during storage and use
S04a	Bury or remove spillages
S04b	Harmful to birds/game and wildlife. Treated seed should not be left on the soil surface. Bury or remove spillages
S04c	Dangerous to birds/game and wildlife. Treated seed should not be left on the soil surface. Bury or remove spillages

S04d To protect birds/wild animals, treated seed should not be left on the soil surface. Bury or remove spillages

S05 Do not reuse sacks or containers that have been used for treated seed for food or feed

S06a Wash hands and exposed skin before meals and after work

S06b Wash hands and exposed skin after cleaning and re-calibrating equipment

S07 Do not apply treated seed from the air

S08 Treated seed should not be broadcast

S09 Label treated seed with the appropriate precautions using printed sacks, labels or bag tags supplied

S10 Only use automated equipment for planting treated seed potatoes

Vertebrate/Rodent control products

V01a Prevent access to baits/powder by children, birds and other animals, particularly cats, dogs, pigs and poultry

V01b Prevent access to bait/gel/dust by children, birds and non-target animals, particularly dogs, cats, pigs, poultry

V02 Do not prepare/use/lay baits/dust/spray where food/feed/water could become contaminated

V03a Remove all remains of bait, tracking powder or bait containers after use and burn or bury

V03b Remove all remains of bait and bait containers/exposed dust/after treatment (except where used in sewers) and dispose of safely (e.g. burn/bury). Do not dispose of in refuse sacks or on open rubbish tips.

V04a Search for and burn or bury all rodent bodies. Do not place in refuse bins or on rubbish tips

V04b Search for rodent bodies (except where used in sewers) and dispose of safely (e.g. burn/bury). Do not dispose of in refuse sacks or on open rubbish tips

V04c Dispose of safely any rodent bodies and remains of bait and bait containers that are recovered after treatment (e.g. burn/bury). Do not dispose of in refuse sacks or on open rubbish tips

V05 Use bait containers clearly marked POISON at all surface baiting points

Medical advice

M01 This product contains an anticholinesterase organophosphorus compound. DO NOT USE if under medical advice NOT to work with such compounds

M02 This product contains an anticholinesterase carbamate compound. DO NOT USE if under medical advice NOT to work with such compounds

M03 If you feel unwell, seek medical advice immediately (show the label where possible)

M04a In case of accident or if you feel unwell, seek medical advice immediately (show the label where possible)

M04b In case of accident by inhalation, remove casualty to fresh air and keep at rest

M05a If swallowed, seek medical advice immediately and show this container or label

M05b If swallowed, do not induce vomiting: seek medical advice immediately and show this container or label

M05c If swallowed induce vomiting if not already occurring and take patient to hospital immediately

M06 This product contains an anticholinesterase carbamoyl triazole compound. DO NOT USE if under medical advice NOT to work with such compounds

SECTION 6

Appendix 5
Key to Abbreviations and Acronyms

The abbreviations of formulation types in the following list are used in Section 2 (Pesticide Profiles) and are derived from the *Catalogue of Pesticide Formulation Types and International Coding System* (CropLife International Technical Monograph 2, 5th edn, March 2002).

1 Formulation Types

AB	Grain bait
AE	Aerosol generator
AL	Other liquids to be applied undiluted
AP	Any other powder
BB	Block bait
BR	Briquette
CB	Bait concentrate
CC	Capsule suspension in a suspension concentrate
CF	Capsule suspension for seed treatment
CG	Encapsulated granule (controlled release)
CL	Contact liquid or gel (for direct application)
CP	Contact powder (for direct application)
CR	Crystals
CS	Capsule suspension
DC	Dispersible concentrate
DP	Dustable powder
DS	Powder for dry seed treatment
EC	Emulsifiable concentrate
EG	Emulsifiable granule
EO	Water in oil emulsion
ES	Emulsion for seed treatment
EW	Oil in water emulsion
FG	Fine granules
FP	Smoke cartridge
FS	Flowable concentrate for seed treatment
FT	Smoke tablet
FU	Smoke generator
FW	Smoke pellets
GA	Gas
GB	Granular bait
GE	Gas-generating product
GG	Macrogranules
GL	Emulsifiable gel
GP	Flo-dust (for pneumatic application)
GR	Granules
GS	Grease
GW	Water soluble gel
HN	Hot fogging concentrate
KK	Combi-pack (solid/liquid)
KL	Combi-pack (liquid/liquid)
KN	Cold-fogging concentrate
KP	Combi-pack (solid/solid)
LA	Lacquer
LI	Liquid, unspecified
LS	Solution for seed treatment
ME	Microemulsion
MG	Microgranules
MS	Microemulsion for seed treatment
OD	Oil dispersion

OL	Oil miscible liquid
PA	Paste
PC	Gel or paste concentrate
PS	Seed coated with a pesticide
PT	Pellet
RB	Ready-to-use bait
RH	Ready-to-use spray in hand-operated sprayer
SA	Sand
SC	Suspension concentrate (= flowable)
SE	Suspo-emulsion
SG	Water soluble granules
SL	Soluble concentrate
SP	Water soluble powder
SS	Water soluble powder for seed treatment
ST	Water soluble tablet
SU	Ultra low-volume suspension
TB	Tablets
TC	Technical material
TP	Tracking powder
UL	Ultra low-volume liquid
VP	Vapour releasing product
WB	Water soluble bags
WG	Water dispersible granules
WP	Wettable powder
WS	Water dispersible powder for slurry treatment of seed
WT	Water dispersible tablet
XX	Other formulations
ZZ	Not Applicable

2 Other Abbreviations and Acronyms

ACP	Advisory Committee on Pesticides
ACTS	Advisory Committee on Toxic Substances
ADAS	Agricultural Development and Advisory Service
a.i.	active ingredient
AIC	Agriculture Industries Confederation
BBPA	British Beer and Pub Association
CDA	Controlled droplet application
CPA	Crop Protection Association
cm	centimetre(s)
COPR	Control of Pesticides Regulations 1986
COSHH	Control of Substances Hazardous to Health Regulations
CRD	Chemicals Regulation Directorate
d	day(s)
Defra	Department for Environment, Food and Rural Affairs
EA	Environment Agency
EBDC	ethylene-bis-dithiocarbamate fungicide
EAMU	Extension of Authorisation for Minor Use
FEPA	Food and Environment Protection Act 1985
g	gram(s)
GCPF	Global Crop Protection Federation (now CropLife International)
GS	growth stage (unless in formulation column)
h	hour(s)
ha	hectare(s)
HBN	hydroxybenzonitrile herbicide
HI	harvest interval
HSE	Health and Safety Executive
ICM	integrated crop management
IPM	integrated pest management
kg	kilogram(s)
l	litre(s)
LERAP	Local Environmental Risk Assessments for Pesticides

SECTION 6

m	metre(s)
MAFF	Ministry of Agriculture, Fisheries and Food (now Defra)
MBC	methyl benzimidazole carbamate fungicide
MEL	maximum exposure limit
min	minute(s)
mm	millimetre(s)
MRL	maximum residue level
mth	month(s)
NA	Notice of Approval
NFU	National Farmers' Union
OES	Occupational Exposure Standard
OLA	off-label approval
PPE	personal protective equipment
PPPR	Plant Protection Products Regulations
SOLA	specific off-label approval
ULV	ultra-low volume
VI	Voluntary Initiative
w/v	weight/volume
w/w	weight/weight
wk	week(s)
yr	year(s)

3 UN Number details

UN No.	Substance	EAC	APP	Hazards Class	Sub risks	HIN
1692	Strychnine or Strychnine salts	2X		6.1		66
1760	Corrosive Liquid, N.O.S., Packing groups II & III	2X	B	8		88
1830	Sulphuric acid, with more than 51% acid	2P		8		80
1993	Flammable Liquid, N.O.S., Packing group III	•3Y		3		30
2011	Magnesium Phosphide	4W[1]		4.3	6.1	
2588	Pesticide, Solid, TOXIC, N.O.S.	2X		6.1		66/60
2757	Carbamate Pesticide, Solid, Toxic	2X		6.1		66/60
2783	Organophosphorus Pesticide, Solid, TOXIC	2X		6.1		66/60
2902	Pesticide, Liquid, TOXIC, N.O.S., Packing Groups I, II & III	2X	B	6.1		66/60
3077	Environmentally hazardous substance, Solid, N.O.S.	2Z		9		90
3082	Environmentally hazardous substance, Solid, N.O.S.	•3Z		9		90
3265	Corrosive Liquid, Acidic, Organic, N.O.S., Packing group I, II or III	2X	B	8		80/88
3351	Pyrethroid Pesticide, Liquid, Toxic, Flammable, Flash Point 23 °C or more	•3W	A(fl)	6.1	3	663/63
3352	Pyrethroid Pesticide, Liquid, TOXIC, Packing groups I, II & III	2X	B	6.1		66/60

Ref: Dangerous Goods Emergency Action Code List 2009 (TSO).
[1] Not applicable to the carriage of dangerous goods under RID or ADR.

Appendix 6
Definitions

The descriptions used in this *Guide* for the crops or situations in which products are approved for use are those used on the approved product labels. These are now standardised in a Crop Hierarchy published by the Pesticides Safety Directorate in which definitions are given. To assist users of this *Guide* the definitions of some of the terminology where misunderstandings can occur are reproduced below.

Rotational grass: Short-term grass crops grown on land that is likely to be growing different crops in future years (*e.g. short-term intensively managed leys for one to three years that may include clover*)

Permanent grassland: Grazed areas that are intended to be permanent in nature (e.g. permanent pasture and moorland that can be grazed).

Ornamental Plant Production: All ornamental plants that are grown for sale or are produced for replanting into their final growing position (*e.g. flowers, house plants, nursery stock, bulbs grown in containers or in the ground*).

Managed Amenity Turf: Areas of frequently mown, intensively managed, turf that is not intended to flower and set seed. It includes areas that may be for intensive public use (*e.g. all types of sports turf*).

Amenity Grassland: Areas of semi-natural or planted grassland subject to minimal management. It includes areas that may be accessed by the public (*e.g. railway and motorway embankments, airfields, and grassland nature reserves*). These areas may be managed for their botanical interest, and the relevant authority should be contacted before using pesticides in such locations.

Amenity Vegetation: Areas of semi-natural or ornamental vegetation, including trees, or bare soil around ornamental plants, or soil intended for ornamental planting. It includes areas to which the public have access. It does NOT include hedgerows around arable fields.

Natural surfaces not intended to bear vegetation: Areas of soil or natural outcroppings of rock that are not intended to bear vegetation, including areas such as sterile strips around fields. It may include areas to which the public have access. It does not include the land between rows of crops.

Hard surfaces: Man-made impermeable surfaces that are not intended to bear vegetation (*e.g. pavements, tennis courts, industrial areas, railway ballast*).

Permeable surfaces overlying soil: Any man-made permeable surface (excluding railway ballast) such as gravel that overlies soil and is not intended to bear vegetation

Green Cover on Land Temporarily Removed from Production: Includes fields covered by natural regeneration or by a planted green cover crop that will not be harvested (*e.g. green cover on setaside*). It does NOT include industrial crops.

Forest Nursery: Areas where young trees are raised outside for subsequent forest planting.

Forest: Groups of trees being grown in their final positions. Covers all woodland grown for whatever objective, including commercial timber production, amenity and recreation, conservation and landscaping, ancient traditional coppice and farm forestry, and trees from natural regeneration, colonisation or coppicing. Also includes restocking of established woodlands and new planting on both improved and unimproved land.

Farm forestry: Groups of trees established on arable land or improved grassland including those planted for short rotation coppicing. It includes mature hedgerows around arable fields.

Indoors (for rodenticide use): Situations where the bait is placed within a building or other enclosed structure, and where the target is living or feeding predominantly within that building or structure.

Herbs: Reference to Herbs or Protected Herbs when used in Section 2 may include any or all of the following. The particular label or EAMU Notice will indicate which species are included in the approval.

Agastache spp.
Angelica
Applemint
Balm
Basil
Bay
Borage (except when grown for oilseed)
Camomile
Caraway
Catnip
Chervil
Clary
Clary sage
Coriander
Curry plant
Dill
Dragonhead
English chamomile
Fennel
Fenugreek
Feverfew
French lavender
Gingermint
Hyssop
Korean mint
Land cress
Lavandin
Lavender
Lemon balm
Lemon peppermint
Lemon thyme
Lemon verbena
Lovage
Marigold
Marjoram
Mint
Mother of thyme
Nasturtium
Nettle
Oregano
Origanum heracleoticum
Parsley root
Peppermint
Pineapplemint
Rocket
Rosemary
Rue
Sage
Salad burnet
Savory
Sorrel
Spearmint
Spike lavender
Tarragon
Thyme
Thymus camphoratus
Violet
Winter savory
Woodruff

Herbs for Medicinal Uses: Reference to Herbs for Medicinal Uses when used in Section 2 may include any or all of the following. The particular label or EAMU Notice will indicate which species are included in the approval

Black cohosh
Burdock
Dandelion
Echinacea
Ginseng
Goldenseal
Liquorice
Nettle
Valerian

Appendix 7
References

The information given in *The UK Pesticide Guide* provides some of the answers needed to assess health risks, including the hazard classification and the level of operator protection required. However, the Guide cannot provide all the details needed for a complete hazard assessment, which must be based on the product label itself and, where necessary, the Health and Safety Data Sheet and other official literature.

Detailed guidance on how to comply with the Regulations is available from several sources.

Pesticides: Code of Practice

Code of Practice for Using Plant Protection Products, 2006 (Defra publication PB 11090)

Known as the 'Green Code', this revised Code of Practice, which came into effect on 15 December 2005, replaced all previous editions of the *Code of Practice for the Safe Use of Pesticides on Farms and Holdings*. It also replaced the voluntary code of practice for the use of pesticides in amenity and industrial areas (the 'Orange Code'). The code explains how plant protection products can be used safely and so meet the legal requirements of the Control of Pesticides Regulations 1986 (COPR) (ISBN 0-110675-10-X), and the Control of Substances Hazardous to Health Regulations 2002 (COSHH) (ISBN 0-110429-19-2). The principal source of information for making a COSHH assessment is the approved product label. In most cases the label provides all the necessary information, but in certain circumstances other sources must be consulted, and these are listed in the Code of Practice.

The code, which applies to England and Wales only, is available in CD-ROM format and electronically from the Chemicals Regulation Directorate and Health and Safety Executive websites. A Welsh language version is available in printed and electronic formats. A Scottish version (approved by the Scottish Parliament) is being produced in printed and electronic formats. Northern Ireland will produce its own version of the code in due course.

Other Codes of Practice

Code of Practice for Suppliers of Pesticides to Agriculture, Horticulture and Forestry (the 'Yellow Code') (Defra Booklet PB 0091)

Code of Good Agricultural Practice for the Protection of Soil (Defra Booklet PB 0617)

Code of Good Agricultural Practice for the Protection of Water (Defra Booklet PB 0587)

Code of Good Agricultural Practice for the Protection of Air (Defra Booklet PB 0618)

Approved Code of Practice for the Control of Substances Hazardous to Health in Fumigation Operations. Health and Safety Commission (ISBN 0-717611-95-7)

Code of Best Practice: Safe use of Sulphuric Acid as an Agricultural Desiccant. National Association of Agricultural Contractors, 2002. (Also available at www.naac.co.uk/?Codes/acidcode.asp)

Safe Use of Pesticides for Non-agricultural Purposes. HSE Approved Code of Practice. HSE L21 (ISBN 0-717624-88-9)

Other Guidance and Practical Advice

HSE (by mail order from HSE Books – See Appendix 2)

COSHH – A brief guide to the Regulations 2003 (INDG136)

A Step by Step Guide to COSHH Assessment, 2004 (HSG97) (ISBN 0-717627-85-3)

Defra (from The Stationery Office – see Appendix 2)

Local Environment Risk Assessments for Pesticides (LERAP): Horizontal Boom Sprayers.

Local Environment Risk Assessments for Pesticides (LERAP): Broadcast Air-assisted Sprayers.

Crop Protection Association and the Voluntary Initiative (see Appendix 2)

Every Drop Counts: Keeping Water Clean

Best Practice Guides. A range of leaflets giving guidance on best practice when dealing with pesticides before, during and after application.

H2OK? Best Practice Advice and Decision Trees - July 2009/10

BCPC (British Crop Production Council – see Appendix 2)

Plantprotection.co.uk (*The UK Pesticide Guide* online) – subscription resource

The Pesticide Manual (16th edition) (ISBN 978-1-901396-86-7)

The Pesticide Manual online (www.bcpcdata.com) – subscription resource

The GM Crop Manual (ISBN 978-1-901396-20-1)

The Manual of Biocontrol Agents (4th edition) (ISBN 978-1-901396-17-1)

IdentiPest PC CD-ROM (ISBN 1-901396-05-3) available via www.plantprotection.co.uk

Garden Detective PC CD-ROM (ISBN 1-901396-32-0)

Small Scale Spraying (ISBN 1-901396-07-X)

Field Scale Spraying (ISBN 1-901396-08-8)

Using Pesticides (ISBN 978-1-901396-10-2)

Safety Equipment Handbook (ISBN 1-901396-06-1)

Spreading Fertilisers and Applying Slug Pellets (ISBN 978-1-901396-15-7)

The Environment Agency (see Appendix 2)

Best Farming Practices: Profiting from a Good Environment

Use of Herbicides in or Near Water

SECTION 7
INDEX

Index of Proprietary Names of Products

The references are to entry numbers, not to pages. Adjuvant names are referred to as 'Adj' and are listed separately in Section 4. Products also registered, i.e. not actively marketed, are entered as *PAR* and are listed by their active ingredient in Section 3.

REFERENCES ARE TO ENTRY NUMBERS NOT PAGES

REFERENCES ARE TO ENTRY NUMBERS NOT PAGES

REFERENCES ARE TO ENTRY NUMBERS NOT PAGES

REFERENCES ARE TO ENTRY NUMBERS NOT PAGES

REFERENCES ARE TO ENTRY NUMBERS NOT PAGES

REFERENCES ARE TO ENTRY NUMBERS NOT PAGES

REFERENCES ARE TO ENTRY NUMBERS NOT PAGES

REFERENCES ARE TO ENTRY NUMBERS NOT PAGES

REFERENCES ARE TO ENTRY NUMBERS NOT PAGES

REFERENCES ARE TO ENTRY NUMBERS NOT PAGES

REFERENCES ARE TO ENTRY NUMBERS NOT PAGES

NOTES

NOTES

NOTES

NOTES

NOTES

NOTES

THE UK PESTICIDE GUIDE 2013

RE-ORDERS

☐ Please send me ____ more copies of *The UK Pesticide Guide 2013* at £47.50 each

Postage: single copy £5.95; orders up to £180 £8.95… orders over £180 FREE

Outside the UK please add £10 per order for delivery.

Name _____ Position_____

Institution _____ Department_____

Address _____

City _____ Region _____ Postcode_____ Country_____

Tel_____ Fax_____ E-mail_____

EU countries except UK – VAT No:_____

Payment (pre-payment is required):

☐ I enclose a cheque/draft for £_____ payable to BCPC. Please send me a receipt.

☐ I wish to pay by credit card: ☐ Visa ☐ Mastercard ☐ Amex ☐ Switch

Please charge to my card £_____ and send me a receipt. Name of issuing bank_____

Card no. ☐☐☐☐ ☐☐☐☐ ☐☐☐☐ ☐☐☐☐

Expiry date ☐☐/☐☐ Security code ☐☐☐   Switch cards only: Start date ☐☐/☐☐ Issue no. ☐

Signature_____ Date_____

Name and address of cardholder if different from above:_____

Please photocopy and return to:
BCPC BCPC Publications Sales, 7 Omni Business Centre, Omega Park, Alton, Hants GU34 2QD, UK
Tel: 01420 593 200, Fax: 01420 593 209, Email: publications@bcpc.org, Web: www.bcpc.org

FUTURE EDITIONS

☐ I wish to take out an annual order for ____ copies of each new edition of *The UK Pesticide Guide*

☐ Please send me advance price details for the 2014 edition of *The UK Pesticide Guide* when available

Order by phone: 01420 593 200 or online: www.bcpc.org/shop

Bookshop orders to:

www.cabi.org
Marston Book Services
160 Milton Park
Abingdon, Oxfordshire, OX14 4SD, UK
Tel +44(0) 1235 465577 Fax: +44(0) 1235 465555
Email: direct.orders@marston.co.uk

Bulk discount:
100+ copies	30%
50–99	25%
10–49	15%
List price £47.50	

Thank you for your order